# Advances in Computational Methods in Sciences and Engineering 2005

Lecture Series on Computer and Computational Sciences
Editor-in-Chief and Founder: Theodore E. Simos

**Volume 4A**

# Advances in Computational Methods in Sciences and Engineering 2005

Selected Papers from the International Conference of Computational Methods in Sciences and Engineering 2005 **(ICCMSE 2005)**

Recognised Conference by the European Society of Computational Methods in Sciences and Engineering (ESCMSE)

Editors:

Theodore Simos and George Maroulis

ISBN: 90 6764 443 9

PRINTED IN THE NETHERLANDS BY RIDDERPRINT BV, RIDDERKERK
COVER DESIGN: ALEXANDER SILBERSTEIN

Brill Academic Publishers
P.O. Box 9000, 2300 PA Leiden
The Netherlands

*Lecture Series on Computer
and Computational Sciences*
Volume 4, 2005, pp. i-iii

# Preface for the Advances in Computational Methods in Sciences and Engineering 2005

## (Selected Papers from the International Conference of Computational Methods in Sciences and Engineering 2005 (ICCMSE 2005)

## Recognised Conference by the European Society of Computational Methods in Sciences and Engineering (ESCMSE)

This volume brings together selected contributed papers presented in the International Conference of Computational methods in Science and Engineering (ICCMSE 2005), held in the Resort Hotel Poseidon, Loutraki, Corinth, Greece between 21st and 26th October 2005.

The aim of the conference is to bring together computational scientists from several disciplines in order to share methods and ideas.

Topics of general interest are: Computational Mathematics, Theoretical Physics and Theoretical Chemistry. Computational Engineering and Mechanics, Computational Biology and Medicine, Computational Geosciences and Meteorology, Computational Economics and Finance, Scientific Computation. High Performance Computing, Parallel and Distributed Computing, Visualization, Problem Solving Environments, Numerical Algorithms, Modelling and Simulation of Complex System, Web-based Simulation and Computing, Grid-based Simulation and Computing, Fuzzy Logic, Hybrid Computational Methods, Data Mining, Information Retrieval and Virtual Reality, Reliable Computing, Image Processing, Computational Science and Education etc.

The International Conference of Computational Methods in Sciences and Engineering (ICCMSE) is unique in its kind. It regroups original contributions from all fields of the traditional Sciences, Mathematics, Physics, Chemistry, Biology, Medicine and all branches of Engineering. It would be perhaps more appropriate to define the ICCMSE as a Conference on Computational Science and its applications to Science and Engineering. Based on the universality of mathematical reasoning the ICCMSE favours the interaction of various fields of Knowledge to the benefit of all. Emphasis on the multidisciplinary character of the Conference and the opening to new forms of interaction was central to the ICCSME 2003 held at Kastoria, in the north of Greece and to the ICCMSE 2004 held at Hotel Armonia, Athens, Greece.

In addition to the general programme the Conference offers an impressive number of Symposia. The purpose of this move is to define more sharply new directions of expansion and progress for Computational Science & Engineering.

We note that for ICCMSE there is a co-sponsorship by American Chemical Society. ICCMSE also is endorsed by American Physical Society.

More than 800 extended abstracts have been submitted for consideration for presentation in ICCMSE 2005. From these extended abstracts we have selected 500 extended abstracts after international peer review by at least two independent reviewers. These accepted papers will be presented at ICCMSE 2005.

After ICCMSE 2005 the participants can send their full papers for consideration for publication in one of the journals that have accepted to publish selected Proceedings of ICCMSE 2005. We would like to thank the Editors-in-Chief and the Publishers of these journals. The full papers will be considered for publication based on the international peer review by at least two independent reviewers.

We would like also to thank:

- The Scientific Committee of ICCMSE 2005 (see in page iv for the Conference Details) for their help and their important support. We must note here that it is a great honour for us that leaders on Computational Sciences and Engineering have accepted to participate the Scientific Committee of ICCMSE 2005.

- The Symposiums' Organisers for their excellent editorial work and their efforts for the success of ICCMSE 2005.

- The invited speakers for their acceptance to give keynote lectures on Computational Sciences and Engineering.

- The Organising Committee for their help and activities for the success of ICCMSE 2005.

- Special thanks for the Secretary of ICCMSE 2005, Mrs Eleni Ralli-Simou (which is also the Administrative Secretary of the European Society of Computational Methods in Sciences and Engineering (ESCMSE)) for her excellent job.

- Special thanks for the student of Professor Simos, Mr. Zacharias Anastassi for his excellent activities for typesetting this document.

**Prof. Theodore Simos**
President of ESCMSE
Department of Computer Science
and Technology
University of the Peloponnese
Tripolis
Greece

**Prof. George Maroulis**
Department of Chemistry
University of Patras
Patras
Greece

August 2005

Brill Academic Publishers
P.O. Box 9000, 2300 PA Leiden
The Netherlands

*Lecture Series on Computer
and Computational Sciences*
Volume 4, 2005, pp. iv-v

# Conference Details

## *International Conference of Computational Methods in Sciences and Engineering 2005 (ICCMSE 2005), Resort Hotel Poseidin, Loutraki, Corinth, Greece, 21-26 October, 2005.*

**Recognised Conference by the European Society of Computational Methods in Sciences and Engineering (ESCMSE)**

### Chairman and Organiser

**Professor T.E. Simos,** President of the European Society of Computational. Methods in Sciences and Engineering (ESCMSE). Active Member of the European Academy of Sciences and Arts, Corresponding Member of the European Academy of Sciences and Corresponding Member of European Academy of Arts, Sciences and Humanities, Department of Computer Science and Technology, Faculty of Sciences and Technology, University of Peloponnese, GR-221 00 Tripolis, Greece.

### Co-Chairman

**Professor George Maroulis,** Department of Chemistry, University of Patras, GR-26500 Patras, Greece.

### Scientific Committee

**Dr. Hamid R. Arabnia, USA**
**Dr. B. Champagne, Belgium**
**Prof. S. Farantos, Greece**
**Prof. I. Gutman, Serbia & Montenegro**
**Prof. P. Mezey, Canada**
**Prof. C. Pouchan, France.**
**Dr. G. Psihoyios, Vice-President ESCMSE**
**Prof. B.M. Rode, Austria.**
**Prof. A. J. Thakkar, Canada**

### Invited Speakers

**Prof. Xavier Assfeld, France**
**Prof. B. Champagne, Belgique.**
**Prof. K. Balasubramanian, USA.**
**Prof. R.J. Bartlett, USA.**
**Prof. D. Clary, UK.**

Prof. E.R. Davidson, USA.
Prof. S. Farantos, Greece.
Prof. W. Goddard, USA
Prof. E. Ghysels, USA
Prof. J.Jellinek, USA
Prof. B. Kirtman, USA.
Prof. M.Kosmas, Greece
Prof. J. Leszczynski, USA.
Prof. M. Meuwly, Switzerland.
Prof. A. Painelli, Italy.
Prof. M.G. Papadopoulos, Greece.
Prof. M. Pedersen, USA.
Prof. C. Pouchan, France.
Prof. V. Renugopalakrishnan, USA
Prof. B.M. Rode, Austria.
Prof. P. Schwerdtfeger, New Zealand.
Prof. G. Scuseria, USA.
Prof. P. Senet, France.
Prof. A.J. Thakkar, Canada.
Prof. A. van der Avoird, The Netherlands.
Prof. P. Weinberger, Austria.
Prof. S. S. Xantheas, USA

Brill Academic Publishers
P.O. Box 9000, 2300 PA Leiden
The Netherlands

*Lecture Series on Computer
and Computational Sciences*
Volume 4, 2005, pp. vi-vii

# European Society of Computational Methods in Sciences and Engineering (ESCMSE)

## Aims and Scope

The *European Society of Computational Methods in Sciences and Engineering (ESCMSE)* is a non-profit organization. The URL address is: http://www.uop.gr/escmse/

The aims and scopes of *ESCMSE* is the construction, development and analysis of computational, numerical and mathematical methods and their application in the sciences and engineering.

In order to achieve this, the *ESCMSE* pursues the following activities:

• Research cooperation between scientists in the above subject.
• Foundation, development and organization of national and international conferences, workshops, seminars, schools, symposiums.
• Special issues of scientific journals.
• Dissemination of the research results.
• Participation and possible representation of Greece and the European Union at the events and activities of international scientific organizations on the same or similar subject.
• Collection of reference material relative to the aims and scope of *ESCMSE*.

Based on the above activities, *ESCMSE* has already developed an international scientific journal called **Applied Numerical Analysis and Computational Mathematics (ANACM)**. This is in cooperation with the international leading publisher, **Wiley-VCH**.

**ANACM** is the official journal of *ESCMSE*. As such, each member of *ESCMSE* will receive the volumes of **ANACM** free of charge.

## Categories of Membership

**European Society of Computational Methods in Sciences and Engineering (ESCMSE)**

Initially the categories of membership will be:

• **Full Member (MESCMSE):** PhD graduates (or equivalent) in computational or numerical or mathematical methods with applications in sciences and engineering, or others who have contributed to the advancement of computational or numerical or mathematical methods with applications in sciences and engineering through research or education. Full Members may use the title MESCMSE.

• **Associate Member (AMESCMSE):** Educators, or others, such as distinguished amateur scientists, who have demonstrated dedication to the advancement of computational or numerical or mathematical methods with applications in sciences and engineering may be elected as Associate Members. Associate Members may use the title AMESCMSE.

• **Student Member (SMESCMSE):** Undergraduate or graduate students working towards a degree in computational or numerical or mathematical methods with applications in sciences and engineering or a related subject may be elected as Student Members as long as they remain students. The Student Members may use the title SMESCMSE

• **Corporate Member:** Any registered company, institution, association or other organization may apply to become a Corporate Member of the Society.

## Remarks:

1. After three years of full membership of the European Society of Computational Methods in Sciences and Engineering, members can request promotion to Fellow of the European Society of Computational Methods in Sciences and Engineering. The election is based on international peer-review. After the election of the initial Fellows of the European Society of Computational Methods in Sciences and Engineering, another requirement for the election to the Category of Fellow will be the nomination of the applicant by at least two (2) Fellows of the European Society of Computational Methods in Sciences and Engineering.

2. All grades of members other than Students are entitled to vote in Society ballots.

3. All grades of membership other than Student Members receive the official journal of the ESCMSE Applied Numerical Analysis and Computational Mathematics (ANACM) as part of their membership. Student Members may purchase a subscription to ANACM at a reduced rate.

**We invite you to become part of this exciting new international project and participate in the promotion and exchange of ideas in your field.**

Brill Academic Publishers
P.O. Box 9000, 2300 PA Leiden
The Netherlands

*Lecture Series on Computer
and Computational Sciences*
Volume 4, 2005, pp. viii-xliii

# Table of Contents

**Field Theory**

# Symposiums of ICCMSE 2005

Brill Academic Publishers
P.O. Box 9000, 2300 PA Leiden,
The Netherlands

*Lecture Series on Computer*
*and Computational Sciences*
Volume 4, 2005, pp. 1-4

# Simulation of Diffraction Efficiency in Oriented Bacteriorhodopsin Films

**P. Acebal[1], L. Carretero, S. Blaya, R.F. Madrigal, A. Murciano, A. Fimia**

Departamento de Ciencia y Tecnología de Materiales. Universidad Miguel Hernández, Av. Ferrocarril s/n Apd o. 03202 Ed. Torrevaillo, Elx (Alicante) Spain

Received 11 July, 2005; accepted in revised form 5 August, 2005

*Abstract:* In this communication we present theoretical calculations of the optical properties for oriented bacteriorhodopsin films. Macroscopic optical properties of the film were related with the microscopic properties of the bacteriorhodopsin, which has been calculated using density functional method. Using the calculated optical properties, we had simulated the diffraction and transmission efficiencies for mixed phase and absorption gratings in oriented bacteriorhodopsin films. The results show that highest diffraction efficiencies than for the randomly oriented material can be obtained for the oriented films. The analysis of the transmission efficiencies shows that the Borrmann effect can be observed for a wide range of wavelengths.

*Keywords:* Bacteriorhodopsin, Optical properties, Density functional, Holography

## 1 Introduction

In this communication we present theoretical calculations of the optical properties for oriented bacteriorhodopsin films. The macroscopic properties were related to the microscopic properties of the Purple Membrane units, which can be estimated taking into account that these units are composed by bacteriorhodopsin trimers[1]. Making use of some approximations, the microscopic properties were calculated using a density functional method. Finally, transmission and diffraction efficiencies for mixed phase and absorption gratings recorded in oriented BR films were simulated.

## 2 Theoretical Background

### 2.1 Macroscopic optical properties

Macroscopic optical properties for an oriented film can be related to the microscopic properties of the different elements of the system. In our case, we divided the lineal optical susceptibility ($\widehat{\chi}$) components into two contributions. On one hand the contribution of the matrix ($\chi_0$), where the zero denotes that it is independent of the electromagnetic field frequency for the range considered. On the other hand, we considered the contribution of the Purple Membrane (PM) units ($\alpha_{ii}^{PMn}$). Therefore, the lineal optical susceptibility components can be written as:

$$\chi_{aa}(\omega;\omega) = \chi_0(1 - w_{PM}) + \sum_n N_n \left( \alpha_{zz}^{PMn}(\omega;\omega) G_{aa}^{zz} + (\alpha_{xx}^{PMn}(\omega;\omega) + \alpha_{yy}^{PMn}(\omega;\omega)) G_{aa}^{yy} \right) \quad (1)$$

---

[1]Corresponding author. E-mail: pablo@dite.umh.es

where $w_{PM}$ is the weight fraction of PM, $N_n$ is the density of Purple Membrane units in the n state, $\alpha_{ii}^{PMn}$ are the components of the polarizability for the PM units in the n state, while the G functions are related to the orientational averages of the PM units and are given by:

$$G_{zz}^{zz} = \langle \vartheta^2 \rangle; \quad G_{zz}^{yy} = G_{yy}^{zz} = \frac{(1 - \langle \vartheta^2 \rangle)}{2}; \quad G_{yy}^{yy} = \frac{(1 + \langle \vartheta^2 \rangle)}{4} \tag{2}$$

where $\langle \vartheta^2 \rangle = 1/3$ for a randomly oriented system and $\langle \vartheta^2 \rangle = 1$ for a totally oriented material. In the same way, the components of the absorption coefficient matrix ($\widehat{\Theta}$) can be expressed as a function of the microscopic properties:

$$\Theta_{aa}(\omega; \omega) = \sum_n N_n \left( \sigma_{zz}^{PMn}(\omega; \omega) G_{aa}^{zz} + (\sigma_{xx}^{PMn}(\omega; \omega) + \sigma_{yy}^{PMn}(\omega; \omega)) G_{aa}^{yy} \right) \tag{3}$$

where $\sigma_{ii}^{PMn}$ are the components of the microscopic absorption coefficient matrix. It is important to note that the microscopic properties have to be corrected with the local field factors [2].

## 2.2 Microscopic optical properties

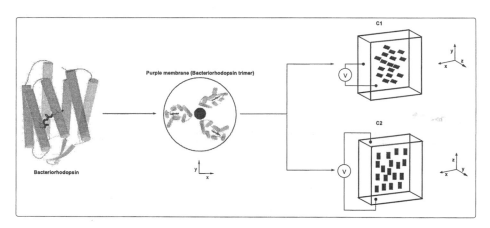

Figure 1: Schematic representation of the bacteriorhodopsin and Purple Membrane units structure.

Since Purple Membrane units are formed by bacteriorhodopsin trimers (see figure fg1)[1], we expressed the microscopic properties of these PM units as a function of the microscopic properties of the bacteriorhodopsin units ($\alpha_0$, $\widehat{\alpha}^{BRn}$ and $\widehat{\sigma}^{BRn}$):

$$\widehat{\alpha}^{PMn} = 3\alpha_0 . \widehat{I} + \sum_{i=1}^{3} \widehat{M}_i^{-1} \widehat{\alpha}^{BRn} \widehat{M}_i \tag{4}$$

$$\widehat{\sigma}^{PMn} = \sum_{i=1}^{3} \widehat{M}_i^{-1} \widehat{\sigma}^{BRn} \widehat{M}_i \tag{5}$$

where $\alpha_0$ denotes the sum of the traces of static polarizability of the 296 aminoacids that form the BR protein, $\widehat{\alpha}^{BRn}$ and $\widehat{\sigma}^{BRn}$ are the polarizability matrix and microscopic absorption coefficient matrix of the light active core of the protein (retinal chromophore and surrounding aminoacids) respectively, while $\widehat{M}$ are the transformation matrices between the frame of the BR and the PM. The properties of light active core are given by[3]:

$$\alpha_{ii}^{BRn}(\omega;\omega) \approx \sum_{e,e\neq 1} \alpha_{ii,e}^{BRn}(0) + \alpha_{ii,1}^{BRn}(0)\Omega_1^{BRn}(\omega;\omega) \tag{6}$$

$$\sigma_{ii}^{BRn}(\omega) = \frac{\omega(\mu_{eg,i}^n)^2}{c\hbar\epsilon_0} \frac{Exp[-\frac{(\omega-\omega_{eg}^n)^2}{2\kappa^2}]}{\sqrt{2\pi}\kappa} \tag{7}$$

where $\mu_{eg}^n$ is the ground to excited state transition dipole moment of the n state of BR, $\omega_{eg}^n$ is the frequency of this transition, $\kappa$ denotes the width of the absorption curve, while $\alpha_{ii,e}^{BRn}(0)$ and $\Omega_1^{BRn}(\omega;\omega)$ are given by equations:

$$\alpha_{ii,e}^{BRn}(0) = \frac{2(\mu_{ge,i}^n)^2}{\hbar\epsilon_0\omega_{eg}} \tag{8}$$

$$\Omega_1^{BRn}(\omega;\omega) = \frac{(\omega_{1g}^n)^2}{(\omega_{1g}^n)^2 - \omega^2} \tag{9}$$

# 3 Modelization procedure

Previous section describes the macroscopic optical properties of oriented BR films as a function of the microscopic properties of the light active core of the BR and the aminoacids chain. The general procedure employed to calculate the microscopic properties involves the geometry optimization of the different elements with a HF/6-31G method, and the evaluation of these microscopic properties using the B3LYP/6-31+G* method. Since the optical properties of the Protonated Schiff Base Retinal are greatly modified by the surrounding medium[4], the modelization procedure of the light active core requires more complex procedure of calculation, which is described in detail in reference [3]. All the calculations were performed with the Gaussian 98 package[5].

# 4 Results and discussion

The model described in the previous sections with the calculated properties allows us to simulate the behavior of the oriented films as a holographic recording material. For these simulations we employ the model of Montemezzani et al. [6] that allows us to evaluate both, diffraction and transmission efficiencies, for mixed phase and absorptions grating in anisotropic media. In the simulations we only considered the B and M states of the bacteriorhodopsin.

The results of the simulations are showed in the figure 2. The simulations were done for three different configurations, C1 and C2 (see figure1) and the randomly oriented material (isotropic). As can be seen, higher diffraction efficiencies can be obtained for the oriented films than for the randomly oriented case in the symmetric setup (except for the vertical polarization in the C2 configuration). In the asymmetric setup, diffraction efficiencies of oriented films are only higher than isotropic film for the cases of vertical polarization in the C1 configuration and horizontal polarization in the C2 configuration.

Other important result of the simulations can be observed if we compare the total transmittance ($\eta$+T) with the transmittance that will have the film without the grating ($T_{ab}$). As is shown, the total transmittance is higher than $T_{ab}$, which in an anomaly known as Borrmann effect. This effect is observed over a wide range of wavelengths and is greater for the asymmetric setup.

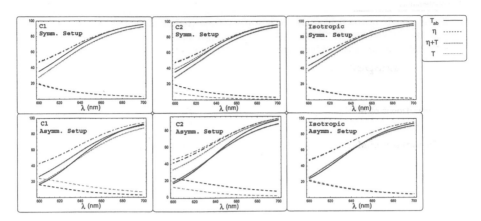

Figure 2: Simulation of the diffraction efficiency ($\eta$), the transmission efficiency (T), the sum of both ($\eta$+T) and the transmission due to absorptive losses ($T_{ab}$) for a BR film (Conditions used: weight fraction of BR of 0.02, thickness of $3\ 10^{-4}$m, conversion efficiency of 0.67, incident angles $-7.5^\circ$ and $40^\circ$ for the symmetric setup the asymmetric setup respectively with difference with the diffracted angle of $15^\circ$). The grey color and the black color denote vertical and horizontal polarizations of the light respectively.

# References

[1] L. O. Essen, R. Siegert, W. Lehmann, and D. Oesterhelt, Proc. Natl. Acad. Sci. USA **95**, 11673 (1998).

[2] R. Wortmann and D. M. Bishop, J. Chem. Phys. **108**, 1001 (1998).

[3] P. Acebal, S. Blaya, and L. Carretero, Phys. Rev. E **72**, 011909 (2005).

[4] R. R. Birge, Annu. Rev. Phys. Chem. **41**, 683 (1990).

[5] M. J. Frisch *et al.*, GAUSSIAN 98, Revision A.7, Gaussian, Inc, Pittsburg PA, 1998.

[6] G. Montemezzani and M. Zgonik, Phys. Rev. E **55**, 1035 (1997).

Brill Academic Publishers
P.O. Box 9000, 2300 PA Leiden,
The Netherlands

*Lecture Series on Computer
and Computational Sciences*
Volume 4, 2005, pp. 5-8

# An Efficient Estimation Scheme for the Space Dependent Dispersion of a Solute Transport Equation

**Jaemin Ahn**[1]

Department of Mathematics, Soonchunhyang University, Asan, 336-745, South Korea

**Chung-Ki Cho**[2]

Department of Mathematics, Soonchunhyang University, Asan, 336-745, South Korea

**Sungkwon Kang**[3]

Department of Mathematics, Chosun University, Gwangju, 501-759, South Korea

Received 8 July, 2005; accepted in revised form 29 July, 2005

*Abstract:* An efficient parameter estimation scheme for the space-dependent dispersion of a solute transport equation in porous media is developed. Using the optimization and the adaptive Laplace transform Galerkin method, the scheme reduces the computational costs significantly compared with those of the conventional methods. The accuracy and efficiency of the scheme are shown.

*Keywords:* Parameter estimation, porous media, inverse Laplace transform

*Mathematics Subject Classification:* 65M06, 65M32, 65Y20

## 1   Introduction

The one-dimensional solute transport equation in a soil is modelled by the following equation[4, 6].

$$\frac{\partial C}{\partial t} - \frac{\partial}{\partial x}\left(D(x)\frac{\partial C}{\partial x}\right) + v(x)\frac{\partial C}{\partial x} = f(x,t), \quad (x,t) \in (0,X) \times \mathbb{R}^+, \tag{1}$$

$$C(x,0) = 0, \ C(0,t) = C_L(t), \ C(X,t) = C_R(t), \tag{2}$$

where $C(x,t)$ is the solute concentration, $v(x)$ is the transport mean velocity, $f$ is a source/sink term, $X$ is the soil depth, $C_L$ and $C_R$ are given functions. In [2], a parameter estimation problem with a constant dispersion coefficient $D$ was considered. In this paper, we consider equations (1)-(2) with a space-dependent dispersion coefficient.

## 2   Parameter estimation

Let $Q$ be a parameter set and $\tilde{Q} \subset Q$ be an admissible parameter subset which embodies any physically motivated constraints on $D$. The solution of (1)-(2) with the parameter $D \in Q$ will be

---

[1]The research of this author was supported by Korean Research Foundation Grant (KRF-2004-037-C00011). E-mail: gom@postech.ac.kr

[2]Corresponding author. E-mail: ckcho@sch.ac.kr

[3]E-mail: sgkang@chosun.ac.kr

denoted by $C(x, t; D)$. Let $T_o$ be a fixed time, and $\{x^\alpha\}_{1 \leq \alpha \leq n}$ a set of observation points in $(0, X)$. For each $\alpha$, let $\Lambda_\alpha$ be the averaged solute concentration in a sufficiently small neighborhood, say, $I_\alpha = [x^\alpha - \frac{1}{2}\Delta x, x^\alpha + \frac{1}{2}\Delta x]$, of the observation point $x^\alpha$. Then our parameter estimation problem becomes

**Problem P.** Find $D^* \in \tilde{Q}$ which minimizes $\mathbf{J}(D) = \sum_{\alpha=1}^{n} \left\{ \left[ \int_{I_\alpha} C(x, T_o; D)\, dx \right] / \Delta x - \Lambda_\alpha \right\}^2$.

Since $\tilde{Q}$ is an infinite-dimensional in general, we consider a sequence of finite dimensional subsets $Q_\mathbf{M}$ of $\tilde{Q}$. Let $C^\mathbf{N}(x, T_o; D_\mathbf{M})$ be the approximated solution for (1)-(2) with the space approximation dimension $\mathbf{N}$ with the parameter $D_\mathbf{M} \in Q_\mathbf{M}$. Consider the following finite dimensional problems.

**Problem $\mathbf{P_M^N}$.**

Find $D_\mathbf{M}^\mathbf{N} \in Q_\mathbf{M}$ which minimizes $\mathbf{J}^\mathbf{N}(D_\mathbf{M}) = \sum_{\alpha=1}^{n} \left\{ \left[ \int_{I_\alpha} C^\mathbf{N}(x, T_o; D_\mathbf{M})\, dx \right] / \Delta x - \Lambda_\alpha \right\}^2$.

Adopting a suitable approximation scheme such as the Crank-Nicolson-Galerkin finite element method, each problem $\mathbf{P_M^N}$ has a solution $D_\mathbf{M}^\mathbf{N}$ and a subsequence of $\{D_\mathbf{M}^\mathbf{N}\}_{\mathbf{M,N}}$ converges to the given true parameter $D$ [3]. To solve Problem $\mathbf{P_M^N}$, the Finite Difference Levenberg Marquardt(FDLM) method[5] is applied. For simplicity, the Jacobian and the Hessian $H$ of $\mathbf{J}^\mathbf{N}$ are approximated by finite differences. Let $[a_{k,l}] = \frac{1}{2}H$ and $[\beta_k] = \frac{1}{2}\nabla \mathbf{J}^\mathbf{N}$. Then

$$\beta_k = \frac{1}{2}\frac{\partial \mathbf{J}^\mathbf{N}}{\partial d_k} \approx \sum_{\alpha=1}^{n} \left[ C^\mathbf{N}(x^\alpha, T_o; D_\mathbf{M}) - \Lambda_\alpha \right] \mathbf{D}_{d_k}[C^\mathbf{N}(x^\alpha, T_o; D_\mathbf{M})], \tag{3}$$

$$a_{k,l} = \frac{1}{2}\frac{\partial^2 \mathbf{J}^\mathbf{N}}{\partial d_k \partial d_l} \approx \sum_{\alpha=1}^{n} \mathbf{D}_{d_k}[C^\mathbf{N}(x^\alpha, T_o; D_\mathbf{M})]\, \mathbf{D}_{d_l}[C^\mathbf{N}(x^\alpha, T_o; D_\mathbf{M})], \tag{4}$$

where $D_\mathbf{M} = (d_1, d_2, \cdots, d_n)$, $\mathbf{D}_{d_k}[C^\mathbf{N}(x^\alpha, T_o; D_\mathbf{M})] = \frac{C^\mathbf{N}(x^\alpha, T_o; D_\mathbf{M} + \epsilon \vec{e}_k) - C^\mathbf{N}(x^\alpha, T_o; D_\mathbf{M})}{\epsilon}$, and $\vec{e}_k$ is the $d_k$-directional unit vector. With an initial guess $D_\mathbf{M}^{(0)} \in Q_\mathbf{M}$, we generate a sequence of iterates $D_\mathbf{M}^{(m+1)} = D_\mathbf{M}^{(m)} + \delta^{(m)}$ until an appropriate stopping criterion is satisfied. Let $\lambda$ be a Marquardt parameter for scaling. Define $a'_{i,i} = a_{i,i}(1 + \lambda)$ and $a'_{i,j} = a_{i,j}$ for $i \neq j$. Then the displacement vector $\delta^{(m)}$ is obtained by solving

$$[a'_{k,l}]\delta^{(m)} = [\beta_k]. \tag{5}$$

**Algorithm (FDLM)**

Step 1. For a given initial guess $D_\mathbf{M}$, evaluate $\mathbf{J}^\mathbf{N}(D_\mathbf{M})$.

Step 2. Compute $[a'_{k,l}]$ and $\vec{\beta}$.

Step 3. Solve (5) and evaluate $\mathbf{J}^\mathbf{N}(D_\mathbf{M} + \delta)$.

Step 4. If $\mathbf{J}^\mathbf{N}(D_\mathbf{M} + \delta) \geq \mathbf{J}^\mathbf{N}(D_\mathbf{M})$, increase $\lambda$ by 10, and go back to Step 2-3.

Step 5. If not, decrease $\lambda$ by 10, update $D_\mathbf{M}$ as $D_\mathbf{M} + \delta$, and go back to Step 2-3.

For a given spatial approximation dimension $\mathbf{N}$, the computational costs for Steps 1-5 will be increased as $\mathbf{M}$ and the number of iterations increase. To overcome these difficulties, we adopt the adaptive Laplace transform Galerkin(ALTG) method. Applying the Laplace transform to (1)-(2), the transformed nodal concentrations are solved by the Galerkin formulation. From the formulation, we obtain the complex-valued linear system

$$(S + pM)\bar{\mathbf{C}} = \bar{\mathbf{b}}, \tag{6}$$

where $\bar{\mathbf{C}} = [\bar{C}_1, \bar{C}_2, ..., \bar{C}_N]^T$ is the transformed vector of nodal concentrations, $p$ is the Laplace transform variable, $\bar{\mathbf{b}}$ is the transformed vector containing source/sink terms, $S$ and $M$ are the "steepness" matrix and "mass" matrix, respectively. Based on the inversion algorithm in [1], the time-dependent nodal concentration $C_j(t)$ is approximated by

$$C_j(t) \approx \frac{1}{T} \exp(\gamma t) \left[ \frac{\bar{C}_j(p_0)}{2} + \sum_{k=1}^{\infty} \text{Re} \left\{ \bar{C}_j(p_k) \exp\left(\frac{ik\pi t}{T}\right) \right\} \right] \approx \frac{1}{T} \exp(\gamma t) \text{Re} \left\{ u_j(z, K_j) \right\}, \quad (7)$$

where $j$ is the nodal index, $i = \sqrt{-1}$, $z = \exp(i\pi t/T)$, $0 < t < 2T$, and the partial fraction $u_j(z, K_j)$ with order $K_j$ approximates the series in (7). Using $\bar{C}_j(p_k)$'s obtained from (6) with $p = p_k$, $u_j(z, K_j)$ are evaluated successively by combining the quotient-difference scheme and recurrence relations. The order $K_j$ is determined adaptively for each node until the criterion $|u_j(z, K_j) - u_j(z, K_j - 1)| < Tol$ is satisfied. For the convection-dominated transport problems, we expect that $K_j$ is large in a steep gradient region, and is small in a smooth region. This adaptive determination of $K_j$'s has the similar effects of the spatial adaptiveness to those in front tracking or multileveling. Since the Laplace transform and its inverse are involved only in the time-integration, our methodology can be applied to multidimensional problems.

## 3 Numerical results

For our simulation, $X = 50\,\text{m}$, $v(x) = 0.1\,\text{m/day}$, $f(x,t) = 0$, $C_L = 1.0\,\text{mg/Lm}$, and $C_R = 0\,\text{mg/L}$ in (1)-(2) were chosen. The true parameter $D^{\text{tr}}(x) \in C[0, X] = Q$ is shown in Figure 1(b). Given the true parameter, the observation data $\{\Lambda^\alpha\}$ were chosen by using the finite element approximation with the Crank-Nicolson time-stepping, which is given by 4096 equally spaced mesh grid and the Courant constraint $C_r = 0.1$, at the final time $T_o = 250\,\text{days}$ and at the uniformly distributed 50 observation points $\{x^\alpha\}$, where $x^\alpha = \alpha X/51$, $\alpha = 1, ..., 50$. $Q_M$ was chosen by the M-dimensional subspace spanned by the linear splines. We started with the constant function 0.01 as an initial guess and with the scaling factor $\lambda = 0.001$ for FDLM. For finite difference formulations (3)-(4) of FDLM, $\epsilon$ was given by 0.001. As a typical conventional method, the Galerkin method with the Crank-Nicolson time-stepping(FEMCN) together with FDLM was used to compare the numerical results of our method(FDLM-ALTG). For FEMCN, the Courant constraint $C_r = 0.1$ was chosen to satisfy the stability condition. The tolerance $Tol$ in ALTG was chosen so that the local errors for ALTG and FEMCN to be similar. Table 1 shows the output-least-squared(OLS) error and the maximum error for the estimated parameters. It is easy to see that both methods have similar accuracy as we expected. However, the computational costs for ALTG are much less than those for FEMCN. Here, the OLS-error means $\{ \sum_{\alpha=1}^{50} [C^{\mathbf{N}}(x^\alpha, T_o; D_{\mathbf{M}}^{\mathbf{N}}) - \Lambda_\alpha]^2 \}^{1/2}$. Figure 1(a) shows the computational costs for both methods and Figure 1(b) shows a typical parameter estimation convergence of FDLM-ALTG.

Table 1: OLS-Error and Maximum error

| M | N | OLS-Error | | $\|D^{tr} - D_M^N\|_\infty$ | |
|---|---|---|---|---|---|
| | | FDLM-ALTG | FDLM-FEMCN | FDLM-ALTG | FDLM-FEMCN |
| 4 | 128 | 1.02948E-03 | 9.97921E-04 | 8.23788E-04 | 1.65662E-03 |
| | 256 | 3.54873E-04 | 3.32514E-04 | 6.65874E-04 | 1.52001E-03 |
| | 1024 | 3.23398E-04 | 3.21105E-04 | 5.90308E-04 | 1.45895E-03 |
| 8 | 128 | 3.44607E-04 | 3.47249E-04 | 9.10550E-04 | 2.74381E-03 |
| | 256 | 5.65146E-05 | 5.51687E-05 | 2.69279E-04 | 6.75562E-04 |
| | 1024 | 4.32608E-05 | 4.27730E-05 | 6.61634E-05 | 1.32481E-04 |
| 10 | 128 | 2.69660E-04 | 2.70394E-04 | 8.36649E-04 | 2.19547E-03 |
| | 256 | 3.22463E-05 | 3.23632E-05 | 2.13832E-04 | 5.40564E-04 |
| | 1024 | 3.15863E-05 | 3.13170E-05 | 4.11794E-05 | 6.68919E-05 |

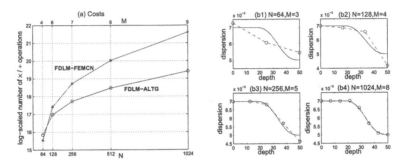

Figure 1: (a) Computational costs. (b) The solid line is the true parameter and the dashed lines with circle are the approximations.

# References

[1] J. Ahn, S. Kang and Y. Kwon, A flexible inverse Laplace transform algorithm and its application, *Computing* **71** 115-131(2003).

[2] J. Ahn, C.-K. Cho, S. Kang and Y. Kwon, An efficient parameter estimation technique for a solute transport equation in porous media, *LNCS* **3045** 847-856(2004).

[3] H.T. Banks and K. Kunisch, *Estimation Techniques for Distributed Parameter Systems.* Birkhäuser, Boston, 1989.

[4] F. Giacobbo, M. Marseguerra and E. Zio, Solving the inverse problem of parameter estimation by genetic algorithms: the case of a groundwater conataminant transport model, *Annals of Nuclear Energy.* **29**, 967-981(2002).

[5] W. Press, S. Teukolsky, W. Vetterling and B. Flannery, *Numerical Recipes in C: The Art of Scientific Computing.* Cambridg Univ. Press, New York, 1992.

[6] T.W.H. Sheu and Y.H. Chen, Finite element analysis of contaminant transport in groundwater, *Appl. Math. Comput.* **127** 23-43(2002).

Brill Academic Publishers
P.O. Box 9000, 2300 PA Leiden,
The Netherlands

*Lecture Series on Computer
and Computational Sciences*
Volume 4, 2005, pp. 9-12

# A Fast Laplace Transform Finite Difference Scheme for the Black-Scholes Equation

**Jaemin Ahn**[1]

Department of Mathematics, Soonchunhyang University, Asan, 336-745, South Korea

**Sungkwon Kang**[2]

Department of Mathematics, Chosun University, Gwangju, 501-759, South Korea

**YongHoon Kwon**[3]

Department of Mathematics, POSTECH, Pohang, 790-784, South Korea

Received 8 July, 2005; accepted in revised form 29 July, 2005

*Abstract:* An accurate and efficient numerical method for solving the Black-Scholes equation with the nonconstant volatility is developed. Based on the inverse Laplace transform, the computational costs for the method is significantly less than those for the conventional ones. The accuracy and efficiency of the method are shown.

*Keywords:* Inverse Laplace transform, time-integration, Black-Scholes equation.

*Mathematics Subject Classification:* 65M06, 65Y20

## 1    Introduction

A European call option with non-constant volatility is modelled by the following modified Black-Scholes equation[3, 6]:

$$\frac{\partial u}{\partial t} - \frac{1}{2}v^2 s^2 \frac{\partial^2 u}{\partial s^2} - (r-d)s\frac{\partial u}{\partial s} + ru + f(s,t) = 0, \ (s,t) \in (0, S_{\max}) \times (0, \mathrm{T}), \tag{1}$$

$$u(s,0) = \max(s-E,0) - (1-E/S_{\max})s, \ s \in [0, S_{\max}], \tag{2}$$

$$u(0,t) = u(S_{\max},t) = 0, \ t \in [0,\mathrm{T}], \tag{3}$$

where $u(s,t) = w(s,t) - w(S_{\max},t)s/S_{\max}$, $w = w(s,t)$ is the European call option price at the underlying asset price $s$ and at the time to expiration $t$, $S_{\max}$ is a sufficiently large underlying asset value, $\mathrm{T}$ is the expiry date, $E$ is the strike price, $r$ is the interest rate, $d$ is the dividend rate, $v = v(s)$ is the volatility of the underlying asset, and $f(s,t) = Es\frac{r-d}{S_{\max}}\exp(-rt)$.

[1]The research of this author was supported by Korean Research Foundation Grant (KRF-2004-037-C00011). E-mail: gom@postech.ac.kr

[2]E-mail: sgkang@chosun.ac.kr

[3]Corresponding Author, This research was supported partially by Com²MaC-KOSEF and by Postech research funds. E-mail: ykwon@postech.ac.kr

## 2   Derivation of algorithm

Applying the Laplace transform to the Black-Scholes equation (1)-(3), we have

$$p\bar{u} - u(s,0) - \frac{1}{2}v^2 s^2 \frac{\partial \bar{u}}{\partial s^2} - (r-d)s\frac{\partial \bar{u}}{\partial s} + r\bar{u} + \bar{f} = 0, \tag{4}$$

$$\bar{u}(0,p) = \bar{u}(S_{\max}, p) = 0, \tag{5}$$

where $p$ is the Laplace transform variable in the complex plane, and $\bar{f}(s,p) = Es\frac{(r-d)}{S_{\max}}\frac{1}{p+r}$.

Let $N$ be a positive integer and let $h = S_{\max}/N$. Assume that $p$ is a fixed complex number. Let $\{\bar{u}_{k,p}\}$ be the finite difference solution on the uniform mesh $\{s_k\}_{1 \le k \le N-1}$, where $s_k = kh$ and $\bar{u}_{k,p}$ approximates $\bar{u}(s_k, p)$. Let $v_k$ and $\bar{f}_{k,p}$ represent $v(s_k)$ and $\bar{f}(s_k, p)$, respectively. Then (4)-(5) can be written as the following system of equations:

$$K(p)\bar{u}(p) = \bar{b}(p), \tag{6}$$

where $K(p) = \{K_{ij}\}$ is the $p$-dependent matrix containing the "steepness" matrix and the "mass" matrix, $\bar{u}(p) = [\bar{u}_{1,p}, \bar{u}_{2,p}, \cdots, \bar{u}_{N-1,p}]^T$ is the transformed vector of nodal prices, and $\bar{b}(p) = [\delta_1, \delta_2, \cdots, \delta_{N-1}]^T$ is the transformed vector containing the effect of forcing terms. After solving the linear system (6), by taking the inverse Laplace transform and applying the Trapezoidal rule to the inverse Laplace transform integration, $u(s_j, t)$ can be obtained as

$$u(s_j, t) \approx \frac{1}{T}\exp(\gamma t)\left[\frac{\bar{u}_{j,p_0}}{2} + \sum_{k=1}^{\infty}\mathrm{Re}\left\{\bar{u}_{j,p_k}\exp\left(\frac{ik\pi t}{T}\right)\right\}\right] = \frac{1}{T}\exp(\gamma t)\mathrm{Re}\left\{\sum_{k=0}^{\infty}a_k^{(j)}z^k\right\} \tag{7}$$

where $i = \sqrt{-1}$, $a_0^{(j)} = \frac{1}{2}\bar{u}_{j,p_0}$, $a_k^{(j)} = \bar{u}_{j,pk}$, $k = 1, 2, \cdots$, $z = \exp(i\pi t/T)$ and $0 < t < 2T$. In general, the power series in equation (7) converges slowly. To accelerate its convergence, we use the quotient-difference(q-d) algorithm[5]. In the q-d scheme, the coefficients $a_k^{(j)}$ in (7) are used to obtain the coefficients of the following corresponding continued fraction[1, 5]

$$v_j(z) = d_0^{(j)}/(1 + d_1^{(j)}z/(1 + d_2^{(j)}z/(1 + \cdots))).$$

The successive convergents $v_j(z, L_j) = d_0^{(j)}/(1 + d_1^{(j)}z/(1 + \cdots + d_{L_j}^{(j)}z))$ can be evaluated by the following recurrence relations:

$$A_k^{(j)} = A_{k-1}^{(j)} + d_k^{(j)}zA_{k-2}^{(j)}, \quad B_k^{(j)} = B_{k-1}^{(j)} + d_k^{(j)}zB_{k-2}^{(j)}, \quad k = 1, 2, \cdots, \tag{8}$$

with $A_{-1}^{(j)} = 0$, $B_{-1}^{(j)} = 1$, $A_0^{(j)} = d_0$, $B_0^{(j)} = 1$. Then $v_j(z, L_j) = A_{L_j}^{(j)}/B_{L_j}^{(j)}$. By a careful investigation of the recurrence relations (8) and the diagonalwise operations[1] of the q-d algorithm, $d^{(j)}$s and $v_j(z)$ can be obtained successively so that the following stopping criterion

$$|v_j(z, L_j^*) - v_j(z, L_j^* - 1)| < Tol \tag{9}$$

can be inserted into the iteration process of the inverse Laplace transform, where $L_j^*$ is the *computational optimal order*. Therefore, the approximation for $u_j(t)$ is given by

$$u(s_j, t) \approx \frac{1}{T}\exp(\gamma t)\,\mathrm{Re}\left\{v_j(z, L_j^*)\right\}. \tag{10}$$

After solving the linear system (6) once, the diagonal evaluation in the inversion process proceeds once for each transformed price $\bar{u}(s_j, p)$, and the stopping criterion (9) is checked for each $j$. If the stopping criterion is satisfied for some $j$, no more evaluations for the inversion are performed for such $j$. Therefore, when the criterion (9) is satisfied for all $j$, our adaptive Laplace transform finite difference method(ALTFD) is completed. The full algorithm is the following.

**Algorithm 2.1(ALTFD)**

Step 1.    For $j = 1, ..., N - 1$, set $fl_j = False$.

Step 2.    Set $flag = N - 1$.

Step 3.    Solve the linear system (6) for $p = p_0, p_1$, and $p_2$.

Step 4.    For $j = 1, ..., N - 1$, do Steps 5-6.

Step 5.      Determine $d_0^{(j)}, d_1^{(j)}$, and $d_2^{(j)}$ by the q-d algorithm.

Step 6.      Evaluate $v_j(z, 1)$ and $v_j(z, 2)$ by the recurrence relations (8).

Step 7.    Set $L = 2$.

Step 8.    While $flag \neq 0$, do Steps 9-17.

Step 9.      Set $L = L + 1$.

Step 10.    Solve the linear system (6) for $p = p_L$.

Step 11.    For $j = 1, ..., N - 1$, do Steps 12-17.

Step 12.      If $fl_j = False$, do Steps 13-17.

Step 13.      Determine $d_L^{(j)}$ by diagonalwise operations in the q-d algorithm.

Step 14.      Evaluate $v_j(z, L)$ by the recurrence relations (8).

Step 15.      If $|v_j(z, L) - v_j(z, L - 1)| < Tol$, do Steps 16-17.

Step 16.      Set $fl_j = True$ and $flag = flag - 1$.

Step 17.      Set $u(s_j, t) = (\exp(\gamma t)/T)\text{Re}\{v_j(z, L)\}$.

# 3   Numerical Results

The backward difference and the Crank-Nicolson methods were chosen for conventional time-marching schemes to compare with our numerical results. We denote FDMBD and FDMCN the finite difference method with the backward difference time-stepping and the finite difference method with the Crank-Nicolson time-stepping, respectively.

For our numerical simulation, $T = 0.5$, $r = 0.05$, $d = 0.01$, $v = 0.35$, $E = 70$, and $S_{\max} = 200$ were chosen. The central difference method was chosen to approximate $\frac{\partial w}{\partial s}$. For the time-marching schemes, the time-step $\Delta t$ was chosen as $T/(N - 1)$ to satisfy the stability condition, where $(N - 1)$ is the approximation dimension for the asset price domain. For ALTFD, $T = 0.8\text{T}$ and $\gamma = -\log 10^{-6}/(1.6\text{T})$ were chosen in (7) which are known to be adequate for general purposes[2]. The tolerance $Tol$ for the stopping criterion (9) was chosen as $Tol = \frac{T}{\exp(\gamma\text{T})} \frac{1}{N^2}$ so that, by [4], the local error for ALTFD becomes

$$|u(s_j, \text{T}) - \frac{e^{\gamma\text{T}}}{T}\text{Re}\{v_j(z, L_j^*)\}| \approx |\frac{e^{\gamma\text{T}}}{T}\text{Re}\{v_j(z, L_j^* - 1) - v_j(z, L_j^*)\}| \leq \frac{1}{N^2}. \quad (11)$$

Therefore, ALTFD and FDMCN have similar accuracy. Table 1 shows the maximum, the minimum, and the average of computational optimal orders $L_j^*$s with respect to the mesh size $N$. It is easy to see that the maximum and the average of $L_j^*$s are approximately proportional to $O(\log N)$. Figure 1 shows the accuracy and efficiency of ALTFD compared with those of FDMCN and FDMBD. From Figures 1 (a) and (b), it can be observed that the accuracy of ALTFD and FDMCN are similar as we expected, but, the computational costs for ALTFD is much less than those for FDMCN or FDMBD for large $N$s, say, $N \geq 256$. Since the matrices in the linear system from the finite difference formulation have the bandwidth 1, the cost of solving the linear system once becomes $O(N)$. In the time-marching schemes, we need to solve the linear system $N - 1$ times. Therefore, the total computational cost for the time-marching schemes FDMBD or FDMCN becomes $O(N^2)$.

On the other hand, the computational costs for the inverse Laplace transforms and for solving the linear system for ALTFD are $O(\sum_{j=1}^{N-1}(L_j^*)^2)$ and $O((\max\{L_j^*\})N))$, respectively, so that the total cost for ALTFD becomes $O(\sum_{j=1}^{N-1}(L_j^*)^2 + \max\{L_j^*\}N)$. Therefore, for large $N$s, ALTFD is much more efficient than the conventional methods such as FDMBD or FDMCN due to the relation $\max(L_j^*) = O(\log N)$.

Table 1: $N$ and $L_j^*$

| $N$ | 16 | 32 | 64 | 128 | 256 | 512 | 1024 |
|---|---|---|---|---|---|---|---|
| $\max(L_j^*)$ | 12 | 13 | 14 | 14 | 15 | 16 | 17 |
| $\min(L_j^*)$ | 11 | 11 | 11 | 12 | 12 | 13 | 13 |
| $avg(L_j^*)$ | 11 | 12 | 13 | 13 | 14 | 15 | 16 |

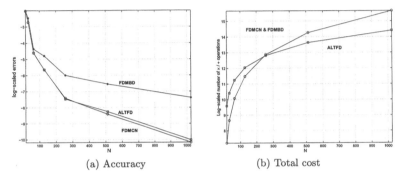

(a) Accuracy        (b) Total cost

Figure 1: Accuracy and computational costs for ALTFD and FDMs

# References

[1] J. Ahn, S. Kang and Y. Kwon, A flexible inverse Laplace transform algorithm and its application, *Computing* **71** 115-131(2003).

[2] J. Ahn, C. Cho, S. Kang and Y. Kwon, An efficient parameter estimation technique for a solute transport equation in porous media *LNCS* **3045** 847-856(2004).

[3] F. Black and M. Scholes, The pricing of options and corporate liabilites, *Journal of Political Economy* **81** 637-659(1973).

[4] G. Blanch, Numerical evaluation of continued fractions, *SIAM Review* **6** 383-421(1964).

[5] F.R. De Hoog, J.H. Knight and A.N. Stokes, An improved method for numerical inversion of Laplace transforms, *SIAM J. Sci. Stat. Comput.* **3**, 357-366(1982).

[6] J. Lishang and T. Youshan, Identifying the volatility of underlying assets form option prices, *Inverse Problems* **17**, 137-155(2001).

Brill Academic Publishers
P.O. Box 9000, 2300 PA Leiden
The Netherlands

*Lecture Series on Computer
and Computational Sciences*
Volume 4, 2005, pp. 13-16

# Adaptive Model Based Control of Optimized Bioprocess Temperature

B. Akay[1]

Department of Chemical Engineering,
Faculty of Engineering,
University of Ankara,
06100 Ankara, Turkey

Received 4 August, 2005; accepted in revised form 15 August, 2005

*Abstract:* Adaptive Internal Model Control was examined when it was applied to control temperature of a jacketed batch bioreactor in which yeast was produced . A software program was used which is capable of receiving a signal from a temperature sensing probe within the bioreactor and regulating an input to that reactor mixture in order to maintain control set point. The control parameters were changed continuously during control operation. Inputs and outputs were recorded and plotted by a the PC using Visual Basic software. Very satisfactory results were obtained for IMC system to track the reactor temperature. All control programs and data acquisition were achieved by using a comprehensive application development tool.

*Keywords:* Adaptive Algorithm, Bioprocess, Temperature control

## 1. Introduction

Yeast productivity and quality are the most important factors in biochemical processes. Factors such as other industrial products, economic conditions, demand for cheap and high quality products and commercial competition require minimum production cost and maximum production yield. Therefore , optimum operating conditions must be determined and the bioreactor must be controlled under these conditions.

*S.cerevisiae* microorganism which is known as Baker's yeast is widely used in food and bioproduct industries [1]. It is produced by using glucose or molasses as carbon sources under aerobic condition. The optimal operating conditions can be found by means of several experimental methods. Bioprocess parameters such as pH, temperature and substrate concentration were determined by using statistical experimental design technique [2] in order to obtain the maximum yeast concentration and specific growth rate and then the reactor temperature was controlled to maintain this optimal condition

In some control studies, more effective and modern control systems such as self tuning PID control system have been applied to chemical processes and bioprocesses [3]. The control technique used in this work is dealing with adaptive control systems. Optimal value of tuning parameter was calculated initially depending on the on-line computer control performance. The control parameters were changed using Recursive Least Square method.

The ISE Iintegral Square Error) criteria was used to the determine the control performance:

$$ISE = \sum_{t=0}^{t} (y - y_{set})^2 \qquad (1)$$

[1] Corresponding author. E-mail: bakay@eng.ankara.edu.tr

## 2  Materials and Method

*Saccharomyces cerevisiae* yeast (NRRL Y-567) was obtained from the ARS culture collection (Northern Regional Research Center, Peoria, IL, U.S.A.).  Stock cultures were maintained on agar slants containing (in g/L): Glucose (20), yeast extract (6), $K_2HPO_4$ (3), $(NH_4)_2SO_4$ (3.35), $NaH_2PO_4$ (3.76),  $MgSO_4.7H_2O$  (0.52), $CaCl_2.4H_2O$ (0.01) and agar (20) (pH 5).

The cells growing on the newly prepared slants were inoculated into the same liquid medium (without agar) and cultivated at 30 °C for 24 h in an incubator-shaker. Cells in the exponential growth phase were inoculated from the seed culture into the growth medium in which optimization experiments were made. Control experiments were carried out in 2 liters  bioreactor (Fig. 1).Adaptive control parameters were changed continuously during control operation. Inputs and outputs were recorded  and plotted by a  the PC using Visual Basic software. All control programs and data acquisition were achieved   by using a comprehensive application development tool. This package provides an icon-based , mouse driven system for designing real time Automation  and Control Strategies, System Monitor Displays and Dynamic Operator Displays.

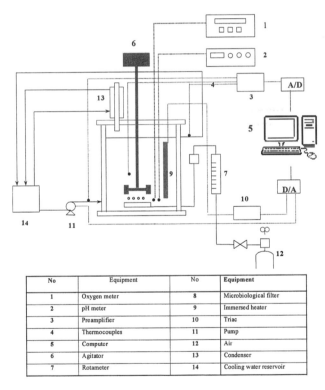

| No | Equipment | No | Equipment |
|----|-----------|----|-----------|
| 1 | Oxygen meter | 8 | Microbiological filter |
| 2 | pH meter | 9 | Immersed heater |
| 3 | Preamplifier | 10 | Triac |
| 4 | Thermocouples | 11 | Pump |
| 5 | Computer | 12 | Air |
| 6 | Agitator | 13 | Condenser |
| 7 | Rotameter | 14 | Cooling water reservoir |

**Figure 1.** Experimental  system

Bioreactor temperature is measured by a platinum resistance thermometer based on resistance change. The accurate and repeatable change in resistance occurs with a change in temperature.  On the Process Control Unit a three wire system was used.  The heating element housed in the process tank is controlled by the computer using  a pulse width modulated technique. The required 'on-time'  pulse width   is determined from data collected in a computer controlled heating mode. Under computer control the microcomputer looks at the difference in temperature between the demand and actual temperature of the bioreactor mixture.

## 3 Results

Experimental results of temperature control by PID algorithm are given in Figure 2. Initially the culture temperature was at the set point, but the response was sluggish and oscillatory. The parameters of PID control system are calculated on Cohen-Coon the basis of method [4]. Relevant PID parameters are found as $K_C = 65.04$, $\tau_I = 1.96$ min, $\tau_D = 0.21$ min.

Experimental results of temperature control by Adaptive control algorithm are given in Figure 3. As can be seen, even if the culture temperature is at the set point at the beginning of the experiment, the adaptive control system brings this temperature quickly to the desired value and then it follows set point smoothly. The ISE values were computed for each control system as $(ISE)_{Adaptive} = 0.67$ and $(ISE)_{PID} = 5.47$. Figure 3 indicates that the performance of the Adaptive control system gives better and more efficient response in following the desired set point than the performance of PID control system.

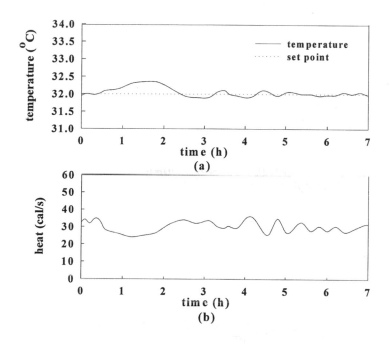

**Figure 2.** PID control of the bioreactor temperature **(a)** Change in the bioreactor temperature with time **(b)** Change in manipulated variable with time

## 4 Discussion

In the present work, the effectivenes of the Adaptive and PID controllers have been examined using ISE performance criterion and off-line measurement of product concentration. Experimental study has been realised to compare the performance of these control systems . Heat input given to the bioreactor mixture was chosen as the manipulated variable and the main control objective is to maintain the bioreactor temperature at the optimal constant set point. In all cases, the controlling heat input changes rapidly to give satisfactory set point tracking. When the process is under the reaction heat effect, Adaptive control has better performance than PID controller experimentally. On-line computer control of PID system doesn't seem to be an acceptable way to produce the desired baker's yeast quality.

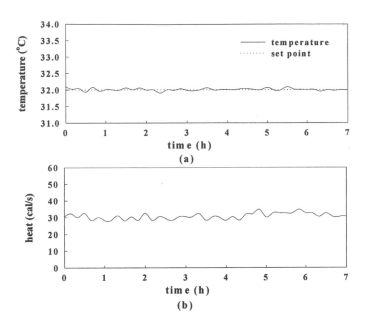

**Figure 3.** Adaptive control of the bioreactor temperature **(a)** Change in the bioreactor temperature with time **(b)** Change in manipulated variable with time

## Acknowledgements

The author wishes to thank Ankara University Biotechnolgy Institue for support.

## References

[1] J.E. Bailey and D.F. Ollis, Biochem.Eng.Fund., Mc.Graw Hill, New York,1986.

[2] M. Alpbaz, N. Bursalı, S. Ertunç and B. Akay, Application of a statistical technique to the production of *Saccharomyces cerevisiae* (baker's yeast), *Biotechnol Appl. BioChem.*, Vol 26 . 91-96. (1997)

[3] B. Akay , S. Ertunç , A. Kahvecioğlu , H.Hapoğlu and M. Alpbaz , Adaptive Control of *S.cerevisiae* Production, *Food and Bioproducts Processing*, Vol 80, 28-38. (2000)

[4] D.R. Omstead, Computer Control of Fermentation Processes, CRC Press , Inc., Boca Raton, Florida, 2000.

Brill Academic Publishers
P.O. Box 9000, 2300 PA Leiden
The Netherlands

*Lecture Series on Computer
and Computational Sciences*
Volume 4, 2005, pp. 17-20

# Immittance Matrix Partition by Hypothetical Capacitor

A. R. Alyyan, M. A. Lahham, M. S. Arni, A. M. O'kool and S. A. Al-Jufout[1]

Department of Electrical Engineering,
Faculty of Engineering,
Tafila Technical University,
66110 Tafila, Jordan

Received 7 February, 2005; accepted in revised form 24 April, 2005

*Abstract:* A mathematical model of a radial power system for transient and steady-state condition calculations has been developed. This model has been represented as a system of differential equations for currents determination and a system of algebraic equations for bus voltages determination. The calculating time mainly depends on the size of the bus immittance matrix inverse, which should be calculated at each step of solution. In order to decrease the size of the immittance matrix, a hypothetical capacitor has been connected to a certain node of the radial power system. This hypothetical capacitor decreased the matrix size and splitted it into two matrices and consequently the calculating time has been decreased by 24%. The results of modelling without and with hypothetical capacitor have been compared, where the maximum error is 1.09%. Decreasing the calculating time allows to model a larger power system using the same resources of the personal computer.

*Keywords:* Power System, Mathematical modelling, Calculating time, Transient, Steady-state.

## 1.    Introduction

Models are essential to the operation of modern power systems, simulation whose is necessary for both planning and operation and depends on appropriate models. In the operation arena models are typically more comprehensive than those used for planning. Operational models support analysis of incoming data as well as simulation of expected and unexpected operational scenarios.

Mathematical modelling and solution on computers is the only practical approach to system analysis and planning studies for a modern power system with its large size, complex and integrated nature. When carrying out technical calculations it is necessary to make use of equivalent circuits for various components, and then combine these circuits in order to represent the interconnection of the components in the actual electrical network [1]. The relatively short lengths of medium and low voltage distribution circuits enable simple modelling techniques to be used for lines. The radial network configurations usually used make it possible to simplify the network model, and large matrix models are seldom necessary. It is usually sufficiently accurate to ignore the capacitance of a distribution circuit and represent it by a series impedance, except when carrying out voltage calculations on a long cable, for example, when a π- or T-equivalent circuit with capacitive shunt branches should be used.

## 2.    Radial Power System Modelling

Fig. 1 shows the modelled radial power system, where its single-phase equivalent circuit is shown in Fig. 2. To decrease the number of differential equations, the mathematical model can be represented in a rectangular system of coordinates $(x, y)$ [2]. Ignoring the capacitor of Fig. 2, the differential equations in $x$ axis are as follows:

$$\frac{di_{sx}}{dt} = \frac{v_{sx} - v_{1x}}{L_T} - \frac{R_T}{L_T} i_{sx} \; ; \; \frac{di_{jkx}}{dt} = \frac{v_{jx} - v_{kx}}{L_{jk}} - \frac{R_{jk}}{L_{jk}} i_{jkx} \; ; \; \frac{di_{lx}}{dt} = \frac{v_{lx}}{L_l} - \frac{R_l}{L_l} i_{lx} \, ,$$

where $j = 1, 2, \ldots, 4$; $k = j + 1$; $l = 1, 2, \ldots, 5$.

---

[1] Corresponding author. Associate Professor of Electrical Engineering Department. E-mail: drjufout@yahoo.com

The equation of the nodal voltages for each axis can be derived from Kirchhoff's law for the current derivatives: the algebraic sum of the current derivatives entering any node is zero:

$$\sum \frac{di_x}{dt} = 0, \quad \sum \frac{di_y}{dt} = 0.$$

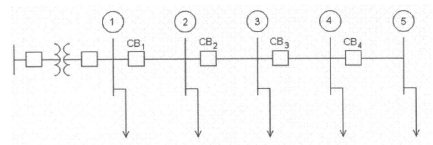

Figure 1: The single-line diagram of a radial power system.

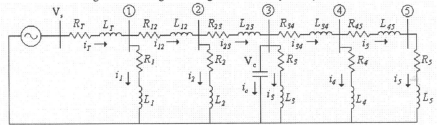

Figure 2: The single-phase equivalent circuit of a radial power system with a hypothetical capacitor on node 3.

Thus the equation for nodal voltages determination can be represented in matrix form as follows:

$$\begin{bmatrix} \mathbf{V_x} \\ \mathbf{V_y} \end{bmatrix} = \begin{bmatrix} \mathbf{L} & \mathbf{0} \\ \mathbf{0} & \mathbf{L} \end{bmatrix}^{-1} \begin{bmatrix} \mathbf{I_x} \\ \mathbf{I_y} \end{bmatrix}, \tag{1}$$

where for $x$ axsis

$$\mathbf{V_x} = \begin{bmatrix} v_{1x} \\ v_{2x} \\ v_{3x} \\ v_{4x} \\ v_{5x} \end{bmatrix}; \quad \mathbf{I_x} = \begin{bmatrix} \alpha_1 i_{1x} + \alpha_{12} i_{12x} - \alpha_T i_{sx} + (1/L_T)v_{sx} \\ \alpha_2 i_{2x} + \alpha_{23} i_{23x} - \alpha_{12} i_{12x} \\ \alpha_3 i_{3x} + \alpha_{34} i_{34x} - \alpha_{23} i_{23x} \\ \alpha_4 i_{4x} + \alpha_{45} i_{45x} - \alpha_{34} i_{34x} \\ \alpha_5 i_{5x} - \alpha_{45} i_{45x} \end{bmatrix};$$

$$\mathbf{L} = \begin{bmatrix} \dfrac{1}{L_1}+\dfrac{1}{L_{12}}+\dfrac{1}{L_T} & -\dfrac{1}{L_{12}} & 0 & 0 & 0 \\ -\dfrac{1}{L_{12}} & \dfrac{1}{L_2}+\dfrac{1}{L_{12}}+\dfrac{1}{L_{23}} & -\dfrac{1}{L_{23}} & 0 & 0 \\ 0 & -\dfrac{1}{L_{23}} & \dfrac{1}{L_3}+\dfrac{1}{L_{23}}+\dfrac{1}{L_{34}} & -\dfrac{1}{L_{34}} & 0 \\ 0 & 0 & -\dfrac{1}{L_{34}} & \dfrac{1}{L_4}+\dfrac{1}{L_{34}}+\dfrac{1}{L_{45}} & -\dfrac{1}{L_{45}} \\ 0 & 0 & 0 & -\dfrac{1}{L_{45}} & \dfrac{1}{L_5}+\dfrac{1}{L_{45}} \end{bmatrix},$$

where the damping coefficients are calculated as follows:

$$\alpha_i = \frac{R_i}{L_i}; \quad \alpha_{ij} = \frac{R_{ij}}{L_{ij}},$$

where

$R_i$ - the load resistance connected to bus *I*;

$R_{ij}$ - the transmission line resistance connected between buses *i* and *j*;

$L_i$ - the load inductance connected to bus *i*;

$L_{ij}$ - the transmission line inductance connected between buses *i* and *j*;

$R_T$, $L_T$ - the resistance and leakage inductance of the power transformer respectively.

### 3.     Radial Power System Modelling with Hypothetical Capacitor

The calculating time depends on the number of the algebraic equations (the bus immittance (impedance or admittance) matrix size) and the number of differential equations. In order to decrease the size of the bus immittance matrix, it is suggested to add a hypothetical capacitor to node 3 (see Fig. 2). Its capacitance is low and consequently its reactance is high and thus the current flowing through it is small. The nodal voltage on bus 3 can be replaced by the capacitor voltage, which is determined in *x* axis by:

$$\frac{dv_{Cx}}{dt} = \frac{1}{C}(i_{23x} - i_{3xx} - i_{3x}),$$

where $C$– the capacitance of the hypothetical capacitor.

To determine the bus voltages, the Kirchhoff's Low for the current derivatives should be applied to each node except node 3. Thus the equation for nodal voltages determination can be represented by Eq. (1) where its components will be as follows:

$$\mathbf{V}_x = \begin{bmatrix} V_{1x} \\ V_{2x} \end{bmatrix}; \text{ where } V_{1x} = \begin{bmatrix} v_{1x} \\ v_{2x} \end{bmatrix} \text{ and } V_{2x} = \begin{bmatrix} v_{4x} \\ v_{5x} \end{bmatrix}.$$

The bus immittance matrix is:

$$L = \begin{bmatrix} L_1 & 0 \\ 0 & L_2 \end{bmatrix},$$

where

$$L_1 = \begin{bmatrix} \dfrac{1}{L_{12}} + \dfrac{1}{L_1} + \dfrac{1}{L_T} & -\dfrac{1}{L_{12}} \\[2mm] -\dfrac{1}{L_{12}} & \dfrac{1}{L_{12}} + \dfrac{1}{L_{23}} + \dfrac{1}{L_2} \end{bmatrix}; \quad L_2 = \begin{bmatrix} \dfrac{1}{L_{34}} + \dfrac{1}{L_{45}} + \dfrac{1}{L_4} & -\dfrac{1}{L_{45}} \\[2mm] -\dfrac{1}{L_{45}} & \dfrac{1}{L_{45}} + \dfrac{1}{L_5} \end{bmatrix}.$$

The current column matrix is:

$$I_x = \begin{bmatrix} I_{x1} \\ I_{x2} \end{bmatrix},$$

where

$$I_{x1} = \begin{bmatrix} \alpha_1 i_{1x} + \alpha_{12} i_{12x} - \alpha_T i_{Tx} + \dfrac{V_{sx}}{L_T} \\[2mm] \alpha_2 i_{2x} + \alpha_{23} i_{23x} - \alpha_{12} i_{12x} + \dfrac{1}{L_{23}} V_{cx} \end{bmatrix}; \quad I_{x2} = \begin{bmatrix} \alpha_4 i_{4x} + \alpha_{45} i_{45x} - \alpha_{34} i_{34x} + \dfrac{1}{L_{34}} V_{cx} \\[2mm] \alpha_5 i_{5x} - \alpha_{45} i_{45x} \end{bmatrix}.$$

We notice that the equation for node 3 is omitted. Thus as shown from the algorithm above, the system of the algebraic equations was decreased by one equation and divided into two systems in each axis. In this algorithm, in order to solve the system the inverse of two 2 ×2 matrices should be found, while in the former algorithm the inverse of 5 ×5 matrix should be calculated for each axis of coordinates. On the other hand the number of differential equations has been increased by one in each axis.

### 4.   Results and Discussion

The differential equations have been solved by Fourth-Order Rung-Kutta method. The system of algebraic equations has been solved by Gauss method. The power transformer is rated 16 MVA, 110/22 kV with leakage reactance of 11% and resistance of 1%. The transmission line resistance and reactance are assumed to be 0.1284 Ω/mi and 0.788 Ω/mi respectively. The value of the hypothetical capacitor is very small and equals to 0.01μF thus the current flowing through it is approximately zero.

The calculating time using the power system model with the hypothetical capacitor is decreased by 24%. This is due to the decrease of the bus immittance matrix size, i.e. instead of finding the inverse of 5x5 matrix, the inverse of two 2x2 matrices should be found. On the other hand, one differential equation is added to determine the voltage on the hypothetical capacitor.

Fig. 3 shows the transmission line (1-2) phase currents during three-phase fault on bus 4. This figure shows the prefault and fault currents from which it is obvious that the developed model can be used for both transient and steady-state condition computation.

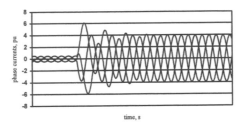

Figure 3: The prefault and fault transmission line (1-2) phase currents during three-phase short-circuit on bus 4 (the base current is 594 A).

This figure consists of six curves of phase currents: three phase currents are calculated without a hypothetical capacitor and three of them are calculated with the capacitor, but due to the complete coincidence, it seems that there are only three curves. Results obtained by the algorithm without hypothetical capacitor and the algorithm with hypothetical capacitor are compared where the maximum percentage error is 1.09%.

## 5.  Conclusion

A mathematical model of a radial power system is developed. The model is used for transient and steady-state condition computation. It consists of two systems of equations: a system of differential equations for currents determination and a system of algebraic equations for bus voltages determination, where the calculating time mainly depends on the bus immittance matrix size and the number of the differential equations. Decreasing the calculating time is achieved by introduction a hypothetical capacitor in a certain node. Its capacitance is small and thus the current flowing through it is also small. Introduction of the hypothetical capacitor decreases the bus immittance matrix by one and splits it into two matrices. On the other hand the system of the differential equations increases by one. The calculating time of the suggested algorithm with a hypothetical capacitor is less by 24% and the maximum percentage error due to the introduction of this capacitor is 1.09%. Thus the suggested algorithm allows to model a larger power system using the same resources of the personal computer. In addition, this capacitor can be used for medium-length transmission line modelling and for studying the problems of power factor correction.

## References

[1] John J. Grainger, William D. Stevenson Jr. *Power System Analysis*. UK: McGraw-Hill, 1994.

[2] Al-Jufout S., 2003. Modelling of the Cage Induction Motor for Symmetrical and Asymmetrical Modes of Operation. *Computers & Electrical Engineering International Journal*, Elsevier Science Ltd., New Mexico, USA. Vol. 29, № 8, pp. 851-860.

Brill Academic Publishers
P.O. Box 9000, 2300 PA Leiden
The Netherlands

*Lecture Series on Computer
and Computational Sciences*
Volume 4, 2005, pp. 21-24

# Theoretical Studies on Solid α,ω-Diamine Salts

Ana M. Amado[a], Sónia Fiuza[a], J.C. Otero[b], M. Paula M. Marques[a] and Luís A.E. Batista de Carvalho[1a]

[a]Química-Física Molecular, Universidade de Coimbra, 3004-535 Coimbra, Portugal
[b]Departamento de Química-Física, Faculdade de Ciências, Univ. Málaga, 29071 Málaga, Spain

Received 1 July, 2005; accepted in revised form 13 July, 2005

*Abstract:* The present study reports a structural analysis of the homologous series of α,ω-diamine dihydrochlorides ($[H_3N(CH_2)_nNH_3]^{2+} \cdot 2Cl^-$) in the solid state, through *ab initio* molecular orbital calculations. Several molecular models were built and assessed for their ability to accurately represent these systems, which displayed distinct conformational behaviours according to the length of their carbon chain. It was verified that the larger the amines the more intermolecular interactions had to be considered by the theoretical models, from (N)H···Cl and (C)H···Cl close-contacts to weak London dispersion forces.

*Keywords: ab initio* calculations, structural analysis, diamine salts, hydrogen bonding

## 1. Introduction

The development of new anticancer drugs is presently a vigorous area of research, which aims at overcoming the limitations displayed by most of the clinically used chemotherapeutic agents, such as cisplatin (*cis*-diamminedichloroplatinum (II), *cis*-$Pt(NH_3)_2Cl_2$) – *e.g.* adverse side-effects, low specificity and acquired resistance. The design of multinuclear platinum(II) complexes, with aliphatic diamines as metal linkers, is one of the most promising strategies for achieving this goal [1]. A particular group of Pt(II) chelates comprise two or three cisplatin-like moieties, their main structural features being: (i) each metal ion is coordinated to two chloride atoms and two amine groups, in a *cis*-orientation (similar to cisplatin); (ii) the metal centres are linked through linear diamines of variable length. The cytotoxic effect of these systems is known to be ruled by structure-activity relationships (SAR's) [2,3] which are thus the object of intense study. Therefore, the knowledge of the structural behaviour of the diamines acting as linkers in these complexes is of the utmost relevance.

In the present work, a conformational analysis of the homologous series of α,ω-diamine dihydrochlorides of general formula $[H_3N(CH_2)_nNH_3]^{2+} \cdot 2Cl^-$ (Figure 1), in the solid state, was carried out by *ab initio* methods. According to the corresponding X-ray structures, these compounds may be divided into three different groups – group 1: n=2; group 2: n=3 and 4; group 3: n≥5 – in order to build suitable molecular models that will accurately represent each group, thus allowing to obtain a reliable theoretical prediction of the experimental data.

## 2. Results and Discussion

All *ab initio* calculations were carried out using the Gaussian 98 program (G98W) [4]. The B3LYP/6-31G* theory level was considered using the integration grid of 75 radial shells and 302 angular points per shell (G98W keyword grid=75302, using a FineGrid). All molecular geometries were fully optimised using the following cuttoffs for forces and step sizes: 0.000015 Hartree/Bohr for maximum force, 0.000010 Hartree/Bohr for root-mean-square force, 0.000060 Bohr for maximum displacement and 0.000040 Bohr for root-mean-square displacement (G98W keyword opt=tight).

---

[1] Corresponding author: E-mail: labc@ci.uc.pt

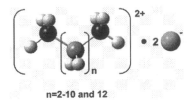

**n=2-10 and 12**

Figure 1: Schematic representation of the homologous series of $\alpha,\omega$-diamine dihydrochlorides ($[H_3N(CH_2)_nNH_3]^{2+} \cdot 2Cl^-$) studied in this work.

For the smallest element of the $[H_3N(CH_2)_nNH_3]^{2+} \cdot 2Cl^-$ series, the 1,2-ethylenediamine dihydrochloride (n=2), it was recently shown that both the number and spatial arrangement of the chloride counterions are essential for an accurate theoretical simulation of both structural and vibrational features [5]. In fact the best theoretical molecular model (***model 5***, Fig. 2) was found to comprise one ethylenediamine cation surrounded by six chloride ions, in accordance with the X-ray core structure reported for this system [6]. By correctly mimetising the intra-layer (N)H$\cdots$Cl intermolecular interactions known to occur in the solid state structure, this molecular model was verified to accurately reproduce both the geometrical parameters and vibrational pattern – wavenumbers and isotopic shifts – thus allowing to clarify previous controversial assignments.

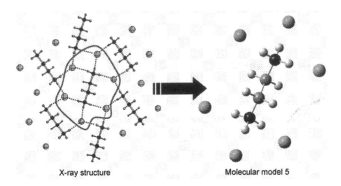

X-ray structure                    Molecular model 5

Figure 2: Representation of the X-ray structure of 1,2-ethylenediamine dihydrochloride [6] and of molecular ***model 5*** [5].

Regarding the amines of the second group (n=3 and 4), however, it was found that molecular ***model 5*** does not reproduce their structural preferences. Actually, calculations based on this model yielded identical distances for the (N)H$_2\cdots$Cl and (N)H$_3\cdots$Cl close-contacts (Fig. 3), in constrast to the X-ray data which indicates that the latter are longer. Moreover, these calculations were unable to reproduce the slight bending observed for the carbon skeleton of n=3 and n=4 diamines.

A careful analysis of the X-ray data reported for these $\alpha,\omega$-diamine dihydrochlorides shows that, for n>2 besides the characteristic (N)H$\cdots$Cl interactions there is a network of weaker (C)H$\cdots$Cl intermolecular close-contacts (Fig. 3). In the light of these crystallographic core structures, the molecular ***model 6*** was used (Fig. 3) and found to correctly mimetise group 2 diamines giving rise to the best fit between calculated and experimental parameters, namely the deviation from linearity and the relative magnitude of the (N)H$\cdots$Cl interactions. The (C)H$\cdots$Cl distances were also accurately reproduced by this model.

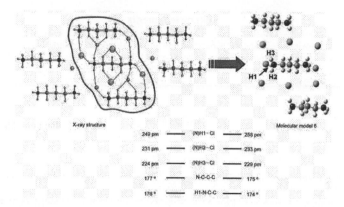

Figure 3: Representation of the X-ray structure of 1,3-propyldiamine dihydrochloride [7] (used as an example of group 2 diamines) and of molecular *model 6*.

Both *models 5 and 6* showed to be unsuitable for a correct mimetisation of group 3 diamines (n≥5), for which the X-ray core geometry comprises two amine units interacting through (N)H···Cl close-contacts as well as by London dispersion forces between the two hydrocarbon chains, leading to a curvature of the molecular skeleton. In the light of this data, the theoretical *model 7* (Fig. 4) was considered, leading to a quite good calculated to experimental fit.

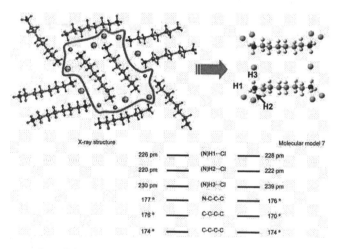

Figure 4: Representation of the X-ray structure of 1,6-hexyldiamine dihydrochloride [8] (used as an example of group 3 diamines) and of molecular *model 7*.

While for 1,2-ethylenediamine only (N)H···Cl intermolecular interactions have to be represented by the theoretical model, for 1,3- and 1,4- diamines both (N)H···Cl and (C)H···Cl close-contacts should be considered. For the larger diamines, in turn, weak dispersion forces must also be accounted for, in order to properly reproduce the experimental solid state structure (*e.g.* distortion of the carbon chain relative to an *all*-trans conformation).

## Acknowledgments

The authors acknowledge financial support from the Portuguese Foundation for Science and Technology – *Unidade de Química-Física Molecular and Research Project POCTI/47256/QUI/2002* (co-financed by the european community fund FEDER). SF also acknowledges a PhD fellowship SFRH/BD/17493/2004.

## References

[1] T.D. McGregor, A. Hegmans, J. Kaspárková, K. Neplechová, O. Nováková, H. Penazová, O. Vrána, V. Brabec and N. Farrell, *J. Biol. Inorg Chem.* **7**, 397 (2002) and refs. therein.

[2] M.P.M. Marques, M.T. Girão da Cruz, M.C. Pedroso de Lima, A.. Gameiro, E. Pereira and P. Garcia, *Biochim. Biophys. Acta (MCR)* **1589**, 63 (2002).

[3] L.J. Teixeira, M. Seabra, E. Reis, M.T. Girão da Cruz, M.C. Pedroso de Lima, E. Pereira, M.A. Miranda and M.P.M. Marques, *J. Med. Chem.,* **47**, 2917 (2004).

[4] **Gaussian 98, Revision A.9**, M. J. Frisch, G. W. Trucks, H. B. Schlegel, G. E. Scuseria, M. A. Robb, J. R. Cheeseman, V. G. Zakrzewski, J. A. Montgomery, Jr., R. E. Stratmann, J. C. Burant, S. Dapprich, J. M. Millam, A. D. Daniels, K. N. Kudin, M. C. Strain, O. Farkas, J. Tomasi, V. Barone, M. Cossi, R. Cammi, B. Mennucci, C. Pomelli, C. Adamo, S. Clifford, J. Ochterski, G. A. Petersson, P. Y. Ayala, Q. Cui, K. Morokuma, D. K. Malick, A. D. Rabuck, K. Raghavachari, J. B. Foresman, J. Cioslowski, J. V. Ortiz, A. G. Baboul, B. B. Stefanov, G. Liu, A. Liashenko, P. Piskorz, I. Komaromi, R. Gomperts, R. L. Martin, D. J. Fox, T. Keith, M. A. Al-Laham, C. Y. Peng, A. Nanayakkara, M. Challacombe, P. M. W. Gill, B. Johnson, W. Chen, M. W. Wong, J. L. Andres, C. Gonzalez, M. Head-Gordon, E. S. Replogle and J. A. Pople, Gaussian, Inc., Pittsburgh PA, 1998.

[5] A.M. Amado, J.C. Otero, M.P.M. Marques and L.A.E. Batista de Carvalho, *ChemPhysChem,* **5**, 1837 (2004).

[6] M. Bujak, L. Sikorska and J. Zalesli, *Z. Anorg.Allg. Chem.* **626**, 2535 (2000).

[7] J. Brisson and F. Brisse, *J. Crystallogr. Spectrosc. Res.* **12**, 39 (1982).

[8] N. Borkakoti, P.F. Lindley, D.S. Moss and R.A. Palmer, *Acta Cryst.* **B34**, 3431 (1978).

Brill Academic Publishers
P.O. Box 9000, 2300 PA Leiden
The Netherlands

*Lecture Series on Computer and Computational Sciences*
Volume 4, 2005, pp. 25-32

# Polarizability and Capacitance of Platonic solids and Kerr constant of Proteins

Sergio R. Aragon[1], and David K. Hahn

Department of Chemistry & Biochemistry,
San Francisco State University,
San Francisco, California

Received 30 June, 2005; accepted in revised form 13 July, 2005

*Abstract:* A precise implementation of the Boundary Element method has been applied to the computation of the polarizability and the capacitance of the five conducting Platonic solids. The method is demonstrated to be accurate and precise by comparison with analytical values for spheroids. Our work agrees with path integral methods to 5 decimal places when adaptive triangulations of the polygon faces are used. The method is then applied to the computation of the Kerr constant of proteins, using a simple ansatz for the relation of the polarizability of the conducting body and the actual dielectric body. Agreement with experiment is excellent for 7 out of 8 proteins when the orientation of the protein dipole moment is optimized with respect to the orientation of the polarizability tensor in the crystallographic frame.

*Keywords:* platonic solids, polarizability, capacitance, Kerr constant, proteins, boundary elements

*PACS SubjectClassification:* Electromagnetism, Biological physics

*PACS:* 41.20Cv, 42.25.Lc, 87.15.Aa

## 1. Introduction

The polarizability of molecules and molecular aggregates is required in order to understand measurements involving electromagnetic scattering, and the dynamics of these systems in the presence of electric fields. In addition, as emphasized in the work of Douglas, Garboczi and co workers[1], the polarizability is also essential in the description of the thermodynamic properties of mixtures since it mediates particle –particle interactions. There has been much work on the numerical computation of these quantities because analytical solutions are obtainable only for simple smooth shapes characterized by an ellipsoid. The Platonic solids have been an important challenge because despite the large amount of symmetry, no analytical solutions exist even for the cube. Two classes of numerical methods have been put forth in the study of these problems: a) Finite element[1] and boundary element methods[2,3], and b) random walk models or path integral methods[1,4]. All of these numerical methods are capable of handling arbitrarily shaped objects, a requirement in the application to biomolecules such as proteins.

Douglas and Garboczi[1,4] have shown that the path integral method is both accurate and fast. However, that method is restricted to problems which have the electrostatic Green function as a propagator in the interaction. Some hydrodynamic problems can be approximated by the so-called pre-averaging approximation in which the Oseen tensor is orientationally averaged to yield a propagator proportional to the electrostatic one. We have shown that this approximation is unsuitable for transport properties except for the case of translational diffusion[5]. Thus, we emphasize the boundary element method in this work and demonstrate that we can also obtain high accuracy for the Platonic solids and for amorphous objects such as proteins. The polarizabilities are extended to arbitrary dielectric constant via an ellipsoidal ansatz and applied to the computation of the Kerr constant of proteins. Our work on the boundary element method applied to hydrodynamics is presented elsewhere[6].

## 2. Theory

Zhou[7] has shown that the surface charge density, $\sigma$, of an arbitrarily shaped conducting particle subjected to an electric field can be determined by formulating Poisson's equation as an integral of Green's function, **G**, over the surface of the molecule:

---

[1]Corresponding author. Email: aragons@sfsu.edu

$$\vec{y} = \int_{sp} G(\vec{x}, \vec{y})(\hat{x}_1 \sigma_1 + \hat{x}_2 \sigma_2 + \hat{x}_3 \sigma_3) dS_x \qquad (1)$$

where the electrostatic Green function is (omitting a factor of $4\pi$):

$$G(\vec{x}, \vec{y}) = \frac{1}{|\vec{x} - \vec{y}|} \qquad (2)$$

and the surface charge density is expressed in terms of three auxiliary densities $\sigma_i(r)$.
An approximate solution is obtained by converting the integral equation into
a matrix equation. This is accomplished by discretizing the surface of the molecule
into boundary elements. The key assumption is that the surface charge density
has a constant value within each of the N boundary elements, then, for each auxiliary density, the $\mu^{th}$
component of the $i^{th}$ surface element satisfies the linear system given by,

$$y_i^\mu = \sum_{j=1}^{N} \int_{\Delta_j} G(\vec{x}, \vec{y}_i) dS_x \cdot \sigma_\mu(\vec{x}_j) \qquad (3)$$

With the same super matrix $\mathbf{G}_{ij}$ of integrals over the Green function, one can solve 3 different linear
systems to obtain the unknown surface charge densities. This is achieved in our program PBEST with
the aid of Lapack routines that call a hardware optimized BLAS library. Computations on a
tetrahedron composed of 17560 triangles take less than 5 minutes on an AMD Opteron 248 machine.
The integrals of the $\mathbf{G}_{ij}$ matrix are computed essentially exactly, as a special case of those for
hydrodynamics as described in Aragon[6].

The computation of the capacitance proceeds in identical fashion except that the total charge $\sigma_c$
satisfies a scalar integral equation analogous to eq. 1, with "1" as the left hand side. The super matrix
is used a total of four times to obtain the quantities of interest.

Once the unknown surface quantities have been computed, we may easily compute the polarizability
tensor and capacitance by an integration over the surface, or equivalently, by a discrete sum over the
surface elements. The expressions are:

$$\alpha^o_{\mu\nu} = 4\pi \int_{sp} x_\mu \sigma_\nu ds \qquad (4)$$

$$C = 4\pi \int_{sp} \sigma_c ds \qquad (5)$$

A convenient set of quantities that can be used to compare the BE methodology to analytical results is
the computation of the depolarization factors $L_j$, for a particle of volume $V_p$, from the eigenvalues of
the polarizability tensor,

$$L_j = \frac{V_p}{4\pi\alpha^o_{jj}} \qquad (6)$$

Increasing the number of boundary elements gives a better approximation to the exact surface, and
therefore the exact solution, at the cost of an increase in computational time and memory requirements.
To obtain the most precise values, the properties are computed for a series of values N boundary
elements, and then extrapolated vs. 1/N to an infinite number of surface triangles.

## 3. Results for Ellipsoids and Polygons

The precision of our results can be observed from the linearity of the extrapolations to infinite number of surface
elements. Our ellipsoids where triangulated by modifying the tessellation produced by Mathematica: every
quadrilateral element was divided into two triangles. For this case, no attempt was made to more finely define the
ends of a prolate or the edges of an oblate ellipsoid. In the case of an ellipsoid of revolution, figure 1 shows that
the extrapolations can be done to high precision.

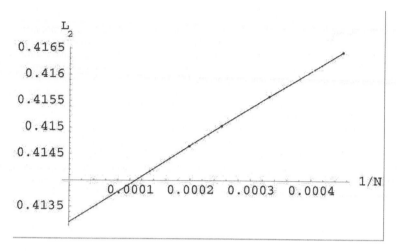

Figure 1: $L_2$ Extrapolation to infinite number of triangles. Oblate ellipsoid with axial ratio p = 2.

The results for ellipsoids are given as a function of axial ratio for prolate and oblate cases in Tables 1. The errors compared to the analytical results[8] are insignificant at a few parts per million, demonstrating that the method is both accurate and precise.

Table 1: Depolarization Factors of Ellipsoids of Revolution

| p | $L_1$(PBEST) | $L_1$ | % Err. | $L_2$, $L_3$(PBEST) | $L_2$, $L_3$ | % Err. |
|---|---|---|---|---|---|---|
| 1/10 | 0.86085 | 0.86080 | 0.005 | 0.069580 | 0.069598 | -0.026 |
| 1/9 | 0.84734 | 0.84729 | 0.006 | 0.076381 | 0.076356 | 0.032 |
| 1/8 | 0.83092 | 0.83087 | 0.006 | 0.084591 | 0.084565 | 0.032 |
| 1/7 | 0.81054 | 0.81051 | 0.004 | 0.094754 | 0.094747 | 0.007 |
| 1/6 | 0.78465 | 0.78459 | 0.008 | 0.10769 | 0.10771 | -0.020 |
| 1/5 | 0.75051 | 0.75048 | 0.003 | 0.12476 | 0.12476 | 0.000 |
| 1/4 | 0.70366 | 0.70364 | 0.003 | 0.14818 | 0.14818 | 0.000 |
| 1/3 | 0.63541 | 0.63539 | 0.004 | 0.18235 | 0.18231 | 0.022 |
| 1/2 | 0.52723 | 0.52720 | 0.006 | 0.23637 | 0.23640 | -0.013 |
| 1 | 0.33332 | 0.33333 | -0.003 | 0.33332 | 0.33333 | -0.003 |
| 2 | 0.17355 | 0.17356 | -0.006 | 0.41321 | 0.41322 | -0.002 |
| 3 | 0.10872 | 0.10871 | 0.009 | 0.44567 | 0.44565 | 0.006 |
| 4 | 0.075434 | 0.075407 | 0.036 | 0.46235 | 0.46230 | 0.012 |
| 5 | 0.055873 | 0.055821 | 0.093 | 0.47211 | 0.47209 | 0.005 |
| 6 | 0.043243 | 0.043230 | 0.030 | 0.47841 | 0.47839 | 0.006 |
| 7 | 0.034623 | 0.034609 | 0.042 | 0.48273 | 0.48270 | 0.007 |
| 8 | 0.028432 | 0.028421 | 0.037 | 0.48580 | 0.48580 | 0.000 |
| 9 | 0.023812 | 0.023816 | -0.016 | 0.48807 | 0.48809 | -0.004 |
| 10 | 0.020291 | 0.020286 | 0.024 | 0.48980 | 0.48986 | -0.012 |

In Table 2 we compare values computed for the polarizability and capacitance (per unit length) of platonic solids with those of the path integral method[1,4] and other implementations of the boundary element method[3].

Table 2: Capacitance and Polarizability of Platonic Solids

| Platonic Solid | C/L | | | | $[\sigma]_\infty$ | | |
|---|---|---|---|---|---|---|---|
| | PBEST | | Path Int. | Read | PBEST | | Path Int. |
| | *Uniform* | *Edged* | | | *Uniform* | *Edged* | |
| tetrahedron | 0.356467 | 0.356806 | 0.35680(3) | 0.35651(4) | 5.01445 | 5.02911 | 5.029(1) |
| cube | 0.66034 | 0.660628 | 0.66069(2) | 0.6606767(4) | 3.63865 | 3.64317 | 3.6437(6) |
| octahedron | 0.509195 | 0.509458 | 0.50945(3) | | 3.54455 | 3.5507 | 3.5509(2) |
| dodecahedron | 1.24624 | 1.24647 | 1.24648(5) | | 3.17618 | 3.1778 | 3.1779(3) |
| icosahedron | 0.815805 | 0.815894 | 0.81584(3) | | 3.12962 | 3.13068 | 3.1305(4) |

First we see that all methods yield values that are quite good and the differences do not arise till the fourth decimal place. The data in Table 2 show that the details of the triangulation method are important. In our work (PBEST), a triangulation method in which the triangles have uniform size everywhere on the face of the polyhedron (Fig 2a) as the refinement continues is shown to be less accurate than the edge enhanced triangulation (*Edged*). This triangulation is shown in Figure 2b. In the *Edged* case, the triangulation is not uniform but is always finer near the edges and corners than everywhere else on the faces. The number of elements required to yield accuracy beyond 4 decimals is large: over 35,000 triangles for the tetrahedron and 17,650 triangles for the cube using PBEST. At the most stringent precision and accuracy requirements we find that the Path Integral method and the BE method are equivalent if the triangulations are done with care, and numbers can be reliably obtained to five decimal places. The tetrahedron work of Read[3], using the BE method, does not agree with ours or with the path integral methods at high precision. This is probably due to the use of a mixture of rectangular (near the edges) and triangular surface elements by this group, making the extrapolation to N infinite less reliable at the highest levels of precision. It is not clear that the precision implied in their work on the cube (with about 20,000 surface elements) is trustworthy as stated, given the problem with the tetrahedron case.

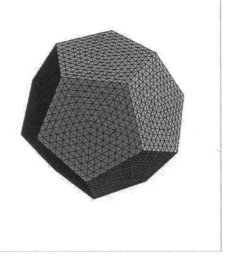

Figure 2a: Uniform triangulation of the dodecahedron into 3840 platelets.

Figure 2b: Edge-enhanced triangulation of the dodecahedron into 4740 platelets.

## 4. Proteins

The triangulation method utilized for proteins has been described previously[6]. Since proteins are not conducting bodies, and we can only compare with experimental measurements, we have to introduce an effective dielectric constant for the body. We are aware that a computation of the polarizability taking into account the dielectric constant is feasible[9]; here we make a simple approximation that nevertheless turns out to work well. The ansatz corresponds to using the $\varepsilon$ dependence of the polarizability of ellipsoids[8] with the actual depolarization tensor computed for the protein using equation 6 for each of the tensor components of $\alpha_\infty$. The polarizability component as a function of the relative dielectric constant $\varepsilon = \varepsilon_b/\varepsilon_m$ (where b = body, m = medium) is given by,

$$\alpha_{ij}(\varepsilon) = \frac{(\varepsilon-1)V_p/4\pi}{1+(\varepsilon-1)L_{ij}} \qquad (7)$$

and the tensor components of the depolarization factors are computed from the conducting body polarizability, *in the crystal frame*, with an extension of eq. 6:

$$L_{ij} = \frac{V_p}{4\pi\alpha_{ij}^\infty} \qquad (8)$$

Eq. (7) is exact for triaxial ellipsoids where the off diagonal depolarization factors are zero. The BE method, taking the ellipsoid as a conducting body, computes the exact depolarization factors as shown in Table 1 above. To produce a rigorous $\alpha_{ij}(\varepsilon)$ tensor we would ideally compute the diagonal terms from the diagonalized $\alpha_\infty$ tensor, and then rotate it to the proper orientation in the crystal frame. However, we only know the orientation of the conducting body tensor in the crystal frame, and these two tensors do not necessarily have the same orientation; assuming so would be an approximation. Since eq. 7 for non-ellipsoidal shapes is an approximation to begin with, we utilize the simple ansatz shown instead to generate the approximate tensor in the crystal frame. A precise computation of $\alpha_{ij}(\varepsilon)$ using the methods of Senior[9] will eliminate all of these approximations –this work is being pursued in our laboratory.

A suitable measurement that we can compare with is the Kerr constant in dilute aqueous solution. The measurement of the electric birefringence is described elsewhere[10]. This quantity, however also depends on the protein dipole moment, and in most of the examples below, this quantity dominates the response by comparison to the dc electric polarizability, but is directly dependent on the polarizability in the optical regime. The specific Kerr constant, $K_{sp}$, can be computed by a formula due to Wegener[11], adapted here to the case of no electro-hydrodynamic coupling, and without the approximation that the relative index of refraction is unity. Then we have,

$$K_{sp} = \left(\frac{n^2+2}{3}\right)\left(\frac{3Tr[\alpha(\varepsilon_{opt})\cdot(\mu\mu + kT\alpha(\varepsilon)] - Tr[\alpha(\varepsilon_{opt})]Tr[\mu\mu + kT\alpha(\varepsilon)]}{60n^2\varepsilon_0 V_p k^2 T^2}\right) \quad (9)$$

The formula depends on the index of refraction of the solvent, n, the relative dielectric constant for protein/water in the optical range (a laser with wavelength in the visible is used to probe the molecular orientation) $\varepsilon_{opt} = 1.2 = (1.45/1.33)^2$, and the equivalent ratio in the dc or zero frequency range, $\varepsilon = 4/80$[12]. A typical value for the index of refraction of proteins was obtained from the work Willner[13]. Eq. 7 is used for both of these ranges, using these values. The contribution of the dc polarizability, $\alpha(\varepsilon)$, is very small compared to the dipole moment in most of these cases. $K_{sp}$ has been evaluated for a series of proteins and the data is shown in Table 4 along with a comparison with experiment. The dipole moment has been either measured, or computed with Amber 8, however, its orientation in the crystallographic frame is not known. The expression for $K_{sp}$ is written in terms of tensors (and the dyadic $\mu\mu$) and is sensitive to the relative orientation of the frames defining the polarizability and the dipole moment. The polarizabilities have been computed in the crystallographic frame. The eigenvalues of the polarizability tensor and the Euler angles required to rotate this tensor into the crystallographic frame are given in Table 4. The dipole moment vector has been represented by $\mu = \mu(\hat{x}Sin\,\theta Cos\,\phi + \hat{y}Sin\,\theta Sin\,\phi + \hat{z}Cos\,\theta)$. Table 3 shows the orientation that the dipole moment must have in that frame in order to have our calculation match experiment precisely. In all cases, the azimuth angle $\phi = 0$. The match can be done in all cases except for β-lactoglobulin. In most cases, we obtain perfect agreement with experiment using a small angular deviation of the dipole moment orientation from the crystallographic z-axis. Given that proteins are globular in shape, that they should pack with their dipole moment closely aligned to the crystallographic axis makes sense. The largest deviation occurs for tubulin, which is not globular but axially extended. In this case, the dipole moment appears to point a little away from the molecular axis. In the case of ncd, the negative theta angle gives rise to the observed negative birefringence.

Table 3: Protein Specific Kerr constants

| Protein | Dipole | | θ | $V_p$ | $K_{sp}$ x $10^{-16}$m$^2$/V$^2$ | | |
|---------|--------|-----------|--------|---------|------|------|-----------|
| | *Debye* | *Reference* | *Degree* | *Å$^3$* | *Calc* | *Exp* | *Reference* |
| tubulin | 1443 | 14 | 11.8 | 85909.7 | 8.14 | 8.13 | 15 |
| tropomyosin | 6300 | 16 | 3.7 | 29672.0 | 31.1 | 31.1 | 16 |
| ovalbumin | 305 | 17 | 4.2 | 41397.5 | 0.516 | 0.515 | 18 |
| lysozyme | 122 | 19 | 3.6 | 13656.7 | 0.121 | 0.121 | 18 |
| ribonuclease a | 350 | 20 | 1.7 | 12993.0 | 0.212 | 0.213 | 18 |
| β-lactoglobulin | 790 | 21 | 0 | 34201.0 | 0.368 | 0.296 | 18 |
| kinesin(1-349) | 1042 | 14 | 0.54 | 36857.4 | 0.364 | 0.36 | 22 |
| ncd(335-700) | 331 | 14 | -5.3 | 35631.0 | -1.65 | -1.65 | 22 |

Table 4: Eigenvalues (Å$^3$) and Eigenvector Matrix Euler Angles[a] (°) of the Protein Polarizabilty Tensor

| Protein | $\omega_1$ | $\omega_2$ | $\omega_3$ | $\alpha$ | $\beta$ | $\gamma$ |
|---|---|---|---|---|---|---|
| tubulin | 65000 | 31000 | 38000 | −81.7 | −9.28 | 72.5 |
| tropomyosin | 10800 | 14100 | 1200000 | −60.9 | −40.8 | 70.4 |
| ovalbumin | 29800 | 14500 | 13900 | −130.0 | 16.5 | 174.0 |
| lysozyme | 3610 | 4280 | 7600 | −17.3 | −43.4 | 22.3 |
| ribonuclease a | 3920 | 7921 | 5099 | −29.0 | 27.0 | −5.66 |
| β-lactoglobulin | 11820 | 12950 | 29300 | −53.7 | 26.1 | 106.0 |
| kinesin(1-349) | 14010 | 27380 | 12890 | −33.8 | −12.5 | 67.8 |
| ncd(335-700) | 25200 | 15800 | 13500 | −105.0 | 48.0 | 60.6 |

[a]Obtained by equating the elements of the eigenvector matrix with those of the euler rotation matrix,

$$A(\alpha,\beta,\gamma) = \begin{pmatrix} \cos\gamma\cos\beta\cos\alpha-\sin\gamma\sin\alpha & \cos\gamma\cos\beta\sin\alpha+\sin\gamma\cos\alpha & -\cos\gamma\sin\beta \\ -\sin\gamma\cos\beta\cos\alpha-\cos\gamma\sin\alpha & -\sin\gamma\cos\beta\sin\alpha+\cos\gamma\cos\alpha & \sin\gamma\sin\beta \\ \sin\beta\cos\alpha & \sin\beta\sin\alpha & \cos\beta \end{pmatrix}$$

where the angle $\alpha$ represents rotation around the crystallographic z-axis, $\beta$ represents rotation around the resulting x-axis, and $\gamma$ represents rotation around the resulting z-axis.

## 5. Conclusions

In conclusion, we observe that our implementation of the BE method is very precise and accurate and that we can obtain excellent agreement with measured Kerr constants of proteins based on measured dipole moments and computed polarizability tensors.

## Acknowledgments

This research was supported through a grant from the National Institutes of Health, MBRS SCORE Program - Grant #S06 GM52588 to SA.

## References

[1] J.F. Douglas and E.J. Garboczi, Intrinsic Viscosity and the polarizability of particles having a wide range of shapes, *Advances in Chemical Physics* (Editors. I. Prigogine and S.R. Rice), John Wiley & Sons, New York, Vol 91, 85-151, 1995.

[2] F.H. Read, Capacitances and singularities of the unit triangle, square, tetrahedron and cube, COMPEL, 23, 572-578, 2003.

[3] E. Goto, Y. Shi, and N. Yoshida, Extrapolated surface charge method for capacity calculation of polygons and polyhedra, *J. Comput. Phys.*, 100, 105-115, 1992.

[4] M.L. Mansfield, J.F. Douglas, and E.J. Garboczi, Intrinsic viscosity and the electrical polarizabililty of arbitrarily shaped objects, *Phys. Rev. E.*, 64, 061401:1-16, 2001.

[5] S.R. Aragon and D.K. Hahn, The Pre-Averaged Hydrodynamic Interaction Revisited via Boundary Element Computations, *J. Chem. Theory Comput.* (submitted, 2005).

[6] S.R. Aragon, A precise boundary element method for macromolecular transport properties, *J. Comput. Chem.*, 25, 1191-1205, 2004.

[7] H-X. Zhou, Calculation of translational friction and intrinsic viscosity. I. General formulation for arbitrarily shaped particles, *Biophys. J.*, 69, 2286-2297, 1995.

[8] C.F. Bohren and D.R. Huffman, *Absorption and scattering of light by small particles*, John Wiley & Sons, New York, 1983.

[9] D.F. Herrick and T.B.A. Senior, The dipole moments of a dielectric cube, *IEEE Transactions on antennas and propagation*, X, 590-592, 1977.

[10] S. Highsmith and D. Eden, Transient electric birefringence of heavy meromyosin, *Biochem,* 24, 4917, 1985.

[11] W. A. Wegener, Transient electric birefringence of dilute rigid-body suspension at low field strengths, *J. Chem. Phys.* 84, 59896004, 1986.

[12] H. –X. Zhou, Boundary element solution of macromolecular electrostatics: interaction energy between two proteins, *Biophys. J.*, 65, 955-963, 1993.

[13] M. Zayats, D. A. Raitman, V. I. Chegel, A. B. Kharitonov, I. Willner, Probing antigen-antibody binding processes by impedance measurements on ion-sensitive field-effect transistor devices and complementary surface plasmon resonance analyses: development of cholera toxin sensors, *Anal. Chem.*, 74, 4763-4773, 2002.

[14] S.R. Aragon and D.K. Hahn, unpublished results.

[15] W. Bras, G. P. Diakun, G. Maret, H. Kramer, J. Bordas, and F. J. Medrano, The susceptibility of pure tubulin to high magnetic fields: a magnetic birefringence and x-ray fiber diffraction study, *Biophys. J.*, 74, 1509-1521, 1998.

[16] C. A. Swenson, N. C. Stellwagen, Electric birefringence studies of rabbit tropomyosin, *Biopolymers*, 27, 1127-1141, 1988.

[17] S. Takashima, A study of proton fluctuation in protein. Experimental study of the Kirkwood-Shumaker theory, *J. Phys. Chem.*, 69, 2281-2286, 1965.

[18] S. Krause and C. T. O'Konski, Electric properties of macromolecules. VIII. Kerr constants and rotational diffusion of some proteins in water and in glycerol–water solutions, *Biopolymers*, 1, 503-515, 1963.

[19] S. Takashima and K. Asami, Calculation and measurement of the dipole moment of small proteins: Use of protein database, *Biopolymers*, 33, 59-68, 1993.

[20] E. H. Grant and S. E. Keefe, Dipole moment and relaxation time of ribonuclease, *Phys. Med. Biol.*, 19, 701-707, 1974.

[21] T. M. Shaw, E. F. Jansen, H. Lineweaver, The dielectric properties of β-lactoglobulin in aqueous glycine solutions and in the liquid crystalline state, *J. Chem. Phys.*, 12, 439-448, 1944.

[22] D. Eden, B. Q. Luu, D. J. Zapata, E. P. Sablin, F. J. Kull, Solution structure of two molecular motors: nonclaret disjunctional and kinesin, *Biophys. J.*, 68, 59s-65s, 1995.

Brill Academic Publishers
P.O. Box 9000, 2300 PA Leiden
The Netherlands

*Lecture Series on Computer
and Computational Sciences*
Volume 4, 2005, pp. 33-36

# Flow Cessation of Viscoplastic Fluids in Poiseuille Flows

Ioannis Argyropaidas, Evan Mitsoulis[1]

School of Mining Engineering and Metallurgy,
National Technical University of Athens,
GR-157 80 Zografou, Greece

Georgios Georgiou

Department of Mathematics and Statistics,
University of Cyprus,
PO Box 20537, CY-1678 Nicosia, Cyprus

Received 15 July, 2005; accepted in revised form 10 August, 2005

*Abstract:* Viscoplastic fluids exhibit both viscous and plastic behaviour having a yield stress. The Bingham model is the most famous viscoplastic model, which describes the behaviour of materials with a yield stress. The cessation of a pressure-driven Poiseuille flow of a Bingham plastic is studied to deduce finite stopping times unlike the infinite times taken for Newtonian fluids. The numerical solution employs finite elements in space and a fully implicit scheme in time. The numerical calculations confirm previous theoretical findings that a fluid with nonzero yield stress will stop after a finite time which is called stopping time. The decay of the volumetric flow rate per unit area, which is exponential in the zero yield-stress case, is accelerated and eventually becomes linear as the yield stress is increased.

*Keywords:* Poiseuille Flow; Bingham Plastic; Yield Stress; Viscoplasticity; Cessation

## 1. Introduction

Viscoplastic fluids have characteristics of viscous fluids and elastic solids due to the presence of a yield stress. Such fluids are polymer gels, slurries, sludge, mud, foodstuff, etc. [1]. In pressure-driven flows, one can bring the fluid to a halt by reducing the applied pressure gradient to zero in Poiseuille flows. In a Newtonian fluid, the corresponding velocity fields decay to zero in an infinite amount of time [2,3]. In a Bingham plastic, the velocity fields go to zero in a finite time, which emphasizes the role of the yield stress [3,4]. The objective of the present work is to compute numerically the stopping times and make comparisons with theoretical upper bounds for the cessation of the axisymmetric Poiseuille flow of a Bingham fluid, including both the case of flow in a tube (Fig. 1a) and in an annulus (Fig. 1b).

## 2. Governing equations

The flow is governed by the usual Navier-Stokes conservation equations of mass and momentum. Because of the simple shear flow, the analysis for the time-dependent problem assumes one-dimensional field in space, with the axial velocity $u_z(r,t)$. We assume that at $t=0$ the velocity $u_z(r,t)$ is given by the steady-state solution and that at $t=0^+$ the steady-state pressure gradient, $(-dp/dz)^s$, is set to zero. The dimensionless form of the $z$-momentum equation is

$$\frac{\partial u_z}{\partial t} = f + \frac{1}{r}\frac{\partial}{\partial r}\left(r\tau_{rz}\right), \tag{1}$$

where lengths are scaled by the gap $H = r_o - r_i$ for annular and by $H = R$ for tube, velocity by the mean steady-state velocity $V$, pressures $p$ and shear stress components $\tau_{rz}$ by $\mu V/H$, and time by $\rho H^2/\mu$, where $\rho$ is the constant density of the fluid and $f$ is the dimensionless pressure

---

[1] Corresponding author. E-mail: mitsouli@metal.ntua.gr

gradient. Instead of the ideal Bingham model [1], we use the regularized constitutive equation proposed by Papanastasiou [5], which in dimensionless form is given by

$$\tau_{rz} = \left\{ \frac{Bn\left[1 - \exp\left(-M\,\dot{\gamma}\right)\right]}{\dot{\gamma}} + 1 \right\} \frac{du_z}{dr}, \tag{2}$$

where $\dot{\gamma} = |du_z/dr|$, while the Bingham number $Bn$ and the stress growth exponent $M$ are given by

$$Bn \equiv \frac{\tau_0 H}{\mu V} \quad \text{and} \quad M \equiv \frac{mV}{H}, \tag{3}$$

where $m$ is a stress growth exponent, $\tau_0$ is the yield stress and $\mu$ is a constant viscosity.

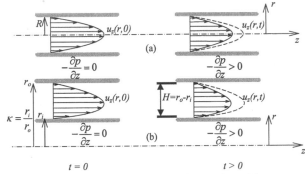

Figure 1: Cessation of Poiseuille flow in a: (a) tube, (b) annulus with aspect ratio $\kappa$.

## 3. Numerical results

Since there are no analytical solutions to the flows under study, in the case of the Bingham plastic or the Papanastasiou model, the Galerkin Finite Element Method (G/FEM) has been used. Rectangular meshes of 9-node Lagrangian two-dimensional elements were used with quadratic basis functions for the velocities and linear basis functions for the pressure ($u$-$v$-$p$ formulation). The element aspect ratio $r/z$ was 5, with 50 elements in the $r$-direction and 5 elements in the $z$-direction. In both directions the nodes were equally spaced. For the time discretization, we used the standard fully implicit Euler backward-difference scheme with a dimensionless time step $dt \leq 10^{-3}$. At each time step, the nonlinear system of discretized equations was solved using Picard iterations (direct substitution). The criteria for convergence of the system of equations were that the norm of the error for the velocities and the norm of the residuals were both less than $10^{-4}$.

The general stopping criterion for the "numerical cessation" of the flow was that the norm of the dimensionless deceleration was less than a small number $\varepsilon$. Here $\varepsilon = (\rho H^2/\mu V)10^{-3}$, where $\mu V/\rho H^2$ is the *characteristic deceleration*. The total dimensionless time, $T_f$, found with the above criterion and with constant $dt = 10^{-3}$, is given in all figures where transient solutions are shown. Finally, in any flow of a Bingham plastic, determination of the yielded ($\tau_{rz} \geq \tau_0$) and unyielded ($\tau_{rz} \leq \tau_0$) regions in the flow field is necessary. The yield point locations are determined accurately by quadratic interpolation in the $r$-direction over the nearest nodes where the difference $\tau_{rz} - \tau_0$ changes sign.

The velocity profiles for tube and for annulus with $\kappa = 0.3$ are given in Figure 2 as a function of the normalized radius

$$\bar{r} = (r + R)/2R \quad \text{(tube)}, \qquad \bar{r} = (r - r_i)/(r_o - r_i) \quad \text{(annulus)}, \qquad \bar{r} \in [0,1]. \tag{4}$$

The yield points are shown as symbols connected with lines. In all figures $M = 200$.

The increasing dimensionless times shown are respectively 0%, 10%, 30%, 50%, 70%, 90% and 100% of $T_f$. The unyielded regions increase with time for the same $Bn$ number and $\kappa$ ratio. They also increase substantially with $Bn$ for the same time and $\kappa$ ratio. The times to reach $T_f$ decrease with increasing $Bn$.

Figure 3 shows the evolution of dimensionless mean velocity, $u_{avg}$, for various Bingham numbers, for tube and for annulus with ratio $\kappa = 0.3$. These curves show the dramatic effect of the yield stress, which accelerates the cessation of the flow. The effect of $\kappa$ is small with smaller values needing more time, all other parameters being equal. In the Newtonian case ($Bn = 0$) and for small Bingham numbers, the decay of the mean velocity is exponential at least initially. At higher Bingham numbers, the decay of $u_{avg}$ becomes polynomial and eventually linear.

The times at which $u_{avg} = (10^{-3}, 10^{-5})$ for a flow in a tube and in an annulus with various $\kappa$-values are plotted as functions of the Bingham number in Fig. 4. The two times coincide for moderate or large Bingham numbers, which indicates that the flow indeed stops at a finite time.

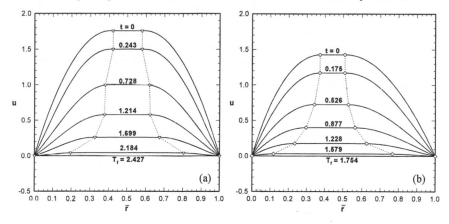

Figure 2: Evolution of the velocity in cessation of Poiseuille flow of a Bingham fluid with $Bn=1$ and $M=200$: (a) tube, (b) annulus with ratio $\kappa=0.3$.

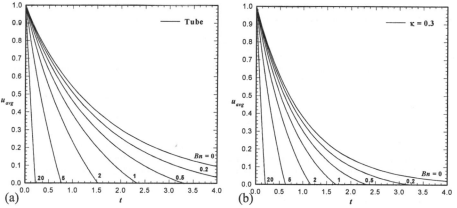

Figure 3: Evolution of the volumetric flow rate per unit area during the cessation of Poiseuille flow (a) in a tube and (b) in an annulus with $\kappa = 0.3$ of a Bingham fluid with $M=200$ and various Bingham numbers.

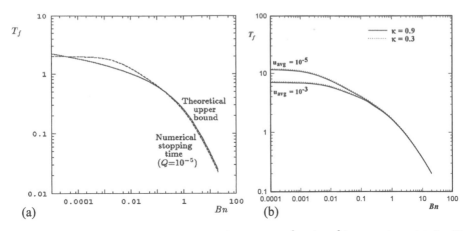

Figure 4: Comparison of calculated times for $u_{avg} = 10^{-3}$ and $10^{-5}$ in cessation of Poiseuille flow with calculated times for $Q=10^{-5}$ (a) in a tube and (b) in an annulus with different $\kappa$-values. Bingham fluid with $M=200$. The theoretical upper bound for the tube is from eq. (5).

For the case of the tube, the theoretical upper bound has been found to be [4]

$$T_f = \frac{1}{a_1^2} \ln \left\{ a_1^2 \frac{\| u_z(r,0) \|}{2Bn - f} + 1 \right\}, \quad f < 2Bn, \tag{5}$$

where $\qquad a_1^2 \cong 5.7831, \quad \| u_z(r,0) \| = \left[ 2 \int_0^1 u_z^2(r,0) r dr \right]^{1/2}$ (6)

For the annular flow, the corresponding bound is the subject of a current investigation.

## Acknowledgments

Part of this research is supported by the "HERAKLEITOS" program of the Ministry of Education and Religious Affairs of Greece (#68/0655). The Project is co-funded by the European Social Fund (75%) and National Resources (25%).

## References

[1]    R.B. Bird, G.C. Dai and B.J. Yarusso, The rheology and flow of viscoplastic materials, *Rev. Chem. Eng.* **35** 1-70 (1983).

[2]    T. Papanastasiou, G. Georgiou and A. Alexandrou, *Viscous Fluid Flow,* CRC Press, Boca Raton, 1999.

[3]    R. Glowinski, *Numerical Methods for Nonlinear Variational Problems*, Springer-Verlag, New York, 1984.

[4]    R.R. Huilgol, B. Mena and J.-M. Piau, Finite stopping time problems and rheometry of Bingham fluids, *J. Non-Newtonian Fluid Mech.* **102**, 97-107 (2002).

[5]    T.C. Papanastasiou, Flows of materials with yield, *J. Rheol.* **31** 385-404 (1987).

Brill Academic Publishers
P.O. Box 9000, 2300 PA Leiden,
The Netherlands

*Lecture Series on Computer
and Computational Sciences*
Volume 4, 2005, pp. 37-43

# Quantum Nuclear Scattering calculations on the BEgrid

**F. Arickx**[1]**, J. Broeckhove, K. Vanmechelen, P. Hellinckx, V. Vasilevsky**

University of Antwerp,
Department of Mathematics and Computer Science,
Middelheimlaan 1, BE-2020 Antwerp, Belgium.

Received 9 July, 2005; accepted in revised form 31 July, 2005

*Abstract:* We report on user experiences gathered with quantum nuclear scattering calculations on the BEgrid, the Belgian EGEE Grid infrastructure. We discuss aspects such as Grid utilization, ease-of-use and porting the application to the BEgrid environment.

*Keywords:* Nuclear scattering, Cluster models, J-Matrix methodology, Grid computation

*PACS:* 21.10., 21.60.Gx, 21.60.Ev, 25.55.ci, 24.30.Gd.

## 1  Introduction

This contribution presents the user experiences gathered in gridifying an application for calculating quantum nuclear scattering. The grid platform is the BEgrid which hosts the Belgian part of the EGEE project. As a member of the BEgrid consortium, our research group has experience in its procedures and maintenance, as well as in gridifying applications.

## 2  The Scattering Problem

The $^5H$ nucleus has a large neutron excess and lies beyond the neutron drip line. It has, in the last five years, been the object of quite a few experimental investigations aimed at finding a clear evidence of the existence of resonance structures in $^5H$.

Different theoretical models and methods have been used to calculate the energy and width of the resonances. In this contribution we focus on an application of the microscopic three-cluster model [1] formulated in the context of the Modified J-Matrix method described in [2], and model $^5H$ with the $^3H + n + n$ three-cluster configuration.

The three-cluster wave function can be written as

$$\Psi_{SL;JM} = \widehat{\mathcal{A}} \left\{ \left[ \Phi_1 \left( t \right) \Phi_2 \left( n \right) \Phi_2 \left( n \right) \right]_S \phi_L \left( \mathbf{q}_1, \mathbf{q}_2 \right) \right\}_{JM} \tag{1}$$

where $\Phi_1 \left( t \right)$ is an antisymmetric shell-model wave function describing the internal structure of $^3H$ with three nucleons in the $s$-shell. The neutron wave function $\Phi_2 \left( n \right)$ only includes spin and isospin variables of the neutron. $\widehat{\mathcal{A}}$ stands for the overall antisymmetrization operator. The relative behavior of clusters can be described by two sets of Jacobi coordinates as shown in Figure 1.

---

[1]Corresponding author. E-mail: Frans.Arickx@ua.ac.be

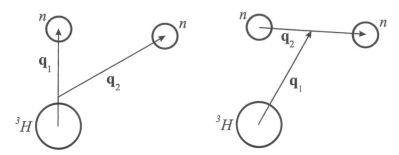

Figure 1: Possible choices of Jacobi coordinates for the $^3H + n + n$ configuration in $^5H$.

The inter-cluster wave function $\phi_L(\mathbf{q}_1, \mathbf{q}_2)$ of relative three-cluster motion is to be determined by solving the Schrödinger equation with the appropriate boundary conditions. We therefore expand the wave function $\phi_L(\mathbf{q}_1, \mathbf{q}_2)$ onto a Hyperspherical Harmonic basis:

$$
\begin{aligned}
\phi_L(\mathbf{q}_1, \mathbf{q}_2) &= \sum_{l_1,l_2} \phi_{l_1,l_2;L}(\rho,\theta) \{Y_{l_1}(\mathbf{q}_1) Y_{l_2}(\mathbf{q}_2)\}_{LM} \\
&= \sum_{K,l_1,l_2} \phi_{K;l_1,l_2;L}(\rho) \chi_K^{(l_1,l_2)}(\theta) \{Y_{l_1}(\mathbf{q}_1) Y_{l_2}(\mathbf{q}_2)\}_{LM}
\end{aligned}
\tag{2}
$$

Hypermomentum $K$ and partial angular momenta $l_1$ (along $\mathbf{q}_1$) and $l_2$ (along $\mathbf{q}_2$) define the three-cluster geometry, and characterize the different scattering channels. These three quantum numbers are collectively denoted as $c = \{K; l_1, l_2\}$. The hyperradial wave function is expanded onto the basis of the 6-dimensional radial oscillator:

$$
\phi_{c;L}(\rho) = \sum_{n_\rho} C_{n_\rho}^{(c;L)} \Phi_{n_\rho}^{(K)}(\rho)
\tag{3}
$$

It should be noted here that all spectroscopic calculations - bound state energies, phase shifts, cross sections etc. - are independent of the specific choice of Jacobi coordinate system (see Figure 1) if one uses a complete set of Hyperspherical Harmonics for given momentum $K$ (i.e. all possible values of partial angular momenta compatible with this $K$).

We solve the Schrödinger equation by substituting (1) as an ansatz with (2), (3) and obtain a matrix equation in the expansion coefficients $C_{n_\rho}^{(c;L)}$ (see [1]). Because of the oscillator expansion, we can obtain the solution through the Modified J-Matrix approach [2]. This is an implementation of the Resonating Group Method in function space. As in coordinate space, we distinguish an interaction and an asymptotic region, but now in the space of expansion coefficients. The latter are split in two subsets: the first set represents the three-cluster wave function in the internal region and is determined by solving the Schrödinger matrix equation; the second set is connected with the asymptotic form of the wave functions and has to represent the appropriate boundary conditions. The asymptotic set for a scattering boundary condition is obtained from the solution of reference hamiltonians describing three non-interacting clusters. These hamiltonians are defined in [1]. There will be $N_c$ (the number of channels) independent solutions, with an asymptotic behavior for large hyperradius $\rho$

$$
\Psi_c^{(c_0)} \rightarrow \Phi_1(A_1) \Phi_2(A_2) \Phi_3(A_3) \sum_c f_c^{(c_0)}(\rho) \chi_c(\Omega_5)
\tag{4}
$$

where $c_0$ refers to the incoming channel, and $f_c$ denotes the asymptotic expansion coefficients, defined by

$$f_c^{(c_0)}(\rho) \to \delta_{c_0,c}\psi_c^{(-)}(k\rho) - S_{c_0,c}\psi_c^{(+)}(k\rho) \tag{5}$$

Here $\psi_c^{(-)}(k\rho)$ $(\psi_c^{(+)}(k\rho))$ is the incoming (outgoing) channel wave function and $S_{c_0,c}$ is the $S$-matrix describing the transition from the initial channel $c_0$ to the final channel $c$.

The resonance parameters can be deduced from the eigenphase shifts, obtained by diagonalizing the $S$-matrix. In this eigenchannel representation one has ($\alpha$ enumerates the uncoupled eigenchannels)

$$S_\alpha = \exp\{2i\delta_\alpha\}, \qquad \alpha = 1, 2, \ldots N_c$$

The relation between the original $\|S_{c,c'}\|$ and diagonal $\|S_\alpha\|$ forms of the $S$- matrix is

$$S_{c,c'} = \sum_\alpha U_\alpha^c S_\alpha U_\alpha^{c'}$$

with $\|U_\alpha^c\|$ an orthogonal matrix. The extraction of resonance position and width is done in the traditional way by

$$\left.\frac{d^2\delta_\alpha}{dE^2}\right|_{E=E_\alpha} = 0, \quad \Gamma = 2\left(\frac{d\delta_\alpha}{dE}\right)^{-1}\bigg|_{E=E_\alpha} \tag{6}$$

The problem to be solved is thus essentially twofold: (1) set up the matrix equation by substituting (1), (2) and (3) for the solution of the schrödinger equation, i.e. calculate the energy matrix, and (2) solve this equation subject to the appropriate boundary conditions. A major advantage of the current approach is that an explicit representation of the wave function is obtained. This allows for a detailed analysis of the resonance wave functions, as well as for the possibility to calculate additional physical quantities at the resonance energies.

In this contribution we will focus on resolving the first problem in a distributed way for an extensive basis. We will present a (limited) solution of the system though to prove the validity of the obtained energy matrix.

The energy matrix to be determined can be denoted by ($c = \{K; l_1, l_2\}$)

$$\left\langle c_i; L, n_i \left| \hat{H} \right| c_j; L, n_j \right\rangle = \left\langle K_i, l_i, n_i \left| \hat{H} \right| K_j, l_j, n_j \right\rangle \tag{7}$$

where $\hat{H}$ is the Hamiltonian, or energy, operator, and $i$ and $j$ distinguish basis states; the right hand side in (7) simplifies the notation, omitting $L$ as an overall constant, and replacing the combination $(l_1 l_2)_i$ by $l_i$

The theory to obtain (7) [1] is well beyond the scope of this paper, but it can be broken down to

$$\left\langle\!\left\langle K_i, l_i \left| \hat{H} \right| K_j, l_j \right\rangle\!\right\rangle = \sum_t \sum_{l_r} \sum_{l_s} R(K_i, l_i, l_r, t) \left\langle\!\left\langle K_i, l_r \left| \hat{H} \right| K_j, l_s \right\rangle\!\right\rangle_t R(K_j, l_s, l_j, t) \tag{8}$$

where $\langle\!\langle\rangle\!\rangle$ stands for a matrix over all $n_i$, $n_j$ indices. The $R$ factors are so-called Raynal-Revai coefficients as discussed in [1], and the nature and range of index $t$ depends on the nucleus ($^5H$) and its cluster decomposition ($^3H + n + n$).

The granularity of the problem is clear from (8), and reduces the problem to a fork and join algorithm by calculating all independent $\left\langle\!\left\langle K_i, l_r \left| \hat{H} \right| K_j, l_s \right\rangle\!\right\rangle_t$ matrices for fixed $K_i, K_j$ and all allowed combinations $l_r, l_s, t$ (the fork), followed by a summation to obtain (8) (the join). In this paper we discuss a calculation for $L = 0$ and $K = 0, 2, 4, .., 16$, a range of $l_1 = l_2 = 0, 1, .., K/2$

values, and 45 $t$ values. All of the computational code components are implemented in Fortran90 using the Intel v7 compiler.

A master-worker distribution model seems an evident choice for this problem, because all of the fork tasks are independent of one another. The tasks are file based, meaning that they get their input from a series of files and write their results into one. All input files except one, a configuration file which contains the particular indices for the current computation, have a constant content for all tasks.

## 3    BEgrid

The BEgrid project involves an infrastructure collaboration between several universities and research institutions in Belgium. It is currently running LCG 2.2 middleware [3] and includes sites from the universities of Antwerp, Ghent, Leuven and Brussels as well as other research oriented institutions such as the Flanders Marine Institute. More institutions are expected to join on a short term basis. BEgrid is part of the North European region of the EGEE project. Many sites are currently in deployment phase, with 205 CPUs operational at the beginning of 2005.

The LCG middleware consists of a number of functional modules that co-operate to provide workload, data and information management services. A *resource broker* matches job requirements, which are expressed in JDL (Job Description Language), to the available resources. The user has the ability to state complex infrastructural and operational job requirements using a standardized information schema called GLUE. The broker also supports user-defined ranking of resources that fulfill the hard job requirements, based on soft QoS requirements such as the average queued time for a job at a compute resource. After a computing resource is selected by the broker, the job is forwarded to the *computing element* (CE) associated with that resource. The computing element virtualizes compute resources such as PBS, LSF or Condor farms, providing uniform access to heterogeneous compute back-ends.

In the same way as the CE provides a virtualization layer for compute resources, a *storage element* (SE) delivers such a layer for heterogeneous storage resources. A *replica management service* (RMS) supports file based data replication and orchestrated access to files stored across SEs. Although jobs may interact with the RMS to access large data files, small I/O requirements can be fulfilled by specifying input and output *sandboxes*. These allow the user to indicate which files should be staged in and out before and after the computation. All interactions with the Grid infrastructure take place from a *user interface* node which supports APIs, command line, and graphical interfaces to BEgrid.

The LCG middleware supports a batch oriented computing model, and is currently in process of providing better support for data dependencies between jobs and more tightly coupled computations. This does not affect the mapping of our problem to the LCG based computing infrastructure as jobs are trivially parallel. A characteristic that does affect our job distribution scheme is the fact that long delays have been observed when submitting jobs and receiving their results on a zero loaded computing resource. The computation time for individual indices also varies widely from a few seconds to a few hours. Therefore it was deemed necessary to group multiple calculations in a single job, in order to mitigate the performance penalty induced by these delays. To determine the size of these groupings we monitored the number of queued jobs over all CE queues during the computation and adjusted the group size accordingly. The idea is that as long as all CE queues are sufficiently loaded to keep the compute resources busy, the submission delay will not affect the total wall-clock time of the computation.

All file staging was performed using the input and output sandboxes since the calculations are compute intensive. After jobs finish, a polling component on the user interface fetches the job output.

## 4   Results

We have used 24 3.2 GHz Pentium 4 nodes at Antwerp and 8 dual-Opteron/244 nodes (1.8 GHz) at Leuven. With application level benchmarks we have determined that this amounts to 40.76 normalized 3.2 CPUs, which is the maximum reachable speedup factor on this setup. Both sites use a 1 Gigabit Ethernet interconnect. The user interface and resource broker were both located at Antwerp.

Table 1 shows the turn-around time of the distributed computations for different values of $K$, and compares these to the pure sequential execution timing. The speedup factor which denotes the ratio between sequential and distributed execution time is also shown for each K. The table depicts low speedup figures for computations with $K < 10$. More time is spent deploying the jobs than processing them. The reason is that most of the tasks with low $K$ have a very small execution time. From $K = 10$ onwards, the picture changes, as the mean execution time for these jobs rises significantly. The low speedup values can be attributed to job submission delays which are in the order of seconds, and result fetching delays which are in the order of minutes. These make it difficult to achieve optimal resource utilization in the beginning and at the end of the computation. The delays were not found to be caused by data transmission overhead but rather by the workload management system itself.

Table 1: Speedup achieved on the available BEgrid resources.

| K | sequential | distributed | speedup | utilization |
|---|---|---|---|---|
| 0 | 1s | 7m57s | 0.00 | 0.00 |
| 2 | 4s | 9m00s | 0.01 | 0.02 |
| 4 | 43s | 16m27s | 0.04 | 0.11 |
| 6 | 6m18s | 16m59s | 0.32 | 0.91 |
| 8 | 40m17s | 18m54s | 1.63 | 5.23 |
| 10 | 3h34m45s | 18m42s | 9.40 | 28.17 |
| 12 | 15h47m39s | 49m36s | 15.27 | 46.87 |
| 14 | 61h07m19s | 2h01m05s | 24.61 | 74.30 |
| 16 | 210h34m54s | 6h07m31s | 34.38 | 84.34 |

Turn-around timing is not the most appropriate measure if we wish to gauge the usefulness of distributing an application, as the amount of available resources varies significantly from one run to another. Therefore, we consider the utilization rate of the participating nodes. It is defined as the percentage of time the processor is performing the actual computation. This measure provides an indication of the overhead introduced by both the middleware and data transfers, and is largely independent of the hardware used. Table 1 lists the utilization as a function of $K$. It indicates strongly decreasing overhead for increasing $K$-values. Job submission and result fetching delays were the main causes for overhead.

In order to validate the results of the calculation we have solved the matrix equation and determined the eigenphases and their resonance properties for a maximal value of $K = 10$ and 25 oscillator shells in each channel. In Table 2 we show the convergence of the resonance properties as a function of $K$. One notices that the results are converging well, but that additional $K$-values are needed above 10 to obtain true stability. It should be noted that the converged results are in fair agreement with experiment.

Table 2: Convergence of resonance properties for $J^\pi = 1/2^+$ as a function of $K$ and 25 oscillator shells

| $K$ | $E$(MeV) | $\Gamma$(MeV) | $E$(MeV) | $\Gamma$(MeV) |
|---|---|---|---|---|
| 0 | - | - | - | - |
| 2 | 1.750 | 3.371 | 3.700 | 15.165 |
| 4 | 1.770 | 2.875 | 3.640 | 16.90 |
| 6 | 1.680 | 2.413 | 4.310 | 9.356 |
| 8 | 1.550 | 1.934 | 4.250 | 8.724 |
| 10 | 1.480 | 1.763 | 4.230 | 7.847 |

In Fig. 2 we show the eigenphases for the case $L = 0, J^\pi = 1/2^+$ where one clearly discerns the first resonance, and the start of the second one.

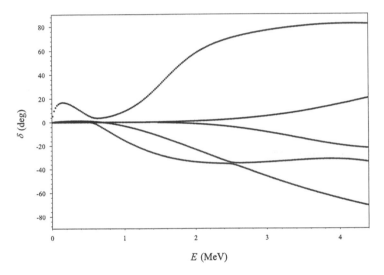

Figure 2: Eigenphase obtained for $L = 0, J^\pi = \frac{1}{2}^+$.

These results indicate that the procedure to calculate the energy matrix in a distributed way is successful and leads to physically relevant results for $^5H$ comparable to experiment.

## 5    Conclusion

One of the main merits of distributing the computation on BEgrid is the efficiency of the implementation process itself: only 230 lines of bash scripting code were needed to capture all job distribution and execution logic. Though the highly parallel and independent nature of the distributed subtasks is certainly of influence, this is also directly related to the fact that core Grid services for workload, data and information management as well as core security provisions, are delivered 'out-of-the-box'. Another important advantage is the fact that coding an application to a model and infrastructure accepted on a European scale, opens the possibility for accessing a large scale infrastructure. On the other hand, one should be aware that gaining access to a general purpose (inter)national Grid resource still requires "social"interactions for the application to be accepted at remote sites.

A potential shortcoming of BEgrid is the high job submission and result fetching delay which may affect applications that have to face a fine grained level of job distribution. Although our approach of wrapping multiple application level jobs into a single Grid job mitigated these effects, it was still difficult to achieve a good load balance at the beginning and near the end of the total computation.

In summary, the BEgrid platform is well suited for gridifying a highly parallel CPU-intensive application. The platform provides standardized well-defined services for distributive processing. This delimited set of services to approach the Grid is a merit of the system, because of its straightforward usage, as well as a shortcoming because of its inherent limitations. With respect to computing resources, it exhibits strong scalability, which is an important asset for large scale distributed applications.

## Acknowledgment

Support from the Fonds voor Wetenschappelijk Onderzoek Vlaanderen (FWO), G0488-03N, and the Afdeling Technologie en Innovatie van het Ministerie van het Vlaams Gewest is gratefully acknowledged. V. Vasilevsky is grateful to the Department of Mathematics and Computer Science of the University of Antwerp (UA) for hospitality.

## References

[1] V. S. Vasilevsky, A. V. Nesterov, F. Arickx and J. Broeckhove: *The Algebraic Model for Scattering in Three-s-cluster Systems: Theoretical Background.* Phys. Rev. C63 (2001) 034606:1–16

[2] J. Broeckhove, F. Arickx, W. Vanroose and V. Vasilevsky: *The Modified J-Matrix method for Short-Range Potentials.* J. Phys. A: Math. Gen 37 (2004) 1-13

[3] A. D. Peris, P. Mendez Lorenzo, F. Donno, A. Sciabà, S. Campana, R. Santinelli, *LCG-2 User Guide.* URL: https://edms.cern.ch/file/454439/LCG-2-Userguide.pdf.

Brill Academic Publishers
P.O. Box 9000, 2300 PA Leiden
The Netherlands

*Lecture Series on Computer
and Computational Sciences*
Volume 4, 2005, pp. 44-47

# A New Approach of Dynamics Model

M. Arshad, Taj M. Khan, Dr. M. A. Choudhry

Department of Electrical and Electronics Engineering,
University of Engineering and Technology
Taxila, Pakistan

Received 10 July, 2005; accepted in revised form 03 August, 2005

*Abstract:* This work is aimed at studying the dynamic behavior of Parallel Robot Manipulator. This mechanism has six degrees of freedom, having a fixed base and a platform which acts as a free body, although, with some constraints. This paper discusses the simulation of dynamic models of the Parallel robot manipulator. Dynamics is the study of motion of a mechanism when torques or external forces that produce that motion are taken into consideration. Robot arm dynamics deal with the mathematical formulations of the equations of robot arm motion, which are a set of mathematical equation describing the dynamic behavior of the manipulator.

*Keywords:* Dynamics model, Parallel robot manipulator, simulation.

## 1. Introduction

The actual dynamics model of a robot arm can be obtained from known physical laws such as the laws of Newtonian mechanics and Legrangian mechanics. These are used for developing the dynamic equations of motion for the joints of the manipulator in terms of specified geometric and inertial parameters of the links. Lagrange-Euler(L-E) and Newton-Euler (N-E)formulations can then be applied to develop the actual robot arm motion equations.

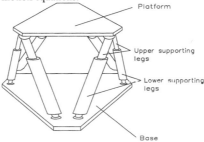

;Fig. 2.1 Six—degree—of freedom Stewart mechanism

As these equations describe the dynamic behavior of the same physical robot manipulator so they are equivalent to each other but their structure is different.

The derivation of the dynamic model of a manipulator based on the L-E[11-13] formulation is simple and systematic. Bejczy, using the 4×4 homogeneous transformation matrix representation of the kinematics chain and the Lagrangian formulation, has shown that the dynamic motion equation for a six-joint Standford robot arm are highly non linear and consist of inertia loading, coupling reaction forces between joints (Coriolis and centrifugal), and gravity loading effects. These torques/forces depend on the manipulator's physical parameters, instantaneous joint configuration, joint velocity and acceleration, and the load it carrying.

The L-E equations are very difficult to utilize for real-time control purposes unless they are simplified.

The Lagrange and Newton-Euler formulation techniques become very much complicated when they are applied to multi-loop mechanisms[3]. The Lagrange formulation gives a closed form set of equations but a large number of derivatives are required to be computed. Similarly in the Newton-Euler formulation each element is treated as a separate object or system and is treated independently. So a large number of unwanted forces are to be calculated. D' Alembert principle for solving the inverse dynamics of closed loop mechanisms is more complicated when applied to spatial mechanisms consisting of multi-degrees-of -freedom joints.

Therefore in this paper an efficient computational scheme has been developed for formulating explicit equations of motion for open loop mechanisms. The advantage of this scheme over the Lagrange and Newton-Euler formulations is that it automatically eliminates the non-contributing constraint forces without calculating lengthy differentials. Using kinematical constraint equations, This scheme can be extended for closed-loop mechanisms containing single-degree-of-freedom joints

## 2. Dynamics Model

Under this scheme the dynamical model of the above mechanism has been constructed in five steps.

1. Partial angular and partial linear velocities

   For the platform $\underline{\omega}_P = \theta_x \underline{n}_1 + \theta_y \underline{n}_2 + \theta_z \underline{n}_3$

   For supporting legs $\underline{\omega}_{Ai} = \underline{\omega}_{Bi} = \underline{\omega}_{Li}$

   $$\underline{\omega}_{L_i} = \sum_{j=1}^{6} \left[ (w_{1j})_{L_i} \underline{n}_1 + (w_{2j})_{L_i} \underline{n}_2 + (w_{3j})_{L_i} \underline{n}_3 \right] \dot{l}_j$$

2. Generalized inertia forces,

   For platform $\overset{*}{F}_P = -m_P \overset{*}{a}_P$

   For lower legs $\overset{*}{F}_{A_i} = -m_{A_i} \overset{*}{a}_{A_i}$

   For upper legs $\overset{*}{F}_{R_i} = -m_{B_i} \overset{*}{a}_{B_i}$

3. Inertia torque

   $$\overset{*}{\underline{T}}_{A_i} = \left[ R_{A_i}(t) \right] \overset{*}{\underline{T}}_{A_i(loc)}$$

   $$\overset{*}{\underline{T}}_{B_i} = \left[ R_{B_i}(t) \right] \overset{*}{\underline{T}}_{B_i(loc)}$$

4. Generalized active forces

   $$\underline{G}_P = -m_P g \underline{n}_3$$
   $$\underline{G}_{A_i} = -m_{A_i} g \underline{n}_3$$
   $$\underline{G}_{B_i} = -m_{B_i} g \underline{n}_3$$

5. Dynamical equation of motion

   $$\overset{*}{\underline{K}} + \overset{..}{\underline{K}} = 0$$

   or $$[m]\ddot{\underline{l}} + \underline{b} + \underline{g} + [a]\underline{\sigma} + \underline{f} = 0$$

The equations were employed for developing the dynamic model. Singular configuration of the mechanism have also been investigated.

Simulation results of the inverse dynamic model has been developed. Complete dynamic models have been developed and simulated to study the behavior of 6 DOF Parallel robot manipulator, using C++ computer language.

## 3. Conclusions

Two sets of results were obtained, one by ignoring the leg masses, and the other by taking leg masses (0.5 kg) into account. Actuator forces were computed in each test for same desired trajectories and have been plotted. From these simulation tests, it can be deducted that when platform is moved along x-axis, actuator forces for each leg are different, as shown (Figure 1). When the manipulator is moved along y-axis, actuator forces on legs pairs are same (Figure 2). While for movement along z-axis, forces on each leg are same (Figure 3). Similarly, when platform was rotated about $\theta_x$, forces on leg 1, 3 and 5 were same and that of 2, 4 and 6 were same. Rotating the manipulator about $\theta_y$, it was observed that forces on each leg were different. Similarly forces on each pair were experienced as same, when platform was moved about $\theta_z$. Similar results are obtained for masses of legs taken into account, as described previously.

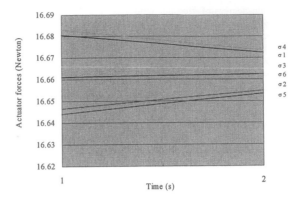

Figure 1: Plot of actuator forces versus time ($\ddot{x} = 1$ m/s$^2$, $m_P$=10 kg and $m_{Ai}$=$m_{Bi}$=0)

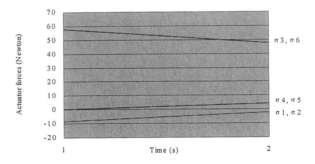

Figure 2: Plot of actuator forces versus time ($\ddot{y}$ =1 m/s$^2$, $m_P$=10 kg and $m_{Ai}$=$m_{Bi}$=0)

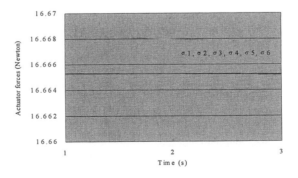

Figure 3: Plot of actuator forces versus time ($\ddot{z}$ =1 m/s$^2$, $m_P$=10 kg and $m_{Ai}$=$m_{Bi}$=0)

## Acknowledgments

The author wishes to thank the anonymous referees for their careful reading of the manuscript and their fruitful comments and suggestions.

## References

[1] K.E. Zanganeh, R. Sinatra and J. Anglels, *Kinematics and dynamics of a six-degree-of-freedom parallel manipulator with revolute legs*, J robotica(1997) volume 15 pp(3385-394)

[2] D.N. Nenchev, S. Bhattacharya and M. Uchiyama, Department of Aeronautics and Space Engineering, Tohoku University Japan, *Dynamic Analysis of Parallel Manipulators under the Singularity-Consistent Parameterization* Robotica July August 1997 Cambridge University press

[3] J.-P. Merlet *Designing a Parallel Manipulator for a Specific Workspace,* The International journal of Robotics Research, Aug 1997.

[4] A Fattah and G. Kasaei *Kinematics and dynamics of a parallel manipulator with a new architecture*, J Robotica (2000) volume 18. pp(535-543)

[5] Xing-Jun Liu, Jinsong Wang, Feng Gao, and Li-Ping Wang *Mechanism design of a simplified 6-DOF 6-RUS parallel manipulator*, Robotica (2002) volume 20

[6] Alain Codourey and Burdet, E. 1997, *A body oriented method for finding a linear form fo the dynamic equation of fully parallel robots,* Robotics And automation Vol. 2 Washington, DC: IEEE (pp. 16612-1618)

[7] Alain Codourey *Dynamic Modeling of parallel Robots for Computed-Torque Control Implementation,* The International journal of Robotics Research, vol 17, No12, December 1998

[8] Tatsuo Arai, Kei Yuasa, Yahushi Mae, Kenji Inoue, Kunio Miyawaki, and Noriho Koyachi *A Hybrid Drive Parallel Arm for Hevy Material Handling,* IEEE Robotics & Automation Magazine (March 2002)

[9] Abdelhamid Liab, *Adaptive output Regulation of Robot Manipulators Under Actuator Constraints,* IEEE Transaction on *Robotics and automation* (Dec 2000)

[10] Robert E. Parkin, *APPLIED ROBOTICS ANALYSIS*, Prentice HAL, Englewood, N.J. (1991)

[11] Kane, T.R., and Faessler, H. (1984). *Dynamics of Robots and manipulators involving closed loops.* 35th CISM-IFTOMM Symposium on Theory and Practice of Robots and Manipulators, Udine, Italy.

[12] Kane, T.R., and Levinson, D.A. (1983) . *The use of Kane's dynamical equations in robotics Int. J.* Robotics Vol. 2 No. 31, 3-21.

[13] Kok-Soon Chai, Ken Young and Ian Tuersley, *A Practical calibration process using partial information for a commercial Stewart Platform*, Robotica (Volume 20 2002) Cambridge University Press.

Brill Academic Publishers
P.O. Box 9000, 2300 PA Leiden
The Netherlands

*Lecture Series on Computer
and Computational Sciences*
Volume 4, 2005, pp. 48-51

# Three Dimensional Modelling of Turbulence

L. Balas[1] and A. İnan

Department of Civil Engineering,
Faculty of Engineering and Architecture,
Gazi University,
06570 Ankara, Turkey

Received 14 February, 2005; accepted in revised form 16 March, 2005

*Abstract:* Effect of turbulence modeling on the simulation of wind induced circulation patterns in coastal waters, has been investigated. Eddy viscosities are calculated by a two equation k- ε turbulence model, and by a two equation k-ω turbulence model, that are commonly applied in modeling of coastal transport processes. Kinetic energy of turbulence is k, dissipation rate of turbulence is ε, and frequency of turbulence is ω. In the modeling of turbulence by k-ε model and by k-ω model, a composite finite element-finite difference method has been used. The governing equations are solved by the Galerkin Weighted Residual Method in the vertical plane and by finite difference approximations in the horizontal plane. The water depths are divided into the same number of layers following the bottom topography. Therefore, the vertical layer thickness is proportional to the local water depth. From the applications, it has been seen that application of two equation k-ω turbulence model in the prediction of wind induced circulation in coastal waters leads to better predictions compared to k- ε model.

*Keywords:* turbulence, modeling, finite difference, finite element, wind induced circulation, transport

*Mathematics Subject Classification:* Turbulent transport, mixing

*PACS:* 76F25

## 1. Introduction

The developed implicit baroclinic three dimensional numerical model (HYDROTAM-3), is capable of computing water levels and water particle velocity distributions in three principal directions by solving the Navier-Stokes equations. The governing hydrodynamic equations in the three dimensional cartesian coordinate system with the z-axis vertically upwards, are [1],[2],[3],[4]:

$$\frac{\partial u}{\partial x}+\frac{\partial v}{\partial y}+\frac{\partial w}{\partial z}=0 \tag{1}$$

$$\frac{\partial u}{\partial t}+u\frac{\partial u}{\partial x}+v\frac{\partial u}{\partial y}+w\frac{\partial u}{\partial z}=fv-\frac{1}{\rho_o}\frac{\partial p}{\partial x}+2\frac{\partial}{\partial x}(v_h\frac{\partial u}{\partial x})+\frac{\partial}{\partial y}(v_h(\frac{\partial u}{\partial y}+\frac{\partial v}{\partial x}))+\frac{\partial}{\partial z}(v_z(\frac{\partial u}{\partial z}+\frac{\partial w}{\partial x})) \tag{2}$$

$$\frac{\partial v}{\partial t}+u\frac{\partial v}{\partial x}+v\frac{\partial v}{\partial y}+w\frac{\partial v}{\partial z}=-fu-\frac{1}{\rho_o}\frac{\partial p}{\partial y}+2\frac{\partial}{\partial y}(v_h\frac{\partial v}{\partial y})+\frac{\partial}{\partial x}(v_h(\frac{\partial v}{\partial x}+\frac{\partial u}{\partial y}))+\frac{\partial}{\partial z}(v_z(\frac{\partial v}{\partial z}+\frac{\partial w}{\partial y})) \tag{3}$$

$$\frac{\partial w}{\partial t}+u\frac{\partial w}{\partial x}+v\frac{\partial w}{\partial y}+w\frac{\partial w}{\partial z}=-\frac{1}{\rho_o}\frac{\partial p}{\partial z}-g+\frac{\partial}{\partial y}(v_h(\frac{\partial w}{\partial y}+\frac{\partial v}{\partial z}))+\frac{\partial}{\partial x}(v_h(\frac{\partial w}{\partial x}+\frac{\partial u}{\partial z}))+\frac{\partial}{\partial z}(v_z\frac{\partial w}{\partial z}) \tag{4}$$

where, x,y:horizontal coordinates, z:vertical coordinate, t:time, u,v,w:velocity components in x,y,z directions at any grid locations in space, $v_z$:eddy viscosity coefficients in z direction, $v_h$:horizontal eddy viscosity coefficient, f:corriolis coefficient, ρ(x,y,z,t):water density, g:gravitational acceleration, p:pressure.

As the turbulence model, firstly, modified k-ω turbulence model is used. Model includes two equations for the turbulent kinetic energy k and for the specific turbulent dissipation rate or the turbulent frequency ω [5]. Equations of k-ω turbulence model are given by the followings.

---

[1] Corresponding author. E-mail: lalebal@gazi.edu.tr

$$\frac{dk}{dt}=\frac{\partial}{\partial z}\left[\sigma^* v_z \frac{\partial k}{\partial z}\right]+P+\frac{\partial}{\partial x}\left[\sigma^* v_h \frac{\partial k}{\partial x}\right]+\frac{\partial}{\partial y}\left[\sigma^* v_h \frac{\partial k}{\partial y}\right]-\beta^* \varpi k$$

(5)

$$\frac{d\varpi}{dt}=\frac{\partial}{\partial z}\left[\sigma^* v_z \frac{\partial \varpi}{\partial z}\right]+\alpha \frac{\varpi}{k} P+\frac{\partial}{\partial x}\left[\sigma^* v_h \frac{\partial \varpi}{\partial x}\right]+\frac{\partial}{\partial y}\left[\sigma^* v_h \frac{\partial \varpi}{\partial y}\right]-\beta \varpi^2$$

(6)

The stress production of the kinetic energy is defined by;

$$P=v_h\left[2\left(\frac{\partial u}{\partial x}\right)^2+2\left(\frac{\partial v}{\partial y}\right)^2+\left(\frac{\partial u}{\partial y}+\frac{\partial v}{\partial x}\right)^2\right]+v_z\left[\left(\frac{\partial u}{\partial z}\right)^2+\left(\frac{\partial v}{\partial z}\right)^2\right]$$

(7)

Eddy viscosity is calculated as;

$$v_z=\frac{k}{\varpi}$$

(8)

At high Reynolds Numbers, the constants are used as; $\alpha=5/9$, $\beta=3/40$, $\beta^*=9/100$, $\sigma=1/2$ and $\sigma^*=1/2$. Whereas at lower Reynolds numbers they are calculated as;

$$\alpha^*=\frac{1/40+R_T/6}{1+R_T/6}, \quad \alpha=\frac{5}{9}\frac{1/10+R_T/2.7}{1+R_T/2.7}(\alpha^*)^{-1}, \quad R_T=\frac{k}{\varpi v}, \quad \beta^*=\frac{9}{100}\frac{5/18+(R_T/8)^4}{1+(R_T/8)^4}, \quad \sigma=\sigma^*=1/2$$

(9)

where, $R_T$ is the Reynolds number of the turbulence.

Secondly, as the turbulence model a two equation k-ε model has been applied. Equations of k-ε turbulence model are given by the followings.

$$\frac{\partial k}{\partial t}+u\frac{\partial k}{\partial x}+v\frac{\partial k}{\partial y}+w\frac{\partial k}{\partial z}=\frac{\partial}{\partial z}\left(\frac{v_z}{\sigma_k}\frac{\partial k}{\partial z}\right)+P-\varepsilon+\frac{\partial}{\partial x}\left(v_h \frac{\partial k}{\partial x}\right)+\frac{\partial}{\partial y}\left(v_h \frac{\partial k}{\partial y}\right)$$

(10)

$$\frac{\partial \varepsilon}{\partial t}+u\frac{\partial \varepsilon}{\partial x}+v\frac{\partial \varepsilon}{\partial y}+w\frac{\partial \varepsilon}{\partial z}=\frac{\partial}{\partial z}\left(\frac{v_z}{\sigma_\varepsilon}\frac{\partial \varepsilon}{\partial z}\right)+C_{1\varepsilon}P\frac{\varepsilon}{k}-C_{2\varepsilon}\frac{\varepsilon^2}{k}+\frac{\partial}{\partial x}\left(v_h \frac{\partial \varepsilon}{\partial x}\right)+\frac{\partial}{\partial y}\left(v_h \frac{\partial \varepsilon}{\partial y}\right)$$

(11)

where, k :Kinetic energy, ε:Rate of dissipation of kinetic energy, P: Stress production of the kinetic energy. The vertical eddy viscosity is calculated by:

$$v_z = C_\mu \frac{k^2}{\varepsilon}$$

(12)

The following universal k-ε turbulence model empirical constants are used $C_\mu=0.09$, $\sigma_\varepsilon=1.3$, $C_{1\varepsilon}=1.44$, $C_{2\varepsilon}=1.92$. To account for large scale turbulence generated by the horizontal shear, horizontal eddy viscosity can be simulated by the Smagorinsky algebraic subgrid scale turbulence model [6];

$$v_h=0,01\Delta x\Delta y\left(\left(\frac{\partial u}{\partial x}\right)^2+\left(\frac{\partial v}{\partial y}\right)^2+\frac{1}{2}\left(\frac{\partial u}{\partial y}+\frac{\partial v}{\partial y}\right)^2\right)^{1/2}$$

(13)

where $\Delta x$ and $\Delta y$ are horizontal mesh sizes. Some other turbulence models have also been widely applied in three dimensional numerical modeling of wind induced currents such as one equation turbulence model and mixing length models. They are also used in the developed model HYROTAM-3, however it is seen that two equation turbulence models give better predictions compared to others. Applied mixing length model is,

$$v_z=l^2\left[\left(\frac{\partial u}{\partial z}\right)^2+\left(\frac{\partial v}{\partial z}\right)^2\right]^{1/2}$$

(14)

Where the mixing length $l$ is defined by,

$$l=\kappa z\cdots ; \kappa z\leq 0.1H$$
$$l=0.1H\cdots ; \kappa z>0.1H$$

(15)

where κ=0.41 Von-Karman constant; H: water depth.

One equation mixing length model has been used as,

$$\frac{dk}{dt}=\frac{\partial}{\partial z}\left(\frac{v_z}{\sigma_k}\frac{\partial k}{\partial z}\right)+v_z\left(\frac{\partial u}{\partial z}\right)^2-\varepsilon$$

(16)

Where $\sigma_k=1$, and dissipation of kinetic energy and vertical eddy viscosity are defined by,

$$\varepsilon=\frac{C_d k^{3/2}}{L_0};v=\frac{C_m k^2}{\varepsilon}$$

(17)

where $C_m=C_d=0.3$. and dissipation length $L_0$ is given by:

$$L_o=\alpha_* \eta^m (1-\eta)^p L_{om} ; 0\leq\eta<1/2$$

$$L_o=L_{om}=b_* H; 1/2\leq\eta<1$$

(18)

where $\eta$ is a nondimensional length changing in between 0 and 1. At the surface $\eta=1$, and at the bottom $\eta=0$'dır. The constants $\alpha_*$, $b_*$, m, p are selected as 4, 0.105, 1 and 1,respectively [7].

Solution method is a composite finite difference-finite element method [1],[2],[3],[4]. Equations are solved numerically by approximating the horizontal gradient terms using a staggered finite difference scheme. In the vertical plane however, the Galerkin Method of finite elements is utilized. A detailed presentation of the finite difference approximations and Galerkin Method of finite elements to the governing equations, and boundary conditions were given in Balas and Özhan [1]. Water depths are divided into the same number of layers following the bottom topography. At all nodal points, the ratio of the length (thickness) of each element (layer) to the total depth is constant. To increase the vertical resolution, wherever necessary, grid clustering can be applied in the vertical plane. Grids can be concentrated near the bottom, surface, or intermediate layers. The mesh size may be varied in the horizontal plane. The system of nonlinear equations is solved by the Crank Nicholson Method which has second order accuracy in time.

## 2. Comparison of Model Applications

Simulated velocity profiles by using k-ε turbulence model, k-ω turbulence model, one-equation turbulence model, a mixing length model and a constant vertical eddy viscosity, have been compared with the experimental results of wind driven turbulent flow of an homogeneous fluid conducted by Tsanis and Leutheusser [8]. Laboratory basin had a length of 2.4 m., a width of 0.72 m. and depth of H=0.05

meters. The Reynolds Number, $R_s=\dfrac{u_s H\rho}{\mu}$ was 3000 ($u_s$ is the surface velocity, H is the depth of

the flow, $\rho$ is the density of water and $\mu$ is the dynamic viscosity). Resulted velocity profiles are given in Figure 1.

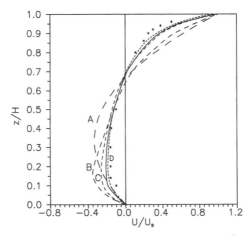

Figure 1. Velocity profiles (solid line: k-ε turbulence model, *: experimental data, dotted line A: constant vertical eddy viscosity, dotted line B: one equation turbulence model, dotted line C:mixing length model, dotted line D: k-ω turbulence model).

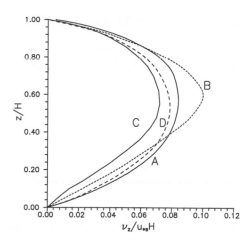

Figure 2. Distribution of vertical eddy viscosity by different turbulence models (solid line A: k-ε turbulence model, dotted line B: one-equation turbulence model, dotted line C:mixing length model, dotted line D: k-ω turbulence model).

## 3. Conclusion

Turbulence models used in three dimensional modeling of wind induced currents in coastal waters have been presented. Generally, two-equation turbulence models give improved estimations compared to other turbulence models. It has been seen that the two equation k-ω turbulence model predictions are better than the two equation k-ε turbulence model.

## Acknowledgments

The author wishes to thank the anonymous referees for their careful reading of the manuscript and their fruitful comments and suggestions.

## References

[1] L. Balas and E. Özhan, An Implicit Three Dimensional Numerical Model to Simulate Transport Processes in Coastal Water Bodies, *International Journal for Numerical Methods in Fluids* **34** 307-339 (2000).

[2] L. Balas and E. Özhan, Three Dimensional Modelling of Stratified Coastal Waters, *Estuarine, Coastal and Shelf Science* **56** 75-87(2002).

[3] L. Balas and E. Özhan Applications of a 3-D Numerical Model to Circulations in Coastal Waters, *Coastal Engineering Journal* **43** 99-120(2001).

[4] L. Balas and E. Özhan, A Baroclinic Three Dimensional Numerical Model Applied to Coastal Lagoons, *Lecture Notes in Computer Science* **2658** 205-212 (2003).

[5] V.S. Neary, F. Sotiropoulos, A.J. Odgaard, Three Dimensional Numerical Model of Lateral Intake Inflows, *Journal of Hyraulic Engineering* **125** 126-140 (1999).

[6] B. Mohammadi and O. Pironneau, *Analysis of the K-Epsilon Turbulence Model*, John Wiley and Sons, London, U.K, 1994.

[7] C. Koutitas and B. O'Connor, B.,Modelling Three-Dimensional Wind-Induced Flows, *Journal of the Hydraulic Division*, ASCE, 1843-1865 (1980).

[8] K.I. Tsanis and H.J. Leutheusser, The Structure of Turbulent Shear-Induced Countercurrent Flow, *Journal of Fluid Mechanics* **189** 531-552 (1988).

Brill Academic Publishers
P.O. Box 9000, 2300 PA Leiden,
The Netherlands

*Lecture Series on Computer
and Computational Sciences*
Volume 4, 2005, pp. 52-55

# Sign-consistent *ab initio* Molecular Wave Functions

**P. Barragán, L. F. Errea, L. Fernández, A. Macías, L. Méndez, I. Rabadán[1] and A. Riera**

Laboratorio Asociado al CIEMAT de Física Atómica y Molecular en Plasmas de Fusión,
Departamento de Química,
Universidad Autónoma de Madrid,
Madrid-28049, Spain

Received 8 July, 2005; accepted in revised form 8 August, 2005

*Abstract:* We present a method to ensure the sign consistency of molecular wave functions and the dynamical couplings that are obtained from them. The method also allows to systematically "diabatize" avoided crossings between two potential energy surfaces, including conical intersections.

*Keywords:* molecular wave functions, sign-consistency, non-adiabatic couplings, diabatization.

*PACS:* 34.70.+e, 34.10.+x

## 1 Introduction

The theoretical treatment of atom (ion)-atom (molecule) collisions, at low energies, is usually carried out by employing a molecular expansion (see e.g. [1]) of the collisional wave function in terms of the electronic wave functions of the (quasi)molecule formed by the colliding systems. This method involves a first step where one solves the clamped-nuclei electronic equation in the Born-Oppenheimer approximation to obtain the adiabatic molecular functions (MFs) and potential energy surfaces (PES). Excitation and charge transfer processes then take place through non-adiabatic transitions between the Born-Oppenheimer states, induced by the dynamical or non-adiabatic coupling terms. Also, non-adiabatic transitions are relevant in some reactive processes.

PES and couplings are the required input of the dynamical calculation. The calculation of MFs may require large configuration interaction (CI) expansions and the evaluation of non-adiabatic couplings is an important practical aspect for which numerical and analytical techniques have been proposed (see [5] and references therein) and implemented [2, 3].

An important practical drawback of the method is the erratic sign of the calculated dynamical couplings. These show, in general, unphysical teeth-saw shapes because the overall sign of the MFs is arbitrary; indeed, it depends on the numerical diagonalization procedure of the Hamiltonian matrix.

In [4], we have proposed a method based on the calculation of the delayed overlap matrix (DOM) to solve this problem. Here we summarize this method and show an application with a calculation of a radial non-adiabatic coupling for the $H^+ + N_2$ collisional system, of relevance in modeling proton precipitation in the Earth's upper atmosphere.

---

[1]Corresponding author. E-mail: ismanuel.rabadan@uam.es

## 2 Basic equations

In a molecular treatment of ion-atom and ion-molecule collisions, the collisional wave function is expanded in terms of MFs, $\Phi_l$, which are (approximate) solutions of the equation:

$$H_{\text{elec}}\Phi_l(q;Q) = E_l(Q)\Phi_l(q;Q) \tag{1}$$

where $H_{\text{elec}}$ is the clamped-nuclei electronic Hamiltonian, $q$ and $Q$ denote electronic and nuclear coordinates, respectively, and $E_l(Q)$ are the PES.

To solve this equation, one often employs the configuration interaction method (CI), in which the system wave function is expanded as a linear combination of configurations, $\psi_j$:

$$\Phi_l(q;Q) = \sum_j c_{jl}\psi_j(q;Q) \tag{2}$$

Here, the coefficients $c_{jl}$ are obtained variationally from the secular equation and the electronic configurations $\psi_j$ are spin- and symmetry- adapted, antisymmetrized products of molecular spin-orbitals.

The collisional wave function is then written as a linear combination of the MFs:

$$\Psi(Q,q) = \sum_l F_l(Q)\Phi_l(q;Q) \tag{3}$$

where $F_l(Q)$ are calculated by substituting the expansion (3) in the corresponding full (dynamical) Schrödinger equation, yielding a set of differential equations that are solved numerically. Transitions between molecular states are induced by non-adiabatic couplings, which are the matrix elements of the nuclear gradient operator $\nabla_Q$:

$$M_{lm} = \langle \Phi_l(q;Q)|\hat{X} \cdot \nabla_Q|\Phi_m(q;Q)\rangle \tag{4}$$

where $\hat{X}$ is a unit vector in any direction of $Q$-space.

In practice, the PES and couplings are evaluated in a set of nuclear geometries $\{Q_i, \ldots, Q_N\}$, and they are then interpolated at the points needed to numerically solve the system of differential equations for the functions $F_l(Q)$. In these interpolation nodes, the couplings can be evaluated numerically to first order in $\delta$, by taking

$$M_{lm}(Q_i) = \delta^{-1}O_{lm}(Q_i) + \mathcal{O}(\delta^2) \tag{5}$$

where $O_{lm}$ is the DOM between two nearby points:

$$O_{lm} = \langle \Phi_l(q;Q_i)|\Phi_m(q;Q_i + \delta\hat{X})\rangle \tag{6}$$

and can be evaluated as explained in [3].

To calculate the dynamical couplings of Eq. (5), we need to solve Eq. (1) at two different grid points: $Q_i$ and $Q_i + \delta\hat{X}$. If one employs a basis of real functions, both $-\Phi_l$ and $\Phi_l$ are solutions of this equation, which means that the sign of $\Phi_l(q;Q)$ is arbitrary, leading to arbitrary signs of the matrix elements $M_{lm}(Q_i)$. A locally sign consistent coupling is obtained from the original coupling by using the sign of the diagonal elements of the DOM:

$$M_{lm}^s(Q_i) = k_m M_{lm}(Q_i) \tag{7}$$

where

$$k_m = \frac{|O_{mm}(Q_i)|}{O_{mm}(Q_i)} \tag{8}$$

# 3   Sign consistency

The procedure explained above only ensures that dynamical couplings between different electronic states are sign consistent at a given nuclear geometry, $Q_i$. Now, we are interested in obtaining sign consistent couplings in a set of nuclear geometries defining a path or a domain of nuclear configurations. For this purpose, we construct a new set of sign-consistent molecular states $\{\Phi_i^c\}$, as explained below.

Let $\{Q_1, \ldots, Q_N\}$ be an ordered grid of points along a given path in the nuclear configuration space, where the set of adiabatic molecular states $\{\Phi_j\}$ are known. We assume that the erratic sign of $\Phi_j$ can only arise in the CI step.

The first step is the evaluation of the DOM:

$$O_{jj}(Q_i) = < \Phi_j(Q_{i-1}) | \Phi_j(Q_i) > \tag{9}$$

with $i = 2, \ldots, N$. Next, we define $n_i(j, j)$ from the following mapping:

$$n_i(j, j) = \begin{cases} 0 & \text{if} \quad |O_{jj}(Q_i) - 1| < \epsilon_1 \\ 1 & \text{if} \quad |O_{jj}(Q_i) + 1| < \epsilon_1 \\ 2 & \text{otherwise} \end{cases} \tag{10}$$

where $\epsilon_1$ is a given threshold (0.2 in our calculations). In these expressions, $n_i(j, j) = 0$ means that the state $\Phi_j$ does not change sign when going from $Q_{i-1}$ to $Q_i$, while $n_i(j, j) = 1$ implies that there was a change of sign. In either case, we define a new MF $\Phi_j^c$ by taking:

$$\Phi_j^c(Q_i) = (-1)^{n_i(j,j)} \Phi_j(Q_i) \tag{11}$$

On the other hand, $n_i(j, j) = 2$ may indicate that there is an avoided crossing between the energy of $\Phi_j$ and some other state, $\Phi_{j'}$, in the path from $Q_{i-1}$ to $Q_i$. To check this possibility, we calculate the delayed overlaps with "energetically-near" states:

$$O_{jj'}(Q_i) = < \Phi_j(Q_{i-1}) | \Phi_{j'}(Q_i) > \tag{12}$$

which leads, for each $j'$, to the mapping $n_i(j, j')$:

$$n_i(j, j') = \begin{cases} 0 & \text{if} \quad |O_{jj'}(Q_i) - 1| < \epsilon_1 \\ 1 & \text{if} \quad |O_{jj'}(Q_i) + 1| < \epsilon_1 \\ 2 & \text{otherwise} \end{cases} \tag{13}$$

If $n_i(j, j') \neq 2$ for any $j'$, we proceed to "diabatize" the crossing:

$$\phi_j^c(Q_i) = (-1)^{n_i(j,j')} \phi_{j'}(Q_i) \tag{14}$$

This means that the energies of MFs $j$ and $j'$ must be exchanged when going from $Q_{i-1}$ to $Q_i$.

The result $n_i(j, j') = 2 \ \forall j'$ indicates that $Q_i$ is too far from $Q_{i-1}$ to obtain a DOM close to 1 for any $j'$. In this case, we add an intermediate point to the grid and repeat the process.

In the new $\{\Phi_j^c\}$ basis, the corrected couplings $M_{lm}^c(Q_i)$, sign-consistent with those at $Q_{i-1}$, are:

$$M_{lm}^c(Q_i) = (-1)^{n_i(l,l') + n_i(m,m')} M_{l'm'}^s(Q_i) \tag{15}$$

where $l'$ and $m'$ are those for which $n_i$ is either 0 or 1.

In this work, we have applied this method to the calculation of a radial coupling in $H^+ + N_2$ collisions. Here we take $H^+$ at the origin, $R$, the position vector of the center of mass of $N_2$, $\rho$, the vector between the two N nuclei and $\alpha$ the angle between $R$ and $\rho$. The radial non-adiabatic

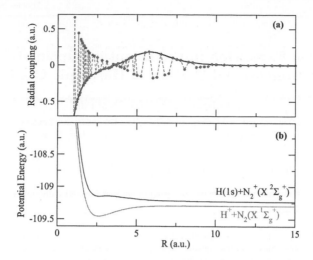

Figure 1: (a): radial coupling between the states of panel (b), in the $\{\Phi\}$ ($\bullet$) and $\{\Phi^c\}$ (solid line) representations. (b) Energies of the first two $^1A'$ MFs of $H^++N_2$ as functions of $R$ for $\rho = 2.08$ $a_0$ and $\alpha = 45°$.

coupling between two electronic states is, then, $< \Phi_l | \partial \Phi_m / \partial R |_{\rho,\alpha} >$. In panel (a) of Fig. 1 we show the raw $M_{12}^s$ coupling of eq. (15) with symbols and the sign-corrected one, $M_{12}^c$, with solid line. We can see that the original couplings (symbols) jump erratically, while those in the $\{\Phi^c\}$ basis (11) (solid line) are perfectly smooth. The potential energy curves of the two electronic states involved (the first two $^1A'$ MFs of $H^++N_2$ with $\rho = 2.08$ $a_0$ and $\alpha = 45°$, for $1 \leq R \leq 15$) are shown in Fig. 1(b).

## Acknowledgment

This work has been partially supported by MEC projects FIS2004-04145 and and ENE2004-06266. IR acknowledges MEC for a "Ramón y Cajal" contract.

## References

[1] B. H. Bransden and M. H C. McDowell. *Charge Exchange and the Theory of Ion-Atom Collisions.* Oxford, Clarendon, 1992.

[2] R. J. Buenker, S. D. Peyerimhoff, and P. J. Bruna. In I. G. Csizmadia and R. Daudel, editors, *Computational Theoretical Organic Chemistry.* Dordrecht, Reidel, 1981.

[3] J. F. Castillo, L. F. Errea, A. Macías, L. Méndez, and A. Riera. *J. Chem. Phys.* **103** 2113–2116(1995).

[4] L. F. Errea, L. Fernández, A. Macías, L. Méndez, I. Rabadán, and A. Riera. *J. Chem. Phys.* **121** 1663(2004).

[5] A. Macías and A. Riera. *Phys. Rep.*, **81** 299(1982).

Brill Academic Publishers
P.O. Box 9000, 2300 PA Leiden,
The Netherlands

*Lecture Series on Computer*
*and Computational Sciences*
Volume 4, 2005, pp. 56-59

# Fluctuation Expansion in the Quantum Optimal Control of One Dimensional Perturbed Harmonic Oscillator

**N.A. Baykara[1] and M. Demiralp[2]**

Department of Mathematics, Faculty of Science and Letters,
Marmara University, Göztepe, 34722, İstanbul, Turkey

Group for Science and Methods of Computing, Informatics Institute,
İstanbul Technical University, Maslak, 34469 İstanbul, Turkey

Received 10 July, 2005, accepted in revised form 31 July, 2005

*Abstract:* This work considers the quantum optimal control of one dimensional harmonic oscillator under linear control agents. The system's potential energy function contains a perturbation term which is bounded everywhere in the space variable's domain. We use the most preferred cost functional to construct the necessary equations. Equations are converted to a boundary value problem for a set of ordinary differential equations containing the expectation values of certain operators and the terms corresponding to the transitions between the states described by the wave and costate functions. Resulting equations involve certain undesired expectation values and transition terms. We use a recently developed scheme called fluctuation expansion to approximate these terms at the sharply localized wave and costate function limits. This enables us to construct n infinite number of ordinary differential equations and accompanying boundary conditions whose both halves are given at the beginning and end of the control. These equations are truncated and then the resulting boundary value problem is solved iteratively.

*Keywords:* Quantum Theory, Fluctuations, Distributions

*Mathematics Subject Classification:* 81Q99, 60E05

## 1 Control Equations in Temporally Varying Entities and Their Solution

In this work we deal with the application of the recently developed Fluctuation Expansion Method [1, 2, 3] to quantum optimal control problems. The governing equations for the optimal control of this system are obtained by setting the first variation of the most widely used cost functional[4] to

---

[1]Corresponding author. E-mail: baki@be.itu.edu.tr
[2]E-mail: demiralp@be.itu.edu.tr

zero. They are given below

$$ih\frac{\partial}{\partial t}\left|\psi(t)\right\rangle = \left[\widehat{H}_0 - \mu E(t)\right]\left|\psi(t)\right\rangle, \qquad \left|\psi(0)\right\rangle = \left|in\right\rangle \tag{1}$$

$$ih\frac{\partial}{\partial t}\left|\lambda(t)\right\rangle = \left[\widehat{H}_0 - \mu E(t)\right]\left|\lambda(t)\right\rangle - W_p(t)\left\langle\psi(t)\left|\widehat{O}'\right|\psi(t)\right\rangle\widehat{O}'\left|\psi(t)\right\rangle,$$
$$\left|\lambda(T)\right\rangle = -i\frac{\eta}{\hbar}\widehat{O}\left|\psi(T)\right\rangle \tag{2}$$

$$E(t) = \frac{2}{W_E(t)}\Re e\left(\left\langle\lambda(t)\left|\mu\right|\psi(t)\right\rangle\right), \qquad \eta \equiv \left\langle\psi(T)\left|\widehat{O}\right|\psi(T)\right\rangle - \tilde{O} \tag{3}$$

where Dirac's bra and ket notation is used and hence $\left|\psi(t)\right\rangle$, $\left|\lambda(t)\right\rangle$ are respectively waveket and costate ket. Their bra counterparts are respectively denoted by $\left\langle\psi(t)\right|$ and $\left\langle\lambda(t)\right|$ although they are not listed in the above equations. Bra counterpart of (1) and (2) can be written by taking Hermitian conjugates equations (1) and (2) and recalling the fact that the Hermitian conjugate of a ket is its bra counterpart and vice versa. Wave and costate functions or their ket and bra counterparts describe forward and backward evolution in time. $\widehat{O}$ and $\tilde{O}$ stand for the **objective operator** and its expectation value's prescribed **target value** respectively. $\widehat{O}'$ symbolizes the **penalty operator** whose expectation value's square should be suppressed during the system's control. $W_p(t)$ and $W_E(t)$ are given weight functions appearing in the cost functional. $\Re e$ means real part whereas $E(t)$ denotes the external field amplitude. $\widehat{H}_0$ and $\mu$ denote the isolated system's Hamiltonian and the dipole function varying solely spatially. The parameter $T$ stands for the interaction time period. $\left|in\right\rangle$ characterizes the prescribed initial form of the wave function. $\eta$ is called the **deviation parameter** and is introduced to facilitate brevity in further analysis. $\eta$ has a deeper physical meaning beyond being a measurer for the deviation of the expectation value of objective operator and its prescribed target value. It is a quite important agent monitoring the realizability and effectivity of the control.

The equations above are rather complicated since they contain partial differential equations. Since we do not directly need the wave and costate functions as long as physical observables are under consideration, it is better to deal with the expectation values and transition matrix elements of certain operators. To this end we can define the following four entities as unknown functions of time

$$p(t) \equiv \left\langle\psi(t)\left|\widehat{p}\right|\psi(t)\right\rangle \qquad q(t) \equiv \left\langle\psi(t)\left|\widehat{x}\right|\psi(t)\right\rangle, \tag{4}$$

$$r(t) \equiv 2\Re e\left(\left\langle\lambda(t)\left|\widehat{p}\right|\psi(t)\right\rangle\right), \qquad s(t) \equiv 2\Re e\left(\left\langle\lambda(t)\left|\widehat{x}\right|\psi(t)\right\rangle\right) \tag{5}$$

These functions form a closed structure under temporal differentiation except the transition term between the states described by the wave and costate function via the derivative of the perturbation function, $V(x)$, appearing in the Hamiltonian as will be given soon. We can also construct four boundary conditions two of which are given at the beginning of the control while the remaining two are given at the final moment of the control. Before giving these equations and accompanying conditions we need to make further specifications as follows

$$\widehat{H}_0 \equiv -\frac{\hbar^2}{2m}\frac{\partial^2}{\partial x^2} + \frac{\kappa}{2}x^2 + V(x) \equiv \frac{1}{2m}\widehat{p}^2 + \frac{\kappa}{2}\widehat{x}^2 + \widehat{V}(\widehat{x}) \tag{6}$$

$$\mu \equiv x \equiv \widehat{x}, \qquad \widehat{O} \equiv \alpha_1\widehat{x} + \alpha_2\widehat{p}, \qquad \widehat{O}' \equiv \alpha_1'\widehat{x} + \alpha_2'\widehat{p} \tag{7}$$

where $m$ and $\kappa$ symbolize the mass and the elastic force constant of the system under consideration whereas the function $V(x)$ stands for the given potential function term perturbing the system from the harmonic oscillator. The parameters $\alpha_1$, $\alpha_2$, $\alpha_1'$, $\alpha_2'$ are given constants.

Differentiation of both sides of each equality in (4) and (5) with respect to $t$ and utilization of the partial differential equations for the wavebra and waveket enable us to write

$$\dot{p}(t) = -\kappa q(t) - \langle \psi(t) | V'(\widehat{x}) | \psi(t) \rangle + E(t), \qquad \dot{q}(t) = \frac{1}{m}p(t), \tag{8}$$

$$\dot{r}(t) = -2\Re e\left(\left\langle \lambda(t) \left| \widehat{V}'(\widehat{x}) \right| \psi(t) \right\rangle\right), \qquad \dot{s}(t) = \frac{1}{m}r(t) - \alpha_1'\alpha_2' W_p(t)q(t) - {\alpha_2'}^2 W_p(t)p(t), \tag{9}$$

which are accompanied by the conditions $p(0) = p_{in}$, $q(0) = q_{in}$, $r(T) = -\eta\alpha_1$, and $s(T) = \eta\alpha_2$. The deviation equation defining $\eta$ becomes

$$\eta = \alpha_1 q(T) + \alpha_2 p(T) - \widetilde{O} \tag{10}$$

Now we can use the first terms of the recently developed Fluctuation Expansions[1] for the approximation of the expectation value in (8) and transition matrix element in (9) as follows

$$\left\langle \psi(t) \left| \widehat{V}'(\widehat{x}) \right| \psi(t) \right\rangle \approx V'(q(t)), \tag{11}$$

$$2\Re e\left(\left\langle \psi(t) \left| \widehat{V}'(\widehat{x}) \right| \psi(t) \right\rangle\right) \approx \frac{s(t)}{q(t)}\left[V'(q(t)) - V'(0)\right] \tag{12}$$

These equations give the zero fluctuation approximations to equations (8 − 9) and present a boundary value problem of four ordinary differential equations in time. If in the fluctuation expansion of the expectation values in (11) and (12), the first two terms were kept then we would have ten first order differential equations for ten unknowns. Six of these conditions are related to the fluctution terms and, in total, half of the conditions is given at the beginning and the other half is specified at the final moment of control. We may write the following approximations in place of (11) and (12) in this latter case

$$\left\langle \psi(t) \left| \widehat{V}'(\widehat{x}) \right| \psi(t) \right\rangle \approx V'(q(t)) - \frac{\varphi_1(t)}{q(t)^2}\left[V'(q(t)) - V'(0) - V''(0)q(t)\right] \tag{13}$$

$$2\Re e\left(\left\langle \lambda(t) \left| \widehat{V}'(\widehat{x}) \right| \psi(t) \right\rangle\right) \approx \frac{s(t)}{q(t)}\left[V'(q(t)) - V'(0)\right]$$

$$-\frac{\theta_1(t)}{q(t)^2}\left[V'(q(t)) - V'(0) - V''(0)q(t)\right] \tag{14}$$

where

$$\varphi_1(t) \equiv \left\langle \psi(t) \left| \widehat{x}\widehat{P}_\psi^{(c)}\widehat{x} \right| \psi(t) \right\rangle, \qquad \theta_1(t) \equiv 2\Re e\left(\left\langle \lambda(t) \left| \widehat{x}\widehat{P}_\psi^{(c)}\widehat{x} \right| \psi(t) \right\rangle\right) \tag{15}$$

These entities are first order fluctuation terms in expectation value and in transition matrix element between the states described by the wave and costate functions. They satisfy first order differential equations containing the four fluctuation terms, $\varphi_{1,1}(t)$, $\varphi_{1,2}(t)$, which are expectation values of the operators $\widehat{p}\widehat{P}_\psi^{(c)}\widehat{x} + \widehat{x}\widehat{P}_\psi^{(c)}\widehat{p}$ and $\widehat{p}\widehat{P}_\psi^{(c)}\widehat{p}$ respectively, and $\theta_{1,1}(t)$, $\theta_{1,2}(t)$, which are transition terms between the states described by the wave and costate functions through the same operators respectively. Here $\widehat{P}_\psi^{(c)}$ projects to the complement of the space spanned by the wave function. These satisfy four differential equations with four boundary conditions such that the equations and conditions do not bring about any extra unknowns. We do not give explicit structures here.

In the case of zero fluctuation we can write the following vector differential equation

$$\dot{\mathbf{z}}(t) = \mathbf{A}(q(t), t)\mathbf{z}(t) - V'(0)\mathbf{e}_1 \tag{16}$$

where $\mathbf{z}(t)$ is the vector whose four elements are $p(t)$, $q(t)$, $r(t)$, $s(t)$ respectively and

$$\mathbf{A}(t) \equiv \begin{bmatrix} \mathbf{A}_1 & \mathbf{A}_2 \\ \mathbf{A}_3 & \mathbf{A}_1 \end{bmatrix}, \qquad \mathbf{A}_1(t) \equiv \begin{bmatrix} -\overline{V}(q(t)) & -\kappa \\ \frac{1}{m} & 0 \end{bmatrix} \qquad \mathbf{A}_2(t) \equiv \begin{bmatrix} 0 & -\frac{1}{W_E(t)} \\ 0 & 0 \end{bmatrix}$$

$$\mathbf{A}_3(t) \equiv \begin{bmatrix} \alpha_1' \alpha_2' W_p(t) & {\alpha_1'}^2 W_p(t) \\ -{\alpha_2'}^2 W_p(t) & -\alpha_1' \alpha_1' W_p(t) \end{bmatrix}, \qquad \overline{V}(q(t)) \equiv \frac{[V'(q(t)) - V'(0)]}{q(t)} \tag{17}$$

The accompanying boundary conditions of (16) can be written by using the boundary conditions on the elements of the vector $\mathbf{z}(t)$ and the explicit definition of $\eta$.

$$\mathbf{I}_{4T}\mathbf{z}(0) = \mathbf{v}_1, \qquad \left[\mathbf{v}_2\mathbf{v}_3^T\mathbf{I}_{4T} - \mathbf{I}_{4B}\right]\mathbf{z}(T) = \widetilde{O}\mathbf{v}_2 \tag{18}$$

$\mathbf{I}_{4T}$ and $\mathbf{I}_{4B}$ are $(2 \times 4)$ matrices which are in fact the top and bottom halves of $(4 \times 4)$ unit matrix, $\mathbf{I}_4$, and $\mathbf{v}_1^T \equiv [p_{in} \; q_{in}]$, $\mathbf{v}_2^T \equiv [-\alpha_1 \; \alpha_2]$, $\mathbf{v}_3^T \equiv [\alpha_2 \; \alpha_1]$.

First and higher order truncated equations also present quite similar equations although they are not given here. Equation (16) can be solved iteratively by using appropriate propagators (evolution matrices). The evaluation of the propagators can be realized by using certain expansions with the assumption that $q(t)$ is known. Then a new $q(t)$ is obtained from (16) to form a cycle for an iterative scheme. Certain implementations to compare zeroth and first order truncations will be presented in the conference and then in the resulting publication.

## 2    Concluding Remarks

Here we have applied recently developed Fluctuation Expansion Method to the quantum control of one dimensional systems which are certain perturbed form of the harmonic oscillator. Our purpose is to show the applicability of the Fluctuation Expansion method and we focus only on the zeroth and first order truncation equations for simplicity. We expect rapid convergence in heavy systems with weak bonds. Results are promising and encourage us for a broader class of problems.

### Acknowledgment

The second author wishes to acknowledge financial support from the State Planning Organization (DPT) of Turkey and thanks Turkish Academy of Science for its partial support.

### References

[1] M. Demiralp, A Fluctuation Expansion in Integration Under Almost Sharply Localized Weight, (submitted), 2005.

[2] M. Demiralp, A Fluctuation Expansion Method for the Evalution of a Function's Expectation Value, *To appear in the Proceedings of ICNAAM2005, Rhodes, Greece*, 2005.

[3] M. Demiralp, Determination of Quantum Expectation Values Via Fluctuation Operator, *To appear in the Proceedings of ICCMSE2005, Loutraki, Greece*, 2005.

[4] M. Demiralp and H. Rabitz, Optimally Controlled Quantum Molecular Dynamics: A Perturbation Formulation and the Existence of Multiple Solutions, *Phys. Rev. A*, **47**, 831, 1993.

Brill Academic Publishers
P.O. Box 9000, 2300 PA Leiden,
The Netherlands

*Lecture Series on Computer
and Computational Sciences*
Volume 4, 2005, pp. 60-64

# A Constrained and Nonsmooth Hydrothermal Problem

**L. Bayón[1]; J.M. Grau; M.M. Ruiz; P.M. Suárez**

Department of Mathematics, University of Oviedo, Spain

Received 25 June, 2005; accepted in revised form 13 July, 2005

*Abstract:* This paper addresses a hydrothermal problem that simultaneously considers non-regular Lagrangian and non-holonomic inequality constraints, obtaining a necessary minimum condition. It is further shown that the discontinuity of the lagrangian does not translate as discontinuity in the derivative of the solution. Finally, a solution algorithm is developed and applied to an example.

*Keywords:* Optimal Control, Clarke's Gradient, Hydrothermal Optimization

*Mathematics Subject Classification:* 49J24, 49A52

## 1 Introduction

This paper deals with the optimization of hydrothermal problems. In a previous paper [1], we considered a hydrothermal system with one hydro-plant and $m$ thermal power plants that had been substituted by their thermal equivalent and addressed the problem of minimizing the cost of fuel $F(P)$ during the optimization interval $[0, T]$

$$F(P) = \int_0^T \Psi(P(t))dt \tag{1.1}$$

$$P(t) + H(t, z(t), z'(t)) = P_d(t), \ \forall t \in [0, T] \tag{1.2}$$

$$z(0) = 0, z(T) = b \tag{1.3}$$

where $\Psi$ is the function of thermal cost of the thermal equivalent and $P(t)$ is the power generated by said plant. The following must be also be verified: the equilibrium equation of active power (1.2), and the boundary conditions (1.3), where $P_d(t)$ is the power demand, $H(t, z(t), z'(t))$ is the power contributed to the system at the instant $t$ by the hydro-plant, $z(t)$ being the volume that is discharged up to the instant $t$ by the plant, $z'(t)$ the rate of water discharge of the plant at the instant $t$, and $b$ the volume of water that must be discharged during the entire optimization interval. In said paper, we likewise considered constraints for the admissible generated power ($P(t) \geq 0$ and $H(t, z(t), z'(t)) \geq 0$). The mathematical problem $(P_1)$ was stated in the following terms:

$$\min_{z \in \Theta_1} F(z) = \min_{z \in \Theta_1} \int_0^T \Psi\left[P_d(t) - H(t, z(t), z'(t))\right] dt = \min_{z \in \Theta_1} \int_0^T L(t, z(t), z'(t))dt$$

$$\Theta_1 = \{z \in \widehat{C}^1[0, T] \mid z(0) = 0, z(T) = b, 0 \leq H(t, z(t), z'(t)) \leq P_d(t), \forall t \in [0, T]\}$$

[1]Corresponding author. EUITI, 33204 Gijón, Asturias (Spain). E-mail: bayon@uniovi.es

where $(\widehat{C}^1)$ is the set of piecewise $C^1$ functions. The problem $(P_1)$ was formulated within the framework of optimal control [2] and

$$\mathbb{Y}_q(t) := -L_{z'}(t, q(t), q'(t)) \cdot \exp\left[-\int_0^t \frac{H_z(s, q(s), q'(s))}{H_{z'}(s, q(s), q'(s))} ds\right] \tag{1.4}$$

was called *the coordination function of* $q \in \Theta_1$, obtaining the following result:

**Theorem 1.** *If $q$ is a solution of $(P_1)$, then $\exists K \in \mathbb{R}^+$ such that:*

$$\mathbb{Y}_q(t) \text{ is } \begin{cases} \leq K & \text{if } H(t, q(t), q'(t)) = 0 \\ = K & \text{if } 0 < H(t, q(t), q'(t)) < P_d(t) \\ \geq K & \text{if } H(t, q(t), q'(t)) = P_d(t) \end{cases}$$

In another previous paper [3], a problem of hydrothermal optimization with pumped-storage plants was addressed, though without considering constraints for the admissible generated power. In this kind of problem, the derivative of $H$ with respect to $z'$ ($H_{z'}$) presents discontinuity at $z' = 0$, which is the border between the power generation zone (positive values of $z'$) and the pumping zone (negative values of $z'$). The mathematical problem $(P_2)$ was stated in the following terms:

$$\min_{z \in \Theta_2} F(z) = \min_{z \in \Theta_2} \int_0^T \Psi\left[P_d(t) - H(t, z(t), z'(t))\right] dt = \min_{z \in \Theta_2} \int_0^T L(t, z(t), z'(t)) dt$$

$$\Theta_2 = \{z \in \widehat{C}^1[0, T] \mid z(0) = 0, z(T) = b\}$$

where $L(\cdot, \cdot, \cdot)$ and $L_z(\cdot, \cdot, \cdot)$ are the class $C^0$ and $L_{z'}(t, z, \cdot)$ is piecewise continuous ($L_{z'}(t, z, \cdot)$ is discontinuous in $z' = 0$). Denoting by $\mathbb{¥}_q(t), q \in \Theta_2$ the function:

$$\mathbb{¥}_q(t) := -L_{z'}(t, q(t), q'(t)) + \int_0^t L_z(s, q(s), q'(s)) ds \tag{1.5}$$

and by $\mathbb{¥}_q^+(t)$ and $\mathbb{¥}_q^-(t)$ the expressions obtained when considering the lateral derivatives with respect to $z'$. The problem $(P_2)$ was formulated within the framework of nonsmooth analysis [4], using the generalized (or Clarke's) gradient, the following result being proven:

**Theorem 2.** *If $q$ is a solution of $(P_2)$, then $\exists K \in \mathbb{R}^+$ such that:*

$$\begin{cases} \mathbb{¥}_q^+(t) = \mathbb{¥}_q^-(t) = K & \text{if } q'(t) \neq 0 \\ \mathbb{¥}_q^+(t) \geq K \geq \mathbb{¥}_q^-(t) & \text{if } q'(t) = 0 \end{cases}$$

This paper merges the two previous studies, simultaneously considering non-regular Lagrangian and non-holonomic inequality constraints (differential inclusions), obtaining a necessary minimum condition. Furthermore, under certain convexity conditions, we shall establish the result (*smooth transition*) that the derivative of the minimum presents a constancy interval, the constant being the value for which $L_{z'}(t, z, \cdot)$ presents discontinuity. Finally, we shall present a solution algorithm and shall apply it to an example.

## 2 Mathematical Statement and Resolution of the Problem

In this paper, we consider a hydrothermal system with one thermal plant (the thermal equivalent) and one pumped-hydro plant, which will have certain constraints in both geneararion and pumping for $H$. We shall take $H_{\min}$ (maximum pumping capacity) as the lower boundary and $H_s(t) = \min\{H_{\max}, P_d(t)\}$ ($H_{\max}$ being maximum generation) as the upper boundary.

The mathematical problem $(P_3)$ may be stated in the following terms:

$$\min_{z\in\Theta} F(z) = \min_{z\in\Theta} \int_0^T \Psi\left[P_d(t) - H(t, z(t), z'(t))\right] dt = \min_{z\in\Theta} \int_0^T L(t, z(t), z'(t)) dt$$

$$\Theta = \{z \in \widehat{C}^1[0, T] \mid z(0) = 0, z(T) = b, H_{\min} \le H(t, z(t), z'(t)) \le H_s(t), \forall t \in [0, T]\}$$

where $L(\cdot, \cdot, \cdot)$ and $L_z(\cdot, \cdot, \cdot)$ are the class $C^0$ and $L_{z'}(t, z, \cdot)$ is piecewise continuous. We shall assume that $\Psi$ is strictly increasing and strictly convex, that $H$ verifies $H_{z'} > 0$, and $H_z(t, z(t), 0) = 0$, and the strictly increasing nature of $L_{z'}(t, z, \cdot)$. We shall establish the necessary minimum condition for this problem with non-regular Langrangian and constraints on the admissible functions, employing to this end the coordination function, $\mathbb{Y}_q(t)$.

We shall denote by $\mathbb{Y}_q^+(t)$ and $\mathbb{Y}_q^-(t)$ the expressions obtained when considering in (1.4) the lateral derivatives with respect to $z'$. We shall prove that these functions also verify Theorem 2 in the same way as $¥_q^+(t)$ and $¥_q^-(t)$, and that for the stated problem, except in those values of $z'$ for which $L_{z'}(t, z, \cdot)$ is not continuous, Theorem 1 will continue to be valid. We thus obtain the following result:

**Theorem 3.** *If $q$ is a solution of $(P_3)$, then $\exists K \in \mathbb{R}^+$ such that:*

i) *If $L_{z'}(t, q(t), \cdot)$ is discontinuous at $q'(t)$* $\implies$ $\mathbb{Y}_q^+(t) \le K \le \mathbb{Y}_q^-(t)$
ii) *If $L_{z'}(t, q(t), \cdot)$ is continuous at $q'(t)$* $\implies$

$$\mathbb{Y}_q(t) \quad is \quad \begin{cases} \le K & if \quad H(t, q(t), q'(t)) = H_{\min} \\ = K & if \quad H_{\min} < H(t, q(t), q'(t)) < H_s(t) \\ \ge K & if \quad H(t, q(t), q'(t)) = H_s(t) \end{cases}$$

## 3   Smooth Transition

In this section, we present a qualitative aspect of the solution of $(P_3)$. We prove that, under certain conditions, the discontinuity of the derivative of the Langrangian does not translate as discontinuity in the derivative of the solution. In fact, it is verifed that the derivative of the extremal where the minimum is reached presents an interval of constancy, the constant being the value for which $L_{z'}(t, z, \cdot)$ presents discontinuity. The character $C^1$ of the solution is thus guaranteed.

**Theorem 4.** *Let $L(\cdot, \cdot, \cdot)$ be the Lagrangian of the functional $F$ in the conditons stated above, and let us assume that the function $L_{z'}(t_0, z(t_0), \cdot)$ is strictly increasing (decreasing) and discontinuous in 0. If $q$ is minimum (maximum) for $F$, then:*

*(i) $t_0$ is not an isolated point of a change in the sign of $q'$.*
*(ii) $q' \equiv 0$ in some interval that contains $t_0$.*
*(iii) $q'$ is continuous in $t_0$.*

Note that this result has a very clear interpretation in terms of pumping plants: under optimum operating conditions, pumping plants never switch brusquely from generating power to pumping water or vice versa, but rather carry out a smooth transition, remaining inactive during a certain period of time.

## 4   Optimization Algorithm

¿From the computational point of view, the construction of $q_K$ can be performed with the use of a discretized version of Theorem 3. The problem will consist in finding for each $K$ the function $q_K$ that satisfies conditions i) and ii) of Theorem 3, and from among these functions, an admissible function $q_K \in \Theta$. In general, the construction of $q_K$ cannot be carried out all at once over the entire interval $[0, T]$. The construction must necessarily be carried out by constructing and successively concatenating the extremal arcs, until completing the interval $[0, T]$, where:

· $H_{\min} < H(t, q(t), q'(t)) < H_s(t)$ (free extremal arcs), or
· $q'(t) = 0$ (the hydro-plant is on shut-down), or
· $H(t, q(t), q'(t)) = H_s(t)$ (the hydro-plant generates all the demanded power or its technical maximum), or
· $H(t, q(t), q'(t)) = H_{\min}$ (the hydro-plant is functioning at its maximum pumping power)

If the values obtained for $q$ and $q'$ do not obey the constraints, we force the solution $q_K$ to belong to the boundary until the moment when the conditions of leaving the domain (established in Theorem 3) are fulfilled.

## 5   Application to a Hydrothermal Problem

A program that resolves the optimization problem was elaborated using the Mathematica package and was then applied to one example of hydrothermal system made up of the thermal equivalent plant and a hydraulic pumped-storage plant.

We use the quadratic model: $\Psi(x) := c_1 + c_2 x + c_3 x^2$, for the fuel cost model of the equivalent thermal plant. The power production $H$ of the hydroplant (variable head) is a function of $z(t)$ and $z'(t)$ and its power consumption during pumping is a lineal function of the amount of water pumped ($M \cdot z'(t)$). Hence the function $H$ is defined piecewise as

$$H(t, z(t), z'(t)) := \begin{cases} A(t) \cdot z'(t) - B \cdot z(t) \cdot z'(t) & \text{if} \quad z'(t) > 0 \\ M \cdot z'(t) & \text{if} \quad z'(t) \le 0 \end{cases}$$

where $A(t) = \frac{B_y}{G}(S_0 + t \cdot i)$, $B = \frac{B_y}{G}$. The parameters are: $G(m^4/h.Mw)$ the efficiency, $i(m^3/h)$ the natural inflow, $S_0(m^3)$ the initial volume, and $B_y(m^{-2})$ a parameter that depends on the geometry of the tanks. $M(h.Mw/m^3)$ is the factor of water-conversion of the pumped-storage plant and we consider two cases: (1) $M_1 = (1,04)A(0)$ and (2) $M_2 = (1,10)A(0)$.

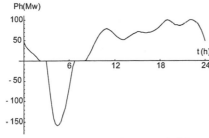

Figure 1. Optimal solution with $M_1$.     Figure 2. Optimal solution with $M_2$.

Figures 1 and 2 presents the optimum solution. It can be seen how the interval of smooth transtion varies when considering two different values of $M$.

## References

[1] L. Bayón, J. Grau, M.M. Ruiz and P.M. Suárez, A Bolza Problem in Hydrothermal Optimization. Lecture Series on Computer and Computational Sciences (Proceedings of the ICCMSE 2004), Eds.: T.E. Simos and G. Maroulis; Vol. I, 57-60, 2004.

[2] E. Lee and L. Markus, *Foundations of optimal control theory*, John Wiley & Sons, New York, 1967.

[3] L. Bayón, J. Grau, M.M. Ruiz and P.M. Suárez, Nonsmooth Optimization of Hydrothermal Problems. Proceedings of the CMMSE 2004, Eds.: E. Brandas, and J. Vigo-Aguiar; Vol. I, 116-125, 2004.

[4] F.H. Clarke, *Optimization and nonsmooth analysis*, John Wiley & Sons, New York, 1983.

Brill Academic Publishers
P.O. Box 9000, 2300 PA Leiden
The Netherlands

*Lecture Series on Computer
and Computational Sciences*
Volume 4, 2005, pp. 65

# Cross-layer Simulations for Performance Evaluation of Self-organized Wireless Networks

Monique BECKER[1a], Alexandre DELYE de CLAUZADE de MAZIEUX[1], Vincent GAUTHIER,
Michel MAROT[1]

Received 25 June, 2005; accepted in revised form 13 July, 2005

*Abstract:* Ad hoc wireless networks are self organizing multi-hop wireless networks where all nodes take part in the process of forwarding packets, hence they are very different from conventional computer networks. First, radio resources are rare and time varying. Second, the network topology is mobile and the connectivity is unpredictable. Third, if we consider, for example, an architecture-based on 802.11 WLAN further problems appear due to, for example, the presence of hidden stations, exposed stations and the capturing phenomena.

Due to the complex interactions between processes and protocols in a traditional layered architecture, the intrinsic dynamic characteristics of resource availability in an ad-hoc wireless environment and the emerging end-users' requirements, we advocate that a simulation environment that takes into account inter-layer interaction can provide a suitable framework to analyse and design wireless networks that need to maintain the quality of the service (QoS) offered to end-users. This approach allows upper layer(s) to consider the performance of lower layers in order to, for example, select a path which provides QoS guarantees.

We present a few results that lead us to believe that a dynamic autonomic simulation environment should be built.

*Keywords*: wireless, networks, performance evaluation, simulations, cross-layer, Ad-hoc, optimization.

## References

[1] Alexandre Delye de Clauzade de Mazieux, Vincent Gauthier, Michel Marot, Monique Becker, "CAAC Mechanism : Cluster Address AutoConfiguration Mechanism" (July 2005). 3rd International Conference on Performance Modelling and Evaluation of Heterogeneous Networks, HET'NET'05, Ilkley( UK).

[2] V. Gauthier, R. de Rasse, M. Marot, M. Becker, "On a Comparison of some Ad-hoc Routing Protocols taking into account the Radio Interferences", (July 2005), 3rd International Conference on Performance Modelling and Evaluation of Heterogeneous Networks, HET'NET'05, Ilkley( UK).

[1] CNRS SAMOVAR (UMR-5157) GET/INT, 9 rue Charles Fourier, 91011 Evry Cedex, France
[a] Corresponding author: monique.becker@int-evry.fr

Brill Academic Publishers
P.O. Box 9000, 2300 PA Leiden,
The Netherlands

*Lecture Series on Computer
and Computational Sciences*
Volume 4, 2005, pp. 66-69

# Tuning a Portfolio Management Model on Historical Data using a Parallel Three-Stage Stochastic Programming Model

S.Benkner[2], L. Halada[3], M.Lucka[1] and I. Melichercik[4]

[1,2]Institute of Scientific Computing, University of Vienna
Nordbergstrasse 15, A-1090 Vienna, Austria

[3]Institute for Informatics, Slovak Academy of Sciences,
Dubravska 9, 842 37 Bratislava, Slovakia

[4]Department of Applied Mathematics and Statistics,
Faculty of Mathematics, Physics and Informatics,
Comenius University, Mlynská Dolina, Bratislava, Slovakia

Received 8 July, 2005; accepted in revised form 30 July, 2005

*Abstract:* A multi-stage model for allocation of financial resources to bond indices in different currencies is presented. The model is tested on historical data of interest rates and exchange rates. We consider a three-stage stochastic programming model solved by the interior point method (IPM). The application of the BQ method to the IPM allows a decomposition of the large linear system to smaller blocks allowing thus solving it in parallel. The designed parallel code is written in the Fortran/MPI programming language using the LAPACK library calls for managing the linear algebra operations on small block matrices and experiments were executed on a IMP 1350 Linux cluster. We outline the possibility of using of the program as a Grid service in the context of the Vienna Grid Environment.

*Keywords:* portfolio management, interior point method, parallel computing, large-scale optimization, stochastic programming

*Mathematics Subject Classification:* 90C15, 82B31, 60H35, 68W10, 90C51, 90C06

## 1 Introduction

We present a model for allocation of financial resources to bond indices in different currencies. Such portfolios are subject to interest rate risk and exchange rate risk. The stochastic properties of possible future development of exchange rates and interest rates are represented in the form of a scenario tree. The objective is to maximize the expected value of the portfolio at defined time horizon. Mathematically such models lead to multi-stage stochastic programming. When one deals with several currencies, the realistic trees are "bushy" and the number of scenarios grows exponentially with the number of stages. Thus, the computation of such problems could be extremely large and computationally intractable. Multi-stage models are recently often used in the bond portfolio management. Concerning the development of such models we refer to successful and valuable contributions of [1], [2] and [3]. In [5] a multi-period dynamic model for fixed-income portfolio management under uncertainty, using multi-stage stochastic programming was developed.

---

[1]Corresponding authors. E-mail: lucka@par.univie.ac.at, igor.melichercik@fmph.uniba.sk.

The scenarios of the term structure of interest rates were generated using Monte Carlo simulations. Their results confirmed that multi-period models outperform classical single-period models. There is a considerable research effort to develop efficient parallel methods for solving this problem on parallel computer architectures. In [4] we have presented a parallel interior point algorithm for solving three-stage stochastic linear problem which comes from three-period models of portfolio management. As a next step, the efficiency of the model is being tested on historical data of interest rates and exchange rates.

## 2 Problem formulation

The stochastic properties are represented in the form of a scenario tree. Denote by $\mathcal{F}_\tau$, $1 \leq \tau \leq T$ the set of nodes at time $\tau$. For any $\omega \in \mathcal{F}_\tau$, $1 < \tau \leq T$, there is a unique element $a(\omega) = \omega' \in \mathcal{F}_{\tau-1}$, which is the unique predecessor of $\omega$.

**Financial instruments.** Denote the decision variables of this process as
$b_j^{(\tau)}(\omega^\tau), s_j^{(\tau)}(\omega^\tau), h_j^{(\tau)}(\omega^\tau), \; \omega^\tau \in \mathcal{F}_\tau$: The amount of index j bought, sold and held in period $\tau$, and constants as
$c^{(0)}, h_j^{(0)}$: Initial cash available and composition of the initial portfolio,
$\xi_j^{(\tau)}(\omega^\tau), \chi_j^{(\tau)}(\omega^\tau), \; \omega^\tau \in \mathcal{F}_\tau$: Bid and ask prices of the $j$-th index in the base currency.

**Constraints.** The amount of bought and sold units of index should be nonnegative. We forbid short positions. Therefore the number of hold units is nonnegative.

**Equations.** *Inventory balance and cash-flow accounting for the Period 1*

$$h_j^{(0)} + b_j^{(1)} - s_j^{(1)} = h_j^{(1)} \; \forall j \quad \text{and} \quad c^{(0)} + \sum_j \xi_j^{(1)} s_j^{(1)} = \sum_j \chi_j^{(1)} b_j^{(1)}. \tag{1}$$

*Inventory balance and cash-flow accounting for the Period $\tau$, where $1 < \tau < T$:*

$$h_j^{(\tau-1)}(a(\omega^\tau)) + b_j^{(\tau)}(\omega^\tau) - s_j^{(\tau)}(\omega^\tau) = h_j^{(\tau)}(\omega^\tau) \quad \forall \omega^\tau \in \mathcal{F}_\tau \; \forall j. \tag{2}$$

$$\sum_j \xi_j^{(\tau)}(\omega^\tau) s_j^{(\tau)}(\omega^\tau) = \sum_j \chi_j^{(\tau)}(\omega^\tau) b_j^{(\tau)}(\omega^\tau) \quad \forall \omega^\tau \in \mathcal{F}_\tau. \tag{3}$$

**Objective function.** The terminal wealth calculation is given by:

$$WT(\omega^T) = \sum_j \xi_j^{(T)}(\omega^T) h_j^{(T-1)}(a(\omega^T)) \quad \forall \omega^T \in \mathcal{F}_T.$$

The objective function maximizes the expected terminal wealth. It can be written as:

$$\text{Maximize } E(WT), \tag{4}$$

where $E(WT) = \sum_{\omega^T \in \mathcal{F}_T} \pi(\omega^T) WT(\omega^T)$ and $\pi(\omega^T)$ is the probability of the scenario $\omega^T$.

## 3 Testing the model using historical data

The three-stage model is being tested using historical data of interest rates and exchange rates. We consider portfolios composed of 4 indices. The indices include USD, CHF, GBP and EUR government bonds. At each time step a scenario tree and the input data for the stochastic program must be generated. The initial composition of the portfolio is represented by the decisions $h_j^{(1)}$, $j = 1, 2, 3, 4$ from the previous calculation. The parameters of the scenario tree are functions of past historical data.

Mathematical formulation of the objective function is expressed in the form $c^T x$, where $x$ is the vector of decision variables and the vector $c$ is derived from (4). In the case of the three-stage problem it leads to solving the equation

$$maximize\ c^T x,\ subj.to\ A^{(3)}x = b. \tag{5}$$

The matrix $A^{(3)}$ is a sparse matrix and has a regular structure.

$$A^{(3)} = \begin{pmatrix} A_0^{(2)} & & & & & \\ T_1^{(2)} & A_1^{(2)} & & & & \\ T_2^{(2)} & & A_2^{(2)} & & & \\ T_3^{(2)} & & & A_3^{(2)} & & \\ \cdots & & & & \cdots & \\ \cdots & & & & & \cdots \\ T_{N^{(3)}}^{(2)} & & & & & A_{N^{(3)}}^{(2)} \end{pmatrix}$$

The regular structure means that every matrix $A_k^{(2)}$ has the same structure as $A^{(3)}$. The matrices $A_k^{(1)}$ (for 4 currencies) have the size 5x12 and have 16 nonzero elements. The matrices $T_k^{(2)}$ are sparse matrices with 4 nonzero elements and the same size as $A_k^{(2)}$. All nonzero elements of the matrix $A^{(3)}$ and the vector $b$ are derived from equations (1) - (3). There are many possibilities how the problem (5) can be solved. We have used the Interior Point Method (IPM) in the frame of the Mehrotra's Predictor Corrector algorithm (MPC) defined in [8], p.198. The MPC method is an iterative algorithm creating a sequence of solutions $(x^{k+1}, y^{k+1}, z^{k+1})$, $k = 0, 1, 2....$ The crucial step by finding them is solving the system of linear equations with the system matrix

$$A^{(3)} D^{(3)} (A^{(3)})^T, \tag{6}$$

where $D^{(3)}$ is a diagonal matrix. A three-stage procedure for solving this problem was designed and summarized in [7]. We have proposed and implemented a parallel algorithm based on the MPC algorithm for solving the problem (5). For implementation details see [6] and [4]. The application of the IPM method to the system (6) decomposes this "large" linear system into many "small" block systems that are solved in parallel with very few communications between processes. For solving linear algebra operations on small blocks the linear algebra package LAPACK library calls are used in the frame of an MPI implementation. There is no need to store the whole matrix in this step. But in the MPC algorithm one need to determine the right-hand side of the equation 5 and to multiplicate the matrix $A^{(3)}$ by vector $x^{(k)}$. For these reasons we have stored it in the compressed row storage form and used a sparse matrix-vector multiplication. The experiments were executed on a new IBM Linux cluster, University of Technology, Vienna, consisting of 144 Pentium 4 (3.6GHz) processors (2 processors per node) communicating over Fast Ethernet node interconnect. The execution time needed for solving one three-stage problem $81 \times 81 \times 16$ (81 possibilities of the price vector at the end of the first period, 81 possibilities of the price vector for each position $\omega^2 \in \mathcal{F}_2$ at the end of the second period and 16 possibilities of the price vector for each position $\omega^3 \in \mathcal{F}_3$ at the end of the third period), where the size of the matrix (6) was 33215x33215, varied in dependence on the number of processors from 10 to 2 seconds. The preliminary testing calculations ([4]) showed that:

1. it makes sense to increase the number of stages,
2. there is a difference between the multi-stage decision and successive single-stage decisions,
3. the optimal decision is sensitive with respect to different generations of the scenario tree.

The time measurements experiments for the whole problem will be presented at the conference and will be the subject of our next paper. The achieved results proved that increasing number of stages improves the quality of the optimal decision.

For facilitating the use of the whole program product (tree generation and stochastic program) we plan to provide the three-stage stochastic program as a Grid service using the Vienna Grid Environment (VGE) [9]. The user can choose the model for the scenario tree generation and give the initial composition of the portfolio. The VGE enables to transparently run the code on an appropriate parallel architecture with respect to the size of the problem and the time needed for the execution. Another possibility is to provide the input data in the prescribed format and to run only the three-stage stochastic program solving the problem (5) as a Grid service. The VGE gives moreover the possibility to many users to run the three-stage stochastic programs on different input data simultaneously. A more detailed description of the Grid service as well as experimental results will be a presented at the conference and will be the subject of our future work.

## Acknowledgment

This work was supported by the Special Research Program SFB F011 "AURORA" of the Austrian Science Fund FWF and VEGA Agency (1/9154/02, 1/1004/04), Slovakia.

## References

[1] A. Beltratti, A. Consiglio and S. A. Zenios, Scenario Modeling for the Management of International Bond Portfolios, *Annals of Oper. Res.* **85**, pp.227–247, 1999.

[2] A. J. Bradley and D. B. Crane, A Dynamic Model for Bond Portfolio Management, *Management Science* **19**, pp.139–151, 1972.

[3] B. Golub, M. Holmer, R. Mc Kendall and S. A. Zenios, Stochastic Programming Models for Money Management, *European Journal of Operations Research* **85**, pp.282–296, 1995.

[4] L. Halada, M. Lucka and I. Melichercik, Optimal Multistage Portfolio Management Using a Parallel Interior Point Method,*Proc. of ALGORITMY'2005*, High Tatras, Slovakia, March 13-18, pp. 359-368, 2005.

[5] M. Holmer, R. Mc Kendall, Ch. Vassiadou-Zeniou and S. A. Zenios, Dynamic Models for Fixed-Income Portfolio Management under Uncertainty, *Journal of Economic Dynamics and Control* **22**, pp.1517–1541, 1998.

[6] L. Halada, S. Benkner, M. Lucka, Parallel Birge and Qi Factorization of Three-stage Stochastic Linear Programs, *Parallel and Distributed Computing Practices* **5**, No.3, Nova Science Publishers, pp. 301–311, 2002.

[7] G. CH. Pflug, L. Halada, A Note on the Recursive and Parallel Structures of the Birge and Qi Factorization for Tree Structured Linear Programs, *Computational Optimization and Applications* **24**, No. 2–3, pp. 251–265, 2003.

[8] Stephen. J. Wright, *Primal-Dual Interior Point Methods*, ISBN 0-89871-382-X, SIAM, 1997.

[9] S. Benkner, I. Brandic, G. Engelbrecht, R. Schmidt, VGE - A Service-Oriented Environment for On-Demand Supercomputing, *Proceedings of the Fifth IEEE/ACM International Workshop on Grid Computing (Grid 2004)*, Pittsburgh, PA, USA, November, 2004.

Brill Academic Publishers
P.O. Box 9000, 2300 PA Leiden
The Netherlands

*Lecture Series on Computer*
*and Computational Sciences*
Volume 4, 2005, pp. 70-73

# Inverse Laplace transforms of luminescence relaxation functions

Mário N. Berberan-Santos[1]

Centro de Química-Física Molecular,
Instituto Superior Técnico,
1049-001 Lisboa, Portugal

Received 14 June, 2005; accepted in revised form 12 July, 2005

*Abstract:* Laplace transforms find application in many fields, including time-resolved luminescence. In this work, relations that allow a direct (i.e., dispensing contour integration) analytical calculation of the original function from its transform are re-derived and used for the determination of the distributions of rate constants of several relaxation functions, including the stretched exponential (Kohlrausch) and the compressed hyperbola (Becquerel) luminescence decay laws, and the asymptotic power law and the Mittag-Leffler relaxation functions. General results concerning the relation between relaxation function and distribution of rate constants are also obtained.

*Keywords:* Laplace transform, luminescence decay kinetics, relaxation function, stretched exponential, asymptotic power law, Mittag-Leffler function.

*Mathematics Subject Classification:* 33E12 Mittag-Leffler functions and generalizations, 44A10 Laplace transform, 60E07 Infinitely divisible distributions; stable distributions

*PACS:* 02.30.Gp, 02.30.Uu

## 1. Introduction

The Laplace transform $F(s)$ of a function $f(t)$ is defined by [1-3]

$$F(s) = \int_0^\infty f(t)e^{-st}dt. \tag{1}$$

The Laplace transform is a powerful tool for solving linear differential equations, ordinary and partial; linear difference equations; and linear equations involving convolutions. For this reason, it finds application in many fields. Furthermore, in relaxation processes, including time-resolved luminescence spectroscopy, the relaxation function is either the transform or the original function of a Laplace transform pair, the other function of the pair being also of physical relevance.

Luminescence decays are widely used in the physical, chemical and biological sciences to get information on the structure and dynamics of molecular, macromolecular, supramolecular, and nano systems [4]. In the simplest cases, luminescence decay curves can be satisfactorily described by a sum of discrete exponentials, and the respective pre-exponential factors and decay times have a clear physical meaning. However, the luminescence decays of inorganic solids are usually complex. Continuous distributions of decay times or rate constants are also necessary to account for the observed fluorescence decays of molecules incorporated in micelles, cyclodextrins, rigid solutions, sol-gel matrices, proteins, vesicles and membranes, biological tissues, molecules adsorbed on surfaces or linked to surfaces, energy transfer in assemblies of like or unlike molecules, etc.

In such cases, the luminescence decay is written in the following form:

$$I(t) = \int_0^\infty H(k)e^{-kt}dk, \tag{2}$$

---

[1] E-mail: berberan@ist.utl.pt

with $I(0)=1$. This relation is always valid because $H(k)$ is the inverse Laplace transform of $I(t)$, which is a well-behaved function. The function $H(k)$, also called the eigenvalue spectrum (of a suitable kinetic matrix), is normalized, as $I(0)=1$ implies that $\int_0^\infty H(k)dk = 1$. In most situations (e.g. in the absence of a rise-time in the decay), the function $H(k)$ is nonnegative for all $k>0$, and $H(k)$ can be understood as a distribution of rate constants (strictly, a probability density function, PDF). This PDF, or distribution of rate constants, gives important information of the dynamics of the luminescent systems [5,6], but is not always easy to infer from the decay law $I(t)$. In the remaining of this work, and in view of the specific application to be considered, the notation of Eq. (2) will be retained.

The more difficult step in the application of Laplace transforms is the inversion of the transform to obtain the desired solution. In many cases, the inversion is accomplished by consulting published tables of Laplace transform pairs [1-3]. More generally, and in the absence of such a pair, the inversion integral can be applied [2,3]. This integral is

$$H(k) = \frac{1}{2\pi i} \int_{c-i\infty}^{c+i\infty} I(t)e^{kt}\, dt,$$

(3)

where $c$ is a real number larger than $c_0$, $c_0$ being such that $I(z)$ has some form of singularity on the line $Re(z) = c_0$ but is analytic in the complex plane to the right of that line, i.e., for $Re(z) > c_0$. Eq. (3) is usually evaluated by contour integration [2,3].

## 2. General inverse Laplace transform

A simple form of the inverse Laplace transform of a relaxation function can be obtained by the method outlined in [7]. Briefly, the three following equations can be used for the direct inversion of a function $I(t)$ to obtain its inverse $H(k)$,

$$H(k) = \frac{e^{ck}}{\pi} \int_0^\infty \left[ Re[I(c+i\omega)]\cos(k\omega) - Im[I(c+i\omega)]\sin(k\omega) \right] d\omega,$$

(4)

$$H(k) = \frac{2e^{ck}}{\pi} \int_0^\infty Re[I(c+i\omega)]\cos(k\omega)\, d\omega \quad k>0,$$

(5)

$$H(k) = -\frac{2e^{ck}}{\pi} \int_0^\infty Im[I(c+i\omega)]\sin(k\omega)\, d\omega \quad k>0.$$

(6)

where $c$ was defined in Eq. (3).

## 3. An example of a continuous distribution: Stretched exponential (Kohlrausch) kinetics

For the stretched exponential (or Kohlrausch) decay law

$$I(t) = \exp\left[ -\left( \frac{t}{\tau_0} \right)^\beta \right],$$

(7)

where $\tau_0$ is a parameter with dimensions of time, one obtains, from Eqs. (4) and (7) [7,8]

$$H_\beta(k) = \frac{\tau_0}{\pi} \int_0^\infty \exp\left[ -u^\beta \cos\left( \frac{\beta\pi}{2} \right) \right] \cos\left[ k\tau_0 u - u^\beta \sin\left( \frac{\beta\pi}{2} \right) \right] du.$$

(8)

From Eqs. (5) and (6) alternative forms are ($k > 0$)

$$H_\beta(k) = \frac{2\tau_0}{\pi} \int_0^\infty \exp\left[ -u^\beta \cos\left( \frac{\beta\pi}{2} \right) \right] \cos\left[ u^\beta \sin\left( \frac{\beta\pi}{2} \right) \right] \cos(k\tau_0 u) du,$$

(9)

and

$$H_\beta(k) = \frac{2\tau_0}{\pi} \int_0^\infty \exp\left[ -u^\beta \cos\left( \frac{\beta\pi}{2} \right) \right] \sin\left[ u^\beta \sin\left( \frac{\beta\pi}{2} \right) \right] \sin(k\tau_0 u) du.$$

(10)

The function $H_\beta(k)$ is plotted in Fig. 1 for several values of the parameter $\beta$.

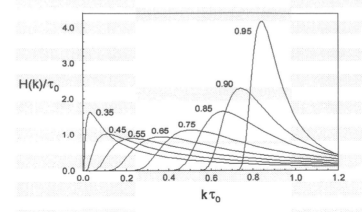

Figure 1: Distribution of rate constants (probability density function) for the Kohlrausch decay law obtained by numerical integration of Eqs. (8) and (13). The number next to each curve is the respective $\beta$.

$H_\beta(k)$ can be expressed by elementary functions only for $\beta=1/2$,

$$H_{1/2}(k) = \frac{e^{-\frac{1}{4k\tau_0}}}{\sqrt{4\pi k^3 \tau_0}} \quad ,$$

(11)

and is variously called Smirnov and Lévy PDF. A form explicitly displaying the asymptotic behavior for large $k$ was recently obtained for $\beta=1/4$ [9],

$$H_{1/4}(k) = \frac{\tau_0}{8\pi (k\tau_0)^{5/4}} \int_0^\infty u^{-3/4} \exp\left[-\frac{1}{4}\left(\frac{1}{\sqrt{k\tau_0 u}}+u\right)\right] du \ .$$

(12)

Pollard's relation [10] which is the only previously known integral representation of $H_\beta(k)$,

$$H_\beta(k) = \frac{\tau_0}{\pi} \int_0^\infty \exp(-k\tau_0 u)\exp\left[-u^\beta \cos(\beta\pi)\right]\sin\left[u^\beta \sin(\beta\pi)\right] du.$$

(13)

was obtained from Eq. (3) by defining a special contour. Eqs. (8)-(10) can also be obtained by contour integration [8], but not so directly as by the procedure described above.

## 4. An example of a discrete distribution: mixed first- and second-order kinetics

For mixed first- and second-order kinetics the decay of delayed fluorescence is [11]

$$I(t) = \left(\frac{\beta}{e^{k_1 t} - 1 + \beta}\right)^2,$$

(14)

where

$$\beta = \frac{k_1}{k_1 + k_2 N_0},$$

(15)

and $N_0$ is the initial number of excited luminophores. Eq. (14) gives, upon expansion

$$I(t) = \beta^2 \sum_{n=1}^\infty n(1-\beta)^{n-1} e^{-(n+1)k_1 t},$$

(16)

hence the rate constant spectrum is infinite but discrete

$$H(k) = \beta^2 \sum_{n=1}^\infty n(1-\beta)^{n-1} \delta\left[k-(n+1)k_1\right].$$

(17)

## References

[1]  D. V. Widder: *Advanced Calculus*. Prentice-Hall, Englewood Cliffs, 1947.

[2]  R. E. Bellman, R. S. Roth: *The Laplace Transform*. World Scientific, Singapore, 1984.

[3]  D. A. McQuarrie: *Mathematical Methods for Scientists and Engineers*. University Science Books, Sausalito, 2003.

[4]  B. Valeur: *Molecular Fluorescence. Principles and applications.* Wiley-VCH, Weinheim, 2002.

[5]  M. N. Berberan-Santos, P. Choppinet, A. Fedorov, L. Jullien, B. Valeur, Multichromophoric cyclodextrins. 6. Investigation of excitation energy hopping by Monte-Carlo simulations and time-resolved fluorescence anisotropy, *Journal of the American Chemical Society* **121** 2526-2533 (1999).

[6]  E.N. Bodunov, M.N. Berberan-Santos, E.J. Nunes Pereira, J.M.G. Martinho, Eigenvalue spectrum of the survival probability of excitation in nonradiative energy transport, *Chemical Physics* **259** 49-61 (2000).

[7]  M.N. Berberan-Santos, Analytical inversion of the Laplace transform without contour integration. Application to luminescence decay laws and other relaxation functions, *Journal of Mathematical Chemistry* **38** 163-171 (2005).

[8]  M.N. Berberan-Santos, E.N. Bodunov, B. Valeur, Mathematical functions for the analysis of luminescence decays with underlying distributions 1. Kohlrausch decay function (stretched exponential), *Chemical Physics* in press (2005).

[9]  M.N. Berberan-Santos, Relation between the inverse Laplace transforms of $I(t^\beta)$ and $I(t)$. Application to the Mittag-Leffler and asymptotic inverse power law relaxation functions, *Journal of Mathematical Chemistry* **38** 263-268 (2005).

[10] H. Pollard, The representation of $e^{-x^\lambda}$ as a Laplace integral, *Bulletin of the American Mathematical Society* **52** 908-910 (1946).

[11] M.N. Berberan-Santos, E.N. Bodunov, B. Valeur, Mathematical functions for the analysis of luminescence decays with underlying distributions 2. Becquerel (compressed hyperbola) and related decay functions, *Chemical Physics* in press (2005).

Brill Academic Publishers
P.O. Box 9000, 2300 PA Leiden
The Netherlands

*Lecture Series on Computer
and Computational Sciences*
Volume 4, 2005, pp. 74-76

# Calculation of Vibrational Levels of Three-Atomic Molecules in the Ground Rotational State on Example of NO₂.

D.S. Bezrukov[1], Yu.V. Novakovskaya

Laboratory of Quantum Mechanics and Molecular Structure,
Faculty of Chemistry, Moscow State University,
119992 Moscow, Russia

A. Bäck, G. Nyman

,Department of Chemistry, Physical Chemistry,
Göteborg University,
SE-412 96 Göteborg, Sweden

Received 9 August, 2005; accepted in revised form 12 August, 2005

*Abstract:* A new method, allowing calculation all bound vibrational states of threeatomic molecules (J=0), was applied to NO2. The advantage of the method as compared to the proceeding one is the high calculation speed for the whole vibrational spectrum (at least 1000 times faster). The obtained values are in a excellent agreement with the values of [1]. Unfortunately, this method doesn't provide eigenfunctions.

*Keywords:* Vibrational levels, hyperspherical coordinates, nitrogen dioxide

## 1. Introduction

Calculations of vibrational eigenvalues have been a subject of the great interest due to its fundamental relevance to experementalists (to obtain molecular structure) and theoreticians (to develop better potential energy surfaces).

For diatomic molecules, which have only one vibrational degree of freedom, composing and diagonalization of the Hamiltonian matrix ($10^4$-$10^6$ elements) is a simple task. The similar task gets much more difficault even in a case of three-atomic molecules, due to a large size of the Hamiltonian matrix constituting as many as $10^7$x$10^7$ elements. The methods used to treat this problem include both time-dependent and time-independent approaches. If one tries to calculate vibrational levels near the dissociation threshold, only time-independent methods can provide good results without significant time concerning.

Nitrogen dioxide plays a great role in atmospheric processes. A lot of investigations were devoted to this molecule, including calculations of all vibrational levels in a frame of time-independent approaches [1]. In this work modified scheme for a rapid calculation of all bound vibrational levels is proposed.

## 2. Mathematical details

We used a slightly modified [2] version of hyperspherical coordinates of Johnson [3-5]. The Hamiltonian for the vibrational motion of a triatomic molecule with zero total angular momentum (J=0) in modified hyperspherical coordinates is given by the following expression:

$$\hat{H} = -\frac{\hbar^2}{2\mu}\frac{\partial^2}{\partial\rho^2} + \frac{8\hbar^2}{\mu\rho^2}\hat{L}^2(\theta,\phi) + \frac{15\hbar^2}{8\mu\rho^2} + V(\rho,\theta,\phi),\tag{1}$$

where $\hat{L}^2(\theta,\phi)$ is the grand angular momentum operator,

---

[1] Corresponding author. E-mail: 20mitya02@mail.ru

$$L^2(\theta,\phi) = -\left( \frac{1}{\sin\theta} \frac{\partial}{\partial\theta} \sin\theta \frac{\partial}{\partial\theta} + \frac{1}{4\sin^2(\theta/2)} \frac{\partial^2}{\partial\phi^2} \right). \qquad (2)$$

For eliminating singularities, the mixed basis-grid representation of a wavefunction was used, namely, grid representation for $\rho$ combined with the hyperspherical harmonics in $\phi$ and $\theta$.

In calculating the vibrational eigenvalues we used a filtering technique taking the Green operator as a filter. The main idea of this approach is in the construction of a filter operator $F(\hat{H}, E)$, which acts on a random wavefunction $\psi_0$ and gives filtered wavefunction $\psi_E$, which in some sense close to eigenfunction corresponding to energy E.

$$F(\hat{H}, E)\psi_0 = \psi_E, \qquad (3)$$

$$F(\hat{H}, E) = \frac{1}{\hat{H} - E}, \qquad (4)$$

$$(\hat{H} - E)\psi_E = \psi_0, \qquad (5)$$

Lanczos basis was used to construct the Hamiltonian matrix which is sparse in this basis so that we can construct such matrix, which contains the whole vibrational spectrum. The program [1] has used an iterative scheme in order to obtain both eigenvalues and eigenfunctions in a narrows window of energy spectrum, but the computational costs have been too high if one tends to calculate all eigenvalues and eigenfunctions.

### 3. Program code

The Hamiltonian representation in hyperspherical coordinates is described in [2]. Fortran program code, performing calculations for the states of even symmetry of ABA molecule was written by Back [1] and Yu [2]. We develop a more general code, which allows one to calculate vibrational spectrum for symmetrical and non-symmetrical molecules. We include multiprocessor support using MPI library. The efficiency of the calculations is demonstrated on Fig. 1.

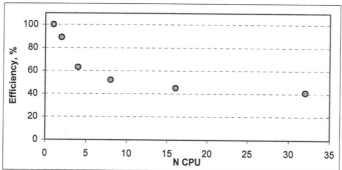

Figure 1: Efficiency of multiprocessor variant.

### 4. Results and discussions

In order to verify the technique we simulated $NO_2$ molecule with the same grid of 168x105x145 points [1] and the same potential energy surface [6]. All calculations were made on 32 Opteron processor cluster within 40 hours. The main result of simulation can be summarized are follows:

- The effective code, supporting multi-processoring has been written for both symmetrical and nonsymmetrical threeatomic molecules.
- The obtained results for nitrogen dioxide showed an excellent agreement with previously calculation of all bound states.
- The convergence of the results is plotted in Fig. 2.
- A new technique has permitted to calculate all bound states by $7 \times 10^4$ evaluations of Hamiltonian operator action onto wavefunction, while the old program demanded more than $16 \times 10^6$ evaluations.

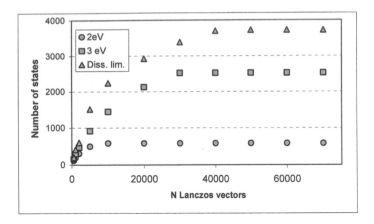

Figure 2: Convergence on number of bound vibrational states.

## Acknowledgments

The authors wish to thank Dr. Sture Nordholm from Göteborg University. The research is supported by the Russian Foundation for Basic Research, project no. 05–03–33153a and INTAS YS project. no. 03-55-2056 . The authors also wish to thank Research Computer Centre of Moscow State University for the computational resources.

## References

[1] A. Back. Vibrational eigenstates of $NO_2$ by a Chebyshev-MINRES spectral filtering technique. *J.Chem. Phys.* **117** 8314-8320 (2002).

[2] H.-G. Yu. An efficient grid calculation of vibrational states for $H_3^+$ with geometric phase in hyperspherical coordinates. *Chem. Phys. Lett.* **281** 312-318 (1997).

[3] B. R. Johnson. On hyperspherical coordinates and mapping the internal configurations of a tree body system. *J. Chem. Phys.* **73** 5051-5058 (1980).

[4] B. R. Johnson. The classical dynamics of three particles in hyperspherical coordinates. *J. Chem. Phys.* **79** 1906-1915 (1983).

[5] B. R. Johnson. The quantum dynamics of three particles in hyperspherical coordinates. *J. Chem. Phys.* **79** 1916-1925 (1983).

[6] E. Leonardy, C. Petrongolo, G. Hirsch and R. J. Buenker. Ab initio study of $NO_2$. *J. Chem. Phys.* **105** 9051-9067 (1996).

Brill Academic Publishers
P.O. Box 9000, 2300 PA Leiden,
The Netherlands

*Lecture Series on Computer
and Computational Sciences*
Volume 4, 2005, pp. 77-80

# Two Approaches for Modelling Hydrate Growth

Trygve Buanes[1], Bjørn Kvamme, Atle Svandal

Department of Physics and Technology, University of Bergen
Allégaten 55, N-5007 Bergen, Norway

Received 12 July, 2005; accepted in revised form 8 August, 2005

*Abstract:* We consider two different approaches to model growth of $CO_2$ hydrate, phase field theory and a model based on cellular automata. We show that the latter model has the benefit of being much more computationally efficient, and still giving results which are consistent with the former model.

*Keywords:* Gas hydrates, $CO_2$, Growth from solutions

*PACS:* 81.10.Aj, 81.10.Dn

## 1 Introduction

Gas hydrates are crystalline structures of water which enclathrates small non-polar "guest" molecules. The presence of these guest molecules can stabilise the ice-like structure at temperatures well above the melting point of pure ice. The kinetics of hydrate formation, as well as the macroscopic structure and surface properties of the formed hydrate, depends on the kinetics of mass transport, heat transport and the free energy changes related to the phase transition. The macroscopic structure of the formed hydrate, and corresponding macroscopic hydrate properties, depends sensitively on the free energy difference for the hydrate formation. These macroscopic properties of the growing hydrate essentially determines the degree of agglomeration of hydrate particles and eventually plugging of flow lines. In this work we therefore investigate two mean field approaches to hydrate growth that model kinetic rates as well as corresponding hydrate crystal structures related to the thermodynamic driving forces. Cellular automata uses a Monte Carlo approach to evaluate the most probable growth paths. The phase field theory is based on the free energy functional related to the phase transition, and involves the integration of a coupled set of differential equations in time and space.

## 2 Phase Field Theory

Mesoscopic modelling of hydrate growth has up to now mainly been done using phase field theory [1, 2]. Although phase field theory primarily has been developed for modelling solidification of metallic melts, it also has potential for providing insight into the kinetics of hydrate growth.

The version of phase field theory we use includes three fields; the phase $\phi$, molar $CO_2$ concentration $c$ and microscopic orientation, $\theta$. Note that for historical reasons $\phi = 0$ corresponds to solid and $\phi = 1$ to liquid in the scope of phase field theory. From the free energy functional

$$ F = \int d^3r \left\{ \frac{\varepsilon_\phi^2 T}{2} |\nabla \phi|^2 + \frac{\varepsilon_c^2 T}{2} |\nabla c|^2 + w(c)Tg(\phi) + [1 - p(\phi)][f_S(c,T) + f_{ori}(|\nabla \theta|)] + p(\phi)f_L(c,T) \right\} \quad (1) $$

---

[1] Corresponding author. E-mail: Trygve.Buanes@ift.uib.no

subject to conservation of the field $c$, the equations of motion are derived [3]:

$$\dot{\phi} = M_\phi \left\{ \nabla \left( \frac{\partial f}{\partial \Delta \phi} \right) - \frac{\partial f}{\partial \phi} \right\} + \zeta_\phi,$$

$$\dot{c} = \nabla \left\{ Dc(1-c) \nabla \left[ \left( \frac{\partial f}{\partial c} \right) - \nabla \left( \frac{\partial f}{\partial \nabla c} \right) \right] \right\} + \zeta_c, \tag{2}$$

$$\dot{\theta} = M_\theta \left\{ \nabla \left( \frac{\partial f}{\partial \nabla \theta} \right) - \frac{\partial f}{\partial \theta} \right\} + \zeta_\theta.$$

$\varepsilon_\theta$, $\varepsilon_c$ and $w(c)$ are related to the interface free energy, interface thickness and melting temperature [4], and can be obtained from experiments or molecular dynamics simulations. The functions $g(\phi)$ and $p(\phi)$ are not completely fixed, but their form is constrained by the requirement of thermodynamical consistency [5]. $f_L$ and $f_S$ are the free energy densities of the aqueous solution and solid hydrate, respectively, and $f_{ori}$ is an extra contribution added to take into account orientational differences in the solid phase. The terms $\zeta_i$ are Langevin noise terms added to model thermal fluctuations in the system.

## 3  Monte Carlo Cellular Automata

In [6] we proposed a model based on cellular automata combined with Monte Carlo for simulations of $CO_2$ hydrate growth. This approach is much simpler than phase field theory, and the primary goal is to have a more computationally efficient tool enabling us to study larger systems and longer time spans.

The basis of our model is Metropolis tests where the change in free energy as response to change of phase, or change in $CO_2$ concentration or temperature is considered. As input we use the free energy parametrised with respect to phase, $\phi$, ($\phi = 1$ corresponds to solid hydrate, and $\phi = 0$ corresponds to liquid water with dissolved $CO_2$), $CO_2$ molar fraction, $x_{CO_2}$, and temperature $T$. In specific, each time step consist of three steps: solidification, $CO_2$ diffusion and temperature diffusion. The criteria for solidification is that a cell should have at least one solid neighbour, and that

$$r < e^{-\beta \Delta f(x_{CO_2}, T) \left[ 1 - \lambda(\Phi_n - 6) \right]}, \tag{3}$$

where $0 \leq r \leq 1$ is a random number with a flat distribution, $\beta$ is the characteristic energy for the solidification process, $\Delta f(x_{CO_2}, T)$ is the change in free energy if the cell with molar $CO_2$ concentration $x_{CO_2}$ and temperature $T$ changes its phase from liquid to solid, $\Phi_n = \sum_n \omega_n \phi_n$ is a weighted sum over solid neighbours with weights $\omega_n = 2$ for nearest neighbours, $\omega_n = 1$ for next nearest neighbours and $\omega_n = 0$ otherwise. The $\Phi_n$-term is included to take surface energy effects into account, and its strength is parametrised with $\lambda$. The diffusion of $CO_2$ is done using a Monte Carlo implementation of Fick's law. At each time step one of the nearest neighbours are drawn at random for each cell. The current

$$j_c = -D_c(\Delta n_{CO_2} + \delta_c), \tag{4}$$

where $n_{CO_2}$ is proportional to the number density of $CO_2$ molecules and $\delta_c$ is a random number with a Gaussian distribution centred at 0, runs if

$$r < e^{-\beta \Delta f(j_c)}, \tag{5}$$

where $0 \leq r \leq 1$ is a random number with a flat distribution, $\beta$ is the characteristic energy for the diffusion, $D_c$ is the diffusion coefficient, and $\Delta f(j_c)$ is the change in free energy due to the current $j_c$. The temperature diffusion is done in the same way, but with $\beta = 0$, such that the current

$$j_T = -D_T(\Delta T + \delta_T), \tag{6}$$

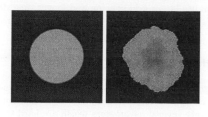

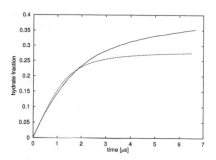

Figure 1: Hydrate particle after $5.6\mu s$. The system is 128×128 nm. *Left*: Phase field theory simulation. *Right*: Cellular automaton simulation.

Figure 2: Fraction of total system converted into hydrate as a function of time. *Dashed line*: Phase field theory simulation. *Solid line*: Cellular automaton simulation.

where $D_T$ is the thermal diffusion coefficient and $\delta_T$ is a random number with a Gaussian distribution centred at 0, is exchanged. In our system the heat transport is much faster than the mass transport. Since equation 6 require $D_T < 0$ to be consistent, we must apparently choose $D_c$ very small. But this would slow down the simulation considerably. To circumvent this we replace $D_T \rightarrow D_T/m$ and run the temperature diffusion $m$ times.

In addition we also need to establish length and time scales since these are not inherently defined in this model. Since the phase can only take the values 0 (liquid) and 1 (solid), but nothing in between, we interpret the size of a cell to be of the same magnitude as the interface thickness between solid hydrate and an aqueous solution. The time scale is connected to the length scale by the diffusion rate.

## 4    Results

We do simulations on a two dimensional system of size 128×128 nm. Since we are presently not able to handle nucleation we let the initial system have a small hydrate particle at the centre of the simulation window. The rest of the simulation window is filled with an aqueous solution with 3.3% $CO_2$. The simulations are done with initial temperature of 274K and with constant pressure at 150 bar.

Both models produces compact hydrate particles, but the cellular automaton approach gives rise to a slightly less regular particle than the perfect circular particle produced by the phase field theory, see Figure 1. As shown in [6] the morphology of the hydrate particle produced by the cellular automaton approach is strongly dependent on the driving forces, large driving forces giving rice to branched structures. With a small modification of $|\nabla\phi|^2$-part of the free energy functional, equation 1, branching can also be seen within the scope of phase field theory [7].

Monitoring the growth rate, measured in terms of fraction of the system converted into hydrate, we find that the models are in good agreement at short times as shown in Figure 2. However, the phase field simulation approaches a lower asymptotic value than the cellular automaton simulation. The reason for this is that the hydrate grown in the cellular automaton simulation has a little smaller $CO_2$ concentration than that of the phase field simulation. This difference becomes increasingly important as the $CO_2$ from the aqueous solution is consumed by the growing hydrate particle.

Considering the computational efficiency of the two approaches, we find that the cellular automaton simulations is nearly a factor 20 faster than the phase field simulation. This is mainly due to the fact that in the cellular automata we can use larger grid spacing and longer time steps. In simulations where temperature effects can be neglected, the speed-up of the cellular automaton approach will be even larger.

## 5  Conclusion

We have considered simulations of $CO_2$ hydrate growth using a phase field theory and a model based on cellular automata. The cellular automaton approach is certainly less rigorous than the phase field theory. The main problem is that there are several parameters which are hard to find correct values for from experiments or from more fundamental considerations. In particular, the characteristic energies used in the Metropolis tests, and the noise terms added to the diffusion currents are hard to quantify. But the simplicity of the model is also its strength; we have seen that using the cellular automaton approach simulations are done much more efficiently and thus allows for larger systems and longer time spans.

Presently the phase field theory is by far the most mature of the two approaches, but the computational efficiency of the cellular automata makes this model an interesting supplement to the more rigorous calculations of the phase field theory.

## Acknowledgements

This work has been supported financially by The Research Council of Norway and Hydro. We would also like to thank Laszlo Gránásy whose contribution in the phase field theory simulations has been invaluable.

## References

[1] B Kvamme, A Graue, E Aspenes, T Kuznetsova, L Gránásy, G Tóth, T Pusztai, and G Tegze. Kinetics of solid hydrate formation by carbon dioxide: Phase field theory of hydrate nucleation and magnetic resonance imaging. *Physical Chemistry Chemical Physics*, 6(9), 2003.

[2] B Kvamme, L Gránásy, T Kuznetsova, A Svandal, and T Buanes. Towards a kinetic model for hydrate sealing of $CO_2$ in reservoirs. In *7th International Conference on Greenhouse Gas Control Technologies*, 2004.

[3] L Gránásy, T Pusztai, and J Warren. Modelling polycrystalline solidification using phase field theory. *Jorurnal of Physics: Condensed Matter*, 16(41):R1205–R1235, 2004.

[4] L Gránásy, T Pusztai, G Tóth, Z Jurek, M Conti, and Kvamme B. Phase field theory of crystal nucleation in hard sphere liquid. *Journal of Chemical Physics*, 119(19), 2003.

[5] S L Wang, Sekera R F, Wheeler A A, Murray B T, Coriell S R, Braun R J, and McFadden G B. Thermodynamically consistent phase-field models for solidification. *Physica D*, 69:189–200, 1993.

[6] T Buanes, B Kvamme, and A Svandal. Computer simulation of $CO_2$ hydrate growth. *Journal of Crystal Growth*. In Press.

[7] A Svandal, B Kvamme, L Gránásy, and T Pusztai. The influence of diffusion on hydrate growth. *Journal of Phase Equlibria and Diffusion*, 26(5), 2005. In press.

Brill Academic Publishers
P.O. Box 9000, 2300 PA Leiden,
The Netherlands

*Lecture Series on Computer
and Computational Sciences*
Volume 4, 2005, pp. 81-84

# An Implicit Finite Difference Scheme for Simulating the Space Fractional-order Riesz Diffusion Processes

$^{a,b}$**X. Cai** and $^{a,c}$**F. Liu** [1]

$^a$School of Mathematics Sciences, Xiamen 361021, China,
$^b$Department of Mathematics,Jimei University, Xiamen 361021, China,
$^c$School of Mathematical Sciences, Queensland University of Technology,
GPO Box 2434, Brisbane, Q 4001, Australia

Received 30 May, 2005; accepted in revised form 13 June, 2005

*Abstract:* In this paper, space fractional-order Riesz diffusion equation is considered. The space fractional-order Riesz diffusion equation is obtained from the standard diffusion equation by replacing the second order derivative in space by a fractional Riesz derivative of order $\alpha$ $(1 < \alpha \leq 2)$. It has been used to model anomalous diffusion or dispersion, where a particle plume spreads at a rate inconsistent with the classical Brownian motion model. An implicit difference scheme is proposed. The stability and convergence results of the implicit difference scheme are analyzed. Finally, the numerical results are presented to demonstrate that the computed result is in good agreement with theoretical analysis.

*Keywords:* Space fractional-order, Riesz diffusion processes, implicit difference scheme, Stability, Convergence.

*Mathematics Subject Classification:* 34D15, 34K26, 65L20.

## 1 Introduction

Fractional-order partial differential equations have recently been used to new applications in engineering, physics, finance, and hydrology [1]. Fractional-order partial differential equations are equation involving derivatives of non-integer order. These equations have been studied by a number of authors [2, 3].

However, numerical methods for the space fractional-order diffusion equation are limited. A different numerical methods for solving fractional partial differential-order equation was studied by Liu et al. in the recent paper [4, 5, 6, 7]. They transformed the partial differential equation into a system of ordinary differential equations (Method of Lines), which was then solved using backward differentiation formulas. Fix and Roop [8] developed a finite element method for a two-point boundary value problem. Meerschaert et al. [9] proposed finite difference approximations for fractional advection-dispersion flow equations.

Nevertheless, all above works were only concerned space fractional derivatives by Caputo definition or Grünwald definition. It can be found that fractional derivatives by Riesz definition is more difficult than above two definitions.

In this paper, we consider a space fractional-order Riesz diffusion equation. The analytical solution of the equation is studied by Cai and Liu [10]. It was pointed out that analytical solution of the

---

[1]Xin Cai, E-mail address: cx85@263.net. Professor Fawang Liu, E-mail address: fwliu@xmu.edu.cn / f.liu@qut.edu.au

problem is difficult to solve and it is important to develop computationally efficient numerical methods. This motivates us to consider implicit difference scheme. As far as author's knowledge, numerical solution for the fractional-order Riesz diffusion equation has never been considered.

## 2    An implicit difference scheme and its properties

**Definition:** Riesz definition of fractional-order derivative is

$$_xD^\alpha u(x,t) = g(\alpha)(_xD^\alpha_{a+} +_x D^\alpha_{b-})u(x,t), \tag{1}$$

where $g(\alpha) = -\frac{1}{2cos(\frac{\pi\alpha}{2})}, \alpha \neq 1$,

$$_xD^\alpha_{a+}u(x,t) = \frac{1}{\Gamma(n-\alpha)} \frac{\partial^n}{\partial x^n} \int_a^x \frac{u(\xi,t)d\xi}{(x-\xi)^{\alpha+1-n}}, n = [\alpha] + 1, \tag{2}$$

$$_xD^\alpha_{b-}u(x,t) = \frac{(-1)^n}{\Gamma(n-\alpha)} \frac{\partial^n}{\partial x^n} \int_x^b \frac{u(\xi,t)d\xi}{(\xi-x)^{\alpha+1-n}}, n = [\alpha] + 1, \tag{3}$$

In this paper, we study the space fractional-order Riesz diffusion processes (SFORDP)

$$\frac{\partial u(x,t)}{\partial t} = d_xD^\alpha u(x,t), \quad x \in (a,b), \quad t \in I\!R^+, \tag{4}$$

$$u(x,0) = \psi(x), \quad x \in [a,b], \tag{5}$$

$$u(a,t) = 0, \quad u(b,t) = 0, \quad t \geq 0, \tag{6}$$

where $_xD^\alpha u(x,t)$ is the Riesz's fractional derivative. $\alpha \in [\beta,2]$, $\beta > 1$, $d > 0$ is a constant. Let $\bar{g} = -\frac{1}{2cos(\frac{\pi\beta}{2})}$, thus $g(\alpha) \in [\frac{1}{2}, \bar{g}]$.

We consider the equidistant mesh partition in space $a = x_0 < x_1 < \ldots < x_M = b$, where we set $h = \frac{b-a}{M}$ be the grid step in space, $M$ is a positive integer. We consider the equidistant mesh partition in time $0 = t_0 < t_1 < \ldots < t_N \leq T$. Let $\tau$ be the grid step in time, $N\tau \leq T$, $T$ is a fixed number. Let $u_i^n$ be the numerical solution of $u(x_i, t_n)$.

Using shifted Grünwald formala to approximate $_xD^\alpha_{a+}u(x_i,t)$ and $_xD^\alpha_{b-}u(x_i,t)$ respectively

$$\overline{_xD^\alpha_{a+}u(x_i,t)} = \frac{1}{h^\alpha} \sum_{k=0}^{i+1} w_k u(x_{i-k+1},t), \tag{7}$$

$$\overline{_xD^\alpha_{b-}u(x_i,t)} = \frac{1}{h^\alpha} \sum_{k=0}^{M-i+1} w_k u(x_{i+k-1},t), \tag{8}$$

where $w_0 = 1, w_1 = -\alpha, w_i = (-1)^i \frac{\alpha(\alpha-1)\cdots(\alpha-i+1)}{i!}, \ for \ i \geq 2$.

We approximate Eq. (4) using an implicit finite-difference scheme (IFDS), for $1(1)M - 1$,

$$u_i^{n+1} = u_i^n + r\{\sum_{k=0}^{i+1} w_k u_{i-k+1}^{n+1} + \sum_{k=0}^{M-i+1} w_k u_{i+k-1}^{n+1}\}, \tag{9}$$

where $r = \frac{dg\tau}{h^\alpha}$.

The initial and boundary condition (5) and (6) can be discretized as

$$u_0^n = u_M^n = 0, n \geq 0, u_i^0 = \psi(x_i), i = 1(1)M - 1. \tag{10}$$

The equations (9) and (10) result in the following linear system of equations

$$BU^{n+1} = U^n, \tag{11}$$

where $U^n = (u_0^n, u_1^n, \cdots, u_{M-1}^n, u_M^n)^T$ and $B = (b_{ij})$ is a $(M+1) \times (M+1)$ matrix, which independent on $t$. These coefficients, for $i = 1(1)M - 1$, are defined as follows

$$b_{ij} = \begin{cases} -rw_{i-j+1}, & when \quad j < i-1, \\ -r(w_0 + w_2), & when \quad j = i-1, \\ 1 - 2rw_1, & when \quad j = i, \\ -r(w_0 + w_2), & when \quad j = i+1, \\ -rw_{j-i+1}, & when \quad j > i+1, \end{cases} \tag{12}$$

while $b_{00} = 1$, $b_{0j} = 0$ for $j = 1(1)M$, $b_{MM} = 1$, $b_{Mj} = 0$ for $j = 1(1)M$.
We have the following main results.

**Theorem 1:** The coefficients matrix $B$ is strictly diagonally dominant and $\|B^{-1}\|_\infty \leq 1$.

**Theorem 2:** The implicit finite-difference scheme IFDS (9) and (10) for (SFORDP) (4) – (6) is unconditionally stable.

**Theorem 3:** The implicit finite-difference scheme IFDS (9) and (10) for (SFORDP)(4) – (6) is unconditionally convergence in order $O(h + \tau)$ by norm $\|\cdot\|_\infty$.

## 3 Numerical results

In this paper, we will consider the space fractional-order Riesz diffusion processes, which is governed by the following equation

$$\frac{\partial u(x,t)}{\partial t} = d_x D^\alpha u(x,t), \quad 0 < x < \pi, \quad t \geq 0, \quad 1 < \alpha \leq 2, \tag{13}$$

$$u(x,0) = \psi(x), \quad 0 < x < \pi, \tag{14}$$

$$u(0,t) = 0, \quad u(\pi,t) = 0, \quad t \geq 0, \tag{15}$$

where $_x D^\alpha u(x,t)$ is the Riesz's fractional derivative, $d = 0.4$, $\psi(x) = x^2 \cos(\frac{x}{2})$.
In order to illustrate the efficiency of our method, the method of lines (MOL) was used for comparison. The MOL was firstly introduced by Liu et al. [4, 5] and has been demonstrated that MOL is a computationally efficient method for solving the space fractional partial differential equations. According to Theorem 2, the IFDS is unconditionally stable. From table 1, it can be seen that IFDS is stable and convergence for the different $\tau$.

## Conclusions

In this paper, an implicit finite difference scheme for simulating the space fractional-order Riesz diffusion processes has been described and demonstrated. The method is based on the shifted Grünwald definition of fractional derivative. The stability and convergence of the IFDS are analyzed. Finally, the numerical results are presented to demonstrate that the computed result is in good agreement with theoretical analysis.

## Acknowledgment

This work has been supported by the National Natural Science Foundation of China under Grant 10271098 and the Natural Science Foundation of Fujian Province, China under Grant A04100021.

Table 1: IFDS for $\alpha = 1.8, h = \frac{\pi}{100}, M = 100, t = 1.0$, and different values of $\tau$.

| $(X, 1.0)$ | $\tau = 0.02$ | $\tau = 0.01$ | $\tau = 0.005$ | MOL |
|---|---|---|---|---|
| 0.314159 | 0.2938302515 | 0.2946240126 | 0.2950231776 | 0.2954232535 |
| 0.628319 | 0.5651782624 | 0.5663308026 | 0.5669116008 | 0.5674951846 |
| 0.942478 | 0.8316316016 | 0.8325998888 | 0.8330894381 | 0.8335829949 |
| 1.256637 | 1.0733132531 | 1.0736148099 | 1.0737689162 | 1.0739256358 |
| 1.570796 | 1.2510249445 | 1.2503559364 | 1.2500201337 | 1.2496833425 |
| 1.884956 | 1.3173783879 | 1.3156883775 | 1.3148374431 | 1.3139818376 |
| 2.199115 | 1.2314649504 | 1.2290475690 | 1.2278311801 | 1.2266090256 |
| 2.513274 | 0.9741731997 | 0.9717054305 | 0.9704655930 | 0.9692222621 |
| 2.827433 | 0.5578451149 | 0.5562198278 | 0.5554043810 | 0.5545881722 |

# References

[1] I. Podlubny: *Fractional Differential Equations*. Academic Press, 1999.

[2] W. Wyss, The fractional diffusion equation, *J. Math. Phys.* **27** 2782-2785(1986).

[3] R. Gorenflo, Yu. Luchko and F. Mainardi, Wright function as scale-invariant solutions of the diffusion-wave equation, *J. Comp. Appl. Math.* **118** 175-191(2000).

[4] F. Liu, V. Anh and I. Turner, Numerical solution of the fractional-order Advection-Dispersion Equation, The Procceding of An International Conference on Boundary and Interior Layers -Computational and Asymptotic Methods, Perth, Australia, 159-164, (2002).

[5] F. Liu, V. Anh, I. Turner, Numerical Solution of the Space Fractional Fokker-Planck Equation, Journal of Computational and Applied Mathematics 166, 209-219 (2004).

[6] F. Liu, S. Shen, V. Anh and I. Turner, Analysis of a discrete non-Markovian random walk approximation for the time fractional diffusion equation, *ANZIAM J.*, (2005), to appear.

[7] S. Shen and F. Liu, Error analysis of an explicit finite difference approximation for the space fractional diffusion, *ANZIAM J.*, (2005), to appear.

[8] G.J. Fix and J.P. Roop, Least squares finite element solution of a fractional order two-point boundary value problem, *Comput. Math. Appl.* (2004), to appear.

[9] M.M. Meerschaert, Charles Tadjeran, Finite difference approximations for fractional advection-dispersion flow equations, *Journal of Computational and Applied Mathematics-sJournal of Computational and Applied Mathematics* **172** 65-77(2004).*Journal of Computational and Applied MathematicsJournal of Computational and Applied Mathematics* **172** 65-77(2004).

[10] X. Cai and F. Liu, The Explicit Finite Difference Scheme for the Space Fractional-order Partial Differential Equation with Initial-boundary Value Problem, Submit to *J. Appl. Math and Computing*. in Korea. (2005).

Brill Academic Publishers
P.O. Box 9000, 2300 PA Leiden
The Netherlands

*Lecture Series on Computer
and Computational Sciences*
Volume 4, 2005, pp. 85-88

# Statistical Mechanical Study of Dimethyl-Sulfoxide/Water Eutectic Mixture by Molecular Dynamics Simulation

M. Chalaris[1], J. Samios[2]

Laboratory of Physical Chemistry,
Department of Chemistry,
National and Kapodistrian University of Athens,
Panepistimiopolis 15771 Athens, Greece

Received 7 March, 2005; accepted in revised form 10 March, 2005

*Abstract: A molecular dynamics simulation study of the local structures and H-bond distribution for water and dimethyl-sulfoxide(DMSO) in the eutectic DMSO/Water mixture, $2H_2O:1DMSO$, is presented. The temperature range studied was from room temperature down to - 68.5°C. The various site-site pair distribution functions (pdfs) were calculated and their temperature dependence was obtained. It is found that over the thermodynamic conditions investigated here DMSO forms hydrogen bonds with water molecules which are longer lived than water-water hydrogen bonds. A further analysis of the hydrogen bonds was undertaken and the degree of aggregation was obtained. Finally, the transport properties of the mixture has been calculated and compared with available experimental data.*
*Keywords: Mechanics, MD simulation; dimethyl-sulfoxide; water; transport properties; hydrogen bonding*

*Mathematics Subject Classification:* 82B05, 82D15, 82C70

## 1. Introduction

Dimethyl-sulfoxide (DMSO - $(CH_3)_2SO$) in aqueous solutions has been studied extensively using a variety of experimental and theoretical techniques. This is due to its wide range of applications in industry, environmental science, as well as in pharmacology and medicine.

The phase diagram of the water-DMSO system shows a superposition of eutectics, with a minimum at 2:1 molar composition of water and DMSO [1]. At this composition, the freezing point of water is depressed down to ~ - 70°C. DMSO is therefore used as a cryoprotector to prevent freezing and thawing damage to living cells and tissues.

The present work describes the use of the Molecular Dynamics (MD) simulation technique to study the properties of the DMSO/Water mixture at the eutectic $2H_2O:1DMSO$ over a wide range of temperatures. The main purpose here is to understand the various thermodynamic, dynamic and structural properties of the system and to explore the intermolecular forces responsible for the formation of hydrogen bonds among the molecules of the system.

## 2. Simulation methods and potentials

In this treatment, the MD simulations of the DMSO/Water mixture were carried at various thermodynamic conditions for which experimental data are available in the literature [1, 2]. All the state points studied here are listed in Table 1. Systems of 500 rigid molecules were simulated using periodic boundary conditions in the microcanonical (NVE) statistical mechanical ensemble. The temperature was maintained using the well-known Nose-Hoover thermostat. The effective pair potentials used were the following: MCS for DMSO [3] and SPC/AG for water [3]. The intermolecular interactions DMSO – DMSO, $H_2O$ - $H_2O$ and DMSO - $H_2O$ were of the site – site Lennard – Jones (LJ) type with electrostatic terms. Also, the Lorentz-Berthelot mixing rules were used to calculate cross interaction parameters. The direct Ewald method was employed to treat the long-range coulombic interactions. The translational and rotational equations of motions are solved using Leapfrog algorithms. The integration time step was $1 \cdot 10^{-15}$ ps. After equilibration, each MD run was extended to additional configurations for about 1 ns.

---

[1] E-mail: mhalaris@cc.uoa.gr
[2] Corresponding Author: E-mail: isamios@cc.uoa.gr

## 3. Results and Discussion

In this paper we will analyze the hydration structure of DMSO aqueous solutions 2H₂O:1DMSO for six different temperatures. Our purpose was to study the effect of temperature on the structure of solution of DMSO in Water. In the following, we will describe the pair correlation functions, which reveal the average structure of the solution. Also, we will analyze the structure of water the solution at different temperatures as well as its hydrogen-bonded network

### 3.1 Thermodynamic and transport properties

A summary of the bulk thermodynamic properties obtained is given in Table 1. As mentioned above, experimental densities were used throughout the simulations studies. The potential energy of the system is in good agreement with available experimental data at 298 K. The potential energy of the system increases as the temperature decreases, as expected.

Table 1: Simulated points of the eutectic 2H₂O:1DMSO mixture and thermodynamic results derived from this NVE-MD study. The numbers in parentheses are experimental measurements taken from Ref. [4] and [5]

| ρ(g/ml) | T (K) | T$^{MD}$(K) | -U$_P$ (KJ/mol) | D$_W$(10$^{-9}$m$^2$s$^{-1}$) | D$_D$(10$^{-9}$m$^2$s$^{-1}$) | η$_B$(Poise) |
|---------|-------|-------------|-----------------|-------------------------------|-------------------------------|--------------|
| 1.167 | 204.5 | 206.2 | 52.057 | 0.0206 | 0.0116 | - |
| 1.159 | 213.0 | 210.0 | 51.292 | 0.0608 | 0.0459 | 349 (359) |
| 1.151 | 223.0 | 226.0 | 51.00 | 0.0451 | 0.0314 | 128 (138.6) |
| 1.130 | 248.0 | 254.9 | 49.831 | 0.1528 | 0.0872 | 21.77 (22.72) |
| 1.109 | 273.0 | 271.9 | 48.226 | 0.3972 | 0.1580 | 5.88 (6.027) |
| 1.088 | 298.0 | 298.3 | 46.400 | 0.9701(0.820) | 0.427 (0.36) | 2.33 (2.394) |

The self –diffusion coefficients and the bulk viscosity of the mixture are also determined. Values for the self –diffusion coefficients are obtained by determining the long time slope of the mean square displacement of the molecular centers of motion with time. Also, the bulk viscosity was estimated using the Einstein – type expression [6]. The values of the properties of each component in the mixture are listed in Table 1 Moreover the self-diffusion coefficients for water and DMSO as a function of the temperature is given in Fig 1.

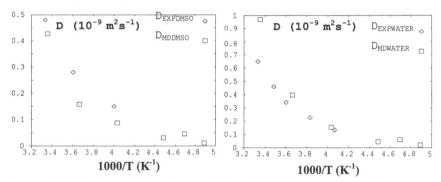

Figure 1: Self-diffusion coefficients for water (right) and DMSO (left) as function of temperature.

The agreement between simulations and experimental results is quite good. Water and DMSO exhibit different diffusion dynamics. The water self-diffusion coefficients, D$_W$, provide Arrhenius type temperature dependence. The Stokes –Einstein equation was found to describe well the relation ship between calculated bulk viscosity and diffusion coefficient.

### 3.2 Solution structure

The pair distribution functions g(r) were computed during the periods of 1ns data production. Fig. 2 shows the amplitudes of the first molecular coordination shell peaks of the radial distributions g(r )[O$_W$—H$_w$], g(r)[O$_D$—H$_w$] and g(r)[S$_D$-H$_W$] as a function of temperature. The results indicate that the nearest water molecules are hydrogen bonded to the oxygen and sulfur in DMSO molecules, respectively.

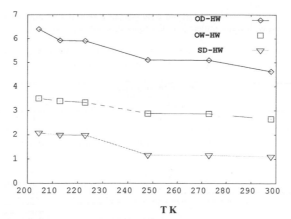

Figure 2: Variation of the first peak height of g(r)'s for $O_W$—$H_w$, $O_D$—$H_w$ and $S_D$-$H_w$ in the eutectic $2H_2O:1DMSO$ mixture as a function of temperature.

The corresponding $g(r)[O_W$—$O_w]$, $g(r)[O_D$—$O_w]$ show similar behavior as the temperature decreases. The peak locations in pdfs are slightly affected with the temperature in Fig 2, whereas the peak amplitudes change significantly. In the case of the $g(r)[O_W$—$H_w]$, we see that as temperature is decreased, water molecules in the solutions becomes more structured, as revealed by the sharpness of the peaks. Furthermore, the heights of the peaks confirm the enhancement of water structure with respect to the bulk water as reported earlier [7]. Also, figure 2 shows the peak amplitudes between the oxygen in the sulfonyl group of DMSO and the hydrogen water sites $g(r)[O_D$—$H_w]$. As before, the decrease of temperature let to a sharpness of the peaks of the radial distribution functions. Importantly, all pair correlations show that water molecules form a linear hydrogen bond to the OS group atoms in DMSO, as the first peak in the $g(r)[O_D$—$H_w]$, is observed at a distance of 1.60 Å, while the first peak in the $g(r)[O_D$—$O_w]$ is observed at a distance of 2.55 Å. At this concentration of DMSO in the solutions at all the temperatures studied, it may be expected that the structure of the solutions corresponds to the existence of predominant $1DMSO:2H_2O$ aggregates.

### 3.3 Water structure
Since water molecules can form an effective hydrogen-bonded network around the DMSO molecule, we decided to explore the tetrahedral order of water in the mixture as a function of temperature. Previous simulation results and neutron diffraction data have shown that for temperatures above 246 K the average tetrahedral coordination of water is preserved for concentrations as high as 35% mole fractions of DMSO. Such tetrahedral order may be characterized by the positions of the first and second peaks of $O_WO_W$ pair distribution functions, which roughly have to satisfy the tetrahedral relation $r_2 = 2(2/3)^{1/2}r_1$. Our $O_WO_W$ functions show that this tetrahedral order is still clearly observed even for temperatures above the 223K, but it doesn't exist at very low temperatures (213 K and 204.5 K)

### 3.4 Hydrogen-bonding structure
As mentioned above, due to the fact that pdfs do not provide detailed information concerning the local order in binary mixtures, we have carry out a detailed analysis of hydrogen bonding network in the mixture to gain deeper insight into the aqueous structures. A basic aspect of the hydrogen bonding network analysis is the probability distribution which gives information on the number and type of hydrogen bonds that a molecule forms with other molecules. For the system we have employed the same geometric criterion used in a previous work [7]. For the DMSO-water mixture, we found that two molecules are regarded to be hydrogen bonded if their separations satisfy the relations: $R_{OO} \leq 3.20$ Å, $R_{OH(w)} \leq 2.40$ Å and the angle $H_wO...O \leq 30°$, where $H_W$ is the hydroxyl proton. The corresponding values for the water-water hydrogen bond criterion are: $R_{O-O} \leq 3.60$ Å, $R_{O-H} \leq 2.60$ Å, and for the angle $H-O...O \leq 30°$. We also computed the average number of hydrogen bonds ($N_{MD}$) for the different kinds in each simulated temperature. The results are presented in the following Table 2 for all the types of hydrogen bonds. We observe that the number of hydrogen bonds donated by water is almost constant with temperature. For the DMSO molecules, we found that most of them exhibit two hydrogen bonds

with water molecules over the entire temperature range, but a remarkable number of molecules form only one hydrogen bond with water molecule in the aqueous solution.

Table 2: Hydrogen bonding analysis: Average percentage $f_i$ of water-water and DMSO- Water molecules with i (0, 1,2,3,4 ...) hydrogen bonds and the total number of H-bonds $N_{MD}$ from this NVE-MD study.

| WATER-WATER | | | | | | |
|---|---|---|---|---|---|
| $f_i$ | 298 K | 273 K | 243 K | 223 K | 213 K | 204.5 K |
| 0 | 1.55±0.76 | 0.97±0.52 | 0.59±0.36 | 0.18±0.21 | 0.57±0.44 | 0.84±0.18 |
| 1 | 16.00±1.95 | 13.95±1.68 | 12.27±1.35 | 9.0±0.97 | 11.73±1.35 | 9.47±0.89 |
| 2 | 40.96±2.47 | 39.91±2.35 | 40.10±1.80 | 40.08±1.66 | 37.56±1.66 | 37.88±1.48 |
| 3 | 33.41±2.41 | 35.25±2.49 | 38.52±2.27 | 40.90±1.72 | 36.88±1.81 | 40.49±1.77 |
| 4 | 7.86±1.30 | 9.75±1.26 | 8.45±1.35 | 9.75±0.90 | 13.10±1.08 | 11.19±1.04 |
| 5 | 0.20±0.23 | 0.17±0.22 | 0.10±0.18 | 0.1±0.16 | 0.18±0.22 | 0.14±0.20 |
| $N_{MD}$ | 403.63±8.0 | 418.90±6.98 | 424.0±5.5 | 439.81±4.3 | 438.74±6.8 | 441.2±3.65 |

| DMSO (O)- WATER (H) | | | | | | |
|---|---|---|---|---|---|
| $f_i$ | 298 K | 273 K | 243 K | 223 K | 213 K | 204.5 K |
| 0 | 1.32±0.9 | 1.65±0.87 | 1.18±0.56 | 0.79±0.28 | 0.10±0.26 | 1.95±0.55 |
| 1 | 49.40±3.37 | 43.33±3.20 | 35.18±2.33 | 38.17±1.85 | 38.48±4.00 | 33.36±1.74 |
| 2 | 46.95±3.54 | 52.47±2.99 | 61.07±2.66 | 58.71±2.29 | 59.40±3.87 | 63.23±1.82 |
| 3 | 2.33±1.15 | 2.55±1.26 | 2.57±1.20 | 2.19±1.19 | 2.01±1.17 | 1.47±0.79 |
| $N_{MD}$ | 225.5±5.56 | 234.0±4.73 | 247.6±3.80 | 243.5±3.10 | 245.0±6.73 | 246.3±2.76 |

| DMSO( S) - WATER(H ) | | | | | | |
|---|---|---|---|---|---|
| $f_i$ | 298 K | 273 K | 243 K | 223 K | 213 K | 204.5 K |
| 0 | 72.12±3.62 | 71.29±3.56 | 71.67±3.23 | 71.71±3.20 | 71.42±3.17 | 69.74±3.23 |
| 1 | 22.94±3.45 | 23.33±3.45 | 22.82±3.21 | 22.10±3.18 | 24.0±3.0 | 24.80±3.31 |
| 2 | 4.42±1.67 | 4.84±1.70 | 4.99±1.72 | 5.53±1.74 | 4.34±1.60 | 5.12±1.62 |
| 3 | 0.48±0.57 | 0.50±0.57 | 0.50±0.57 | 0.62±0.64 | 0.25±0.41 | 0.33±0.47 |
| $N_{MD}$ | 50.1±7.0 | 51.97±6.88 | 51.59±6.40 | 52.83±6.56 | 50.19±6.02 | 54.12±6.04 |

The ratio between the number of hydrogen bonding with two and the number of hydrogen bonding with one water molecules increase as the temperature decreases. Therefore the trimmer DMSO· $(H_2O)_2$ seems to be more stable in the solution than the corresponding dimmer DMSO·$H_2O$.

## References

[1] D. Martin and H.G. Hauthal, *Dimethyl Sulfoxide*, Wiley, New York, 1975; J.L. Lindberg and C. Mijari, *Acta Chem. Scand,* **17**, 1447 (1963).

[2] J.M.G. Cowie and P.M. Toporowski: *Association in the binary liquid system Dimethyl Sulphoxide – Water, Can. J. Chem.* **39** 2240 -2244 (1961).

[3] M.E. Chalaris: *Statistical mechanical studies via molecular dynamics simulation techniques of liquid systems consisted of protic (MeOH, $H_2O$) and aprotic (DMSO, DMF) solvents*: Doctoral Dissertation, National and Kapodistrian University of Athens, Greece, 1999 (in Greek).

[4] K. J. Parker and D. J. Tomlison, *Trans. Faraday Soc.,* **67**, 1302 (1971); S. A. Schichman, R. L. Amey, *J. Phys. Chem.,* **75**, 98 (1971); W. M. Madigosky and R.W. Warfield, *J. Chem. Phys.,* **78**, 1912 (1983).

[5] J. T. Cabral, A. Luzar, J. Texeira and M-C. Bellissent -Funnel: *Single-particle dynamics in dimethyl-sulfoxide/water eutectic mixture by neutron scattering, J. Chem. Phys.,***113(19)**, 8736-8745 (2000).

[6] H.J.V. Tyrell and K.R. Harris, Diffusion in Liquids, Butterworth & Co, London, 1984; M.P. Allen and D.J. Tildesley, Computer Simulation of Liquids, Clarendon, Oxford, 1987.

[7] T R. L. Mancera, M. Chalaris, K. Refson and J. Samios: *Molecular dynamics simulation of dilute aqueous DMSO solutions. A temperature-dependence study of the hydrophobic and hydrophilic behaviour around DMSO, Phys. Chem. Chem. Phys* **6** 94-102(2004).

Brill Academic Publishers
P.O. Box 9000, 2300 PA Leiden
The Netherlands

*Lecture Series on Computer
and Computational Sciences*
Volume 4, 2005, pp. 89-93

# Adaptive Repetitive Control for an Eccentricity Compensation of Optical Disk Drivers

K.B. Chang[1], I.J. Shim and G.T. Park

ISRL, Korea University, 1, 5-ka Anam-dong Sungbuk-ku
136-701, Seoul, South Korea

Received 18 July, 2005; accepted in revised form 10 August, 2005

*Abstract:* This paper present an adaptive repetitive control scheme for optical disk drivers to track a periodic reference signal with dynamic change in period. Periodic disturbances can be adequately attenuated using the concept of repetitive control, provided the known period. Optical disk drivers support various speeds. So optical disk drivers have the varying periodic disturbance. To deal with time varying periodic disturbances, a proposed repetitive controller is turned based on repetitive control to change sampling frequency to follow the change of reference period. The proposed adaptive repetitive control consists of two portions, the repetitive controller and the frequency multiplier, where the former uses a varying sampler operating at a variable sampling rate maintained at fixed multiple times of the disturbance frequency and the latter generate the vary sampling frequency based on the disturbance frequency. An adaptive repetitive control scheme is proposed, implemented on experimental set of an optical disk driver, and demonstrate the effectiveness of the proposed methods and the improvement of the random access time.

*Keywords:* Optical, Disc, Servo, Storage, Programmable, Filter

## 1. Introduction

In track-following servo to read data in the optical disk drive, the laser beam spot should be positioned on the track within a specified tracking error bound in the presence of external disturbance. The track-following servo system has disturbances to come from track geometry and eccentric rotation of the disk. The disturbances contain significant periodic components appearing at known fundamental frequency corresponding to the disk rotational velocity. Therefore, the disturbances should be effectively rejected in order to achieve the required tracking accuracy. To effectively reject such disturbances, repetitive control has been employed. And the stability and the design methods of the repetitive controller has been analyzed and proposed [3], [7].

However, in many situations such disturbances are varying periodic components. In optical disk driver, the periodic disturbances have a time varying period, too. In the case of CLV (Constant Linear Velocity), the period of outer is 2.5 times as longest as the period of inner. So the period varies continuously. In the case of CAV (Constant Angular Velocity), when we change the speed of disk drivers, the period has a region to vary the period. A number of adaptive repetitive control [1], [2], [4] have been developed for time varying periodic signal, of which the sampling rate of the repetitive controller is adjusted to multiples of the signal frequency. However, the existing repetitive controllers need to estimate the varying period from an analog signal and modify a characteristic of the repetitive controller or the number of the sample data. To fine the changes of the disturbance period, we use the FG signal of the motor which indicates the period of the disturbance. It is difficult and complicate to find the new characteristic of the pre-compensation filter. To synchronize samples required [1], the adaptive repetitive controller needs the complicated function to find synchronized samples. These

---

[1] Corresponding author. Electrical Engineering, Korea University, E-mail: lslove@korea.ac.kr

controllers has deal with a few varying periodic signal, but in the optical disk driver, the changes of the disturbance period is the various period and is the continuous change.

To solve this problem, this paper presents a new adaptive repetitive control, which deals with varying sampling frequency. We choose the band limited repetitive controller [5], [6]. The proposed controller is added on the track-following servo controller. The proposed controller consists of the repetitive controller and the frequency multiplier. The track-following servo controller uses the fixed sampling rate and provides compensation on non-periodic signals and stabilizing the plant. The track-following servo controller has the higher sampling frequency than the rotation frequency. So if we use this high sampling frequency, we can achieve better accuracy of the measurement and the multiple frequency of the varying periodic disturbance. The new sampling frequency based on the disturbance period is used for sampling the data of the band limited filter and the repetitive controller. So the characteristic of the band limited filter changes in proportion to the varying periodic disturbance frequency which is based on the rotation frequency.

The remainder of the paper is organized as follows. Section II gives a brief introduction to the optical disk driver tracking following servo system and discusses the problem. Section III reviews the existing repetitive controllers. Section IV presents the proposed adaptive repetitive controller. In Section V, the experimental plant and the proposed adaptive repetitive controller are implemented in real optical disk driver. Additionally, the digital implementation of the servo system is described experimental result is presented. Concluding remarks are given in Section VI.

## 2. System Description

In a track-following servo mechanism, a compound actuator is composed of a small fine actuator mounted on top of large coarse actuator. While the fine actuator with fast response time follows the track in the track-following servo, the coarse actuator with relatively slow response time moves the fine actuator slightly over the operating range. Therefore, tracking performance is almost entirely dependent on how accurately the fine actuator is controlled. So only the fine actuator is considered as the tracking actuator in this paper. The main purpose of track-following servo is to position the laser spot on the exact center of a track. However, the disk eccentricity shakes the track depending on the period of disk rotation. Because the displacement of tracking actuator determines the position of laser spot, the tracking actuator should be periodically shaken at the same rate with track oscillation in order to preserve the relative distance between the laser spot and the center of the track within an allowable limit. The eccentric rotation of disk can be considered as a periodic disturbance. The tracking actuator is usually represented as linear system model. There exist unavoidable modeling errors caused by changes in the operating environment of allowable error margins of components. Nevertheless, it is reasonable to assume that all the parameters of the actuator are within known intervals. Suppose that, in consideration of parameter uncertainty, the tracking actuator is described as a second-order linear transfer function as following as:

$$P(s) = \frac{k_n w_n^2}{s^2 + 2\xi w_n s + w_n^2} \tag{1}$$

where $k_n$ is the DC sensitivity, $w_n$ is the resonance period, $\xi$ is the damping ratio. The tracking error e is detected by the optical sensor and is then fed to the feedback compensator C via the sensor gain K. The feedback compensation C is designed so as to satisfy the desired closed-loop bandwidth, loop gain, and so on determined by design specifications such as maximum allowable tracking error and information on extraneous disturbance. It is noted that the more the eccentricity and the rotation speed of disk increase, the more the bandwidth and the magnitude of periodic disturbance also do. To solve this problem, C should make loop gain larger and the closed-loop bandwidth wider. However, this attempt has some demerits that control input unnecessarily increases and the overall system becomes unstable. This limitation of the existing system affects the performance of the track-following and the random access time. To effectively reject the variable periodic disturbance, the adaptive repetitive controller is added on the existing track-following controller. The design of such a feedback compensator has been fully dealt with in the existing control literature and it is not of concern in this paper.

## 3. Experimental Results

The existing paper introduced the simulation methodology [3], [7]. We use that simulation methodology. In the case of the frequency multiplier, it is implemented on the digital logic. So we use

NC-Verilog simulator which is the logic simulation tool and verify the function of the frequency multiplier.

In the experimental setup, all optical disk driver system (e.g., track seeking system, focus servo, cd-encoder, ATAPI) was implemented on chip and the adaptive repetitive controller was implemented on the MCU. We experiment on real production. In experimental optical disk drivers, the behavior of tracking actuator is well approximated by the second-order system at low frequency. The equation was previously described. The DC sensitivity (kn) is 1mm/V. The resonance frequency (1/wn) and the damping ratio ($\xi$) are approximately 57Hz and 0.033. The control objective is to maintain the tracking error within ±0.05um against various disturbances. To achieve this spec, in the 8x speed of the CLV mode, a servo bandwidth of 2.5 kHz is required, together with a loop gain exceeding 67.6dB in the low frequency region. The existing digital compensator that meets these requirements and provides the sufficient gain and phase margin is a lead-lag compensator expressed by

$$C(z) = \left[ \begin{array}{l} \dfrac{k_{10}+k_{11}z^{-1}}{1-k_{12}z^{-1}} \times \dfrac{1+k_{13}z^{-1}}{1-k_{14}z^{-1}} \times \dfrac{1+k_{15}z^{-1}}{1-k_{16}z^{-1}} \times (k_{10}+k_{10}z^{-1}) \\[2mm] + \left( \dfrac{1+k_{20}z^{-1}}{1-k_{21}z^{-1}} \times \dfrac{k_{22}(k_{24}+k_{25}z^{-1})}{1-k_{23}z^{-1}} \right) \end{array} \right] \times ktg \times 2^6 \tag{2}$$

The adaptive repetitive controller was implemented on 32bit –microprocessor running at 33.8688MHz. The existing digital equalizer was implemented on full digital filter, and is described in synthesized Verilog-HDL code.

The block diagram of the experimental system is shown in Fig. 1. This consisted of optical disk drivers' board, IBM PC, JTAG emulator with serial interface. Optical disk drivers' board consists of the actuator unit, analog circuits to drive the actuator, the digital servo controller, and MCU.

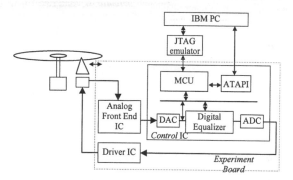

Figure 1: Block diagram of the experimental optical disk driver

All programs are written in C language. The interrupt sub-routine executed at the sampling time from the frequency multiplier. This routine is used to apply the control to the actuator, to read the tracking error with varying periodic disturbance and to calculate the F1 and F2 filter. In the search mode, it also controls the switch sw1 and sw2. When the search mode is the rough track search, both sw1 and sw2 are off. When the search mode is the fine track search, sw1 is off and sw2 is on.

We use the 140um and 210um eccentric test disk. The experiment starts at the 24x speed. In that speed, the frequency of the disturbance is 80Hz and the rotation number is 4800 rpm. Then the adaptive controller is turned on. So the frequency multiplier generates the sampling frequency and the adaptive controller work on the rejection of the disturbance.

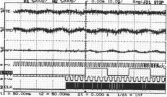

Figure 2: The signal and sampling frequency of the start point

Fig. 2 shows the signal and the frequency multiplier. There are three analog signals. The first signal is tracking error, the second is the control signal and the third is the output of the adaptive repetitive controller. There are four digital signals. There are FG signal, the detected one rotation signal, the rising edge signal of the FG signal, and the varying sampling frequency.

Next the speed changes into the 48x speed. In that speed, the frequency of the disturbance is 160Hz and the rotation number is 9600 rpm. According to change the frequency, the frequency multiplier and the band pass filter follow the change of the frequency.

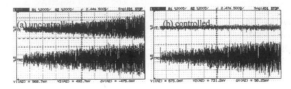

Figure 3: when the speed is changed from 24x to 48x, the measured error signal before (a) and after (b) with adaptive repetitive controller

Fig. 3 shows the tracking error and the control signal of the system between 24x and 48x. Fig. 3(a) shows the tracking error and the control signal of the existing servo controller. The upper and the lower signal are the tracking error signal and control signal. The error and the control signal increase in amplitude and frequency. Because the control system has the limitation of the filter's characteristic, the tracking error is not compensated, even though the signal increases. Fig. 3(b) shows the tracking error and the control signal of the servo controller with the adaptive controller. The adaptive repetitive controller compensates the tracking error for the varying periodic disturbance. So the control signal increases, but the tracking error signal does not increase.

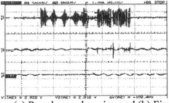

Figure 4: Track search waveforms, (a) Rough search region and (b) Fine search region

Fig. 4 shows the tracking error, control signal, and repetitive controller's output signal. Fig. 4(a) shows the signals of the rough search mode. In the rough search mode, when the sled motor moves to the target track, the control signal is controlled by the center servo. So the adaptive repetitive controller generates the reference voltage. Internally, it remembers the disturbance amplitude and the period. Fig. 4(b) shows the signals of the fine search mode. In the fine search mode, the actuator is moved to the target track by the track seeking system, but this system is not enough to compensate the periodic disturbance. So we add the control signal of the seeking system to the output of the adaptive repetitive controller.

To estimate the performance of the optical disk driver with the adaptive repetitive controller, we estimate the random access time in IBM PC environment. The comparison is provided to demonstrate the significant performance improvement of the proposed controller over the existing controller. Both systems used the same track seeking system. We experiment at 32x and 40x speed. The random access time of the adaptive controller is 20% faster than one of the existing servo controller. The random access time is estimated in 32x speed and 40x speeds. In 32x of the uncontrolled system, the random access of the optical disk driver is over the spec. In 40x speed, it is impossible to access the track. While in the optical disk driver with the adaptive repetitive controller, both 32x and 40x speed achieve the good result. Table 1 shows the average access time of the optical disk driver.

Table 1: Average access time of optical disk driver

| 210um DISK | Uncontrolled | | controlled | |
|---|---|---|---|---|
| Speed | (32X) | (40X) | (32X) | (40X) |
| Average time(mSec) | 109.84 | Fail | 85.4 | 90.8 |
| Total number | 10000 | 1000 | 10000 | 10000 |
| 0~99ms | 5109 | | 6712 | 6015 |

| ~199ms | 4297 | | 3259 | 3925 |
|---|---|---|---|---|
| ~299ms | 503 | | 28 | 59 |
| ~399ms | 78 | | 0 | 0 |
| ~499ms | 9 | | 0 | 0 |
| ~999ms | 4 | | 0 | 0 |
| over 1 sec | 0 | | 0 | 0 |

## 4. Conclusion

This paper considered the control of the varying periodic disturbance for a track-following servo system of an optical disk driver. An adaptive repetitive control is proposed and implemented on the real production of an optical disk driver. The frequency multiplier is used to generate the adaptive sampling frequency and to synchronize the disturbance and control period exactly. The experiments demonstrate the effectiveness of the proposed adaptive repetitive controller.

We develop for discrete system to track a variable periodic signal with varying sampling time. We apply the proposed adaptive repetitive control to the real production.

## References

[1] Z.Cao and Gerard F. Ledwich. "Adaptive repetitive control to track variable periodic signals with fixed sampling rate", IEEE ASME Transactions on Mechatronics. Vol.7, No.3 Sep, 2002, pp378-384.

[2] G.M Dotsch and Henk T. Smakman, "Adaptive Repetitive Control of a Compact Disc Mechanism", the 34th Conference on Decision & Control, IEEE 1995, pp1720-1725.

[3] T.Y.Doh, J.R.Ryoo, M.J.Chung "Repetitive controller design for track-following servo system of Optical Disk Driver", IEEE AMC2002, 2002.

[4] Gerard Ledwich, A. Bolton, "Tracking periodic inputs using sampled compensators" Proc. Inst. Elect. Eng. Pt. D, vol. 138 no. 3, 1991.

Brill Academic Publishers
P.O. Box 9000, 2300 PA Leiden
The Netherlands

*Lecture Series on Computer
and Computational Sciences*
Volume 4, 2005, pp. 94-98

# Intelligent Building Network System Integration using TCP/IP

Kyungbae Chang[1], Iljoo Shim and Gwitae Park

Intelligent System Research Laboratory,
Korea University,
1, 5-ka, Anam-dong, Sungbuk-ku,
136-701 Seoul, South Korea

Received 27 July, 2005; accepted in revised form 10 August, 2005

*Abstract:* In this paper, we suggest the possibility of integration method adopting TCP/IP in building network system. The importance of integration in building control network system cannot be over-emphasized, for integrative system management is essential in energy saving and building management. We discuss some problems in practice and propose a solution for these problems, by organically connecting TCP/IP and Ethernet which are well distributed already.

*Keywords*: IBS, TCP/IP, Integration, Building Network

## 1. Introduction

In building control network system, solution for integration which guarantees the interoperability among the systems is important for interoperability has merits in terms of building maintenance, convenience in management, and energy saving.

Recently Building Control Systems have been often used not only to control building equipment such as HVAC, lighting efficiently but also to make living rooms com-fort. As Building System have developed, they have been various and complex. Building Automation Systems have also benefited from management, labor and energy cost. Thus integrative management is a requisite for maximized energy saving and the enhancement of network efficiency, and the control open protocol improves the effectiveness in integration.

But it is impossible to integrate Building Control Systems as one system, because subsystem in Building Control Systems is generally supplied by different vendors and they use each other open protocols. There are researches about open protocol to improve interoperability. The two most common protocols are LonWorks and BACnet.

The core of the LonWorks technology served by Echelon Corporation lies in a communication protocol called LonTalk.

LonWorks use Neuron Chip in order to develop a special type of communications, LonTalk. Controller that makes use of LonWorks can communicate with each other through "Standard Network Variable Types" or SNVTs. Although LonWorks has been applied in various markets for a few some years, it is a defect that some vendors have used the innate ability of neuron chip to contain applications in addition to communications at modest cost. More performance comes at a higher cost.

BACnet is an American National Standard. This is literally a book which describes in great detail how to create automation and controls system which may interoperate with other BACnet systems. BACnet has two important advantages which are vendor-independence and forward-compatibility. This is accomplished using an object-oriented approach for representing all information within each controller. If different vendors properly applied BACnet, systems of them can interoperate with each other. But BACnet is not a guarantee that forces all systems to interoperate. And implementation is difficult, because of great detail.

Generally, Building Network Systems have been used together with more than two open protocols. Top of Building Network Systems have integrated to use TCP/IP, one of open protocols. Building system

---

[1] Corresponding author. Kyungbae Chang. E-mail: lslove@korea.ac.kr

integration using many open protocols is set with limits in comparison with it using one protocol. Because network equipment which is connected each other protocols is inevitably needed and each other protocols are connected physically.

In this paper, we propose the scheme using only TCP/IP for advanced integration(for the problems proposed at the beginning) and We will verify the possibility of integration that uses TCP/IP ,which gets expansion property of the system, independence property from certain company, and interoperability when we integrate it by the proposed way.

## 2. Unification of the Building Automation System

The control of the various facility systems in the building is important because there are convenience of management and the reduction of utilization. In order to control efficiently, each of the facility systems is integrated and the open protocol is developed. Because the utilization of the open protocol can guarantee interoperability between different manufacturing companies, even if each system is installed in a different manufacturing company, the system can be used as a unified system. Accordingly, the new open protocols-the LonWorks and BACnet are developed for the control. Also, the integration method is invested by the new things. The following thing shows the integration component of a real building: [1].

### 2.1 The Construction to Combine the Company Protocol into the TCP/IP

The Fig.1 is the method that general buildings usually use. Each vendor protocol is used in the subordinate systems. In order to unify those systems, facility servers and the SI server are connected by the TCP/IP and the Ethernet. Because the unification with the TCP/IP is just the integration as the data network, it is possible to monitor the control information. However, it is impossible to control the DDC directly in the SI server.

The unified monitoring is materialized by the method. However, it is difficult to keep, to repair, and to extend the system because the innate protocol is not opened from a supply company. Also, because a different protocol is used in each other systems, the system will need a little more flexibility. Moreover, it causes the lack of the various connected scenario as well as the consumption for the maintenance, repairing and reduction of energy.

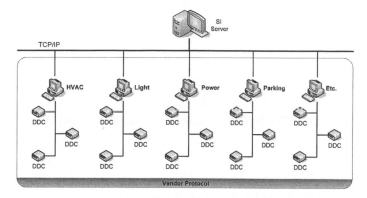

Figure 1: the Construction to Combine the Company Protocol into the TCP/IP

### 2.2 The Construction to Combine the Company Protocol or Open Protocol into the TCP/IP

The Fig.2 shows that not only certain sections of open protocols in the devices and the existing vendor protocol are used but also the TCP/IP is used from the each device server to the unified server. A lot of buildings which will like to unify the building automation system use this method. The method uses the gateway in each facility in order to control the DDC in the SI server where it is impossible to control the DDC with the simple unification such as the Fig. 1. Due to the method, it is possible not only to monitor but also to control all information of the DDC as a request from the development institution.

As using certain sections of open protocol, the maintenance and the extension of the systems are improved. Also, the interoperability can make connected scenario possible. However, the maintenance and the extension without the open protocol make a company subordinated. Also, it makes the discrepancy of efficiency large as the character of the gateway because the hardware and the software are unified by different protocols between the top and the bottom point. The unification with the gateway makes the innate protocol connected with TCP/IP of the top unstable.

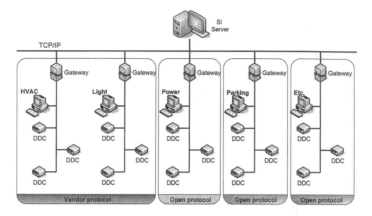

Figure 2: the Construction to Combine the Company Protocol and Open Protocol into the TCP/IP

### 2.3 The Construction to Combine Many Kinds of Open Protocols into the TCP/IP

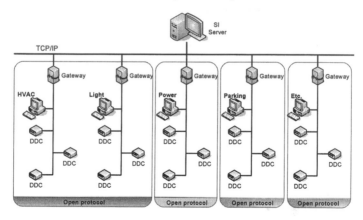

Figure 3: the Construction to Combine Many Kinds of Open Protocols into the TCP/IP

The Fig.3 shows that many kinds of open protocols in the bottom and TCP/IP in the top are used in order to make up for a few weak points. Because the utilization of open protocol in each system guarantees the interoperability in each other devices, it makes the various connected scenario possible. Also, it will cut down on expenses for energy, maintenance, mending, and management. However, in order to unify the TCP/IP from the top, the other devices are requested because the open protocol should be converted into the TCP/IP protocol by the gateway. Due to the devices, it is difficult to combine into the physical protocol precisely.

The neuron chip should be used to embody LonWorks in the open protocols. Due to the neuron chip, it is comfortable to apply the products and the hardware into the building network system. Moreover, the neuron chip makes the organization of the hardware convenient. However, because the neuron chip of echelon which is used in the LonWorks needs special software, there is a defect to rely on just one company.

Also, if we use BACnet, it is not easy to develop the system at the development company. Moreover, because there are not many development companies of each system, there are a small number of products and it is difficult to embody building a network system.

## 3. A Unification Plan Using TCP/IP

I suggest a method that unifies whole building systems as TCP/IP and Ethernet to fix up some problems at system building status.

The Fig. 4 is a constituent illustration, which unify whole building systems as TCP/IP: [2], [3]. TCP/IP has been widely used as protocol that is most active in re-search development and is well developed in the Internet among open protocol. In addition, embodiment of hardware is easy because there are many skills and developers and there are many products. At present, because the Internet is indispensable at a building, a unification that uses the Internet line that is already installed can save in-stall the fee. And because it doesn't need Gateway the unification that is guaranteed interoperability is possible.

When whole building is unified with open protocol, TCP/IP, maintenance and re-pair are east, expansion of system is guaranteed and energy fee can be saved. More-over, because TCP/IP is not being dependent on special processor and software, the problem of the existing method is improved and more efficient operation and management would be possible.

If whole building is unified construction of network would be various by a special character of TCP/IP. The Fig.5 shows topology that is variously changed when whole building unifies TCP/IP.

Unlikely topology of an existing building, it can communicate SI Server to DDC directly, without via facility server. This topology can solve a problem. It is that when an existing facility server is frozen, it cannot be geared with DDC of different facility. And free-network topology can install flexibly by considering distance between facilities when it install systems.

I expect that these constructions would save operation, management and energy fee with various connected scenario and freedom of installation.

This method's unification solved above all problems, but new problems would hap-pen if we use data network protocol, TCP/IP as control network: [4]. If we use data network protocol, TCP/IP as building control network, we have to consider problems that come from difference of data network and control network.

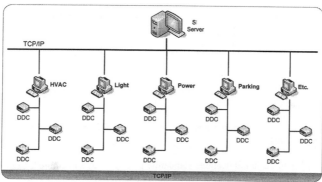

Figure 4: A Constituent Illustration of a Unification using TCP/IP

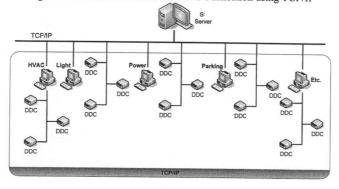

Figure 5: A New Construction of a Unification Plan of TCP/IP

Control network has to be solved with delay from few data transmission and high transmission frequency, packet loss when it transmits an important data, priority in collision confusion and a choice of transmission media: [5][6]. However, using a limited place, building can solve these problems. I will verify through next experiment.

## 4. Conclusion

In order to use open protocol in integrating buildings, the standards must be adequate to the network transmission delay, shouldn't have any packet losses, and must guarantee interoperability. Under this precondition, we suggest using TCP/IP, which is normally used to integrating open protocol.

Through many experiments under different scenarios, we proved that not only sys-tem construct cost declines, because of the reduction of the system integrating equipment using switch, but also it can have various kinds of network structure.

Looking through the weak points of various open protocols used for the integration of building-automating-system, we can overcome this weakness by using TCP/IP. Further we have suggested the possibility of control network.

## References

[1] IBS Korea, The study on Smart Building Systems, 2003.

[2] "Understanding and evaluation Ethernet and TCP/IP technologies for industrial automation", OPTO 22

[3] Dave VanGompel, Peter Cleaveland, "Moving popular industrial protocols to Ethernet TCP/IP", Control Solutions, 2001

[4] Max felser, "Ethernet TCP/IP in automation a short introduction to real-time requirement", IEEE 2001

[5] Feng-Li Lian, James R.Moyne, and Dawn M. Tibury, "Performance evaluation of control networks", IEEE Control System Magazine, 2001

[6] Wei Zhang, Michael S. Branicky, and Stephen M. Philips, "Stability of networked control system", IEEE Control System Magazine, 2001

[7] Mike Donlon, "Using standard internet protocols in building automation", Network Controls White Paper, 2001

[8] Chris Chapman, "Communications standards: past, present and future", Control & Automation Systems Technologies Professional Network

[9] Jhon Petze, "Powering smart energy networks now", AutomatedBuilding.com Article, 2000

[10] Kees van Grieken, "TCP/IP in building management systems", AutomatedBuildings.com Article, 2002

Brill Academic Publishers
P.O. Box 9000, 2300 PA Leiden
The Netherlands

*Lecture Series on Computer
and Computational Sciences*
Volume 4, 2005, pp. 99-104

# DSCM (Dual Switching Control Module) Design to Heighten Reliability of System

Kyungbae Chang[1], Iljoo Shim and Gwitae Park

School of Electrical Engineering
Korea University,
1, 5-ka, Anam-dong, Seongbuk-gu, Seoul,
136-701, South Korea

Received 30 July, 2005; accepted in revised form 14 August, 2005

*Abstract:* As computer system has developed, the break down rate of system has been increased. So, engineers have been interested in reliability of system. General method that is used to heighten reliability of system is Redundancy of System by dual switching control. In this paper, we propose DSCM which can give Redundancy structure in existent Computer (Microsoft Window 2000 OS) base system and we experiment redundancy structure that is applied DSCM to currently using train control system in IFC (Interface Computer) system.

*Keywords:* Reliability, Switching Control, Redundancy

## 1. Introduction

In recent years, the computer industry has developed so that the computer gets diversified and complicated and the system was applied to various areas with using computers. Computer based systems have been developed a lot, being combined systematization and its system in the electricity field. However, the reliability of the system is quite low. That's because. The stability of system has not been considered compared to the functional improvement of the computer using system. If the low reliability of electricity system brings about system fault, it'll suffer not only great financial loss, but also the toll of lives, thus a lot of system developers started to have an interest in reliability of system. The methods to strengthen the reliability of system are fault avoidance which reduces the breakdown of demagnetization by using high quality demagnetization and fault tolerant control that doesn't influence the function of system even if something's wrong[1][2]. The breakdown by the unknown reason of change of managing environment and by software bug is hard to handle with fault avoidance as system gets complicated and larger. Thus, fault tolerant control is being researched a lot nowadays, most used and studied way among these fault tolerant control methods is redundancy of system by dual switching control.

This method makes two systems as duplicated structures so that another prepared system can take over its role and work right if the error happens.

We proposed and realized module which enables established Microsoft Window OS based PC system like Fig. 1 concept of DSCM to redundancy by using dual switching control. This paper suggests one method to censor the process of system based on Microsoft Window OS and another one to exchange information each other effectively for two PCs. We estimated the function of module developed from a sham experiment and applied it to IFC system used in railroad control now in 2004.

Next session 2 gives you detail on the structure and method of dual switching control module and session 3 will experiment dual switching control by redundancy arbitrary process made for performance comparison and IFC system used in railroad control now. The last session 4 draws a conclusion and tells you the future task.

---

[1] Corresponding author. Kyungbae Chang. E-mail: lslove@korea.ac.kr

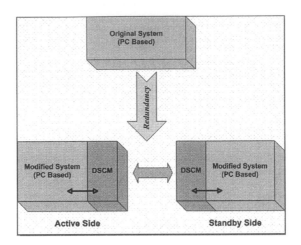

Figure 1: Concept of DSCM.

## 2. Dual Switching Control Module

In this session we explain about DSCM. DSCM is the process for redundancy established PC based system. ESCM is the process for redundancy the established PC based system, which watches if it's ok or not by adding DSCM module to the established system and if it finds out something is wrong. It hands over the information to the other system and it enables switching.

To realize the function of redundancy, first of all, you need routine that watches established movement process, that exchange information with the other person and that determines active or standby by using collected information.

We account for the structure of DSCM, the way to censor your process, how to ex-change information with others, how to decide the movement mode and lastly, the parts established process should be amended.

### 2.1 The structure of DSCM

You need the two same PC systems in order to get redundancy structure of HOT standby form [4]. These two systems provide you the same input and output and determined Active side system by the priority order outputs, Also, DSCM performs within the same PC which has the parallel with established movement processor as module form, Fig. 2 shows you the system structure by using DSCM.

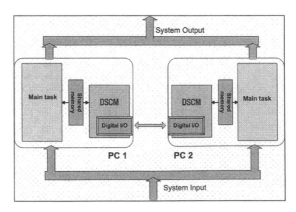

Figure 2: Architecture of DSCM.

In Fig. 2 Main Task and DSCM are the processor that drive on Microsoft Window OS and here, mail task was rectified so that the established movement processor can be applied to DSCM. Shared memory and Digital I/O were used as bride to notice Main Task or opponent system. The DSCM access the certain data of Main Task to watch the state of Main Task.

The way separate process shares data on Windows OS is to use IPC (Intercrosses Communication) in MAPI (Messaging Application Programming Interface) provided by Windows [5]. IPC uses mechanism such as Clipboard, Com, WM COPYDATA, FILE MAPPING, DSCM continues to sharing datum at regular intervals, so it is advantageous to File MAPPING mechanism which only uses a physical memory.

Therefore, DSCM uses shared Memory that puts to use File Mapping mechanism. Interface methods are really various for two PC Systems to exchange datum information, Among them, the method that uses Printer Port (SPP, EPP, ECP), Serial Port (RS-232) and LAN (Ethernet) are most used interfaces, But, these interfaces are slow in transmission speed since all of them need controller and driver for data trans-mission and DSCM uses Digital I/O Board which is most simple and stable interface for fear that they might have errors in controller and driver.

### 2.2 How to censor the movement process

DSCM exchanges the information and watch the state by using Shared memory. Above all, we define the state of Main Task like TABLE 1.

Table 1: State of Main Task

| State | Description | Code |
|---|---|---|
| Normal | Active state without something wrong | 0x04 |
| Warning | Unstable state in warning state, when lose track of signal once | 0x02 |
| NoResponse | Irresponsive state, when lose track of signal | 0x03 |
| Error | Error state, main task judged that operating is impossible state | 0x01 |

We decide the state of Main Task by using two ways. First method is to take performing process name and handle and then check the movement. Second one is to ex-change information reading and writing the data structure by turns on Shared Memory. The structure used this time shows TABLE 2.

RW member is the data that judges whether or not Main Task and DSCM update the information. Code member has intelligence to deliver for each other. Checksum member finds errors in using memory or errors by miscalculation. And, sampling time member is used for setting the best time by conveying cycle time figure to exchange information each other, since two processes perform respectively. You should go to trouble to set the same period in reading and writing the information once.

Table 2: Data Structure

| Member | Description | Type |
|---|---|---|
| Handle | Shared Memory's handle, value was decided when memory creates | Char |
| Code | Include information of each other | Char |
| Sampling Time | Include sampling cycle time | Char |
| RW | 0x01 - When DSCM wrote data<br>0x02 - When Master Task wrote data | Char |

You should try setting the same period by given the 2/5 delay of sampling time if you miss the signal like Fig. 3. If you don't get any responses, it recognizes No Response State like Fig. 4.

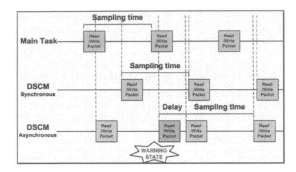

Figure 3: Synchronization.

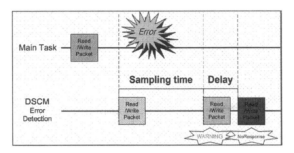

Figure 4: No Response.

### 2.3 The way to exchange information with others

DSCM realizes redundancy with the perfectly same system like itself. I've already mentioned that these two same systems use DIO Board in sharing each other's information. The information that must be shared for redundancy is only two modes like movement possibility of system and working mode at present. DSCM shares the in-formation by using the second contingent for quick switching.

In the first place, the possibility of movement gives a state signal per DIO Period time by Heart Response Signal. DSCM understands that the counterpart system cannot work when there's no response for timer period. Current movement mode is Current Mode Signal, it indicates Active mode in the state of 'H' and stand by or error in 'L' state.

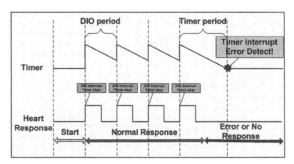

Figure 5: Heart Response Signal.

### 2.4 Mode determination of DSCM

DSCM makes a definition there modes and work after watching the state of Main Task and its counterpart. First, Active mode is a movement mode of Active Side. Standby mode is a movable mode which is a movement mode of Standby Side system. Lastly, it changes to Emergency mode. It gives priority to the counterpart system by not out-putting Heart Response Signal. If you realize the system as redundancy, two systems have the equal priority so you have to decide which system should be Active Side when they become Normal State simultaneously.

Table 3: Operating Mode

| Mode | Description | Code |
|---|---|---|
| Active | Operating mode on Active Side | 0x01 |
| Standby | Mode that system operating is possible on Standby Side | 0x02 |
| Emergency | Mode when DSCM judge itself error witch is system operating is impossible | 0x03 |

DSCM suggests you two choices. Master/Slave method that gives a priority to either system and Random Select method that gives a random priority.

Master/Slave method is that the Master input system has got the top priority after inputting Master and Slave on DSCM process. Random Select method is to give priority to the previous working system if two systems behave the same mode and if they convert the same mode at the same time, take the random figure between Time Slot(1~10)* DIO Period[ms]. In other words, the system which has less random figure gets the priority, however, if the same random figure is settled and converted to the same mode, the priority is chosen according to the random figure again.

### 2.5 Amendment from established system

DSCM is module for redundancy of system based on PC. At that time, system needs to be revised. PC system needs installed DIO Board for two systems to exchange information, if the established PC system doesn't have DIO Board. It needs an additional Board installation and Device Driver and Library the program requires. And you should installed DSCM movement program and make it work before Main Task drives, established system hands over the state information of system to DSCM, and it needs revision to decide the movement and to receive from DSCM. Data exchange and data Synchronization which is DLL (Dynamic Link Library) form library offered by DSCM use Shared memory which Main Task will provide, at this moment. Main Task program must add routine that calls four functions provided by DLL and must compile and make the program. TABLE 4 shows you the functions provided by DLL.

Table 4: Function for Main Task

| Function | Description |
|---|---|
| DSCMOn | Function for start process |
| DSCMOff | Function for end process |
| DSCMGet-Mode | This function gets the mode of DSCM |
| DSCMSet-State | In case of Main Task has itself diagnosis function, this function informs own state to DSCM |

## 3. Conclusion

Earlier, we presented redundancy method of module form which enables redundancy for system based on PC. Presented ways are simple but the amendment from the established system, which enables redundancy of the system. We've figured out capacity of redundancy system through experiments.

From now on, we're studying the way to find out the state of movement program with-out any revision to support the perfect module form and adding and amending a few of algorithm for exchanging information with the counterpart system.

## References

[1] Sae-Hwa Park, "A Study on the Control system with Dual Structure to Enhance its reliability", Journal of Control. Automation and Systems Engineering, 1990. 10.

[2] K. S. Song, H. K. Yeo, "Hot standby Computer Structure: An Availability Analysis of Switching Control System with Hot Standby Fault Tolerant Architecture", Jounal of Korea Information Processing Society, 1995. 11.

[3] Barry W. Johnson, "Reliability & Safety Analysis of a Fault-Tolerant Controller", IEEE MICRO, p22-p33, 1984.

[4] Hwan-Geun Yeo, "The Implementation Method of Synchronization Unit in Hot Standby Dual Switching Control System", Jounal of Electronics and Telecommunications Research Institute, 1995.

[5] http://msdn.microsoft.com/.

Brill Academic Publishers
P.O. Box 9000, 2300 PA Leiden
The Netherlands

*Lecture Series on Computer
and Computational Sciences*
Volume 4, 2005, pp. 105-111

# Performance Evaluation of the Smart Airbag ECU

Kyungbae Chang[1] and Gwitae Park

Department of Electrical Engineering,
Korea University,
Seoul 136-701, South Korea

Received 1 August, 2005; accepted in revised form 12 August, 2005

*Abstract:* This paper is about the method of evaluation of the performance of the Smart Airbag ECU, and the development of it's automated measuring device. Because of the smart airbag's many modules and surrounding devices, sensitive measuring is difficult and it takes a lot of time. But by using the automated performance evaluation device, reliable and precise data can be measured. It can also save the measuring time. The test program has been made by LabVIEW. By the measurement and statistic analysis of the data, the reliability can be analyzed.

*Keywords:* Smart Airbag ECU(SAE), Occupant Classification System(OCS), Distributed System Interface(DSI), Electric Control Unit(ECU), Function Test Equipment(FTE)

## 1. Introduction

The airbag started with a single driver side airbag (1 Fire Circuit), but recently the smart airbag has at least 12 airbags installed on a single car. The twelve are Driver side Dual Stage Airbag (2 Fire Circuit), Passenger side Dual Stage Airbag (2 Fire Circuit), Driver/Passenger side Belt Pretensioner (2 Fire Circuit), Front Driver/Passenger side Side Airbag (2 Fire Circuit), Rear Driver/Passenger side Side Airbag (2 Fire Circuit) and Driver/Passenger side Curtain Airbag (2 Fire Airbag).

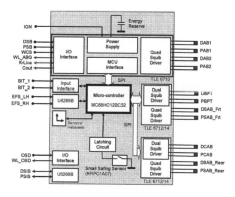

Figure 1: Block diagram of SAE.

In order to accurately control the many airbags in a crash situation, the control device is considered to be very important. Figure 1 shows an example of the recently developed SAE (12 Firing Circuit). The SAE uses various sensor modules in order to minimize passenger casualties, and to inhibit unnecessary deployments. OC sensors or PPD sensors are used to monitor, with a 1 sec interval, the presence of a passenger in the passenger seat, if it's a child of an adult, of if there is a child seat. Seat Track sensors

---

[1] Corresponding author. Kyungbae Chang. E-mail: lslove@korea.ac.kr

are used to measure the distance between the Passenger side seat and the passenger airbag module. If the distance is too short, airbag deployment is prohibited, because airbag expansions in these situations are even more dangerous.

To estimate if the Seat-belt is used, seat-belt Buckle Switch sensors are used. If the Seat-belt is being used, the airbag deployment algorithm activates slower, and if it isn't being used, airbag deployment algorithm becomes more rapid because the passenger's movements are more faster under this situation. The Ball in Tube sensor is the device capable of estimating the degree of the impact, before the actual crash impact force data is transmitted to the airbag's ECU acceleration sensor. As side way crashes delivers the impact to the passenger more rapidly than a front side crash, it is more dangerous. In order to indicate these side way crash situations, Driver/Passenger side Side Impact sensors are used. When side way crashes occur, the Side Impact Sensor activates the airbag deployment algorithm more rapidly, transmitting the airbag deployment decision results to the ECU. The use of many sensors like these, are obstacles for the evaluation of the reliability and performances of the airbag ECU.

Measuring the performances and securing the reliability of an SAE developed product is even more difficult. The number of evaluation items are increased, and various sub devices that elevate the airbag ECU's abilities must be tested and verified also.

## 2. Performance Evaluation Items

The SAE is consisted of many functions, and each function surveys and diagnoses the situation for any faults with the each of its cycles. The results are used to determine the crash, and to show the current condition of the SAE. The following are the diagnosis items, of which the SAE performs.

### 2.1 Ignition Voltage Check

In order to inspect any ignition voltages outside of the normal bounds, the SAE measures the input voltage and performs defect diagnosis, under the criterion shown in Figure 2. If "Vbatt too High" or "Vbatt too Low" appears for more than 10 seconds, it is recognized as a defect and lights up the warning lamps. If it recovers normal ignition voltage for over 10 seconds, the defects are considered fixed and the warning lamp goes off.

Figure 2: Ignition Voltage Check Criterion.

By using the variable power supply device, the FTE changes the input voltage over/under the normal value, and by doing this it finds the point where the fault has occurred.

### 2.2 Squib Leakage to Vbatt/GND Detection

In order to detect short circuits of the GND/Vbatt, which interferes with the deployment of the airbag or the measurement of the squib resistance, the firing circuit is observed. The range of leakage resistance, which is used to diagnose "Leakage to Vbatt" and "Leakage to GND" is shown in Table 1. The time required for the detection is 4 seconds, and if there is no defect for 8 continuous seconds, it is recorded as a fault of the past.

Table 1: Criterion of firing circuit Leakage to Vbatt/GND decision.

| Short Resistance Range | Content | Trouble |
|---|---|---|
| Rsc-<2kΩ | Vbatt/GND leakage | Clear Detection |
| Rsc->10kΩ | Normal state | No Detection |
| 2kΩ ≤ Rsc- ≤ 10kΩ | Tolerence | Detection or NOT |

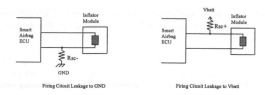

Figure 3: Firing Circuit Leakage to Vbatt/GND.

In order to generate a Leakage fault, the Decade Box is installed in the Rsc, and the resistance is varied. By using Table 1's leakage resistance range, we know if a fault were detected. The Decade Box's resistance values are changed automatically to the values sent from the GPIB card.

### 2.3 Firing Circuit too High/Low Resistance Check

In a constantly vibrating automobile, wire leakage and disconnections will happen. Also it weakens the connectors bonding, instantly opening the circuit. These kinds of defects can result in unneeded airbag expansions or no expansions when needed. Resistance can be shown by the constant equivalence between the 1'st stage airbag module, which is in front of the driver's side, and SAE. The driver side resistance is higher than other resistance ranges. This is because clock springs are used to provide electricity to the steering wheel module, and at this part resistance increases. The loop resistance of the Front driver side airbag firing circuit is shown in Figure 4. Thus,

$$R_{DAB} = R_S + R_T + R_W + R_C$$

where, $R_S$ = Squib Resistor
$R_T$ = Connector Terminal Contact Resistor
$R_W$ = Wiring Harness Resistor
$R_C$ = Contact Coil Resistor

The passenger side airbag's equivalence resistance is equal, to the driver side equivalence model without the clock spring resistance

$$R_{PAB} = R_S + R_T + R_W$$

where, $R_S$ = Squib Resistor
$R_T$ = Connector Terminal Contact Resistor
$R_W$ = Wiring Harness Resistor

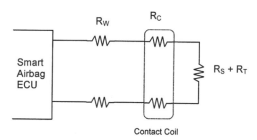

Figure 4: Equivalence resistance of the Driver side Front Airbag.

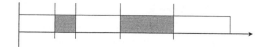

Figure 5: Criterion of drivers side Front Airbag Resistance Decision(1'st stage).

The Passenger side airbag resistance decision is shown in Figure 6. The equivalent resistance decision criterion of the other 10 airbag modules, which does not include the driver side airbag resistance decision criteria, is equal to the resistance decision criterion of the passenger side airbag.

Figure 6: Criterion of passenger side Front Airbag Resistance Decision (1'st Stage).

In order to measure the equivalent resistance, a high precision Decade Box is used. The Decade Box changes the resistance value down to two decimal places, by grip communication. To do this in a short time, the average of the upper limit and the lower limit is used. For example, if you want to measure the boundary value of "Rs too High", we start with the average of "Rs OK(2.8)" and "Rs too High(5.4)". The recognition of "Rs too High" must be confirmed, and if it hasn't been confirmed, measurement starts over again, beginning with the average of "Rs too High(5.4)" and the former average. By this method, the boundary value can be found in short time.

### 2.4 Firing Circuit Resistance Open Check

If the connector's bond becomes weak by the vibration of the car, the firing circuit's resistance instantly becomes open. To survey these situations, the resistance decision criterion in Figure 7 is used, which is the same method with the firing circuit's too High/Low resistance check.

Figure 7: Criterion of Firing Circuit Resistance Open Resistance Decision (1'st Stage).

### 2.5 Seat-belt Buckle Switchk

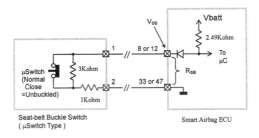

Figure 8: Interface between the SAE and the Seat-belt Buckle Switch (Micro Switch Type).

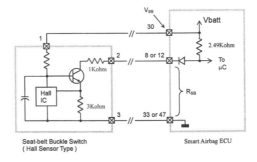

Figure 9: Interface between the SAE and the Seat-belt Buckle Switch (Hall Sensor Type).

Figure 8 and Figure 9 shows the two types of interface methods between Seat-belt Buckle Switch and SAE. The SAE diagnoses problems, and checks whether if the seat-belt is used, by surveying the Seat-belt Buckle Switch equivalence resistance and RSB. Debouncing logic is used when the switch is detected, in order to prohibit the errors caused from mechanical chattering when the switch is On/Off. Seat-belt Buckle Switch equivalent resistance, voltage measured from the RSB and High-side (Pin #8 of #12), and the diagnosis of the VSB is shown in Table 2.

Table 2: Criterion of Seat-belt Buckle Switch decision.

| Seat-belt buckle switch Input resistance($R_{SB}$) Range | Voltage at Pin #8 or #12($R_{SB}$) | Diagnosis |
|---|---|---|
| $10.0k\Omega \leq R_{SB}$ | $0.82 \times Vbatt \leq V_{SB}$ | Short Power and Open Circuit |
| $4.4k\Omega < R_{SB} < 10.0k\Omega$ | $0.56 \times Vbatt < V_{SB} < 0.82 \times Vbatt$ | Tolerance |
| $3.6k\Omega \leq R_{SB} \leq 4.4k\Omega$ | $0.48 \times Vbatt \leq V_{SB} \leq 0.56 \times Vbatt$ | Belt Buckled(Switch Open) |
| $1.1k\Omega < R_{SB} < 3.6k\Omega$ | $0.28 \times Vbatt < V_{SB} < 0.48 \times Vbatt$ | Tolerance |
| $0.9k\Omega \leq R_{SB} \leq 1.1k\Omega$ | $0.21 \times Vbatt \leq V_{SB} \leq 0.28 \times Vbatt$ | Belt Unbuckled(Switch Close) |
| $0.3k\Omega < R_{SB} < 0.9k\Omega$ | $0.10 \times Vbatt < V_{SB} < 0.21 \times Vbatt$ | Tolerance |
| $R_{SB} \leq 0.3k\Omega$ | $V_{SB} \leq 0.10 \times Vbatt$ | Short GND and Short Circuit |

In order to diagnose whether if the Seat-belt Buckle Switch is working normally, fault detection is surveyed by using a low precision Decade Box, instead of the Seat-belt Buckle Switch, to vary the resistance. In order to diagnose whether if the Seat-belt Buckle Switch is working normally, fault detection is surveyed by using a low precision Decade Box, instead of the Seat-belt Buckle Switch, to vary the resistance. If the measured value is equal to the diagnosis boundaries of Table 2, it is recorded as normal.

### 2.6 Seat Track Sensor Test

In order to measure the distance between the passenger seat and the passenger airbag module, the Seat Track sensor is used. The device's formation is equal to the Seat-belt Buckle Switch. The status of the Seat Track sensor is divided into three categories, "Open", "Rearward", and "Forward and Short". In a "Short" situation, airbag expansion is prohibited. In a "Forward" situation the rapid crash algorithm is performed. Resistance range and diagnosis is equal in the Seat-belt Buckle Switch's "Belt Buckled" situation and the "Rearward'" situation. Also, "Belt Unbuckled" situations and "Forward" situations have the same resistance range and diagnosis too. Defect diagnosis methods are the same with the Seat-belt Buckle Switch also.

### 2.7 Ball In Tube

The Ball In Tube calculates the impact force before the actual impact force from a frontal crash is delivered to the ECU's acceleration sensor. The decision of whether or not to deploy the airbag is made with this calculation. Predicting the crash situation helps to make the airbags more accurate.
The device formation is similar to the Seat-belt Buckle Switch. But the state changes are based on the voltage change by the movement of mass. The Ball In Tube can be divided into four situations. "STB(Open)", "Switch Open", "Switch Close" and "STG". In the "Switch Close" situation, the airbags are ready to deploy, and in the "Switch Open" situation, airbags motion normally. In the other two situations, a warning lamp is switched on to indicate the defect. In order to check the Ball In Tube's performances, movement of mass is required. But in this FTE, as the voltage changes with the movement of mass, performances are evaluated by modifying resistance. Resistance is modified by low precision Decade Box.

### 2.8 First Warning Lamp Test (Bulb Type)

The ECU uses the warning lamp to tell its conditions or defects to the driver. If there is a defect in the inner or surrounding devices, the warning lamp lights up. In order to check if the actual condition is the same with the command it made to turn on/off the warning lamp, the ECU measures the warning lamps output port voltage. The criterion for the voltage range of defect is shown in Table 3.

Table 3: Criterion for Warning lamp Fault check.

| Warning Lamp Status | Light Power Range | Diagnosis | Trouble |
|---|---|---|---|
| On | $V_{WL} \leq 3.5V$ | NORMAL STATE | NO DETECTION |
| | $V_{WL} \geq 4.5V$ | Short power or warning lamp short | Clear Detection |
| | $3.5V < V_{WL} < 4.5V$ | Tolerance | Detection or NOT |
| Off | $V_{WL} \geq 0.8V*Vbatt$ | Normal state | No Detection |
| | $V_{WL} \leq 0.4V*Vbatt$ | Short GND or Light open | Clear Detection |
| | $0.4V*Vbatt < V_{WL} < 0.8V*Vbatt$ | Tolerance | Detection or NOT |

Defect diagnosis items that can be performed with the warning lamp On, can be performed when the early bulb is being examined of when the warning lamp is turned on by activation defects.

By the use of DMM(Digital Multi Meter) in the warning lamp output, which is on the wire harness, voltage is read through GPIP communication. The measured voltage is verified by the criterion shown in Table 3. Performance survey items of the Warning lamp, varies by its on/off situations. This fact must be considered in the process of evaluation.

## 2.9 PPD/OC

For the passenger detection of the passenger side seat, PPD and OC are used. The PPD is connected with the Force Sensitive Resistor(FSR) in the passenger seat. It measures the resistance changed by the presence of a passenger, and it performs self-diagnosis every 1.2 seconds and sends these information to the ECU. The transferred information is "Passenger Not Found", "PPD Fault" and "Passenger Found".

With more advanced technology than PPD, the OC transfers the information as "Class0", "Class1" and "Class2". "Class2" means a passenger is present. In order to measure the interface functions between the two devices, an Emulator, which can generate various situations by the PPD and OC, has been developed. The Emulator generates signals based on commands given by the PC, and surveys if the ECU recognizes it.

## 2.10 ECU Autonomous Time Test

Even with the power supply being cut, the ECU must survive for more than 150msec, saving the past and present data. To perform this function, Reserve Capacitor is used. In order to verify if the data storing process is functioning normally, the Reserve Capacitor opens the voltage input pin by force and observes the ECU. FTE cannot take apart and analyze the ECU. Therefore in order to survey the existence of the inner MICOM, the warning lamp pin is watched. If the MICOM does not function the warning lamp lights up, and by measuring the time past, approximate ECU Autonomous Time can be measured. By impressing the power and reading fault items, monitoring whether or not the data is recorded on the memory is possible.

## 2.11 Crash Output

The Crash Output is an open collector type switch output with a 500mA sink current capacity, with an Output Impedance of 100w. This switches to GND for 200ms after a crash is indicated.

In order to evaluate the interface performance between SAE and the Door Control Module, the ECU outputs the Crash Output signal by force, and measures this with DMM.

## 2.12 Et Cetera

By measuring the time used to minimally turn off the lights of the warning lamp, when provided with power, the ECU inner program's time related defects can be distinguished. The many items that are related to time use the same subroutine.

Minus Option Test, which is divided by option, is measured to verify wrongly manufactured ECU's errors, which is caused by ECU assemblers with different car type parts insertion, program errors of the Auto-inset machine, or by insertion of a wrong parameter.

By the use of ageing equipments, the wire resistance increases and errors may occur. For the calibration of these incorrect measurement values, the wire resistance value is repaired periodically by the 4-Wire Measurement method.

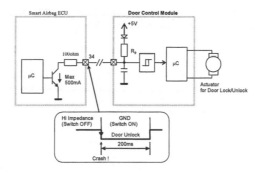

Figure 10: Interface between SAE and Door Control Module.

## 3. Conclusion

This paper describes the development of a device capable of evaluating the SAE, which is a next generation airbag, automatically. By accurately controlling the many airbag modules, and by communicating with its many surrounding sensor devices, the SAE must decide the airbag deployment time. Because of these many functions, the SAE has more items to test than the existing devices. By the use of the FTE, which has been developed by this study, errors made by test engineering decreases, and therefore the product's performances are measured precisely, increasing the product's reliability. Development time can be reduced also, due to its shortened performance evaluation time.

## References

[1] http://www.ni.com/labview.

[2] FMVSS 208, "Occupant Crash Pretension", U.S. Department of Transportation, National Highway Traffic Safety Administrations, 1988

[3] J. Hou, J.Tomas, "Optimization of Driver Side Airbag and Restraint System by Occupant Dynamics Simulation", SAE 952703, 1995.

[4] GT. Narwani, T. Subbian, "Optimization of Passenger Airbag using Occupant Simulation", SAE 930240, 1993.

[5] Ching-Yao Chan, "A Treatise on Crash Sensing for Automotive Air Bag Systems", IEEE/ASME Trans. On Mechatronics, Vol. 7, No. 2, June 2002.

[6] Yan Lu, Christian Marschner, Lutz Eisenmann and Sivart Sauer, "The New Generation of BMW Child Seat and Occupant Detection System SBE 2", IJAT, Vol. 3, No. 2, pp.53-56(2002).

[7] Robert Barnard and Miloslav Riener, "Vehicular Air-Bag Control Based on Energy, Momentum, and Semometric Spaces", IEEE Trans. On Vehicular Technology, Vol. 49, No. 5, Sep. 2000.

Brill Academic Publishers
P.O. Box 9000, 2300 PA Leiden
The Netherlands

*Lecture Series on Computer
and Computational Sciences*
Volume 4, 2005, pp. 112-116

# Global and Local Electrophilicities as Descriptors of Toxicity

P. K. Chattaraj[1,*], U. Sarkar[1], D. R. Roy[1], R. Parthasarathi[2], J. Padmanabhan[2], V. Subramanian[2]

[1]Department of Chemistry, Indian Institute of Technology, Kharagpur 721302, INDIA
[2]Chemical Laboratory, Central Leather Research Institute, Adyar, Chennai 600 020, INDIA

Received 2 July, 2005; accepted in revised form 11 July, 2005

*Abstract*: Global and local electrophilicities of several polychlorinated dibenzofurans and aliphatic amines are calculated at the B3LYP/6-31G* level, in the gas and solvent phases, using different types of population analysis schemes. While the former toxins act as electron acceptors latter toxins behave as electron donors when they interact with biosystems. Charge transfer is shown to be one of the major causes of toxicity of those molecules. Experimental toxicity correlates well with that calculated using the global and local electrophilicities.

*Keywords*: Electrophilicity, Philicity, Charge transfer, Toxicity

*Mathematics SubjectClassification:* PACS: 31.10, 31.15.A, 31.15.E, 33.15, 87.15.B

*PACS:* 31.10, 31.15.A, 31.15.E, 33.15, 87.15.B

Density functional theory [1-3] is quite successful in providing theoretical foundations of qualitative chemical concepts. Global electrophilicity index ($\omega$) is defined [4] in terms of the chemical potential ($\mu$) and hardness ($\eta$) as

$$\omega = \frac{\mu^2}{2\eta} \tag{1}$$

A generalized concept of philicity containing electrophilic, nucleophilic and radical reactions has been proposed [5]. The condensed- to- atom variants for the atomic site k have been written as

$$\omega_k^\alpha = \omega f_k^\alpha \tag{2}$$

where $\alpha$ = +, - and 0 refer to nucleophilic, electrophilic and radical attacks respectively and $f_k^\alpha$ are the respective Fukui functions. The $\omega_k^\alpha$ will vary from point to point in a molecule but the sum of any $\omega_k^\alpha$ over all atoms is conserved. The changes in the global and local reactivity profiles of reacting systems are important to understand the reactivity of the chemical system. In this context some exact conditions of extremals of electrophilicity index along an arbitrary reaction coordinate have been examined [6] and its variation during molecular vibrations and internal rotations are also analyzed [6]. The toxicity of polychlorinated biphenyls (PCB), benzidine and dibenzofuran [7] has been quantified with the help of electrophilicity. Effects of electric field [8] and solvent [9] on the global and local electrophilicities have been analyzed. Reactivity trends of some more molecules and reactions in terms of philicity have also been analyzed [10].

The main objective of the present study is to analyze the potential of global and local electrophilicities in estimating toxicity of various systems. Development of quantitative structure activity relationship (QSAR) in the *de novo* drug design and prediction of toxicity of various molecules have attracted wide attention [11]. The successes of electrophilicity and local electrophilicity indices, as opposed to several disjoint descriptors, in the prediction of biological activity and toxicity of various molecules have been illustrated [12, 13]. It is shown that the electrophilicity can be used as a descriptor of biological activity/toxicity and it is unbelievable that a single descriptor can provide such a beautiful correlation. In the present study the usefulness of electrophilicity descriptors in the prediction of

---

[1,*]Corresponding author. E-mail: pkc@chem.iitkgp.ernet.in
* Presenting Author: e- mail pkc@chem.iitkgp.ernet.in

toxicity is demonstrated by considering the polychlorinated dibenzofurans (PCDFs) and a set of aliphatic amines as examples. All the previously determined biological activity data [14], that is the negative of the log of molar concentration of chemical necessary to displace 50% of radiolabeled TCDD from Ah receptor ($pIC_{50}$), are utilized for this purpose. Experimental biological activity ($pIC_{50}$) for PCDFs are correlated with their corresponding calculated $pIC_{50}$ values determined using two parameter multiple regression analysis in gas and solvent phases using MPA, NPA and HPA schemes. A set of aliphatic amines, which act as electron donors in their interaction with biomolecules are also studied for their log ($IGC_{50}^{-1}$) activity [14].

The geometries of polychlorinated dibenzofurans (PCDFs) and aliphatic amines have been optimized using 6–31G* basis set in the framework of B3LYP theory. All calculations have been performed using the G98W & G03W suites of programs [15]. Hirschfeld population scheme has been used to calculate FF values as implemented in the DMOL package [16] employing BLYP/DND method. Solvent phase optimization has been carried out for other selected systems using polarizable continuum model (PCM) [17]. The atomic charges for all the above molecules have been obtained in the framework of B3LYP theory using Mulliken Population Analysis (MPA) [18]. Natural population analysis (NPA) [19] are also used to derive atomic charges.

Electronegativity, joint hardness and fraction of charge transferred between a toxin (PCDF/amine) and a biosystem simulated by nucleic acid bases (adenine, guanine, cytosine, thymine and uracil)/ DNA base pairs (GCWC and ATH) indicated that PCDFs act as electron acceptors and aliphatic amines as electron donors. Hence, for QSAR analysis of PCDFs, the atom with maximum value of local electrophilic power ($\omega^{+}_{max}$) in a molecule alongwith the electrophilicity index ($\omega$) are considered as independent variables. For aliphatic amines the $\omega^{-}$ of the atom with the maximum value of local philicity ($\omega^{-}_{max}$) in a molecule alongwith the electrophilicity index ($\omega$) are considered as independent variables. In this presentation only some representative cases would be reported although the results are encouraging in almost all cases.

Experimental biological activities ($pIC_{50}$) of polychlorinated dibenzofurans (PCDF) which act as electron acceptors as well as that ($pIGC_{50}$) of aliphatic amines which act as electron donors when they interact with biosystems are correlated with their corresponding activities ($pIC_{50}/pIGC_{50}$) calculated using the electrophilicity index and the local electrophilic power through regression analysis in both gas and solvent phases. A good correlation has been obtained for all the systems showing the significance of the selected conceptual DFT based descriptors in the prediction of toxicity in gas and solution phases with similar trends originating from different population analysis schemes. Reasonably good correlation of the amount of charge transfer with toxicity implies that the charge transfer plays a crucial role in the observed toxic behavior of PCDFs and amines.

The global interactions between the constituents of the selected system namely, PCDFs, and NA bases/base pairs have been determined using the parameter $\Delta N$, which represents the fractional number of electrons, transferred from a system A to a system B. Generally, electron flows from less electronegative system to more electronegative and this fact along with the definition of $\Delta N$ clearly shows that charge transfer values (Figure 1) are positive for PCDFs in most cases representing them as electron acceptors. Also maximum amount of charge flows from Guanine and GCWC to PCDFs among the selected bases/base pairs.

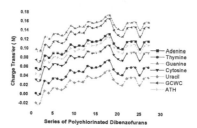

Figure 1: Charge transfer ($\Delta N$) between different polychlorinated dibenzofurans and nucleic acid bases/base pairs in gas phase.

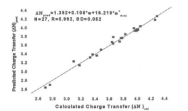

Figure 2: Variation of calculated Charge transfer ($\Delta N_{cal}$) with the predicted charge transfer ($\Delta N_{pred}$) during interaction of polychlorinated dibenzofuran with GCWC base pair.

In the case of interaction with PCDFs, Guanine and GCWC donate maximum charge among the selected bases and base pairs and Uracil and ATH, the minimum. The variation of calculated charge transfer with the charge transfer predicted ($\Delta N_{pred}$) using $\omega$ and NPA derived local philicity ($\omega^{+}_{max}$), during the interaction of polychlorinated dibenzofuran with GCWC base pairs (Figure 2) is reported here and they give a correlation coefficient of 0.992. A good correlation shows the importance of the selected descriptors in the prediction of charge transfer and hence the toxicity of the system.

The gas phase data of the selected set of 27 PCDFs are considered for evaluating the correlation between the observed and calculated values of $pIC_{50}$ for NPA derived charges. A plot between the observed $pIC_{50}$ values and that calculated using $\omega$ and NPA derived local philicity ($\omega^{+}_{max}$) for a set of 27 PCDFs (Figure 3) shows a correlation of 0.999 representing the significance of the philicity indices in toxicity prediction.

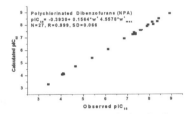

Figure 3: Observed versus calculated values of $pIC_{50}$ using NPA derived charges for PCDFs in gas phase.

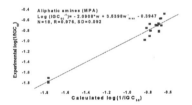

Figure 4: Observed versus calculated values of log ($1/IGC_{50}$) using MPA derived charges for the set of aliphatic amines in gas phase.

Aliphatic amines are known to be electron donors in their interaction with biomolecules. Hence for the regression analysis, the $\omega^{-}_{max}$ of the atom with the maximum value of local philicity ($\omega^{-}_{max}$) in a molecule along with electrophilicity index ($\omega$) has been considered as independent variables. The gas phase data of the selected set of 18 aliphatic amines including the observed and calculated values of the activity $\log(IGC_{50}^{-1})$ for MPA derived charges are presented in Figure 4. $\omega$ and MPA derived $\omega^{-}_{max}$ are capable of providing a correlation of 0.976 exhibiting the power of these descriptors in the toxicity prediction.

**Conclusions:** It is demonstrated through the DFT level calculations of global and local electrophilicities, using various population analysis schemes in the gas and solvent phases, that the toxicity in both electron donors like aliphatic amines and acceptors like polychlorinated dibenzofurans stems from the charge transfer between the toxin and the biosystem. These two descriptors are capable of providing the toxicity values which match very well with the corresponding experimental ($pIC_{50}$ and $pIGC_{50}$) values.

### Acknowledgment

We are thankful to CSIR, New Delhi for financial assistance.

### References

[1] R. G. Parr and W. Yang, *Density Functional Theory of Atoms and Molecules* (Oxford University Press: New York), 1989.

[2] P. Geerlings, F. De Proft and W. Langenaeker, Conceptual Density Functional Theory. *Chem. Rev.* **103** 1793-1874(2003).

[3] P. K. Chattaraj, S. Nath and B. Maiti, *"Reactivity Descriptors" in Computational Medicinal Chemistry and Drug Discovery.* eds., J. Tollenaere, P. Bultinck, H. D. Winter and W. Langenaeker (Marcel Dekker: New York) Chapter 11, pp. 295- 322, 2003.

[4] R. G. Parr, L. V. Szentpaly and S. Liu, Electrophilicity index, *J. Am. Chem. Soc.* **121** 1922-1924(1999).

[5] P. K. Chattaraj, B. Maiti and Sarkar, U. Philicity: A unified treatment of chemical reactivity and selectivity, *J. Phys. Chem. A* **107** 4973-4975(2003).

[6] E. Chamorro, P. K. Chattaraj and P. Fuentealba, Variation of the Electrophilicity Index along the Reaction Path, *J Phys Chem A* **107** 7068- 7072(2003). R. Parthasarathi, M. Elango, V. Subramanian and P. K. Chattaraj, Variation of electrophilicity during molecular vibrations and internal rotations, *Theor. Cem. Acc.*, 1-9(2005)

[7] R. Parthasarathi, J. Padmanabhan, V. Subramanian, B. Maiti and P. K. Chattaraj, Chemical Reactivity Profiles of Two Selected Polychlorinated Biphenyls, *J. Phys. Chem. A* **107** 10346-10352(2003). R. Parthasarathi, J. Padmanabhan, V. Subramanian, B. Maiti and P. K. Chattaraj, Toxicity Analysis of 33'44'5 - Pentachloro Biphenyl Through Chemical Reactivity and Selectivity Profiles, *Current Sci.* **86** 535-542(2004). R. Parthasarathi, J. Padmanabhan, V. Subramanian, U. Sarkar, B. Maiti and P. K. Chattaraj, Toxicity Analysis of Benzidine Through Chemical Reactivity and Selectivity Profiles: A DFT Approach, *Internet Electron J. Mol. Des.* **2** 798-813(2003). R. Parthasarathi, M. Elango, J. Padmanabhan, V. Subramanian, D. R. Roy, U. Sarkar and P. K. Chattaraj, Application of Quantum Chemical Descriptors in Computational Medicinal Chemistry for Chemoinformatics, *Ind. J. Chem. A* (In Press)

[8] R. Parthasarathi, V. Subramanian and P. K. Chattaraj, Effect of electric field on the global and local reactivity indices, *Chem. Phys. Lett.* **382** 48- 56(2003).

[9] J. Padmanabhan, R. Parthasarathi, U. Sarkar, V. Subramanian and P. K. Chattaraj, Effect of solvation on the condensed Fukui function and the generalized philicity index, *Chem. Phys. Lett.* **383** 122- 128(2004).

[10] R. Parthasarathi, J. Padmanabhan, M. Elango, V. Subramanian and P. K. Chattaraj, Intermolecular reactivity through the generalized philicity concept, *Chem. Phys. Lett.* **394** 225-230(2004). P. K. Chattaraj, U. Sarkar, R. Parthasarathi and V. Subramanian, DFT study of some aliphatic amines using generalized philicity concept, *Int. J. Quant. Chem.* **101** 690-702(2005). P. K. Chattaraj and D. R. Roy, Local Descriptors around a Transition State: A Link between Chemical Bonding and Reactivity, *J. Phys. Chem. A* **109** 3771-3772(2005).

[11] C. Hansch, D. Hoekman, A. Leo, D. Weininger and C. Selassie, Chem-Bioinformatics: Comparative QSAR at the Interface between Chemistry and Biology, *Chem. Rev.,* **102** 783-812(2002).

[12] R. Parthasarathi, V. Subramanian, D. R. Roy and P. K. Chattaraj, Electrophilicity Index as a Possible Descriptor of Biological Activity, *Bioorg. Med. Chem.* **12** 5533-5543(2004).

[13] D. R. Roy, R. Parthasarathi, B. Maiti, V. Subramanian and P. K. Chattaraj, Electrophilicity as a possible descriptor for toxicity prediction, *Bioorg. Med. Chem.* **13** 3405-3412 (2005).

[14] C. L. Waller and J. D. McKinney, Three-Dimensional Quantitative Structure-Activity Relationships of Dioxins and Dioxin-like Compounds: Model Validation and Ah Receptor Characterization, *Chem. Res. Toxicol.* **8** 847-858(1995). T. W. Schultz, TETRATOX: *Tetrahymena pyriformis* population growth impairment endpoint-A surrogate for fish lethality, *Toxicol. Methods* **7** 289-309(1997)

[15] *Gaussian 03,* & *Gaussian 98,* Revision B.03; Gaussian, Inc.: Pittsburgh, PA.

[16] *DMOL³*, Accelrys, Inc. San Diego, California.

[17] V. Barone, M. Cossi and J. Tomasi, A new definition of cavities for the computation of solvation free energies by the polarizable continuum model, *J. Chem. Phys.* **107** 3210-3221(1997).

[18] R. S. Mulliken, Electron Population Analysis on LCAO-MO Molecular Wave Functions. I. *J. Chem. Phys.* **23** 1833-1840(1955).

[19] A. E. Reed and F. Weinhold, Natural bond orbital analysis of near-Hartree-Fock water dimmer, *J. Chem. Phys.* **78** 4066-4073(1983). A. E. Reed, R. B. Weinstock and F. Weinhold, Natural population analysis, *J. Chem. Phys.* **83** 735-746(1985).

Brill Academic Publishers
P.O. Box 9000, 2300 PA Leiden,
The Netherlands

*Lecture Series on Computer
and Computational Sciences*
Volume 4, 2005, pp. 117-120

# Estimation of Spatially Varying Diffusivity in an Infiltration Problem

**Chung-Ki Cho**[1]

Department of Mathematics, Soonchunhyang University, Asan 336-745, Korea

**Sungkwon Kang**

Department of Mathematics, Chosun University, Gwangju 501-759, Korea

**YongHoon Kwon**

Department of Mathematics, POSTECH, Pohang 790-784, Korea

Received 8 July, 2005; accepted in revised form 29 July, 2005

*Abstract:* A parameter estimation scheme for the spatially varying soil water diffusivity modelled by a nonlinear convection-diffusion equation is developed, and its convergence is proved. A numerical simulation is performed.

*Keywords:* parameter estimation, convection-diffusion equation, approximation, infiltration

*Mathematics Subject Classification:* 65K10, 86A22, 93A30

## 1   Introduction

The rainfall infiltration in the soil of finite depth is modelled by the following form of modified quasilinear convection-diffusion equation[1]

$$v_t = (d(z)v_z)_z - 2a(v + u_L + b)v_z + f(z,t),$$
$$v(z,0) = 0, \tag{1}$$
$$[a(v + u_L + b)^2 - dv_z](0,t) = f_0(t), \quad v(L,t) = 0,$$

where $v = u - u_L$, $u$ is the volumetric water content, $t$ is the time, $z$ is the depth, $u_L$ is a given constant, $f_0$ and $f$ are given functions, and $d$, $a$, and $b$ are geophysical parameters. The function parameter $d$ is the soil water diffusivity. Here, all the parameters and the constants in (1) are assumed to be chosen that the infiltration rates are less than the saturated hydraulic conductivity so that the soil remains unsaturated. The existence and uniqueness of the classical solution for (1) are known[2, 3]. In [4, 5, 6], Cho et. al. considered the related parameter estimation problems with a constant diffusion coefficient $d$. However, when $d$ is space dependent, the analysis becomes more complex.

Based on geophysical considerations, assume that the diffusivity is positive and bounded, i.e., let $Q = \{d \in C^{1+\alpha}[0,L] \mid 0 < d_0 \leq d \leq d_1, \|d\|_{C^{1+\alpha}[0,L]} \leq d_2\}$ be the parameter space equipped

---

[1]Corresponding author. E-mail: ckcho@sch.ac.kr

with the usual supremum norm. Then the parameter estimation problem is to determine the parameter $d \in Q$ from a given set of observations.

Let $T_o$ be a fixed observation time, $\{z_j\}_{j=1}^m$ a fixed set of points in $[0, L]$, and $\Delta z$ a fixed positive number. Suppose we have observed the water contents $\omega_j$'s in $\frac{1}{2}\Delta z$-neighborhoods of $z_j$'s at time $T_o$. Consider the parameter-to-output mapping $\Gamma : Q \to \mathbb{R}^m$, $d \mapsto (U_1(d), \dots, U_m(d))$, where

$$U_i(d) = \int\limits_{z_i - \frac{1}{2}\Delta z}^{z_i + \frac{1}{2}\Delta z} v(z, T_o; d)\, dz + u_L \Delta z$$

and the following optimization problem:

**Problem (P)** Let $\tilde{Q}$ be an admissible parameter subset of the parameter space $Q$. Given a set of measurements $\omega = (\omega_1, \dots, \omega_m)$, find $d^* \in \tilde{Q}$ that minimizes the cost functional $J : \tilde{Q} \to \mathbb{R}$ defined by

$$J(d) := \|\Gamma(d) - \omega\|_2^2 = \sum_{j=1}^m [U_j(d) - \omega_j]^2.$$

## 2 Forward problem

To establish the continuity properties of the analytical and numerical solutions for (1) with respect to the parameter $d$, let $S = \{\Phi \in H^1(0, L) \mid \Phi(L) = 0\}$. For spatial discretizations, define $S^N = \{\Phi \in C[0, L] \mid \Phi|_{[\frac{j}{N}L, \frac{j+1}{N}L]}$ is linear for each $j = 0, 1, \dots, N-1$, and $\Phi(L) = 0\}$, where $N \in \mathbb{N}$. Let $I_N$ be the corresponding linear spline interpolating operator, $I_N g$ be the element of $S^N$ satisfying $(I_N g - g)(\frac{j}{N}L) = 0$, $j = 0, 1, \cdots, N$. Define a projection operator $P_N : S \to S^N$ by $(d\,[P_N g - g]_z, \Phi_z) = 0$ for all $\Phi \in S^N$.

We now consider the time discretization. For $M \in \mathbb{N}$, let $k = T/M$ be the time step size. For $t \in \{0, k, 2k, \cdots, T\}$, $R_k(t) = \{k, 2k, \cdots, t-k, t\}$, $\bar{R}_k(t) = \{0\} \cup R_k(t) = \{0, k, 2k, \cdots, t-k, t\}$. Let $\Phi_j$ be the standard hat function at the $j$-th spatial node $jL/N$. Then, for each $t \in \bar{R}_k(T)$, let $v^N$ is of the following form $v^N(z, t) = \sum_{j=0}^{N-1} w_j(t)\Phi_j(z)$. Using the Sobolev inequality, Green's Theorem, the Hölder inequality, Young's inequality, and the Poincare inequality, we have the following convergence and continuity results.

**Theorem 2.1** There exists a positive constant $C$ such that $\|v(d) - v(\tilde{d})\|_{L^\infty[0,T;L^2(0,L)]} \leq C\|d - \tilde{d}\|_{L^\infty[0,L]}$ for all $d, \tilde{d} \in Q$.

**Theorem 2.2** There exists a positive constant $C$ such that $\max_{t \in \bar{R}_k(T)} \|v(d) - v^N(d)\|(t) \leq CN^{-1}$ holds for all $d \in Q$ and for all sufficiently large $N$.

**Theorem 2.3** For each $N$, there exists a positive constant $C$ depending only on $N$ such that for any $d, \tilde{d} \in Q$, $\max_{t \in \bar{R}_k(T)} \|v^N(d) - v^N(\tilde{d})\|(t) \leq C(N)\|d - \tilde{d}\|_{L^\infty[0,L]}$.

## 3 Parameter estimation

Let $d_3$ be a sufficiently small positive constant, and let $\tilde{Q}$ to be the closure of $Q^0$ in the $C^{1+\alpha}$-norm, where $Q^0 = \{d \in H^3(0, L) \mid 0 < d_0 \leq d, \|d\|_{H^3} \leq d_3\}$. Then, by the Rellich-Kondrachov compactness theorem[7], $Q^0$ is a precompact subset of $C^{1+\alpha}[0, L]$, $\tilde{Q}$ is a compact subset of $C^{1+\alpha}[0, L]$, and, hence, $\tilde{Q}$ becomes a compact subset of the parameter space $Q$ defined in Section 1. Then, with $\tilde{Q}$ as an admissible parameter subset, the cost functional $J$ in Problem (P) is continuous. Therefore, by the compactness of $\tilde{Q}$, $J$ has a minimum which is a solution of Problem (P).

Now, we construct a finite dimensional approximation scheme for Problem (P). For a $K \in \mathbb{N}$, let $Y^K = \{g \in C^2[0, L] \mid g|_{[j\frac{L}{K}, (j+1)\frac{L}{K}]}$ is a cubic polynomial for each $j = 0, \ldots, K-1\}$ and let $\tilde{I}_K$ be the corresponding clamped cubic spline interpolating operator. That is, for $g \in C^1[0, L]$, $\tilde{I}_K g$ is the element of $Y^K$ satisfying $(\tilde{I}_K g)(jL/K) = g(jL/K)$, $0 \leq j \leq K$, $(\tilde{I}_K g)'(0) = g'(0)$, $(\tilde{I}_K g)'(L) = g'(L)$. Define $Q^K = \tilde{I}_K Q^0$ and consider the following finite-dimensional approximation problems.

**Problem $(P^{N,K})$** Given a set of measurements $\omega = (\omega_1, \ldots, \omega_m)$, find $d^* \in Q^K$ that minimizes

$$J^N(d) = \|\Gamma^N(d) - \omega\|_2^2 = \sum_{j=1}^m \left[U_j^N(d) - \omega_j\right]^2, \text{ where } U_j^N(d) = \left[\int_{z_j - \frac{1}{2}\Delta z}^{z_j + \frac{1}{2}\Delta z} v^N(z, T_o; d)\, dz + u_L \Delta z\right].$$

From the continuity results obtained in Section 2, we have the following theorem so called *function space parameter estimation convergence*(FSPEC)[8].

**Theorem 3.1** (i) For each $N, K \in \mathbb{N}$, there exists a solution $d^{N,K} \in Q^K$ of the problem $(P^{N,K})$. (ii) There exist increasing sequences $\{N_s\}$, $\{K_s\}$ in $\mathbb{N}$ such that the resulting subsequence $\{d^{N_s,K_s}\}$ of $\{d^{N,K}\}$ converges to an element in $\tilde{Q}$. (iii) Suppose that $\{N_s\}$ and $\{K_s\}$ are increasing sequences in $\mathbb{N}$. If the corresponding subsequence $\{d^{N_s,K_s}\}$ of $\{d^{N,K}\}$ converges to $d^* \in \tilde{Q}$, then $d^*$ is a solution to the original problem (P).

**Example.** Let $L = 25\,cm$, $u_L = 0.03$, $a = 0.5928$, $b = -0.0065$, and assume that the true parameter is given by

$$d^{tr}(z) = 0.25 + 0.15 \cos(\frac{\pi}{25} z),$$

and let f and $f_0$ be defined so that the solution $u^{tr}$ of the corresponding model problem is descibed as in Figure 1-(A).

The observation time and the observation regions were chosen as $T_o = 120$ minutes and $\{[z_j - \frac{1}{2}\Delta z, z_j + \frac{1}{2}\Delta z]\}_{0 \leq j \leq 10}$, respectively, where $z_j = 0.1jL$ and $\Delta z = 0.01L$. Measured water contents in the observation intervals were made from the true solution.

Finite-Difference-Levenberg-Marquardt(FDLM) algorithm[10] was adopted for the optimization process. We started with the constant function 0.125 as an initial guess. Table 1 and Figure 1-(B) show the function space parameter estimation convergence(FSPEC) property of our approximation scheme. It is easy to observe that the approximated diffusivity functions converge to the true one as the approximation dimensions N and K increase.

Table 1: FSPEC

| K | N | OLS-Error | $\|d^{tr} - d^{N,K}\|_\infty$ | K | N | OLS-Error | $\|d^{tr} - d^{N,K}\|_\infty$ |
|---|---|---|---|---|---|---|---|
| 1 | 32 | 7.5444e−07 | 9.8669e−02 | 2 | 32 | 6.1451e−08 | 3.7514e−02 |
|   | 64 | 5.1721e−07 | 9.0986e−02 |   | 64 | 1.6887e−08 | 1.7563e−02 |
|   | 128 | 4.2650e−07 | 8.7172e−02 |   | 128 | 5.3230e−09 | 8.3880e−03 |
|   | 256 | 3.8725e−07 | 8.5228e−02 |   | 256 | 2.0648e−09 | 4.4025e−03 |
| 4 | 32 | 1.6329e−08 | 5.9323e−02 | 8 | 32 | 6.6037e−11 | 1.8437e−01 |
|   | 64 | 3.6470e−09 | 2.8766e−02 |   | 64 | 2.6104e−11 | 6.8616e−02 |
|   | 128 | 7.8561e−10 | 1.4726e−02 |   | 128 | 6.8426e−12 | 2.8496e−02 |
|   | 256 | 1.7757e−10 | 7.9747e−03 |   | 256 | 1.7795e−12 | 1.3468e−02 |

# Acknowledgment

The work by Chung-Ki Cho was supported by Korea Research Foundation Grant (KRF-2003-015-C00073).

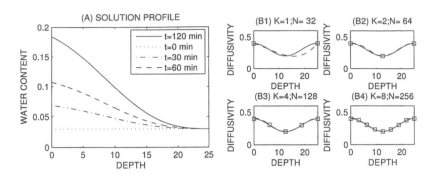

Figure 1: (A) Solution profiles at various times. (B) The solid lines represent the true diffusivity $d^{tr}$ and the dashed lines with square show the approximations.

# References

[1] R. Hills and A. Warrick, Burgers' equation: A solution for soil water flow in a finite length, *Water Resour. Res.* **29** 1179-1184(1993).

[2] R. Bürger and W.L. Wendland Existence, uniqueness, and stability of generalized solutions of an initial-boundary value problem for a degenerating quasilinear parabolic equation, *J. Math. Anal. Appl.* **218** 207-239(1998).

[3] O.A. Ladyzenskaja, V.A. Solonnikov and N.N. Ural'ceva, *Linear and Quasilinear Equations of Parabolic Type.* AMS, Providence, RI, 1968.

[4] C.-K. Cho, Y. Kwon and S. Kang, Parameter estimation for an infiltration problem, *Computers Math. Applic.* **33** 53-67(1997).

[5] C.-K. Cho, S. Kang and Y. Kwon, Estimation of soil water distribution: identifiability and observation design, *Computers Math. Applic.* **34** 105-120(1997).

[6] C.-K. Cho, S. Kang and Y. Kwon, Parameter estimation problem for nonhysteretic infiltration in soil, *J. KSIAM* **4** 11-22(2000).

[7] A. Quarteroni and A. Valli, *Numerical Approximation of Partial Differential Equations.* Springer-Verlag, Berlin Heidelberg, 1994.

[8] H.T. Banks and K. Kunish, *Estimation Techniques for Distributed Parameter Systems.* Birkhäuser, Boston, 1989.

[9] B. Clothier, J. Knight and I. White, Burgers' equation : Application to field constant-flux infiltration, *Soil Sci.* **132** 255-261(1981).

[10] W. Press, S. Teukolsky, W. Vetterling and B. Flannery, *Numerical Recipies in C : The Art of Scientific Computing.* Cambridge Univ. Press, New York, 1992.

Brill Academic Publishers
P.O. Box 9000, 2300 PA Leiden
The Netherlands

Lecture Series on Computer[1]
and Computational Sciences
Volume 4, 2005, pp. 121-125

# A Simplistic Technique to Optimize Reactive Power Management in the Electrical Installation for Ensuring Maximum Savings and Reducing Cost of Electricity Consumed

Muhammad Ahmad Choudhry[1], Wazir Zada, Aamir Hanif and Tahir Mehmood

Department of Electrical Engineering
University of Engineering and Technology
Taxila, Pakistan

Received 15 July, 2005: accepted in revised form 9 August, 2005

*Abstract:* There exist many techniques to optimize reactive power in the electrical systems and reduce cost of providing electricity to the consumer. In this paper, a simplistic algorithm for reactive power capacity calculation in electric distribution system is presented. Various loads of similar reactive power requirement are categorized and their demand factor is selected in a way that it has engineering justifications.

The proposed software works on principle of reactive power fulfillment at low voltage level, which substantially reduces cost of electricity and increases electrical system efficiency. It allows commercial and industrial consumers to save appreciably on their electric utility bill.

This work is augmented by application software, which takes a few inputs and produces commonly required distribution systems parameters. Application of this work helps user in computing reactive power requirements, system KVA (size of transformer) and cost of compensation. Another feature of this system is the facility to predict poor reactive power management in the system. The technique also calculates the payback period for the recommended compensation capacity.

*Keywords:* Reactive power, simultaneity factor, compensation capacity, power factor improvement,

*Mathematics Subject Classification:* Programming Languages

*PACS:* 681N15

## 1. Introduction

A conventional method used for reactive power compensation depends upon approximate calculation. It is based on the assumption of a lagging power factor (PF) value before compensation and a multiplying factor is selected for a target PF at intersection point from a standard table of capacitor manufacturer [1]. This factor so selected is then multiplied with load KW for suggesting total KVAr compensation. However, this method has inherent disadvantages of over/under calculation of compensation capacity and selecting transformer size. Thus resulting in higher initial capital investment. In this paper, the proposed technique is based on a comprehensive load data of the equipment to be installed thus resulting in a very realistic approach at planning stage thus bringing about a much optimized cost of electricity bill as well as equipment cost in the transmission system.

## 2. Power Loading of Distribution System

In order to design an installation, the actual maximum load demand likely to be imposed on the power supply system must be assessed [2]. To base the design simply on the arithmetic sum of all the loads existing in the installation would be extravagantly uneconomical, and bad engineering practice. The aim is to show how all-existing and projected loads can be assigned various factors to account for diversity. Demand factor $D_f$ and simultaneity factor $K_{s\ are}$ used which is the ratio of Maximum demand

[1] Muhammad Ahmad Choudhry, Member IEEE. Email: drahmad@uettaxila.edu.pk

to connected load. The values given to the factors are based on experience and on records taken from actual installations [3].

The determination of these factors is the responsibility of the designer, since it requires a detailed knowledge of the installation and the conditions in which the individual circuits are to be exploited [4] that depends upon the nature of activities of the consumer and to some extent upon the production load of the industry. For this reason, it is not possible to give precise values for general application.

## 3. Reactive Power Management

In this method table from manufacture of electrical equipment is used, that is based on the formula
$$KVAr = kW (\tan\varphi_1 - \tan\varphi_2)$$
Where $\varphi_1$ & $\varphi_2$ are the angles of initial PF1 and Target PF2 respectively.
$KVAr = kW$ x multiplying factor from table provided by the manufacturer.

This scheme has got many associated problems. The main disadvantage is that you have to deal with assumptions i.e. you don't have exact known values for computations. An in-accurate assumption can lead to non-precise and in-accurate results (because values of initial PF and expected load are unknown unless assumed suitably or ascertained from electricity bill if exists [5]). But for the corporate customers having their own grid system (with meters installed at one/two points) and the customers own 11KV distribution system comprising of multiple substations we don't have the values of PF and load. A non-accurate assumption may lead to power quality problem.

Finally this technique cannot precisely optimize the PF and load therefore compensation will also be non-precise [5]. Further the size of transformer and other allied equipment will also be disproportionate. It will negatively affect capital cost of substation equipment.

## 4. Software Design for Compensation Capacity Calculation

To start with the type of customer's end-use devices and their power consumption behavior is studied. Different loads are aggregated in categories for example induction motor, lighting, electronic, air-conditioning and induction furnaces loads etc Nominal PF values are selected , because, it is difficult to forecast individual loads PF.

In this paper Demand factor is used for individual load category and Simultaneity factor is used as an overall factor for the installation although both are the same terms by definitions. Practical approach will be used for selecting values of these factors for each installation so that it has engineering justifications and a reduced (optimized) investment cost of static condensers and their controls. The compensation will be worked out on LT side of the supply transformer.

### 4.1. General Manipulation

**Application Toolbar:** - To facilitate the user general tasks functions, which user may need to perform from time to time like insert or a Save a Record, Clear, Close a screen etc. is provided on mouse-click through Application Toolbar. This toolbar will be available to user throughout the application.

Each of the Buttons on application toolbar is explained below:
*Save:* The **Save** button is used to update changes in the database after a new record is inserted or deleted.
*Clear:* The **Clear** button is used to clear contents of the current window.
*Close:* The **Close** button is used to close the current window.
*Refresh:* The **Refresh** button is used to fill latest values from database into the lists and drop downs in the  current window.
*Print:* The **Print** button is used to print report.
*Cut:* The **Cut** button is used to cut the seleted text.
*Copy:* The **Copy** button is used to copy the seleted text.
*Paste:* The **Paste** button is used to paste the cut/copied text.

## 4.2. Input Screen

### 4.2.1 Categories Information

| Categories Information | | | |
|---|---|---|---|
| Name: | | Categories Information | |
| Purpose: | | To enter a new category of load | |
| Description: | | This screen is used to enter all detail about the load categories | |
| **Typical Course of Events** | | | |
| **User Action** | | **System Response** | |
| 1 | Open the screen | 2 | System displays list of categories. |
| 3 | User selects category and double clicks the mouse | 4 | The selected category is displayed in the appropriate field. |
| 5 | User enters new category name | | -- |
| 6 | User presses the save button | 7 | System saves the data in database after validation. |
| 8 | User presses the clear button | 9 | System clear the current screen |
| 10 | User presses the Refresh button | 11 | All values of categories are filled into the list |
| 12 | User presses the Close button | 13 | System closes the current form |

### 4.2.2 Compensation Capacity Calculation

| Module: Compensation Capacity Calculation | | | |
|---|---|---|---|
| Name: | | Compensation Capacity Calculation | |
| Purpose: | | To calculate the following:<br>• Compensation Capacity value<br>• System KVA<br>• Total cost of compensation ( Rs)<br>• Low PF penalty per Month ( Before compensation)<br>• Payback period. | |
| Description: | | This screen is used to enter all detail to calculate compensation | |
| **Typical Course of Events** | | | |
| **User Action** | | **System Response** | |
| 1 | Open the screen | 2 | System fills list of ref no and categories. |
| 3 | User selects the desired category from the list and double clicks the mouse | 4 | The selected category is displayed in the appropriate field. |
| 5 | User enters rated power | 6 | -- |
| 7 | User enters demand factor | 8 | -- |
| 9 | User enters power factor | 10 | System calculates and displays Active & Reactive power |
| 11 | User enters simultancity factor | 12 | System calculates and displays sum of rated power, power factor, active & reactive power of the system. |
| 13 | User enters Rate (Rs) per KW of fixed charges | 14 | --- |
| 15 | User enters Cost per KVAr | 16 | --- |
| 17 | User enters Target power factor | 18 | System calculates and display: Compensation Capacity, System KVA, Cost of compensation ( Rs), Low PF penalty , and Payback period. |
| 19 | User presses the save button | 20 | System saves the data after validation. |
| 21 | User presses the clear button | 22 | System clear the current screen |
| 23 | User presses the Refresh button | 24 | All values of categories are filled into the list |
| 25 | User presses the Close button | 26 | System closes the current form |

## 4.3. Database Design

| Table Name | | Categories | | | |
|---|---|---|---|---|---|
| Table Description | | To store Categories information | | | |
| Primary Key | | Code | | | |

| Columns | Description | Data type | Nulls ? | PK ? | FK ? |
|---|---|---|---|---|---|
| Code | Category | Integer | NOT NULL | Yes | No |
| Description | Category Name | VARCHAR | NOT NULL | No | No |

### 4.4. Users

| Table Name | | Users | | | |
|---|---|---|---|---|---|
| Table Description | | Users table to secure access | | | |
| **Columns** | **Description** | **Data type** | **Nulls ?** | **PK ?** | **FK ?** |
| Code | House No | VARCHAR2 | NOT NULL | No | Yes |
| Description | Room No | VARCHAR2 | NULL | No | No |
| Password | ID Number of Employee | VARCHAR2 | NOT NULL | No | Yes |

### 4.5.  Compensation Capacity Calculation

| Table Name | | PI | | | |
|---|---|---|---|---|---|
| Table Description | | To store Compensation Capacity Information | | | |
| Primary Key | | Code | | | |
| **Columns** | **Description** | **Data type** | **Nulls ?** | **PK ?** | **FK ?** |
| Code | Category | Number | NOT NULL | Yes | Yes |
| Ref No | Ref No of Case | Text | NOT NULL | Yes | No |
| Total Active Power | Total Active Power | Number | NULL | No | No |
| Total Reactive Power | Total Reactive Power | Number | NULL | No | No |
| Power Factor | Power Factor | Number | NOT NULL | No | No |
| SF | Simultaneity Factor | Number | NOT NULL | No | No |
| SF Active Power | Active Power applying Simultaneity Factor | Number | NOT NULL | No | No |
| SF Reactive Power | Reactive Power  applying Simultaneity Factor | Number | NOT NULL | No | No |
| Rate Per KW | Rate Per Kilo Watt | Number | NOT NULL | No | No |
| Cost Per KVAr | Cost Per KVAr | Number | NOT NULL | No | No |
| Target Power Factor | Target Power Factor | Number | NOT NULL | No | No |
| CCC | Compensation Capacity Calculation | Number | NOT NULL | No | No |
| System KVA | System KVA | Number | NOT NULL | No | No |
| Total Cost Comp | Total Cost Comp | Number | NOT NULL | No | No |
| Penalty | Penalty | Number | NULL | No | No |
| Payback Period | Payback Period | Number | NULL | No | No |

### 4.6.  Compensation Capacity Detail

| Table Name | PI_Detail | | | | |
|---|---|---|---|---|---|
| Table Descrip. | To store Compensation Capacity detailed information | | | | |
| Primary Key | Code | | | | |
| **Columns** | **Description** | **Data type** | **Nulls ?** | **PK ?** | **FK ?** |
| Code | Computer Generated Code | Number | NOT NULL | Yes | No |
| PI Code | PI Master Table Primary Key used here as Foreign Key | Number | NOT NULL | No | Yes |
| Ref No | Ref No of Master Table | Text | NOT NULL | No | Yes |
| Category Code | Category Code from Categories Table | Number | NOT NULL | No | Yes |
| Rated Power | Calculated value of Rated Power | Number | NULL | No | No |
| Demand Factor | Calculated value of Demand Factor | Number | NULL | No | No |
| Power Factor | Power Factor | Number | NULL | No | No |
| Calc. Power | Calculated Power | Number | NULL | No | No |
| Reactive Power | Reactive Power | Number | NULL | No | No |
| Angle | Angle | Number | NULL | No | No |
| Cost Angle | Cosine of Angle | Number | NULL | No | No |
| Tan Angle | Tangent of Angle | Number | NULL | No | No |

## 5.  Discussion of Results

To utilize the system in an actual working environment, a case of Bilal Engg Industry was taken as an example with total connected load of 387kW. The industry has nine different categories of load with known rated power, nominal PF, demand factor, simultaneity factor, Rate of fixed charges, cost per KVAr and Target PF.

**Bilal Engineering Industries**

| Category | Rated Power | Demand Factor | Power Factor | Active Power | Reactive Power |
|---|---|---|---|---|---|
| Injection-Moulding m/c, s | 50 | 0.40 | 0.45 | 20 | 39.69 |
| Computers & Accessories | 10 | 0.75 | 0.7 | 7.5 | 7.651 |
| Hoist | 12 | 0.4 | 0.75 | 16 | 14.11 |
| Lighting Load | 15 | 0.9 | 0.76 | 13.5 | 11.544 |

| Electronic loads | 30 | 0.5 | 0.77 | 15 | 12.429 |
|---|---|---|---|---|---|
| Power drill m/c, s | 20 | 0.5 | 0.8 | 10 | 7.50 |
| Induction furnace | 100 | 0.45 | 0.8 | 45 | 33.75 |
| Air conditioners | 100 | 0.9 | 0.8 | 90 | 67.50 |
| Air compressor | 50 | 0.7 | 0.8 | 35 | 26.25 |

For Simultaneity Factor = 0.66
Active power=166.32 KW and Reactive Power = 145.47984KVAr
Rate / kW of Fixed Charges = 520.00 Pak Rupees
Power Factor =0.75
Target Power Factor = 0.95
Cost/ KVAr = 1250.00 Pak Rupees
System KVA = 175.07368
Compensation Capacity = 90.813 KVA
Power factor penalty/ month = 19690.04134 Pak Rupees
Total cost of Compensation =113516 Pak Rupees
Payable Period =5.76 months

## 6. Conclusions

This technique of compensation capacity calculation enable the user to achieve optimum Reactive power management in the electrical installation for ensuring maximum savings and reducing cost of electricity consumed under given load conditions.

## Acknowledgements

The authors wish to thank the anonymous referees for their careful reading of the manuscript and their fruitful comments and suggestions.

## References

[1] Robert B. Morgan," Improving Power Factor for Greater Efficiency-Part 1", pp. 64-76, September 1994 EC&M.

[2] Donald Beeman, "Industrial power system hand-book", 1st Edition, The McGraw Hill Book Company, INC 1955.

[3] Low Voltage Capacitor and Capacitor Banks, ALPES Technologies France 2000.

[4] Mr. Edwin COEY C "Electrical Installation Guide "According to IEC International Standards Schneider Group France, pp. E1-E25, 1996.

[5] M. Abdul Aziz, "Power Factor and your Utility bill in Egypt." IEEE 2003.

Brill Academic Publishers
P.O. Box 9000, 2300 PA Leiden,
The Netherlands

*Lecture Series on Computer*
*and Computational Sciences*
Volume 4, 2005, pp. 126-130

# An Elementary Speculative Strategy in the case of Informations about the Future Behaviour of the Market

## A.P.L. Cipollini[1]

Department of Mathematics,
Faculty of Mathematics,
University of Milan-Bicocca,
U5 via Cozzi 53, 20125, Milan, Italy

Received 10 July, 2005; accepted in revised form 3 August, 2005

*Abstract:* In reality there are cases in which the Long/Short Seller has at disposal some informations about the future behaviour of the market. In detail there are situations in which, observed along a financial horizon a stock price, is known either precisely the price at maturity of the stock or, more generally, a possible interval of values among which the stock price at maturity is included. It comes natural to wonder if it is possible in this context to build an arbitrage strategy. Main theorical contribution of the article is to construct an elementary speculative strategy that brings to a sure profit without risk. In particular here is modelled a 0/1 strategy, that corresponds to the situation in which the Long/Short Seller owns some units of the stock on which he has informations on the behaviour if its price at maturity and takes into account two possible decisions: either to sell the unit of stock owned at the maturity date or to sell them at a date before. Crucial result of the article is that under suitable conditions that the values known at maturity must satisfy, there exists an optimal stock price in correspondence to a date before maturity, at which the Long/Short seller maximizes its profit. It is derived the exact explicit formula of the optimal stock price and of the arbitrage profit. Moreover, in order to test the validity of the speculative strategy, it has been computed a complete asympotic analysis of the behaviour of both the optimal price and the arbitrage profit, as the maturity date grows in time. The model used in order to describe the stock price fluctuations is the usual Black and Sholes with as white noise part a Random Walk for the discrete time setting and a Brownian motion for the continuous time context.

*Keywords:* Asset Price Forecast, Speculative Strategies, Insider Trading, Random Walks, Brownian motions, etc

*Mathematics Subject Classification:* Primary 60G51; Secondary 91B28

## 1   Introduction

One of the most intriguing topic in financial research, during the past decades, is the study, motivated by the insider trading phaenomenon, of the influence of events and informations on the market's value of a society. Examples of contributes in that direction of research are, from the financial point of view, the famous papers of Fama-Fisher-Jensen-Roll [1] and MacKinlay [2] and, from the mathematical point of view, the mathematical machinery formulated for the first time by Jeulin, Yor [3] and Wu [4].

---

[1]Corresponding author. E-mail: alessandro.cipollini@unimib.it, alessandro.cipollini@fastwebnet.it

At the same time the denvelopment of new statistical, probabilistical and econometrical tecniques, through which it is possible to analyse the past behaviour of the market, improves the ability to do statistically significative predictions about the future behaviour of prices. Forecasts based on rigorous statistical methods, even if not necessarily able to give exact informations, achieve more and more a role of importance as guidelines for the investor.

A question that consequently arises is how the informations about the future behaviour of the market, either generated by sure exogenous factors (insider trading, etc) or by statistically rigorous quantitative methods or by both, should be used.

Consider a market that consists in an investor that owns a unit share of a stock and has informations about the future behaviour of the market. In detail it is possible to distinguish into two different types of informations: given a financial horizon and a price process of a stock, we say that an investor owns a *first type* information when he knows explicitly the price $s \in [0, \infty)$ that a unit share of a stock has at maturity and a *second type* information when he knows a possible interval of prices $[s_1, s_2] \subseteq [0, \infty)$ ($s_1 < s_2$) among which the price of a unit share of stock is at maturity.

The presence of an information about the future behaviour of the market, permits to build an elementary speculative *strategy*: either the insider decides to keep the unit share of the stock and to sell it at maturity, at the known price/prices, or prefers to sell the unit share of stock before maturity at a price that must be at least higher than the maturity price known. Main aim of the article is to formalize and study this 0/1 elementary strategy and in particular to analyse the corresponding profit of the insider in order to show that, under suitable conditions that the first/second type information must satisfy, an arbitrage profit is ensured.

## 2   The construction of the speculative strategy and the main results

Given $T > 0$ maturity date assume, according to [5], [6] and [7], that $S$ stock price process is the strong solution of the following Stochastic Differential Equation:

$$\begin{cases} dS_\tau = \mu S_\tau d\tau + \sigma S_\tau dW_\tau, & \tau \in (t, T] \\ \quad\quad S_t = x, & a.s., \end{cases} \tag{1}$$

for $W : \Omega \times [0, T] \to \mathcal{R}$ Random Walk/Brownian motion defined on some probability space $(\Omega, \mathcal{F}, \mathcal{P})$ and adapted with respect to a complete filtration $\{\mathcal{F}_\tau\}_{\tau \in [0,T]}$.

Let us define formally the 0/1 strategy in the case of a private information of the first kind, since the other case follows in the same manner. Introduce the partition $\eta$ of $\Omega$ into the subsets $A_\eta$ and $B_\eta$, where

$$A_\eta = \{\omega \in \Omega : S_T = s\}, \qquad B_\eta = \{\omega \in \Omega : S_T \neq s\}.$$

Note that working in the probability space $(\Omega|_A, \mathcal{F}_A, P_A)$, with $\Omega_A = \Omega \cap A_\eta$, $\mathcal{F}_A$ corresponding $\sigma$−field and $P_A = P(.|A_\eta)$ conditional distribution, it is equivalent to assume that at maturity the trajectories of $S$ assume exactly the value $s$.

Call $D$ the set of all admissible prices and given any possible price $P \in D$ attained by the stock along the financial horizon $[0, T]$, define as a speculative strategy:

$$g(P, \omega) = s 1_{S_0 e^{(\mu - \sigma^2/2)T + \sigma v_T} < P} + P 1_{S_0 e^{(\mu - \sigma^2/2)T + \sigma v_T} \geq P}, \qquad \omega \in \Omega_A, \qquad P \in D, \tag{2}$$

for $v_T = \max_{\tau \in [0,T]} W_\tau$.

For any $P \in D$ price, the profit that the insider performs selling the unit share of stock at the first time at which the stock price process hits the price $P$, i.e. stops the game at the level $P$, is given by:

$$\pi_{T,s}(P) = E^{P_A}_{\Omega_A}(g) = E^P_\Omega(g|A_\eta) = s + (P - s)P(v_T \geq P|A_\eta), \qquad P \in D. \tag{3}$$

The investor is interested in the price $\overline{P} > s$ solution of the following optimization problem:

$$\begin{cases} \max_{P \in [0,\infty)} \pi_{T,s}(P), \\ P \in D, \end{cases} \tag{4}$$

The existence of such a point of maximum depends from the price known $s$; consequently define the function

$$\Pi_T(s) = \max_{P \in E} \pi_{T,s}(P), \qquad s \in [0, \infty). \tag{5}$$

Clearly if $s$ known price is such that the optimal value $\overline{P}$ exists and in particular $\Pi_T(s) > s$, than it will be convenient for the insider to sell the unit share of stock exactly when the stock price process will attain the value $\overline{P}$. Note that in this case $\overline{P}$ is an arbitrage price. Conversely if such an $\overline{P}$ does not exist than the optimal value will be the known price $s$ to which corresponds the value $\Pi_T(s) = s$.

In the case of a private information of the first kind in the discrete time context:

**Theorem 2.1** *Assume that $S : \Omega \times \{0, ..., n\} \to \{-n, ..., +n\}$ is the stock price process whose fluctuations are modelled throught a symmetric Random Walk $W$ defined on $(\Omega, \mathcal{F}, \mathcal{P})$ and call $s \in [0, \infty)$ the known price at maturity. The optimal price at which the insider has to sell the unit share of stock in order to maximize its profit, is*

$$\overline{P} = S_0 \exp\left\{ \left( \mu - \frac{\sigma^2}{2} \right) n + \sigma \overline{h} \right\}, \tag{6}$$

*where*

$$\overline{h} = \left[ \left| \frac{3k + 2 + \left( k^2 + 4n + 4 \right)^{1/2}}{4} \right| \right],$$

*and $k = (1/\sigma)\{\ln(s/S_0) - (\mu - \sigma^2/2)\}$. Moreover the profit of the insider that corresponds to the optimal price $\overline{P}$ is:*

$$\Pi_n(s) = se^{\sigma H_n(k)}, \tag{7}$$

*where*

$$H_n(k) = \left[ \left| \frac{2 - k + \left( k^2 + 4n + 4 \right)^{1/2}}{4} \right| \right] \frac{\left( \frac{n+k}{2} - \left[ \left| \frac{3k+2+(k^2+4n+4)^{1/2}}{4} \right| \right] \right)}{\left( \frac{n+k}{2} \right)}.$$

Note that both formulas can be computed explicitly. A priori the optimal price $\overline{P}$ exists always but it can happen that to that price corresponds a profit not higher than the one obtained selling the unit share of stock at maturity. It is useful, from this point of view, to study the behaviour of the function $\Pi_n(.)$, whenever the maturity date increases in time:

**Theorem 2.2** *Set $S : \Omega \times \{0, ..., n\}$ stock price process whose fluctuations are modelled throught a symmetric Random Walk and suppose that, given any $s \in [0, \infty)$ known price at maturity $\overline{P} = e^{\mu n + \sigma \overline{h}}$ is the optimal price with $\Pi_n(.)$ corresponding profit. Assume that the known price at maturity, evolves in time according to:*

$$\lim_{n \to \infty} \frac{1}{\sigma} \left( \frac{\ln\left(s(n)\right) - \left( \mu - \frac{\sigma^2}{2} \right) n}{n^\beta} \right) < \infty, \qquad \beta \in (0, 1/2]. \tag{8}$$

*The following statements are true:*

- *(i) If $\beta \in (0, 1/2)$, than:*

$$\Pi_n\left(s(n)\right) \sim se^{\sigma e^{-1} n^{1/2}/2}. \tag{9}$$

- *(ii) If $\beta = 1/2$, than:*

$$\Pi_n\big(s(n)\big) \sim s \exp\left\{\sigma n^{1/2}(C_x - 4x)e^{-C_x(2x-C_x)/4}/4\right\}, \tag{10}$$

*with $C_x = (x^2 + 4)^{\frac{1}{2}} + 3x$.*

- *(iii) If $\beta \in (1/2, 1)$, than:*

$$\Pi_n\big(s(n)\big) \sim s. \tag{11}$$

Observe that if the condition of the theorem is satisfied, than an optimal price $\overline{P}$ exists and it is convenient for the insider to sell the unit share of the stock at the first time at which the stock price process attains such a value. Conversely the optimal price $\overline{P}$ defined as into the theorem 2.1 exists but it is useless, since the corresponding profit is asymptotically the same than the one obtained selling at the maturity date.

Similarly in the case of a private information of the second type:

**Theorem 2.3** *Assume that $S : \Omega \times \{0, ..., n\} \rightarrow \{-n, ..., +n\}$ is the stock price process whose fluctuations are modelled throught a symmetric Random Walk $W$ defined on $(\Omega, \mathcal{F}, \mathcal{P})$ and call $[s_1, s_2] \in [0, \infty)$ the known prices at maturity. The optimal price at which the insider has to sell the unit share of stock in order to maximize its profit, is*

$$\overline{P} = S_0 \exp\left\{\left(\mu - \frac{\sigma^2}{2}\right)n + \sigma\overline{h}\right\}, \tag{12}$$

*where the value $\overline{h}$ is defined by the system*

$$\begin{cases} \left(\frac{h-j_2}{2h-j_2}\right)e^{-(2h-j_2)^2/2n} - \left(\frac{h-j_1}{2h-j_1}\right)e^{-(2h-j_1)^2/2n} = 0, \\ \left(\frac{h-j_2+1}{2h-j_2+2}\right)e^{-(2h-j_2+2)^2/2n} - \left(\frac{h-j_1+1}{2h-j_1+2}\right)e^{-(2h-j_1+2)^2/2n} \\ + \left(\frac{h-j_1}{2h-j_1}\right)e^{-(2h-j_1)^2/2n} - \left(\frac{h-j_2}{2h-j_2+2}\right)e^{-(2h-j_2)^2/2n} \geq 0. \end{cases} \tag{13}$$

Moreover:

**Theorem 2.4** *Set $S : \Omega \times \{0, ..., n\}$ stock price process, whose random fluctuations are modelled through an asymmetric Random Walk and assume that $\overline{P}$ is the optimal price to which corresponds a profit of $\Pi_n(.)$. Assume that the interval of known prices $I = [s_1, s_2]$ evolve in time according to the law:*

$$\lim_{n \to \infty} \frac{1}{\sigma}\left(\frac{\ln\big(s_i(n)\big) - (\mu - \frac{\sigma^2}{2})n}{n^\beta}\right) < \infty, \qquad \beta \in (0, 1/2), \tag{14}$$

*for all $i = 1, 2$ and $n \geq 0$. Than the following statements are true:*

- *(i) If $\beta \in (0, 1/2)$, than the optimal price $\overline{P}$ satisfies, as $n \to \infty$*

$$\overline{P} = S_0 \exp\left\{\left(\mu - \frac{\sigma^2}{2}\right)n + \sigma\overline{h}\right\} \sim \exp\left\{\left(\mu - \frac{\sigma^2}{2}\right)n - \frac{c\sigma n^\beta}{2}\right\} \tag{15}$$

*to which corresponds a profit $\Pi_n(.)$ that satisfies as $n \to \infty$*

$$\Pi_n([s_1(n), s_2(n)]) \sim S_0 e^{(\mu-\sigma^2/2)n+\sigma E(W_n|W_n \in [-n^\beta, +n^\beta])} \exp\left\{\sigma\left(\frac{cn^{1/2}}{2}\right)e^{-c^2/2}\right\}, \tag{16}$$

*for $c > 1$.*

- *(ii) If $\beta \in (1/2, 1)$, than the optimal price $\overline{P}$ is completely determined by the expression*

$$\overline{P} = S_0 \exp\left\{\left(\mu - \frac{\sigma^2}{2}\right)n + \sigma E(W_n | W_n \in [-n^\beta, +n^\beta])\right\}, \tag{17}$$

*to which corresponds a profit $\Pi_n(.)$ that satisfies as $n \to \infty$*

$$\Pi_n([s_1(n), s_2(n)]) \sim S_0 \exp\left\{\left(\mu - \frac{\sigma^2}{2}\right)n + \sigma E(W_n | W_n \in [-n^\beta, +n^\beta])\right\}. \tag{18}$$

It is possible to show that in the continuous time context and in particular using a Brownian motion in order to interpret the randomness of the stock price fluctuations, the same formulas proved are still valid.

Finally observe that the conditions (8) and (14) are reasonable, since, by the Central Limit theorem, the distance of a Random Walk running from the origin up to the time $n$ has an order at infinity of $\sqrt{n}$.

## 3 Conclusions

In the case of a first/second type information about the behaviour of the stock price at maturity, if either the condition (8) or (14) are satisfied than there exists an optimal price specified by the explicit formulas (6) and (12), at which it is possible to sell the stock owned, realizing an arbitrage profit. ¿From an applicative point of view the investor, verified the conditions (8) and (14) has to sell the stock owned in correspondence to the first time at which the stock price attains the optimal price computed. Finally has to be remarked that the analysis may be extended in a continuous time context and in particular brings to analogous results and conclusions.

## Acknowledgment

The author wishes to thank Professor Luigi Santamaria and the anonymous referees for their careful reading of the manuscript and their fruitful comments and suggestions.

## References

[1] E. F. Fama, L. Fisher, M. Jensen and R. Roll, The adjustments of Stock Prices to New Informations, *International Economic Review*, 1969.

[2] A. C. MacKinlay, Event studies in Economics and Finance, *Journal of Economic Literature*, 13-39, 1997.

[3] T. Jeulin, M. Yor, Grossissement de Filtrations, examples et applications, *Lecture Notes in Mathematics*, Springer Verlag, 1985.

[4] C. T. Wu, *Construction of Brownian Motions in Enlarged Filtrations and their role in Mathematical Model of Insider Trading*, Phd thesis, Humboldt-Universitat zu Berlin.

[5] F. Black, M. Sholes, The Pricing of Options and Corporate Liabilities, *Journal of Political Economy*, 81, 637-659, 1973.

[6] P. Wilmott, J. Dewynne, S. Howison, *Option pricing: mathematical models and computation* Oxford Financial Press, 1993.

[7] M. Baxter, Rennie, *Option pricing: mathematical models and computation* Oxford Financial Press, 1993.

Brill Academic Publishers
P.O. Box 9000, 2300 PA Leiden,
The Netherlands

*Lecture Series on Computer*
*and Computational Sciences*
Volume 4, 2005, pp. 131-134

# A Mathematica package for the solution of ordinary differential equations using Taylor series method

**F. Costabile, A. Napoli**[1]

Department of Mathematics,
Università della Calabria,
87036 Rende (Cs) ITALY

Received 9 July, 2005; accepted in revised form 29 July, 2005

*Abstract:* In the present paper we will present a new Mathematica package for the numerical solution of initial-value problems by means of Taylor-series method.

*Keywords:* Taylor series, Initial value problem, Mathematica package

*Mathematics Subject Classification:* 65L05, 41A58, 65K05

## 1 Extended abstract

Mathematica has two build-in functions, DSolve and NDSolve, to solve ordinary and partial differential equations. The first one finds symbolic solutions to both linear and nonlinear ordinary differential equations, the other one finds numerical solutions to ordinary differential equations. In the present paper we will present a new Mathematica package, TaylorMet for the numerical solution of initial-value problems by means of Taylor-series method. Numerical examples provide favorable comparisons with these Mathematica solvers.

Taylor series method is one of the oldest analytic-numeric method for approximate solution of ODE-IVPs. It was given up when Runge-Kutta methods were introduced and studied, due to the huge amount of work required by computation of higher order derivatives. In fact, instead of dealing with the one function $f$, one has to operate simultaneously with the $n$ functions $f$, $f', \ldots, f^{(n-1)}$. Mathematica has the capabilities that we need, i.e. it automates the computation of the higher analytical derivatives.

Let us consider the initial value problem

$$
\begin{cases}
\mathbf{y}'(x) = \mathbf{f}(x, \mathbf{y}(x)) & x \in [x_0, b] \\
\mathbf{y}(x_0) = \mathbf{y}_0
\end{cases}
\tag{1}
$$

where $\mathbf{f}(x, \mathbf{y}(x))$ is a vector-valued function defined for $x \in [x_0, b]$ and arbitrary vector $\mathbf{y} \in \mathbb{R}^n$, and $\mathbf{y}_0 \in \mathbb{R}^n$ is a preassigned vector.

We assume that a unique solution of this problem exists. Moreover, we assume that $\mathbf{f}(x, \mathbf{y})$ is analytic in the neighborhood of the initial values $x_0, y_0$, so that the series

$$
\mathbf{y}(x_0 + h) = T_n(x_0, \mathbf{y}; h) + R_n(\mathbf{y})
$$

converges to the exact solution of (1). $T_n(x_0, \mathbf{y}; h)$ is the Taylor polynomial of $\mathbf{y}$ about $x_0$ and $R_n(\mathbf{y})$ is the remainder term, which, under the previous hypothesis, tends to zero when $h \to 0$.

---

[1] E-mail: costabil@unical.it, a.napoli@unical.it

Taylor method of order $n$ has the property that the accuracy can be increased arbitrarily by increasing $n$. For this reason this method is suitable for computations requiring high accuracy, as the determination of physical constants or initial conditions for periodic problems.

When $\mathbf{y}'(x)$ involves not just $x$ but the unknown $\mathbf{y}$ as well, the higher derivatives may not be easy to come by, but they get very messy. Therefore the problem is reduced to the determination of the Taylor coefficients, that is to the computation of the derivatives $\mathbf{y}^{(k)}(x)$ up to a suitable order.

Taylor series method presents several peculiarities. One of them is the easy formulation as a variable-step method, which permits to automatize the control of the error.

In the global formulation the error increases quickly for $x$ greater than the initial point. A remedy can be giving approximate solution as a piecewise polynomial defined in subintervals of the whole interval. In both cases we directly have a dense output.

The idea of the method is quite simple: the interval $[x_0, b]$ is divided in subintervals $[x_i, x_{i+1}]$ with $x_{i+1} = x_i + h$ where $h$ is the stepsize of the integration procedure. Given the initial condition $\mathbf{y}(x_i) = \mathbf{y}_i$, the value of the solution at $x_{i+1}$ is approximated by the $N$–th degree Taylor polynomial of $\mathbf{y}(x)$ at $x = x_i$. If $N$ is fixed, it is theoretically possible to a priori determine at each step the value of $h$ so that the local error will be as small as requested.

Several formulations of variable-stepsize Taylor methods can be found in the literature. Here we use the approach given in [4]. Once we have obtained an approximation to the solution of the differential equation as a power series, we compute the interval into which the approximation is within the allowed tolerance $\varepsilon$.

Let us denote $Y_j = 1/j! d^j y(x_{i-1})/dx^j$. If the root criterion for the convergence of the series is applicable, we have

$$(\|Y_n\|_\infty h^n)^{1/n} < k < 1 .$$

If $N$ is the order of the method, to obtain a maximum stepsize within the tolerance $\varepsilon$, we choose $h$ such that

$$h = (k - \varepsilon)\|Y_N\|_\infty^{-1/N}.$$

From the root criterion of convergence, for the local error $\mathbf{z}_N$ we have

$$\mathbf{z}_N < k^{N+1} + k^{N+2} + \cdots = \frac{k^{N+1}}{1 - k}.$$

Now we impose that $\mathbf{z}_N$ is less than $\varepsilon$, thus the value of $k$ is the solution of the following implicit equation

$$\frac{k^{N+1}}{1 - k} = \varepsilon . \tag{2}$$

We observe that $k$ depends only on $\varepsilon$ and $N$, thus it does not change during all the integrations.

Then we report the results obtained by applying the Taylor series method to find numerical approximations of the solutions to some test problems. For some problems the Mathematica solver `DSolve` is not able to find solutions. Hence we have compared the package `TaylorMeth` with `NDSolve`. All executions have been carried out on a Pentium based computer. As we know the exact solutions, we have computed the true error.

The main drawback of Taylor series methods is that they are explicit, so they have all the limitations of this kind of schemes, among which they are not suitable, in general, for stiff problems. The examples show that the proposed package works well also in these cases.

**Example 1.**

$$\begin{cases} y_1' (x) = -2y_1 + y_2 + 2 \sin x \\ y_2' (x) = -1.999 y_1 + 0.999 y_2 + 0.999 (\sin x - \cos x) \\ y_1 (0) = 0 \\ y_2 (0) = \frac{1}{2} \end{cases} \tag{3}$$

The solutions are $y_1(x) = k_1 e^{-x} + k_2 e^{-0.001x} + \sin x$, and $y_2(x) = k_1 e^{-x} + 1.999 k_2 e^{-0.001x} + \cos x$, with $k_1 = \frac{1}{1.998}$ and $k_2 = -k_1$.

The Mathematica call is

TaylorMet[{-2y1[x]+y2[x]+2Sin[x],-1.999y1[x]+0.999y2[x]+0.999(Sin[x]-Cos[x])},
{y1[x],y2[x]},{0,0.5},{x,0,1},n,toll];

Here we report the work-precision diagrams. As usual, on the $y$-axis there are the execution times (in seconds), and on the $x$-axis the number of significant digits.

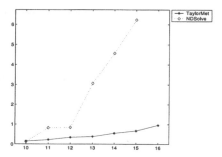

Figure 1: Computed work precision diagram (problem 3).

In this case we cannot have an error less than $10^{-15}$ by NDSolve.

The real power of Taylor series is for very stringent tolerances. Thus, if we need to track a solution whose error comes close to zero, then we just have to fix a little step or a high order.

The Mathematica function TaylorMeth takes as input: f, the function or the list of functions $f(x, y)$; yy, the list of dependent variables of f; y0, the list of the initial values of the variables specified in yy; {t,t0,t1}, where t is the independent time variable, t0 the initial value of t, t1 the final value of t; m, the order of the Taylor Method; toll, the requested tolerance.

```
TaylorMeth[f_, yy_, y0_, {t_, t0_, t1_}, m_, toll_] := Module[{k, dd, a},
  k = Part[FindRoot[x(m + 1)/(1 - x) == toll, {x, toll}], 1, 2];
  dd = D[yy, t];
  a = NestList[(D[#, t]/.Thread[dd->f])&, f, m];
  NestWhileList[TaylorStep[a, yy, #, t, toll, m, k]&, N[{t0, y0}], #[[1]] < t1&]];
```

```
TaylorStep[a_, yy_, {t0_, y0_}, t_, toll_, n_, k_] := Module[{h},
  h = (k-toll) Norm[(a[n]/.Thread[yy -> y0]/.t->t0), \[Infinity]]^(-1/n);
  {t0 + h, y0 + Sum[h^j a[j]/.Thread[yy->y0]/.t->t0, {j,1,n}]} ]
```

# References

[1] G.F.Corliss, Y.F.Chang, Solving Ordinary Differential Equations Using Taylor Series, ACM Trans. Math. Software, 8(2) (1982), 114-144.

[2] E.Hairer, S.P.Norsett, G.Wanner, *Solving Ordinary Differential Equations I*, Springer-Verlag, Berlin (1992).

[3] J.Hoefkens, *Rigorous numerical analysis with high order Taylor methods*, PhD thesis, Michigan State University (2001).

[4] M. Lara, A. Elipe, M. Palacios, Automatic programming of recurrent power series, Math. Comp. Simul. 49 (1999), 351-362.

Brill Academic Publishers
P.O. Box 9000, 2300 PA Leiden
The Netherlands

*Lecture Series on Computer*
*and Computational Sciences*
Volume 4, 2005, pp. 135-138

# A New Quantitative Structure-Property Relationship (QPRS) Model based on Topological Distances of Non-Isomorphic Subgraphs

Manuel Urbano Cuadrado, Irene Luque Ruiz, Gonzalo Cerruela García, Miguel Ángel Gómez-Nieto [1]

Department of Computing and Numerical Analysis. University of Córdoba
Campus de Rabanales. Albert Einstein Building
E-14071 Córdoba, Spain

Received 20 July, 2005; accepted in revised form 12 August, 2005

*Abstract:* A Quantitative Structure-Property Relationship (QSPR) model has been developed using a new method proposed in this paper, which is aimed at overcoming disadvantages related to the use of similarity calculations in quantitative approaches. The method uses the concept of topological descriptor (*td*) but applied to non-isomorphic subgraphs. A symmetrical matrix comprising Euclidean distances according to differences between the non-isomorphic subgraphs is built. This symmetrical matrix is used as input of partial least squares regression (PLSR) processes for predicting sublimation enthalpies of PolyChlorinated Biphenyls. Statistical results ($R^2$ in full cross validation, SECV —Standard Error in Cross Validation—, slope and bias) of our model were obtained and compared with the use of topological descriptors in univariate calibration and with those results from the literature.

*Keywords:* Graph theory, Topological descriptors, Similarity and distance, QSPR, PLSR.

*Mathematics Subject Classification:* 05C12, 58K65, 62H20, 92E10, 94C15.

## 1. Introduction

Similarity measurements have been widely used in computational chemistry. Thus, similarity calculation algorithms support many approaches for both screening of chemical databases and searching and prediction of physical-chemical properties and biological activities [1]. In recent years, methods that correlate the structural similarity between molecules with their properties have been proposed based on the following chemical principle: *"structurally similar molecules show similar properties and biological activities"* [2,3].

Computational resources involved in the structural similarity calculation are great owing to the previous detection of isomorphism. With the aim of overcoming this disadvantage, different methods for graph isomorphism calculation have been developed [4]. For this purpose, the use of binary fingerprints constitutes the basis of many similarity approaches. However, this similarity calculation based on the transformation of chemical structures to fingerprints has shown problems in Quantitative Structure-Property Relationship (QSPR) models. Non-consideration of kind, size and number of substructures produce low correlations between properties and fingerprint similarity values.

In addition to the computational cost commented upon above, similarity measurements yield inconsistencies when different molecules show equal isomorphism, as can be seen in Figure 1.

When we calculate structural similarity between the *A*, *B* and *C* graphs, the isomorphism consisting of 11 vertexes and 12 edges is equal for the three graphs. Similarity between any pair of the three graphs is therefore equal ($S_{A,B}=S_{A,C}=S_{B,C}$). Nevertheless, properties of the molecules represented by the *A*, *B* and *C* graphs are different. This fact explains the low correlation achieved using structural similarity.

We propose a new method for obtaining similarity / distance measurements between molecular graphs based on the consideration of non-isomorphic subgraphs. In this paper, the method is applied to

---

[1] Corresponding author. Email: mangel@uco.es. Phone: +34-957-212082. Fax: +34-957-218630

the construction of a QSPR model for sublimation enthalpy prediction of polychlorinated biphenyls (PCBs).

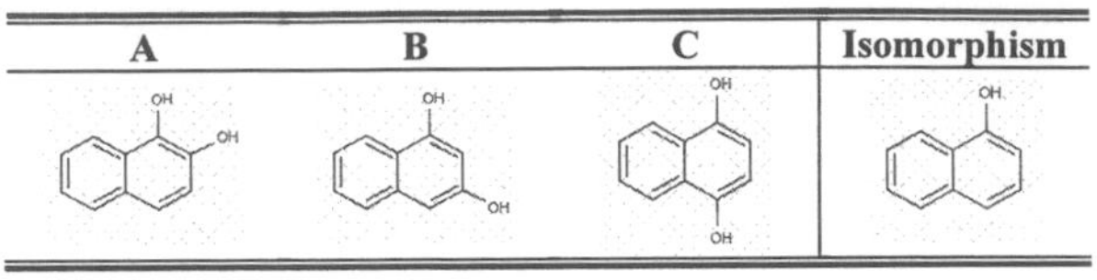

| A | B | C | Isomorphism |
|---|---|---|---|
| | | | |

Figure 1: Graphs showing equal isomorphism.

## 2. Topological Distances between Non-Isomorphic Subgraphs

Similarity between two graphs $G_A$ and $G_B$ that represent the molecules $A$ and $B$ is expressed as follows:

$$S_{A,B} = f(I_{A,B}) \tag{1}$$

where $S_{A,B}$ is a value within the range [0,1] that shows the similarity between the graphs $G_A$ and $G_B$; $I_{A,B}$ is the isomorphism in the graphs $G_A$ and $G_B$; and $f$ is a function (algorithm) or criterion that matches $S$ and $I$. Thus, different similarity values can be obtained depending on the method employed for the calculating of isomorphism between molecular graphs, namely: *MCES* (Maximum Common Edges Subgraph), *MCS* (Maximum Common Subgraph) or *AMCS* (All Maximum Common Subgraphs) [4]. When methods based on the transformation of graphs into fingerprints are used, different similarity values are also obtained.

As above stated, similarity measurements can lead to deviations in the correlation between molecular topology and properties (*QSPR*). Our proposal takes into account the characteristics between subgraphs that do not form the isomorphism $I_{A,B}$. We intend to search for relationships between variations of properties of molecules and their differences according to structural topology.

Thus, we express the structural difference between two molecular graphs $G_A$ and $G_B$ as follows:

$$\Gamma_{A,B} = g(td[A - f(I_{A,B})], \, td[B - f(I_{A,B})]) \tag{2}$$

where $f(I_{A,B})$ has equal meaning to that shown in expression (1); $A-f(I_{A,B})$ and $B-f(I_{A,B})$ represent the subgraphs of $A$ and $B$, respectively, that do not form the isomorphism $I_{A,B}$; $g()$ is a function aimed at obtaining a distance value between $td[A-f(I_{A,B})]$ and $td[B-f(I_{A,B})]$; and $\Gamma_{A,B}$ is a metric technique that calculates the structural difference between the molecules $A$ and $B$.

The method is therefore open with respect to several factors, namely: kind of isomorphism (functions $f$), different descriptors or topological variables accounting for the non-isomorphic subgraphs in $A$ and $B$, and the technique employed for measuring of distances between $td()$ values. Thus, different models can be developed aimed at correlating $\Gamma_{A,B}$ values with physical-chemical properties and/or biological activities.

In this paper we have developed a QSPR model for predicting sublimation enthalpies of PCBs using the method proposed. In this approach, the following considerations have been taken into account:

- Isomorphism calculation (function $f$) is based on the *MCS* (Maximum Common Subgraph).
- Topological descriptors ($td$) used in this model are the Wiener, HyperWiener and Valence Overall Wiener (*VOW*) indexes [3,5]. The later descriptor is supported by the Wiener calculation from the distances matrix ($D$) of a molecular graph. Elements $D(i,j)=1$ are replaced by elements $D(i,j)=x$, where $x$ is the relative bond distance between the vertices (atoms) $i$ and $j$ with respect to the reference value corresponding to the C-C bond distance.
- Euclidean distance is the function $g$ used as a measure of $td[A-f(I_{A,B})]$ y $td[B-f(I_{A,B})]$.

## 3. QSPR model for the prediction of sublimation enthalpy of PCBs

PCBs have attracted the attention of the scientific community owing to the environmental problems related to organohalogen compounds. Non-flammability and chemical stability of these compounds, in addition to their lipophilicity, are responsible for their widespread problems. The compounds studied consisted of 210 molecules —from biphenyl to decachlorobiphenyl, considering structural isomers for intermediate substituted biphenyls—.

For this set of compounds, a structural distance matrix is built. Each element $(i, j)$ of this symmetrical matrix (called dissimilarity matrix) stores the $\Gamma_{i,j}$ for each pair of the compounds set studied.

The sublimation enthalpy $\Delta_{sub}H_m$ *(298.15 K)* is a molecular property that provides information about the intermolecular forces that lead to the packing observed in the solid state. Experimental values of this parameter for biphenyl and sixteen PCBs were obtained from bibliography [6]. As Table 1 shows these values were used in learning and testing of the model.

The development of a QSPR model for predictions of sublimation enthalpy and the study of its efficiency is an exploratory analysis to pre-evaluate potential approaches. Partial least squares regression (PLSR) was the statistical technique employed for the construction of the models. In this technique, a reduction of the $\Gamma$ space into latent variables, which are linear combinations of the original space, is carried out. Both high correlation between predictors and consideration of variances of predictors and properties justify the selection of PLSR for multivariate calibration.

Table 1: Biphenyl and PCBs used in this study. Experimental data have been obtained from the bibliography. Predicted results (full cross validation) and residuals obtained with the proposed model are also shown.

| Compound | Experimental | Predicted | Residual | Compound | Experimental | Predicted | Residual |
|---|---|---|---|---|---|---|---|
| (structure) | 82.1 | 84.5 | -2.4 | | | | |
| (structure) | 86.3 | 88.2 | -1.9 | (structure) | 101.0 | 100.6 | 0.4 |
| (structure) | 82.4 | 93.0 | -10.6 | (structure) | 101.0 | 101.8 | -0.8 |
| (structure) | 96.9 | 91.2 | 5.7 | (structure) | 109.8 | 110.9 | -1.1 |
| (structure) | 105.1 | 95.0 | 10.1 | (structure) | 122.7 | 122.1 | 0.6 |
| (structure) | 99.5 | 97.6 | 1.9 | (structure) | 101.5 | 104.4 | -2.9 |
| (structure) | 103.6 | 99.9 | 3.7 | (structure) | 122.7 | 127.0 | -4.3 |
| (structure) | 95.6 | 100.3 | -4.7 | (structure) | 114.2 | 111.6 | 2.6 |
| (structure) | 108.3 | 107.4 | 0.9 | (structure) | 119.1 | 118.3 | 0.8 |

Full cross validation was the methodology used in the development of the model due to the low number of objects with the reference value known (17 compounds). $N$ cycles, where $N$ is the number of objects, are realised and final regression equation is the average of the individual fittings. $N-1$ and $1$ objects compose the learning and testing sets, respectively, with a composition assuring that all the objects have been used in testing (also called leave One Out, LOO). Statistical parameters used in the model evaluation were $R^2$ (full cross validation), SECV (Standard Error in Cross Validation), slope and bias.

Univariate analysis using topological descriptors (Wiener, HyperWiener and VOW indexes) as predictors and multivariate analysis using similarity matrix of the set of studied compounds have also been carried out in order to study the usefulness of considering both non-isomorphic subgraphs and Euclidean distance as the function $g$. Descriptor values used in the univariate analysis were those calculated using the entire topology of the graph and similarity matrix was obtained considering the cosine index and the *MCES* graph isomorphism approach. The statistical parameters relevant to this study were obtained after a full cross validation process.

The VOW descriptor yielded the best univariate approach and, this was then used for multivariate analysis. The reason for using a dissimilarity space lead to better results than those achieved when *TD(VOW)* values and multivariate analysis using the similarity matrix were considered.

As Table 2 shows, PLSR applied to $\Gamma$ matrixes increases $R^2$ and slope values, and decreases SECV and bias numbers. Thus, accuracy and precision are improved when the method proposed is employed. In addition, according to chemometric criteria, the $R^2$ values obtained permits the use of the model for quantitative prediction tasks, thus overcoming the screening role of the applications with $R^2 < 0.70$. Predicted values and residuals using the model proposed are shown in Table 1.

The method was also compared with the computational approach from which experimental values of sublimation enthalpy were obtained [6]. This approach is based on the *Comparative Molecular Field*

*Analysis* (CoMFA) technique. Table 2 shows the values for the parameters used in our study and available from the above commented approach. These statistic values support the higher efficiency of our method.

Table 2: Statistical results for the proposed method regarding with other studied approaches

| Method | $R^2$ (Full Cross Validation) | Standard Error in Cross Validation (kJ/mol) | Slope | Bias (kJ/mol) |
|---|---|---|---|---|
| Univariate (*TD*) | 0.56 | 8.62 | 0.81 | 19.42 |
| Multivariate (*Similarity*) | 0.72 | 6.42 | 0.92 | 7.90 |
| CoMFA model | 0.75 | — | — | 2.27 |
| **Multivariate ($\Gamma$)** | **0.87** | **4.51** | **0.98** | **1.79** |

## 4. Remarks

In this paper we propose a new QSPR model based on the consideration of graph isomorphism and the measuring of distances between the non-isomorphic subgraphs existing between a set of molecular graphs. Different kinds of isomorphism, topological invariants and distance approach can be considered in order to obtain a distance (dissimilarity) matrix for property prediction. Thus, different models can be developed aimed at correlating $\Gamma_{A,B}$ values with physical-chemical properties and/or biological activities.

Besides this, distance measurements might be used to calculate "*fine similarity*" values between molecules. These corrected similarity values account not only for the structural similarity between two molecular graphs (subgraphs isomorphism) but also for the approximate similarity between the remaining non-isomorphic subgraphs. We are using distances and approximate similarity values for the development of new QSPR models and screening methods.

## Acknowledgments

We wish to thank The *Comisión Interministerial de Ciencia y Tecnología* (CICyT) for financial support (Project TIN2004-04114-C02-01).

## References

[1] G.,M. Downs; J.M. Barnard. Clustering and Their Uses in Computational Chemistry. In *Reviews in Computational Chemistry*. Lipkowitz, K.B., Boyd, D.B. (Eds.) Wiley-VCH. New York. 2003, 18, 1-39.

[2] O. Ivanciuc; A.T. Balaban. The Graph Description of Chemical Structures. In *Topological Indices and Related Descriptors in QSAR and QSPR*. Devillers, J., Balaban, A. T. (Eds.) Gordon and Breach Science Publishers. The Netherlands. 1999, 59-167.

[3] D.H. Rouvray.; A.T. Balaban. Chemical Applications of Graph Theory. In *Applications of Graph Theory*. R.J. Wilson and L.W. Beineke (Eds.) Academic Press. 1979, 177-221.

[4] G. Cerruela García;, I. Luque Ruiz; M.A. Gómez-Nieto. Step-by-Step Calculation of All Maximum Common Substructures through a Constraint Satisfaction Based Algorithm. *J. Chem. Inf. Comput. Sci.* **44** 30-41(2004).

[5] X. Li; J. Lin. The Valence Overall Wiener Index for Unsaturated Hydrocarbons. *J. Chem. Inform. Comput. Sci.* **42** 1358-1362 (2002).

[6] S. Swati Puri; J.S. Chickos; W.J. Welsh. Three-Dimensional Quantitative Structure-Property Relationship (3D-QSPR) Models for Prediction of Thermodynamic Properties of Polychlorinated Biphenyls (PCBs): Enthalpy of Sublimation. *J. Chem. Inf. Comput. Sci.* **42** 109-116 (2002).

Brill Academic Publishers
P.O. Box 9000, 2300 PA Leiden
The Netherlands

*Lecture Series on Computer*
*and Computational Sciences*
Volume 4, 2005, pp. 139-142

# A MP2 Study of Low Energy Structures of the Fluorobenzene Dimer

J. Czernek[1]

Department of Bioanalogous and Special Polymers,
Institute of Macromolecular Chemistry,
Academy of Sciences of the Czech Republic,
Heyrovsky Square 2,
CZ-162 06 Prague, The Czech Republic

Received 10 July, 2005; accepted in revised form 4 August, 2005

*Abstract:* The region of the potential energy surface of the fluorobenzene dimer exhibiting planar and near-planar arrangements of the monomers was explored using the MP2/aug-cc-pVDZ approach. Three minima were located: the first one being the $C_{2h}$ symmetry structure with CH/F interactions, the second was a nonplanar, nonsymmetrical structure with two different CH/F contacts, and the third one featured CH/CH interactions and the $C_s$ symmetry. Their RI-MP2 interaction energies extrapolated to the complete basis set limit amounted to $-10.43$, $-10.61$, and $-3.39$ kJ/mol, respectively.

*Keywords:* ab initio, electron correlation, MP2, interaction energy, fluorobenzene

*PACS:* 31.15.Ar, 31.15.Md, 31.25.Qm

## 1. Introduction

Noncovalent interactions affect a variety of chemical, physical, and biochemical processes and are a focus of intense experimental and theoretical scrutiny [1]. The CH/F [2] and CH/CH [3] interactions (that is, the attraction between C–H bond and fluorine atom and between two C–H bonds, respectively) are important manifestations of noncovalent forces. In some biomolecular systems, both these types of weak interactions are present and may thus be relevant to highly important phenomena including protein folding and inhibition of enzymes [4]. Moreover, unnatural nucleotides with the ability to form the CH/F and CH/CH contacts were prepared and incorporated into DNA oligonuclotides [5], [6]. Significantly, the pairs of fluorine-substituted phenyl nucleobase analogues were shown not only to stabilize the duplex DNA, but to be enzymatically replicated with reasonable efficiency and selectivity as well [6]. Fluorobenzene-containing skeletons are thus promising unnatural base candidates for an expansion of the genetic alphabet [7]. Consequently, there is a need for an understanding the origin and the magnitude of various intermolecular interactions in clusters containing fluorine-substituted phenyl groups, with the fluorobenzene dimer as the primary model. Ab initio molecular orbital calculations can be usefully employed to study both the qualitative and quantitative aspects of these interactions [8]. Interestingly, to the best of our knowledge, the fluorobenzene dimer was not studied using high-level quantum chemical methods. This is quite surprising, because, first, it is an important model, relevant to a number of phenomena including CH/F and CH/CH interactions (see below), and, second, the dimers of fluorobenzene and aromatic compounds, and of fluorobenzene and rare-gas atoms, were the subject of numerous investigations (see [9] and [10], respectively, for recent examples). To fill this gap, a highly accurate second-order Møller–Plesset perturbational theory (MP2) treatment of the low-energy region of the fluorobenzene dimer is presented here. Thus, the planar and nearly-planar minima are located applying the MP2 method and the aug-cc-pVDZ (augmented correlation-consistent polarized-valence double zeta) basis set of atomic orbitals and their MP2 energies are extrapolated to the complete basis set limit employing the family of aug-cc-pVXZ basis sets (see Methods for details). Based on a thorough analysis of the energetic and geometrical data, issues related to the mechanisms of stabilization of the fluorobenzene dimer are discussed.

---

[1] Corresponding author. Phone: +420-296809290. Fax: +420-296809410. E-mail: czernek@imc.cas.cz

## 2. Methods

It was repeatedly shown in the literature that for highly accurate studies of weak intermolecular interactions, the currently available density functional theory approaches, while computationally much cheaper, are generally inferior to the MP2 method (see [11] for a recent reference). On the other hand, the fairly large size of the system investigated currently precludes an application of the method(s) treating the electron correlation at some higher level (in particular, an exceedingly accurate CCSD(T) strategy, the coupled cluster singles and doubles method with noniterative inclusion of triple electron excitations) together with sufficiently large basis set(s). Thus, various conformers were initially generated by using interactive computer graphics (program Insight II (2000), Accelrys Inc., San Diego, California) and subjected to the full geometry optimization in the resolution of the identity (RI) MP2 approximation [12] employing the aug-cc-pVDZ basis set; the TURBOMOLE V5-6 program package [13] was used. Three stationary points on the potential energy surface of the fluorobenzene dimer were considered further (see below). Namely, they were fully optimized at the MP2(Frozen Core)/aug-cc-pVDZ level and verified to be minima by calculating the harmonic vibrational frequencies (all real for each structure); the Gaussian 03 suite of programs [14] was adopted. Their interaction energies were then calculated using the RI-MP2/aug-cc-pV$X$Z approach with $X = 2$ (DZ), 3 (TZ), and 4 (QZ), where $X$ is the cardinal number associated with each basis set, and corrected for the basis set superposition error (BSSE) by applying the counterpoise method [15]. The Hartree–Fock $\Delta E_X^{HF}$ and the electron correlation $\Delta E_X^{corr}$ portions of the BSSE-corrected interaction energy $\Delta E$ ( $\Delta E = \Delta E^{HF} + \Delta E^{corr}$ ) were extrapolated to their complete basis set (CBS) estimates by performing the fits to the form which assumes the components of the energy to approach their basis set limits by power laws [16]:

$$\Delta E_X^{HF} = \Delta E_{CBS}^{HF} + A^{HF} X^{-\alpha} \tag{1}$$

$$\Delta E_X^{corr} = \Delta E_{CBS}^{corr} + A^{corr} X^{-\beta} \tag{2}$$

Using the aug-cc-pVDZ basis set, it was found that the RI-MP2 interaction energies are almost the same (within one hundredth of kJ/mol), and their computation is more than ten times faster, as compared to the canonical MP2 method. As a consequence, quite sizeable calculations of the RI-MP2/aug-cc-pVQZ interaction energies (1580 basis functions) could also be carried out.

## 3. Results and Discussion

Figures 1 and 2 show the planar MP2/aug-cc-pVDZ minima featuring the CH/F and CH/CH interactions, respectively. The third MP2/aug-cc-pVDZ minimum found is a nonplanar structure with the CH/F interactions, exhibiting the twist of 30.08° (as expressed by the C–F/H–C dihedral angle). Interestingly, in the case of the two structures featuring CH/F interactions, all the C–F and C–H bond lengths, where F and H are atoms making F/H contact, are the same (when rounded off to five significant digits): 1.3743 and 1.0920 Å accordingly. For the planar structure, the H/F distance is 2.3250 Å and the C–F/H and C–H/F angles are 117.21° and 176.32°. In the nonplanar structure there are two different H/F contacts, one with the H/F separation of 2.3298 Å and the C–F/H and C–H/F angles of 131.78° and 168.67°, respectively, and another with the values of 2.3261 Å, 119.61°, and 146.49° accordingly.

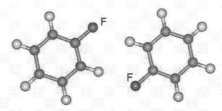

Figure 1: The Planar, $C_{2h}$ Symmetric MP2/aug-cc-pVDZ Minimum of the Fluorobenzene Dimer.

In the case of the "CH/CH structure", the C–H1, H1/H2, and C–H2 distances (see Figure 2 for numbering) are 1.0925, 2.2913, and 1.0935 Å, respectively, while the C–H1/H2 and C–H2/H1 angles are 177.81° and 116.72° accordingly. Additional structural details may be derived from the atomic coordinates, which can be obtained from the author upon request.

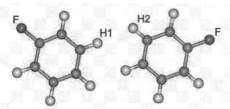

Figure 2: The Planar, $C_s$ Symmetric MP2/aug-cc-pVDZ Minimum of the Fluorobenzene Dimer.

Table 1 summarizes the BSSE-corrected RI-MP2 interaction energies together with their Hartree–Fock (HF) and electron correlation components calculated using the family of the aug-cc-pVXZ basis sets; the corresponding CBS estimates are also given. As expected [8], the HF portion of the dimerization energy converges much faster to its CBS limit than the correlation contribution. Significantly, the aug-cc-pVQZ results are fairly close (within ca. 0.2 kJ/mol) to the CBS counterparts (see Table 1).

For all the arrangements, the interactions energies are quite small, pointing to the van der Waals character of attraction in the equilibrium structures of the fluorobenzene dimer. Rather surprisingly, the CBS-extrapolated dimerization energies of the two "CH/F" complexes differ by only ca. 0.2 kJ/mol. However, their bonding characteristics vary significantly, which is also reflected in the differences of the HF contribution to the interaction energy (*cf.* Table 1). The overall stabilization of these complexes is comparable to, for example, that of the most stable configuration of the fluoroform dimer (−6.8 kJ/mol at the MP2/aug(df,pd)–6-311G** level [2]).

The "CH/CH" complex exhibits even smaller stabilization and, unlike in the case of the "CH/F" structures, the HF and correlation contributions to the interaction energy are of similar magnitude (Table 1). Its overall stabilization is similar to, for example, that found for the $D_{3d}$ structure of the methane dimer (−1.8 kJ/mol at the MP2/aug(df,pd)–6-311G** level [3]).

Table 1: Energy characteristics (in kJ/mol) for the investigated structures. See the text for details.

| structure | $\Delta E_{DZ}^{HF}$ | $\Delta E_{DZ}^{corr}$ | $\Delta E_{DZ}$ | $\Delta E_{TZ}^{HF}$ | $\Delta E_{TZ}^{corr}$ | $\Delta E_{TZ}$ |
|---|---|---|---|---|---|---|
| planar CH/F | −0.51 | −8.33 | −8.84 | −0.17 | −9.79 | −9.96 |
| nonplanar CH/F | −0.19 | −8.80 | −8.99 | +0.16 | −10.29 | −10.13 |
| CH/CH | +9.75 | −11.52 | −1.78 | +10.09 | −13.00 | −2.91 |

| structure | $\Delta E_{QZ}^{HF}$ | $\Delta E_{QZ}^{corr}$ | $\Delta E_{QZ}$ | $\Delta E_{CBS}^{HF}$ | $\Delta E_{CBS}^{corr}$ | $\Delta E_{CBS}^{corr}$ |
|---|---|---|---|---|---|---|
| planar CH/F | −0.07 | −10.22 | −10.29 | −0.03 | −10.40 | −10.43 |
| nonplanar CH/F | +0.26 | −10.73 | −10.47 | +0.30 | −10.91 | −10.61 |
| CH/CH | +10.19 | −13.43 | −3.24 | +10.23 | −13.62 | −3.39 |

## 4. Conclusions

The region of the potential energy surface of an important but neglected model system, the fluorobenzene dimer, was explored using a reliable MP2/aug-cc-pVDZ approach. Three minima were located: the first one being the $C_{2h}$ symmetry structure with (two equivalent) CH/F interactions, the second was a nonplanar, $C_1$ symmetry structure with two different CH/F contacts, and the third one featured CH/CH interactions and the $C_s$ symmetry. Based on a demanding RI-MP2/aug-cc-pVXZ (X=2, 3, and 4) treatment, their CBS interaction energies were estimated to be −10.43, −10.61, and −3.39 kJ/mol, respectively. These results reveal the intrinsic (that is, due to the assumed structure of isolated molecules only, the environment absent) preferences for formation of fluorobenzene-containing fragments. Currently an analysis of the electrostatic, repulsive, and correlation contributions to the interaction energies is being carried out.

## Acknowledgments

This research has been supported by the Academy of Sciences of the Czech Republic (Grant KJB4050311 and Project T400500402 in the program "Information Society"). Time allocation in the Czech Academic Supercomputer Centre and in the Mississippi Center for Supercomputing Research is gratefully acknowledged.

## References

[1] B. Brutschy and P. Hobza, Van der Waals molecules III: Introduction, *Chemical Reviews* 100 3861-3862(2000).

[2] S. Tsuzuki et al., Ab initio calculations of intermolecular interaction of $CHF_3$ dimer: Origin of attraction and magnitude of CH/F interaction, *Journal of Physical Chemistry A* 107 7962-7968 (2003).

[3] S. Tsuzuki et al., Magnitude of interaction between n-alkane chains and its anisotropy: High-level ab initio calculations of n-butane, n-pentane, and n-hexane dimers, *Journal of Physical Chemistry A* 108 1031-1036(2004).

[4] A. DerHovanessian et al., Models of F•H contacts relevant to the binding of fluoroaromatic inhibitors to carbonic anhydrase II, *Organic Letters* 1 1359-1362(1999).

[5] A. A. Henry et al., Efforts to expand the genetic alphabet: Identification of a replicable unnatural DNA self-pair, *Journal of the American Chemical Society* 126 6923-6923 (2004).

[6] S. Matsuda and F. E. Romesberg, Optimization of interstrand hydrophobic packing interactions within unnatural DNA base pairs, *Journal of the American Chemical Society* 126 14419-14427 (2004).

[7] S. A. Benner, Redesigning genetics, *Science* 306 625-626(2004).

[8] G. Chałasiński and M. M. Szczęśniak, State of the art and challenges of the ab initio theory of intermolecular interactions, *Chemical Reviews* 100 4227-4252 (2000).

[9] M. O. Sinnokrot and C. D. Sherill, Substituent effects in pi-pi interactions: Sandwich and T-shaped configurations, *Journal of the American Chemical Society* 126 7690-7697 (2004).

[10] J. L. C. Fajin et al., Fluorobenzene-argon ground-state potential energy surface, *Journal of Chemical Physics* 120 8582-8586(2004).

[11] S. M. Cybulski and C. E. Seversen, Critical examination of the supermolecule density functional theory calculations of intermolecular interactions, *Journal of Chemical Physics* 122 014117(2005).

[12] F. Weigend and M. Häser, RI-MP2: First derivatives and global consistency, *Theoretical Chemistry Accounts* 97 331-340(1997).

[13] R. Ahlrichs et al., Electronic structure calculations on workstation computers: The program system TURBOMOLE, *Chemical Physics Letter* 162 165-169(1989).

[14] M. J. Frisch et al., Gaussian 03, Revision C.02. Gaussian, Inc., Wallingford CT, 2004.

[15] S. F. Boys and F. Bernardi, The calculation of small molecular interactions by the difference of separate total energies: Some procedures with reduced errors, *Molecular Physics* 19 553-557(1970).

[16] D. G. Truhlar, Basis-set extrapolation, *Chemical Physics Letters* 294 45-48(1998).

Brill Academic Publishers
P.O. Box 9000, 2300 PA Leiden,
The Netherlands

*Lecture Series on Computer
and Computational Sciences*
Volume 4, 2005, pp. 143-145

# Computation of Switching Surfaces using Grobner Basis

## A. Delavarkhalafi[1]

Department of Mathematics,
Faculty of Sciences ,
University of Shahre-Kord,
P. O. Box: 115, Iran

Received 31 March, 2005; accepted in revised form 13 June, 2005

*Abstract:* Optimal control is one of the most widely used and studied methodologies in modern systems theory. In this paper we consider the classical problem of time optimal control with at most $n$ switchings and we apply the Grobner basis approach to find effective solutions.

*Keywords:* Optimal control, switching surfaces, Grobner basis, real algebraic geometry

*Mathematics Subject Classification:* 49J15

## 1 Introduction

The Grobner basis method is widely used in optimization theory such as optimal control [1], integer programming [2], polynomial programming [3] and so on. As it is well known, time optimal problems lead to switching surfaces by polynomial equations [4], [5]. We consider a general form of the classical time optimal control as follows:

$$\frac{d^n x}{dt^n} = u(t), \quad x(0) = 0, \quad -\alpha \le u(t) \le \beta, \quad \alpha, \ \beta > 0 \tag{1}$$

The traditional approach usually requires explicit determination of the switching surfaces where the sign of the control input changes. Explicit expressions for switching strategy in the simplest cases are extremely complicated [4]. A fundamental technique in computational algebraic geometry in order to solve a system of polynomial equations is Grobner basis [6], [7]. In the next section we compute switching surfaces of problem (1.1), using Grobner basis.

## 2 Computational aspects of switching surface

We consider the problem (1.1). Our objective is to drive the system from an initial condition $x(0)$ to target $x(t_m)$, in minimum time $t_m$. Using new variables $x_1, ..., x_n$, we can convert the problem (1.1) as follows:

$$\dot{x}_1 = x_2, ..., \dot{x}_{n-1} = x_n, \ \dot{x}_n = u, \quad -\alpha \le u(t) \le \beta \tag{2}$$

The Hamiltonian is $\mathcal{H} = 1 + \sum_{i=1}^{n-1} \psi_i(t)x_{i+1} + \psi_n(t)u(t)$ and the costate equations are $\dot{\psi}_i(t) = -\frac{\partial \mathcal{H}}{\partial x_i}$ i.e.

$$\dot{\psi}_1(t) = 0, \ \dot{\psi}_2(t) = -\psi_1(t), ..., \dot{\psi}_n(t) = -\psi_{n-1}(t) \tag{3}$$

---

[1]delavar@sku.ir, adelavarkh@yahoo.com

A closed form expression for the optimal $u(t)$ tests the location of the state vector with regard to a switching surfaces. Considering Hamiltonian, the optimal $u(t)$ is given by $u(t) = -sign(\psi_n(t))$. Since $\psi_n(t)$ is a polynomial with degree $n-1$, therefore it has at most $n-1$ distinct roots. Designate by $t_1, ..., t_n$, the length of the successive intervals where $u(t)$ stays constant, and hence $t_m = \sum_{i=1}^n t_i$.

**lemma 1**: If $s(t)$ is unit step function, we have

$$\int (p-a)^n s(p-a) dp = \frac{1}{n+1}(p-a)^{n+1} s(p-a) \tag{4}$$

Proof: we can prove the above lemma by the principle of mathematical induction.

The particular choice of $u(t)$ is

$$u(t) = \beta s(t) - (\alpha+\beta)s(t-t_1) + ... + (-1)^{n-1}(\alpha+\beta)s(t - \sum_{i=1}^{n-1} t_i), \quad n \geq 2 \tag{5}$$

By lemma (2.1), w can compute $x_1(t_m), ..., x_n(t_m)$ as follows:

$$x_n(t_m) = \beta \sum_{i=1}^n t_i - (\alpha+\beta) \sum_{i=2}^n t_i + ... + (-1)^{n-1}(\alpha+\beta)t_n \tag{6}$$

$$x_{n-1}(t_m) = \frac{\beta\{\sum_{i=1}^n t_i\}^2}{2!} - \frac{(\alpha+\beta)\{\sum_{i=2}^n t_i\}^2}{2!} + ... + \frac{(-1)^{n-1}(\alpha+\beta)t_n^2}{2!}$$

$$x_1(t_m) = \frac{\beta\{\sum_{i=1}^n t_i\}^n}{n!} - \frac{(\alpha+\beta)\{\sum_{i=2}^n t_i\}^n}{n!} + ... + \frac{(-1)^{n-1}(\alpha+\beta)t_n^n}{n!}$$

We read these equations as polynomials in $t_1, ..., t_n$ with parametric coefficients that depend on $x_1(t_m), ..., x_n(t_m)$. In [6], we will see how Grobner basis techniques may be used to solve polynomial equations system (2.5). We will study the systematic methods for eliminating variables from systems of polynomial equations. The basic strategy of elimination theory will be given in two main theorems:

The Elimination Theorem and the Extension Theorem. We solve the complex version of the switching problem, namely in system (2.5) we are interested in solving the following question: if $x_i(t_m) \in \mathcal{C}$, $i = 1, ..., n$, does the system have complex solutions $t_1, ..., t_n$ ?. The answer will be positive. Now that we have known the existence of complex solutions $t_1, ..., t_n$ for any parameter set $(x_1(t_m), ..., x_n(t_m))$ let us search for the existence of real nonnegative solutions for real parameters . This will solve our switching control problem. If $x_i(t_m) \in \mathbb{R}$, $i = 1, ..., n$, does the system have nonnegative real solutions $t_1, ..., t_n$?. If there is a positive solution $t_1, ..., t_n$, then the value of the optimal control $u$ assumes the values $\beta$, $-\alpha, ..., \beta(or -\alpha)$, with $u(0) = \beta$. If no positive solution exists, then the present value of the optimal control is $u(0) = -\alpha$. We will use Grobner basis together with an algorithm from real algebraic geometry called Sturm sequences [7] for desirable solution of polynomial equations system (2.5).

## 3    Conclusion

In this paper we consider a general form of the classical time optimal control. Using Grobner basis technique and Sturm sequences, we can find , a complete solution of the problem of identifying switching surfaces, in the sense that provide a symbolic computer program which solve the problem for a reasonable number of variables.

# References

[1] U. Walther and T. T Georgiou and A.Tannenbaum *On The Computation of Switching Surfaces in Optimal Control: A Grobner Basis Approach"*,IEEE Transactions on Automatic Control 46, 4, 534-540, 2001.

[2] A. Delavarkhalafi *Solving Integer Programming Using Algebraic Method*, 35 Annual Iranian Mathematics Coference 2005.

[3] Y. J. Chang and B. W. Wah *Polynomial Programming Using Grobner Bases*, Proc. Computer Software and Applications Conference 1994.

[4] L. S. Pontryagin and V. G.Boltyaynkij and R. v. Gamkrelidze and E. F. Mishchenko *The Mathematical Theory of Optimal Processes*, Naoka (Russian) 1976.

[5] A. A. Agrachev and Yu. L.Sachkov *Control Theory from the Geometric Viewpoint*, Moscow Pereslav-Zalessky-Trieste 2003.

[6] D. Cox and J. O. Little and D. O'shea *Ideals, Varieties, and Algorithms*, Springer 1997.

[7] J. Bocchnak and M. Coste and M. F. Roy *Real Algebraic Geometry*, Springer 1998.

Brill Academic Publishers
P.O. Box 9000, 2300 PA Leiden,
The Netherlands

*Lecture Series on Computer
and Computational Sciences*
Volume 4, 2005, pp. 146-149

# Determination of Quantum Expectation Values Via Fluctuation Expansion

M. Demiralp[1]

Group for Science and Methods of Computing, Informatics Institute,
İstanbul Technical University, Maslak, 34469 İstanbul, Turkey

Received 10 July, 2005, accepted in revised form 31 July, 2005

*Abstract:* The goal of this work is, the utilization of a method recently developed by the author, in quantum dynamical problems. Main idea is based on the relation between quantum and classical dynamics such that quantum dynamical problems become their classical dynamical counterparts when they are formulated via expectation values of certain operators and the probability density tends to be sharply localized. Expectation values can be expanded into a series in ascending appearence multiplicity of the complement of the projection operator which projects the space spanned by the wave function. This expansion contains fluctuation terms as unknowns besides the expectation values. We construct an infinite set of differential equations. The finite truncations of this set enables us to approximate the quantum dynamics of the system under consideration.

*Keywords:* Quantum Theory, Fluctuations, Distributions

*Mathematics Subject Classification:* 81Q99, 60E05

## 1  Introduction

In this work we consider an isolated one dimensional quantum system whose equation of motion can be written as follows

$$i\hbar\frac{\partial}{\partial t}\left|\psi(t)\right\rangle = H\left|\psi(t)\right\rangle \qquad \left[-\frac{\hbar^2}{2m}\frac{\partial^2}{\partial x^2}+V(x)\right]\left|\psi(t)\right\rangle, \qquad \left|\psi(0)\right\rangle = \left|in\right\rangle \qquad (1)$$

where H, $\hbar$, $m$, $V(x)$, $\left|in\right\rangle$ denote respectively the Hamiltonian, reduced Planck constant, mass, potential energy function, and the initial form of the waveket of the system under consideration. We have used Dirac's bra and ket notation although we have not given the equation satisfied by wavebra since it is in fact the Hermitian conjugate of the above equation. We will not attempt to solve (1) for the waveket since we do not want to use this entity directly. Instead, we are going to deal with the expectation value of the position operator, $\widehat{x}$, whose action on its operand is simply multiplication by the independent variable $x$ and the momentum operator $\widehat{p}$ which differentiates and multiplies by $-i\hbar$ its operand. We define

$$p(t) \equiv \left\langle\psi(t)\left|\widehat{p}\right|\psi(t)\right\rangle, \qquad q(t) \equiv \left\langle\psi(t)\left|\widehat{x}\right|\psi(t)\right\rangle \qquad (2)$$

The differentiation of the equations in (2) enables us to write

$$\dot{p}(t) = -\left\langle\psi(t)\left|\{H,\widehat{p}\}\right|\psi(t)\right\rangle, \qquad \dot{q}(t) = -\left\langle\psi(t)\left|\{H,\widehat{x}\}\right|\psi(t)\right\rangle \qquad (3)$$

---

[1]E-mail: demiralp@be.itu.edu.tr

where we have used the Poisson Bracket Notation defined as follows for any couple of operators, $\hat{O}_1$ and $\hat{O}_2$, whose domains and ranges are the same

$$\left\{\hat{O}_1, \hat{O}_2\right\} \equiv \frac{1}{i\hbar}\left(\hat{O}_1\hat{O}_2 - \hat{O}_2\hat{O}_1\right) \tag{4}$$

This allows us to rewrite (2) as follows

$$\dot{p}(t) = -\langle \psi(t)|V'(x)|\psi(t)\rangle, \qquad p(0) = \langle in|\hat{p}|in\rangle, \tag{5}$$

$$\dot{q}(t) = \frac{1}{m}q(t), \qquad q(0) = \langle in|\hat{x}|in\rangle \tag{6}$$

The expectation value of $V'(x)$ in (5) is an unknown at the moment. To deal with it we can use the Fluctuation Expansion Method recently developed by the author[1, 2]. It brings an infinite number of unknowns which are called fluctuation terms to the stage. The fluctuation expansion for a function $f(x)$ can be written as

$$\left\langle \hat{f}(\hat{x})\right\rangle = f(\langle\hat{x}\rangle) + \sum_{n=1}^{\infty}\sum_{\ell=0}^{n}\sum_{k=0}^{n-\ell} C_{k,\ell}^{(n)}(q)\,\varphi_\ell(q)\,\frac{1}{(2n+1-k-\ell)!}f^{(2n+1-k-\ell)}(q),$$

$$1 \le n < \infty \tag{7}$$

where the fluctuation terms are defined as

$$\varphi_n(q) \equiv \left\langle \left[\hat{x} - q\hat{I}\right]\left\{\left[\hat{I} - \hat{P}_\psi\right]\left[\hat{x} - q\hat{I}\right]\right\}^n\right\rangle, \qquad 0 \le n < \infty \tag{8}$$

Here, in fact, $\varphi_0(q)$ is not a true fluctuation term but is included for the sake of formulation simplicity. The $C_{k,\ell}^{(n)}(q)$ terms are defined by the following recursions

$$C_{n,0}^{(n)}(q) = 0, \qquad\qquad 1 \le n < \infty; \tag{9}$$

$$C_{k,0}^{(n)}(q) = \sum_{k_1=0}^{k}\varphi_{k_1+1}(c)\,C_{k-k_1,k_1}^{(n-1)}(c), \qquad 1 \le n < \infty, \quad 0 \le k \le n-1; \tag{10}$$

$$C_{k,\ell}^{(n)}(q) = C_{k,\ell-1}^{(n-1)}(c), \qquad\qquad 1 \le n < \infty, \quad 0 \le k \le n-1, \quad 1 \le \ell, k+\ell \le n \tag{11}$$

with the initial conditions

$$C_{0,0}^{(1)}(q) = \varphi_1(q), \qquad C_{1,0}^{(1)}(q) = 0, \qquad C_{0,1}^{(1)}(q) = 1 \tag{12}$$

Here we have not shown the dependence on time since the original paper[1] was not written for quantum mechanics but for univariate integration. Now all these urge us to replace (5) with the following equation

$$\dot{p}(t) = -V'(q(t)) - \sum_{n=1}^{\infty}\sum_{\ell=0}^{n}\sum_{k=0}^{n-\ell} C_{k,\ell}^{(n)}(q(t))\,\varphi_\ell(t)\,\frac{1}{(2n+1-k-\ell)!}V^{(2n+2-k-\ell)}(q(t)), \tag{13}$$

where we have embedded $q(t)$ dependence of the fluctuation term to global $t$ dependence. This equation contains an infinite number of unknown functions of time. They are the fluctuation terms defined above. As can be easily seen $\varphi_0\left(q(t)\right)$ vanishes due to the definition of $q(t)$. We need to construct a differential equation and an accompanying initial condition for each fluctuation term. This can be done by simple differentiation with respect to time and brings about the necessity of defining certain new momentum operator dependent fluctuation terms we call "Associate Fluctuation Terms". The differentiation of these terms construct all the necessary equations whose right hand sides contain nothing other then terms expressed in terms of these unknowns.

## 2 Truncated Set of Equations Based on Fluctuation Expansion

Due to the existence of an infinite number of unknowns in (13) and in the fluctuation term related differential equations not given here explicitly, it is hard to get analytical solutions to these equations. However, we can truncate these infinite set of equations by omitting all fluctuation terms whose order is greater than a prescribed integer say $N$. Here the order of a fluctuation term is defined as the product of its power by its index. We are not going to attempt to give the explicit structure of the general, $N$–th Order Truncated Set of Equations. Instead we are going to deal with only zeroth and first order truncations here. In zeroth order case, the right hand side of (13) becomes just $-V'\left(q(t)\right)$. Resulting equation together with (6) forms two differential equations with two initial conditions and can be solved at least numerically when $V(x)$ is specified. Of course the solvability of these equations depends on the structure and especially on the singularities of $V(x)$. Here we assume that $V(x)$ has a sufficiently regular structure not to lead problems of unsolvability. This level of truncation corresponds to the classical mechanical limit of the system's behavior. Therefore the quality of this level approximation increases as the localization sharpness of the system's wave function goes to Dirac's delta function.

The differential equations for the case of first order truncation can be written as follows, after various intermediate manipulations

$$\dot{p}(t) \;=\; -V'\left(q(t)\right) - \frac{1}{2}V'''\left(q(t)\right)\varphi_1(t) \qquad p(0) = \langle in \,|\widehat{p}|\, in\rangle, \qquad (14)$$

$$\dot{q}(t) \;=\; \frac{1}{m}q(t), \qquad q(0) = \langle in \,|\widehat{x}|\, in\rangle, \qquad (15)$$

$$\dot{\varphi}_1(t) \;=\; \frac{1}{m}\varphi_{1,1}(t), \qquad \varphi_1(0) = \varphi_{1,in}, \qquad (16)$$

$$\dot{\varphi}_{1,1}(t) \;=\; -2V''\left(q(t)\right)\varphi_1(t) + \frac{2}{m}\varphi_{1,2}(t), \qquad \varphi_{1,1}(0) = \varphi_{1,1,in}, \qquad (17)$$

$$\dot{\varphi}_{1,2}(t) \;=\; -V''\left(q(t)\right)\varphi_{1,1}(0), \qquad \varphi_{1,2}(t) = \varphi_{1,2,in}, \qquad (18)$$

where

$$\widehat{P}_\psi(t) \equiv |\psi(t)\rangle \langle\psi(t)| \tag{19}$$

$$\varphi_1(t) \equiv \left\langle\psi(t)\left|\left[\widehat{x}-q(t)\widehat{I}\right]\left[\widehat{I}-\widehat{P}_\psi(t)\right]\left[\widehat{x}-q(t)\widehat{I}\right]\right|\psi(t)\right\rangle \tag{20}$$

$$\varphi_{1,1}(t) \equiv \left\langle\psi(t)\left|\left[\widehat{p}-p(t)\widehat{I}\right]\left[\widehat{I}-\widehat{P}_\psi(t)\right]\left[\widehat{x}-q(t)\widehat{I}\right]\right|\psi(t)\right\rangle$$
$$+\left\langle\psi(t)\left|\left[\widehat{x}-q(t)\widehat{I}\right]\left[\widehat{I}-\widehat{P}_\psi(t)\right]\left[\widehat{p}-p(t)\widehat{I}\right]\right|\psi(t)\right\rangle \tag{21}$$

$$\varphi_{1,2}(t) \equiv \left\langle\psi(t)\left|\left[\widehat{p}-p(t)\widehat{I}\right]\left[\widehat{I}-\widehat{P}_\psi(t)\right]\left[\widehat{p}-p(t)\widehat{I}\right]\right|\psi(t)\right\rangle \tag{22}$$

and the initial values $\varphi_{1,in}$, $\varphi_{1,1,in}$, and $\varphi_{1,2,in}$ are evaluated by replacing $t$ by 0 and $\psi(t)$ by *in* in the equations from (20) to (21).

The ordinary differential equations and the accompanying boundary conditions from (14) to (18) can be solved at least numerically and the differences of the solutions from the solutions of zero fluctuation case can be inspected. Although we do not explicitly report the results here we can comfortably say that this difference diminishes as the localization of the wave function increases.

## 3 Concluding Remarks

We have presented the application of a recently developed method called Fluctuation Expansion to quantum dynamical problems. Although we have given all the information to construct the full infinite set of equations we have focused on only zeroth and first order truncations here. What we have proposed here can be extended to most generalized truncation of the infinite set of equations. Towards this goal symbolic and/or numerical codes can be developed in future applications. The method developed here is more efficient in the cases where the considered system's behavior is close to the classical limit. That is, heavy molecular systems with rather weak internal forces are more convenient for the application of this method.

## Acknowledgment

The author wishes to acknowledge financial support from the State Planning Organization (DPT) of Turkey and thanks Turkish Academy of Science for its partial support. He also thanks Professor N. A. Baykara for his careful reading of the manuscript and very useful comments.

## References

[1] M. Demiralp, A Fluctuation Expansion in Integration Under Almost Sharply Localized Weight, (submitted), 2005.

[2] M. Demiralp, A Fluctuation Expansion Method for the Evalution of a Function's Expectation Value, *To appear in the Proceedings of ICNAAM2005, Rhodes, Greece*, 2005.

Brill Academic Publishers
P.O. Box 9000, 2300 PA Leiden,
The Netherlands

*Lecture Series on Computer
and Computational Sciences*
Volume 4, 2005, pp. 150-153

# Structure analysis of second order reduced density matrix expanded in geminal basis

**G. Dezső**[a,b,1] **I. Bálint**[a,c,d], **I. Gyémánt**[a]

a) Department of Theoretical Physics,
University of Szeged,
H-6720 Szeged, Hungary
b) Department of Technology and Product Engineering,
College of Nyíregyháza,
H-4401 Nyíregyháza, Hungary
c) Department of Pharmaceutical Analysis,
University of Szeged,
H-6720 Szeged, Hungary
d) Department of Natural Sciences,
Dunaújváros Polytechnic,
Dunaújváros, H-2400 Táncsics M. u. 1.

Received 8 July, 2005; accepted in revised form 29 July, 2005

*Abstract:* The structure of second-order reduced density matrix (2RDM) of N-electron systems is studied. The wave function corresponding to the 2RDM is assumed to be linear combination of Slater-deteminants (a FCI-type wave function) constructed over a finite dimensional orthonormal one-particle basis set. Our considerations are independent from the size and the quality of the one-particle basis. Using the definition of second-order reduced density matrix, and properties of Slater-determinants a set of inequalities are derived for the 2RDM. Inequalities can be formulated without the wave function, and without higher order density matrices, involving only the density matrix, and can be considered as necessary N-representability conditions.

*Keywords:* second order reduced density matrix, 2RDM, N-representability, geminal basis

## 1 Introduction

It is well known, that the wave function is not the sole mathematical object for describe a quantum-mechanical system. Among others, density matrices, especially second order reduced density matrices (2RDM) are one of the most promising possibilities, providing fuull description of the system in the most compact form. Today 2RDM however can not take over the role of wave function in quantum-mechanics. The reason for this is the so-called N-representability problem, what proved to be a really complicated, and hard mathematical task in the last half century, and remained unsolved up till now.

The development of the concept of Contracted Schrödinger Equation opened a new method in quantum chemistry, serving results comparable with those of truncated CI methods (SDCI). At the same time the exact N-representability is not ensured by this method. Really one could almost

---

[1]Corresponding author. E-mail: dezsog@nyf.hu

say, that the error in energy caused by not taking into account completely of electron correlation by approximate wave functions, is replaced to the error in CSE theory due to not ensuring perfect N-representability. The reader can find a good summary on density matrices and many-electron densities in reference [1].

Authors developed a method, wich avoids the N-representability problem by indirect optimization of 2RDM. This method has surprisingly good convergence properties, but parameters of the wave function needed to ensure the exact N-representability [4, 5, 6, 7].

Recently perturbation theory corrections were published for correcting the error of the energy obtained by minimizing the density matrix directly applying the known conditions of N-representability [2], and variational calculations were performed for density matrices of bosonic systems including harmonic interactions showing that two-positivity may be exact for them [3].

In this paper we add remarks on N-representability problem. The second-order reduced density matrix is expanded in geminal basis, and inequalities are derived, what can be considered as necessary N-representability conditions.

## 2 Derivation of inequalities

The most general approximation of the wave function of a system consisted of N electrons, in a finite dimensional subspace of the Hilbert-space is the so-called full-CI wave function:

$$\Psi = \sum_{\alpha=1}^{\mu} c_\alpha \Phi_\alpha, \quad \mu = \binom{M}{N}, \tag{1}$$

where $\Psi$ denotes the N-electron wave function, and $\{\Phi_\alpha\}_{\alpha=1...\mu}$ are Slater-deteminants forming a basis of the subspace. We assume, that Slater-deteminants are constructed over an M-dimensional orthonormal one-particle basis $\{\varphi_i(1)\}_{i=1...M}$. The number of determinants is $\binom{M}{N}$, determinants are orthogonal, and each of them has the norm of $N!$ . The two-electron density matrix ($\Gamma^{(2)}$) is defined by the formula

$$\Gamma^{(2)}(1, 2; 1', 2') = \int \Psi(1, 2, 3, \ldots, N) \Psi^*(1', 2', 3, \ldots, N) \, d3 \ldots dN. \tag{2}$$

Although the nomination "density matrix" is widely used for this quantity, really it is the kernel of an integral operator, the so-called density operator. In the followings we call it density matrix, as it is usual. Inserting the linear combination (1) into (2), after the Laplace-expansion of the determinants the density matrix has the form:

$$\Gamma^{(2)}(1, 2; 1', 2') = \sum_{i<j, k<l}^{M} g_{i,j}(1, 2) \, \Theta_{i,j;k,l} \, g_{k,l}^*(1', 2'), \tag{3}$$

where $g_{i,j}(1, 2) = \psi_i(1)\psi_j(2) - \psi_i(2)\psi_j(1), \quad i < j$ is an anti-symmetrized two-electron function (geminal), and

$$\Theta_{i,j;k,l} = \frac{1}{N!} \sum_{\alpha,\beta=1}^{\mu} c_\alpha c_\beta^* \int {}^\alpha g_{i,j}^+(3, \ldots, N) \, {}^\beta g_{k,l}^{+*}(3, \ldots, N) \, d3 \ldots dN. \tag{4}$$

The ${}^\alpha g_{i,j}^+(3, \ldots, N)$ denotes the adjoint geminal of $g_{i,j}(1, 2)$ in Slater-determinant $\Phi_\alpha$ . The second-order reduced density matrix is given in geminal basis by formula (3).

Let us rearrange the position of coefficients $c_\alpha$ and $c_\beta$ in (4):

$$\Theta_{i,j;k,l} = \frac{1}{N!} \sum_{\alpha,\beta=1}^{\mu} \int c_\alpha \, {}^\alpha g_{i,j}^+(3,\dots,N) \; c_\beta^* \, {}^\beta g_{k,l}^{+*}(3,\dots,N) \, d3\dots dN. \tag{5}$$

While determinants $\Phi_\alpha$ are N-electron functions, we can identify them by an index-set $I_\alpha$ containing ordered indices of one-electron functions from which the determinant is built up. In the set $I_\alpha$ there are N piece of integers between 1 and M, $I_\alpha = \{\alpha_1, \alpha_2 \dots \alpha_N\}$. Similarly, the adjoint geminals can be identified with index sets $I_\lambda$ containing N-2 piece of integers sequentially. Really, the denomination "adjoint geminal" is for the relationship with the N-electron Slater determinant, but one can handle adjoint geminals as N-2-electron functions. We number adjoint geminals, and in the followings they will be noted as $\{g_\lambda^+\}_{\lambda=1\dots\binom{M}{N-2}}$. Eventually, geminal $g_{i,j}$ can be considered as a "mini-determinant" (2 by 2 determinant) identified by the index-set $I_\kappa = \{i,j\}$ so we will write $g_\kappa$ instead of $g_{i,j}$.

If we have determinant $\Phi_\alpha$, then $I_\kappa \subset I_\alpha$ and $I_\lambda \subset I_\alpha$ means, that $g_\kappa \cdot g_\lambda$ appears in the Laplace-expansion (according to geminals) of $\Phi_\alpha$ multiplied by + or - sign given by expansion rules of determinants. In the followings these signs are denoted as: $\omega(\kappa,\lambda)$. For mathematical consistency: $\omega(\kappa,\lambda) = 0$ if $I_\kappa \cap I_\lambda$ is not the empty set.

The fact, that $I_\alpha = I_\kappa \cup I_\lambda$ will be expressed by writing $\alpha(\kappa,\lambda)$.

Now we introduce the following notations:

$$D_\lambda^\kappa = c_{\alpha(\kappa,\lambda)} \, \omega(\kappa,\lambda) \tag{6}$$

$$D_{\lambda'}^{\kappa'} = c_{\alpha(\kappa',\lambda')} \, \omega(\kappa',\lambda') \tag{7}$$

Applying notations above in formula (??) we can write for $\Theta$:

$$\Theta_{\kappa,\kappa'} = \frac{1}{N!} \sum_{\lambda,\lambda'=1}^{\nu} D_\lambda^\kappa D_{\lambda'}^{\kappa'} \underbrace{\int g_\lambda^+(3,\dots,N) g_{\lambda'}^+(3,\dots,N) d3\dots dN}_{\delta_{\lambda,\lambda'}\cdot(N-2)!} \tag{8}$$

Because of the orthogonality of N-2 electron functions (N-2 by N-2 determinant functions constructed over an orthonormal one-electron basis set!) the summation indices $\lambda$ and $\lambda'$ can be set to be equal:

$$\Theta_{\kappa,\kappa'} = \frac{(N-2)!}{N!} \sum_\lambda D_\lambda^\kappa D_\lambda^{\kappa'}. \tag{9}$$

Quantities $D_\lambda^\kappa$ can be considered as the $\lambda$th element of vector $\vec{D}^\kappa$, so we can write the element of two-electron density matrix $\Theta$ as a scalar product of two vectors:

$$\Theta_{\kappa,\kappa'} = \frac{(N-2)!}{N!} \vec{D}^\kappa \cdot \vec{D}^{\kappa'}. \tag{10}$$

This result can be summarized in one sentence as: for an N-particle fermion system, using M-dimensional one-particle function basis, the elements of the second-order reduced density matrix in geminal basis are scalar products of $\binom{M}{2}$ piece of $\binom{M}{N-2}$ dimensional vectors.

The expression (10) of $\Theta_{\kappa,\kappa'}$ gives a straightforward oppotunity for continuing our considerations. We can use the most basic features of vectors, and rules of vector operations.

The norm of a vector is positive definite, so the sum of k piece of vectors $\vec{D}_\kappa$ is positive or zero:

$$\left(\sum_{i=1}^k \vec{D}^{\kappa_i}\right)^2 \geq 0. \tag{11}$$

Here $\kappa_i$ is an element of an index set containing k integer of possible values of geminal indices $\kappa$.

$$\{\kappa_1, \kappa_2 \ldots \kappa_k\} \subseteq \{1, 2 \ldots \binom{M}{2}\}$$ (12)

Evaluating the square of the sum of k piece of vectors for all possible value of k, a hierarchy of inequalities can be formulated:

$$\frac{1}{2} \sum_{i=1}^{k} \Theta_{\kappa_i \kappa_i} \geq \sum_{i,j=1, i<j}^{k} |\Theta_{\kappa_i \kappa_j}| \quad \forall k \ and \ \forall \{\kappa_1, \kappa_2 \ldots \kappa_k\}.$$ (13)

The number of inequalities (13) can be calculated as:

$$\binom{\binom{M}{2}}{2} + \binom{\binom{M}{2}}{3} + \ldots + \binom{\binom{M}{2}}{\binom{M}{2}} = \sum_{k=2}^{\binom{M}{2}} \binom{\binom{M}{2}}{k} = 2^{\binom{M}{2}}.$$ (14)

These inequalities are independent from each other, this can be easily checked numerically for small values for k.

This hierarchy of inequalities is a set of necessary N-representability conditions, in the sense, that an N-representable second-order density matrix must fulfill all of them.

## Acknowledgements

G.D. acknowledges the Ministry of Education of Hungary for supporting by the Békési György fellowship (BÖ 92/2003), and College of Nyíregyháza for financial support.

## References

[1] Jerzy Cioslowski (ed.): *Many electron densities and reduced density matrices* (Mathematical and computational chemistry 1, series editor Paul G. Mezey), Kluwer Academic/Plenum Publishers, New York, 2000

[2] T. Juhász, D. A. Mazziotti Perturbation theory corrections to the two-particle reduced density matrix variational method *Journal of Chemical Physics*: **121/3** 1201-1205 (2004)

[3] G. Gidofalvi, D. A. Mazziotti: Boson correlation energies via variational minimiyation with the two-particle reduced density matrix: Exact N-representability conditions for harmonic interactions *Physical Review A* **09** 042511 (2004)

[4] I.Bálint, G.Dezső, I.Gyémánt, Journal of Molecular Structure, THEOCHEM **501-502** 125-132 (2000)

[5] I.Bálint, G.Dezső, I.Gyémánt, J.Chem.Inf.Comp.Sci. **41** 806-810 (2001)

[6] G.Dezső, I.Bálint, I.Gyémánt, Journal of Molecular Structure, THEOCHEM **542** 21-23 (2001)

[7] I.Bálint, G.Dezső, I.Gyémánt, Int.J.Quant.Chem **84**(1) 32-38 (2001)

Brill Academic Publishers
P.O. Box 9000, 2300 PA Leiden
The Netherlands

*Lecture Series on Computer*
*and Computational Sciences*
Volume 4, 2005, pp. 154-158

# Numerical analysis of pressure field on curved self-weighted metallic roofs due to wind effect by the Finite Volume Method

J. J. del Coz Díaz[(1)], J.A. González Pérez[(2)] and P.J. García Nieto[(3)]

(1) Department of Construction Engineering
High Polytechnic School,
University of Oviedo,
Edificio Departamental N° 7 – 33204 – Gijón (Spain)

(2) Department of Construction
Faculty of Sciences,
University of Seville,
33007 – Seville (Spain)

(3) Department of Mathematics
Faculty of Sciences,
University of Oviedo,
C/ Calvo Sotelo s/n – 33007 – Oviedo (Spain)

Received 3 August, 2005; accepted in revised form 15 August, 2005

*Abstract:* In this paper, an evaluation of distribution of the air pressure is determined throughout the curved and open self-weighted metallic roof due to the wind effect by the finite volume method (FVM) [7, 9]. Data from experimental tests carried out in a wind tunnel involving a reduced scale model of a roof was used for comparison. The nonlinearity is due to Navier-Stokes equations [2-6] that govern the turbulent flow. The calculation has been carried out keeping in mind the possibility of turbulent flow in the vicinities of the walls, and speeds of wind have been analyzed understood between 30 and 40 m/s. Finally the forces and moments are determined on the cover, as well as the distribution of pressures on the same one, comparing the results obtained with the Spanish and European Standards rules, given place to the conclusions that are exposed in the study.

*Keywords:* Computational fluid dynamics, finite volume modeling, numerical and experimental methods, steady incompressible flow, curved self-weighted metallic roofs

*Mathematics Subject Classification:* Finite volume methods, Navier-Stokes equations, Numerical methods
*PACS:* 76M10, 76D05, 74S10

## 1. Introduction

In order to obtain the flow solution we have employed the *finite volume method* (FVM) [7] which has the following characteristics: (1) the equations are discretized by the *finite volume method* (FVM) [9], (2) the space discretization is carried out by means of *non-structured* meshs, (3) a *direct simulation* of Navier-Stokes equations is used, (4) the time integration scheme used is *implicit,* and (5) the communication scheme used will be the *level IV* [5]. The final results are compared with experimental measurements and the finite element solution.

The finite volume method is a robust and cheap method for discretization of conservation laws. By robust, we mean a scheme which behaves well even for particularly difficult equations, nonlinear systems of hyperbolic equations. It is in general more adequate than the finite difference method [3], which requires a simple geometry and does not allow discontinuities of the parameters in the equations. The principle of the finite difference method is, given a number of discretization points which may be defined by a mesh, to assign one discrete unknown per discretization point, and to write one equation per discretization point. At each discretization point, the derivatives of the unknown are replaced by the finite difference through the use of Taylor's expansion. Note that this method is not so clear to implement in the case of discontinuities of the coefficients of the equation (which may appear in the case of nonhomogeneous media, for instance). In the case of the finite volume method, discontinuities of the coefficients will not be any problem if the mesh is chosen such that the discontinuities of the coefficients occur over the boundaries of the control volumes. Note that the finite volume has often called 'finite difference scheme' or 'cell centered difference scheme' because of the finite difference approach which is in general used for the approximation of the flux on the boundary of the control volumes. However, the finite volume scheme differs from the finite difference scheme in that the finite difference approximation is used for the flux rather than for the operator itself.

The finite element method [1, 4, 10] is based on the multiplication of the equation by a 'shape function'. The unknown is then expanded as a linear combination of shape functions (this is the so-called 'Galerkin expansion'), and the resulting equation is integrated over the domain. The finite volume method is sometimes called a 'discontinuous finite element method' since the original equation is multiplied by the characteristic function of each grid cell (i.e. $1_p(x) = 1$ si $x \in p$, $1_p(x) = 0$ si $x \notin p$). However there is no expansion of the unknown with respect to shape functions in the finite volume method (Galerkin expansion). The finite element method (FEM) can be much more precise than the finite volume method (FVM) when using higher order polynomials, but it requires an adequate functional framework, which is not always possible to obtain.

The system of self-weighted metallic roofs constitutes an original alternative in the construction field (see figure 1). The Blocotelha shells carry out a double function based on the principle that the element of roof has to work like resistant element too: on one hand they act like beam and on the other hand like casing. In the model the air is divided into three-dimensional finite volumes, of this kind incompressible fluid with the characteristic properties of air. The dimensions of the roof are indicated in Figure 1.

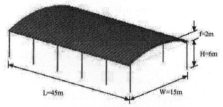

Figure 1: Dimensions of the roof.

-Length: $L = 45$ meters.

-Width: $W = 15$ meters.

-Height: $H = 6$ meters.

-Arrow: $f = 2$ meters.

-Curvature radius: $R = 15$ meters.

On the other hand, the Blocotelha company orders a report with the results given by test carried out in a wind tunnel [8]. The test was performed on a reduced scale model (1:100) of a roof with dimensions like the above described.

## 2. Mathematical model

For this problem of external flow around a body, the domain is an air volume that contains to the roof. This air box is divided in finite volumes with the characteristic properties of the air. This process is called meshing. All the volume of the box is meshed excepting but the space occupied by the roof (see Figure 2).

Figure 2: Finite element model.

The Navier-Stokes equations can be written in complete integral form for a control volume $\Omega$ contained in a surface $\Gamma$ and neglecting the volume (bulk) forces in the following way [7, 9]:

$$\frac{\partial}{\partial t}\int_{\Omega}\vec{Q}\,dV + \oint_{\Gamma}\left(\vec{F}\cdot\vec{n}\right)dS - \oint_{\Gamma}\left(\vec{F}_v\cdot\vec{n}\right)dS = 0 \tag{1}$$

where $\vec{Q}$ is the flow's variables vector,

$$\vec{Q} = \left[\rho, \rho u, \rho v, \rho w, \rho e_0\right]^T \tag{2}$$

$\vec{F}$ is the inviscid flow vector including the pressure variations and $\vec{F}_v$ is the viscous flow vector. Besides $\rho$ is the density of fluid, $\vec{v} = \left(u, v, w\right)^T$ are the cartesian components of velocity, $p$ is the pressure, $e_0$ the total energy, $\tau_{ij}$ is the viscous stress tensor which is function of velocity gradients and $\mu$ is the molecular viscosity that is function of temperature following the Sutherland law [2, 6].

## 3. Analysis of results

Since the case of transverse wind stream is the most unfavorable, by means of finite volume method will be studied only this case. It is also checked if it is necessary to take into account the pressure variations taking place in the lower face of the roof.

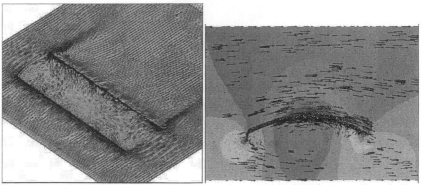

Figure 3: Numerical results on the roof by Finite Volume Method (FVM): stream lines (left) and velocity field (right).

From the $C_P$ distribution the dimensionless coefficients of sustentation $C_L$ and resistance $C_D$ can be calculated (see figure 4). The measures obtained in the central measurement sections are fitted taking place the following mathematical expression:

$$C_P = -0.5\,\sin\left[4\cdot\left(\theta+\frac{\pi}{6}\right)\right] - 0.6 \tag{3}$$

where $\theta$ is the position angle in polar coordinates, varying from $-30^o$ in the inlet edge to $+30^o$ in the outlet edge.

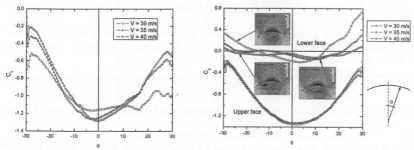

Figure 4: Sustentation coefficient $C_L$ as a function of the position angle θ in polar coordinates (upper) and pressure coefficient $C_P$ as a function of the position angle θ in polar coordinates (lower), both obtained with finite volume method (FVM) for the velocities $V = 30$, $V = 35$ and $V = 40$ m/s.

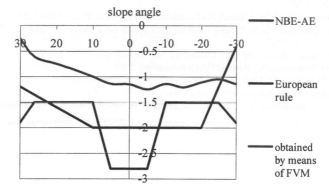

Figure 5: Coefficients $C_P$ obtained with FVM [1] and the estimated one by the Standards rules.

Comparing the pressure coefficients estimated by the Standard NBE-AE and by the European Standard with those obtained by the finite volume method (FVM) [1, 4], Figure 5 is obtained. As it is noted, the Standard NBE-AE is the only that considers the case of curved roof and open nave, and therefore it is approximated more to the reality. The European Standard only considers flat roofs when the nave is open. This Standard increases the suction in the ridge and the ends of roof to avoid the separation of stream in these zones. Due to the huge quantity of constructive possibilities, it is very difficult to elaborate an entire Standard about this subject. Anyway, the finite volume method (FVM) [7, 9] and finite element method (FEM) [1, 4, 10] are shown as useful and cheap tools to study this phenomenon.

## 4. Conclusions

A computational procedure has been developed based on the general-purpose finite volume method code, for modeling and simulating the air pressure on the self-weighted metallic roofs due to the wind effect. Finite volume method along with data from experimental tests carried out in a wind tunnel [8] involving a reduced scale model of a roof is used for comparison purposes. The findings of this study suggest that it may be possible to devise a practical procedure for establishing a self-weighted metallic roof model by using a combined experimental/computational approach.

From the experimental measurements it is devised that the suction on the upper face of the roof is the most unfavorable case in case of lateral stream. When the direction of the stream is longitudinal or oblique the zones of suction are considerably smallest. In case of two roofs the effects of the wind on the second roof are smaller than the first roof.

## Acknowledgments

The authors express deep gratitude to Construction Department and Department of Mathematics at Oviedo University and Department of Construction at Sevilla University for useful assistance. Helpful comments and discussion are gratefully acknowledged..

## References

[1] ANSYS User's Manual: *Procedures, Commands and Elements* Vols. I, II and III. Swanson Analysis Systems, 2004.

[2] I.M. Cohen, P.K. Kundu, *Fluid Mechanics*, Academic Press, New York, 2004.

[3] C.A.J. Fletcher, *Computational Techniques for Fluid Dynamics*, Springer-Verlag, Berlin, 1991.

[4] V. Girault, P.R. Raviart, *Finite Element Methods for Navier-Stokes Equations: Theory and Algorithms*, Springer-Verlag, New York, 1986.

[5] C. Hirsch, *Numerical Computation of Internal and External Flows*, John Wiley & Sons, New York, 1997.

[6] L.D. Landau, E.M. Lifshitz, *Fluid Mechanics*, Pergamon Press, New York, 1991.

[7] R.J. Leveque, *Finite Volume Methods for Hyperbolic Problems*, Cambridge University Press, New York, 2002.

[8] Technical Report of the Laboratory of Aerodynamic Buildings, "Bundesversuchs und Forschungsanstalt", Arsenal (Vienna-Austria), 2000.

[9] H. Versteeg, W. Malalasekra, *An Introduction to Computational Fluid Dynamics: The Finite Volume Method Approach*, Prentice Hall, New York, 1996.

[10] O.C. Zienkiewicz, R.L. Taylor, *The Finite Element Method: Solid and Fluid Mechanics and Non-linearity*, McGraw-Hill Book Company, London, 1991.

Brill Academic Publishers
P.O. Box 9000, 2300 PA Leiden
The Netherlands

*Lecture Series on Computer
and Computational Sciences*
Volume 4, 2005, pp. 159-162

# Non-linear buckling analysis of a self-weighted metallic roof by the finite element method

J. J. del Coz Díaz[1], P. J. García Nieto[2], J. A. Vilán Vilán[3], A. Martín Rodríguez[1] and A. Lozano Martínez-Luengas[1]

(1) Department of Construction Engineering
High Polytechnic School, University of Oviedo,
Edificio Departamental Nº 7 – 33204 – Gijón (Spain)

(2) Department of Mathematics
Faculty of Sciences, University of Oviedo,
C/ Calvo Sotelo s/n – 33007 – Oviedo (Spain)

(3) Department of Mechanical Engineering
High University School of Industrial Engineering, University of Vigo
Campus de Lagoas-Marcosende - 36200 – Vigo (Spain)

Received 11 July, 2005; accepted in revised form 19 July, 2005

*Abstract:* This paper aims to describe the development of a numerical model to accurately simulate the non linear buckling of self-weighted metallic roofs by the finite element method (FEM), which have different span lengths, ranging from 22 to 30 meters, and the same cross section. In this way, the collapse buckling load was calculated in two steps: firstly a linear buckling was carried out and secondly, an initial imperfection was added to the geometrical model and the non-linear analysis was performed. Finally the results and conclusions reached in this work are shown.

*Keywords:* Finite element modeling, non-linear buckling analysis, plasticity, large displacements.

*Mathematics Subject Classification:* Finite element methods, buckling, large strain, rate-independent theories

*PACS:* 74S05, 74G60, 74C15

## 1. Introduction

The use of finite element method [2-4, 5-6] shows innumerable advantages of economical and practical order due, on the one hand, to the cost that plays the realization of real tests, and in the other hand, to the technical difficulty of the same. Because it is based on the variational formulation of differential equation, it is much more flexible than finite difference and finite volume methods, and can thus be applied to more complicated problems. In general, engineering problems are mathematical models of physical situations. Mathematical models are differential equations with a set of corresponding boundary and initial conditions. The differential equations are derived by applying the fundamental laws and principles of nature to a system or a control volume. These governing equations represent balance of mass, force, or energy [4]. The main objective of this paper is to determine, by the finite element method [3, 5], the ultimate buckling load of different curved and open self-weighted metallic roof on which a very large snow load is applied. The portuguese company Blocotelha Coberturas Metálicas Autoportantes S.A., applicant of the present work, manufactures and installs self-weighted metallic roofs for industrial naves, agricultural buildings and sporting installations.

The system of self-weighted roofs constitutes an original alternative in the construction field. The Blocotelha / Intertelha shells carry out a double function based on the principle that the element of roof has to work like resistant element too: on one hand they act like beam and on the other hand like casing.

This system furnishes the following advantages:

- Innovating designs.
- Simplicity and economy of structure.
- Quickness in the assembling.
- Shorter number of joints among shells.
- Absence of maintenance. Belts and porchs are not necessary.
- It allows the use of isolation and translucents.
- The structure of support may be of steel, concrete, wood, etc.

The self-weighted system lets to carry curves by means of arcs which can reach 30 meters span and straight roofs that can go as high as 15 meters. In both cases, there is a total lack of belts, porchs, beams or any other intermediate resistant element (See Figure 1).

Figure 1: Self-weighted metallic roof.

However, the self-weighted roofs may be more vulnerable to the snow loads. It is well known that a vertical snow loads on the roof produces a compressive stresses in the unstiffened element and local buckling or global buckling can appear. The lightness of the shells constituting the self-weighted roofs and the fact that the shells are not supported by any resistant element make this phenomenon more problematical. This problem is specially problematic when the slope is small and the span is large, since the snow load can remain in the roof and to freeze, given place to the collapse of the structure.

## 2. Mathematical model

The resolution of this problem implies the simultaneous study of two non-linearities: (1) material non-linearity (plastic behaviour in this case), (2) geometric non-linearity or large displacements. We will describe the equations that govern the behaviour of these non-linearities next shortly.

### 2.1. Plasticity

'Plastic' behaviour of solids is characterized by a non-unique stress-strain relationship as opposed to that of non-linear elasticity. Indeed, one definition of plasticity may be the presence of irrecoverable strains on load removal. If uniaxial behaviour of a material is considered, a non-linear relationship on loading alone does not determine whether non-linear elastic or plastic behaviour is exhibited. Unloading will immediately discover the difference, with the elastic material following the same path and the plastic material showing a *history-dependent*, different, path [2, 5-6]. Many materials show an ideal plastic behaviour in which a limiting yield stress, $\sigma_y$, exists at which the strains are indeterminate. For all stresses below such yield a linear (or non-linear) elasticity relationship is assumed. A further refinement of this model is one of hardening/softening plastic material in which the yield stress depends on some parameter $\kappa$ (such as plastic strain $\vec{\varepsilon}_p$).

### 2.2. Large displacements

Whether the displacements (or strains) are large or small, equilibrium conditions between internal and external 'forces' have to be satisfied. Thus, if the displacements are prescribed in the usual manner by a finite number of nodal parameters $\vec{a}$, we can obtain the necessary equilibrium equations using the *virtual work principle* [6]:

$$\Psi(\vec{a}) = \int_V \overline{B}^T \vec{\sigma} \, dV - \vec{f} = 0 \tag{1}$$

where $\Psi$ once again represents the sum of external and internal generalized forces, and in which $\overline{B}$ is defined from the strain definition $\vec{\varepsilon}$ as:

$$d\vec{\varepsilon} = \overline{B} \, d\vec{a} \tag{2}$$

The bar suffix has now been added for, if displacements are large, the strains depend non-linearly on displacement, and the matrix $\overline{B}$ is now dependent on $\vec{a}$.

Figure 2: Finite element model.

## 3. Analysis of results

The analysis was carried out in two phases [1-2, 3]. First, a pre-buckling analysis was accomplished and then, a nonlinear analysis was performed updating the geometry of the finite element model to the deformed shape for the buckling's first mode.

*3.1 Pre-Buckling analysis*

A buckling analysis was carried out with the purpose of determining the eventual influence of geometric imperfections in the behavior of the model. Thus a compressive snow load on the roof, $q_u$, was applied. The method chosen for mode extraction was subspace iteration method [1-2, 4], with ten modes.

Once the mode shapes were reviewed, the geometry of the finite element model was updated according to the displacement results of the previous analysis for the first mode shape. A displacement multiplier of span length divided by 500 was applied, corresponding to the value of permissible variation for the tile geometry.

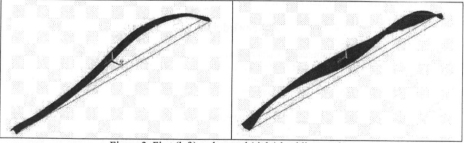

Figure 3: First (left) and second (right) buckling modes.

*3.2 Nonlinear Analysis*

The following step consisted of applying on the model a compressive solicitation, as it is indicated above and to introduce the material non-linearities as a multilinear stress-strain relationship.

The compressive load was applied ramped, starting with a 0.1% snow load and the solution controls were also adjusted to improve convergence. Thus the parameter "time" varied in the range from 0.1 to 0.0001% snow load, the geometric non-linearities were activated and the inertial effects were not included, the number of equilibrium iterations was specified and the tolerance convergence values of out-of-balance forces were delimited to 0.005.

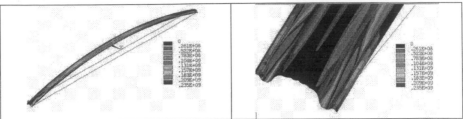

Figure 4: Stresses in the metallic roof for the collapse buckling load (left) and detail of the tile in the support (right).

## 4. Conclusions

The method of the finite elements has been shown as a suitable tool in the modelling and analysis of singular structures, such as the non-linear buckling analysis of a self-weighted metallic roof. The BLOCOTELHA curved self-weighted metallic roofs represent a robust option against to the traditional roofs, both its capacity to comply with the job of design and its assembly simplicity. Besides these self-weighted metallic roofs are light, have a reduced maintenance and offer a pleasant aesthetic result. It is important to emphasize that the structure collapse takes place due to two reasons: (a) Roof spar breaking; (b) Roof tile buckling. It is also important to denote that, in general, it is better the behaviour of the roofs with a 1.5 mm thickness against to those 1.25 mm thickness roofs, since the first ones resist larger snow load values, irrespective of kind of spars used in the structure. It is also important to observe the relevance of the roof thickness, since in the 1.25 mm thickness roofs the collapse takes place due to failure of tail in the joint zone with the spars, while in the 1.5 mm thickness roofs the collapse, from spans greater than 23 meters, is due to the breaking of the roof spars. The 1.5 mm thickness roofs suffer collapse due to the breaking of the 12 mm spars. Thus the behaviour of the roofs is studied if the spar is modified, replacing the previous one by 16 mm diameter spars, the 1.5 mm thickness roofs are not only, but also for the 1.25 mm thickness roofs. For this change of spar diameter the following conclusions were obtained:

- In 1.5 mm thickness roofs, larger snow load values supported for them are obtained, so that the spars do not cause problems and therefore, the collapse takes place due to the tail breaking, in the joint zone to the spar.
- Greater snow load values supported by the structure are obtained in the 1.25 mm thickness roofs. This is due to that the spars will absorb the stresses that arise in the joint zones between tail and spars. Thus larger snow load values can be supported on the roofs.

Therefore we can determine that in certain load situations would be suitable to change from the 12 mm diameter cable to the 16 mm diameter cable.

## Acknowledgments

The authors express deep gratitude to Construction Department and Department of Mathematics at Oviedo University and Department of Mechanical Engineering at Vigo University for useful assistance. Helpful comments and discussion are gratefully acknowledged. We thank to Swanson Analysis Inc. for the use of ANSYS University Intermediate program.

## References

[1] ANSYS User`s Manual: *Procedures, Commands and Elements* Vols. I, II and III. Swanson Analysis Systems, 2004.
[2] K. Bathe, *Finite Element Procedures*, Englewood Cliffs, Prentice-Hall, New York, 1996.
[3] T. Belytschko, *Nonlinear Finite Elements for Continua and Structures*, John Wiley & Sons, New York, 2000.
[4] D. Braess, *Finite Elements: Theory, fast solvers, and applications in solid mechanics*, Cambridge University Press, New York, 2001.
[5] J.N. Reddy, *An Introduction to Nonlinear Finite Element Analysis*, Oxford University Press, New York, 2004.
[6] O.C. Zienkiewicz, R.L. Taylor, *The Finite Element Method: Solid and Fluid Mechanics and Non-linearity*, McGraw-Hill Book Company, London, 1991.

Brill Academic Publishers
P.O. Box 9000, 2300 PA Leiden,
The Netherlands

*Lecture Series on Computer*
*and Computational Sciences*
Volume 4, 2005, pp. 163-166

# Cooperative Ant Colonies Solve the Vehicle Routing Problem

Karl F. Doerner[1], Richard F. Hartl[2] and M.Lucka[3]

[1,2]Institute for Management Science
Bruenner Strasse 72, A-1210 Vienna, Austria

[3]Institute for Software Science, University of Vienna
Nordbergstrasse 15, A-1090 Vienna, Austria

Received 10 July, 2005; accepted in revised form 5 August, 2005

*Abstract:* We discuss possibilities of parallelization of the Ant Colony Optimization (ACO) method for solving the Vehicle Routing Problems. Achieved results are compared from the point of view performance and the quality of solution. We show how the use of the pheromone information exchange within independent populations improves the solution quality. Moreover we will present the differences in the solution quality of the asynchronous variant in comparison to the synchronous variant.

*Keywords:* optimization problems, transportation, parallel programming

*Mathematics Subject Classification:* 90C08,90B06,90C27,68W10,65Y05

## 1 Introduction

The Vehicle Routing Problem (VRP) involves the construction of a set of vehicle tours starting and ending at a single depot and satisfying the demands of a set of customers, where each customer is served by exactly one vehicle and neither vehicle capacities nor maximum tour lengths are violated. Therefore no efficient exact solution methods are available, and the existing solution approaches are of heuristic nature. In this article we focus on solving the VRP using different parallel versions of Ant Colony Optimization (ACO). Based on the observation of real ant's foraging behavior ACO was developed as a graph-based, iterative, constructive meta-heuristic by Dorigo et al. [5]. The main idea of ACO is that a population of computational ants repeatedly builds and improves solutions to a given instance of a combinatorial optimization problem. From one generation to the next a joint memory is updated that guides the work of the successive populations. The memory update is based on the solutions found by the ants and more or less biased by their associated quality.

Recently some possible parallelization strategies for ACO have been proposed,which can be classified into *fine-grained* and *coarse-grained* strategies [6]. In fine-grained parallelization strategies usually several artificial ants of a colony are assigned to each processor and therefore frequent information exchange between the small sub-colonies of ants (i.e. an information exchange between the processors) takes place. Coarse-grained parallelization schemes run several colonies in parallel.

---

[1]E-mail: karl.doerner@univie.ac.at
[2]E-mail:richard.hartl@univie.ac.at
[3]E-mail:lucka@par.univie.ac.at

This strategy is also referred to as *multi colony approach*. The information exchange among colonies is done at certain intervals or numbers of iterations [1]. The important questions in implementing the multi colony approach are when, which and how information should be exchanged among the colonies.

In our work we consider both fine-grained and coarse-grained parallelization strategies, as well as a combination of these two strategies. We consider two different variants of the mixed parallelization strategy. The first variant is that independent colonies cooperate by exchanging some information on good solutions found so far. Preliminary results were reported in ([1]). The second variant is that the problem is divided into several subproblems and each colony works on a partition of the problem. After a certain amount of runtime the partial solutions are combined to one solution and the problem is partitioned in a different way. Preliminary results were reported in ([4]).

Moreover, we study the differences between synchronous and asynchronous communication within the different strategies on the fine-grained level. In the synchronous fine-grained variant the different artificial ants wait until the all the solutions within a population are constructed. Afterwards, the global best and the best solutions of the iteration is determined. The construction phase of the ants can vary substantially in runtime. Therefore, it can be more efficient to alter the original behavior of the algorithm in order to exploit the possibilities the parallel architecture provides. We modified our algorithm in the following way. We introduced a data repository where the global best solutions and some of the elite solutions are stored. Whenever an ant finishes the construction phase the solution is sent to the data repository, and immediately the ant continues in constructing the next solution. After a certain time period a pheromone update out of the best solutions in the repository is performed.

Our goal in parallelization of the Savings based ACO algorithm is on the one hand to speed up its execution and on the other hand to improve the solution quality by exploiting the possibilities provided by parallel architectures.

## 2 Savings based ACO algorithms for the VRP

The Savings based ACO algorithm published in [9] and repeated here mainly consists of the iteration of three steps: (1) generation of solutions by ants according to private and pheromone information; (2) application of a local search to the ants' solutions, and (3) update of the pheromone information.

Solutions are constructed based on the well known Savings Algorithm due to Clarke and Wright [2]. In this algorithm the initial solution consists of the assignment of each customer to a separate tour. After that for each pair of customers $i$ and $j$ the following savings values are calculated:

$$s_{ij} = d_{i0} + d_{0j} - d_{ij}, \tag{1}$$

where $d_{ij}$ denotes the distance between locations $i$ and $j$, the index 0 denotes the depot, and $s_{ij}$ represent the savings of combining two customers $i$ and $j$ on one tour contrary to serving them on two different tours. In the iterative phase, customers or partial tours are combined by sequentially choosing feasible entries from the list of saving values. A combination is infeasible if it violates either the capacity or the tour length constraints. The decision making about combining customers is based on a probabilistic rule taking into account both savings values and the pheromone information. Let $\tau_{ij}$ denote the pheromone concentration on the arc connecting customers $i$ and $j$ telling us how good the combination of these two customers $i$ and $j$ was in previous iterations. In each decision step of an ant, we consider the $k$ best combinations still available, where $k$ is a parameter of the algorithm which we will refer to as 'neighborhood' below. Let $\Omega_k$ denote the set of $k$ neighbors, i.e. the $k$ feasible combinations $(i, j)$ yielding the largest savings, considered in a given decision step, then the decision rule is given by equation (2).

$$\mathcal{P}_{ij} = \begin{cases} \dfrac{s_{ij}^{\beta}\tau_{ij}^{\alpha}}{\sum_{(h,l)\in\Omega_k} s_{hl}^{\beta}\tau_{hl}^{\alpha}} & \text{if } (i,j)\in\Omega_k \\ \\ 0 & \text{otherwise.} \end{cases} \tag{2}$$

In (2), $\mathcal{P}_{ij}$ is the probability of choosing to combine customers $i$ and $j$ on one tour, while $\alpha$ and $\beta$ bias the relative influence of the pheromone trails and the savings values, respectively. This algorithm results in a (sub-)optimal set of tours through all customers, once no more feasible savings values are available.

The pheromone management centers around two concepts borrowed from Genetic Algorithms, namely ranking and elitism to deal with the trade-off between exploration and exploitation.

## 3  Decomposing the VRP

In the standard Savings based Ant System [9] one population of Ants for solving the whole problem instance is used. This is reasonable for only small problem instances. The runtime for large problem instances is prohibitive and the solution quality decreases. Therefore, the main idea is to split up large problems into a number of smaller problems that can be solved both more effectively and more efficiently. We use an own population of ants for each sub-problem. This algorithm works better than the standard Savings based Ant System. Extensive computational results on the sequential version were published in [10]. The design of the algorithm is well suited for the development of a parallel variant.

The D-Ant algorithm is repeated in the following section. The idea is based on Taillard's decomposition algorithm.

Taillard ([7]) realized that good solutions to most VRP instances of moderate to large size feature some spatial characteristics that allow to exploit problem decomposition. For uniform problems, a partition into sectors is suggested, while for non-uniform problems Taillard uses a partitioning method based on arborescences and associated shortest paths. The core of the algorithm consists of the iteration of the following steps. First, the problem is partitioned with the appropriate method (as mentioned above). Each partition is then solved using a standard Tabu Search approach. After a certain number of iterations the sub-problems are re-joined. Using these methods, Taillard was able to find all the best known solutions for the classic benchmark instances for the VRP ([8]). We adapted this mechanism within our Savings base Ant algorithm. We denote this algorithm as D-Ant algorithm ([10]).

## 4  Computational Results

First we report some results on independent populations. The experiments with the parallelized version were run on a Beowulf cluster with 16 compute nodes, each equipped with four 700 Mhz Pentium III Xeon processors and connected via Myrinet.

By using pheromone information exchange within independent populations we can easily improve the solution quality. The solution quality without communication is 1389.8. We can increase the solution quality to 1381.56 if we exchange the global best and the elite solutions every 50 iterations between the two colonies and reset the pheromone matrix. By communicating only the global best solution we achieve almost the same improved solution quality. The runtime is increased by resetting the pheromone matrix from about 150 seconds to over 200 seconds. A good tradeoff between solution quality and runtime can be the strategy with no reset of the matrix and the communication of the global best and elite solutions (sol. qual. 1384.09, runtime 149.27 sec.).

The communication of the whole pheromone matrix leads to a decrease in solution quality as well as worse runtime behavior.

Second some results on the D-ant algorithm are reported. It is obvious that for all the different settings we get about the same solution quality as in the sequential D-Ant algorithm. When we use 8 processes we have a slightly better speed up (3.85) and efficiency (0.48) by execution the subproblems in parallel than executing the subproblems sequentially and using all the 8 processors in a parallel execution of the population of ants (speedup 3.30 - efficiency 0.41).

The differences in the solution quality of the asynchronous variant in comparison to the synchronous variant are also presented.

# References

[1] Benkner, S., Doerner, K. F., Hartl, R. F., Kiechle, G. and Lucka, M (2005): Communication Strategies for Parallel Cooperative Ant Colony Optimization on Clusters and Grids. to appear: Lecture Notes on Computer Science

[2] G. Clarke, and J. W. Wright. Scheduling of Vehicles from a Central Depot to a Number of Delivery Points. *Operations Research*, 12:568–558, 1964.

[3] K. F. Doerner, R. F. Hartl, G. Kiechle, M. Lucka, M. Reimann. Parallel Ant Systems for the Capacitated Vehicle Routing Problem. In *Evolutionary Computation in Combinatorial Optimization: 4th European Conference, EvoCOP 2004*, number LLNS 3004 in Lecture Notes in Computer Science, pages 72–83, Coimbra, Portugal, April 5-7, 2004. Springer.

[4] Doerner, K. F, Hartl, R. F., Lucka, M., 2005. A parallel Version of the D-Ant Algorithm for the Vehicle Routing Problem. R.Trobec, P.Zinterhof, M.Vajtersic, A.Uhl (Eds.): *Proceedings of the International Workshop on Parallel Numerics 2005*, ISBN 961-6303-67-8, April 20-23, Portoroz, Slovenia, pp. 109-118.

[5] M. Dorigo, and L. M. Gambardella. Ant Colony System: A Cooperative Learning Approach to the Travelling Salesman Problem. *IEEE Transactions on Evolutionary Computation*, 1(1):53–66, 1997.

[6] M. Dorigo, T. Stuetzle. The Ant Colony Optimization Metaheuristic: Algorihtms, Applications, and Advances. In F. Glover, and G. A. Kochenberger, editors, *Handbook of Metaheurisitcs*, pages 251–285, January 2003. Kluwer.

[7] Taillard, E. D. (1993): Parallel iterative search methods for vehicle routing problems. Networks **23** 661–673.

[8] Christofides, N.; Mingozzi, A. and Toth, P.: The vehicle routing problem. In: Christofides, N., Mingozzi, A., Toth, P. and Sandi, C. (Eds.): Combinatorial Optimization. Wiley, Chicester (1979).

[9] Reimann, M., Stummer, M. and Doerner, K. (2002): A Savings based Ant System for the Vehicle Routing Problem. in: Langdon, W. B. et al. (eds.): Proceedings of the Genetic and Evolutionary Computation Conference (GECCO 2002), Morgan Kaufmann, San Francisco, 1317–1325.

[10] Reimann, M., Doerner, K. F. and Hartl, R. F. D-Ants: Savings Based Ants divide and conquer the vehicle routing problem Computers & Operations Research, vol. 31 (4). pp. 563–591.

Brill Academic Publishers
P.O. Box 9000, 2300 PA Leiden
The Netherlands

*Lecture Series on Computer
and Computational Sciences*
Volume 4, 2005, pp. 167-170

# A Perspective on Computational Science and Its Status in Academia

Denis Donnelly

Department of Physics,
Siena College,
Loudonville, NY, 12211, USA
donnelly@siena.edu

Received 10 July, 2005; accepted in revised form 13 July, 2005

*Abstract:* The origins of computational science (CLS) can reasonably be associated with the seminal work of Enrico Fermi, John Pasta, and Stanislaw Ulam which took place in the mid-1950' s. The CLS field has developed enormously since then and has reached the stage where verification and validation are key to future developments. The lag time for the transfer, of this interdisciplinary research approach toward problem solving, to the educational system has been considerable, but the rate of program development in the academy is now increasing. The general outline of that development and some example programs are discussed.

*Keywords:* Computational Science, Computational Science Education, Fermi, Pasta, Ulam,

*Mathematics SubjectClassification:* 01.50.Ht,01.40.Di,01.40.Gm,97B40

## 1. Origins of Computational Science

One essential moment in the history of computational science took place in the mid-fifties at Los Alamos. Enrico Fermi, a physicist, John Pasta, a physicist, turned computer scientist, and Stanislaw Ulam, a mathematician, the three forming perhaps the first computational science group, attacked the following problem (FPU). Given a series of masses connected by springs with a non-linear response, from a given normal mode, how will the energy be dispersed over time? The results were not what they expected and it took 20 years for an explanation[1].

## II. Computational Science in Academia

The FPU calculations were performed in 1954. Fifty years later, the computational science field has developed substantially with greater maturity limited by the problems of verification and validation. However, in terms of formal academic programs, computational science is still in its early stages, which are characterized by rapid growth in number of programs and in the variety of their content.

If we look at a traditional undergraduate science programs such as the physics major, we find that the educational process does a reasonable job at teaching physics but at least until recently physics programs have done a poor job teaching anything to do with computing. Data from a survey indicate that the physics itself was well taught, laboratory and research ability were taught in a satisfactory manner, and software and computing skills were inadequately treated.

In a survey with a different emphasis, workers in science, mathematics, and engineering who had graduated five to seven years earlier were asked what skills they needed most in their employment. The results: Scientific problem solving – 85 %, Synthesizing Information – 66 %, Mathematical Skills – 58 %, Physics Principles – 52%, Lab skills – 50 %, Software skills – 49 %, and Modeling and Simulation – 47 %.

Computational science does address these needs. The skills required in the survey are the goals of computational science programs are all addressed in computational science programs: scientific problem solving (the overarching goal of computational science), synthesizing information (often implies a multidisciplinary approach), mathematical skills (more applied than abstract), software skills and modeling and simulation (an essential part of computational science).

### III. Examples of computational science programs

Computational science degree programs, unlike other disciplines which have developed over long periods of time, do not yet have a standard curriculum. Briefly outlined here are two bachelor programs, two master programs, and one PhD program. The descriptions are to provide a glimpse at the curricula and will be for one track at a given institution. They will *not* describe the entire program in any one of the cases. Justice cannot be done to a program by describing it as a list of requirements; faculty, resources, level of rigor, student body, ambience, and other factors are essential to judging the quality and nature of a program. Consult the institutions directly for more encompassing information.

At the bachelor's level, there is a zeroth-order consensus as to what should be included. The following outline provides an approximation.

General Outline of Topics to be Covered in a Bachelor's Level Program

1. Programming/tools
   A. Numerical Environment (Matlab, Mathcad)
   B. Symbolic Environment (Mathematica, Maple)
   C. Languages (Java, C++, Fortran 90, LaTeX, …)
   D. Operating Systems (Unix (Linux), Windows, OS-X)
   E. Libraries (LAPACK, BLAS, Netsolve, … )
2. Applied and Computational Mathematics
   A. Calculus – differential, integral, vector
   B. Differential Equations
   C. Linear Algebra
   D. Numerical Analysis
   E. Discrete Mathematics
   F. Probability/ Statistics
3. Application Science
   A. Need more than the equivalent of a minor in the particular science
   B. Must attain a modest level of theoretical understanding and problem solving for that discipline.
4. Visualization
   A. Tools should be able to handle large data sets
   B. 2-D and 3-D plots, as well as slicing and dicing for higher dimensional data sets
   C. Produce publication quality figures
5. Modeling and Simulation
   course should include applications associated with –
   A. Ordinary differential equations
   B. Coupled equations
   C. Linear and non-linear systems
   D. Random processes
   E. Numerical Integration / Monte Carlo methods
   F. Partial differential equations
   G. Fast Fourier transform / wavelets
6. High Performance Computing
   A. Programming on clusters or supercomputers
   B. Familiarity with MPI or equivalent
   C. Hardware and performance issues

The computational science program leading to a bachelor's degree at Siena College[2] has a curriculum comprising computer science, mathematics, and physics in roughly equal parts as shown in the table below. In addition, there is a capstone course. This is a project oriented course with an emphasis on high performance computing. This CLS program requires 73 credit hours out of the 120 required for the typical bachelor's degree in the USA.

In a highly abbreviated form, one example of the bachelor's level program at the University of Erlangen-Nurnberg[3] is shown in the table below. Although the number of courses is approximately the same for the two bachelor programs, the Erlangen program is at a more advanced level.

| Required courses For Siena's bachelor's program | Computer Science | Mathematics | Physics |
|---|---|---|---|
| Introductory Courses | Intro to CS | Calculus I | General Physics I |
| | Intro to (Java) Programming | Calculus II | General Physics II |
| | Data Structures | Calculus III | Modern Physics I |
| Upper level Courses | Numerical Methods | Differential Equations | Modern Physics II |
| | Analysis of Algorithms | Linear Algebra | Computational Physics |
| | Plus two of the Following | Computer Algebra Systems | Classical Mechanics |
| | Graphics | Discrete Math | Simulation and Modeling |
| | Bioinformatics | | |
| | Object-Oriented Design and Prog | | |
| | Software Engineering | | |
| Total – 21 + capstone | 7 | 7 | 7 |

| Required courses for Erlangen bachelor's program – number of courses | Computer Science and other courses | Mathematics | Thermodynamics and Fluid Mechanics |
|---|---|---|---|
| | 2 technical CS / computer architec | 4 engineering math | 1 experimental physics |
| | 2 Algorithms | 2 numerical analysis | 2 thermodynamics |
| | 1 software systems | | 3 fluid mechanics |
| | 1 theoretical CS | | |
| | 2 scientific computing | | |
| | 1 seminar | | |
| | 1 bachelor thesis | | |
| Total - 22 | 10 | 6 | 6 |

The University of Utah offers both a master's degree[4] and a Ph.D. in Computational Science and Engineering[5]. The requirements for the master's degree are in the first year
- Intro to Scientific Computing
- Advanced Scientific Computing
- Scientific Visualization
- Mathematical Modeling
- Case Studies in Computational Engineering and Science
- An elective course
- CES seminar

In the second year students should take at least two graduate level elective courses, when in the Thesis program. The thesis program requires 25 hours of course work and at least 6 hours of research.

The master's program at the University of Erlangen-Nurnberg[3] takes two years and allots the final six months solely to the writing of a thesis. One example of course requirements (of many possible), where each of the listed courses is four-credit, could be

- Simulation and modeling
- Performance modeling of computer systems
- Numerical Simulation of Fluids
- Scientific Visualization.

The Erlangen thesis requirement is quite demanding. Despite similar names, the degrees are not quite comparable and one needs to be aware of the local culture, for example, of the previous Diploma program in Germany and the current transition throughout Europe as directed by the Bologna agreement.

At the University of Utah, one approach to the PhD[5] is the Scientific Computing Track within the Computing Degree Program. Overall a student's work must comprise at least 50 semester hours of graduate course work and dissertation research. Before the fifth semester, a student must complete two courses in advanced scientific computing, and one course each in high performance computing and scientific visualization. Further, a student must take four elective courses (twelve hours) which involve the themes of scientific computing which are directly applicable to the student's dissertation research.

These all too brief summaries are not intended as detailed explanations of the programs. Rather they are presented here to indicate those subject areas where the programs direct the student. For true details of the programs consult the degree granting institution.

## IV Final Comments

In the few examples of CLS curricula discussed here, we find a consistency in the areas to be studied. How the themes are implemented is surely institution dependent but a general thematic approach exists; scientific computing, visualization, modeling and simulation, and high performance computing are consistent subjects of study at all levels.

Many of the most interesting real world problems are no longer within the realm of a single discipline. Despite the fact that academic institutions are built around individual disciplines, the future may well be in various multi-disciplinary groupings. For complex scientific problem solving, computational science is the wave of the future.

## Acknowledgments

The author wishes to thank Siena College for support of this work.

## References

[1] For an interesting history of the FPU problem see this reference and references therein. "The Genesis of Simulation in Dynamics," Thomas P. Weissert, New York, 1997.

[2] http://www.siena.edu/physics/computationalscience

[3] For information about the computing program at the Universitaet Erlangen-Nurnberg contact Dr. Ulrich.Ruede@informatik.uni-erlangen.de . A 40 page document is available describing their program in detail. The university website is http://www.uni-erlangen.de.

[4] http://www.ces.utah.edu/

[5] http://www.cs.utah.edu/research/areas/computation/PhDdegree.html

Brill Academic Publishers
P.O. Box 9000, 2300 PA Leiden,
The Netherlands

*Lecture Series on Computer
and Computational Sciences*
Volume 4, 2005, pp. 171-174

# Time–Space Least–Squares Spectral Element Method for Population Balance Equations

## C.A. Dorao, H.A. Jakobsen[1]

Department of Chemical Engineering,
Norwegian University of Science and Technology,
N-7491 Trondheim, Norway

Received 6 August, 2005; accepted in revised form 16 August, 2005

*Abstract:* In multiphase chemical reactor analysis the prediction of the dispersed phase distribution plays a major role in obtaining reliable results. The Population Balance Equation is a well stablished equation for describing the evolution of the dispersed phase. However, the numerical treatment of such equation is computationally intensive.
In this work, a Time–Space Least Squares Spectral method is presented for solving the Population Balance Equation including breakage and coalescence processes.

*Keywords:* Population Balance Equation, Least Squares Spectral Method

*Mathematics Subject Classification:* 47.55.Kf; 02.70.Hm; 02.60.Nm

*PACS:* 47.55.Kf; 02.70.Hm; 02.60.Nm

## 1 Introduction

Population Balance Modeling (PBM) is a well established method for describing the evolution of population of entities such as bubbles, droplets or particles. In particular, in multiphase flow problems the PBM is used to include information about the dispersed phase distribution into the multifluid model.

Based on the population balance approach the dispersed phase is treated by using a probability density function, PDF, for instance $f(\mathbf{r}, \xi, t)$ where $\mathbf{r}$ is the spatial vector position, $\xi$ is the property of interest of the dispersed phase, and $t$ the time. This PDF, $f(\mathbf{r}, \xi, t)$, can represent the average amount of number of particles per unit volume around the point $\mathbf{r}$, with the property $\xi + d\xi$ in the instant $t + dt$. The evolution of this PDF must take into account the different processes that control the PDF such as breakage, coalescence, growth and advective transport of the particles. The resulting equation is a nonlinear partial integro–differential equation which requires to be solved by a suitable numerical method. In this regard, the Time–Space Least–Squares Spectral Element Method (LS–SEM) is applied in this work for the numerical treatment of this type of problems. The basic idea in the LS–SEM is to minimize the integral of the square of the residual over the computational domain [1, 6, 9, 8]. In the case when the exact solutions are sufficiently smooth the convergence rate is exponential. For time dependent problems, the space–time formulation, i.e. time is treated as an additional dimension, allows high order accuracy both in space and in time [7]. In this way, space–time can be solved at once, or per time–step on a space–time slab in a kind of semi–discrete formulation. In this work, the time–space LS–SEM is applied for solving a time dependent Population Balance Equation.

---

[1]Corresponding author. Tel. +47-73-594132; fax: +47-73-594080. E-mail: hugo.jakobsen@chemeng.ntnu.no

## 2 Population Balance Equation

The Population Balance Equation (PBE) can be written like

$$\mathcal{L}f(\xi,t) = g(\xi,t), \qquad\qquad \text{in } \Omega \qquad (1)$$
$$\mathcal{B}_0 f(\xi,t) = f_0(\xi), \qquad\qquad \text{on } \Gamma_0 \qquad (2)$$

with $\Omega = [\xi_{min}, \xi_{max}] \times [0,T]$ where $\xi_{min}$ and $\xi_{max}$ are for instance the minimum and maximum particle size, and $T$ the final simulation time. The initial condition is applied on $\Gamma_0 = \{(\xi,t) \in \partial\Omega : t = 0\}$. $\mathcal{L}$ is a non-linear first order partial integro–differential operator,

$$\mathcal{L}f(\xi,t) \equiv \frac{\partial f(\xi,t)}{\partial t} + \mathcal{L}_{PB}f(\xi,t) \qquad (3)$$

and $\mathcal{B}_0$ the identity operator.

The changes in the population are given by the $0^{th}$ moment conservative Population Balance operator,

$$\mathcal{L}_{PB}f(\xi,t) = -b(\xi)f(\xi,t) + \int_\xi^{\xi_{max}} h(\xi,s)\,b(s)f(s,t)ds -$$
$$f(\xi,t) \int_{\xi_{min}}^{\xi_{max}+\xi_{min}-\xi} c(\xi,s)f(s,t)ds + \int_{\xi_{min}}^\xi c(\xi-s,s)f(\xi-s,t)f(s,t)ds \qquad (4)$$

Detailed discussion about the meaning of each term can be found in [5].

## 3 The Least Squares Method

The Least–Squares formulation is based on the minimization of a norm–equivalent functional. This consists in finding the minimizer of the residual in a certain norm. The norm–equivalent functional is given by

$$\mathcal{J}(f;g,f_0) \equiv \frac{1}{2} \parallel \mathcal{L}f - g \parallel_{Y(\Omega)}^2 + \frac{1}{2} \parallel \mathcal{B}_0 f - f_0 \parallel_{Y(\Gamma_0)}^2$$

with the norms defined like

$$\parallel \bullet \parallel_{Y(\Omega)}^2 = \langle \bullet, \bullet \rangle_{Y(\Omega)} = \int_\Omega \bullet \bullet \, d\Omega, \qquad \parallel \bullet \parallel_{Y(\Gamma_0)}^2 = \langle \bullet, \bullet \rangle_{Y(\Gamma_0)} = \int_{\Gamma_0} \bullet \bullet \, ds \qquad (5)$$

Based on variational analysis, the minimization statement is equivalent to:
    Find $f \in X(\Omega)$ such that

$$\lim_{\epsilon \to 0} \frac{d\mathcal{J}(f + \epsilon\, v; g, f_0)}{d\epsilon} = 0 \qquad\qquad \forall v \in X(\Omega) \qquad (6)$$

Consequently, the necessary condition can be written as:
    Find $f \in X(\Omega)$ such that

$$\mathcal{A}(f,v) = \mathcal{F}(v) \qquad\qquad \forall v \in X(\Omega) \qquad (7)$$

with

$$\mathcal{A}(f,v) = \langle \mathcal{L}f, \mathcal{L}v \rangle_{Y(\Omega)} + \langle \mathcal{B}_0 f, \mathcal{B}_0 v \rangle_{Y(\Gamma_0)} \qquad (8)$$
$$\mathcal{F}(v) = \langle g, \mathcal{L}v \rangle_{Y(\Omega)} + \langle f_0, \mathcal{B}_0 v \rangle_{Y(\Gamma_0)} \qquad (9)$$

where $\mathcal{A} : X \times X \to \mathbb{R}$ is a symmetric, continuous bilinear form, and $\mathcal{F} : X \to \mathbb{R}$ a continuous linear form. Finally, the discretization statement consists in searching the solution in a reduced subspace, i.e. $f_N(r, \xi) \in X_N(\Omega) \subset X(\Omega)$. Hence, $f_N$ can be expressed like

$$f_N(\xi, t) = \sum_{i=0}^{N_1} \sum_{j=0}^{N_2} f_{ij} \; \varphi_i(\xi) \; \varphi_j(t), \quad \text{with } f_{ij} = f(\xi_i, t_j) \tag{10}$$

where $\varphi_i(\xi)$ and $\varphi_j(t)$ are the 1D Lagrangean interpolants through the Gauss–Legendre collocation points and Gauss–Lobatto–Legendre collocation points respectively [3, 2].

It is important to mention that the non–linear terms require a linearization step previous to the application of the previous framework which was previously discussed in [4].

## 4  Numerical Example

To show the properties of the method the following problem with analytical solution is used

$$b(\xi) = \xi^2, \quad h(\xi, \tilde{\xi}) = 1/\tilde{\xi}, \quad c(\xi, \tilde{\xi}) = 1$$

$$g(\xi, t) = \frac{1}{750} e^{-5-2t-5\xi} \Big( 222 \; e^{t+5\xi} + 30 \; e^{5\xi} \; \xi(-6 + 5\xi) - \tag{11}$$

$$5e^5 \xi(-6 + 25\xi^2) + 6 \; e^{5+t}(-2 - 135\xi - 25\xi^2 + 125\xi^3) \Big);$$

$$f_{exact}(\xi, t) = \xi \; e^{-5\xi-t} \tag{12}$$

where $(\xi, t) \in \Omega = [0, 1] \times [0, T]$. The behaviour of the moment percentual error $\mu_k$ is studied which is defined like

$$\mu_k = 100 \left( \int_{\xi_{min}}^{\xi_{max}} \left( \xi^k \; f(\xi, T) - \xi^k \; f_{exact}(\xi, T) \right) d\xi \right) \Big/ \int_{\xi_{min}}^{\xi_{max}} \xi^k \; f_{exact}(\xi, T) \; d\xi \tag{13}$$

In figure (1) on the LHS, the h–convergence of $\mu_k$ with respect to $N_2$, i.e. the polynomial order used for expanding $t$, is shown keeping $N_1 = 10$ fix. As expected, an exponential convergence rate is observed. On the RHS of figure (1), the $\Delta t$–convergence of $\mu_k$ is ploted keeping $N_1 = 10$ and $N_2 = 3$ fixed. The convergence rate is linear with a slope depending on $N_2$.

## 5  Conclusions

A time–space Least Squares Spectral method was presented for solving the Population Balance Equation including breakage and coalescence processes. The use of a time–space approach allows to increase the temporal accuracy or for a given accuracy increase the time step reducing the final computational cost. Besides, if the exact solution is sufficiently smooth, the convergence rate is exponential.

Further work is required for coupling such type of methods with available multifluid solvers.

## Acknowledgment

The PhD fellowship (Dorao, C. A.) financed by the Research Council of Norway through a Strategic University Program (CARPET) is gratefully appreciated.

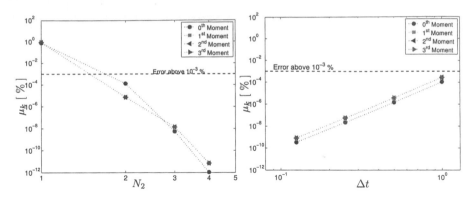

Figure 1: p–convergence of the percentual moments ($N_1 = 10$, $T = 1$, $\Delta t = 0.1$) on the RHS. $\Delta t$–convergence of the percentual moments $\mu_k$ ($N_1 = 10$, $N_2 = 3$, $T = 1$) on the LHS.

# References

[1] P. Bochev: *Finite element methods based on least–squares and modified variational principles*, Technical report, POSTECH, 2001.

[2] M.O. Deville, P.F. Fischer and E.H. Mund: *High-order methods for incompressible fluid flow*. Cambridge University Press, 2002.

[3] C.A. Dorao and H.A. Jakobsen: Spatially dependent Population Balance Problems. *submitted to Chemical Engineering Science*

[4] C.A. Dorao and H.A. Jakobsen: Application of the Least Square Method to Population Balance Problems. *submitted to Computer & Chemical Engineering.*

[5] H.A. Jakobsen, H. Lindborg and C.A. Dorao: Modeling of Bubble Column Reactors: Progress and Limitations. *Industrial & Engineering Chemistry Research.* **44**(14) 5107-5151 (2005).

[6] B. N. Jiang: *The Least–Squares Finite Element Method, Theory and Applications in Computational Fluid Dynamics and Electromagnetics*. Springer Verlag, 1998.

[7] B. De Maerschalck: *Space–Time least–squares spectral element method for unsteady flows - application and evaluation for linear and non–linear hyperbolic scalar equations*, Master Thesis Report, Delft University of Technology, Dept. of Aerospace Engineering, The Netherlands, 2003.

[8] J. P. Pontaza and J. N. Reddy: Space–time coupled spectral/hp least squares finite element formulation for the incompressible Navier–Stokes equation. *Journal of Computational Physics.* **190**(2) 523-549 (2003).

[9] M.M.J.Proot and M.I. Gerritsma: Least–Squares Spectral Elements Applied to the Stokes Problem. *Journal of Computational Physics.* **181**(2) 454-477 (2002).

Brill Academic Publishers
P.O. Box 9000, 2300 PA Leiden,
The Netherlands

*Lecture Series on Computer*
*and Computational Sciences*
Volume 4, 2005, pp. 175-178

# MPI and JavaRMI Skeletons for the Branch-and-Bound Technique

## I. Dorta, C. León[1]

Departamento de Estadística, I.O. y Computación,
University of La Laguna, E-38271 La Laguna, Tenerife, Spain
http://nereida.deioc.ull.es

Received 9 July, 2005; accepted in revised form 3 August, 2004

*Abstract:* This article proposes a comparison between the MPI and JavaRMI implementation of skeletons for the Branch-and-Bound technique. Concretely, sequential, parallel and distributed generic skeletons to implement this algorithmic technique are presented. The skeletons have been implemented using three different programming languages: C, C++ and Java. Some preliminary computational results for the C and C++ sequential implementation are presented.

*Keywords:* Branch-and-Bound Technique, Skeletons, Parallelism, Distributed algorithms

*Mathematics Subject Classification:* 68R99

## 1 Introduction

In this paper, we present skeletons to solve Optimization Problems using the Branch-and-Bound technique. The skeletons proposed could be used on as much of networks of computers under Linux, like of high performance computers. The user will be provided to the possibility to solve its problems, as much of sequential form as of parallel and distributed forms without having to modify its code. The skeletons has been implemented using three different programming languages: C, C++ and Java. The first approximation uses the imperative language C. However, to have modular system where fundamental features like data structure easily can be changed and inherited by different modules Object-Oriented programming will be called for using the C++ and Java languages. There are two different parts in our proposal. The first one compare the sequential behaviour of the skeletons with different languages implementations. And the second part consists of a comparison between the parallel and distributed tools to implement the parallel and the distributed versions. An algorithm for the resolution of the classic 0-1 Knapsack Problem [3] has been implemented using the three implementations proposed. The parallel implementations will be made using MPI [4] (Message Passing Interface) and the distributed implementation uses Java RMI [5] (Remote Method Invocation).

## 2 The user interface of the skeletons

In our skeletons two principal parts are distinguished: One that implements the *resolution pattern* provided by the library and a part which the user has to complete with the particular characteristics

---

[1]E-mail: {isadorta|cleon}@ull.es

of the *problem to solve* and that will be used by the resolution pattern [1]. There is an intuitive relationship between the participating entities in the resolution pattern and the classes or structures which will be implemented by the user. The part provided by the skeleton, that is, the resolution pattern, is implemented through classes (structures in the C language case). These classes are denominated *provided classes*. The part which the user fills in with their particular problem will be named *required* classes.

The adjustment that has been accomplished by the user consists of two steps. First, the problem has to be represented through data structures and then, using them, the user has to implement the required functionalities of the classes. These functionalities will be invoked from the particular resolution pattern (because the interface for such classes is known) so that, when the application has been completed, the expected functionalities applied to the particular problem are obtained.

## 3   The algorithm of the skeletons

Branch-and-Bound algorithms are general methods applied to various combinatorial optimization problems that belong to the NP-hard problems class. We consider a combinatorial optimization problem as a tuple $\Pi = (I, S, f, g)$ where:

- $I$ is the set of instances of $\Pi$. If $x \in I$ we say that $x$ is an *instance* (or an input) of $\Pi$.

- Given an instance $x \in I$, $S(x)$ denotes the *set of feasible solutions* of $x$.

- For any instance $x \in I$ and any feasible solution $\sigma \in S(x)$, $f(x, \sigma)$ represents a real value, the measure (or cost) of $\sigma$ with respect to $\Pi$ and $x$. The function $f$ is called the *objective function*.

- $g \in \{\max, \min\}$. The goal of $\Pi$ is to find a feasible solution that optimizes $f$ according to $g$: given an input $x \in I$, determine an *optimal solution* $\sigma' \in S(x)$ such that $f(x, \sigma') = g\{f(x, \sigma) \mid \sigma \in S(x)\}$.

- A *subproblem* $\Pi_i$ is a tuple $\Pi_i = (I, S_i, f, g)$ where $S_i(x)$ is a subset of the underlying space I.

Given a *subproblem* $\Pi_i$, it can be broken down into $\Pi_{i1}, \Pi_{i2}, ..., \Pi_{ik}$ by a branching operation where $S_i = \bigcup_{j=1}^{k} S_{ij}$. Thus any feasible solutions $\sigma \in S_i$ belong to some $S_{ij}$ and conversely any $\sigma \in S_{ij}$ belong to $S_i$. Let $\mathcal{Q}$ denote the set of *subproblems* currently generated. A *subproblem* $\Pi_i \in \mathcal{Q}$ that is neither broken down nor tested yet is called *live*. The set of *live subproblems* are denoted by $\mathcal{L}$. For each tested subproblem in $\mathcal{Q}$ its *lower bound* and *upper bound* are computed. The greatest lower bound so far obtained is called the *best solution value* or *incumbent value* and is denoted by $bs$. The solution $bs$ is called the *best solution* and is stored in $\mathcal{T}$. The algorithm proceeds by repeating the test of lives subproblems. The selection of a live subproblem for the next test is performed by a *search function* $s$, such that, $s(\mathcal{L}) = \mathcal{L}$.

The skeletons proposed apply the Branch-and-Bound method to a combinatorial optimization (i.e. maximization) problem following next steps:

**Step 1** (initialization): $\mathcal{L} := \{\Pi_0\}$, $\mathcal{Q} := \{\Pi_0\}$, $bs := -\infty$ and $\mathcal{T} := \emptyset$ (empty set).

**Step 2** (search): Go to Step 9 if $\mathcal{L} = \emptyset$. Otherwise go to Step 3 after selecting $\Pi_i := s(\mathcal{L})$ for the test.

**Step 3** (update): If $lower\_bound(\Pi_i) > bs$, let $bs := lower\_bound(\Pi_i)$ and $\mathcal{T} := \{\sigma\}$, where $\sigma$ is a feasible solution of $\Pi_i$ carrying out $f(x, \sigma) = lower\_bound(\Pi_i)$. Go to Step 4.

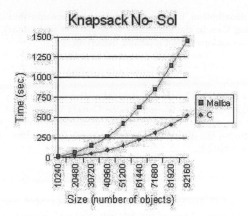

Figure 1: Comparison between C and C++ sequential implementations

**Step 4**: Go to Step 8 if $\Pi_i \in G$ (the set of partial problems $\Pi_i$ solved in the course of computing the *upper_bound*$(\Pi_i)$). Otherwise go to Step 5.

**Step 5** (upper bound test): Go to Step 8 if *upper_bound*$(\Pi_i) \leq bs$. Otherwise go to Step 6.

**Step 6**: Go to Step 8 if there exists a $\Pi_k(\neq \Pi_i) \in \mathcal{Q}$ such that $f(\Pi_k) \geq f(\Pi_i)$. Otherwise go to Step 7.

**Step 7** (branch): Break down $\Pi_i$ into $\Pi_{i1}, \Pi_{i2}, ..., \Pi_{ik}$ and let
$\mathcal{L} := \mathcal{L} \cup \{\Pi_{i1}, \Pi_{i2}, ..., \Pi_{ik}\} - \{\Pi_i\}$, $\mathcal{Q} := \mathcal{Q} \cup \{\Pi_{i1}, \Pi_{i2}, ..., \Pi_{ik}\}$. Return to Step 2.

**Step 8** (terminate $\Pi_i$): Return to Step 2 after letting $\mathcal{L} := \mathcal{L} - \Pi_i$.

**Step 9** (termination): Stop. $bs$ is equal to $f(\Pi_0)$, and $\mathcal{T}$ stores an optimal solution of $\Pi_0$, if $bs > -\infty$. If $bs = -\infty$, $\Pi_0$ is unfeasible.

## 4    Conclusions

Firstly, we accomplished a study of the time required by the sequential Branch and-Bound skeleton implemented, using the imperative language C and an the Object Oriented language C++. The C++ skeleton implementation used is the one provided by the MaLLBa library [2]. The experiments were executed on a Pentium III/600 Mhz with 256 Mb of memory. Figure 1 shows the times obtained executing the C++ implementation versus the Ansi C implementation, for different sizes of the 0-1 Knapsack Problem; these problems were generated for sizes between [10240..92160]. As can be appreciated, the computational results obtained with C++ were good, but lightly less efficient than those obtained using C language. In spire of this, an oriented object (OO) methodology provides the advantages: modularity, re-usable, modifiable, and so on, and also is added the interpretation facility of the skeleton, mainly because there is quite an intuitive relationship between the entities participating in the resolution pattern and the classes implemented by the skeleton.

Other accomplished comparison was between different Object Oriented languages. The language Java offers greater advantages for the development of the components software associated with the geographical distribution, but the current Java implementations do not improve the performance obtained using C++ in the sequential version. The preliminary results for the parallel and distributed implementations shown a similar behaviour.

## Acknowledgement

This work was partially supported by the Spanish "Ministerio de Ciencia y Tecnología" and FEDER under the project TIC02-04498-C05-05, and by the Canary Government Project COF2003/022. We also have received the support of the European Community-Research Infrastructure Action under the FP6 "Structuring the European Research Area" Programme, under the Project HPC-EUROPA (RII3-CT-2003-506079).

## References

[1] I. Dorta, C. León and C. Rodríguez, Comparison between MPI and OpenMP Branch-and-Bound Skeletons, *In Proceedings of the 8th International Workshop on High-Level Parallel Programming Models and Supportive Environments (HIPS'03)*, Nice, France, 66-73 (2003).

[2] I. Dorta, C. León, and C. Rodríguez, Parallel Branch-and-Bound Skeletons: Message Passing and Shared Memory implementations, *Fifth International Conference on Parallel Processing and Applied Mathematics (PPAM2003) Proceedings Published in the Springer Verlag Series LNCS* **3019** 286-291, (2004).

[3] S. Martello, P. Toth, Knapsack Problems: Algorithms and Computer Implementations, *John Wiley & Sons Ltd*, (1990).

[4] M. Snir, S. W. Otto, S. Huss-Lederman, D.W. Walker and J. Dongarra, MPI – the Complete Reference, 2nd Edition, MIT Press, (1998).

[5] Javasoft. RMI Specification. See `http://java.sun.com/products/jdk/rmi/` (2004).

Brill Academic Publishers
P.O. Box 9000, 2300 PA Leiden,
The Netherlands

*Lecture Series on Computer*
*and Computational Sciences*
Volume 4, 2005, pp. 179-182

# An Approach for Simultaneous Structure Determination and Parameter Estimation Using MINLP Techniques

Stefan Emet[1]

Department of Mathematics
University of Turku, FIN-20014 Turku, Finland

Received 8 July, 2005; accepted in revised form 29 July, 2005

*Abstract:* The present paper presents an approach of simultaneously solving a structure determination and parameter estimation problem using Mixed Integer Nonlinear Programming (MINLP) techniques. It is shown that the minimization of Akaike's Information Criterion (AIC), the Bayesian Information Criterion (BIC), and the Final Prediction Error (FPE) can efficiently be modeled as MINLP problems. The problems are solved using the Extended Cutting Plane (ECP) method. The presented techniques are applied in determining the structure and the parameters of some illustrative Autoregressive Moving Average (ARMA) time series. The described techniques can also be applied on dynamical systems.

*Keywords:* Akaike's Information Criterion, Mixed Integer Nonlinear Programming, Extended Cutting Plane Method

*Mathematics Subject Classification:* 90C11, 90C90

## 1   Introduction

Given a set of observations $y_i$, made at time points $t_i$, $i = 1, \ldots, N$, the classical least squares problem of estimating the parameters $\theta_i$, $i = 1, \ldots, p$, of a function $f$, describing the underlying system, can be formulated as follows:

$$\min_{\theta_1, \ldots, \theta_p} \left\{ \sum_{i=1}^{N} (f(t_i, \theta) - y_i)^2 \right\} \tag{1}$$

In many practical applications [1], [2] it is challenging to find out the parameters and/or what function should be used in order to descibe the underlying system as well as possible. A tool that can be applied for this purpose is the Akaike's Information Criterion (AIC) [3]:

$$-2 \ln L + 2p \tag{2}$$

where $L$ denotes the maximum likelihood function and $p$ is the number of parameters. The logarithm of $L$, for $N$ normally distributed, independent, random variables $e_k$, and a variance $\sigma^2$, is given by [4]:

$$\ln L = -\frac{N}{2} \ln(2\pi) + \frac{N}{2} \ln(|\frac{1}{\sigma^2}|) - \frac{1}{2\sigma^2} \sum_{k=1}^{N} e_k^2 \tag{3}$$

---

[1]Corresponding author. E-mail: stefan.emet@utu.fi

Substituting (3) in (2) yields

$$N \ln(2\pi) + N \ln(|\sigma^2|) + \frac{1}{\sigma^2} \sum_{k=1}^{N} (f(t_k, \theta) - y_k)^2 + 2p \tag{4}$$

If the variance $\sigma^2$ is known, or predefined [5], the criterion to be minimized can be formulated as

$$\min_{\theta_1, \ldots, \theta_p} \left\{ \frac{1}{\sigma^2} \sum_{k=1}^{N} (f(t_k, \theta) - y_k)^2 + 2p \right\} \tag{5}$$

In (5) it is assumed that $p$, which gives the structure of $f$, is known, but in many cases it is difficult to choose $p$ in advance. One possibility is to solve (5) for different values of $p$ and observe which value gives the best solution [6], [7]. In the following, an alternative approach which enables the solving of both the structure determination and parameter estimation simultaneously, is presented.

## 2    MINLP Formulation

Introducing a binary variable, $\beta_i$, for the existence of each parameter $\theta_i$, and lower and upper bounds, $\theta_{i,\min}$ and $\theta_{i,\max}$, the optimization problem (5) can be formulated as follows:

$$\min \left\{ \frac{1}{\sigma^2} \sum_{i=1}^{N} (f(t_i, \theta) - y_i)^2 + 2 \sum_{i=1}^{n} \beta_i \right\}$$
$$\theta_i - \theta_{i,\max} \cdot \beta_i \le 0$$
$$-\theta_i + \theta_{i,\min} \cdot \beta_i \le 0 \tag{6}$$
$$\beta_i \in \{0, 1\}, \ i = 1, \ldots, n.$$

The minimization of Akaike's criterion might, however, in some applications result in redundant parameters [7]. In order to keep the number of parameters of moderate size it is possible to use the following Bayesian Information Criterion (BIC) [3]:

$$N \ln(2\pi) + N \ln(|\sigma^2|) + \frac{1}{\sigma^2} \sum_{i=1}^{N} (f(t_i, \theta) - y_i)^2 + (1 + \ln N)p \tag{7}$$

In (7) the number of observations, that is, the sample size, $N$, is damping the number of parameters. The corresponding MINLP formulation can be written as follows:

$$\min \left\{ \frac{1}{\sigma^2} \sum_{i=1}^{N} (f(t_i, \theta) - y_i)^2 + (1 + \ln N) \sum_{i=1}^{n} \beta_i \right\}$$
$$\theta_i - \theta_{i,\max} \cdot \beta_i \le 0$$
$$-\theta_i + \theta_{i,\min} \cdot \beta_i \le 0 \tag{8}$$
$$\beta_i \in \{0, 1\}, i = 1, \ldots, n.$$

Note, that the objectives in (6) and (8) are of a similar form; the sum of squared residuals and a penalty term. The penalty term can easily be modified in such a way that lower or higher priority is put on the number of parameters, the sample size, costs etc. Assume that $f$ is convex and that $\sigma^2$ is defined [5], then the objectives in (6) and (8) are also convex, which is a nice property because many solvers are able to guarantee global optimality for such problems.

Another approach which also takes the sample size into consideration is given by the Final Prediction Error (FPE) [8]:

$$\sum_{i=1}^{N}(f(t_i, \theta) - y_i)^2 \frac{N+p}{N-p} \tag{9}$$

The (FPE) in (9) is nonconvex, with respect to $p$, which might induce difficulties in the solving.

## 3   Numerical Examples

An Autoregressive Moving Average (ARMA) series of the order $(p, q)$ can be written as follows:

$$X_t = \alpha_0 + \alpha_1 X_{t-1} + \cdots + \alpha_p X_{t-p} + \epsilon_t + \beta_1 \epsilon_{t-1} + \cdots + \beta_q \epsilon_{t-q} \tag{10}$$

The MINLP formulations in (6) and (8) were applied on the following set of series:

$$X_t = \epsilon_t + \frac{1}{3}X_{t-1} \tag{11}$$

$$X_t = \epsilon_t + 0.25X_{t-1} - 0.75X_{t-2} \tag{12}$$

$$X_t = \epsilon_t + 0.7X_{t-1} - 0.2X_{t-2} + 0.5X_{t-3} \tag{13}$$

$$X_t = \epsilon_t + 0.9\epsilon_{t-1} \tag{14}$$

$$X_t = \epsilon_t - 0.7\epsilon_{t-1} + 0.5\epsilon_{t-2} \tag{15}$$

$$X_t = \epsilon_t + 0.75\epsilon_{t-1} + 0.75X_{t-1} \tag{16}$$

$$X_t = \epsilon_t - 0.8\epsilon_{t-1} + 0.5X_{t-2} \tag{17}$$

where $\epsilon_t$ are normally distributed random variables with the mean 0 and the variance 1. The corresponding MINLP problems were solved using the Extended Cutting Plane method (ECP), [9], which is an extension of Kelley's method [10]. The ECP method is a general purpose MINLP method with applicability to a large variety of problems [9], [2]. The residuals and variances were calculated as described in [11] and [6], respectively. The results of determining the order and of estimating the parameters of data generated from series (11)-(17) are presented in Table 1.

Table 1: Results of series (11)-(17).

| series | criterion | estimated model | $\hat{\sigma}_\epsilon^2$ |
|--------|-----------|-----------------|---------------------------|
| (11) | $AIC = 126.35$ | $X_t = \epsilon_t + 0.339X_{t-1}$ | 1.321 |
|       | $BIC = 129.90$ | $X_t = \epsilon_t + 0.339X_{t-1}$ | 1.321 |
| (12) | $AIC = 93.49$ | $X_t = \epsilon_t + 0.1233X_{t-1} - 0.7562X_{t-2}$ | 0.936 |
|       | $BIC = 97.63$ | $X_t = \epsilon_t - 0.7571X_{t-2}$ | 0.961 |
| (13) | $AIC = 115.76$ | $X_t = \epsilon_t + 0.776X_{t-1} - 0.242X_{t-2} + 0.434X_{t-3}$ | 1.166 |
|       | $BIC = 125.36$ | $X_t = \epsilon_t + 0.660X_{t-1} + 0.309X_{t-3}$ | 1.214 |
| (14) | $AIC = 113.38$ | $X_t = \epsilon_t + 0.904\epsilon_{t-1}$ | 1.122 |
|       | $BIC = 116.92$ | $X_t = \epsilon_t + 0.916\epsilon_{t-1}$ | 1.122 |
| (15) | $AIC = 93.82$ | $X_t = \epsilon_t - 0.794\epsilon_{t-1} + 0.450\epsilon_{t-2}$ | 0.965 |
|       | $BIC = 101.01$ | $X_t = \epsilon_t - 0.822\epsilon_{t-1} + 0.468\epsilon_{t-2}$ | 0.968 |
| (16) | $AIC = 100.80$ | $X_t = \epsilon_t + 0.623\epsilon_{t-1} + 0.745X_{t-1}$ | 1.004 |
|       | $BIC = 107.93$ | $X_t = \epsilon_t + 0.620\epsilon_{t-1} + 0.746X_{t-1}$ | 1.003 |
| (17) | $AIC = 81.20$ | $X_t = \epsilon_t - 0.872\epsilon_{t-1} - 0.371X_{t-2}$ | 0.802 |
|       | $BIC = 88.36$ | $X_t = \epsilon_t - 0.854\epsilon_{t-1} - 0.347X_{t-2}$ | 0.803 |

## 4   Summary

In the present paper, techniques for solving structure determination and parameter estimation problems were presented. It was shown that such problems can be solved in a simultaneous manner by minimizing, for instance, Akaike's information criterion.

Examples of joint model structure determination and parameter estimation of some time series were finally illustrated. The examples were numerically solved using the ECP-method that has been proven efficient on many complex engineering problems. The presented methods can in a similar way also be applied on industrial problems.

## References

[1] A. Sundberg, S. Emet, P. Rehn, L. Vähäsalo B. and Holmbom, Estimation of proportions of different pulp types in recycled paper and deinked pulp by acid methanolysis, *Proc.* $13^{th}$ *Intern. Symp. Wood Pulping Chem.*, Auckland, New Zealand, 361-365 (2005).

[2] H. Skrifvars, S. Leyffer and T. Westerlund, Comparison of certain MINLP algorithms when applied to a model structure determination and parameter estimation problem. *Computers chem. Engng*, **22** 1829-1835 (1998).

[3] H. Akaike, A New Look at Statistical Model Identification. *IEEE Trans. Automat. Contr.*, **19** 716-722 (1974).

[4] A. M. Goodwin and R. L. Payne: *Dynamic System Identification. Experimental Design and Data Analysis.* Academic Press, 1977.

[5] S. Karrila and T. Westerlund, An Elementary Derivation of the Maximum Likelihood Estimator of the Covariance Matrix, and an Illustrative Determinant Inequality, *AUTOMATICA*, **27** 425-426 (1990).

[6] M. B. Priestley: *Spectral Analysis an Time Series.* Academic Press Inc., 1981.

[7] C. Chatfield: *The Analysis of Time Series, An Introduction.* Chapman and Hall, 1989.

[8] H. Akaike, Fitting autoregressive models for prediction. *Ann. Inst. Stat. Math.*, **20** 425-439 (1969).

[9] T. Westerlund and R. Pörn, Solving Pseudo-Convex Mixed Integer Problems by Cutting Plane Techniques. *Optimization and Engineering*, **3** 253-280 (2002).

[10] J. Kelley, The cutting-plane method for solving convex programs. *Journal of SIAM*, **8** 703-712 (1960).

[11] G. E. P. Box and G. M. Jenkins: *Time Series Analysis, Forecasting and Control.* Holden-Day, San Francisco, 1970.

Brill Academic Publishers
P.O. Box 9000, 2300 PA Leiden
The Netherlands

*Lecture Series on Computer
and Computational Sciences*
Volume 4, 2005, pp. 183-186

# Using Shape Function in Cell Centred Finite Volume Formulation for Two Dimensional Stress Analysis

N. Fallah[1]

Department of Civil Engineering,
University of Guilan,
P. O. Box 3756, Rasht, Iran

Received 7 July, 2005; accepted in revised form 28 July, 2005

*Abstract:* A novel finite volume formulation for the stress analysis of two dimensional elastic solids is presented. Mesh element is used as control volume or cell and the centre of the cells are considered as the computational nodes. A new approach is adopted to calculate stresses on the faces of the cell by using interim elements that surround cell faces. The interim elements are isoparametric. Shape functions are used to describe the displacement variations in the interim elements, and hence across the enclosed faces. The equilibrium equations of a cell are approximated at the integration points, which are located in the interim elements. In this approach stress continuity will be guaranteed on common faces of the adjacent cells. The formulation is used to predict the stress field in a large thin plate containing a small circular hole subjected to tensile stresses at the ends. The comparison between the predicted results and the analytical results reveals the capability of the presented formulation in prediction of accurate results.

*Keywords:* Finite volume, Cell centred, Shape function, Thin plate, Stress concentration

## 1. Introduction

Although the initial purpose for the development of finite volume technique in the realm of solid mechanics was to provide a uniform discretization technique to analyze problems of coupled systems of fluid flow and solid body interaction, but the interesting performance of the technique in pure solid mechanics problems directed some research works to explore more features of the finite volume technique in solid mechanics applications as its own. The applications of the technique include stress analysis of linear elastic structures [1-4] and analysis of materially nonlinear problems [5]. Works were also directed to the modelling of geometric nonlinear problems [6], modelling of bending behavior of thin and thick plates [7] and two-dimensional solid analysis incorporating rotational degrees of freedom [8].

Some features of various formulations that were developed to model the foregoing applications may seem different. However, they converge to satisfy the relevant conservation principle, the equilibrium of small part of the domain as control volume, and ensuring that stress varies continuously across the common face of adjacent control volumes. In this paper, a cell centred based finite volume formulation is developed for prediction of displacement, strain and stress in two-dimensional solids. A 4-node isoparametric element is formed to surround an individual face of the control volume. The element vertices are the centres of the two adjacent cells lying on either side of the face and the nodes at each end of the face. Bilinear shape function is used to describe the displacement variation in the element. Using this approach guarantees the stress continuity across cell faces and also provides an effective and simple tool for deformation gradient calculations.

## 2. Formulation

Figure 1 shows a part of the geometry of a two dimensional solid domain. The solid domain is discretized by mesh of elements with arbitrary shapes. Each element is now considered as a control volume or a cell. The control volume's centre coincides with the element centre. For a static analysis, in the absence of body force, the equilibrium equation governing the conservation of force for a given internal cell **P**, which is surrounded by neighbouring cells, can be written in the form

---

[1] Corresponding author. E-mail: fallah@guilan.ac.ir

$$\int_A \mathbf{T}\boldsymbol{\sigma}dA = 0 \tag{1}$$

Where $\boldsymbol{\sigma}$ is stress vector, matrix $\mathbf{T}$ includes components of the outward unit normal to the boundaries of the cell. The integral is a surface integral over the faces bounding the cell, denoted by $A$. In 2-D problems for the cell with constant thickness, we have

$$\int_\Gamma \mathbf{T}\boldsymbol{\sigma}dL = 0 \quad , \quad \boldsymbol{\sigma} = \begin{bmatrix} \sigma_x & \sigma_y & \sigma_{xy} \end{bmatrix}^\mathrm{T} \quad , \quad \mathbf{T} = \begin{bmatrix} n_x & 0 & n_y \\ 0 & n_y & n_x \end{bmatrix} \tag{2}$$

Where the integral is a line integral. Substituting the linear elastic constitutive equation in Eq. (2), for a cell enclosed with $k$ faces, gives

$$\sum_{i=1}^{k} \int_{\Gamma_i} \mathbf{T}\mathbf{D}\boldsymbol{\varepsilon}dL = 0 \tag{3}$$

Where matrix $\mathbf{D}$ includes the material elastic constants and $\boldsymbol{\varepsilon}$ is the elastic strain vector. The equilibrium equation (3) is exact and no approximation has been introduced so far. In order to evaluate the line integrals in Eq. (3) a distribution of displacement has to be assumed. Here, a bilinear distribution of displacement is employed in an imaginary 4-noded element, which surrounds each face of the cell. This imaginary element is referred to as interim element here.

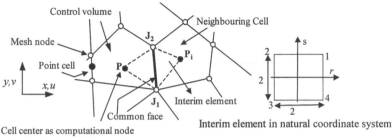

Figure 1: Control volume on a mesh of elements and an interim element on control volume's face

The interim element is considered as an isoparametric element and the interpolation functions are defined in the natural coordinate system of the element. Displacement vector, $\mathbf{u}$, can be approximated anywhere within the interim element using shape functions. In order to evaluate the first order derivative of displacement components with respect to global coordinate system, $xy$, which are presented in the equilibrium equation (3), the chain rule of differentiation is applied. Expressing strain on each face of the cell in terms of nodal displacements of interim element that surrounds the face yields strain-displacement matrix, $\mathbf{B}$ that can be used in Eq. (3) as

$$\sum_{i=1}^{K} \int_{\Gamma_i} \mathbf{T}\mathbf{D}\mathbf{B}\tilde{\mathbf{u}}dL = 0 \tag{4}$$

Where $\tilde{\mathbf{u}}$ is a vector that contains nodal displacements of the associated interim element with each face of the cell. Each integral in Eq. (4) can be evaluated at the centre of interim element as location of integration point. The integrals are evaluated for all the faces bounding the cell and therefore Eq. (4) may be represented in matrix form as

$$\begin{bmatrix} \overline{\mathbf{R}}_x & \overline{\mathbf{R}}_y \end{bmatrix}^\mathrm{T} [\tilde{\mathbf{u}}] = 0 \tag{5}$$

Where the entries of the first matrix include coefficients that are dependent on the elastic properties and geometry of the cell. The second term in Eq. (5) includes nodal displacement components of all interim elements that surround the cell. To express Eq. (5) only in terms of displacements at the centres of a given cell and all neighbouring cells, the corner nodal displacements must be eliminated. A bilinear variation in displacement across the region that surrounds a given corner node can be assumed. By using this assumption the displacement of that node can be expressed in terms of the displacements at the centres of each of the cell that meet at the node. After the elimination of displacements of all corner nodes, Eq. (5) becomes

$$\begin{bmatrix} \mathbf{R}_x & \mathbf{R}_y \end{bmatrix}^\mathrm{T} [\mathbf{u}] = 0 \tag{6}$$

In which each individual equation represents the relation of displacement components at the centre of cell $\mathbf{P}$ to those at the centres of all surrounding cells.

To incorporate the boundary conditions a point cell is used to transmit the boundary conditions onto the adjacent internal cell. The point cell is considered on the boundary next to the internal cell and at the middle of the face lying on the boundary. If displacement boundary conditions are applied, the applied displacement, $\mathbf{u}'$, can be transformed to global coordinate system as $\mathbf{u}$ which provides two equations. In case of stress boundary condition, the applied stress components, $\boldsymbol{\sigma}'$, can be transformed to global coordinate system as

$$\mathbf{T}^*\boldsymbol{\sigma} = \boldsymbol{\sigma}' \tag{7}$$

Using constitutive equation in Eq. (7) and approximating the strain on point cell using interim element whose vertices are the center of the internal cell, point cell and the nodes lying at either side of the point cell provides two equations. Each of the foregoing equation includes the displacement components of the end nodes of the boundary face, which must be eliminated. The elimination procedure is similar to that used for the internal cell. At the end of this procedure the equations that relate the unknown displacements of the point cell to those in neighbouring cells and to the applied boundary stresses will be obtained. These equations are in the form of Eq. (6) with nonzero right hand side. When the mixed boundary conditions are applied, a proper combination of equations mentioned above should be used. Equations associated with the internal cells and boundary conditions provide a system of simultaneous linear equations, which relates the unknown displacements to one and other and can be expressed in matrix form as

$$\mathbf{RX} = \mathbf{C} \tag{8}$$

Where $\mathbf{R}$ contains the coefficients relating the unknown displacements associated with the cells. $\mathbf{X}$ is a vector including the unknown variables and vector $\mathbf{C}$ represents the known values on the boundaries. The matrix $\mathbf{R}$ has sparse nature and non-symmetric, however, Eq. (8) can be solved by any appropriate solver technique to yield the displacements in the field.

## 3. Numerical example

The formulation presented in this paper has been applied to a large thin plate containing a small circular hole of radius $a$ and subjected to uniform tensile stress $\sigma_0$ at the ends, as shown in figure 2. The analytical solutions for stress field are known in the literature [9].

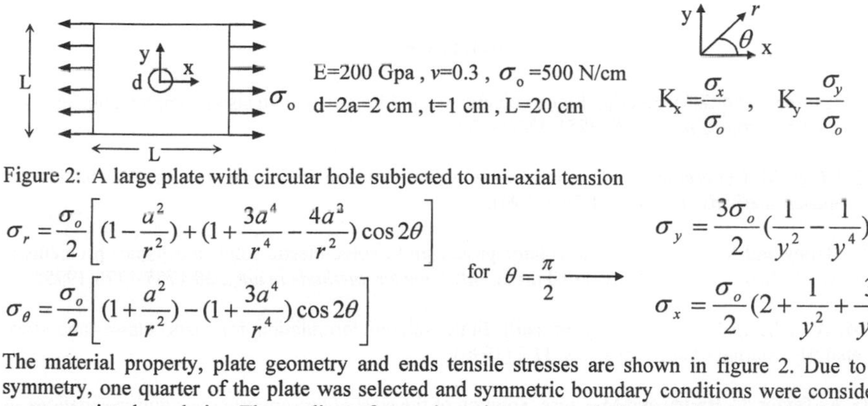

E=200 Gpa , $v$=0.3 , $\sigma_0$ =500 N/cm

d=2a=2 cm , t=1 cm , L=20 cm

$K_x = \dfrac{\sigma_x}{\sigma_o}$ , $K_y = \dfrac{\sigma_y}{\sigma_o}$

Figure 2: A large plate with circular hole subjected to uni-axial tension

$$\sigma_r = \frac{\sigma_o}{2}\left[(1-\frac{a^2}{r^2})+(1+\frac{3a^4}{r^4}-\frac{4a^2}{r^2})\cos 2\theta\right]$$

$$\sigma_\theta = \frac{\sigma_o}{2}\left[(1+\frac{a^2}{r^2})-(1+\frac{3a^4}{r^4})\cos 2\theta\right]$$

for $\theta = \dfrac{\pi}{2} \longrightarrow$

$$\sigma_y = \frac{3\sigma_o}{2}(\frac{1}{y^2}-\frac{1}{y^4})$$

$$\sigma_x = \frac{\sigma_o}{2}(2+\frac{1}{y^2}+\frac{3}{y^4})$$

The material property, plate geometry and ends tensile stresses are shown in figure 2. Due to the symmetry, one quarter of the plate was selected and symmetric boundary conditions were considered on appropriate boundaries. The gradient of stress in regions near to the hole is high; as such a fine mesh was used in these regions. Plate was meshed in three levels, as shown in figure 3. The stresses in x and y directions were computed at the middle of the faces lying on vertical axial line, $x$=0, of the plate. The stress values were normalized with respect to the ends tensile stress as $K_x$ and $K_y$. The normalized stress values or stress concentration factors were compared with the analytical solutions. Figure 3 shows the comparisons between the predicted results and the analytical results by applying different mesh. As figure shows, using coarse mesh results in discrepancies particularly in regions near to the hole for factor, $K_x$. By using fine mesh these differences will become negligible in distances far from the hole but in the regions near to the hole the predicted stress concentration factors are slightly higher than the analytical values but became smaller. Due to the fact that in the formulation the stresses on cell faces were considered uniform, it is expected that the differences should become smaller by using finer mesh in regions near to the hole. The results achieved from this test depict the capability of the presented formulation to provide accurate stress field.

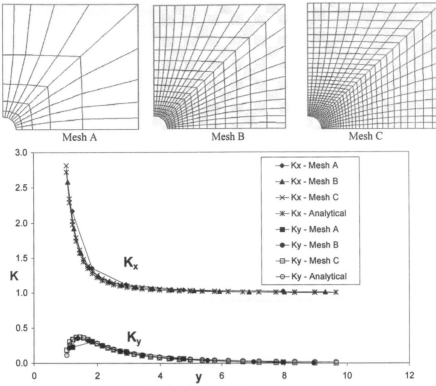

Figure 3: Variation of stress concentration factors along the vertical central axis of the plate

## References

[1] I. Demirdzic and S. Muzaferija, Finite volume method for stress analysis in complex domains', *Int. J. numer. methods in eng.,* **37** 3751-3766(1994).

[2] E. Oñate, M. Cervera and O. C. Zienkiewicz, A finite volume format for structural mechanics, *Int. J. numer. methods eng.,* **37** 181-201(1994).

[3] C. Bailey and M. Cross, A finite volume procedure to solve elastic solid mechanics problems in three dimensions on an unstructured mesh, *Int. J. numer. methods in eng.,* **38** 1757-1776(1995).

[4] M. A. Wheel, A geometrically versatile finite volume formulation for plane elastostatic stress analysis, Journal of strain analysis, **31** 2 (1996).

[5] G. A. Taylor, C. Bailey and M. Cross, Solution of the elastic/visco-plastic constitutive equations: A finite volume approach, *Appl. Math. Modelling,* **19** 743-760(1995).

[6] N. A. Fallah, C. bailey, M. Cross and G. A. Taylor, Comparison of finite element and finite volume methods application in geometrically nonlinear stress analysis, *Applied Mathematical Modelling,* **24** 439-455(2000).

[7] N. Fallah, Cell vertex and Cell centred finite volume methods for plate bending analysis, *Computational methods in applied Mechanics and Engineering,* **193** 3457-3470(2004).

[8] P. Wenke, M. A. Wheel, A finite volume method for solid mechanics incorporating rotational degrees of freedom, *Computers & Structures,* **81**, 321-329(2003).

[9] A. C. Ugural, S. K. Fenster, *Advanced strength and applied elasticity,* Prentice Hall, 1993

Brill Academic Publishers
P.O. Box 9000, 2300 PA Leiden
The Netherlands

*Lecture Series on Computer
and Computational Sciences*
Volume 4, 2005, pp. 187-190

# A New Approach in Cell Centred Finite Volume Formulation for Plate Bending Analysis

N. Fallah[1]

Department of Civil Engineering,
University of Guilan,
P. O. Box 3756, Rasht, Iran

Received 7 July, 2005; accepted in revised form 22 July, 2005

*Abstract:* In this paper a novel approach is developed in the application of cell centred finite volume method for plate bending analysis. The essence of this new approach lies in the use of interim elements and evaluating derivatives of unknown variables using the natural coordinate system of the interim elements. Mindlin-Reissner plate theory is applied in which the lateral shear effects are taken into account. The plate is meshed by elements that have arbitrary number of sides. These multi-faced elements are considered as control volumes or cells. The conservation of resultant forces, equilibrium equations, is written in the discretized form for the each cell. To evaluate the resultant forces on the faces of the cell, in the equilibrium equations, a 4-node interim element is used that enclosed the face. The interim element is isoparametric and its vertices are the centres of the two adjacent cells lying on either side of the face and the nodes at each end of the face. Shape functions are used to interpolate the unknown variables in the interim elements, and hence across the enclosed faces. The interpolation functions are defined in the natural coordinate system of the interim element. The derivative of unknown variables is evaluated in the natural coordinate of the interim element and then mapped back to global coordinate system. The equilibrium equations of a cell are approximated at the integration points, which are located in the interim elements. In this approach stress continuity will be guaranteed on common faces of the adjacent cells, which is a prominent feature of the finite volume method. To incorporate the boundary conditions point cells are used to transfer the boundary conditions to the adjacent cells. To demonstrate the capability of the present method in the predictions of accurate results a thin plate is analyzed and the results are compared with the analytical predictions. Further studies of the present method show the formulation is capable to analyze thin and thick plates. It is noticeable that the formulation does not show shear locking problem in the thin plate analysis, which occurs in the Mindlin based finite element formulation for the thin plate analysis. This extended studies cannot be included in this paper regards to the paper length.

*Keywords:* Finite volume, Cell centred, Plate bending, Thick plate, Thin plate.

## 1. Plate formulation

The mid-plane of the plate is meshed to a number of elements that each of them is referred to as a control volume or a cell [1], figure 1. A control volume is bounded by arbitrary number of faces and its centre is considered as computational node. By applying Mindlin-Reissner assumptions in the small displacement plate bending theory, the displacement components of a point of coordinates $x, y, z$ are

$$u = z\beta_x(x, y) \qquad , \qquad v = z\beta_y(x, y) \qquad , \qquad w = w(x, y) \tag{1}$$

Where $w$ is the transverse displacement, $\beta_x$ and $\beta_y$ are the rotations of the outward normal of the un-deformed mid-plane ($z=0$ plane) in the $xz$ and $yz$ planes respectively, figure 2a. Regardless of in-plane forces, the governing equations of a portion of the plate, which lies in the $xy$ plane, can be written as the equilibrium relations of applied forces and stress resultant forces acting on the mid-plane of the plate as

$$\sum M_x = 0 \; , \quad \sum M_y = 0 \; , \quad \sum F_z = 0 \tag{2}$$

Where the first two equations express equilibrium of moments about $x$ and $y$ axes respectively and the last equation expresses equilibrium state in the $z$ direction. For a given control volume surrounded by $k$ faces and assuming a uniform distribution of moments and shear forces along the bounding faces, the equilibrium Eq. (2) can be discretized as

---

[1] Corresponding author. E-mail: fallah@guilan.ac.ir

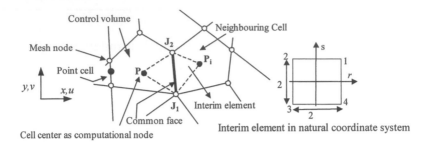

Figure 1: Control volume on a mesh of elements and an interim element on control volume's face

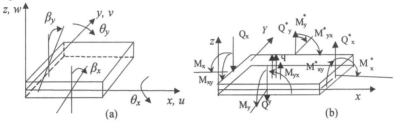

Figure 2: The positive convention of moments, shear forces and section rotations

$$\sum_{i=1}^{k}\left\{\begin{bmatrix} n_x & 0 & n_y \\ n_y & 0 & n_x \\ 0 & 0 & 0 \end{bmatrix}\begin{bmatrix} M_x \\ M_y \\ M_{xy} \end{bmatrix} - \begin{bmatrix} (y_m - y_c)n_x & (y_m - y_c)n_y \\ (x_m - x_c)n_x & (x_m - x_c)n_y \\ n_x & n_y \end{bmatrix}\begin{bmatrix} Q_x \\ Q_y \end{bmatrix}_i\right\} l_i = \begin{bmatrix} 0 \\ 0 \\ qA_c \end{bmatrix}_i \tag{3}$$

Which can be expressed more concisely as

$$\sum_{i=1}^{k} \mathbf{C}_i \mathbf{M}_i l_i - \sum_{i=1}^{k} \mathbf{C}_i^* \mathbf{Q}_i l_i = \mathbf{P}_c \tag{4}$$

Where illustrated sign convention in figure 2b was used and for a given face: $n_x$ and $n_y$ are cosine directions of outward normal of the face, $x_m$ and $y_m$ are the coordinates of midpoint of the face, $x_c$ and $y_c$ are the coordinates of cell centre (computational node), $q$ is the uniform distributed load applies upon the cell with the mid-surface area of $A_c$ and $l_i$ is the length of the face. All moments and shear forces are measured per unit length. By applying the Mindlin-Reissner plate theory and substituting linear constitutive equation into Eq. (4) we have

$$\sum_{i=1}^{k} \mathbf{C}_i \mathbf{D} \varepsilon_i^* l_i - \sum_{i=1}^{k} \mathbf{C}_i^* G^* \gamma_i l_i = \mathbf{P}_c \tag{5}$$

Where $G^*=kGt$, in which $t$ is the plate thickness, $G$ is the shear modulus and $k$ is the shear correction factor. Matrix $\mathbf{D}$ contains elastic coefficients. For a given face, $\varepsilon^*$ includes curvatures and $\gamma$ includes lateral shear strains as

$$\varepsilon^* = \left[\frac{\partial \beta_x}{\partial x} \quad \frac{\partial \beta_y}{\partial y} \quad \frac{\partial \beta_x}{\partial y} + \frac{\partial \beta_y}{\partial x}\right]^T , \quad \gamma = \begin{bmatrix} \gamma_{xz} & \gamma_{yz} \end{bmatrix}^T = \left[\frac{\partial w}{\partial x} + \beta_x \quad \frac{\partial w}{\partial y} + \beta_y\right]^T \tag{6}$$

For calculation of the curvatures and the lateral strains corresponding to a given face, a 4-noded interim element which enclosing the face is used. As shown in figure 1, the vertices of the interim element are the centres of the two adjacent cells lying on either side of the face and the nodes at each end of the face. The interim element is considered as an isoparametric element in which a bilinear distribution of displacement and rotations are assumed where

$$w = \sum_{j=1}^{4} N_j w_j , \quad \beta_x = \theta_y = \sum_{j=1}^{4} N_j \theta_y^j , \quad -\beta_y = \theta_x = \sum_{j=1}^{4} N_j \theta_x^j \tag{7}$$

Substituting Eq (6) into Eq. (7), the curvatures and the transverse shear strains can be expressed in terms of nodal displacements and rotations of interim element as

$$\boldsymbol{\varepsilon}^* = \mathbf{B}_b \tilde{\mathbf{u}} \quad , \quad \boldsymbol{\gamma} = \mathbf{B}_s \tilde{\mathbf{u}} \tag{8}$$

Where $\tilde{\mathbf{u}}$ includes the nodal displacements and rotations of interim element. Substituting Eq. (8) into Eq. (5) gives

$$(\sum_{i=1}^{k} \mathbf{C}_i \mathbf{DB}_{b}^{i} l_i - \sum_{i=1}^{k} \mathbf{C}_{i}^{*} G^* \mathbf{B}_{s}^{i} l_i)\bar{\mathbf{u}} = \mathbf{P}_c \tag{9}$$

Where $\bar{\mathbf{u}}$ includes the nodal unknowns corresponding to the interim elements all around the cell. In order to express Eq. (9) in terms of the unknown values just associated with the centres of given cell and all neighbouring cells, the unknowns corresponding to the corner nodes of the cell should be eliminated. It can be achieved by assuming a bilinear variation of unknowns across the region surrounding a given corner node. This assumption enables to express the corner nodal unknowns in terms of the unknowns at the centers of each of the cell that meet at the node. After the elimination of the unknowns corresponding to the all corner nodes, Eq. (9) can be expressed more concisely as

$$\mathbf{R}_c \mathbf{u} = \mathbf{P}_c \tag{10}$$

Where $\mathbf{u}$ includes the unknowns associated with the centre of the cell and the neighbouring cells. Eq. (10) represents the relation of unknowns at the centre of the given cell to those at the centres of all surrounding cells. To incorporate the boundary conditions a point cell is used to transmit the boundary conditions onto the adjacent internal cell. The point cell is considered on the boundary next to the internal cell and at the middle of the face lying on the boundary. If displacement boundary conditions are applied, the applied displacement, $\mathbf{u}'$, can be transformed to global coordinate system as $\mathbf{u}$ which provides three equations. In case of force boundary condition, the applied forces including moments and shear forces, $\mathbf{F}'$, can be transformed to global coordinate system as

$$\mathbf{T}^* \mathbf{F} = \mathbf{F}' \tag{11}$$

Where $\mathbf{T}^*$ represents the transformation matrix. Using constitutive equation in Eq. (11) and approximating the curvatures and strains on the point cell provides three equations. To approximate the curvatures and strains on point cell an isoparametric interim element was used, whose vertices are the center of internal cell, point cell and the nodes lying at either side of the point cell, figure 1. Each foregoing equation includes the unknowns corresponding to the end nodes of the boundary face, which must be eliminated. The elimination procedure is similar to that used for the internal cell. At the end of this procedure the equations that relate the unknowns of the point cell to those in the neighbouring cells and to the applied boundary forces will be obtained. These equations are in the form of Eq. (10). When the mixed boundary conditions are applied, a proper combination of equations mentioned above should be used. Equations associated with the internal cells and boundary conditions provide a system of simultaneous linear equations, which relates the unknowns to one and other and can be expressed in matrix form as

$$\mathbf{RX} = \mathbf{H} \tag{12}$$

Where $\mathbf{R}$ contains the coefficients relating unknowns associated with the cells. $\mathbf{X}$ is a vector includes the unknown variables and vector $\mathbf{H}$ represents the known values on the boundaries and the applied loads upon the plate. The matrix $\mathbf{R}$ has sparse nature and non-symmetric, however, Eq. (12) can be solved by any appropriate solver technique to yield the solution.

## 4. Numerical example

To demonstrate the accuracy of the predictions of the present formulation a simply supported square plate under the uniform load was analyzed. The geometry and material property of the plate is shown in figure 3. The hard boundary condition [1] was assumed where lateral displacement of boundaries and rotations about the outward normal to each of the plate boundaries were constrained. The plate was meshed to 16 divisions in $x$ direction and 15 divisions in $y$ direction. The divisions were so considered to place the plate center on the face of a cell, where the stresses were calculated. The plate was analyzed and the variation in bending moments and shear forces along the central axis of the plate are given in figure 4 and compared with the analytical solutions obtained from Reference [2]. As figure shows, the predicted results are in agreement with the analytical results. Extended studies, which are not presented here, show the capability of the present method for the analysis of plate with different geometries and different boundary conditions. These studies also show the method is shear locking free in the analysis of thin plates. It is well known that the shear locking problem occurs in the displacement based finite element formulation, using Mindlin assumptions, for the bending analysis of thin plates. The shear locking of the finite element formulation is not easy to resolve and needs specific treatments like reduced integration method [3-4], selective integration method [5] and mixed interpolation method [6].

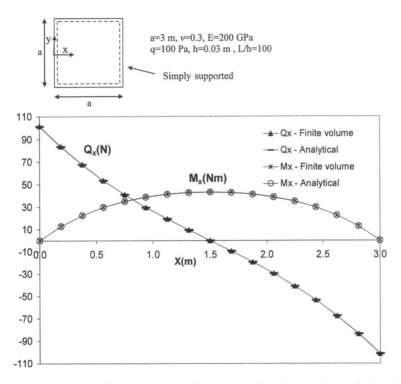

Figure 3: Variation of shear force and bending moment along the central axes of a simply supported plate under the uniform loading.

# References

[1] N. Fallah, Cell vertex and Cell centred finite volume methods for plate bending analysis, *Computational methods in applied Mechanics and Engineering,* **193** 3457-3470(2004).

[2] S. P. Timoshenko and S. Woinowsky-Krieger, *Theory of plates and shells*, McGraw Hill, Singapore, 1970.

[3] O. C. Zienkiewicz and R. L. Taylor, *The Finite Element Method,* **2,** 4th edition, McGraw-Hill.

[4] E. Hinton and D. R. J. Owen, *Finite element software for plate and shells*, Pineridge Press Limited,1984.

[5] T. j. R. Hughes, M. cohen and M. Haroun, Reduced and selective integration    techniques in the finite element analysis of plates, *Nuclear engineering and design,* **46** 203-222(1978).

[6] K. J. Bathe and E. N. Dvorkin, Short communication, A four-node plate bending element based on Mindlin-Reissner plate theory and a mixed interpolation, *Int. J. Numer. Methods Engrg,* **21** 367-383(1985).

Brill Academic Publishers
P.O. Box 9000, 2300 PA Leiden
The Netherlands

*Lecture Series on Computer*
*and Computational Sciences*
Volume 4, 2005, pp. 191-194

# Performance of Different DFT/ECP Combinations in the Study of Platinum Anticancer Drugs

Sónia Fiuza, Ana M. Amado, Luís A.E. Batista de Carvalho and M. Paula M. Marques[1]

Química-Física Molecular,
Universidade de Coimbra,
3004-535 Coimbra, Portugal

Received 1 July, 2005; accepted in revised form 11 July, 2005

*Abstract:* In the present study the performance of several combinations of *ab initio* molecular orbital methods was evaluated, aiming at a thorough conformational analysis of the well-known anticancer agent cisplatin (*cis*-diamminedichloroplatinum (II), cDDP). Different DFT/ECP approaches were tested, coupling the relativistic ECP's CEP-4G/CEP-31G and LANL2MB/LANL2DZ to the DFT approximations B3LYP and mPW1PW. From an energetic point of view, the results obtained were compared with the ones yielded by the AE basis sets 6-31G* and 6-311G** (at both the HF and MP2 levels). The type of ECP considered and inclusion of an f-polarisation function at the Pt valence shell were found to be determinant for an accurate reproduction of the AE results for this Pt(II) chelate, while the effect of the DFT protocol used proved to be of little significance.

*Keywords*: cisplatin, *ab initio* calculations, conformational analysis, density functional theory, effective core potentials

## 1. Introduction

Since its discovery in the late 1960's [1], cisplatin (*cis*-diamminedichloroplatinum (II), cDDP) has been among the most used drugs in cancer chemotherapy. It displays a significant activity against several types of neoplastic cells, with an exceptional efficacy against testicular cancers. Despite this remarkable biological efficacy, cDDP is also responsible for severe toxic side effects and development of resistance. Therefore, numerous studies have been carried out aiming at the design of novel anticancer agents, coupling a lower toxicity to a higher efficacy and tissue specificity. Thousands of new metal drugs have lately been synthesised and tested, including mono- and polynuclear Pt(II) chelates [2-5]. The knowledge of the conformational preferences of this kind of platinum compounds is essential for the understanding of the structure-activity relationships (SAR's) ruling their antitumour activity.

Thanks to the continuously increasing number of new theoretical methods, quantum chemistry has become an essential tool in inorganic chemistry, namely for the prediction and understanding of the structural behaviour and vibrational spectra of coordination compounds. However, when transition metal complexes are concerned, the huge number of electrons from the metal ion render most of the commonly used theoretical levels, such as the Møller-Plesset (MP) correlation energy approximation and some all-electron (AE) basis sets, prohibitively computer-demanding. This is enhanced by the complexity of the system, namely in multinuclear metal chelates. Thus, it is particularly important to perform an accurate and systematic evaluation of different theoretical approximations, which will hopefully enable to select the *"best method"* for the mimetisation of this kind of complex systems. Some of these approximations may be combinations between density functionals (DFT's, accounting for the electron correlation effects) and effective core potentials (ECP's, representing the inert metal core electrons).

---

[1] Corresponding author: E-mail: pmc@ci.uc.pt

The present report describes an *ab initio* conformational analysis performed for cDDP (in the Gaussian 98W program [6]), using different standard relativistic ECP's combined with two distinct DFT protocols. The ECP's tested were the ones of Stevens [7] (CEP-4G and CEP-31G) and of Hay and Wadt [8] (LANL2MB and LANL2DZ), with and without polarisation functions at the heavy atoms. The DFT's used correspond to the G98W keywords B3LYP [9] and mPW1PW [10]. The predicted number of conformational minima and their relative energies were compared with the results obtained with two widely used AE basis sets – 6-31G** [11] and 6-311G** [12] – using both the HF (Hartree-Foch) [13] and MP2 [14] methodologies.

## 2. Results and Discussion

Figure 1 comprises the three geometries obtained for cDDP after full optimisation, for all theory levels used. Concerning the DFT calculations, they were found to lead to only one real minimum in the potential energy surface, cDDP1, regardless of the AE basis set or the type of ECP (presence or absence of polarisation functions at the Pt and/or the N and Cl atoms). Geometries cDDP2 and cDDP3 yield negative eigenvalues in the harmonic frequency calculations, thus corresponding to first- and second order saddle-points, respectively. The results obtained with the HF and MP2 methods, in turn, show to be strongly dependent on both the AE basis set and the ECP's: calculations with 6-31G** lead to only one real minimum (cDDP1, as for the DFT calculations), while those with 6-311G** predict three close-lying minima (cDDP1, cDDP2 and cDDP3), cDDP1 being the most stable one if zero-point vibrational energy (zpve) correction is considered. As to the ECP calculations, LANL2MB and LANL2DZ predict a single minimum (cDDP1), while CEP-4G and CEP-31G lead to two distinct situations: the minimal splitting scheme points to two conformers (cDDP1 and cDDP2), while splitting of the valence shell yields three close-lying minima similarly to those calculated with the AE 6-311G** basis set (again, cDDP1 only becomes the most stable one when zpve correction is performed).

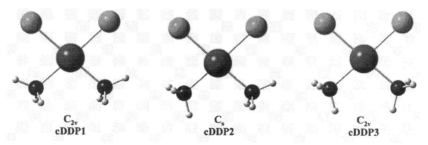

$C_{2v}$
cDDP1                    $C_s$
cDDP2                    $C_{2v}$
cDDP3

Figure 1: Schematic representation of the geometries yielded by full geometry optimisation of cDDP. (The symmetry point group is included).

Figure 2 is a graphical representation of the relative energies between optimised geometries, for different AE theory levels, with and without a polarisation function at the platinum valence shell (both minima and saddle-points are included). The corresponding plot for the ECP calculations, using the two DFT protocols B3LYP and mPW1PW, is presented in Figure 3.

From the AE results (Fig. 2) it was possible to conclude that, regardless of the basis set considered, the DFT calculations yield larger relative energy values than the HF and MP2 methods, the former giving rise to the lowest values.

The HF and MP2 results displayed a very slight effect of the f-polarisation function at the Pt atom, which is almost negligible for HF. The splitting of the of the N and Cl valence shells (6-31G** → 6-311G**), in turn, led to a decrease of the relative stability energy range. However, the magnitude of this effect was found to depend on the presence or absence of the polarisation function at the metal atom, the largest decrease having been detected for MP2 when no polarisation was considered. The smallest effects due to this metal polarisation, in turn, were observed for the HF method.

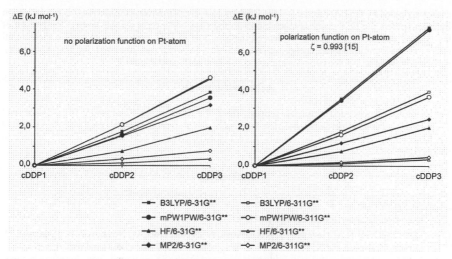

Figure 2: Graphical plots of the relative energies *vs* optimised geometries for cDDP, yielded by full optimisation at different AE theoretical levels, with (right) and without (left) an f-polarization function at the Pt atom.

When using the DFT methodology without an f-polarization function at the metal, the splitting of the valence shell of the N and Cl atoms is responsible for a stability increase of cDDP1 relative to the other two conformations. The opposite effect was verified when the Pt valence shell is augmented with an f-polarisation function. Actually, when this function is added, the calculations predicted a slight decrease of the conformational relative energies at the 6-311G** level, and a significant stabilisation of cDDP1 relative to cDDP2 and cDDP3 when the 6-31G** basis set is used. Finally, the type of DFT approach considered did not prove to have a significant effect on the conformational energy values.

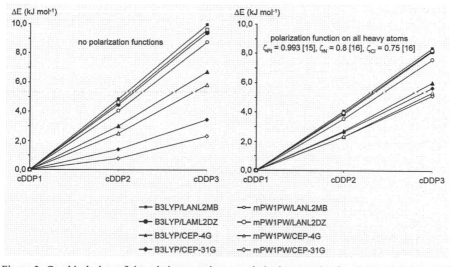

Figure 3: Graphical plots of the relative energies *vs* optimised geometries for cDDP, yielded by full optimisation at different ECP theoretical levels, using the two DFT protocols B3LYP and mPW1PW, with (right) and without (left) an f-polarization function at the Pt atom.

Concerning the ECP results (Fig. 3), it may be concluded that the ECP's of Hay and Wadt (LANL2MB and LANL2DZ) yield higher energy differences between conformers than the ones of Stevens (CEP-4G and CEP-31G), the largest values being obtained for the minimal splitting schemes. Inclusion of an f-polarisation function at the metal valence shell was found to reduce the gap between the relative energies yielded by the LANL-type and the CEP-type ECP's. Moreover, considering this f-polarization function renders the ECP energies much closer to the ones obtained at the AE levels. Comparison of the B3LYP and mPW1PW results evidenced that the type of DFT protocol used had no meaningful effect on the conformational energy values, similarly to what was observed for the AE results, particularly when an f-polarisation function is considered at the metal atom.

## Acknowledgments

The authors acknowledge financial support from the Portuguese Foundation for Science and Technology – *Unidade de Química-Física Molecular and Research Project POCTI/47256/QUI/2002* (co-financed by the european community fund FEDER). SF also acknowledges a PhD fellowship SFRH/BD/17493/2004.

## References

[1]    B. Rosenberg, L. Van Camp, J.E. Trosko and V.H. Mansour, *Nature* **222** 385-386 (1969).

[2]    N. Farrell, *Cancer Invest.* **11**, 578-589 (1993).

[3]    T.D. McGregor, A. Hegmans, J. Kaspárková, K. Neplechová, O. Nováková, H. Penazová, O. Vrána, V. Brabec and N. Farrell, *J. Biol. Inorg. Chem.* **7**, 397-404 (2002) and refs. therein.

[4]    M.P.M. Marques, M.T. Girão da Cruz, M.C. Pedroso de Lima, A.. Gameiro, E. Pereira and P. Garcia, *Biochim. Biophys. Acta (MCR)* **1589**, 63-70 (2002).

[5]    L.J. Teixeira, M. Seabra, E. Reis, M.T. Girão da Cruz, M.C. Pedroso de Lima, E. Pereira, M.A. Miranda and M.P.M. Marques, *J. Med. Chem.*, **47**, 2917-2925 (2004).

[6]    **Gaussian 98, Revision A.9**, M. J. Frisch, G. W. Trucks, H. B. Schlegel, G. E. Scuseria, M. A. Robb, J. R. Cheeseman, V. G. Zakrzewski, J. A. Montgomery, Jr., R. E. Stratmann, J. C. Burant, S. Dapprich, J. M. Millam, A. D. Daniels, K. N. Kudin, M. C. Strain, O. Farkas, J. Tomasi, V. Barone, M. Cossi, R. Cammi, B. Mennucci, C. Pomelli, C. Adamo, S. Clifford, J. Ochterski, G. A. Petersson, P. Y. Ayala, Q. Cui, K. Morokuma, D. K. Malick, A. D. Rabuck, K. Raghavachari, J. B. Foresman, J. Cioslowski, J. V. Ortiz, A. G. Baboul, B. B. Stefanov, G. Liu, A. Liashenko, P. Piskorz, I. Komaromi, R. Gomperts, R. L. Martin, D. J. Fox, T. Keith, M. A. Al-Laham, C. Y. Peng, A. Nanayakkara, M. Challacombe, P. M. W. Gill, B. Johnson, W. Chen, M. W. Wong, J. L. Andres, C. Gonzalez, M. Head-Gordon, E. S. Replogle and J. A. Pople, Gaussian, Inc., Pittsburgh PA, 1998.

[7]    W. Stevens, H. Basch and J. Krauss, *J. Chem. Phys.* **81**, 6026 (1984); W. J. Stevens, M. Krauss, H. Basch and P. G. Jasien, *Can. J. Chem.* **70**, 612 (1992); T. R. Cundari and W. J. Stevens, *J. Chem. Phys.* **98**, 5555 (1993).

[8]    P. J. Hay and W. R. Wadt, *J. Chem. Phys.* **82**, 270 (1985); W. R. Wadt and P. J. Hay, *J. Chem. Phys.* **82**, 284 (1985); P. J. Hay and W. R. Wadt, *J. Chem. Phys.* **82**, 299 (1985).

[9]    A. D. Becke, *J. Chem. Phys.* **98**, 5648 (1993).

[10]   C. Adamo and V. Barone, *J. Chem. Phys.* **108**, 664 (1998).

[11]   W. J. Hehre, R. Ditchfield and J. A. Pople, *J. Chem. Phys.* **56**, 2257 (1972); P. C. Hariharan and J. A. Pople, *Mol. Phys.* **27**, 209 (1974); M. S. Gordon, *Chem. Phys. Lett.* 76, 163 (1980); P. C. Hariharan and J. A. Pople, *Theo. Chim. Acta.* **28**, 213 (1973).

[12]   M. J. S. Dewar and C. H. Reynolds, *J. Comp. Chem.* **2**, 140 (1986); K. Raghavachari, J. A. Pople, E. S. Replogle and M. Head-Gordon, *J. Phys. Chem.* **94**, 5579 (1990); A. D. McLean and G. S. Chandler, *J. Chem. Phys.* **72**, 5639 (1980); R. Krishnan, J. S. Binkley, R. Seeger and J. A. Pople, *J. Chem. Phys.* **72**, 650 (1980).

[13]   J. A. Pople and R. K. Nesbet, *J. Chem. Phys.* **22**, 571 (1954); R. McWeeny and G. Dierksen, *J. Chem. Phys.* **49**, 4852 (1968).

[14]   M. J. Frisch, M. Head-Gordon and J. A. Pople, Chem. *Phys. Lett.* **166**, 275 (1990); M. J. Frisch, M. Head-Gordon and J. A. Pople, *Chem. Phys. Lett.* **166**, 281 (1990); M. Head-Gordon and T. Head-Gordon, *Chem. Phys. Lett.* **220**, 122 (1994).

[15]   A.W. Ehlers, M. Böhme, S. Dapprich, A.Gobbi, A. Höllwarth, V. Jonas, K.L. Köhler, R. Stegmann, A. Veldkamp and G. Frenking, *Chem. Phys. Lett.*, **208**, 111-114 (1993).

[16]   W.J. Heyre, L. Radom, P.V.R. Schleyer and J.A. Pople, *Ab Initio Molecular Orbital Theory*, Wiley & Sons, New York, 1986.

Brill Academic Publishers
P.O. Box 9000, 2300 PA Leiden,
The Netherlands

Lecture Series on Computer
and Computational Sciences
Volume 4, 2005, pp. 195-197

# The Sound of a Different Bell: Density Functional Theory as a Tool for the Study of Classical Liquids

M. Franchin and E. Smargiassi[1]

Department of Theoretical Physics,
University of Trieste,
Strada Costiera 11,
34014 Trieste, Italy

Received 10 July, 2005; accepted in revised form 3 August, 2005

Density-functional methods have become the cornerstone of state-of-the-art ab-initio calculations of materials properties, and recently a vast interest has arisen about the possibility of exploiting the full power of the Hohenberg-Kohn theorem by avoiding the use of wavefunctions altogether.

A similar approach (with the energy replaced by the Helmholtz free-energy) can be used to study complex classical fluids at finite temperature and in particular highly-asymmetric, charged fluids[1]. In these systems, in fact, we are faced with a similar problem to the one we face while trying to compute the properties of a system composed by ions and electrons; namely, we have two species of particles, the "big" ones — macromolecules (macroions) instead of ions — and the "small" ones — ions (microions) instead of electrons. Again, in analogy with the quantum case, our aim is to exploit the fact that the small particles move typically much faster than the big ones by assuming that the former follow adiabatically the latter. Thus, we may expect that the techniques needed for the solution of the classical problem can be of help for the solution of the quantum problem and vice versa.

Of course, there is normally no need to treat the microions quantum mechanically and in fact they are always treated as classical particles using classical DFT[2]. This makes the problem easier than in the quantum case, because there are no problems in finding the "ideal" part of the free energy functional (the kinetic energy part in the QM case): in fact it becomes simply the free-energy functional for an ideal gas:

$$F_0[\rho] = k_B T \int d^3r\, \rho(r) \left[ \ln\left(\Lambda^3 \rho(r)\right) - 1 \right] \qquad (1)$$

where $\Lambda$ is an arbitrary length, often taken to be equal to the De Broglie length of the microions. The full free-energy functional is the sum of this term and of the interaction terms between all the particles:

$$F[\rho] \quad = \quad F_0[\rho] + F_{mm}[\rho] + F_{Mm}[\rho] + F_{MM} \qquad (2)$$

$$F_{mm}[\rho] \quad = \quad \frac{1}{2} \int_\Omega d^3r' d^3r'' \rho(r')\rho(r'') V_{int}(r' - r'') \qquad (3)$$

$$F_{Mm}[\rho] \quad = \quad \int_\Omega d^3r' \rho(r') V_{ext}(r') \qquad (4)$$

---

[1]Corresponding author. E-mail: smargiassi@ts.infn.it

where the subscript $m$ indicates microions and $M$ macroions, $F_{MM}$ is the direct macro-macro interaction, $V_{int}$ the coulombic interaction between microions and $V_{ext}$ the interaction between one macroion and one microion. (The solvent is ignored except in that it modifies the static dielectric constant of the medium). This so-called the Poisson-Boltzmann model is formally equivalent to the Hartree approximation in the quantum case, but fortunately in the classical case it works very satisfactorily, and it is in fact the standard approach to the study of many systems.

This approach, which can be easily generalized to the case in which there are two species of microions: "coions", having the same charge of the macroions, and "counterions", having opposite charge, has already been used to treat a few system, e.g. spherical micelles[3]. We aim to study another system of theoretical and practical importance, the synthetic clay known as Laponite. Natural clays are formed by rigid, charged platelets of typical sizes of 100-300 Å in diameter and very thin (about 1 Å thick); Laponite, being artificial, has the big advantage of being monodisperse (i.e., all the platelets are almost identical) and thus easier to study.

We have studied, as a preliminary step, the case of one and two platelets surrounded by a sea of counterions or of both counter- and coions. The first problem is to find a suitable and fast algorithm to minimize the free-energy of the system of microions at fixed macroions configuration. We tested the simple steepest descent method and, not surprisingly, found it wanting. Better results are obtained using a conjugate-gradient algorithm. Even in this case we found that troubles arise due to the fact that the coions are strongly repelled by the macroions and thus their density becomes very small at the macroions location. It is therefore easy to obtain an unphysically negative density, for which the free-energy functional is not even defined.

Fortunately, a simple trick allows one to get round the difficulty. If one assumes that the macro-counterion interaction is exactly the opposite of the macro-coion interaction — indeed, this is exactly the case if the interaction is purely Coulombic — it is easy to show that the product of the two densities of microions is a constant. Hence, whenever the density of coions becomes negative it can be simply replaced by the reciprocal of the density of counterions, which is always positive.

The speed of convergence however is still not high enough. We tried a few different preconditioning functions in order to speed up convergence, but it it difficult to go much beyond the standard uniform-gas forms. These are of dubious utility since the counterions condensate on close proximity of the platelets, having opposite charge, and this creates a sharp peak in their density; it is obvious that treating this case as similar to an uniform-gas case cannot be of much help.

Another problem is that this peak takes quite a long time to converge. Moreover, a reasonable description of this peak requires a large number of basis functions (we use plane waves), with all the problems it entails. We have thus explored the possibility of "pseudising" the platelet by adding to it a suitable "frozen" distribution of counterions that mimics the peak as accurately as possible. This leaves one with a smoother density of counterions as a dynamical variable, again in analogy with the QM case where the only electrons explicitly treated in the pseudopotential approximation are the valence electrons and not the core electrons.

The results of this procedure so far look quite encouraging; we can show that for typical cases, especially at moderate-to-high concentrations of macroions, the standard linear-response (Debye-Hückel) theory is in its original form is utterly inadequate to describe the behaviour of laponite. It is, however, possible to improve it substantially by a suitable adjustment of some parameters.

Our next step will be to try and perform a Montecarlo simulation on an ensemble of platelets in various points of the phase diagram.

## Acknowledgment

The authors wish to thank Prof. M. Dijkstra and G. Pastore for very helpful discussions and suggestions.

## References

[1] J.-P. Hansen and E. Smargiassi,in *Conference Proceedings— vol 49*, edited by K. Binder and G. Ciccotti (SIF, Bologna, 1996).

[2] R. Evans, *Adv. in Physics* **28**, 143 (1979).

[3] H. Löwen, J.-P. Hansen and P.A. Madden, *J. Chem. Phys.* **98** 3275 (1993).

Brill Academic Publishers
P.O. Box 9000, 2300 PA Leiden
The Netherlands

*Lecture Series on Computer
and Computational Sciences*
Volume 4, 2005, pp. 198-203

# An Advanced Numerical Method for Recovering Image Velocity Vectors Field

E. Francomano[1], C. Macaluso, A. Tortorici, E. Toscano

Dipartimento di Ingegneria Informatica
Facoltà di Ingegneria
Università degli Studi di Palermo
Viale delle Scienze, I-90128, Palermo, Italia

Received 8 July, 2005; accepted in revised form 30 July, 2005

*Abstract:* This paper addresses an innovative method devoted to estimate the velocity vectors field of the apparent motion of brightness patterns. This task plays a fundamental role in many real-time computer vision applications. In this paper the flow field is performed by adopting a bivariate quasi-interpolant operator based on centered cardinal *B*-spline functions onto a non linear minimizing algorithm. The solving model involves massive computational amount so that high-performance computing need to rapidly proceed in the computations. Therefore an analysis of the process has been carried out on distributed multiprocessors platforms. The process has shown to be synchronous, with good tasks balancing and requiring few amount of data transfer.

*Keywords:* Image velocity vectors field, *B*-spline functions, Quasi-interpolant operator, Distributed multiprocessors.

*Mathematics SubjectClassification:* 47A58, 65Y05, 65Y20, 65D05, 65D07 .

*PACS:* 07.05.Pj.

## 1. Introduction

The most common approach for the analysis of visual motion is based on the computation of the velocity field composed of vectors describing the instantaneous velocity of image elements. This computation is currently motivated by many applications in computer vision and it is an important prerequisite when visual tasks, such those involving a robot autonomous locomotion and its physical interaction with the environment, are required [1]-[5]. In this paper a non linear minimizing technique [6] involving a bivariate quasi-interpolant operator has been used to perform velocity vectors field estimates. The quasi-interpolant operator adopted composed with a timely scaling process gives rise to a multiresolution coarse-to-fine process [7], [8]. Improvements in the solution can be performed by increasing the granularity. Furthermore, at each level of the multigrid process the non linear minimizing technique involves iterations to better generate velocity values at the current grid and more computations than those implicated at the previous level are required. Moreover, the mainly processing tasks, involved in the model, require regular and repetitive estimations computationally expensive. As a consequence it is advisable to proceed in the computation by employing highly performing architectures to rapidly obtain accurate results. In this paper notes on data distribution, communications requirement and computational demanding of the solving model are provided.

---

[1] Corresponding author. E-mail: e.francomano@unipa.it

## 2. Quasi-Interpolant Operator

In this paper a quasi-interpolant operator:

$$(Qf)(x) = \sum_{l \in Z} (\Lambda f)(l) N^d \left(x + \frac{d}{2} - l\right) \tag{1}$$

has been taken into account involving the convolution operator [7], [8]:

$$(\Lambda f)(l) = \sum_{j \in Z} \lambda_{l-j} f_j, \quad l \in Z \tag{2}$$

and the $d^{th}$ order cardinal $B$-spline function $N^d(x)$ [9] with support $(0,d)$. The (1) is a bounded linear and local operator which supplies simple and computationally efficient schemes for constructing cardinal spline approximations. Moreover, a multiresolution process can be carried out: by considering the data sequence in the spline space $V_0^d$ a data sampling in the spline space $V_s^d$ can be provided by fixing the sampling step as small as $h = 2^{-s}$.

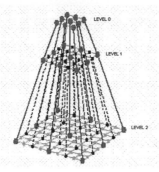

Figure 1: Three levels of the coarse-to-fine process.

By considering $d = 4$ and $s = 1$ the bivariate formula can be written as:

$$
\begin{aligned}
(Qf)(x,y) = \sum_{l \in Z} \sum_{k \in Z} &\left\{ \left[ \frac{1}{36} f(l+1,k+1) - \frac{2}{9} f(l+1,k) + \frac{1}{36} f(l+1,k-1) - \frac{2}{9} f(l,k+1) + \right.\right. \\
&\left. + \frac{16}{9} f(l,k) - \frac{2}{9} f(l,k-1) + \frac{1}{36} f(l-1,k+1) - \frac{2}{9} f(l-1,k) + \frac{1}{36} f(l-1,k-1) \right] \\
&\left. N^4(x+2-l) N^4(y+2-k) \right\}.
\end{aligned}
\tag{3}
$$

By fixing $x = q$ and $y = r$, $q,r \in Z$, the only terms in the summations are related to the indexes $l$ and $k$ verifying the following conditions: $q-1 \le l \le q+1$ and $r-1 \le k \le r+1$. The formula (3) can be rewritten requiring only data information in even points giving the evaluation in odd points and implicitly performing one resolution level that is:

$$
\begin{aligned}
(Qf)(x,y) = \sum_{j \in Z} \sum_{k \in Z} &\left[ \sum_i \sum_j c_{ij} f(2i,2j) N^4(x-2l+2) N^4(y-2k+2) + d_{ij} f(2i+1,2j+1) \right. \\
&N^4(x-2l+3) N^4(y-2k+3) + g_{ij} f(2i,2j+1) N^4(x-2l+2) N^4(y-2k+3) + \\
&\left. + h_{ij} f(2i+1,2j) N^4(x-2l+3) N^4(y-2k+2) \right]
\end{aligned}
\tag{4}
$$

where

| | |
|---|---|
| $c_{-11} = c_{11} = c_{-1-1} = c_{1-1} = \frac{1}{576}$<br><br>$c_{01} = c_{10} = c_{-10} = c_{0-1} = -\frac{13}{288}$<br><br>$c_{00} = \frac{169}{144}$<br><br>$c_{ij} = 0$    otherwise | $d_{-21} = d_{11} = d_{-2-2} = d_{1-2} = \frac{1}{144}$<br><br>$d_{-11} = d_{01} = d_{10} = d_{1-1} = d_{0-2} = d_{-1-2} = d_{-2-1} = d_{20} = -\frac{7}{144}$<br><br>$d_{00} = d_{0-1} = d_{-1-1} = d_{-10} = \frac{49}{144}$<br><br>$d_{ij} = 0$    otherwise |
| $g_{1-1} = g_{11} = g_{-21} = g_{-2-1} = \frac{1}{288}$<br><br>$g_{0-1} = g_{01} = g_{-1-1} = g_{-11} = -\frac{13}{144}$<br><br>$g_{10} = g_{-20} = -\frac{7}{288}$<br><br>$g_{00} = g_{-10} = \frac{91}{144}$<br><br>$g_{ij} = 0$    otherwise | $h_{1-2} = h_{11} = h_{-1-2} = h_{-11} = \frac{1}{288}$<br><br>$h_{1-1} = h_{10} = h_{-1-1} = h_{-10} = -\frac{13}{144}$<br><br>$h_{0-2} = h_{01} = -\frac{7}{288}$<br><br>$h_{0-1} = h_{00} = \frac{91}{144}$<br><br>$h_{ij} = 0$    otherwise |

## 3.   Image Velocity Vectors Field Estimator

Given a sequence of images, the function $I_t(x,y)$ is the image brightness over time computed at pixel $(x,y)$. By considering the displacement vector $(u_t(x,y), v_t(x,y))$ resulting from the apparent motion of brightness patterns in the images, the aim is to recover, at each pixel, the components $u_t$ and $v_t$. A solution to this problem consists in minimizing a Sum of Squared Differences (SSD) involving $I_t(x,y)$ [10], [11]. By taking into account a sequence composed by only two frames the SSD relation can be expressed as:

$$E = \sum_i \sum_j \left[ I_1(x_i + u_{ij}, y_j + v_{ij}) - I_0(x_i, y_j) \right]^2 \tag{5}$$

where $u_{ij} = u(x_i, y_j)$ and $v_{ij} = v(x_i, y_j)$ are computed in all image pixels. The idea is to represent the displacement field by using a local approximation model such as a quasi-interpolant based on the operator $Qf$ previously described.

Let $\{\hat{u}_{mn}, \hat{v}_{mn}\}$ a set of initial values fixed in some image pixels. The (5) must be minimized with respect to $\hat{u}_{mn}$ and $\hat{v}_{mn}$ and the modified Levenberg-Marquardt iterative non linear minimization technique can be adopted [6] in order to correct the initial values $\{\hat{u}_{mn}, \hat{v}_{mn}\}$.

By indicating with:

$$e_{ij} = I_1(x_i + u_{ij}, y_j + v_{ij}) - I_0(x_i, y_j) \tag{6}$$

where $u_{ij}$ and $v_{ij}$ are the values computed with the expression (4), and with:

$$(G_{ij}^x, G_{ij}^y) = \nabla I_1(x_i + u_{ij}, y_j + v_{ij}) \tag{7}$$

the intensity gradient at $(x_i, y_j)$, the gradient of $E$ can be computed as:

$$\gamma_{mn}^u = \frac{\partial E}{\partial \hat{u}_{mn}} = 2 \sum_i \sum_j e_{ij} G_{ij}^x R_{ij}^{(mn)}, \quad \gamma_{mn}^v = \frac{\partial E}{\partial \hat{v}_{mn}} = 2 \sum_i \sum_j e_{ij} G_{ij}^y R_{ij}^{(mn)} \tag{8}$$

where

$$R_{ij}^{(mn)} = \sum_l \sum_k c_{\tilde{m}\tilde{n}} N^4(x_i - 2l + 2)N^4(y_j - 2k + 2) + d_{\tilde{m}\tilde{n}} N^4(x_i - 2l + 3)N^4(y_j - 2k + 3) +$$
$$+ g_{\tilde{m}\tilde{n}} N^4(x_i - 2l + 2)N^4(y_j - 2k + 3) + h_{\tilde{m}\tilde{n}} N^4(x_i - 2l + 3)N^4(y_j - 2k + 2) \tag{9}$$

where $\tilde{m}$ and $\tilde{n}$ depend on $m, n, l, k$ and $-2 \le \tilde{m}, \tilde{n} \le 2$.

The computational process generates an Hessian block matrix $H$ which entries are:

$$h_{(mn)(pq)}^{uu} = \frac{\partial^2 E}{\partial \hat{u}_{mn} \partial \hat{u}_{pq}} = 2 \sum_i \sum_j (G_{ij}^x)^2 R_{ij}^{(mn)} R_{ij}^{(pq)} \tag{10}$$

$$h_{(mn)(pq)}^{uv} = \frac{\partial^2 E}{\partial \hat{u}_{mn} \partial \hat{v}_{pq}} = 2 \sum_i \sum_j G_{ij}^x G_{ij}^y R_{ij}^{(mn)} R_{ij}^{(pq)} \tag{11}$$

$$h_{(mn)(pq)}^{vv} = \frac{\partial^2 E}{\partial \hat{v}_{mn} \partial \hat{v}_{pq}} = 2 \sum_i \sum_j (G_{ij}^y)^2 R_{ij}^{(mn)} R_{ij}^{(pq)} \tag{12}$$

The matrix $H$ is used to compute an update flow vector by setting the system: $(H + \lambda D)\Delta\varepsilon = -\gamma$, where $\lambda$ is a stabilization term [6], $D$=diag($H$), $\Delta\varepsilon$ is an increment of the current displacement estimates $(\hat{u}_{mn}, \hat{v}_{mn})$ and $\gamma$ is the block gradient which entries are in (8).

At each step of the Levenberg-Marquardt algorithm a sparse multibanded linear system must be solved [12].

## 4. Notes On Parallel Features

A parallel algorithm is penalized from the communication overhead mainly due to the data distribution, data exchange and collection of local results. Therefore, it is important to determine the best data partitioning by considering that the solving scheme involves a local operator requiring information from neighbour values for each application point. When a processor needs to compute operations on boundary points, the local operator will also requires values relative to points belonging to adjacent processors. Consequently, a communication overhead depending on the length of the boundaries will occur besides the initial data distribution. Two data partitions, denoted as *block* and *strip* partition, are considered by supposing to work with a multiprocessor system equipped with *NP* processors.

### 4.1 Data distribution

*a) Block partition*

The data are divided into blocks amongst the *NP* processors arranged along a mesh of size $P_x \times P_y$. Let $a \cdot b$ the number of the image pixels and $a_p \cdot b_p$ those belonging to a generic processor where $a_p = \text{int}_{\text{sup}}(a/P_x)$, $b_p = \text{int}_{\text{sup}}(b/P_y)$. When local operators are applied during the process, each processor is involved in two data transfer with each of neighbouring processors. The amount of data exchanged by each processor is $4w(a_p + b_p)$ at most where $w$ is the width of set values required by the quasi-interpolant operator in use. Therefore, the total amount of data to be transferred is $4w[(a_p + b_p)NP - (a + b)]$.

*b) Strip partition*

The data are divided into strips among *NP* processors. In this case, each processor is involved at most in 4 data transfer. Hence, the total data transfer is $4(NP - 1)$ and by indicating with $\beta$ the lowest value between $a$ and $b$, the total amount of data to be exchanged is $4\beta w(NP - 1)$.

### 4.2 Computations and communications

The input data $I_0$ and $I_1$ are involved in a different manner in the solving scheme. $I_0$ is not modified during the overall process and it can be partitioned at the beginning of the computation among all the available processors. The values of the image brightness $I_1$ are interested in numerical derivatives computed with difference schemes and must be updated. Hence, a suitable data exchange must be performed for $I_1$ between adjacent processors. Let $m^{(0)} \cdot n^{(0)}$ be the initial velocity guesses number on each processor. In the following the analysis is performed by considering the strip data partition along the horizontal direction with $m^{(0)}$ guesses.

The velocity components are partitioned among the processors as $I_1$ but with a larger amount of data to be exchanged. In fact, for each value $u_{ij}$ to be computed, a processor uses the values in the set $\left\{ u_{l_i l_j} \middle| i-1 \le l_i \le i+2, j-1 \le l_j \le j+2 \right\}$ . Analogous considerations are extended to the $v_{ij}$ computations. For each communication the total amount of transferred data for computing $u_{ij}^{(k)}$ and $v_{ij}^{(k)}$, at a generic level $k$ is upper bounded by $2^{(k+3)} m^{(0)}$. The quasi-interpolant operator is applied a number of times equal to $2^{(k+3)} [2^{(k+1)} 3(m^{(0)} - 1)(n^{(0)} - 1) + m^{(0)} + n^{(0)} - 2]$.

Regarding the numerical derivatives $I_1^x$ and $I_1^y$, data transfer are required only for the $I_1^y$, i.e. each processor performs at most $4m^{(0)}$ exchanges. Then at the level $k$, each data transfer involves $2^{(k+1)} (m^{(0)} - 1) + 1$ values and the total number of derivatives which must be computed is $2^{(k+3)} (m^{(0)} \cdot n^{(0)} + 1) + 2$ . Similar operative modalities must be performed for computing the derivatives $G_{ij}^x$ and $G_{ij}^y$ subsequently involved in the computational scheme. The values obtained up to now allow to build a linear system. Each processor can compute the known vector $\gamma$ and the matrix entries regarding velocity values belonging to itself. To this end, it is necessary to exchange data regarding the $G_{ij}^x$, $G_{ij}^y$ and $e_{ij}$ evaluated at the boundary points. The amount of data involved for each data transfer is the same as the one required by the quasi-interpolant operator. Each processor can compute the values of the formula (9) without any data transfer. Therefore, the known vector and the Hessian matrix are already distributed among all processors. The Hessian matrix is a sparse multibanded matrix [12], [13] and by applying a suitable $\lambda$ parameter [6], an iterative solver can be successfully employed. In order to compute the matrix-vector products which are involved at each time step of the iterative solver, only few values of the updated vector must be exchanged between adjacent processors.

As a conclusion, the computational process is suitable to run on distributed multiprocessors platforms resulting synchronous, with a good tasks balancing and requiring an exiguous amount of data transfer.

## Acknowledgments

This work has been supported by the Italian *Ministero dell'Istruzione, dell'Università e della Ricerca* (MIUR).

## References

[1] G. Avid, Inherent ambiguities in recovering 3-D motion and structure from noisy flow field, *IEEE Transactions on Pattern Analysis and Machine Intelligence* **11** 477-489(1999).

[2] K. Horn, *Robot Vision*. MA MIT Press, Cambridge, 1986.

[3] S. Negahdaripour and S. Lee, Motion recovery from image sequences using only first order optical flow, *International Journal of Computer Vision* **9** 163-184(1992).

[4] K. Prazdny, On the information in optical flow, *Computer Vision Graphics* **23** 239-259(1983).

[5]  A. Bab-Hadiashar and J. Suter, Robust optic flow computation, *International Journal of Computer Vision* **29** 59-77(1998).

[6]  W.H.Press, B.P.Flannery, S.A.Teukolsky  and W.T.Vetterling, *Numerical Recipes in C: The Art of Scientific Computing*, Cambridge University Press 2$^{nd}$ ed., 1992.

[7]  C. K. Chui, *An introduction to wavelets*. Academic Press, 1992.

[8]  C. K. Chui, *Multivariate splines*. CBMS-NSF Regional Conference Series in Applied SIAM 54, 1988.

[9]  L. Schumaker, *Spline function: Basic theory*. J.Wiley & Sons, 1981.

[10] E. De Micheli, V. Torre and S. Uras, The analysis of time varying image sequences, *Proceeding First Euro. Conf. Comp. Vision* 45-50(1990).

[11] R. Szeliski and H. Shum, Motion estimation with quadtree splines , *IEEE Transactions on Pattern Analysis and Machine Intelligence* **18** 1199-1210(1996).

[12] G. H. Golub and C. F. Van Loan, *Matrix Computation*. The Johns Hopkins University Press, 1990.

[13] D. P. Bertesekas and J. N. Tsitsiklis, *Parallel and distributed computation – Numerical methods*. Prentice Hall, 1989.

Brill Academic Publishers
P.O. Box 9000, 2300 PA Leiden,
The Netherlands

*Lecture Series on Computer*
*and Computational Sciences*
Volume 4, 2005, pp. 204-206

# Remark on Testing Irreducibility of Polynomials over Finite Fields [1]

S.B.Gashkov, I.B.Gashkov

Department Mechanics and Mathematics
Moscow State University, 119899 Moscow, Russia
Department of Engineering Sciences,Physics and Mathematics,
Karlstad University, Sweden

Received 11 February, 2005; accepted in revised form 12 June, 2005

*Abstract:* We prove some improvement for well-known upper bound of complexity of testing irreducibility of polynomials over finite fields.

*Keywords:* Irreducible polynomial, finite fields

## 1 Introduction

To test irreducibility of a polynomial $f$ over given finite field $\mathbb{F}_q$ with given degree $n$, following well-known algoritm from [1], [2] can be used.

We compute the polynomial sequence $f_{k+1}(x) = f_k^q(x) \ rem \ f(x)$, starting from $f_0(x) = x$, where we denote by $g \ rem \ f$ the remainder of $g(x)$ on division by $f(x)$: $g \ rem \ f = g \bmod f$ and $\deg(g \ rem \ f) < \deg f$. It is clear, that $f_k = x^{q^k} \ rem \ f(x)$. The polynom $f$ is irreducible iff $f_n(x) = x$ and for any prime $s \mid n \ gcd(f_{n/s}(x) - x, f(x)) = 1$, where the $gcd$ of two polinomials (not both zero) is the monic polynomial of largest degree dividing both polynomials. Naturally, if $f_{n/s}(x) = x$ for some $s > 1$, then polynomial $f(x)$ is reducible.

In [3] (see also [2]) was obtained the following upper bound of complexity of testing irreducibility

$$O(M(n)\log_2 q + (n^{(\omega+1)/2} + n^{1/2}M(n))d(n)\log_2 n),$$

where $d(n)$ is the number of distinct prime divisors of $n$, $M(n)$ is the complexity of polynomial multiplication over $\mathbb{F}_q$ (i.e., the total number of field operations), and $\omega$ is matrix multiplication exponent (world record $\omega < 2.376$ is due to Winograd-Coppersmith, see for example [2]). It is known that $d(n) \lesssim \ln n / \ln\ln n$ and $d(n) \sim \ln\ln n$ on average (see [4], [5]).

The proof [2] of given bound for complexity of testing irreducibility is based on fast modular composition of polynomials, suggested in [6]. The operation of modular composition $g(h)rem f$ of polynomials $g, h$ with $\deg g, \deg h < \deg f$, can be performed with complexity of

$$O(n^{(\omega+1)/2}) + 6n^{1/2}(M(n) + O(n))$$

field operartions, where $n = \deg f$ (see[6], [2]).

---

[1] The research was support by The Royal Swedish akademy of Sciences

## 2    Improved bound of complexity of testing irreducibility

Here we prove

**Theorem 1** *Algorithm for testing irreducibility off $\in \mathbb{F}_p[x]$ of degree $n$ can be implemented so as to use*

$$O\left( M(n)\log_2 q + \left( n^{(\omega+1)/2} + n^{1/2}M(n) \right) \left( \log_2 n + d(n)\frac{\log_2 n}{\log_2 \log_2 n} \right) \right)$$

*operations in* $\mathbb{F}_q$.

Proof.

Let's consider defined above sequence $f_k(x)$. It is clear that $f_k = x^{q^k}$ *rem* $f$ has degree less than $n$. As was pointed above, $f$ is irreducible iff $f_n(x) = x$ and for any prime $s \mid n$ $gcd(f_{n/s}(x) - x, f(x)) = 1$. Complexity of computing of the gcd of two polinomials is $O(M(n)\log n)$ (see [2]). Hence the cost of computing of all polinomials $(f_{n/s} - x, f)$ is $O(d(n)M(n)\log_2 n) = O(M(n)\log_2^2 n)$.

We have to estimate the cost for computing of all polynomials $f_{n/s}$. The cost for $f_1 = x^q$ *rem* $f$ is $((l(n) + 1)(3M(n) + O(n)) + 6M(n) = O(M(n)\log_2 q)$, where $l(n)$ is shortest length of addition chain for $n$; the first bound can be obtained by P.Montgomery's method (see [2]), and second bound can be obtained by binary method for constructing of addition chains (see [7]). Since

$$f_{i+j}(x) = x^{p^{i+j}} \ rem \ f = (x^{p^i} \bmod f)^{p^j} \bmod f = f_i^{p^j}(x) \bmod f =$$

$$= f_i\left( x^{p^j} \right) \bmod f = f_i\left( x^{p^j} \bmod f \right) \bmod f = f_i(f_j) \bmod f = f_i(f_j) \ rem \ f,$$

the polynom $f_{i+j}$ is computed using given $f_i, f_j$ by modular composition $f_{i+j} = f_i(f_j) \ rem \ f$ with complexity $O(n^{(\omega+1)/2} + n^{1/2}M(n))$ (see [2]). In generally, modular composition is not commutative operation, but in considered case polynomials $f_i, f_j$ may be transposed. Thus the polynomial system $\{f_{n/s}\} = \{f_{n_1}, \ldots, f_{n_d}, \}, n_1 = n, d = d(n)$ can be computed using

$$((l(q) + 1)(3M(n) + O(n)) + 6M(n) + l(n_1, \ldots, n_d)O(n^{(\omega+1)/2} + n^{1/2}M(n))$$

field operations, where $l(n_1, \ldots, n_d)$ is shortest length of addition chain that contains all numbers $n_1, \ldots, n_d$ (this notation see in [7]). Using Yao's theorem (see [7], part 4.6.3, ex.37), we see that

$$l(n_1, \ldots, n_d) = \log_2 n + O(d)\frac{\log_2 n}{\log_2 \log_2 n} = O(\frac{\log_2^2 n}{(\log_2 \log_2 n)^2}).$$

For almost all $n$ $d(n) = O(\ln \ln n)$, hence $l(n_1, \ldots, n_d) = O(\log_2 n)$. Using some result from [8], we have that

$$l(n_1, \ldots, n_d) = \log_2 n + O(d) + \frac{R}{\log_2 R}\left( 1 + O\left( \left( \frac{\log_2 \log_2 R}{\log_2 R} \right)^{1/2} \right) \right),$$

where $R = n^d/m$, $m$ is the product of all prime divisors of $n$.

Finally, we have that complexity of testing is

$$O\left( M(n)\log_2 q + \left( n^{(\omega+1)/2} + n^{1/2}M(n) \right) \left( \log_2 n + d(n)\frac{\log_2 n}{\log_2 \log_2 n} \right) \right),$$

because the second additive term in this estimation is great than the cost $O(M(n)\ln^2 n)$ for the gcd computing.

If $q = 2$, then the estimation from theorem is

$$O\left(\left(n^{(\omega+1)/2} + n^{1/2}M(n)\right)\left(\log_2 n + d(n)\frac{\log_2 n}{\log_2 \log_2 n}\right)\right).$$

If we use fast polynomial multiplication, for example $M(n) = O(n^{\omega/2})$, then we have the estimation

$$O\left(n^{(\omega+1)/2}\left(\log_2 n + d(n)\frac{\log_2 n}{\log_2 \log_2 n}\right)\right).$$

## References

[1] R. Lidl and H. Niderreiter, Finite Fields, Addison-Wesley, 1983.

[2] J. von zur Gathen and J. Gerhard, Modern computer algebra, Cambridge University Press, 1999.

[3] J. von zur Gathen and V. Shoup, Computing Frobeius maps and factoring of polynomials, Comput. complexity 2(1992), 187-224.

[4] K. Prachar, Primzahlverteilung, Springer-Verlag, 1957.

[5] J. von zur Gathen and M. Giesbrecht, Constructing normal bases in finite fields. J. Symbolic Computation 10(1990), 547-579.

[6] R.P.Brent and H.T.Kung, Fast algoritms for manipulating formal power series. J. of the ACM v.25, N 4(1978), 581-595.

[7] D.Knut, The Art of programming, v.2, Addison-Wesley Longman, 1998.

[8] S.B.Gashkov and V.V.Kochergin, On addition chains of vectors, gate circuits, and the complexity of computation of power, Syberian Advances in Mathematics v.4, No 4(1994), ,1-16.

Brill Academic Publishers
P.O. Box 9000, 2300 PA Leiden
The Netherlands

*Lecture Series on Computer
and Computational Sciences*
Volume 4, 2005, pp. 207-210

# A Generic Design of Ladder Algorithm for Automation of Backwash Water Treatment Plant

Aamir Hanif, M. A. Choudhry[1] and Tahir Mehmood

Department of Electrical Engineering
University of Engineering and Technology
Taxila, Pakistan.

Received 03 June 2005: accepted in revised form 27 June, 2005

*Abstract:* This paper is about development of PLC software for control and automation of a water treatment system which is called "Backwash Water Treatment Process". The algorithm developed in this paper may be used to develop PLC based software for any control process.

Developed Software is based on the filter backwash sequence which determines the time to backwash and the time to stop backwashing the filter and the quantity of Aluminum sulfate $Al_2 (SO_4)_3$ to be added. The decision is based on data acquired from limit switches, level switches/sensors, counters and timers. The logic sequence is designed to minimize production down time when a unit was off line due to backwashing and ensured that the primary sludge tank had sufficient capacity to accept the dirty backwash water.

The algorithm is developed by considering filter backwash sequence, the I/O table developed for all input and output devices obtained by assigning the corresponding INPUT and OUTPUT number to these devices considering the particular programmable controller, developed timing diagram showing the length of occurrence of process (I/O) signals, flow chart developed for the control process and using CX-Programmer software supporting predominant ladder programming language for CQM1 PLC used as Master PLC and CPM2A-60 CDR PLCs used as slave PLCs for control of a water treatment system "backwash water treatment process. CQM1 is meant for advanced control system operations requiring 16 to 192 I/O points per PLC. CX-Simulator software was used for simulating software developed by CX-Programmer.

*Keywords:* Ladder programming, Automatic control, PLC, Backwashing sequence.

*Mathematics Subject Classification:* Logic Programming

*PACS:* 68N17

## 1. Introduction

Unfortunately the lack of standardization coupled with continually changing technology has made PLC common nightmare of incompatible protocols and physical networks [1]. During 1980s an attempt was made to standardize communications with General Motors manufacturing automotive protocol. It was a time for reducing the size of PLC and making them programmable through symbolic programming on personal computers, instead of dedicated programming terminals [2].

IEC has recently standardized programmable controllers and their associated peripherals to facilitate automation industry. The commission has developed a series of standards (IEC-61131 series) to address salient issues and problems confronting the automation industry. It also provides guidelines for the implementation of programming languages in programmable controller and their programming support environment [4].

## 2. Systematic Approach

The concept of controlling a plant is a simplistic in nature. It involves a systematic approach by following the operation procedure as shown in Figure. 1. The algorithm involves the following steps.

---

[1] Muhammad Ahmed Choudhry, Member IEEE. E-mail: drahmad@uettaxila.edu.pk

*A. Determine the Sequence of Operation*

We have to identify the equipment or the system to be controlled. The ultimate purpose of the programmable controller will be to control an external system [3]. In simple, one needs to determine the sequence of operation in steps form for onward development of flow chart by completely understanding the system detailed operation.

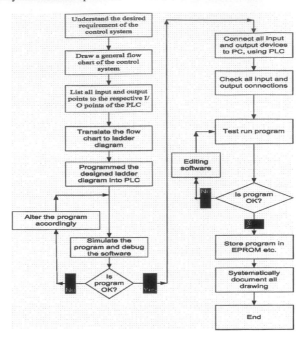

Figure 1:    A systematic Approach to Develop PLC Program

*B. Assignment of Inputs and Outputs*

After identifying all the INPUT and OUTPUT devices, the numbers corresponding to the INPUT and OUTPUT of the particular programmable controller will be assigned. The actual wiring will follow the numbers of the programmable controller.

*C. Develop Timing Diagram*

Timing diagram is developed to   show   the length of occurrence of signals (I/O signals) of control process.

*D. Develop Flow Chart*

Flow chart is drawn considering the sequence of operation of control process.

*E. Writing of the Program*

Write ladder program by following the control system sequence of operation as determined by step A.

## 3. Chain of Operation

After fulfilling the pre-requisites for automatic operation and the start up sequence, we converted detailed theoretical description of sequence of operation for backwashing a filter in steps, so that flow chart and hence software can be easily developed.

### A. Initiation of a Backwash

Three methods for initiating backwashing sequence will be available.

*1)    Timed wash/automatic   initiation:*   The filter will be washed in sequence by a timer, which will be adjusted from 12 to 36 hours.

*2)    Manual initiation:*   it will be initiated from   the   key operated pushbuttons on the front of the filter consoles.

*3)    Semi-Automatic initiation:* This method is explained in Figure 2.

### B.  Filter Backwash Sequence

The wash sequence will consists of the following steps as described below. The condition which must be met before jumping from one step to next is listed in each step.

Step 1.   i)   Signal the appropriate "filter washing" lamp on   the filter console.
  ii)   Signal   the   inlet penstock (V202/X) to   close    and   confirm.
  iii)   Start the filter drain down period, which is 30 minute from closing the filter inlet penstock
  Inlet penstock closed
  End of Step 1.

Step 2.   Signal the filter outlet valve (V-204/X) to closed and confirm.
  Inlet penstock closed
  Outlet valve closed
  End of step 2.

Similarly our sequence of operation will be defined until Step 19 arises, which is

Step 19. i)   Remove the signal "filter washing" lamp on the filter console.
  ii)   Initiate the "time since last wash" timer for that filter.

### C.   Backwash Sequence Failure

Any failure in the backwash sequence will initiate the "backwash sequence fail" lamp on the front of the filter control consoles (FCC). It will also initiate a common klaxon located in the filter building.

## 4.  Software Design

After defining the backwash water treatment process, we determined the inputs and outputs from the sequence of operation for backwash water treatment plant. These are for the CQM1 PLC (Master PLC in our control process) and the CPM2 PLCs (Slave PLCs in our control process).  Corresponding address numbers for each input and output for these PLCs were also given. Timing diagram for filter backwash sequence was developed. Flow chart was made by considering the sequence of operation for backwash water treatment plant.

Ladder program was written after completing above tasks by following the steps of control system sequence of operation as written in section 3, I/O tables defined and flow chart developed. Now we applied power to the programmable controller. Depending on the type of programmable controller, an I/O generation to prepare the system configuration was done. After that, we entered our program in the memory by computer aided ladder software tool i.e. OMRON CX-programmer and checked for any coding errors by means OMRON CX-Simulator software. PLCs self diagnostic functions enabled for us easy and fast troubleshooting of the system.

Test run was done thoroughly until it was safe to operate by anyone.

## 5. Conclusions and Discussions

The main contribution to this research is to demonstrate how the PLC based software's can be developed for automatic control process, because PLC can control from a simple pick and place system to a much complex servo positioning system. The algorithm developed in this software may be used to develop PLC based software for any control process.

Although the software which we have developed for backwash water treatment plant yield the required results, but this software can be further improved /shortened by using different control logical techniques/strategies using PLC instruction set and different functions depending upon the PLC programmer experience and logical approach.

## Acknowledgements

The authors wish to thank the anonymous referees for their careful reading of the manuscript and their fruitful comments and suggestions.

## References

[1] Alan J. Crispin, "Programmable logic controllers and their Engineering applications", 2nd edition, The McGraw-Hill Company (U.K) 1997.

[2] Ian G.Warnak, "Programmable Controller Operation and Application", Prentice Hall International (U.K) Ltd. 1998.

[3] Richard A. Cox, "Technicians Guide to Programmable Logic Controllers,"4th edition, Delmar Thomson Learning (Canada) Inc. 2001.

[4] Programmable controllers –part 3: Programming languages, International Electro technical Commission, IEC 61131-3, 2003.

Brill Academic Publishers
P.O. Box 9000, 2300 PA Leiden,
The Netherlands

*Lecture Series on Computer*
*and Computational Sciences*
Volume 4, 2005, pp. 211-214

# Embedding Wave Function Theory in Density Functional Theory

Thomas M. Henderson[1]

Tyndall National Institute,
University College,
Lee Maltings,
Prospect Row,
Cork, Ireland

Received 9 July, 2005; accepted in revised form 2 August, 2005

*Keywords:* Multi-scale, Computational Chemistry

*PACS:* 31.15.-p

## 1   Embedded Wave Function Theory

Because *ab initio* wave function theory (WFT) proceeds by making systematic approximations to the Schrödinger equation, it reliably makes predictions of the physical and chemical properties of atoms and molecules. By systematically relaxing these approximations (*i.e.* increasing the level of correlation or the size of the computational basis), one can obtain arbitrarily accurate solutions of the Schrödinger equation. Unfortunately, this accuracy and reliability require such exorbitant computational effort that high-level calculations can be applied only to relatively small systems.

Density functional theory (DFT) is an attractive reformulation of quantum mechanics since it provides usually reasonable results at a fraction of the cost of traditional wave funtion methods. But as we do not know the exact exchange-correlation functional, we must resort to approximate functionals which may or may not prove reliable in any given system. And if these functionals give poor results, it is usually not possible to systematically improve the quality of the calculation within the DFT framework.

The computational chemist thus has two choices: On the one hand, there is an inexpensive method which may not yield meaningful results, and on the other there is a much more reliable method which may be prohibitively expensive. In general, we do not have an ideal solution to this dilemma. But at times, we will be interested primarily in a small part of a larger system. In cases like these, one would like to use WFT to focus on the chemically interesting part of the problem without going to the expense of applying WFT everywhere. In other words, one might wish to employ a multi-scale technique which uses WFT for one part of a system and DFT for the rest. Though this is not a new idea, we see prospects for much additional work towards this goal and herein present one framework.

Suppose we have a system $Z$ which we wish to conceptually divide into two subsystems $A$ and $B$. We carry out such a division by partitioning the molecular orbitals in $Z$ into two sets; usually this will require some form of spatial localization, but this is not strictly required and we may

---

[1]Corresponding author. E-mail: tom.henderson@tyndall.ie

choose to instead define the partitioning in any other way. For example, we may choose subsystem $A$ to consist of the valence orbitals and subsystem $B$ to consist of the rest.

In any event, having partitioned $Z$ into $A$ and $B$, let us further suppose that we wish to treat $A$ with some accurate *ab initio* wave function technique and are willing to use Kohn-Sham to treat $B$ and the explicit interactions between $A$ and $B$. The total energy would then be

$$E_Z = E_Z^{DFT} + E_A^{WFT} - E_A^{DFT}. \tag{1}$$

Provided that we use the same orbitals in both the DFT and the WFT calculation and that we use exact exchange DFT, we ave

$$E_A^{WFT} - E_A^{DFT} = E_{c,A}^{WFT} - E_{c,A}^{DFT}, \tag{2}$$

*i.e.*, the difference between the WFT and DFT energies in region $A$ is simply the difference in the WFT and DFT correlation energies in that region.

The DFT correlation energy in a region $A$ is given explicitly by

$$E_{c,A}^{DFT} = E_c^{DFT}[n_A] \tag{3}$$

where $n_A$ is the density in that region, while the WFT correlation energy in region $A$ is

$$E_{c,A}^{WFT} = E_c^{WFT}[T_A], \tag{4}$$

where $T_A$ represents the wave function amplitudes for determinants which differ from the reference only by excitation from orbitals in region $A$ to other orbitals in region $A$. Thus, the total energy becomes

$$E_Z = E_Z^{DFT} + E_c^{WFT}[T_A] - E_c^{DFT}[n_A]. \tag{5}$$

If the exact $E_c^{DFT}[n]$ were used, then we would expect $E_c^{WFT}[T_A]$ and $E_c^{DFT}[n_A]$ to cancel exactly, and $E_Z$ would reduce to the exact Kohn-Sham result. Of course, we do not have the exact $E_c[n]$ and we expect that the energy of (5) will be an improvement over the approximate DFT result.

However, we must still evaluate the wave function amplitudes required, which is where all the complications arise. Our amplitude equations will take the generic form

$$\mathcal{H}T = G[T] \tag{6}$$

with different choices for $\mathcal{H}$ and $G[T]$ corresponding to different wave function theories. Partitioning the Slater determinants into $A$ and $B$ sets, we could solve for the amplitudes in $A$ via

$$\left(\mathcal{H}_{AA} - \mathcal{H}_{AB}\mathcal{H}_{BB}^{-1}\mathcal{H}_{BA}\right)T_A = G_A - \mathcal{H}_{AB}\mathcal{H}_{BB}^{-1}G_B, \tag{7}$$

but for our purposes such a manipulation will generally be of limited utility, since we will rarely be able to construct $\mathcal{H}_{BB}^{-1}$ efficiently and since $G_A$ and $G_B$ will generally depend upon $T_B$.

To make progress, we presume that we have a reasonable approximation for $T_B$, so that we may solve instead

$$\mathcal{H}_{AA}T_A = G_A - \mathcal{H}_{AB}T_B, \tag{8}$$

which is surely simpler. Perturbation theory is the obvious tool for the construction of an approximate $T_B$, but even there we will often be faced with a problem: even solving perturbatively for $T_B$ will tend to be rather expensive computationally, and we do not wish to limit ourselves to systems which we can treat with perturbation theory.

Consider, however, a further subdivision of the determinants in $B$ into sets $B_1$ and $B_2$, respectively "close to" and "far from" $A$.[2] We presume that the coupling between $B_2$ and $A$ is

────────────────────────

[2]Our use of the terms "near" and "far" is intentionally vague, as their precise meaning will depend on how the original partitioning is carried out.

sufficiently weak that we can neglect it, so that we need solve the perturbation theory equations only for $T_{B_1}$. This will in general require yet further consideration, as we may not have negligible coupling between $B_1$ and $B_2$, and various methods to handle this problem are being considered.

Regardless of the precise details, though, a multi-scale calculation of the type we envision would proceed along the following lines:

- Perform an initial Kohn-Sham calculation on the whole system $Z$, obtaining a reference energy and a set of orbitals.

- Partition the orbitals into subsets $A$ and $B$.

- Calculate the Kohn-Sham portion of the correlation energy in region $A$, *i.e.* $E_c^{DFT}[n_A]$.

- Partition the excited Slater determinants into subsets $A$, $B_1$, and $B_2$.

- Calculate $T_{B_1}^{(1)}$ with coupling to $B_2$ handled in one way or another (*e.g.* with the aid of repeated partitionings or via effective potentials).

- Calculate $T_A$ from (8).

- Calculate the *ab initio* portion of the correlation energy, $E_c^{WFT}[T_A]$.

These ideas have been implemented and we will present simple applications.

## 2   Conclusions

We have sketched one approach to embedding an accurate wave function calculation in a simple density functional calculation. Our approach would cost only marginally more than performing a wave function calculation with no coupling between the wave function region and the rest of the system, but with genuine two-body coupling between the wave function region and its environment should provide real advantages in accuracy for those cases in which the wave function part of the system is not well-isolated from its surroundings. We anticipate using such an approach to study a wide variety of problems, such as, for example, defects in solids or reactive sites in large molecules.

## Acknowledgment

This work was supported by Science Foundation Ireland. The author wishes to thank Dr. Jim Greer, Dr. Paul Delaney, and Dr. Giorgos Fagas for stimulating discussions.

## References

[1] J.L. Whitten and T.A. Pakkanen, Chemisorption theory for metallic surfaces: Electron localization and the description of surface interactions, *Physical Review B* **21** 4357-4367 (1980).

[2] M. Svensson, S. Humbel, R.D.J. Froese, T. Matsubara, S. Sieber, and K. Morokuma, ONIOM: A Multilayered Integrated MO+MM Method for Geometry Optimizations and Single Point Energy Predictions. A Test for Diels-Alder Reactions and Pt(P($t$-Bu)$_3$)$_2$ + H$_2$ Oxidative Addition, *Journal of Physical Chemistry* **100** 19357-19363 (1996).

[3] N. Govind, Y.A. Wang, and E.A. Carter, Electronic-Structure Calculations by First-Principles Density-Based Embedding of Explicitly Correlated Systems, *Journal of Chemical Physics* **110** 7677-7688 (1999).

[4] T. Klüner, N. Govind, Y.A.Wang, and E.A.Carter, Periodic Density Functional Embedding Theory for Complete Active Space Self-Consistent Field and Configuration Interaction Calculations: Ground and Excited States, *Journal of Chemical Physics* **116** 42-54 (2002).

[5] Q. Cui, H. Guo, and M. Karplus, Combining *Ab Initio* and Density Functional Theories with Semiempirical Methods, *Journal of Chemical Physics* **117** 5617-5631 (2002).

[6] B.G. Janesko and D. Yaron, Explicitly Correlated Divide-and-Conquer-Type Electronic Structure Calculations Based on Two-Electron Reduced Density Matrices, *Journal of Chemical Physics* **119** 1320-1328 (2003).

Brill Academic Publishers
P.O. Box 9000, 2300 PA Leiden
The Netherlands

*Lecture Series on Computer
and Computational Sciences*
Volume 4, 2005, pp. 215-218

# Convexification Strategies for Solving the Nonlinear Sum of Ratios Problem

C.H. Huang[1], H.L. Li

Institute of Information Management,
National Chiao Tung University,
Management Building 2,
1001 Ta Hsueh Road, Hsinchu, Taiwan 300

Received 6 July, 2005; accepted in revised form 29 July, 2005

*Abstract:* This paper presents a new method for solving nonlinear sum of ratios problem (NSRP) by the convexification strategies to obtain a global optimum. First we transform the fractional functions into the signomial terms with introducing positive variables. Then by referring to the proposed convexification strategies, all signomial terms are converted into convex and concave terms. The concave terms are then represented by piecewise linearization method. A numerical example is illustrated to show the usefulness of the proposed method.

*Keywords:* Nonlinear sum of ratios problem, fractional function, convexification, piecewise linearization.

*Mathematics Subject Classification:* 90C32

*PACS:* 90C32

## 1. Introduction

Consider the following nonlinear sum of ratios problem (NSRP):

$$Min \quad \sum_{i=1}^{m} \frac{n_i(x)}{d_i(x)} \tag{1a}$$

$$s.t.$$

$$x \in X \tag{1b}$$

where $n_i(x)$, $d_i(x) \in R^m$, $i = 1, 2, ..., m$, and $X$ is a compact, convex set in $R^m$ defined by a system of linear equalities and inequalities.

There have been many studies on nonlinear sum of ratios problem. Benson [1] has presented two branching processes using concave envelops for solving the nonlinear sum of ratios problem. Quesada and Grossmann [4] propose an algorithm for linear fractional and bilinear programs.

In this paper we propose an algorithm for solving the nonlinear sum of ratios problem. The paper is organized as follow. In Section 2, the proposed methods including the convexification strategies and piecewise linearization method are presented. In order to illustrate the usefulness of the proposed method, a numerical example is to be demonstrated in Section 3. Finally, Section 4 contains some concluding remarks.

---

[1] Corresponding author. Tel.: +886-3-571-2121 ~ 57430. E-mail: leohuang.iim92g@nctu.edu.tw

## 2. Proposed Methods

Introducing positive variables $y_1$, $y_2$, ..., $y_m$ such that $d_i(x)y_i = 1$, for $i = 1,\ 2,\ ...,m$, then the NSRP results in the following signomial problem (SP):

$$Min \ \sum_{i=1}^{m} n_i(x)y_i \tag{2a}$$

s.t.

$$d_i(x)y_i = 1, \ i = 1,\ 2,\ \cdots,\ m \tag{2b}$$

$$x \in X \tag{2c}$$

**Proposition 1.** The optimal solution of the SP in (2) is an optimal solution of the NSRP in (1). □

It is clear that each fractional term $\dfrac{n_i(x)}{d_i(x)}$, for $i = 1,\ 2,\ ...,m$, in (1a) can be expressed as a signomial term with introducing a positive variable $y_i$. The rest of this section proposes the convexification strategies for treating the signomial terms. The advantage of the strategies is to reduce the number of concave terms significantly.

**Lemma 1.** Consider a twice-differentiable signomial function $f(\mathbf{x}) = c\prod_{i=1}^{n}x_i^{\alpha_i}$, $\mathbf{x} = (x_1,\ x_2,\ ...,\ x_n)$, $c,\ x_i,\ \alpha_i \in R$, $\forall i$, denote $H(\mathbf{x})$ is the Hessian matrix of $f(\mathbf{x})$. The determinant of $H(\mathbf{x})$ can be expressed as $\det H(\mathbf{x}) = (-c)^n (\prod_{i=1}^{n}\alpha_i x_i^{n\alpha_i - 2})(1 - \sum_{i=1}^{n}\alpha_i)$ [2]. □

**Proposition 2.** A twice-differentiable signomial function $f(\mathbf{x}) = c\prod_{i=1}^{n}x_i^{\alpha_i}$ is convex for $c, x_i > 0$, and $\alpha_i \le 0$, $\forall i$ [2]. □

**Proposition 3.** A twice-differentiable signomial function $f(\mathbf{x}) = c\prod_{i=1}^{n}x_i^{\alpha_i}$ is convex for $c < 0$, $x_i, \alpha_i > 0$, and $1 - \sum_{i=1}^{n}\alpha_i \ge 0$, $\forall i$ [2]. □

By referring to the proposed convexification strategies, all signomial terms are converted into convex and concave terms. All concave terms in the SP problem can be represented by piecewise linearization method as follows.

**Proposition 4.** Let $L(f(x))$ be the piecewise linearization function of $f(x)$ as shown in Figure 1, where $a_i$, $i = 1,\ 2,\ ...,m$, are the break points of $f(x)$, and $s_i$, $i = 1,\ 2,\ ...,m$, are the slopes of line segments between $a_i$ and $a_{i+1}$, $s_i = \dfrac{f(a_{i+1}) - f(a_i)}{a_{i+1} - a_i}$, for $i = 1,\ 2,\ ...,m-1$. Let $x = \sum_{j=1}^{k}2^{j-1}u_j$, where $0 \le x \le x^U$, $k = \lceil \log_2(x^U + 1) \rceil$, and $u_j \in \{0,\ 1\}$. Then $f(x)$ can be expressed as the piecewise linearization function $L(f(x))$ below.

$$L(f(x)) \ge f(a_i) + s_i(x - a_i) - M(w + \sum_{j=1}^{k}c_{ij}u_j), \tag{3a}$$

$$L(f(x)) \le f(a_i) + s_i(x - a_i) + M(w + \sum_{j=1}^{k}c_{ij}u_j), \tag{3b}$$

$$a_i - M(w + \sum_{j=1}^{k}c_{ij}u_j) \le x \le a_{i+1} + M(w + \sum_{j=1}^{k}c_{ij}u_j), \tag{3c}$$

where $i = 1, 2, ..., m-1$, $j = 1, 2, ..., k$, $k = \left\lceil \log_2(x^U + 1) \right\rceil$, $u_j \in \{0, 1\}$, and $M$ is a big positive value. If $\left\lceil \dfrac{i-1}{2^{j-1}} \right\rceil$ is an even value, then $c_{ij} = 1$, otherwise $c_{ij} = -1$. $w$ is the number where $c_{ij} = -1$. $\square$

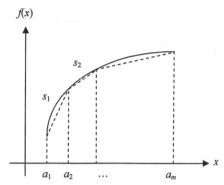

Figure 1. Piecewise linearization function of $f(x)$

## 3. Illustrative Example

Consider the following nonlinear sum of ratios problem.

**Example 1.** [1]

$$\min \quad \frac{x_1^2 - 3x_1 + x_2^2 - 3x_2 - 3.5}{x_1 + 1} - \frac{x_2}{x_1^2 - 2x_1 + x_2^2 - 8x_2 + 20} \tag{4}$$

s.t.

$$2x_1 + x_2 \le 6, \tag{5}$$
$$3x_1 + x_2 \le 8, \tag{6}$$
$$x_1 - x_2 \le 1, \tag{7}$$
$$x_1, \; x_2 \ge 1. \tag{8}$$

Based on Proposition 1, Example 1 can be formulated as the following signomial problem with introducing positive variables $y_1$, and $y_2$.

$$\min \quad (x_1^2 - 3x_1 + x_2^2 - 3x_2 - 3.5)y_1 - x_2 y_2 \tag{9}$$

s.t.

$$(x_1 + 1)y_1 = 1, \tag{10}$$
$$(x_1^2 - 2x_1 + x_2^2 - 8x_2 + 20)y_2 = 1, \tag{11}$$
$$(5) - (8).$$

Based on the convexification strategies, the nonlinear terms $x_1^2 y_1$, $-3x_1 y_1$, $x_2^2 y_1$, $-3x_2 y_1$, $-x_2 y_2$ in (9), $x_1 y_1$ in (10), and $x_1^2 y_2$, $-2x_1 y_2$, $x_2^2 y_2$, $-8x_2 y_2$ in (11) are non-convex which can be transformed into convex terms $z_{11}^{-2} w_{11}^{-1}$, $-3z_{12}^{\frac{1}{2}} w_{12}^{\frac{1}{2}}$, $z_{21}^{-2} w_{11}^{-1}$, $-3z_{22}^{\frac{1}{2}} w_{12}^{\frac{1}{2}}$, $-z_{22}^{\frac{1}{2}} w_{22}^{\frac{1}{2}}$, $z_{11}^{-1} w_{11}^{-1}$, $z_{11}^{-2} w_{21}^{-1}$, $-2z_{12}^{\frac{1}{2}} w_{22}^{\frac{1}{2}}$, $z_{21}^{-2} w_{21}^{-2}$, and $-8z_{22}^{\frac{1}{2}} w_{22}^{\frac{1}{2}}$ respectively where $z_{11} = x_1^{-1}$, $z_{12} = x_1^2$, $z_{21} = x_2^{-1}$, $z_{22} = x_2^2$, $w_{11} = y_1^{-1}$, $w_{12} = y_1^2$, $w_{21} = y_2^{-1}$, and $w_{22} = y_2^2$.

The problem then becomes:

$$min \quad z_{11}^{-2}w_{11}^{-1} - 3z_{12}^{\frac{1}{2}}w_{12}^{\frac{1}{2}} + z_{21}^{-2}w_{11}^{-2} - 3z_{22}^{\frac{1}{2}}w_{12}^{\frac{1}{2}} - 3.5y_1 - z_{22}^{\frac{1}{2}}w_{22}^{\frac{1}{2}}$$ (12)

$s.t.$

$$z_{11}^{-1}w_{11}^{-1} + y_1 = 1,$$ (13)

$$z_{11}^{-2}w_{21}^{-1} - 2z_{12}^{\frac{1}{2}}w_{22}^{\frac{1}{2}} + z_{21}^{-2}w_{21}^{-2} - 8z_{22}^{\frac{1}{2}}w_{22}^{\frac{1}{2}} + 20y_2 = 1,$$ (14)

$$z_{11} = x_1^{-1},$$ (15)

$$z_{12} = x_1^2,$$ (16)

$$z_{21} = x_2^{-1},$$ (17)

$$z_{22} = x_2^2,$$ (18)

$$w_{11} = y_1^{-1},$$ (19)

$$w_{12} = y_1^2,$$ (20)

$$w_{21} = y_2^{-1},$$ (21)

$$w_{22} = y_2^2,$$ (22)

$(5)-(8).$

The equality constraints $(15)-(22)$ which are non-convex terms are represented by piecewise linearization method based on Proposition 4. The transformed program is then a convex program. Solve this problem using LINGO [3] to obtain the optimal solution as $(x_1, x_2) = (1.00, 1.74)$ and the objective value = 4.061 which is a global optimum with tolerance error 0.001.

## 4. Conclusions

In this paper an algorithm is proposed for solving the nonlinear sum of ratios problem. The fractional functions are transformed into the signomial terms with introducing positive variables. The convexification strategies are to convert the signomial terms into convex and concave terms. The concave terms are then piecewise linearized. The nonlinear sum of ratios problem can easily be solved by the proposed method.

## Acknowledgments

The authors wish to thank the anonymous referees for their careful reading of the manuscript and their fruitful comments and suggestions.

## References

[1] H. P. Benson, Using concave envelops to globally solve the nonlinear sum of ratios problem, *Journal of Global Optimization* **22** 343-364(2002).

[2] H.L. Li and J.F. Tsai, Global optimization for generalized geometric programming problems, *A dissertation for the degree of doctor of philosophy in institute of information management*, National Chiao Tung University, Hsinchu, Taiwan (2003).

[3] LINGO Release 8.0, LINDO System Inc. 2002.

[4] I. Quesada and I. Grossmann, A global optimization algorithm for linear fractional and bilinear programs, *Journal of Global Optimization* **6** 39-76(1995).

Brill Academic Publishers
P.O. Box 9000, 2300 PA Leiden,
The Netherlands

*Lecture Series on Computer*
*and Computational Sciences*
Volume 4, 2005, pp. 219-222

# Information Theoretical Analysis of Diagrams Represented by Cutting Segments

## N. Ikeda[1]

Junior College Division, Tohoku Seikatsu Bunka College,
981-8585, Niji-no-oka 1-18-2, Izumi-ku, Sendai, Miyagi, Japan

Received 5 July, 2005; accepted in revised form 20 July, 2005

*Abstract:* The boundary of two-dimensional shapes is investigated from the viewpoint of information theory. A study was carried out on the adequate lattice size for approximating the boundary as a set of cutting segments by examining the convergence behavior of information entropy. The calculation of conditional information entropy and KL divergence derives a complicated behavior of the dependence of data representing a line on its slope.

*Keywords:* Information entropy, Kullback-Liebler divergence, Boundary representation

*Mathematics Subject Classification:* 68P05, 68P30, 68U10, 68U05

## 1 Introduction

The method of detailed representation of the boundary of shapes has significance in computer graphics. For example, the marching cube method [1] is extensively used in imaging processes and the Kitta cube [2] was proposed for development of the Volume CAD [3]. In many cases, the data needed for boundary representation is assigned to voxels or pixels, so the description of the material depends on the size of the voxel. However, the size of voxels for maintaining detailed description of the material without producing redundant data has not been considered adequately.

We have tried to treat this problem by using information entropy which plays the central role in information theory. In our previous works, we obtained information entropy of a relative frequency of a kind of cutting segments for simple shapes such as lines and circles [4]. It is however only natural that the focus be on the sequence of cutting segments rather than a kind of cutting segments in order to investigate the boundary of objects.

In this paper, the information entropy is calculated to discuss an adequate size of pixels to describe the boundary of two-dimensional shapes. Conditional information entropy and KL (Kullback-Liebler) divergence are introduced later to consider the dependence of the cutting segments made from a line on its slope. This information is expected to be useful for condensing the data of cutting segments made from a curve.

## 2 The information entropy, conditional information entropy and KL divergence of cutting segments

The cutting segments can be constructed from an original two-dimensional shape as follows. Let a shape be located on a two-dimensional square lattice and regard the boundary of the shape as a

---

[1] Corresponding author. E-mail: nobikeda@mishima.ac.jp

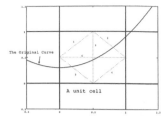 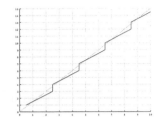

Figure 1: The dotted lines present the six types of cutting segments in one unit cell. The cutting segment numbered 6 is adopted to approximate the curve shown here.

Figure 2: An example of cutting segments deduced from a line. The sequence of cutting segments is $434314343143431\cdots$.

straight line as far as one unit cell is concerned. As a rule, the straight line links two points on the middle of one side of the unit cell. We call these lines obtained like this as cutting segments. Then six types of cutting segments can be constructed in one unit cell, each of which must be assigned to different data.(see Figure 1) If sequences with length 3 is attention, the possible number of sequences is $6^3 = 216$. But the continuous condition of the boundary reduces the number 216 to $6 \times 3 \times 3 = 54$. For example, a sequence $434314343143431$ which is shown in Figure 2 produces the sequence with length 3 like $434, 343, 431, 314, 143, 434, 343, 314, \cdots$.

The definition of the information entropy is

$$H^{(l)} = \sum_i P_i^{(l)} \log P_i^{(l)}$$

where $P_i^{(l)}$ means a relative frequency of appearance of $i$−th kind of sequence of cutting segments with length $l$. Similarly, the conditional information entropy $H_n = H(X_{n+1}|X_n, X_{n-1}, \cdots, X_1)$ can be systematically defined as

$$H_n = \sum_{X_n, X_{n-1}, \cdots, X_1} P(X_n, X_{n-1}, \cdots, X_1) \sum_{X_{n+1}} P(X_{n+1}|X_n, X_{n-1}, \cdots, X_1) \log P(X_{n+1}|X_n, X_{n-1}, \cdots, X_1),$$

where $P(X_{n+1}|X_n, X_{n-1}, \cdots, X_1)$ is the conditional probability of the appearance of a cutting segment $X_{n+1}$ under the condition that the previous sequence of the cutting segments is $X_n, X_{n-1}, \cdots, X_1$. This quantity indicates a predictability of the next segment from previously known cutting segments with length $n$. Later, it will be shown that the information entropy made from lines strongly depends on its slope.

KL divergence, which is going to be used to discuss the difference between sequences of cutting segments made from lines with different slopes, is defined as

$$D(\vec{P}, \vec{Q}) = \sum_i P_i \log P_i - \sum_i P_i \log Q_i,$$

where relative frequencies $P_i$ and $Q_i$ are, respectively, calculated for a line under consideration and for a reference line with a specific slope.

## 3   Results of calculation

Various slopes are undertaken here. Lines are worth considering because they are localized approximation of continuous and smooth curves.

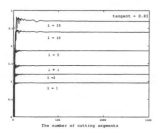 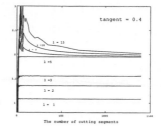

Figure 3: The dependence of the entropy on the number representing a finite line with slope 0.81. Stable value can be easily obtained for any $l$.

Figure 4: An exceptional case indicating a slow convergence of $H^{(l)}$ for large $l$. An invariant limit value of $H^{(l)}$ for $l \geq 5$ can be found.

Examples of the evaluated information entropy $H^{(l)}$ of a line with finite length are shown in Figure 3 and Figure 4 for some $l$, where the calculation is carried out for various cutting segments needed to represent the line. Of course, the smaller the size of the unit cell of the square lattice becomes, the larger the number of cutting segments becomes, thus deriving convergence of the information entropy. In many cases, $H^{(l)}$ indicates fast approach to nearly a limit value regardless of the value of $l$ when the number of cutting segments is about several ten times and stable behavior can be found when the number of cutting segments is larger than several hundred. This behavior is peculiar to a line where $H^{(l)}$ calculated by using curves indicates slow convergence when $l$ is large. There are exceptional cases at some particular slopes of the line, one of which is shown in Figure 4. In the figure, a slow convergence of $H^{(l)}$ for large $l$ and an invariant limit value of $H^{(l)}$ with $l$ can be found for $l \geq 7$ .

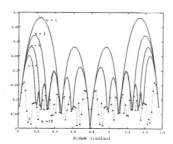

Figure 5: The dependence of the conditional information entropy $H_n$ on a slope of the line for various $n$.

It is entirely true of the conditional information entropy that the behavior strongly depends on the slope of the line, which is shown in Figure 5. In the figure dependence of $H_n$ on a slope of the line is presented for various $n$. $H_n$ is nearly 0 when the line is nearly horizontal (or vertical) because of the evident reason that only the same kind of cutting segment can appear in such a situation. But $H_n$ in the figure indicates very complicated behavior where there are some points where $H_n$ is invariant as $n$ increases and other points where $H_n$ becomes rapidly or slowly small as $n$ increases.

The results of KL divergence are shown in Figure 6 and Figure 7. There are some divergence

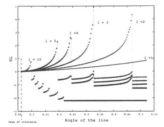 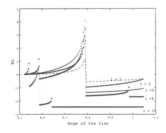

Figure 6: KL divergence between a line and reference line. The reference line has an angle $\pi/50$ with respect to horizontal.

Figure 7: KL divergence between a line and a reference line. The reference line has an angle $\pi/6$ with respect to horizontal.

points, the origin of which is a null value of $Q_i$ in the definition of $D(\vec{P}, \vec{Q})$. In the figure, calculation is carried out except infinite terms in which $Q_i = 0$. The frequency of the appearance of divergence points differs according to the reference slope. The change in the sequence of cutting segments is dependent not only on the changes of a slope of the line but the absolute value of the referring slope.

## 4 Concluding Remarks

We adopt information entropy to realize quantitative evaluation the sufficient number of cutting segments representing a finite line. The use of information entropy is significant in that it has the definite meaning of quantification of the uncertainty of data and it enables discussion of the sequence of cutting segments in terms of other quantities in information theory.

Another remarkable point found in this article is the complicated dependence of the entropy of lines on its slope. Similarly the KL divergence calculated from two lines with different slopes does not depend only on the angle formed where the two lines cross. It is natural that KL divergence can detect changes in the slope of a line. However, compared to the slope of the original line, the value of KL divergence should have an important role from the viewpoint of examining the data of cutting segments. Further consideration of data treatment should provide ideas on data condensation, for example, by gathering the sequence of cutting segments with small KL divergence value. The calculation in this paper should serve as preliminary knowledge for future attempts.

## References

[1] W.E. Lorensen and H.E. Cline, A High-Resolution 3D surface construction algorithm, *SIG-GRAPH*, vol. 21-4,(1987),163-169.

[2] Y. Teshima, et al.,Enumeration of Cutting Points Configuration in Cube Cutting, *Human and Artificial Intelligence Systems (Proceedings of HART2004)*, Advanced Knowledge of International, (2004), 407-414.

[3] K. Kase, et al., Volume CAD, *Volume Graphics 2003 Eurographics / IEEE TCVG Workshop Proceedings*, pp145-150,pp173, (2003).

[4] N. Ikeda, Y. Teshima, Information measure for diagrams described by cutting segments, *Human and Artificial Intelligence Systems (Proceedings of HART2004)*,Advanced Knowledge of International, (2004), 485-489.

Brill Academic Publishers
P.O. Box 9000, 2300 PA Leiden
The Netherlands

*Lecture Series on Computer
and Computational Sciences*
Volume 4, 2005, pp. 223-226

# Adenosine Triphosphate ATP, a Cellular Key Molecule, is Affected by Non-Resonant Optical Frequencies

A. Amat [1,2,3,4], J. Rigau [2], R. W. Waynant [3], I. K. Ilev [3] and J. J. Anders [4]

[2] Histology and Neurobiology Unit, Faculty of Medicine and Health Sciences, Rovira i Virgili University, C. Sant Llorenç 21, 43201 Reus, Spain

[3] Division of Physics, Centre for Devices and Radiological Health, Food and Drug Administration, 12725 Twinbrook Parkway Rockville, MD 20857, United States.

[4] Department of Anatomy, Physiology and Genetics, Uniformed Services University of the Health Sciences, 4301 Jones Bridge Road Bethesda MD 20814, United States.

Received 30 July, 2005; accepted 20 August, 2005

*Abstract:* Adenosine triphosphate (ATP) is a cellular key molecule because it provides free energy to biochemical processes by coupling endergonic and exergonic reactions. The molecule is synthesized in the mitochondria, a cellular organelle, after the complete oxidation of foodstuffs occurs. ATP is constituted by two phosphate bonds, which are the structures that store the chemical energy and are cleaved by the enzymes to release its energetic content to cellular reactions. By inducing an electric field using laser light in the visible and near-infrared optical frequencies, which are non-resonant for ATP, a charge displacement in the phosphate structure may be expected resulting in instability, and as a consequence enzymatic cleavage will be more efficient, powering the biochemical reactions in which ATP is a substrate. Since charge displacement will modify the dipole moment of ATP, the measure of the refractive index of an ATP solution before an after irradiation with non-resonant optical frequencies will be an indicator of how the molecule is altered by non-absorbed light.

*Keywords*: ATP (adenosine triphosphate), laser light, non-resonant optical frequencies, dipole moment, refractive index.

## 1. ATP is the cellular energy currency

Living things require a continuous input of free energy for three major purposes: the performance of mechanical work in muscle contraction and other cellular movements, the active transport of molecules and ions, and the synthesis of macromolecules and other biomolecules from simple precursors. The free energy used in these processes, which maintain an organism in a state that is far from equilibrium, is derived from the environment. *Chemotrophs* such as mammalian cells obtain this energy via the oxidation of foodstuffs, whereas *phototrophs* such as plants obtain it by trapping light energy. The free-energy donor in most energy-requiring processes is adenosine triphosphate (ATP). The central role of ATP in energy exchanges in biological systems was perceived in 1941 by Fritz Lipmann and Hermann Kalckar [1].

ATP is a nucleotide consisting of an adenine plus a ribose (adenosine) and a triphosphate unit (Fig. 1). To consider the role of ATP as an energy carrier, we can focus on its triphosphate unit. ATP is an energy-rich molecule because its unit contains two phosphor-oxygen bonds (the phosphate linked to the adenosine to form AMP has different characteristics because it binds a $CH_2O^-$ group and not another

---

1 Corresponding author. E-mail: albert.amat@urv.net

PO⁻ group). A large amount of energy is liberated when ATP is broken down (in a biochemical reaction catalyzed by an enzyme) into ADP plus orthophosphate ($P_i$) or into AMP plus pyrophosphate ($PP_i$): 7.3 kcal/ mol of free energy for each bond [1]. Since the energy that ATP stores in the phosphates can be used by any metabolic reaction, ATP is said to be the cellular energy currency that fuels the cell biochemistry [2, 3]. It can be also considered as a molecule that translates different kinds of energy into a common language. In this sense ATP is a key molecule.

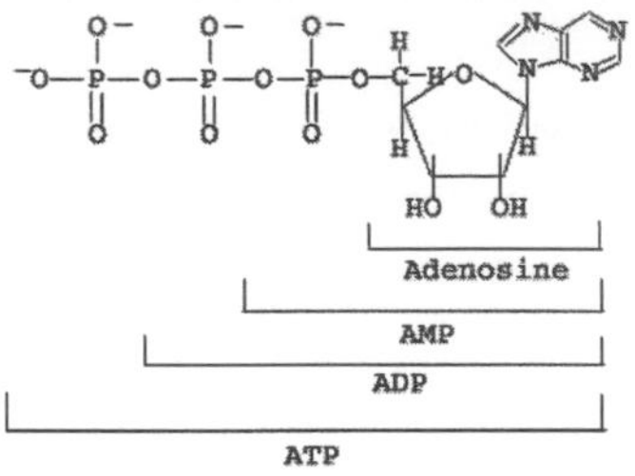

Figure 1: ATP has an adenine attached to a ribose that is called adenosine. The number of phosphates names the molecule and determines its biological function.

## 2.    ATP is synthesized in the mitochondria, the cellular power station

ATP is mostly synthesized in the mitochondria, the energy-supply cellular organelle of a chemotroph organism, after a series of redox reactions in the electronic transport chain located in the inner membrane of the organelle [4]. In aerobic conditions, 36-38 molecules of ATP are synthesized from one molecule of glucose, the most important foodstuff [5]. In anaerobic conditions, when the oxygen levels are insufficient for cell metabolism and ATP synthesis by mitochondrial respiration, a molecule of glucose is fully oxidized in the cytoplasm to produce two molecules of ATP and pyruvate in a process called glycolysis [6].

## 3.    ATP responds to electromagnetic energy

ATP has a strong absorption in the UV, with two important peaks at 210 and 260 nm. [7]. If we consider visible light and near-infrared light, ATP is not colored and has no absorption in that range of the spectrum. ATP absorbs in the middle-infrared part of the electromagnetic spectrum, at 8000 to 11000 nm, which corresponds to mid-infrared frequencies [8].

ATP in solution shows an unexpected response when irradiated with visible and near-infrared wavelengths, which are not absorbed. Light-exposed ATP (635, 655 and 830 nm laser light) increased the kinetic parameters of various cellular and non-cellular reactions in which the molecule is a substrate [9-11]. The mechanism for this response to non-resonant optical frequencies can be related to the electric field that light induces in the medium, which is an insulator medium because it is constituted by dipoles, i.e. water an ATP. ATP is a dipolar molecule with a measured dipole moment of 30 Debyes and a relaxation frequency of 50 MHz [12]. At the end of the enzymatic process the last phosphate of the molecule is split to release the energy that it contains, it is necessary to analyze how light can affect this structure. The last phosphate bond of ATP has different interconvertable resonance isoforms of similar but not equal energy (Fig. 2) [5]. Other resonance forms of the orthophosphate are chemically highly improbable and do not contribute significantly to the total resonance state of the molecule, because two like charges are adjacent, which increases the instability of the structure (Fig. 3) [13]. Bonds with different resonance forms usually have electrons that are less tightly bound to the nucleus and, thus, light produces larger responses when interacting with them [14]. Therefore, it is possible that the displacement of the bound charges that the induced electric field produces in the molecule results in the formation of a new resonant state of ATP, which will be easily cleaved by the enzyme because of the repulsion force produced by two adjacent like electric charges.

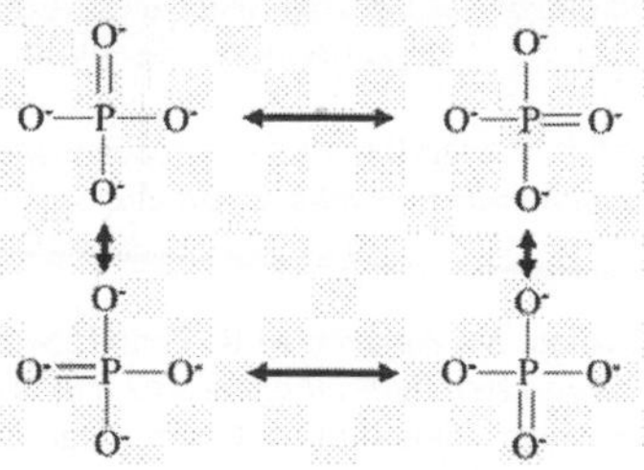

Figure 2: Probable resonance structures of orthophosphate, where two negative charges are not in adjacent atoms.

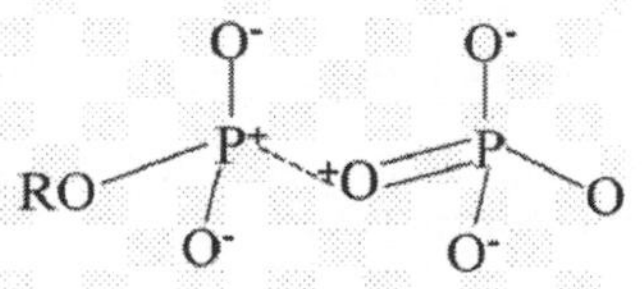

Figure 3: Improbable resonance structure of orthophosphate, unstable because of two adjacent like electrical charges, which contributes to the breakdown of ATP's terminal bond.

## 4.     Measurement of the refractive index of an ATP solution

The hypothesis of a charge redistribution produced by the light-induced electric field can be tested experimentally by measuring the refractive index of an ATP solution and checking whether it is altered after exposing the molecule to non-absorbed light. A different charge distribution will change the dipole moment of ATP and, therefore, the refractive index of an ATP solution will be modified.

## References

[1] H.M. Kalckar, *50 years of biological research-from oxidative phosphorylation to energy requiring transport regulation.* Ann. Rev. Biochem., 60 (1991), 1-37.

[2] D.C. Rees and J.B. Howar, *Structural bioenergetics and energy transduction mechanisms.* J. Mol. Biol., 293 (1999), 343-350.

[3] C.H. Schilling, D. Letscher and B.O. Palsson, *Theory of the systemic definition of metabolic pathways and their use in interpreting metabolic function from a pathway-oriented perspective.* J. Theor. Biol., 203 (2000), 229-248.

[4] T. Friedrich and B. Bottcher, *The gross structure of the respiratory complex I: a Lego System.* Biochim. Biophys. Acta, 1608 (1) (2004), 1-9.

[5] M.A Bianchet, P.L Pederse and L.M. Amzel, *Notes on the mechanism of ATP synthesis.* J Bioenerg. Biomembr., 32 (5) (2000), 517-521.

[6] J.M Berg, J.L Tymoczko and L. Stryer, *Biochemistry* (5th Ed.). W.H. Freeman and Company, New York, 2002.

[7] A. Amat, *The effect of visible and near-infrared light on adenosine triphosphate (ATP).* Doctoral Thesis. Rovira i Virgili University, Reus-Spain, 2005.

[8]   C.J. Pouchert, *The Aldrich Library of Infrared Spectra* (3$^{rd}$ Ed.). Aldrich Chemical Company, Milwaukee ,1981

[9]   S. Gagliardi, A. Atlante and S. Passarella, *A novel property of adenine nucleotides: sensitivity to helium-neon laser in mitochondrial reactions.* Biochem Mol Biol Int., Mar; 41(3) (1997), 449-460.

[10]  A. Amat, J. Rigau, R. Nicolau, M. Aalders, M.R. Fenoll, M.J.C. van Gemert and J. Tomàs, *Effect of red and near-infrared laser light on adenosine triphosphate (ATP) in the luciferine–luciferase reaction.* J.  Photochem. Photobiol. A: Chem., 168 (2004), 59–65.

[11]  A. Amat, J. Rigau, R.W. Waynant, I.K. Ilev, J.M. Tomàs and J.J. Anders, *Modification of the intrinsic fluorescence and biochemical behavior of adenosine triphosphate ATP after irradiation with visible and near-infrared laser light.* J. Photochem. Photobiol. B: Biol., 81 (2005), 26-32.

[12]  E.H. Grant, R.J. Sheppard and G.P. South, *Dielectric behaviour of biological molecules in solution.* Clarendon Press, Oxford, 1978.

[13]  C.K. Mathews, K.E. van Holde and K.G. Ahern, *Biochemistry* (3$^{rd}$ Ed.). Benjamin/Cummings, San Francisco, 2000.

[14]  R.W. Boyd, *Non-linear optics* (2$^{nd}$ Ed.). Academic Press, Amsterdam, 2003.

Brill Academic Publishers
P.O. Box 9000, 2300 PA Leiden
The Netherlands

*Lecture Series on Computer
and Computational Sciences*
Volume 4, 2005, pp. 227-230

# Systems with Combined Damping:
# Finite Element Modelling and Dynamic Analysis

E. Barkanov[1], W. Hufenbach[2] and L. Kroll[2]

[1]Institute of Materials and Structures, Riga Technical University
Kalku St. 1, LV-1658, Riga, Latvia

[2]Institute of Lightweight Structures and Polymer Technology, Technical University of Dresden
Dürerstrasse 26, D-01062, Dresden, Germany

Received 8 July, 2005, accepted in revised form 30 July, 2005

*Abstract:* The objective of the present study is to develop the universal methods and algorithms for dynamic analysis of systems with combined damping using the finite element method. As an example, systems with viscous and viscoelastic, structural and external damping are examined. Two forms of viscous damping are available in the present implementation: Raleigh damping and element damping. The viscoelastic material behaviour is represented by the complex modulus model preserving exactly the frequency dependence of the storage and loss moduli. Dynamic characteristics of systems with combined damping are evaluated by the method of complex eigenvalues, from the resonant peaks of the frequency response function and using the steady state vibrations. Numerical examples are given to demonstrate the validity and application of the approaches developed.

*Keywords:* systems with viscous and viscoelastic, structural and external damping; finite element method; free vibration, frequency and transient response analyses.

## 1. Introduction

Vibration analysis of systems made from viscous and viscoelastic damping materials and discrete damping devices receives a considerable spreading in recent years [1,2]. This is connected with further improvements of commercially available damping composite materials and increasing their applications in various fields of engineering: aerospace technology, shipbuilding, automotive industry, mechanical engineering, etc. A big attention is paid also to the analysis of systems including different types of damping that gives the possibility to investigate many complex problems like a seismic response of inelastic structures, vibrations of machine foundations, dynamics of rotor-bearing systems, vibrations arising in railway carriages and cars, etc. The objective of the present study is to develop the universal methods and algorithms for dynamic analysis of systems with combined damping using the finite element method.

## 2. Modelling of Damping

Two forms of viscous damping are available in the present implementation: Raleigh damping and element damping. In this case the damping matrix has the following form:

$$\mathbf{C} = \alpha \mathbf{M} + \beta \mathbf{K} + \sum_{j=1}^{n} \mathbf{C}_{el}$$

where $\mathbf{M}$ is the mass matrix of a structure, $\mathbf{K}$ is the stiffness matrix of a structure, $\alpha$ and $\beta$ are constants to be determined from two given damping ratios that corresponds to two unequal frequencies of vibrations, $\mathbf{C}_{el}$ is the element damping matrix and $n$ is a number of elements with specified damping. In many practical structural problems, the mass damping or alpha damping may be ignored ($\alpha=0$). The

---

[1] Corresponding author.
E-mail: barkanov@latnet.lv

element damping involves using element types having viscous damping characteristics. There is possibility to specify more than one form of viscous damping in a model.

To describe the rheological behaviour of viscoelastic materials, the complex modulus representation is used. Using this model, the constitutive relations will be expressed in the frequency domain as follows

$$\sigma_0 = E^*(\omega)\varepsilon_0 = E(\omega)[1 + i\eta(\omega)]\varepsilon_0, \quad \eta(\omega) = \frac{E''(\omega)}{E(\omega)}$$

where $\sigma_0$ and $\varepsilon_0$ are an amplitude of the harmonically time-dependent stress and strain respectively, $E^*$ is the complex modulus of elasticity, $E, E''$ are the real and imaginary parts of the complex modulus of elasticity, $\eta$ is a loss factor and $\omega$ is a frequency. It is necessary to note that the storage and loss moduli in this case are defined in the frequency domain by experimental technique for each material and can be used directly in the numerical analysis after application of the curve fitting procedure. In the case, when the storage and loss moduli are constant values we have to deal with so called the hysteretic damping model.

## 3. Dynamic Analysis

The forced vibration equation of a structure with viscous and viscoelastic damping presented by the complex modulus model appears as follows in a matrix form:

$$\mathbf{M}\ddot{\mathbf{X}}^* + \mathbf{C}\dot{\mathbf{X}}^* + \mathbf{K}^*(\omega)\mathbf{X}^* = \mathbf{F}(t)$$

where $\mathbf{M}$ is the mass matrix of a structure, $\mathbf{C}$ is the damping matrix of a structure depending on the damping model used, $\mathbf{K}^*(\omega) = \mathbf{K}(\omega) + i\mathbf{K}''(\omega)$ is the complex stiffness matrix of a structure. $\mathbf{K}(\omega)$ is determined using the elastic $E(\omega)$ and shear $G(\omega)$ moduli, while $\mathbf{K}''(\omega)$ is found using the imaginary parts of the complex moduli $E''(\omega) = \eta_E(\omega)E(\omega)$ and $G''(\omega) = \eta_G(\omega)G(\omega)$, where $\eta_E(\omega), \eta_G(\omega)$ are the material loss factors in tension and shear, respectively, and $\omega$ is a frequency. $\mathbf{X}^*, \dot{\mathbf{X}}^*, \ddot{\mathbf{X}}^*$ are the complex vectors of displacements, velocities and accelerations, $\mathbf{F}(t)$ is the load vector. In the present paper three methods are used in order to investigate the damping properties of different systems. There are the method of complex eigenvalues, the method evaluating the modal loss factors from the frequency response and the transient response analysis giving the possibility to estimate the damping properties of structures under certain loads using the steady state vibrations.

### 3.1 Free Vibration Analysis

Damped eigenfrequencies and corresponding loss factors can be determined from the free vibration analysis of a structure. In this case the right part of the forced vibration equation is zero, i.e. $\mathbf{F}(t) = 0$.

By substitution $\mathbf{X}^* = \overline{\mathbf{X}}^* e^{i\omega^* t}$, the characteristic equation becomes

$$\left(-\omega^{*2}\mathbf{M} + i\omega^*\mathbf{C} + \mathbf{K}^*(\omega)\right)\overline{\mathbf{X}}^* = 0$$

where $\overline{\mathbf{X}}^*$ is the complex eigenvector and $\omega^* = \omega + i\omega''$ is the complex eigenfrequency. The real part $\omega$ represents the damped eigenfrequency of a structure and the imaginary part $\omega''$ specifies the rate of decay of the dynamic process. This quadratic characteristic problem can be reduced to a linear one by doubling the order of the system as follows

$$\omega^* \begin{bmatrix} i\mathbf{C} & -\mathbf{M} \\ -\mathbf{M} & 0 \end{bmatrix} \begin{bmatrix} \overline{\mathbf{X}}^* \\ \omega^*\overline{\mathbf{X}}^* \end{bmatrix} = \begin{bmatrix} -\mathbf{K}^*(\omega) & 0 \\ 0 & -\mathbf{M} \end{bmatrix} \begin{bmatrix} \overline{\mathbf{X}}^* \\ \omega^*\overline{\mathbf{X}}^* \end{bmatrix}$$

which can be written as the non-linear generalised eigenvalue problem

$$\mathbf{A}^*(\omega)\overline{\mathbf{Y}}^* = \lambda^* \mathbf{B}^* \overline{\mathbf{Y}}^*$$

with $\lambda^* = \omega^*$, $\overline{\mathbf{Y}}^* = \begin{bmatrix} \overline{\mathbf{X}}^* \\ \omega^*\overline{\mathbf{X}}^* \end{bmatrix}$, $\mathbf{A}^*(\omega) = \begin{bmatrix} \mathbf{K}^*(\omega) & 0 \\ 0 & \mathbf{M} \end{bmatrix}$, $\mathbf{B}^* = \begin{bmatrix} -i\mathbf{C} & \mathbf{M} \\ \mathbf{M} & 0 \end{bmatrix}$

where $\lambda^* = \lambda + i\lambda''$ is the complex eigenvalue, $\overline{\mathbf{Y}}^*$ is the complex eigenvector, $\mathbf{A}^*(\omega)$ and $\mathbf{B}^*$ are the complex symmetrical matrices.

A solution of the non-linear generalised eigenvalue problem starts with a constant frequency ( $\omega = $ const ). Then at each step the linear generalised eigenvalue problem with $\mathbf{A}^*(\omega) = $ const is solved by the Lanczos method [3], which is programmed in a truncated version, where the generalised eigenvalue problem is transformed into a standard eigenvalue problem with a reduced order symmetric tridiagonal matrix. Orthogonal projection operations are employed with greater economy and elegance using elementary reflection matrices. The iteration process terminates, when the following condition is satisfied

$$\frac{|\omega_{i+1} - \omega_i|}{\omega_i} * 100\% \leq \xi$$

where $\xi$ is a desired precision and $\omega_{i+1}$ is the real part of eigenfrequency of a structure calculated from the linear generalised eigenvalue problem with the storage and loss moduli for the frequency $\omega_i$, which was obtained from the same equation in the previous step. The modal loss factor of a structure for each vibration mode are determined by the following relation

$$\eta_n = 2\frac{\lambda_n''}{\lambda_n}$$

This approach gives the possibility to preserve an exact mathematical formulation for the damping models examined and to calculate structures with high damping.

### 3.2 Frequency Response Analysis

In the case of harmonic vibrations $\mathbf{F}(t) = \overline{\mathbf{F}}e^{i\omega t}$, a solution of the forced vibration equation is found in the form $\mathbf{X}^*(t) = \overline{\mathbf{X}}^* e^{i\omega t}$ and the system of complex linear equations is obtained

$$\left(-\omega^2 \mathbf{M} + i\omega\mathbf{C} + \mathbf{K}^*(\omega)\right)\overline{\mathbf{X}}^* = \overline{\mathbf{F}}$$

where $\omega$ is the frequency for which the response of a structure is calculated, $\overline{\mathbf{X}}^*$ is the complex amplitude of the displacements and $\overline{\mathbf{F}}$ is the amplitude of applied force. A solution of the system of complex linear equations for each given frequency is performed by the Gauss algorithm [4].

Dynamic characteristics of a structure (eigenfrequencies and corresponding loss factors) can be easily obtained from the frequency response. The eigenfrequencies $f_n = \omega_n / 2\pi$ of a structure present the points of the real part of the response spectrum, where the amplitude of the displacements is zero, but the corresponding loss factors can be determined by analysing the resonant peaks at frequencies $f_a$ and $f_b$ for a particular mode as follows

$$\eta_n = \frac{1 - (f_b / f_a)^2}{1 + (f_b / f_a)^2}$$

This method takes a considerable computing time, since the the dynamic stiffness matrix $\left(-\omega^2 \mathbf{M} + i\omega\mathbf{C} + \mathbf{K}^*(\omega)\right)$ must be recalculated, decomposed and stored at each of the numerous frequency steps. On this reason for systems modelled by a great number of degrees of freedom and in the case of a great number of desired dynamic characteristics to be calculated, it is more efficient to use the results of the free vibration analysis. The frequency response analysis may be successfully applied in the case, when it is necessary to determine a small number of desired dynamic characteristics, or when the eigenfrequency of the undamped structure is already known and only it recalculation and determination of the corresponding loss factor for the damped structure is necessary. This approach also gives the possibility to preserve an exact mathematical formulation for the damping models examined and to calculate structures with high damping.

### 3.3 Transient Response Analysis

The transient response of the system, described above, cannot be obtained effectively applying direct integration methods or modal superposition method, because in this case it is not possible to determine a variation of the material properties $E^*(\omega)$ and $G^*(\omega)$ with respect to time. The time domain behaviour of a structure may be obtained from the frequency domain response by the Fourier transform technique.

The method proposed is based on the assumption that any complex input signal can be interpolated by trigonometric polynomials. It is more convenient for this purpose to use the Fourier transform to find the frequency spectra of excitation

$$\mathbf{F}^*(\omega_j) = \mathsf{F}\big[\mathbf{F}(t_k)\big]$$

where $t_k$ is a set of discrete times for the excitation $\mathbf{F}(t)$ and for the response $\mathbf{X}^*(t)$, $\omega_j$ is a set of discrete frequencies for the frequency spectra of excitation $\mathbf{F}^*(\omega)$ and for the frequency response $\mathbf{X}^*(\omega)$. The response of a structure for each trigonometric component is calculated exactly using the matrix of transfer functions. Incidentally, it is necessary to solve the following system of complex linear equations:

$$\big[-\omega_j^2 \mathbf{M} + i\omega_j \mathbf{C} + \mathbf{K}^*(\omega_j)\big]\mathbf{X}^*(\omega_j) = \mathbf{F}^*(\omega_j)$$

The displacements of a structure in the time domain can be obtained by the inverse Fourier transform

$$\mathbf{X}^*(t_k) = \mathsf{F}^{-1}\big[\mathbf{X}^*(\omega_j)\big]$$

To characterise the transient response, the value of maximum displacement in a time of load action and logarithmic decrement showing the velocity of vibration fading are taken into consideration. Numerical realisation of the Fourier transform is performed by the routine using a variant of the fast Fourier transform algorithm [5] known as the Stockham self-sorting algorithm, which takes advantage of the cyclic repetition of the complex exponentials in the discrete Fourier transform and drastically reduces the number of calculations required.

## 4. Numerical Examples

As numerical examples demonstrating an application of the approaches described above, the free vibration, frequency and transient response analyses of systems with viscous and viscoelastic, structural and external damping presented by homogeneous and sandwich beams, viscoelastic spring and dashpot have been carried out. As results the damped eigenfrequencies, corresponding loss factors, and displacements in the frequency and time domains have been determined. A good agreement between dynamic characteristics obtained by different methods is observed.

## 5. Conclusions

The present approaches were developed with the aim to use them as universal tools in the free vibration, frequency and transient response finite element analysis of systems with combined damping including viscous and viscoelastic, structural and external damping. A good coincidence of the dynamic characteristics obtained by different methods demonstrates a validity of the technique developed.

## Acknowledgments

The authors gratefully acknowledge support of this research work from the "Deutscher Akademischer Austauschdienst" (DAAD) and from the Latvian Council of Science under grant No. 04.1180.

## References

[1] R. Chandra, S.P. Singh and K. Gupta, Damping studies in fiber-reinforced composites – a review, *Composite Structures* **46** 41-51 (1999).
[2] D.V. Balandin, N.N. Bolotnik and W.D. Pilkey, Review: optimal shock and vibration isolation, *Shock and Vibration* **5** 73-87 (1998).
[3] E.N. Barkanov, Method of complex eigenvalues for studying the damping properties of sandwich type structures, *Mechanics of Composite Materials* **1** 90-94 (1993).
[4] K. J. Bathe and E. L. Wilson, *Numerical Methods in Finite Element Analysis*, Prentice-Hall, Englewood Cliffs, New Jersey, 1976.
[5] E. O. Brigham, *The Fast Fourier Transform*, Prentice-Hall, 1974.

Brill Academic Publishers
P.O. Box 9000, 2300 PA Leiden
The Netherlands

*Lecture Series on Computer*
*and Computational Sciences*
Volume 4, 2005, pp. 231-234

# Optimal Design of Large Wheel Loaded Vehicle Decks

E. Barkanov[1], A. Chate[1], E. Skukis[1] and T. Gosch[2]

[1]Institute of Materials and Structures, Riga Technical University
Kalku St. 1, LV-1658, Riga, Latvia

[2]Flensburger Schiffbau-Gesellschaft mbH & Co. KG
Batteriestrasse 52, D-24939, Flensburg, Germany

Received 8 July, 2005; accepted in revised form 30 July, 2005

*Abstract:* The methodology based on experimental design and response surface technique is developed for the optimal design of large wheel loaded vehicle decks using the finite element method. The weight optimisation and parametric study are carried out for a sandwich-grillage construction representing the wheel loaded vehicle deck.

*Keywords:* optimisation, plans of experiments, response surfaces, finite element method, sandwich-grillage construction, wheel loaded vehicle deck.

## 1. Introduction

The purpose of the present study is the methodology development for the optimal design of large wheel loaded vehicle decks widely applied in ship building industry. Due to large dimension of the numerical problem to be solved, an optimisation methodology is developed employing the method of experimental design and response surface technique [1]. This methodology is a collection of mathematical and statistical techniques that are useful for the modelling and analysis of problems in which a response of interest is influenced by several variables and the objective is to optimise this response. Optimisation procedure based on experimental design is not only an effective tool for the optimum design of different systems and processes requiring computationally expensive analyses, but it also easily combines modelling and optimisation stages and requires less intervention from an analyst in comparison with other approaches. Moreover, this approach is general in the sense that it permits to optimise any systems or processes under arbitrary conditions with respect to any objective function (e.g. performance, durability, integrity, reliability, cost) taking into account all practical requirements [2].

## 2. Theoretical Background

An engineering approach of optimisation based on experimental design and response surface technique is presented in Figure 1. It is necessary to note, that in each of these stages, it is possible to solve a problem by different methods.

### 2.1 Plans of Experiments

Let us consider a criterion for elaboration of the plans of experiments independent on a mathematical model of the designing object or process. The initial information for elaboration of the plan is number of factors $n$ and number of experiments $k$. The points of experiments in the domain of factors are distributed as regular as possible. For this reason the following criterion is used

$$\Phi = \sum_{i=1}^{k} \sum_{j=i+1}^{k} \frac{1}{l_{ij}^2} \Rightarrow \min$$

where $l_{ij}$ is the distance between the points having numbers $i$ and $j$ ( $i \neq j$ ). Physically it is equal to the minimum of potential energy of repulsive forces for the points with unity mass if the magnitude of these repulsive forces is inversely proportional to the distance between the points.

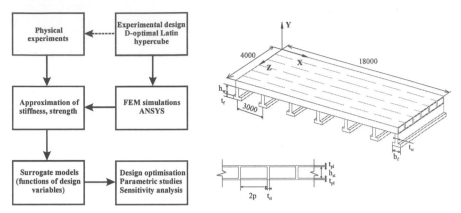

Figure 1: Optimisation process.                    Figure 2: Sandwich-grillage construction.

For each number of factors $n$ and number of experiments $k$ it is possible to elaborate a plan of experiments. But it needs much computer time, therefore each plan of experiment is elaborated only once and it can be used for various designing cases. The plan of experiments is characterised by the matrix of plan $B_{ij}$. When the domain of factors is determined as $x_j \in [x_j^{\min}, x_j^{\max}]$, the points of experiments are calculated by the following expression

$$x_j^{(i)} = x_j^{\min} + \frac{1}{k-1}\left(x_j^{\max} - x_j^{\min}\right)\left(B_{ij} - 1\right), \quad i = 1,2,...,k, \quad j = 1,2,...n$$

Then the numerical calculations and/or physical experiments are carried out in these points.

## 2.2 Approximation Technique

In this approach, the form of the equation of regression is unknown previously. There are two requirements for the equation of regression: accuracy and reliability. Accuracy is characterised as minimum of standard deviation of the table data from the values given by equation of regression. Increasing the number of terms in the equation of regression it is possible to obtain a complete agreement between the table data and the values given by equation of regression. However, it is necessary to note that prediction at the intervals between the table points can be not so good. For improvement of prediction, it is necessary to decrease of distance between the points of experiments by increasing of the number of experiments or by decreasing of the domain of factors. Reliability of the equation of regression can be characterised by affirmation that standard deviations for the table points and for any other points are approximately the same. Obviously the reliability is greater for a smaller number of terms of the equation of regression.

The equation of regression can be written in the following form

$$y = \sum_{i=1}^{p} A_i f_i(x_j)$$

where $A_i$ are the coefficients of the equation of regression, $f_i(x_j)$ are the functions from the bank of simple functions $\varphi_1, \varphi_2, ..., \varphi_m$ which are assumed as,

$$\varphi_m(x_j) = \prod_{i=1}^{s} x_j^{\alpha_{mi}}$$

where $\alpha_{mi}$ is a positive or negative integer including zero. Synthesis of the equation from the bank of simple functions is carried out in two stages: selection of perspective functions from the bank and then step by step elimination of the selected functions.

## 3.3 Non-Linear Optimisation

Constrained non-linear optimisation problem can be written in the following form

$$\min F(x); H_i(x) \geq 0; G_j(x) = 0$$
$$i = 1,2,...,I; \ j = 1,2,...,J$$

where $I$ and $J$ are the numbers of inequality and equality constraints. This problem is replaced to the unconstrained minimisation problem in which the constraints are taken into account with the penalty functions. New version of random search method is used for solving of the formulated optimisation problem.

## 3. Finite Element Modelling

Wheel loaded vehicle deck presents the homogeneous steel plate supported by the grillage construction consisting of transverse webs. In the optimisation study the homogeneous steel plate is change by the I-core laser-welded sandwich panel without foam material inside with the purpose to decrease the weight of wheel loaded vehicle deck. Geometry of sandwich-grillage construction is given in Figure 2. Sandwich-grillage construction is under bending load presented by the wheel loads of 2+3 axes road trailer vehicle with total weight 60t and self-weight load. The acceleration factor for the self-weight and wheel loads is 1.25. Simply supported boundary conditions are applied at flanges of every transverse girder. All components of sandwich-grillage construction are made from steel.

Sandwich-grillage construction is modelled and analysed by ANSYS. To decrease considerably dimension of the numerical problem, the I-core laser-welded sandwich panel is changed by the homogeneous orthotropic plate with equivalent properties [3,4] – stiffness, thickness and density. In this case the finite element model of sandwich-grillage construction are built using 4-node shell element SHELL 63. It is necessary to note that von Mises stresses in sandwich plate cannot be used for an evaluation of strength. For sandwich plate the stress resultants and moments should be employed to calculate stresses in the top and bottom face sheets, and in the sandwich I-core. In the present investigation this analysis is not performed.

## 4. Optimal Design

Optimisation problem is formulated as a minimum weight design subjected to stiffness and strength constraints:

Objective function:
$W(\mathbf{x}) \rightarrow \min$

$W(\mathbf{x}) = W_s(\mathbf{x}_1) + W_g(\mathbf{x}_2)$

Design parameters:
$\mathbf{x} = \{\mathbf{x}_1, \mathbf{x}_2\}$

$\mathbf{x}_1 = \{2p, h_{st}, t_{st}, t_{pl}\}$

$\mathbf{x}_2 = \{t_w, h_w, t_f, b_f\}$

Constraints:

$w_l(\mathbf{x}) \le [w_l] = \dfrac{B}{200} + \dfrac{e}{200} = 35$ (mm)

$w_{gl}(\mathbf{x}) \le [w_{gl}] = \dfrac{B}{200} = 20$ (mm)

$\sigma(\mathbf{x}) \le [\sigma] = 180$ (N/mm$^2$)

where $W(\mathbf{x})$ is the weight of sandwich-grillage construction consisting of sandwich plate weight $W_s(\mathbf{x}_1)$ and grillage weight $W_g(\mathbf{x}_2)$, $\mathbf{x} = \{\mathbf{x}_1, \mathbf{x}_2\}$ is the vector of design parameters consisting of sandwich plate parameters $\mathbf{x}_1 = \{2p, h_{st}, t_{st}, t_{pl}\}$ and grillage parameters $\mathbf{x}_2 = \{t_w, h_w, t_f, b_f\}$. As the design parameters the following values are taken: $80 \le 2p \le 250$ (mm) - stiffener spacing in sandwich panel, $20 \le h_{st} \le 80$ (mm) - stiffener height, $3 \le t_{st} \le 6$ (mm) - stiffener thickness, $3 \le t_{pl} \le 6$ - thickness of face plates, $4 \le t_w \le 10$ (mm) - web thickness, $450 \le h_w \le 750$ (mm) - web height, $4 \le t_f \le 30$ (mm) - flange thickness, $50 \le b_f \le 250$ (mm) - flange width. It is necessary to note that all thicknesses are examined as discrete value design parameters with the following steps: $\Delta t_{st} = \Delta t_{pl} = \Delta t_w = \Delta t_f = 0.5$ (mm). As the stiffness and strength constraints the following values are taken: $[w_l]$ - maximum deflection of sandwich panel, $[w_{gl}]$ - maximum deflection of girder, $[\sigma]$ -

Table 1

Optimal weight of sandwich-grillage construction.

| $h_w$ [mm] | 700 | 750 | 720 | 490 | 450 | 450 | 450 |
|---|---|---|---|---|---|---|---|
| $b_f$ [mm] | 50 | 100 | 50 | 50 | 50 | 170 | 130 |
| $t_f$ [mm] | 30 | 4 | 30 | 30 | 30 | 4 | 4 |
| $t_w$ [mm] | 7 | 8.5 | 6.5 | 7 | 6 | 8 | 8 |
| $t_{pl}$ [mm] | 5.5 | 3.5 | 3 | 3 | 3 | 3 | 3 |
| $h_{st}$ [mm] | 20 | 30 | 40 | 50 | 60 | 70 | 80 |
| $t_{st}$ [mm] | 6 | 6 | 3 | 3 | 3 | 3 | 3 |
| 2p [mm] | 250 | 230 | 250 | 250 | 250 | 250 | 250 |
| $\sigma$ [MPa] | 180 | 178 | 180 | 180 | 177 | 179 | 179 |
| $w_l$ [mm] | 34 | 35 | 33 | 18 | 10 | 9 | 9 |
| $w_{gl}$ [mm] | 0.9 | 2.3 | 0.7 | 1.0 | 1.2 | 1.8 | 1 |
| Weight [kg] | 7845 | 5850 | 4989 | 4783 | 4691 | 4776 | 4809 |

maximum equivalent stress, where $B = 4000$ (mm) is the width of sandwich-grillage construction and $e = 3000$ (mm) is the span between girders in grillage.

The experimental design is formulated for 8 design parameters and 135 experiments. For global approximations it is planned to use the second order polynomial functions. It is well known that for such approximating functions the $D$-optimal experimental design is the most suitable. Since we have discrete and continuous design variables, the combined Latin Hypercube and $D$-optimal experimental design is chosen. To get better description of behaviour functions, the second order polynomials with some eliminated points have been chosen for optimisation. The minimization problem has been solved by the random search method. The results of weight optimisation of the sandwich-grillage construction with different stiffener height are given in Table 1.

## 5. Conclusions

The methodology based on experimental design and response surface technique has been developed for the optimal design of large wheel loaded vehicle decks. To describe the behaviour of sandwich-grillage construction under bending loads, the finite element method has been applied in the sample points of experimental design. Approximations of the original functions for behavioural constraints have been obtained using the second order polynomials with some eliminated points. Minimization problem has been solved for the real construction by the method of random search employing the approximating functions instead of original functions. Parametric study has been carried out additionally for a designer convenience.

## Acknowledgments

This work was supported by the European Commission, FRAMEWORK6 Program, project SAND.CORe, Contract No. TCA3-CT-2004-506330.

## References

[1] R.H. Myers and D.C. Montgomery, *Response Surface Methodology. Process and Product Optimisation Using Designed Experiments*, Wiley: New York, 2002.
[2] E. Barkanov, R. Rikards and A. Chate, Numerical optimisation of sandwich and laminated composite structures, In: *Structural Optimization*, Ed. S. Hernández, M. El-Sayed and C.A. Brebbia (Computational Mechanics Publications: Southampton, Boston) 311-318 (1995).
[3] D. Zenkert, *The Handbook of Sandwich Construction*, EMAS Publishing, 1997.
[4] T.-S. Lok and Q.-H. Cheng, Elastic deflection of thin-walled sandwich panels, *Journal of Sandwich Structures and Materials* **1** 279-298 (1999).

Brill Academic Publishers
P.O. Box 9000, 2300 PA Leiden,
The Netherlands

*Lecture Series on Computer
and Computational Sciences*
Volume 4, 2005, pp. 235-238

# Heuristic Functions, Primitive Recursion and Recursivity

A.Garrido[1]

Departamento de Matemáticas Fundamentales. Facultad de Ciencias, UNED, Madrid, Spain.

Received 8 June, 2005; accepted in revised form 28 June, 2005

*Abstract:* When we approach to a problem of Artificial Intelligence, we can distinguish between Classification, Searching and Representation Methods. Within Searching Procedures, we have the *Blind Search* (without knowledge of the Domain) and the *Heuristic Search* (with knowledge of the Domain).
The Theory of Functions provide us with very useful tools (such as the *heuristic function*) for the representation and solution of such type of problems. Such function show, for instance, the estimation of distance to the final node, or equivalently, the number of steps to reach the solution.
Also, in the problems with Fuzzy Logic and Non-Monotonic reasoning appears a very interesting tool: the *membership function*, with the information about the degree of completion of the condition defining the respective Fuzzy Set or Fuzzy Relation.
Finally, we analyze the interesting Ackermann function, famous counterexample, created to prove that the $\mu-recursivity$ does not imply the primitive recursion, with its consequences in Automata Theory.

*Keywords:* Recursive Functions, Mathematical Analysis, Automata Theory, A. I.

## 1   Heuristic Search

Relative to the search with knowledge of the domain, in an initial phase, it was generally thought that all paths can be explored by the computer. But this is too optimistic: such exploration can be often very difficult, because of the phenomenon of "combinatory explosion" of the ramifications, when we advance in the expansion. Their spatial and temporal complexity advise us against their realization. For this reason, we need to choose, first, the most promising trajectories. In this way, we can not obtain the best solution, but an efficient approach to it.

Now, we introduce a new mathematical tool: the *heuristic evaluation function, f*. By such function, we assign to each node, $n$, the value $f(n)$, which give us the estimation of the real distance (unknown), from the current node, $n$, to the final node, $m$.

There are *critics on the Heuristic Search*, because its *unpredictability. It finds good solutions, but not necessarily the best*. This makes convenient the introduction of the *algorithm* $A^*$, with its useful properties of *completeness* and *admissibility*.

## 2   Fuzzy Logic, Set and Relations

When we solve problems in A. I., their representation will be through the Fuzzy Logic techniques, a very useful procedure. We need Fuzzy Sets, Fuzzy Relations and so on, to describe the uncertainty of our world. It is showed through a new function, the *membership function*, which describes the degree of fulfilment for each element of the property defining the set.

---

[1] E-mail: algbmv@telefonica.net

# 3 Recursivity and Primitive Recursion

We comment now some of Recursive Theory of Functions. Our purpose is the study of computability. As basic element, the function $f$. Its domain will be composed by $m$-uples of non-negative integers, and its range must be $n$-uples of the same type of numbers. So: $f : (N^*)^m \to (N^*)^n$, where: $N^* = N \cup \{0\}$

The first distinction will be between partial and total functions, according to that the original set coincides or not with their domain. Obviously, the function:

$$div : (N^*)^2 \to N^*$$

$$(x, \ y) \mapsto div \ (x, \ y) = \left[ \frac{x}{y} \right]$$

is a *partial function*, so defined, because it does not map the pairs of the form $(x, 0)$. Whereas the function:

$$plus : (N^*)^2 \to N^*$$

$$(x, \ y) \mapsto plus \ (x, \ y) = x + y$$

is a *total function*.

We construct $f$, *primitive recursive function*, from the set of initial functions, clearly computable, by standard operations, such as combination, composition and primitive recursion, applied in finite number. Into the set of *initial functions*, we have the subsequent: *zero* function ($\zeta$), *successor* function ($\sigma$) and *projection* functions, $\pi_j^i$ ($i \succeq j$).

We can combine two functions in this way:

$$f : (N^*)^k \to (N^*)^m$$

$$(x_1, x_2, ...x_k) \mapsto f \ (x_1, x_2, ...x_k) = (y_1, y_2, ..., y_m)$$

$$g : (N^*)^k \to (N^*)^n$$

$$(x_1, x_2, ...x_k) \mapsto g \ (x_1, x_2, ...x_k) = (z_1, z_2, ..., z_n)$$

$$f \times g : (N^*)^k \to (N^*)^{m+n}$$

$$(x_1, x_2, ...x_k) \mapsto [f \times g] \ (x_1, x_2, ...x_k) = (f \ (x_1, x_2, ...x_k), \ g \ (x_1, x_2, ...x_k))$$

$$= (y_1, y_2, ..., y_m, z_1, z_2, ..., z_n)$$

so, we take first the images of $f$ and then, the images of $g$, constructing a $(m+n) - uple$. This combination of functions is usually denoted by $\times$ and does not coincide, obviously, with the usual composition of functions $(f \circ g)$, of similar name.

The *recursivity* can be introduced, from $f$ and $g$, by:

$$f \ (x, \ 0) = g \ (x)$$

$$f \ (x, \ y+1) = h \ (x, \ y, \ f \ (x, y))$$

Generalizing, we can consider the primitive recursion as a technique that gives the step from the functions:

$$g : (N^*)^k \to (N^*)^m$$

$$h : (N^*)^{k+m+1} \to (N^*)^m$$

to:

$$f : (N^*)^{k+1} \to (N^*)^m$$

in this way:

$$f(\mathbf{x}, 0) = g(\mathbf{x})$$
$$f(\mathbf{x}, y+1) = h(\mathbf{x}, y, f(\mathbf{x}, y))$$

where $\mathbf{x}$ is the k-uple $(x_1, x_2, ...x_k) \in (N^*)^k$ .

*Primitive recursive functions* are, for instance: *predecessor, mult, exp, monus, eq*, their negation ($\neg eq$), the *tabular* functions, *coc* (not *div*) and so on.

In the classification into the set of functions, according to more restricted conditions each time , we have:

$$\{initial\ functions\} \subset \{recursive\ primitive\ functions\} \subset \{computable\ functions\}$$

Also we know that if $f$ is recursive primitive, then $f$ is total. That is:

$$\{recursive\ primitive\ functions\} \subset \{total\ functions\}$$

For instance, the initial functions. But the converse of the inclusion above is not true: *There exist computable functions no recursive primitives*. Because they are partial, no total functions.

So, the function *div* (different to the previous *coc*) is not recursive primitive. Remember that:

$$coc : (N^*)^2 \to N^*$$
$$(x, y) \mapsto coc(x, y) = \begin{array}{l} [x \div y], \ if\ y \neq 0 \\ \quad 0, \ if\ y = 0 \end{array}$$

At first, it may be thought that there is a coincidence between primitive recursive function and computable total (also called $\mu - recursive$) function classes. But this is false. There exists a famous *counterexample*, of *W. Ackerman*: the function $A$, named after him.

But it is very easy to proof the existence of a $\mu - recursive$ (total and computable) function: $f : N^* \to N^*$, which is not recursive primitive.

*Abridged proof:* Any recursive primitive function, $f$, can be obtained from initial functions, combining, composing and by primitive recursion method, applied a finite number of times. Therefore, $f$ can be defined by a finite string of symbols. Such recursive primitive functions, according to this, can be ordered by their length, and those of the same length, by lexicographic ordering. So, we reach an ordered collection:

$$F = \{f_i\}_{i \in N^*} = \{f_0,\ f_1,\ f_2...\}$$

But it is possible to construct a $\mu - recursive$ function such that does not belong to $F$. Such function can be:

$$f(i) \equiv f_i(i) + 1, \forall i \in N^*$$

Although it can also be defined as:

$$\text{fixed } k, \ f(i) \equiv f_i(i) + k, \forall i \in N^*$$

Our $f$ is total, because for each $i \in N^*$, $f(i)$ exists. And $f$ is computable, because we can find, always, the value of $f(i)$. All the initial functions are computable, and also their composition, combination or primitive recursion, in finite number. By definition, suffices to take the successor of $f_i(i)$.

But the primitive recursion fails: If $f$ is so, then: $f \equiv f_n$, for some $n \in N^*$. However, in such case: $f(n) = f_n(n)$, specifically in $n$.

But this falls in total contradiction with its definition. In effect, we have simultaneously: $f(n) \equiv f_n(n) + 1$ and the aforementioned equality. Therefore,

$$f_n(n) = f_n(n) + 1$$

for some $n \in N^*$. But this implies obviously that: $0 = 1\ (!)$. So, $\mu - recursivity$ does not imply primitive recursivity.

The *Ackermann's function:* $A : (N^*)^2 \to N^*$ can be defined through the equations:

$$A(0, \ y) = y + 1$$
$$A(x + 1, \ 0) = A(x, \ 1)$$
$$A(x + 1, \ y + 1) = A(x, \ A(x + 1, \ y))$$

This function was not actually written in this form by Wilhelm Ackermann. Instead, he found that the z-fold exponentiation of $x$ with $y$ can result a good example of recursive function which was not primitive recursive. Later, a logician of Hungary, Rosza Péter, expressed this as a function of two variables, very similar to the aforementioned.

We can prove easily its $\mu - recursivity$ (that is, its computability and total character). The *proof* of its total character is reached by induction on the pairs of entries, lexicographily ordered. And $A$ is obviously computable, because $A$ is a total function, according to the definition. But $A$ *fails in the primitive recursion.*

So, finally, the classification will be established in this sense:

$$\{initial \ f.\} \subset \{recursive \ primitive \ f.\} \subset \{\mu - recursive \ f.\} \subset \{computable \ f.\}$$

improving the initial sequence by a new class of functions.

## References

[1] W.J. Clancey,: Notes on heuristic classification. *Artificial Intelligence*,59 (1993)191-196.

[2] Clancey, W. J.: Heuristic Classification. *Artificial Intelligence*27 (1985), 289-350.

[3] Garrido, Angel: Complexity of Algorithms. EpsMsO'05. Intern. Conf. held in Athens, Greece. July 2005. Published in their Proceedings, CD-Rom and paper.

[4] Garrido, Angel: Logics in A. I. Conference *Logic in Hungary*. Uni Corvinus, Budapest. Janos Bolyai Mathematical Society. August 2005.

[5] Mira et al.: *Aspectos Básicos de la Inteligencia Artificial.* Ed. Sanz y Torres. Madrid, 1995.

Brill Academic Publishers
P.O. Box 9000, 2300 PA Leiden
The Netherlands

*Lecture Series on Computer
and Computational Sciences*
Volume 4, 2005, pp. 239-243

# Prediction of thawing time of frozen foodstuffs with a one dimensional finite difference model

Y. Onur Devres*'**, Kenan Gürsoy**

Istanbul Technical University, Faculty of Chemical and Metallurgical Engineering, Food Engineering Department, Maslak, 34469, Istanbul, Turkey*

Istanbul Technical University, Informatics Institute, Computational Science and Engineering Program, Maslak, 34469, Istanbul, Turkey**

Received 17 August, 2005; accepted in revised form 22 August, 2005

*Abstract:* In this study, a description of the thawing process is developed using a model based on the freezing point depression method. In previous studies, phase change heat was taken as a constant and the thermophysical properties of frozen foodstuffs were taken from related tables whereas in the present study, phase change heat is taken as a temperature dependent variable and the thermophysical properties of frozen foodstuffs are calculated by the model itself. This has made the model gain considerable freedom. When the thawing times obtained by operating the model and the published experimental studies were compared, an average difference of -0.59% was observed. However, the differences between the results from the model and the thawing experiments carried out in this study were found to average only -0.49%. The effects of certain assumptions and errors that were made during both the development of the model and the conduction of experiments were investigated.

## 1. Introduction

Prediction of freezing and thawing times is important for the design and operation of food processing plants to assure high end-product quality and efficient use of production capacity. For these purposes, many studies have been done in this field [1-3]. General approaches to the solution can be gathered in three groups as follows [1,2]:
- Theoretical methods
- Semi-theoretical methods
- Experimental methods

Prior to computers being commonly available, all solutions to theoretical methods were analytic. As is well-known however, the analytical solutions are availed only for some simple geometries such as rectangular slabs, cylinders and spheres, which are rarely used in practical life. On the other hand, the problems can be solved with constant thermophysical properties in the analytical methods. However, all the properties are dependent on temperature and the structure of the foodstuffs is heterogeneous. Consequently, all the above assumptions reduce the accuracy of the results. In the last 20-25 years, computer technology has been improved and high speed low price computers are now available everywhere. Freezing/thawing problems are now usually solved numerically and so to a greater accuracy.

In this study, the thawing problem is solved numerically with a novel approach. A description of the thawing process is developed using a model based on the freezing point depression method [4-12]. In the previous studies, phase change heat was taken as a constant and the thermophysical properties of frozen foodstuffs were taken from related tables whereas in the present study, phase change heat is taken as a temperature dependent variable and the thermophysical properties of frozen foodstuffs are calculated by the model itself. This has made the model gain considerable freedom.

## 2. Materials and Method

The thawing experiments were conducted in a cabinet with dimensions of 1m x 1m x 2m using Tylose slabs according to DIN 8953 [13-15]. Tylose slabs (0.1m x 0.1m x 0.05m) were placed inside the 0.1 m thick polystyrene block, leaving only one side exposed to the surrounding medium to allow one dimensional heat transfer. As a result, at the positions x=0 and x=L boundary conditions of the third and second kind respectively occurred. The upper side of the block was covered with a sheet of aluminum foil to prevent the loss of moisture to the surrounding air. Five Cu/CuNi pin point thermocouples were placed inside the sample at regular intervals. Slabs were frozen in a -30°C cold store and before thawing, were kept in a constant-temperature freezer for three days. During the thawing process, the polystyrene block was also placed in a bigger packing case with very thick glass wool insulation at each of its five surfaces to minimize heat gain. The thawing cabinet was specially designed for humid air thawing processes. The design was undertaken in TUBITAK (Scientific and Technical Research Council of Turkey) Marmara Research Centre. In this cabinet, air temperature was controlled by a double contact thermostat (ELIMKO 4111, Turkey) and air speed was changed using a three phase AC motor speed controller (SAM-EL, Turkey) to regulate the speed of a fan motor. For heating and cooling, two separate coils, one for steam and the other for water, were employed. The temperature within the slabs was recorded at 5 separate points and the mean air temperature from three readings was recorded at three minute intervals using a Temperature Micro-processor (ELLAB CMC 821, Denmark). During experimental study two different air velocities were used. The experiments were terminated after the coldest point (at x=L) rose above 0°C. After the experiments, the data were transferred to a computer for graphical analysis.

For measuring the surface heat transfer coefficient, 0.1m x 0.1m x 0.01m aluminum slabs were employed with their upper surfaces polished to minimize radiation heat gain. The surface heat transfer coefficient is strictly dependent on the surface temperature so it changes during the experimental period. Therefore, the heating period (from -20°C to 10°C) of the aluminum slab was divided into three intervals and the surface heat transfer coefficient was calculated separately for each interval. The mean of the three intervals was calculated for later use. During calculation of heat transfer coefficient, the following equation was used [16] :

$$h_c = \frac{m_{al} \; c_{p-al}}{A \; \Delta t} \ln\left(\frac{T - T_\infty}{T_{init} - T_\infty}\right) \tag{1}$$

## 3. Mathematical Modeling

In capillary-porous bodies such as foodstuffs, "the substance bound with the capillary-porous field in a region of negative temperatures can be in the form of a solid (ice), sub-cooled liquid, vapour or gas [17]. In this study a small finite region, the control volume (CV) is taken into account during heat transfer analysis. It is also assumed that the CV has three imaginary parts, ice, sub-cooled water and dry matter The effects of vapour or gas in the CV are neglected.

During the thawing process, the sub-cooled water fraction is increased until all the ice turns into water. Within this period the temperature of the CV changes with time. Consequently both water fraction and temperature are time-dependent variables. At temperatures below freezing point of the foodstuff, unfrozen water is known as a temperature dependent function as follows [10-12]:

$$\%UFW = \frac{G - H\,T}{K - M\,T} \qquad\qquad T \le T_f \tag{2a}$$

$$G = E \; (F - \%BW) + \%BW \tag{2b}$$
$$H = D \; (F - \%BW) \tag{2c}$$
$$K = - \; (E - 1) \; \%WC \tag{2d}$$
$$M = -D \; \%WC \tag{2e}$$
$$D = -0.9809994 \; 10^{-2} \tag{2f}$$
$$E = 1.000134 \tag{2g}$$
$$F = \frac{n^* \; \%S \; M_w}{M_s} \tag{2h}$$

Mass and enthalpy variables given below:

$$m_{ice} = (1 - \%UFW) \; \%WC \; m_{cv} \tag{3a}$$
$$h_{ice} = c_{p-ice} \; (T - T_{init}) \tag{3b}$$

$$m_w = \%UFW \, \%WC \, m_{cv} \tag{3c}$$

$$h_w = c_{p-w} \, (T - T_{init}) \tag{3d}$$

$$m_{dm} = (1 - \%WC) \, m_{cv} \tag{3e}$$

$$h_{dm} = c_{p-dm} \, (T - T_{init}) \tag{3f}$$

$$\Delta q = \left( D_1 + D_2 \, \frac{G - H \, T}{K - M \, T} + A_1 \, \frac{D_2 \, T + D_3}{(K - M \, T)^2} \right) \frac{dT}{dt} \tag{4a}$$

where

$$D_1 = \%WC \, m_{cv} \, c_{p-ice} + (1 - \%WC) \, m_{cv} \, c_{p-dm} \tag{4b}$$

$$D_2 = \%WC \, m_{cv} \, c_{p-w} + \%WC \, m_{cv} \, c_{p-ice} \tag{4c}$$

$$D_3 = -T_{init} \, D_2 + \%WC \, m_{cv} \, h_m \tag{4d}$$

$$A_1 = G \, M + H \, K \tag{4e}$$

Finally heat conduction equation becomes:

$$\frac{d}{dx}\left( k \frac{dT}{dx} \right) dx \, dy \, dz = \left( D_1 + D_2 \, \frac{G - H \, T}{K - M \, T} + A_1 \, \frac{D_2 \, T + D_3}{(K - M \, T)^2} \right) \frac{dT}{dt} \tag{5}$$

The numerical solution (in this study it is finite difference) to Equation (5) gives the temperature distribution inside the material.

## 4. Computer Program

The program was written in the FORTRAN 77 and MATLAB 6.5. In previous studies, the volumetric enthalpy and thermal conductivity of the food over an appropriate temperature range had to be input to the computer. Therefore, at every step, the thermophysical properties of foodstuffs had to be calculated individually from relevant tables. These tables were derived or measured assuming special conditions such as constant water content and freezing point. This situation caused some difficulties when working with a foodstuff with thermophysical properties that were not defined in the tables. In this study, however, as shown in a previous section, water content, bound water and freezing temperature can be chosen individually. Consequently the model itself offers considerable freedom for cases where water contents and freezing points differ from the table values and also those cases that are not covered by the tables. In the computer program, a parallel model approach [18-20] with second order polynomial is used [19, 21-25] when calculating thermophysical properties of foodstuff and pure substances, e.g. water and ice.

## 5. Results and Discussions

Cleland et al. [26, 27] made some thawing experiments for the verification of the accuracy of any method used to predict thawing times. They indicated that the comparison must be made with reliable experimental data. For their experiments they used Tylose and minced lean beef formed into the shapes of slabs, infinite cylinders and spheres. They also mentioned that the overall error in temperature measurement and control was estimated to be less than 0.5 K; the thickness of the slabs was measured to an accuracy of 0.5 mm and the overall 95% confidence bound for the experimental error of the surface heat transfer coefficient was estimated to be 8.0%. The percentage difference of 10.0% at the 95% level of confidence is assumed during the comparison of experimental and predicted thawing times.

Cleland and Earle [28] gave the freezing point of Tylose containing 77% water as -0.6°C and total phase change heat as 209 MJ/m³. In the previous study [11, 12], using Equation (2a) with the parallel model approach, thermophysical properties were defined and compared with Cleland and Earle's data [28]. A good fit was observed with 77% water content, 12.5% bound water and -0.666°C freezing point. These properties were assumed for comparison between the thawing times of this study and those of Cleland et al. [26]. The differences between the experimental and predicted thawing times vary between -14.33% and 6.86% with an arithmetical mean of -0.59%. This value is given as -0.8% by Cleland et al. [26].

## Acknowledgements

The authors wish to acknowledge the financial support from the State Planning Organization (DPT) of Turkey.

## References

[1] Anon. Freezing and defrosting time of food *ASHRAE Fundamentals Handbook* (1985) Chapter 30.

[2] Hayakawa, K., Scott, K. R., Succar, J. Theoretical and semi-theoretical methods for estimating freezing or thawing time *ASHRAE Transactions* (1985) **91** Part 2B 371-384.

[3] Cleland, D. J., Cleland, A. C., Earle, R. L. Prediction of freezing and thawing times for foods - a review *Int J Refrig* **9** (1986) 182.

[4] Heldman, D. R. Predicting the relationship between unfrozen water fraction and temperature during food freezing using freezing point depression *Transactions of the ASAE* (1974) 63-66.

[5] Heldman, D. R., Gorby, D. P. Prediction of thermal conductivity in frozen foods *Transactions of the ASAE* (1975) 740-744.

[6] Fontan, C. F., Chirife, J. The evaluation of water activity in aqueous solutions from freezing point depression *J Food Tech* **16** (1981) 21-30.

[7] Larkin, J. L., Heldman, D. R., Steffe, J. F. An analysis of factors influencing precision of thermal property values during freezing *Int J Refrig* **7** (1984) 86-92.

[8] Chen, C. S. Thermodynamic analysis of the freezing and thawing of foods : enthalpy and apparent specific heat *J Food Sci* **50** (1985) 1158-1162.

[9] Chen, C. S. Thermodynamic analysis of the freezing and thawing of foods : ice content and mollier diagram *J Food Sci* **50** (1985) 1163-1166.

[10] Devres, Y. O. *Depression method for predicting unfrozenwater fraction in frozen foodstuffs* Research Memorandum No.115 South Bank Polytechnic - Institute of Environmental Engineering London (1989) 34p.

[11] Devres, Y. O. *Mathematical modeling of thawing processes of frozen foodstuffs and reduction of the thawing losses* Ph.D. Thesis Yildiz University-Istanbul (1990) 184p.

[12] Devres, Y. O. An analytical model for thermophysical properties of frozen foodstuffs *Proc XVIII$^{th}$ Int Cong Refrig* **IV** (1991) 1908-1912.

[13] Anon. *DIN 8953 Household frozen food cabinets and household food freezer requirements, marking, testing* Deutsche Normen Sheet 2 Alleinverkauf der Normblätter durch Beuth Verlag GmbH, Berlin 30 (1973).

[14] Anon. *DIN 8954 Offene Verkaufskühlmöbel, allgemeine Prüfbedingungen* Deutsche Normen Teil 3 Alleinverkauf der Norm blätter durch Beuth Verlag GmbH, Berlin 30 (1983).

[15] Gutschmidt, J. *Über das Herstellen und Verpacken der Karlsruher Prüffmasse Kältetechnik* **8** (1960) 226-229.

[16] Succar, J., Hayakawa, K. A response surface method for the estimation of convective and radiative heat transfer coefficients during freezing and thawing of foods *J Food Sci* **51** (1986) 1314-1322.

[17] Luikov, A. V. *Heat and Mass Transfer in Capillary-Porous Bodies* Pergamon Press (1966).

[18] Pham Q. T., Willix, J. Thermal conductivity of fresh lamb meat, offals and fat in the range -40 to +30°C : measurements and correlations *J Food Sci* **54** (1989) 508-515.

[19] Comini, G., Bonacina, C., Barina, S. Thermal properties of foodstuffs *Proc Int Ins Refrig B1, C1/2 Com Meeting on the Thermophysical Properties of Foodstuffs* Bressanone (1974) 163-172.

[20] Cleland, D. J. Personal communication, Massey University Department of Biotechnology, New Zealand (1989).

[21] Anon. Thermal properties of foods *ASHRAE Fundamentals Handbook* (1985) Chapter 31.

[22] Wilson, H. A., Singh, R. P. Numerical simulation of individual quick freezing of spherical foods *Int .J Refrig* **10** (1987) 149-155.

[23] Devres, Y. O. *En küçük kareler yöntemi ile eğri yaklaştirmasi (regresyon analizi) VAX paket bilgisayar programý* (curve fitting (regression analyses) VAX package computer programme using least squares method) Food and Refrigeration Technology Department Publications No.121 Scientific and Technical Research Council of Turkey, Marmara Research Centre (1989) 40p.

[24] Holman, J. P. *Heat Transfer* McGraw-Hill Kogakusha Ltd Fourth Edition (1976).

[25] Perry, R. H., Green, D. W., Maloney, J. O. Perry's Chemical Engineers' Handbook McGraw Hill Book Company 6th Edition (1984).

[26] Cleland, D. J., Cleland, A. C., Earle, R. L., Byrne, S. J. Prediction of thawing times for foods of simple shape *Int J Refrig* **9** (1986) 220-228.

[27] Cleland, D. J., Cleland, A. C., Earle, R. L., Byrne, S. J. Experimental data for freezing and thawing of multi-dimensional objects *Int J Refrig* **10** (1987) 22-31.

[28] Cleland, A. C., Earle, R. L. Assessment of freezing time prediction methods *J Food Sci* **49** (1984) 1034-1042.

Brill Academic Publishers
P.O. Box 9000, 2300 PA Leiden
The Netherlands

*Lecture Series on Computer
and Computational Sciences*
Volume 4, 2005, pp. 244-248

# Prediction and Study of Passive Pollutant Dispersion in a Street-Canyon in London Using Computational Fluid Dynamics Techniques

A. Galani[*,1], P. Neofytou[1], A. Venetsanos[1], J. Bartzis[1,2], S. Neville[3]

[1]*Environmental Research Lab., INT-RP, NCSR Demokritos, Aghia Paraskevi, 15310 Athens, Greece*
[2]*University of West Macedonia, Department of [1]Energy Resources Rngineering and Management,
Kastorias and Fleming Str., Kozani, Greece*
[3]*Westminster Council House, Marylebone Road, London, NW1 5PT, United Kingdom*

Received 6 August, 2005; accepted in revised form 16 August, 2005

*Abstract:* Pollution levels in an urban street-canyon area are determined numerically as part of the European research project OSCAR using the ADREA-HF code. Aim of the modelling is to investigate the flow-field and CO concentrations in the area and compare with measurements. Results show a tendency of overprediction of concentration by the code attributed mostly to the uncertainty of the meteorological data and emission levels within the studied time frames. The concentration distribution and flow field within the canyon are shown to be highly correlated whereas the in-canyon induced vortex plays a prominent role in the concentration dispersion.

*Keywords:* Passive pollutant, Dispersion, Street-Canyon, Computational Fluid Dynamics, OSCAR

## 1. Introduction

Dispersion of pollutants originating from traffic is related to the geometry of the urban area and traffic conditions. Computational Fluid Dynamics (CFD) is becoming increasingly important as a tool to assist with modeling the airflow and dispersion of pollutants among complex urban geometries. This tool, which takes account of meteorological scenarios, building designs and proximity of roadways to pollutant sources, enables more accurate predictions combined with cost effectiveness. Furthermore, the limit of pollutant levels set by the World Health Organisation has led to an increased research activity as to the specification of the influence of car emissions on the air quality in urban street canyons. Urban street-canyons consist of uniform parallel building complexes on either side of the street and induce flow recirculations and/or stagnant conditions thus prohibiting the dispersion of pollutants away from inhabited areas.

With respect to CFD applications on environmental flows, a review was carried out by Vardoulakis *et al.*[1] that includes evaluation of several CFD methods applied in meteorological, wind-tunnel and street canyon studies. In addition, Walton *et al.*[2] pursued LES (Large-Eddy Simulation) for the problem of mean flow and turbulence in cubic street canyons. Their results show good agreement between simulations and experimental data. Finally, CFD computations using the ADREA-HF code have been carried out by Neofytou *et al.*[3] in order to parametrically study the pollution in a street canyon by assuming different wind directions.

The current study is carried out in the framework of the Optimised Expert System for Conducting Environmental Assessment of Urban Road Traffic (OSCAR) project. This project aspires to assess the environmental impact of road traffic in terms of traffic flows, emissions and air pollution. A combined emission measurement and model-prediction campaign was carried out in Marylebone Road in London, United Kingdom in 2003-2004. CO, NO, $NO_2$, $NO_x$ and $SO_2$ concentrations at selected time periods are provided by measurements whereas the wind directions and wind speeds are measured at rooftop level. Furthermore, emission data during the time periods under investigation are also provided. The numerical predictions are accomplished using the CFD code ADREA-HF[4], which also has been used in the past for environmental flow predictions[5,5,6].

*Corresponding Author
email: galani@ipta.demokritos.gr

## 2.   Methodology

### 2.1 Measurements

The monitoring campaign in Marylebone Road was conducted in 2003 - 2004 (01.01.2003 - 31.12.2004), in order to provide comprehensive air quality databases. The street-canyon has an aspect ratio (Average Height/Width) approximately equal to 1:2. The monitoring site is located in a purpose built cabin on Marylebone Road opposite Mme Tussauds. The sampling point is located at a height of 3m, around 1m from the kerbside. Traffic flows of over 80,000 vehicles per day pass the site on six lanes. Street level air quality measurements and on-site electronic traffic counts were conducted throughout the campaign. CO concentrations, wind-direction, wind-speed and emission data were available with the time resolution of one hour. The emission factors were assessed by the Department for the Environment, Food and Rural Affairs.

### 2.2 Numerical Method

The methodology consisted in solving the transient, Reynolds averaged, mass and momentum 3D conservation equations for the mean flow and the mass fraction conservation equation for the pollutant dispersion, until steady state conditions were reached. Boundary conditions were zero gradient and given value for the inflow boundaries, zero gradient for the outflow boundaries, wall functions for velocities at the buildings surfaces and ground and finally zero vertical velocity at the top of domain.

The computational domain that includes all buildings in the area surrounding Marylebone Road is constructed using actual coordinates provided by the Westminster Council House and is shown in Figure 1. It consists of a 1000x940x200m area discretised as a 70x70x35 grid which is refined near the measurement location in order to more accurately capture the wind field and concentration distribution. As regards to the comparison with measurements, two different datasets were selected for the pollutant under consideration, both covering a 2 hour time period. Each dataset provides values for wind speed, wind direction and concentration at a resolution of one hour. Higher accuracy purposes led to the development of a Fortran 90 code in order to select the proper datasets. According to this code, a dataset will be accepted, if the data standard deviation does not exceed the limited values of 5.0, 0.2, 12 and 2.0 for wind direction, wind speed, traffic count and traffic speed correspondingly. Hence, the parameters from each dataset do not substantially fluctuate within the selected time periods and therefore a mean value was derived for each parameter. The aforementioned values for wind speed and direction were used as input data for calculations in order to compare with the value of concentration from the corresponding dataset. In addition, the source of pollution which is principally the emissions from the cars passing Marylebone Road (Fig. 1) was modelled for each dataset as an area source along the street emitting homogenously and with constant rate.

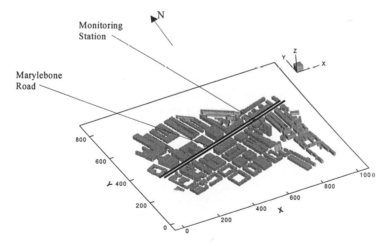

Figure 1. Computational domain of the Marylebone Road area.

## 3.   Results And Discussion

The datasets selected and the corresponding mean values for the wind speed (WS), wind direction (WD) and street level concentration of CO (SLC in mg/m³) correspond to weekdays and are shown in

Table 1. The background concentration values of CO (BC in mg/m³) correspond to reported annual mean values. The point where the measurements are taken corresponds to $x$=727.3m, $y$=508.7m, $z$=3m of the computational domain. It can be seen that although ADREA-HF has a tendency to overpredict the measurements, it follows their trend i.e. the ratio of the predicted to the observed in the measurements (SLC minus BC) values is approximately the same.

The most important factor of uncertainty in the computations is the definition of the source of emissions, the wind speed and wind direction mean values. Due to the fact that the aforementioned data are given on hourly basis, it is most possible their corresponding values to exhibit substantial fluctuations within this large time frame. Also, the considered background concentration values are not the most appropriate, since the latter are given on annually basis.

| Dataset/ Case | Date | Time frame | WS (m/s) | WD (deg) | BC (mg/m³) | SLC (mg/m³) | SLC-BC (mg/m³) | SLC$_{prediction}$ |
|---|---|---|---|---|---|---|---|---|
| 1 | 17.2.2003 | 11:00-13:00 | 1.37 | 211.6 | 0.44 | 0.90 | 0.46 | 0.83 |
| 2 | 19.4.2004 | 15:00-17:00 | 3.53 | 92.8 | 0.44 | 0.53 | 0.09 | 0.14 |

Table 1. Overview of datasets and numerical predictions for CO.

It is very interesting though to see how the concentration distribution is behaving with respect to the flow field. The comparison is carried out between datasets 1 and 2, for which the same source of emissions is assumed. For the height of the measurement point ($z$=3m) the concentration distribution is illustrated in Figure 2. In order to represent the concentration field, a limit of 0.18mg CO/m³ of air was set, below which concentration values are not shown. First, it can be seen that Marylebone Road plays the most important role to the pollution in its vicinity and that can be attributed to the fact that this street has the most traffic compared to its neighbouring ones. Thus, the pollution from the latter is more prominent.

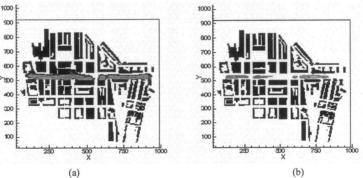

(a)                                                            (b)

Figure 2. Concentration distribution for CO at $z$=3m for cases (a) 1; (b) 2.

Furthermore, it can be seen that the concentration levels are higher for case 1 compared to case 2 at $z$=3m despite the fact that emission rates in case 2 are a bit higher ($3.0 \cdot 10^{-8}$ m/s) than those in case 1 ($2.5 \cdot 10^{-8}$ m/s). This can be explained from the wind field for that height (Fig. 3) where the wind speeds within the street-canyon are relatively higher for case 2 and therefore the dispersion mechanism is more intense. The latter is caused not only from the higher freestream velocity for case 2 that causes more marked street-canyon effects but also from the fact that the wind direction, which is for this case almost aligned with Marylebone Road, forces wind to move towards the west side of the street thus causing pollutant attenuation along x-axis.

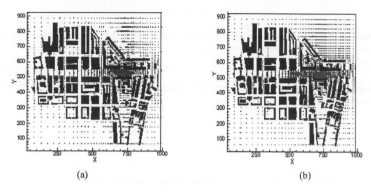

Figure 3. Wind-velocity field at $z$=3m for cases: (a) 1; (b) 2.

In both cases 1 and 2, the existence of a street-canyon vortex can explain the upwind accumulation of the concentration within the canyon observed in Fig. 2. It is worthwhile pointing out that although velocities within the canyon are higher than at the crossing of Marylebone Road with Baker Street (Fig. 2), the street-canyon vortex is responsible for maintaining higher in-canyon concentrations than at the crossing, where the concentration is much lower, despite of the fact that the emissions at both points are the same. Therefore higher velocities do not necessarily mean higher dispersion rates. Figures 4 and 5 present the concentration distribution and the velocity field for dataset 1 at three lateral planes along the z-axis, correspondingly. The first height is at $z$=1.70m, which is the population average height and it is shown that concentration levels there are much higher than the corresponding ones at $z$=19.87m (buildings average height) and at $z$=70.55m (maximum height of buildings). It is observed that as the height increases, the wind speed increases too, thus causing a more intense dispersion mechanism and consequently a decrease at pollutant concentration levels. In figure 4c, the pollutant accumulates substantially in a small region in front of the two highest buildings of the complex. The latter implies that the height of the buildings is another important factor that influences the dispersion mechanism. A similar behaviour is also observed for dataset 2.

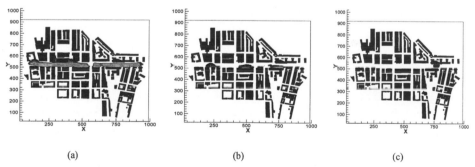

Figure 4. Concentration distribution for dataset 1 at (a) z=1.70m; (b) z=19.87m; (c) z=70.55m.

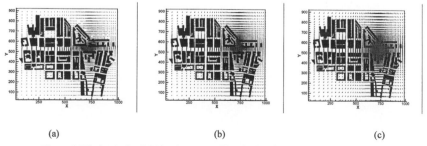

Figure 5. Wind-velocity field for dataset 1 at (a) z=1.70m; (b) z=19.87m; (c) z=70.55m.

## 4. Conclusions

Pollution-dispersion modelling in an urban area was carried using ADREA-HF code. It seems that the code tends to overpredict the CO concentrations mainly because of the uncertainty involving both the meteorological data and the emissions for the specific time frames considered here. The presence of a street-canyon vortex can explain the upwind accumulation of the concentration within the canyon observed. Furthermore, the wind speed and direction as well as the height of the buildings determine in a major degree the dispersion mechanism. Finally, along z-axis, the wind velocity increases causing a decrease at the pollutant concentration levels.

## Acknowledgements

The financial support from the European Union under the contract EVK4-CT-2002-00083 is gratefully acknowledged.

## References

[1]  S. Vardoulakis, B. E. A. Fisher, K. Pericleous, N. Gonzalez-Flesca, Modelling air quality in street canyons: a review, *Atmos. Environ.* **37** 155-182 (2003)

[2]  A. Walton, A.Y.S. Cheng, W.C. Yeung, Large-eddy simulation of pollution dispersion in an urban street canyon-Part I: comparison with field data, *Atmos. Environ.* **36** 3601-3613 (2002).

[3]  P. Neofytou, A. G. Venetsanos, S. Rafailidis, J. G. Bartzis, Numerical Investigation of the Pollution Dispersion in an Urban Street-Canyon. *In press*, Environ. Modell. Softw (2005).

[4]  Bartzis, J.G., "ADREA-HF: A three-dimensional finite volume code for vapor cloud dispersion in complex terrain", EUR report 13580 EN (1991).

[5]  P. Neofytou, A. G. Venetsanos, D. Vlachogiannis, J. G Bartzis, A. Scaperdas, CFD Simulations of the Wind Environment around an Airport Terminal Building. *In press*, Environ. Modell. Softw (2005).

[6]  A.G. Venetsanos, J.G. Bartzis, J. Würtz, D.D. Papailiou, Comparative modeling of a passive release from an L-shaped building using one, two and three-dimensional dispersion models, *Int. J. Environ. Pollut.* **14** 324-333 (2000).

Brill Academic Publishers
P.O. Box 9000, 2300 PA Leiden
The Netherlands

*Lecture Series on Computer
and Computational Sciences*
Volume 4, 2005, pp. 249-253

# Combined heat and fluid flow in a double wall infant incubator

M. K. Ginalski , A. J. Nowak , L. C. Wrobel [1]

Institute of Thermal Technology,
Silesian University of Technology
Konarskiego 22, 44-100 Gliwice, Poland

Received 9 July, 2005; accepted in revised form 31 July, 2005

*Abstract:* The main objective of this study is to investigate the major physical processes taking place inside an infant incubator, after modifications have been made to the interior chamber. Additional screens (double walls) are commonly used in modern incubators to decrease radiation heat losses from the infant child. However, these improvements affect the flow and temperature fields inside the incubator. The present study investigates the effect of these modifications on the convective heat flux from the infant's body to the surrounding environment inside the incubator. A combined analysis of airflow and heat transfer due to conduction, convection, radiation and evaporation has been performed, in order to calculate the temperature and velocity fields inside the incubator before and after the design modification. Due to the geometrical complexity of the model, CAD applications were used to generate a computer-based model. All numerical calculations have been performed using the commercial CFD package Fluent, together with in-house routines used for managing purposes and User-Defined Functions which extend the basic solver capabilities. The results obtained were numerically verified as well as compared with results published in the literature.

*Keywords:* CFD, biomedical engineering, coupled heat and mass transfer, convection, radiation, evaporation

## 1. Introduction

Computational Fluid Dynamics (CFD) has been greatly developed over recent years, mostly due to the rapid advances in computer technology. It is now possible to calculate scientific problems in domains with complex geometries, including several combined processes taking place simultaneously. The present study investigates the major mechanisms of heat transfer between a newborn infant and the closed environment inside an infant incubator, in order to validate the usefulness of a design modification made to the incubator by fitting additional over-head screens. The mass flow and heat transfer mechanisms are analysed before and after the proposed design modification to decrease radiation heat losses from the infant's body. Velocity and temperature fields have been calculated in order to investigate the impact of the proposed modification on the thermal comfort of the baby. Convection and radiation heat losses from the baby, due to the surrounding environment, were obtained and compared in both cases.

## 2. Solution procedure

In order to make the calculation process fully automatic, a managing program was created to deal with all the essential stages of the numerical simulation, step by step. After the creation of the model geometry in the CAD package CATIA, the pre-processor GAMBIT is used to generate the numerical grid. In the next step, the managing in-house program reads information regarding boundary conditions, material properties and initial parameters. Following this, the FLUENT main solver is launched and calculations are performed. Once a grid has been read into Fluent, all remaining operations are performed within the solver.

## 3. Mathematical Model

Basic assumptions:
- Steady-state model of the combined heat and mass transfer processes.
- Airflow through the incubator is laminar and viscous.
- Original shapes and general dimensions of the infant incubator and the newborn baby have been set according to available data.

- The amount of heat lost by a premature baby due to conduction is very small compared to the total heat exchange between the child and the surrounding environment. For this reason, this process was not included in this study.
- The impact of the water evaporation process from the newborn's body on the internal incubator thermal conditions was examined. The Discrete Phase Model was used to investigate this process.
- Heat exchange due to respiration has been included and subtracted from the total amount of heat generated by the newborn's body.
- The Discrete Ordinate Model was used to model the radiation process between the infant and the surrounding environment.

Mathematical model components:

The heat transfer analysis is based on the energy equation [1]

$$\nabla(k\nabla T) = \rho c \frac{DT}{Dt} \tag{1}$$

where T is the temperature (K), k is the thermal conductivity (W/mK), $\rho$ is the density (kg/m3), c is the specific heat (W/kgK), and t is time (s). The derivative on the right-hand side of equation (1) is the total derivative:

$$\frac{DT}{Dt} = \frac{\partial T}{\partial t} + \mathbf{u}_x \frac{\partial T}{\partial x} + \mathbf{u}_y \frac{\partial T}{\partial y} + \mathbf{u}_z \frac{\partial T}{\partial z} = \frac{\partial T}{\partial t} + \mathbf{u} \cdot \nabla T \tag{2}$$

where ux, uy, uz, are the velocity components of vector u in the x-, y-, z-direction, respectively (m/s). Since only steady-state problems are studied in this work, the first term on the right-hand side of equation (2) vanishes.

The above equations are complemented by the continuity and momentum equations [1], namely

$$\nabla \cdot \mathbf{u} = 0 \tag{3}$$

$$\rho \frac{D\mathbf{u}}{Dt} = \mathbf{F} - \nabla p + \mu \nabla^2 \mathbf{u} \tag{4}$$

where p is the pressure (N/m2), F represents the body force term which in the present case has only a vertical component    Fz = $\rho$g in the z-direction (N/m2), g is gravity acceleration (m/s2) and $\mu$ is the dynamic viscosity (Ns/m2).

The Boussinesq approximation was adopted for the buoyancy term in equation (4). Thus, density takes the usual form

$$\rho = \rho_0 (1 - \beta(T - T_0)) \tag{5}$$

where $\beta$ is the thermal expansion coefficient (1/K), T0 and $\rho$0 represent the operating parameters.

Essential contributions to the total heat transfer between a newborn infant and its environment are given by the thermal radiation and water evaporation. In this paper, only surface radiation described by the brightness of the relevant surface elements, i.e. the infant's skin and walls of the incubator, has been considered.

The last section of the mathematical model includes the equations which describe the evaporation of sweat particles from the surface of the body. As the trajectory of a particle is computed, Fluent keeps track of the heat, mass and momentum gained or lost by the particle stream that follows that trajectory [2]. These quantities can be incorporated in the subsequent continuous phase calculations. Therefore, while the continuous phase always impacts on the discrete phase, it is also possible to incorporate the effect of the discrete phase trajectories on the continuous phase.

The heat transfer rate from the newborn child to the surrounding environment is directly related to the temperature and air velocity within the incubator. Moreover, based on this information, Fluent calculates the heat losses from the infant's body and provides results of heat losses due to convection, radiation and evaporation.

## 4. Results and discussion

The simulation confirms the air movement conforms to the basic working principles of the Caleo™ Infant Incubator [3]. Figures 1-2 and 7-8 show clear differences in flow and temperature fields between the two cases analysed. In the incubator with the internal screen, the main air stream from the inlets separates into two. One of these streams starts circulating and warms up the baby. The second stream warms up only the internal screen and after mixing with the stream from the inlets located on the opposite side of the incubator, it moves directly to the outlets. The analysis also shows that the circulating air creates a zone around the baby where the air velocity is almost zero. Due to this fact, convection heat losses from the child to the surrounding air in this region are greatly reduced. This process can be seen in Figures 3-4, which shows the distribution of the convective heat flux on the infant's body (negative values on the scale indicates heat flux provided to the baby). The upper part of the chest and the head of the child are in direct contact with the circulating air. Therefore, heat transfer due to convection is larger in those areas.

The distribution of the radiative heat flux on the newborn's skin is presented in Figures 5-6. A reduction of the heat loss due to radiation can be seen in Figure 6. This fact indicates that incubators with internal screens (double walls incubators) significantly decrease the radiative heat losses from the infant's body.

The last stage of the analysis examines the evaporation process from the infant's skin. Evaporation of moisture greatly decreases the temperature of the skin. Therefore, the impact of this energy balance component is very important for the thermal comfort of the infant. Relevant numerical values of evaporation heat losses are presented in Table 1. In extremely low birth weight infants, evaporation is a major source of heat loss. Evaporative heat flux from the infant's body is steadily distributed and the influence of radiation and convection is almost unnoticeable. Using a high ambient humidity in the incubator reduces the loss of heat by evaporation. However, in this study, this process was only examined for air with ambient humidity of 50%.

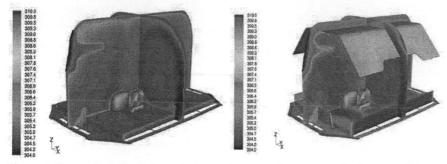

Figure 1. Temperature distribution within an infant incubator without the over-head screen. (K)

Figure 2. Temperature distribution within an infant incubator with the over-head screen. (K)

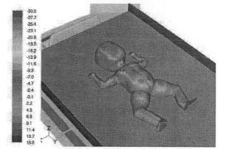

Figure 3. Contours of the convective heat flux plotted on the newborn's skin nursed inside the incubator without the over-head screen. (W/m2)

Figure 4. Contours of the convective heat flux plotted on the newborn's skin nursed inside the incubator with the over-head screen. (W/m2)

Figure 5. Contours of the radiative heat flux plotted on the newborn's skin  nursed inside the incubator without the over-head screen. (W/m2)

Figure 6. Contours of the radiative heat flux plotted on the newborn's skin  nursed inside the incubator with the over-head screen. (W/m2)

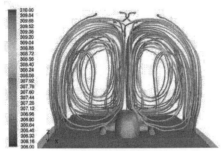

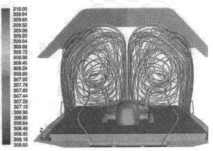

Figure 7. Path lines colored by total temperature inside the incubator without over-head screen. (K)

Figure 8. Path lines colored by total temperature inside the incubator with over-head screen. (K)

Table 1: Solution of the system of equations produced based on the requirement of the minimal phase-lag. Phase-Lag (PL) and interval of periodicity (IP) of the produced methods

| Incubator with over-head screen | Incubator without over-head screen |
|---|---|
| Heat transfer rate due to convection | |
| 0.02 W | 0.015 W |
| Heat transfer rate due to radiation | |
| 0.356 W | 0.451 W |
| Heat transfer rate due to evaporation | |
| 1.720 W | 1.714 W |
| Total heat transfer rate from the newborn's body | |
| 2.08 W | 2.18 W |
| Heat generated by the newborn's body | |
| 1.25 W | 1.25 W |
| Deficit of heat | |
| -0.62 W | -0.715 W |

## 6. Conclusions

As shown in this work, modern numerical methods can successfully be used in biomedical engineering to solve complex heat and mass flow problems with high accuracy. The procedure described in this study is very efficient and flexible. New designs, materials and technologies can be examined first by using numerical models, which will save time and money in the complete design cycle of new products. The possibility of interfacing two different types of software, such as CAD and CFD, was also examined in this paper. In order to generate the complex incubator and human body geometries, a powerful CAD application (CATIA) was employed. A STEP file was then used to transfer the geometry created by the CAD application to the CFD software. Models transferred from CAD applications are much more accurate compared with models created directly in CFD packages. Therefore, a heat transfer analysis which is strongly dependent on the model geometry can be performed with higher accuracy. Unlike previous studies, using the procedures presented in this paper, individual cases can be examined. This includes identical twin infants placed in the same incubator or some extremely ill premature newborns. It is also possible to examine several special cases and use the results to create a system of graphs or equations, which can then be used for the determination of the heat transfer rate from the infant's body to the surrounding environment within an infant incubator.

From the thermal point of view, the Caleo™ infant incubator is a well-designed neonatal unit. The temperature and velocity fields show that the characteristic geometry and the specific way the air flows within the incubator are unique and very efficient. An additional over-head screen not only reduces radiative heat flux, but also changes the airflow inside the incubator in such a way that convective heat fluxes are reduced. Convective heat flux should be minimized in order to reduce the total heat rate that is lost by the child. Moreover, in this type of incubator, there is a possibility of providing an additional amount of heat to the child (by convection only) that can be useful for significantly reducing the total heat lost to the environment or to increase the temperature of the newborn's body, if necessary.

The results obtained in this study are in good agreement with previous calculations based on simpler methods [4]. However, much more detailed information can be generated via CFD simulations. Therefore, it can be assumed that the algorithm presented here is efficient and possesses the necessary capabilities for calculating the thermal factors present during the neonatal care within an infant incubator. By using a numerical method, we can save time and money during the design process of new infant incubators, while maintaining levels of accuracy which are at least equivalent to those achieved in previous studies using simplified models [5, 6]. Additional measurements in real hospital conditions are currently being undertaken to further validate the models.

## 7. References

[1]. Fleur L. *Strand. Physiology - A Regulatory System Approach.* Collier Macmillan Publishers, London, 1980..

[2]. H.K. Versteeg, W. Malalasekera. *Computational Fluid Dynamics. The Finite Volume Method.* Longman, NY, 1995.

[3]. *Technical documentation of the Caleo™ Infant Incubator.* Dragwerk AG, Lubeck, Germany, 1994.

[4]. A.J. Lyon, C. Oxley, *HeatBalance, a computer program to determine optimum incubator air temperature and humidity. A comparsion aginst nurse settings for infants less than 29 weeks gestation.* Early Human Development 62, 33-41,2001.

[5]. J. Wasner. *Heat Balance of Premature Infants.* Dragwerk AG, Lubeck, Germany, 1994.

[6]. J.M. Rennie, N.R.C. Robertom, *A Manual of Neonatal Intensive Care.* Arnold Hodder, London, 2002.

Brill Academic Publishers
P.O. Box 9000, 2300 PA Leiden
The Netherlands

*Lecture Series on Computer
and Computational Sciences*
Volume 4, 2005, pp. 254-258

# The effect of process parameters on thawing time of frozen foodstuffs with a one dimensional finite difference model

Kenan Gürsoy*, Y. Onur Devres*,**

Istanbul Technical University, Informatics Institute, Computational Science and Engineering Program,
Maslak, 34469, Istanbul, Turkey*

Istanbul Technical University, Faculty of Chemical and Metallurgical Engineering,
Food Engineering Department, Maslak, 34469, Istanbul, Turkey**

Received 17 August, 2005; accepted in revised form 23 August, 2005

*Abstract:* During thawing and freezing of a foodstuff several parameters influence its processing time. These are ambient temperature, thickness of product, heat transfer coefficient at the surface, thermophysical properties of product, initial temperature of product, water and bound water contents of product. During definition of mathematical model of these processes, all these parameters are used to predict thawing or freezing time. In reality, however, ambient temperature is measured in certain accuracy; each system has a different precision, sensitivity and resolution. During measurement of thickness, one millimeter of mismeasurement can be done very easily. Convection and radiation heat transfer coefficients are calculated from equations with certain assumptions or experimentally measured with some restrictions. Therefore for evaluation of a model and discussing its validation, such parameters should be also considered. In this study effects of small changes in process parameters due to imprecise measurement to thawing time are investigated and their counter effects to each other are discussed.

## 1. Introduction

Thawing is a phase change process, which is necessary when frozen foodstuffs need to be consumed. It is described as all ice crystal content of food must be turned into liquid phase. Thawing problem in catering industry generally occurs in thawing of frozen poultry. Insufficient cooking of food due to unthawed part of food causes food poisoning. In industrial size of thawing process, it is important to determine size and amount of frozen food to plan an optimal thawing process. During thawing process surface temperature is higher then its interior temperature. This situation presents favorable medium for growing of microorganisms on the surface. Therefore thawing process should also be planned to have minimal microorganism growth on food surface during process [1].

Hot air is generally used in industrial thawing system designs. This type thawing systems' performance depends on following parameters [2]

- Size and shape of food
- Thermal properties of food
- Enthalpy change
- Initial and final temperature of food
- Thawing medium fluid's thermal properties
- Thawing medium fluid's temperature, humidity and speed
- Packaging properties of food

Natural or forced convection systems can be used during thawing process. Natural convection system can be used for foods in small sizes and in small amounts. It causes long thawing time in thick foods because of its lower heat transfer rate then forced convection. For this reason there is a risk of microbial growth on surface of foods when natural convection is used in thawing process. Advantage of natural convection is only its low cost.

Most of the industrial thawing system uses forced convection and air is a thawing medium fluid. Increasing speed of air results in effective thawing process since it also causes higher heat transfer coefficient. Relative humidity of air is also an important parameter because it affects weight loss of product during thawing and product quality after thawing. In general it is recommended to use air that has 85-95% relative humidity for thawing process of meat.

In some systems water is used as a thawing medium for its better thermophysical properties than air. On the other hand water should be changed frequently in order to prevent microbial growth. Water as a thawing medium systems are used for big amount but small in size foodstuffs. In these systems frozen foodstuff should be packed to restrict migration of water soluble content of foodstuffs [1-4].

In this paper, effects of thawing process' parameters were examined. One dimensional finite difference method is used in mathematical modeling. It is focused on effects of small changes in parameters on thawing time calculations.

## 2. Method

A one dimensional finite difference based mathematical model was prepared in this study to predict thawing time. The model is written in FORTRAN 77 and Matlab 6.5. This model calculates thermophysical properties at every time step with respect to temperature.

In the model, thermophysical properties of foodstuff and pure substances were calculated with parallel model approach [5] with second order polynomial is used. [6, 8-12]

Three-dimensional heat conduction equation in Cartesian coordinates can be written as follows:

$$q_x + q_y + q_z = q_{x+dx} + q_{y+dy} + q_{z+dz} + \frac{dE_t}{dt} \tag{1}$$

For the cases where $(dT/dy)=0$ and $(dT/dz)=0$, it is one-dimensional heat transfer, and with the phase change, Equation (1) can be simplified as given below :

$$\frac{d}{dx}\left(k\frac{dT}{dx}\right)dx\, dy\, dz = \frac{d(m\, h)}{dt} \tag{2}$$

The boundary conditions are as follow:

$$h_c\left(T - T_\infty\right) = -k\frac{dT}{dx} \qquad x = 0, \quad t\rangle 0 \tag{3a}$$

$$\frac{dT}{dx} = 0 \qquad x = L, \quad t\rangle 0 \tag{3b}$$

$$T = T_{init} \qquad 0 \le x \le L, \quad t = 0 \tag{3c}$$

Finally Equation (2) becomes:

$$\frac{d}{dx}\left(k\frac{dT}{dx}\right)dx\, dy\, dz = \left(D_1 + D_2\frac{G - H\,T}{K - M\,T} + A_1\frac{D_2\,T + D_3}{\left(K - M\,T\right)^2}\right)\frac{dT}{dt} \tag{4}$$

The numerical solution (in this study it is finite difference) to Equation (4) gives the temperature distribution inside the material [1].

The mathematical model's results had been compared with published and experimental data. Four experimental conditions are used in this study. It is necessary to use parameters to run the model such as ambient temperature, thickness of product, heat transfer coefficient at the surface, initial temperature of product, water and bound water contents. Parameters are changed in acceptable small ranges (±5%) and searched the change in thawing times. It is also analyzed multiple changed parameters' effect on thawing time calculations [1].

## 3. Results and Discussion

Hayakawa et al [13] developed simplified procedures for predicting the freezing times of food of a two dimensional rectangular shape or a finitely cylindrical shape by using a computer model. Using these formulae, they investigated the relative error of estimated freezing times arisen from frequently observed values of physical or operational constants. In their analyses, a combination of 1 K difference in operational (initial and final) temperatures, 2 K difference in surrounding medium temperature, 0.1 K difference in initial freezing point, 5% difference in specific heat parameters, 10% difference in thermal diffusivity, 10% difference in Biot number and 0.001m difference in food dimension, resulted in 57.3% relative error where 10.9% comes from operational temperature, 7.6% from food dimension, 30.4% from thermophysical property, 8.4% from boundary property. They also observed that the thermal radiation and moisture loss from the food surfaces had no effect on the above parameters.

Cleland et al. [14, 15] published some thawing experiments data for the verification of the accuracy of any method used to predict thawing times. They used Tylose and minced lean beef formed into the shapes of slabs, infinite cylinders and spheres. They also mentioned that the overall error in temperature measurement and control was estimated to be less than 0.5 K; the thickness of the slabs was measured to an accuracy of 0.5 mm and the overall 95% confidence bound for the experimental error of the surface heat transfer coefficient was estimated to be 8.0%. The percentage difference of 10.0% at the 95% level of confidence is assumed during the comparison of experimental and predicted thawing times.

In this study, a similar approach was used during analyses of the changes in process parameters in thawing time. For the calculations, Cleland et al. [14]'s results were used. The thawing time of a slab was observed to be 16 560 s and 16 200 s for two different runs under the same conditions. The model run with these parameters resulted in 16 306 s where the percentage differences are -1.53% and 0.65% respectively.

The effects of small changes in process parameters on the prediction of thawing time were investigated and the results are shown in Table 1. For this purpose, every parameter was changed individually and the resulting predicted thawing time compared with the ideal case.

Table 1: The effects of deviation in process parameters on predicted thawing time

| Run | $\Delta x$ | $h_c$ | $T_\infty$ | $T_{init}$ | $t_{exp}$ | $t_{pre}$ | Difference | Remarks |
|---|---|---|---|---|---|---|---|---|
| | m | W m$^{-2}$ °C$^{-1}$ | °C | °C | s | s | % | |
| | 0.02625 | 50.40 | 13.4 | -23.6 | 16357 | | | The results of the model with experimental data |
| 1 | 0.02625 | 50.40 | 13.9 | -23.6 | 16357 | 15848 | -2,81 | 0.5°C increase in surrounding medium temperature |
| 2 | 0.02625 | 50.40 | 12.9 | -23.6 | 16357 | 16795 | 3,00 | 0.5°C decrease in surrounding medium temperature |
| 3 | 0.02625 | 50.40 | 13.4 | -24.1 | 16357 | 16288 | 0,11 | 0.5°C decrease in initial temperature |
| 4 | 0.02625 | 50.40 | 13.4 | -23.1 | 16357 | 16324 | -0,11 | 0.5°C increase in initial temperature |
| 5 | 0.02625 | 55.44 | 13.4 | -23.6 | 16357 | 15589 | -4,40 | 10% increase in heat transfer coefficient |
| 6 | 0.02625 | 45.36 | 13.4 | -23.6 | 16357 | 17185 | 5,39 | 10% decrease in heat transfer coefficient |
| 7 | 0.02525 | 50.40 | 13.4 | -23.6 | 16357 | 15378 | -5,69 | 0.001 m decrease in thickness |
| 8 | 0.02725 | 50.40 | 13.4 | -23.6 | 16357 | 17259 | 5,84 | 0.001 m increase in thickness |
| 9 | 0.02725 | 55.44 | 13.9 | -24.1 | 16357 | 16071 | -1,44 | 1, 4, 5, 8 conditions are together |
| 10 | 0.02725 | 45.36 | 12.9 | -24.1 | 16357 | 18736 | 14,90 | 2, 4, 6, 8 conditions are together |
| 11 | 0.02525 | 55.44 | 13.9 | -23.1 | 16357 | 14260 | -12,55 | 1, 3, 5, 7 conditions are together |
| 12 | 0.02625 | 50.40 | 13.4 | -23.6 | 16357 | 16102 | -1,25 | 75.46% water content (2% decrease in water content) |

| 13 | 0.02625 | 50.40 | 13.4 | -23.6 | 16357 | 15976 | -2,02 | Bound water is taken as 15% instead of 12.5% |

Table 1(continued)

| 14 | 0.02625 | 50.40 | 13.4 | -23.6 | 16357 | 16635 | 2,02 | Bound water is taken as 10% instead of 12.5% |
| 15 | 0.02625 | 50.40 | 13.4 | -23.6 | 16357 | 17195 | 5,45 | Freezing point is taken as -0.5°C instead of -0.666°C |
| 16 | 0.02625 | 50.40 | 13.4 | -23.6 | 16357 | 15750 | -3,41 | Freezing point is taken as -1.0°C instead of -0.666°C |
| 17 | 0.02625 | 50.40 | 13.4 | -23.6 | 16357 | 16647 | 2,09 | 90% of calculated conduction heat transfer coefficient |
| 18 | 0.02625 | 50.40 | 13.4 | -23.6 | 16357 | 16174 | -0,81 | 95% of calculated conduction heat transfer coefficient |
| 19 | 0.02625 | 50.40 | 13.4 | -23.6 | 16357 | 15366 | -5,76 | 105% of calculated conduction heat transfer coefficient |
| 20 | 0.02625 | 50.40 | 13.4 | -23.6 | 16357 | 15018 | -7,90 | 110% of calculated conduction heat transfer coefficient |

# References

[1] Devres, Y. O. *Mathematical modeling of thawing processes of frozen foodstuffs and reduction of the thawing losses* Ph.D. Thesis Yildiz University-Istanbul (1990) 184p.

[2] James,S.J. and Bailey, C., "The Theory and Practice of Food Thawing", Proceedings of the COST 91 Seminar, Athens, 1983, eds. P. Zeuthen et al., Thermal Processing and Quality of Foods, 1st Edition, Elsevier Applied Science Publishers, (1984), 566-578

[3] Delgado , Adriana E., Sun, Da-Wen,"One-dimensional finite difference modelling of heat and mass transfer during thawing of cooked cured meat", Journal of Food Engineering 57 (2003) 383–389

[4] Hoke, K., Houska, M., Kyhos, K., Landfeld, A., "Use of a computer program for parameter sensitivity studies during thawing of foods", Journal of Food Engineering 52 (2002) 219–225

[5] Pham Q. T., Willix, J. Thermal conductivity of fresh lamb meat, offals and fat in the range -40 to +30°C : measurements and correlations *J Food Sci* **54** (1989) 508-515.

[6] Comini, G., Bonacina, C., Barina, S. Thermal properties of foodstuffs *Proc Int Ins Refrig B1, C1/2 Com Meeting on the Thermophysical Properties of Foodstuffs* Bressanone (1974) 163-172.

[7] Cleland, D. J. Personal communication, Massey University Department of Biotechnology, New Zealand (1989).

[8] Anon. Thermal properties of foods *ASHRAE Fundamentals Handbook* (1985) Chapter 31.

[9] Wilson, H. A., Singh, R. P. Numerical simulation of individual quick freezing of spherical foods *Int J Refrig* **10** (1987) 149-155.

[10] Devres, Y. O. *En küçük kareler yöntemi ile eğri yaklaştirmasi (regresyon analizi) VAX paket bilgisayar programý* (curve fitting (regression analyses) VAX package computer programme using least squares method) Food and Refrigeration Technology Department Publications

No.121 Scientific and Technical Research Council of Turkey, Marmara Research Centre (1989) 40p.

[11] Holman, J. P. *Heat Transfer* McGraw-Hill Kogakusha Ltd Fourth Edition (1976).

[12] Perry, R. H., Green, D. W., Maloney, J. O. Perry's Chemical Engineers' Handbook McGraw Hill Book Company 6th Edition (1984).

[13] Hayakawa, K., Nonino, C., Succar, J. Two dimensional heat conduction in food undergoing freezing : predicting freezing time of rectangular or finitely cylindrical food *J Food Sci* **48** (1983) 1841-1848.

[14] Cleland, D. J., Cleland, A. C., Earle, R. L., Byrne, S. J. Prediction of thawing times for foods of simple shape *Int J Refrig* **9** (1986) 220-228.

[15] Cleland, D. J., Cleland, A. C., Earle, R. L., Byrne, S. J. Experimental data for freezing and thawing of multi-dimensional objects *Int J Refrig* **10** (1987) 22-31.

Brill Academic Publishers
P.O. Box 9000, 2300 PA Leiden
The Netherlands

*Lecture Series on Computer
and Computational Sciences*
Volume 4, 2005, pp. 259-261

# DFT Study and NBO (Natural Bond Orbital) Analysis of the Mutual Interconversion of Cumulene Compounds

Saeed Jameh-Bozorghi[1], Davood Nori-Shargh[1,2†], , Romina Shakibazadeh[1] and Farzad Deyhimi[3]

1) Chemistry Department, Graduate Faculty, Arak branch, Islamic Azad University, Arak, Iran
2) Chemistry Department, Science and Research Campus, Islamic Azad University, Hesarak, Poonak, Tehran, Iran
Chemistry Department, Shahid Beheshti University, Evin-Tehran 19839, Iran

Received 11 June, 2005; accepted in revised form 22 June, 2005

*Abstract:* The Becke, Lee, Yang and Parr density functional (B3LYP) method was used to investigate the configurational properties of allene (1,2-propadiene) (**1**), 1,2,3-butatriene (**2**), 1,2,3,4-pentateriene (**3**), 1,2,3,4,5-hexapentaene (**4**), 1,2,3,4,5,6-heptahexaene (**5**), 1,2,3,4,5,6,7-octaheptaene (**6**), 1,2,3,4,5,6,7,8-nonaoctaene (**7**) and 1,2,3,4,5,6,7,8,9-decanonaene (**8**). B3LYP/6-31G* level of theory showed that the mutual interconversion energy barrier in compounds **1-8** are 50.14, 31.49, 28.75, 20.37, 19.32, 14.31, 14.00 and 10.49 kcal mol$^{-1}$, respectively. The results showed a linear relationship between the corresponding B3LYP/6-31G* mutual interconversion energy barriers and the average C=C double bond lengths $\left(\overline{d}\right)$, in the ground state geometries of cumulene compounds. The results showed also that the difference between the $\left(\overline{d}\right)$ values in cumulene compounds **1** and **2** is larger than those between **7** and **8**, which suggest that with large n (number of carbon atoms in cumulene chain), the $\overline{d}$ values approach to a limiting value. Also, NBO//B3LYP/6-31G* results revealed that the π-bond occupancies decrease from **1** to **8**, and inversely, the π-antibond occupancies increase from compound **1** to compound **8**. The difference between the average of π-bonds and π-antibond occupancies ($\Delta = \overline{\pi}_{occupancy} - \overline{\pi}^{*}_{occupancy}$) could be considered as a criteria for the mutual interconversion in compounds **1-8**. These Δ values decrease from compound **1** to compound **8** series as following: 1.92087, 1.80788, 1.75314, 1.70637, 1.67144, 1.64337, 1.62065 and 1.60170, respectively, as calculated by B3LYP/6-31G* level of theory. The decrease of Δ values for compounds **1-8**, follow the same trend as the barrier heights of mutual interconversion in compounds **1-8**. Accordingly, besides the allylic resonant stabilization effect in the transition state structures, the results revealed, particularly, that the Δ value could be considered as significant criteria for the mutual interconversion in cummulene compounds **1-8**.

*Keywords:* cumulene, linear chain, molecular modeling, DFT calculations, configurational properties

Zero point (*ZPE*) and total electronic ($E_{el}$) energies ($E_o = E_{el} + ZPE$) for various configurations of compounds **1-8**, as calculated by the B3LYP/6-31G*[1-3] level of theory are given in Table 1. These results reveal that the mutual interconversion energy barriers for compounds **1-8** decrease by increasing of the number of C atoms (chain length) in these compounds (see Table 1).

It can be seen that the chain length in the transition state geometries is longer than those of the corresponding ground state structures. The average bond length $\left(\overline{d}\right)$ of C=C in cumulene chain can be defined by:

$$\overline{d} = \frac{\sum_{i=1}^{n-1} d_i}{n-1} , \quad (n \geq 3)$$

---

† Corresponding author: E-mail address: nori_ir@yahoo.com, Fax: +98 21 4817175, Tel: +98 21 4817170

Using B3LYP/6-31G* calculated bond length $(d_i)$, the average bond length $(\bar{d})$ of C=C double bonds in the ground state geometry of cumulene compounds were calculated by above Eq., and reported in Table 2. It is interesting to note that there is linear relationship between the corresponding B3LYP/6-31G* mutual interconversion energy barriers and these average bond length $(\bar{d})$ in cumulene compounds **1-8** (see Fig. 1). As it can also be seen from Table 2, the difference between the average bond lengths $(\bar{d})$ in cumulene compounds **1** and **2** is larger than those between **7** and **8**. Therefore, it could be expected that with large n (number of carbon atoms in cumulene chain) the differences between the corresponding $\bar{d}$ would disappear. Note that Mölder *et al.*[4] have already reported that in cumulene compounds "the bond lengths have no remarkable alternation within the chain and converge to the asymptotic limit as the value of *n* is getting larger". Similarly, our calculation confirm that the average bond length $(\bar{d})$ in cumulene compounds **1-8** converge to an asymptotic limit value (see Table 2). In addition, our NBO[5,6] analysis reveal that the mean occupancies of π bonding orbitals (e.g. $\bar{\pi}$ ) decrease from compound **1** to compound **8**, while the π* antibonding orbital occupancies ( $\bar{\pi}^*$ ) increase for these compounds (e.g. **1** to compound **8**) (see Table 2 and Fig. 1).

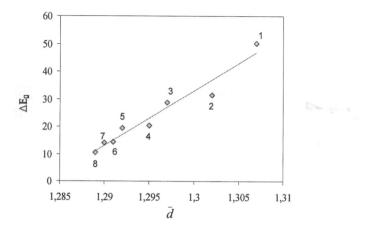

Fig.1. Calculated $\Delta E_0^a$ versus $\bar{d}$ (the average bond length) for compounds **1-8**.

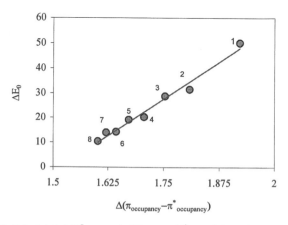

Fig.2. Calculated $\Delta E_0^a$ versus $\Delta=(\bar{\pi}_{occupancy} - \bar{\pi}^*_{occupancy})$ for compounds **1-8**.

| method | GS | | | | TS | | | |
|---|---|---|---|---|---|---|---|---|
| compounds | ZPE | $E_{el}$ | $E_0$ | $\Delta E_0^a$ | ZPE | $E_{el}$ | $E_0$ | $\Delta E_0^a$ |
| 1 | 0.054406 | -116.657676 | -116.603270 | 0.000000 | 0.050576 | -116.573945 | -116.523368 | 0.079901 |
| | | | | (0.000000) | | | | (50.138860) |
| 2 | 0.059430 | -154.730753 | -154.671323 | 0.000000 | 0.056057 | -154.677199 | -154.621142 | 0.050181 |
| | | | | (0.000000) | | | | (31.488830) |
| 3 | 0.064160 | -192.806627 | -192.742466 | 0.000000 | 0.060949 | -192.757598 | -192.696648 | 0.045818 |
| | | | | (0.000000) | | | | (28.751100) |
| 4 | 0.069193 | -230.883114 | -230.813921 | 0.000000 | 0.066362 | -230.847816 | -230.781454 | 0.032468 |
| | | | | (0.000000) | | | | (20.370000) |
| 5 | 0.074224 | -268.960901 | -268.960901 | 0.000000 | 0.071104 | -268.926993 | -268.855888 | 0.030788 |
| | | | | (0.000000) | | | | (19.320086) |
| 6 | 0.097326 | -307.038658 | -306.959332 | 0.000000 | 076567 | -307.013093 | -306.936526 | 0.022806 |
| | | | | (0.000000) | | | | (14.311042) |
| 7 | 0.084277 | -345.117189 | -345.032911 | 0.000000 | 0.081129 | -345.091728 | -345.010599 | 0.022312 |
| | | | | (0.000000) | | | | (14.000860) |
| 8 | 0.089359 | -383.195557 | -383.106198 | 0.000000 | 0.086670 | -383.176165 | -383.089486 | 0.016712 |
| | | | | (0.000000) | | | | (10.486900) |

Table 1. Calculated total energies E, zero-point energies ZPE (from B3LYP/6-31G*), and relative energies E (Eh), (in hartree), for the energy-minimum and energy-maximum geometries of compounds 1-8. (Numbers in parenthesis are the corresponding $\Delta E$ values in kcal mol-1.
  a Relative to the minimum)

| | 1 | 2 | 3 | 4 | 5 | 6 | 7 | 8 |
|---|---|---|---|---|---|---|---|---|
| $\overline{\pi}_{occupancy}$ | 1.97234 | 1.90752 | 1.88172 | 1.85594 | 1.83857 | 1.82328 | 1.81200 | 1.80180 |
| $\overline{\pi}^{*}_{occupancy}$ | 0.05147 | 0.09964 | 0.12858 | 0.14957 | 0.16713 | 0.17911 | 0.19135 | 0.20010 |
| $\Delta=(\overline{\pi}_{occupancy} - \overline{\pi}^{*}_{occupancy})$ | 1.92087 | 1.80782 | 1.75314 | 1.70637 | 1.67144 | 1.64337 | 1.62065 | 1.60170 |
| $\overline{r}$ | 1.307 | 1.302 | 1.297 | 1.295 | 1.292 | 1.291 | 1.290 | 1.289 |

Table 2. B3LYP/6-31G* calculated mean of  bonding and  * anti bonding orbital occupancies, the difference between  and  values (i.e.  = - *) and the average bond length ( ) for compounds 1-8.

Using the obtained occupancy values, a "Δ" parameter could be defined as: $\Delta = \overline{\pi}_{occupancy} - \overline{\pi}^{*}_{occupancy}$ . The plot of $\Delta E_0$ *vs* $\Delta$, shown in Fig. 2, reveal a linear proportionality between them. These results indicate that with the increase of $\Delta$ values, the corresponding $\Delta E_0$ (e.g. mutual interconversion energy barrier) decreases. Consequently, the $\Delta$ parameter (i.e. $\Delta = \overline{\pi}_{occupancy} - \overline{\pi}^{*}_{occupancy}$ ) could be proposed as criteria for evaluation of the easiness of mutual interconversion in cumulene compounds.

# References

[1]    M. J. Frisch, *et al.* GAUSSIAN 98 (Revision A.3) Gaussian Inc. Pittsburgh, PA, USA, 1998.

[2]    A. D. Becke, *J. Chem. Phys.*, **98**, 5648 (1993).

[3]    C. Lee, W. Yang and R. G. Parr, *Phys. Rev.* B, **37**, 785 (1988).

[4]    U. Mölder, P. Burk and I. A. Koppel, *J. Mol. Struct. (THEOCHEM)*, **712**, 81 (2004).

[5]    E. D. Glendening, A. E. Reed, J. E. Carpener, F. Weinhold, NBO Version 3.1.

[6]    A. E. Reed, L. A. Curtiss, and F. Winhold, *Chem. Rev.*, **88**, 899 (1988).

Brill Academic Publishers
P.O. Box 9000, 2300 PA Leiden
The Netherlands

*Lecture Series on Computer*
*and Computational Sciences*
Volume 4, 2005, pp. 262-265

# Digital Hearing Aid DSP Chip Parameter Fitting Optimization

Soon Suck Jarng[1], You Jung Kwon[1], Je Hyoung Lee[1]

Department of Information, Control & Instrumentation,
Chosun University, South Korea

Received 8 June, 2005; accepted in revised form 29 June, 2005

*Abstract:* DSP chip parameters of a digital hearing aid (HA) should be optimally selected or fitted for hearing impaired persons. The more precise parameter fitting guarantees the better compensation of the hearing loss (HL). Digital HAs adopt DSP chips for more precise fitting of various HL threshold curve patterns. A specific DSP chip such as Gennum GB3211 was designed and manufactured in order to match up to about 4.7 billion different possible HL cases with combination of 7 limited parameters. This paper deals with a digital HA fitting program which is developed for optimal fitting of GB3211 DSP chip parameters. The fitting program has completed features from audiogram input to DSP chip interface. The compensation effects of the microphone and the receiver are also included. The paper shows some application examples.

*Keywords:* Digital Hearing Aid, GB3211 DSP Chip, Parameter Fitting, Optimization, Hearing Loss, Audiogram

*Mathematics SubjectClassification:* 07.05.Tp

*PACS:* 07.05.Wr

## 1. Introduction

Hearing impaired persons have increased HL threshold curves over audio frequency band, so that their hearing impairments may be partly compensated by HAs. Before 1990s analog HAs dominated HA markets even though analog HAs are limited in the HL compensation because they are not modifiable. In analog ITE(In-The-Ear) type HAs a logarithmic volume control switch is the only way of modification. There were some conventional fitting formulas developed for better choice of amplification and compression [1]. The advent of the sophisticated semiconductor technology as well as the better understanding of the hearing physiology opened the age of digital HAs last decade. Digital HAs are modifiable, that is, each different type of hearing impairment can be precisely compensated with the same class of the digital HA. These multi-purposed digital HAs are possible by adopting DSP based IC chip design and manufacturing. DSP chips for digital HAs are divided into two categories; specific DSP chip and general DSP chip. The specific HA DSP chip was produced earlier because the miniature size of the chip package was required for ITE type HA fabrication [2]. Recently, general DSP chips are getting down in packaging size with lower power consumption, so as to be applied to ITE type HAs [3, 4]. As digital HAs support more functions and flexibilities, the fitting method of the digital HAs becomes important issues in the hearing aid market. This paper applies a specific DSP chip such as Gennum GB3211 (Fig. 1) to digital HA fabrication and shows the result of the optimal parameter fitting program development for the chip.

The main features of GB3211 are 4 channel nonlinear compressive active filtering and 4 extra linear biquad filtering. Those 8 active digital filters are used for fitting of various patterns of HL threshold curves. In this paper a GB3211 chip fitting program was developed for digital HAs in which the most optimal DSP chip parameters were selected from 4.7 billion combinations of possible parameters.

## 2. DSP parameter fitting procedures and results

The whole procedure of the parameter fitting is charted in Fig. 2.

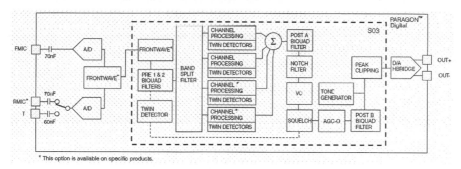

Figure 1: Circuit diagram of GB3211(GB3210) digital HA specific DSP chip [2].

### 2.1 Audiogram Hearing Threshold

The first step of the parameter fitting is to read the HL threshold of the hearing impaired person .

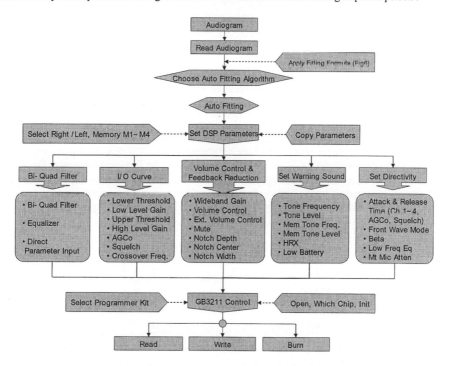

Figure 2 : The flow chart of the parameter fitting.

### 2.2 Fitting Formula

The second step is to calculate the HA amplification as a function of frequency by a conventional fitting formula such as 'FIG6' Formula [1].

### 2.3 DSP chip parameters' fitting

The third step is to calculate DSP chip parameters for fitting to the HA amplification function. The word, fitting, is sometimes confused. In general 'fitting' means fitting formula as in the second step. The parameter fitting in the third step means the proper adjustment of the DSP chip parameters in order for the DSP chip to resemble to the HA amplification function derived by the second step. Fig. 3 shows the results of the fitting formula (three black thin lines) and the parameter fitting (three colored thick lines) by a fitting program supplied by Gennum Co. The blue line is for 40dB input sound level, and the green and the violet lines are for 60dB and 80dB input sound levels respectively. The parameter fitting seems to be all right at low frequency bands, but the parameter fitting is not well resemble to the HA amplification functions at high frequency bands. Therefore this paper tried better parameter fitting than Gennum's.

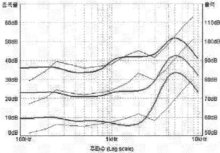

Figure 3 : The results of the fitting formula (thin lines) and the parameter fitting (thick lines) by a fitting program supplied by Gennum Co. Blue: 40dB Input Level, Green: 60dB Input Level, Violet: 80dB Input Level. (Left ear)

If we look at the operational principle of the Gennum 3211 DSP chip, some better optimal parameter fitting method could be resulted. The Gennum 3211 chip divides the frequency band into four channels. Each channel boundary is defined as cross frequency, CO. Each channel is controlled by each digital filter; CH1=Low Pass (LP) Filter, CH2=Band Pass (BP) Filter, CH3= Band Pass (BP) Filter, CH4= High Pass (HP) Filter [5, 6]. The four digital filter responses are summed to resemble to the HA amplification function. The channel digital filter is Butterworth type. The first channel is composed by the third order low pass filters. $cf$ is cut-off frequency and $co1$ is the first cross frequency. The second channel is composed by second order high pass filters as well as third/fourth order low pass filters. The third channel is composed by third/fifth order high pass filters as well as second/third order low pass filters. The fourth channel is composed by second/fifth order high pass filters. (Equation (1))

$$
\begin{aligned}
&\left(LP_{3rd}\big|_{cf=co1}\right)^2 \\
&\left(HP_{2nd}\big|_{cf=co1} \times LP_{4th}\big|_{cf=co2}\right) \times \left(HP_{2nd}\big|_{cf=co1} \times LP_{3rd}\big|_{cf=co2}\right) \\
&\left(HP_{3rd}\big|_{cf=co2} \times LP_{2nd}\big|_{cf=co3}\right) \times \left(HP_{5th}\big|_{cf=co2} \times LP_{3rd}\big|_{cf=co3}\right) \\
&\left(HP_{2nd}\big|_{cf=co3}\right) \times \left(HP_{5th}\big|_{cf=co3}\right)
\end{aligned}
\tag{1}
$$

Table 1 : Butterworth digital filter formula. N is a number of order [5].

| N | $N^{th}$ -order Low-Pass Filter | $N^{th}$ -order High-Pass Filter |
|---|---|---|
| Even | $H(s)=\prod_{k=1}^{N/2} \dfrac{\omega_c^2}{s^2+2\omega_c \cos\phi_k s+\omega_c^2}$ | $H(s)=\prod_{k=1}^{N/2} \dfrac{s^2}{s^2+2\omega_c \cos\phi_k s+\omega_c^2}$ |
| Odd | $H(s)=\dfrac{\omega_c}{s+\omega_c}\prod_{k=1}^{(N-1)/2} \dfrac{\omega_c^2}{s^2+2\omega_c \cos\phi_k s+\omega_c^2}$ | $H(s)=\dfrac{s}{s+\omega_c}\prod_{k=1}^{(N-1)/2} \dfrac{s^2}{s^2+2\omega_c \cos\phi_k s+\omega_c^2}$ |

Fig. 4 shows the results of the fitting formula (three black thin lines) and the parameter fitting (three colored thick lines) by a fitting program developed by authors. In comparison with Fig. 3 the optimal parameter fitting seems to be better at all frequency bands than Gennum's

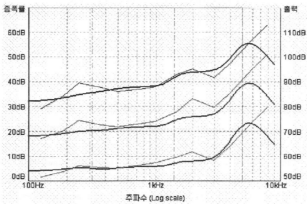

Figure 4 : The results of the fitting formula (thin lines) and the optimal parameter fitting (thick lines) by a fitting program developed by authors. Blue: 40dB Input Level, Green: 60dB Input Level, Violet: 80dB Input Level. (Left ear)

## 3. Conclusion

This paper applies a specific DSP chip such as Gennum GB3211 to digital ITE type HA fabrication and shows the result of the optimal parameter fitting program development for the chip. Details of chip parameters are explained. The fitting program has completed features from audiogram input to DSP chip interface. The compensation effects of the microphone and the receiver are also included. The comparison between Fig. 3 and Fig. 4 shows that the DSP chip parameters such as 4.7 billion possible combinations of parameters should be optimally chosen to resemble to the amount of hearing threshold compensation derived by the fitting formula. Extra digital filters need to be added for the complete fitting to the HA amplification function. Even though the present DSP chip parameter fitting program provides the best fitting for the fitting formula, the final precise fitting should be manually adjusted by the user, that is, the hearing impaired person [7,8].

## References

[1] Harvey Dillon, "Hearing Aids", Printed by Thieme, 2001.

[2] http://www.gennum.com/hip/dproducts/paragon/gb3211.htm

[3] http://www.gennum.com/hip/dproducts/duet/gb3212.htm

[4] http://www.dspfactory.com/technology/signaklaratechnology.html

[5] Chaewook Lee, "Recent digital signal processing", published by BooksHill Co., 2002.

[6] Gennum Co., "Biquad filters in PARAGON digital hybrid", Doc. No. 20205-1, 2001.

[7] B. Kollmeier and V. Hohmann, "Loudness estimation and compensation employing a categorical scale," in *Advances in Hearing Research*, G. A. Manley, G. M. Klump, C. Köppl, H. Fastl, and H. Oeckinghaus, Eds. Singapore: World Scientific, pp. 441-451. 1995.

[8] S. Launer and B. C. J. Moore, "Use of a loudness model for hearing aid fitting. V. On-line gain control in a digital hearing aid," *Int. J. Audiol.*, vol. 42, pp. 262-273, 2003.

Brill Academic Publishers
P.O. Box 9000, 2300 PA Leiden
The Netherlands

*Lecture Series on Computer
and Computational Sciences*
Volume 4, 2005, pp. 266-269

# Model and Algorithm Research for Seeking Efficient Monitor-Nodes Measuring Network Traffic

JIANG Hong-Yan[1], LIN Ya-Ping[1], HUANG Sheng-Ye[1], LIU Xiao-Fan[1]
1.College of Computer and Communication,
Hunan University,
CN-410082 Changsha, China

Received 12 June, 2005; accepted in revised form 26 July, 2005

*Abstract:* The problem of seeking for monitor-nodes to measure the network traffic is abstractly regarded as the problem of finding out the minimum weak vertex cover of a graph which is NP-hard. First, an approximation algorithm based on the concept of incidence matrix is proposed in this paper. Also, we analyze the complexity of the algorithm. Furthermore, the algorithm is expanded to seek for the minimum weak vertex cover of a graph that has weights on its nodes. According to the theoretical analysis and the simulation results, our novel algorithm is comparatively more patulous than the traditional algorithms, and more effective in searching for smaller weak vertex cover as well.

*Keywords:* Weak vertex cover, nodes with weights, flow conservation, NP-hard, incidence matrix

## 1. Introduction

Effective monitoring of network Traffic is a key verifying Qos guarantees. However, frequently inquires by massive equipments to measure the network traffic lead to extra pressure on the net, reducing the performance of the routers. For the sake of optimizing the capability of the net, it is important to establish highly efficient and reasonable system used to monitor the traffic[1]. In recent years, many scholars have done some research into this issue and usually transform the problem of measuring the network traffic into the problem of finding out the minimum vertex cover or weak vertex cover of an undirected graph. Since these are all NP-hard, approximation algorithms are usually adopted. In [2], the author proposed an effective technique and used three approximation algorithms to find the minimum vertex cover of a graph. In [3], the author theoretically analyzed the greedy algorithm. Without exception, all these algorithms can not be applied to the situation in which graph has weights on its nodes. In this paper, we put forward an new approximate algorithm. Furthermore, we apply this method to the situation in which graph has weights on its nodes. We also analyze the time Complexity of this algorithm, compare it with the the traditional algorithms through simulation. As the results show, our algorithm can find smaller vertex cover and is more expansible.

## 2. Model and algorithm of setting effective monitoring nodes

Our abstract model of a data network is an undirected graph $G=(V,E)$, where $V=(v_1,v_2,...,v_n)$ denotes the set of network nodes which can be regarded as a router in IP web. $E=(e_1,e_2,...,e_m)$ represents the set of edges(i.e.,physical links)connecting the routers, we let $n=|V|$ and $m=|E|$ denotes the the number of G's nodes and edges, further, for a node $v_i$ $E$, $e_k=(v_i,v_j)$ denotes the shortest path from $v_i$ to $v_j$ $degree(v)$ presents the number of lines belongs to node $v$. Two restrictive conditions should be satisfied: $Degree(v) \geq 2$ and the *law of flow-conservation* should be obeyed.

---

[1] JIANG Hong-Yan, a Ph.D. candidate of Hunan University,senior, engineer, China,,Email: ipaddress@hntelecom.net.cn or jhy@hnpta.net.cn

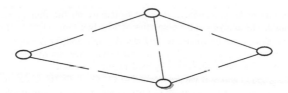

Fig1.network graph G

According to the analysis above, the problem of minimizing the number of effective monitoring nodes becomes equivalent to determining a minimum VC for G. In this paper,by using the conception of incidence matrix, we put forward an algorithm named SMN selecting the minimum vertex cover set:
Begin
1) Write out the incidence matrix $A_1 = A = (a_{ij})$ of $G$, where $i \in \{1,2,...,n\}, j \in \{1,2,...,m\}$ ;

2) let $S = \varnothing, k = 1$ ; 3)while ( $A_k \neq 0$ )  Begin

4) $\forall\ i \in (1,2,...,n)$ , calculate the sums of elements in each row $\sum_{j=1}^{m} a_{ij}$ of $A_k$ , choose the row with

the maximum sum, marked as $\max_{i} \sum_{j=1}^{m} a_{ij} = a_k$ , else if in $A_k$ ,there are two rows $i_1, i_2$ whose sums

are the maximum value $a_k$ and if $i_1 \leq i_2$, then we choose $k = i_1$, and the vertex $v_k$ corresponding to

this row can be obtained . 5)  $S=S+\{ v_k \}$; 6)  In the incidence matrix $A_k$ , cancel the row corresponding

to $v_k$ and the columns which have elements 1 in the row.  And then , in the left incidence matrix, cancel all the other rows in which the total of all elements in the row is no more than 1 and the corresponding columns with the elements valued 1 in these rows, do so until no new row or column can be deleted, let $A_{k-1}$ be the incidence matrix which are obtained at last. 7) $k = k+1$ ;
    End // end of while ()
    End // end of Algrithm
Take fig 1 as an example,we illuminate the algorithm *SMN*.
1) The incidence matrix of network G in fig.1 is:

$$A_1 = \begin{array}{c} \\ v_1 \\ v_2 \\ v_3 \\ v_4 \end{array} \begin{array}{ccccc} e_1 & e_2 & e_3 & e_4 & e_5 \\ \left( 1 & 1 & 0 & 0 & 0 \right. \\ 0 & 1 & 1 & 1 & 0 \\ 0 & 0 & 0 & 1 & 1 \\ \left. 1 & 0 & 1 & 0 & 1 \right)_{4 \times 5} \end{array}$$

2) when i=1, $\sum_{j=1}^{m} a_{1j} = 2$ ; when i=2, $\sum_{j=1}^{m} a_{2j} = 3$ ; when i=3, $\sum_{j=1}^{m} a_{3j} = 2$ ;when i=4, $\sum_{j=1}^{m} a_{4j} = 3$ ;
Obviously, when $i=2$, the sum of the row is the most, so $v_2$ is chosen.
3)After deleting the row corresponding to $v_2$ in the incidence matrix and the columns with elements valued 1(etc. $e_2$, $e_3$ and $e_4$ )in the row,we can obtain the left incidence matrix

$$\begin{array}{c} \\ v_1 \\ v_3 \\ v_4 \end{array} \begin{array}{cc} e_1 & e_5 \\ \left( 1 & 0 \right. \\ 0 & 1 \\ \left. 1 & 1 \right)_{3 \times 2} \end{array}$$

Then  delete other rows consecutively in which the total of the elements is no more than 1,such as $v_1$, $v_3$ ,and the columns relevant to $e_1$ and $e_5$ ( both values are 1), we can get $A_2$=0. So we obtain the vertex cover set S by the algrithm *SMN* S={ $v_2$}. Alternatively, the traffic in each link can be measured after only one inquiry on one node. Commonly, we have to inquire all the four nodes by means of the regular measure method. The analysis of the properties of this algorithm is listed below:

**Theorem 1:** If $G=(V,E)$ is a simple connected undirected graph, and for $\forall\ v\ V$, $Degree(v) \geq 2$, then the set S of nodes selected by the algrithm *SMN is a weak* vertex cover set of graph G.

It is not difficult for one to prove theorem 1 by mathematic induction.

**Theorem 2:** For the graph $G$ and $S$, if we use $G'=G-S=\{\ G_1, G_2, \ldots\ G_j\ \}$ to represent the set in which $S$ and its incident edges are removed from graph $G_l$ ($1 \leq l \leq j$)denotes some connected subnet graph of $G'$, $N_i$ and $M_i$ denote the number of nodes and links in graph $G_l$ *respectively*, then for any given $G_l$, $N_i \geq M_i$

To prove the theorem, one needs to remember that flow conservation is assumed.

**Theorem3:** Given an undirected graph $G=(V,E)$, where $V=(v_1, v_2, \ldots, v_n)$denotes the set of network nodes, $E=(e_1, e_2, \ldots, e_m)$ represents the set of edges, $n=|V|$ and $m=|E|$ denotes the the number of G's nodes and edges, we let $t=max(m,\ n)$, then the time complexity of algorithm SMN is $O(t^2)$.

Limited by text length, the proof of the theorem is omitted.

Now we expand this algorithm to the situation in which graph has weights on its node. Granted that the cost of setting an effective monitor on node $v_i$ is $f_i$ Obviously, the larger $f_i$ is, the less value of setting a monitor on node $v_i$ would be ,and less possible that $v_i$ could be chosen. On the other side, the profit of setting a monitor on node $v_i$ can be measured by the traffic and the access frequency on it. We use $b_i$ to represent it and it can be measured by statistic software, $b_i$ has a certain extent of stability, but could be effected by varied factors to change dynamically . Thus, in order to rectify $b_i$ dynamically, we can modify the established mathematical model dynamically, along with the optimal solution. Obviously, the bigger $b_i$ is, the more valuable will it be to set a monitor on node $v_i$.let $c_i = f_i/b_i$, we can see that $c_i$

gives us a quantitative description, let $d_i = c_i / \sum_{j=1}^{m} a_{ij}$, it provides the value on each link connected to node $v_i$, which is also the major parameter in the next algorithm.

According to the analysis above, We expand SMN to get a generalized SMN algorithm aiming at finding the minimum weak vertex cover of a graph that has weights on its nodes., the algorithm is:

1. Choose a node with the least value of $d_i$ denote it by $v_{i1}$;
2. In the incidence matrix $A_k$, cancel the rows corresponding to $v_{i1}$ and the columns in which the elements are valued 1 in the row , and then , in the rest incidence matrix, cancel all the other rows the total of all elements of which are no more than 1 and the corresponding columns with the elements valued 1 in these rows, until no new row or column can be deleted. Let $A_{k+1}$ be the incidence matrix which are obtained at last.
3. With regard to the rest nodes, repeat the step 1),2), until the last incidence matrix becomes null.

Then,the selected node set $\{\ v_{i1}, v_{i2}, \ldots, v_{in}\}$ is the minimum vertex cover set。

We still consider Fig. 1. as an example to specifically explain the idea of generalized SMN algorithm, suppose $(f_1, f_2, f_3, f_4)=(20,30,20,54)$, $(b_1, b_2, b_3, b_4)=(2,5,4,6)$ , then $(c_1, c_2, c_3, c_4)=(10,6,5,9)$ , $(d_1, d_2, d_3, d_4)=(5,2,2.5,3)$, because $d_2$ is minimum, select $v_2$; delete the row $v_2$ and column $e_2$, $e_3$ and $e_4$ from the incidence matrix, then recompute $d_i$.Take its turn to delete the other rows in which all the sum of the line element is not more than 1,for example $v_1$, $v_3$ ,and delete the correspond column $e_1$, $e_5$ in which the element in the related rows are 1, we obtain $A_2=0$, thus the vertex set is $S=\{\ v_2\}$

## 3. Simulation Experiment

Refering to [2], we do a simulation experiment, comparing the performance of the four algorithms: the maximal matching heuristic for simple VC[2,6], Weak VC formulation – a variant of the maximal matching heuristic[2,6], GREEDYRANK algorithm[2], and the performance of the SMN algorithm in this paper. The simulations are based on network topologies generated using the Waxman Model [5], which is a popular topology model for networking research, Different network topologies are generated by varying three parameters: 1) $n$, the number of nodes in the network graph; 2) $\alpha$, a parameter that controls the density of short edges in the network; 3)ß, a parameter that controls the average node degree,4) $N_{match}^{vc}$, $N_{match}^{wvc}$, $N_{rank}^{wvc}$ and $N_{smn}^{wvc}$ to represent the number of monitor nodes used in the four different algorithms above,5). Avg.Degre is used to represent the average degree of nodes and it grows with the growth of $\beta$ value. 6) $N_{smn}^{wvc} / n$ represents the ratio of the number of monitor nodes , obtained by SMN algorithm, and the total number of the nodes. Simulation results are showed below:

TABLE1. comparing four different algorithms ($n$=500, $\alpha$=0.4, $\beta$={0.02,...,0.08})

| *Avg.Degree* | $N^{vc}_{match}$ | $N^{wvc}_{match}$ | $N^{wvc}_{rank}$ | $N^{wvc}_{smn}$ | $N^{wvc}_{smn}/n$ |
|---|---|---|---|---|---|
| 4.4 | 387 | 255 | 165 | 129 | 0.26 |
| 8.6 | 441 | 372 | 254 | 221 | 0.44 |
| 12.6 | 453 | 408 | 307 | 296 | 0.59 |
| 16.9 | 466 | 431 | 334 | 327 | 0.65 |

From the result listed in table one we know the nodes selected by SMN algorithm is less than those by three other algorithms.

## 4. Conclusions

In this paper, we propose an approximate algorithm SMN to solve the problem of placing monitor-nodes to measure the network traffic. Compared with the algorithms presented in [2,6], our algorithm performs better. As for the problem of efficiently monitoring traffic of a large complex network, we can greatly play down the scale of the problem by choosing one more node, assisting in optimizing the performance of the network. The algorithm is based on central control, it may confront challenges in extremely large-scale networks. Further study is to develop a distributed version of SMN.

## References

[1] Y. Breitbart, Chee-Yong Chan, M. Garofalakis, R. Rastogi, and A. Silberschatz, *Efficiently monitoring bandwidth and latency in IP networks*, Murrary Hill,NJ:Bell Laboratories,2000.

[2] Y. Breitbart, Chee-Yong Chan, M. Garofalakis, R. Rastogi, and A. Silberschatz, *Efficiently monitoring bandwidth and latency in IP networks*, INFOCOM 2001, vol.2: 933-942.

[3] Liu XH, Yin JP, Tang LL, Zhao JM. Analysis on methods of efficient monitoring of network traffic, Chinese Journal of Software, 2003, 14(2) :300-304. In Chinese.

[4] R. Caceres, N.G. Duffield, A. Feldmann, J. Friedmann, A. Greenberg, R. Greer, T. Johnson, C. Kalmanek, B. Krishnamurthy, D. Lavelle, P.P. Mishra, K.K. Ramakrishnan, J. Rexford,F.True, and J.E. van der Memle, *Measurement and analysis of IP network usage and behavior*, IEEE Communications Magazine,May 2000, 38(5):144-151.

[5] B.M. Waxman, Routing of Multipoint Connections, *IEEE Jrnl. On Selected Areas in Communications*, December 1988,6(9): 1617-1622.

[6] V.V.Vaziram,*Approximation Algorithms,*Springer-Verlag,2000.

Brill Academic Publishers
P.O. Box 9000, 2300 PA Leiden,
The Netherlands

*Lecture Series on Computer
and Computational Sciences*
Volume 4, 2005, pp. 270-274

# Numerical Solution of the two-dimensional time independent Schrödinger Equation by symplectic schemes based on Magnus Expansion

**Z. Kalogiratou**

Department of International Trade,
Technological Educational Institute of Western Macedonia at Kastoria,
P.O. Box 30, GR-521 00, Kastoria, Greece

**Th. Monovasilis**

Department of Computer Science and Technology,
Faculty of Sciences and Technology,
University of Peloponnese,
GR-221 00 Tripolis, Greece
Department of International Trade,
Technological Educational Institute of Western Macedonia at Kastoria,
P.O. Box 30, GR-521 00, Kastoria, Greece

**T.E. Simos**[1]

Department of Computer Science and Technology,
Faculty of Sciences and Technology,
University of Peloponnese,
GR-221 00 Tripolis, Greece

Received 16 August, 2005; accepted in revised form 17 August, 2005

*Abstract:* The solution of the two-dimensional time-independent Schrödinger equation is considered by partial discretisation. The discretized problem is treated as an ordinary differential equation problem and solved numerically by symplectic methods based on Magnus expansion. The problem is then transformed into an algebraic eigenvalue problem involving real, symmetric matrices.

*Keywords:* Schrödinger equation, Magnus expansion,symplectic schemes, partial discretisation,

*Mathematics Subject Classification:* 65L05;

## 1  Introduction

The time-independent Schrödinger equation is one of the basic equations in quantum mechanics. Plenty of methods have been developed for the solution of the one-dimensional time-independent

---

[1]Corresponding author. Active Member of the European Academy of Sciences and Arts. E-mail: simos-editor@uop.gr, tsimos@mail.ariadne-t.gr

Schrödinger equation. In the literature this problem has been treated by means of discretization of both variables $x$ and $y$. Here we use partial discretization only on the variable $y$, then we have an ordinary differential equation problem. Symplectic integrators were proven to be suitable integrators for the numerical solution of the one-dimensional Schrödinger equation. Recently in their work Liu, ct. al. [3] and the authors [2] developed a numerical method for the numerical solution of the two-dimensional time independent Schrödinger equation. Here the problem will be treated by symplectic methods based on Magnus Expansion. The methods will be tested on the two-dimensional harmonic oscillator and the two-dimensional Henon-Heils potential.

## 2  Partial discretisation of the two-dimensional equation

The two-dimensional time-independent Schrödinger equation can be written in the form

$$\frac{\partial^2 \psi}{\partial x^2} + \frac{\partial^2 \psi}{\partial y^2} + (2E - 2V(x,y))\psi(x,y) = 0, \tag{1}$$

$$\psi(x, \pm\infty) = 0, \quad -\infty < x < \infty,$$

$$\psi(\pm\infty, y) = 0, \quad -\infty < y < \infty$$

where $E$ is the energy eigenvalue, $V(x,y)$ is the potential and $\psi(x,y)$ the wave function. The wave functions $\psi(x,y)$ assymtotically approaches infinity away from the origin. We consider $\psi(x,y)$ for $y$ in the finite interval $[-R_y, R_y]$ and

$$\psi(x, -R_y) = 0 \quad \text{and} \quad \psi(x, R_y) = 0$$

the boundary conditions. We also consider partition of the interval $[-R_y, R_y]$

$$-R_y = y_{-N}, \; y_{-N+1}, \; \ldots, \; y_{-1}, \; y_0, \; y_1, \; \ldots, \; y_{N-1}, \; y_N = R_y$$

where $y_{j+1} - y_j = h_y = R_y/N$.
We approximate the partial derivative with respect to $y$ with the difference quotient

$$\frac{\partial^2 \psi}{\partial y^2} = \frac{\psi(x, y_{j+1}) - 2\psi(x, y_j) + \psi(x, y_{j-1})}{h_y^2}$$

and substitute into the original equation

$$\frac{\partial^2 \psi}{\partial x^2} = -\frac{1}{h_y^2}\psi(x, y_{j+1}) - B(x, y_j)\psi(x, y_j) - \frac{1}{h_y^2}\psi(x, y_{j-1})$$

where

$$B(x, y_j) = 2\left(E - V(x, y_j) - \frac{1}{h_y^2}\right)$$

We also define the $2N - 1$ length vector

$$\Psi(x) = (\psi(x, y_{-N+1}), \psi(x, y_{-N+2}), \ldots, \psi(x, y_0), \ldots, \psi(x, y_{N-2}), \psi(x, y_{N-1}))^T$$

which we can write as

$$\Psi(x) = (\psi_1(x), \psi_2(x), \ldots, \psi_N(x), \ldots, \psi_{2N-2}(x), \psi_{2N-1}(x))^T$$

or

$$\Psi(x) = (\psi_1(x), \psi_2(x), \dots, \psi_k(x))^T$$

where $k = 2N - 1$. Also consider the functions $\dot\phi_j(x) = \psi_j(x)$ for $j = 1, 2, \dots, k$, and

$$\Phi(x) = (\phi_1(x), \phi_2(x), \dots, \phi_k(x))^T$$

Equation (1) can be written as

$$\begin{pmatrix} \dot\Phi \\ \dot\Psi \end{pmatrix} = \begin{pmatrix} 0 & -S(x) \\ I & 0 \end{pmatrix} \begin{pmatrix} \Phi \\ \Psi \end{pmatrix} \tag{2}$$

where $S(x)$ is a $k \times k$ matrix

$$S(x) = \begin{pmatrix} B(x, y_{-N+1}) & 1/h_y^2 & & & \\ 1/h_y^2 & B(x, y_{-N+2}) & 1/h_y^2 & & \\ \ddots & & \ddots & & \ddots \\ & & 1/h_y^2 & B(x, y_{N-2}) & 1/h_y^2 \\ & & & 1/h_y^2 & B(x, y_{N-1}) \end{pmatrix}$$

The matrix $S(x)$ can be written in terms of three matrices the identity matrix I, the diagonal matrix $V$ which contains the potential at the mesh points $y_{-N+1}, \dots, y_{N-1}$ and the tridiagonal matrix $M$ with diagonal elements $-2$ and off diagonal elements 1.

$$S(x) = 2EI - 2V(x) + \frac{1}{h_y^2}M$$

## 3  Numerical methods based on Magnus Expansion

We notice that equations (2) have a linear structure, this is of the general form

$$\dot Y = A(x)Y \tag{3}$$

where

$$Y(x) = \begin{pmatrix} \Phi(x) \\ \Psi(x) \end{pmatrix} \quad \text{and} \quad A(x) = \begin{pmatrix} 0 & -S(x) \\ I & 0 \end{pmatrix}$$

For the scalar case and in the case that the matrices $A(x)$ and $\int_0^x A(t)dt$ commute the solution of (3) with initial condition $Y(0) = Y_0$ is given by

$$Y(x) = \exp\left(\int_0^x A(t)dt\right) Y_0$$

In our case matrices $A(x)$ and $\int_0^x A(t)dt$ do not commute. Following the approach of Magnus there is a matrix function $\Omega(t)$ such that

$$Y(x) = \exp\left(\Omega(x)\right) Y_0$$

and $\Omega(x)$ can be approximated by the Magnus expansion (Hairer [1],p.118)

$$\Omega(x) = \int_0^x A(t)dt \; - \; \frac{1}{2}\int_0^x \left[\int_0^t A(\tau)d\tau, A(t)\right]dt$$

$$+ \; \frac{1}{4}\int_0^x \left[\int_0^t \left[\int_0^\tau A(\sigma)d\sigma, A(\tau)\right]d\tau, A(t)\right]dt$$

$$+ \; \frac{1}{12}\int_0^x \left[\int_0^t A(\tau)d\tau, \left[\int_0^t A(\sigma)d\sigma, A(t)\right]\right]dt + \cdots$$

Zanna [4] constructed numerical methods of the solution of (3) by replacing $A(x)$ locally by an interpolation polynomial and truncating accordingly the above series.

The second order method is based on the mid point rule

$$Y_{n+1} = \exp\left(hA(x_n + h/2)\right)Y_n \tag{4}$$

The fourth order method

$$Y_{n+1} = \exp\left(\frac{h}{2}(A_1 + A_2) + \frac{\sqrt{3}h^2}{12}[A_2, A_1]\right)Y_n \tag{5}$$

is based on the two-stage Gauss quadrature on the points

$$c_{1,2} = \frac{1}{2} \mp \frac{\sqrt{3}}{6}, \quad A_1 = A(x_n + c_1 h) \quad \text{and} \quad A_2 = (x_n + c_2 h).$$

We can also write

$$Y_{n+1} = \exp\left(M_k\right)Y_n, \quad k = 2 \quad \text{or} \quad 4$$

for the second and fourth order method.

where

$$M_2 = \begin{pmatrix} 0 & -hS\left(x_n + \frac{h}{2}\right) \\ h & 0 \end{pmatrix}$$

and

$$M_4 = \begin{pmatrix} \frac{\sqrt{3}}{12}h^2(S_1 - S_2) & \frac{-h}{2}(S_1 + S_2) \\ h & \frac{-\sqrt{3}}{12}h^2(S_1 - S_2) \end{pmatrix}, \quad S_1 = S(x_n + c_1 h), \quad S_2 = S(x_n + c_2 h)$$

In order to implement the above methods we shall apply Taylor expansion to the matrix exponential.

## 4   Numerical Results

We shall apply both numerical methods developed above to the calculation of the eigenvalues of the two-dimensional harmonic oscillator and the two-dimensional Henon-Heiles potential.

### 4.1   Two-dimensional harmonic oscillator

The potential of the two-dimensional harmonic oscillator is

$$V(x,y) = \frac{1}{2}(x^2 + y^2)$$

The exact eigenvalues are given by

$$E_n = n + 1, \quad n = n_x + n_y, \quad n_x, n_y = 0, 1, 2, \ldots$$

### 4.2   Two-dimensional Henon-Heiles potential

The potential is

$$V(x,y) = \frac{1}{2}(x^2 + y^2) + (0.0125)^{1/2}\left(x^2 y - \frac{y^3}{3}\right)$$

## Acknowledgment

This research was co-funded by 75% from E.E. and 25% from the Greek Government under the framework of the Education and Initial Vocational Training Program - Archimedes, Technological Educational Institution (T.E.I.) Chalkis.

## References

[1] Hairer E., Lubich Ch., Wanner G., *Geometric Numerical Integration*, Springer-Verlag, 2002.

[2] Monovasilis Th., Kalogiratou, Z, Simos T.E., Numerical Solution of the two-dimensional time independent Schrödinger Equation by symplectic schemes, *Applied Numerical Analysis and Computational Mathematics* 1(2004), 195-204

[3] Liu, X.S., Liu X.Y, Zhou, Z.Y., Ding, P.Z., Numerical Solution of the Two-Dimensional Time-Independent Schrödinger equation by using symplectic schemes, *International Journal of Quantum Chemistry* 38(2001), 303-309.

[4] Zanna,A., Collocation and relaxed collocation for the Fer and Magnus expansions, *SIAM J. Numer. Anal.* 36(1999),1145-1182.

Brill Academic Publishers
P.O. Box 9000, 2300 PA Leiden
The Netherlands

*Lecture Series on Computer
and Computational Sciences*
Volume 4, 2005, pp. 275-278

# Modeling Knowledge in Traditional Chinese Medicine with Influence Diagrams and Possibility Distributions

H.-Y. Kao[1]

[1] Department of Marketing and Distribution Management,
Husan Chuang University,
48 Hsuan Chuang Road, Hsinchu, Taiwan. R.O.C.

Received 20 June, 2005; accepted in revised form 13 July, 2005

*Abstract:* Traditional Chinese medicine (TCM) originated from China and has been developed for more than four thousand years. The domain knowledge of TCM has been accumulated from a great deal of experiences and dialectical methods. However, the knowledge of TCM is mostly qualitative and lacks the mechanism of quantitative inference. Besides, most descriptions of the diseases and symptoms are linguistically vague in TCM, which makes systematization of this medical discipline more difficult. This study proposes a novel approach for knowledge modeling in TCM, influence diagrams. Influence diagrams have been widely used in western medical informatics and industrial knowledge based systems. In conventional influence diagrams, the numerical models are probability distributions associated with the random variables. However, when incomplete knowledge or linguistic vagueness is involved in the medical decisions, the suitability of probability theory and distributions is questioned. This study uses fuzzy sets and possibility distributions to model the uncertainty in influence diagrams. In the influence diagrams, each chance node is associated with a possibility distribution function, which expresses the imprecise relationships among the variables. This study also proposes an algorithm for fuzzy reasoning from the influence diagrams, which answers various problems in Chinese medicine, such as prediction, diagnosis, and so on

*Keywords:* Traditional Chinese medicine, knowledge bases, influence diagrams, possibility distributions, Fuzzy reasoning.

*Mathematics Subject Classification: 03B52, 68T30*

## 1. Influence diagrams in Traditional Chinese Medicine

In Traditional Chinese Medicine (TCM), the diagnosis and decision on a disease can be structured in four levels: diseases, patterns, body signs and symptoms, and treatments. Body signs include tongue fur, pulse, etc; symptoms normally comprise headache, fever, cough, retching, and so on [16-17]. This study proposes a graphical decision model, influence diagrams [5, 7, 9, 10], to model the knowledge in TCM. An influence diagram is a directed acyclic graph (DAG) with three types of nodes: decision nodes, chance nodes, and value nodes. Decision nodes, shown as squares, stand for the alternative actions available to the decision-makers. Chance nodes, or random nodes, shown as circles, represent random variables in the domain. Value nodes, or utility nodes, shown as diamonds, stand for the objectives or utilities to be optimized. A general medical decision problem in TCM can be modeled with the influence diagram in Figure 1.

For example, the definitions of the nodes in *Yang Edema* or *Wind Edema* (acute nephritis) [17] are demonstrated in Table 1. This study uses the uppercase letters to represent the variables and lowercase letters for the value of a node. That is, $s_1$ stands for the value of $S_1$ and $t_i$ represents the alternatives in treatments $T$.

---

[1] Corresponding author. E-mail: teresak@wmail.hcu.edu.tw, teresak_hk@yahoo.com.tw.

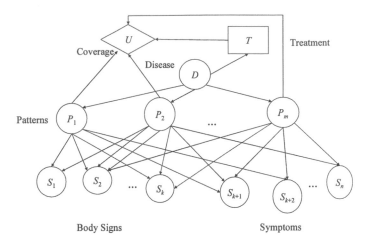

Figure 1: The General Influence Diagram of a Medical Decision in TCM.

Table 1: Definitions of the nodes in a Pediatric Disease, *Yang Edema* (*acute nephritis*) [13, 17-18]

| Disease | D: *Yang Edema* |
|---|---|
| Pattern | $P_1$: Wind edema overflow;<br>$P_2$: Retention of internal water-damp;<br>$P_3$: Retention of damp heat in the interior. |
| Sign/<br>Symptom | $S_1$: Swelling;  $S_2$: Body;  $S_3$: Urine;  $S_4$: Stool;  $S_5$: Fever<br>$S_6$: Throat;  $S_7$: Cough;  $S_8$: Head;  $S_9$: Tongue fur;  $S_{10}$: Pulse |
| Treatment | $T = \{ t_1, t_2, t_3 \}$<br>$t_1$: Dispersing lung-qi and disinhibiting water;<br>$t_2$: Normalizing the flow of Yang qi and disinhibiting water;<br>$t_3$: Clearing heat, disinhibiting dampness and cooling the blood. |

The purposes of medical expert systems include:

a. Based on a set of observed signs and symptoms, compute the belief distribution of every pattern and the disease.

b. Maximize the utility function (coverage) based on the observed signs and symptoms, which can be expressed as (1).

$$Max \quad E[U = u \mid p_i, t_i] = E[f(p_i, t_i)]$$
$$= \sum_{p_i} f(p_i, t_i) \times BEL(p_i \mid s_1, s_2, ..., s_k, s_{k+1}, ..., s_n) \; , \tag{1}$$

where $BEL(p_i \mid s_1, s_2, ..., s_k, s_{k+1}, ..., s_n)$ is the (posterior) belief of the pattern $p_i$ given the observations $s_1, s_2, ..., s_k, s_{k+1}, ..., s_n$. The set of observed nodes is $\mathbf{E} = \{ s_1, s_2, ..., s_k, s_{k+1}, ..., s_n \}$

The possibility information and the value table of *Yang Edema* can be expressed in Table 2 and Table 3. Usually the numerical models for influence diagrams are probability distributions of the chance nodes and the value table for the utility nodes. However, In TCM most descriptions of symptoms, body signs, and estimated values are qualitative and linguistic vague, which makes the adequacy of conventional approaches questioned. This study intends to define the nodes in influence diagrams as fuzzy variables and use possibility distribution functions [1-3, 6, 12] as the numerical models for TCM. Every chance node in the diagram is assigned a (conditional) possibility distribution and every outcome of the utility node is expressed as a fuzzy number.

Table 2: Possibility information of *Yang Edema* (*acute nephritis*)

---

*Pos* ($S_1$ is "puffy swelling eyelid" | $P_1$ = 1) is HIGH; *Pos* ($S_1$ is "puffy swelling limbs" | $P_2$ = 1) is HIGH; *Pos* ($S_1$ is "light puffy swelling limbs" | $P_3$ = 1) is HIGH.

*Pos* ($S_2$ is "aching and heavy limbs" | $P_1$ = 1) is HIGH; *Pos* ($S_2$ is "heavy cumbersome body" | $P_2$ = 1) is HIGH; *Pos* ($S_2$ is "Glomus and fidget chest" | $P_3$ = 1) is HIGH.

*Pos* ($S_3$ is "scant urine" | $P_1$ = 1) is HIGH; *Pos* ($S_3$ is "short voiding of scant urine" | $P_2$ = 1) is HIGH; *Pos* ($S_3$ is "short voiding of reddish urine" | $P_3$ = 1) is HIGH.

*Pos* ($S_4$ is "Dry bound stool" | $P_3$ = 1) is HIGH.

*Pos* ($S_5$ is "averse to wind and fever" | $P_1$ = 1) is MEDIUM.

*Pos* ($S_6$ is "sore swollen throat" | $P_1$ = 1) is MEDIUM.

*Pos* ($S_7$ is "cough and pant" | $P_1$ = 1) is HIGH.

*Pos* ($S_8$ is "Headache or dizziness" | $P_1$ = 1) is MEDIUM.

*Pos* ($S_9$ is "thin and white" | $P_1$ = 1) is HIGH; *Pos* ($S_9$ is "white and slimy" | $P_2$ = 1) is HIGH; *Pos* ($S_9$ is "yellow and slimy" | $P_3$ = 1) is HIGH.

*Pos* ($S_{10}$ is "floating, rapid, or stringlike" | $P_1$ = 1) is HIGH; *Pos* ($S_{10}$ is "deep and slow" | $P_2$ = 1) is HIGH; *Pos* ($S_{10}$ is "deep and rapid" | $P_3$ = 1) is HIGH.

---

Table 3: Estimated Value Table of *Yang Edema* in Figure 1.

| Treatment \ Pattern | $P_1$ = 1 | $P_2$ = 1 | $P_3$ = 1 |
|---|---|---|---|
| $t_1$ | $v_{11}$ | $v_{21}$ | $v_{31}$ |
| $t_2$ | $v_{12}$ | $v_{22}$ | $v_{32}$ |
| $t_3$ | $v_{13}$ | $v_{23}$ | $v_{33}$ |

## 2. Medical reasoning from influence diagrams and possibility distributions

This study proposes a fuzzy simulation algorithm [1-3, 10, 11] to solve the problems mentioned above.

Step 1: Choose a chance variable $x$ from an ordered list **L** of chance nodes whose value is unknown in the specified sequence. In this problem, **L** = {$D, P_1, P_2, \dots P_m$}.

Step 2: Sample for the chance variable $x$ using its conditional possibility distribution at $\alpha$-level $Pos(x|W_X)^\alpha$. For example, if we denote by $W_D$ the state of all variables except $D$, then the value of $D$ will be chosen by sampling with the distribution $Pos(d|W_D)^\alpha$. If all nodes in **L** are processed for the specified iterations, go to Step 3, else go to step 1.

Step 3: Generate the posterior belief distributions of all nodes in **L**.

Step 4: Compute the expected utility determined by the specified treatment $t_i$ from $T$ and the belief distributions of the patterns.

Step 5: Specify the optimal treatment based on maximizing the expected utility computed in Step 4.

## 3. Conclusions

This study proposes a novel approach for modeling the medical decision in TCM, influence diagrams. Considering the linguistic vagueness of the knowledge in TCM, this study uses possibility to model the uncertain relationships among the variables in the diagrams. The influence diagrams and possibility distributions provide a scientific knowledge model for intelligent systems in TCM. This paper also develops a fuzzy simulation algorithm for answering various queries on the influence diagrams of Chinese medicine, such as diagnosis, prediction, treatment planning.

## Acknowledgments

The author is indebted to the anonymous referees for their careful reading of the manuscript and their fruitful comments. Also special thanks to Tai-Chi Kao in Shanghai University of TCM for her suggestions in translating the terminology of TCM into English.

## References

[1] A.C. III Miller, and T.R. Rice, "Discrete Approximations of Probability Distributions." *Management Science*, 29:352-362, 1983.

[2] B. Bouchon Fuzzy inferences and conditional possibility distributions", Fuzzy sets and systems, 23, 33-41, 1987.

[3] D.L Keefer, "Certainty Equivalents for Three-Point Discrete-Distribution Approximations." *Management Science*, 40(6):760-773, 1994.

[4] E. Castillo, J.M. Gutoerre, and A.S. Hadi, "A New Method for Symbolic Inference in Bayesian Networks." *Networks*, 28:31-43, 1996.

[5] E. Castillo, J.M. Gutoerre, and A.S. Hadi, Expert Systems and Probabilistic Network Models. Springer-Verlag Inc., New York, 1997.

[6] G. Coletti and R.Scozzafava, "Conditional probability, fuzzy sets, and possibility: a unifying view", Fuzzy sets and systems, 144, 227-249, 2004.

[7] J. Pearl, Probabilistic Reasoning in Intelligent Systems: Networks of Plausible Inference. Morgan Kaufmann Publishers, Inc, 1997.

[8] J. Pearl, Causality-Models, Reasoning, and Inference. Cambridge University Press, 13-16, 21-26, 2000.

[9] J.A. Tatman, and R.D. Shachter. "Dynamic Programming and Influence Diagrams." *IEEE Transactions on Systems, Man, and Cybernetics*. 20(2):365-379, 1990.

[10] J. M. Charnes and P. P. Shenoy, "Multistage Monte Carlo method for solving influence diagrams using local computation", *Management Science*, 50(3), 405-418, 2004.

[11] K. Yamada, "Diagnosis under compound effects and multiple causes by means of the conditional causal possibility approach", Fuzzy sets and systems, 145, 183-212, 2004.

[12] L. A. Zadeh, "Fuzzy sets as a basis for a theory of possibility", Fuzzy sets and systems, 1, 3-28, 1978.

[13] N. Wiseman and Y. Feng, Introduction to English Terminology of Chinese Medicine, Ho-Chi Publishing Co., Taipei, Taiwan, 2003. (ISBN: 957-666-538-8)

[14] R. E. Neapolitan, Learning Bayesian networks, Pearson Education Inc., 2004.

[15] S. L. Ji, M. Z. Yu, G. S. Chen and K. R. Jan, Diagnostics in Traditional Chinese Medicine, Grand East Enterprise Ltd., 2003. (ISBN: 957-19-2704-X) (in Chinese)

[16] S. L. Ji, M. Z. Yu, G. S. Chen and J. P. Lee, Theoretical Basics of Traditional Chinese Medicine, Grand East Enterprise Ltd., 2002. (ISBN: 957-19-2703-1) (in Chinese)

[17] T. R. Chen, Pediatrics in Traditional Chinese Medicine, Cheng Chung Book Co., Taipei, Taiwan, 1987 (in Chinese).

[18] Y. M. Zhang, P. Jia and C. D. Liao, Chinese-English Pocket Dictionary of Traditional Chinese Medicine, People's Medical Publishing House, Beijing, 2003. (ISBN: 7-117-05266-X)

Brill Academic Publishers
P.O. Box 9000, 2300 PA Leiden,
The Netherlands

*Lecture Series on Computer*
*and Computational Sciences*
Volume 4, 2004, pp. 279-282

# Bayesian Sample Size Calculations with Imperfect Diagnostic Tests

**Athanassios Katsis[1]**

Department of Social and Education Policy.
University of Peloponnese
20100, Korinthos, Greece

**Hector E. Nistazakis**

Department of Telecommunications Science and Technology.
University of Peloponnese
Tripolis 22100, Greece

Received 23 June, 2005; accepted in revised form 13 July, 2005

*Abstract:* In the planning stages of a medical study involving diagnostic tests, it is common to expect some degree misclassification in the tests results. Sample size calculations must take that fact into account, regardless of the aim of the study which can be either to estimate the prevalence of a disease or to examine the properties of a diagnostic test. We investigate the problem of deriving the optimal sample size in a medical setup when misclassification is present. The optimality criteria are based on posterior credible interval widths with the help of Bayesian computational techniques.

*Keywords:* Sample size, Sensitivity, Specificity, Prevalence, Misclassification, Bayesian point of view.

*Mathematics Subject Classification:* 62K05; 62F15

## 1 Introduction

Let us assume that a researcher is interested in deriving the required sample size to accurately estimate the prevalence of a disease with a diagnostic test. Using a gold standard (or error-free) test, estimating a $(1 - \alpha)\%$ confidence interval of length $w$ for the prevalence, $\pi$, the required sample size, $n$, is given by

$$n = \frac{4Z_{1-\alpha/2}^2 \pi(1 - \pi)}{w^2} \qquad (1)$$

where $Z_{1-\alpha/2}$ is the $1 - \alpha/2$ corresponding value of the standard normal distribution.

However, it is virtually impossible to expect an error-free diagnostic test because it either does not exist or, even if it does, its use is financially prohibitive. Thus, the sensitivity and the specificity of a diagnostic test are used to denote the accuracy of a diagnostic test and are defined as follows:

For sensitivity, $s = \text{P(positive test | person with the disease)}$

---

[1] Corresponding author. E-mail: katsis@uop.gr

and for specificity, $c = $ P(negative test | person without the disease)

Thus, (1) is now modified to account for misclassification , yielding

$$n = \frac{4Z_{1-\alpha/2}^2 p(1-p)}{(w(s+c-1))^2} \qquad (2)$$

where the probability of a positive test is given by $p = \pi s + (1-\pi)(1-c)$.

Both sample size equations require point estimates of all parameters involved. This is a serious problem to the researcher since mostly range of values are usually available from previous studies. Moreover, in some situations there exist more than one diagnostic tests (see Dendukuri *et al.*, 2004). Therefore, (2) is not adequate to handle the occurrence of misclassification suggesting the need for a Bayesian approach where the researcher's prior beliefs are combined with the data.

The literature on sample size specification adjusting for misclassification has been growing in the recent years. From the Bayesian side, Erkanli, Soyer and Angold (1998) discussed two-phase designs using a fallible and an error free device while in the medical context Rahme, Joseph and Gyorkos (2000) derived the required sample size in order to estimate the prevalence of a specified disease when all the prior parameters are independent. Furthermore, Katsis and Toman (2004) proposed a double sampling scheme combining information from a fallible and an error-free device. Finally, in the same context, Gustafson, Le and Saskin (2001) analyze nonidentifiability issues arising in similar studies.

In this article, we examine the most widely used Bayesian sample size criteria to address the problem of misclassification in a medical setup. We investigate specific aspects of the problem such as computational efficiency, the dependence among the prior parameters is examined and the case of more than one diagnostic test.

## 2   Bayesian calculations

Typically, there are two broad categories of Bayesian sample size criteria. The first one involves posterior variances or credible interval widths (Wang and Gelfand, 2002) and the other focuses on the maximization of an expected utility function (Gittins and Pezeshk, 2000). Practical experience suggests that the latter approach is not easily applicable in most sampling situations. Hence, we shall concentrate on the most important credible interval criteria, namely the average coverage criterion (ACC), the average length criterion (ALC) and the worst outcome criterion (WOC).

In the interval-based method, all the available information over the unknown parameter $\theta$ is expressed through the prior distribution $f(\theta)$. Combining this distribution with the likelihood function $l(x|\theta)$ of the observed data $x$ from the data space $X$, we obtain the posterior distribution $f(\theta|x)$.

More specifically, on ACC we require that $n$ be the smallest sample size such that the average coverage of the posterior credible interval $R(x)$ is at least $1 - \alpha$. The expectation is taken over the marginal distribution of the data $x$. In particular, we are seeking the smallest value of $n$ that satisfies the following inequality:

$$\int_X \{ \int_{R(x)} f(\theta|x)d\theta \} m(x)dx \geq 1 - \alpha$$

where

$$m(x) = \int l(x|\theta)f(\theta)d(\theta).$$

The interval $R(x)$ is of fixed length and may be either symmetric around the posterior mean or have the highest posterior density. If there is more than one prior parameter, we integrate out the other parameters, obtaining the marginal posterior density of the parameter of interest. Further computational details are provided in and Adcock (1997).

Conversely, Joseph, du Berger and Belisle (1997) defined the ALC where the coverage is fixed and we average over all the highest posterior density intervals. Therefore, we are looking for the smallest $n$ such that the average length of the $100(1 - \alpha)$ posterior credible intervals over the marginal distribution does not exceed a pre-specified length $w$, that is,

$$\int_X w'(x)m(x)dx \leq w$$

where $w'$ is the length of the posterior credible interval $R(x)$ derived by solving the following equation for each value of $x \in X$:

$$\int_{R(x)} f(\theta|x)d(\theta) = 1 - \alpha$$

Finally, according to the conservative WOC the smallest sample size that satisfies the equation

$$\inf_x \{ \int_{R(x)} f(\theta|x)d(\theta) \} \geq 1 - \alpha$$

is the desired one.

Regarding the prior information for the parameters, many authors suggest the use of independent beta distributions for each parameter. This is done mainly for convenience since the Beta distribution is a rich class of densities. However, it is realistic to assume that the frequency of the disease is independent of the accuracy of the diagnostic test. Thus, $\theta$ is independently distributed of both $s$ and $c$. However, any information about the results of the test among the healthy, say, part of the population will undoubtedly provide a rough idea how the test fares among the people with the disease. The concept of statistical dependence between error probabilities in various settings has been discussed in, among others, Galen and Gambino (1975) and Gunel (1984). Thus, the joint prior distribution of $\theta$, $s$ and $c$ may be thought of as a product of a Beta distribution (for $\theta$) with parameters $a$ and $b$ and a Dirichlet distribution (for $s$ and $c$) with parameters $(\lambda_1, \lambda_2, \lambda_3)$. Therefore we have that

$$f(\theta, s, c) \propto \theta^{a-1}(1 - \theta)^{b-1}s^{\lambda_1-1}c^{\lambda_2-1}(1 - s - c)^{\lambda_3-1}$$

where $a$, $b$, $\lambda_i$, $(i = 1, 2, 3)$ are non-negative numbers.

## 3   Numerical methodology

Since it is very difficult to obtain analytic expressions for almost all of the above mentioned criteria, the non-iterative Monte Carlo method discussed by Ross (1996) is usually very helpful. In this, $k$ random sample points from the joint prior distribution are generated and a specific weight function is defined for each point. This function is used to approximate the posterior mean and variance of the prevalence as well as the average coverage.

Furthermore, Dendukuri *et al* (2004) have suggested an approximate method for the marginal posterior distribution of the prevalence. More specifically, they recommend that a posterior mixture of Beta densities be approximated by a single Beta distribution where its parameters are found by matching the first two moments of the sample to the mean and the variance of a Beta distribution. Subsequently, a Newton-Raphson type algorithm was employed to find Highest Posterior Density intervals. This works best for the ACC and the ALC cases.

Although the above mentioned methodology is developed in a medical context, misclassification can occur in a wide range of applications such as engineering, quality control and auditing. The researcher must choose the most appropriate optimality criterion, decide on the nature of the prior information (i.e., independent or not) and resort to the most helpful numerical technique.

# References

[1] C. A. Adcock, Sample size determination: a review, *Statistician* **46**, 261-283(1997).

[2] N. Dendukuri, E. Rahme, P. Belisle and L. Joseph, Bayesian Sample Size Determination for Prevalence and Diagnostic Test Studies in the Absence of a Gold Standard Test, *Biometrics*, **60**, 388-397(2004)

[3] A. Erkanli, R. Soyer and A. Angold, Optimal Bayesian Two-Phase designs for Prevalence Estimation, *Journal of Statistical Planning and Inference*, **66**, 175-191(1998)

[4] R. S. Galen and S. R. Gambino, *Beyond Normality: The Predictive Value and Efficiency of Medical Diagnoses*, John Wiley & Sons, 1975.

[5] J. Gittins and H. Pezeshk, A behavioral Bayes method for determining the size of a clinical trial , *Drug Information Journal*, **34**, 355-363(2000)

[6] E. Gunel, A Bayesian analysis of the multinomial model for a dichotomous response with nonrespondents, *Communications in Statistics, Theory and Methods*, **13**, 737-751(1984)

[7] P. Gustafson, N. D. Le and R. Saskin, Case-control analysis with partial knowledge of exposure misclassification probabilities, *Biometrics*, **57**, 598-609(2001).

[8] L. Joseph, R. du Berger and P. Belisle, Bayesian and mixed bayesian/likelihood criteria for sample size determination, *Statistics in Medicine*, **16** 769-781(1997)

[9] A. Katsis and B. Toman, A Bayesian double sampling scheme for classifying binomial data, *The Mathematical Scientist*, **81**, 49-53(2004)

[10] E. Rahme, L. Joseph and T. W. Gyorkos, Bayesian sample size determination for estimating binomial parameters from data subject to misclassification, *Applied Statistics*, **49**, 119-128(2000)

[11] S. M. Ross, Bayesians should not resample a prior sample to learn about the posterior, *American Statistician*, **50**, 116(1996)

[12] F. Wang and A. E. Gelfand, A simulation-based approach to Bayesian sample size determination for performance under a given model and for separating models, *Statistical Science*, **17**, 193-208(2002)

Brill Academic Publishers
P.O. Box 9000, 2300 PA Leiden
The Netherlands

*Lecture Series on Computer*
*and Computational Sciences*
Volume 4, 2005, pp. 283-287

# Practical Aspect of Refactoring Patterns for a XP Project

A. Kaya[1], D. Öktem[2], A. Önal[3]

Department of Computer Technologies and Programming
Tire Kutsan Post Secondary Vocational School
Ege University
35900 Tire İzmir, Turkey

Received 14 June, 2005; accepted in revised form 12 July, 2005

*Abstract:* Extreme Programming is a new and very practical software development technique, which is appropriate to apply for small projects with small software development teams. Refactoring is one of the activities of XP which encourages the developer to find the most effective source code to execute. It helps the code to be more simple and flexible. To apply refactoring on a source code the software team must be aware of refactoring patterns. There are some tools which can apply some of the refactoring patterns. The tools generated for refactoring is able to apply very few patterns. This paper states that extreme programming world needs more developed tools, especially some tools so-called "intelligent".

*Keywords:* Extreme Programming, Intelligent Refactoring, Refactoring Patterns.

## I. Introduction

EXTREME Programming is quite new methodology in object oriented software development, which has a different philosophy than the classical software development techniques. Unlike the traditional waterfall development styles, XP has a different iterative and incremental way of developing source code. This means that analysis, design and testing are done in each small step of the development, but in different order.

One of the basic principles of XP is the tight communication with the customers and obtaining rapid feedback to interfere on time to change the software. With XP technique, developers can interfere any time because each step of the source code is under control and every development step is saved for the flashbacks of the program code. This paper involves with the experiences of producing software by using the XP techniques, with most basic activities like Refactoring and Unit Testing.

## II. What Is Refactoring?

The word "refactoring", in fact, has two meanings: one as a noun and one as a verb. As a noun it means: "a change made to the internal structure of software to make it easier to understand and cheaper to modify without changing its observable behavior". And as a verb it means: "to restructure software by applying a series of refactorings without changing its observable behavior" [1].

The definitions tell that refactoring is an activity done to make the code easier and simpler to understand. Since XP always aims to provide a flexible structure to the source code, it is an important step to keep the code simple in any phase of the software project. So by refactoring the code continuously, it is possible to add more features to the source code without making it more complex.

The act of refactoring centers on testing to validate that the code still functions. The activity of testing gives the software development team the courage to refactor and keep the code simple [2]. In order to understand the mission of refactoring, another important activity of XP, testing, must be exercised.

### *A. Testing*

The most important test activity of XP is "Unit Testing". In unit testing, each single atomic line of source code is tested before it is implemented. By this way, the testing process of software does not consist of entering a value and expecting the program to return the right result. Unit tests are done internally, to control the behavior of the program. Lots of unit tests must be written to prove that the code is on the right way.

After the source code passes the unit tests successfully, this means that the code is secure to accept new feature additions. At that moment, refactoring must be done while making the necessary feature addition. After that, the code must be retested, if tests fail, appropriate corrections must be done and retested again. If the tests are successful, that this means that refactoring is successful, and the code has improved in a correct manner without affecting the external behavior of the program.

The definition of the "simplest design", which is aimed in a program, can be defined as follows:

1.  All unit tests are running.

2.  The code communicates all of its design concepts.

3.  The code does not contain redundancy.

4.  Under the above rules, the code contains the smallest possible number of classes and methods [3].

The activity of writing tests and refactoring must be put forward together to obtain a simple, flexible, understandable program code. In order to do that, the unit tests must be automated, and the refactorings must be applied by using some refactoring patterns.

## III.  Refactoring Patterns

For the word pattern, a definition given here can be suitable: "Problem solution formula that codifies exemplary design principles". According to another definition, pattern means: "a named description of a problem and solution that can be applied to new contexts with advice on how to apply it in novel situations and discussion of its trade-offs" [4].

As the definitions say, refactoring patterns are somehow code samples, in which the developer should decide where to use any of them while refactoring. Refactoring patterns obtain lots of solutions to clarify the source code.

All patterns have names, expressing the job that the pattern does when applied to the code. For example the pattern name "Rename Method" tells the procedure of renaming a method, when this change of name is necessary. Each pattern has a rule for applying, that is, the series of steps that must be passed to realize the change.

In this paper, some refactoring patterns will be explained, which are used in the project OBS. But the point that we want to emphasize is not the patterns used. The point is how those patterns can behave as "intelligent refactorings".

## IV.  About the XP project - OBS

OBS is the Turkish abbreviation of the words "Ogrenci Bilgi Sistemi", which means "Student Information System". This project is implemented as the application phase of the Master thesis, "Object Oriented Software Development with Extreme Programming" [5]. In this project, XP techniques are applied and tested. The problems faced and experiences are explained in detail. So, as an activity of XP, refactoring and refactoring patterns are used also.

According to the XP style, classes and their methods are created one by one, by writing the unit tests for each statement. So, after the creation of a class and a method, a refactoring activity will be necessary to integrate a new feature on the existing parts of the program.

The program code has three main classes, these are "Ders" (Lectures), "Ogrenci" (Students) and "Not" (Grades). While these classes are created, the order followed was to create "Ders" first, than "Ogrenci". "Not" was an attribute in the class "Ogrenci". Later on, this design seemed insufficient and "Not" became a class as "Ders" and "Ogrenci".

The processes like explained above all need some refactorings. The list of refactoring patterns that were applied on the project is given below.

## V. Patterns Used In The Project

1. **Replace Data Value with Object:** the "Not" field has changed from being an attribute in the object "Ogrenci" to a class because "Not" was insufficient as a field of the object.

2. **Change Unidirectional Association to Bidirectional:** The object "Ders" was able to keep the list of the students attending at that lecture, where the list of lectures that a student is assigned was not available. The direction of the association became bidirectional when this pattern is applied.

3. **Rename Method:** When the function of a method changes, it is necessary to change its name too. For example, the method "testAddDers()" has changed to "testAddandReturnDers()" after the returning value of the method is added.

4. **Remove Parameter:** The method deleteDers() had used the classes "Ders" and "Ogrenci" as parameters at the beginning. But it is understood that it would be enough to keep "dkod" and "no" fields as parameters representing the code of the lecture and number of the student.

5. **Extract Method:** To make any of the transactions in OBSDatabase class, it needed to be connected to the database. To do that, a method named connect() is created and called before all other processes.

These patterns mentioned above are just a small part of all the refactoring patterns stated in literature. There are lots of patterns which are advised to be used. But the important point is to know which pattern can be used in which case.

## VI. Problems Faced in practice

While developing a program code with XP style, the developers must always be on the alert to find the bad smell in the code. These bad smells can be because of anything that takes the code away from being simple and flexible. In practice, the most important fact is to find the appropriate refactoring pattern for that code sample.

The developer always perceives that the code needs a change. But the developer must also have deep knowledge and experience about the refactoring patterns to decide where to use the suitable pattern.

When this project started, we were too inexperienced about the refactoring patterns. Even in a small renaming event, we understood that we had to rename, but we weren't aware that we were already refactoring the code and this refactoring had a name.

The software developers must be very experienced in object oriented programming concept. They must be able to think in objects because many of the patterns are built on the object orientation logic. Lack of experience was the major disadvantage for the team while development. The most important point in refactoring is to feel the right pattern to refactor.

## VII. Intelligent Refactoring?

For inexperienced developers, it is quite difficult to seize the pattern suitable for the refactoring activity. And sometimes, the pattern to be applied is so obvious that it is a burden on the developer to apply it because it might be time consuming to deal with such easy-to-find refactorings. In such a case, the tools supporting the extreme programming environment must help the developer.

Not only for refactoring, but also for the other activities of extreme programming, tool support has a great importance. In our project, the development environment used is Eclipse 2.1 [6]. The Eclipse environment can help developer with some of the transformations stated in Martin Fowler's book [1]. Renaming an element is the most popular and easy transformation supported with Eclipse. Renaming can be made on packages, units, methods and method parameters. A compilation unit may be moved to another package. Addition to those, some part of a method can be extracted to form another method. Eclipse prevents the code from being more complex by applying these refactorings automatically. For more complex ones, the developer must be careful to catch and apply them.

Except Eclipse, there are a few more refactoring tools. One of the most popular one is RefactorIT. It has more features than Eclipse; it can make the transformations of the following refactorings:

- Rename (Method, field, type, package prefix)
- Move (Class, method/field, push up/push down in hierarchy)
- Extract (Method, superclass/interface, introduce explaining variable)
- Create (Factory method, constructor, add delegate methods, override/implement methods)
- Inline (variable, method) [7].

For any of the extreme programming tools, it is not enough to let the tool to rename elements because refactoring is not just renaming. There are many refactoring patterns stated but there is no specific tool to "decide" that a refactoring is necessary for the code. Even it doesn't decide, a tool with the refactoring templates would be very useful. Refactoring activity needs a tool, where the names of the elements of a program are given, and when the appropriate refactoring pattern is found, the tool itself applies the pattern without any objection to the developer.

Renaming or moving an element in the project by the help of the existing refactoring tools may only prevent the developer from being confused because of the previous and next locations and names of the methods or classes. This is also an important fact, but it is not the entire problem. The refactoring tools must be "intelligent" enough to make the developer feel more comfortable in finding the appropriate pattern to apply on the source code. Since XP is a whole with its philosophy and its tools, a tool supporting intelligent refactoring activity would increase the self-confidence of the software developers.

## VIII.  Future Work

Our suggestion for the future work on the refactoring subject is more developed tools which can perceive the need of refactoring in the code must be implemented.

Such a tool must have the ability of recognizing the source code with its class, method and variable names. After analyzing the code, the refactoring tool must propose a list of possible refactorings to be applied on the code. If the developer approves the possible refactorings that the tool offers, then the tool must make the needed transformations to apply the proposed patterns.

Also the tool must keep the track of the transformations on the code done throughout the life cycle of the source code. So it then can be possible for the developer to reach the history of the program's implementation.

## IX.  Conclusion

Our experience on the master thesis about the application of extreme programming proved that tools are a great necessity for doing XP. Development environments can afford that need in testing, integrating or version controlling, but for refactoring the case is a bit different. Because refactoring is something that the developer feels that he has to do. So by claiming a tool to do this for the developer, we are hoping to see an intelligent tool to "feel" that the code must be refactored.

The next step in developing software tools for extreme programming should be a complete tool for refactoring as explained above. Intelligence level in the software tools being used currently is not enough to meet the needs of developing software effectively. So more intelligent tools are needed for this aspect of software development in the future.

## References

[1] M. Fowler, "Refactoring: Improving the Design of Existing Code" Massachusetts: Addison-Wesley, 1999, pp. 53–54.
[2] R. Hightower, N. Lesiecki. "Java Tools for Extreme Programming". Wiley Computer Publishing, 2002. pp. 8.
[3] J. Link, "Unit Testing in Java", Morgan Kaufmann Publishers, 2003, pp. 8.
[4] C. Larman, "Applying UML and Patterns", Prentice Hall PTR, 2001, pp. 4, 218.

[5] D. Öktem, Master thesis called "Object Oriented Software Development with Extreme Programming", Ege University, 2003, pp. 1.

[6] [Online] www.eclipse.org

[7] [Online] http://www.refactorit.com/index.php?id=1370

Brill Academic Publishers
P.O. Box 9000, 2300 PA Leiden,
The Netherlands

Lecture Series on Computer
and Computational Sciences
Volume 4, 2005, pp. 288-291

# Vertical temperature in the atmoshere during surface inversions

## H. Koshigoe[1], T. Shiraishi and M. Ehara

Department of Urban Environment Systems,
Faculty of Engineering, Chiba University,
263-8522, Chiba, Japan

Received 5 August, 2005; accepted in revised form 16 August, 2005

*Abstract:* In this article we present a mathematical model for the surface inversion in the meteorology and propose a numerical method which is based on the finite difference approximation and the Schwartz distribution. Through our discussion, a numerical simulation which is characterized by the inversion of the temperature gradient is presented.

*Keywords:* direct method, Schwartz distribution , surface inversion

*Mathematics Subject Classification:* 65F05, 65F22

## 1  Introduction

The purpose of this article is to present a mathematical model and a numerical simulation for a singular temperature ($u$) which is called the surface inversion in the meteorology. The atmospheric temperature usually decreases in proportion to the altitude but when the weather of the surface inversion appears it does increase near the ground and the inversion layer appears at the altitude where the inversion of the temperature gradient occurs ( cf. Figure 1).

¿From the viewpoint of the meteorology, one of the reasons that the surface inversion appears is the radiative cooling on the ground by the difference of the temperature in the daytime and night. This fact is introduced in our mathematical model. Here we state our motivation. The inversion layer appears at the altitude of about 200 meters and it covers a town. Hence the town loses a function of diffusions and the environmental degradation is observed (for example, the fumigation of pollution and the white smog).

In order to analyze the character of the singular temperature in the surface inversion, we shall propose a mathematical model with the transmission conditions (refs. [1] and [5]) on the ground which describes the relative cooling and show its numerical simulations( cf. Figure 2).

The content of this article is as follows: In Section 2 we shall formulate a mathematical model for the surface inversion by the Schwartz distribution. In Section 3 we shall describe a finite difference approximation of the above formulation and finally, in Section 4, we shall show the numerical result

[1]Corresponding author.E-mail: koshigoe@faculty.chiba-u.jp
Research partially supported by Japanese Grant-in-Aid for Scientific Research, No.14540103

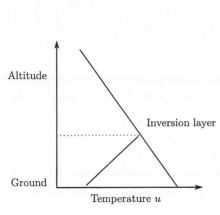

**Figure 1** Surface Inversion

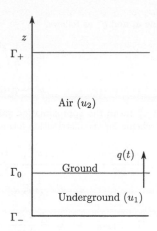

**Figure 2** Radiative Cooling q(t)

by the direct method we propose here (refs. [3] and [4]).

## 2 Distribution formulation

Let $\chi_i$ $(i = 0, 1, 2)$ be a characteristic function in $\Omega_i$ $(i = 0, 1, 2)$ and set $a = \epsilon_0\chi_0 + \epsilon_1\chi_1 + \epsilon_2\chi_2$. Then the distribution theorem (refs. [2] and [6]) shows that Problem I is equivalent to

**Distribution formula** Let $V = H^1(\Pi)$.
Find u $\in L^2(0, T; V)$ satisfying

$$\begin{cases} \dfrac{\partial u}{\partial t} = \dfrac{\partial}{\partial z}\left(a(z)\dfrac{\partial u}{\partial z}\right) + q(t)\,\delta_{\Gamma_0} & \text{in } (0, T) \times (\Gamma_-, \Gamma_+) \\[2mm] \lambda_1 u_1 + (1 - \lambda_1)\dfrac{\partial u_1}{\partial z} = g_1 & \text{on } (0, T) \times \Gamma_- \\[2mm] \lambda_2 u_2 + (1 - \lambda_2)\dfrac{\partial u_2}{\partial z} = g_2 & \text{on } (0, T) \times \Gamma_+ \\[2mm] u(0, z) = f_1(z)\chi_{(\Gamma_-, \Gamma_0)} + f_2(z)\chi_{(\Gamma_0, \Gamma_+)} & \text{in } (\Gamma_-, \Gamma_+). \end{cases} \tag{1}$$

Hereafter let $V'$ denote the dual space of $H^1(\Pi)$.

**Theorem 1** *There is a unique solution* $u \in L^2(0, T; V)$ *satisfying (1).*

## 3 Finite difference scheme with the Dirac distribution

In this section we consider the finite difference equation for the distribution formula (1). We divide the interval $(\Gamma_-, \Gamma_+)$ into N subintervals, and let mesh size $\Delta z = 1/N$ and $\Delta t$ denote time mesh size. Set $z_i = i \, \Delta z$ $(i = 0, 1, \cdots, N)$, $t_n = n \, \Delta t$ $(n = 0, 1, \cdots)$; $z_0 = \Gamma_-$, $z_{N_1} = \Gamma_0$ and $z_N = \Gamma_+$.

We also define $a_i$ and $q^n$ as follows.

$$a_i = \frac{1}{h} \int_{(i-1/2)\Delta z}^{(i+1/2)\Delta z} a(z) \, dz,$$

$$q^n = q(n\Delta t).$$

Moreover let $u_i^n$ mean the approximation value of $u(i\Delta x, n\Delta t)$. Using these notation, we define the implicit scheme for the distribution formula (1) as follows.

$$\frac{u_i^n - u_i^{n-1}}{\Delta t} = \frac{a_{i-\frac{1}{2}} u_{i-1}^n - \left(a_{i-\frac{1}{2}} + a_{i+\frac{1}{2}}\right) u_i^n + a_{i+\frac{1}{2}} u_{i+1}^n}{(\Delta z)^2}$$

$$+ \quad q^n \cdot \frac{1}{\Delta z} \delta_{i,N_1} \quad \text{for } i = 0, 1, 2, \ldots, N_1, \ldots, N, \quad n \geq 0 \,, \tag{2}$$

$$(1 - \lambda_1) u_0^n \quad + \quad \lambda_1 \frac{u_1^n - u_{-1}^n}{\Delta z} = g_1^n \quad \text{for } n \geq 0 \,,$$

$$(1 - \lambda_2) u_N^n \quad + \quad \lambda_2 \frac{u_{N+1}^n - u_{N-1}^n}{\Delta z} = g_2^n \quad \text{for } n \geq 0$$

where $\delta_{i,N_1}$ denotes Kronecker's delta and $q(t) \, \delta_{\Gamma_0}$ is approximated by a function $q^n \frac{1}{h} \chi_{((N_1 - \frac{1}{2})h, (N_1 + \frac{1}{2})h)}$.

## 4    Numerical simulation

We now go back to the numerical scheme of (2) and compute it for the case where $\epsilon_1 = 0.5$, $\epsilon_2 = 0.1$, $q(t) = 1 - 5t$, $\lambda_1 = 1$, $\lambda_2 = 0.5$, $g_1 = 10$ and $g_2 = 0$. We have taken $\Delta z = 1/50$ and $\Delta t = 0.001$. The numerical result shows the surface inversion appears as follows.

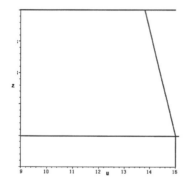
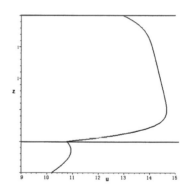

**Figure 3**  t=0                    **Figure 4**  t=0.2

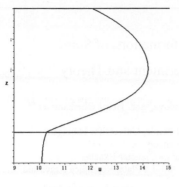

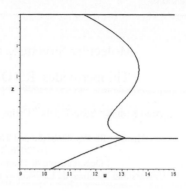

**Figure 5** t=0.6                    **Figure 6** t=1.0

Figure 3 is the initial temperature. Figure 4 shows that the temperature near the ground decreases rapidly and that the inversion layer appears. After that, the altutude of the inversion layer becomes higher from Figure 5 and 6. Therefore these numerical result shows that the surface inversion occurs.

Through our discussion, we showed the efficiency of a direct method by use of the Schwartz distribution. The corresponding results in 2-dimension space will be reported in a forthcoming publication.

## Acknowledgment

The authors wish to thank Prof.K.Kitahara of Kogakuin University and Dr.A.Ohi for significant discussion.

## References

[1] R. Glowinski, J.L. Lions and R. Tremolieres, Numerical analysis of variational inequalities, Studies in Mathematics and its applications Vol.8, North-Holland (1981).

[2] H. Kawarada, Free boundary problem - theory and numerical method - , Tokyo University Press (1989)(in Japanese) .

[3] H. Koshigoe, Direct Solver on FFT and SEL for Diffraction Problems with Distribution, Lecture Notes in Computer Science, 3037, Springer,(2004), 105-112.

[4] H. Koshigoe, Direct Method for Solving a Transmission Problem with a Discontinuous Coefficient and the Dirac Distribution, Lecture Notes in Computer Science,2659,Springer,(2003), 388-398.

[5] J.L. Lions, Optimal control of systems governed by partial differential equations, 170, Springer-Verlarg (1971).

[6] S. Mizohata, The theory of partial differential equations, Cambridge at the University Press(1973).

Brill Academic Publishers
P.O. Box 9000, 2300 PA Leiden
The Netherlands

*Lecture Series on Computer
and Computational Sciences*
Volume 4, 2005, pp. 292-294

# Molecular Structures and Conformations of Some

# Thioperoxides RS-OR': Experiment and Theory

Areti Kosma,[a] Sonia E. Ulic,[b] Carlos O. Della Vedova,[b] and Heinz Oberhammer.[a]*

[a] Institut für Physikalische und Theoretische Chemie, Universität Tübingen,
D-72076, Tübingen, Germany.
[b] Departamento de Química. Facultad de Ciencias Exactas.
Universidad Nacional de La Plata, 47 esq. 115, (1900) La Plata, República Argentina

Received July 20, 2005; accepted July 24, 2005

*Abstract*: All peroxides RO-OR' and disulfanes RS-SR', whose gas phase structures are known, possess gauche structures and most of them possess dihedral angles between 80° and 120°. Thioperoxides of the type RS-OR' are much less stable than peroxides and disulfanes and very little is known about their structural properties. Sulfenic acid, HSOH, dimethoxysulfane, $CH_3OSOCH_3$ and dimethoxydisulfane, $CH_3OSSOCH_3$, possess gauche structures with dihedral angles between 75° and 90°. A combination of vibrational spectroscopy, gas electron diffraction and quantum chemical calculations results for trifluoromethanesulfenyl acetate, $CF_3S-OC(O)CH_3$, and for trifluoromethanesulfenyl trifluoroacetate, $CF_3S-OC(O)CF_3$, in a mixture of two conformers with the prevailing component possessing gauche structure. The minor form (18(5)% and 11(5)%, respectively) possesses an unexpected trans structure around the S-O bond. The C=O bond of the acetyl group is oriented syn with respect to the S-O bond in both conformers. A similar experimental and theoretical study of S-(fluoroformyl)O-(trifluoroacetyl)thioperoxide, $FC(O)S-OC(O)CF_3$, results also in a mixture of two conformers. In this thioperoxide, however, both conformers possess gauche structure around the S-O bond and differ only by the orientation of the FC(O) group (C=O bond syn or anti with respect to the S-O bond).

*Keywords*: Thioperoxides, Conformational Properties, Matrix-Infrared Spectroscpy, Gas Electron Diffraction, Quantum Chemical Calculations

## 1. Introduction

Experimental studies of a large number of gaseous non-cyclic peroxides of the type ROOR' using gas electron diffraction or microwave spectroscopy resulted in skew or gauche structures with dihedral angles $\phi$(ROOR') between 80° and 120°. Only a few peroxides with bulky substituents possess larger dihedral angles. Similarly, disulfanes of the type RSSR' possess gauche structures with in general smaller dihedral angles around 90°. Thioperoxides of the type RSOR', which are derivatives of the sulfenic acid HSOH, are much less table than peroxides or disulfanes and so far gas phase structures of only three such compounds have been studied, sulfenic acid, dimethoxysulfane, MeOSOMe, and dimethoxydisulfane, MeOSSOMe. These compounds possess gauche structures with dihedral angles between 75° and 90°. All these experimental structures are reproduced satisfactorily by high-level quantum chemical calculations.

Typical calculated potential functions for internal rotation around the central bond in the parent species HOOH, HSSH and HSOH possess minima for gauche conformations and maxima for cis and trans orientation. Whereas the trans maxima for peroxides are only about 0.5 kcal/mol above the gauche minima, these maxima are about 4 to 5 kcal/mol above the gauche minima in disulfanes and thioperoxides.

## 2. Trifluoromethanesulfenyl Acetates CF3S-OC(O)CX3, X = H and F

Recently we studied the structures and conformational properties of two derivatives of the sulfenic acid, trifluoromethanesulfenyl acetate and trifluoromethanesulenyl trifluoroacetate. In the infrared spectra of these compound in an Ar matrix two band are observed in the carbonyl stretching region, which are split by 58 cm$^{-1}$ in the case of the acetate. This might suggest the presence of two conformers with syn and anti orientation of the C=O bond relative to the S-O bond. However, quantum chemical calculations performed with different methods predict for these two conformers a splitting of the C=O vibrations of only 0 to 6 cm$^{-1}$ and, furthermore, a relative free energy of the anti form which is 5-6 kcal/mol higher than that of the cis conformer. These results make the assignment of the vibrational spectra to these two conformers unlikely. In the next step the potential function for internal rotation around the S-O bond was calculated with different computational methods for the energetically favored structure with syn orientation of the C=O bond relative to the S-O bond. All calculations result in functions which possess a minimum for trans orientation ($\phi$(CSOC) = 180°) in addition to the global minima for gauche conformers. The calculations predict a splitting of the C=O vibrations between gauche and trans conformers of 45 – 61 cm$^{-1}$, in very good agreement with the experimental value and free energy differences of 0.4 to 1.2 kcal/mol in good agreement with the relative intensities of the two bands in the vibrational spectra. The results for the trifluoroacetate are very similar.

In the next step the structures of both compounds were determined by gas electron diffraction. Comparison between experimental and calculated radial distribution curves demonstrates that the gauche conformer with syn orientation of the C=O bond is the prevailing form. The agreement with the experimental curve improves slightly if a small amount of trans conformer is included. For the acetate a contribution of 15(5)% of trans conformer is derived from the matrix spectra and 8(12)% from the GED experiment. Quantum chemical calculations predict contributions between 11 and 34%. For the trifluoroacetate derivative 11(5) % trans conformer result from IR spectra, 18(9)% from GED and 15 - 57% from quantum chemical calculations.

Thus, experiments and theory result in a mixture of two conformers for these thioperoxides with gauche and trans orientation around the S-O bond. The C=O bond is oriented synperiplanar with respect to the S-O bond in both conformers. These two thioperoxides are the first examples in this class of compounds – peroxides, disulfanes and thioperoxides – for which a stable trans form is observed. We are unable to offer a convincing explanation for this unexpected result. One might suspect that conjugation between the $\pi$ system of the acetyl group C(O)CX$_3$ and the p-shaped electron lone pairs of oxygen and sulfur stabilize the trans structure. If this were the case, the trans form should be stabilized even more, if the CF$_3$ group bonded to sulfur is substituted by a second carbonyl group. In order to proof this assumption a new thioperoxide, fluoroformyl trifluoroacetyl thioperoxide was synthesized.

## 3. Fluoroformyl Trifluoroacetyl Thioperoxide, FC(O)S-OC(O)CF3

For this compound the conformational properties are more complicated. Independent of the orientation around the S-O bond, gauche or trans, four conformations are feasible, depending on the orientation of the two C=O bonds of the FC(O) and CF$_3$C(O) group. Both can be oriented synperiplanar (sp-sp) or antiperiplanar (ap-ap) or one can be oriented antiperiplanar and the other synperiplanar (ap-sp or sp-ap). In the infrared matrix spectrum four prominent bands are observed in the C=O stretching region, split over a range of 32 cm$^{-1}$. This indicates a mixture of two conformers. It was attempted to determine the kind of conformers in combination with quantum chemical calculations.

In the first step the potential function for internal rotation around the S-O bond was calculated with different computational methods. B3LYP and MP2 methods with standard basis sets (6-31G*) predict functions with a shallow minimum for trans orientation in addition to the global minima for gauche structures. If diffuse functions are added to the basis sets, the trans minimum predicted by the B3LYP method becomes even more shallow and it disappears in the potential function predicted with the MP2 approximation. Futhermore, if zero-point vibrational energies are added to the electronic energies, the trans minimum disappears in all cases except in the B3LYP curve derived with standard basis sets. Further increase of the basis sets has no effect on the overall shape of the potential function. Thus, these calculated potential functions make the existence of a stable trans structure unlikely, but the results are not unambiguous. Predicted relative free energies are between 1 and 2 kcal/mol and would be compatible with the experimental vibrational spectrum. On the other hand, the predicted splitting of the C=O stretching vibrations for a mixture of gauche and trans conformer of more than 74 cm$^{-1}$ is not compatible with the observed splitting of 32 cm$^{-1}$.

Next we concentrated on the various conformers with gauche orientation around the S-O bond. All calculations predict the sp-sp conformer to be the most stable form and the conformer with the C=O bond of the FC(O) group rotated to antiperiplanar orientation (ap-sp) to be about 1 kcal/mol higher in energy. Both structures with the C=O bond of the trifluoroacetyl group in antiperiplanar orientation are predicted to be 6 to 9 kcal/mole higher in energy and are not observable in our experiments. Thus, from the calculations we expect a mixture of two conformers with gauche orientation around the S-O bond which differ by the orientation of the C=O bond of the FC(O) group. These two conformers are the only ones, for which the calculated C=O vibrations reproduce the experimetnal pattern.

This result is confirmed by the GED experiment. The calculated radial distribution curves for sp-sp and ap-sp conformers differ strongly in the distance range between 2.5 and 3 Å. Only a mixture of these two structures leads to a satisfactory fit of the experimental curve. The GED experiment results in a mixture of 82(7)% sp-sp conformer and 18(7)% ap-sp conformer. Both conformers possess gauche structures around the S-O bond. This result is in contrast to our expectation that two carbonyl groups bonded to oxygen and sulfur would stabilize the trans structure of a thioperoxide.

The most interesting geomtric parameters in these thioperoxides are the S-O bond length and the dihedral angle of the gauche forms. The S-O bond lengths (1.659(4) Å and 1.663(5) Å in $CF_3S$-$OC(O)CH_3$ and $CF_3S$-$OC(O)CF_3$, respectively) and the dihedral angles (100(4)° and 101(3)°, respectively) in the prevailing gauche forms of the trifluoromethanesulfenyl acetates are equal within their experimental uncertainties. The S-O bond in fluoroformyl trifluoroacetyl thioperoxide is slightly shorter (1.647(5) Å) and the dihedral angle is considerably smaller (75(3)°). All experimental geometric parameters are reproduced satisfactorily by the quantum chemical calulations, except for the S-O bond lengths and the dihedral angles. The S-O bond lengths in all three compounds are predictet too long by 0.04 to 0.05 Å and the calculated dihedral angles in the trifluoromethanesulfenyl acetates are about 10° smaller than the experimental values.

## Acknowledgements

Financial support by the Deutsche Forschungsgemeinschaft is gratefully acknowledged.

## References

[1] Winnewisser, G.; Lewen, F.; Thoswirth, S.; Behnke, M.; Hahn, J.; Gauss, J.; Herbst. E.; *Chem Eur. J.* **2003**, *9*, 5501.

[2] Baumeister, E.; Oberhammer, H.; Schmidt, H.; Steudel, R. *Heteroatom Chem.* **1991**, *2*, 633.

[3] Steudel, R.; Schmidt, H.; Baumeister, E. Oberhammer, H.; Koritsanszky, T. *J. Phys. Chem.* **1995**, *99*, 8987.

Brill Academic Publishers
P.O. Box 9000, 2300 PA Leiden,
The Netherlands

*Lecture Series on Computer
and Computational Sciences*
Volume 4, 2005, pp. 295-298

# ClONO Loss Mechanism in The Presence of NO$_2$: A Quantum-Mechanical Study

**S. Kovačič[1], A. Lesar**

Department of Physical and Organic chemistry,
Institute 'Jožef Stefan',
SI-1000 Ljubljana, Slovenia

**M. Hodošček**

Centre for Molecular Modeling,
National Institute of Chemistry,
Hajdrihova 19, SI-1000 Ljubljana, Slovenia

Received 8 July, 2005; accepted in revised form 30 July, 2005

*Abstract:* Possible reaction pathways of ClONO loss in the presence of NO$_2$ have been calculated using the B3LYP/6-311+G(3df) level of theory. Depending on the mutual orientation of the approaching molecules, ClNO$_2$+NO$_2$ or ClONO$_2$+NO products might be formed. The energy barriers are 4.1 and 20.6 kcal/mol, respectively. The first reaction, with exothermicity of 13.3 kcal/mol, is preferred over the reaction producing ClONO$_2$+NO, which is endothermic by 8.5 kcal/mol.

*Keywords:* bimolecular reaction, chlorine nitrite, nitryl chloride, chlorine nitrate, density functional calculations

## 1 Introduction

Over the past two decades, it has become evident that nitryl chloride, ClNO$_2$, and its isomer chlorine nitrite, ClONO, play an important role in the chemistry of the stratosphere. These species are formed through the reaction of Cl with NO$_2$ in the presence of a third body and thus couple the ClO$_x$ and NO$_x$ ozone depletion cycles. The cis isomer of ClONO was found to be the major product (yield $\geq$ 80%) of the gas phase reaction between atomic chlorine and NO$_2$ [1], although it is calculated to be thermochemically less favorable than ClNO$_2$ [2].

The reaction between ClONO and NO$_2$ presents possible atmospheric loss process of ClONO. In the present work, two reaction channels yielding either nitryl chloride or chlorine nitrate

$$ClONO + NO_2 \rightarrow ClNO_2 + NO_2 \qquad (1)$$
$$\rightarrow ClONO_2 + NO \qquad (2)$$

were investigated using ab initio methods, where the cis conformer, ClONO$_c$, has been considered.

---

[1]Corresponding author. e-mail: sasa.kovacic@ijs.si

## 2  Computational methods

The electronic molecular orbital calculations were carried out using GAUSSIAN 03 program [3]. We carefully selected the level of theory to obtain reliable results and to reduce the cost of the calculation at the same time. Geometries of reactants, adducts and transition states were fully optimized using the Becke three-parameter nonlocal exchange functional [4] with the nonlocal correlation of Lee, Yang, and Parr (B3LYP) [5, 6] in conjunction with the 6-311+G(3df) basis set. This level of theory provide reasonable structural parameters that are comparable both with the experimental values [7, 8] and with the values obtained with the CCSD(T) method [2] for the Cl-NO$_2$ system. The harmonic frequencies of all species were computed from analytical derivatives at the B3LYP/6-311+G(3df) level using geometries calculated at the same level of theory, in order to characterize the nature of the stationary points and to determine zero point energies. IRC procedure was used to follow the reaction path in both directions from transition states, to the corresponding reactant and product structures, with a step size of 0.1 amu$^{1/2}$ bohr. The energetics of the reaction was calculated at the B3LYP/6-311+G(3df) level of theory.

## 3  Results

We have considered the reaction channels when nitrogen atom of NO$_2$ molecule attacks either chlorine atom or non-terminal oxygen atom of ClONO species. Transition state TS1 was found to connect reactants and products of reaction (1), and TS2 accompanying the reaction (2). Optimized geometrical parameters of TS1 and TS2 calculated at the B3LYP/6-311+G(3df) level of theory are presented in Figure 1. Both transition states are found to be planar. TS1 represents a loose transition state, where the NO$_2$ moiety is only weakly bonded to Cl atom, with N2-Cl bond length being 2.493 Å. The structures of ClONO and NO$_2$ moieties of TS1 are similar to those of the separate molecules. Moving from TS1 to the products the N2-Cl bond is formed and Cl-O2 bond is dissociated. When the NO$_2$ moiety attacks the ClONO species from the oxygen side, the ClONO$_2$ and NO species might be formed trough TS2. In this process a new N2-O2 bond is formed and a N1-O2 bond of ClONO is dissociated.

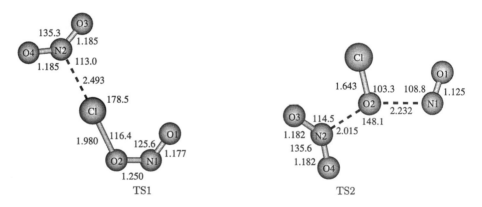

Figure 1: Calculated B3LYP/6-311+G(3df) optimized geometrical parameters of transition states, TS1 and TS2, accompanying the reaction ClONO+NO$_2$.

Figure 2 represents potential energy diagram of the $ClONO_c+NO_2$ reaction calculated at the B3LYP/6-311+G(3df) level of theory. The energy barriers for the pathways passing through TS1 and TS2 are calculated to be 4.1 and 20.6 kcal/mol, respectively. The reaction producing $ClNO_2$ and $NO_2$ radicals is therefore more likely to proceed, with the products being for 13.3 kcal/mol lower in energy than the reactants, $ClONO_c+NO_2$. On the other hand, the $ClONO_2+NO$ channel is calculated to be endothermic by 8.5 kcal/mol. Thus, the present results indicate clearly that the later reaction channel is much less probable. The former reaction channel could be considered as $ClONO_c \leftrightarrow ClNO_2$ isomerization in the presence of the $NO_2$.

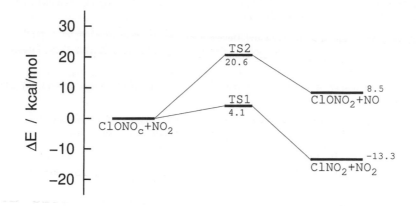

Figure 2: Potential energy diagram for the $ClONO_c + NO_2$ reaction calculated at the B3LYP/6-311+G(3df) level of theory. Relative energies are given in kcal/mol.

## 4  Conclusions

We have found two reaction pathways of ClONO loss in the presence of $NO_2$ using the B3LYP/6-311+G(3df) level of theory. Depending on the mutual orientation of the approaching molecules, $ClNO_2+NO_2$ or $ClONO_2+NO$ products might be formed. The energy barriers are calculated to be 4.1 and 20.6 kcal/mol, respectively. The former reaction yielding nitryl chloride with exothermicity of 13.3 kcal/mol, is preferred over the reaction producing chlorine nitrate, which is endothermic by 8.5 kcal/mol. The former reaction channel could be considered as $ClONO_c \leftrightarrow ClNO_2$ isomerization in the presence of the $NO_2$.

## 5  Acknowledgment

This research was funded by the Ministry of Higher Education, Science, and Technology of Slovenia, programme grant number P2-0148.

# References

[1] H. Niki, P.D. Maker, C.M. Savage and L.P. Breitenbach, Fourier-Transform IR spectroscopic observation of chlorine nitrite, clono, formed via Cl + $NO_2$(+M)→ClONO(+M), *Chemical Physics Letters* **59** 78-79(1978).

[2] T. J. Lee, Ab-initio characterization of $ClNO_2$, cis-ClONO, and trans-ClONO, *Journal of Physical Chemistry* **98** 111-115(1994)

[3] M.J. Frisch et al., Gaussian 03, Revision B.03, Gaussian, Inc., Wallingford, CT, 2004

[4] A.D. Becke, Density-Functional Thermochemistry .3. The role of exact exchange, *Journal of Chemical Physics* **98** 5648-5652(1993).

[5] C. Lee, W. Yang amd R. G. Parr, Development of the Colle-Salvetti Correlation-Energy formula into a functional of the electron-density, *Physical Review B* **37** 785-789 (1988).

[6] B. Miehlich, A. Savin, H. Stoll and H. Preuss, Results obtained with the correlation-energy density functionals of Becke and Lee, Yang and Parr, *Chemical Physics Letters* **157** 200-206 (1989).

[7] J.R. Durig, Y.H. Kim, G.A. Guirgis and J.K. McDonald, *Spectrochimica Acta Part A-Molecular and Biomolecular Spectroscopy* **50** 463-472(1994)

[8] Y. Kawashima, H. Takeo and C. Matsumura, Microwave-spectrum of cis chlorine nitrite, ClONO, *Chemical Physics Letters* **63** 119-122(1979).

Brill Academic Publishers
P.O. Box 9000, 2300 PA Leiden
The Netherlands

*Lecture Series on Computer*
*and Computational Sciences*
Volume 4, 2005, pp. 299-302

# Input File of Mathematical Phantom for MCNP-4B Code

D. Krstić[1], D. Nikezić and N. Stevanović

Department of Physics,
Faculty of Sciences,
R. Domanovica 12 Kragujevac34000,
Serbia and Montenegro

Received 9 June, 2005; accepted in revised form 30 June, 2005

*Abstract:* In this paper we prepared mathematical model of human body as it described in ICRP23, ICRP89 and ICRU48 publications. Computacional models or phantoms are mathematical representations of human anatomy designed to use in dosimetry calculations. Calculations have been done by means of software package MCNP-4B.

*Keywords:* mathematical phantom, ORNL, MCNP-4B code, input file

*PACS:* 02.50.Ng, 07.05.Tp, 87.53.Bn, 87.53.Dq

## 1. Introduction

Computacional models or phantoms are mathematical representations of human anatomy designed to use in dosimetry calculations. They have been used in dosimetry calculations for radiography, radiotherapy, nuclear medicine, radiation protection and investigate the effects of low frequency electromagnetic fields. The Medical Internal Radiation Dose (MIRD) Committe phantom is a heterogenous mathematical representation of the human body that was first designed to calculate absorbed doses from internal radiation [1-2]. This model has been modified over several years to enable its use in calculation of absorbed doses from external radiation. The original MIRD phantom was hermaphrodite and included the gonads of both sexes, and female breasts.

The MIRD phantom has been the basis for various derivations representing infants and children of various ages [3] , gender-specific adult phantoms, called ADAM and EVA [4]. The most recent developments (CHILD and BABY) belong to a new generation of models compiled from computer tomographic (CT) data of real patients [5]. The first paediatric models developed at Oak Ridge National Laboratory (ORNL) are merely scaled-down versions of the adult, later modifications take advantage of the knowledge of pediatric anatomy and human growth.

## 2. Methodology: mathematical phantom

Mathematical phantoms defined in [6-8 ] have been adopted and used in this paper. An age-dependent mathematical phantom series, developed by Eckerman, Cristy and Ryman in Oak Ridge National Laboratory (ORN) [9]. The ORNL mathematical phantom series is an antropomorphic MIRD-type phantom series covering both genders. It contains newborn and individuals of ages 1, 5, 10, 15 and adult. At age 15 the phantom represents both a 15 year old male and an adult female. The heights and weights of the age –dependent ORNL phantom series are listed in Table 2.

Table 2. Characteristics of the ORNL mathematical phantom series

| Phantom age (y) | Height (cm) | Weight (kg) |
|---|---|---|
| Newborn | 51.04 | 3.60 |
| 1 | 74.57 | 9.30 |

---

[1] Corresponding author. Dragana Krstić E-mail: dragana@kg.ac.yu

| 5 | 109.29 | 19.19 |
|---|---|---|
| 10 | 139.97 | 32.54 |
| 15 | 168.41 | 56.86 |
| Adult | 179.00 | 74.03 |

The phantom consists of three major sections: (1) the trunk and arms represented by elliptical cylinder; (2) the legs and feet represented by two truncated circular cones; and (3) the head and neck represented by an elliptical cylinder capped by half an ellipsoid.

In the further text the details about ORNL phantom are given. The origin is in the centre of the trunk base. z- axis is vertically up, x – axis is phantom left and y –axis is directed toward posterior side of the phantom.

Trunk is elliptical cylinder given with inequality

$$\left(\frac{x}{A_T}\right)^2 + \left(\frac{y}{B_T}\right)^2 \leq 1 \tag{1}$$

and

$$0 \leq z \leq C_T, \tag{2}$$

with the parameters $A_T$, $B_T$ and $C_T$ and some other data given in Table 4.

Table 4. Physical dimensions of the trunk section

| Phantom age (y) | Length (cm) | | | Volume (cm$^3$) | Mass (g) |
|---|---|---|---|---|---|
| | $A_T$ | $B_T$ | $C_T$ | | |
| Newborn | 6.35 | 4.90 | 21.60 | 2 050 | 2 030 |
| 1 | 8.80 | 6.50 | 30.70 | 5 350 | 5 350 |
| 5 | 11.45 | 7.50 | 40.80 | 10 660 | 10 650 |
| 19 | 13.90 | 8.40 | 50.80 | 18 050 | 18 130 |
| 15 | 17.25 | 9.80 | 63.10 | 32 920 | 33 500 |
| Adult | 20.00 | 10.00 | 70.00 | 43 090 | 43 470 |

The trunk includes arms, while for female phantom breast are appended to the outside of trunk section. ORNL phantom includes neck in head section, that is represented as right circular cylinder. The head is an right elliptical cylinder topped by half an ellipsoid. The neck is specified by inequality

$$x^2 + y^2 \leq R_H^2 \tag{3}$$

and

$$C_T \leq z \leq C_T + C_{H0} \tag{4}$$

The head is given by the following inequalities

$$\left(\frac{x}{A_H}\right)^2 + \left(\frac{y}{B_H}\right)^2 \leq 1 \tag{5}$$

and

$$C_T + C_{H0} \leq z \leq C_T + C_{H0} + C_{H1} \tag{6}$$

$$\left(\frac{x}{A_H}\right)^2 + \left(\frac{y}{B_H}\right)^2 + \left(\frac{z - (C_T + C_{H0} + C_{H1})}{C_{H2}}\right)^2 \leq 1 \tag{7}$$

and

$$z \geq C_T + C_{H0} + C_{H1} \tag{8}$$

Parameters of the neck and head equations are given in Table 5.

Table 5. *Physical dimensions of the head section*

| Phantom age (y) | Lenght (cm) | | | | | | Volume (cm$^3$) | Mass (g) |
|---|---|---|---|---|---|---|---|---|
| | $R_H$ | $A_H$ | $B_H$ | $C_{H0}$ | $C_{H1}$ | $C_{H2}$ | | |
| Newborn | 2.8 | 4.52 | 5.78 | 1.50 | 7.01 | 3.99 | 831 | 876 |
| 1 | 3.6 | 6.13 | 7.84 | 2.30 | 9.50 | 5.41 | 2 070 | 2 220 |
| 5 | 3.8 | 7.13 | 9.05 | 3.30 | 10.70 | 6.31 | 3 170 | 3 480 |
| 0 | 4.4 | 7.43 | 9.40 | 4.70 | 11.68 | 6.59 | 3 810 | 4 210 |
| 15 | 5.2 | 7.77 | 9.76 | 7.70 | 12.35 | 6.92 | 4 700 | 5 220 |
| Adult | 5.4 | 8.00 | 10.00 | 8.40 | 13.05 | 7.15 | 5 250 | 5 870 |

$C_T$ is given in the table for trunk.
The legs region consists of the part of two circular cones specified by

$$x^2 + y^2 \le \pm x \left( A_T + \frac{A_T}{C_L} z \right)$$
(9)

and

$$-C_L \le z \le 0,$$
(10)

where the sign + is for the left leg and sign − is for the right leg.
Parameters for the legs region are given in Table 6.

Table 6. Physical dimensions of the legs region

| Phantom age (y) | Lenght (cm) | | Volume (cm$^3$) | Mass (g) |
|---|---|---|---|---|
| | $C_L$ | $C_L^{'}$ | | |
| Newborn | 16.8 | 21.6 | 451 | 480 |
| 1 | 26.5 | 37.1 | 1 470 | 1 600 |
| 5 | 48.0 | 65.0 | 4 380 | 4 780 |
| 10 | 66.0 | 90.0 | 8 930 | 9 740 |
| 15 | 78.0 | 100.0 | 15 400 | 16 800 |
| Adult | 80.0 | 100.0 | 20 800 | 22 600 |

All equations for organs with elemental composition, volumes, masses etc, were programmed in input files for MCNP-4B code [10]. Totally 66 cells and 180 surfaces were used for a male phantom in INPUT file. Female phantom is little bit more complicated with 68 cells and 188 surfaces. One option of MCNP gives possibility to plot geometry, i.e. cross of the defined body with a chosen plane surface. Some examples of longitudinal crosses of ORNL phantom, obtained with our input MCNP file are given in Figs. 1 a, b, c.

Fig. 1. Longitudinal cross section of ORNL phantom given for MCNP
## 3. Conclusion

In this paper absorbed doses in organs of human body per one photon from $^{137}$Cs in soil have been calculated as a function of the source depth. By using of MCNP-4B code, absorbed doses were calculated per one photon for all main organs and organs of remainder. Geometry considered here is bottom geometry; conversion coefficients (given in aGy/photon) for all organs of human body are calculated for only one energy i.e. 661.6 keV. $^{137}$Cs is cause of increased occurence of cancer.

## Acknowledgments

The authors would like to thank the Serbian Ministry of Science and Environment Protection, which supported this work through project No 1425.

## References

[1] W.S. Snyder, Estimation of absorbed fraction of energy from photon sources in body
    Organs, *Proceedings of the International Summer School on Radiation Protection, Radiation Dosimetry 1* (Editor: I. Mirić), Cavtat, Yugoslavia, 1970.

[2] W. S. Snyder, M.R. Ford.and G.G. Warner, Estimates of specific absorbed fractions for photon sources uniformly distributed in various organs of heterogeneous phantom, MIRD Pamphlet No. 5, Revised. *Society of Nuclear Medicine*, New York. 1978.

[3] M. Cristy, Mathematical phantoms representing children at various ages for use in estimates of internal dose, Report ORNL/NUREG/TM-367 (Oak Ridge, TN: Oak Ridge National Laboratory), 1980.

[4] R. Kramer, M. Zankl, G. Williams and G. Drexler, The calculation of dose from external photon exposures using reference human phantoms and Monte Carlo metods, Part I: The male (ADAM) and female (EVA) adult matematical phantoms. GSF Bericht S-885, 1982.

[5] M. Zankl, R. Veit, G. Williams, K. Schneider, H. Fendel, N. Petoussi, and G. Drexler, The construction of computer tomographic phantoms and their application in radiology and radiation protection, *Radiation and Environmental Biophysics* 27 153-164(1988).

[6] International Commission on Radiation Units and Measurements, ICRU REPORT 48, Phantoms and Computational Models in Therapy, Diagnosics and Protection, 1992.

[7] International Commission on Radiological Protection, Reference man: Anatomical, physiological and metabolic characteristics, Oxford: Pergamon Press, ICRP Publication 23; 1975.

[8] International Commission on Radiological Protection. Basic anatomical and physiological data for use in radiological protection: reference values. Annals of ICRP23. Oxford: Pergamon Press, ICRP Publication 89. 2002.

[9] K. F. Eckerman, M. Cristy, and , J. C. Ryman, The ORNL mathematical phantom series. http//ats.ornl.gov/documents/mird2.pdf (1996).

[10] J. F. Briesmeister (Ed.): MCNP - A General Monte Carlo N- Particle Transport Code, Version 4B, LA-12625- M, Los Alamos National Laboratory, Los Alamos, New Mexico, 1997.

VSP International
Science Publishers
P.O. Box 346, 3700 AH Zeist
The Netherlands

*Lecture Series on Computer
and Computational Sciences*
Volume 4, 2005, pp. 303-306

# Modeling excess surface energy in dry and wetted calcite systems

Bjørn Kvamme[1], Tatyana Kuznetsova, Daniel Uppstad

[a]Department of Physics, University of Bergen
Allégt. 55, 5007 Bergen, Norway

Received 20 May, 2005; accepted in revised form 13 June, 2005

*Abstract:* We combined the calcite force field of Hwang et el [1] (2001) with the F3C water model and a hybrid Lennard-Jones/van der Waals 3-site potential for $CO_2$ to investigate the $(10\bar{1}4)$ and $(10\bar{1}0)$ cleaving surfaces of calcite under dry and wetted conditions. The wetting fluid included both pure water and water-carbon dioxide mixture. Excess surface energies and structural features of the calcite-fluid interface were analyzed, with the simulation results for the relaxed surfaces confirming the experimentally observed morphology and supporting our conclusion that the relative stability order of calcite cleaving surfaces under investigation will remain unchanged in the presence of water-carbon dioxide mixture as well.

*Keywords:* calcite; carbon dioxide; water; molecular modeling

*PACS:* 61.25.Em; 64.70.Dv; 68.08.De

## 1. Introduction

Calcium and magnesium carbonates are the most abundant within the rhombohedral carbonate mineral family, which comprises 4% of the earth's crust. A detailed understanding of mechanisms and phenomena governing the precipitation and dissolution of calcium carbonate is therefore essential to predict the processes taking place when this mineral comes into a contact with either pure water or aqueous solutions. Molecular simulations are a promising complimentary tool offering molecular-level insights into the separate components of the complex experimental processes and often capable of providing explanations and predictions at observational scales. We present molecular dynamics investigation into two most stable calcite surfaces under dry and wetted conditions, with the wetting fluid including both pure water and water-carbon dioxide mixture.

## 2. Simulations

### 2.1 Details

The molecular dynamics used constant-temperature, constant-pressure algorithm from MDynaMix package of Lyubartsev and Laaksonen [2]. The starting interfacial system was constructed from slabs of bulk water-carbon dioxide mixture and appropriately cleaved calcite crystal, thermalized initially at 298 K and set side by side, with the periodic boundary conditions applied. The resulting systems (described in Table 1) were subsequently equilibrated for several tens of picoseconds before the average collection began. The production time amounted to 200 picoseconds. Time step was set to $10^{-16}$ s to allow for accurate integration of internal degrees of freedom. The system was kept at constant temperature of 298 K. Electrostatic interactions were handled by the Ewald summation technique with a variable number of reciprocal vectors. Linux-based Message Passing Interface (MPI) was used to implement parallel computation on a cluster dual-processor machines. The number of processors ranged from 2 to 12. Cut-off radius for the Lennard-Jones potential and electrostatic forces was 12Å. Six different systems were simulated; system details are stated in table 8.1 below. The calcite crystal was cleaved along the $(10\bar{1}0)$ plane in systems 3, 4 and 5. System 6, on the other hand, had the $(10\bar{1}4)$ plane facing the water-$CO_2$ interface.

---

[1] Corresponding author. E-mail: bjorn.kvamme@ift.uib.no

| System | Box lengths (Å) | Number of molecules |
|--------|-----------------|---------------------|
| 1 | 24.4x25.0x29.0 | 588 $H_2O$, 19 $CO_2$ |
| 2 | 17.1x20.0x29.3 | 330 $H_2O$, 11 $CO_2$ |
| 3 | 17.1x20.0x49.3 | 96 $Ca^{2+}$, 96 $CO_3^{2-}$, 330 $H_2O$, 11 $CO_2$ |
| 4 | 17.1x20.0x49.3 | 96 $Ca^{2+}$, 96 $CO_3^{2-}$, 330 $H_2O$, 11 $CO_2$ |
| 5 | 17.1x20.0x49.3 | 96 $Ca^{2+}$, 96 $CO_3^{2-}$, 330 $H_2O$, 11 $CO_2$ |
| 6 | 24.4x25.0x49.0 | 240 $Ca^{2+}$, 240 $CO_3^{2-}$, 588 $H_2O$, 19 $CO_2$ |

Table 1 Simulated systems

The density profiles of $x$-$y$ averaged quantities of interest were generated by partitioning the simulation box into discrete bins in the $z$ direction. The atomic density profile for atom of type $i$ was thus obtained by

$$\langle \rho_i(z) \rangle = \frac{\langle m_i N_i(z) \rangle}{A \Delta z} \tag{1}$$

where $N_i$ is the number of $i$-type atoms in a slab located between $z$ and $z+\Delta z$, $A$ is the cross-section area, $A = L_x L_y$, $m_i$ is the atom's mass, $\Delta z$, the slab's width. Varying the binning width in 0.095 - 0.475 Å range (1800-360 bins) did not affect the profiles in any significant way.

## 2.2 Molecular force fields

The 3-site F3C water model of Levitt et al.(1997) [3] (with SPC/E charges since we found that the original ones gave a worryingly flat RDF) was used in the simulations. The model is 3-site, and differ from the majority of water force fields in that its hydrogens possess a small amount of van der Waals interaction useful to offset the otherwise unshielded Coulombic attraction between positively charged water hydrogens and calcite anions.

For modeling calcite, we used a combination of two molecule models; a 1-site model for the $Ca^{2+}$ ion and a 4-site model for the $CO_3^{2-}$ ion. Both models were adopted from Hwang et al. (2001)[1]. The $\zeta$, $R_0$, $D_0$ exp-6 parameters were fitted Mayo et al., 1990, [4] to yield accurate lattice parameters.

The water-$CO_2$ interactions used a 3-site Lennard-Jones model carbon dioxide model of Panhuis et al. (1998) [5], well suited to describe the behavior of lone $CO_2$ molecules in water. As for $CO_2$-calcite and $CO_2$-$CO_2$ interactions, we modified the original 5-site $CO_2$ model due to Tsuzuki et al [6] by merging the van der Waals and Coulombic interaction sites to make it into a 3-site one. The charges were also exchanged for the those of Panhuis et al. (1998). We tested the validity of combining these force fields by comparing radial distribution functions (RDFs) of systems 1 and 2 (Table 1) for water oxygen-$CO_2$ oxygen and water oxygen-$CO_2$ carbon with the findings of Panhuis et al [5]. Our RDFs matched those of Panhuis et al quite well.

## 2.3 Simulational results and discussion

Findings reached in this work are not directly comparable to either experimental or modeling results, since nobody (to our knowledge) has performed either experiments or numerical simulations involving calcite in contact with water-$CO_2$ mixture. This is why the validity of our approach was tested on calcite-water systems. Two calcite surfaces ($10\bar{1}4$ and $10\bar{1}0$) were investigated to see which surface had the lowest surface energy. Both experimental results and simulations state that the ($10\bar{1}4$) surface is the most stable one under both dry and wet conditions (corresponds to the lowest surface energy). Our results for pure water and calcite and calcite in vacuum showed a good qualitative agreement with (wildly varying) results of previous numerical treatment of other authors, as well a good quantitative agreement with the work of Hwang et al. (2001). The best agreement was with Hwang et al. because their model was used and their approach followed, though with modifications (cross-interactions). The differences between wet and dry surface energies were also consistent. If the mineral is water wet, the excess energy should be significantly lower for the hydrated surface than for the dry surface. This can be stated as the second point of agreement.

The next step was to investigate how adding $CO_2$ to the water would affect the excess surface energies of the two calcite surfaces ($10\bar{1}4$ and $10\bar{1}0$) under study.

Excess surface energy, $\gamma$, a good indicator of the relative stability of surfaces, as well as the strength of hydration, is presented in Table 2.

$$\Delta E = E_{int\,erface} - E_{bulk} - E_{water}, \tag{2}$$

$$\gamma_{int\,erface} = \frac{\Delta E}{(2 A_s N_A)}, \tag{3}$$

where E is system's total potential energy, $A_s$ is surface area, $N_A$, the Avogadro number.

| System | Box lenghts (Å) | | | Potential energy per 'particle' (kJ/mol) | Excess surface energy (J/m$^2$) |
|--------|------|------|------|---------------------|------------|
|        | x    | y    | z    |                     |            |
| 1      | 24,4 | 25,9 | 29,0 | -43,0770 ±0,0009    | -          |
| 2      | 17,1 | 20,0 | 29,3 | -42.9560 ±0,0018    | -          |
| 3      | 17,1 | 20,0 | 49,3 | -544,1338 ±0,0006   | 0,79(4)    |
| 4      | 17,1 | 20,0 | 49,3 | -544,1621 ±0,0008   | 0,79(0)    |
| 5      | 17,1 | 20,0 | 49,3 | -544,1293± 0,0006   | 0,79(4)    |
| 6      | 24,4 | 25,9 | 49,0 | -662,2044 ±0,0005   | 0,41(9)    |

Table 2 Excess surface energies and potential energies for calcite-water-$CO_2$ systems

The difference in surface energy, $\Delta\gamma$, between the most stable and the next most stable calcite surfaces amounted to

$$\Delta\gamma = \gamma_{(10\bar{1}0)} - \gamma_{(10\bar{1}4)} = (0,794 - 0,419)\frac{J}{m^2} = 0,375\left(\frac{J}{m^2}\right) \tag{4}$$

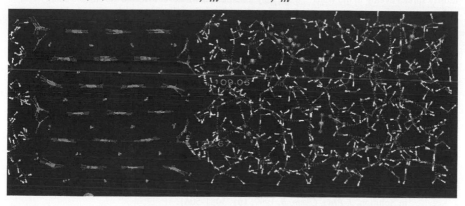

Figure 1: VMD package [7] -generated presentation of simulation system 4 ( $\{10\bar{1}0\}$ cleaving plane. Water-carbon dioxide mixture on the right-hand side.

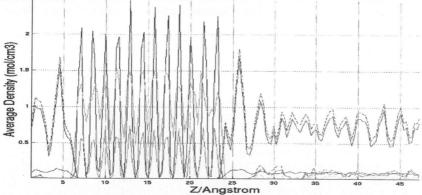

Figure 2: Density profiles of simulation system 4. Full red line is water oxygen; blue - water hydrogen; dashed red line indicates total water density profile; yellow is the carbon (C) in $CO_2$; purple is the oxygen (O) in $CO_2$; black is calcite calcium cation; cyan - oxygen (O) of the calcite carbonate anion; green - carbon (C) of the calcite carbonate anion.

The excess surface energies of surfaces wetted by the mixture of water and $CO_2$ retain the relative stability order. I.e. the ($10\bar{1}4$) surface still clearly has the lowest excess energy (0,794 J/m$^2$). It is by 0,375 J/m$^2$ lower than in the ($10\bar{1}0$) surface (0,419 J/m$^2$), indicating that the ($10\bar{1}4$) surface is still the most stable one energetically. However, it does appear that the increase in excess surface energy associated with the addition of carbon dioxide is larger for the stable ($10\bar{1}4$) surface.

There are several instructive features to be observed in the density profiles. First, the average water densities have higher peaks at each side of the crystal, by the surface. Not the first layer, but the second and third water layer indicates the waters affinity for calcite. This indicates different forces on interface water than on bulk water. Figure 2 might explain the low first peak as a result of the characteristic "ridges" on the calcite surface.

Second, a closer look at the calcite crystal shows that density peaks corresponding to different calcite components are aligned in the bulk, but shifted relative to each other in the interfacial layers. In the layer closest to the interface, one can see that the oxygen and calcium peaks clearly lie closer to the interface than that of carbon. This is probably due to the positively charged calcium ion interacting significantly with the oxygen in water, and the carbonate oxygen attracted by hydrogen in water. This is the origin of the excess energy. In the second layer from the interface, the calcium cation is still closer to the interface than the carbon, but the oxygen peak is shifted to the other side of the carbon peak. This is probably a consequence of the disorder in the first layer.

Third, it appears from the plot that the carbon dioxide has no particular affinity for the calcite interface, mostly spread evenly in the water. So one may assume that whatever affected the surface energy was water restructuring to accommodate the carbon dioxide. Carbon dioxide is a significantly larger molecule than water so it does not fit into the calcite surface as water does. Figures 1 and 2 illustrate this effect quite well. Both VMD-generated views and density profiles suggest that the carbon dioxide does not have any affinity for the interface. They do not cluster either, but position themselves nearer to each other than to the interface.

## References

[1]   Hwang, S., Blanco, M., Demiralp, E., Cagin, T., and Goddard W.A. The MS-Q Force Field for Clay Minerals: Application to Oil Production. *J. Phys. Chem. B*, 2001, **105**, 4122-4127.

[2]   Lyubartsev, A.P., and Laaksonen, A. M.DynaMix - a scalable portable parallel MD simulation package for arbitrary molecular mixtures. *Computer Physics Communications* 2000, **128**, 565-589.

[3]   Levitt, M., Hirshberg, M., Sharon, R., Ladig, K.E., and Dagett, V. Calibration and Testing of a Water Model for Simulation of the Molecular Dynamics of Proteins and Nucleci Acids in Solution. *J. Phys. Chem. B*, 1997, **101**, 5051-5061.

[4]   Mayo, S.L., Olafson, B.D., and Goddard, W.A. Dreiding - A Generic Force-Field for Molecular Simulations. *J. Phys. Chem.*, 1990, **94**, 8897-8909.

[5]   Panhuis, M.I.H., Patterson, C.H., and Lynden-Bell, R.M. A Molecular Dynamics Study of Carbon Dioxide in Water: Diffusion, Structure and Thermodynamics. *Molecular Physics*, 1998, **94**, 963-972.

[6]   Tsuzuki, S., and Tanabe, K. Molecular Dynamics Simulations of Fluid Carbon Dioxide Using the Model Potential Based on Ab Initio MO Calculation. *Computational Material Science*, 1999, **14**, 220-226.

[7]   Humphrey, W., Dalke, A. and Schulten, K. VMD - Visual Molecular Dynamics. *J. Molec. Graphics*, 1996, **14**, 33-38.

Brill Academic Publishers
P.O. Box 9000, 2300 PA Leiden
The Netherlands

*Lecture Series on Computer
and Computational Sciences*
Volume 4, 2005, pp. 307-310

# Investigation into Stability and Interfacial Properties of $CO_2$ Hydrate – Aqueous Fluid System

Bjørn Kvamme[1], Tatyana Kuznetsova

Department of Physics, University of Bergen
Allègt. 55, 5007 Bergen, Norway

Received 21 May, 2005; accepted in revised form 13 June, 2005

*Abstract:* We applied the techniques of Molecular Dynamics (MD) to study the structural and dynamic properties of a stable interface between $CO_2$ hydrate and aqueous solution. The steady-state interface thickness was evaluated from a set of criteria, with the decay of hydrogen signature being the leading one. Applying the criteria has yielded interface width of about 10 Å.

*Keywords:* carbon dioxide hydrate; carbon dioxide; water; molecular modeling

*PACS:* 61.25.Em; 64.70.Dv; 68.08.De

## 1. Introduction

Clathrate hydrates of natural gases are ice-like structures composed of water molecules encaging guest gas molecules. The list of gas molecules capable of forming clathrate hydrates is extensive and includes a number of abundant compounds, from light hydrocarbons to refrigerants to sour gases like carbon dioxide and hydrogen sulfide. Under conditions typical for oil and gas production and transport in colder climates, clathrate hydrates can easily form in pipelines and production equipment.

## 2. Simulation Details

The hydrate part of our system comprised a block of structure I [1,2] carbon dioxide hydrate made of 2x2x11 unit hydrate cells (2024 SPC/E water [3] and 264 three-site $CO_2$ [4] molecules. The hydrate structure was generated as a perfect eqilibrated periodic hydrate crystal arbitrary broken in two (to facilitate possible melting) and brought into contact with a pre-quilibrated 40-Å long slab of 795 SPC/E water molecules. After being stacked together, the translational degrees of motion were switched off for all the molecules to overcome perturbations due to possible overlaps at the interface. The resulting interfacial system ranged about 170 Å in length. Periodic Boundary Conditions were applied in all three directions.

Our MD routine used the MDynaMix package of Lyubartsev and Laaksonen [5] with explicit reversible integrator for NPT-dynamics of Martyna *et al* [6], modified by us to implement implicit quaternion treatment of rigid molecules with Nosé-Hoover thermostat for temperature and pressure [7,8,9]). The time step was set to 1 femtosecond, with the simulation run totaling upward of 5 million steps. The system was kept at constant temperature of 240 K and pressure of 20 MPa by means of Nosé-Hoover thermo- and barostat, with the MD box allowed to fluctuate in all three directions. Both experimental and modeling data indicate a substantial surface tension in the hydrate-liquid system. Therefore, only tangential components of pressure tensor were used to evaluate and control the pressure. The electrostatic interactions were handled by an Ewald summation techniques with a variable number of reciprocal vectors. Linux-based Message Passing Interface (MPI) was used to implement parallel computation on a cluster dual-processor machines.

Our particular choice of temperature was dictated by a desire to obtain a hydrate-liquid system reproducing the real-life conditions characteristic for $CO_2$ storage hydrate on the ocean floor, given the specifics of the water model. Bryk and Haymet [10] found that the stable ice-water interface of SPC/E water lies in the vicinity of 225+/-10K. Thus the chosen value of 240K represented a temperature about 5 degrees above the model-specific freezing point and should insure the liquidity of water. On the other hand, the point of +5 degrees Celcius and 20 Mpa is known to fall within the stability limits of the $CO_2$ hydrate[11], and we expected the simulations to yield a relatively stable hydrate-liquid interface.

---

[1] Corresponding author. E-mail: bjorn.kvamme@ift.uib.no

## 3. Results and Discussion
### 3.1. Density Profiles and Hydrate Stability

The density profiles of $x$-$y$ averaged quantities of interest were generated by partitioning the simulation box into discrete bins in the $z$ direction. The atomic density profile for atom of type $i$ was thus obtained by

$$\langle \rho_i(z) \rangle = \frac{\langle m_i N_i(z) \rangle}{A \Delta z} \tag{1}$$

where $N_i$ is the number of $i$-type atoms in a slab located between $z$ and $z+\Delta z$, $A$ is the cross-section area, $A=L_x L_y$, $m_i$ is the atom's mass, $\Delta z$, the slab's width. Varying the binning width in 0.095 - 0.475 Å range did not affect the profiles in any significant way. On the other hand, results obtained using coarser grids show a rather dramatic binning dependence. For instance, a definite additional long-range *hydrate* structure starts to emerge in case of 0.634 Å spacing (270 bins) and wider ones. This structure becomes quite obvious in case of 90 bins (1.9 Å-wide bin). The "sine" behavior of the envelopes was persistant but definitely not stationary.

A sequence of time-averaged liquid-hydrate density profiles was obtained using 450 and 90-bins sectioning. Figures 1 and 2 present unfiltered 450-bin density profiles taken throughout the production run. Each figure is an average over a hundred snapshots taken every 0.1 picosecond. The starting averaging points are 10 picoseconds apart. The total production run used to evaluate hydrate-liquid system behavior amounted to 4.0-plus nanosecond. Comparison between the 450- and 90-bin pictures shows some substantial differences even in the behavior of "envelopes" themselves. It is not the only example of the fact that neither coarse-grain nor fine-grain profile envelope are capable of describing (and even defining for that matter) the interface on their own.

As noted, we wanted to study a steady-state interface between carbon dioxide hydrate and an aqueous solution. Our previous simulations led us to conclude that temperatures above 260 K lie outside the model-specific stability limit of $CO_2$ hydrate. As for the chosen temperature of 240 K, the analysis of hydrate inerface dynamics (Figures 1 and 2) has revealed a quick destruction of disrupted cages, followed by the dissolution of two high peaks belonging to the next hydrate layer. This process took roughly 700 picoseconds of the simulation time (Figure 1), and the remaining hydrate proved to be stable throughout the rest of the run (4 nanosecond). This stable hydrate structure comprised 18 high and 18 low peaks of "hydrate-quality" and one each low and high interfacial peak. The interfacial peaks are characterized by significantly decreased amplitude and weakened hydrogen signature.

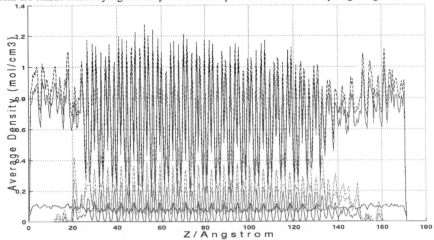

Figure1. Density profiles. 0.73 ns elapsed after the start of the simulation. Full lines -- fine-scale densities for individual atoms. Dashed lines -- summed up fine-scale densities of water and CO2 molecules. Blue -- hydrogen in either liquid or hydrate. Cyan -- water oxygen in either liquid or hydrate. Red -- CO2 oxygen. Magenda -- CO2 carbon.

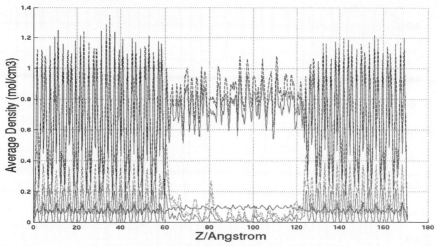

Figure2. Density profiles.3.73 ns elapsed after the start of the simulation. See Figure 1 for conventions.

## 3.2. Interface Width

Our estimate of the interface width was based on the analysis of all of the datasets. It is our belief belief that the density profiles envelopes are not completely reliable since the hydrate profiles reflect not only the interfacial phenomena but long-range oscillations as well, thus causing a perceived broadening of the interface. We have used the stuctural differences between the liquid phase and the hydrate to isolate the transitional region where the crystal ordering makes way for the liquid disorder. And even though the analysis of plain density profiles is complicated by the $CO_2$ diffusing away from the interface and into the water bulk, applying a composite criteria enables one to distinguish between the features characteristic for water molecules encaging a $CO_2$ guest molecule and those of carbon dioxide molecule dissolved in water. Figure 3 presents several of structural details specific to hydrates only. The hydrogen density proved to be an especially useful hydrate marker, since its hydrate profile features a distinctive triple-peak structure, with the central hydrogen peak coinciding with the lower carbon density peak and hydrate-oxygen valley. This hydrogen signature is totally absent in both liquid water and water-CO2 mixtures, so one can define the interface as the transition length along which the triple-peak feature decays from the one charateristic for bulk to non-existant (in combination with other criteria).

Another hydrate-structure marker is supplied by the positioning of water-oxygen valleys relative to the carbon peaks. In general, we found that the density profiles of $CO_2$'s carbon provided the better idea of the guest behavior, therefore we will concentrate on carbon rather than oxygen guest profiles. The regular succession of alternating high and low peaks evident in Figure 3 is a typical feature of Structure I carbon dioxide hydrate when it's projected on a given direction; the water-oxygen valleys flank the high peaks and coincide with the low ones. The latter valleys vary in depth throughout the whole of the hydrate crystal; these long-range envelope oscillations can be clearly seen in coarse-grain density profiles.

The steady-state interface thickness was evaluated from a set of criteria, with the decay of hydrogen signature being the leading one, so it's not a straightforward "10-90" value of the classical vapor-liquid interface evaluations. The other criteria included the correct allignment of the lower carbon peak and its corresponding hydrogen maximum, as well as the roughly equivalent depth of water-oxygen valleys that flank the higher carbon peak. A good example of these features as in their undisturbed hydrate crystal state are present as Items 1 to 3 in Figure 3, while Items 4 and 5 show their deterioration in the interfacial area. Applying the criteria has yielded interface width of about **10** Å. This estimate is close to the value deduced from the envelopes of fine-scale density profiles, so even though the interface profiles (especially the coarse-grained ones) might look broader, our analysis of structural details points to a narrower interface.

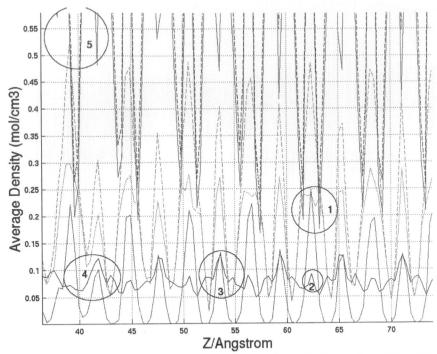

Figure3. Items 1 to 5 point out hydrate structural markers (see text for further details and Figure 1 for conventions).

## References

[1]  von Stackelberg, M., and Müller, H.R. Feste Gashydrate II: Structur und Raumchemie. *Z. Elektrochem.* 1954, **58**, 25-39.

[2]  von Stackelberg, M., and Jahns, W. Feste Gashydrate. *Z. Elektrochem.* 1954, **58**, 162-164.

[3]  Berendsen, H. J. C., Postma, J. P. M., van Gunsteren, W. F., and Hermans, J. in *Intermolecular Forces*, edited by B. Pullman, Reidel, Dordrecht, 1981.

[4]  Harris, J.G., and Yung, K.H. Carbon dioxides liquid-vapor coexistence curve and critical properties as predicted by a simple molecular-model. *J. Phys. Chem.* 1995, **99**, 12021-12024.

[5]  Lyubartsev, A.P., and Laaksonen, A. M.DynaMix - a scalable portable parallel MD simulation package for arbitrary molecular mixtures. *Computer Physics Communications* 2000, **128**, 565-589.

[6]  Martyna, G.J., Tobias, D.J., and Klein M.L. Constant-Pressure Molecular-Dynamics Algorithms. *J. Chem. Phys.* 1994, **101**, 4177-4189.

[7]  Nosé, S. Constant Temperature Molecular-Dynamics Methods. *Progr. Theor. Phys. Suppl.* 1991, **103**, 1-46.

[8]  Hoover, W.G. Canonical Dynamics - Equilibrium Phase-Space Distributions. *Phys Rev A* 1985, **31**, 1695-1697.

[9]  Fincham, D. Leapfrog rotational algorithms. *Molecular Simulation* 1992, **8**, 165-178.

[10] Bryk, T., and Haymet, A.D.J. Ice 1h/water interface of the SPC/E model: Molecular dynamics simulations of the equilibrium basal and prism interfaces. *J. Chem. Phys.* 2002, **117**, 10258-10268.

[11] Sloan, E.D. *Clathrate Hydrates of Natural Gases*, 2nd ed. Marcel Dekker, New York, 1998.

Brill Academic Publishers
P.O. Box 9000, 2300 PA Leiden
The Netherlands

*Lecture Series on Computer
and Computational Sciences*
Volume 4, 2005, pp. 311-314

# Constrained free form deformation on subdivision surfaces

S. Lanquetin[1a], R. Raffin[b], M. Neveu[a]

a) LE2I, UMR CNRS 5158, University of Burgundy, France
b) LSIS, UMR CNRS 6168, University of Provence, France

Received 10 July, 2005; accepted in revised form 5 August, 2005

*Abstract:* This paper proposes to define a deformation method on a subdivision surface. It combines an "easy-to-use" free-form deformation with a Loop subdivision algorithm. The deformation method controls the resulting shape, defining the range (i.e. the impact) of the deformation on an object before applying it. The deformation takes into account the Loop properties to follow the subdivision scheme, allowing the user to fix some constraints at the subdivision-level he works on and to render the final object at the level he wants to. We finally propose an adaptive subdivision of the object driven by the deformation influence.

*Keywords:* geometric modeling, subdivision surface, Loop scheme, free-form deformation, adaptive subdivision

*Mathematics Subject Classification:* 65D17, 65D18

## 1. Introduction

Subdivision surfaces are now widely used in computer graphics. It allows the generation of smooth surfaces, as well in geometric modelers, CAD applications or animation movies. However, handling such surfaces is not easy, due to the subdivision process which can modify the object from one level to another. Large amounts of modeling actions are made by the user, implying up and down changeover from the modeling subdivision level to the smooth rendering one. We present in this paper a tool to define deformations on a subdivision surface, guaranteeing the respect of the deformation whatever the level computed.

In the existing methods to modify or deform subdivision surfaces, Schweitzer [13] proposes a formulation for the subdivision surfaces and the displacement of the mesh vertices. This method is non intuitive and non interactive. Lee defines a deformation of the surface normals, Ehmann and Khodakovsky use respectively methods based on weight linked to the surface vertices or a perturbation curve which modifies the successive subdivisions process ([10], [5], [6]). However, these methods are based on the mathematical definition of the subdivision surfaces which is not trivial for a common user.

On another hand, Free Form Deformation (FFD, [14]) is a well-known method to model objects. It is based on an embedding grid which surrounds the object to be deformed. Once the points of the object are expressed in the grid, moving a vertex of the grid implies the deformation of the embedded object. Several methods propose a better interaction with the object to be deformed, hiding the embedding grid [7][8]. Agron [1] works on Free Form Deformation of subdivision surface. He gives an easy way to interact with the subdivision mesh since the FFD hides the description of the surface. Other methods based on FFD can be used to facilitate the interaction of the user ([3], [12]). Instead of hiding the embedding grid, these methods use displacement constraints on object points and the deformation model insures the satisfaction of these displacements. With these methods the user chooses a point to be displaced, eventually an area of influence of the deformation or a displacement path, and the deformation model insures the satisfaction of these constraints (see figure 2). The *a priori* perception of the resulting object is very intuitive and its computation is done at interactive rate.

The first part of this paper is a quick review of the subdivision surfaces and of the deformation model we used in our method. Then we define the principle and the process to deform a subdivision surface with the free form deformation method (including adaptive subdivision linked to the influence area of the deformation). The last section will present some significant results.

---

[1] Corresponding author. E-mail: sandrine.lanquetin@u-bourgogne.fr

## 2. Subdivision surfaces and deformation methods

### 2.1 Subdivision surfaces

Subdivision surfaces are among the easiest way to generate smooth surfaces. They preserve both advantages of NURBS and polygonal meshes. We choose Loop scheme [11] to apply our results because most of meshes are currently triangular (provided by geometric modelers or reconstructed from laser range images ...). An example of Loop subdivision is shown at Figure 1. From left to right of the figure, every face is split into 4 at each subdivision step.

Figure 1. Example of subdivision surface with a Loop scheme.

### 2.2 Deformation model

If a wide range of deformation methods exists, the majority of them are linked to the representation of the object to be modified. The Free Form Deformations model on which we based our work wish to be generic. It consists in a space deformation, acting only on points without taking into account the topology, geometry or neighborhood information. The same general principle gives the DOGME model [2] where punctual deformation constraints are defined. This method is made more practical with the definition of a radius of influence surrounding a constraint point [3] (Figure 2). Any point of an object lying at the constraint point location is displaced according to the deformation constraint vector. Conversely any point outside the influence area is not moved. The authors use a B-Spline function to model the decreasing deformation from the constraint point to the influence boundary.

Figure 2. The *Scodef* method. Left : a constraint C with an influence radius R is applied on a plane. Center : the influence area. Right : the points belonging to the influence area are displaced.

Raffin [12] extends this deformation model with the definition of constraint curves, various influence sets (star-shaped solids) called influence hulls and paths of deformation that replace the initial vector between the constraint point and its deformed location. The following work is based on this method and applied on subdivision surfaces.

## 3. Principles of our method

Using the deformation seen above, the user defines a constraint, an influence zone and a path of deformation. To compute the deformation fixed by the user for any subdivision level, we need to follow any point up from a level to the next one. As the Loop scheme is approximating, the deformation computed at one level will progressively be eroded after some subdivision steps (see Figure 3).

Figure 3. Successive subdivisions of the original deformed mesh (left) : the constraint is not satisfied.

The resulting deformation does not satisfy the constraint and thus does not correspond to the user's goal. Indeed after one step of subdivision, due to the Loop subdivision, the constraint point will not lie on the mesh surface anymore. The mesh moves away from the constraint point and fewer vertices are influenced (Figure 4b). To avoid this, an interpolating scheme (Butterfly [4] for instance) could be used, but well known artifacts (bulging, waves) appear. Our solution keeps the deformation settings (constraint point, displacement, influence) along the subdivision levels. This will preserve the displacement amplitude but not the constraint point location. For this purpose, we compute the image of the constraint point at any level of the subdivided mesh [9]. Then we use this image as the new constraint point, with the same displacement than the primary constraint as shown on Figure 4.c.

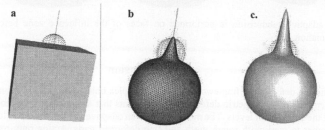

Figure 4. a) the initial control mesh and the deformation constraint. b) deformation of the mesh defined in a. at the $5^{th}$ level of subdivision, keeping the same deformation settings. c) the deformation is performed after the computation of the constraint point image after five subdivisions.

Another improvement consists in subdividing only the faces in the influence area with an adaptive Loop scheme. As the influence of the deformation is fixed by the user, we know the points of the object that will be displaced. The object surface can be refined to improve the quality of the deformation, since it acts on more points. The set of images at Figure 5 shows a deformation process. On the left, the deformation constraint is put on the bunny model, with a spherical influence. The middle image shows the resulting mesh when the whole mesh is subdivided. The right image presents the resulting bunny with a local adaptive subdivision in the influence zone.

Figure 5. From left to right: definition of the deformation, resulting mesh using a global subdivision or a local adaptive subdivision.

## 4. Results

Once we have defined the process to follow the constraint point from a level to another, we have a basic tool to deform a subdivision surface, whatever the level we work on. A more complex deformation can be set, while benefiting from our extended deformation model [12]. In Figure 6, a path of deformation is defined, where the displaced points of the mesh follow a 3D curve and non-isotropic influences allow a better selection of mesh points. The adaptive subdivision permits a fine resulting object that would be impossible with classical methods.

Figure 6. An adaptive subdivision is performed on faces of the influence zone before applying a complex deformation.

## 5. Conclusion

This paper defines a deformation framework to model complex objects with subdivision surfaces. The deformation permits to fix geometric deformation constraints that are satisfied after computation and exactly kept in the subdivision levels. The method allows an adaptive subdivision of influenced parts of the object in order to work with rough mesh and fine deformed parts. Acting only on points of the mesh, the deformation has a very low time cost comparing to the subdivision process.

## References

[1] P. Agron. Free Form Deformation of Subdivision Surfaces. *Course paper; Spring 2002*, SUNY Stony Brook.

[2] P. Borrel and D. Bechmann. Deformations of n-dimensional objects. *Research Report*, IBM Research Division, 1990.

[3] P. Borrel and A. Rappoport. Simple Constrained Deformations for Geometric Modeling and Interactive Design. *ACM Transactions on Graphics*, vol. 13, no. 2, pp . 137-155, 1994.

[4] N. Dyn, D. Levin and J. A. Gregory (1990). A butterfly subdivision scheme for surface interpolation with tension control. *ACM Transactions on Graphics*, vol 9, p. 160-169.

[5] S. Ehmann, A. Gregory, and M. Lin. A Touch-Enabled System for Multiresolution Modeling and 3D Painting. *Journal of Visualization and Computer Animation*, 2001.

[6] A. Khodakovsky and P. Schröder. Fine level feature editing for subdivision surfaces. *Symposium on Solid Modeling and Applications 1999*, pp. 203-211.

[7] WM. Hsu, JF. Hugues and H. Kaufman. Direct manipulation of free form deformation. *Computer Graphics1992*, vol. 26(2), pp. 177-184.

[8] S-M. Hu, H. Zhang, C-L. Tai and J-G. Sun. Direct manipulation of FFD: efficient explicit solutions and decomposible multiple point constraints. *The visual computer 2001*, vol. 17 , pp. 370-379.

[9] S. Lanquetin. Study of subdivision surfaces: intersecting, subdivision accuracy and depth. *PhD thesis*, University of Burgundy, France, 2004 *(in french)*.

[10] A. Lee, H. Moreton and H. Hoppe. Displaced Subdivision Surfaces. *Proceedings of SIGGRAPH 2000*, pp. 85–94, 2000.

[11] C. Loop. Smooth Subdivision Surfaces Based on Triangles. *Department of Mathematics: Master's thesis*, University of Utah, 1987.

[12] R. Raffin, M. Neveu, F. Jaar. Curvilinear Displacement of free form based deformation. *The Visual Computer*, Springer-Verlag , 16 (1), pp. 38-47, 2000.

[13] J.E. Schweitzer. Analysis and Application of Subdivision Surfaces. *PhD thesis*, Department of Computer Science and Engineering, University of Washington, 1996.

[14] T.W. Sederberg and S.R. Parry. Free-Form Deformation of Solid Geometric Models. *Proceedings of Siggraph'86*, vol. 20, no. 4, 1986.

Brill Academic Publishers
P.O. Box 9000, 2300 PA Leiden
The Netherlands

*Lecture Series on Computer
and Computational Sciences*
Volume 4, 2005, pp. 315-318

# The Predictive Behavior of The Phase Transition Temperatures of Imidazolium Based Ionic Liquids

Beáta Lemli,[a] László Kollár,[b] Géza Nagy,[a] Gellért Molnár,[a] Sándor Kunsági-Máté[a,1]

[a] Department of General and Physical Chemistry
University of Pécs
H-7624 Pécs, Hungary
[b] Department of Inorganic Chemistry
University of Pécs
H-7624 Pécs, Hungary

Received 9 July, 2005; accepted in revised form 3 August, 2005

*Abstract:* The applicability of ionic liquids in synthetic or separation science highly depends on the temperature range within they are really liquid. Therefore, the phase transitions are the most important property of these materials. Accordingly, the solid-solid and solid-liquid phase transitions of 1-methyl-3-tetradecylimidazolium hexafluorophosphate ($[C_{14}mim]^+[PF_6]^-$) ionic liquid was studied by density functional calculations and semiempirical AM1 molecular dynamic analysis. Our results show that the recently discovered anomalous structural change observed experimentally during crystalline to crystalline phase transition can be explained at molecular level with a conversion of two stable conformation of the $[C_{14}mim]^+[PF_6]^-$ pair. The solid-liquid phase transition temperature of the imidazolium cation-based ionic liquids was found to be affected by the length of alkyl side chain of the imidazol ring. In our present work a close relationship was found between the melting point of imidazolium cation-related ionic liquids and the dynamics of atomic vibrations in the side alkyl chain of different length of the cation.

*Keywords:* ionic liquid, structure analysis, quantum-chemical calculations, phase transitions

*PACS:* 31.15.Ar, 31.15.Ew, 31.15.Qg, 64.70.Dv, 64.70.Kb

## 1. Introduction

Room temperature ionic liquids (RT-ILs) belong to the class of molten salts. These materials have focused global attention in the last few years, when an intensive search for green alternative solvents has begun. Ionic liquids are usually made of an organic cation and organic or inorganic anion. The nature of these ions determines the physicochemical properties of the melts. Structural changes can be easily introduced in both cationic and anionic parts, so ionic liquids can be designed for particular applications and sets of their properties can be tuned.

To optimize the practical performance of RT-ILs, detailed knowledge about the physical and chemical properties is required; especially structural parameters are needed to describe the interesting processes at the molecular level. Although quantum chemistry is one of the fruitful testing techniques for molecular-scale analysis, only a few computational or theoretical studies on RT-ILs can be found in the literature (e.g. [1,2])

One of the most popular class of ionic liquids is the imidazolium ion containing one. In a recent experimental work [3] a wide range of analytical techniques were applied to characterize the phase behavior of $[C_{14}mim]^+[PF_6]^-$. The known typical phase transitions, such as crystalline to smectic, smectic to nematic, and nematic to isotropic, were identified in agreement with those reported by

---

[1] Corresponding author. E-mail: kunsagi@ttk.pte.hu

Gordon et al. [4]. A low temperature phase transition was also found with an unexpected change in the $a$, $b$ and $c$ unit cell dimensions: the longest dimension $c$ and the shortest dimension $a$ was lengthened, but the third dimension $b$ was shortened with increasing temperature. These experimental investigations of the $[C_{14}mim]^+[PF_6]^-$ suggested the existence of the crystalline polymorphism in case of this ionic liquid.

Further increasing the temperature results in melting. The solid-liquid phase transition temperature of the imidazolium cation-based ionic liquids showed dependency on the length of alkyl side chain of the imidazol ring [4].

Recently we started investigating this crystalline to crystalline phase transition at molecular level by quantum chemical methods. As a result of this work a very close connection between the melting point of imidazolium related ionic liquids and the dynamic movement of atoms in the alkyl side chain was found. Here we report shortly about our results obtained.

## 2. Results

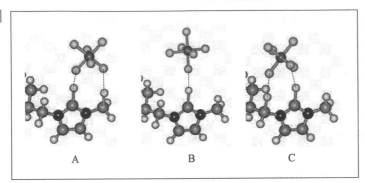

A        B        C

Figure 1: Stable conformations of the $[C_{14}mim]^+[PF_6]^-$ pair (**A** and **C**) and the transition - state conformation (**B**) related to the transition between these two stable conformations. The dotted lines show the H bonds, which stabilize the conformations.

In order to see the credible stable conformers of the $[C_{14}mim]^+[PF_6]^-$ ion pair calculations were performed as follows [1]: The $PF_6^-$ ion was positioned in various locations selected at four actual directions around the imidazol ring and using these starting conformations, geometry optimizations was determined by the AM1 method. After starting from this semiempirical structure, the conformation was calculated with the DFT/BLYP/6-31G* method. We found two stable conformations (Fig. 1 **A** and **C**) and a transition-state conformation (Fig. 1 **B**) between them. The stabilities of the $[C_{14}mim]^+[PF_6]^-$ in the two stable conformations are different: -367 kJ/mol in the **A** and -353 kJ/mol in the **C** conformation. The transition between **A** and **C** state is inhibited by an energy barrier of 43kJ/mol.

Figure 2 shows the variation of the unit cell dimensions $a$ and $b$ in accordance with the two stable conformations of the $[C_{14}mim]^+[PF_6]^-$ pair. It can be seen clearly that the change from the lower energy conformation **A** to conformation **C**, which has a higher energy, results in an increase and in a decrease of the $a$ and $b$ unit cell dimensions, respectively. Figure 2 also shows that this effect is essentially based on the tilted orientation of the imidazol ring relating to its connected alkyl chain (Fig. 2, bottom).

Our results show a change in the basic structure of this RT-IL in a very good agreement with the conclusion of the crystalline to crystalline phase transition of $[C_{14}mim]^+[PF_6]^-$ ionic liquid published earlier [3].

A        C

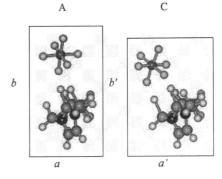

$b$        $b'$

$a$        $a'$

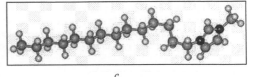

$c$

Figure 2: The consequences of the conformational changes on the unit cell dimensions of the $[C_{14}mim]^+[PF_6]^-$ RT-IL.

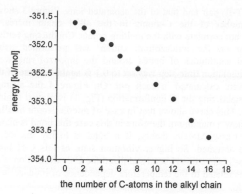

Figure 3: The cation-anion interaction energies (T=0K)

To obtain the phase transition temperature and the mechanistic details of the atomic movements during the transitions from **A** to **C** conformations, a molecular dynamic calculation was done using the AMBER forcefield. The temperature-dependent molecular dynamic simulations were performed at the semiempirical AM1 (Austin Model) level. The simulation time step was 0.1 fs because of the fast C-H vibrations. Ten thousand points were calculated in each run. Finding an appropriate initial condition for molecular dynamics a 'heating' algorithm implemented in HyperChem was used. This procedure heats up the molecular system smoothly from lower temperatures to the temperature T at which molecular dynamics simulation is desired to perform. The starting geometry for this heating phase is a static initial structure. We used the optimized AM1 geometry as an initial structure, and the temperature step and the time step in the heating phase were set to 2K and 0.1 fs, respectively. For molecular dynamics performed on the larger system periodic boundary with AMBER forcefield was used. In this case a periodic box imposes periodic boundary conditions on calculations. Periodic boundary conditions provide identical molecular systems whose are virtual images, identical to the one in the periodic box, surrounding the periodic box. For the purposes of calculations, molecules can move in a constant-density environment. All types of calculations were carried out with the HyperChem Professional 7 (HyperCube) program package.

The transition temperatures determined with the above method are varied from about 255 to 285 K. These theoretical values are close to the experimental value determined as the crystalline to crystalline phase transition temperature (278 K). The agreement is especially good if we consider that the calculated data were obtained from the force field calculation.

In the crystalline phase stable at higher temperature the anion is located closer to alkyl side chain of the cation (**C** conformation). Increasing the temperature, the crystal melts. Due to the practical importance of the melting point of the ionic liquids extensive search was done to find possible structural background that determines the melting temperatures. Excellent experimental works related, showed, that in both cases when the anion of the imidazolium cation is hexafluorophosphate[4] or tetrafluoroborate,[5] the melting point shows the same dependence on the length of alkyl chain of the cation. When the melting point –alkyl chain length dependence was investigated a "flat" minimum was found for two imidazolium ion based RT-ILs independently from the type of anionic components. That

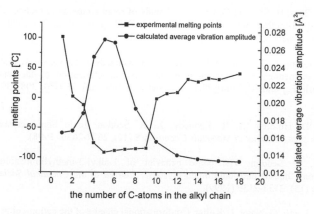

means that in the range, when the alkyl chain consists of $6 \leq n \leq 10$ carbon atoms the melting temperatures are lower [4,5] than in other ranges. To obtain data of the above effects at molecular level, the interaction energy between the cation and anion was calculated first, since the related conductivity measurements show the dissociation of the ion pair during the melting process. The interaction energy between cation and anion was defined as follows: it was calculated

Figure 4: The change of the experimental melting points and the calculated average vibration amplitude of the C-H bond against the number of C-atoms in the alkyl chain.

as the difference of the total energy of the RT-IL pair and that of the separated ions. Figure 3 shows that the interaction energy against the number of the C-atoms in the alkyl chain increases monotonously *i.e.* the interaction energy does not correlate with the melting points. Conducting further this search, molecular dynamic calculation on the imidazolium cation was performed using semiempirical AM1 method. The vibrational amplitudes of bonds around the imidazol ring were collected during 10 ps simulation time. The simulation time step was set to 0.1 fs again because of the fast C-H vibrations. Ten thousand points were calculated in each run. On Figure 4 the average amplitude of C-H bond, which takes part in stabilizing the C conformation (Fig. 1) is plotted against the number of the C-atoms in the alkyl chain. This curve shows that in case of middle length of alkyl chain the vibrational amplitude of this bond show a maximum, therefore in this case this bond is highly excited. According to the Rice-Ramsperger-Cassel-Marcus theory, if a bond is highly excited, a reaction within braking of this bond can be assumed. So higher vibration state of this C-H bond supports the dissociation of the cation-anion pair, since this H bond stabilized the C conformation (Fig. 1). This finding seems to answer the questions related to the temperature dependence of imidazolium-based RT-ILs on length of their side alkyl chain.

## 3. Conclusions

The recently discovered low temperature crystalline to crystalline phase transition of $[C_{14}mim]^+[PF_6]^-$ was examined by *ab initio* density functional method and molecular dynamics calculations at semiempirical level. Our results show that this RT-IL has two stable conformations with different energies. The transition between these two conformations can explain the RT-IL polymorphic crystalline character [6]. The solid-liquid phase transition temperature of the imidazolium cation-based ionic liquids was found experimentally being affected by the length of alkyl side chain of the imidazol ring. Our molecular dynamic calculations showed that the dependence of the melting point of imidazolium type ionic liquids on the length of alkyl chain of their cation is preferably based on the vibration state of the C-H bond taking part in the stabilization of the anion to the imidazolium cation.

## Acknowledgements

Calculations were performed on SunFire 15000 supercomputer located in the Supercomputer Center of the Hungarian National Infrastructure Development Program Office. The financial support of the Hungarian Scientific Research Fund (OTKA Grant TS044800) and that of the Hungarian PRCH Student Science Foundation (regarding participation on the ICCMSE 2005) are highly appreciated.

## References

[1] Z. Meng, A. Dölle, W. R. Carper: Gas phase model of an ionic liquid: semiempirical and *ab initio* bonding and molecular structure, *Journal of Molecular Structure: THEOCHEM* 585, 119-128 (2002).

[2] J. de Andrade, E. S. Böes, H. Stassen: Computational study of room temperature molten salts composed by 1-alkyl-3-methylimidazolium cations-force-field proposal and validation, *Journal of Physical Chemistry B* 106 (51), 13344-13351 (2002).

[3] J. De Roche, C. M. Gordon, C. T. Imrie, M. D. Ingram, A. R. Kennedy, F. Lo Celso, A. Triolo: Application of complementary experimental techniques to characterization of the phase behavior of $[C_{16}mim][PF_6]$ and $[C_{14}mim][PF_6]$, *Chemistry of Materials* 15 (16), 3089-3097 (2003).

[4] C. M. Gordon, J. D. Holbrey, A. R. Kennedy, K. R. Seddon: Ionic liquid crystals: hexafluorophosphate salts, *Journal of Materials Chemistry* 8 (12), 2627-2636 (1998).

[5] J. D. Holbrey, K. R. Seddon: The phase behavior of 1-alkyl-3-methylimidazolium tetraflouroborates; ionic liquids and ionic liquid crystals, *Journal of the Chemical Society, Dalton Transactions* (13) 2133-2139 (1999).

[6] S. Kunsági-Máté, B. Lemli, G. Nagy, L. Kollár: Conformational change of the cation-anion pair of an ionic liquid related to its low-temperature solid-state phase transitions, *Journal of Physical Chemistry B*, 108 (26), 9246-9250 (2004).

Brill Academic Publishers
P.O. Box 9000, 2300 PA Leiden
The Netherlands

*Lecture Series on Computer
and Computational Sciences*
Volume 4, 2005, pp. 319-322

# A New Fast Algorithm for Exact Calculation of the Discrete 2-D and 3-D X-Ray Transform

O. Levi, B.A. Efros[1]

Department of Industrial Engineering and Management,
Faculty of Engineering Sciences,
Ben-Gurion University of the Negev,
Beer-Sheva 84105, Israel

Received 21 June, 2005; accepted in revised form 2 August, 2005

*Abstract:* A new algorithm for fast calculation of line integrals is proposed. The algorithm is applicable for both 2-D and 3-D digital data. We focus our discussion in the case of line integration over 3-D voxel arrays where speed is much more crucial than in the 2-D case. The fundamental idea is simultaneous calculation of a 2D array of line integrals using serial memory access in order to significantly reduce the computational cost. We compare our algorithm to the Fast Slant-Stack algorithm [1] with respect to computation time and accuracy.

*Keywords:* X-Ray Transform, Radon Transform, Line integrals, Fast Slant-Stack.

## 1. Definition of the proposed discrete X-Ray transform

To implement a discrete X-ray transform one needs to define a discrete set of digital lines and a definition of line integration over a discrete voxel array. It is natural to choose a discrete set of lines that intersect with the voxel array and is uniformly sampled with respect to orientation and location. For our proposed algorithm we use a discrete system of lines which we name the slope-intercept system, it is the same system that is used for the fast slant stack algorithm [1] which we will use as reference for performance analysis in the next sections. Alternative viewpoints on 'digital geometry' and 'discrete lines' can be found in [8]. We will next define the slope-intercept system for the 3-D case. It is convenient to translate the 3-D cube's center of mass to the origin (0,0,0). Hence, for an $n$ by $n$ by $n$ voxel array we can define the cube's interior by the set $(x, y, z)$ satisfying $|x|, |y|, |z| \leq n/2$. We consider 3 types of lines: X-driven, Y-driven, Z-driven, depends which axis provides the shallowest slopes. A Z-driven line takes the form:

$$x = s_x z + t_x, \quad y = s_y z + t_y \tag{1}$$

with slopes $|s_x|, |s_y| \leq 1$, and intercepts $|t_x|, |t_y| \leq n$. X and Y-driven lines are defined similarly with an interchange of the roles between X, Y and Z. We consider the family of lines generated this way, where the slopes and intercepts run through the equispaced family:

$$s_x, s_y, s_z \in \{2l/n : l = -n/2, ..., n/2 - 1\}, \quad t_x, t_y, t_z \in \{2l/n : l = -n+1, ..., n-1\} \tag{2}$$

One particularly important property of this line system is that the intersection of any line and the cube starts and ends on the cube's boundary grid points (we consider the set of grid points as the set of voxel corners for all voxels in the cube). Our discrete X-Ray transform performs exact line integration along each of the lines defined above when the 3-D voxel array is treated as a 3-D piece-wise constant function.

---

[1] Corresponding author. E-mail: efros@bgumail.bgu.ac.il

## 2. Overall strategy and the Algorithm

Consider a zero-intercept Z-driven line $L(h,t)$ when $h$ and $t$ are its two intersection points on the cube's boundary. Denote the grid points coordinates of h and t by $(h_X, h_Y, h_Z)$ and $(t_X, t_Y, t_Z)$ respectively. It is easy to show that since the line passes through the cube's center of mass the following holds:

$$
\begin{aligned}
h_Z &= -n/2 & t_Z &= n/2 \\
h_X &= -t_X & h_Y &= -t_Y
\end{aligned}
$$

(3)

We define as *X-grid plane* a plane containing the 2-D set of grid points all sharing the same X-coordinate and similarly, Y-grid plane and Z-grid plane. Obviously, for an $n$ by $n$ by $n$ voxel array there are $n+1$ X-grid planes and the same number of Y-grid planes and Z-grid planes and the intersection of these three sets of planes generates the cube's grid.

If the length of the line segment $L(h,t)$ is $l$ then it is obvious that the line segment intersects with X-grid plane each $l/|\Delta x|$ units, with Y-grid plane each $l/|\Delta y|$ and with Z-grid plane each $l/|\Delta z|$ units, where $\Delta x = x_t - x_h$   $\Delta y = y_t - y_h$   $\Delta z = z_t - z_h$

We use the term "slice" to define the 3D region bounded by two consequent Z-grid-planes. We will show shortly that it is possible to calculate a line integral by finding the line's intersection points with the different cube's slices. Consider now the set of Z-driven lines with the same slope, notice that due to the fact that these lines have an integer intercept difference, the set of intersection points with the cube's slices for any of these lines can be obtained by an integer shift of the set belongs to any other line, where the shift size is exactly the integer intercept difference. Our algorithm takes advantage of the last observation in order to calculate in batch a 2-D set of line integrals over a corresponding set of parallel lines by serial processing of the cube's slices and summing the contribution of each slice in a single step to the whole 2-D set of line integrals.

We will next present some notations that will be used in order to define the algorithm steps in details.

Let $\Psi = \{(\Psi_\Delta(i), \Psi_X(i), \Psi_Y(i), \Psi_Z(i)); i = 1, ..., m\}$ be a data structure associated with a line traveling through the data cube and contains information about the line's intersections with the different cube's slices. In the definition above m is the number of intersections of the line with the different slices, $\Psi_\Delta(i)$ is the i-th intersection's length, $\Psi_X(i)$, $\Psi_Y(i)$ and $\Psi_Z(i)$ are binary vectors for which a 'true' value indicates that the i-th intersection is generated by the corresponding grid plane. We denote by $\Psi^{s_x s_y}$ the structure associated to a line with slopes $s_x, s_y$ and zero intercepts. It should be clear from the above discussion that the structures for each family of parallel lines with slopes $s_x, s_y$ can be all generated by applying shifts to the X and Y binary vectors of $\Psi^{s_x s_y}$ when the shifts sizes are equal to the intercepts difference, we will show how it can be done below. Consider again the zero-intercept Z-driven line $L(h,t)$ from the above discussion, the integral over this line can be exactly computed using the following ideas:

$\Psi_\Delta(i)$ is a multiplier of the i-th sum-member (a voxel's intensity), whereas $\Psi_X(i)$, $\Psi_Y(i)$ and $\Psi_Z(i)$ are used to easily increment indices. Generally, we can express this as:

$$
Xf_{L(h,t)} = \|(h,t)\|_2 \cdot \sum_{i=1}^{m} \psi_\Delta(i) \cdot f(\hat{\psi}_X(i), \hat{\psi}_Y(i), \hat{\psi}_Z(i))
$$

(4)

where

$$
\hat{\psi}_\eta(i) = h_\eta + sign(h_\eta - t_\eta) \cdot \sum_{j=2}^{i} \psi_\eta(j) \quad , \eta = X, Y, Z
$$

(5)

$$
\|(h,t)\|_2 = \sqrt{(h_X - t_X)^2 + (h_Y - t_Y)^2 + (h_Z - t_Z)^2} = n\sqrt{s_x^2 + s_y^2 + 1}
$$

(6)

Traversing the $\Psi^{s_x s_y}$ structure provides us an easy and coherent way for summing all of the involved voxels, properly weighted for an exact computation of a line integral. Until now we discussed only the

Z-driven lines with zero intercepts. We will next show how to simultaneously compute a set of integrals over a family of lines with the same slopes and varying intercepts and doing so with serial data access. As we mentioned earlier, the structures $\Psi$ for 2 Z-driven lines with the same slopes and different intercepts differ only by integer shifts to the X and Y binary vectors when the shifts sizes are exactly the intercepts difference. By looking at (4) it is easy to show that a simultaneous calculation of a complete set of line integrals corresponding to a family of parallel lines can be computed by applying integer shifts to complete Z slices and computing an appropriate weighted sum of the shifted slices. The serial memory access is obvious since in each iteration of the algorithm the voxel data is accessed slice by slice in a sequential order. We call our algorithm Shift-And-Sum (or SHAS). Since these two operations are the most dominant ones in the algorithm.

The algorithm input is a 3-D voxel array $f(x, y, z) \in \Re^{n \times n \times n}$ . The output is a 5-D coefficients array:

$Xf(\eta, s_1, s_2, t_1, t_2) \in \Re^{3 \times n \times n \times 2n \times 2n}$ . The parameter $\eta$ is used to index the line's type (X, Y or Z-driven), $s1, s2$ are the line's slopes and $t1, t2$ are the line's intercepts.

We are using the notation $\Psi^{s_x s_y}$ - for an Array of Z-driven lines with zero intercepts and slopes sx, sy. Let E1 be the operator of extension, padding an array n wide and n tall to be n wide and 2n tall by adding extra rows of zeros symmetrically above and below the input argument. Let E2 be the operator of extension, padding an array n wide and n tall to be 2n wide and n tall by adding extra columns of zeros symmetrically left and right of the input array. Let $\tilde{I}_1 = E_1 I$ and $\tilde{I}_2 = E_2 I$ .

Let S be the circular shift operator over arrays of size $2n \times 2n$ . Then given any array $A \in \Re^{2n \times 2n}$ , the result of $S^d A$ for any integer d is an array $B \in \Re^{2n \times 2n}$ , where the i-th column of A is identical to the column at position $\mathrm{mod}_{2n}(i + d)$ in B.

The procedure for implementing our proposed 3-D discrete X-ray transform is given below:

*for each line type* $\eta \in \{X\text{-}driven, Y\text{-}driven, Z\text{-}driven\}$

   *for each pair of slopes sx, sy*

     - *Derive* $\Psi = \Psi_{s_x s_y}$

$$Xf(\eta, s_1, s_2, :, :) = 1 \cdot \sum_{i=1}^{m} \psi_\Delta(i) \cdot S^{-\hat{\psi}_X(i)} \left( \tilde{I}_1 f(\hat{\psi}_Z(i), :, :) \tilde{I}_2 \right) \left( S^{-\hat{\psi}_Y(i)} \right)'$$

   *end for*

     - *Change grids roles:* $Z \to Y, Y \to X, X \to Z$

*end for*

## 3. Performance analysis and Conclusions

We consider two key measures of performance of our proposed algorithm: accuracy and timing. The SHAS algorithm is exact by it's definition when viewing a digital image as a piecewise constant function, i.e. each volume element associated with constant intensity. In this case each line integral is a weighted sum of the voxels it intersects with. The Fast Slant-Stack algorithm [1] uses a different definition for the line integrals and computed the line integrals over a trigonometric interpolation of the image done in the frequency domain, even though there might be good justifications for such definitions it sometimes results in non physical values for the line integrals, for example negative coefficients for positive data, due to the interpolation and for some applications it might be a major drawback. The number of operations required for the calculation of a single line integral in an image of size $n^3$ is $O(n)$ in our algorithm and only $O(\log n)$ for the Fast Slant-Stack algorithm. Therefore we are not expecting these algorithms to compete for very large image sizes. However, the Slant-Stack algorithm involves the application of non-stardard FFT's that despite of being $O(n \cdot \log n)$ are quite time consuming compare to standard FFT. In fact, the operations count constant multiplier of the SHAS algorithm is about 2 and all operations are only summation and multiplications of real numbers. Therefore we expect that the SHAS algorithm will be faster for moderate images size. Figure 5 presents results of the comparison between the two algorithms when both of them are implemented in C++ programming language with Microsoft Visual C++ .NET compiler and work in double precision. Tests were performed on a Pentium 4, 2.4GHz, 512Mb of RAM. These experimental results verify that

SHAS algorithm is significantly more effective than Slant Stack for voxel arrays smaller than 256 by 256 by 256.

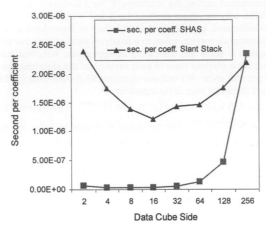

Figure 1: Timing Comparison

We have focused our discussion in the case of 3-D digital images but the same ideas exactly can be applied in 2-D even with more ease where the X-Ray transform is then referred as the Radon Transform. We find the 3-D case more challenging due to the natural high complexity of discrete set of lines in 3-D and the relatively big sizes of data. We have given timing measures of our algorithm and compared it with the Fast Slant-Stack algorithm, despite of the fact that theoretically our algorithm is slower than the Fast Slant-Stack we have shown that for practical 3-D image sizes our algorithm is much faster. One of the most important advantages of our proposed method is its cache-awareness, i.e. it is very well-organized for use with modern hierarchical memory computers [7]. Another advantage over the Fast Slant-Stack data is that our algorithm computes exact line integrals when treating a digital image as a piece-wise constant function and no interpolation is used.

## References

[1]     A. Averbuch, R. R. Coifman, D. L. Donoho, M. Israeli, and J. Walden, Fast slant stack: A notion of Radon transform for data in a Cartesian grid which is rapidly computable, algebraically exact, geometrically faithful and invertible, *Tech. Rep.* (preprint), httt://ww.math.tau.ac.il/amirl/, 2001.

[2]     M. L. Brady, A Fast Discrete Approximation Algorithm For The Radon Transform, *SIAM J. Computing*, Vol. 27, No 1, pp. 107-119, 1998.

[3]     A. Brandt and J. Dym, Fast Calculation of Multiple Line Integrals, *SIAM Journal of Scientific Computing*, Vol. 20, No 4, pp. 1417–1429, 1999.

[4]     D. Donoho and O. Levi, Fast X-Ray and Beamlet Transforms for Three-Dimensional Data, *Technical Report, Statistics,* Stanford University, 1999.

[5]     J. Dym, Multilevel Methods for Early Vision, *Ph.D. thesis*, Weizmann Institute of Science, Rehovot, Israel, 1994.

[6]     W. A. Götze and H. J. Druckmüller. A fast digital Radon transform – an Efficient Means for Evaluating the Hough Transform, *Pattern Recognition*, Vol. 28(12), pp.1985–1992, 1995.

[7]     J. L. Hennessy and D. A. Patterson, Computer Architecture: A Quantitative Approach, *Morgan Kaufmann Publishers*, 2nd edition, San Francisco, CA, 1996.

[8]     G. Herman, Geometry of Digital Spaces, *Birkhauser*, Boston, 1998.

Brill Academic Publishers
P.O. Box 9000, 2300 PA Leiden
The Netherlands

*Lecture Series on Computer*
*and Computational Sciences*
Volume 4, 2005, pp. 323-329

# An Integration Technology of Checkpointing for DMR with Energy-Aware

Li Zhong-wen[1,2], Jiang Ye-peng[1], Yu Shui[3]

[1]Information Science and Technology College,
Xiamen University,
Xiamen 361005, China
[2]Zhongshan Institute of UESTC,
Zhongshan, 528402, China
[3]School of Computing and Mathematics,
Deakin University, Australia

Received 21 June, 2005; accepted in revised 15 August, 2005

*Abstract:* We present an integrated approach that provides fault tolerance, signature and DVS (Dynamic voltage scaling) for DMR (double modular redundancy). Fault tolerance is achieved through tuning the checkpointing schemes to a given architecture, namely, inserting additional CCPs (compare-checkpoints) or SCPs (store-checkpoints) between CSCPs (store-and-compare-checkpoints). Fault tolerance is then combined with DVS. Introducing the overheads of comparison, storage, DVS, the average execution times to complete a task for proposed schemes are obtained, using renewal equation. In addition, an optimal checkpointing interval that minimizes the average times is analytically derived and is numerically computed. Four schemes are compared as numerical examples and results show that compared to previous approaches, the proposed approach significantly reduces task average execution time. Further, we show how signatures can be used to reduce the overhead of checkpoints in our schemes. Our research results can be applied to the other task duplication systems, such as TMR-F, DMR-F-1 and RFCS, etc..

*Keywords:* Fault-tolerant computing, Checkpointing intervals, Signature, DMR, DVS, Renewal equation

## 1. Introduction

Checkpointing is an important method of providing fault-tolerance in real-time control systems that operate in harsh environmental conditions. The following three types of checkpoints are well known: CSCP, SCP and CCP [1-3]. CCPs are used to compare the states of the processors without storing them, while, in SCPs, the processors store their states without comparison. If the two operations are used together in the same checkpoint, we call them CSCPs. Using CCP and SCP, Ziv and Bruck have shown numerically that the task execution time is significantly reduced [1,4]. And with additional CCPs and SCPs, Nakagawa and Fukumoto have used a triple modular redundancy and double modular redundancy to analyze the optimal checkpointing interval that minimizes a task execution time, respectively [5,6].

In addition, many real-time control systems are often energy-constrained since system lifetime is determined to a large extent by the battery lifetime [2]. Examples include autonomous airborne and sea-borne systems working on limited battery supply, space systems working on a limited combination of solar and battery power supply, or time-sensitive systems deployed in remote locations where a steady power supply is not available [3,7]. DVS has emerged as a popular solution to the problem of reducing power consumption during system operation. DVS is made possible by the availability of embedded processors that can dynamically scale the frequency by adjusting the operation voltage [2,3]. Many embedded processors are now equipped with the ability to dynamically scale the operation voltage. Examples include the mobile processors available from Intel with its SpeedStep [8] technology and the Transmeta Crusoe processor with LongRun [9]. In the realm of real-time systems, DVS techniques

2 Corresponding author. E-mail: lizw@xmu.edu.cn. This work is supported partly by Fujian natural science foundation (A0410004), Fujian young science & technology innovation foundation (2003J020), NCETXMU 2004 program, and Xiamen University research foundation (0630-E23011).

focused on minimizing energy consumption of the system, while still meeting the deadlines [10,11]. DVS and fault tolerance for real-time control systems have largely been studied as separate problems. It is only recently that an attempt has been made to combine fault tolerance with DVS [3]. Signatures are used to detect and correct faults in many applications, such as communication channels. In systems with high comparison time, signatures can significantly reduce the checkpoint overhead, and, hence, reduce the execution time of a task.

A combination of DVS, CSCPs (CCPs or SCPs) and signatures can be used to satisfy system's DVS requirement and improve the performance of real-time systems. However, none of the above papers addresses these issues in conjunction. In this paper, we present an integrated approach that facilitates fault tolerance through additional CCPs checkpoints and power management through DVS. Further, we show how signatures can be used to reduce the overhead of checkpoints in our schemes. To the best of our knowledge, this is the first approach that addresses these issues in conjunction. The analysis of the DMR scheme with SCPs is similar to the analysis of the schemes with additional CCPs, and, therefore, it is not included in the paper.

## 2. Computing Models

Assume that a real-time task $\tau$ has a period $T$, a deadline $D$, a fixed quantity of computation cycles $N$ in the fault-free condition $N$ ($D \leqslant T$). Because the variable voltage CPUs are available, the time to execute task $\tau$ depends on the processor speed. We therefore characterize $\tau$ by a fixed quantity $N$, namely, its worst-case number of CPU cycles, needed to execute the task at the minimum processor speed. For the rest of this paper, we normalize the units of $N$ such that the minimum processor speed is 1. That is, if the minimum processor speed is $S$ cycles per second, then we express the number of cycles in units of $S$ cycles and thus normalize the minimum processor speed to $S_{min}=1$ [7]. Of course, $T$ is expressed in terms of the number of CPU cycles at the minimum processor speed.

To simplify analysis and to allow for the derivation of analytical formulas, we would like to assume that a single processor with two speeds $f_1$ and $f_2$, and $f_1$ is the minimum processor speed, namely, $f_1 = S_{min}=1$. Moreover, the processor can switch its speed in a negligible amount of time. In checkpointing scheme, task $\tau$ takes some checkpoints to enable recovery from failure. In this paper, if errors in a checkpoint interval are detected and $\tau$ rolls back to the most recent checkpoint, we let processors to reexecute this checkpoint interval at the maximum speed $f_2$, and reduce processors speed of executing the remainder of $\tau$ to $f_1$.

There are various models proposed to characterize the failure process. The most commonly used assume that the time between failures follows an exponential distribution, Weibull distribution or variations on them (Castillo, McConnel and Siewiorek) [12]. For simplicity of illustration, we assume that the time for a failure to occur is exponentially distributed with constant failure rate $\lambda$. We execute $\tau$ on two independent processors and assume that faults can occur while the processors execute the task, but not during checkpoints. Some notation used in our paper is as following:

$t_s$: the time to store the states of processors.
$t_{cp}$: the time to compare processors' states.
$t_r$: the time to roll back the processors to a consistent state.
$t_{sig}$ : the time to calculate and compare signatures.
$\varepsilon$ : the probability of misdetections of signatures.
$C$: the checkpointing cost.

## 3. Checkpointing schemes with DVS

### 3.1 CSCPs

To analyze the average execution time of a task using the DMR and DVS scheme, we assume that task $\tau$ is divided equally into $m$ intervals of length $T_1 = \dfrac{N}{m}$, and at the end of each interval, CSCP is always placed. At the end of each interval, the states of two processors are compared and stored at CSCP. If two states agree with each other, this execution is recognized to be correct and the processors continue the execution of the next interval. If they do not agree, it is judged that some errors have occurred. Then, two processors are rolled back to the previous CSCP, and repeat the execution of this interval. In figure 1, two processors are rolled back to $(i-1)T_1$ because some errors have occurred during $((i-1)T_1, iT_1)$, and repeat the execution from $(i-1)T_1$.

Task $\tau$ is completed when all executions are succeeded in every checkpointing interval. The average execution time $R_1(1)$ for a CSCP interval $((i-1)T_1, iT_1]$ is consisted of two parts ($i=2,3,\ldots,m$). The first one denotes the average execution time $R_1(1)$ under the condition that there is no error occurrence in each processor and the processor's speed is $f_1$. The second one denotes the average

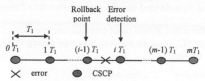

Figure 1: Task execution with CSCP

execution time $R_1(1)$ in the presence of failures and we increase the speed of processors' to $f_2$. The probability that no error occurred between CSCPs is $e^{-2\lambda T_1}$.

$R_1(1)$ is given by a renewal-equation[6, 13]:

$$R_1(1) = (T_1 + t_s + t_{cp})e^{-2\lambda T_1} + [T_1 + t_s + t_{cp} + t_r + kR_1(1)](1 - e^{-2\lambda T_1}) \tag{1}$$

Here $k$ ($k<1$) is the proportionality constant. In the absence of errors, let $R_1(1)$ express the average execution time of $\tau$ corresponding to speed $f_1$, then $kR_1(1)$ is the average execution time of $\tau$ corresponding to speed $f_2$. Solving (1), we have (2)

$$R_1(1) = \frac{(T_1 + t_s + t_{cp})e^{2\lambda T_1} + t_r(e^{2\lambda T_1} - 1)}{e^{2\lambda T_1}(1-k) + k} \tag{2}$$

Therefore, the average execution time $R_1(m)$ is (3).

$$R_1(m) = mR_1(1) = m\frac{(T_1 + t_s + t_{cp})e^{2\lambda T_1} + t_r(e^{2\lambda T_1} - 1)}{e^{2\lambda T_1}(1-k) + k} \tag{3}$$

Since $T_1 = N/m$, we have

$$R_1(m) = \frac{\left[N + m(t_s + t_{cp} + t_r)\right]e^{2\lambda N/m} - mt_r}{e^{2\lambda N/m}(1-k) + k} \tag{4}$$

Of course, $m$ has an upper-border. Checkpointing increases task execution time and in the absence of errors, it might cause a missed deadline for a task that completes in time without checkpointing. This implies that the effective number of checkpoints $m$ must satisfy $m \le \frac{D-N}{C}$. Thus, there exists a finite

number $\tilde{m}(1 \le \tilde{m} \le \frac{D-N}{C})$ which minimizes $R_1(m)$. Putting $T_1 = N/m$ in equation (4), we have

$$R_1(T_1) = \frac{N}{e^{2\lambda T_1}(1-k) + k}\left[\left(1 + \frac{t_s + t_{cp} + t_r}{T_1}\right)e^{2\lambda T_1} - \frac{t_r}{T_1}\right] \tag{5}$$

Since $T_1 = N/m$, there exists an optimal $\tilde{T}_1(\frac{CN}{D-N} \le \tilde{T}_1 \le N)$ which minimizes $R_1(T_1)$ in (5).
Differentiating $R_1(T_1)$ with respect to $T_1$ and setting it equal to zero, we have

$$k \cdot 2\lambda T_1(T_1 + t_s + t_{cp}) + 2\lambda T_1 \cdot t_r$$
$$-[e^{2\lambda T_1}(1-k) + k][(t_s + t_{cp}) + t_r(1 - e^{-2\lambda T_1})] + (t_s + t_{cp}) = t_s + t_{cp} \tag{6}$$

Solving equation (6), we can get $\tilde{T}_1(\frac{CN}{D-N} \le \tilde{T}_1 \le N)$.

Procedure num_CSCPs($N$) is used to calculate $\tilde{m}$ which minimize $R_1(\tilde{m})$.

```
Procedure num_CSCPs(N) {
  Find T̃₁ which minimizes R₁(T₁); /*according to equation (6)*/
  if (T̃₁<N) {
    m = ⌊N / T̃₁⌋;
    if (R₁(m) ≤R₁(m+1)) then  /*according to equation (4)*/
    m̃ = m;
      else m̃ = m+1;
  } else m̃ = 1;
```

```
return m̄ ; }
```

### 3.2 Additional CCPs

Now, each CSCP interval $T_1$ is divided equally into $n$ intervals of length $T_2 = \dfrac{T_1}{n}$. Now, CCPs are placed between CSCPs, and the states of two processors are compared at $iT_2$ ($i$=1,2,..., $n$-1) and $jT_1$. If two states do not agree at $iT_2$ or $jT_1$, some errors have occurred during this interval, and two processors are rolled back to $(j-1)T_1$ (see Figure2).

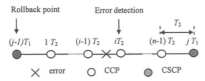

Figure2: Task execution with CCPs

The average execution time $R_2(n)$ for an interval $((j-1)T_1, jT_1)$ is given by a renewal-equation[6, 13].

$$R_2(n) = (nT_2 + nt_{cp} + t_s)e^{-2\lambda nT_2} + \sum_{i=1}^{n} \int_{(i-1)T_2}^{T_2} \left[ iT_2 + it_{cp} + t_r + k \cdot R_2(n) \right] 2\lambda e^{-2\lambda t} dt \tag{7}$$

$$+ \int_{(n-1)T_2}^{nT_2} t_s \times 2\lambda e^{-2\lambda t} dt$$

Solving it, we have

$$R_2(n) = \frac{1}{e^{2\lambda nT_2}(1-k)+k} \left[ t_s e^{2\lambda T_2} + \left( \frac{T_2 + t_{cp}}{1 - e^{-2\lambda T_2}} + t_r \right)\left( e^{2\lambda nT_2} - 1 \right) \right] \tag{8}$$

Since $T_2 = T_1/n$, we have

$$R_2(n) = \frac{1}{e^{2\lambda T_1}(1-k)+k} \left[ t_s e^{2\lambda T_1/n} + \left( \frac{T_1/n + t_{cp}}{1 - e^{-2\lambda T_1/n}} + t_r \right)\left( e^{2\lambda T_1} - 1 \right) \right] \tag{9}$$

Using the similar reason in section 3.1, we can proof that there exists a finite $\tilde{n}$ which minimizes $R_2(n)$ in equation (9). Putting $T_2 = T_1/n$ and rewrite (8), we have equation (10).

$$R_2(T_2) = \frac{1}{e^{2\lambda T_1}(1-k)+k} \left[ t_s e^{2\lambda T_2} + \left( \frac{T_2 + t_{cp}}{1 - e^{-2\lambda T_2}} + t_r \right)\left( e^{2\lambda T_1} - 1 \right) \right] \tag{10}$$

There exists an optimal $\tilde{T}_2$ ($0 < \tilde{T}_2 \leq T_1$) which minimizes $R_2(T_2)$ in (10). Differentiating $R_2(T_2)$ with respect to $T_2$ and setting it equal to zero, we have

$$2\lambda t_s (e^{2\lambda T_2} - 1)^2 + (e^{2\lambda T_1} - 1)\left[ e^{2\lambda T_2} - (1 + 2\lambda T_2) \right] = 2\lambda t_{cp} (e^{2\lambda T_1} - 1) \tag{11}$$

Solving equation (11), we can get $\tilde{T}_2$ ($0 < \tilde{T}_2 \leq T_1$) which minimizes $R_2(T_2)$ and we can calculate $\tilde{n}$ by the similar approach described in procedure num_CSCPs($N$).

## 4. Signatures

### 4.1 CSCPs

Here, each CSCP interval $T_1$ is divided equally into $\xi$ intervals of length $T_3 = \dfrac{T_1}{\xi}$, and signatures are calculated and compared at scheduled time $T_3$. Assume an error has occurred between $(i-1) T_3$ and $iT_3$, and is detected at $(i+j-1) T_3$ ($i<j$). Then processors rollback to the last CSCP and make a retry. The average execution time $R_3(\xi)$ for one CSCP interval is given by a renewal equation[13,14].

$$R_3(\xi) = [\xi(T_3 + t_{sig}) + t_{cp} + t_s]e^{-2\lambda\xi T_3}$$

$$+ \sum_{i=1}^{\xi} \int_{(i-1)T_3}^{iT_3} 2\lambda e^{-2\lambda t} dt \{\varepsilon^{\xi-i+1}[\xi(T_3 + t_{sig}) + t_{cp} + t_s + t_r + k \cdot R_3(\xi)]$$

$$+ \sum_{j=1}^{\xi-i+1} \varepsilon^{j-1}(1-\varepsilon)[(j+i-1)(T_3 + t_{sig}) + t_r + k \cdot R_3(\xi)]\} \tag{12}$$

Solving equation (12), we have

$$R_3(\xi) = (T_3 + t_{sig})\{\xi e^{-2\lambda\xi T_3} + \sum_{i=1}^{\xi} \int_{(i-1)T_3}^{T_3} 2\lambda e^{-2\lambda t} dt [\xi \varepsilon^{\xi-i+1} + \sum_{j=1}^{\xi-i+1} \varepsilon^j (1-\varepsilon)(j+i-1)]\}$$

$$+ (t_{cp} + t_s)[e^{-2\lambda\xi T_3} + \sum_{i=1}^{\xi} \int_{(i-1)T_3}^{T_3} 2\lambda e^{-2\lambda t} dt \cdot \varepsilon^{\xi-i+1}] + t_r \sum_{i=1}^{\xi} \int_{(i-1)T_3}^{T_3} 2\lambda e^{-2\lambda t} dt [\varepsilon^{\xi-i+1} + \sum_{j=1}^{\xi-i+1} \varepsilon^{j-1}(1-\varepsilon)] \tag{13}$$

$$+ k \cdot R_3(\xi) \cdot \sum_{i=1}^{\xi} \int_{(i-1)T_3}^{T_3} 2\lambda e^{-2\lambda t} dt [\varepsilon^{\xi-i+1} + \sum_{j=1}^{\xi-i+1} \varepsilon^{j-1}(1-\varepsilon)]$$

From equation (13), we get

$$R_3(\xi)[1 - k(1-e^{-2\lambda\xi T_3})] = (T_3 + t_{sig})\{\xi e^{-2\lambda\xi T_3} + \sum_{i=1}^{\xi}(e^{-2\lambda(i-1)T_3} - e^{-2\lambda i T_3})[(i-1) + \frac{1-\varepsilon^{\xi-i+1}}{1-\varepsilon}]\}$$

$$+ (t_{cp} + t_s)[e^{-2\lambda\xi T_3} + \sum_{i=1}^{\xi}(e^{-2\lambda(i-1)T_3} - e^{-2\lambda i T_3})\varepsilon^{\xi-i+1}] + t_r \cdot \sum_{i=1}^{\xi}(e^{-2\lambda(i-1)T_3} - e^{-2\lambda i T_3}) \tag{14}$$

Putting $T_3 = \dfrac{T_1}{\xi}$ in equation (14), we have

$$R_3(\xi) = \{(\frac{T_1}{\xi} + t_{sig})[\frac{1-\varepsilon^{\xi}}{1-\varepsilon}e^{2\lambda T_1} + \sum_{i=0}^{\xi-1}(1-\varepsilon^i)e^{2\lambda i \frac{T_1}{\xi}}]$$

$$+ (t_{cp} + t_s)[\varepsilon^{\xi} e^{2\lambda T_1} + (1-\varepsilon)\sum_{i=1}^{\xi-1} \varepsilon^i e^{2\lambda i \frac{T_1}{\xi}}] + t_r(e^{2\lambda T_1} - 1)\}/[(1-k)e^{2\lambda T_1} + k] \tag{15}$$

Obviously, $\xi$ has an upper-border, thus, there exists a finite number $\tilde{\xi}$ which minimizes equation (15). And we can calculate it by the similar approach described in procedure num_CSCPs($N$).

## 4.2 CCPs

Like 3.2, each CSCP interval $T_1$ is divided equally into $\omega$ intervals of length $T_4 = \dfrac{T_1}{\omega}$, CCPs are placed between CSCPs. And signatures are calculated and compared at scheduled time $T_4$. Of course, if an error has occurred between $(i-1)T_4$ and $iT_4$, an error is always detected at $iT_4$. The average execution time $R_4(\omega)$ for one CSCP interval is given by a renewal equation[13,14].

$$R_4(\omega) = e^{-2\lambda\omega T_4}(\omega T_4 + \omega t_{sig} + \omega t_{cp} + t_s)$$

$$+ \sum_{i=1}^{\omega-1} \int_{(i-1)T_4}^{iT_4} 2\lambda e^{-2\lambda t} dt \{\varepsilon[iT_4 + it_{sig} + it_{cp} + t_r + kR_4(\omega)]$$

$$+ (1-\varepsilon)[iT_4 + it_{sig} + (i-1)t_{cp} + t_r + kR_4(\omega)]\} \tag{16}$$

$$+ \int_{(\omega-1)T_4}^{\omega T_4} 2\lambda e^{-2\lambda t} dt \{\varepsilon[\omega T_4 + \omega t_{sig} + \omega t_{cp} + t_s + t_r + kR_4(\omega)]$$

$$+ (1-\varepsilon)[\omega T_4 + \omega t_{sig} + (\omega-1)t_{cp} + t_r + kR_4(\omega)]\}$$

Solving equation (16), we have

$$R_4(\omega) = (T_4 + t_{sig})\{\omega e^{-2\lambda\omega T_4} + \sum_{i=1}^{\omega} \int_{(i-1)T_4}^{iT_4} 2\lambda e^{-2\lambda t} dt [\varepsilon i + (1-\varepsilon)i]\}$$

$$+ t_{cp}\{\omega e^{-2\lambda\omega T_4} + \sum_{i=1}^{\omega} \int_{(i-1)T_4}^{iT_4} 2\lambda e^{-2\lambda t} dt [\varepsilon i + (1-\varepsilon)(i-1)]\} \tag{17}$$

$$+ t_s[e^{-2\lambda\omega T_4} + \int_{(\omega-1)T_4}^{\omega T_4} 2\lambda e^{-2\lambda t} dt \cdot \varepsilon] + t_r \cdot \sum_{i=1}^{\omega} \int_{(i-1)T_4}^{iT_4} 2\lambda e^{-2\lambda t} dt [\varepsilon + (1-\varepsilon)]$$

$$+ kR_4(\omega) \cdot \sum_{i=1}^{\omega} \int_{(i-1)T_4}^{iT_4} 2\lambda e^{-2\lambda t} dt [\varepsilon + (1-\varepsilon)]$$

Putting $T_4 = \frac{T_1}{\omega}$ in equation (17), we have

$$R_4(\omega) = \{(\frac{T_1}{\omega} + t_{sig})\sum_{i=1}^{\omega} e^{2\lambda i \frac{T_1}{\omega}} + t_{cp}[\sum_{i=1}^{\omega} e^{2\lambda i \frac{T_1}{\omega}} - (1-\varepsilon)(e^{2\lambda T_1} - 1)] \tag{18}$$

$$+ t_s[1 + \varepsilon(e^{2\lambda \frac{T_1}{\omega}} - 1)] + t_r(e^{2\lambda T_1} - 1)\} / [(1-k)e^{2\lambda T_1} + k]$$

Obviously, $\omega$ has an upper-border, thus, there exists a finite number $\tilde{\omega}$ which minimizes equation (18). And we can calculate it by the similar approach described in procedure num_CSCPs($N$).

## 5. Numerical examples

Let sch1, sch2, sch3, sch4 represents scheme with DVS & Signature, scheme with DVS and without Signature, scheme without DVS and with Signature, scheme without DVS & Signature, repectively. We carried out a set of experiments to evaluate our checkpointing schemes (sch1) and to compare them with checkpointing scheme with DVS and with Signature[6,14] (sch2, sch3). Without loss of generality, we let $f_2 = 2f_1$[3], it implies $k=0.5$. Assume that $N=1$, $t_{cp} = 2.5 \times 10^{-5}$, $t_s = 5 \times 10^{-4}, t_r = 5.0 \times 10^{-4}$ [6], $\varepsilon = 1.0 \times 10^{-3}$ and $t_{sig} = 1.25 \times 10^{-7}$ [14].

Figure 3 shows the average execution time of a task in these four schemes with CSCPs and

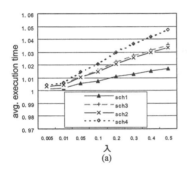
(a)

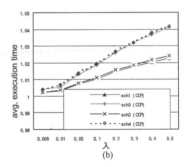
(b)

Figure 3: Comparison of average execution time between schemes with and without DVS/Signature:(a) CSCPs without additional CCPs, (b) CSCPs with additional CCPs

additional CCPs. For each value of $\lambda$, the values of $m$, $n$, $\xi$ and $\omega$, were chosen such that the average execution time is minimized for these four schemes, and optimal $m$, $n$, $\xi$ and $\omega$ are shown in table 1. Figure 3.a shows that signature and DVS can significantly reduce the execution time of a task with CSCPs scheme. It can be seen from the figure 3.b that using DVS gives a significant reduction in the overhead of the execution time and signature can also be used to reduce the execution time of a task.

Table 1: Optimal CSCP number $m$, optimal signature number $\xi$ and $\omega$, optimal CCP number $n$.

| $\lambda$ | CSCPs without additional CCPs | | | | CSCPs with additional CCPs | | | |
|---|---|---|---|---|---|---|---|---|
| | sch1 | sch2 | sch3 | sch4 | sch1 | sch2 | sch3 | sch4 |
| | $m$ $\xi$ | $m$ | $m$ $\xi$ | $m$ | $m$ $\omega$ | $m$ $n$ | $m$ $\omega$ | $m$ $n$ |
| 0.005 | 3  67 | 3 | 4  50 | 4 | 3  5 | 3  5 | 4  4 | 4  4 |
| 0.010 | 4  71 | 4 | 6  47 | 6 | 4  5 | 4  5 | 6  3 | 6  3 |
| 0.050 | 10  63 | 10 | 14  45 | 14 | 10  4 | 10  5 | 14  3 | 14  3 |
| 0.100 | 14  64 | 14 | 20  45 | 20 | 14  5 | 14  5 | 20  3 | 20  3 |
| 0.200 | 20  63 | 20 | 28  45 | 28 | 20  4 | 20  5 | 28  3 | 28  3 |
| 0.300 | 24  65 | 24 | 34  46 | 34 | 24  5 | 24  5 | 34  3 | 34  3 |
| 0.400 | 28  64 | 28 | 39  46 | 39 | 28  5 | 28  5 | 39  3 | 39  3 |
| 0.500 | 31  65 | 31 | 44  46 | 44 | 31  5 | 31  5 | 44  3 | 44  3 |

# 6. Conclusion

In this paper, we have presented a unified approach to improve the performance by unifying checkpointing, DVS and signatures for a real-time task, and have obtained the average execution times to complete a task for each scheme, using renewal equation. Further, we have discussed analytically the optimal numbers of checkpoints that minimize the average execution times. Numerical examples results shows that the proposed approach significantly reduce task average execution time.

## Acknowledgments

The author wishes to thank the anonymous referees for their careful reading of the manuscript.

## References

[1] Ziv A. Analysis of checkpointing schemes with task duplication, IEEE Trans. Computers, 1998,47(2):222-227

[2] Ying Z, Crishnendu C. Task feasibility analysis and dynamic voltage scaling in fault-tolerant real-time embedded systems, Proc. Of the design, automation and test in Europe conference and exhibition (DATE'04)

[3] Ying Z, Crishnendu C. Energy-Aware Adaptive Checkpointing in Embedded Real-Time Systems, Proc. of the design, automation and test in Europe conference and exhibition (DATE'03), 2003

[4] Ziv A, Bruck J. Performance Optimization of Checkpointing Schemes with Task Duplication IEEE Transactions on Computers, 1997, 46(2):1381-1386

[5] Nakagawa S, Fukumoto S, Ishii N. Optimal checkpoint interval for redundant error detection and masking systems , In Proc. First Euro-Japanese Workshop on Stochastic Risk Modeling for Finance, Insurance, Production and Reliability, 1998

[6] Sayori N, Satoshi F, Naohiro I. Optimal Checkpointing Intervals of Three Error Detection Schemes by a Double Modular Redundancy, Mathematical and Computer Modelling, 2003,38:1357-1363

[7] Melhem R, Mosse D, Elnozahy E. The interplay of power management and fault recovery in real-time systems, IEEE Tran. On computers, 2004, 53(2):217-231

[8] Intel Corp, speedstep, http://developer.inte.com/mobile/pentiumIII, 2003

[9] Fleischmann M. Crusoe power management: cutting x86 operating power through LongRun , Embedded processor forum, 2000, 6.

[10] Lee H, Shin H, Min S. Worst case timing requirement of real-time tasks with time redundancy, Proc. Real-Time Computing Systems and Applications, 1999: 410-414, 1999

[11] Hong I, Potkonjak M, Srivastava M. On-line scheduling of hard real-time tasks on variable voltage processor, Proc. Computer-Aided design (ICCAD'98), 1998:653-656

[12] Pummekkat S, Burns A, Davis R. analysis of checkpointing for Real-Time Systems, Time-Critical Computing Systems, 2001, 20:83-102

[13] Osaki S. Applied stochastic system modeling, Springer-Verlag, 1992

[14] Nakagawa S, Fukumoto S, Ishii N. Optimal checkpointing intervals for a double modular redundancy with signatures, Computers and mathematics with applications, 2003,46:1089-1094

Brill Academic Publishers
P.O. Box 9000, 2300 PA Leiden,
The Netherlands

*Lecture Series on Computer*
*and Computational Sciences*
Volume 4, 2005, pp. 330-333

# Open Mathematical Engine Interface and Its Application

W. Liao[1] , D. Lin[2] , P. Wang, Y. Wu[3]
Institute of Computational Mathematics
Kent State University
Kent, Ohio 44242, U.S.A.

Received 9 July, 2005; accepted in revised form 2 August, 2005

*Abstract:* The *Open Mathematical Engine Interface* (OMEI) aims to establish a uniform application programming interface (API) for heterogeneous mathematical computation engines. A common API such as OMEI can make mathematical engines easily accessible by front-ends, tools, and servers. It also enables the development of individual applications that can serve or utilize different engines. The motivation, application framework, specification, and usage scenarios for OMEI are described. The OMEI interface is applied to implement an API for ELIMINO, a symbolic system specializing in polynomial characteristic sets computations. With the OMEI API, ELIMINO becomes a compute engine that is compliant to the *Internet Accessible Mathematical Computation* (IAMC) framework. Thus, ELLIMINO gains the Dragonfly GUI for local and remote users as well as the ability to provide computational services over the Web/Internet.

*Keywords:* Mathematical Engine, Programming Interface, OMEI, IAMC

*Mathematics Subject Classification:* AMS-MOS 68N19, 68N30

*PACS:* 89.20.Ff, 89.20.Hh

## 1 Introduction

The *Open Mathematical Engine Interface* (OMEI) is an application programming interface (API) specification for mathematical compute engines. Following a standard interface specification such as OMEI, applications can access any compliant mathematical engines and vice versa. The concept of OMEI is similar to that of Microsoft's *Open Database Connectivity* (ODBC) which has become the standard for PCs and LANs. Be becoming OMEI compliant, distributed mathematical components, such as front ends, servers, tools can easily interoperate with mathematical engines.

The motivation for this work comes from several areas. First, it is evident that the demand to make mathematical computing accessible over the Internet is increasing. Cooperating with other institutions worldwide, the Institute for Computational Mathematics (ICM) at Kent State University initiated the *Internet Accessible Mathematical Computation* (IAMC) framework project [7, 8, 12, 13, 14] to provide an infrastructure for bringing mathematical computational and educational services over the Internet. The IAMC framework includes an IAMC client, an IAMC server, and a layered protocol model for connecting IAMC clients and servers effectively

---

[1]Department of Computer and Information Sciences, Shepherd University, Shepherdstown, WV 25443, USA. Email: wliao@shepherd.edu.
[2]SKLOIS, Institute of Software, Chinese Academy of Sciences, Beijing 100080, China.
[3]Department of Computer Science and Technology, Tsinghua University, Beijing, China

and efficiently over the Internet. The computation powers of an existing mathematical compute engine can be made available on the Web/Internet by an IAMC server or other similar programs. Consequently, it is essential to have an easy and systematic way for network servers to interface with compute engines.

The second motivation comes from *distributed mathematical computation* (DMC), an important research area in symbolic and numeric computation. The goal of DMC is to make mathematical computation accessible and interoperable remotely. The DMC architecture generally consists of a user interface, a programming interface, and a mathematical data encoding on top of the communication network/protocol layers. Contributions can be easily seen in the user interface (e.g. CAS/PI [6], SUI [2] and GI/S [16]), and the encoding (e.g. OpenMath [1], MathML [4], etc.) , and the protocol (e.g. MCP [13, 14] etc.) levels. Nevertheless, the programming interface area requires more investigation. With a well-defined application programming interface, distributed mathematical components, such as front ends, servers, and GUIs, can interoperate with different mathematical engines. OMEI is an effort in this direction.

The third motivation for OMEI comes from the development of new mathematical systems. Generally, a mathematical system contains two main parts: a computation kernel and a user interface. The same kernel can be served by a number of user interfaces designed for different end users—in industry, education, or scientific research, for example. Depending on its purpose, a user interface can be simple and straight-forward or sophisticated and complex. A standard such as OMEI can separate the development of mathematical engines from user interfaces. Therefore, the two can be developed independently and both will be usable with any OMEI compliant components.

The last but not least motivation is from main-stream applications that can use mathematical enhancements. For example, the word processing system Microsoft Word comes with an Equation Editor to embed mathematical expressions into documents. It would be nice to be able to also evaluate such expressions and to plot their graphs. OMEI compliant mathematical engines will be easier to integrate into such systems either as an embedded component or an external process.

APIs for specific mathematical engines, such as Math/Link [10] by Wolfram Research Inc. OpenMaple from MapleSoft, and Matlab External Interface [11] from MathWorks, exist. They are vendor/engine specific and programming language dependent. Nevertheless, these interfaces provide valuable input and excellent reference for the OMEI effort.

Several other efforts for distributed mathematical computation environment are also noted. JavaMath [5] is proposed as a standard Java API for client-server mathematical computation over Java and Java RMI. OpenXM provides a communication protocol and also a programming paradigm over its protocol layer. Different from JavaMath and OpenXM, OMEI is an abstract programming interface specification that works with any protocol and mathematical encoding. It is language, platform, communication protocol, and mathematical engine independent. Implementations of OMEI are written in specific programming languages for their target mathematical engines.

OMEI has been applied in the IAMC framework to build APIs for the general symbolic computation systems MAXIMA and Mathematica, making them IAMC compute engines. As a result, these systems are accessible through the IAMC front-end Dragonfly on the Internet. Here, we focus on describing the OMEI specification and its application for the ELIMINO symbolic computation system.

## 2   Why OMEI?

OMEI can help mathematical software developers achieve the following goals:

- **Engine-Application Interoperability**
  Compute engines following the OMEI specification will be interoperable with any user in-

terface, tools, and applications that employ the interface. Conversely, an application using OMEI to access one compute engine can access any other engine that is OMEI compliant. With help of the OMEI specification, compute engines, user interfaces, and applications can be developed separately and independently and still work together.

- **Network Accessibility**
  An OMEI compliant mathematical engine can readily interface to Starfish, an IAMC server prototype. This means the engine can be used from the Internet via the IAMC framework. The IAMC framework follows a 3-tier architecture that consists of IAMC client, IAMC server, and back-end compute engine. IAMC uses OMEI to connect IAMC servers, such as Starfish, to back end engines.

- **Integration of Heterogeneous Mathematical Systems**
  An application can access and integrate multiple engines by loading multiple OMEI drivers. An integrated compute engine can combine capabilities from other engines using OMEI. A parallel/distributed problem solving environment can be much easier to achieve over the OMEI programming paradigm.

- **3-tier/multi-tier mathematical systems**
  In general, OMEI can make building 3-tier/multi-tier distributed mathematical systems easier by standardizing the engine API.

## 3  Applications: Accessing ELIMINO/OMEI

Two approaches are available to access our computing environment built over OMEI and ELIMINO. As OMEI has been adopted in IAMC framework as the unified interface between IAMC server and compute engine, accessing ELIMINO/OMEI through an IAMC client would be a natural accessing approach. For end users, any generic IAMC client software can be used as an interactive utility to access the computational services from ELIMINO engine. For application developers, the MCP protocol library can be integrated into any application to send MCP requests to an IAMC server to obtain computational services programmatically.

As the OMEI interface is XML compliant, the OMEI/ELIMINO integrated environment can also be used as the service provider for any XML compliant application, programmatically or interactively. OMEI can be used to deploy a Web service engine and/or used in Grid computing environment.

## 4  Conclusion and Future Work

The OMEI effort investigates ways to establish a uniform API for mathematical compute engines and how such a uniform API can make distributed mathematical computation easier. We have applied the OMEI specification in ELIMINO, new computer-mathematics research system developed in China. The OMEI interface entitles several new features for ELIMINO, a new computer-mathematics research system, with minimal efforts.

Based on OMEI specification, we are currently working on Java Mathematical Engine Interface (JMEI), and are about ready to release its official specification and reference implementation.

## Acknowledgment

Work reported herein has been supported in part by the USA National Science Foundation under Grant CCR-9721343 and INT-9722919, and the Ohio Board of Regents Computer Science Enhancement Funds, the Shepherd University Professional Development Funds, the Chinese National 973 Project NKBRSF G1998030609, Chinese 863 Project 2001AA144030.

# References

[1] ABBOTT, J., DIAZ., A., AND SUTOR, R. S. *Report on OpenMath.* ACM SIGSAM Bulletin (Mar. 1996), 21-24.

[2] DOLEH, Y., AND WANG, P. S. *SUI: A System Independent User Interface for an Integrated Scientific Computing Environment.* In Proc. ISSAC 90 (Aug. 1990), Addison-Wesley (ISBN 0-201-54892-5), pp. 88-95.

[3] I. Foster, C. Kesselman, S. Tuecke, The Anatomy of the Grid: Enabling Scalable Virtual Organization, International J. Supercomputer Applications, 15(3), 2001

[4] ION, P., MINER, R., BUSWELL, S., S. DEVITT, A. D., POPPELIER, N., SMITH, B., SOIFFER, N., SUTOR, R., AND WATT,S. *Mathematical Markup Language (MathML) 1.0 Specification.* (www.w3.org/TR/1998/REC-MathML-19980407), Apr. 1998.

[5] JavaMath. *http://javamath.sourceforge.net/.*

[6] KAJLER, N. *CAS/PI: a Portable and Extensible Interface for Computer Algebra Systems.* Proceedings of ISSAC'92, ACM Press 1992

[7] LIAO, W. and WANG, P. S. *Building IAMC: A Layered Approach.* Proc. International Conference on Parallel and Distributed Processing Techniques and Applications (PDPTA'00). pp. 1509-1516.

[8] LIAO, W. and WANG, P. S. *Dragonfly: A Java-based IAMC Client Prototype.* Lecture Notes on Computing Vol.8. (Proceedings of ASCM 2000.) World Scientific Press, pp. 281-290.

[9] LIAO W., LIN D. and WANG P. S. *OMEI: Open Mathematical Engine Interface.* Proceedings of ASCM'2001, pp 83-91, Matsuyama, Japan, September 26-28, 2001. Lecture Notes Series on Computing Vol. 9, World Scientific.

[10] MathLink Home Page. http://www.wolfram.com/solutions/mathlink.

[11] MATLAB External Interface. http://www.mathworks.com/access/helpdesk/help/ techdoc/matlab.shtml.

[12] WANG, P. S. *Internet Accessible Mathematical Computation.* In the 3rd Asian Symp. on Computer Mathematics (ASCM'98), pp 1-13, Lanzhou Univ., China, 1998.

[13] WANG, P. S., GRAY, S., KAJLER N., LIN D., LIAO W. etc. *IAMC Architecture and Prototyping: A Progress Report.* Proceedings of ACM ISSAC'01, University of Western Ontario, London, Ontario, Canada, July 22-25, 2001.

[14] WANG, P. S. *Design and Protocol for Internet Accessible Mathematical Computation.* In Proc. ISSAC'99 (1999), ACM Press, pp. 291-298.

[15] Web Services Description Language (WSDL) 1.1. W3C Note. 15 March 2001. http://www.w3.org/TR/wsdl.

[16] YOUNG, D. and WANG, P. S. *GI/S: A Graphical User Interface for Symbolic Computation Systems.* In Journal of Symbolic Computation. Vol 4. No. 3. 1987. pp. 365-380.

Brill Academic Publishers
P.O. Box 9000, 2300 PA Leiden
The Netherlands

*Lecture Series on Computer
and Computational Sciences*
Volume 4, 2005, pp. 334-336

# Theoretical and experimental methods applied to development of new medicines. The arise of new era of drug delivery systems based upon cyclodextrins molecular structure

A.C. S. Lino[1,2] and C. Jaime[2]

[1]Departamento de Físico-Química, Instituto de Química, Universidade Estadual de Campinas
CP. 6154 CEP: 13083-340 Campinas SP Brasil
[2]Department de Quimica, Facultat de Ciències, Universidad Autonoma de Barcelona,
E 08193 Bellaterra (Barcelona) Spain

Received 20 August, 2005; accepted in revised form 20 August, 2005

*Abstract:* The interpretation of molecular interactions in both point of view Theoretical and Experimental is the objective of this session. To exemplify how molecular modeling can be a very interesting tool to the interpretation and comprehension of experimental data. The use of simultaneous analysis is one novel and multidisciplinary aspect that was showed as the future of some areas in pure and applied chemistry.
Especially in Cyclodextrins (CDs) chemistry and technology this methodology to be growth and growth. The use of Nuclear Magnetic Resonance (NMR) and Molecular Mechanics (MM) / Molecular Dynamics (MD) is described and applied to development of new medicines using the CDs technology. To improve the potential action of the drug in human body the molecular encapsulation with CDs was already knowledge as a promise changes in Pharmacology. These aspects are because the molecular geometry of CDs with a toroidal shape to form inclusion compounds with numerous molecules. Benzene ring dock sharp in the CDs cavity. Depends on the high of the molecule 1 or 2 CDs molecules can be fit as the 1:1 and 1:2 stoichometry. The CDs geometry has another interesting point. The internal protons $-H_3$ and $-H_5$ is like one anchor to prove the inclusion compounds formation. Experiments of diffusion coefficients can be used to prove the formation and the stoichometry of the inclusion compounds. NMR is the most important experiment to look for the inclusion compounds formation because have a lot of different experiments that can be applied, as for example, ROESY, NOESY, DOSY, 2D NMR, Job Plot, $H^1$ and $C^{13}$. The MM and MD calculations usually are in the perfect harmony with the NMR experiments.

*Keywords :Theoretical, Experimental, Cyclodextrins.*
*Mathematics SubjectClassification:* 49M
*PACS:* 90CXX

## Theoretical and experimental methods

First paper that was writing talk about one compound that after know to be one cyclodextrin (CDs) was made by Villier[1] in 1891. In these times Villier working with the enzymatic degradation of starch with *Baccilus Amylobacter* isolate one compound that first was called Celulosina. The reason was attributed to the similar chemical aspect between these novel compounds and Celulose as inert to oxidation reaction and with one pKa very strong. Probably Villier had one mixture of natives CDs in conditions that impossible to know what exactly was. More than a hundred years after the development of CDs chemistry growthing in exponential scale[2]. The possibility of applications of CDs in industrial scale to Pharmaceutical, Food, Cosmetic and New materials (nanomachines) to be done as one of the most interesting field of pure and applied chemistry. CDs are aligosacharides formed by glucose units linked by a -1,4 C-O-C bonds. The toroidal shape of CDs shows a hydrophilic (out) and hydrophobic (inside) chemical environment. This molecular geometry concerns to CDs the possibility to form stable inclusion compounds with several types of molecules. Most sophisticate part of CDs chemistry is the modify CDs. A lot of CDs are already knowledge and is possible to do others. In this session we are talk about natives and Metyled CDs[3], Isopropyl CDs[4], Large-Rings CDs[5], Carcerands CDs[6] and lypophilic CDs[7]. Nowadays the most important aspects of the CDs technology are the possibility to use:

- Solubilization of nonpolar drugs and on their parenteral formulations;
- Inhibition of side effects related to the toxicity of the drug, with decrease of rejection of the body and increase of the drug effective effect;

- Protection and stabilization of the drug against hydrolysis/oxidation reaction;
- Drug delivery, keeping fixed the concentration of a drug, and increasing its effect along time.

The diffusion coefficients measurements are very important in the CDs chemistry. Lino and Loh[8] using the Taylor-Aris technique showed the molecular encapsulation of the homologue series of p-hydroxy benzoates. With $C_1$ to $C_4$ (methyl, ethyl, propyl and butyl) was possible to distinguee the different stoichometry between butyl and the others. Was stabilised that only butyl p-hydroxy benzoate do inclusion compounds with -CD in one stoichometry 1:2 (one molecule of butyl p-hydroxy benzoate to two of -CD). In this work the question was why only butyl p-hydroxy benzoate forms 1:2 stoichometry? In a recent work, Lino *et. al*[9]. Discovery the answer. In this work was calculated by MM the stability of the inclusion compounds between -CD and the same homologue series of p-hydroxy benzoate and parallel another homologue series, the series of n-alkyl carboxylic acids with -CD. In a previous work Castruonovo *et. al.* dome micro calorimetric experiments with $C_1$ to $C_{13}$ and showed than $n_C = 9$ that H obtained growth to the double of numeric data. The curious is that the n-alkyl carboxylic acids with $n_C = 9$ have the same length of n-alkyl p-hydroxy benzoates with $n_C = 4$. With these simple calculations was stabilised the molecular ruler.

Methyled CDs is good studied. Differential of the native CDs structure pass from the formation of tosyl CDs intermediary and after that one attack with the nucleophilic group the same synthetic strategy to do Isopropyl CDs. These derivatives are interesting but less usual because the block of hydrogen bond in these cases made CDs less soluble that in some cases this is no good.

In these days we have a special attention to the Large-Rings CDs (LRCDs). LRCDs are CDs with a higher number of glucose molecules, 26, 30, 55, 70, 85 and 100. Experimentally the works of Professor UEDA in Hoshi University, Tokyo, Japan, is the most important in the world. The original idea is use the LRCDs in association with native CDs, for example the -CD to do one double inclusion compounds system host-host-guest. If one CD molecule can be see like one spaceship this system do one mother spaceship.

Carcerands CDs is the molecular container type of molecule done with two conjugated native CDs molecules linked from final –OX with derivatives with carbon atoms. This molecular structure is promise to do very stable inclusion compounds.

Lypophilic CDs are CDs based upon the interaction of one molecule, like one steroid, linked in the CDs structure anchored in the cellular membrane. This is very interesting because the promise applications of this kind of CDs are closer to the biochemistry. Is possible to if we can develop this system it is applicable to the cleaner the body and, otherwise, the cells from poisons of the real day life like legal and illegal drugs, tobacco, cholesterol, alcohol, etc., very useful for the society.

I believe that these projects point us in the frontier of chemistry. In a brief future we goanna be more and more stay looking the intimae interactions between molecules, interpretation real data about the true of molecular interactions. The multidisciplinary focus of chemistry can be useful in a large number of theoretical and experimental associations in the fields of pure and applied chemistry. In us case the final product of this strategy is one patent of one new drug with commercial interest. In these days Lino et al have 3 patents based upon only in natives CDs in the INPI Brazil and developing these same patents in the new CDs based structure new hosts. The first was the master of Leite[10] in the Departamento de Engenharia de Alimentos Unicamp Brasil. Using the addiction of ‥ and -CDs to iron II lactate Leite *et. al*[11]. was design a new potent agent for malnutrition and fortification with iron II. The question in this project was the iron II in water is quickly oxidation to iron III and the biodisponibility of iron II is very difficult. In the system of molecular encapsulation of ‥ and -CDs iron II shows a good stability in the time scale studied. Diffusion coefficients[12] were already confirms the formation of inclusion compounds between iron II and ‥ and -CDs. MM calculation were performed to evaluation of the stability of the inclusion compounds formed. HyperChem 6.03v[13] was used with the Forced Fields MM+, AMBER and OPLS in vacuum and with addition of 250 water molecules in one box with dimensions 20 x 20 x 20 Å. Was observed only the formation of 1:1 host-guest inclusion compounds formation in this case.

Another example is the formation of inclusion compounds between Local Anaesthetics (LA) with -CD[14]. Bupivacaine (BVC) and Lidocaine (LDC) were studied by experimental as well by theoretical methods. We use the Differential Scanning Calorimeter (DSC) for experimental measurements. MM calculation was performed to evaluation of the stability of the inclusion compounds formed. HyperChem 6.03v was used to design the appropriated start points calculated with the Forced Fields MM+, after that this primary structure was used to the MacroModel 7.0v[15] calculation with the Forced Field Operator Parameter Liquid Systems (OPLS) with General Bohr Surface Ares (GB/SA) system for solvent. Was used the dielectric constant of water. The OPLS GB/SA results shows that the difference between the enthalpy changes between 1:1 to 1:2 stoichometry is ca. ± 1 Kcal.mol$^{-1}$ for BVC and 10 Kcal.mol$^{-1}$ for the same association for LDC. These guest molecules have an interesting geometry. Two sides on one same molecule. First one aromatic side and another cyclohexanoid side. We denote the type of docking as "aromatic in" and "aromatic out" for the strategy of study. In the case of BVC the "aromatic out" have the most stable structure denote that the two possibilities coexist in a dynamic equilibrium. The both 1:2 stoichometries shows an annealing Tail-to-Tail that the primary –OHs groups of both two    -CDs molecules. This last result is interesting that the Head-to-Head annealing is the most stable to CDs alone[16]. Therefore, this result indicates a sensibility to this strategy to evaluate the intermolecular hydrogen bonds.

The last example that will show here is the inclusion compounds between Praziquentel and    -CD[17]. Praziquantel (PZQ) is a broadly effective anthelminthic drug available for human and veterinary use, being the drug of choice for the treatment of all forms of schistosomiasis. The aqueous solubility of PZQ is rather low. To improve it, without loss of the drug, we prepared complexes of PZQ included in    -CD as a controlled-release system. The inclusion complexes between PZQ/    -CD were studied at 1:1 and 1:2 stoichiometries. Molecular mechanics (MM) calculations were used to foresee the better stoichiometry for the complex as well as the possible orientations of PZQ inside the    -CD cavity. Supporting evidences for the 1:1 aromatic in tail complexation of PZQ in    -CD obtained both by MM through by 1H-NMR experiments (2D-ROESY).

## Acknowledgments

The author wishes to thank the anonymous referees for their careful reading of the manuscript and their fruitful comments and suggestions. A. C. S. L. a pos-doctoral fellowship (Proc. BEX1556/04-5) from CAPES.

## References

[1] A. C. Villier, *TR. Acad. Sci.*, 112, **1891**, 536.

[2] J Szejtli, *Proc. of 1$^{st}$ International Symposium on Cyclodextrins, Budapeste*, Kluwer Acad., **1981**.

[3] A. R. Khan, P. Forgo, K. J. Stine and V. D`Souza: Chem. Rev., 98, **1998**, 1977.

[4] Yiming Fine Chemicals Co. LtD. Benjin, China and CycloLab, Budapeste, Hungary.

[5] P. M. Ivanov and C. Jaime, *J. Phys. Chem. B*, 108, **2004**, 6261.

[6] M. Pons and O Millet *Progr in N M R Spectroscopy*, 38, **2001**, 267.

[7] M. Raoux, R Alzely-Velty, F Djedaine-Pilardi and B Perly, Biophical J., 82, **2002**, 813.

[8] A. C. S. Lino and W. Loh, *J Incl Phen and Macr Chem.*, 36, **2000**, 267.

[9] A. C. S. Lino, Y. Takahata and C. Jaime, *J. of Mol. Str.* THEOCHEM, 594, **2002**, 207.

[10] R A Leite and A C S Lino, Thesis of Master Degree, FEA, Unicamp, **1999**.

[11] R A Leite, A C S Lino and Y Takahata, J of Mol. Struc. THEOCHEM, 644(1-3), **2003**, 49.

[12] A C S Lino and W Loh, Thesis of Master Degree, IQ, Unicamp, **1997**.

[13] http://www.hyperchem.com

[14] L. M. A. Pinto, M. B. de Jesus, E. de Paula, A. C. S. Lino, J. B. Alderete, H. A. Duarte and Y. Takahata, *J. of Mol. Struc.* THEOCHEM, 678, **2004**, 63.

[15] http://www.schrodinger.com/maestro

[16] P. Bonnet, C. Jaime and L. Morrin-Allory, *J of Org. Chem.*, 66, **2001**, 689.

[17] M.B. de Jesus, L.F. Fraceto, L.M.A. Pinto, A.C.S. Lino, Y. Takahata, T. A. Pertinhez, E. de Paula. *Braz. J. Pharm. Sci.*, 39-2 ,**2003**, 254.

VSP International
Science Publishers
P.O. Box 346, 3700 AH Zeist
The Netherlands

*Lecture Series on Computer
and Computational Sciences*
Volume 4, 2004, pp. 337-339

# Monte Carlo Simulation of Hydrogen Adsorption in Single Walled Silicon Nanotubes

George P. Lithoxoos and Jannis Samios

Laboratory of Physical Chemistry,
Department of Chemistry,
University of Athens,
Panepistimiopolis 15771, Zographou Athens,
Greece

Received 18 July, 2005; accepted in revised form 10 August, 2005

*Abstract:* The Grand Canonical Monte Carlo simulation technique was employed to study the adsorption of hydrogen in hypothetical Single Walled Silicon Nanotubes (SWSiN) system of armchair structure. The results show that such systems might be appropriate for hydrogen storage.

*Keywords:* Monte Carlo, Carbon Nanotubes, Silicon Nanotubes, Hydrogen

*Mathematics Subject Classification:* 82O8, 08B80, 82D99

## 1. Introduction

After the discovering of carbon nanotubes by Iijima [1] in 1991 a great number of scientists showed special interest in these new nanostructured materials and tried to explore their unique properties. One of the main topics that has concentrated theoretical and experimental efforts [2] is that of hydrogen storage in nanotubes, because molecular hydrogen is considered to be one of the purest energy sources and its use could contribute to the solution of the energy and environmental pollution problem.

More recently, the scientific community showed increased interest in the use of silicon as a new material able to form nanostructures similar to carbon. In 2002 the successful synthesis of large scale silicon nanotubes took place [3] and showed that it is possible for silicon to exist in the form of nanotubes. Sha et al. [3] reported on the synthesis of silicon nanotubes by a Chemical Vapor Deposition (CVD) process and using a nanochannel $Al_2O_3$ (NCA) substrate. More recently, Jeong et al. [4] synthesized silicon nanotubes on porous alumina using molecular beam epitaxy. M.Zhang et al [5] used density functional theory calculations to study the structural and electronic properties of armchair and zigzag silicon nanotubes. They calculated the average binding energy of silicon atom at 3.311 eV, a much bigger value than that of crystal silicon. Their estimation is that efficient overlap of p orbitals and delocalization of $\pi$ electrons do take place in armchair SWSiN. However Zhang et al. pointed out that such an electronic structure does not happen in zigzag nanotubes and they are less stable than the armchair ones.

In the present study, we constructed a theoretical model of a hypothetical single walled silicon nanotube of armchair structure, based on Zhang's calculations. Our main purpose here was to study the hydrogen storage capacity of SWSiN and also the density profiles of hydrogen in the system by performing Grand Canonical Monte Carlo simulation.

## 2. Model system calculations.

We consider an ideal simulation model of single walled armchair silicon nanotubes based on ab initio calculations [3]. The simulation box is constructed by three rows each of them containing three tubes. So there exist 9 nanotubes arranged on a two-dimensional squared lattice. All tubes have equal diameters and heights and are obtained by rolling up a basal graphitic plane of 148 atoms, forming a cylinder consisting of 140 atoms. The bond length between neighboring silicon atoms is 2.245 A°, which is less than the bulk silicon bond length of 2.352 A°.

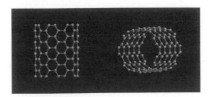

Figure 1. Single Walled Silicon Nanotube model used in simulations.

The site-site Lennard-Jones potential was used to describe the interactions between hydrogen molecules and also between hydrogen molecules and silicon atoms,

$$\upsilon_{LJ}(r) = 4\varepsilon_{H-H}\left[\left(\frac{\sigma_{H-H}}{r}\right)^{12} - \left(\frac{\sigma_{H-H}}{r}\right)^{6}\right]$$

with $\sigma_{HH}$ = 2.958 A°, $\varepsilon_{HH}/k_B$ = 36.7 K, $\sigma_{HSi}$ = 3.179 A° and $\varepsilon_{HSi}/k_B$ = 85.33 K. The quadrupolar moment of the hydrogen molecule is represented by localized point charges q = 0.466|e| on both the hydrogen atoms and a charge $-2q$ on the center of mass.

For the calculation of the chemical potential of the bulk hydrogen, we performed NVT Monte Carlo simulation using Widom's umbrella sampling method. Then, by performing Grand Canonical Monte Carlo simulation, we obtained the gravimetric and volumetric densities of the adsorbed hydrogen and also the density profiles of hydrogen inside and outside the silicon nanotubes. We compared the theoretical results with those of our previous simulations [6] considering carbon nanotubes of similar geometric characteristics and found that model silicon nanotubes adsorb more hydrogen molecules than carbon nanotubes.

Table 1. Gravimetric densities of hydrogen in SWSiN for the corresponding values of pressure and temperature.

| P(Mpa)/T(K) | 293.5 | 274 | 175 |
|---|---|---|---|
| wt% | SiN | SiN | SiN |
| 10 | 3.4 | 3.8 | 6.8 |
| 5 | 2.5 | 2.9 | 6.5 |
| 1 | 0.8 | 1.0 | 4.9 |
| 0.1 | 0.09 | 0.13 | 1.7 |

The hydrogen density profiles obtained from the simulation are presented in figure 2.

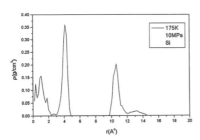 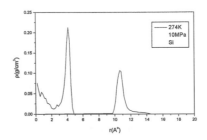

Figure 2. Density profiles of hydrogen in SWSiN for 175K (left) and 274K (right).

Langmuir isotherms were obtained by fitting the theoretical function to the simulation results and are presented in figure3.

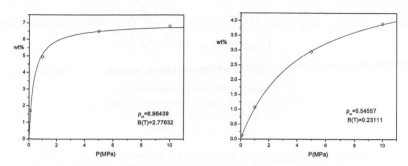

Figure 3. Langmuir isotherms at 175K (left) and 274K (right) of the adsorbed hydrogen in SWSiN.

From the gravimetric densities obtained, we may observe that there is a significant amount of hydrogen adsorbed in silicon nanotubes. It also seems that the international standards for hydrogen adsorption [7] are satisfied at some thermodynamic states. We think that these results are of particular importance for the scientific community and generally for the research on the field of hydrogen storage in nanomaterials.

## References

[1] S. Iijima, *Nature* **354** 56 (1991)

[2] Darkrim F.L., Malbrunot P. and Tartaglia G.P., *Int. J. Hydrogen Energy*, **27**, 193, (2002)

[3] Jian Sha, Junjie Niu, Xiangyang Ma, Jin Xu, Xiaobing Zhang, Qin Yang, Deren Yang *Advanced Materials* **14** 17 1219 (2002)

[4] Seung Yol Jeong, Jae Yon Kim, Hyun Duk Yang, Bin Nal Yoon, Suk-Ho Choi, Hee Kwang Kang, Cheol Woong Yang, Young Hee Lee, *Advanced Materials* **15** 14 1172 (2003).

[5] M. Zhang, Y.H Kan, Q.J. Zang, Z.M. Su, R.S. Wang *Chemical Physics Letters* **379** (2003) 81

[6] George P Lithoxoos and Jannis Samios, *Lecture Series on Computer and Computational Sciences* (VSP International), **1**, 314 (2004)

[7] Scott Hynek, Ware Fuller, Jeffrey Bentley, *Int. J. Hydrogen Energy* **22** 6 601 (1997)

Brill Academic Publishers
P.O. Box 9000, 2300 PA Leiden,
The Netherlands

*Lecture Series on Computer
and Computational Sciences*
Volume 4, 2005, pp. 340-343

# Algorithms for the Construction of the Semi-analytical Planetary Theories

**J.A.López[1]; M. Barreda[2]; J. Artés[3]**

Departamento de Matemáticas.
Escuela Superior de Tecnología y Ciencias Experimentales,
University Jaume I of Castellon,
12071 Castellón, Spain

Received 5 August, 2005; accepted in revised form 17 August, 2005

*Abstract:* The aim of this paper is the construction of package of software necessary for the construction of the semi-analytical planetary theories.
This package is a set of computational tools in celestial mechanics. The main parts of this package are, in first place, a set of methods for the expansion of the inverse of the distance between two planets according to trues, eccentrics, elliptic and mean anomalies; second an iterative algorithm to integrate the Lagrange planetary equations, appropriate to use several kinds of temporal variables in the Fourier ( or Poisson) series expansions, has been developed; and in third place, a Poisson series processor necessary for the construction of the solutions of order second or higher order.

*Keywords:* Celestial Mechanics. Planetary Theories. Algorithms. Orbital Mechanics. Perturbation Theory. Computational Algebra.

*Mathematics Subject Classification:* 70F05, 70F10, 70F15,70M20

*PACS:* 95.10.C, 45.50.P, 96.35.F

## 1   Introduction

One of main problems in celestial mechanics is the construction of the theories of the perturbed motion of the celestial bodies. The main methods used are based in the the perturbation theory. The initial approximation is the two bodies problem $(Sun - Planet_i)$, it is a well known problem, and their solution is completely defined by means the orbital elements $\vec{\sigma}_i$.
The perturbed motion solution can be arranged ussing the Lagrange planetary equations

$$\dot{\vec{\sigma}} = \vec{f}(\vec{\sigma}) = L\frac{\partial R}{\partial \vec{\sigma}} \tag{1}$$

---

[1]Corresponding author. E-mail: lopez@mat.uji.es
[2]E-mail: barreda@mat.uji.es
[3]E-mail: artes@mat.uji.es

where L is a $6 \times 6$ matrix and $R$ is the disturbing potential defined as Levallois [5]:

$$R = \sum_{k=1}^{N} Gm_k \left[ \left( \frac{1}{\Delta_k} \right) + \frac{xx_k * yy_k * zz_k}{r_k^3} \right] \tag{2}$$

where $\Delta_k$ is the distance between the secondary and disturbing body and $r_k$ the distance between the primary and the disturbing body.

Let $\vec{\sigma}_i$ be the solution of the unperturbed two bodies problem for the couple Sun $body_i$. Let $\delta_1 \vec{\sigma}_i$ be the solution of the equation $\delta_1 \vec{\sigma}_i = \vec{f}(\vec{\sigma}_i)$. The quantities $\delta_1 \vec{\sigma}_i$ are of first order in the masses $m_1, .., m_N$, and from them we have in second order order of perturbation the relations

$$\delta_2 \vec{\sigma}_i = \vec{f}_i(\vec{\sigma}_1 + \delta_1 \vec{\sigma}_1, \cdots, \vec{\sigma}_N + \delta_1 \vec{\sigma}_N) \tag{3}$$

and in general

$$\delta_{l+1} \vec{\sigma}_i = \vec{f}_i(\vec{\sigma}_1 + \delta_1 \vec{\sigma}_1 + \cdots + \delta_2 \vec{\sigma}_l, \cdots, \vec{\sigma}_N + \delta_1 \vec{\sigma}_N + \cdots + \delta_N \vec{\sigma}_N) \tag{4}$$

The quantities $\delta_l \vec{\sigma}_i$ are of $l$ order in masses, so to integrate (2) we proceed expanding the second member of the equation (2)until $l$ order in $m_1, .., m_N$. Analytical methods to expand these functions can be obtained from Brower & Clemence [1], Hagihara [3], Simon[7], Tisserand[8], etc.

To evaluate the second member of the equation (1) we can also proceed by mean of semi-analytical expansion (Chapront [2], Kovalewsky [4], López [6]), and for them it is necessary to calculate the inverse of the distance between the disturbing and disturbed bodies as a Fourier series of a appropriate temporal variables.

## 2 Expansion of the inverse of the distance

Let $\vec{r}$, $\vec{r'}$ be the radii vector of the secondary and the disturbing bodies, and let $S$ be the angle between the vectors $\vec{r}$, $\vec{r'}$. The distance between the secondary an the disturbing bodies is given as

$$\Delta^2 = r + r'^2 - 2rr' \cos S = \sum_{j_1, j_2} A_{j_1, j_2} \cos(j_1 \Psi_1 + j_2 \Psi_2 + \varphi_{j_1, j_2}) \tag{5}$$

where $\Psi_1, \Psi_2$ represent appropriate temporal variables (mean anomaly, eccentric anomaly, true anomaly, or elliptic anomaly).

To expand the inverse of the distance, we use the iterative Kovalewsky Algorithm

$$\left( \frac{1}{\Delta} \right)_{k+1} = \frac{3}{2} \left( \frac{1}{\Delta} \right)_k - \frac{1}{2} \left( \frac{1}{\Delta} \right)_k^3 \Delta^2 \tag{6}$$

where $k$ represent the iteration number. An appropriate first approximation (Tisserand [8]) can be taken as

$$\left( \frac{1}{\Delta} \right)_0 = \frac{1}{a'} \left[ b_{1/2}^{(0)} + \sum_{j=1}^{\infty} b_{1/2}^{(j)}(\alpha) \cos jS \right] \tag{7}$$

where $b_s^{(j)}$ are the Laplace coefficients.

## 3    Integration Method

To integrate the Lagrange Planetary equations, we proceed by means of perturbation method. For a generic element $\sigma$ we have in first order the differential equation

$$\dot{\sigma} = \sum_{j_1,j_2} C_{j_1,j_2} \cos(j_1\Psi_1 + j_2\Psi_2 + \varphi_{j_1,j_2}) \qquad (8)$$

Taking the mean anomalies $M_1, M_2$ as temporal variables the integration is immediate. To use another temporal variables, the integration procedure can be modified, for this purpose we can use the algorithm described in López [6]

$$\int_{t_0}^t \cos(j_1\Psi_1 + j_2\Psi_2 + \varphi_{j_1,j_2})dt = \left[\frac{\sin(j_1\Psi_1 + j_2\Psi_2 + \varphi_{j_1,j_2})}{j_1 n_1 + j_2 n_2}\right]_{t_0}^t +$$

$$+ \int_{t_0}^t \cos(j_1\Psi_1 + j_2\Psi_2 + \varphi_{j_1,j_2}) \left[g^{(1)} + g^{(2)}\right] dt \qquad (9)$$

where $n_1$, $n_2$ are the mean motion of bodies.
The extension of this method to integrate the second, third and higher orders of perturbation is direct by using the integration by parts formula.

## 4    Poisson Series Processor

To obtain the differential equation in second and higher order of perturbation, it is necessary the management of Poisson series. For a generic element $\sigma$ we have

$$\sigma(t) = \sigma_0 + \sum_{j_1,j_2} C_{j_1,j_2}(t) \cos(j_1\Psi_1 + j_2\Psi_2 + \varphi_{j_1,j_2}) \qquad (10)$$

where $C_{j_1,j_2}(t)$ are polynomials in $t$.
Usually, the quantities $\sigma(t) - \sigma_0$ are little quantities, so we can approximate a generic function $f(\sigma(t))$ until order $k$ in $m_1, .., m_N$ by means Taylor expansion.
The Poisson processor is a package of software for the management of binomial expansions, elementary function (SIN, COS, EXP,..) expansions, Taylor expansions, series inversion, etc.
The implementation of these procedures in C, Fortran or MATHEMATICA languages presents no difficulties.

## 5    Concluding Remarks

The package of software developed is appropriate to development of the semi-analytical planetary theories using as temporal variables the mean anomalies and other anomalies.
The use of true or elliptic anomalies as temporal variables allows the use of shorter developments than if other anomalies are used.
The integration process can de automated for the Fourier and for the Poisson series by using an appropriated algorithm.

## Acknowledgments

This work has been partially supported by Generalitat Valenciana 05I197.01/1 and Fundación Bancaja 03I284.01/1 grants.

## References

[1] D. BROWER, G. M. CLEMENCE, *Celestial Mechanics*, Ed Academic Press, New York, 1965.

[2] J. CHAPRONT, P. BRETAGNON, M. MEHL, "Une Formulaire pour le calcul des perturbations d'ordres élevés dans les problemès planétaires", *Celestial Mechanics*. **11** (1975) 379–399

[3] Y. HAGIHARA, *Celestial Mechanics*, Ed MIT Press, Cambridge MA, 1970.

[4] J. KOVALEWSKY, *Introduction to Celestial Mechanics*, Ed D. Reidel Publishing Company, DoDrecht-Holland, 1967.

[5] L.L LEVALLOIS, J. KOVALEWSKY, *Geodesie Generale Vol 4*, Ed Eyrolles, Paris, 1971.

[6] J.A.LÓPEZ, M. BARREDA, "A Formulation to Obtain Semi-analytical Planetary Theories Using True Anomalies as Temporal Variable", *proceedings of the International Conference on Computational and Mathematical Methods in Science ans Engineering, CMMSE-2005 Alicante, June 2005, pp. 319-328*

[7] J.L. SIMON, "Computation of the first and second derivatives of the Lagrange equations by harmonic analysis", *Astron&Astrophys.* **17** (1982) 661–692.

[8] F. F. TISSERAND, *Traité de Mecanique Celeste*, Ed Gauthier-Villars, Paris, 1896.

Brill Academic Publishers
P.O. Box 9000, 2300 PA Leiden,
The Netherlands

*Lecture Series on Computer*
*and Computational Sciences*
Volume 4, 2005, pp. 344-347

# Finite Element Methods for Self-adjoint Singular Perturbation Problems

Jean M.-S. Lubuma[1] and Kailash C. Patidar[2]

Department of Mathematics and Applied Mathematics
University of Pretoria, Pretoria 0002
South Africa

Received 24 June, 2005; accepted in revised form 13 July, 2005

We consider the self-adjoint singularly perturbed two-point boundary-value problem

$$-\varepsilon \left(p(x)y'\right)' + q(x)y = f(x) \quad \text{on} \quad (0,1); \; y(0) = 0, \; y(1) = 0, \tag{1}$$

where $\varepsilon > 0$ is a small parameter while $f$, $p$ and $q$ are sufficiently smooth functions satisfying the positivity conditions

$$\widetilde{p} \geq p(x) \geq p^* > 0 \quad \text{and} \quad \widetilde{q} \geq q(x) \geq q^* > 0. \tag{2}$$

Due to condition (2) and to the Lax-Milgram lemma, the problem (1) has a unique variational solution $y_\varepsilon \equiv y \in H_0^1 \equiv H_0^1(0,1)$ such that

$$a\left(y, v\right) = \left(f, v\right), \quad \forall\, v \in H_0^1. \tag{3}$$

where the bilinear and the linear forms are given by

$$a\left(y, v\right) := \int_0^1 \left(\varepsilon p(x)y'(x)v'(x) + q(x)y(x)v(x)\right) dx$$

and

$$\left(f, v\right) := \int_0^1 f(x)v(x)dx.$$

Consider a uniform partition $0 = x_0 < x_1 < \ldots < x_{n-1} < x_n = 1$ and denote by $h$ the mesh width of a single element, i.e., $h = x_{j+1} - x_j = 1/n$, for all $j = 0(1)n - 1$. Let $V_h \subset H_0^1$ be the classical finite element space defined by

$$V_h = \left\{ y_h \in C[0,1], \; y_h|_{[x_i, x_{i+1}]} \text{ is linear }, \; y_h(0) = y_h(1) - 0 \right\}.$$

The finite element method (FEM) for the problem (3) reads as follows: seek $y_h \in V_h$ such that

$$a\left(y_h, v_h\right) = \left(f, v_h\right), \quad \forall\, v_h \in V_h. \tag{4}$$

By Céa's lemma

$$\|y - y_h\|_a^2 = \inf_{v_h \in V_h} \|y - v_h\|_a^2 \tag{5}$$

---

[1]Corresponding author. E-mail: jean.lubuma@up.ac.za, Fax.: +27-12-4203893.
[2]E-mail: kailash.patidar@up.ac.za.

and from the interpolation theory, we have

$$\|y - y_h\|_a \leq Ch/\varepsilon, \tag{6}$$

where

$$\|v\|_a := \sqrt{\int_0^1 (\varepsilon p(x)(v'(x))^2 + q(x)(v(x))^2)\, dx}$$

is the energy norm and $C$ denotes here and after various constants independent of $h$ and $\varepsilon$. The book [2] is our standard reference for the finite element method.

The estimate (6) shows that the classical FEM (4) is not reliable in solving the above singular perturbation problem since the parameters $h$ and $\varepsilon$ cannot vary independently. Thus, we need to design more suitable finite element methods.

Our analysis is based on the following theorem on the singular behavior of the solution $y$ ([1]):

**Theorem 1** *Assume that $f \in C^0[0,1]$ and $q \in C^2[0,1]$. Then the solution $y$ admits the decomposition*

$$y(x) = \gamma_0 S_0(x) + \gamma_1 S_1(x) + g(x)$$

*where: $\gamma_0$ and $\gamma_1$ are real numbers; $g$ is the regular part which satisfies the relation*

$$\left| g^{(k)}(x) \right| \leq C\left(1 + \varepsilon^{1-\frac{k}{2}}\right), \quad \text{for} \quad k = 0, 1, 2;$$

$$S_0(x) = \exp\left(-x\sqrt{\frac{q(0)}{\varepsilon}}\right) - 1 + x - x\exp\left(-\sqrt{\frac{q(0)}{\varepsilon}}\right)$$

*and*

$$S_1(x) = \exp\left(-(1-x)\sqrt{\frac{q(1)}{\varepsilon}}\right) - x - (1-x)\exp\left(-\sqrt{\frac{q(1)}{\varepsilon}}\right)$$

*are the singular functions in the sense that*

$$\sup_{0 < \varepsilon \leq 1} \sup_{x \in [0,1]} \left| S_i^{(k)}(x) \right| = \infty \text{ for } k = 1, 2 \text{ and } i = 0, 1.$$

In order to capture the singular nature of the solution $y$ into the numerical method, we follow an idea which was first introduced by Strang and Fix [3] in the context of boundary-value problems with corner singularities. More precisely, we replace the space $V_h$ by the augmented finite dimensional subspace of $H_0^1$ defined by

$$V_h^+ = V_h \oplus span\{S_0, S_1\}.$$

The singular function method (SFM) reads as follows: find $y_h^+ \in V_h^+$ such that

$$a\left(y_h^+, v_h^+\right) = \left(f, v_h^+\right) \quad \forall\, v_h^+ \in V_h^+.$$

Using (5) and interpolation theory, we have the following result:

**Theorem 2** *The SFM is $\varepsilon$-uniformly convergent of order one and two in the energy and $L^2$-norms, respectively. That is*

$$\|y - y_h^+\|_a \leq Ch \text{ and } \|y - y_h^+\|_{L^2} \leq Ch^2.$$

Our second modification of the FEM is the Mesh Refinement Method (MRM). Its point of departure is (5), which yields the series of relations

$$\|y - y_h\|_a^2 \leq \|y - \Pi_h y\|_a^2$$

$$\leq C \sum_{i=0}^{n-1} \left( \varepsilon |y - \Pi_i y|_{H^1(x_i, x_{i+1})}^2 + \|y - \Pi_i y\|_{L^2(x_i, x_{i+1})}^2 \right)$$

where $\Pi_h$ is the $V_h$-global interpolation operator and $\Pi_i$ is the local interpolation operator on the space of polynomials of degree $\leq 1$ restricted to the interval $[x_i, x_{i+1}]$. Thus, estimating the error $\|y - y_h\|_a^2$ is reduced to estimating each local interpolation error $\varepsilon |y - \Pi_i y|_{H^1(x_i, x_{i+1})}^2$ and $\|y - \Pi_i y\|_{L^2(x_i, x_{i+1})}^2$. It is convenient to deal with the end point $x_0 = 0$, the situation of the node $x_n = 1$ being dealt with by symmetry. We consider the following two cases:

**First case:** $i \neq 0$. We are far away from the layer region $[0, x_1]$. Using the decomposition in Theorem 1 and interpolation theory, we have the estimates

$$\|y - \Pi_i y\|_{L^2(x_i, x_{i+1})}^2 \leq C(x_{i+1} - x_i)^4 \left[ \||g''|\|_{L^2}^2 + x_i^{-3} \right]$$

and

$$\varepsilon |y - \Pi_i y|_{H^1(x_i, x_{i+1})}^2 \leq C(x_{i+1} - x_i)^2 \left[ \||g''|\|_{L^2}^2 + \varepsilon x_i^{-3} \right].$$

**Second case:** $i = 0$. We are in the layer region $[0, x_1]$ and we obtain, as above, the estimates

$$\|y - \Pi_0 y\|_{L^2(0, x_1)}^2 \leq C x_1^4 \left[ \||g''|\|_{L^2}^2 + \varepsilon^{-3/2} \right]$$

and

$$\varepsilon |y - \Pi_0 y|_{H^1(0, x_1)}^2 \leq C x_1^2 \left[ \||g''|\|_{L^2}^2 + \varepsilon^{-1/2} \right].$$

On the nodes $(x_i)$, we impose the mesh refinement condition

$$(x_{i+1} - x_i)^2 \varepsilon x_i^{-3} \leq C h^2 \tag{7}$$

for $i \neq 0$, whereas we require

$$x_1^2 \varepsilon^{-3/2} \leq C h^2 \tag{8}$$

whenever $i = 0$.

**Theorem 3** *If the finite element space $V_h$ is constructed according to the mesh refinement conditions (7) and (8), then the FEM (4) is $\varepsilon$-uniformly convergent of order one and two in the energy and $L^2$-norms, respectively:*

$$\|y - y_h\|_a \leq C h \quad \text{and} \quad \|y - y_h\|_{L^2} \leq C h^2.$$

**Remark 4** If we come back to the initial situation where both end-points $x_0 = 0$ and $x_n = 1$ must be incorporated, a typical example that meets the mesh refinement requirement consists of the nodes defined by

$$x_i := \begin{cases} \frac{1}{2}(\frac{i}{n})^\nu & \text{if } 0 \leq i \leq n \\ 1 - \frac{1}{2}(\frac{2n-i}{n})^\nu & \text{if } n \leq i \leq 2n \end{cases} \quad \text{for a suitable } \nu \geq 1.$$

This example will be exploited in the numerical experiment.

**Remark 5** The SFM has a uniform mesh like the standard FEM. However, the involved stiffness matrix is no longer tridiagonal. The stiffness matrix in the MRM is tridiagonal, though the mesh is not uniform. The reliability of the two methods will also be demonstrated on numerical simulations.

# References

[1] E.P. Doolan, J.J.H. Miller and W.H.A. Schilders, *Uniform Numerical Methods for Problems with Initial and Boundary Layers*, Boole Press, Dublin, 1980.

[2] P.G. Ciarlet, *The Finite Element Method for Elliptic Problems*, North-Holland, Amsterdam, 1978.

[3] G. Strang and G.J. Fix, *An Analysis of the Finite Element Method*, Prentice-Hall, Englewood Cliffs, N.J., 1973.

Brill Academic Publishers
P.O. Box 9000, 2300 PA Leiden
The Netherlands

*Lecture Series on Computer*
*and Computational Sciences*
Volume 4, 2005, pp. 348-352

# Fast Searching and Blind Detecting LSB Steganographic Images on the Internet

Luo Xiangyang[1]  Luo Junyong  Liu Fenlin

Institute of Information Engineering, Information Engineering University.
Zhengzhou, Henan, China

Received 9 July, 2005; accepted in revised form 3 August, 2005

*Abstract:* This thesis discusses the problem of searching and detecting LSB (Least Significant Bits) steganographic images on the Internet, and a full scheme on fast images searching and blind steganalysis is investigated. In this scheme, doubtful images will be detected with RS, SPA and LSM method and associated decision will be given depending on the results of all these three methods. By adopting heuristic method, once the detector captures an image, same images in searching spaces will be found quickly. The simulation results show that the scheme is effective reliable.
*Keywords:* Internet; LSB Steganography; Steganalysis; Image search; Associated decision

## 1. Introduction

As a new art of covert communication, the main purpose of steganography is to convey messages secretly by concealing the very existence of messages under digital media files, such as images, audio, or video files. A popular digital steganography technique is so-called least significant bit (LSB) embedding. Because LSB steganography has some advantages such as conceal capability is excellent and its realization is easy, it has been used widely in the Internet. The research of steganography fall into two parts: steganography and steganalysis. Steganography is to research embedding message into host-object, and steganalysis is to research detecting, extracting and breaking down the message. Generally, If there exists an algorithm that can guess whether or not a given image contains a secret message with a success rate better than random guessing, the steganographic system is considered broken (see [1]). Recently, there are a lot of methods for detection of LSB steganography.

Fridrich et. al. [2] presented a powerful RS method (regular and singular groups method) for detection of LSB embedding that utilizes sensitive dual statistics derived from spatial correlations in images. This method counts the numbers of the regular group and the singular one respectively, and constructs a quadratic equation. The length of message embedded in image is then estimated by solving the equation. Sorina Dumitrescu et. al. [3] proposed SPA, a method to detect LSB steganography via sample pair analysis. When the embedding ratio is more than 3%, this method can estimate it with relatively high precision, the probability of error of average estimated value is 0.023. In some sense, RS and SPA are the two most reliable detectors of thinly-spread LSB steganography until now [4].

Being enlightened by SPA steganalysis method, we develop a LSM (the Least Square Method) algorithm. This algorithm constructs a cubic equation via the least square method, and estimates the message embedding ratio of the steg-image blindly. Experimental results show that this algorithm has less probability of false alarm and less miss probability than SPA.

A full scheme on searching and detecting LSB steganographic images on the Internet is also investigated in this thesis. In the scheme, doubtful images on the Internet are searched with multithread parallel processing technology firstly, irrespective images are ridded off via filtration measure, and doubtful images link database is built, then detect doubtful images with RS, SPA and LSM method respectively. Associated decision is made based on results gained by these three methods finally, and by adopting heuristic method, once the detector captures an image, same images in searching spaces can be found quickly. Experimental results show that the scheme in this thesis is effective and reliable for LSB steganographic images searching and detecting on the Internet.

---

[1] Corresponding author, e-mail: xiangyangluo@126.com

## 2. Steganalysis Algorithm

### A. RS Steganalysis

Let $C$ be the test image, which has $M \times N$ pixels and with pixel values from the set $V$. For example, for an 8-bit grayscale image, $V = \{0,1,\cdots,255\}$. Then divide $C$ into disjoint groups $G$ of $n$ adjacent pixels $G = (v_1,\cdots,v_n) \in C$. The discrimination function is defined as follows:

$$f(v_1, v_2, \cdots, v_n) = \sum_{i=1}^{n-1} |v_{i+1} - v_i| \qquad (1)$$

Generally, the noisier the group $G = (v_1,\cdots,v_n)$ is, the larger the value of the discrimination function becomes. Similarly, the invertible operation $F$ on $x$ called flipping is also defined.

$$F_1 : 0 \leftrightarrow 1, 2 \leftrightarrow 3, \cdots, 254 \leftrightarrow 255,$$

$$F_{-1} : -1 \leftrightarrow 0, 1 \leftrightarrow 2, \cdots, 253 \leftrightarrow 254, 255 \leftrightarrow 256 \qquad (2)$$

$$F_{-1}(v) = F_1(v+1) - 1 \quad \text{for all } v$$

Then the group $G$ is dependent on one of the three types of pixel group. Regular groups: $G \in R$ $f(F(G)) > f(G)$ ; Singular groups: $G \in S$ $f(F(G)) < f(G)$ ; Unusable groups: $G \in U$ $f(F(G)) = f(G)$

For any mask $M$, $F_M(F_{M(1)}(v_1), F_{M(2)}(v_2), \cdots, F_{M(n)}(v_n))$ is also dependent on one of types in the $R$, $S$ and $U$. Fridrich experimentally verified the following two statistical assumptions with a large database of images.

$$R_M \cong R_{-M} \text{ and } S_M \cong S_{-M} \qquad (3)$$

$$R_M(1/2) = S_M(1/2) \qquad (4)$$

The assumptions (3) and (4) make it possible to derive a ratio-estimate equation (5) to calculate the embedding ratio $p$

$$2(d_1 + d_0)z^2 + (d_{-0} - d_{-1} - d_1 - 3d_0)z + d_0 - d_{-0} = 0 \qquad (5)$$

where,

$$d_0 = R_M(p/2) - S_M(p/2), \quad d_1 = R_M(1 - p/2) - S_M(1 - p/2),$$

$$d_{-0} = R_{-M}(p/2) - S_{-M}(p/2), \quad d_{-1} = R_{-M}(1 - p/2) - S_{-M}(1 - p/2),$$

$p$ is calculated from the root $z$ with a smaller absolute value by $p = z/(z - 1/2)$. $\qquad (6)$

### B. SPA Steganalysis

SPA steganalysis is based on a finite state machine whose states are selected multisets of sample pairs called *trace multisets*. Some of the trace multisets are equal in their expected cardinalities if the sample pairs are drawn from a digitized continuous signal. Random LSB flipping causes transitions between these trace multisets with given probabilities and consequently alters the statistical relations between the cardinalities of trace multisets.

Assuming that the digital signal is represented by the succession of samples $s_1, s_2, \cdots, s_N$ (the index represents the location of a sample in a discrete waveform), a sample pair means a two-tuple $(s_i, s_j), 1 \leq i, j \leq N$. Let $P$ be a set of sample pairs drawn from a digitized continuous signal. Regard $P$ as a multiset of two-tuple $(u, v)$, where $u$ and $v$ are the values of two samples. Denote by $D_n$ the submultiset of $P$ that consists of sample pairs of the form $(u, u+n)$ or $(u+n, u)$, where $n$ is a fixed integer, $0 \leq n \leq 2^b - 1$, and $b$ is the number of bits to represent each sample value. For each integer $m$, $0 \leq m \leq 2^{b-1} - 1$, denote by $C_m$ the submultiset of $P$, we denote $C_m$

$$C_m = \left\{ (u, v) \middle\| \left\lfloor \frac{u}{2} \right\rfloor - \left\lfloor \frac{v}{2} \right\rfloor \middle| = m, 0 \leq u \leq 2^b - 1, 0 \leq v \leq 2^b - 1 \right\} \qquad (7)$$

where $\lfloor x \rfloor$ represents the maximal integer of the integers which isn't larger than $x$

It is clear that the multisets $D_n$ form a partition of $P$, the multisets $C_m$ form another partition of $P$, and $D_{2m}$ is contained in $C_m$. we partition $D_{2m+1}$ into two submultisets $X_{2m+1}$ and $Y_{2m+1}$, where $X_{2m+1} = D_{2m+1} \cap C_{m+1}$ and $Y_{2m+1} = D_{2m+1} \cap C_m$, for $0 \leq m \leq 2^{b-1} - 2$, and $X_{2^b - 1} = \Phi$, $Y_{2^b - 1} = D_{2^b - 1}$. $X_{2m+1}$ and $Y_{2m+1}$ contain pairs $(u, v)$ where $|u - v| = 2m + 1$, and those pairs in which the even component is larger are in $X_{2m+1}$, whereas those pairs in which the odd component is larger are in $Y_{2m+1}$.

Based on the mutual transition relation among these multisets and the presupposition $E\left\{\left|\bigcup_{m=i}^{j}X_{2m+1}\right|\right\}=$

$E\left\{\left|\bigcup_{m=i}^{j}Y_{2m+1}\right|\right\}$, paper [3] establishes quadratic equation:

$$\frac{(|C_i|-|C_{j+1}|)p^2}{4}-\frac{(|D'_{2i}|-|D'_{2j+2}|+2\sum_{m=i}^{j}(|Y'_{2m+1}|-|X'_{2m+1}|))p}{2}+\sum_{m=i}^{j}(|Y'_{2m+1}|-|X'_{2m+1}|)=0, \qquad i\geq 1 \qquad (8)$$

and

$$\frac{(2|C_0|-|C_{j+1}|)p^2}{4}-\frac{(2|D_0|-|D'_{2j+2}|+2\sum_{m=0}^{j}(|Y'_{2m+1}|-|X'_{2m+1}|))p}{2}+\sum_{m=0}^{j}(|Y'_{2m+1}|-|X'_{2m+1}|)=0, \qquad i=0 \quad (9)$$

where all parameters except $p$ can be gained by statistical approach, and $0\leq i\leq j\leq 2^{b-1}-1$, empirical values are given in [3], $i=0, j=30$. The smaller absolute value of root of quadratic equation (8) [or (9)] is the estimated length of hidden message.

### C. LSM Steganalysis

In practice, $E\{X_{2m+1}|\}$ isn't absolutely equal to $E\{Y_{2m+1}|\}$, and $E\left\{\left|\bigcup_{m=i}^{j}X_{2m+1}\right|\right\}$ isn't absolutely equal to

$E\left\{\left|\bigcup_{m=i}^{j}Y_{2m+1}\right|\right\}$. Denote $\varepsilon_m=|Y_{2m+1}|-|X_{2m+1}|$, where $\varepsilon_m$ is less than $|X_{2m+1}|$ and $|Y_{2m+1}|$. Then, obtain the following equations:

$$\frac{(|C_m|-|C_{m+1}|)p^2}{4}-\frac{(|D'_{2m}|-|D'_{2m+2}|+2|Y'_{2m+1}|-2|X'_{2m+1}|)p}{2}+|Y'_{2m+1}|-|X'_{2m+1}|=\varepsilon_m(1-p)^2, \quad m\geq 1 \quad (10)$$

and

$$\frac{(2|C_0|-|C_1|)p^2}{4}-\frac{(2|D'_0|-|D'_2|+2|Y'_1|-2|X'_1|)p}{2}+|Y'_1|-|X'_1|=\varepsilon_0(1-p)^2, \qquad m=0 \quad (11)$$

Because $m$ can take value over $2^{b-1}-1$ different values, we can establish $2^{b-1}-1$ different equations.

Considering excellent precision and robustness of the Least Square Method on parameter estimation, we estimate the $2^{b-1}-1$ equations parameters in the Least Square Method. When the absolute value of right of each equation is least, the value of $p$ is the estimated length of hidden message. Practical process is as follows.

Let $A_m=\dfrac{|C_m|-|C_{m+1}|}{4}$, $E_m=|Y'_{2m+1}|-|X'_{2m+1}|$, and $B_m=-\dfrac{(|D'_{2m}|-|D'_{2m+2}|+2|Y'_{2m+1}|-2|X'_{2m+1}|)}{2}$,

then left of equation (10) is $A_m p^2+B_m p+E_m$.

Let $S(i,j,p)=\sum_{m=i}^{j}(A_m p^2+B_m p+E_m)^2$, where $0\leq i<j\leq 2^{b-1}-2$, then $S(i,j,p)=(1-p)^4\sum_{m=i}^{j}\varepsilon_m^2$.

Suppose $q=\dfrac{1}{1-p}$, and $T(i,j,q)=\dfrac{S(i,j,p)}{(1-p)^4}=\sum_{m=i}^{j}\varepsilon_m^2$, then

$$T(i,j,q)=q^4\sum_{m=i}^{j}(A_m(1-\frac{1}{q})^2+B_n(1-\frac{1}{q})+E_m)^2=\sum_{m=i}^{j}((A_m+B_m+C_m)^2 q^4-2(A_m+B_m+C_m)(2A_m+B_m)q^3+$$

$$+(2A_m(A_m+B_m+C_m)+(2A_m+B_m)^2)q^2-2A_m(2A_m+B_m)q+A_m^2).$$

Gain minimum of $T(i,j,q)$ and find $q$ by solving equation $\dfrac{\partial T(i,j,q)}{\partial q}=0$. Namely solving

$$\sum_{m=i}^{j}(2(A_m+B_m+C_m)^2 q^3-3(A_m+B_m+C_m)(2A_m+B_m)q^2+(2A_m(A_m+B_m+C_m)+(2A_m+B_m)^2)q-A_m(2A_m+B_m))=0 \ (12)$$

Generally, there are three different roots $q_1,q_2,q_3$ for this equation, and we select a $q$ to be appropriate solution, where $q\in\{q_1,q_2,q_3\}$, and $T(i,j,q)\leq T(i,j,q_i), i=1,2,3$. Compute corresponding ratio of hidden message $p$, and make the decision whether the image is embedded message by comparing $p$ with the threshold 0.018.

## 3. Searching and Detecting

We call a certain selected web site as target-site, and its pages space composes a limited searching space with homepages, stations which link with it limited times. In this thesis, we mainly discuss how to search and detect the images in the limited searching space. We show below the following four parts respectively: first web page location and the primary selection of the images, image detecting and associated decision, results clustering and database building, and heuristic information searching and quickly location the same images.

### A. First Web Page Location and Primary Selection of the Images

The universal images searching systems on the Internet mainly resolve the following three problem: how to capture images from Internet; how to build corresponding index for captured images and how to search in the images database based on the requirement of users. Unlike the above three problems, our goal here is to search the existence of the hidden information in large numbers of images on the Internet, where the extension of searching can be entire Internet or a certain web site and correlative links. Hence, the starting searching page can be random URL, or one special page selected based on practical detecting scope.

Once we locate the starting page, we can capture images. There are varied images existing on the Internet, we need to capture the images with some basic characteristics to be employed in detecting, hence, must select from images. We adopt the first selection approach to filter some irrespective images. First of all, we collect the images in the selected station automatically, then store all images on pages in the corresponding database in link manner. All pages in the station are sent to pages analyzer, simultaneity, considering the requirement of the hidden data, such as the size of host image, we can use some heuristic information, such as image size, file type, file name, and the color histogram of image, to classify images simply and throw off the images such as advertisement bar, background, icon, button. Generally, we select images whose format is *.jpg,*.bmp,*.png or *.gif, size ranges from 10k to 2M to detect farther.

We adopt multithread parallel processing technology in searching process, which can process multi-operation at one time. The maximum of threads can be adjusted based on practical requirement. Experimental results show that it can improve the efficiency of searching greatly by adopting these technologies.

### B. Image Detect and Associated Decision

The images have some basic characteristics after the primary selection, and then these images will be detected with RS, SPA and LSM method introduced in Section 2.

We consider the three embedding estimate ratios of these three methods as $p_R, p_S, p_L$, and compute their average $\bar{p} = (p_R + p_S + p_L)/3$. We compare $\bar{p}$ with the decision threshold, and make the decision whether there are hidden data in the images. When making the decision, we can adjust the decision threshold based on the limit of the probability of false alarm and that of miss alarm.

Clearly, we can detect with searching, and can store the searching result in the database in order to make selective analyzing later. In the case that emphasizes precision of decision but not high speed, we can carry out associated analysis with more approaches on doubtful images.

### C. Results Clustering and Database Building

Since of the relative stability of web pages, the content of web pages shouldn't change largely within some time, we can build a database to store searching hyperlink result in order to analyze the object farther. In the database built process, we store the gained image links in hiberarchy clustering method. We search images by the breadth-first searching arithmetic, then classify the searching result based on searching link level. On the same levels, storing images by classes depends on web pages' visited catalog, and we commit images to users in hyperlink hiberarchy manner.

After results clustering, processing plentiful searching results set and quickly searching become convenience. Simultaneity, it provides convenience for entirely detecting for searching results.

### D. Heuristic Information Searching and Quickly Location the Same Images

When host images are transmitted on the Internet, sender usually sends more than one same images with hidden message to assure the receiver can receive the message, Thus, after capturing a certain image with hidden message, we need to search and locate all the same images as it in the searching space quickly.

Since we have selected a certain image, we can deploy heuristic searching and detecting in the corresponding link address in the database based on the heuristic conditions such as image name, format, size, veins, link address and so on. Obviously, the heuristic searching performs better than the breadth-first searching and the depth-first searching.

## 4. Simulation Results and Performance Analysis

We carry out the searching and detecting scheme, and make some experiments on the LAN of our university.

We insert 100 steg-images with LSB hidden message in the web pages randomly, where the numbers of 128×128, 256×256, 512×384 and 512×512 images are 25 pieces respective and the amounts of host images with 10%, 20%, ⋯ 100% embedding rate are 10 pieces respective. There are two groups same images, one is Lena image group and the other is Peppers image group (see Figure.1), each group containing 10 same images. We set the searching link depth 4 levels for searching. We embedded message in the LSB of host images randomly.

Figure.1 The sample images (left: Lena, right: Peppers)

The experimental results show that the amount of images searched and stored in the database to be detected after primary selection is 2031, and 97 steg-images are stored. When we set the size of image is between 10K and 2M, and get rid of those images such as advertisement bars, icons and buttons, the amount of images searched is 436,and 97 steg-images are still stored. It is obvious that the amount of doubtful images decreases greatly, thus, the amount of images to be detected decreases greatly.

We detect the 436 images passing through the primary selection with RS, SPA and LSM method, and make a associated decision. It is needed to emphasize that the probability of false alarm is more important than that of miss alarm on searching and detecting steganographic images. This means that we must detect steg-images with high precision with missing some steg-images. Therefore, it is acceptable that the miss probability even reach 50%, but the probability of false alarm must less than 10%. We must take this into consideration when we select threshold. By associated analysis, there are 86 images inserted in the web pages (84 images come from the 100 steg-images) are assured there are LSB embedding message within them. More experiments indicate that the result of searching and detecting is relative to the parameters such as the extension of searching space and the threshold of detecting method.

## 5. Conclusions

The problem of searching and detecting LSB steganographic images on the Internet is discussed in this thesis. A full scheme on images searching and steganalysis is investigated, and an improved algorithm of SPA is proposed. Some techniques such as multithread parallel processing technology, filtration method and heuristic searching are adopted in searching process, and decision are made with RS, SPA and LSM method in detecting process. Experimental results show that the scheme can search doubtful images in the target site quickly, and can make decision whether they are steg-images with LSB embedding quickly.

## References

[1] J. Fridrich, M. Goljan, "Practical Steganalysis of Digital Images – State of the Art", In Delp III, E.J., Wong, P.W., eds.: *Security and Watermarking of Multimedia Contents* IV. Volume 4675 of Proc. SPIE. (2002) 1-13.

[2] A. Westfeld and A. Pfitzmann, Attacks on Steganographic Systems[C], Lecture Notes in Computer Science, vol. 1768, Springer-Verlag, Berlin, 2000, pp. 61-75.

[3] Dumitrescu, S., Wu, X., Wang, Z., Detection of LSB Steganography via Sample Pair Analysis[J], IEEE Transactions on Signal Processing, VOL.51, NO.7, July 2003.

[4] Ker, A.: Quantitive evaluation of pairs and rs steganalysis[C]. In Delp III, E.J., Wong, P.W., eds.: Security, Steganography, and Watermarking of Multimedia Contents VI. Volume 5306 of Proc. SPIE. (2004) 83–97.

Brill Academic Publishers
P.O. Box 9000, 2300 PA Leiden
The Netherlands

*Lecture Series on Computer
and Computational Sciences*
Volume 4, 2005, pp. 353-360

# A Remote Sensing Application Workflow and its Implementation in Remote Sensing Service Grid Node

Ying Luo[1], Chaolin Wu[1], Yong Xue[1,2*], Guoyin Cai[1], Yingcui Hu[1], and Zhengfang Wang[3]

[1]State Key Laboratory of Remote Sensing Science, Jointly Sponsored by the Institute of Remote Sensing Applications of Chinese Academy of Sciences and Beijing Normal University, Institute of Remote Sensing Applications, Chinese Academy of Sciences, P.O. Box 9718, Beijing 100101, China
[2]Department of Computing, London Metropolitan University, 166-220 Holloway Road, London N7 8DB, UK
[3]PuTian Information and Technology Academy, 2 Shangdi 2nd Street, Beijing, China
jennyjordan@hotmail.com, y.xue@londonmet.ac.uk

Received 27 July, 2005; accepted in revised form 11 August, 2005

*Abstract:* In this article we describe a remote sensing application workflow in building a Remote Sensing Information Analysis and Service Grid Node at Institute of Remote Sensing Applications based on the Condor platform. The goal of the Node is to make good use of physically distributed resources in the field of remote sensing science such as data, models and algorithms, and computing resource left unused on Internet. Implementing it we use workflow technology to manage the node, control resources, and make traditional algorithms as a Grid service. We use web service technology to communicate with Spatial Information Grid (SIG) and other Grid systems. We use JSP technology to provide an independent portal. Finally, the current status of this ongoing work is described.

*Keywords:* Grid Computing, Workflow, Remote Sensing

## 1. Introduction

Grid has been proposed as the next generation computing platform for solving large-scale problems in science, engineering, and commerce (Foster and Kesselman 1997, 1999). There are many famous Grid projects today: DataGrid, Access Grid, SpaceGRID, European Data Grid, Grid Physics Network (GriPhyN), Earth System Grid (ESG), Information Power Grid, TeraGrid, U.K. National Grid, etc. Of particular interest are SpaceGRID and Earth System Grid (ESG), which focus on the integration of spatial information science and Grid.

The SpaceGRID project aims to assess how GRID technology can serve requirements across a large variety of space disciplines, such as space science, Earth observation, space weather and spacecraft engineering, sketch the design of an ESA-wide Grid infrastructure, foster collaboration and enable shared efforts across space applications. It will analyse the highly complicated technical aspects of managing, accessing, exploiting and distributing large amounts of data, and set up test projects to see how well the Grid performs at carrying out specific tasks in Earth observation, space weather, space science and spacecraft engineering.

The Earth System Grid (ESG) is funded by the US Department of Energy (DOE). ESG integrates supercomputers with large-scale data and analysis servers located at numerous national labs and research centers to create a powerful environment for next generation climate research. This portal is the primary point of entry into the ESG.

---

* Corresponding author

In this paper we present remote sensing application workflow in building a Remote Sensing Information Analysis and Service Grid Node at Institute of Remote Sensing applications (IRSA) based on the Condor platform. The node is a special node of Spatial Information Grid (SIG) in China.

Grid workflows consist of a number of components, including computational models, distributed files, scientific instruments and special hardware platforms. Abramson et al. (2005) described an atmospheric science workflow implemented by web service. Their workflow integrated several atmosphere models physically distributed. Hai (2003) studied the development of component-based workflow system, and he also studied the management of cognitive flow for distributed team cooperation.

The Condor Project has performed research in distributed high-throughput computing for the past 18 years, and maintains the Condor High Throughput Computing resource and job management software originally designed to harness idle CPU cycles on heterogeneous pool of computers (Basney, 1997). In essence a workload management system for compute intensive jobs, it provides means for users to submit jobs to a local scheduler and manage the remote execution of these jobs on suitably selected resources in a pool. Condor differs from traditional batch scheduling systems in that it does not require the underlying resources to be dedicated: Condor will match jobs (*matchmaking*) to suited machines in a pool according to job requirements and community, resource owner and workload distribution policies and may vacate or migrate jobs when a machine is required. Boasting features such as check-pointing (state of a remote job is regularly saved on the client machine), file transfer and I/O redirection (i.e. remote system calls performed by the job can be captured and performed on the client machine, hence ensuring that there is no need for a shared file system), and fair share priority management (users are guaranteed a fair share of the resources according to pre-assigned priorities), Condor proves to be a very complete and sophisticated package. While providing functionality similar to that of any traditional batch queuing system, Condor's architecture allows it to succeed in areas where traditional scheduling systems fail. As a result, Condor can be used to combine seamlessly all the computational power in a community.

We describe the Grid workflow in remote sensing science and show how it can be implemented using Web Services on a Grid platform in this paper. The workflow supports the coupling of a number of pre-existing legacy computational models across distributed computers. An important aspect of the work is that we do not require source modification of the codes. In fact, we do not even require access to the source codes. In order to implement the workflow we overload the normal file Input/Output (IO) operations to allow them to work in the Grid pool. We also leverage existing Grid middleware layers like Condor to provide access to control of the underlying computing resources.

In this paper, we present the remote sensing information analysis and service Grid node, which is developed in Institute of Remote Sensing Applications, Chinese Academy of Sciences, China. The node will be introduced in Section 2. Several middleware developed in the node and the remote sensing workflow, with some detail of the functions of the various components will be demonstrated in Section 3. Finally, the conclusion and further development will be addressed in Section 4.

## 2. Remote Sensing Information Analysis and Service Grid Node

### 2.1 Spatial Information Grid (SIG)

SIG is the infrastructure that manages and processes spatial data according to users' demand. The goal of SIG is to build an application Grid platform for spatial information community, which can simplify and shield many complex technology and settings, facilitate SIG users, and make good use of resources physically distributed. There are many reasons why one might wish to have the SIG. First, the amount of spatial data is increasing amazingly. So that real time or almost real time processing needed by applications confronts much more difficulties in one single computer. Second, data, algorithm, and/or computing resources are physically distributed. Third, the resources may be "owned" by different organizations. Fourth, the use frequency of some resources is rather low.

A SIG at least contains:
- A remote sensing Remote Sensing Information Analysis and Service Grid Node:
- A data service node: traditional data base to a web service
- A management centre: resource register, find, trade, and management

● A portal: an entry to SIG user

## 2.2 Remote Sensing Information Analysis and Service Grid Node

Remotely sensed data is one of the most important spatial information sources, so the research on architectures and technical supports of remote sensing information analysis and service grid node is the significant part of the research on SIG.

The aim of the node is to integrate data, traditional algorithm and software, and computing resource distributed, provide one-stop service to everyone on Internet, and make good use of everything pre-existing.

The node can be very large, which contains many personal computers, supercomputers or other nodes. It also can be as small as just on personal computer.

Figure 1 describes the architecture of Remote sensing information analysis and service Grid node. The node is part of the SIG, but it also can provide services independently. There're two entries to the node:
1. A portal implemented by JSP. User can visit it through Internet browses, such as Internet Explorer and others.
2. A portal implemented by web service and workflow technology. It is special for SIG. Other Grid resources and Grid systems can integrate with our node through this portal.

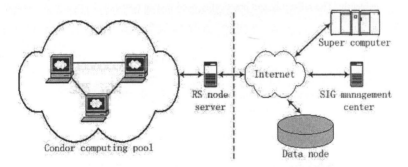

Figure 1 Architecture of remote sensing information analysis and service Grid node

The node provides the application services such as aerosol optical depth retrieval, land surface temperature retrieval, soil moisture retrieval, surface reflectance retrieval, and vegetation index applications by MODIS data.

The remote sensing node server (Figure 2) responds for all the corresponding events about the services on it. It contains:
● Node container: provide the necessary grid work environment;
● Service/workflow: some functional entities or workflows will be encapsulated to services for Grid. Others functional entities called within the node will be encapsulated to workflows. By adding new services into the node container, the capability of the node can be augmented and in turn the function of the whole system can be improved incrementally.
● Middleware: traditional applications cannot be used on condor directly. So middleware is needed to provide an access to condor. The middleware stands for the division and mergence of sub-tasks, and monitor the condor computing pool and the process of sub-tasks running in the pool.
● Node management tool: it responds for issuance, register, and update to SIG management centre. It's up to the node management tool to trigger service and monitor the status of the service.

The node issues and registers its services to SIG manage centre periodically, responses calls of SIG, triggers services, and reports status. Receiving require from SIG manager, the node will find data form either local or remote data server according to the user's requirements, organize computing resource dynamically, trigger services, and monitor their running status. To decrease the total processing time of a task, the node will divide a task into several sub-tasks. The exact number and size of the sub-tasks is according to the current PC number and configure in the Condor computing pool and super computer. Only when the task is large enough, or on the user's request, the node trigger the super computer to do a large sub-task. The method we trigger super computer is different with that of Condor computing pool. The Grid Service Spread (GSS) described by Wang *et al.* (2005) was used to implement it.

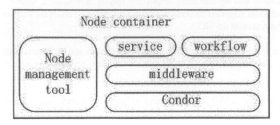

Figure 2. Remote sensing node server

Implementing the node, we refer to use web service technology rather than Globus technology. The reasons are:

1) Globus is a new technology which progressing fast. From the issuance of Globus toolkit version 1.0 at 1999 to version 4.0 at 2005, the protocol and technology in it changed evidently. The reliability and practicality need textual research. Therefore, it's venturesome to follow it.

2) Web service is a mature technology, which has been tested out in the field of industry for decades of years. Globus is adjusting its steps to web service. So long as our node supports web service, our node will be compatible with Globus. Web service ensures the compatibility and expansibility of our node.

3) It's more complex and difficult to develop using Globus than using web service. There're some convenient tools available for web service. But few for Globus.

## 3. Workflow implementation with web service and Grid

Traditional method for link functional components is to program them into a static course, or involve human control. People have to realize all the permutation and combination of the components in order to deal with all possible cases. It need large amount of repeated work, and is discommodious to modify and extend. The workflow technology overcomes these shortages. Using the concept STEP, ACTION, STATE, and TRANSITION in finite state machine (FSM) for reference, workflow can skip among the components flexibly. So it's facility to organize components dynamically according the user request and the environment of the system.

As the temperature retrial example for demonstrating the implement of our node, which will be described in details in sections 3.1 and 3.2, we can cut-out any pre-processing components by configuring the workflow conditions and attributes. For example, if the primary data has been rectified outside our node, the workflow will skip the rectification component and turn into the next one.

The components called by workflows can be reused. So that people were set free from repeated work. For the loose coupling of the components and workflows, it's convenient for developers to modify the components themselves without harm to the workflows, as long as the interface of the components is unchanged. Furthermore, it is convenient to modify the workflows by adding or decreasing components in them.

We also benefit from the termination function of workflow technology. It is particular useful to some remote sensing applications which involve many steps and need long processing time. A workflow instance can be terminate by sending terminal signal artificially, if we don't want the instance execute anymore. An instance can stop itself when it finds out the components in it meet trouble or the required environment is not satisfied. But the workflow cannot stop a running component. It only can stop itself before or after the running component is finished.

Using workflow's status function, we can monitor the running status of a workflow and status of components in it. STATE has four meaning in the remote sensing node:

- State of the node: it describes the attributes of the node, such as busy, idle, current workload, current processing ability, service instance number in queue, etc.
- State of application/workflow instance: it presents whether the instance is finished, or which step is finished.
- State of components in an application: it has only two status: finished and underway.
- State of condor pool: it contains the detail configuration (operation system, IP address, architecture, CPU, and RAM) and running status (IDLE or BUSY) of the PCs in the condor computing pool, the number of the tasks submitted to condor.

### 3.1 An implementation with Web Services on the Condor platform

This implementation is for the connection to the SIG or other Grid application systems. Each application is implemented using web service technology such as SOAP, XML, and WSDL (Sun Microsystems, 2002).

In our node, the workflow engine plays the role of a global coordinator to control different components to fulfil the functions requested by users according to a pre-defined process model. For example, Figure 3 presents the pre-defined process model of land surface temperature retrial. It involved the functional component of data format transfer, rectification, region selection, division, land surface temperature retrial, mergence, and return result to caller. The workflow will skip the functional components in front of the component whose input requirement has been satisfied by the primary data. Unfortunately, the system cannot recognize whether the data has been transferred or rectified without further information. It's necessary for the data node or someone else to describe these characters in a data metafile. The workflow will decide which components should be skipped by the metafile.

The division component calls the middleware shown in Figure 2. The middleware stands for checking the current status of condor computing pool, dividing data into pieces according to the number of the PCs in condor, generating all the files needed by Condor, packaging them, and transferring them to the location where condor can find them. The workflow submits the temperature component and data pieces to Condor. Then Condor stands for manage these sub-tasks. When Condor returns the results of sub-tasks, the mergence component will integrate them into the final result. And return component transfer this final result to caller.

In this test, there were 5 PCs running operation system of Windows 2000 professional. The details of the PCs in condor computing pool are shown in Table 1.

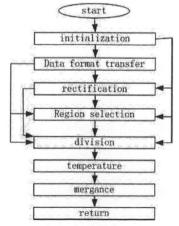

Figure 3. Process model of land surface temperature retrieval

Table 1. Configuration of 5 PCs in the Condor computing pool

| Host name | IP | Arch | CPU | RAM |
|---|---|---|---|---|
| Tgp2.tgp2 | 192.168.0.200 | Intel | 1.7GHz | 256M |
| Cai.tgp2 | 192.168.0.110 | Intel | 2.6GHz | 512M |
| Hu.tgp2 | 192.168.0.119 | Intel | 2.6GHz | 512M |
| Zhong.tgp2 | 192.168.0.11 | Intel | 2.6GHz | 512M |
| Jennyjordan.tgp2 | 192.168.0.120 | Intel | 2.0GHz | 512M |

### 1.2  An implementation with JSP on a Condor platform

This implementation is for users to visit our node directly. The difference between the JSP implementation and web service one is:

1) It is not a web service application. There are no WSDL files and WSDD files for the JSP implementation. It only can be triggered and initialized artificially. It cannot be interact by other machines.

2) The implement with JSP can be man-machine interactive. Of course it can auto run, too. An application can be intervened when it is running.

3) It provides a data up load function before data format transfer component. So that users can process their data using the functions on our node.

Figures 4 to 6 show the execution course of the land surface temperature retrieval application.

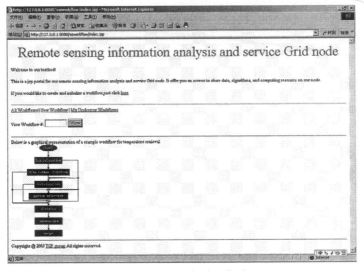

Figure 4 The welcome page of land surface temperature retrieval application

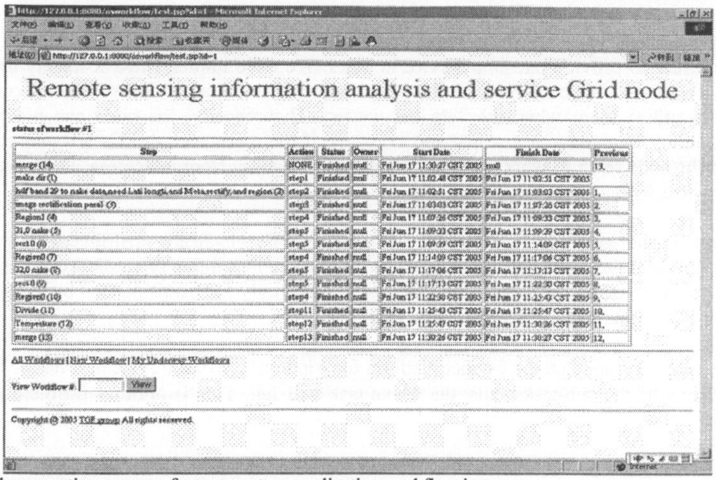

Figure 5. The execution course of a temperature application workflow instance

Figure 6. The feedback information of Tomcat

## 4. Conclusions and future work

In this paper, we introduced our ongoing research on remote sensing information analysis and service Grid node. The node is service-oriented. The whole node is composed of many workflows. Each workflow stands for certain service. Some of the workflows were implemented with web service technology, so that they can be called by the SIG or other systems which follow web service protocol. Web service technology endows our node with compatibility and machine-machine interaction. Other workflows are for JSP web site, so that user can visit our node directly. They only can be triggered artificially. The JSP technology endows our node with man-machine interaction. The execution course of the JSP workflows can be intervened by users. Workflow technology endows our node with easy modification and extensibility. SIG and Condor endows our node with the power of remote cooperation, resource share, and management physically distributed.

Our node is a demonstration to how to integrate data, traditional algorithms and models, and computing resource in order to provide one-stop service to users. We have implemented it mainly by workflow technology. We try to make a large virtual super computer by integrating cheap personal computers on Internet, and utilize them when they are left unused. Theoretically, the node can provide enough processing ability to anyone at anytime and anywhere if there are enough PCs left unused in it. The services on the node seem to work well, but actually it is far away from intactness. There are still many problems to deal with.

One problem is the long file transfer time between the PCs physically distributed. It weakens the decrease of the total processing time improved by using more PCs doing one task. To some multi-band image data, a feasible approach to deal with it is to transmit only the bands needed by the applications. The amount of the data may decrease a scalar grade.

We implemented our node on the Condor platform for windows operation system. It has no break-point function. If one of the sub-tasks fails, the whole task will fail. This is another problem need to be solved.

## Acknowledgement

This work is support by the research projects "Grid platform based aerosol fast monitoring modeling using MODIS data and middlewares development" (40471091) funded by National Natural Science Foundation of China (NSFC), "CAS Hundred Talents Program" (BRJH0101) funded by Chinese Academy of Sciences, and "Remote Sensing Information Processing and Service Node" and "Dynamic Monitoring of Beijing Olympic Environment Using Remote Sensing" (2002BA904B07-2) funded by the Ministry of Science and Technology, China.

## References

[1] Abramson, D., Kommineni, J., McGregor J. L., and Katzfey, J., *1 January 2005,* An Atmospheric Sciences Workflow and its implementation with Web Services. *Future Generation Computer Systems,* Vol.21, Issue 1, pp.69-78

[2] Basney, J., Livny, M., and Tannenbaum, T., 1997, High Throughput Computing with Condor, *HPCU news,* Volume 1(2)

[3] Foster, I., and Kesselman, C., 1997, Globus: a metacomputing infrastructure toolkit, Int. J. Supercomputer Application, 11 (2), 115–128.

[4] Foster, I., and Kesselman, C. (Eds.), 1999, The Grid: Blueprint for a New Computing Infrastructure, Morgan Kaufmann Publishers, USA.

[5] Cai, G. Y., Xue, Y., Tang, J. K., Wang, J. Q., Wang, Y. G., Luo, Y., Hu, Y. C., Zhong, S. B., and Sun, X. S., 2004, Experience of Remote Sensing Information Modelling with Grid computing. *Lecture Notes in Computer Science,* Vol.3519, pp. 989-999

[6] Hai Z. G., 2003, Component-based workflow systems development, *Decision Support Systems,* Volume 35, Issue 4, Pages 517-536

[7] Sun Microsystems, 2002, Web Services made easier: the Java APIs & architectures for XML, http://java.sun.com/webservices/white/ and http://java.sun.com/webservices/webscrvicespack.html.

[8] Wang, Y. G., Xue, Y., Jianqian Wang, Chaolin Wu, Yincui Hu, Ying Luo, Shaobo Zhong, Jiakui Tang, and Guoyin Cai, 2005, Java-based Grid Service Spread and Implementation in Remote Sensing Applications. *Lecture Notes in Computer Science,* Vol.3516, pp.496-503.

Brill Academic Publishers
P.O. Box 9000, 2300 PA Leiden,
The Netherlands

*Lecture Series on Computer*
*and Computational Sciences*
Volume 4, 2005, pp. 361-364

# Learning Characteristics of Multi-Layered Higher Order Neural Networks

**L. Ma**

Shanghai University of Electric Power, 2103 Pingliang Rd Shanghai, China

malxaii@shiep.edu.cn

**H. Miyajima, N. Shigei**

Kagoshima University, 1-21-40 Korimoto, Kagoshima 890-0065, Japan

Received 7 June, 2005; accepted in revised form 28 June, 2005

*Abstract:* In order to approximate any nonlinear system well, the conventional neural networks need a large number of units. However, the BP method by neural networks of large size ensures that the converged point is only a local minimum or even a saddle point. Therefore, effective neural networks of small size are desired. This paper describes that multi-layered higher order neural networks are more effective than the conventional ones in the both cases where BP algorithm and BP algorithm with pruning process are performed.

## 1 Introduction

It is well known that multi-layered neural networks (NNs) using the gradient descent method perform pattern recognition and nonlinear function approximation well and have been successfully applied to many applications[1, 2]. When the number of hidden units in the network is not restricted, it has been theoretically proven that such a network has been an ability to perform approximation any continuous input-output relation with any accuracy[1]. But, in order to perform it, NNs of large size are needed. On the other hand, the gradient descent method based on the BP algorithm ensures that the converged point is only a local minimum or even a saddle point. Therefore, NNs of small size and of good efficient are desired[2]. This paper describes that multi-layered higher order neural networks (MHONNs) are superior to the conventional ones in the both cases where BP algorithm and BP algorithm with pruning process are used. Specifically, three-layered second order NNs (SONNs) as MHONNs are used without loss of generality. The efficiency of BP algorithms for two systems, SONNs and NNs, with the same number of parameters are compared in some simulations such as functional approximation, pattern recognition, generalization ability for uncertain data and the prediction of chaotic series. Further, the superiority of SONNs is also shown even when BP algorithms with pruning process are used.

## 2 Higher Order Neural Networks and Back Propagation Algorithm

First, we consider SONNs each of which consists of $N_m$ ($m = 1, 2, 3$) units, where the 1-th, 2-th and 3-th layers are called input, hidden and output layers, respectively. Each value of input takes any real number and output takes any real numbers from 0 to 1. The state of a unit (the output) is represented as a function of potential $u$. The potential $u$ increases in proportion to the weighted sum of product terms for input variables. That is, the output of the $i$-th unit of the $m$-th layer, $y_i^{(m)}(i = 1, \cdots, N_m, m = 2, 3)$, is determined as follows: $y_i^{(m)} =$

$F[\sum_{l_1=1}^{N_{m-1}-1} \sum_{l_2=l_1+1}^{N_{m-1}} W_{il_1l_2}^{(m-1)} y_{l_1}^{(m-1)} y_{l_2}^{(m-1)} + \sum_{l=1}^{N_m-1} W_{il}^{(m-1)} y_l^{(m-1)} + s_i^{(m)}]$, where $F[x] = \frac{1}{1+e^{-x}}$, $W_{i[L]}^{(m-1)}$ and $s_i^{(m)}$ means the weight and threshold. For data $\boldsymbol{y}^{(o)}$ to input layer, let the desirable value of output layer be represented by $\boldsymbol{d}$, where $\boldsymbol{y}^{(0)} = (y_1^{(0)}, \cdots, y_{N_1}^{(0)})$, $\boldsymbol{d} = (d_1, \cdots, d_{N_3})$. Let $P$ be the number of learning data. The square error $E$ between the output $y_i^{(3),r}$ and the desirable output $d_i^r$ for $r = 1, \cdots, P$ is defined as follows.

$$E = \frac{1}{2} \sum_{r=1}^{P} \sum_{i=1}^{N_3} (y_i^{(3,r)} - d_i^r)^2 \qquad (1)$$

To minimize the Eq.(1), weights are adjusted based on the descend method by $\Delta W_{i[L]} = \eta \frac{\partial E}{\partial W_{i[L]}}$, where $W_{i[L]} = W_{il_1l_2}$ or $W_{il}$ and $\eta$ is called the learning constant. The method is called BP algorithm for neural networks. We have already shown BP algorithm for MHONNs in the previous paper[3].

Next, we explain a pruning algorithm for HONNs. In order to determine the optimal number of units for the hidden layer, the pruning algorithm is introduced. The pruning procedure is based on the error calculating method. This method consists of calculating the changes of mean square error by eliminating one of weights, and is shown as follows.

**step 1** Initialize all the weight values and threshold values.
**step 2** Learn using the BP algorithm during certain iterations.
**step 3** Calculate the initial error $E_0$ by the Eq.(1).
**step 4** Error-calculating-based pruning procedure
  **begin**
      $i := 0$; $S_0 := \{\boldsymbol{w}_1, \cdots, \boldsymbol{w}_k\}$;
      **repeat**
          $i := i + 1$; $S_i := S_0 - \{\boldsymbol{w}_i\}$; calculate $E_i$;
      **until** $i = k$;
      identify $S_j$ with the minimal $E_j$;
      $S_0 := S_j$; **end**,
    where $S_0$ is a set of all weights at the step and $E_i$ is the error for $S_i$.
**step 5** Learn using BP algorithm during certain iterations.
**step 6** Go to step 3 if necessary.
**step 7** Terminated.

The algorithm is repeated until a target number of weights or a definite threshold of error is attained.

## 3 Numerical Simulations

In order to compare SONNs with the conventional NNs, two kinds of simulations are performed. The first simulations are ones about BP learning with the fixed number of weights, and the second simulations are ones about pruning algorithm.

### 3.1 BP learning

**Function approximation:** The functions are given by

$$z = \frac{(x_1 x_2)^2 + 1}{2} \qquad (2)$$

$$z = \frac{4 \sin(\pi x_1) + 2 \cos(\pi x_2)}{12} + 0.5 \qquad (3)$$

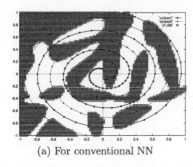
(a) For conventional NN

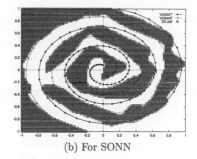
(b) For SONN

Figure 1: Learning results for two spiral problem

Table 1: Results on function approximation by BP learning

| Func. | Method | MSE($\times 10^{-4}$) | | |
|---|---|---|---|---|
| | | Avg. | Min. | Max. |
| Eq.(2) | SONN | 0.1575 | 0.2720 | 1.8930 |
| | Conventional | 2.4965 | 1.0307 | 5.7698 |
| Eq.(3) | SONN | 2.0131 | 0.8310 | 4.9346 |
| | Conventional | 3.4212 | 1.6617 | 5.9953 |

Table 2: Results on chaotic time series prediction

| Method | # of weights | MSE |
|---|---|---|
| SONN | 88→71 | 0.00053 |
| Conventional | 85→71 | 0.00144 |

In each function, input variables are $x_1$, $x_2$ and output variable is z. The number of learning data sets are 100, where the input data are randomly chosen from $[-1, 1]$ and output data are normalized within $[0, 1]$. The number of test data is 529, where the input data are uniformly chosen from $[-1, 1]$. Table 1 shows the results of learning for functions (2) and (3). It shows that SONNs are superior in almost cases to the conventional NNs under the same condition on the numbers of weights.

**Pattern classification–two spirals problem:** This classification problem is discriminating between two sets of learning points which lie on two distinct spirals in the $x$–$y$ plain. These spirals coil three times around the original and around one another. The number of learning data denoted by "o" or "+" in the Fig. 1 for each spiral is 97. It is known that learning of 100% for learning data can not be performed in the conventional model. In our simulation, 97% learning in the conventional model and 100% learning in the SONN are obtained. The Fig.1 shows the result for each model for test data. In this case, SONNs are superior to the conventional model.

**Function approximation with uncertain or vague data:** The system with five data $(-3, 7)$, $(-1, 11)$, $(0, 26)$, $(3, 56)$, $(4, 29)$ for learning is identified as neural networks with one input and one output. The numbers of hidden units are 4 in the conventional model and 2 in SONN, respectively. After learning of 5000 times, the result of the Fig.2 is obtained for test data. The target function is illustrated by using the spline interpolation[4]. The result by SONN is identical with one interpolated by the spline functions. In some other functional simulations, the same results as this case are shown.

### 3.2 Pruning algorithm

**Function approximation:** Two pruning algorithms for two functions of Eqs. (2) and (3) are performed under the condition that the number of learning and test data are 100 and 2000, respectively. Fig.3 shows the result for test data when the weights are pruned from 80 or 85 to 2

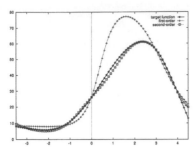
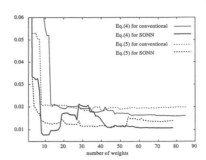

Figure 2: Results for function approximation with a small number of data

Figure 3: Results on function approximation by pruning algorithms

sequentially, where the vertical line means the root mean square error. It shows that SONNs are superior to the conventional model.

**Chaotic time series prediction:** Mackey-glass chaotic time series data are derived from the following equation: $\dfrac{dy_t}{dt} = \dfrac{ay_{t-\tau}}{1 + y_{t-\tau}^{10}} - by_t$. In the problem, given four data points in the time series, $t - 18$, $t - 12$, $t - 6$ and $t$, the point at $t + 85$ is tried to be predicted. Thus, networks comprising 4 input units and an output unit are needed. Learning data sets consist of 3000 data sets from $t = 200$ to $t = 3200$ and test data sets consist of 500 data sets from $t = 5000$ to $t = 5500$. Two pruning algorithms for the problem are performed under the condition that the number of initial weights is 85 or 88, the number of final ones is 71 and the number of learning times is 1000. Table 2 shows that SONNs are superior to the conventional ones.

## 4 Conclusion

This paper describes that SONNs are more effective than the conventional NNs in the both cases where BP algorithm and BP algorithm with pruning process are used. As a result, it was shown that MHONNs were effective even when the small number of units was used. In order to show the effectiveness, some simulations such as functional approximation, pattern recognition, generalization ability for ill-posed problem and the prediction of chaotic series, were performed by the proposed and conventional models with the same number of parameters.

## Acknowledgement

This work was supported by the ShangHai Education Natural Science Foundation (04LB09).

## References

[1] M.M. Gupta et al., "Static and Dynamic Neural Networks", IEEE Press, 2002.

[2] R. Reed, "Pruning Algorithms–A Survey", IEEE Trans. on Neural Networks, vol.4, no.5, pp.740–747, 1993.

[3] L. Ma, H. Miyajima and N. Shigei, "Neural Networks Determining Injury by Salt Water for Distribution Lines", LNCS 3174, pp.436–441, 2004.

[4] J.H. Ahlberg et al., "The Theory of Splines and Their Applications", Academic Press, New York, 1967.

Brill Academic Publishers
P.O. Box 9000, 2300 PA Leiden
The Netherlands

*Lecture Series on Computer
and Computational Sciences*
Volume 4, 2005, pp. 365-368

# The Remeshing Problem in the Multi Scale Strain Localization CAFE Approach

L. Madej [1], P.D. Hodgson[2], M. Pietrzyk[1]

[1]Department of Computer Methods in Metallurgy,
Akademia Gorniczo Hutnicza AGH, 30-059, Krakow, Poland
[2]School of Engineering and Technology
Deakin University, 3217, Victoria, Australia

Received 7 July, 2005; accepted in revised form 29 July, 2005

*Abstract:* A study of the possibilities given by the multi scale CAFE method including the remeshing operation in the FE scheme for the parallel modeling of microshear and shear band propagation in steels during deformation is presented. Modeling of microshear bands development in microscale, and mezoscale and material response based on those processes at the macroscopic scale is possible using this multi scale computational technique. The general idea of the model and solution of the problem of the remeshing is described. The results of simulations of the compression in channel dies are presented.

*Keywords:* strain localization, multi scale modeling, smoothed particle hydrodynamic.

## 1. Introduction

Recent experiments have shown that the initiation and development of micro shear bands and shear bands are crucial for the development of plastic deformation in a variety of metals [1,2]. The traditional way to model material behavior during the deformation process is through application of the Finite Element technique (FE), which is capable of predicting material instabilities such as fracture and strain localization.

However, when using traditional FE modeling, it is very difficult to predict phenomena that take place on different scales in the material during the deformation process. The search for an alternative computational method has been undertaken for several years. The coupled Cellular Automata (CA) – Finite Element (FE) multi scale model is one possible solution. The CAFE approach has already been successfully used to model the microstructure evolution during hot rolling [3] as well as the ductile-brittle fracture [4] during the Charpy test. In the CAFE approach the material behavior can be completely separated from the structural response because calculations are coupled between the cellular automata technique and the finite element method.

The main objective of this work was development of the 3D multi scale CAFE model for the prediction of micro shear bands and shear bands development during the deformation process. The inspiration was the CAFE model developed at The University of Sheffield [4].

## 2. CAFE model for strain localization

The problem of strain localization in metallic materials during deformation has been investigated for over thirty years. Experimental and theoretical analysis of the shear band phenomena during various kinds of deformation has been performed [1,5]. However, there is still a lack of an efficient numerical model that accounts for the influence of the formation of shear bands, and which can adequately describe the material behavior in the FE simulation. This problem has been investigated by several researchers [6,7,8] but the major disadvantage of these models is lack of flexibility and lack of the possibility for generalization, what leads to difficulties with accurate description of various forming operations.

---

[1] Corresponding author, E-mail: lmadej@metal.agh.edu.pl

All of the above are the reason for the ongoing search for an alternative approach to describe strain localization phenomena. Authors have shown in [9] that the multi scale CAFE method should be an efficient method in this field. CAFE is a multi scale approach taking care of phenomena that take place at different scales in the material. Initiation and development of the micro and macro shear bands during various forming processes is one example of such processes. Microshear bands initiate in the microscale, while shear bands appear at the mezoscale. According to these two scales following the approach proposed in [4], two cellular automata spaces representing the material behavior in the micro- and mezoscale are introduced and attached to the finite element code.

In the CAFE model both CA spaces, microshear band space (MSB space) and shear band space (SB space), are defined by several state variables that describe each particular cell, as well as by a set of transition rules defined respectively for those spaces. Transition rules controlling changes between states in MSB and SB space are defined based on experimental knowledge [1,5]. All of the details describing the assumed cell's state and proper transition rules are presented in [9].

Information about the occurrence of microshear and shear bands is exchanged between the CA spaces during each time step, according to the defined mapping operations. Flow of the information between the scales goes in both directions, from macroscale to mezoscale and microscale as well as from microscale and mezoscale to macroscale. In each time increment, information about the stress tensor is sent from the finite element solver to the MSB space, where the development of microshear bands is calculated according to the transition rules. After exchange of information between CA spaces, transition rules for the SB space are introduced, and propagation of the shear bands is modeled. Based on the information supplied by the CA spaces, an equivalent stress $\sigma_p^{CA}$ is calculated and is used to obtain the correction coefficient $\xi$. This coefficient is sent to the FE program and used to modify the flow curve in the next step of FE calculations.

The conventional CAFE approach described in this section is capable of describing the initiation and evolution of microshear bands for the deformation simulations excluding the remeshing problem in the FE mesh during calculation. Whenever a remeshing operation is performed the major problem connected with the lack of the information from the material appears. An aspect of the meshfree Smoothed Particle Hydrodynamics (SPH) [10,11] was used to overcome this problem, as described below.

## 3. Remeshing problem

The change between states using the predefined transition rules [9] is due to the state in the neighbors of a selected cell and the cell itself in the previous time step. According to this essential definition of the CA method a number of CA spaces in the CAFE approach should remain constant during the calculation. This leads to a major disadvantages whenever a complicated scheme of deformation or sample shapes are introduced and remeshing appears in the defined mesh. As a consequence there is a change in the number of the mesh nodes as well as in the number of gauss points and a problem with attaching the CA spaces to the FE nodes appear. This may led to a lack of information from the microscale in some regions of the FE mesh, and eventually cause inaccurate results. An alternative set of points is introduced in the sample area to overcome this problem. The number of so called CA points remains constant during the deformation process and a number of underlying CA spaces attached to each CA point also remains constant. This approach enables the remeshing process and a change in the nodes mesh in the initial mesh.

In each time step an exchange of information between FE points and CA points using the approximation technique from the SPH method is performed. The essential basis of the SPH method is briefly described below.

SPH is a meshless Lagrangian method that uses a pseudo particle method to calculate the field variables. The concept of the integral representation of a function $f(x)$ at the location $x$ in the SPH method is given by of the multiplication of the product of the function and an appropriate kernel function $W_{ij}$:

$$f(x) \cong \int f(x_j) W_{ij}(x - x_j, h) dx \tag{1}$$

If we assume that a value of the $f(x)$ function is known only in a finite set of discrete points, an equation can be written by a summation:

$$\langle f(x) \rangle \cong \sum_{j=1}^{N} f(x_j) W_{ij}(x - x_j, h) V_j \tag{2}$$

where: $\langle \ \rangle$ – kernel approximation, $W_{ij}$ – kernel function, $h$ – smoothing length of the support domain, $V_j$ – volume associated with the $j$ particle.

Equation (2) is a basic equation used in the SPH method. Value of the function at a point $x$ is calculated by summation of the contribution from a set of the neighboring particles ($j$ subscript) in the support domain of the $x$ particle (Figure 1).

During the calculation of the kernel functions a correction suggested in [10] was introduced to achieve a better consistency. A quintic spline function proposed in **Error! Reference source not found.** was used this work:

$$W(R, h) = \alpha_d \begin{cases} (3-R)^5 - 6(2-R)^5 + 15(1-R)^5 & 0 \le R < 1 \\ (3-R)^5 - 6(2-R)^5 & 1 \le R < 2 \\ (3-R)^5 & 2 \le R < 3 \\ 0 & R > 3 \end{cases} \tag{3}$$

where $\alpha_d = 7/478\pi h^2$ in the two dimensional space.

The SPH approximation procedure in the CAFE approach is a two step process in each time increment. In the first step an approximation of the displacement field between the FE and CA points is performed to achieve a proper shape of the CA points in the space (Figure 2), which is a key point to obtain correct results. When a position of CA points is updated, an approximation of the stress field from the FE points is performed. Information about stress values from the CA points is later on sent to the underlying CA spaces and calculation of the

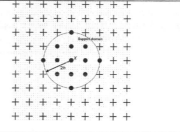

Figure 1. Schematical illustration of the neighboring particles in the support domain of the $x$ point.

initiation and development of micro and shear bands is performed according to the CAFE approach described previously.

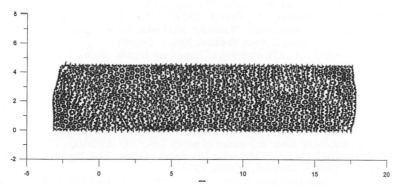

Figure 3. Strain distribution during the channel die tests calculated by the conventional FE model (top) and the CAFE model (bottom).

Developed CAFE model with the approximation is able to perform calculations for different complicated deformation processes. This creates opportunities for a proper validation of the achieved results with the experimental results and will lead to further development of this multi scale approach.

## 4. Conclusion

Results achieved from the strain localization CAFE model replicate qualitatively realistic material behavior including the stochastic phenomena that take place during deformation. The general form of the developed model, which does not constrain application of the model to particular tests or particular processes and allows simulations of any metal forming process, is the main advantage of this approach and makes it a powerful investigative tool. This capability is possible due to the discrete idea of the CA method itself, as well as due to the application of the SPH approximation method to overcome the remeshing problem. The SPH part of the CAFE model also creates opportunities for a proper validation of the achieved results with the experimental data, and will lead to further development of this multi scale approach.

## Acknowledgments

Financial assistance of MNiI, project 11.11.110.643, is acknowledged. One of the authors (LM) thanks Deakin University for financial assistance while Professor Hodgson is grateful to the Australian Research Council who support aspects of this work.

## References

[1] A. Korbel, Perspectives of the Control of Mechanical Performance of Metals during Forming Operations, J. Mater. Proc. Techn, 34, 1992, s. 41-50.
[2] W. Bochniak, A. Korbel, KOBO Type Forming forging of metals under complex conditions of the process, J. Mater. Proc. Techn., 134, 2003, s. 120-134.
[3] S. Das, E.J. Palmiere, I.C. Howard, CAFE: a Tool for Modeling Thermomechanical Processes, Proc. Thermomech. Processing: Mechanics, Microstructure & Control, ed., E.J. Palmiere, M. Mahfouf, C. Pinna, Sheffield, 2002, s. 296-301
[4] A. Shterenlikht, 3D CAFE Modeling of Transitional Ductile – Brittle Fracture in Steels, PhD thesis, Univeristy of Sheffiled, Sheffiled 2003
[5] P. Cizek, Characteristics of Shear Bands in an Austenitic Stainless Steel during Hot Deformation, Mat. Sci. Eng. **A324**, 2002, s. 214-218
[6] W. Wajda, Modelowanie procesów przeróbki plastycznej z uwzględnieniem efektów mikropasm ścinania, PhD thesis, AGH, Kraków, 2005 (in Polish).
[7] M. Pietrzyk, V. Pidvysotskyy, M. Packo, Flow stress model accounting for the strain localization during plastic deformation of metals, Ann. CIRP, 53, 2004, s. 235-238
[8] R.B. Pęcherski, Continuum Mechanics Description of Plastic Flow Produced by Micro-Shear Bands, Technische Mechanik, **18**, 1998, 107-115.
[9] Madej L., Talamantes-Silva J., Howard I.C., Pietrzyk M., Modeling of the initiation and propagation of the shear band using the coupled CAFE model, Archives of Metallurgy and Materials, 2005, to be published.
[10] Bonet J., Kulasegaram S., Correction and stabilization of smooth particle hydrodynamics method with applications in metal forming simulation, Int. J Numer. Meth. Engng., 2000, **47**, 1189-1214.
[11] Liu G.R., Liu M.B., Smoothed Particle Hydrodynamic a meshfree particle method, world scientific publishing, 2003.

Brill Academic Publishers
P.O. Box 9000, 2300 PA Leiden
The Netherlands

*Lecture Series on Computer
and Computational Sciences*
Volume 4, 2005, pp. 369-372

# Methodology to Modify Existing Electromechanical Disc-type Energy Meters to AMR

Tahir Mahmood, M. A. Choudhry[1], Salman Amin, Aamir Hanif

Department of Electrical Engineering,
University of Engineering and Technology
Taxila, Pakistan.

Received 15 July, 2005; accepted in revised form 9 August, 2005

*Abstract:* This paper presents a prototype design and implementation of automated meter reading (AMR) system using Frequency Modulation (FM) and Micro Controller based embedded system. A trend towards replacing the electromechanical meters with new designs, that can transmit their reading without any human intervention is in progress. In spite of developing new systems, this paper describes methodology that can modify existing electromechanical disc-type energy meters to AMR. Visual Basic (VB) programming language is used to interface our prototype design with personal computer at company's end. All the programming of microcontroller is done using Keil software. One such designed hardware unit can convert sixteen different meters data to AMR.

*Keywords:* Energy meter, Embedded system, Energy storage, Energy Pricing, AMR

*Mathematics Subject Classification:* Modulation and demodulation.

*AMS-MOS Number:* 94A14

## 1. Introduction

Since the time that electric power meters were introduced in the 1870s, the basic function of the meters has remained more or less unchanged. Although electronic, computer and communication technologies have advanced greatly; many developed countries however are still using the same technology for residential electricity services that has existed for more than a century. While automatic meter reading (AMR) has gradually been introduced in many places, the cost involves in retrofitting the existing systems may not be justified if they are used merely for meter reading.

From the operator's viewpoint, the existing situation should be improved because of the following reasons:

a) To reduce the time lag between energy supply and actual revenue collection.
b) To minimize the non-payment of bills by customers.
c) To economize the costs and overheads incurred in meter reading, invoicing and revenue collection.
d) To account for the differences between power generated and revenue collected due to energy losses and thefts.
e) To improve the accuracy in meter reading and to eliminate possible mistakes in data entries.
f) To obtain real-time information on the actual energy consumption by the end-users.
g) To enable implementation of flexible or innovative tariffs.

While part of the above issues such as (a), (b), and (c) can be addressed by utilizing Prepaid Meters, the other problems remain unsolved. Discussion on AMR systems is available in [1-5].

From the customer's viewpoint, one of the main reasons for accepting any changes to the existing metering system must be based on much improved customer services. AMR meter will provide

---

[1] Muhammad Ahmed Choudhry, Member IEEE. E-mail: drahmad@uettaxila.edu.pk

solutions to the above issues. The new meters will be able to provide many additional services to the customers as well as gathering vital information for the utility companies.

The automatic meter reading (AMR) system has become a necessity for most utilities as deregulation, free customer choice and open market competition occur in the energy supply sector. But these new meters are costly. Distribution companies (Discos) of developing and third world countries are not in a position to buy and replace thousands of already installed disc-type meters. One solution is that these meters are modified to AMR system using cost effective hardware and simple communication media. Our proposed prototype design is an implementation of this concept.

## 2. Design Overview

There are two options to get reading from electromechanical disc type energy meter. First one is to get reading of meter from dial directly. The other possibility is to count the revolution of disk with the help of a sensor. The developed prototype system uses second option. In the following sections, we will discuss the Designing of sending/receiving end hardware/software, communication, programming, interfacing, feasibility and benefits of prototype energy meter. This prototype system hardware is divided into two main blocks i.e. hardware at user end and hardware at Power Distribution Company's end as shown in Figure 1.

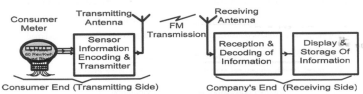

Figure 1: Block Diagram of overall System

*A. Hardware Design at Consumer End*

Figure 2 shows diagram of hardware design at the consumer's end. The blocks labeled sensor 1, 2, 3 ..., 16 comprise meter interfaced with an Opto-electronic sensor. This Opto-electronic sensor is the only thing that needs to be placed inside the electromechanical-meter to count disc revolution. The sensor has three wires. Two wires supply DC power to it and the third gives output pulse. This output pulse is given to one of the pins of microcontroller. The rest of the hardware can be placed at any suitable place e.g. in distribution box located in street. ATMEL microcontroller has 32 Input/Output pins divided into 4 ports each of 8 pin.

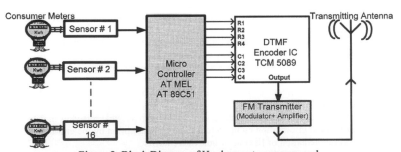

Figure 2: Block Diagram of Hardware at consumer end.

The microcontroller treats 5 volt pulse from the sensor as high input state and programmed to count one revolution of disk as it passes through sensor and so on. Microcontroller is so programmed that revolutions of disk are converted to energy-units consumed. The FM transmitter is used to transmit amount of energy. The transmitter modulates the signal with a fixed carrier frequency. This means that we can only send information of one sensor. This will not make system cost effective, so we have utilized a DTMF (Dual tone multi frequency) encoder IC. It generates 16 predefined frequency combinations depending upon combination of inputs at its 8 input pins. Microcontroller is programmed using Keil software, to give information of a maximum of 16 sensors against these 8 lines of DTMF encoder in time division multiplexing. Microcontroller gave out data of one sensor to DTMF encoder. For each combination given by Microcontroller, DTMF gave an output at a predefined frequency which is used as a modulating signal to FM transmitter which modulates it with a fixed frequency carrier and finally transmits it. Figure 4 shows the sequence of operation within hardware at consumer side.

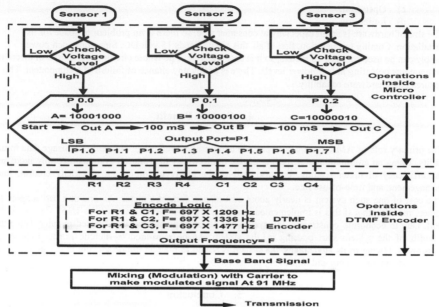

Figure 4: Sequence of Operations performed inside the Hardware on Consumer End.

### B. Hardware Design at Company End

Figure 5 shows block diagram of hardware design at the company's end. The Microcontroller at receiving end programmed with same time interval for getting data of one sensor from DTMF decoder. The output of FM receiver is fed to DTMF decoder IC (CM-8870) and then fed to Microcontroller. Hence we obtain information of each meter at user side. The microcontroller displays the units consumed on a seven segment LED display. An EEPROM (Electrically Erasable Programmable Read Only Memory) is connected externally to the Microcontroller at Company's end. This EEPROM does not lose its data on power failure or resetting of Microcontroller. Microcontroller is programmed in such a way that it reads data stored in EEPROM, every time it is reset. Then it starts counting from that value and also displays it on seven segment LED displays units. In this way, it can count usage of energy for the whole month. Visual Basic programming language is use to interface this prototype design with personal computer.

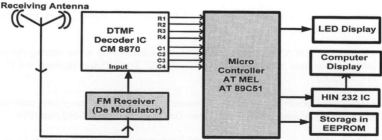

Figure 5: Block Diagram of Hardware at Power Company's End.

## 3.  Feasibility of System

In order to evaluate the feasibility of a system following factors were analyzed;
   a)   Installation of system.
   b)   Space requirement
   c)   Operation.
   d)   Reliability.
The size of hardware is small especially at consumer side, so there is no problem of space for its installation. Coming to operation, first of all, this circuit needs 15 volt DC. Since one such power supply can be used to serve 16 meters, so it is cost effective. The whole circuit is composed of discrete components operating at low power levels. There is hardly any chance of failure of a component. This ensures good hardware reliability.

## 4.  Cost vs. Benefit

The primary role of AMR system is to provide daily usage, total usage and real time usage data to the utility. Using an automatic meter reading device in electricity distribution systems, not only increases the efficiency of meter reading, but also enables the utility to implement electricity control, load management, and time-of-use rate.
The cost of one such system is nearly about cost of one domestic meter. As one unit can support 16 meters meaning that if this is implemented in a large area the overall cost will be $1/16^{th}$ for one AMR unit. Due to economic constraints, the system has to be implemented in a phased manner. Once the benefits of the system start pouring in, the system can be expanded to include greater number of consumers. Owing to its advantages, the system is bound to play an important role in streamlining the power distribution system and improving power sector.

## 5.  Conclusion

In this paper, we have reviewed the technical issues related to the effective construction of AMR system using Microcontroller, while enhancing our understanding about the challenges of designing, wireless communication, programming, interfacing a system. We have developed a prototype  system for testing its capability and performance for long time to the varying extent, with a view to effectively consolidate and empirically examine the sending and receiving end hardware and software.

### Acknowledgments

The author wishes to thank the anonymous referees for their careful reading of the manuscript and their fruitful comments and suggestions.

### References

[1]  I. Hakki Cavdar, Member IEEE "*A solution to remote detection of illegal Electricity usage via power line carrier*," IEEE Trans. Power Delivery Vol.19, pp 1663-1667. Oct, 2004.

[2]  Mark Gutowski "*Strategies for wireless infrastructure*," 14th annual Distibu Tech conference. Jan 20-22, Orlando.

[3]  T.Y Lim and T.W. Chan "*Experimenting remote kilowatt-hour meter reading through Low-voltage power lines at dense housing estates*," IEEE Trans. Power Delivery, vol.17, pp.708-711, July2002.

[4]  S. Mark and D.Radford, "*Design considerations for implementations of large scale automatic meter reading systems*," IEEE Trans. Power Delivery, vol.10, pp.97-103, Jan.1995.

[5]  N.Miura, H.Sato, H. Narita and M.Takaki, "Automatic meter reading system by power line carrier communications," Proc. Inst.Elect.Eng., *vol.137, pp.25-31, 1990.*

Brill Academic Publishers
P.O. Box 9000, 2300 PA Leiden
The Netherlands

Lecture Series on Computer
and Computational Sciences
Volume 4, 2005, pp. 373-376

# Interplay of Charge Distribution and Conformation in Polypeptides. Comparison of Theory and Experiment

Joanna Makowska,[1,2] Katarzyna Bagińska,[1] Franciszek Kasprzykowski,[1] Jorge A. Vila,[2,3] Anna Jagielska,[2] Adam Liwo,[1,2] Lech Chmurzyński,[1±] and Harold A. Scheraga[2]

[1]Faculty of Chemistry, University of Gdańsk, Sobieskiego 18, 80-952 Gdańsk, Poland
[2]Baker Laboratory of Chemistry and Chemical Biology, Cornell University, Ithaca, NY 14853-1301, USA
[3]Universidad Nacional de San Luis, Facultad de Ciencias Físico Matemáticas y Naturales, Instituto de Matemática Aplicada San Luis, Consejo Nacional de Investigaciones Científicas y Técnicas, Ejército de los Andes 950-5700 San Luis, Argentina.

Received 1 July, 2005; accepted in revised form 13 July, 2005

*Abstract:* We assessed the correlation between charge distribution and conformation of flexible peptide by comparing the theoretically calculated potentiometric-titration curves of one model of peptide Ac-Lys-Ala$_{11}$-Lys-Gly$_2$-Tyr-NH$_2$ (P1) in water and methanol, with the experimental curves. The calculation procedure consisted of three steps: (i) global conformational search of the peptide under study using the electrostatically-driven Monte Carlo (EDMC) method with the ECEPP/3 force field plus the SRFOPT or the GBSA solvation model as well as a molecular dynamics method with the AMBER 99/GBSA force field; (ii) re-evaluation of the energy in the pH range considered by using the modified Poisson-Boltzmann approach and taking into account all possible protonation microstates of each conformation, and (iii) calculation of the average degree of protonation of the peptide at a given pH value by Boltzmann averaging over conformations. The experimental titration curves of peptide P1 in water and methanol indicate a remarkable downshift of the first pK$_a$ value compared to the values for reference compounds (n-butylamine and phenol, respectively), suggesting the presence of a hydrogen bond between the tyrosine hydroxyl oxygen and the H$^\varepsilon$ proton of a protonated lysine side chain. The theoretical titration curves agree well with the experimental curves, if conformations with such hydrogen bonds constitute a significant part of the ensemble; otherwise the theory predicts too small a downward pH shift.

*Key words:*peptide conformation; protonation state; potentiometric titration; Poisson-Boltzmann equation; global optimization

## 1. Methods

One part of theory of the method implemented in this work was described in Ref. 1-5. To determine the effect of the reaction field on the ionization of the functional groups, we compared the $\langle x(pH) \rangle$ curves calculated for the peptides under study with those calculated for a mixture of isolated reference model compounds at the same stoichiometric ratio as in the peptide.

For P1 (two types of functional groups) the formula is given by Eq(1):

$$\langle x(pH) \rangle = \frac{2 \times 10^{-pH}}{K_a^{n-BuNH_3^+} + 10^{-pH}} - \frac{K_a^{PhOH}}{K_a^{PhOH} + 10^{-pH}} \qquad (1)$$

where $K_a^{n-BuNH_3^+}$ and $K_a^{PhOH}$ are the acid-dissociation constants of the n-butylammonium cation and phenol, respectively, with the neutral peptide being taken as reference ($\langle x \rangle = 0$).

± Corresponding author: lech@chem.univ.gda.pl

To compare the theoretical titration curves of peptide P1 with the corresponding experimental titration curves, we converted $\langle x(pH)\rangle$ into volume of added titrant ($V(pH)$) as given by Eq. (2).

$$V(pH) = \frac{(2 - \langle x(pH)\rangle)C_p + K_{aw}C_w 10^{pH} - 10^{-pH}}{C^o - K_{aw} 10^{pH} + 10^{-pH}} V_p \qquad (2)$$

where $C_p$ is the concentration of the peptide in the titrated solution, $V_p$ is the volume of the titrated solution, $C^o$ is the concentration of the titrant, $C_w$ is the concentration of water in the solution, and $K_{aw}$ is the acid-dissociation constant of water. Obviously, for aqueous solution we have $K_{aw}C_w = K_w$, where $K_w$ is the ionic product of water.

Table 1: Physicochemical constants of solvents and pK$_a$ values of reference model compounds at 298 K used in calculations

| Solvent | $\sigma$ [dyne/cm][a] | $\varepsilon^b$ | $pK_a^{n-BuNH_3^+}$ | $pK_a^{PhOH}$ |
|---|---|---|---|---|
| Water | $72.0^c$ | $80.0^e$ | $10.5^f$ | $10.1^h$ |
| Methanol | $22.6^d$ | $32.6^e$ | $11.22^g$ | $14.33^i$ |

[a]Surface tension; [b]Dielectric constant; [c]Ref. 6; [d]Ref. 7; [e]Ref. 8; [f]Ref. 9; [g]Ref. 10; [h]Ref. 11; [i]Ref. 8

## 2. Results

Table 2: The pK$_a$ values of peptide P1 in water and methanol at 298.1 K obtained by fitting the equilibrium model of three-stage dissociation to the potentiometric-titration curves. The numbers in parentheses are the standard deviations. (The models of acid-base equilibria were fitted to the resulting titration curves by using the program STOICHIO[12,13] which is based on a nonlinear confluence analysis. This program can treat any model of chemical equilibria and takes into account all possible sources of experimental error (the error in EMF, titrant and titrated solution volume, reagent impurities, etc.).

| | Water | Methanol |
|---|---|---|
| pK$_{a1}$ | 9.36(0.05) | 8.91(0.14) |
| pK$_{a2}$ | 9.94(0.06) | 10.59(0.13) |
| pK$_{a3}$ | 11.32(0.06) | 13.74(0.12) |
| pK$_{aw}$ | $-^a$ | 13.74(0.12) |

[a]In water, pK$_{aw}$ was not an adjustable parameter and the value of 14 was assumed for the negative of the logarithm of the ionic product of water (pK$_w$).

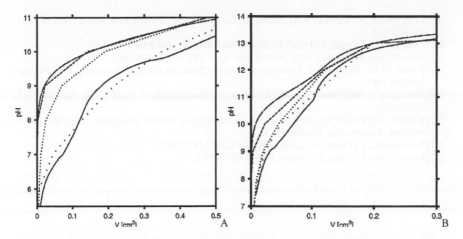

Figure 1: Comparison of theoretically calculated curves for potentiometric titration of peptide P1 in water (A) and methanol (B) with the calculated curves for the potentiometric titration of a stoichiometric mixture of reference model compounds and experimental potentiometric titration curves of peptide P1. Solid line: experimental curve; long-dashed line: stoichiometric mixture of reference model compounds; short-dashed line: curve calculated with the Poisson-Boltzmann approach and EDMC-generated ensemble of conformations; dotted line with dense dots: curve calculated with the Poisson-Boltzmann approach and the ensemble of 2,000 lowest-energy conformations from the MD trajectories; dotted line with sparse dots: curve calculated with the Poisson-Boltzmann approach and the ensemble.

## 3. Discussion

The first $pK_a$ in Table 2 corresponds to the tyrosine hydroxyl engaged in a $(Lys)NH_3^{+\cdots}OH(Tyr)$ hydrogen bond with its oxygen lone-pair serving as a proton acceptor and one of the lysines serving as a donor; this creates a strong positive electrostatic field at the tyrosine hydroxyl group, which facilitates loss of the proton from the OH group, after which a favorable $(Lys)NH_3^{+\cdots}O^-$ (Tyr) salt bridge is formed. Next the protonated ε-amino group of the lysine residue which is not engaged as the proton donor in the salt bridge loses its proton. Because the salt bridge has no net charge, this lysine residue experiences primarily the electrostatic field of the polarized solvent and, consequently, its $pK_a$ value is not much different from that of the reference compound. The last $pK_a$ corresponds to the dissociation of the protonated ε-amino group of the lysine residue engaged in the salt bridge. This $pK_a$ is shifted upward because the group experiences a strong negative electrostatic field from the ionized tyrosine hydroxyl.

For curves (Figure 1) calculated with the ECEPP/3-based ensemble, the theoretically predicted pH shift appears to be too small for peptide P1. The reason for the underestimated pH shift for peptide P1 can be explained in terms of the absence of conformations with the $(Lys)NH_3^{+\cdots}OH(Tyr)$ hydrogen bonds in the ECEPP/3-based ensembleFor the AMBER-based ensemble which contains a significant amount of conformations with the $(Lys^1)NH_3^{+\cdots}OH(Tyr^{16})$ hydrogen bonds, the agreement between the theoretical and experimental titration curves in both solvents improved significantly and, for conformational ensembles composed of conformations that contained such hydrogen bonds, the agreement between theoretical and experimental titration curves is very good.

## References

[1]  Ripoll, D. R.; Vorobjev, Y. N.; Liwo, A.; Vila, J. A.; Scheraga, H. A. *J. Mol. Biol.* **264** 770-783(1996).
[2]  Vorobjev, Y. N.; Scheraga, H. A.; Hitz, B.; Honig, B. *J. Phys. Chem.* **98** 10940-10948(1994).
[3]  Vorobjev, Y. N.; Grant, J. A.; Scheraga, H. A. *J. Am. Chem. Soc.***114**, 3189-3196(1992).
[4]  Vorobjev, Y. N.; Scheraga, H. A. *J. Phys. Chem.* **97** 4855-4864(1993).
[5]  Vorobjev, Y. N.; Scheraga, H. A. *J. Comput. Chem.* **18** 569-583(1997).

[6]  Compostizo, A.; Cancho, S. M.; Rubio, R. G.; Colin, A. C. *Phys. Chem. Chem. Phys.* **3** 1861-1866(2001).
[7]  Wu, M. M.; Cubaud, T.; Ho, C. M. *Phys. Fluid.* **16** L51-L54(2004).
[8]  Bos, M.; van der Linden, W. E. Anal. Chim. Acta **332** 201-212(1996).
[9]  Perrin, D. D., Dissociation Constants of Organic Bases in Aqueous Solutions, Butterworths, London 1965.
[10] Zielińska, J.; Makowski, M.; Maj, K.; Liwo, A.; Chmurzyński, *L. Anal. Chim. Acta* **401** 317-321(1999).
[11] Kortüm, G.; Vogel, W.; Andrussow, K., *Dissociation constants of organic acids in aqueous solution, Butterworths*, London 1961.
[12] Kostrowicki, J.; Liwo, *A. Comput. Chem.* **11**, 195-210(1987).
[13] Kostrowicki, J.; Liwo, A. *Talanta* **37**, 645-650(1990).

Brill Academic Publishers
P.O. Box 9000, 2300 PA Leiden
The Netherlands

*Lecture Series on Computer*
*and Computational Sciences*
Volume 4, 2005, pp. 377-380

# Theoretical Calculations of Homoconjugation Equilibrium Constants in Systems Modeling Acid-Base Interactions the in Side-Chain of Biomolecules Using the Potential of Mean Force

Mariusz Makowski, Joanna Makowska, Lech Chmurzyński[1],

Faculty of Chemistry, University of Gdańsk, Sobieskiego 18, 80-952 Gdańsk, Poland

Received 1 July, 2005; accepted in revised form 11 July, 2005

*Abstract:* The potentials of mean force (PMF's) were determined for systems forming cationic and anionic homocomplexes composed of acetic acid, phenol, isopropylamine, n-butylamine, imidazole and 4(5)-methylimidazole and their conjugated bases or acids, respectively, in three solvents with different polarity and hydrogen-bonding propensity: acetonitrile (AN), dimethyl sulfoxide (DMSO), and water ($H_2O$). For each pair and each solvent a series of umbrella-sampling molecular dynamics simulations with the AMBER force field, explicit solvent, and counter-ions added to maintain a zero net charge of a system were carried out and the PMF was calculated by using the Weighted Histogram Analysis Method (WHAM). Subsequently, homoconjugation equilibrium constants were calculated by numerical integration of the respective PMF profiles. In all cases but imidazole stable homocomplexes were found to form in solution, which was manifested as the presence of contact minima corresponding to hydrogen-bonded species in the PMF curves. The calculated homoconjugation constants were found to be greater for complexes with the OHO bridge (acetic acid and phenol) than with the NHN bridge and they were found to decrease with increasing polarity and hydrogen-bonding propensity of the solvent (i.e., in the series AN > DMSO > $H_2O$), both facts being in agreement with the available experimental data. It was also found that interactions with counter-ions are manifested as the broadening of the contact minimum or appearance of additional minima in the PMF profiles of the acetic acid-acetate, phenol/phenolate system in acetonitrile and the 4(5)-methylimidazole/4(5)-methylimidzole cation conjugated base system in dimethyl sulfoxide.

*Keywords*: homoconjugation, umbrella sampling, potential of mean force, acid-base equilibria

## 1. Methods

Molecular dynamics simulations were carried out with the AMBER[1] suite of programs, using the AMBER 5.0 force field.[2] Each system was placed in a 30 x 27 x 24 Å³ box containing explicit solvent molecules with amount corresponding to experimental solvent density at 298 K (the solvents being water, acetonitrile or DMSO) and simulations were carried out at 298 K in the NVT scheme (constant number of particles, volume, and temperature). A 10 Å cut-off for all non-bonded interactions, including electrostatic interactions, was imposed. Because recent studies[3] showed that neglecting long-range electrostatic interactions could influence the PMF of charged systems, we also carried out some trial calculation with including the Ewald summation[4] to compute the electrostatic energy. The simulation time was 500 ps (picoseconds) and the integration step was 0.001 ps. A total number of 500 000 configurations were generated for each system.

Restraining potentials were imposed on distances between the oxygen (for the acetic acid/acetate and the phenol/phenolate system) or nitrogen (for the remaining systems) atoms of the interacting molecules that are involved in hydrogen bonding to keep them within the derived distance range.

The charges on the atoms of solute molecules needed for the AMBER 5.0 force field were determined by using a standard procedure[6] by fitting the point-charge electrostatic potential to the molecular electrostatic potential computed using the electronic wave function calculated at the restricted Hartree-Fock (RHF) level with the 6-31G* basis set. The program GAMESS[5] was used to carry out quantum-mechanical calculations, while the program RESP[6] of the AMBER 5.0 package was

---

[1] Corresponding author: lech@chem.univ.gda.pl

used to compute the fitted charges. Because the systems studied bear a net charge, chloride or sodium counter ions, respectively, were added to neutralize a system.

Force-field parameters of the solvent molecules were taken from Ref. 7 for water (the TIP3P model), Ref. 8 for DMSO, and Ref. 9 for acetonitrile, respectively. For neutral N-methylguanidine and 4(5)-methylimidazole we assumed the "standard" tautomeric forms corresponding to the neutral forms of histidine and arginine side chains. To determine the potentials of mean force of the systems studied, we carried out umbrella-sampling MD simulations and processed the results by the weighted histogram analysis (WHAM)[11,12] method.

The homoconjugation equilibrium constants, $K_{homo}$, were calculated by integration of the PMF profiles from equation 1, following Ref. 13:

$$K_{homo} = \frac{\int_{r_0}^{r_1} \exp(-\frac{W(r)}{RT}) 4\pi r^2 dr}{\frac{1}{V} \int_{r_1}^{r} \exp(-\frac{W(r)}{RT}) 4\pi r^2 dr} \tag{1}$$

where $r_0$ is the shortest distance found in simulations distance, $r_1$ is the distance at which the first maximum in the PMF occurs, R is the gas constant (8.314 J/(mol K), T is the absolute temperature (298 K) and $V = \frac{4}{3} \cdot \pi \cdot r^3$. Integration was carried out numerically.

## 2. Results

Table 1: Logarithms of the anionic and cationic homoconjugation constants, $\log_{10} K_{homo}$, determined by molecular dynamics calculations with the AMBER force field and explicit solvent models in acetonitrile (AN), dimethyl sulfoxide (DMSO) and water (H$_2$O). For comparison, experimental homoconjugation constant values (exp) determined in acetonitrile[14] are included.

| System[a] | $\log_{10} K_{homo}$ | | | |
|---|---|---|---|---|
| | AN | exp.[e] | DMSO | H$_2$O |
| AcO$^-$/AcOH [b] | 5.12 | 3.21 | 3.82 | 2.63 |
| PhO$^-$/PhOH [b] | 4.43 | 4.80 | 6.19 | 2.67 |
| MeGuaH$^+$/MeGua [c] | 2.73 | f | 3.13 | 2.44 |
| MeImidH$^+$/MeImid [c] | 0.93 | 1.49 | -0.46 | 0.77 |
| ImidH$^+$/Imid [c] | d | d | d | d |
| iso-PropH$^+$/iso-Prop [c] | 4.25 | f | 3.15 | 2.25 |
| n-ButH$^+$/n-But [c] | 4.61 | 1.30 | 3.70 | 2.33 |

[a] Abbreviations for compounds: AcOH – acetic acid; PhOH – phenol; MeGua – methylguanidine; MeImd – methylimidazole; Imid – imidazole; iso-Prop – isopropylamine; *n*-But – *n*-butylamine. [b] Anionic homoconjugation;. [c] Cationic homoconjugation. [d] Equilibrium does not establish. [e] From Ref. 14. [f] Experimental data not available.

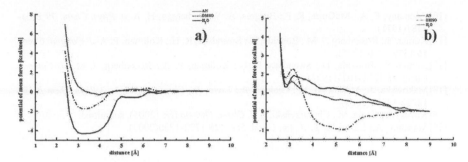

Figure 1: The PMF profiles of the AcO⁻/AcOH (a), MeImidH⁺/MeImid (b) in acetonitrile, (solid lines) dimethyl sulfoxide (dot-dashed lines), and water (dashed lines).

## 3. Discussion

The PMF curves for the acetic acid − acetate (figure 1a) and 4(5)-methylimidazol − 4(5)-methylimidazol (figure 1a) cation pairs and all solvents studied are plotted as functions of the distances between the oxygen (nitrogen) atoms involved in hydrogen bonding. The logarithms of the calculated anionic and cationic homoconjugation constant values [calculated from the PMF profiles following Eq. (1)] are summarized and compared with the available experimental homoconjugation constants determined in acetonitrile in Table 1.

The PMF's of the acetic acid − acetate (figure 1a) and 4(5)-methylimidazol − 4(5)-methylimidazol (figure 1a) cation pairs obtained in our study can be compared to the PMF's of the acetic acid − acetate (Figure 14 of Ref. 15) and 4(5)-methylimidazol − 4(5)-methylimidazol cation pairs (Fig. 8 of Ref. 15), respectively, reported in the study of Masunov and Lazaridis.[15] It should be noted, however, that our results pertain to averaging over all possible orientations, while in their study the orientation was restricted to head-to-head. For the propionic acid − propionate pair Masunov and Lazaridis found a depth of a contact minimum of -2 kcal/mol, while our value for the AcO⁻...AcOH pair (averaged over all orientations) is −0.30 kcal/mol (Figure 1a − dashed lines). For the 4(5)-methylimidazole − 4(5)-methylimidazole cation pair they found a contact minimum with depth of -2 kcal/mol at a distance 3 Å, while our value for the MeImidH⁺-MeImid pair (averaged over all orientations) has the PMF of 0.95 kcal/mol at minimum (Figure 1a − dashed lines). These differences can be attributed to the fact that the orientation of the interacting species in their work was constrained, while in our calculations was not.

## References

[1] Pearlman, D. A.; Case, D. A.; Caldwell, J. W.; Ross, W. S.; Cheatham III, T. E.; DeBolt, S.; Ferguson, D.; Seibel, G.; Kollman, P. A. *Comp. Phys. Commun*, **91** 1-41(1995).

[2] Case, D.A.; Pearlman, D. A.; Caldwell, J. W.; Cheatham III, T. E.; Ross, W. S.; Simmerling, C. L.; Darden, T. A.; Merz, K. M.; Stanton, R. V.; Cheng, A. L.; Vincent, J. J.; Crowley, M.; Ferguson, D. M.; Radmer, R. J.; Seibel, G. L.; Singh, U. C.; Weiner, P. K.; Kollman, P. A. AMBER 5 1997, University of California, San Francisco.

[3] Maksimiak, K.; Rodziewicz-Motowidło, S.; Czaplewski, C.; Liwo, A.; Scheraga, H. A. *J. Phys. Chem. B* **107** 13496-13504(2003).

[4] Darden, T.; York, D.; Pedersen, L. *J. Chem. Phys.* **98** 10089-10092(1993).

[5] Schmidt, M. W.; Baldridge, K. K.; Boatz, J. A.; Elbert, S. T.; Gordon, M. S.; Jensen, J. A.; Koseki, S.; Matsunaga, N.; Nguyen, K. A.; Su, S.; Windus, T. L.; Dupuis, M.; Montgomery, J. A. *J. Comput. Chem.* **14** 1347-1363(1993).

[6] Bayly, C. I.; Cieplak, P.; Cornell, W. D.; Kollman, P. A. *J. Phys. Chem.* **97** 10269-10280(1993).

[7] Jorgensen, W. L.; Chandrasekhar, J.; Madura, J. D.; Impey, R. W.; Klein, M. L.; *J. Chem. Phys.* **79** 926-935(1983).

[8] Liu, H.; Müller-Plethe, F.; van Gunsteren, W. F. *J. Am. Chem. Soc.* **117** 4363-4366(1995).

[9] Jorgensen, W. L.; Briggs, J. M. *Mol. Phys.* **63** 547-558(1988).

[10] Momany, F. A.; McGuire, R. F.; Burgess, A. W.; Scheraga, H. A. *J. Phys. Chem.* **79** 2361-2381(1975).

[11] Kumar, S.; Rosenberg, J. M.; Bouzida, D.; Swendsen, R. H.; Kollman, P. A. *J. Comput. Chem.* **16** 1339-1350(1995).

[12] Kumar, S.; Bouzida, D.; Swendsen, R. H.; Kollman, P. A.; Rosenberg, J. M.; *J. Comput. Chem.* **13** 1011-1021(1992).

[13] Makowska, J.; Makowski, M.; Giełdoń, A.; Liwo, A.; Chmurzyński, L. *J Phys Chem B* **108** 12222-12230(2004).

[14] Kozak, A.; Czaja, M.; Chmurzyński, L. *J, Chem. Thermodyn.* (2005), submitted.

[15] Masunov, A.; Lazaridis, T. *J. Am. Chem. Soc.* **125** 1722-1730(2003).

Brill Academic Publishers
P.O. Box 9000, 2300 PA Leiden
The Netherlands

*Lecture Series on Computer
and Computational Sciences*
Volume 4, 2005, pp. 381-383

# Decomposition method in comparison with Numerical Solutions of Burgers Equation

Ch. Mamaloukas[1], S.I. Spartalis[2] and Z. Manussaridis[3]

[1]Department of Informatics,
University of Economics and Business
76 Patision Str., GR-10 434 Athens, Greece

[2]Department of Production Engineering & Management
School of Engineering, Democritus University of Thrace
University Library Building, Kimeria, GR-671 00 Xanthi, Greece

[3]Department of Planning and Regional Development
Polytechnic School, University of Thessaly
GR-38 221 Volos, Greece

Received 20 May, 2005; accepted in revised form 13 June, 2005

*Abstract*: This paper presents a solution of the one-dimension Burgers equation using the Decomposition Method and compares this solution to the analytic solution [Cole] and solutions obtained with other numerical methods. Even though decomposition method is a non-numerical method, it can be adapted for solving nonlinear differential equations. The advantage of this methodology is that it leads to an analytical continuous approximated solution that is very rapidly convergent [2,7,8]. This method does not take any help of linearization or any other simplifications for handling the non-linear terms. Since the decomposition parameter, in general, is not a perturbation parameter, it follows that the non-linearities in the operator equation can be handled easily, and accurate solution may be obtained for any physical problem.

*Keywords:* Burgers equation, Adomian Decomposition Method, Adomian polynomials.
*Mathematics SubjectClassification:* 65C20, 65M99, 65N55

## 1. Introduction

Many problems in Fluid Mechanics and in Physics are governed generally by the Navier-Stokes equations. These equations can show the behaviour of a certain attribute (e.g. momentum, heat) in space and time. The one-dimension non-linear differential equation which is used as a model for these problems is Burgers equation. This equation is applied to laminar and turbulence flows as well. The Burgers equation which is the one-dimension nonlinear Diffusion Equation is similar to the one dimension Navier-Stokes equation without the stress term. Many researchers tried to find analytic and numerical solutions of this equation using the appropriate initial and boundary conditions. Characteristically in Benton and Platzman [10] are mentioned almost 35 distinct solutions of Burger equation but only the half of them are having physical interest. Agas [9] tried to get approximate solution of Burger equation using a new numerical solution which is called Group of Explicit Method. He also tried the method of Finite Differences and the method of Lines in Finite Elements. The problem he faced was that these methods could not give solutions for big values of the Reynolds number. He also found some problems in convergence.

In this paper, a solution obtained by the Adomian's Decomposition method, which is described briefly in this paper and was used by Mamaloukas [12] for the numerical solution of the one-dimensional

---

[1] Corresponding author: mamkris@aueb.gr
[2] Second author: sspart@pme.duth.gr
[3] Third author: manzac@gen.auth.gr

Kortweg-de Vries equation, is compared graphically to the analytic and to some others numerical methods. As it is shown in the diagrams at the end of this paper this method gives a computable and accurate solution of the problem using only a small number of terms.

## 2. Formulation of the Problem

Consider the Burgers equation with the following form

$$\frac{\partial u}{\partial t} + u\frac{\partial u}{\partial x} = v\frac{\partial^2 u}{\partial x^2}$$

with boundary conditions    $u(0,t) = u(1,t) = 0$      *for*    $t \geq 0$

and initial condition:        $u(x,0) = 4x(1-x)$  *or*  $\sin \pi x$

Using the Adomian's Decomposition Method we finally obtained

$$u = u_0 + \frac{1}{2}\left[v^{-1}L_x^{-1}\left(L_t u + Nu\right) + L_t^{-1}\left(v \ L_x u - Nu\right)\right]$$

where $L_t = \dfrac{\partial}{\partial t}$    and    $L_x = \dfrac{\partial^2}{\partial x^2}$ . The operators $L_t^{-1}$    and    $L_x^{-1}$ are the one and twofold right-inverse operators of $L_t$    and    $L_x$ respectively, given by the form

$$L_t^{-1} = \int(\cdot)\,dt \quad \text{and} \quad L_x^{-1} \iint(\cdot)\,dx\,dx$$

and $u_0$ is to be determined from the initial conditions, so, is

$$u_0 = 4x(1-x) \ \ or \ \ \sin \pi x .$$

If we suggest as a solution of u an approximation of only two terms then we have the solution

$$u = u_0 + u_1 .$$

## 3. Results, Diagrams and Discussions

For the solution of this equation we used the initial conditions $u(x,0) = 4x(1-x)$ and $u(x,0) = \sin \pi x$ without restricting generality. The boundary conditions were $u(0,t) = u(1,t) = 0$    *for*    $t \geq 0$ . For comparison reasons with other published papers we used interspaces $\Delta x = 0.25$ and time amplitude $0.01 \leq t \leq 0.25$

In the above diagrams numerical results of Burger equation are registered for different values of v. For comparison reasons we used values of viscosity $v = 1$, $v = 0.1$, $v = 0.01$, $v = 0.001$

The analytic solution as it is described by Cole [13] is liable to restrictions concerning the values of the coefficient $v = \dfrac{1}{R_e}$ . For example, if the value of the Reynolds number is greater than 1000 then we can not find any solution because Fourier series do not converge. For this reason we try numerical approaches, like finite differences and finite elements.

Concerning finite differences the explicit method give us adequate results if and only if $\lambda \leq 1/2$ . Otherwise results did not converge. With the implicit method we do not need the covenant $\lambda \leq 1/2$, but we need a large number of calculations. Finally, the group of explicit methods gives us adequate results with few calculations and the method is more stable.

Concerning finite elements, the method of lines with Gauss and Hermite was used with initial and boundary conditions from Madsen and Sincovec [14]. These methods, using great values of Reynolds number, gave us adequate results without the limitations for $\Delta x$ and $\Delta t$ , with small number of repetitions and without stability limitations. However, the errors depend first on the choice of the polynomial and second on the choice of $\Delta x$ and $\Delta t$

Concerning the decomposition method from the above diagrams it is obvious how powerful this method is. Using only two terms we can obtain similar results with the other numerical methods and the analytic solution. Of course, in some cases the present solutions deviate from the solutions given in the table. The decomposition solution can be further improved if more-term approximations of the solution are obtained.

As far as accurate results are concerned, computational experience has shown that they can be obtained easily by taking half a dozen terms. In case we do not have a sufficiently high precision by using a few of the $A_n$, then accordingly to Rach R. [15] there are two alternatives. One is to compute additional terms by any of the available procedures. The second approach is to use the Adomian-Malakian "convergence acceleration" procedure [16]. This unique approach conveniently yields the error-damping effect of calculating many more terms of the $A_n$ to determine whether further calculation is required.

The advantage of this method is to avoid simplifications and restrictions which change the non-linear problem to a mathematically tractable one, whose solution is not consistent to physical solution.

Further study on the stability and the convergence of the solutions will prove the accuracy of the above conclusions.

## 4. Conclusions

The great advantage of the decomposition method is that of avoiding simplifications and restrictions which change the non-linear problem into a mathematically tractable one, whose solution is not consistent to physical solution. Further study on the stability and the convergence of the solutions will prove the accuracy of the above method.

## Acknowledgments

The author wishes to thank the anonymous referees for their careful reading of the manuscript and their fruitful comments and suggestions.

## References

[1] Adomian G., Nonlinear Stochastic Operator Equations, Academic Press, (1986)

[2] Adomian G., Nonlinear Stochastic Systems Theory and Applicat. to Physics, Kluwer Acad. Publishers, (1989)

[3] Adomian G., J. Math. Anal. Appl., 119, pp 340-360, (1986)

[4] Adomian G., Appld. Math. Lett., 6, No5, pp 35-36, (1993)

[5] Adomian G., Rach R., On the Solution of Nonlinear Differential Equations with Convolution Product Non-linearities, J. Math. Anal. Appl., 114, pp 171-175, (1986)

[6] Adomian G., Solving Frontier Problems of Physics: The Decomposition Method, Kluwer AcademicPublishers, (1994)

[7] Cherruault Y., Kybernetes, 18, No2, pp 31-39, (1989)

[8] Cherruault Y., Math. Comp. Modeling, 16, No2, pp 85-93, (1992)

[9] Agas C., The Effect of Kinematic Viscosity in the Numerical Solution of Burger Equation, Thessalloniki,1998.

[10] Benton E.R. & Platzman G.W., A Table of Solutions of the one-dimensional Burgers Equation, Quart. Appl. Math., (1972)

[11] Burgers J. M., 'The Nonlinear Diffusion Equation', D. Reidel Publishing Company, Univ. of Maryland, USA (1974)

[12] Mamaloukas C., Numerical Solution of one dimensional Kortweg-de Vries Equation, BSG Proceedings 6, Global Analysis, Differential Geometry and Lie Algebras, pp 130-140, (2001)

[13] Cole J.D., On a Quasilinear Parabolic Equation Occurring in Aerodynamics, A.Appl. Maths, 9, pp 225-236, (1951).

[14] Madsen N. K. and Sincovec R. F., General Software for Partial Differential Equations in Numerical Methods for Differential System, Ed. Lapidus L., and Schiesser W. E., Academic Press, Inc., (1976).

[15] Rach R., A Convenient Computational Form of the Adomian Polynomials, J. Math. Anal. Appl., V 102, pp 415-419, (1984).

[16] Adomian G. and Malakian, Self-correcting approximate solutions by the iterative method for nonlinear Stochastic Differential Equations, J. Math. Anal. Appl., V76, pp 309-327, (1980).

[17] Adomian G. and Malakian, Inversion of Stochastic Partial Differential Operators-The Linear Case, J. Math. Anal. Appl., V 77, pp 505-512, (1980).

[18] Adomian G. and Malakian, Existence of the Inverse of a Linear Stochastic Operator, J. Math. Anal. Appl.,114, pp 55-56, (1986).

[19] Adomian G. and Rach R., Inversion of Nonlinear Stochastic Operators, J. Math. Anal. Appl., 91, pp 39-46, (1983).

Brill Academic Publishers
P.O. Box 9000, 2300 PA Leiden,
The Netherlands

*Lecture Series on Computer
and Computational Sciences*
Volume 4, 2005, pp. 384-387

# Speech Frame Extraction Using Neural Networks and Message Passing Techniques

**A.Margaris[1], E.Kotsialos and M.Roumeliotis**

Department of Applied Informatics, University of Macedonia,
GR-540 06 Thessaloniki, Greece

Received 5 August, 2005; accepted in revised form 15 August, 2005

*Abstract:* The objective of this project is the implementation of a parallel application, capable of extracting speech frames associated with Greek words pronounced by two speakers (one male and one female). The speech recognition paradigm is the Neural Networks approach, and the parallel programming model is the Message Passing Interface (MPI).

## 1 Introduction

Speech recognition is one of the most popular and important types of signal processing. It can be performed in a variety of different ways. Two standard modelling techniques are the Dynamic Time Warping (DTW) [3], and the Hidden Markov Models (HMM) [1][7]. However, in this project, the extraction of speech frames is based on the use of feed-forward neural networks (FFNNs) trained by the back propagation algorithm. There are many research projects studying the usage of neural networks for the speech recognition problem [6], such as Kohonen's Phonetic Typewriter [4] and the Time-Delay Neural Networks (TDNNs) [5].

## 2 Speech Processing

The objective of speech processing is the extraction of feature vectors capable of identifying isolated word frames. In this project this technique was applied to speech samples recorded by an ordinary microphone - using the Sound Recorder Windows application - and a standard on-board PC sound card. The sampling frequency of the recording was 11.025 KHz, with each speech sample to be recorded with a precision of 8 bits. The following operations were applied to the speech signal [2]:

- Pre-emphasis: the speech signal $s(n)$ that represents a single word is passed though a first order FIR (Finite Impulse Response) filter with a transfer function $H(z) = 1 - \alpha z^{-1}$ where $\alpha = 0.950$ in order to be spectrally flattened. By applying this filter, the $DC$ component of the signal is removed.

- Frame Blocking: the pre-emphasized signal is divided into overlapping frames of $N$ samples with the size of the overlap region to be $M$ samples. In this project the values of these parameters were $N = 256$ and $M = 64$.

- Windowing: each one of the speech frames is windowed, in order to minimize the signal discontinuities at the borders of the frame. The standard Hamming window was used in this project.

---

[1]Corresponding author. E-mail: amarg@uom.gr

- LPC analysis [1] : for each windowed speech frame, two vectors containing the LPC (Linear Predictive Coding) and the Cepstral Coefficients, respectively, were calculated by applying the autocorrelation technique and the Levinson-Durbin algorithm [3]. The number of these coefficients - known as LPC order and Cepstral order, respectively - were set equal to 10.

## 3    Training Set Implementation

The speech processing technique as described above, was applied to the Hellenic words [ένα, δύο, τρία, ..., δέκα] that correspond to the English words [one, two, three, ... ten], respectively. In order to achieve a good representation of each word sample, 15 occurrences of each word (10 for training and 5 for recall) have been recorded. For each occurrence, the feature vectors for each word frame were calculated, and an average vector per frame was put in the training set of the neural network. Regarding the training operation, it has been applied to 10 different training sets containing the feature vectors of the average word sample, for each one of the ten words. The average number of training vector pairs was about 60, while the number of inputs was equal to 10 (since 10 LPC coefficients were used to model each speech frame). For each input vector, the corresponding desired output vector was the vector of Cepstral coefficients, calculated by the application of the previously described method.

## 4    The algorithm and the experimental results

The speech recognition algorithm used in this project is based on the Message Passing approach [8], and combines the usage of point to point as well as collective operations. In the first step the training set is created for each word template, by following the rules described above. Then, a feed-forward neural network was constructed and trained by means of the back propagation algorithm. The results of the training operation for the first three words templates are shown in Table 1.

Table 1: Back propagation training results for the learning of the neural network for the first three word templates.

| WORD TEMPLATE | ένα | δύο | τρία |
|---|---|---|---|
| Network File Name | 1.net | 2.net | 3.net |
| Network Structure | 10-15-10 | 10-15-10 | 10-15-10 |
| Learning Rate | 0.8 | 0.7 | 0.4 |
| Sigmoidal Slope | 1.0 | 1.0 | 1.0 |
| Momentum | 0.5 | 0.4 | 0.8 |
| Iterations | 4877 | 1571 | 4344 |
| RMS Error | 0.000024419 | 0.000035843 | 0.00002208 |
| Global Error | 0.119150 | 0.166849 | 0.176932 |

After the construction of the trained neural network, the parallel application was started with a process number $N = 6$. These processes belong to the main process group associated with the default communicator MPI_COMM_WORLD and the process with a rank value equal to zero, was marked as the root process. This process acts as a speech recognition server. It first loads the three data files used by the algorithm, i.e. (a) the neural network configuration file that contains the structure of the trained neural network, (b) the training set file that contains the feature vectors recognized by the network, and (c) the configuration file that contains the parameter values of the speech processing stage. These four values are broadcasted to the remaining processes; each one of them reads the speech data of its own speech file, and performs the speech processing steps to calculate the LPC coefficients of the associated word frames. The LPC vectors produced in this way are then sent to the root process and they are submitted by it to the trained neural network that produces the corresponding output vectors that are supposed to represent the Cepstral vector of the corresponding input frame. The calculation of the total number of messages received by the root process is based to the usage of control messages that contain the number of sent frames per process. This algorithm is shown graphically in figure 1.

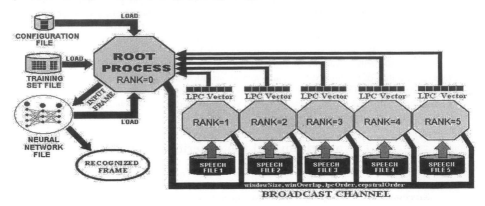

Figure 1: The structure of the parallel application for the speech recognition project.

The recognition accuracy of the above algorithm is shown in figure 2, in a histogram fashion. For each one of the incoming frames, the closest frame of the training set was identified, and a histogram that shows the number of the incoming frames mapped to each Cepstral vector was constructed. The histogram for the first word template is shown in figure 2. Regarding the performance of the parallel application for the first three word templates, it is shown in table 2. It can be seen that, due to the very small training set size, the processing time of the serial case is equal - or even smaller - to the processing time of the parallel case, where the time delay of the network traffic associated with the message passing operations has to be added to the CPU execution time. However, in a real application with hundreds of speech files, each one of them containing hundreds of frames, the processing time of the parallel application is expected to be quite smaller than the processing time of the serial case.

Table 2: Execution time of the speech frame extraction application for the first three word templates and for the serial as well as the parallel case. Due to hardware limitations, the parallel application run on a small home network of 2 PC's with a 2.5GHz CPU and 256 MB RAM.

|  | ένα | δύο | τρία |
|---|---|---|---|
| Serial Case (N=1, P=1) | 0.492854221 | 0.640342528 | 0.644194388 |
| Parallel Case (N=6, P=1) | 0.672103354 | 0.817163466 | 0.832702871 |
| Parallel Case (N=6, P=2) | 0.846682693 | 0.956584556 | 0.924264254 |

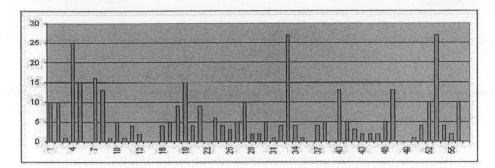

Figure 2: Histogram showing the number of mapped incoming frames for each training set vector pair and for the word template ένα (first histogram).The horizontal axis represents the frame position inside the word template, while the vertical axis is associated with the occurrence frequency of each speech frame.

## 5 Conclusions and Future Work

The objective of this project was the demonstration of the capabilities of the message-passing parallel programming model, as it is applied to speech processing applications. This is a very simple application that uses a simple process group, but it can be improved - by using the LAM-MPI implementation [9] - to support the parallel speech processing task in a client - server fashion.

## References

[1] L. Rabiner, B-H Juang *Fundamentals of Speech Recognition*, Prentice Hall, 1993, ISBN 0-130-151572.

[2] L. Rabiner, R. Schafer *Digital Processing of Speech Signals*, Pearson Education, 1978, ISBN 0-132-136031.

[3] J. Deller, J. Hansen, J. Proakis *Discrete Time Processing of Speech Signals*, Wiley - IEEE Press, 1999, ISBN 0-780-353862.

[4] T. Kohonen, *The Neural Phonetic Typewriter*, Computer, Volume 21, No. 3, 1988.

[5] K. Lang, Waibel A, *A Time-Delay Neural network Architecture for Isolated Word Recognition*, Neural networks, Volume 3, 1990.

[6] J. Tebelskis, *Speech Recognition Using Neural Networks*, PHD Thesis, School of Computer Science Carnegie Mellon University Pittsburgh, Pennsylvania 15213-3890, 1995.

[7] J. Hennebert, *Hidden Markov Models and Artificial Neural Networks for Speech and Speaker Recognition*, PHD Thesis, School of Computer Science Carnegie Mellon University Pittsburgh, Pennsylvania 15213-3890, 1995.

[8] P. Pacheco, *Parallel Programming with MPI*, Morgan Kaufmann Publishers Inc, San Francisco, California, 1997, ISBN 1-55860-339-5.

[9] *LAM-MPI Documentation*, available from URL *http://www.lam-mpi.org*.

Brill Academic Publishers
P.O. Box 9000, 2300 PA Leiden
The Netherlands

*Lecture Series on Computer
and Computational Sciences*
Volume 4, 2005, pp. 388-389

# Photoionisation of Nitric Oxide in the Upper Atmosphere.
# A MQDO Study

I.Martín[†]; E. Mayor; A. M. Velasco
Department of Physical Chemistry
University of Valladolid
E-47005 Valladolid (SPAIN)

Received 9 August, 2005; accepted in revised form 17 August, 2005

In recent years, nitric oxide (NO) has attracted increased attention as an atmospheric pollutant, and also as an important reagent in astrophysical and atmospheric chemistry. For instance, on the one hand, NO has a direct contribution to the destruction of Ozone in the Stratosphere, and, on the other, is involved in the photochemical production of $O_3$ in the Troposphere. It also plays a key role in combustion processes, being a major product of internal combustion engines and combustion power plants. Recently, nitric oxide has also been found to be of relevance in biology and physiology.

The role played by NO in the formation of the Ionospheric D-region was also realized, whilst the study of nitric oxide in the Mesosphere was stimulated by in situ measurements of the gas conditions. The photodissociation of NO has long been recognized as an important process in the Mesosphere and Stratosphere.

Crutzen brought attention to the role played by this compound as a major participant in the Stratospheric photochemical balance, given the existence of a substantial source of NO.

Nitric oxide arises from the reaction of nitrous oxide with atomic oxygen in its first excited level, following chemical reaction,

$$N_2O + O(^1D) \rightarrow 2\, NO$$

The photodissociation of NO in the Mesosphere and Stratosphere takes place through predissociation in the $\delta(0\text{-}0)$ and $\delta(1\text{-}0)$ **bands** of the $C^2\Pi\text{-}X^2\Pi$ transition, known as "delta bands". The entire process may be illustrated as follows,

$$NO(X^2\Pi) + h\nu \rightarrow NO(C^2\Pi) \rightarrow N(^4S) + O(^3P)$$

Particular attention must be paid to the details of light absorption in the 189.4-191.6 nm and 181.3-181.5 nm spectral ranges, respectively. This requires an accurate knowledge of the rotational structure of the bands.

For a good account of the above processes, calculations of the photoabsorption cross section must be performed on a line-by line basis. Accordingly, we have computed the integrated photoabsorption cross section of the rotational lines belonging to the $\delta(0\text{-}0)$ and $\delta(1\text{-}0)$ bands of the $C^2\Pi$ - $X^2\Pi$ electronic transition, as a first step in our study of the Mesospheric and Stratospheric

---

[†] Corresponding author. E-mail*: imartin@qf.uva.es

photodissociation of NO.

Numerical values of the relevant photoabsorption cross sections can enable one to calculate the rates at which atmospheric molecules are photodissociated. In order to calculate the photodissociation coefficient of NO we assume that the predissociation occurring in the $\delta(0\text{-}0)$ and $\delta(1\text{-}0)$ bands is mainly responsible for the photodissociation of NO, as previous works have demostrated. Particular attention must be devoted to the penetration of solar radiation in the upper atmosphere and its crucial effects on the predissociation of NO.

In this paper the effects of the dissociation of NO in the Mesosphere and Stratosphere at altitudes between 20 and 90 Km will be sought.

The Molecular Quantum Defect Orbital (MQDO) approach [1] has been employed in the calculations. This methodology has proven to yield quantitative intensities for transition intensities involving molecular Rydberg states, sometimes with vibrational and rotational resolution, in a variety of molecular species. This methodology also has been applied to the photolysis rate constant calculations [2- 5] for the atmospherically important molecules $CF_3Cl$ and $CF_2Cl_2$. In this communication we will present results that can have direct application to the study of models for the Stratosphere and Mesosphere.

# References

[1] I. Martín, C. Lavin, A. M. Velasco, J. Karwwoski, and G. H. F. Dierksen, Chem. Phys. 202, 307 (1996).
[2] I. Martín, C. Lavin, and J. Karwwoski, Chem. Phys. Lett. 255, 89 (1996).
[3] E. Bustos, A. M. Velasco, I. Martin, and C. Lavin, J. Phys.Chem. A 106, 35 (2002).

[4] A. M. Velasco, I. Martin, and C. Lavin, J. Phys.Chem. A 106, 6401 (2002).
[5] E. Mayor, A. M. Velasco, and I. Martín, J. Phys. Chem. A 108, 5699 (2004).

Brill Academic Publishers
P.O. Box 9000, 2300 PA Leiden,
The Netherlands

*Lecture Series on Computer*
*and Computational Sciences*
Volume 4, 2005, pp. 390-393

# Partially Linear SVMs used to Predict Mine Slope Movements

## J. M. Matías[1], J. Taboada[2], C. Ordóñez[2]

[1]Department of Statistics, University of Vigo, Spain
[2]Department of Natural Resources, University of Vigo, Spain

Received 9 July, 2005; accepted in revised form 30 July, 2005

*Abstract:* Several autoregressive models were used with a view to predicting monthly movement in a mine slope with an impact on the safety of the mining operation. These models were: ARMAX models, local linear regression models –LLR), and support vector machines (SVMs). We also propose a partially linear version of the SVMs, PL-SVM, which unlike standard SVMs, uses a linear kernel, composed of a linear kernel in which a number of variables participate, and a non-linear kernel which affects the other variables. In our problem, the PL-SVM improves on the results of the other approaches, including those for the classical non-parametric partially linear models.

*Keywords:* Partially Linear Models, SVM, Prediction, Mining Safety.

*Mathematics Subject Classification:* 62P30-62G08-62M10-68T99-93E99

## 1 Introduction

The support vector machine (SVM) approach [3] is one of the most relevant non-parametric curve-smoothing methods of recent years. However, as used to date the model implemented by the SVMs generally makes difficult to distinguish between the different covariables that influence the response. At best, and if, for example, a Gaussian kernel is used, the influence of the different covariables can be distinguished through the use of a variance-covariance matrix, possibly non-diagonal, and with different values in the diagonal. But this approach implies estimation problems (a great deal of data is required to estimate the large number of parameters), as well as computational problems.

To date, partially linear models [2] that use SVMs have not as yet been implemented that would permit the postulation of a linear relationship between the response and a group of covariables and of a more complex relationship between the response and the other covariables.

One such endeavor was the PL-LSSVM, developed by [1]. However, despite including SVM in its name, Espinoza's PL-LSSVM is basically a spline, to which a linear term has been added. The PL-LSSVM is not, therefore, an authentic SVM, as it does not permit regularization of the coefficients of the linear term nor does it permit use of the Vapnik $\varepsilon$-insensitive loss [3] (below we will assume a regression problem).

In this article, we describe a method for constructing a simple and intuitive partially linear SVM that can be easily extended to other partially parametric models (e.g., to partially polynomial models), and apply it, using an autoregressive approach, to the problem of predicting mine slope movement.

---

[1]Corresponding author. E-mail: jmmatias@uvigo.es

Open-pit mining requires mining slopes to be created, whose instability often slow down operations and affect a mine's safety. The instabilities arise in the exploitation process, which involves the extraction of materials that formerly provided support to the slopes, but which, due to gravity and interstitial pressure from the water within the slopes, end up becoming unstable.

From the data provided by a robotic topographic total station, which measures displacements in real time in three dimensions at different topographic key points located in mine slopes, we endeavor to predict slope movement which would enable short-term mining operation planning to be adjusted.

## 2 Partially Linear SVM

Let the model be as follows:

$$Y_i = h(\mathbf{Z}_i; \beta) + g(\mathbf{T}_i) + \varepsilon_i, \ i = 1, ..., n \tag{1}$$

$$= \psi(\mathbf{X}_i; \beta) + \varepsilon_i, \ i = 1, ..., n \tag{2}$$

where for $i = 1, ..., n$, $\varepsilon_i$ is i.i.d. zero-mean random noise, $\mathbf{Z}_i$ is the ith observation of the random vector $\mathbf{Z} \in \mathbb{R}^{d_1}$, $h(\cdot; \beta) : \mathbb{R}^{d_1} \to \mathbb{R}$ is a known and predetermined function (which will serve as the parametric term in the model), parameterized by a vector $\beta \in \mathbb{R}^m$ to be estimated, $g : \mathbb{R}^{d_2} \to \mathbb{R}$ is an unknown function, and $\mathbf{T}_i$ is the ith observation of the random vector $\mathbf{T} \in \mathbb{R}^{d_2}$. Likewise, we denote as $\mathbf{X} = (\mathbf{Z} \ \mathbf{T})$ the vector for all the covariables, with observations $\mathbf{X}_i = (\mathbf{Z}_i \ \mathbf{T}_i) \in \mathbb{R}^d$, with $d = d_1 + d_2$, $i = 1, ..., n$, and with $\psi$ as the regression function $\mathbb{E}(Y|\mathbf{X} = \mathbf{x})$. The model in (1) includes as a particular case the partially linear model – when $h(\mathbf{Z}_i; \beta) = \mathbf{Z}_i^t \beta$, – but also other parametric models (e.g. higher degree polynomials).

The model in (2) is a typical application of SVM to regression. Thus, by means of a transformation $\phi : \mathcal{S} \subset \mathbb{R}^d \to \mathbb{R}^r$ in which $r$ may be infinite, a feature space is defined $\mathcal{F} = \{\phi(\mathbf{x}) : \mathbf{x} \in \mathcal{S} \subset \mathbb{R}^d\}$ as equipped with an internal product defined in turn by means of a positive definite function (kernel) $\langle \phi(\mathbf{x}_i), \phi(\mathbf{x}_j) \rangle_{\mathcal{F}} = k(\mathbf{x}_i, \mathbf{x}_j)$ in which the problem is posed of determining the optimum hyperplane $f_{\mathbf{w},b}(\mathbf{x}) = \langle \mathbf{w}, \phi(\mathbf{x}) \rangle + b$. The solution takes the form $\mathbf{w} = \sum_{\text{s.v.}} \alpha_i \phi(\mathbf{x}_i)$, and so:

$$f_{\mathbf{w},b}(\mathbf{x}) = \sum_{i \in SV} \alpha_i \langle \phi(\mathbf{x}_i), \phi(\mathbf{x}) \rangle + b = \sum_{i \in SV} \alpha_i k(\mathbf{x}_i, \mathbf{x}) + b \tag{3}$$

where the abbreviation SV represents the set of support vectors (for which $\alpha_i \neq 0$).

The solution in (3) is difficult to interpret in terms of the covariables and, therefore, it is not very useful for problems of the type in (1). Our approach to dealing with this drawback is to make use of the properties of the positive definite functions: if the functions $\bar{k}_i$, $i = 1, 2$ defined in $\mathbb{R}^{d_i} \times \mathbb{R}^{d_i}$, $i = 1, 2$ are positive definite functions, and the transformations $\varphi_1(\mathbf{x}) = \mathbf{z} \in \mathbb{R}^{d_1}$ and $\varphi_2(\mathbf{x}) = \mathbf{t} \in \mathbb{R}^{d_2}$ are used, where $\mathbf{x} = (\mathbf{z} \ \mathbf{t})$, then the function $k : \mathbb{R}^d \times \mathbb{R}^d \to \mathbb{R}$ defined as:

$$k(\mathbf{x}, \mathbf{x}') = \bar{k}_1(\varphi_1(\mathbf{x}), \varphi_1(\mathbf{x}')) + \bar{k}_2(\varphi_2(\mathbf{x}), \varphi_2(\mathbf{x}')) = k_1(\mathbf{z}, \mathbf{z}') + k_2(\mathbf{t}, \mathbf{t}')$$

is a positive definite function and can be used as the kernel in an SVM.

As for the feature space and its inner product, the above can be interpreted as the choice of a transformation $\phi : \mathbb{R}^d \to \mathbb{R}^r$, with two components $\phi^{(i)} : \mathbb{R}^d \to \mathbb{R}^{r_i}$, with $\mathcal{F}_i = \{\phi^{(i)}(\mathbf{x}) : \mathbf{x} \in \mathcal{S} \subset \mathbb{R}^d\}$, $i = 1, 2$, such that $\mathcal{F} = \mathcal{F}_1 \oplus \mathcal{F}_2$, where $\phi^{(1)}(\mathbf{x}) = \phi_1(\mathbf{z})$ and $\phi^{(2)}(\mathbf{x}) = \phi_2(\mathbf{t})$, with $\phi_i : \mathbb{R}^{d_i} \to \mathbb{R}^{r_i}$, $i = 1, 2$ such that:

$$\langle \phi(\mathbf{x}), \phi(\mathbf{x}') \rangle_{\mathcal{F}} = \langle \phi_1(\mathbf{z}), \phi_1(\mathbf{z}') \rangle_{\mathcal{F}_1} + \langle \phi_2(\mathbf{t}), \phi_2(\mathbf{t}') \rangle_{\mathcal{F}_2} = k_1(\mathbf{z}, \mathbf{z}') + k_2(\mathbf{t}, \mathbf{t}')$$

In order to deal with partially linear problems, it is sufficient to select $\phi^{(1)}(\mathbf{x}) = \phi_1(\mathbf{z}) = \mathbf{z}$ with an inner product: $\langle \phi_1(\mathbf{z}), \phi_1(\mathbf{z}') \rangle_{\mathcal{F}_1} = k_1(\mathbf{z}, \mathbf{z}') = \langle \mathbf{z}, \mathbf{z}' \rangle$ and $\phi^{(2)}(\mathbf{x}) = \phi_2(\mathbf{t})$ with an inner product in $\mathcal{F}_2$ defined by a kernel $k_2(\mathbf{t}, \mathbf{t}')$ with sufficient approximating properties to represent the complexity of the function $g$.

The above approach can be generalized to other functions $h(\cdot; \beta) : \mathbb{R}^{d_1} \to \mathbb{R}$ parameterized by $\beta$. For example, if $h$ is a polynomial of degree $a$ in $\mathbf{z}$, suffice it to choose: $\langle \phi_1(\mathbf{z}), \phi_1(\mathbf{z}') \rangle_{\mathcal{F}_1} = k_1(\mathbf{z}, \mathbf{z}') = (\langle \mathbf{z}, \mathbf{z}' \rangle + c)^a$. Consequently, the solution in (3) becomes:

$$f_{\mathbf{w},b}(\mathbf{x}) = \sum_{i \in SV} \alpha_i \langle \phi(\mathbf{x}_i), \phi(\mathbf{x}) \rangle_{\mathcal{F}} + b = \sum_{i \in SV} \alpha_i (\mathbf{z}^t \mathbf{z}_i + c)^a + \sum_{i \in SV} \alpha_i k_2(\mathbf{t}_i, \mathbf{t}) + b \qquad (4)$$

And if, for example, $a = 1$ and $c = 0$, then it takes the form:

$$f_{\mathbf{w},b}(\mathbf{x}) = \mathbf{z}^t \sum_{i \in SV} \alpha_i \mathbf{z}_i + \sum_{i \in SV} \alpha_i k_2(\mathbf{t}_i, \mathbf{t}) + b = \mathbf{z}^t \hat{\beta} + \hat{g}(\mathbf{t})$$

where $\hat{\beta} = \sum_{i \in SV} \alpha_i \mathbf{z}_i$ and $\hat{g}(\mathbf{t}) = \sum_{i \in SV} \alpha_i k_2(\mathbf{t}_i, \mathbf{t}) + b$.

## 3 Application to the problem of predicting mine slope movement

The valley of Meirama – located in NW Spain at about 200 meters above sea level and measuring some 3.5 kms long by 1 km wide – contains a brown lignite open-pit mine. Almost from the commencement of mining operations, the NE slope has shown signs of instability.

Given the difficulty of predicting the behavior of the slope, a supervision and control system was installed with the aims of furnishing data for the creation of the geological, geotechnical and mathematical models, and providing information on both drainage efficacy and displacements in the granite mass.

With a view to developing a mine planning tool that will enable production goals to be achieved in acceptably stable and safe conditions, we set one month as the period for the prediction. A total of 31 slope displacement measurements were obtained for the period up to April 2004, as also rainfall data (liters/m$^2$) for the periods corresponding to these measurements.

The following techniques were used to predict slope displacement: a linear autorregressive model (AR), local linear regression (LLR) and support vector machines (SVMs).

With a view to evaluating the influence of rainfall on slope displacement, these techniques were applied without and with the inclusion of rainfall data in the model. In the latter case, the partially linear versions of the different methods were used (except, logically, for the ARMAX model, which is linear), using the rainfall variable in the linear term.

In the displacement data, no autocorrelation or partial autocorrelation of significance was observed beyond the first lag, and so a single lag was used in the models (models with a greater number of lags failed to improve on the results obtained using one lag).

Five-fold-cross-validation was used to select the model for both the non-parametric techniques and the SVMs. Both the mean squared error (MSEval) and the mean absolute error (MAEval) were recorded for the data excluded in each training run.

Table 1 shows the results obtained using the different techniques, with the data rescaled to $[0, 1]$ in order to facilitate model selection. The SVMs whose results are shown were trained using quadratic loss (SVM_2) and $\varepsilon = 0$ (equivalent to a spline), and linear loss (SVM_1) and $\varepsilon = 0.075$. The partially linear models are identified using PL prefixed to the abbreviation for each technique.

The following comments can be made in relation to the table data:

1. Inclusion of rainfall data in the model clearly improves predictive capacity.

2. The partially linear models improve significantly on the results of the models without a linear term. This would suggest that the assumption of a linear influence by rainfall on slope

| Model | MSEval | MAEval | $\beta$ |
|---|---|---|---|
| **Without rainfall** | | | |
| ARIMA | 0.2895 | 0.9028 | |
| LLR | 0.2887 | 0.8757 | |
| SVM_2 $\varepsilon = 0$ | 0.2837 | 0.8664 | |
| SVM_1 $\varepsilon = 0.075$ | 0.2865 | 0.9607 | |
| **With rainfall** | | | |
| ARIMAX | 0.2300 | 0.7933 | 0.4787 |
| LLR | 0.1999 | 0.7241 | |
| SVM_2 $\varepsilon = 0$ | 0.1984 | 0.7108 | |
| SVM_1 $\varepsilon = 0.075$ | 0.1938 | 0.7420 | |
| **PLM with rainfall** | | | |
| PL-LLR | 0.1728 | 0.7270 | 0.6136 |
| PL-SVM_2 $\varepsilon = 0$ | 0.1727 | 0.7202 | 0.5752 |
| PL-SVM_1 $\varepsilon = 0.075$ | 0.1685 | 0.7251 | 0.4548 |

Table 1: Results of the different techniques used. 1) without rainfall data; 2) with rainfall data; and 3) with rainfall in the linear term in partially linear models. The means for quadratic error and for absolute error are shown for the data successively excluded from training in the five-fold cross-validation process, as also - for partially linear models - the estimation of $\beta$ rainfall variable coefficient in the linear term.

movements is by no means exaggerated and facilitates interpretation of the influence of this variable.

3. The partially linear model that behaves most satisfactorily is the PL-SVM described in this article, which surpasses the classical PL-LLR models.

## 4   Conclusions

In this work we have described a simple and intuitive method for constructing partially linear support vector machines. The proposed method likewise enables partially parametric SVMs to be constructed (e.g. partially polynomial models).

The linear version of the method was used to predict mine slope movement, and results were compared to standard support vector machines and classical partially linear non-parametric techniques.

For our problem, the partially linear SVM produced the best results; this would indicate its potential for modeling and interpreting the relationship between rainfall and monthly displacements in a mine slope.

## References

[1] M. Espinoza, K. Pelckmans, L. Hoegaerts, J. Suykens, B. De Moor, A comparative study of LS-SVMs applied to the Silver box identification problem. *Proceedings of the 6th IFAC Nonlinear Control Systems (NOLCOS)*, Stutgartt, Germany, 2004.

[2] P. Speckman, Kernel smoothing in partial linear models. *Journal of the Royal Statistical Society, Series B*, **50**, 413-436 (1988).

[3] V. Vapnik, *Statistical Learning Theory*. John Wiley & Sons, 1998.

Brill Academic Publishers
P.O. Box 9000, 2300 PA Leiden
The Netherlands

*Lecture Series on Computer
and Computational Sciences*
Volume 4, 2005, pp. 394-397

# Estimation of Chemical Potential in Molten Rare Earth and Alkali Chlorides by Molecular Dynamics Simulation

M. Matsumiya[1] and M. Shimoike

Department of Chemical Science and Engineering,
Miyakonojo National College of Technology,
473-1, Yoshio, Miyakonojo, Miyazaki, Japan

Received 6 July, 2005; accepted in revised form 27 July, 2005

*Abstract:* It is very important to develop the pyrochemical process using the molten salts in order to recover the rare earth as a recycling system and necessary to investigate the possibility for recovery of the rare earth in the molten salt baths. The molecular dynamics methods enabled us to estimate the possibility for the recovery of the rare earth as for the thermodynamic properties such as the chemical potential for the alkali metal in the molten salt baths. In this work, we performed a successful estimation of the chemical potential and the activity of the molten binary mixtures containing the rare earth and the alkali chlorides from the viewpoint of an energy distribution function. These thermodynamic parameters allowed us to estimate the consumption energy, which was needed to the development of the pyrochemical treatment process.

*Keywords:* Activity, Chemical potential, MD simulation, Molten salts, Pyrochemical Process, Rare earth

## 1. Introduction

It is necessary to develop the pyrochemical recycling process for the rare earth metals with high purity as an industrial technology. We have been demonstrated that the pyrochemical process [1] combined the electromigration [2] and the electrowinning [3] processes in the molten salts was a promising method for the production of high-purity rare earth metals. In addition, we also proposed that the effectiveness of the molecular dynamics (MD) methods [4] enabled us to indicate that the limitation of the enrichment degree of the alkali-chloride mixtures as a theoretical approach. Then it is important to investigate the thermodynamic properties for the rare earth elements in the molten salts.

Widom [5] demonstrated that the canonical ensemble average of $\exp(-u_f/kT)$ provides the excess chemical potential as $\mu^{ex} = -RT\ln\langle \exp(-u_f/kT)\rangle = -RT\left\{\int_{-\infty}^{\infty} f(u_f)\exp(-u_f/kT)du\right\}$, where $R$, $T$, $k$ and $f(u)$ are the gas constant, an absolute temperature, the Boltzman constant and an energy distribution of $u_f$, which is the energy between the original system and the system with the inserted fictitious particles, respectively. On the other hand, Shing and Gubbins [6] showed that the canonical ensemble average of $\exp(u_r/kT)$ yields the excess chemical potential as $\mu^{ex} = RT\ln\langle \exp(u_r/kT)\rangle = RT\left\{\int_{-\infty}^{\infty} g(u_r)\exp(u_r/kT)du\right\}$, where $g(u)$ is an energy distribution of $u_r$, which is the energy difference between the original system and the system with the particles removed.

[1] Corresponding author. Tel/+81-986-47-1218, Fax/+81-986-47-1231, E-mail: mmatsumi@cc.miyakonojo-nct.ac.jp

Although these two methods require only the positional data at the temperature in the equation, the chemical potentials estimated by these methods largely depend on the lower and upper limits of the energy distributions of $f(u)$ and $g(u)$, respectively. Powles et al. [7] then showed that the intercept of a linear plot of $L(u)$ versus $u$ gives the chemical potential, i.e.,

$$L(u) \equiv \ln\left(\frac{f(u)}{g(u)}\right) = \beta u - \beta \mu^{ex}, \quad \beta = \frac{1}{kT} \tag{1}$$

In order to verify the effectiveness of this method, an MD simulation of the molten alkali chlorides and the rare earth binary mixture was executed in this work.

## 2. Computation

Our applied MD simulation proposed by Nose [8] with the 7 points predictor-corrector method using the periodic boundaries. The parameter of a heat reservoir was determined carefully after several attempts. The Born-Mayer-Huggins pair potential was applied in our molten system:

$$\Phi_{ij} = \frac{z_i z_j e^2}{r} + A_{ij} b \exp\left[(\sigma_i + \sigma_j - r)/\rho\right] - \frac{c_{ij}}{r^6} - \frac{d_{ij}}{r^8} \tag{2}$$

Coulomb interaction between two ions was evaluated by the Ewald method. The potential parameters of the alkali metal and the halogen elements demonstrated by Tosi and Fumi [9] were employed. The parameter $c_{ij}$ was estimated from the ionic polarizability [10]. The corresponding parameters for the mixture were determined by the combination rule represented by Larsen et al. [11]. At the beginning, the initial cell, in which the ions were arranged in the crystalline structure, were first annealed with the constant temperature method postulated by Woodcock [12]. From the runs during more than $10^4$ time steps by isothermal-choric (NVT) MD simulations after attainment of the equilibrium, the structural and the dynamical properties were obtained. The excess chemical potential was estimated from the positional data of the ions recorded every 20 MD time steps.

## 3. Results and Discussion

The total structure factor of the molten NdCl$_3$-NaCl (20:80) system is shown in **Fig. 1**, compared with the experimental results. As for the NdCl$_3$ pure melt, the ratio of $r_{Cl-Cl}$=0.347nm to $r_{Nd-Cl}$=0.275nm in the calculated pair correlation functions is roughly equal to $\sqrt{2}$ as calculated from the geometry of the octahedron. Thus, it has been believed that the local structure of the trichloride melts is simply 6-fold octahedral coordination [NdCl$_6$]$^{3-}$ because the vibration modes corresponding to the 6-fold octahedral coordination was observed for some rare earth trichloride melts in the photochemical spectroscopic studies. Moreover, The network structure of [NdCl$_6$]$^{3-}$ units could be considered as a distorted corner-sharings of 6-fold complex ions according to the correlation, $r_{Cl-Cl} < r_{Nd-Nd} < 2r_{Cl-Cl}$.

In order to calculate the chemical potential of NaCl in the molten NdCl$_3$-NaCl system, we made MD simulations at the liquids temperatures of the phase diagram. In addition, in order to estimate the activity of NaCl, we made an MD simulation of NaCl in the super-cooled state at this temperature. In order to obtain the energy distributions, $f(u)$ and $g(u)$, 500 combinations of pairs of Na$^+$ and Cl$^-$ ions were removed from or added to the system randomly. The energy distribution functions of NaCl in the molten NdCl$_3$-NaCl (10:90) system, $f(u)$ and $g(u)$, are depicted in **Fig. 2**. The excess chemical potential of NaCl is estimated from the intercept of this linear line. The activity of NaCl in the molten NdCl$_3$-NaCl system is calculated by $RT \ln a_2 = \mu_I(x_2) - \mu_I(2)$, where $T$ and $a_2$ are the liquidus temperature and the activity of NaCl. $\mu_I(x_2)$ is the chemical potential of NaCl in the molten binary system and $\mu_I(2)$ is the chemical potential of the pure super-cooled NaCl at the liquidus temperature. The chemical potentials of NaCl at the liquidus temperature estimated by MD simulation $\mu_I(2)$ and the activity of NaCl $a_2$ are tabulated in **Table 1**. The activity of NaCl in the molten NdCl$_3$-NaCl system estimated by MD simulation is close to that in the molten DyCl$_3$-NaCl [13]. Finally, for the practical application of the pyrochemical process, it is important to estimate the chemical potential in the range of the whole concentrations, because the detailed thermodynamic information would lead to better and higher separation efficiency.

## 4. Conclusion

It is of importance to obtain the thermodynamic properties for the rare earth elements in the molten salt baths because it is essential to develop the pyrochemical process for the recovery of rare earth elements. Based on this purpose, we performed the determination of the activity of the alkali metals in the molten rare earth binary mixtures from the theoretical estimation. As a result, it is demonstrated that the chemical potential and the activity of the alkali metal were consistent with the similar rare earth ions. These parameters would be contributed to the development of the pyrochemical process using the molten salts. These results allow us to estimate the detailed calculation of the consumption energy in the further pyrochemical process.

Table 1: The chemical potentials and the activity coefficients of NaCl in the molten NdCl$_3$-NaCl system estimated by MD simulation.

| $(x_1:x_2)$ | $\mu_I(2)$ (kJ/mol) | $\mu_I(x_2)$ (kJ/mol) | $RT \ln a_2$ (kJ/mol) | $a_2$ |
|---|---|---|---|---|
| (10:90) | -891.28 | -887.52 | -3.76 | 0.67 |
| (20:80) | -864.84 | -858.63 | -6.21 | 0.51 |
| (30:70) | -842.32 | -833.48 | -8.84 | 0.35 |
| (40:60) | -823.46 | -814.23 | -9.23 | 0.33 |

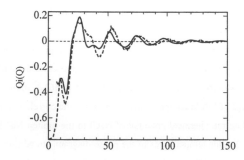

Figure 1: Total structure factor of the molten NdCl$_3$-NaCl (20:80) system
Solid line: the calculated results from MD simulation, dotted line: the experimental results

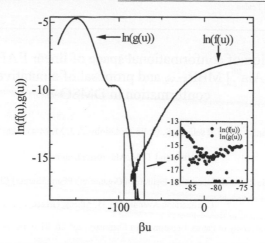

Figure 2: The energy distribution functions $f(u)$ and $g(u)$ of NaCl
in the molten $NdCl_3$-NaCl (10:90) system

## Acknowledgments

We are also especially indebted to the financial support for Kyushu Industrial Technology Center (KITEC) in attendance of this international conference.

## References

[1] M. Matsumiya and H. Matsuura, Pyrochemical process for electrochemically base metals in molten halides combined countercurrent electromigration and electrowinning methods, *J. Electroanal. Chem.* **579** 329-336 (2005).

[2] M. Matsumiya and S. Matsumoto, Enrichment of negative elements in molten quasi-binary nitrates, *Electrochem. Commun.*, **7(2)**, 147-150 (2005).

[3] M. Matsumiya and S. Matsumoto, Electrochemical studies on lanthanum ions in molten LiCl-KCl-eutectic mixture, *Z. Naturforsch.*, **59a**, 711-714 (2004).

[4] M. Matsumiya and K. Seo, A Molecular dynamics simulation of the transport properties of molten $(La_{1/3}, K)Cl$, *Z. Naturforsch.*, **60a**, 187-192 (2005).

[5] B. Widom, Some topics in the theory of fluids, *J. Chem. Phys.* **39**, 2808-2812 (1963).

[6] K. S. Shing and K. E. Gubbins, The chemical potential in dense fluids and fluid mixtures via computer simulation, *Mol. Phys.* **46**, 1109-1128 (1982).

[7] J. G. Powles, W. A. B. Evans and N. Quirke, non-destructive molecular-dynamics simulation of the chemical potential of a fluid, *Mol. Phys.* **46**, 1347-1370 (1982).

[8] S. Nose, A molecular dynamics method for simulations in the canonical ensemble, *Mol. Phys.*, 52, 255 (1984).

[9] M. P. Tosi and F. G. Fumi, Ionic sizes and born repulsive parameters in the NaCl-type alkali halides- II, *J. Phys. Chem. Solids* **25**, 45-52 (1964).

[10] Y. Iwadate, J. Mochinaga, and K. Kawamura, Refractive Indexes of Ionic Melts, *J. Phys. Chem.*, **85**, 3708-3712 (1981).

[11] B. Larsen, T. Forland and K. Singer, A Monte Carlo calculation of thermodynamic properties for the liquid NaCl+KCl mixture, *Mol. Phys.* **26**, 1521-1532 (1973).

[12] L. V. Woodcock, Isothermal molecular dynamics calculations for liquid salts, *Chem. Phys. Lett.* **10**, 257-261 (1971).

[13] M. Sakurai, R. Takagi, A. K. Adya, and M. Gaune-Escard, Estimation of chemical potential and the activity of NaCl in molten $DyCl_3$-NaCl by molecular dynamics simulation, *Z. Naturforsch.* **53a**, 655-658 (1998).

Brill Academic Publishers
P.O. Box 9000, 2300 PA Leiden
The Netherlands

*Lecture Series on Computer*
*and Computational Sciences*
Volume 4, 2005, pp. 398-401

# Exploration of conformational space of linear EAE antagonist [Arg$^{91}$, Ala$^{96}$] MBP$_{87-99}$ and proposal of a putative bioactive conformation in DMSO-$d_6$.

E. D. Mantzourani[1,2], T. V. Tselios[2], S. Golič-Grdadolnik[3], J. M. Matsoukas[2], A. Brancale[4], J. A.

Platts[5], and T. M. Mavromoustakos[1*]

[1] National Hellenic Research Foundation, Institute of Organic and Pharmaceutical Chemistry, 48 Vassileos

Constantinou Avenue, 116 35, Athens, Greece

[2] University of Patras, Department of Chemistry, 265 00, Rion, Patras, Greece

[3] Laboratory for Molecular Modeling and NMR Spectroscopy, National Institute of Chemistry, Hajdrihova 19, SI-

1001 Ljubljana, Slovenia

[4] Welsh School of Pharmacy, Cardiff Univ., Redwood Build., King Edward VII Av., CF1 3XF, Cardiff, Wales, UK

[5] Department of Chemistry, Cardiff University, P.O. Box 912, CF10 3TB, Cardiff, Wales, UK

Received 20 August, 2005; accepted in revised form 20 August, 2005

*Abstract:* [Arg$^{91}$, Ala$^{96}$] MBP$_{87-99}$ is an altered peptide ligand (APL) of myelin basic protein (MBP), shown to actively inhibit experimental autoimmune encephalomyelitis (EAE) that is studied as a model of multiple sclerosis. The APL has been rationally designed by substituting two of the critical residues for recognition by the T-cell receptor. A conformational analysis of the APL has been sought, both experimentally and theoretically, in order to comprehend the stereoelectronic requirements for antagonistic activity and to propose a putative bioactive conformation based on spatial proximities of the native peptide in the crystal structure. Amino acid fragments of the peptide have been found to form elements of secondary structure, in accordance to the NMR data, representing the true flexibility of the peptide. All derived conformations were scanned for compliance with criteria for biological activity and a match was found. This proposed bioactive conformation complies with criteria for HLA-DR2b recognition, as primary and secondary anchors occupy the same region in space. On the contrary, amino acids that serve as TCR contacts are orientated differently. In this structure, interactions of the side chains which are largely populated in the conformational ensemble are present, accenting the combination of NMR and molecular dynamics as a valuable tool for the proposal of bioactive conformations.

*Keywords:* [Arg$^{91}$, Ala$^{96}$] MBP$_{87-99}$, NMR, molecular dynamics, bioactive conformation

# 1. Introduction

Multiple sclerosis (MS) is an inflammatory, demyelinating disease involving the white matter of the central nervous system (CNS), prevalent in Northern Europeans and North Americans. Its etiology remains unknown, but both environmental and genetic factors seem to play an important role. Previous studies have demonstrated that the HLA-DR2b (DRB1*1501) haplotype is present at an increased frequency in northern European Caucasoid patients with MS. Modern approaches towards the therapeutic management of MS involve the design and use of peptide analogues of disease-associated myelin epitopes. The idea is to block the formation of the trimolecular complex MHC – Altered Peptide Ligand (APL) - T cell receptor (TCR) and therefore actively inhibit the disease. The best animal model of MS known so far is Experimental Autoimmune Encephalomyelitis (EAE), a relapsing experimental demyelinating disease characterized by focal areas of inflammation and demyelination throughout the CNS. An epitope of myelin basic protein (MBP), $MBP_{87-99}$ is shown to be immunodominant. In this study an investigation of the 3D structure of an APL that was shown to be an EAE antagonist has been performed, $[Arg^{91}, Ala^{96}]$ $MBP_{87-99}$ (Figure 1). The molecule has been rationally designed by substituting two of the critical residues for TCR recognition; the amino acids $Lys^{91}$ and $Pro^{96}$ were replaced by Arg and Ala respectively. $Lys^{91}$ and $Pro^{96}$ are two anchor residues forming H-bonds with the TCR. The three-dimensional structure of the APL is essential to understand better its physicochemical properties for antagonistic biologic activity, as it governs the binding to MHC and TCR.

Within this context, a combination of unrestrained and backbone restrained molecular dynamics simulations, and experimental 1D and 2D NMR spectroscopic methods has been applied to establish a putative bioactive conformation for the APL. This is defined as a conformation that enables binding with the MHC, but prevents binding with the TCR and therefore T-cell activation. The proposed bioactive conformation is indeed a low energy structure, in which a range of the experimental NOE distance restraints is in accordance with the conformation of $MBP_{87-99}$ provided by crystallographic data. This work establishes the necessity of using both experimental NOE data and theoretical modelling studies for exploring and understanding the stereo-electronic requirements for EAE inhibition in order to rationally design a peptide mimetic molecule.

# 2. Experimental

The high-resolution NMR spectra were recorded on a Varian INOVA 600-MHz spectrometer at 298 K. Data were processed and analyzed with FELIX software package from Accelrys Software Inc. Cross-peak volumes in NOESY spectra were calculated by integration routine within the FELIX software. For calculation of distances the two-spin approximation was used and an integrated intensity of the geminal pair of protons $\gamma_1$ and $\gamma_2$ of $Ile^{93}$ was assumed to have a distance of 1.78 Å. Based on these distances and direct observation of peaks, a set of strong, medium, and weak NOEs was established.

Computer calculations were performed on a RM 3GHz Pentium IV workstation using MOE 2004.03 by Chemical Computing Inc. Molecular dynamics simulations were performed and the derived conformations were examined for consistency with experimental distance information designated by the obtained NOEs. Thus, populations of various conformers that represent local minima at the potential energy surface are identified.

## 3. Discussion-Conclusions

The natural ligand was isolated from the complex, and its conformation was used as a basis from which to establish criteria, in order to identify a putative bioactive conformation among all the low energy conformers that were generated. A bioactive conformation is defined in this study as one that theoretically exhibits optimal HLA binding affinity, having canonical MHC binding motifs, but fails to be recognized by the TCR and therefore to trigger an immune response. Thus, a set of criteria was used: a) The structure should adopt a general backbone conformation very similar to that of the natural peptide, with an RMSD value of less than 2 Å when backbone Ca, C, and N atoms are superimposed. b) Distances between amino acids that serve as primary (Val[87], Phe[89]) and as secondary anchors (Asn[92], Ile[93], and Thr[95]) for binding to HLA-DR2b, should have a deviation smaller 20% when compared to the x-ray structure. c) In conformations that fulfill a) and b), residues that were previously found to be important for TCR recognition (His[88], Phe[89], and Lys[91]) should be at a different position than on the natural peptide, and should present different spatial relations with the residues that bind to HLA-DR2b. RMSD of this conformation when superimposed to MBP[85-99] from the x-ray complex is 1.59 Å.

In Figure 1 it is superimposed with MBP[87-97] obtained from the crystal structure, by isolating the molecule and deleting the first two amino acids. The crystal coordinates were subjected to side-chain energy minimization, for removal of unfavorable atomic interactions. The C-terminus is different due to the replacement of Pro[96] in the native peptide with an Ala in the APL. As can be observed, though, the backbone of the N-terminus is almost identical. This is very important because it is known that the N-terminus is important for binding. MHC contacts lie within the same region in space. Phe[90] has the correct orientation to occupy the large hydrophobic pocket in HLA, whereas TCR contact Phe[89] is no longer prominent and solvent exposed. Therefore it is not accessible for interaction with the TCR.

The combination of NMR and MD studies led to a proposed bioactive conformation that when superimposed to the structure obtained by crystallography with a 2.6 Å resolution, presents a striking similarity in the peptide binding sequence (backbone RMSD 1.23 Å). Thus, this approach rises as a valuable tool in the rational design of molecules with desired biologic activity.

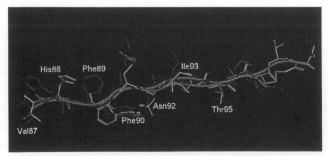

**Figure 1:** Superimposition between MBP[87-97] obtained from the crystal structure of the complex with HLA (blue) and the proposed bioactive conformation of [Arg[91], Ala[96]] MBP[87-99] (yellow). It is evident that primary MHC anchors (Val[87], Phe[90]) and secondary MHC anchors (Asn[92], Ile[93], Thr[95]) occupy the same region in space, whereas TCR contacts (His[88], Phe[89]) have different orientation between the two structures.

## Acknowledgments

Present work is supported by the Ministry of Development, Secretariat of Research and Technology of Greece (Grant EPAN YB/76).

## References

[1]. Kenealy S. J., Perical-Vance M. A., and Haines J. L. The Genetic Epidemiology of MS *J. Neuroimmunol.* 2003, **143**, 7-12.

[2]. Prat E., and Martin R. The immunopathogenesis of MS *J. Rehab. Res. Devel.* 2002, **39**, 187-200

[3]. Coles A. J., and Wing M., Pulsed Monoclonal Antibody Treatment and Autoimmune Thyroid Disease in MS *Lancet* 1999, **354**, 1691-1695,

[4]. Vergelli M., Hemmer B., Utz U., Vogt A., Kalbus M., Tranquill L., Conlon P., Ling P., Steinman L., McFarland H., and Martin R. Differential Activation of Human Autoreactive T cell Clones By Altered Peptide Ligands Derived from MBP$_{87-99}$ *Eur. J. Immunol.*, 1996, **26**, 2624-2634

[5]. Vogt, A.B., Kropshofer, H., Kalbacher, H., Kalbus, M., Rammensee, H.G., Coligan, J.E., and Martin, R. Ligand motifs of HLA-DRB5*0101 and DRB1*1501 molecules delineated from self-peptides *J. Immunol.* 1994, **153**, 1665

[6]. Karin, N., Binah, O., Grabie, N., Mitchell, D.J., Felzen, B., Solomon, M.D., Conlon, P., Gaur, A., Ling, N., and Steinman, L. Short peptide-based tolerance without self-antigenic or pathogenic activity reverse autoimmune disease. *J Immunol.* 1998 **15;160**(10):5188-94

[7]. Chemical Computing Group Inc. 1010, Sherbrooke Street W, Suite 910; Montreal, Quebec; Canada H3A 2R7

[8]. Dyson, H. J., and Wright, P. E. Defining solution conformations of small linear peptides. *Annu. Rev. Biophys. Biophys. Chem.* 1991, **20**, 519-538

[9]. Smith, K.J., Pyrdol, J., Gauthier, L., Wiley, D.C. and Wucherpfennig, K.W. Crystal structure of HLA-DR2 (DRA*0101, DRB1*1501) complexed with a peptide from human myelin basic protein *J Exp Med* 1998, **188**(8), 1511

Brill Academic Publishers
P.O. Box 9000, 2300 PA Leiden
The Netherlands

*Lecture Series on Computer*
*and Computational Sciences*
Volume 4, 2005, pp. 402-405

# Theoretical Calculations constitute a Valuable Tool to Understand the Molecular Basis of Hypertension

T. Mavromoustakos, P. Zoumpoulakis, M. Zervou, N. Benetis

Institute of Organic and Pharmaceutical Chemistry,
National Hellenic Research Foundation, Vas. Constantinou 48, 11635 Athens Greece

Received 20 August, 2005; accepted in revised form 20 August, 2005

*Abstract:* In the last decade, several efforts have been made to comprehend the molecular basis of hypertension through the study of Angiotensin II antagonists (AT1 antagonists, SARTANs). The structural features which determine the pharmacophoric segments of SARTANs are examined using a combination of NMR and conformational analysis. It has been proposed that AT1 antagonists bind on the transmembrane segments of the AT1 receptor through a two step mechanism. This fact has motivated us to the study of interactions between AT1 antagonists and membranes using several experimental techniques and theoretical simulations. Meanwhile, computer programs dedicated to docking small molecules into protein binding pockets *in silico* have been applied in order to explore the interactions between the pharmacophore groups of the antagonists and the crucial for binding aminoacids. Such a study may reveal the stereoelectronic requirements for drug activity (molecular basis of hypertension) and aids in the design and synthesis of novel antihypertensive drugs.

*Keywords:* membrane bilayers, theoretical calculations, AT1 antagonists

## 1. Introduction

AT1 antagonists constitute the last generation of drugs for the treatment of hypertension, a growing undesired symptom which damages health of about 1/5 of Greek population. These pharmaceutical molecules are designed and synthesized to mimic the C-terminal segment of the vasoconstrictive hormone Angiotensin II ( Ang II). They exert their action by blocking the binding of Ang II on the AT1 receptor. It is known that AT1 antagonists bind on the transmembrane segments of the AT1 receptor. A two step mechanism for their action including interaction with membrane bilayers and lateral diffusion to the active site of the receptor was proposed (Figure 1).

## 2. Results

To provide evidence on the proposed two step mechanism the following methodologies are applied. (a) Conformational Analysis on AT1 antagonists; (b) Drug interactions with membrane bilayers; (c) theoretical docking study on the receptor site and (d) biophysical experimental techniques. Biophysical techniques aim in understanding the molecular basis of AT1 antagonism and include: Solid State NMR and Raman Spectroscopies, X-ray Diffraction and Differential Scanning Calorimetry. They provide information on the dynamic properties of membrane bilayers in the absence and presence of drug molecules as well their topography.

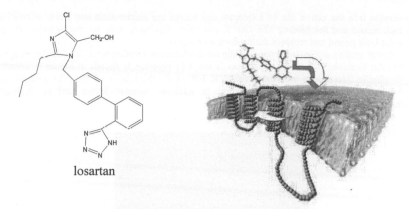

losartan

*Figure 1: (left) chemical structure of losartan (the first SARTAN in the market); (right) two step mechanism of action for the AT1 antagonists including interaction with membrane bilayers and lateral diffusion to the active site of the receptor.*

*Conformational analysis:* The coupling of theoretical calculations and experimental results provide possible bioactive conformations which will be used in the drug interactions with membrane bilayers and receptor site. Figure 2 shows the steps utilized for the conformational analysis of another AT1 antagonist, eprosartan. A putative bioactive conformation of eprosartan is shown in the same figure.

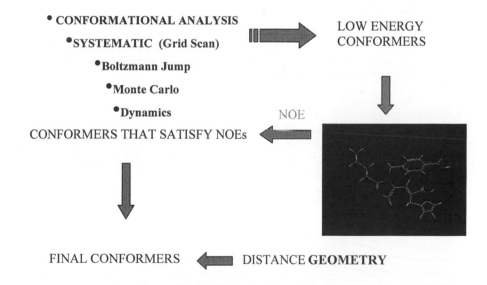

*Figure 2: Steps utilized for deriving bioactive conformations of eprosartan.*

*Docking experiments:* FlexX and DOCK algorithms have been used to identify the drug:receptor site interactions. The first algorithm places a base fragment of the ligand in the pocket and then constructs fragment by fragment the remainder while searches for favorable interactions between the ligand and the protein trying to avoid steric overlaps at each step. During this procedure, the conformation of the protein is kept stable. The docking and subsequent scoring were performed using the formal charges and the default parameters of the FlexX program. In addition, the second algorithm, DOCK (SYBYL), docks molecules into the active site of a receptor and scores the electrostatic and steric interactions between each ligand and the binding site. DOCK can allow flexibility only for the ligand or for the receptor or for both ligand and receptor during docking process.

These theoretical studies aid in understanding the stereoelectronic requirements for docking drugs in the receptor. An example of docking of losartan in the AT1 receptor is shown in Figure 3. Losartan exerts hydrophobic interactions with aminoacid Val 108 of the third helix of the AT1 receptor and other hydrophobic aminoacids in spatial vicinity. In addition, losartan favors multiple hydrogen bondings between its tetrazole with Lys199 and His256.

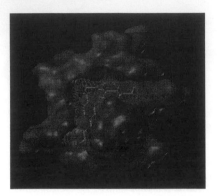

*Figure 3: Losartan   docked in the active site of the AT1 receptor.*

*Simulations of Solid State $^{31}P$ NMR Spectroscopy:* The above docking interactions point out that drug:receptor interactions must be localized in the vicinity of head group of phospholipids. Solid State $^{31}P$ NMR Spectroscopy when it is coupled with theoretical calculations can provide more detail molecular information for drug interactions in the vicinity of the phosphorous area.

A mathematical algorithm is developed in our laboratory in which simulation of experimental spectra can be achieved. The simulation of experimental spectra provides parameters that can give valuable quantitative information on the dynamic interactions of drugs incorporated in membrane bilayers versus the mesomorphic changes of lipid bilayers.

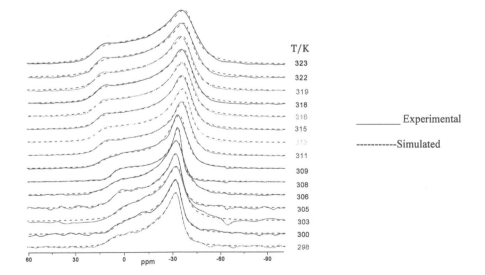

*Figure 4: Experimental (solid lines) and simulated spectral (broken lines) of solid state $^{31}P$ NMR spectra of lipid bilayers in the presence of AT1 antagonist candesartan.*

# 3. Conclusions

In conclusion, experimental and theoretical calculations are in harmony and support the proposed two step mechanism. More specifically the insertion of the amphiphilic AT1 antagonist losartan and its lateral diffusion in the active site, takes place in the vicinity of the phospholipid head group. Moreover, theoretical docking studies have indicated several specific interactions with the aminoacids of the transmembrane part of the receptor.

## Acknowledgments

The authors acknowledge GSRT for funding this research activity through EPAN GSRT YB/76 (2004-2007). *Production of new double action (angiotensin II inhibitors with adrenergic inhibitors) anti-hypertensive drugs and overlapping of intra-intentions with angiotensin II inhibitors for the reduction of reconstruction.* ]

## References

[1] P. Zoumpoulakis, I. Daliani, M. Zervou, I. Kyrikou, E. Siapi, G. Lamprinidis E. Mikros, T. Mavromoustakos. Losartan;s molecular basis of interaction with membranes and AT1 receptor. Chem. Phys. Lipids **125**, 13-25 (2003).

[2] T. Mavromoustakos, M. Zervou, P. Zoumpoulakis, I. Kyrikou, N.P. Benetis, L. Polevaya, P. Roumelioti N. Giatas, A. Zoga, P. Moutevelis Minakakis, A. Kolocouris, D. Vlahakos, S. Golic Grdadolnik, J. Matsoukas. Conformation and Bioactivity. Design and Discovery of Novel Antihypertensive Drugs. Current Topics in Medicinal Chemistry, **4**, 385-401 (2004).

[3] T. Mavromoustakos, P. Zoumpoulakis, I. Kyrikou, A. Zoga, E. Siapi, M. Zervou, I. Daliani, D. Dimitriou, A. Pitsas, C. Kamoutsis, P. Laggner. Efforts to Understand the Molecular Basis of Hypertension Through Drug:Membrane Interactions. Current Topics in Medicinal Chemistry, **4**, 445-459 (2004).

[4] Nikolas-P. Benetis, I. Kyrikou, T. Mavromoustakos, M. Zervou. Static $^{31}$P CP NMR Multilamellar Bilayer Broadlines in the Absence and presence of the Bioactive Dipeptide beta-Ala-Tyr or Glu 314(1-3) 57-72 (2005).

Brill Academic Publishers
P.O. Box 9000, 2300 PA Leiden,
The Netherlands

*Lecture Series on Computer
and Computational Sciences*
Volume 4, 2005, pp. 406-409

# Phase Space Consideration in Quantum Dynamical Fluctuation Expansion

**E. Meral and M. Demiralp[1]**

Group for Science and Methods of Computing, Informatics Institute,
İstanbul Technical University, Maslak, 34469 İstanbul, Turkey

Received 10 July, 2005; accepted in revised form 4 August, 2005

*Abstract:* A new method called "Fluctuation Expansion" has been recently proposed for solving quantum dynamical problems . This method is succesfully applied to the dynamics of expectation values such that not the wave function itself but the expectation values of the position and momentum operator are taken into consideration and then the expectation value of certain entities depending on position and momentum operators are expressed via fluctuation expansion. This brings an infinite number of new unknown, fundamental and associate fluctuation terms. The derivatives of expectation values of the position and momentum operators and fluctuation operators are expressed in terms of these terms again. The resulting infinite set of differential equations and accompanying initial conditions are truncated at certain fluctuation order and then tried to be solved. Here we take a one dimensional time independent Hamiltonian and consider the zeroth and first order truncations. Equations obtained in this way are investigated in the sense of phase space which is defined by the expextation values of the position and momentum operator and the fluctuations entering the truncated equations. We find certain subspaces of the phase space such that there, certain expressions remain constant . This helps us to reduce the dimension of the problem.

*Keywords:* Quantum Theory, Fluctuations, Distributions

*Mathematics Subject Classification:* 81Q99, 60E05

## 1   Introduction

Consider the following Schrödinger's equation written for the waveket of an isolated one dimensional quantum system

$$i\hbar\frac{\partial}{\partial t}\,|\psi(t)\rangle \;=II\,|\psi(t)\rangle \quad \left[-\frac{\hbar^2}{2m}\frac{\partial^2}{\partial x^2}+V(x)\right]|\psi(t)\rangle\,, \qquad |\psi(0)\rangle=|in\rangle \tag{1}$$

where H, $\hbar$, $m$, respectively denote the Hamiltonian, reduced Planck constant, and the mass of the system while $V(x)$, and $|in\rangle$ stand for the potential energy function of the system, and the initial form of the waveket respectively. Dirac's bra and ket notation has been used here although the equation satisfied by wavebra is not given since it is in fact the Hermitian conjugate of the above equation. We are not concerned here with the wave function directly. Instead the expectation value

[1] Corresponding author:E-mail: demiralp@be.itu.edu.tr

of the position operator, $\widehat{x}$, which multiplies its operand simply with the independent variable, $x$, and the momentum operator, $\widehat{p}$, which multiplies its operands derivative with $-i\hbar$. We define

$$p(t) \equiv \langle \psi(t) | \widehat{p} | \psi(t) \rangle, \qquad q(t) \equiv \langle \psi(t) | \widehat{x} | \psi(t) \rangle \tag{2}$$

These entities satisfy the following equations as can be shown by differentiating the equations in (2) and then using certain properties of Poisson's Bracket[1]

$$\dot{p}(t) = -\langle \psi(t) | V'(x) | \psi(t) \rangle, \qquad p(0) = \langle in | \widehat{p} | in \rangle, \tag{3}$$

$$\dot{q}(t) = \frac{1}{m} q(t), \qquad q(0) = \langle in | \widehat{x} | in \rangle \tag{4}$$

We need to express the expectation value of $V'(x)$ in (3) more explicitly. We can use the recently developed Fluctuation Expansion Method[2, 3]. It requires the determination of an infinite number of unknowns which are called fundamental and associate fluctuation terms. The fluctuation expansion for a function $f(x)$ can be written as

$$\left\langle \widehat{f}(\widehat{x}) \right\rangle = f(\langle \widehat{x} \rangle) + \sum_{n=1}^{\infty} \sum_{\ell=0}^{n} \sum_{k=0}^{n-\ell} C_{k,\ell}^{(n)}(q) \, \varphi_\ell(q) \, \frac{1}{(2n+1-k-\ell)!} f^{(2n+1-k-\ell)}(q),$$

$$1 \leq n < \infty \tag{5}$$

which requires the fluctuation terms defined as

$$\varphi_n(q) \equiv \left\langle \left[ \widehat{x} - q\widehat{I} \right] \left\{ \left[ \widehat{I} - \widehat{P}_\psi \right] \left[ \widehat{x} - q\widehat{I} \right] \right\}^n \right\rangle, \qquad 0 \leq n < \infty \tag{6}$$

Here $\varphi_0(q)$ does not really represent a fluctuation. It is included in fluctuation terms for the sake of generalization. The $C_{k,\ell}^{(n)}(q)$ terms appearing in (5) are defined by the following recursions

$$C_{n,0}^{(n)}(q) = 0, \qquad 1 \leq n < \infty; \tag{7}$$

$$C_{k,0}^{(n)}(q) = \sum_{k_1=0}^{k} \varphi_{k_1+1}(c) \, C_{k-k_1,k_1}^{(n-1)}(c), \qquad 1 \leq n < \infty, \quad 0 \leq k \leq n-1; \tag{8}$$

$$C_{k,\ell}^{(n)}(q) = C_{k,\ell-1}^{(n-1)}(c), \qquad 1 \leq n < \infty, \quad 0 \leq k \leq n-1, \quad 1 \leq \ell, k+\ell \leq n \tag{9}$$

with the initial conditions

$$C_{0,0}^{(1)}(q) = \varphi_1(q), \qquad C_{1,0}^{(1)}(q) = 0, \qquad C_{0,1}^{(1)}(q) = 1 \tag{10}$$

These formulae do not contain time explicitly since they are taken from the source paper as they are.

Now all these urge us to replace (3) with the following equation

$$\dot{p}(t) = -V'(q(t)) - \sum_{n=1}^{\infty} \sum_{\ell=0}^{n} \sum_{k=0}^{n-\ell} C_{k,\ell}^{(n)}(q(t)) \, \varphi_\ell(t) \, \frac{1}{(2n+1-k-\ell)!} V^{(2n+2-k-\ell)}(q(t)), \tag{11}$$

where we have not explicitly shown $q(t)$ dependence of the fluctuation term, instead, we have shown global $t$ dependence only.

## 2   Truncated Set of Equations Based on Fluctuation Expansion

Equation (11) contains an infinite number of fluctuation terms called "Fundamental Fluctuation Terms" (11) and (4) are not sufficient to determine everything. We need to construct an infinite number of equations for this purpose. This construction can be realized simply by differentiating the definition equation of each fluctuation term. However this procedure produces new type of fluctuation terms containing momentum operator and its expectation value in the definition. This situation urges us to produce more ordinary differential equations for these new terms by differentiating their definition equations. When all these are done we obtain a uniquely compatible infinite set of ordinary differential equations and accompanying initial conditions.

It is hard to explicitly write this infinite set of ordinary differential equations and accompanying boundary conditions. However it is possible to give the explicit expressions of the truncations from this set. Truncations are done by omitting all fluctuation terms whose orders are greater than a prsecribed integer value which is called "Truncation Order". The fluctuation order mentioned here for a monomial is defined as the sum of the products of each individual factor's index by its power. The zeroth order truncated equations are given as

$$\dot{p}(t) \;=\; -V'\left(q(t)\right) \qquad p(0) = \langle in\,|\widehat{p}|\,in\rangle\,, \tag{12}$$

$$\dot{q}(t) \;=\; \frac{1}{m}q(t), \qquad q(0) = \langle in\,|\widehat{x}|\,in\rangle\,, \tag{13}$$

These contain no fluctuation and in fact corresponds to the classical limit behavior of the system.

The first order truncation of the equations are given below

$$\dot{p}(t) \;=\; -V'\left(q(t)\right) - \frac{1}{2}V'''\left(q(t)\right)\varphi_1(t) \qquad p(0) = \langle in\,|\widehat{p}|\,in\rangle \tag{14}$$

$$\dot{q}(t) \;=\; \frac{1}{m}q(t), \qquad q(0) = \langle in\,|\widehat{x}|\,in\rangle \tag{15}$$

$$\dot{\varphi}_1(t) \;=\; \frac{1}{m}\varphi_{1,1}(t), \qquad \varphi_1(0) = \varphi_{1,in} \tag{16}$$

$$\dot{\varphi}_{1,1}(t) \;=\; -2V''\left(q(t)\right)\varphi_1(t) + \frac{2}{m}\varphi_{1,2}(t), \qquad \varphi_{1,1}(0) = \varphi_{1,1,in} \tag{17}$$

$$\dot{\varphi}_{1,2}(t) \;=\; -V''\left(q(t)\right)\varphi_{1,1}(0), \qquad \varphi_{1,2}(t) = \varphi_{1,2,in} \tag{18}$$

where three new differential equations and accompanying initial conditions for three new entities are added to the zeroth order case equations. These entities contain all possible first order fluctuation terms over the position and momentum operators. The fluctuations we call "Fundamental Fluctuation Terms" contain only position operator whereas the fluctuations defined to contain momentum operator alone or together with the position operator are called "Associated Fluctuation Terms". These terms describe the corrections to the Classical Limit Behavior and hence originate from the quantum dynamical nature of the system.

## 3   Phase Space Considerations

If we consider zeroth order truncation equations then the phase space of these equations is spanned by $p$ and $q$. Since the Hamiltonian of the system is time independent it is quite easy to show that the expectation value of the Hamiltonian is constant in time. Within zero fluctuation approximation, this means

$$\frac{1}{2m}p(t)^2 + V\left(q(t)\right) = K_1 \tag{19}$$

where $K_1$ is constant to be determined from the initial values of $p(t)$ and $q(t)$. This equation defines a curve inside the phase space and enables us to solve the zeroth order equations first for $q(t)$ (an inverse function evaluation is required) and then for $q(t)$.

## 4   Concluding Remarks

We have presented an application to quantum dynamical problems of a recently developed method called Fluctuation Expansion Method. Although we have given all the necessary information to construct the full infinite set of equations we have focused on only zeroth and first order truncations here. What we have proposed here can be extended to most generalized truncation of the infinite set of equations. Towards this goal symbolic and/or numerical codes can be developed in future applications. The method developed here is more efficient in the cases where the considered system's behavior is close to the classical limit. That is, heavy molecular systems with rather weak internal forces are more convenient for the application of this method.

## Acknowledgment

The second author wishes to acknowledge the financial support from the State Planning Organization (DPT) of Turkey and thanks Turkish Academy of Science for its partial support of this work. Both authors thank Professor N. A. Baykara for his careful reading of the manuscript and very useful comments.

## References

[1] L.D. Landau and F.M. Lifshitz: *Quantum Mechanics*. Pergamon, New York, 1965.

[2] M. Demiralp, A Fluctuation Expansion in Integration Under Almost Sharply Localized Weight, (submitted), 2005.

[3] M. Demiralp, A Fluctuation Expansion Method for the Evaluation of a Function's Expectation Value, *To appear in the Proceedings of ICNAAM2005, Rhodes, Greece*, 2005.

Brill Academic Publishers
P.O. Box 9000, 2300 PA Leiden
The Netherlands

*Lecture Series on Computer
and Computational Sciences*
Volume 4, 2005, pp. 410-413

# The Gas-Phase Proton Affinity: Basis Set And Correlation Effects

A. Mohajeri[1]

Department of Chemistry, College of Sciences

Shiraz University, Shiraz, 71454, Iran

Received 9 June, 2005; accepted in revised form 29 June, 2005

*Abstract*: High level ab initio and density functional theory calculations have been performed for evaluation of gas-phase proton affinities in nine simple molecules. Investigation on the effects of basis set and electron correlation which are essential to get reliable results is the main goal of this research. A series of computations were carried out with different basis sets and at different levels of theory and a reasonable comparison have been made with the experimental values. On the basis of this investigation, it is found that the role of polarization functions in predicting the proton affinities is more important than that of diffuse functions. The proton affinities were also calculated at G2 and CBS-Q methods in which high correlation level and extrapolation to the complete basis set are used. The correlation between the results obtained at G2 and CBS-Q methods and the experimental values indicated excellent agreement and the average absolute errors were less than 1 kcal/mol.

*Keywords*: Gas-phase proton affinity, polarization, diffuse functions, correlation, G2, CBS-Q

## 1. Introduction

Proton transfer reactions have a crucial role in chemistry and biomolecular process of living organisms [1]. So, many attempts have been carried out to obtain the accurate values of proton affinity in recent years [2-4]. The proton affinity is the quantity describes the ability of a molecule to accept proton. This value links the ion thermochemistry to that of neutral molecule. The importance of gas-phase proton affinity is due to two reasons. First, the majority of chemical reactions deal with acid-base concepts [5]. Therefore knowledge of basicity of a compound in the gas-phase which is directly connected with its proton affinity is helpful for evaluation of its reactivity. Second, investigation on most atmospheric and space reactions in the gas phase requires complete information about their involved chemical compounds.

The proton affinity of a compound B is defined as negative of enthalpy change in the following reaction at 298.15 K.

$$B + H^+ \rightarrow BH^+ \tag{1}$$

Although there are experimental techniques for determining the proton affinity, it is difficult to measure its value for biologically interesting molecules and kinetically unstable intermediates. Recent developments in computational methods enable chemist to provide accurate values for thermodynamic properties. Theoretical calculations can be carried out much faster and less cost than experiment. Moreover, there is less limitation in size and stability of the molecules under investigation in theoretical methods. However, proton affinity can be obtained absolutely rather than relative by using computational approaches.

In this research systematic calculations of proton affinities have been performed for verifications of the reliability of different methods and basis sets.

The proton affinities (PA) were calculated for 9 different halides of three neighboring groups and the obtained results were compared with their corresponding experimental values. In order to study the basis set effect the calculations were carried out at HF, MP2 and B3LYP levels using 10 different basis sets.

---

[1] Corresponding author. E-mail: mohajeri@chem.susc.ac.ir, admin@mohajeri.info

## 2. Computational methods

A full geometry optimization was carried out for all considered molecules and their corresponding protonated species. Total energies, zero pint energies and enthalpies at 298.15 K are computed for all species.

The proton affinities were obtained as negative of the enthalpy change for the reaction (1). All calculations were performed by Gaussian 98 program at HF, MP2 and B3LYP levels using 10 different basis sets. The basis set dependence of the obtained results is an important factor that is considered in this research. In order to check how polarization and diffuse functions impress the evaluated PAs, these functions are added step by step to the smallest 6-31G basis set. In this work G2 [6] and CBS-Q [7] were also used as highly accurate methods for studying the correlation between the computed proton affinities and their corresponding experimental values [8].

## 3. Results and discussion

The deviation of computed proton affinities from their corresponding experimental values ($PA_{calc.}$-$PA_{exp.}$) are collected in Table 1 at HF, MP2 and B3LYP levels, respectively. We started from 6-31G basis set without the polarization and diffuse functions. It can be seen in Table 1 that the obtained results with this basis set are the worst among all studied basis sets. The deviations are the least for the first member of each group, i.e. $NH_3$, $H_2O$ and HF, while for the molecules with heavier atoms the errors become more in all groups. Another important point is the dependence of computed proton affinities on the applied method. HF theory severely underestimates the calculated proton affinities while considering the correlation terms in MP2 and B3LYP reduces the average absolute errors about 50%.

Further improvement of basis set is achieved by adding polarization functions. The 6-31G* and 6-31G** are slightly larger than the previously discussed 6-31G basis set as they include a d-type polarization function on the heavy (non-hydrogen) atoms and a p-type function on hydrogen atoms, respectively. So, it can be expected that considerable improvement are seen in 6-31G* and some more improvement in 6-31G** results in comparison to 6-31G. As the results in Table 1 show adding polarization functions is more effective in the heavier atoms. Therefore, the less deviations can be seen in the third member of each group ($AsH_3$, $H_2Se$ and HBr).

To provide more accurate description the diffuse functions are added to the 6-31G** basis set. In 6-31+G** a set of diffuse s- and p-orbitals are added to non-hydrogen atoms and in 6-31++G** a set of s-functions are added to hydrogen atom. As it is obvious from the results in Table 1 no significant improvement is seen in the predication of proton affinities with these basis sets in comparison with the results obtained by the previous 6-31G** basis set. The comparison of mean absolute errors reveals that the calculated PAs have not been improved by going from 6-31+G** to 6-31++G**, while addition of polarization function on hydrogen atoms shows significant improvement on the obtained PAs.

Correlation effect is crucial in determination of thermodynamic properties. In comparison with HF results, MP2 and B3LYP have the results in better agreements with experiment. Also the results indicate that the correlation effect is more vital for the heavier molecules, while in the first member of each group the deviations are in the range of MP2 and B3LYP results. In the final step the calculations were performed using G2 and CBS-Q methods. A reasonable straightforward way to compare the experiment and theory is to calculate the linear regression between the calculated and experimental values. The agreement between these two is then reflected by the slope and intercept of the correlation line. A slope different from unity and a non-zero intercept imply systematic deviation between theory and experiment. For G2 slope and intercept of the correlation line are 1.005 and 0.1627. The obtained values for the slope and intercept of the correlation line in CBS-Q are 1.01 and 2.5031. The average absolute errors for G2 and CBS-Q are 0.82 and 0.92 kcal/mol, respectively. Therefore G2 is a reliable method in predicting thermodynamic properties at chemical accuracy.

Table 1. Deviations of computed proton affinities from experimental values (The results at HF, MP2 and B3LYP levels are in the first, second and third rows for each basis set)*

| | Basis set | $NH_3$ | $PH_3$ | $AsH_3$ | $H_2O$ | $H_2S$ | $H_2Se$ | HF | HCl | HBr |
|---|---|---|---|---|---|---|---|---|---|---|
| 1 | 6-31G | -4.09 | -12.94 | -23.50 | 15.72 | -23.79 | -14.35 | -8.00 | -27.19 | -23.53 |
| | | -0.76 | -10.31 | -3.97 | 1.42 | -19.30 | -7.97 | -5.04 | -22.36 | -15.93 |
| | | -0.78 | -10.88 | -5.86 | -2.64 | -17.37 | -5.77 | -2.68 | -18.92 | -12.51 |
| 2 | 6-31G* | -5.54 | -5.19 | -2.27 | -5.25 | -10.49 | -3.84 | -4.91 | 0.93 | -9.40 |
| | | -4.57 | -7.11 | -5.96 | -5.71 | -9.38 | -4.32 | -1.72 | -11.28 | -9.07 |
| | | -4.67 | -8.29 | -6.29 | -4.12 | -8.00 | -0.02 | 0.73 | -9.18 | -3.52 |
| 3 | 6-31G** | -3.19 | -3.02 | 1.91 | 0.19 | -8.13 | -1.34 | 0.05 | -10.22 | -6.40 |
| | | -1.97 | -3.40 | -0.30 | -0.51 | -4.89 | 0.07 | 3.16 | -4.98 | -4.02 |
| | | -3.16 | -6.84 | -2.79 | 0.05 | -6.33 | 1.86 | 4.62 | -6.19 | -1.02 |
| 4 | 6-31+G** | -6.54 | -3.22 | 1.39 | -4.17 | -7.89 | -3.81 | -6.76 | -10.07 | -8.79 |
| | | -7.32 | -3.82 | -1.04 | -7.51 | -5.07 | -2.66 | -7.35 | -5.08 | -6.59 |
| | | -8.88 | -7.78 | -3.98 | -7.40 | -7.21 | -1.46 | -7.26 | -7.22 | -4.38 |
| 5 | 6-31++G** | -6.51 | -3.08 | 2.86 | -4.11 | -7.96 | -3.36 | -6.75 | -10.02 | -7.72 |
| | | -7.34 | -3.69 | 0.43 | -7.52 | -4.97 | -2.09 | -7.32 | -4.98 | -5.50 |
| | | -8.91 | -7.72 | -2.70 | -7.41 | -3.87 | -0.90 | -7.24 | -7.18 | -3.33 |
| 6 | 6-311G | -4.23 | -11.43 | -4.85 | 1.72 | -21.31 | -15.22 | -9.68 | -25.16 | -21.03 |
| | | -2.71 | -9.10 | -2.55 | 0.78 | -17.05 | -9.91 | -8.78 | -20.69 | -14.68 |
| | | -3.16 | -11.37 | -6.81 | 1.30 | -17.08 | -10.28 | -7.26 | -19.52 | -13.83 |
| 7 | 6-311G* | -5.20 | -4.45 | -2.81 | -2.22 | -8.98 | -7.02 | -8.59 | -12.99 | -11.86 |
| | | -4.79 | -6.82 | -7.02 | -2.03 | -8.40 | -0.05 | -6.44 | -10.69 | -10.15 |
| | | -5.83 | -9.14 | -8.15 | -2.74 | -8.40 | -4.81 | -5.55 | -10.39 | -7.67 |
| 8 | 6-311G** | -4.21 | -1.66 | 0.28 | -1.07 | -6.36 | -5.48 | -3.86 | -8.67 | -8.90 |
| | | -4.73 | -3.09 | -2.30 | -1.60 | -3.35 | -4.07 | -0.19 | -10.15 | -4.25 |
| | | -5.67 | -7.47 | -5.85 | -2.05 | -6.64 | -3.98 | -1.43 | -6.81 | -5.30 |
| 9 | 6-311+G** | -6.79 | -1.75 | 0.40 | -4.56 | -6.42 | -5.41 | -7.32 | -8.25 | -8.93 |
| | | -8.73 | -3.05 | -2.19 | -7.50 | -3.25 | -4.04 | -6.23 | -2.63 | -4.34 |
| | | -9.75 | -7.35 | -5.81 | -8.07 | -6.53 | -4.06 | -8.21 | -6.30 | -5.55 |
| 10 | 6-311++G** | -6.78 | -1.74 | 0.41 | -4.64 | -6.43 | -5.40 | -7.30 | -8.19 | -8.89 |
| | | -8.24 | -3.06 | -2.17 | -7.43 | -3.24 | -4.02 | -6.20 | -2.57 | -4.30 |
| | | -9.73 | -7.36 | -5.69 | -8.03 | -6.54 | -4.07 | -8.19 | -6.26 | -5.54 |

*Values in kcal/mol

## 4. Conclusion

Two influential factors such as basis set dependence and correlation effect were investigated in the calculation of gas-phase proton affinity. The results obtained at different levels of theory were compared with their experimental data. It is found that the accurate values for proton affinities can be obtained by using flexible basis sets and explicit recovery of electron correlation. Considering the basis set dependence of deviations indicates that the polarization functions have an important role in calculation of PAs. This role becomes more significant in the heavier atoms. It is also found that the size of basis set is not an important determinant factor for the basis set quality. The results also show that less deviation can be seen by using more correlated method.

## References

[1] A. Onufriev, A. Smondyrev and D. Bashford, Proton affinity changes driving unidirectional proton transport in the bacteriorhodopsin photocycle, *Journal of Molecular Biology* **332** 1183-1193 (2003).

[2] R. R. Sauers, A computational study of proton and electron affinities, *Tetrahedron* **55** 10013-10026 (1999).

[3] J. K. Wolken and F. Turecek, Proton affinity of uracil. A computational study of protonation site, *Journal of American Society for Mass spectrometry* **11**1065-1071 (2000).

[4] C. A. Deakyne, Proton affinities and gas phase basicities: theoretical methods and structure effects, *International Journal of Mass Spectrometry* **227** 601-616 (2003).

[5] G. F. Cerofolini and N. Re, Studies in the acid-base theory. The base strength of amines, *Chemical Physics Letters* **339** 375-379 (2001).

[6] L. A. Curtiss, K. Raghavachari, J. A. Pople, Gaussian-2 theory Using reduced Moller-Plesset orders *Journal of Chemical Physics* **98** 1293-1298 (1993).

[7] A. L. L. East, B. J. Smith and L.Radom, Entropies and free energies of protonation and proton-transfer reactions, *Journal of American Chemical Society* **119** 9014-9020(1997) .

[8] E. P. L. Hunter and S. G. Lias, Evaluated gas phase basicities and proton affinities of molecules, *Journal of Physical Chemistry Reference Data.* **27** 413-656 (1998).

Brill Academic Publishers
P.O. Box 9000, 2300 PA Leiden,
The Netherlands

*Lecture Series on Computer
and Computational Sciences*
Volume 4, 2005, pp. 414-416

# Surface Magnetism in $YCo_2$

**P. Mohn[1], S. Khmelevskyi, and J. Redinger**

Center for Computational Materials Science,
Institute for general Physics,
Vienna University of Technology,
A-1060 Vienna, Austria

Received 5 July, 2005; accepted in revised form 6 July, 2005

*Abstract:* Using full-potential electronic structure calculations, we predict that the (111) surface of the cubic Laves phase Pauli paramagnet $YCo_2$ is ferromagnetic. The magnetism of the (111) surface is independent of the termination of the surface, does not extend beyond two Co layers. $YCo_2$ appears to be a prominent candidate to demonstrate the phenomenon of surface-induced itinerant magnetism localized in two dimensions. The top Co layer appears to be intrinsically magnetic and it induces magnetism in the second layer thus causing a meta-magnetic-like transition localized in two dimensions.

*Keywords:* Itinerant electron magnetism, surface magnetism, metamagnetism

*PACS:* 71.20.Eh,75.30.Kz,75.70.Rf

## 1

During recent years materials whose electronic structure is close to the onset of the magnetic order have attracted a huge interest in the field of the strongly correlated electron systems. The reason for this is the wide recognition of the fact that critical fluctuations, which develop in the region of the magnetic instability often result in very unusual or exotic phenomena, which are absent in normal solids. During the course of this development more traditional studies of itinerant electron meta-magnetic (IEM) [1] compounds and weakly magnetic (WM) materials [2], which can be easily be driven to the instability region by applying a magnetic field or pressure, have gained a new vista since they provide a natural choice for studying the development of critical spin-fluctuations. One of the important issues in the theory of quantum critical phenomena is related to the understanding of the effects of the reduced dimensionality on the properties of the quantum fluctuations. In this paper we argue that two dimensional (2D) strongly spin-fluctuating system may be found at the surface of the bulk strongly exchange enhanced Pauli paramagnet $YCo_2$, which for a long time has been the most studied, though canonical, IEM system [3] with a critical field induced IEM transition at $\approx 70T$ [4]. We have calculated the electronic structure of the $YCo_2$ surface within the framework of the Local Spin Density Approximation using a full potential Linear Augmented Plane Wave method as embodied in the FLAIR code[5]. The surface was modeled by a slab with Y and Co terminated $YCo_2(111)$ surfaces and a sufficient thickness to guarantee that the central layers show the electronic structure of the bulk material. Further details of the calculations can be found in Ref.[6]. We have found a stabilization of the intrinsically magnetic state at the top Co surface layer and a fast decay of the magnetic moment deeper into the bulk [6] almost independently on

---

[1]Corresponding author. E-mail: phm@cms.tuwien.ac.at

the termination and the size of the surface lattice relaxation. In this paper we provide more insight concerning the details of the electronic structure of the YCo₂ surface with particular emphasis to the peculiar situation in the second Co layer. In Fig. 1 we show the calculated atom projected DOS of the topmost Co layer for the unrelaxed Co terminated and relaxed Co and Y terminated geometry. It can be seen that in all cases the bands are magnetically split and corresponding atomic moments are found to be 1.33, 1.05 and $0.79\mu_B$, respectively. These values are close to the Co moments in bulk YCo₂ in magnetic fields above the IEM [3, 4]. In the case of the Y surface termination the moment has a smaller value since some Co bonds are more saturated than in case of Co termination. Taking into account also that the same calculations give a paramagnetic ground state in the bulk calculations and no Co moment deeper in the bulk in slab calculations proofs that the appearance of the local moments on the YCo₂ surface is independent on the termination as well as on the surface relaxation.

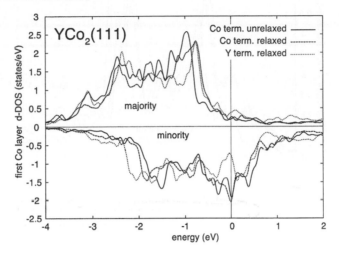

Figure 1: Ground state, atom projected DOS of the top Co surface layer calculated for different termination of the YCo₂(111) surface.

In the case of Co termination the magnetic moment of the second Co layer is equal to $0.6\mu_B$/atom whereas deeper in the bulk it remains at values $\leq 0.01\mu_B$. It appears that in the second Co layer the moment is not intrinsic but induced by the magnetic interaction with top Co layer. This can be seen from the Fig. 2 where the atomic DOS of the non-magnetic (non-spin-polarized, NM) calculations are presented for different Co layers. For the surface Co layer, the Fermi energy $E_F$ intersects at the top of the Co d-band with a DOS large enough to fulfill the Stoner criterion for the onset of magnetism. This increase of the surface DOS at $E_F$ is due to a reduced coordination number, which results in the well known band narrowing at the surface and a subsequent shift of the d-states to maintain charge neutrality. For the almost filled Co d-band, the shift is towards higher energy. In the second and third Co layer this effects is very weak since the reduction of the coordination number occurs only in the second nearest neighbor shell. The values of the NM DOS at the Fermi level (Fig. 2) are almost equal for the 2nd and 3rd Co layers. Since now the 2nd Co layer is magnetic and 3rd not it can be concluded that magnetism of the 2nd Co layer is induced in a sort of IEM transition caused by the interaction with the intrinsically magnetic top surface layer. The comparison of the DOS features of bulk YCo₂ with those of the 2nd and 3rd

Co layers also suggests that these layers are much closer to a magnetic instability than the bulk material. A development of strong spin-fluctuations in the surface region, which would effectively be localized in two dimensions can thus be expected. To conclude we note, that recent progress in the preparations of Laves Phase thin films and multi-layers [7] gives evidence that studies of the magnetism in $YCo_2$ on the surface is possible. Here we have shown that such studies may be highly interesting in the context of developments in the field of quantum critical phenomena and itinerant magnetism in general.

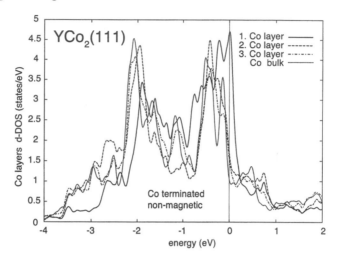

Figure 2: Non-magnetic atom-projected DOS for the first three Co-layers in comparison with the bulk $YCo_2$ DOS.

# References

[1] E. P. Wohlfarth and P. Rhodes, Phil. Mag. 7, 1817 (1962).

[2] T. Moriya, Spin Fluctuations in Itinerant Electron Magnetism, (Springer-Verlag, Berlin, 1985).

[3] N. H. Duc and T. Goto, in Handbook on Physics and Chemistry of Rare-Earths, (ed. by K. A Gscheider and L. Eyring, North-Holland, Amsterdam, 1999) Vol. 29, p.177.

[4] T. Goto, T. Sakakibara, K. Murata and H. Komatsu, Solid State Comm. 72, 945 (1989).

[5] http://www.uwm.edu/~weinert/flair.html

[6] S. Khmeleskyi, P. Mohn, J. Redinger and M. Weinert, Phys. Rev. Lett. 94, 146403 (2005).

[7] F. Robaut, S. Jaren, N. Cherief-Bandbrahim, and G. Meyer, Appl. Phys. Lett. 69, 1643 (1996).

Brill Academic Publishers
P.O. Box 9000, 2300 PA Leiden,
The Netherlands

*Lecture Series on Computer
and Computational Sciences*
Volume 4, 2005, pp. 417-420

# Application of Symplectic Partitioned Runge-Kutta Methods to Hamiltonian Problems

## Th. Monovasilis

Department of Computer Science and Technology,
Faculty of Sciences and Technology,
University of Peloponnese,
GR-221 00 Tripolis, Greece
Department of International Trade,
Technological Educational Institute of Western Macedonia at Kastoria,
P.O. Box 30, GR-521 00, Kastoria, Greece

## Z. Kalogiratou

Department of International Trade,
Technological Educational Institute of Western Macedonia at Kastoria,
P.O. Box 30, GR-521 00, Kastoria, Greece

## T.E. Simos[1]

Department of Computer Science and Technology,
Faculty of Sciences and Technology,
University of Peloponnese,
GR-221 00 Tripolis, Greece

*Abstract:* In this paper we present second, third and fourth order symplectic partitioned Runge-Kutta (PRK) methods with several stages. We apply all methods to the numerical integration of the harmonic oscillator, the pendulum and Kepler problems.

*Keywords:* partitioned Runge-Kutta methods, Symplectic Integrators, Hamiltonian systems

*Mathematics Subject Classification:* 65L05;

## 1   Introduction

For the separable Hamiltonian $H(p,q) = T(p) + U(q)$ the Hamiltonian system is given by

$$p' = -\frac{\partial U}{\partial q}(q), \quad q' = \frac{\partial T}{\partial p}(p) \tag{1}$$

---

[1]Corresponding author. Active Member of the European Academy of Sciences and Arts. E-mail: simos-editor@uop.gr, tsimos@mail.ariadne-t.gr

Any explicit symplectic PRK method for the system (1) can be written in the form:

$$k_i = -\frac{\partial U}{\partial q}(Q_i), \qquad l_i = \frac{\partial T}{\partial p}(P_i)$$

$$P_i = p_n + h\sum_{j=1}^{i} b_j k_j, \qquad Q_i = q_n + h\sum_{j=1}^{i-1} B_j l_j$$

$$p_{n+1} = p_n + h\sum_{i=1}^{s} b_i k_i \qquad q_{n+1} = q_n + h\sum_{i=1}^{s} B_i l_i$$

Since 1983 many authors Ruth [5], Yoshida [8],McLachlan [3] developed symplectic integrators based on the idea of PRK methods. In the next tables the coefficients of PRK methods of orders 2,3 and 4 are given. This methods can be found in  citeMcLachlan1,[4], Ruth [5], Yoshida [8].

Table 1: Coefficients of Second Order Symplectic Methods

| Stages | Coefficients |
|--------|--------------|
| 2 | $b_1 = \frac{1}{2}$ , $\quad b_2 = \frac{1}{2}$, $\quad B_1 = 1$, $\quad B_2 = 0$ |
| 2 | $b_1 = \frac{2-\sqrt{2}}{2}$, $\quad b_2 = \frac{\sqrt{2}}{2}$, $\quad B_1 = \frac{\sqrt{2}}{2}$, $\quad B_2 = \frac{2-\sqrt{2}}{2}$ |
| 3 | $b_1 = z$, $\quad b_2 = 1 - 2z$, $\quad b_3 = b_1$, $\quad B_1 = \frac{1}{2}$ , $\quad B_2 = \frac{1}{2}$ $\quad B_3 = 0$ <br> $z = \frac{a^2 + 6a - 2}{12a} \approx 0.193183$ $\quad a = (2\sqrt{326} - 36)^{\frac{1}{3}}$ |

Table 2: Coefficients of Third Order Symplectic Methods

| Stages | Coefficients |
|--------|--------------|
| 3 | $b_1 = \frac{7}{24}$ , $\quad b_2 = \frac{3}{4}$, $\quad b_3 = -\frac{1}{24}$ $\quad B_1 = \frac{2}{3}$, $\quad B_2 = -\frac{2}{3}$, $\quad B_3 = 1$ |
| 3 | $b_1 = B_3$, $\quad b_2 = B_2$ , $\quad b_3 = B_1$, $\quad B_3 = 1 - B_1 - B_2$, <br> $B_1 = \left(\frac{1}{9y} - \frac{w}{2} + \sqrt{y}\right)^{\frac{1}{2}} - \frac{1}{3\sqrt{y}}$, $\quad B_2 = \frac{1}{4B_1} - \frac{B_1}{2}$, <br> $z = -\left(\frac{2}{27} - \frac{1}{9\sqrt{3}}\right)^{\frac{1}{3}}$ $\quad w = -\frac{2}{3} + \frac{1}{9z} + z$, $\quad y = \frac{1}{4}(1 + w^2)$ |

Table 3: Coefficients of Fourth Order Symplectic Methods

| Stages | Coefficients |
|---|---|
| 4 | $b_1 = b_4 = \frac{x_1}{2}$, $\quad b_2 = b_3 = \frac{x_0 + x_1}{2}$, $\quad B_1 = B_3 = x_1$, $\quad B_2 = x_0$, $\quad B_4 = 0$, $$x_0 = -\frac{2^{1/3}}{2 - 2^{1/3}} \quad x_1 = \frac{1}{2 - 2^{1/5}}$$ |
| 7 4 | $B_1 = 0.5153528374311229364$, $\quad B_2 = -0.085782019412973646$ $B_3 = 0.4415830236164665242$, $\quad B_4 = 0.1288461583653841854$ $b_1 = 0.1344961992774310892$, $\quad b_2 = -0.2248198030794208058$ $b_3 = 0.7563200005156682911$, $\quad b_4 = 0.3340036032863214255$ |
| 5 | $b_1 = \frac{1}{2} - z$, $\quad b_2 = -\frac{1}{3} + z$, $\quad b_3 = \frac{2}{3}$, $\quad b_4 = b_2$, $\quad b_5 = b_1$ $B_1 = 1$, $\quad B_2 = -\frac{1}{2}$, $\quad B_3 = B_2$, $\quad B_4 = B_1$, $\quad B_5 = 0$, $\quad z = \frac{\sqrt{\frac{7}{8}}}{3}$ |
| 6 | $b_1 = \frac{642 + \sqrt{471}}{3924}$, $\quad b_2 = \frac{121(12 - \sqrt{471})}{3924}$, $\quad b_3 = \frac{-11 + 10\sqrt{471}}{327}$, $\quad B_1 = \frac{6}{11}$, $\quad B_2 = -\frac{1}{22}$ $b_4 = b_3$, $\quad b_5 = b_2$, $\quad b_6 = b_1$, $\quad B_3 = B_6 = 0$, $\quad B_4 = B_2$, $\quad B_5 = B_1$ |
| 6 | $b_1 = \frac{14 - \sqrt{19}}{108}$, $\quad b_2 = \frac{20 - 7\sqrt{19}}{108}$, $\quad b_3 = \frac{5 + 2\sqrt{19}}{27}$, $\quad B_1 = \frac{2}{5}$, $\quad B_2 = -\frac{1}{10}$, $\quad B_3 = \frac{2}{5}$ $b_4 = b_3$, $\quad b_5 = b_2$, $\quad b_6 = b_1$, $\quad B_4 = B_2$, $\quad B_5 = B_1$, $\quad B_6 = 0$ |
| 6 | $b_1 = 0.40518861839525227722$, $\quad b_2 = -0.287144040816524089$, $b_3 = 0.38195542242127181$ $B_1 = -\frac{3}{73}$, $\quad B_2 = \frac{17}{59}$, $\quad B_3 = \frac{2179}{4307}$, $b_4 = b_3$, $\quad b_5 = b_2$, $\quad b_6 = b_1$, $\quad B_5 = B_1$, $\quad B_4 = B_2$, $\quad B_6 = 0$ |
| 6 | $b_1 = \frac{x_1}{2}$, $\quad b_2 = \frac{x_1 + x_2}{2}$, $\quad b_3 = \frac{x_2 + x_3}{2}$, $\quad B_1 = x_1$, $\quad B_2 = x_2$, $\quad B_3 = x_3$, $b_4 = b_3$, $\quad b_5 = b_2$, $\quad b_6 = b_1$, $\quad B_5 = B_1$, $\quad B_4 = B_2$, $\quad B_6 = 0$ $$x_1 = \frac{1}{4 - \sqrt[3]{4}}, \quad x_2 = x_1, \quad x_3 = -\frac{\sqrt[3]{4}}{4 - \sqrt[3]{4}}$$ |
| 6 | $b_1 = \frac{x_1}{2}$, $\quad b_2 = \frac{x_1 + x_2}{2}$, $\quad b_3 = \frac{x_2 + x_3}{2}$, $\quad B_1 = x_1$, $\quad B_2 = x_2$, $\quad B_3 = x_3$, $b_4 = b_3$, $\quad b_5 = b_2$, $\quad b_6 = b_1$, $\quad B_5 = B_1$, $\quad B_4 = B_2$, $\quad B_6 = 0$ $x_1 = 0.28$, $\quad x_2 = 0.62546642846767004501$, $\quad x_3 = -0.810932856935$ |

## 2 Numerical Results

We apply our methods to the following problems

**Harmonic Oscillator**

$$p' = -k^2 q, \quad q' = p$$

the Hamiltonian of this problem is

$$H(p,q) = \frac{1}{2}(p^2 + k^2 q^2).$$

**Pendulum**

$$p' = -\sin q, \quad q' = p$$

the Hamiltonian of this problem is

$$H(p,q) = \frac{1}{2}p^2 - \cos q.$$

**Two body problem**

$$p_1' = -\frac{q_1}{(q_1^2 + q_2^2)^{3/2}}, \quad q_1' = p_1 \quad p_2' = -\frac{q_2}{(q_1^2 + q_2^2)^{3/2}}, \quad q_2' = p_2$$

the Hamiltonian of this problem is

$$H(p_1, p_2, q_1, q_2) = \frac{1}{2}(p_1^2 + p_2^2) - \frac{1}{\sqrt{(q_1^2 + q_2^2)}}.$$

# References

[1] Hairer E., Lubich Ch., Wanner G., *Geometric Numerical Integration*, Springer-Verlag, 2002.

[2] Liu, X.S., Liu X.Y, Zhou, Z.Y., Ding, P.Z., Pan, S.F., Numerical Solution of the One-Dimensional Time-Independent Schrödinger equation by using symplectic schemes, *International Journal of Quantum Chemistry* 79(2000),343-349.

[3] McLachlan R., Atela P., The accuracy of symplectic integrators, *Nonlinearity*, 5(1992) 541-562.

[4] McLachlan R.,On the numerical integration of ordinary differential equations by symmetric composition methods, *SIAM J. Sci. Comput*, 16(1995) 151-168.

[5] Ruth R.D., A canonical integration technique, *IEEE Transactions on Nuclear Science*, NS 30 (1983),2669-2671.

[6] Sanz-Serna, J.M., Calvo, M.P., *Numerical Hamiltonian Problem*, Chapman and Hall, London, 1994.

[7] Suzuki M., General theory of higher-order decomposition of exponential operators and symplectic integrators , *Physics Letters A* 165(1992),387-395.

[8] Yoshida H., Construction of higher order symplectic integrators, *Physics Letters A* 150(1990),262-268

Brill Academic Publishers
P.O. Box 9000, 2300 PA Leiden
The Netherlands

*Lecture Series on Computer
and Computational Sciences*
Volume 4, 2005, pp. 421-424

# Binding States and Structural Analysis of Hydroxyl on Ag(111)

Alejandro Montoya and Brian S. Haynes

Department of Chemical Engineering
The University of Sydney
N.S.W. 2006
Australia

Received 18 July, 2005; accepted in revised form 10 August, 2005

*Abstract:* Periodic density functional total-energy calculations are used to examine the site preference for adsorption, ordered structures and binding energies of OH adsorption to the Ag(111) surface. We find on the Ag(111) facet that the three-hollow adsorption sites are energetically the most favorable for OH binding. The binding energy decreases slightly when the surface coverage is increased from 0.11 up to half monolayer and then drops rapidly on increasing the surface coverage up to full monolayer. Surface reconstruction is determined to be small after OH chemisorption but considerable changes to the OH geometry are observed. The tilt of the O-H axis out of the normal plane of the Ag(111) surface increases from 0.1 degrees at a surface coverage of 0.11 to 63.14 degrees at full monolayer.

*Keywords:* Periodic DFT-GGA, Hydroxyl, Ag(111), Binding states, Structural analysis

*Subject Classification:* Modelling surface reactions

## 1. Introduction

The chemical nature of hydroxyl groups (OH) chemisorbed to silver surfaces has attracted attention since it is believed to act as a reaction intermediate in a number of important catalytic processes such as oxidation of hydrocarbons and alcohols, electrochemical processes and in the water gas shift reaction.

The chemical properties of OH on metal surfaces have been extensively studied [1]. However, atomic level understanding of OH binding states, local structures and reaction channels in the Ag(111) is limited. It has been observed that silver forms pinholes in the outer surface when it is reacted with oxygen-containing hydrogen species over 600°C [2-4]. The chemical mechanism of this process is not well understood. It has been postulated that desorption of water from sub-surface sites are responsible for the pinholes creation. OH recombination and reduction with adsorbed hydrogen atoms may also play a role in the surface reconstruction of the silver catalyst.

Surface OH species have been experimentally identified after reaction of hydrogen, water, and methanol with oxygen doped Ag(111). HREEL peaks at 300 cm$^{-1}$ and 3640 cm$^{-1}$ [5], and Raman peaks shifts at 554 cm$^{-1}$ and 985 cm$^{-1}$ have been assigned to the surface HO-Ag(111) and $\delta$(OH), respectively [6]. Linear and perpendicular orientations of OH above the plane of the Ag(111) surface have been suggested from experimental studies. Klaua and Madey, using electron stimulated desorption ion angular distribution (ESDIAD) techniques, have found a perpendicular orientation in the temperature range of 80K-293K [7]. Carley *et al* using photoelectron and vibrational spectroscopy studies have observed a change in the OH orientation from bent configuration at low temperature (160K) to a linear configuration after desorbing H$_2$O molecules (220 K) [5]. These studies suggest that OH configuration is surface dependent. However, systematic studies of the structure coverage-dependence of OH on the Ag(111) facet is not known to us.

In this paper we report results for the interaction of OH on the Ag(111) at different coverages and provide insight into the site-specific and structure-reactivity relationships in connection with the hydrogen-oxygen interaction with silver.

Corresponding authors e-mail: montoya@chem.eng.usyd.edu.au, haynes@chem.eng.usyd.edu.au

## 2.  Computational technique

Molecular calculations were performed using the plane-wave density functional theory (DFT) code Dacapo with Vanderbilt ultrasoft pseudo-potentials [8]. The Ag surface is represented by periodic slabs with 5 layers repeated in a supercell geometry with 15 Å of vacuum between successive metal slabs. The 3 lower layers are maintained at DFT-bulk geometry (DFT calculated Ag lattice constant is 4.14 Å vs 4.09 Å experimentally) and the 2 upper Ag atom layers are allowed to relax. OH is placed on one side of the slab and the induced dipole moment is taken into account by applying a dipole correction [9]. Three different unit cells have been used, (1 x 1), (2 x 2) and (3 x 3), to represent surface coverages of 1.0, 0.25 and 0.11 monolayer, respectively. The (2 x 2) unit cell is also used to represent surface coverages of 0.5 and 0.75 using two and three surface species. The self-consistent field was carried out using an electronic temperature of $K_BT = 0.1$ and $K_BT = 0.0001$ for the Fermi distribution of the slabs and gas phase molecules, respectively, with Pulay mixing of the resulting electronic density. Total energies have been extrapolated to $K_BT = 0.0$ eV.

The Perdew and Wang nonlocal corrections to the exchange-correlation energy were included self-consistently [10]. The Kohn-Sham equations were solved using a plane-wave basis set of kinetic energy of 400 eV. A Monkhorst-Pack mesh [11] of (6 x 6 x 1) k-points was used for the (2 x 2) unit cell and the k-points density was maintained as constant as possible for the various unit cell sizes. Optimization is achieved when the root-mean-square force on the atomic nuclei was less than 0.05 eV/Å. Test calculations have been performed to ensure convergence of the binding energy. The total energy of isolated OH is calculated in a cubic cell of side length 12 Å where the spin-polarization correction is included during minimization.

## 3.  Energetics of Hydroxyl Chemisorption

The binding energy of OH to the top-most layer of Ag(111) is calculated and reported with respect to the energy of an isolated OH according to the equation:

$$E_b^{OH_{top}/Ag(111)} = -[E(Ag(111) + nOH_{top}) - E(Ag(111)) - nE(OH)]/n$$

where *n* is the number of OH species that are adsorbed per unit cell and *E(OH)*, *E(Ag(111))* and *E(Ag(111) + nOH<sub>top</sub>)* are the energies of hydroxyl, silver slab and OH-silver system, respectively. By definition, a positive binding energy indicates an exothermic adsorption (stable) and a negative number indicates an endothermic adsorption (unstable) with respect to OH in the gas phase.

The binding energy of OH to the most stable adsorption sites of the Ag(111) is shown in Figure 1. The inset in the figure illustrates the fcc site (above an Ag atom of the third layer) and the hcp site (above an Ag atom of the second layer) that are used to established the coverage-dependence binding energy profile. We find that these sites are energetically the most favorable for OH adsorption. Thus, we will discuss only the two three-hollow sites hereafter. It can be seen that the binding energy of OH to the Ag(111) surface is coverage dependent. The binding energy decreases slightly as the surface coverage (Θ) is increased up to Θ = 0.5 and then it drops rapidly by increasing the surface coverage up to Θ = 1.0. This trend shows a small OH-OH interaction at low coverage and significant OH-OH interaction at high coverage. The predicted binding energy is in the range of 44.5 – 57.7 kcal mol⁻¹ below monolayer. The stability of OH is slightly favored on the fcc site at the lowest surface coverage studied and switches to slightly prefer the hcp site at high surface coverage. The largest energy difference between the two three-hollow sites is calculated to be 0.80 kcal mol⁻¹. The difference is small and should be considered as a guide only.

The binding energy of OH to Ag(111) is smaller compared to the DFT binding energy of H in the $H_2O$ molecules. The low binding energy of OH below monolayer and the instability of H chemisorbed to the Ag(111) surface [12] are consistent with the non dissociation of $H_2O$ on Ag(111) below 100 K [7], apparently due to a high energy barrier for dissociation.

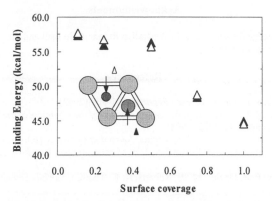

Figure 1. Coverage dependence of the binding energy of OH to silver. The inset shows the fcc and hcp adsorption sites of the Ag(111) surface. The larger gray circles represent the Ag atoms of the first row. Binding energy to the fcc site is shown with open triangles and to hcp sites with filled triangles.

## 4. Structural Parameters of OH Chemisorption

We found that structure changes to the silver slab are small after OH chemisorption in the surface coverage studied. The first and second interlayer Ag-Ag distances decreased by ~ 0.8 % and 0.01% after chemisorption. However, we found that the tilt of the O-H axis out of the normal plane of the Ag(111) surface is surface dependent. The tilt is defined as the degrees out of the normal plane where O is directly bound to the surface and H points to the gas phase. The tilt increases from 0.1 degrees at a surface coverage of $\Theta = 0.11$ to 63.14 degrees at a surface coverage of $\Theta = 1.0$ in the fcc site. The stability of OH is higher when it is linear at low coverage and bent at high coverage. This trend is consistent with the HREEL analysis of Carley et al.[5], who observed a bent OH group and linear OH group before and after $H_2O$ desorption from the Ag(111) surface.

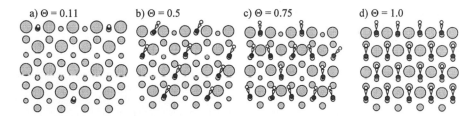

Figure 2. Geometrical representation of OH binding states to fcc sites of Ag(111). Silver atoms are shown in grey where the largest circles represent the first Ag layer. Oxygen atoms are shown in black and hydrogen in white circles.

The direction of the tilted H atom is dependent on the surface coverage as shown in Figure 2. The hydrogen atom is tilted towards the next OH group to maximize the hydrogen-oxygen interaction and to minimize the HO-OH repulsive interaction. As can be seen in Figure 2, hydrogen atoms of the tilted OH groups are above the hcp sites. Conversely, hydrogen atoms of the tilted OH groups in the hcp positions (not shown in Figure 2) are above the fcc sites. The O-H bond length and the distance above the center of mass of the first layer at $\Theta = 0.11$ are 0.9807 Å and 1.4496 Å, respectively. The O-H bond length and the distance to the surface increase gradually up to 0.9926 Å and 1.6470 Å at full monolayer, respectively. It indicates a decrease in the surface-hydroxyl interaction due to a significant attraction at long range of O and H atoms. Reaction paths towards the formation of $H_2O$ from hydroxyl and hydrogen species will be presented.

## Acknowledgments

The authors acknowledge the support of the Australian Research Council and the Australian Partnership for Advanced Computing.

## References

[1]     M. A. Henderson. *Surface Science Reports* **46,** 1-308 (2002).
[2]     G. J. Millar, M. L. Nelson, P. J. R. Uwins. *Journal of the Chemical Society, Faraday Transactions* **94,** 2015-2023 (1998).
[3]     G. J. Millar, M. L. Nelson, P. J. R. Uwins. *Journal of Catalysis* **169,** 143-156 (1997).
[4]     G. I. N. Waterhouse, G. A. Bowmaker, J. B. Metson. *Applied Catalysis, A: General* **265,** 85-101 (2004).
[5]     A. F. Carley, P. R. Davies, M. W. Roberts, K. K. Thomas. *Surface Science* **238,** L467-L472 (1990).
[6]     X. Bao, M. Muhler, B. Pettinger, Y. Uchida, G. Lehmpfuhl, R. Schloegl, et al. *Catalysis Letters* **32,** 171-183 (1995).
[7]     M. Klaua, T. E. Madey. *Surface Science* **136,** L42-L50 (1984).
[8]     B. Hammer, L. B. Hansen, J. K. Nørskov. *Phys. Rev. B* **59,** 7413–7421 (1999).
[9]     J. Neugebauer, M. Scheffler. *Phys. Rev. B* **46,** 16067–16080 (1992).
[10]    J. P. Perdew, J. A. Chevary, S. H. Vosko, K. A. Jackson, M. R. Pederson, D. J. Singh. *Phys. Rev. B* **46,** 6671–6687 (1992).
[11]    H. J. Monkhorst, J. D. Pack. *Phys. Rev. B* **13,** 5188–5192 (1976).
[12]    A. Montoya, A. Schlunke, B. S. Haynes. *In preparation* (2005).

Brill Academic Publishers
P.O. Box 9000, 2300 PA Leiden,
The Netherlands

*Lecture Series on Computer
and Computational Sciences*
Volume 4, 2005, pp. 425-428

# A Generalized False Position Numerical Method for Finding Zeros and Extrema of a Real Function

José F.M. Morgado and Abel J.P. Gomes[1]

Dept. Informatics,
University of Beira Interior,
6200-001 Covilhã, Portugal

Received 5 August, 2005; accepted in revised form 12 August, 2005

*Abstract:* Root-finding numerical methods are based on the Intermediate Value Theorem. It states that a root of a real function $f : \mathbb{R} \to \mathbb{R}$ is bracketed in a given interval $[A, B] \subset \mathbb{R}$ if $f(A)$ and $f(B)$ have opposite signs, i.e. $f(A).f(B) < 0$. But, some roots, say local minima or maxima, cannot be bracketed this way because the condition $f(A).f(B) < 0$ is not satisfied. In this case, we normally use a specific numerical method for bracketing an extremum, checking then whether it is a zero of $f$ or not. In contrast, this paper introduces a single numerical method, called *generalised false position method*, that is capable of computing not only zeros but also extrema of a function. Thus, it determines any zero in a given interval $[A, B] \subset \mathbb{R}$ even when the Intermediate Value Theorem is not satisfied. This is particularly important in sampling implicit curves and surfaces that evaluate either positive or negative everywhere, a computer graphics problem that has remained unsolved for so long.

*Keywords:* Zeros and extrema of real functions, numerical methods without derivatives, sampling of implicit curves and surfaces

*Mathematics Subject Classification:* 65D05, 65D17, 65Y20.

## 1 Introduction

Numerical methods such as the *bisection method, false position method, Newton-Raphson method* amongst others are root-finding algorithms, i.e. they are used for bracketing a root. Recall that a root is *bracketed* in the interval $[A, B]$ if $f(A)$ and $f(B)$ have opposite signs, i.e. $f(A).f(B) < 0$. In this paper, a root of this sort is called a *crossing zero*, as illustrated in Figure 1(a).

Although this class of methods satisfying the Intermediate Value Theorem succeeds in finding most zeros (i.e. crossing zeros), they fail to bracket zeros that are also extrema, here called *extremal zeros*. We say that an extremum is *bracketed* in the interval $[A, B]$ if $f(A)$ and $f(B)$ have identical signs, i.e. $f(A).f(B) > 0$. If $f(A) > 0$ and $f(B) > 0$, "go downhill, taking steps of increasing size, until your function starts back uphill" [4]. At this point, we have only to check that such a minimum is a zero. Analogously, if $f(A) < 0$ and $f(B) < 0$, "go uphill, taking steps of increasing size, until your function starts back downhill" [4]. Then, we have only to check whether such a maximum is a zero.

Despite the existence of some algorithms for finding extrema (maxima and minima) of functions in the numerical analysis literature, there is no single method for catching up both crossing and

---

[1]Corresponding author. Department of Informatics, Univ. Beira Interior, 6200-001 Covilhã, Portugal. E-mail: agomes@di.ubi.pt

extremal zeros. Such a method is described in the next section and is called *generalized false position method* (GFP). It has linear convergence and has been successfully used in sampling planar implicit curves against some space partitioning of a rectangular subspace $\Omega \subset \mathbb{R}^2$ [3].

## 2   GFP Method

The GFP method is based on the ratio of similar triangles. Its main novelty is that it uses a single interpolation formula for determining both zeros and extrema.

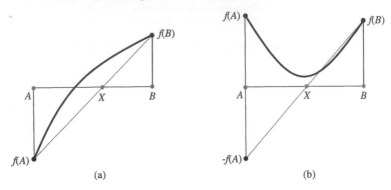

(a)                                                                          (b)

Figure 1: Illustration of the GFP method for (a) zeros and (b) extrema.

Let us consider the interval $[A, B]$, as illustrated in Figure 1(a) and (b), being $|f(A)|$ and $|f(B)|$ the absolute values of the function at $A$ and $B$, respectively. The endpoints of the interval $[A, B]$ work as two distinct estimates for a root of $f(x) = 0$. The next estimate $X$ is the intersection point of the straight-line segments $\overline{(A, 0)(B, 0)}$ and $\overline{(A, |f(A)|)(B, |f(B)|)}$, which is given by

$$X = A + \frac{|f(A)|}{|f(A)| + |f(B)|}.(B - A) \tag{1}$$

In fact, by using the vectorial equation of the segment defined by $A$ and $B$ as follows:

$$X(t) = A + t.(B - A) \tag{2}$$

we easily see that $A = X(0)$, $B = X(1)$, and the next estimate $X$ occurs at $t = \frac{|f(A)|}{|f(A)|+|f(B)|}$, a result that is given by the following equality of triangles:

$$\frac{|f(A)|}{t} = \frac{|f(B)|}{1 - t} \tag{3}$$

Remarkbly, as explained below, the equation (1) works for both zeros and extrema.

### 2.1   Zeros

The computation of a zero is illustrated in Figure 1(a). If $f(A).f(B) < 0$, assuming that $f(A) < 0$, and substituting $|f(A)| = -f(A)$ and $|f(B)| = f(B)$ into (1), we obtain

$$X = A + \frac{f(A)}{f(B) - f(A)}.(B - A) \tag{4}$$

that is the interpolation formula of the well-known false position method [2]. It is used to determine crossing zeros.

## 2.2   Extrema

The computation of an extremum is exactly the same for minima and maxima. The procedure can be described as follows. If $f(A).f(B) > 0$, assuming that $f(A) > 0$, then substituting $|f(A)| = f(A)$ and $|f(B)| = f(B)$ into (1), we obtain

$$X = A + \frac{f(A)}{f(B) + f(A)}.(B - A) \qquad (5)$$

that is used to compute the next estimate of an extremum. Similarly, if $f(A).f(B) > 0$, with $f(A) < 0$, and replacing $|f(A)| = -f(A)$ and $|f(B)| = -f(B)$ into (1), we also get the next estimate $X$ through (5).

Remarkably, the formula (5) works equally well for minima and maxima, independently of whether $f(A)$ and $f(B)$ are both positive or negative. In fact, looking at the equations (4) and (5), we see that it is enough to change the sign of either $f(A)$ to determine the next estimate of an extremum through the false position technique. This is illustrated in Figure 1(b)) for a minimum.

But, the condition $f(A).f(B) > 0$ only indicates that there may exist an extremum between $A$ and $B$. We have to evaluate the finite difference $d(x) = \frac{f(x+h)-f(x)}{h}$, with an infinitesimal $h$, at $A$ and $B$ and test if they have opposite signs, i.e. $d(A).d(B) < 0$, to guarantee that an extremum exists in between. We use the finite difference —the discrete counterpart of the derivative— of a function at a given point for two main reasons. First, the false position method does not use derivatives. So, the generalized false position method does not use them for *consistency* either. Second, there are zeros and extrema at which derivatives do not exist. For example, $f(x) = |x|^{1/3}$ has a cusp at $x = 0$ that is also a minimum. Derivatives cannot be used in this case because the program would break down.

Then, we have to check whether the finite difference $d(X)$ of the estimate $X$ is very close to zero. If so, we get an extremum at $X$ approximately. Besides, if $f(X) \approx 0$, with an accuracy of $10^{-6}$, then $X$ is a zero. Therefore, formula (1) can be used for computing the next estimate of either a zero or an extremum.

## 3   GFP implementation

The C function that implements the generalized false position method returns either a zero or an extremum bracketed in a given interval $[A, B]$:

```
#include <math.h>
#define ACY 1.0E-6          accuracy for estimate, function, and finite difference values
#define h 1.0E-7            infinitesimal for finite diferences
```

```
double GFP(float (*func)(double), double A, double B, int n)
```
Using the generalized false position method, find the root or extremum of a function func known to lie between A and B. The root is refined until its accuracy is ±ACY.

```
{
        if (n==0)
            return;
        if (fabs(B-A)< ACY)
            return (A+B)/2;
        double fA=(*func)(A);
        double fB=(*func)(B);
        double dA=((*func)(A+h)-(*func)(A))/h;
```

```
double dB=((*func)(B+h)-(*func)(B))/h;
double X=A+fabs(fA)/(fabs(fA)+fabs(fB))*(B-A);
double fX=(*func)(X);
      if (fabs(fX)< ACY)
      return X;
if (fA*fX<0)
      return GFP(*func,A,X,n-1);
if (fX*fB<0)
      return GFP(*func,X,B,n-1);
double dX=((*func)(X+h)-(*func)(X))/h;
if (fabs(dX)< ACY)
      return X;
if (dA*dX<0)
      return GFP(*func,A,X,n-1);
if (dX*dB<0)
      return GFP(*func,X,B,n-1);
}
```

## 4   Conclusions

In the literature there are distinct numerical methods for computing zeros and extrema. In contrast, the method introduced in this paper computes both zeros and extrema of a real function through a single iteration formula. This algorithm has been designed for sampling implicit curves and surfaces independently of the sign of function in the neighborhood of each point of them. In particular, our algorithm can detect and sample implicit curves and surfaces whose describing functions evaluate either positive or negative everywhere. For example, the circle $C$ described by function $f(x, y) = (x^2 + y^2 - 1)^2$ in $\mathbb{R}^2$ is a set of points of minimal zeros such that $f$ evaluates positive everywhere in $\mathbb{R}^2$ $C$. It can be sampled by the GFP algorithm, but not by algorithms (e.g. those embedded in well-known mathematical packages such as Mathematica, MatLab, and Maple) that use the Intermediate Value Theorem.

## References

[1] Brent RP. An algorithm with guaranteed convergence for finding a zero of a function. *The Computer Journal* 1971; **14**(4):422-425.

[2] Conte S, de Boor C. *Elementary Numerical Analysis: An Algorithmic Approach* (3rd edn). McGraw-Hill Book Co., 1981.

[3] Morgado JF, Gomes AJ. A BSP-based algorithm for dimensionally non-homogeneous planar implicit curves with topological guarantees. (submitted for publication) 2005.

[4] Press W, Flannery B, Teukolsky S, Vetterling W. *Numerical Recipes in C* (2nd ed.). Cambridge University Press, 2nd edition, 1992.

[5] Ridders CF. A new algorithm for computing a single root of a real continuous function. *IEEE Transactions on Circuits and Systema* 1991; **CAS-36**(11):979–980.

Brill Academic Publishers
P.O. Box 9000, 2300 PA Leiden
The Netherlands

*Lecture Series on Computer
and Computational Sciences*
Volume 4, 2005, pp. 429-432

# Large-Scale Water Flow Simulation in *Mathematica* Computing Environment with a PC Cluster

**Panjit Musik**[1,2,*] and **Krisanadej Jaroensutasinee**[2]
Science and Technology[1], Nakhon Si Thammarat Rajabhat University
School of Science[2], Walailak University, Thasala, Nakhon Si Thammarat 80160 Thailand.

Received 18 August, 2005; accepted in revised form 18 August, 2005

*Abstract:* This work presents a large scale water flow simulation using the lattice Boltzmann method (LBM) with a parallel computation extension for the two dimensional cavity flow simulation. A nine-velocity square lattice model (D2Q9 Model) is used in the LBM simulation. The LBM is an extension of the lattice gas algorithms and approximates the Navier-Stoke equations on a Cellular Automata (CA). A CA was programmed in *Mathematica* computing environment defining with the 6 CA rules, which were in terms of the particle distribution function. Each of the rules operates on the cells using the information in a Moore neighborhood, which consist of the surrounding nine cells. Parallelization is achieved by using horizontal domain decomposition while each decomposed domain is computed on each nodes of our WAC16P4 Cluster composed on 16 Pentium 4 personal computers. The parallel program is written in *Mathematica* and parallel computation is done with the Parallel Computing Toolkit. We also invent a new way of communicating necessary information with adjacent slave processors via the Server Message Block (SMB) protocol. Exact simulation results as a small scale simulation, which are comparable to those of previous work, are obtained. With our current tools, it is possible to model large scale computation (e.g. 1000×1000 or 2000×2000 lattice sizes) with less calculation time through our new parallelisation scheme.

*Keywords:* Lattice Gas (cellular) Automata, Lattice Boltzmann method, Parallel lattice Boltzmann method, Parallel computations, Cavity flow, D2Q9 Model, *Mathematica*

## 1. Modeling method

The method we used in the simulation is Lattice Boltzman Method with BGK model. The lattice Boltzmann BGK (LBGK) equation (in lattice unit) is

$$f_{\sigma i}(\bar{x} + \bar{e}_{\sigma i}, t+1) - f_{\sigma i}(\bar{x}, t) = \frac{1}{\tau}\left[f_{\sigma i}(\bar{x}, t) - f_{\sigma i}^{(0)}(\bar{x}, t)\right] \quad (1)$$

$f_{\sigma i}(\bar{x}, t)$ is the probability of finding a particle at $\bar{x}$ and time t with discrete velocity $\bar{e}_{\sigma i}$. $\sigma$ is the type of particle ($\sigma = 0,1,2$). $\tau$ is the single relaxation time. The equilibrium distribution functions, $f_{\sigma i}^{(0)}$ is defined as

$$f_{0i}^{(0)} = \frac{4}{9}\rho\left(1 - \frac{3}{2}|\bar{u}|^2\right), f_{1i}^{(0)} - \frac{1}{9}\rho\left(1 + 3(\bar{e}_{1i}\cdot\bar{u}) + \frac{9}{2}(\bar{e}_{1i}\cdot\bar{u})^2 - \frac{3}{2}|\bar{u}|^2\right), f_{2i}^{(0)} = \frac{1}{36}\rho\left(1 + 3(\bar{e}_{2i}\cdot\bar{u}) + \frac{9}{2}(\bar{e}_{2i}\cdot\bar{u})^2 - \frac{3}{2}|\bar{u}|^2\right)(2)$$

The relaxation time is related to the viscosity by $\tau = (6v+1)/2$ where $v$ is the kinematic viscosity and. The macroscopic density $\rho$ and velocity vector $\bar{u}$ are defined in terms of the particle distribution function by

$$\rho = \sum_\sigma \sum_i f_{\sigma i} \quad (4), \quad \bar{u} = \frac{1}{\rho}\sum_\sigma \sum_i f_{\sigma i}\bar{e}_{\sigma i} \quad (2)$$

## 2. Parallelisation of Water Flow Simulation in a Cavity

The parallel machine we used is a distributed memory computer. One front end node and 16 Slave nodes with Intel Pentium 4, 2.4 GHz CPU, 1.00 GB RAM, 2 Fast Ethernet Switch 10/100/1000 mbps

---

* Corresponding author: Email address: 44300036@wu.ac.th
Second author: cxresearch@gmail.com

and 8-Station KVM Switch are used. The operating system is Linux Redhat 9, and Windows XP Profession. The parallel program is written in *Mathematica* 5 and parallel computation is done with the Parallel Computing Toolkit. We also invent a new way of communicating necessary information with adjacent slave processors via the Server Message Block (SMB) protocol. A horizontal domain-decomposition are configured according to [2] which will give the highest speedup time (Figure 1c).

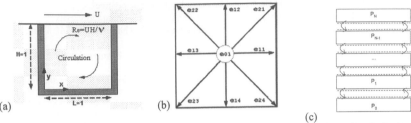

(a)   (b)   (c)

**Figure1.** (a) Cavity flow configuration (b) D2Q9 Model (c) Ghost point configuration for horizontal domain decomposition. Dashed lines represent extra rows of ghost point.

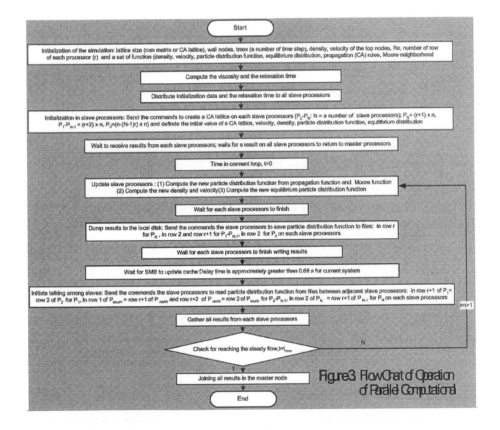

### 3. Results and Discussion

Testing result from the WAC16P4 Cluster from 1 node to 16 nodes (Figure 2a) shows the relationship between computing time for a single step and lattice size. Exponential increases are noted. The computing time is reduced to half from 1 to 2 nodes and keeps decreasing (Figure 2 a, b). The linear speedup increases as computation lattice sizes increase (Figure 3 a, b). The efficiency of 16 processors with 2000×2000 lattice size is about 92.93 % (Figure 3b).

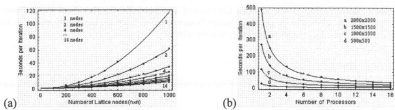

(a)                 (b)

**Figure 2.** Computing of WAC16P4 Cluster. (a) Number of Lattice Nodes, (b) Number of Processors. Dots represent computing time for one step from actual simulation and the line is the result of an exponential fitted function.

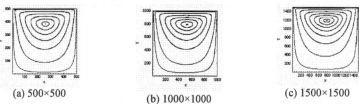

(a)                 (b)

**Figure 3.** Number of Processors: (a) Speedup $(T_1/NT_N)$, (b) Parallel Efficiency (%)

Figure 4a-c show fluid streamlines when reaching steady state with Re=10 for different lattice sizes. The better resolutions can be observed as the number of lattice nodes increases. The observed circulation in the cavity is clockwise. The centre of the primary vortex is located at (0.5160,0.7680) for 500×500 lattice size, (0.5100,0.7880) for 1000×1000 lattice size, (0.5067,0.7933) for 1500×1500 lattice size for present work and (0.5165,0.7650) for 256×256 lattice size in [1]. The results of [1] and the present work are similar when Re=10.

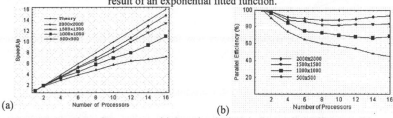

(a) 500×500         (b) 1000×1000         (c) 1500×1500

**Figure 4.** Fluid streamlines when reaching steady state with Re=10. (a) 500x500, (b) 1000x1000, (c) 1500x1500

Fluid streamlines are shown in Figure 5 when the system reaches steady state with 256×256 lattice sizes for different Re. We observe the primary vortex moving towards centre when Re grows higher e.g. for Re=100 the vortex is at (0.6132,0.7344) for the present work, and at (0.6196,0.7373) in [1]. For Re=400, the vortex is at (0.5664,0.6093) for the present work, and at (0.5608,0.6078) in [1] and at (0.5547,0.6094) in [3]. For Re=1000, the vortex is at (0.5313,0.5664) for the present work, and at (0.5333,0.5647) in [1] and at (0.5313,0.5625) in [3]. From the plots, circular streamlines for high Re are noted. The secondary vortex is observed at the bottom left and right when Re increases.

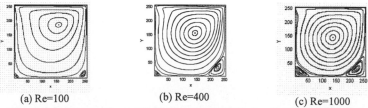

(a) Re=100         (b) Re=400         (c) Re=1000

**Figure 5.** Fluid streamlines of fluid flow in 256×256 lattice size for different Re.

The velocity components, $u$ and $v$, along the vertical and horizontal centre lines for different Re are shown in Figure 6. The velocity profiles change from curved at lower values of Re to linear for higher Re values. These results also show good agreement with [1,3].

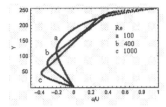 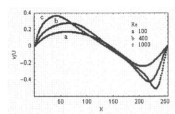

**Figure 6.** The velocity components of cavity flow for different Re. (a) The velocity profiles for $u$ through the geometric center of the cavity, (b) The velocity profiles for $v$ through the geometric center of the cavity.

## 4. Conclusions

Our parallelisation of LBM could be optimised further for the best speedup. The linear speedup for 2D cavity flow is independent of the computation lattice size. A bout 90% efficiency has been obtained with large data size on WAC16P4 Cluster. The program also has to be improved in order to reduce communicated time between nodes with the Parallel Computing Toolkit and SMB protocol. We would like to study further on 3D cavity and 3D in other shapes to observe flow patterns for CA program in large scale.

## Acknowledgements

This paper is partial fulfillment of requirement for the PhD at Walailak University. We would like to thank Walailak University for financial supports to this work under pre-doctoral fellowship and the Complex System Key University Research Unit of Excellence CXKURUE).

## References

[1] Hou S. L., Zou Q., Chen S. Y., Doolen G., and Cogley A. C. (1995). *Simulation of cavity flow by the lattice Boltzmann method*, Journal of Computational Physics, Vol. 118, pp. 329–347.
[2] Satofuka N.,Nishioka T. (1999). *Parallelization of lattice Boltzmann method for incompressible flow computations*, Computational Mechanics, Vol. 23, No.2, pp. 164-171.
[3] Guo.Z.,Shi.B.,Wang.N. (2000). *Lattice BGK Model for Incompressible Navier-Stokes Equation*, Journal of Computational Physics, Vol. 165, pp. 288-306.

Brill Academic Publishers
P.O. Box 9000, 2300 PA Leiden
The Netherlands

*Lecture Series on Computer
and Computational Sciences*
Volume 4, 2005, pp. 433-435

# Natural Bond Orbital (NBO) Analysis of the Metallotropic Shifts in Cyclopentadienyl(trimethyl)silane, -germane and -stannane

Davood Nori-Shargh*[1,2], Fahimeh Roohi[1], Farzad Deyhimi[3] and Reza Naeem-Abyaneh[1]

1) Chemistry Department, Graduate Faculty, Arak branch, Islamic Azad University, Arak, Iran
2) Chemistry Department, Science and Research Campus, Islamic Azad University,
Hesarak, Poonak, Tehran, Iran
3) Chemistry Department, Shahid Beheshti University, Evin-Tehran 19839, Iran

Received 27 June, 2005; accepted in revised form 1 July, 2005

*Abstract:* NBO//B3LYP/3-21G, NBO//B3LYP/3-21G* and NBO//B3LYP/LANL2DZ* analysis was used to investigate the aptitude of the alkyl and metallotropic 1,2-shifts in 5-*tert*-butylcyclopentadiene (1), cyclopentadienyl(trimethyl)silane (2), cyclopentadienyl (trimethyl) germane (3) and cyclopentadienyl(trimethyl)stannane (4). B3LYP/3-21G, B3LYP/3-21G* and B3LYP/LANL2DZ* results show that the MMe$_3$ (M=Si (2), Ge (3) and Sn (4)) migration barrier heights around cyclopentadienyl rings in compounds 1-4 decrease in following order: 4<3<2<1. The NBO analysis of donor-acceptor (bond-antibond) interactions revealed that the hyperconjugation between $\sigma_{Ccyclpentadienyl-M}$ (M=Si (2), Ge (3) and Sn (4)) and $\pi^*_{C=C}$ of cyclopentadienyl ring ($\sigma \rightarrow \pi^*$) facilitate the metal migration around cyclopentadienyl ring. The donor-acceptor interactions resulted in decreasing of occupancy of $\sigma_{Cyclopentadienyl-M}$ bonding orbital of C5-M bonds of the idealized Lewis structure in following order: 4<3<2<1, and also increasing occupancy of $\pi^*_{cyclopntadienyl}$ anti-bonding orbital of C1-C2 and C3-C4 bonds of cyclopentadienyl rings in following order: 4>3>2>1, as calculated by NBO//B3LYP/3-21G, NBO//B3LYP/3-21G* and NBO//B3LYP/LANL2DZ*. The results suggest that in compounds 1-4, the metallotropic shifts is controlled by $\sigma \rightarrow \pi^*$ energetic stabilization. The delocalization energies from $\sigma_{C-M}$ bonding orbital to $\pi^*_{C=C}$ antibonding orbital of cyclopentadienyl ring in compounds 1-4 are 1.45, 7.24, 7.63 and 12.49 kcal.mol$^{-1}$, respectively, as calculated by B3LYP/3-21G level of theory. Also, B3LYP/3-21G* results show that the delocalization energies from $\sigma_{C-M}$ bonding orbital to $\pi^*_{C=C}$ antibonding orbital of cyclopentadienyl ring in compounds 1-4 are 1.45, 6.58, 9.18 and 14.11 kcal.mol$^{-1}$. Further, B3LYP/LANL2DZ* results show that the above delocalization energies in compounds 1-4 are 1.26, 6.08, 8.86 and 13.47 kcal.mol$^{-1}$, respectively.

*Keywords:* molecular modeling, ab initio calculations, natural bond orbital, cyclopentadienyl

## Results and Discussion

An interesting correlation of ionization potentials with the fluxional behavior of compounds 2-4 indicated that the $\sigma$-$\pi$ hyperconjugation of the relevant C$_{cylopebntadienyl}$-M bonds with the ring diene systems in the ground state structure of cyclopentadienyl compounds appears to be an important factor in controlling the rate of sigmatropic rearrangements.[1-3] However, it has to be noted that there is no quantitative data which show the quantitative relationship between $\sigma$-$\pi$ ($\sigma \rightarrow \pi^*$ or $\pi \rightarrow \sigma^*$) hyperconjugation and metallotropic rearrangements in compounds 1-4 (see scheme 1):

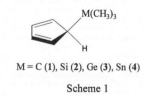

o      M = C (1), Si (2), Ge (3), Sn (4)

Scheme 1

Zero-point (*ZPE*) and total electronic ($E_{el}$) energies ($E_o = E_{el} + ZPE$) for ground state and transition state structures of compounds **1-4**, as calculated by B3LYP/3-21G, B3LYP/3-21G* and B3LYP/LANL2DZ* levels of theory,[4] are given in Table 1. NBO analysis was performed to calculate the $\sigma_{Ccyclopentadienyl-M}$ (M=C (**1**), Si (**2**), Ge (**3**) and Sn (**4**)) bonding and antibonding orbital occupancies and the stabilization energy ($E_2$) associated with $\sigma \rightarrow \pi^*$ or $\pi \rightarrow \sigma^*$ delocalizations (see Table 2). The metallotropic 1,2-shift energy barriers in compounds **2** and **3**, as calculated by B3LYP/3-21G* level of theory, are in good agreement with the reported dynamic [1]H-NMR data, while for compound **4**, the calculated energy barrier using B3LYP/LANL2DZ* method is close to the experimental data (see Table 1). These results show that the energy barriers for the Me$_3$M shifts decrease with the following order: $E_a(\mathbf{4}) < E_a(\mathbf{3}) < E_a(\mathbf{2}) < E_a(\mathbf{1})$. From the results presented in Table 1, the following trends in delocalization energies of compounds **1-4** can be observed: a) $\sigma \rightarrow \pi^*$ delocalization energies ($E_2$) increase as **1<2<3<4**; b) $\pi \rightarrow \sigma^*$ delocalization energies decrease as **1>2>3>4**, as calculated by B3LYP/3-21G, B3LYP/3-21G* and B3LYP/LANL2DZ* levels of theory. The NBO analysis of donor-acceptor (bond-antibond) interactions show that the stabilization energy for $\sigma \rightarrow \pi^*$ ($\sigma_{Ccyclpentadienyl-M}$ (M=C (**1**), Si (**2**), Ge (**3**) and Sn (**4**) and $\pi^*_{Csp2=Csp2}$ of cyclopentadienyl ring)) is 1.45, 7.24, 7.63 and 12.49 kcal mol$^{-1}$, respectively. These results revealed that the electron transferring from metal fragment to the cyclopentadienyl ring in compound **4** is greater than compound **3**, in compound **3** is greater than compound **2**, and in compound **2** is greater than compound **1**. Therefore, the metal migration around cyclopentadienyl ring in compound **4** should require lower energy than compound **3**, and in compound **3** lower than compound **2**.

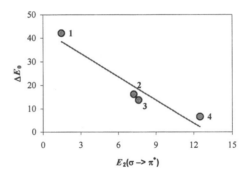

Fig.1. Calculated $\Delta E_o$ (in kcal mol$^{-1}$) for **1-4**, plotted as a function of $E_2(\sigma_{C5-M} \rightarrow \pi_{C1-C2}^*)$.

Table 1. Calculated total electronic energies $E_{el}$, zero-point energies *ZPE* (from B3LYP/3-21G, B3LYP/3-21G* and B3LYP/LANL2DZ*), and relative energies $\Delta E_o$ ($E_h$), (in hartree), for the energy-minimum and energy-maximum geometries of compounds **1-4**.

| | B3LYP/3-21G | | | | B3LYP/3-21G* | | | | B3LYP/LANL2DZ* | | | |
|---|---|---|---|---|---|---|---|---|---|---|---|---|
| | ZPE | $E_{el}$ | $E_o$ | $\Delta E_o^a$ | ZPE | $E_{el}$ | $E_o$ | $\Delta E_o^a$ | ZPE | $E_{el}$ | $E_o$ | $\Delta E_o^a$ |
| **1**, $C_s$ | 0.207601 | -349.446059 | -349.238458 | 0.000000 | 0.207600 | -349.446059 | -349.238459 | 0.0000000 | 0.206225 | -351.308364 | 351.102139 | 0.000000 |
| | | | | (0.000000) | | | | (0.000000) | | | | (0.000000) |
| [**1** →**1'**]$^{a}$, $C_s$ | 0.206118 | -349.377432 | -349.171315 | 0.067143 | 0.206118 | -349.377432 | -349.171315 | 0.067144 | 0.204305 | -351.237689 | 351.033384 | 0.068755 |
| | | | | (42.132904) | | | | (42.133531) | | | | (43.144450) |
| **2**, $C_s$ | 0.196666 | -599.568930 | -599.372264 | 0.000000 | 0.196300 | -599.653608 | -599.457307 | 0.000000 | 0.196332 | -317.149547 | 316.953215 | 0.000000 |
| | | | | (0.000000) | | | | (0.000000) | | | | (0.000000) |
| [**2** →**2'**]$^{a}$, $C_s$ | 0.196922 | -599.543414 | -599.346492 | 0.025772 | 0.196342 | -599.628211 | -599.431869 | 0.025438 | 0.196403 | 317.116921 | 316.920526 | 0.032689 |
| | | | | (16.172188) | | | | (15.962599) | | | | (20.5122674) |
| **3**, $C_s$ | 0.195349 | 2378.618141 | 2378.422792 | 0.000000 | 0.194882 | 2378.792768 | 2378.797887 | 0.000000 | 0.194486 | -317.005955 | 316.811469 | 0.000000 |
| | | | | (0.000000) | | | | (0.000000) | | | | (0.000000) |

| | B3LYP/3-21G | | | | B3LYP/3-21G* | | | | B3LYP/LANL2DZ* | | | |
|---|---|---|---|---|---|---|---|---|---|---|---|---|
| [3→3']†, $C_s$ | 0.196168 | 2378.596178 | 2378.400928 | 0.021864 | 0.194819 | 2378.774118 | 2378.579299 | 0.018588 | 0.194513 | -316.981873 | 316.787360 | 0.024109 |
| | | | | (13.719879 | | | | 11.664156 | | | | (15.128638) |
| 4, $C_s$ | 0.193096 | 6310.839906 | 6310.646804 | 0.000000 | 0.192552 | 6310.914014 | 6310.721462 | 0.000000 | 0.192455 | 316.597267 | 316.404812 | 0.000000 |
| | | | | (0.000000) | | | | (0.0000000) | | | | (0.000000) |
| [4→4']†, $C_s$ | 0.192594 | 6310.828971 | 6310.636378 | 0.010426 | 0.192079 | 6310.905787 | 6310.713709 | 0.007753 | 0.192220 | -316.583640 | 316.391420 | 0.013392 |
| | | | | (6.542419) | | | | (4.865085) | | | | (8.403614) |

Table 2. B3LYP/3-21G, B3LYP/3-21G* and B3LYP/LANL2DZ* calculated stabilization energies ($E_2$) for $\sigma_{C5\text{-}M}\to\pi_{C1\text{-}C2}^*$ and $\pi_{C1\text{-}C2}\to\sigma_{C5\text{-}M}^*$; occupancies for $\sigma_{C5\text{-}M}$, $\pi_{C1\text{-}C2}$ bonding and $\pi_{C1\text{-}C2}^*$, and $\sigma_{C5\text{-}M}^*$ anti-bonding orbitals in the ground state structures of compounds **1-4**.

| Method | B3LYP/3-21G | | | | B3LYP/3-21G* | | | | B3LYP/LANL2DZ* | | | |
|---|---|---|---|---|---|---|---|---|---|---|---|---|
| Compound | 1 | 2 | 3 | 4 | 1 | 2 | 3 | 4 | 1 | 2 | 3 | 4 |
| Stabilization energy ($E_2$) | | | | | | | | | | | | |
| $\sigma_{C5\text{-}M}\to\pi_{C1\text{-}C2}^*$ | 1.45 | 7.24 | 7.63 | 12.49 | 1.45 | 6.58 | 9.18 | 14.11 | 1.26 | 6.08 | 8.86 | 13.46 |
| $\pi_{C1\text{-}C2}\to\sigma_{C5\text{-}M}^*$ | 3.20 | 1.84 | 1.71 | 1.66 | 3.20 | 1.91 | 1.89 | 1.56 | 3.06 | 2.59 | 2.39 | 2.26 |
| | | | | | | | | | | | | |
| Occupancies | | | | | | | | | | | | |
| $\sigma_{C5\text{-}M}$ | 1.95197 | 1.85700 | 1.83624 | 1.75516 | 1.95197 | 1.86164 | 1.80794 | 1.72296 | 1.94273 | 1.88096 | 1.83823 | 1.77299 |
| $\pi_{C1\text{-}C2}$ | 1.91983 | 1.89653 | 1.89078 | 1.87290 | 1.91983 | 1.89528 | 1.88449 | 1.86389 | 1.91201 | 1.89640 | 1.88916 | 1.87771 |
| $\sigma_{C5\text{-}M}^*$ | 0.03559 | 0.05605 | 0.07132 | 0.08055 | 0.03589 | 0.06018 | 0.07933 | 0.09326 | 0.04565 | 0.04550 | 0.05432 | 0.05609 |
| $\pi_{C1\text{-}C2}^*$ | 0.10446 | 0.16560 | 0.17924 | 0.23053 | 0.10445 | 0.16491 | 0.19242 | 0.24825 | 0.11216 | 0.15690 | 0.17818 | 0.21868 |
| | | | | | | | | | | | | |

## Acknowledgement

The author wishes to thank the anonymous referees for their careful reading of the manuscript and Dr D. Tahmassebi for fruitful discussions.

## References

[1]. Yu. A. Ustynyuk, P. I. Zakhraov, A. A. Azizov, V. K. Potapov and I. M. Pribytkova, Mass spectra of monohapto-cyclopentadienyl derivatives of Group IVB elements, *J. Organomet Chem.*, **88** 37-64 (1975).

[2]. S. Cradock, R. H. Findlay and M. H. Palmer, Bonding in methyl- and silylcyclopentadiene compounds: a study by photoelectron spectroscopy and ab initio molecular orbital calculations, *J. Chem. Soc., Dalton Trans.*, 1650-1654 (1974).

[3]. S. Cradock, E. A. V. Ebsworth, H. Moretto, and D. W. H. Rankin, Photoelectron spectra and fluxional behavior in σ-cyclopentadienes, *J. Chem. Soc., Dalton Trans.*, (5) 390-392 (1975).

[4]. M. J. Frisch, *et al.*, GAUSSIAN 98 (Revision A.3), Gaussian Inc. Pittsburgh, PA, USA 1998.

Brill Academic Publishers
P.O. Box 9000, 2300 PA Leiden
The Netherlands

*Lecture Series on Computer*
*and Computational Sciences*
Volume 4, 2005, pp. 436-437

# Kinetic Control of the Production of Fullerene Derivatives: Illustrative Computations on Thermodynamics and Kinetics of C60O and C60O2

Z. Slanina and S. Nagase

Department of Theoretical Molecular Science, Institute for Molecular Science,
Myodaiji, Okazaki 444-8585, Japan

Received 5 August, 2005; accepted in revised form 12 August, 2005

*Abstract:* C60O has represented a long standing case of experiment-theory disagreement, and the problem becomes even more serious with C60O2. The computations cannot identify the calculated ground state with the observed species. The puzzle is treated here from a kinetic point of view. Once activated complexes are exactly located, theory and observation agree in the terms of the computed activation parameters. Both mono- and di-oxide formation is considered as a case where activation-energy order is reversed compared to the standard thermodynamic terms.

*Keywords:* Fullerenes and heterofullerenes; Quantum-chemical calculations; Kinetic control.

*PACS:* 31.15.Ew; 36.40.Cg; 74.70.Wz.

## 1. Computational Outline

Fullerene oxides were the first observed fullerene derivatives and they have naturally attracted attention of both experiment [1] and theory [2]. The oxides C60On have been observed up to n=4 while computations have so far treated only mono- and di-oxides (in fact, for the C60O2 aggregates only a very few, special selected structures were treated). With the mono-oxide, two structures have in particular been considered, originated in bridging a 5/6 or 6/6 bond in the pristine C60 - 5/6 or 6/6 isomers. Moreover, C60O has consistently been computed in a disagreement with observations - the ground state of the system is calculated, regardless the method applied, as the 5/6 species while in the experiment only the 6/6 structure can be observed. The theory-experiment disagreement cannot be explained through temperature (i.e., entropy) factors. The problem becomes even more serious with C60O2. In experiment, only 6/6 6/6 structures have been reported. However, in computations the higher thermodynamic stability of the 5/6 mono-oxide is somehow conserved even with di-oxides. Consequently, the computations are unable to identify the computed ground state with the observed 6/6 6/6 species [3] as among the computed bridged structures a 5/6 5/6 isomer is the lowest in energy. The disturbing problem can be still studied as a possible kinetic control [4,5] of the relative stabilities (i.e., not only temperature but also time factor should be reflected).

In this report, the full geometry optimization and also vibrational harmonic analysis is performed not only for local energy minima but also for related activated complexes. Using the activation parameters, computed at the semiempirical, HF, and DFT levels of theory, we indeed can put theory and observation in agreement for the first time. As already mentioned, there are two principal isomers of the C60O itself, either created by bridging a 5/6 bond between a pentagon and a hexagon or a 6/6 bond between two hexagons. The 5/6 mono-adduct is lower in energy and its critical 5/6 bond is broken; the AM1 computed stabilization energy is about 52 kJ/mol. Still, experiment reveals only the 6/6 mono-adduct. In order to study the problem further, we allocated activated complexes for formations of the 5/6 and 6/6 mono-oxides. It turns out that they are in fact not of epoxo- but rather of keto-forms. We were not able to allocate an activated complex with two C-O bonds, but only structures where the oxygen atom is bonded to just one carbon atom. The keto-forms exhibit just one imaginary frequency. Moreover, the vibrational eigenvector associated with the imaginary frequency has a clear meaning. In one direction it represents a motion towards dissociation of the C-O bond. In the opposite direction it leads to a formation of the second C-O bond, i.e., to a bridging of a C-C bond by the oxygen atom. Hence, we can speak on the 5/6 and 6/6 activated complexes or transition states.

In our calculations the electronic-state multiplicity was always 1, i.e. a singlet state. In real experimental situation we can deal with triplet rather than singlet oxygen atom. Hence, at some point we should expect a jump from the singlet to a triplet hypersurface. This should happen at quite long C-O separations. Thus, we can expect that the related transition probabilities will be about the same for the 5/6 and 6/6 reaction path. Hence, it should not bring to some critical difference in reaction velocities. The activation energy for the 6/6 path is lover by about 52 kJ/mol than that for the 5/6 path. It means that, for example, at room temperature the 6/6 isomer must be almost exclusively formed during the initial stages of the reaction. Although we also computed the activation entropy, its role turns out to be negligible in this case. Only relatively very long reaction times can allow for a thermodynamic equilibrium. At longer times also reverse (backward) reactions come into the play, and eventually the interplay of the four rate processes creates the equilibrium between both isomers. Hence, we suggest that the formation of the mono-oxide is, under the considered experimental conditions, controlled by kinetics rather than by thermodynamics.

Moreover, this finding also plays a key role in understanding the di-oxo derivative formation. Let us suppose that the dioxo-fullerene is created in a two-step mechanism - formation of a mono-oxide and addition of the second oxygen atom. This in fact, within the kinetic control, eliminates the 5/6 5/6 di-oxides in spite of the fact they are quite low-lying in thermodynamic stability scale. In other words, if the kinetic control works for the di-oxide formation, it should produce either 6/6 6/6 species or at most some 6/6 5/6 structures. From a combinatorial point of view the di-oxides are quite numerous. The 5/6 mono-oxide has Cs symmetry and this allows for 48 possibilities in bridging by the second oxygen atom. In the more symmetric 6/6 isomer (C2v symmetry) we have only 28 possibilities for the second bridging. For example, if we place both oxygen atoms on a common ring, the three lowest-energy structures are of the 5/6 5/6 type and the structure highest in energy exhibits the 6/6 6/6 pattern. Moreover, other structures (di-keto, vertical, peroxo) should be considered, too. This may suggest quite an extensive computational treatment. However, observations conclude that the most important di-oxide (labeled (I)) is a structure where both oxygen atoms are placed over two 6/6 bonds in one common hexagon. There is still another structure, less populated and labeled (II), where the two 6/6 bonds belong to two different, but closely related hexagons. Hence, we limited our calculations to the (I) and (II) species, and allocated the related local minima and activated complexes. We also checked an activated complex for one 6/6 5/6 structure and confirmed that it is considerably higher in energy in comparison with the (I) activated complex.

The activated complexes for the addition of the second oxygen atom again have a keto-form. For a given product, one can in fact consider two different activated complexes, originated in an approach to one or the other carbon atom of the C-C bond being bridged over; unless there is a symmetry, the two activated complexes differ in energy. The (I) species is preferred kinetically by some 8 kJ/mol, and also more stable thermodynamically. This represents a fair theory-experiment agreement. We also checked activated complexes for formation of all possible 6/6 6/6 C60O2. No one is located lower than the (I) activated complex (though the differences are modest and a full treatment of the entropic effects should be considered). This finding further supports the consistency of the finally established good theory-experiment agreement.

## Acknowledgments

The reported research has been supported by a Grant-in-aid for NAREGI Nanoscience Project, and for Scientific Research on Priority Area (A) from the Ministry of Education, Culture, Sports, Science and Technology of Japan.

## References

1   W. A. Kalsbeck and H. H. Thorp, *J. Electroanal. Chem.* 314, 363 (1991).
2   K. Raghavachari, *Chem. Phys. Lett.* 195, 221 (1992).
3   K. Winkler, D. A. Costa, A. L. Balch, and W. R. Fawcett, *J. Phys. Chem.* 99, 17431 (1995).
4   Z. Slanina, F. Uhlik, L. Stobinski, H.-M. Lin, and L. Adamowicz, *Int. J. Nanosci.* 1, 303 (2002).
5   Z. Slanina, L. Stobinski, P. Tomasik, H.-M. Lin, and L. Adamowicz, *J. Nanosci. Nanotech.* 3, 193 (2003).

Brill Academic Publishers
P.O. Box 9000, 2300 PA Leiden
The Netherlands

*Lecture Series on Computer
and Computational Sciences*
Volume 4, 2005, pp. 438-443

# Virtual Plant Evolution with Genetic Algorithms

D. Öktem[1], A. Kaya[2], Ş. Çakır[3]

Department of Computer Technologies and Programming
Tire Kutsan Post Secondary Vocational School
Ege University
35900 Tire İzmir, Turkey

Received 14 June, 2005; accepted in revised form 11 July, 2005

*Abstract* – Genetic Algorithms (GA) have a wide range of application areas. GAs can be used for information retrieval and to increase the efficiency of the information gathered. The techniques, which are used to obtain good offspring results by using GAs in reproduction, may be applied to a very popular plant, cotton. Cotton is a valuable plant, which people make use of it in many areas for centuries. In this study, it is suggested that the life cycle of a plant can be obtained by using genetic algorithms for each phase of the growth.

*Keywords* – Cotton, Genetic Algorithms, Plant Evolution.

## I. INTRODUCTION

Genetic Algorithms (GAs) are probabilistic search methods that have been developed by John Holland in 1975 [1][2]. GA's applied natural selection and natural genetics in artificial intelligence to find the globally optimal solution to the optimization problem from the feasible solutions. GAs have been applied to various domains, including timetable, scheduling, robot control, signature verification, image processing, packing, routing, pipeline control systems, machine learning, and information retrieval [3][4].

Genetic algorithms can also be used for the simulation of a system. This system may be either a working plant or a living organism. If the proper data about the system can be converted to the input data of a genetic algorithm, the algorithm can reach to the expected result about the system. For the plant cotton, the result aimed to be obtained is the growth phases of the plant. For each phase, the environmental conditions that the organism needs will change. For each phase, different input data representing the sunlight, moisture, vitamins and minerals in the soil must be changed.

## II. WHAT IS A GENETIC ALGORITHM?

GA is defined as a "detailed sequence of actions to perform to accomplish some task [11]. One branch of algorithm theory, genetic programming, is currently receiving much attention. This is a technique for getting software to solve a task by 'mating' random programs and selecting the fittest in millions of generations [12].

Elaborates: 'Genetic algorithms use natural selection, mutating and crossbreeding within a pool of sub-optimal. Scenarios: Better solutions live and worse ones die -allowing the program to discover the best option without trying every possible combination along the way.

GAs are general – purpose search algorithms that use principles inspired by natural population genetics to evolve solutions to problems [5]. The basic idea is to maintain a population of knowledge structures that evolves over time through a process of competition and controlled variation. Each structure in the population represents a candidate solution to the concrete problem and has an associated fitness to determine the process of competition which structures are used to form new ones [6]. The algorithm uses the rules which are applied in nature for the production of new generations.

## III.  THE PRINCIPLES OF A GENETIC ALGORITHM

In genetic algorithms, a population is simply a collection of chromosomes representing possible solutions to the specific problem at hand. These chromosomes are altered or modified using the genetic operators in order to create a new generation. This evolutionary process is repeated a predetermined number of times or until no improvement in the solution to the problem is found.

The chromosomes in the population can be represented as strings of binary digits. Bit string representations are used in many real-world applications of genetic algorithms. This kind of representation has several advantages: they are simple to create and manipulate, many types of information can be easily encoded, and the genetic operators are easy to apply.

The evaluation of a chromosome is done to test its "fitness" as a solution, and is achieved, typically, by making use of a mathematical formula known as an objective function. The objective function plays the role of the environment in natural evolution by rating individuals in terms of their fitness. Choosing and formulating an appropriate objective function is crucial to the efficient solution of any given genetic algorithm problem [10].

There are three basic rules to pass the genetic information from parents to offspring. These rules are "reproduction", "crossover" and "mutation".

The reproduction operator selects a subset of individuals from the population, such that individuals with a higher fitness have a greater probability of being chosen. Fitness is determined by an objective function, which must be written by a domain expert. Individuals who have been selected for reproduction create two offspring (per couple) by exchanging genetic material.

The exchange of genetic information is via single point crossover. This means that a point is chosen, at random, along the genome, and the first child receives the genetic information from the first parent before the crossover point, and receives the remainder from the second parent. The second child is created from the opposite information. Crossover attempts to propagate high quality 'building blocks' of genetic information to successive generations, thereby eventually creating individuals with many good qualities.

Mutation is randomly applied to the children to prevent premature convergence of the population. Mutation is an attempt to balance exploration of the unknown areas of the search space and exploitation of the areas that are known to be better than the average of the current generation [7].

Selection is often implemented with a 'roulette wheel' technique, where the slots in the roulette wheel are in proportion to the fitness of the individual [8].

For the plant cotton case, searching for different paths and different values of fitness based on the characteristics of each growth phase will be appropriate.

Genetic algorithms are implemented through the following algorithm described by algorithm 1, wherein parameters popSize, fit and genNum are the population maximum size, the expected fitness of the returned individual and the maximum number of generation allowed respectively.

*Algorithm 1. GA(popsize, fit, genNum) : Individual;*
*generation = 0;*
*population = initialPopulation ( );*
*fitness [popsize]= evaluate(population) ;*
*do*
   *parents    = select (population) ;*
   *population  = reproduce (parents) ;*
   *fitness     = evaluate (population );*
   *generation  = generation + 1;*
*while(fitness[i]<fit, $\forall$ i ∈population)&(generation<genNum);*

*return fittestIndividual (population) ;*
*end algorithm.*

In Algorithm 1, function initialPopulation returns a valid random set of individuals that would compose the population of first generation, function evaluate returns the fitness of a given population storing the result into fitness. Function select chooses according to some random criterion that privilege fitter individuals, the individuals that should be used to generate the population of the next generation and function reproduction implements the crossover and the mutation process to actually yield the new population [15].

## IV. THE CHARACTERISTICS OF COTTON

Just like other plants, cotton gathers nutrients from gases in the air and nutrients in the soil. While maturating, the plant cotton needs more sunlight and less moisture because much moisture in soil or rain decreases the quality of cotton in this phase.

**Figure 1** – Different Phases of Cotton Plant

When cotton is only a seed, it already includes all the parts of a mature plant. These parts are two cotyledons, which will form the seed leaves, the epicotyls that will form the stem, the hypocotyls to form the crook, and the radicle, which will spring the roots. This seed strongly needs the moisture in soil and the path of the water follows the tissue around the embryo to the radicle cap at the narrow end of the seed. To grow up to a mature plant, root, stem and leaf systems must be specialized and developed [16].

There are four phases in a plant's life cycle. These phases are germination, sprouting, maturation and reproduction. For each of these phases, the needs of a plant will obviously differ. Watering the plant is not needed till the sprouting phase because the moisture in soil is enough for the plant at the beginning. Watering is critical between the sprouting and maturation phases, that corresponds to a period of 45-50 days. Watering must be done approximately once in 15 days [17].

Nitrogen (N), phosphorus (P) and potassium (K) are three basic elements that cotton needs for living. The ratio of P need throughout the life cycle of the plant is mostly uniform, but N and K need for the plant changes. The amounts needed for N and K are low till the beginning of the sprouting phase, but from the beginning of sprouting, the ratio of N and K amounts needed increase. The growth cycle of a cotton plant with ranges of days is as shown in Table 1.

Table 1: Growth Cycle of the Cotton Plant [16]

|  | Number of days | |
| --- | --- | --- |
| Stage of Growth | Range | Average |
| Planting to emergence | 5 – 20 | 10 |
| Emergence to square | 27 – 38 | 32 |
| Square to first bloom | 20 – 27 | 23 |
| First bloom to peak bloom | 26 – 45 | 34 |
| Bloom to open boll | | |
| • Early and midseason bloom | 45 – 55 | 50 |
| • Late season bloom | 55 – 70 | 60 |
| Growing season | 120 – 150 | 140 |

## V. THE NUMERICAL INFORMATION NEEDED FOR VIRTUAL EVOLUTION

In order to simulate a living organism as a computer program, we strongly need to obtain some numerical information about the organism. The gathered numerical information for cotton is as follows:

- The time needed for the growth process of cotton is 180 – 200 days [17].
- Sunlight is an important factor for the plant's growth. Throughout the life of the plant, it needs more than 60% percent of sunny days [17].
- For 1 decare of soil, the optimum number is 7140 plants [17]. (This means that we have to divide all nutrient values given per decare by 7140.)
- Nutrient percentages of a cotton plant: 11.97% N (nitrogen), 2.11% P (phosphorus) and 8.27% K (potassium).
- The minerals that the plant takes from soil are: N, $P_2O_5$, $K_2O$, MgO, and CaO.
- Watering period of the plant: Once in 15 days and 80 $m^3$/decare water for each watering.
- Nutrient amounts that the plant takes from soil:

Table 2: Nutrient Amounts per Decare

| Nutrient | Amount (kg/decare) |
|----------|--------------------|
| N | 10.5 |
| $P_2O_5$ | 4.1 |
| $K_2O$ | 7.9 |
| MgO | 2.7 |
| CaO | 8.0 |

- The dry weight for the plant in a decare is measured as 530 kg [18].

Since it is planned to simulate the evolution of just one cotton plant, the values given above must be calculated to find the amounts of nutrients for one plant. By using this numerical information, the needed bit strings can be produced from the following data:

- N amount for one plant: 10500 gr / 7140 = 1,470588235 gr.
- $P_2O_5$ amount for one plant: 4100 gr / 7140 = 0,574229691 gr.
- $K_2O$ amount for one plant: 7900 gr / 7140 = 1,106442577 gr.
- MgO amount for one plant: 2700 gr / 7140 = 0,378151261 gr.
- CaO amount for one plant: 8000 gr / 7140 = 1,120448179 gr.
- The amount of water for each plant in each watering: $80m^3$ / 7140 = 11,20448179 lt.

## VI. BIT STRINGS NEEDED FOR VIRTUAL PLANT EVOLUTION

To grow a plant virtually, the growth conditions of the plant must be imitated by the help of a computer program. All the data considered must be in the form of bit strings for the genetic algorithm to work on. The data, which will be represented as bit strings are about the plant's environment. The important factors for a plant to grow are sunlight, moisture, watering and nutrients in soil.

The important point for the virtual evolution is to determine which phases of the evolution will have which inputs. Each input of the plant will be organized as a separate

bit string because usages of the inputs are independent from each other.

Nutrient amounts will be arranged as bit strings with the following method:

- Each amount will be taken with 6 digits of precision (ex: N amount = 1,470588)
- The amounts will be multiplied by 1000000 (ex: 1470588).
- The obtained result will be converted to binary. This number will be considered as the amount to be reached at the end of the evolution (ex: 101100111000001111100).
- In order to use these bit strings, it is determined as each input to have 24 digits (3 bytes for each).

According to this method, the values of the bit strings that are aimed to be reached for the evolution are as follows:

Table 2 – Nutrient Amounts per Decare

| Input | Previous Value | x 1000000 | Final Bit Strings |
|-------|----------------|-----------|-------------------|
| N | 1,470588 gr | 1470588 | 000101100111000001111100 |
| P$_2$O$_5$ | 0,574230 gr | 0574230 | 000010001100001100010110 |
| K$_2$O | 1,106443 gr | 1106443 | 000100001110001000001011 |
| MgO | 0,378151 gr | 0378151 | 000001011100010100100111 |
| CaO | 1,120448 gr | 1120448 | 000100010001100011000000 |
| Water | 11,204482 lt | 11204482 | 010101010111011110000010 |

These bit strings represent the final values of the plant after 200 days of growth. Some of the inputs must be obtained regularly throughout the 200 days of evolution, but some must be improved with different intervals and ratios. So, at the beginning of the virtual evolution, these values will be divided by 200 roughly. For each cycle of evolution corresponding to a single day of growth, the values will be increased with the determined steps. This will be done by choosing the suitable fitness functions separately for each input bit string. While doing this, since this study is so new, unpredictable and unexpected conditions will be neglected. These unexpected conditions can be classified as:

- Drought or too much rain for the summer season,
- Soil's fatigue because of too many sowing and lack of minerals in soil,
- Too many disinfection or bug diseases because of lack of disinfection, etc.

The future aim of this study is to simulate a living organism, by the help of the input data given above.

## VII. CONCLUSION

Artificial intelligence, from the beginning of its existence, tries to understand the principles of life and tries to simulate it. Genetic algorithm is one of the techniques to understand, simulate and animate both the living and non-living systems we can see around.

Simulation of a life cycle of the cotton may help us to understand the behavior and evolution of this plant. Here, we tried to state that, if all the needed information about a plant can be written in bit strings, the total evolution cycle of the plant can be simulated. This study may be a contribution to the application areas of genetic algorithms.

## REFERENCES

[1] David, L. *Handbook of Genetic Algorithms*. New York: Van Nostrand Reinhold. 1991.
[2] Goldberg, D.E. *Genetic Algorithms: in Search, Optimization, and Machine Learning*. New York: Addison-Wesley Publishing Co. Inc. 1989.
[3] Kraft, D.H. et. al. "The Use of Genetic Programming to Build Queries for Information Retrieval." in *Proceedings of the First IEEE Conference on Evolutional Computation*. New York: IEEE Press. 1994. PP. 468-473.
[4] Martin-Bautista, M.J. et. al. "An Approach to An Adaptive Information Retrieval Agent using Genetic Algorithms with Fuzzy Set Genes." In *Proceeding of the Sixth International Conference on Fuzzy Systems*. New York: IEEE Press. 1997. PP.1227-1232.
[5] Holland J.H., Adaptation in Natural and Artificial Systems. Ann Arbor: The University of Michigan Pres (1975) (The MIT Pres, London, 1992)
[6] Cordon O., Herrera F. Genetic Algorithms and Fuzzy Logic in Control Processes, DECSAI-95109, 1995.
[7] Leonard J., Interactive Game Scheduling With Genetic Algorithms, Royal Melbourne Institute of Technology University, 1998.
[8] Goldberg D., Genetic Algorithms in Search, Optimization and Machine Learning, Addison Wesley, 1989.
[9] [Online] Available: http://www.botany.org/bsa/misc/carn.html
[10] Potter W.D., Robinson R.W., Using the Genetic Algorithm to Find Snake-In-The-Box Codes, University of Georgia.
[11] Foldoc (2003). Free online dictionary of computing [online] http://www.wornbat.doc.ic.ac.uk/folded/

[12] Khan, J. (2002) It's alive' Wired, 10(3) [Online]
http://www.wired.com/wired/archieve/10.03/everywhere.html

[13] Pal, S. K. and Majumder, D.D. (1986) Fuzzy Mathematical approach to pattern recognition, John Wiley, New York.

[14] Sameer Singh, (1998) "Fuzzy Nearest Neighbour Method for Time Series Forecasting", Proc. 6th European Congress on Intelligent Techniques and Soft Computing (EUFIT'98). Aachen, Germany, vol. 3 pp. 1901-1905.

[15] Nadia Nedjah and Luiza de Macedo Mourelle (2002) Minimal Addition-Subtraction Chains Using Genetic Algorithms, ADVIS 2002, LNCS 2457, pp. 303-313, 2002.

[16] Del Deterling and Dr. Kamal M. El-Zik, 1982, Progressive Farmer, Progressive Farmer Editorial Offices, Dallas, Texas.

[17] Atilla Şahin, İsmail Ekşi (1998), Cotton Agriculture, Turkish Republic Agricultural Research Headquarters Nazilli Cotton Research Institute, Publication No:50.

[18] Atilla Şahin (1985), Cotton Agriculture, Turkish Republic Agricultural Research Headquarters Nazilli Cotton Research Institute, Publication No:40.

Brill Academic Publishers
P.O. Box 9000, 2300 PA Leiden,
The Netherlands

*Lecture Series on Computer*
*and Computational Sciences*
Volume 4, 2005, pp. 444-448

# Runge - Kutta Methods for Fuzzy Differential Equations

## S. Ch. Palligkinis[1], G. Papageorgiou [2] and I. Th. Famelis[3]

Received 10 July, 2005; accepted 3 August, 2005

*Abstract:* Fuzzy Differential Equations generalize the concept of crisp Initial Value Problems. In this article, the numerical solution of these equations is dealt with. The notion of convergence of a numerical method is defined and a category of problems which is more general than the one already found in Numerical Analysis literature is solved. Efficient s-stage Runge - Kutta methods are used for the numerical solution of these problems and the convergence of the methods is proved. Several examples comparing these methods with the previously developed Euler method are displayed.

*Keywords:* fuzzy numbers, fuzzy differential equations, numerical solution, Runge-Kutta Methods, convergence of numerical methods

## 1   Fuzzy Differential Equations (FDEs)

Fuzzy Set Theory is a quite new branch of Mathematics which deals with sets whose margins are not strictly defined (see [4] or [14] for a comprehensive introduction). They are characterized by their *membership function*, which a generalized characteristic function with values in $[0,1]$ intead of $\{0,1\}$. A special category of fuzzy sets are *fuzzy numbers*. We should consider these these sets as realizations of the concept "approximately $c$", where $c \in \mathbb{R}$. They are defined as followed:

**Definition 1** *A fuzzy number is a normalized fuzzy set $\widetilde{M}$ of $\mathbb{R}$, for which membership function $\mu_{\widetilde{M}}$ is upper semicontinuous, $\widetilde{M}$ is convex and sets $\{x \in \mathbb{R} : \mu_{\widetilde{M}}(x) = a\}$ are compact for $a \in (0,1]$. The set fuzzy numbers is denoted as $F(\mathbb{R})$.*

Function $\mu$ is a generalization of characteristic function of classical (or *crisp*) sets. It defines the *grade of membership* of each number in the fuzzy number. Let $\widetilde{P}$ be a fuzzy number and let also the following sets:

$$\{x \in \mathbb{R} : \mu_{\widetilde{P}}(x) = a\} = [(P_1^a)(x), (P_2^a)(x)], \ 0 < a \le 1$$

These (crisp) sets are called *a-cuts* of the fuzzy number and it turns out that they characterize the fuzzy number. Furthermore, arithmetic operations of fuzzy numbers can be defined through their a-cuts:

**Proposition 1** *If $\widetilde{P}, \widetilde{Q} \in F(\mathbb{R})$, then for $a \in (0,1]$,*

$$[\widetilde{P} + \widetilde{Q}]_a = [P_a^1 + Q_a^1, P_a^2 + Q_a^2],$$

$$[\widetilde{P} \cdot \widetilde{Q}]_a = [min\{P_a^1 Q_a^1, P_a^1 Q_a^2, P_a^2 Q_a^1, P_a^2 Q_a^2\}, max\{P_a^1 Q_a^1, P_a^1 Q_a^2, P_a^2 Q_a^1, P_a^2 Q_a^2\}].$$

[1]Department of Applied Sciences, TEI of Chalkis, GR 34400 Psahna, Greece.
[2]Deparment of Applied Mathematics and Physical Sciences, NTUA
[3]Department of Mathematics, Technological Educational Institution (T.E.I.) of Athens. Corresponding author. E-mail: ifamelis@math.ntua.gr

Let $\widetilde{P}, \widetilde{Q} \in F(\mathbb{R})$. If there exists a fuzzy number $\widetilde{R}$ such that $\widetilde{P} + \widetilde{R} = \widetilde{Q}$, then this number is unique and it is called *Hukuhara difference* of $\widetilde{P}, \widetilde{Q}$ and is denoted by $\widetilde{Q} - \widetilde{P}$, [12].

Let $A, B$ two nonempty bounded subsets of $\mathbb{R}$. The *Hausdorff distance* between $A$ and $B$ is

$$d_H(A, B) = \max[\sup_{a \in A} \inf_{b \in B} |a - b|, \sup_{b \in B} \inf_{a \in A} |a - b|]$$

If $\widetilde{P}, \widetilde{Q} \in F(\mathbb{R})$ the *distance* $D$ between $\widetilde{P}$ and $\widetilde{Q}$ is defined as:

$$D(\widetilde{P}, \widetilde{Q}) = \sup_{a > 0} d_H([\widetilde{P}]_a, [\widetilde{Q}]_a) \tag{1}$$

The above distance is the tool we use in order to define numerical convergence for our methods.

We now concentrate on *fuzzy functions*, i.e. functions $f : \mathbb{R} \to F(\mathbb{R})$. We use the following definition for the derivative:

**Definition 2** *Let $U$ be an open interval in $\mathbb{R}.A$ fuzzy function $f : \mathbb{R} \to F(\mathbb{R})$ is called to be H-differentiable in $x_0 \in U$ if there exists $\widetilde{f}'(x_0) \in F(\mathbb{R})$ such that limits $\lim_{h \to 0+}[\widetilde{f}(x_0 + h) - \widetilde{f}(x_0)/h]$ and $\lim_{h \to 0+}[\widetilde{f}(x_0) - \widetilde{f}(x_0 - h)/h]$ both exist and they are equal to $\widetilde{f}'(x_0)$ [12].*

When this derivative exists, it is also written as:

$$[f'(t)]_a = [(f_1^a)'(t), (f_2^a)'(t)], \qquad 0 < a \leq 1, \tag{2}$$

There exist various definitions for the derivative of fuzzy functions (see [5], [6], [11], [12] and [13]). However, if $(f_1^a)', (f_2^a)'$ are continuous functions with reference to both $t$ and $a$ (this property is called *continuity condition*), the following proposition is proved (see [2]):

**Proposition 2** *Assume the continuity condition holds. If one of the derivatives defined in [5], [6], [11], [12] or [13] exists and it is a fuzzy number, then so do the others and they are all equal. Their value is provided from (2).*

The continuity condition is assumed to hold for all fuzzy functions in the rest of the paper. The *Fuzzy Initial Value Problem (FIVP)* we face in this paper is defined as:

$$\widetilde{x}'(t) = f(t, \widetilde{x}, \overline{C}), \qquad \widetilde{x}(t_0) = \widetilde{x}_0 \in F(\mathbb{R}) \tag{3}$$

where $\widetilde{x}$ is the unknown fuzzy function, $t \in [t_0, T] \subseteq \mathbb{R}$ and $\overline{C}$ is a vector of triangular fuzzy numbers (i.e. fuzzy numbers whose membership function has a triangular graph). We point out that $f$ is a continuous fuzzy function whose fuzziness is because of $\overline{C}$. This means that if we replace $\overline{C}$ with a vector of real numbers, $f$ will become a crisp function.

Let $[\widetilde{x}(t)]_a = [x_1(t, a), x_2(t, a)], a \in (0, 1]$ the a-cuts of the desired solution. If $\overline{C} = (\overline{C}_1, \overline{C}_2, \cdots, \overline{C}_m)$ where $\overline{C}_1, \overline{C}_2, \cdots, \overline{C}_m \in F(\mathbb{R})$, then problem (3), in its parametric form, is written for $a \in (0, 1]$ as follows [2]:

$$x_1'(t, a) = F(t, x_1^a(t), x_2^a(t), \underline{c}, \overline{c}) = \min\{f(t, x, c_1, \cdots, c_m) | x \in [x_1(t, a), x_2(t, a)],$$
$$c_j \in [C_i^1(a), C_i^2(a)]\} \tag{4}$$
$$x_2'(t, a) = G(t, x_1^a(t), x_2^a(t), \underline{c}, \overline{c}) = \max\{f(t, x, c_1, \cdots, c_m) | x \in [x_1(t, a), x_2(t, a)],$$
$$c_j \in [C_i^1(a), C_i^2(a)]\}$$

**Proposition 3** *Let $\widetilde{f} : [t_0, T] \times F(\mathbb{R}) \to F(\mathbb{R})$ be continuous and assume that there exists a $L > 0$ such that $D(\widetilde{f}(t, \widetilde{x}), \widetilde{f}(t, \widetilde{y})) \leq L \cdot D(\widetilde{x}, \widetilde{y})$ for all $t \in [t_0, T], \widetilde{x}, \widetilde{y} \in F(\mathbb{R})$. then the problem (3) has a unique solution in $[t_0, T]$ [8].*

## 2   Runge - Kutta Methods for FDEs

Let $P = \{t_0, t_1, ..., t_N = T\}$ and $0 < a_0 < a_1 < \cdots < a_m = 1$ be discrete sets of points in $[t_0, T]$ and $[0, 1]$ correspondingly. When solving FDEs numerically, we try to find approximately the limits of every a-cut of the form $[\widetilde{x}(t_i)]_{a_j}$, $i = 0, ..., N, j = 0, ..., m$. We know that these a-cuts are intervals and, as a result, we solve for every $a_j$ a system of two equations that give us the the the two limits of the interval. It is known [9] that for every $P \subseteq \mathbb{R}, diam(P) = sup\{|p - q|, p, q \in P\}$ and consider that $P$ has the property that $\lim_{n \to \infty} diam(P) \to 0$. We can now give the following definitions:

**Definition 3** *A numerical method giving approximations $x_i$ of the solutions $x(t_i)$ of FIVP (3) is said to be* convergent *if (3) follows the conditions of Proposition 3 and*

$$\max_{0 \leq i \leq N} D(x(t_n), x_n) \to 0 \qquad as \; diam(P) \to 0 \tag{5}$$

**Definition 4** *The* global truncation error $E_i$ *of a numerical method giving approximations $x_i$ of the solutions $x(t_i)$ of FIVP (3) at the i-th step is given by:*

$$E_i = D(x(t_i), x_i)$$

**Definition 5** *A numerical method is said to be* pth-order *if, for given step h, we have* $\max_{i=0,...N} E_i(h) = O(h^p)$.

Using results presented in [3],[7] and [10], we define s-stage Runge - Kutta methods for FDEs (3) through the following equations:

$$\widetilde{x}_{n+1} = \widetilde{x}_n + \Phi(t_n, \widetilde{x}_n, h) \tag{6}$$

where

$$\Phi(t_n, \widetilde{x}_n, \overline{C}, h) = \sum_{i=1}^{s} b_i k_i$$

$$k_1 = f(t_n, \widetilde{x}_n, \overline{C})$$

$$k_i = f(t_n + c_i h, \widetilde{x}_n + h \sum_{j=1}^{i-1} a_{ij} k_j, \overline{C}) \qquad i = 2, ..., s.$$

We assume that the following conditions hold:

$$\sum_{i=1}^{s} b_i = 1, \qquad c_i = \sum_{j=1}^{i-1} a_{ij}, i = 1, 2, ..., s$$

It is obvious that constants $k_i$ are fuzzy numbers. We also denote $x^a(t) = [x_1^a(t) \; x_2^a(t)]'$, $x_n^a = [x_1^n(a) \; x_1^n(a)]'$ and $k_i^a = [k_{i1}^a \; k_{i2}^a]'$.

Let $Q = \{(t, v, w, u_1^1, u_1^2, u_2^1, u_1^2, ..., u_m^1, u_m^2) : t \in [t_0, T], w \in \mathbb{R}, v \in (-\infty, w], u_i^1 \in [C_i^1(a), C_i^1(1)], u_i^2 \in [C_i^2(1), C_i^2(a)]\}$. It can be shown that the following Proposition holds:

**Proposition 4** *Let $F, G$ of problem (4) with $F, G \in C^p(Q)$ and let their partial derivatives be bounded over $Q$. Then, for arbitrary fixed $a : 0 \leq a \leq 1$, the Runge - Kutta approximates of equation (6)converge to the exact solutions $x_1^a(t)$ and $x_2^a(t)$ uniformly in t.*

# 3  Numerical Results

Using Runge-Kutta method defined in the previous section, we solved the following problem:

$$\widetilde{x}'(t) = \overline{C}\widetilde{x}(t), \quad \widetilde{x}(0) = \widetilde{x}_0, \qquad \widetilde{x}_0 = (8.0/8.5/9.0) \tag{7}$$

where $\overline{C}$ is a triangular fuzzy number. We used two different numbers for our calculations: $\overline{C} = (1.0/2.0/3.0)$ (which means that the value of $\mu_{\overline{C}}(x) = 0$ for $x \notin [1,3]$ and the summit of the triangular graph is for $x = 2$) and $\overline{C} = (-4.0, -3.0, -2.0)$.

Problem (7) can only be solved with our generalized method and the numerical results showed that a 4-step Runge-Kutta method using classical coefficients retains order 4, with our definition of order.

Furthermore, we solved the following problem that was first solved in [11]:

$$\widetilde{x}'(t) = t \cdot \widetilde{x}(t), \qquad \widetilde{x}(-1) = \widetilde{x}_0 \in F(\mathbb{R}), \tag{8}$$

where $[\widetilde{x}_0]_a = [\widetilde{x}(-1)]_a = [\sqrt{e} - 0.5(1 - a), \sqrt{e} + 0.5(1 - a)]$ and $t \in [-1, 1]$.

For problem (8), Euler method was also applicable. We found out that Euler retains order 1, while our method solves the particular problem with order 4.

## Acknowledgment

This research was co-funded by 75% from E.E. and 25%from Greek government, under the framework of the Education and Initial Vocational Training Program - Archmides of the TEI of Chalkis.

## References

[1] S. Abbasbandy and T. Allah Viranloo, Numerical Solution of Fuzzy Differential Equation by Runge-Kutta Method, Nonlinear Studies, Vol. 11, No. 1 (2004), pp. 117-129

[2] J. J. Buckley and T. Feuring, Fuzzy Differential Equations, Fuzzy Sets and Systems 110 (2000), p.43-54

[3] J. C. Butcher, The Numerical Analysis of Ordinary Differential Equations (J.Wiley & Sons, 1987)

[4] D. Dubois and H. Prade, Fuzzy Sets and Systems: Theory and Applications (Academic Press, London,1980)

[5] D. Dubois and H. Prade, Towards Fuzzy Differential Calculus Part 3: Differentiation, Fuzzy Sets and Systems 8 (1982), p.225-233

[6] R. Goetschel and W. Voxman, Elementary Fuzzy Calculus, Fuzzy Sets and Systems 18 (1986) p.31-43

[7] E. Hairer, S. P. Norsett and G. Wanner, Solving Ordinary Differential Equations I, (Springer-Verlag, Berlin, 1987).

[8] O. Kaleva, Fuzzy Differential Equations, Fuzzy Sets and Systems 24 (1987) p.301-317

[9] A. N. Kolmogorov and S. V. Fomin, Introductory Real Analysis(Prentice-Hall, 1970)

[10] J. D. Lambert, Numerical Methods for Ordinary Differential Systems (J. Wiley & Sons, 1990)

[11] M. Ma, M. Friedman and A. Kandel, Numerical Solutions of Fuzzy Differential Equations, Fuzzy Sets and Systems 105 (1999) p.133-138

[12] M. L. Puri and D. A. Ralescu, Differentials of Fuzzy Functions, Journal of Mathematical Analysis and Applications 91 (1983) p.552-558

[13] S. Seikkala, On the Fuzzy Initial Value Problem, Fuzzy Sets and Systems 24 (1987) p.319-330

[14] H.-J. Zimmermann, Fuzzy Set Theory and its Applications(3rd edition, Kluwer, Dordrecht, 1996)

Brill Academic Publishers
P.O. Box 9000, 2300 PA Leiden,
The Netherlands

*Lecture Series on Computer
and Computational Sciences*
Volume 4, 2005, pp. 449-454

# Improving On-the-fly Race Detection for Message-Passing Programs [1]

**Mi-Young Park, Young-Cheol Kim, Moon-Hye Kang, and Yong-Kee Jun** [2]

Computer Science, Gyeongsang National University
Computer Engineering, Jinju International University
Jinju, 660-701 South Korea

Received 6 August, 2005; accepted in revised form 16 August, 2005

*Abstract:* Detecting the first race to occur in each process is important to debugging message-passing programs effectively, because such a message race may affect other races in the same process. The previous techniques to detect such races require more than two monitored runs of a program or analyze a trace file which size is proportional to the number of messages. This paper introduces an new on-the-fly technique to detect the first race to occur in each process without generating a trace file. Detecting first races helps avoid overwhelming the programmer with too much debugging information.

*Keywords:* message-passing program, debugging, message race, on-the-fly detection, first races

## 1 Introduction

In asynchronous message-passing programs, *a message race* [3, 5] occurs toward a receive event, if two or more messages are sent over communication channels on which the receive listens, and they are simultaneously in transit without guaranteeing the order of their arrivals. Message races should be detected for debugging a large class of message-passing programs [11] effectively, because nondeterministic order of arrivals of the racing messages causes unintended nondeterminism of programs [5, 8, 10].

Especially, it is important to detect the *first race* [8, 9] occurred in each process, because the first race can never be affected from the other races which occurred in the same process [8, 9]. The previous techniques [3, 5, 8, 9] which can detect the first races is inefficient because they require more than two monitored runs of a program [3, 8, 9] or analyze a trace file which size is proportional to the number of messages [5].

This paper introduces an on-the-fly technique to detect the first races occurred in each process without generating a trace file. This technique presents a new data structure, called as a *message history* which consists of two vectors, to store the events to be involved in the first race in each process. One vector contains the send events which sent a message most lately to each process. This information used to know the latest event happened before the event receiving the message and to know the start point of concurrency between processes. The other vector contains the receive events which occurred for the first time after the latest send events of the process. This

---

[1]This work was supported in part by Research Intern Program of Korea Research Foundation and by Grant No. R05-2003-000-12345-0 from the Basic Research Program of the Korea Science and Engineering Foundation.

[2]Corresponding author. Also involved in Research Institute of Computer and Information Communication, Gyeongsang National University. E-mail: jun@gsnu.ac.kr

vector used to find the first racing receive event of each process. Using the message history, our technique detects the first race to occur in each process during only one execution.

## 2 Background

There are several tools for detecting message races such as MAD [7], MARMOT [4, 6], and MPVisualizer [1, 2]. MAD offers a variety of debugging features such as placement of breakpoints on multiple processes, inspection of variables, an event manipulation feature, and a record&replay mechanism [7]. MARMOT is to verify the standard conformance of an MPI program automatically during runtime and help to debug the program in case of problems such as deadlocks, and race condition [4, 6]. MPVisualizer includes a trace/reply mechanism and a graphical interface and the engine of the tool detects and notifies the occurrence of race conditions [1, 2].

These tools detect message races just by locating the use of wild card receives as sources of race conditions. In this case the programmers are overwhelmed by the vast information of the results of race detection and they are embarrassed in debugging. Because the vast information may include the affected races which do not need to be debugged [8]. Therefore we need to detect only the races to be debugged with accuracy. Especially, it is important to detect a *locally-first race* [9] which is the first race to occur in a process. Because the first race occurred in a process can never be affected from the other races which occurred in the same process.

Previous techniques [3, 5] to detect locally-first races require more than two monitored executions [3, 8, 9] or analyze a trace file which size is proportional to the number of messages [5]. The efficient techniques [8, 9] among the previous techniques present a two-pass hybrid technique. The first pass determines whether any race occurs and traces an approximate location where a locally-first race occurs in each process. The second pass locates exactly the first racing receive event of the locally-first race to occur in each process, and detects racing messages. However the techniques still require two monitored runs of a program so that they are not efficient for long-run programs.

## 3 Locally-first Race Detection

The *message history* consists of two vectors to store the events to be involved in the first race in each process. One vector contains the send events which sent a message most lately to each process. This information used to know the latest event happened before the event receiving the message and to know the start point of concurrency between processes. The other vector contains the receive events occurred for the first time after the latest send events of each process. This vector used to find the first racing receive event.

The data structure of a message history consists of a pair of vectors named as *LastSend* and *FirstRecv*. Both of vectors are a $P$-length array ($P$ is the number of processes) and each element of the array corresponds to each process. Each element of the vectors of a process has a *local clock* of the process. The local clock is incremented after each send/receive event and uniquely identifies events by numbering them sequentially within a process. *LastSend* of the message history contains the local clocks of the send events which sent a message most lately to the other processes. *FirstRecv* of the message history contains the local clocks of the first receive event occurred for the first time after the latest send event of the process. *LastSend* is updated in each send event and *FirstRecv* is updated in each receive event. When *FirstRecv* is updated, each value should be larger than each element of *LastSend*.

Figure 1 shows the algorithm for maintaining a message history during an execution. Figure 1.a shows the algorithm for initializing each element of *LastSend* and *FirstRecv* of a message history. It is called only one when a program starts. Figure 1.b shows the algorithm for updating *LastSend*

```
0   CheckHistoryInit()
1   for i from 1 to size do
2       LastSend[i] := 0            0   CheckHistoryRecv()
3       FirstRecv[i] := 0          1   for all i in size do
4   end for                        2       if FirstRecv[i] <= LastSend[i] then
            (a)                    3           FirstRecv[i] := localclock
0   CheckHistorySend()             4       end if
1   LastSend[dest] = localclock    5   end for
            (b)                               (c)
```

Figure 1: Message History Algorithm

of a message history. In the algorithm *dest* is an integer specifying the process which will receive a message. Whenever a send event occurs, the *LastSend* is updated to the current local clock so that it always keeps the latest send event for each process.

Figure 1.c shows the algorithm for updating *FirstRecv*. In this algorithm *size* is an integer specifying the number of processes. In the line 2 it determines whether each element of *FirstRecv* is equal or smaller than each element of *LastSend*. if any element of *FirstRecv* is equal or smaller than the corresponding element of *LastSend*, the element of *FirstRecv* is updated to the current local clock to keep a larger value. Because *FirstRecv* should contain the local clock of the receive event which occurred for the first time after the latest send event of each process.

Figure 2 shows the message history of each event in $P_2$ when the algorithms are applied in $P_2$. When a program starts, each message history is initialized with zero. When a send event *a* sends a message to $P_4$, the forth index of *LastSend* of $P_2$ is updated to the current local clock 1. Because each element of *LastSend* corresponds to each process and the forth index of the *LastSend* corresponds to $P_4$. When the receive event *d* occurs, each element of *FirstRecv* is updated to the current local clock 2. Because each element of *FirstRecv* < 00000 > is equal or smaller than the corresponding element of *LastSend* < 00010 >. So *FirstRecv* changed to < 22222 > to keep a larger value than each element of *LastSend*.

When a send event *e* sends a message to $P_1$, the first element of *LastSend* changed to the current local clock 3. When a receive event *j* occurs, the elements of *FirstRecv* is updated to the current local clock 5 only if it is equal or smaller than the corresponding element of *LastSend*. In this case the first and the third element of *FirstRecv* is updated to 5 so that *FirstRecv* is < 52522 >. When a receive event *l* occurs, *FirstRecv* does not change because there is not any element which is equal or smaller than the corresponding element of *LastSend*.

Figure 3 shows the algorithm for detecting locally-first races using message history. The algorithm performs race detection at each receive event in a process. Line 3 locates the most recent receive event in the same process that listened over the channel where the current receive event listens over a channel, *thisChan*. Line 4 checks the ordering between *PrevRecv* and *Send*. If *PrevRecv* is concurrent with *Send* based on timestamp vector, this message must be racing toward *PrevRecv*. And it checks if *FirstRecv[source]* is smaller than *firstRace*. *firstRace* is initialized with a max integer and has the local clock of the first racing receive event of the process. *source* is an integer specifying the process which sent the message to the current process. So *FirstRecv[source]* indicates the first racing receive event where the received message is racing.

If *FirstRecv[source]* is smaller than *firstRace*, it means that currently detected race happens before the previous race and that a new locally-first race is detected. Therefore the information of a new locally-first race is stored in lines from 5 to 7; the first racing receive event is stored in *firstRace*, the channel where *Recv* listened is stored in *firstChan*, and the message is stored in *RacingMessage*. Line 8 checks if the received message is racing and if *FirstRecv[source]* is equal to

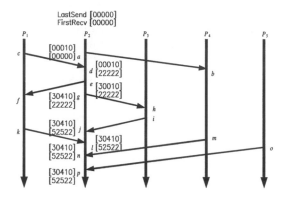

Figure 2: Message History of $P_2$

```
0  CheckFirstRace()
1  Send = a event that sent a Message to this receive event
2  Recv = this receive event
3  prevRecv := PrevBuf[thisChan]
4  if (prevRecv ↛ Send) and FirstRecv[source] < firstRace then
5        firstRace := FirstRecv[source]
6        firstChan := thisChan
7        RacingMessage := Message
8  else if ((prevRecv ↛ Send) and FirstRecv[source] = firstRace) then
9        RacingMessage = RacingMessage ∪ Message
10 end if
11 for all i in Channels do
12       PrevBuf[i] = Recv
13 end for
```

Figure 3: Detection Algorithm

*firstRace*. If two conditions are satisfied, it means that the received message is racing toward the current *firstRace*. So *RacingMessage* includes the received message.

Let us apply the algorithm to the Figure 2. In the receive event $d$ and $j$, any race does not be detected and the *firstRace* does not change. In the receive event $l$, it detects a race because the previous receive event $j$ and the send event $k$ are concurrent. Then *firstRace* changed from a max integer to 5. Because the message $msg(k)$ is received from $P_1$ and the first element of *FirstRecv* is 5. When a receive event $n$ occurs, it detects a race and *firstRace* changed to 2. Because the message $msg(m)$ is received from $P_4$ and the fourth element of *FirstRecv* is 2 which is smaller than the current *firstRace* 5. When a receive event $p$ occurs, the message is just included in *RacingMessage*. Because the message $msg(o)$ is received from $P_5$ and the fifth element of *FirstRecv* is 2 which is equal to the current *firstRace* 2. When this execution terminates, *firstRace* is 2 and *RacingMessage* includes $\{msg(m), msg(o)\}$. Therefore this algorithm detects accurately the locally-first race in $P_2$ as $\langle d, \{msg(m), msg(o)\}\rangle$.

# 4 Conclusion

In this paper we present an on-the-fly technique to detect the first race to occur in each process without generating a trace file. For this, we present a new data structure, called as message history, and a new detection algorithm using the message history. Detecting locally-first races helps avoid overwhelming the programmer with too much information when a programmer debugs message-passing programs. Therefore our technique is very useful for the tools of debugging message-passing programs.

# References

[1] Cláudio, A.P., J.D. Cunha, and M.B. Carmo, "MPVisualizer: A General Tool to Debug Message Passing Parallel Applications," *7th High Performace Computing and Networking Europe*, Lecture Notes in Computer Science, 1593:1199-1202, Springer-Verlag, April 1999.

[2] Cláudio, A.P., J.D. Cunha, and M.B. Carmo, "Monitoring and Debugging Message Passing Applications with MPVisualizer," *8th Euromicro Workshop on Parallel and Distributed Processing*, pp.376-382, IEEE, Jan. 2000.

[3] Damodaran-Kamal, S. K., and J. M. Francioni, "Testing Races in Parallel Programs with an OtOt Strategy," *Int'l Symp. on Software Testing and Analysis*, pp. 216-227, ACM, Aug. 1994.

[4] Krammer, B., K. Bidmon, M.S. Müller, and M.M. Resch, "MARMOT: An MPI Analysis and Checking Tool," *In proceedings of PARCO'03*, 13:493-500, Elsevier, Sept. 2003.

[5] Kilgore, R., and C. Chase, "Re-execution of Distributed Programs to Detect Bugs Hidden by Racing Messages," *30th Annual Hawaii Int'l. Conf. on System Sciences*, Vol. 1, pp. 423-432, Jan. 1997.

[6] Krammer, B., M.S. Müller, and M.M. Resch, "MPI Application Development Using the Analysis Tool MARMOT," *4th International Conference on Computational Science*, Lecture Notes in Computer Science, 3038:464-471, Springer-Verlag, june 2004.

[7] Kranzlmüller D., C. Schaubschläger, and J. Volkert, "A Brief Overview of the MAD Debugging Activities," *4th International Workshop on Automated Debugging* (AADEBUG 2000), Aug. 2000.

[8] Netzer, R. H. B., T. W. Brennan, and S. K. Damodaran-Kamal, "Debugging Race Conditions in Message-Passing Programs," *SIGMETRICS Symp. on Parallel and Distributed Tools*, pp. 31-40, ACM, May 1996.

[9] Park, M., and Y. Jun, "Detecting Unaffected Race Conditions in Message-Passing Programs," *11th European PVM/MPI User's Group Meeting*, Lecture Notes in Computer Science, 3241:268-276, Springer-Verlag, Sept. 2004.

[10] Rimnac, A., and D. Kranzlmüller, "Nondeterminism in Parallel Programs: Experiences with Real-World Applications," *the International Conference on Parallel and Distributed Processing Techniques and Applications* (PDPTA'04), pp.108-113, CSREA Press, June, 2004.

[11] Snir, M., S. Otto, S. Huss-Lederman, D. Walker, and J. Dongarra, *MPI: The Complete Reference*, MIT Press, 1996.

Brill Academic Publishers
P.O. Box 9000, 2300 PA Leiden,
The Netherlands

*Lecture Series on Computer*
*and Computational Sciences*
Volume 4, 2005, pp. 455-458

# Estimating the Regularization Parameter for the Linear Sampling Method in Acoustics

## G. Pelekanos[1]

Department of Mathematics and Statistics,
Southern Illinois University
Edwardsville, IL 62025, USA

## V. Sevroglou

Department of Mathematics
University of Ioannina,
GR-45110 Ioannina, Greece

Received 17 June, 2005; accepted in revised form 13 July, 2005

*Abstract:* In real world applications, the noise level is often unknown, so there has been an interest in methods for selecting the regularization parameter without prior knowledge of the noise level. One of the two most popular approaches is the L-curve method. In this work we use the linear sampling method in combination with the L-curve approach for the visualization of scattering objects from noisy far field data.

*Keywords:* far-field operator, L curve, regularization parameter

*Mathematics Subject Classification:* 45A05, 65F22, 45Q05

## 1   Introduction

For inverse acoustic scattering the original linear sampling method was introduced by Colton and Kirsch [2] and mathematically clarified in [3]. Two of its attractive features are that no low-or high frequency approximation is needed and that a priori knowledge of the boundary condition(s) is not required. Its main drawback however is that it requires complete far-field data.

Recently Kirsch [8] improved the original version of the linear sampling method, leading to the so-called $(F^\star F)^{1/4}$ -method.

This method along with its previous version is known to give only an explicit characterization of the scattering obstacle (i.e. it only determines the support of the refractive index) and it is also very easy to implement, since it just involves the solution of a linear integral equation whose kernel consists of the measured data.

In this work we consider a two dimensional cylindrical scatterer represented by an impenetrable obstacle. In [2] and [3] it has been shown that if $k$ is the wavenumber of a TM polarized electromagnetic wave and $\mathbf{y_o}$ is a point inside the scatterer, then the support of the scatterer can be identified by solving the far-field equation

$$(Fg)(\widehat{\mathbf{r}}) = \frac{e^{i\pi/4}}{\sqrt{8\pi k}} e^{-ik\widehat{\mathbf{r}}\cdot\mathbf{y_o}} \tag{1}$$

---

[1]Corresponding author. E-mail: gpeleka@siue.edu

where $F : L^2(\Omega) \to L^2(\Omega)$ with $\Omega = \{x \in R^2, |x| = 1\}$, such that

$$(Fg)(\hat{\mathbf{r}}) = \int_\Omega u_\infty(\hat{\mathbf{r}}; \hat{\mathbf{d}}) \, g(\hat{\mathbf{d}}) \, ds(\hat{\mathbf{d}}) \tag{2}$$

is the far-field operator, and $u_\infty(\hat{\mathbf{r}}; \hat{\mathbf{d}})$ denotes the far-field data that correspond to incident direction $\mathbf{d}$ and observation direction $\hat{\mathbf{r}}$, obtained using Nyström's method [2] . In [2] it has been shown that $\|g\|_{L^2(\Omega)}$ approaches infinity when $\mathbf{y_o}$ approaches the boundary of the scatterer from the inside. The major drawback of the procedure above is that it is not clear what happens in the case when $\mathbf{y_o}$ approaches the boundary of the scatterer from the inside. In order to answer this question, Kirsch [8] proposed solving the following far field equation instead

$$(F^\star F)^{1/4} g(\hat{\mathbf{r}}) = \frac{e^{i\pi/4}}{\sqrt{8\pi k}} e^{-ik\hat{\mathbf{r}}\cdot\mathbf{y_o}} \tag{3}$$

In particular he showed that for (3), $\|g\|_{L^2(\Omega)}$ completely characterizes the shape of the scattering obstacle. It is well known that inverse problems are ill-posed in the sense of Hadamard and hence any numerical implementation for their solution requires regularization. Apparently the problem above is not an exception. Tikhonov regularization with Morozov's discrepancy principle for the selection of the regularization constant were traditionally used by many researchers, see [3] and [8]. However, discrepancy principles in general require a priori knowledge of the noise level and hence they may be of limited practical importance since in most real world applications the noise level is unknown. One of the most popular parameter selection methods that don't require prior knowledge of the noise level, is the L-curve criterion. The L-curve criterion was introduced by Hansen [5] and is a log-log plot of the norm of a regularized solution versus the norm of the corresponding residual norm. In addition, in contrast with the discrepancy principle this method is "error-free", meaning the error-norm is not required.

## 2 Regularization

Our goal in this section is to formulate a regularization algorithm for the solution of the far-field equation

$$(F^\star F)^{1/4} g(\hat{\mathbf{r}}) = f \tag{4}$$

with

$$f = \frac{e^{i\pi/4}}{\sqrt{8\pi k}} e^{-ik\hat{\mathbf{r}}\cdot\mathbf{y_o}} \tag{5}$$

Existence and uniqueness for the solution of (4) is presented in [8]. It is worthwhile to mention here that the left hand side of (4) is contaminated with noise due to far-field measurements $u_\infty$, while the right hand side is known exactly.

Discretizing the Fredholm integral equation (4) we arrive to a problem formulation in matrix-vector format:

$$A\mathbf{x} = \mathbf{b}$$

where $A$ is an $n \times n$ matrix. To keep things simple we assume that in the above matrix-vector form the errors in the right-hand side are dominant. This assumption does not put any limitations on the applicability of the work as errors $E$ in the coefficient matrix $A$ can be mapped to the right hand side $\mathbf{b}$. If $A = A_{\text{exact}} + E$ then the noisy right-hand side $\mathbf{b}$ can be written as: $\mathbf{b} = \mathbf{b}_{\text{exact}} + \mathbf{e}$ where $\mathbf{e} = (A - A_{\text{exact}})\mathbf{x}$ is the term that maps the errors in the coefficient matrix to the right hand side.

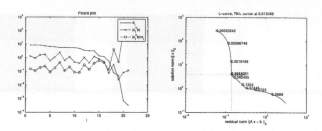

Figure 1: SVD coefficients (left) and L-curve (right)

We now compute $\mathbf{x}^{(\gamma)}$ by minimizing the Tikhonov functional

$$\|A\mathbf{x}^{(\gamma)} - \mathbf{b}\|^2 + \gamma\|\mathbf{x}^{(\gamma)}\|^2 \tag{6}$$

where the regularization parameter $\gamma$ is real and positive. Following [7], the Tikhonov solution $\mathbf{x}^{(\gamma)}$ can be written conveniently in terms of the SVD of $A$

$$\mathbf{x}^{(\gamma)} = \sum_{j=1}^{n} \frac{\sigma_j^2}{\gamma + \sigma_j^2} \frac{\mathbf{u}_j^T \mathbf{b}}{\sigma_j} \mathbf{v}_j \tag{7}$$

The selection of $n$ as an upper bound for the sum above is done through the Picard condition [6]:

*The exact SVD coefficients $|\mathbf{u}_j^T \mathbf{b}|$ decay faster than the $\sigma_j$*

as this will ensure that the solution for the ill-posed problem exists.

## 3   Numerical Validation

In this section we present a reconstruction of an impenetrable peanut shape object from a finite number of $u_\infty$ measurements. In particular the far field is produced using Nystrom method [1] in the case of 21 incident waves with $k = 4$ measured at 21 observation points. In addition, a 5% Gaussian noise is pointwise added to each element of the far-field matrix. The left graph in fig.1 above shows that the right-hand side that satisfies the Picard condition is determined by the indices below 17 (i.e. $n = 17$ in (7)) where the solution coefficients $|\frac{\mathbf{u}_j^T \mathbf{b}}{\sigma_j}|$ begin to increase as a result of errors in the data vector. In that region, the Fourier coefficients $|\mathbf{u}_j^T \mathbf{b}|$ decay faster than the singular values. For indices above 17, we are dealing with lost information due to dominating errors. Hence, the components corresponding to these latter indices should be disregarded in the regularized solution by choice of a suitable regularization parameter $\gamma$. We now proceed with the determination of the index where the solution coefficients start to increase without having to rely on the Picard plot for every problem. Thus, we need to determine the optimal regularization parameter $\gamma$ based on a well-defined criterion.

One very popular method is the so-called *L-curve criterion* due to Hansen [5]. The L-curve is a plot of the two-norm of the regularized solution versus the residual two-norm in a log-log scale for all valid regularization parameters. The L-curve for our test problem is the right plot in fig.1 above. The flat horizontal part of the L-curve is characterized by over-regularized solutions where the norm of the solution is small and does not change much as the regularized solution is very smooth. The vertical part is characterized by under-regularized solutions that are dominated by

Figure 2: Visualization of a peanut (left), without regularization (middle), and with regularization (right)

the effect of errors, hence the solution norm grows fast. Choosing the regularization parameter using the L-curve is the trade-off between the size of the solution norm and the residual norm. The L-curve criterion defines the *optimal* value of the regularization parameter to be at the corner of the L-curve. At the corner, we find the solution that has the smallest residual and a limited size. Note this is only possible in a plot in log-log scale. In addition the faster the Fourier coefficients decay, the sharper and more distinct will the corner of the L-curve be. Algorithms for computing the corner of the L-curve seek to find the point on the curve associated with maximum curvature. In this work we have used the algorithms appear at [4]. Using $\lambda = 0.0155$ as the regularization parameter chosen by the L-curve (see right plot in fig. 1) we reconstruct our object using the Tikhonov regularization described in the previous section. Fig.2 shows reconstructions without (middle) and with (right) regularization. It appears that regularization was necessary as expected, and that the selection of the regularization parameter via the L-curve was successful.

## References

[1] D. Colton, D. and R. Kress, *Inverse Acoustic and Electromagnetic Scattering Theory*. Springer-Verlag, New York, 1992

[2] D. Colton, D. and A. Kirsch, A simple method for solving the inverse scattering problems in the resonance region, *Inverse Problems* **12** 383-393 (1996).

[3] D. Colton , M. Piana and R. Potthast A Simple Method using Morozov's Discrepancy Principle for Solving Inverse Scattering Problems, *Inverse Problems* **13** 1477-93(1999)

[4] P. C. Hansen Regularization Tools: A Matlab Package for Analysis and Solution of Discreet Ill-Posed Problems. *Numer. Algo.* **6** 1-35 (1994)

[5] P. C. Hansen *Rank-Deficient and Discrete Ill-Posed Problems*. SIAM, Philadelphia, 1998.

[6] P. C. Hansen *The L-Curve and its use in the numerical treatment of inverse problems*. Computational Inverse Problems in Electrocardiology. *WIT Press* 119-142 (2001)

[7] A. Kirsch , *An Introduction to the mathematical theory of inverse problems*. Berlin: Springer, 1996.

[8] A. Kirsch Characterization of the Shape of a Scattering Obstacle using the Spectral Data of the Far Field Operator *Inverse Problems* **14** 1489-1512 (1998).

Brill Academic Publishers
P.O. Box 9000, 2300 PA Leiden,
The Netherlands

Lecture Series on Computer
and Computational Sciences
Volume 4, 2005, pp. 459-462

# Fixed Point Iterative Method to solve 1D wave equation. Application to 1D photonic crystals

## Manuel Perez, L. Carretero[1] , P. Acebal

Universidad Miguel Hernandez. Dpto. de Ciencia y Tecnologia de los Materiales.
Avda. Ferrocarril. 3202.Elche. Alicante. (Spain)

Received 11 July, 2005; accepted in revised form 5 August, 200

*Abstract:* In this paper we introduce an iterative method based on the classical fixed point theorem to solve the one-dimensional linear wave equation in the frecuency domain. One of the biggest interests of this method is that it gives both numerical and analytical approximated wave equation solutions. The field of application of this method are general non-magnetic 1D linear media. In particular, we apply it to concrete one-dimensional linear photonic crystals to obtain the electric field profile inside the crystal.

*Keywords:* Fixed point theorem, photonic crystals

## 1    The Fixed Point Iterative Method

In this section we introduce the *Fixed Point Iterative Method* (**FPIM**). The **FPIM** is a method to solve linear[2] differential ecuations both numerically and analytically and it is based on the *Fixed Point Theorem*, a classical result in Mathematical Analysis [1].

Let us consider the following complex second-order differential equation:

$$y''(z) - h(z) y(z) = 0, \ with \ z \in [a, b], \ y(a) = e_1 \ and \ y'(a) = e_2; \ e_1, e_2 \in \mathbb{C} \qquad (1)$$

where $h : [a, b] \rightarrow \mathbb{C}$ is a continuos aplication. We can rewrite (1) **in its integral equivalent form** integrating both members in the real variable $z$ twice. Then equation (1) becomes:

$$y(z) = e_1 + e_2 \cdot (z - a) + \int_a^z \left( \int_a^x h(\tau) y(\tau) d\tau \right) dx, \ with \ z \in [a, b] \qquad (2)$$

The **FPIM to solve equation** (1) (alternatively equation (2)) is described as follows:

1. We choose any (as simple as possible) continuos aplication $y_0 : [a, b] \rightarrow \mathbb{C}$.

2. We calculate, either numerically or analytically[3] the following term of the series given by the recurrency:

$$y_{n+1}(z) = e_1 + e_2 \cdot (z - a) + \int_a^z \left( \int_a^x h(\tau) y_n(\tau) d\tau \right) dx, \ with \ z \in [a, b], \ n \ integer \qquad (3)$$

---

[1]Corresponding author. E-mail:luis@dite.umh.es

[2]It could eventually be used for nonlinear equations under certain conditions.

[3]In order to carry out the FPIM analytically we should choose the first term $y_0$ of the sequence as simple as possible to simplify the integrals

3. We compare the terms $y_n$ and $y_{n+1}$ by evaluating the difference $e_n(z) = |y_{n+1}(z) - y_n(z)|$, with $z \in [a, b]$. If $e_n(z)$ is small enough *for every* $z \in [a, b]$, then $y_n(z)$ is a good aproach to the solution, so we can finish. Otherwise we need more iterations, so we come back to step 2.

The minimum number of iterations required to reach a good aproach of the solution depends on the maximum value of $|h(z)|$ with $z \in [a, b]$ and the size of the interval $b - a$. As they increase, the **FPIM** requires a greater number of iterations. Hoewer, thanks to Mathematical Analysis we know that if $h$ is continuos in $[a, b]$, the **FPIM** always converges to the solution, no matter how many iterations are needed.

In order to carry out the **FPIM analytically** it is suitable to choose $y_0(z) = C$ for every $z \in [a, b]$, where $C$ is a **constant**. In this case, since $h : [a, b] \to \mathbb{C}$ is continuos, we know that $h$ admits Fourier series in the form:

$$h(z) = \sum_{k=-\infty}^{+\infty} a_k e^{i\frac{2\pi k}{b-a}z}, \; for \; z \in [a, b] \tag{4}$$

Taking into account (3) and (4) we obtain a very specific analytical expression for $y_n(z)$ with the FPIM, which is the following:

$$y_n(z) = \sum_{p=0}^{p_n} \sum_{q=-r_n}^{q_n} c_{pqn} z^p e^{i\frac{2\pi q}{b-a}z}, \; with \; c_{pqn} \in \mathbb{C} \; for \; each \; p, q, n \tag{5}$$

## 2 Solving the 1D frecuency-domain wave equation with the FPIM

The one-dimensional linear wave equation for the electric field $E$ in the frecuency domain in a certain material in $[a, b]$ has the following form:

$$\frac{\partial^2 E(z)}{\partial z^2} + \varepsilon_r(z) \left(\frac{2\pi}{\lambda}\right)^2 E(z) = 0, \; with \; z \in [a, b] \tag{6}$$

where $\varepsilon_r(z)$ is the *relative permitivity*[4] along the $z$ axis and $\lambda$ is the *wavelength*. Comparing ecuations (1) and (6) it is easy to deduce that equation (6) **is a particular case of equation** (1), so it can be solved by the **FPIM**. In fact, if we choose:

$$h(z) = -\left(\frac{2\pi}{\lambda}\right)^2 \varepsilon_r(z) \tag{7}$$

then equation (6) becomes equation (1).

### 2.1 Application to photonic crystals

In this section we are going to apply the **FPIM** to a one-dimensional photonic crystal of $10\mu m$ whose relative permitivity is given by:

$$\varepsilon_r(z) = \varepsilon_1 + \varepsilon_2 \cos\left(\frac{2\pi z}{a}\right) \tag{8}$$

---

[4]In this paper we will assume that $\varepsilon_r(z)$ is continuos and that $\varepsilon_r(z)$ is the complex permitivity, i.e. $\varepsilon_r(z) = \varepsilon_{rr}(z) - i\varepsilon_{ri}(z)$.

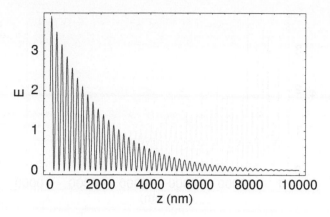

Figure 1: $|E|^2$ as a function of z position ($\lambda = 633nm$)

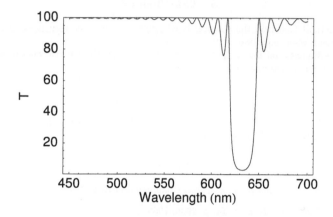

Figure 2: Transmission spectra as a function o as function of $\lambda$

where $\varepsilon_1 = 1.54^2, \varepsilon_2 = 0.03 \cdot 2\sqrt{\varepsilon_1}$ and $a = \frac{0.633}{2\sqrt{\varepsilon_1}}$. The relative permitivity of the outside medium is $\varepsilon_c = 1.54$. In this case, equation 6 has been solved by different authors. In particular, Kogelnik [2] solved it using coupled wave theory also used recently by Morozov [4], and rigorous methods [3] have been also applied. In fact, exact solution for equation 6 can be obtained as a linear combination of Mathieu functions. Figure 1 shows the local power $|E|^2$ (for a wavelength of $\lambda = 633nm$) obtained using the **FPIM** method, the local power $|E(0)|^2 > |E(10\mu m)|^2$ due to the multiple reflections inside the periodic dielectric material, with the result that transmission for this wavelength the transmission is near to 0 % as can be seen in figure 2 . As can be seen in figure 3, the maximum error defined as the difference between the exact Mathieu solution and the obtained from **FPIM** method is lower that 4 %.

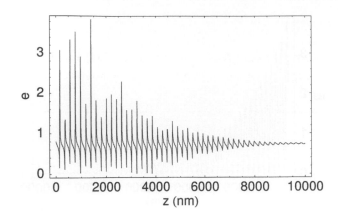

Figure 3: Error $e(\%)$ for $|E|^2$ as a function of z position ($\lambda = 633nm$)

## 3   Conclusions

An iterative method based on the classical fixed point theorem has been used to solve the one-dimensional linear wave equation in the frecuency domain.  In particular, we have applyed it to concrete one-dimensional linear photonic crystals to obtain the electric field profile inside the crystal being the numerical results in agreement with the exact Mathieu solutions.

## Acknowledgment

This work was financially supported by the Comision Interministerial de Ciencia y Tecnología (CICYT) of Spain (projects MAT2004-04643-C03-03)

## References

[1] G. Jameson, *Topology and normed spaces.* Chapman and Hall, 1974.

[2] H. Kogelnik *Bell. Syst. Tech. J.* **48**, p. 2909, 1969.

[3] J. M. Jarem and P. Banerjee , *Computational methods for electromagnetic and optics systems.* Marcell Dekker , 2000.

[4] G. V Morozov and D. Sprung and J. Martorell *Phys. Review E* **69**, p. 016612, 2004.

Brill Academic Publishers
P.O. Box 9000, 2300 PA Leiden
The Netherlands

*Lecture Series on Computer
and Computational Sciences*
Volume 4, 2005, pp. 463-466

# A Movement Tracking Management Model with Kalman Filtering, Global Optimization Techniques and Mahalanobis Distance

Raquel R. Pinho[1], João Manuel R. S. Tavares and Miguel V. Correia

Universidade do Porto, Faculdade de Engenharia
Rua Dr. Roberto Frias, s/n, 4200-465 Porto
Portugal

Received 11 July, 2005; accepted in revised form 8 August, 2005

*Abstract:* In this paper we address the problem of tracking feature points along image sequences. To analyze the undergoing movement we use a common approach based on Kalman filtering which performs the estimation and correction of the feature point's movement in every image frame. The criterion proposed to establish correspondences, between the group of estimates in each image and the new data to include, minimizes the global matching cost based on the Mahalanobis distance. In this paper, along with the movement tracking, we use a management model which is able to deal with the occlusion and appearance of feature points and allows objects tracking in long sequences. We also present some experimental results obtained that validate our approach.

*Keywords:* Kalman filter, Mahalanobis distance, Movement Analysis, Features Tracking, Image Analysis.

*Mathematics Subject Classification:* 62H35, 91B70.

## 1. Introduction

Features' tracking along image sequences is a complex problem whose automatic detection and execution evolved dramatically in the past decade. The applications of movement tracking can almost be extended indefinitely. For instance, in the case of human movement analysis it can be used in medical diagnosis, physical therapy or sports. Tracking may also be used in other areas such as surveillance systems, weather analysis, object deformation, traffic monitorization, etc [1-3].

Many possible applications require tracking of several objects simultaneously, and involve complex problems as their appearance and disappearance of the scene. The complexity of the tracked objects and the interactions of the various parameters involved have stimulated new technologies such as high-speed cameras and computerized movement analysis laboratories, which have allowed new insights into tracking [1]. Automated movement observation systems can often provide very significant advantages. In particular, events are recorded more reliably because the computer algorithm always works in the same way, and the system does not suffer from observer fatigue or drift, so observations can continue almost indefinitely.

Movement analysis with motion capture video systems and interactive modelling systems can help the analysis, diagnosis, and documentation of movements with tools that may be useful in each application area.

To track the movement we used a widespread tracking technique based on a stochastic method, the Kalman filter [4, 5], but combined it with optimization techniques in order to increment robustness to occlusion and non-linear movement, and also to assure the best global matching in each time instance. Within this approach we integrate a management model which can deal with the appearance, disappearance and occlusion of tracked features, especially useful in long image sequences. The Kalman filter estimates a process by using a form of feedback control: the filter estimates the process state at each time instance and then obtains feedback in the form of (noisy) measurements. The drawbacks of this method are related to its relatively high restrictive assumptions [5]. In our previous

---

[1] Corresponding author. E-mail: rpinho@fe.up.pt

works, we have used optimization techniques and Mahalanobis distance to reduce some of the effects of the Kalman's filter when those restrictions are not satisfied, which is the case of many tracking applications (see for example [6]).

This paper is organized as follows. In the next section a brief introduction is made to the Kalman filter. In section 3, we describe our solution to the correspondence problem by using optimization techniques with Mahalanobis distance. In section 4 we explain how the management model deals with the tracked features, as well as their appearance and disappearance from the image scene. Then some experimental results are shown on synthetic and real movement image sequences. In the last section some conclusions will be held and perspectives on future work will be given.

## 2. The *Kalman* Filter

The Kalman filter is a widespread optimal recursive Bayesian stochastic method [4]. It provides "optimal" estimates that minimize the mean of squared error. In this work, the system state in each time step is the position, velocity and acceleration of each tracked feature, and new measurements are incorporated whenever a new image frame is considered.

The equations for the Kalman filter fall into two groups: time update equations and measurement update equations. The time update equations are responsible for projecting forward the current state and error covariance estimates to obtain the a priori estimates for the next time step. The measurement update equations are responsible for the feedback [4]. One of the drawbacks of the Kalman filter is the restrictive assumption of Gaussian posterior density functions at every time step and many tracking problems involve non-linear movement (human gait is just an example). To solve this problem, several techniques have been proposed such as the Extended Kalman Filter, the Unscented Kalman Filter or other stochastic methods, for example particle filters [5], but their computational expense may be questioned.

## 3. Correspondence with Optimization and Mahalanobis Distance

To introduce new measurement data in each step of the Kalman filter a criterion of correspondence between the set of measurements and the given estimates must be used, so that for each features position estimate there may exist at most one new measurement to match.

By Kalman's default approach the search area for each feature position in the image plane is an ellipse. If the filter converges, better estimates will be given and the search areas will successively decrease to a minimum value, so the computational cost due to the points correspondence decreases [7, 8]. However, this default approach may present some problems: there may not be any point in the search area, or there might be several points instead. And, even if there is only one correspondence for each point, there is no guarantee that the best set of correspondences has been achieved.

To overcome those ambiguities, optimization techniques may be used to obtain the best global set of correspondences between the previous predictions and the actual set of measurements [6]; the cost of each correspondence can be given by the normalized Mahalanobis distance [6]. The Mahalanobis distance between two points is scaled by the statistical variation in each component of the point. The Mahalanobis distance values will be inversely proportional to the quality of the prediction/measurement correspondence, so to optimize correspondences we minimize the associated cost function.

## 4. Tracked Features Management Model

During tracking new features can rise at any time of the sequence, but they may also disappear for some instances or even definitively. So, in real sequences one has to decide if it still is necessary to keep on tracking a feature because it might be occluded, or if it has disappeared definitively from the scene. This decision is of greater importance if many features are being tracked, if the image sequence is long, if real-time results are needed, etc. To manage the tracking of features we use a management model which associates a confidence value to each tracked feature in a frame [8]. In each tracking instance, all the features that are visible will have their confidence value increased one level until a maximum value and decreased if not. Whenever a tracked feature reaches a determined minimum confidence value, it will be as if the feature definitively disappears and so its tracking will cease. In the future, if the same feature reappears it will be considered as a new feature to track.

In the results presented in this paper, all confidence values are integers between 0 (zero) and 5 (five), and by default all features are initialized with a confidence value of 3 (three).

## 5. Experimental Results

In this paper we will exemplify the above approach on some movement sequences. For the first example, figure 1, consider a synthetic sequence of five frames which initially contains three blobs to follow (B, C and E), in the second frame two new blobs appear (A and E) and in the next frames the tracked blobs appear and disappear randomly. In this example we can notice that while a point is visible its search area decreases, but when it disappears then its uncertainty increases. The confidence values associated to the tracked features using the management model are displayed in table 1.

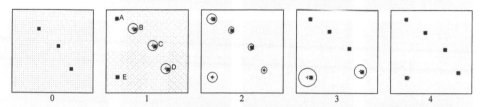

Figure 1: Tracking blobs A-E in a 5 frame image sequence: 0 - original 1$^{st}$. frame; 1..4 – last 4 frames with Kalman's search areas represented by ellipses (in this case by circles as the radius are equal), predicted positions marked by +, and positions after the correction phase indicated by x.

Table 1. Confidence value of each tracked feature with the used management model.

| "Blobs" \ Steps | 0 | 1 | 2 | 3 | 4 |
|---|---|---|---|---|---|
| A | - | 3 | 4 | 5 | 5 |
| B | 3 | 4 | 5 | 5 | 5 |
| C | 3 | 4 | 5 | 5 | 5 |
| D | 3 | 4 | 3 | 4 | 5 |
| E | - | 3 | 2 | 3 | 4 |

For the next example consider the tracking of people in a surveillance system of a shopping centre, figure 2 (images obtained from [9]). In the first height frames three people are tracked. In the 10$^{th}$ frame another person starts to be followed. In the 12$^{th}$ frame one of the previously tracked persons starts to enter a store and so the proposed management model will cease tracking him in the 16$^{th}$ frame. If he reapers along the next five frames, the management model would maintain the previous tracking; on the other hand, if he comes out after a longer period of time our approach would initialize his tracking as a new feature.

## 6. Conclusions and Future Work

We have addressed the problem of tracking feature points along image sequences to analyse their movement. To do so, we used the Kalman filter which is able to predict and correct the tracked features position, as well as their velocity and acceleration. To incorporate the data captured in each image frame we used optimization techniques and Mahalanobis distance. This approach allows the incorporation of measured data even if it would be out of the default Kalman search area, which happens for example in the case of a movement abrupt change [6], and presents a global matching criterion.

In this paper we use a tracked features management model which associates to each feature state a confidence value that may distinguish occlusion and disappearance cases. By doing so, continuous image sequences may be considered with real-time results as this approach reduces the number of simultaneously tracked features.

This work should be continued by comparing the obtained results to those given by other stochastic methods used for tracking such as the unscented Kalman filter, or particle filters. In the case of human gait analysis, this work can be considered in the tracking of the used anatomical features.

## Acknowledgments

This work was partially done in the scope of the project "Segmentation, Tracking and Motion Analysis of Deformable (2D/3D) Objects using Physical Principles", reference POSC/EEA-SRI/55386/2004, financially supported by FCT - Fundação para a Ciência e a Tecnologia from Portugal.

The first author would like to thank the support of the PhD grant SFRH / BD / 12834 / 2003 of the FCT.

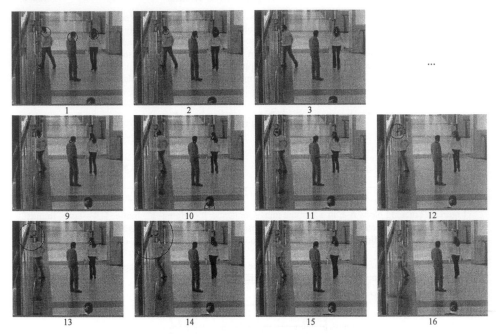

Figure 4: Tracking persons in a shopping centre: the used management model allows the tracking of features during long image sequences.

## References

[1]   A. Azarbayejani, C. Wren, A. Pentland, Real-Time 3D Tracking of the Human Body, *IMAGE'COM 96*, Bordeaux, France, (1996).

[2]   J. Aggarwal, Q. Cai, Human Motion Analysis: A Review, *Computer Vision and Image Understanding*, **73** 428/440(1999).

[3]   O. Masoud, N. Papanikolopoulos, Recognizing Human Activities, *Proceedings IEEE International Conference on Advanced Video and Signal Based Surveillance AVSS2003*, Miami, 157/162(2003).

[4]   G. Welch, G. Bishop, An Introduction to Kalman Filter, T. Report, Ed.: University of North Carolina at Chapel Hill (1995).

[5]   M. Arulampalam, S. Maskell, N. Gordon, T. Clapp, A Tutorial on Particle Filters for Online Nonlinear/Non-Gaussian Bayesian Tracking, *IEEE Transactions on Signal Processing*, **50** 174/188(2002).

[6]   R. Pinho, J. Tavares, M. Correia, Human Movement Tracking and Analysis with Kalman Filtering and Global Optimization Techniques, *II International Conference On Computational Bioengineering*, Lisbon, Portugal, (2005).

[7]   M. Correia, A. Campilho, A. Padilha, Motion Estimation for Traffic Analysis, *7° Congresso Português de Reconhecimento de Padrões*, Portugal, 831/838(1995).

[8]   J. Tavares, A. Padilha, Matching Lines in Image Sequences with Geometric Constraints, *7° Congresso Português de Reconhecimento de Padrões*, Portugal, (1995).

[9]   EC-Funded-CAVIAR-project, I. 2001-37540, http://homepages.inf.ed.ac.uk/rbf/CAVIAR/ OneShopOneWait1cor.mpg, (2004).

Brill Academic Publishers
P.O. Box 9000, 2300 PA Leiden
The Netherlands

*Lecture Series on Computer
and Computational Sciences*
Volume 4, 2005, pp. 467-470

# Application of Nonlinear Dynamics in WLANs

G. Plitsis

Aachen University of Technology,
Germany

Received 1 July, 2005; accepted in revised form 8 August, 2005

*Abstract:* Chaos engineering for wireless communications is an emerging field of research. Different modulation schemes based on chaos have been proposed such as Chaos Shift Keying (CSK) and Differential Chaos Shift Keying (DCSK). There are certain advantages that nonlinear dynamics could offer to conventional wireless access systems such as increased security and encryption, low-power communication, and hardware simplicity by the application of deterministic chaos in conventional communication systems. Chaos could also be used in broadband wireless access systems offering high encryption and abundance of digital codes by simply changing the initial conditions of the chaotic map. In multipath environments, chaotic modulation schemes could offer a solution to cope with noise.

*Keywords:* Nonlinear dynamics; wireless communications; WLANs; chaos.

## 1. Introduction

Chaos-based mobile communications is an emerging field of research that has as its aim to integrate chaos-based modulation schemes into conventional ones. It tries to take advantage of chaotic oscillations for spread spectrum communications. The idea of using chaotic signals for digital communications has been triggered and inspired by L.M. Pecora (Pecora and Carroll 1990) that observed that two identical chaotic electronic circuits that start with different initial conditions can synchronize. Different modulation schemes based on chaos have been proposed during the last years. Some of them such as the Differential Chaos Shift Keying (DCSK) do not use the property of synchronization and transmit reference as well as modulated signals. Others such as Chaos Shift Keying rely on the property of synchronization in order to transmit information but are susceptible to channel noise.

The IEEE 802 committee has established different standards for Local Area Networks (LAN), Wireless Local Area Networks (WLAN), and Wireless Personal Area Networks (WPAN). The 802.11b specification for WLANs affects the physical layer with high data bit rates. It defines the infrastructure-based mode and the ad hoc mode. The physical layers defined include two spread spectrum based layers and a diffuse infrared specification.

## 2. Background of chaos engineering

As an example of synchronization we consider a system driven by a chaotic signal. The Lorenz system is given by,

$$x = \sigma \cdot (y - x)$$
$$\dot{y} = -xz + rx - y \qquad \qquad (1)$$
$$\dot{z} = xy - by$$

The driven system is specified by,

$$\dot{y}' = -xz' + rx - y'$$
$$\dot{z}' = xy' - bz' \qquad \qquad (2)$$

which is a coupled system with the Lorenz being a five dimensional system specified by $x$, $y$, $z$, $y'$, and $z'$.

Civilizations of ancient times were trying to produce structure from chaos. They were trying to explain phenomena that seemed to be governed by no rule, which is to be of chaotic nature.

*"The very first of all, chaos came into being."* Hesiod,
Theogeny 116

Centuries later, Poincaré (1854-1912) has described long before what has later been identified as chaotic behavior,

*"...quite negligible reason, escaping our attention due to its smallness, can cause a considerable effect, which we cannot foresee, and such a case we say that the phenomenon is a result of chance".*

The 1970s have been the years that chaos research has officially started. The physicist Feigenbaum has proceeded in a significant discovery concerning chaos and nonlinear dynamics. By studying different nonlinear systems, he produced periodic oscillations as well as chaos. The problem of controlling chaos has attracted attention from the early 1990's. Multiple applications have aroused such as in ecology, biology, mechanical engineering, music, plasma and lasers, space engineering, earthquakes, medicine, etc.

Some decades ago, scientists were trying to eliminate the chaotic disturbances to produce structured communications. One of the most important problems in chaotic phenomena has been the synchronization between different systems. After it has been found that chaotic systems can synchronize, chaotic behavior has received much attention. There can be different synchronization methods such as synchronization by linear feedback, synchronization of the inverse system, synchronization by decomposition into subsystem, etc.

We consider as an example the Chua's circuit [5] that shows chaotic behavior and is described by,

$$C_1 \frac{du_1}{dt} = \frac{1}{R} \cdot (u_2 - u_1) - g(u_1)$$

$$C_2 \frac{du_2}{dt} = \frac{1}{R} \cdot (u_2 - u_1) + i_L \tag{3}$$

$$L \frac{di_L}{dt} = -u_2$$

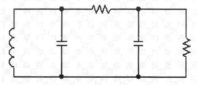

Figure 1: Chua's circuit.

## 3. Nonlinear dynamics and WLANs

### 3.1. A simplified CSK-based wireless system

The simplified simulation model [2], [3] consists of a Base Station (BS) as well as Mobile Terminals (MTs) that have chaos generators. Data is broadcast to MTs and once the MTs are synchronized, they can communicate with each other and with the BS. For synchronization purposes, VCOs and PLLs are used. The channel is modeled as AWGN.

The chaos generator can use different maps for different users such as cubic map, skew tent map, logistic map, and bernoulli-shift map given by,

$$x_{k+1} = 4x_k^3 - 3x_k \tag{4}$$

$$x_{k+1} = \begin{cases} \frac{x_k}{0.6} & 0 \le x_k \le 0.6 \\ \frac{1 - x_k}{0.4} & 0.6 \le x_k \le 1 \end{cases} \tag{5}$$

$$x_{k+1} = 1 - 2x_k^2 \tag{6}$$

$$x_{k+1} = \begin{cases} 1.2x_k + 1 & x_k < 0 \\ 1.2x_k - 1 & x_k > 0 \end{cases} \tag{7}$$

The model is simplified as no special scramblers and interleavers were used.

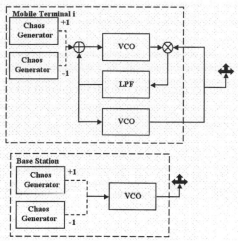

Figure 2: The simulation model with chaos generators.

The results that appear in the following graphs present the spectrum provided by link level simulations. As far as the advantages and disadvantages of a system based on the link level simulator with chaos generators is concerned, its hardware implementation is easy as well as cheap and it provides high encryption but the synchronization of CSK is still an open question and the Bit Error Ratio (BER) of such a system is inferior to that of the conventional. On the other hand, DCSK provides easy synchronization but with a lower data bit-rate when compared to CSK.

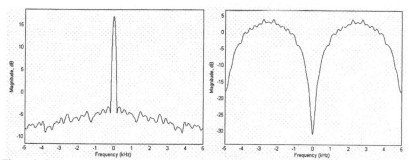

Figure 3: Spectrum when skew tent map / Bernoulli-shift map and AWGN channel is applied.

### 3.2. WiFi and control of jamming effects with the use of deterministic chaos

Wireless communication systems [1] are subject to different forms of distortions and denial of service attacks. Jamming is one common PHY/MAC layer attack and be either malicious with an aim to prevent communication in environments such as the military or unintentional because for instance of the coexistence of different wireless devices functioning in different frequency bands and with different protocols.

An adversary could brake down a communication link with low energy cost especially in the case of omni-directional communication. Non-malicious jamming can occur in both military and commercial frequencies, for instance the 2.4 GHz Industrial, Scientific and Medical band is full of non-interoperable standards such as IEEE 802.11 protocols, Bluetooth, and HiperLAN II.

The strength of the jamming signal can be measured through the jamming-to-signal ration given by,

$$\frac{J}{S} = \frac{P_j G_{jr} G_{rj} R_{tr}^2 L_r B_r}{P_t G_{tr} G_{rt} R_{jr}^2 L_j B_j} \tag{8}$$

where $P_j$ the jammer power, $G_{xy}$ the different gains, $R_{xy}$ the different distances, $L_r$ the communication signal loss, $B_r$ the receiver bandwidth, $P_t$ the transmitter power, $L_j$ the jammer signal loss, and $B_j$ the bandwidth of the jammer of the transmitter.

One of the usual techniques being used to prevent jamming is the spread-spectrum technique that forces the jammer to use more power than the sender and one common type is the Direct Sequence Spread Spectrum (DS-SS) technique. The results for a reference narrowband sinusoidal jammer and IEEE 802.11b are being presented in the following figures. It seems that the BER is being improved when chaos is being applied in the system.

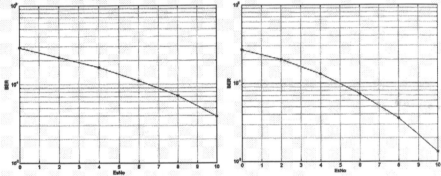

Figure 4: BER of IEEE 802.11b for 1Mbps DBPSK under narrowband sinusoidal jammer and after the replacement of the Barker sequence by a chaotic one (Ikeda map).

## 4. Conclusions

This paper described the application of chaos engineering in broadband wireless access systems and specifically in IEEE 802.11b and how it can influence the narrowband sinusoidal jammer. Wireless LAN security issues have been described. Results based on link level simulations have been provided. Finally, a simplified CSK-based wireless access system in a form of a link-level simulator was investigated.

## References

[1] G. Plitsis, "Chaos Engineering for IEEE 802.11b Security Improvement", *The 13th International Workshop on Nonlinear Dynamics of Electronic Systems (NDES'05)*, Potsdam, Germany, Sept. 2005.

[2] G. Plitsis, "Chaos Shift Keying and IEEE 802.11a", *The 10th IFIP International Conference on Personal Wireless Communications,* Colmar, France, Aug. 2005.

[3] A. Dmitriev, B. Kyarginsky, A. Panas, and S. Starkov, "Direct Chaotic Communication System Experiments," *Proc. of NDES'01*, Netherlands, 2001 157-160.

[4] T. Yang and L.O. Chua, "Chaotic digital code-division multiple access (CDMA) communication systems," *Int. J. Bifurcation and Chaos,* vol. 7, no. 12, pp. 2789–2805, Dec. 1997.

[5] H. Dedieu, M. P. Kennedy, and M. Hasler, "Chaos shift keying: Modulation and demodulation of a chaotic carrier using self-synchronizing Chua's circuit," *IEEE Transactions on Circuits and Systems I,* pp. 634- 642, 1993.

Brill Academic Publishers
P.O. Box 9000, 2300 PA Leiden
The Netherlands

*Lecture Series on Computer
and Computational Sciences*
Volume 4, 2005, pp. 471-474

# Chaos Engineering against Conflicts between WiFi and Bluetooth Coexisting Devices

G. Plitsis

Aachen University of Technology,
Germany

Received 1 July, 2005; accepted in revised form 8 August, 2005

*Abstract:* A common problem caused because of different simultaneously operating wireless systems is the problem of the interference caused by this coexistence. The physical layer of Bluetooth (IEEE 802.15.1) uses Frequency Hopping Spread Spectrum (FHSS) and that of Wi-Fi (IEEE 802.11b) uses Direct Sequence Spread Spectrum (DSSS). They both operate in the industrial, scientific, and medical (ISM) band of 2.4 GHz and thus can cause considerable interference to each another. Chaos engineering is an emerging field of research in wireless communications aiming to offer increased security, hardware simplicity of transmitters, or greater tolerance in multipath or mixed environments. In the case of coexistence of different wireless RF systems it could offer greater tolerance to the interference caused. This paper proposes ways of exploiting chaos to overcome the interference caused by the IEEE 802.11b and Bluetooth coexistence. The results are based on link-level simulations.

*Keywords:* Chaos engineering; Bluetooth; WiFi; IEEE 802.15.1; IEEE 802.11b; coexistence.

## 1. Introduction

The idea of using chaotic signals for digital communications has been triggered and inspired by L.M. Pecora that observed that two identical chaotic electronic circuits that start with different initial conditions can synchronize. Communication systems enable us to transmit / disseminate information. In their basic form include an information source, a transmitter, a channel, and a receiver. Their aim is to produce a structured sequence of data that can be transmitted over a channel and survive the chaotic phenomena existing there. As an example of synchronization we consider a system driven by a chaotic signal. The Lorenz system [2] is given by,

$$x = \sigma \cdot (y - x)$$
$$\dot{y} = -xz + rx - y \tag{1}$$
$$\dot{z} = xy - by$$

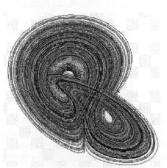

Figure 1: Lorenz attractor.

## 2. WiFi (IEEE 802.11b) and Bluetooth (IEEE 802.15.1) coexistence

The IEEE 802 committee has established different standards for Local Area Networks (LAN), Wireless Local Area Networks (WLAN), and Wireless Personal Area Networks (WPAN). IEEE 802.11b [1], [4] defines the infrastructure-based mode and the ad hoc mode. The physical layers defined include two spread spectrum based layers and a diffuse infrared specification. Direct Sequence Spread Spectrum (DSSS) is usually employed with data rates of 1 Mbps and 2 Mbps. IEEE 802.15.1 [3] protocol for WPANs or else Bluetooth has been standardized by IEEE with an aim to offer wireless broadband access in an area of up to 10m. It uses Frequency Hopping Spread Spectrum (FHSS) and functions in the same band (2.4 GHz ISM) that IEEE 802.11b and thus cause interference to one another.

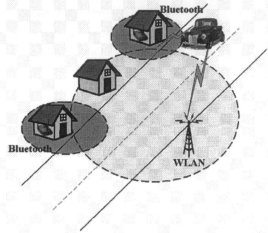

Figure 2: Mobile user receiving WLAN data as well as interference because of the WLAN-Bluetooth coexistence.

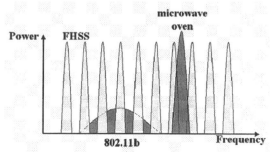

Figure 3: Frequency Hopping Spread Spectrum (FSSS) power against other coexisting systems.

A WiFi device occupies about a quarter of the 83.5 MHz bandwidth available in the ISM band and the devices share the channel in a Time Division Multiple Access (TDMA) basis whereas Bluetooth devices do not have a channel fixed in a frequency as frequency hopping is being used. They hop in a pseudorandom manner among 79 channels at 1 MHz distances. Bluetooth hops to a new channel at 2.4 GHz every 1,600 times a second. The power level of 802.11b is about 100 mW and at this rate can support data rate of about 11 Mbps at a maximum distance of about 100m. In contrast, Bluetooth supports lower data rates of about 1 Mbps with power level of about 1 mW. As Bluetooth devices occupy the entire band, it is usual that also Bluetooth devices collide with each other, not just WiFi and Bluetooth devices. Bluetooth devices degrade substantially the WiFi performance when they are close to the WiFi stations. When the Bluetooth devices are moved 10m away, then this phenomenon almost vanishes. This is so as the power of the signal is stronger when the devices are close to each other and thus the interference is stronger. Other devices / applications functioning in the same environment could also cause interference, devices such as microwave ovens and WLAN devices of other standards (e.g. HiperLAN/2).

## 3. Results

The results present the case of IEEE 802.11b functioning terminals under Bluetooth interference and the improvement in performance when Ikeda map is used for spreading instead of the Barker sequence.

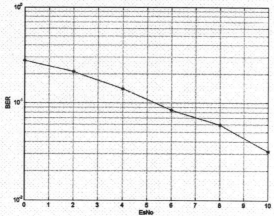

Figure 4: BER of IEEE 802.11b for 1Mbps DBPSK under strong Bluetooth interference.

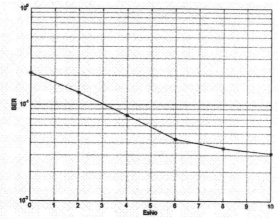

Figure 5: BER of IEEE 802.11b for 1Mbps DBPSK under strong Bluetooth interference when Ikeda map is used for spreading instead of Barker.

## 4. Conclusions

Coexistence is a common interference problem in wireless communications. The coexistence of Bluetooth for WPANs and Wi-Fi for WLANs has been described. The way chaos engineering could improve the performance of such a system has been briefly described. There are many other benefits of such an application such as high encryption, lower power consumption, etc. for wireless communications.

## References

[1] IEEE, "Part 11: Wireless LAN Medium Access Control (MAC) and Physical Layer (PHY) specifications: Higher-Speed Physical Layer Extension in the 2.4 GHz Band", 1999.

[2] O. E. Rossler, "An equation for continuous chaos," *Phys. Lett.*, A57, 397, 1976.

[3] Bluetooth Special Interest Group, "Specifications of the Bluetooth System, vol. 1, v.1.0B 'Core' and vol. 2 v1.0B 'Profiles", 1999.

[4] IEEE Std. 802-11, "IEEE Standard for Wireless LAN Medium Access Control (MAC) and Physical Layer (PHY) Specification ", 1997.

[5] C. Chiasserini and R. Rao, " Coexistence mechanisms for interference mitigation between IEEE 802.11 WLANs and bluetooth ," in *Proceedings of INFOCOM 2002*, pp. 590–598, 2002.

[6] N. Golmie, "Bluetooth Adaptive Frequency Hopping and Scheduling," in *Proceedings of MILCOM'03*, Boston, MA, 2003.

[7] F.C.M. Lau and C.K. Tse, "Coexistence of chaos-based communication systems and conventional spread-spectrum communication systems," *Int. Symp. Nonl. Theory & Appl.*, Xian, China, 2002.

Brill Academic Publishers
P.O. Box 9000, 2300 PA Leiden,
The Netherlands

*Lecture Series on Computer
and Computational Sciences*
Volume 4, 2005, pp. 475-478

# Formalization of a Biological Domain with Beta Binders

**Katerina Pokozy-Korenblat, Federica Ciocchetta, and Corrado Priami**

Department of Informatics and Telecommunications,
University of Trento,
Povo, Italy

Received 10 July, 2005; accepted in revised form 4 August, 2005

*Keywords:* biology, modelling, SBML, formal methods, process algebras

## 1 Introduction

The aim of this work is to reserve for biologists the whole power of formal methods analysis. Traditionally in construction of biological models semi-formal languages are used. These languages are characterized by a strong-defined syntax and a very indistinct semantics. When the model is specified in such a way, its interpretation depends on the intuition of concrete specialist, or of the intuition that is implicitly supposed in a used simulation tool. Since semi-formal specification is not all-sufficient, the result of its analysis is biased and based on nonformalized knowledge. For using formal methods we have to fix semantics of any syntactic construction that can be used for description of the system. In this case an analysis of the system would be objective. On the other hand side, explicitly defined semantics, with which a user can compare his own intuition, helps to understand better a described biological phenomenon.

We imagine that no existing formal languages can be really used by biologists directly. The first reason of this fact is the level of abstraction of formal languages. Usually a formal language have a very limited set of primitives. It means that to specify formally some phenomenon intuitive understandable for biologists we have to use a complex construction in a formal language. This fact decreases readability of formal specifications and a biological intuition of the formal language. The second reason is the fact that biological formal languages appear as adaptation to biology of one of existing formal languages originally created for description of computer systems. Surely, in such adaptation not all biological phenomena could be expressed in a simple and natural form.

Thus, to use formal methods effectively in biology we need a readable and intuitive understandable for biologists intermediate specification language. We chose as such a language SBML [2], the language which becomes one of widely used standard in specification of biological phenomena. Note that SBML is a semi-formal language, it only structures the biological knowledge but does not fix any semantics under its constructions. However, for each biological domain SBML specification assumes some determine intuitive semantics. Our task here is to formalize this intuition.

To apply formal methods for analysis of our SBML specification we have to make some assumptions sufficient for fixing an intuition of the specification.

We chose metabolic pathways as the biological domain which we are going to formalize. We distinguish the following types of enzymatic reactions: single substrate reactions, multiple substrate reactions, reactions with inhibitors, reactions with co-enzymes/co-factors, and simplified reactions. This list of reactions is powerful enough to cover most of phenomena of the domain. Rarely exceptions (e.g. special cases of inhibition) are asked to be expressed in more detail in terms of complex formation/splitting. For each type of reactions we fix its semantics. The choosing semantics is a formalization of biological intuition about the corresponding phenomenon.

Note that the way of translation depends on the granularity of our model. For metabolic pathways we suppose to consider molecule level and reactant level models. The first one describes behavior of separate molecules which form complexes and are transformed to other molecules. The reactant level model describes general direction of reactions and changing of concentration of reactants. This approach is adequate for qualitative models, or models operating only with changing of concentrations of some critical reactants. When we have no enough information about atomic reactions, we talk about more abstract model in which some stable states of the system are essential.

As a target formalism we choose Beta Binders [1]. This formalism allows to express in a readable and intuitive clear form composition and splitting of molecular complexes. Furthermore, since Beta Binders formalism can support a compositional property, it could be available for specifying models composing from different domains elements.

In this paper we present two ways of formalization of the chosen biological domain which are sufficient to apply to an arbitrary metabolic network all the power of formal analysis.

## 2  SBML

SBML [2] is a XML-based format for the representation of quantitative and qualitative models of biochemical reaction networks.

An SBML model is characterized by a list of species with a distinguished subset of initially available elements and a list of reactions. A reaction in SBML model is described by fields of reactants, products, modifiers and reversibility.

## 3  Beta Binders

We consider here the restricted version of Beta Binders [1] which is sufficient for description of the chosen biological domain.

A Beta Binders specification can be represented by a set of boxes and a set of join/split functions controlling interactions of boxes. Each box is characterized by a set of binders which control a possible behavior of the box. Binders can be labelled as active or hidden (written as $<name>$ and $<name>^h$, correspondingly). The evolution of boxes is described by two macro-operations: join two boxes together, and split a box in two. This operations are defined by join/split functions whose view is very flexible depending on the task. For our purpose we will use the following schemes of join and split functions:

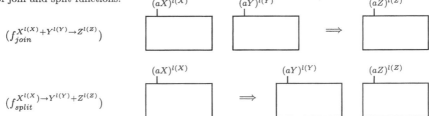

where $l() \in \{' ', 'h'\}$ is a label indicating an active or a hidden state of the corresponding binder.

## 4  Molecule Level Translation

We suppose that our biological model is defined by a set of intersected pathways and a set of initially available species. For any enzyme we suppose its amount to be sufficient for execution of the corresponding reaction. Then we can present all enzymes as initially available species. In this section we talk about molecule level model, and any reaction is considered as interactions of molecules.

The Beta Binders formalism allows to express directly formation and splitting of molecular complexes. In our translation we represent each enzymatic reaction as formation of some intermediate complexes and modifying its reactants inside these complexes.

The Beta Binders specification of an SBML model is constructed from boxes corresponding to each molecule of initially available species and a set of join/split functions describing all possible reactions. A molecule of an available species $S$ is represented as a box with the unique binder $aS$. Each reaction is represented by interaction of several boxes. During this interaction some of the boxes are transformed to new ones corresponding to the products of the reaction. Bellow we use the following name convention: the name of complex composed of two elements $A$ and $B$ (species or their complexes) is denoted by $AB$.

Now we describe schemes of complex formation/splitting corresponding to the distinguished types of reactions. Enzymes are protein molecules that are tailored to recognize and bind specific reactants and catalyze their chemical transformation to products. So each enzymatic reaction can be considered as forming of enzyme+substrate complex and further transformation of the reactant to a product. Note that the intermediate enzyme+substrate complex also can be splitted into the original reactant and enzyme. Thus, a **single substrate reaction** $r : S \xrightarrow{E} P$ is represented by the following set of join/split functions is $F^{j/s}(r) = \{f_{join}^{S+E \to SE},\ f_{split}^{SE \to S+E},\ f_{split}^{SE \to P+E}\}$.

Enzymatic reactions with two substrates usually involve transfer of an atom or a functional group from one substrate to the other. The schemes of two-substrate reactions can be different. In some cases, both substrates are bound to the enzyme at some point in the course of the reaction, forming a ternary complex. In other cases, the first substrate is converted to product and dissociates before the second substrate binds, so no ternary complex is formed. However, if we have no information about the concrete mechanism of intermediate complexes formation, we choose some arbitrary scheme and label all intermediate complexes as hidden to constraint their using only for this reaction. The external effect of the reaction would be the same independently of our choice. In our translation we present **two substrate reaction** in the following way: the first substrate form a complex with enzyme, and then the second substrate is bounded with this complex. In such a case the set of join/split functions of the reaction $r : S_1 + S_2 \xrightarrow{E} P_1 + P_2$ is $F^{j/s}(r) = \{f_{join}^{S_1+E \to S_1E^h},\ f_{split}^{S_1E^h \to S_1+E},\ f_{join}^{S_1E^h+S_2 \to S_1S_2E^h},\ f_{split}^{S_1S_2E^h \to S_1E^h+S_2},\ f_{split}^{S_1S_2E^h \to S_1E^h+P_2},\ f_{split}^{S_1E^h \to P_1+E}\}$.

Similar to two-substrate reactions, we consider any reaction with more then two income or outcome elements (including enzymes) as an abstract representation of a sequence of more simple reactions. A reaction without enzymes is considered as enzymatic reaction with unknown enzyme.

A competitive inhibitor completes with the substrate for the active site of an enzyme. While the inhibitor occupies the active site it prevents binding of the substrate to the enzyme. So we can obtain enzyme+substrate complex, or enzyme+inhibitor complex, alternatively. The set of join/split functions of the **reaction with competitive inhibition** $r : S \xrightarrow{E,I} P$ would be $F^{j/s}(r) = \{f_{join}^{S+E \to SE},\ f_{split}^{SE \to S+E},\ f_{join}^{I+E \to IE},\ f_{split}^{IE \to I+E},\ f_{split}^{SE \to P+E}\}$.

Noncompetitive inhibitors bind at a site distinct from the substrate active site, and unlike a competitive inhibitor, binds only to the enzyme+substrate complex. The set of join/split functions of the **reaction with noncompetitive inhibition** $r : S \xrightarrow{E,I} P$ would be $F^{j/s}(r) = \{f_{join}^{S+E \to SE},\ f_{split}^{SE \to S+E},\ f_{join}^{I+SE \to ISE},\ f_{split}^{ISE \to I+SE},\ f_{split}^{SE \to P+E}\}$.

Some enzymes require an additional chemical components (co-factors or co-enzymes) for their activity. A co-enzyme is bound to the enzyme protein forming a catalytically active group that bind a substrate. So, this type of reactions is translated as multiple substrate reaction with a defined order of binding of substrates.

For reversible reaction we have to consider additionally a possible backward transformation of products of the reaction. For this purpose we add a backward operation to each splitting of an intermediate complex.

## 5 Particular features of reactant level translation

In this section we consider a specifics of formalization of reactant level models. Such type of formalization is important, for example, when we analyze dependencies between different parts of the model.

In this approach we correspond unique box to each initially available species. The main difference from the molecular approach is that we are interested only in the main direction of a reaction but not in all possible transformations of each molecule. So we does not use backward splitting of intermediate complexes and back direction of reversible reactions. Note that an information about presence of all reactants of the translated complex reaction have to be available already on a step of formation of the first intermediate complex. To solve this problem we add conditions on the presence of all reactant of the complex reaction to join functions corresponding to forming of intermediate complexes.

The type of inhibition is not essential on a reactant level translation because we are interested only in a proportional part of bound/active enzyme in the output of reaction but not in intermediate complexes produced during the reaction. So, we can consider inhibition as producing of bound enzyme by binding an active one with the inhibitor. The reactant level inhibition is formalized as one-directional transformation of $E$ to $E^h$ or, alternatively, of $S$ to $P$ through a hidden intermediate $SE$ complex.

The approach was approbated on a number of models from KEGG database [3]. One of considered models is a set of metabolic pathways related to the Citrate Cycle which is a typical example in Metabolic Pathways domain. It demonstrates the simplicity of composing of pathways and usability for the analysis of correlation between species belonging to different pathways.

## 6 Conclusion

In this paper the problem of formal specification of biological phenomena was considered. The presented formalization of metabolic pathways domain is the first step toward specification and formal analysis of multi-domain biological models.

The proposed method of formalization of biological domains does not depend essentially on the original language of a biological specification. It can be easily adapted to a format of one of existing databases. So the proposed approach gives wide possibilities for automatic analysis of already collected biological knowledge.

Since our approach support a compositional property of specifications, complex models consisting of arbitrary amount of metabolic pathways can be analyzed using formal methods techniques.

## References

[1] C. Priami and P. Quaglia: Beta binders for biological interactions. *Proc. 2nd Int. Workshop on Computational Methods in Systems Biology*, Lecture Notes in Bioinformatics, V. 3082, p. 21-34, 2005

[2] A. Finney and M. Hucka: Systems Biology Markup Language (SBML) Level 2: Structures and Facilities for Model Definitions. http://sbml.org/documents/ 2003

[3] KEGG Database website. http://www.genome.jp/kegg/

Brill Academic Publishers
P.O. Box 9000, 2300 PA Leiden,
The Netherlands

*Lecture Series on Computer*
*and Computational Sciences*
Volume 4, 2005, pp. 479-482

# Potential Energy Surfaces For Triatomic And Tetratomic van der Waals Complexes

**R. Prosmiti[1], A. Valdés, P. Villarreal, G. Delgado-Barrio**

Instituto de Matemáticas y Física Fundamental, C.S.I.C.,
Serrano 123, 28006 Madrid, Spain

Received 4 July, 2005; accepted in revised form 14 July, 2005

*Abstract:* We report on interaction potential surfaces calculations of $He_n$–ICl (n=1 and 2) van der Waals (vdW) complexes. For constructing such surfaces *ab initio* technology is used at the coupled-cluster (CCSD(T)) level of theory. Relativistic effects are included with the use of large-core pseudopotentials for the iodine atom, while efficient augmented correlation consistent polarized basis sets are employed for the He and Cl ones, to ensure saturation in interaction energies in the highest level of electron correlation treatment. First, the interaction potential for the HeICl molecule is calculated. The CCSD(T) results predict the existence of three minima for both linear He–ICl and He–ClI and near T-shaped configurations, in agreement with the available experimental data. In turn, CCSD(T) calculations are performed for the ground $He_2ICl$ intermolecular potential. The surface is characterized by four minima corresponding to the 'police-nightstick(1)', linear He–ICl–He, 'police-nightstick(2)' and tetrahedral configurations, and the minimum energy pathways through them. It is found that results obtained by summing three-body parameterized HeICl interactions and the He–He interaction are in very good accord with the corresponding CSSD(T) configuration energies of the $He_2ICl$. Variational calculations using a sum of three-body interactions are presented to study the bound states of the vdW $He_2ICl$ complex. The binding energies $D_0$ and the corresponding vibrationally averaged structures are determined for different isomers of the clusters and their comparison with the recent experimental data is discussed.

*Keywords:* potential energy surface, electronic structure calculations, van der Waals clusters

*PACS:* 31.15.Ar, 31.50.Bc, 31.50.Df, 33.15.Fm, 33.20.Vq, 36.40.Mr

## 1 Introduction

Weak interactions and intramolecular dynamics in van der Waals (vdW) complexes, particularly of rare gas atoms with halogen and interhalogen molecules, have attracted strong attention in recent years (see recent review Ref.[1] and references therein).

Until recently, most models of vdW interactions were based on additive atom-atom forces, for describing some general features of their spectroscopy and dynamics, In general, dynamic simulations are quite sensitive to fine details in the potential, and therefore, an accurate representation of the interaction energies is very important. During last years, *ab initio* methods have progressed sufficiently and interaction potentials for triatomic vdW systems, formed by a rare-gas atom and dihalogen molecule, have been computed with high accuracy [2]. Studies of larger species are more

---

[1] Corresponding author. E-mail: rita@imaff.cfmac.csic.es

complex and the difficulty in the evaluation of their potential surfaces increases with their size. Up to now accurate potentials have been obtained by inversion of spectroscopic data [3] or through high level *ab initio* calculations [4, 5] for several triatomic vdW systems. Thus, the interactions for such clusters are available with satisfactory accuracy, which permits the testing of various models of nonadditivity for their ability to reproduce a number of experimental observations. In general, such model surfaces are useful for studying the relaxation dynamics of impurities embedded in He nanodroplets. These facts made complexes composed of two rare-gas atoms and a dihalogen molecule especially attractive targets for the study of nonadditive forces.

More recently, experimental data are available [6] on the structure and dynamics of heavier triatomic and tetratomic rare-gas dihalogen systems, and thus the purpose of the this study is to present a reliable modeling of the potential energy surfaces (PESs) reproducing the available experimental data for HeICl and He$_2$ICl complexes.

## 2  Methods, Results and Conclusions

The *ab initio* calculations are performed also using the *Gaussian 98* package [7]. Computations are carried out at the CCSD(T) levels of theory. We use Jacobi coordinates $(r, R, \theta)$ for the PES of HeICl complex, while the He$_2$–ICl system is described using the $(r, R_1, R_2, \theta_1, \theta_2, \gamma)$ coordinate system. $r$ is the bond length of ICl; $R_1, R_2$ are the intermolecular distances of each He atom from the center of mass of I–Cl, $\theta_1$ is the angle between the $\mathbf{R_1}$ and $\mathbf{r}$ vectors, while $\theta_2$ is the one between $\mathbf{R_2}$ and $\mathbf{r}$, and $\gamma$ is the angle between the $\mathbf{R_1}$ and $\mathbf{R_2}$ vectors. For I atom the Stuttgart-Dresden-Bonn (SDB) large-core energy-consistent pseudopotential in conjunction with the augmented correlation consistent triple zeta (SDB-aug-cc-pVTZ) valence basis set is employed, while for the Cl atom the aug-cc-pVTZ basis and for the He atoms the aug-cc-pV5Z are used from EMSL library [8]. For a better description of the long range interactions we use an additional set $(3s3p2d2f1g)$ of midbond functions.

The supermolecular approach is used for the determination of the intermolecular energies, $\Delta E$:

$$\Delta E = E_{\mathrm{He_nICl}} - E_{\mathrm{BSSE}} - E_{\mathrm{He_n}} - E_{\mathrm{ICl}}, \tag{1}$$

where $E_{\mathrm{He_nICl}}$, $E_{\mathrm{He_n}}$, and $E_{\mathrm{ICl}}$ are the energies of He$_n$–ICl, He$_n$ and ICl, respectively, with $n = 1, 2$. The correction, $(E_{\mathrm{BSSE}})$, for the basis-set superposition error is calculated using the standard counterpoise method [9]. Intermolecular energies are calculated for several configurations of the triatomic and tetratomic complexes fixing the ICl bondlength at its equilibrium value $r_e = 2.321$ Å.

In order to represent the PES for the He–ICl complex we used an analytical functional form to fit the CCSD(T) *ab initio* points. We used an expansion in Legendre polynomials, $P_\lambda(\cos\theta)$, to describe the two-dimensional He...ICl interaction potential,

$$V(R, \theta; r_k) = \sum_\lambda V_\lambda(R) P_\lambda(\cos\theta), \tag{2}$$

with $\lambda = 0 - 10$. The $V_\lambda(R)$ coefficients are obtained by a collocation method applying the following procedure. For each of $\theta_i)_{i=1-11}$, we fitted the CCSD(T) data to a Morse-vdW function,

$$V(R; \theta_i) = \alpha_0^i(\exp(-2\alpha_1^i(R - \alpha_2^i)) - 2\exp(-\alpha_1^i(R - \alpha_2^i))) - \frac{\alpha_3^i}{R^6} - \frac{\alpha_4^i}{R^8}, \tag{3}$$

with parameters $\alpha_0^i$, $\alpha_1^i$, $\alpha_2^i$, $\alpha_3^i$ and $\alpha_4^i$, where $i = 1 - 11$. These parameters are fitted to the *ab initio* points using a nonlinear least square technique. The He–ICl potential minima and the corresponding barrier are displayed in Figure 1a, where the minimum energy path, $V_m$, of the minimum energy for $r=2.321$ Å are plotted as a function of the angle $\theta$, together with the $n=0$

(collinear), $n=1$ (asymmetric T-shaped) and $n=2$ (antilinear) angular probability distributions for $J = 0$ vdW levels obtained from the variational bound state calculations. The calculated binding energies for He-ICl vdW cluster are [10] $D_0^L = 18.29$ cm$^{-1}$ and $D_0^T = 15.15$ cm$^{-1}$ for the collinear and near T-shaped isomers in excellent agreement with recent experimental estimates [11].

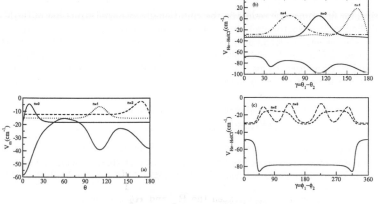

Figure 1: (a) Minimum energy path, $V_m$ in cm$^{-1}$ as a function of angle $\theta$ for HeICl. The probability $\int |\Psi|^2 \sin\theta dR$ distributions for the indicated vdW levels of HeICl are also depicted. Minimum energy path, $V_m$ in cm$^{-1}$ as a function of angle $\gamma$, (b) for planar $\gamma=(\theta_1-\theta_2)$ and (c) for no-planar with $\theta_1 = \theta_2 = 90°$, $\gamma=(\phi_1-\phi_2)$ with $\theta_1 = \theta_2 = 90°$ configurations. The probability $\int |\Psi|^2 \sin\gamma dR$ distributions for the indicated vdW levels of He$_2$ICl are also depicted.

In order to extract information on nonadditive interactions in He$_2$ICl we examine the its equilibrium structures based on the *ab initio* calculations [12]. By partitioning the interaction energy into components, we found a similar nature of binding in triatomic and tetratomic complexes of such type, and thus information on intermolecular interactions available for triatomic species might serve to study larger systems. Thus, the He$_2$–ICl potential energy function is represented as a sum of three-body HeICl interactions and the He–He one.

$$V(r_e, R_1, R_2, \theta_1, \theta_2, \gamma) = \sum_i V_{He_i ICl}(r_e, R_i, \theta_i) + V_{HeHe}(R_1, R_2, \gamma) \tag{4}$$

where the corresponding $V_{He_i ICl}(r_e, R_i, \theta_i)$ terms with $i = 1$ and 2, are the CCSD(T) parameterized potential of the HeICl complex [10] and $V_{HeHe}(R_1, R_2, \gamma)$ term is the potential function for He$_2$. One-dimensional representations of the potential are shown in Figures 1b and 1c, where minimum energy paths are plotted as a function of the angle $\gamma=\theta_1-\theta_2$ for planar (see Figure 1b) and no-planar with $\theta_1 = \theta_2 = 90°$ and $\gamma = \phi_1 - \phi_2$ (see Figure 1c) configurations. The existence of four ('police-nightstick(1)', linear, 'police-nightstick(2)' and tetrahedral) minima is established for the He$_2$ICl ground PES. To our knowledge, for first time such results on the vibrationally averaged structures of He$_2$ICl are presented. This finding may contribute to fit the rotationally resolved excitation spectrum of He$_2$Cl$_2$ or similar species, where the traditional tetrahedral structural models, based on pairwise additive potentials, have failed.

Variational bound state calculation is carried out for the above surface and vdW energy levels and eigenfunctions for $J = 0$ are evaluated for He$_2$ICl. Radial and angular distributions are calculated for the three lower vdW states. All of them are well localized in configuration space, with an exception of the broad distribution of the angle $\gamma$ for the $n = 2, 3$ states, due to the weak

He-He interaction (see Fig. 1c). The ground state corresponds to a 'police-nightstick' isomer, while the next two excited vdW levels are assigned to linear and tetrahedral ones. The binding energies and the average structures for these species are determined to be $D_0=33.51$ cm$^{-1}$ with $R_{1,2}^0=4.38$ Å, $D_0=31.60$ cm$^{-1}$ with $R_{1,2}^0=4.86$ Å, and $D_0=30.46$ cm$^{-1}$ with $R_{1,2}^0=4.38$ Å, respectively. The above values are in very good agreement with recent experimental data available by Loomis's group [6].

To our knowledge, for first time such results on the vibrationally averaged structures of He$_2$ICl are presented. This finding may contribute to fit the rotationally resolved excitation spectrum of He$_2$Cl$_2$ or similar species, where the traditional tetrahedral structural models, based on pairwise additive potentials, have failed. Such model should be applicable to a broad class of Rg$_2$XY, with Rg=rare-gas and X,Y=halogen atoms, vdW clusters. Whether the properties of the weak bonding in such systems can be predicted by the sum of atom-diatom interactions deserve further investigation.

## Acknowledgment

The authors thank to Centro de Calculo (IMAFF), CTI (CSIC), CESGA and GSC (CIEMAT) for allocation of computer time. This work has been supported by DGICYT, Spain, Grant No. FIS2004-02461. R.P. acknowledges a contract from the Comunidad Autónoma de Madrid, Spain.

## References

[1] A. Rohrbacher, N. Halberstadt and K.C. Janda, *Annu. Rev. Phys. Chem.*, **51**, 405 (2000).

[2] K. Higgins, F.-M. Tao and W. Klemperer, *J. Chem. Phys.*, **109**, 3048 (1998); R. Prosmiti, C. Cunha, P. Villarreal and G. Delgado-Barrio, *J. Chem. Phys.*, **119** 4216 (2003).

[3] J.M. Hutson, *J. Chem. Phys.*, **89**, 4550 (1988); **96**, 6752 (1992).

[4] G. Chalasinski, M.M. Szczesniak, *Chem. Rev.* **100**, 4227 (2000).

[5] R. Prosmiti, P. Villarreal, and G. Delgado-Barrio, *Chem. Phys. Lett.*, **359**, 473 (2002); R. Prosmiti, A. Valdés, P. Villarreal, and G. Delgado-Barrio, *J. Phys. Chem. A*, **108** 6065 (2004).

[6] R.A. Loomis (private communication).

[7] Gaussian 98, Revision A.7, M. J. Frisch, *et al.*, Gaussian, Inc., Pittsburgh PA, (1998).

[8] Enviromental Molecular Sciences Laboratory, *http://www.emsl.pnl.gov/*.

[9] S.F. Boys and F. Bernardi, *Mol. Phys.* **19**, 553 (1970).

[10] R. Prosmiti, C. Cunha, P. Villarreal and G. Delgado-Barrio, *J. Chem. Phys.*, **117** 7017 (2002).

[11] M.D.Bradke and R.A. Loomis *J. Chem. Phys.* **118**, 7233 (2003).

[12] R. Prosmiti, A. Valdés P. Villarreal and G. Delgado-Barrio, *J. Chem. Phys.*, submitted (2005).

Brill Academic Publishers
P.O. Box 9000, 2300 PA Leiden,
The Netherlands

*Lecture Series on Computer*
*and Computational Sciences*
Volume 4, 2005, pp. 483-486

# An Efficient Facial Animation Method
# for Mobile Agent Display

H.W. Pyun

Broadcasting Technical Research Institute,
Korean Broadcasting System,
Seoul, Korea

Received 8 July, 2005; accepted in revised form 29 July, 2005

*Abstract:* We present more efficient facial animation method for mobile agent display on mobile devices than existing methods on desktop platforms. Using a technique known as the wire curve [1], a formulation is provided to extract wire curves and deformation parameters from a facial model. We show how to make facial animation efficiently through the extracted wire curves. In addition, our proposed scheme facilitates both local deformation and non-uniform blending by making use of the power of wire deformation.

*Keywords:* real-time facial animation, multiple face models, local deformation, wire deformation.

## 1  Introduction

Recently, there is an explosion of mobile devices, such as mobile phones or PDAs(Personal Digital Assistant). Mobile agent is a conversational character running on such devices to provide personal e-mail, message and news reporting service. The implementation of the mobile agent requires various techniques including information retrieval, knowledge inference and display. Particulary, mobile agent display is motivated to support familiar interaction with users. However, it is difficult to display plausible mobile agent on the mobile devices due to its limited computational power and resource allocation.

Multiple face models, called templates, are widely used for facial animation of the conversational characters on desktop platforms.[2, 3, 4, 5, 6, 7, 8, 9, 10]. Each of these models reflects both a facial expression of different type and designer's insight to be a good guideline for animation. Those templates comprise a facial expression database from which we select appropriate face models to blend and to deform. However, a template consists of a set of vertices with few handles to control such geometric operations except the vertices themselves. It is almost impossible to keep all the vertices of multiple face models on mobile devices with limited resources. Furthermore, to achieve smooth local deformation and non-uniform blending without any tools, one would have to deal with every vertex of face models involved in these operations.

Due to its abstract representation and capability of local control on facial features such as eyes and a mouth, "wires"[1] have become a popular tool for facial animation. The basic idea of wire deformation is to locally deform the vertices of a facial feature according to the displacements of wire curves from their references together with deformation parameters (see Section 2.) Therefore, by deforming a wire curve, the features near the curve are also deformed accordingly. These features may be refined further with deformation parameters. Recently, wire deformation has been incorporated in a well-known animation software called Maya$^{TM}$ as a standard deformation tool.

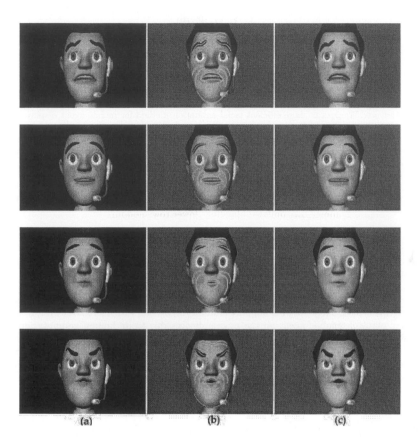

Figure 1: The original face templates and their corresponding reconstructed face templates. (a) The original face templates with sad, happy, surprised, and angry expression (b) The extracted wire curves (c) The reconstructed face templates

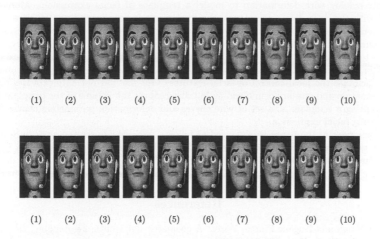

Figure 2: upper row: uniform blending. lower row: non-uniform blending.

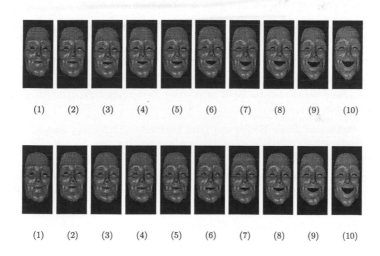

Figure 3: upper row: uniform blending. lower row: non-uniform blending.

However, with a variety of modeling tools available, we cannot expect that a designer does necessarily employ wire deformation to model a template of facial expression. Moreover, even modeled with the wire deformation tool, the final result is represented not by wire curves and deformation parameters applied to a reference face model, but by the vertices themselves displaced by those curves and parameters. In order to abstractly represent the template geometry, we present a method to extract a set of wire curves and deformation parameters from a template regardless of its history, that is, how it has been created. This can increase the usability on mobile devices and also facilitate the local control of facial features. Our method not only provides handles for local deformation and non-uniform blending but also reduces the volume of the template database, by representing a template as a set of wire curves and deformation parameters characterizing its corresponding facial expression.

The remainder of the paper is organized as follows. We provide related work in Section 2. In Section 3, we give an introduction to the wire deformation technique. We present a formulation for extracting wire curves and deformation parameters in Section 4. In Section 5, we demonstrate how our technique can be used for facial animation. Finally, we conclude this paper in Section 6.

## References

[1] Karan Singh and Eugene Fiume, Wires: A geometric deformation technique, *SIGGRAPH 98 Conference Proceedings*, 299-308(1998).

[2] Volker Blanz and Thomas Vetter, A morphable model for the synthesis of 3D faces, *SIG-GRAPH 1999 Conference Proceedings*, 187-194(1999).

[3] Ian Buck and Adam Finkelstein and Charles Jacobs and Allison Klein and David H. Salesin and Joshua Seims and Richard Szeliski and Kentaro Toyama, Performance-driven hand-drawn animation, *Symposium on Non Photorealistic Animation and Rendering*, 101-108(2000).

[4] Erika Chuang and Chris Bregler, Performance driven facial animation using blendshape interpolation, *Stanford University Computer Science Technical Report, CS-TR-2002-02*, 2002.

[5] Adele Hars, Masters of motion capture, *Computer Graphics World*, 27-34(1996).

[6] Pushkar Joshi and Wen C. Tien and Mathieu Desbrun and Frederic Pighin, Learning controls for blend shape based realistic facial animation, *Eurographics/SIGGRAPH Symposium on Computer Animation*, 2003.

[7] Frederic Pighin and Jamie Hecker and Dani Lischinski and Richard Szeliski and David H. Salesin, Synthesizing realistic facial expressions from photographs, *SIGGRAPH 98 Conference Proceedings*, 75-84(1998).

[8] Frederic Pighin and Richard Szeliski and David Salesin, Resynthesizing facial animation through 3D model-based tracking, *International Conference on Computer Vision*, 143-150(1999).

[9] Hyewon Pyun and Yejin Kim and Wonseok Chae and Hyung Woo Kang and Sung Yong Shin, An example-based approach for facial expression cloning, *Eurographics/SIGGRAPH Symposium on Computer Animation*, 2003.

[10] Qingshan Zhang and Zicheng Liu and Baining Guo and Harry Shum, Geometry-driven photorealistic facial expression synthesis, *Eurographics/SIGGRAPH Symposium on Computer Animation*, 2003.

Brill Academic Publishers
P.O. Box 9000, 2300 PA Leiden
The Netherlands

*Lecture Series on Computer
and Computational Sciences*
Volume 4, 2005, pp. 487-491

# Computational Methods in Pursuit-Evasion Problems

J. Reimann[1] and G. Vachtsevanos

Department of Electrical and Computer Engineering,
Georgia Institute of Technology
Atlanta, GA 30332 USA

Received 18 July, 2005; accepted in revised form 10 August, 2005

*Abstract:* In this paper, an approach to reduce the computational complexity of multi-player pursuit-evasion differential games is presented. The approach consists of a prediction stage, a target assignment stage, and an execution stage. Numerical complexity considerations will be discussed briefly followed by a simple multiplayer differential game example to highlight some numerical implementation issues.
*Keywords:* Multiplayer Differential Games, Level Set Methods
*Mathematics SubjectClassification:* 91A12
*PACS:* 91A12

## 1. Introduction

The problem of coordinating a swarm of multiple robots to achieve a common objective has in recent years received much attention. By allowing several robots to consolidate their resources, they are able to solve more demanding problems. However, from a computational perspective, the coordination of multiple robots is very difficult, especially if each robot in the swarm has to make real-time control decisions.

In this paper, we present an approach to solving the problem of having a swarm of pursuers intercept several evading targets. The approach is based on the multiplayer pursuit-evasion differential game framework; however, due to the computational requirements of solving the game in its complete form a decomposition technique will be used to arrive at a real-time solution scheme.

## 2. Multiplayer Pursuit-Evasion Differential Games

The multiplayer pursuit-evasion game consists of a set of pursuing and evading players. The pursuing and the evading players' movements are governed by the following dynamics,

$$\dot{x}_{pi} = f_{pi}(x, u_{pi})$$
$$\dot{x}_e = f_e(x, u_e) \tag{1}$$

where $u_{pi}$ and $u_e$ are the control input of pursuer $i$ and the evader respectively. The functions $f_e()$ and $f_{pi}()$ are considered to be smooth but potentially nonlinear functions. Most pursuit-evasion games consists of only two coalitions, that is, only two groups of players with similar goals; consequently, we will in this paper focus only on pursuit-evasion games of this type.

In order to completely describe the differential game, an objective function will have to be prescribed. As an example, the minimum time problem will be used to highlight possible computational issues in differential pursuit-evasion games, that is, the cost functional considered will be of the form:

$$\min_{u_p} \max_{u_e} J(x, u_p, u_e, T) = \min_{u_p} \max_{u_e} \int_0^T dt \tag{2}$$

where the parameter T is a part of the functional to be minimized. To simplify the problem, it is assumed that the differential game considered is a so-called complete information game, that is, as the game is being played all the players in the game are aware of the other players' current states.
In order to determine the time-optimal trajectory, the following equality has to hold,

---

[1] Corresponding author. E-mail: gtg221d@mail.gatech.edu

$$\min_{u_p} \max_{u_e} \left\langle \frac{\partial V(x)}{\partial x}, f(x, u_p, u_e) \right\rangle + 1 = 0, \tag{3}$$

where $f(x,u_p,u_e)=f_p(x,u_p) - f_e(x,u_e)$, and $V(x)$ is the value function. The condition expressed in (3) is the Hamilton-Jacobi-Bellman equation, which is a sufficient condition used to generate a time optimal solution given that the value function $V(x)$ is zero upon game termination and positive otherwise [3]. Finally, a termination condition will have to be introduced,

$$\Psi(x(T)) = 0 \tag{4}$$

For the game involving multiple pursuers, the collection of possible termination points becomes quite substantial, since in many instances capture is achieved by only a small subset of the pursuers. Hence, for some of the pursuers the final position is not specified. However, it should be noted that even though some of the pursuers will not intercept the target, they are still able to influence the evading target's escape strategy.

## 3. Computational Complexity

To solve multiplayer differential pursuit-evasion games without utilizing a simplification scheme requires significant computational resources. In multiplayer differential games, two sources of complexity combine to create an extremely computationally demanding problem. First, each player will have to optimize its actions over its entire state space, and since the state space in most cases increases exponentially with the number of states considered, the optimization is in itself computationally demanding. Secondly, the cooperation between the players is also a problem of exponential complexity, that is, for every move that one player is considering, all the possible moves of the other players will have to be considered. Hence, the dimension of the space in which the multiplayer differential game is being played is n·m, where n is the size of the state space of each player and m is the number of players. Naturally, the dimensionality of the multiplayer differential games will in most cases exclude brute force numerical schemes if a real-time solution scheme is required.

## 4. Problem Decomposition

To avoid the exponential complexity, which naturally arises when considering the cooperation between the players in the pursuing and evading coalitions, the control effort of each player will have to be decoupled from the other players. Naturally, the decoupling technique will create solutions in which cooperation is not explicitly considered. Hence, the algorithm will have to incorporate the cooperative aspects of the game after the decomposition has been performed.

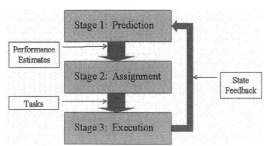

Figure 1: The problem simplification process

A decomposition approach to solve the multiplayer pursuit-evasion game consisting of three stages is shown in Figure 1.

### Stage 1. Prediction

The prediction algorithm is used to estimate the cost of capturing each of the evaders. To arrive at the estimate, the multiplayer pursuit-evasion game is decomposed into multiple two-player pursuit-evasion differential games by considering all combinations of pursuers and evaders. Naturally, the number of combinations can be reduced significantly if the information obtained from the two-player pursuit evasion games is utilized in an insightful manner. For instance, consider a scenario in which three pursuers are guaranteed to capture an evader within τ seconds. However, assuming that the third pursuer's interception time is much greater than $\tau$, then it may be reasonable to disregard the action of the third pursuer and reassign it to another task.

### Stage 2. Assignment

When decomposing the problem, the pursuers will have to be assigned a target. The assignment can be based on several factors, such as cost of interception or target importance. Since pursuit-evasion games in most cases only consist of two coalitions, the assignment process can be viewed as a bipartite graph-matching problem. Algorithms such as the Hungarian Method ($O(n^3)$) or simple Greedy-type algorithms may enable very fast implementation of the assignment process.

### Stage 3. Execution

The execution stage determines the control used to solve the decomposed multiplayer differential game. However, if the state of the game changes significantly from the predicted behavior, the first and second stages will have to be executed again with the new state information of the players. Considering the current state of the game generates the pursuers' control, and the appropriate action is taken based on an estimate of the least favorable maneuver the evading target can perform.

Once the initial assignment has taken place, the execution stage can solve the problem without receiving data from the other stages. However, the consequence of not performing target reassignment regularly throughout the game is that interception of the evading targets will in most cases become more costly.

## 5. Example Problem and Simulation Results

As an example of the decomposition approach, consider the game played in 2-dimensions in which three pursuers are attempting to capture one evader. The motion of each vehicle is described by the following set of differential equations,

$$
\begin{aligned}
\dot{x}_{pi} &= V_{pi} \cdot \cos(H_{pi}) \\
\dot{y}_{pi} &= V_{pi} \cdot \sin(H_{pi}) \\
\dot{H}_{pi} &= W_p \\
\dot{x}_e &= V_e \cdot \cos(H_e) \\
\dot{y}_e &= V_e \cdot \sin(H_e) \\
\dot{H}_e &= W_e
\end{aligned}
\tag{5}
$$

where $V_{pi}, W_{pi}, V_e$ and $W_e$ are the players' controls and $i = 1, 2$ and $3$. The termination condition will be,

$$
\Psi(x(T)) = (r_1^2 - 1) \cdot (r_2^2 - 1) \cdot (r_3^2 - 1)
\tag{6}
$$

where $r_i$ is the distance from pursuer $i$ to the evader. The cost to be minimized is given in Eq. 2, and the minimum time solution for each pursuer is determined by using the Marching Method outlined in [1].

By only considering a single evader, the assignment process is in this example trivial. The decomposition is done by dividing the game space into regions based on the capabilities of each player. The evading target will be located in at least one of the regions, and it is then the responsibility of the pursuer in the region to intercept the evading target. The other players determine the point at which it most likely that the evading target will enter their region, and they will head toward that point.

An example of the output produced by the Marching Method is shown in Figure 2. Naturally, the solution is only an approximation to the actual minimum time solution, and since the error is accumulative in nature, the error increases the further the solution is extended out into the future.

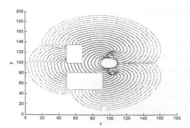

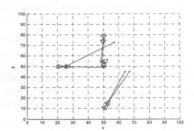

Figure 2: Minimum time contours generated using
the fast marching method

Figure 3: A Three Pursuer and One Evader
Simulation Example after Two Iterations

One of the advantages of using the Marching Method is its ability to incorporate obstacles into the solution as indicated by the two square boxes shown in Figure 2.

In Figure 3, the three pursuers are attempting to intercept the evading target located at (50,50). The x's mark the points the players are headed toward after each iteration of the algorithm. Notice that the pursuer located at (50,10) is not headed directly toward the evader, but rather toward the position at which it expects it may be able to intercept the evader and thereby play an active role in the game. The pursuer originally located at (20,50) has after the second iteration of the algorithm also decided not to intercept the evader directly, but rather attempt to ensure that the evader does not escape by passing through the turning radius of the pursuer located in the top of the picture.

In Figure 4 and Figure 5 the subdivision of the game space used to determine the possible points of interception is shown. Notice, that two of the pursuers originally attempted to directly intercept the evader since it was located on the boundary between the two players regions of responsibility. Once the evader moves into one of the regions, the responsibility to intercept the target is handed off to the pursuer responsible for the particular region in which the evader is located.

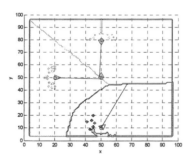

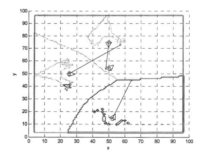

Figure 4: Game Space Division at First Iteration    Figure 5: Game Space Division at Second Iteration

The processing time per iteration was about 10 – 15 seconds in Matlab, which is significant improvement compared to the complete solution implemented in [9] which required several days to arrive at a solution. It should be noted that the exponential nature of the problem has not been completely eliminated by decomposing the problem, since the problem of determining the minimum time solution for each of the vehicles is still of exponential complexity with respect to the dimension of the players' state space.

## 6. Conclusion

An approach to reducing the computational complexity of the multiplayer differential game was presented. Implementation issues such as complexity and dimensionality of the numerical scheme were discussed. A Multiplayer differential game example was provided to highlight some implementation issues such as determining the optimal control of the players and obstacles avoidance. Finally, the approach proved to run significantly faster than the current iterative solution schemes to the multiplayer differential games.

## Acknowledgments

The authors wish to thank the anonymous referees for their careful reading of the manuscript and their fruitful comments and suggestions.

## References

[1] M. Branicky, Ravi Hebbar and Gang Zhang. *A Fast Marching Algorithm for Hybrid Systems.* Proceedings of the 38th Conference on Decision and Control, pp 4897-4902, December 1999.

[2] P. Smereka. *Semi-Implicit Level Set Methods for Curvature and Surface Diffusion Motion.* Journal of Scientific Computing, Vol. 19, Nos. 1-3, December 2003.

[3] S. Sundar, Z. Shiller. *A Generalized Sufficient Condition for Time-Optimal Control.* ASME Journal of Dynamic Systems, Measurement and Control, Vol 118 No. 2, pp. 393-396, June 1996.

[4] R. Isaacs. *Differential Games.* Krieger, New York, 1965.

[5] Bardi, M., Falcone, M. *Numerical methods for pursuit-evasion games in Stochastic and differential games: theory and numerical methods.* Advances in Dynamic Games. vol. 4 pp. 105-175, Birkhauser, 1999.

[6] H. J. Kushner. *Numerical approximations for stochastic differential games.* SIAM J. Control & Optimization, vol. 41, no. 2, pp. 457-486, 2002.

[7] J. Hespanha and M. Prandini. *Optimal pursuit under partial information.* Proceedings of the 10th Mediterranean Conference on Control and Automation, July 2002.

[8] R. Vidal, O. Shakernia, H. Kim, D. Shim and S. Sastry. *Probabilistic Pursuit-Evasion Games: Theory, Implementation, and Experimental Evaluation.* IEEE Transactions on Robotics and Automation, Vol. 18, No. 5. 2002.

[9] D. M. Stipanovic, Sriram, and C. J. Tomlin. *Strategies for agents in multi-player pursuit evasion games.* Proceedings of the Eleventh international Symposium on Dynamic Games and Applications, Tucson, Arizona, December 18-21, 2004.

Brill Academic Publishers
P.O. Box 9000, 2300 PA Leiden
The Netherlands

*Lecture Series on Computer
and Computational Sciences*
Volume 4, 2005, pp. 492-495

# Synchronizing Finite Automata with Short Reset Words

A. Roman[1]

Institute of Computer Science,
Faculty of Mathematics and Computer Science, Jagiellonian University,
Nawojki 11, 30-072 Cracow, Poland

Received 9 July, 2005; accepted in revised form 3 August, 2005

*Abstract:* Finding synchronizing sequences for the finite automata is a very important problem in many practical applications (part orienters in industry, reset problem in biocomputing theory, network issues etc). Problem of finding the shortest synchronizing sequence is NP-hard, so polynomial algorithms probably can work only as heuristic ones. In this paper we propose two versions of polynomial algorithms which work better than well-known Eppstein's Greedy and Cycle algorithms.

This work is supported by the State Committee for Scientific Research – Grant No. 3 T11C 010 27.

*Keywords:* synchronizing words, reset sequences, Černý Conjecture

*Mathematics Subject Classification:* 68Q25, 68Q45

## 1. Introduction

We define the finite automaton without initial or final states as a triple $\mathbf{A} = (Q, A, \delta)$, where Q is the finite set of states, A is the finite alphabet, and $\delta$ is the transition function from Q×A into Q. The free monoid $A^*$ is the set of all words over A. It contains the empty word $\varepsilon$. We use the notation $|w|$ for the length of $w$, i.e. the number of letters in $w$. The length of an empty word is 0. If $u$ and $v$ are two words, then $u.v$ is a word $uv$ which is the concatenation of them. If $w=w_1w_2...w_k$, we say that $w_i$ is a subword of $w$. We extend the transition function on the whole free monoid $A^*$ in a natural way:

$$\delta(q, aw) = \delta(\delta(q, a), w), \quad q \in Q, a \in A, w \in A^* \tag{1}$$

Let $\mathbf{A}$ be the finite automaton. We say that the word $w$ synchronizes $\mathbf{A}$ iff

$$\exists w \in A^* : \forall p, q \in Q \quad \delta(p, w) = \delta(q, w). \tag{2}$$

If such a word exists for $\mathbf{A}$, we say that $\mathbf{A}$ is synchronizing. If $w$ is the synchronizing word for $\mathbf{A}$ and there is no shorter one, we say that $w$ is the minimal synchronizing word (MSW) for $\mathbf{A}$. It is easy to find a synchronizing word for a given automaton, but the problem of finding *minimal* synchronizing word is NP-complete [1]. In 1964 Černý stated the following conjecture:

**Conjecture (Černý).** If the *n*-state automaton is synchronizing, then the length of its MSW is not greater than $(n-1)^2$.

The Conjecture turned out to be true for some special cases ([1], [4], [7], [8]) but in general the problem is still open. The problem of finding the minimal synchronizing word for a given automaton seems to be just a nice combinatoric puzzle, but in fact there is a deep connection between the problem and applications (see for example the pioneer works of Natarajan [2], [3]). Synchronizing sequences are used in engineering, e.g. in part orienters (see the very good example in [4]), biocomputing ([5],

---

[1] E-mail: roman@ii.uj.edu.pl

[6]), network theory etc. Kari [7] gives other examples for possible applications: simple error recovery in finite automata, leader identification in processor networks, road map problem.

## 2. The pair automaton

Let $A=(Q, A, \delta)$ be the finite automaton. For a given automaton $A$ we define the pair automaton $A^2$ as a triple $(Q', A', \delta')$ where:

- $Q'$ is a set of states. Each element of $Q'$ is either the 2-element subset of $Q$: $\{p,q\}$, where $p$ and $q$ belong to $Q$ $(p \neq q)$ or a special state $q_0$.
- $A'=A$
- $\delta'$ is a transition function defined in the following way:

$$\delta'(\{p,q\},a) = \begin{cases} q_0 & if \quad \delta(p,a)=\delta(q,a), \\ \{\delta(p,a),\delta(q,a)\} & otherwise, \end{cases} \tag{3}$$

where $\{p, q\} \in Q'$, $a \in A'$. We also define $\delta(q_0, a)=q_0$ for all $a \in A'$. The following Lemma establishes the relation between the synchronization of $A$ and some property of its pair automaton.

**Lemma 1.** Let $A$ be the finite automaton and $A^2$ its pair automaton. Then $A$ is synchronizing iff the following condition holds:

$$\forall s \in Q' \; \exists w \in A'^* : \delta'(s,w)=q_0. \tag{4}$$

The proof comes directly from the definition of synchronizing word. We also define $d(q) = \min_{w \in A^*}\{|w|: \delta(q,w)=q_0\}$ as the minimal distance in the pair automaton from $q$ to $q_0$. This distance is defined in the terms of the proper word's length. By $W(q)$ we denote the word which realizes this minimum. If for a given word $w$ we have $\delta(Q,w)=P$, then the states which at this moment belong to P are called the active states. States from Q\P are called inactive states. Of course, at the beginning of the synchronization process each state is active. If $w$ is the synchronizing word for A and $\delta(Q,w)=q_0$, only $q_0$ is an active state at the end of the synchronization process.

## 3. Algorithms

Now we will describe two well-known heuristic algorithms which find possibly the shortest synchronizing word for a given automaton A. They were introduced by Eppstein in [1]. The Greedy Algorithm finds a pair of states such that the word synchronizing them is the shortest one and transforms all active states with this word. The Cycle Algorithm does the same, but one state in the pair (in which the synchronization takes place) is fixed.

We will now introduce new algorithms based on $d$ function for the pair automata. In each step the algorithms will find the sequence which synchronizes at least two states of the pair automaton, but now we will also look one step forward, that is – we will be checking how the choice of a particular state $q$ (and, automatically, the synchronizing subword $W(q)$) in the pair automaton will affect the positions of active states. The estimation of how "good" is a given distribution of active states among all states of the pair automaton requires introducing a measure for it. We will use the $d$ function and the following heuristics:

Suppose that at some stage of our procedure $p$ is an active state. We choose a word $w=W(q)$ and we look how the value of $d$ for $p$ changes before and after the word $w$ is applied. The considered difference is defined as follows:

$$\Delta_q(p,w) = \begin{cases} d(\delta(p,w))-d(p) & if \quad p \neq q \\ 0 & if \quad p=q. \end{cases} \tag{5}$$

For a given word $w=W(q)$ we can compute $\Delta_q$ for all active states and summarize this values:

$$\Phi_1(w,q) = \sum_{p \in Act} \Delta_q(p,w), \tag{6}$$

where *Act* is the set of all active states and $q$ is the currently considered state. We compute $\Phi_1$ for all words $W(r)$, where $r$ is an active state. We also assume that if for two words $u$ and $v$ $\Phi_1(u) < \Phi_1(v)$, then it is better to apply $u$ than $v$ at the stage because after applying $u$ to all active states, all transformed active states are closer to the synchronizing state than if $v$ is applied. The $\Phi_1$ function is our measure described above and it is the base measure in the first algorithm.

In the second algorithm we add the reward function which equals the length of the chosen word. The $\Phi_2$ function with "reward" factor is defined as follows:

$$\Phi_2(w,q) = (\sum_{p \in Act} \Delta_q(p,w)) + |w|. \tag{7}$$

The procedure `SynchroPL` is almost the same as `SynchroP` – the only difference is that $\Phi_2$ stands for $\Phi_1$. One can prove the following facts:

**Proposition 2.** The time complexity of greedy and cycle algorithm is $O(n^3)$ .

**Proposition 3.** The time complexity of SynchroP and SynchroPL is $O(\dfrac{241}{1920} n^5) \approx O(\dfrac{1}{8} n^5)$.

## 4. Numercial results

It is very difficult to analyse the synchronizing algorithms because still little is known about the property of "being synchronizable". For example, we don't even know any simple property of the synchronization. Although there are many properties (for example, see Lemma 1.), they are all defined in algorithmic rather than in theoretical way and therefore are worthless in theoretical analysis.

That is why the only way to compare algorithms is to do a computer experiment: generate all $n$-state synchronizing automata, find the synchronizing words for them using all four algorithms, and compare the lengths of words being the algorithms output. The best algorithm should find the shortest synchronizing words. We did the experiment for $n=2,3,4,5$. We generated all synchronizing automata over binary alphabet with transition functions $\delta=(a_1...a_n)(b_1...b_n)$, where $t_i=\delta(i,t)$ and $(a_1...a_n) \le (b_1...b_n)$ in lexicographic order. Now, let us use two methods for estimating the quality of our algorithms. The first one is some kind of a "global method" – we take into consideration all results returned by the algorithm. The second one is focused only on cases in which the algorithm works optimal, i.e. the returned word is exactly a MSW.

Method 1. We define $m(n,ALG)$ in the following way:

$$m(n, ALG) = \frac{\sum\limits_{A \in Syn(n)} (ALG(A) - MSW(A))}{|Syn(n)|}, \tag{8}$$

where A is an automaton, n is the (fixed) number of states, ALG(A) is the length of synchronizing word for A found with algorithm ALG and Syn(n) is the set of all synchronizing $n$-state automata.

The value $m(n,ALG)$ says how much longer is a synchronizing word found by algorithm ALG than MSW length. For example, if ALG is the exponential, optimal algorithm which always finds the shortest synchronizing word, then $m(n,ALG)=0$. If for a given $n$ and two algorithms A1 and A2 we have $m(n,A1)<m(n,A2)$, we say that A1 works better in finding synchronizing words for $n$-state automata.

The result of our experiment, in terms of $m$ value from the Method 1., is presented in Fig. 1. We can see that the Cycle Algorithm is the least efficient one. Our two new algorithms work better than Eppstein's greedy algorithms. The second one (with reward factor) is the most efficient one: for example, for 5-state automata it finds the synchronizing word of length |MSW|+0.17, whereas the Cycle Algorithm finds the word of length |MSW|+0.66.

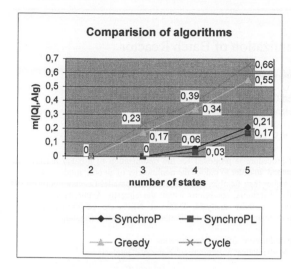

Fig. 1 Comparision of algorithms

Method 2. This is a very simple method; we just compute the ratio of optimal results returned by a given algorithm. Let us define the value $k$:

$$k(n, ALG) = \frac{\sum\limits_{A \in Syn(n)} [ALG(A) = MSW(A)]}{|Syn(n)|}, \quad (9)$$

where $[expression] - 1$ iff *expression* is true and 0 otherwise. If for two algorithms A1 and A2 we have $k(n,A1) < k(n,A2)$, we can say that for $n$-state automata algorithm A2 works better because it finds more optimal synchronizing words (MSWs). The results of the experiment are presented in Table 1. The cell in $n$-th row and "ALG" column contains the value $k(n,ALG)$. This method gives the same order of algorithms quality: Cycle Algorithm is the least efficient one, SynchroPL is the most efficient one. For example, SynchroPL gives the optimal result cases (i.e. it returns MSW) for 5-state automata in 87% of. Cycle algorithm does it only in 55% and Greedy – in 60% of cases.

The weakness of this algorithms is their complexity – one can try to reduce it with, for example, introducing another $\Phi$ function computable in $O(n)$ time. Then the complexity of the whole algorithm would reduce from $O(n^5)$ to $O(n^4)$.

Table 1
Optimal Results Ratio

| n | Number of synchr. automata | Cycle (%) | Greedy (%) | SynchroP (%) | SynchroPL (%) |
|---|---|---|---|---|---|
| 2 | 5 | 5 (100) | 5 (100) | 5 (100) | 5 (100) |
| 3 | 270 | 208 (77) | 225 (83) | 270 (100) | 270 (100) |
| 4 | 25728 | 17674 (69) | 18465 (72) | 24341 (95) | 24910 (97) |
| 5 | 4031380 | 2221524 (55) | 2423148 (60) | 3428673 (85) | 3510181 (87) |

## References

[1]  D. Eppstein, Reset Sequences for Monotonic Automata, *SIAM J. Comput.* 19(1990), 500-510.
[2]  B. K. Natarajan, An algorithmic Approach to the Automated Design of Parts Orienters, *Proc. 27th Annual Symp. Foundations of Computer Science, IEEE* (1986), 132-142.
[3]  B. K. Natarajan, Some paradigms for the automated design of parts feeders, *Internat. J. Robotics Research* 8(1989) 6, 89-109.
[4]  D. S. Ananichev, M. V. Volkov, Synchronizing Monotonic Automata, *Lecture Notes in Computer Science*, 2710(2003), 111--121.
[5]  Y. Benenson, T. Paz-Elizur, R. Adar, E. Keinan, Z. Livneh, E. Shapiro, Programmable and autonomous computing machine made of biomolecules, *Nature* 414(2001).
[6]  Y. Benenson, R. Adar, T. Paz-Elizur, L. Livneh, E. Shapiro, DNA molecule provides a computing machine with both data and fuel, *Proc. National Acad. Sci. USA* 100(2003) 2191-2196.
[7]  L. Dubuc, Les automates circulaires et la conjecture de Černý, *Inform. Theor. Appl.* 32(1998), 21-34.
[8]  A. N. Trahtman, The existence of synchronizing word and Cerny Conjecture for some finite automata, *Second Haifa Workshop on Graph Theory, Combinatorics and Algorithms*, Haifa (2002).
[9]  A. Salomaa, Compositions over a Finite Domain: from Completeness to Synchronizable Automata, *Turku Centre for Computer Science, Technical Report* No 350(2000).

Brill Academic Publishers
P.O. Box 9000, 2300 PA Leiden
The Netherlands

*Lecture Series on Computer*
*and Computational Sciences*
Volume 4, 2005, pp. 496-499

# Profit optimization of Batch Reactor

M. Ropotar, Z. Kravanja [1]

Faculty of Chemistry and Chemical Engineering, University of Maribor,
Smetanova 17 Maribor, Slovenia

Received 8 July, 2005; accepted in revised form 30 July, 2005

*Abstract:* A systematic and robust procedure is proposed for the optimization of a batch reactor with known kinetics under changeable demand. In order to reduce the nonlinearity of an NLP model, orthogonal collocation is performed on a fixed, rather than flexible, finite element and parallel Legendre polynomials are used to define continuously optimal terminal conditions within the element. A one-parametric NLP optimization is applied to perform sensitivity analysis. Different NLPs are performed to obtain concentration and control profiles for different scenarios, with limited or unrestricted demand.

*Keywords:* Batch reactor, orthogonal collocation, off-line optimization
*Mathematics Subject Classification:* Mathematical programming, Numerical analysis
*PACS:* 90 C39, 90 C31, 65 D30

## 1. Introduction

Over the last decade, research in to the optimization of batch reactors could be classified into three interconnected categories: i) modeling, ii) dynamic optimization and iii) on-line optimization, the latter being the current prevailing activity. The modeling category is typically oriented towards a more realistic description of a batch reactor (Zaldivar et al., 1996) and towards the use of special modeling techniques and strategies in the cases where imperfect knowledge of kinetic studies is involved, e.g. the use of tendency models (Fotopoulos et al., 1998) or a sequential experiment design strategy based on reinforcement learning (Martinez, 2000). The second category is related to more advanced aspects for the dynamic optimization of batch reactors, e.g. robust optimization of models, characterized by parametric uncertainty (Ruppen et al., 1995), or stochastic optimization of multimodal batch reactors (Carrasco and Banga, 1997). And finally in work, related to the on-line optimization, different control schemes were proposed, e.g. feedforward/state feedback laws in the presence of disturbances and nonlinear state feedback laws for batch processes with multiple manipulated inputs were developed (Raman and Palanki, 1996, 1998). Loeblein et al. (1998) proposed a method that estimates the likely economic performance of the on-line optimizer, and Abel and Marquard (2003) proposed scenario-integrated on-line optimization to construct model-predictive control scheme capable of normal operation and failure situations.

This paper describes a robust procedure for the dynamic off-line optimization of batch reactors. The objective is to maximize profit under changeable product demand.

## 2. Robust optimization procedure

The procedure consists of the following steps:
1) Simulation is useful for a preliminary analysis of the behavior of a given kinetic system, and to provide a good initial point for nonlinear programming (NLP).
2) Development of a robust NLP model: a differential-algebraic optimization problem (DAOP) model is converted into a robust NLP model by the use of Orthogonal Collocation on fixed Finite Element (OCFE), where the inner optimal point is modeled continuously by the use of a parallel Legendre polynomial representation. In this way, the nonlinearity of the model is significantly reduced.
3) NLP optimization:
   - Analysis of the DAOP model: one-parametric NLP optimization, with production rate (demand) as a variating parameter, is applied to define the sensitivity of batch reactor

---

[1] Corresponding author. E-mail: zdravko.kravanja@uni-mb.si

parameters vs. product demand. Reaction temperature and other model parameters can also be taken as a variating parameter.

- NLP optimization of different scenarios with unrestricted or fixed demand, penalty for by-product removal, etc.
- NLP optimization using global optimizers.

In our case Mathcad Professional was used for the simulation, GAMS/CONOPT for NLP and OQNLP for global optimization.

## 3.  Development of a robust NLP model

*DAOP:* Let us consider the following general DAOP model, consisting of profit objective function (eq. 1) for a given period of $N_b$ batches, subjected to component (eq. 2) and heat balances (eq. 3):

$$\max P = N_b \cdot \left[ \sum_{p \in prod} C_p \cdot c_p V - \sum_{r \in react} C_r \cdot c_r V - C_{heat/cool} \left( \Phi_{preheat/precool} + \int_0^{t^{opt}} d\Phi_{heat/cool} dt \right) \right] \quad (1)$$

s.t. $\quad \dfrac{dc_r}{dt} = -r_r, \quad \forall r \in react, \quad \dfrac{dc_p}{dt} = r_p \quad \forall p \in prod, \quad t \in [0, t^{opt}] \quad (2)$ DAOP

$$\dfrac{dT}{dt} = -\sum_{p \in prod} \dfrac{\Delta_r H_p}{\rho \cdot c_p} \cdot (r_p) + \dfrac{\Phi_s(t)}{V \cdot \rho \cdot c_p}, \quad \Phi_{preheat/precool} = (+/-)\rho \cdot V \cdot c_p \cdot (T_0 - T_b) \quad (3)$$

where $c_p$ and $c_r$ are concentrations of products and reactants; $V$ is reactor volume; $\Phi_{preheat/precool}$ and $\Phi_s$ are heat flows of preheating/precooling and heating/cooling respectively; $T_0$ is temperature at the beginning of reaction, $T_b$ is temperature before preheating/precooling (i.e. 293 K); $t^{opt}$ is optimal termination time; $\Delta_r H$ is enthalpy of reaction; $\rho$ is mixture density; $c_p$ is mass heat capacity and $C_p$, $C_r$, $C_{heat/cool}$ are cost coefficients for products, reactants, utility and preheating/precooling respectively.

*Example problem:* The conversion of a (DAOP) model into an NLP model is illustrated by a simple example problem of endothermic consecutive reaction A $\rightarrow$ B $\rightarrow$ C, B being the desired product. Data are given in Table 1.

Table 1: Data for example problem

| Data | $R$ | $k_{0,A}$ | $k_{0,B}$ | $\Delta_r H_A$ | $\Delta_r H_B$ | $\rho$ | $E_A$ | $E_B$ | $c_p$ | $V$ |
|------|-----|-----------|-----------|----------------|----------------|--------|-------|-------|-------|-----|
| Unit | J/mol·K | l/mol·s | l/mol·s | kJ/mol | kJ/mol | kg/m$^3$ | J/mol | J/mol | kJ/kg·K | m$^3$ |
| Value | 8.314 | 32500 | 32500 | 50 | 50 | 700 | 46000 | 53000 | 1.5 | 0.8 |

*Robust NLP model:*

$$\min_{t^{opt}, c_p(t), c_r(t), \Phi_{preheat}, \Phi_s} Z = \dfrac{28800}{t^{opt} + 600} \cdot \left( \begin{array}{c} C_1 c_B^{opt} V - C_2 \cdot c_A^0 V - C_3 \cdot c_C^{opt} V - C_4 \cdot \Phi_{preheat} - \\[4pt] C_5 \cdot \dfrac{t^{opt}}{2} \sum_{n=1}^{N} A_n \sum_{j=0}^{K} \Phi_{S,j} \cdot \prod_{k=0, k \neq j}^{K} \dfrac{\dfrac{t^{opt}}{2}(x_n + 1) - t_k}{t_j - t_k} \end{array} \right) \quad (4)$$

Mass balances:

$$R_B(t_i) = \sum_{j=0}^{K} c_{Bj} \cdot \prod_{k=0, k \neq j}^{K} \dfrac{t_i - t_k}{t_j - t_k} - k_0 \cdot e^{\frac{-E_{AA}}{R \cdot T_i}} \cdot c_{Ai} + k_0 \cdot e^{\frac{-E_{AB}}{R \cdot T_i}} \cdot c_{Bi} = 0 \quad (5)$$

$$R_C(t_i) = \sum_{j=0}^{K} c_{Cj} \cdot \prod_{k=0, k \neq j}^{K} \dfrac{t_i - t_k}{t_j - t_k} - k_0 \cdot e^{\frac{-E_{AB}}{R \cdot T_i}} \cdot c_{Bi} = 0 \quad (6) \text{ R-NLP}$$

$$(c_{A0} - c_{Ai}) = (c_{Bi} - c_{B0}) + (c_{Ci} - c_{C0}) \quad (7)$$

$$\forall i = 1, 2, \dots K$$

Energy balance:

$$R_T(t_i) = \sum_{i=0}^{K} T_i \cdot \prod_{k=0,k\neq j}^{K} \frac{t_i - t_k}{t_j - t_k} + \frac{\Delta_r H_B}{\rho \cdot c_p} \cdot \left( k_0 \cdot e^{\frac{-E_{aA}}{R \cdot T_i}} \cdot c_{Ai} - k_0 \cdot e^{\frac{-E_{aB}}{R \cdot T_i}} \cdot c_{Bi} \right) +$$

$$\frac{\Delta_r H_C}{\rho \cdot c_p} \cdot k_0 \cdot e^{\frac{-E_{aB}}{R \cdot T_i}} \cdot c_{Bi} - \frac{\Phi_{S,i}}{V \cdot \rho \cdot c_p} = 0$$

(8)

Initial conditions:

$$c_A(0) = c_{A0}, \quad c_B(0) = c_{B0}, \quad c_C(0) = c_{C0}$$

Optimal outlet concentrations by parallel Legendre polynomials:

$$c_B^{opt}(t^{opt}) = \sum_{j=0}^{K} c_{Bj} \cdot \prod_{k=0,k\neq j}^{K} \frac{t^{opt} - t_k}{t_j - t_k}$$

(9)

$$c_C^{opt}(t^{opt}) = \sum_{j=0}^{K} c_{Cj} \cdot \prod_{k=0,k\neq j}^{K} \frac{t^{opt} - t_k}{t_j - t_k}$$

(10)

It should be noted that profit and number of batches are defined for production of 8 hours and 600 sec of non-operating period between batches. Thus, number of batches is $28880/(t^{opt}+600)$.

## 4. One-parametric NLP optimization

One-parametric NLP optimization is applied to perform sensitivity analysis of the (R-NLP) problem. Figure 1 indicates that the sensitivity of product B, $S_B$, decreases while the production rate of product B, $F_B$, increases with the reaction temperature. Therefore, an appropriate trade-off between the selectivity and the production rate has to be achieved by optimization. When demand and, hence, the production rate was taken as a variating parameter, a sensitivity curve of profit vs. production rate was obtained (Figure 2).

The OCFE method has been applied to convert the DAOP model. A system of differential equations is transformed into an algebraic system of equations. Residuals (R) are included directly into the NLP model as constraints, with coefficients that become decision variables. Collocation points correspond to the shifted roots of an orthogonal Legendre polynomial and residuals are enforced at the collocation points. Initial and collocation coefficients $c_{Ai}$, $c_{Bi}$, $c_{Ci}$ are then used in parallel Legendre polynomials with $t^{opt}$ as a degree of freedom, in order to define optimal terminal concentrations $c_B^{opt}$, $c_C^{opt}$ which define the revenue in the objective function. For approximating the integral of heat consumption in the objective function, the Gaussian integration formula $\int_a^b f(x)dx \approx \frac{b-a}{2} \sum_{n=1}^{N} A_n f(x_n)$ was used. Note that the initial and collocation coefficients for heat flow $\Phi_S$ are used in the Gaussian integration.

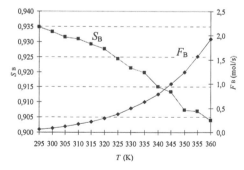

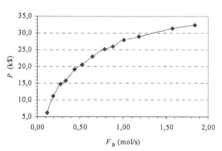

Figure 1: A trade-off between selectivity and production rate.

Figure 2: Profit vs. demand

## 5. Global NLP optimization of different scenarios

A different situation occurs when demand is restricted, e.g. to 0,60 mol/s, or when it is unlimited: terminal time is 42 minutes vs. 33 minutes, number of batches within 8 hours is 10 vs. 12 and profit is 14,16 k$ vs. 15,92 k$. For any demand, restricted or unrestricted, NLP optimization produces different profiles. Concentration and control profiles for unrestricted demand are shown in Figure 3. The solution was obtained in 226 sec of CPU time on a 2,8 GHz Pentium processor by the OQNLP global solver and the GAMS/CONOPT as NLP solver.

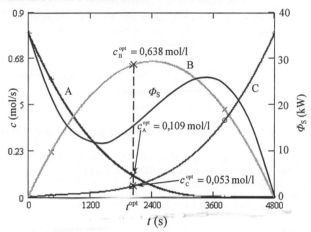

Figure 3: Concentration profiles and control variable profile.

## 6. Conclusions

A robust procedure and a robust NLP model for the optimization of a batch reactor, with known kinetics, is outlined in the contribution. Since orthogonal collocation is defined as a fixed rather than flexible finite element, nonlinearities and probable nonconvexities are significantly reduced, even if some additional nonlinearities are introduced because of parallel Legendre polynomials, in order to define continuously optimal terminal reactor conditions. The development of an MINLP model is under way to apply orthogonal collocation on several fixed finite elements, in order to decrease any errors relating to the approximation of the numerical integration.

## References

[1] O. Abel, W. Marquardt, Scenario-integrated on-line optimisation of batch reactors. *Journal of Process Control*, 13 (8) 703-715, 2003.

[2] E. F. Carrasco, J. R. Banga, Dynamic optimization of batch reactors using adaptive stochastic algorithms. *Industrial & Engineering Chemistry Research*, 36 (6) 2252-2261, 1997.

[3] J. Fotopoulos, C. Georgakis and H. G. Stenger, Use of tendency models and their uncertainty in the design of state estimators for batch reactors. *Chem. Eng. & Proc.*, 37 (6) 545-558, 1998.

[4] C. Loeblein, J. D. Perkins, D. Bonvin, Economic performance analysis in the design of on-line batch optimization systems. *Journal of Process Control*, 9 (1) 61-78, 1999.

[5] E. C. Martinez, Batch process modeling for optimization using reinforcement learning. *Comp. Chem. Eng.*, 24 (2-7) 1187-1193, 2000.

[6] S. Rahman, S. Palanki, On-line optimization of batch process in the presence of measurable disturbances. *AICHE Journal*, 42 (10) 2869-2882, 1996.

[7] S. Rahman, S. Palanki, State feedback synthesis for on-line optimization of batch reactors with multiple manipulated inputs. *Comp. Chem. Eng.*, 22 (10) 1429-1439, 1998.

[8] D. Ruppen, C. Benthack, D. Bonvin, Optimization of batch reactor operation under parametric uncertainty – computational aspects. *Journal of Process Control*, 5 (4) 235-240, 1995.

[9] J. M. Zaldivar, H. Hernandez and C. Barcons, Development of a mathematical model and a simulator for the analysis and optimisation of batch reactors: Experimental model characterisation using a reaction calorimeter. *Thermochimica Acta*, 289 (2) 267-302, 1996.

Brill Academic Publishers
P.O. Box 9000, 2300 PA Leiden,
The Netherlands

*Lecture Series on Computer*
*and Computational Sciences*
Volume 4, 2005, pp. 500-503

# Reduced Dimensionality Approaches in Quantum Dynamics. Ozone Photodissociation including Nonadiabatic Effects

**I.G. Ryabinkin[1], N.F. Stepanov**

Laboratory of Quantum Mechanics and Molecular Structure,
Faculty of Chemistry, Moscow State University,
119992, Moscow, Russia

**G. Nyman**

Department of Chemistry, Physical Chemistry
Göteborg University,
SE-412 96 Göteborg, Sweden

Received 9 August, 2005; accepted in revised form 12 August, 2005

*Abstract:* The singlet–to–triplet ratio in the ozone UV–photodissociation process was estimated using the 1D effective Hamiltonian. Effective Hamiltonian was very similar to the Reaction Path Hamiltonian proposed by Miller, Handy and Adams. Energies for eight singlet electronic states of ozone along a reaction path were calculated by multistate quasidegenerate perturbation theory (MCQDPT) with state–averaged multiconfigurational self–consistent field's wavefunctions. Diabatic states were built by smooth interpolation of the *ab initio* points in the vicinity of quasi–crossings. Subsequent quantum dynamics simulations were made in the wave–packet approach using computer program written by authors. The influence of nonadiabatic interactions onto absorption spectrum was estimated.

*Keywords:* ozone photodissociation, reduced dimensionality, effective Hamiltonian, nonadiabatic effects, quantum dynamics, singlet–to–triplet ratio, absorption spectrum

## Introduction

Ozone plays an important role in UV-protection of the human life. It has a strong absorption band in the 230-290 nm region (Hartley band) which is well studied experimentally [1]. Nevertheless, a theoretical description of the absorption process faces with huge difficulties owing to a complex structure of the excited states of an ozone molecule. Multiple attempts in a computer simulation of the absorption spectrum was undertaken [2, 3, 4], but until recently nobody represented a comprehensive picture.

Full–dimensional simulations require full–dimensional potential energy surfaces for several low–lying excited states of an ozone molecule. Additionally, Born–Oppenheimer approximation breakdown is observed when the excited molecule is dissociating upon light absorption. In this situation one has to calculate nonadiabatic coupling matrix elements of the kinetic energy or, alternatively, use a diabatic representation. Both solutions require additional computational resources.

---

[1]Corresponding author. E-mail: ilya@phys016-amd3.chem.msu.ru

In this paper another approach was implemented. The next sections organised as follows: Firstly, the 1D effective Hamiltonian for the photodissociation of triatomic molecule will be derived. After that, *ab initio* electronic structure calculations will be briefly discussed. In the next section some attention will be paid to the coming beyond Born–Oppenheimer approximation and diabatic state construction. The details of dynamics simulations and the results will be presented in the last section.

## 1    1D Hamiltonian derivation

Derivation started from the full–dimensional (3D) Hamiltonian of the non–rotating ($J = 0$) ABA molecule written in the bonds–angle (natural) coordinates ($r_1$ and $r_2$ are the distances from the central nucleus B to the A nuclei and $\theta$ is the valence angle):

$$\hat{H}_0 = \hat{T}_1 + \hat{T}_2 + \hat{V}_0, \tag{1}$$

where

$$\hat{T}_1 = -\frac{1}{2}\left[\frac{1}{\mu_1}\frac{\partial^2}{\partial r_1^2} + \frac{1}{\mu_2}\frac{\partial^2}{\partial r_2^2} + \left(\frac{1}{\mu_1 r_1^2} + \frac{1}{\mu_2 r_2^2}\right)\frac{1}{\sin\theta}\frac{\partial}{\partial\theta}\sin\theta\frac{\partial}{\partial\theta}\right], \tag{2}$$

$$
\begin{aligned}
\hat{T}_2 = {} & \frac{1}{\mu_3}\left[-\cos\theta\frac{\partial^2}{\partial r_1 \partial r_2} + \left(\frac{1}{r_1}\frac{\partial}{\partial r_2} + \frac{1}{r_2}\frac{\partial}{\partial r_1} - \frac{1}{r_1 r_2}\right)\right. \\
& \left. \times\left(\sin\theta\frac{\partial}{\partial\theta} + \cos\theta\right) + \frac{\cos\theta}{r_1 r_2}\left(\frac{1}{\sin\theta}\frac{\partial}{\partial\theta}\sin\theta\frac{\partial}{\partial\theta}\right)\right].
\end{aligned}
\tag{3}
$$

The reduced masses are $\mu_1 = \mu_2 = m_B m_A/(m_A + m_B)$ and $\mu_3 = m_B$, and the volume element is $\sin\theta\, d\theta\, dr_1\, dr_2$.

The bending motion of the molecule was frozen leading to the intermediate 2D Hamiltonian. This assumption works well at early stages of the photodissocation process but at the last stage so called adiabatic freezing [5] should preferably be used. It has been proven by authors that the similar 2D Hamiltonian can be obtained [6]. The differences occured in the effective mass multiplier which affects only spatial scale. In the case of relatively fast processes these changes introduce only small errors to the results.

In order to switch over to the one–dimensional case we introduced a new curvilinear nonorthogonal coordinate system which depends on the reaction path. It was assumed that the reaction path on the diabatic hypersurface of an excited state is "similar" the the reaction path for the ground state. This assumption can be lately verified by inspection of the first–order derivative of the energy in the direction other than the reaction path. The effects of the motion of a system in this direction was naturally included in the effective potential energy. Final form of the 1D effective Hamiltonian is given by Eqn. 4 and 5:

$$\hat{H}_{final} = -\frac{1}{2\mu}\frac{\partial^2}{\partial\eta^2} + W, \tag{4}$$

$$W = \left(1 + \frac{f'(\eta)^2}{2}\right)\frac{\omega(\eta)}{2} + \frac{1}{16\mu}\left(\frac{\omega'(\eta)}{\omega(\eta)}\right)^2 + V\left(f(\eta), \eta, \theta_{eq}\right), \tag{5}$$

where $\mu = \mu_1$, $r_1 = f(r_2)$ is a reaction path parametrization, $\omega(\eta)$ is a harmonic frequency in the "orthogonal" direction, $V(f(\eta), \eta, \theta_{eq})$ — electronic energy along reaction path, and $\theta_{eq} = 116.8°$ is the equilibrium angle in the non excited ozone molecule.

## 2    Electronic structure calculations details

Multistate quasidegenerate perturbation theory (MCQDPT) based on the multiconfigurational self–consistent field with complete active space (CASSCF) wavefunctions was used to calculate electronic energy of eight low–lying singlet electronic states of an ozone molecule. Augmented correlation–consisted triple zeta (aug-cc-pVTZ) basis set and full–valence (18 electrons over 12 orbitals) active space was involved in state–averaged CASSCF calculations followed by a perturbation correction. The first and the second derivatives in the "orthogonal" to the reaction path directions was approximated by three–points finite–differences formulas. Totally 160 single–point energies calculations were made which took approximately 1000h on the Athlon 3200+ system. All calculation were made using PC-GAMESS suite of programs [7].

## 3    Multistate effective Hamiltonian and diabatic states

Multistate generalization of the effective Hamiltonian (4) can be made straightforwardly: the kinetic energy part is assumed to be a diagonal matrix and off–diagonal elements of the potential energy matrix are due to the diabatic representation of the electronic potential $V(f(\eta), \eta, \Theta_{eq})$.

   This approach implies the knowledge of the electronic potential in the diabatic representation. Quasidiabatic states were built by a smooth interpolation of adiabatic points in the regions far from quasi–crossings [8]. Specially designed orthogonal system of functions was used for this task. The off–diagonal elements can be approximated easily if one knows both diabatic and adiabatic energies. In our case only one pair of states had noticeable interaction and the other off–diagonal elements assumed to be zero.

## 4    Dynamics simulations details

The solution of the time–dependent Schrödinger equation was obtained using wavepacket approach. Time–dependent wavepacket was represented by its values on the even–spaced spatial grid. The grid–step size was chosen to obtain converged results. The edges of the grid were placed in the region unavailable to the main part of the wavepacket during the whole time of propagation. Due to the relatively cheap calculations no absorbing potential was used. An evolution operator $U(t)$ was estimated using split–operator approximation [9]

$$U(t) = \exp\left(-\frac{i}{\hbar}t\hat{H}\right) \approx \exp\left(-\frac{i}{2\hbar}t\hat{V}\right) \exp\left(-\frac{i}{\hbar}t\hat{T}\right) \exp\left(-\frac{i}{2\hbar}t\hat{V}\right),$$

where $\hat{H}$ is the Hamiltonian operator, $\hat{T}$, $\hat{V}$ are the kinetic and potential energy operators, respectively.

   Singlet–to–triplet ratio was found to be 94:4 which is in excellent agreement with experimental one (from 95:5 to 91:9 [10]). Also it should be noted that the main part of the triplet products formed very quickly which justified our preliminary assumptions.

The absorption band simulations were performed for two cases:

- complete neglecting of the nonadiabatic coupling matrix elements, and

- using couplings discussed in the previous section.

No significant changes in the intensity of the calculated absorption band were found. This observation strongly supports the idea that the shape of the absorption band insensitive to the quality of the potential energy surfaces far from Frank–Condon region.

## Acknowledgements

The research is supported by the Russian Foundation for Basic Research, project no. 05-03-33153a and INTAS YS project no. 03-55-2058. Authors wish to thanks Department of Chemistry (Physical Chemistry) of Göteborg University (Sweden) for the given computer resources.

## References

[1] N.J Manson, J.M. Gingell, J.A. Davies, I.C. Walker, and M.R.F. Siggel, VUV optical absorption and energy–loss spectroscopy of ozone, *Journal of Physics B* **29** 3075–3089(1996).

[2] M.G. Sheppard, R.B. Walker, Wigner method studies of ozone photodissociation, *Journal of Chemical Physics* **78** 7191–7199(1983).

[3] N. Balakrishnan, G.D. Billing, Three-dimensional wave packet studies of ozone photodissociation in the Hartley band: Converged autocorrelation functions and absorption spectra, *Journal of Chemical Physics***101** 2968–2977(1994).

[4] G. Barinovs, N. Marković, G. Nyman, 3D wavepacket calculations of ozone photodissociation in the Hartley band: convergence of the autocorrelation function, *Chemical Physics Letters* **315** 282–286(1999).

[5] J.M. Bowman J. M, Reduced Dimensionality Theory of Quantum Reactive Scattering, *Journal of Physical Chemistry* **95** 4960–4968(1991).

[6] I.G. Ryabinkin, G. Nyman, N.F. Stepanov, Nonadiabatic effects in ozone photodissociation, To be published.

[7] Alex A. Granovsky, http://classic.chem.msu.su/gran/gamess/index.html

[8] F.T. Smith, Diabatic and Adiabatic Representation for Atomic Collision Problem, *Physical Review* **179** 111–123(1986).

[9] N. Balakrishnan N, C. Kalyanaraman, N. Sathyamurthy, Time–dependent quantum mechanical approach to reactive scattering and related process, *Physics Reports* **280** 79–144(1997).

[10] N. Taniguchi N, S. Hayashida, K. Takahashi, and Y. Matsumi, Sensitivity studies of the recent new data on $O(^1D)$ quantum yields in $O_3$ Hartley band photolysis in the stratosphere, *Atmospheric Chemical Physics Discussion* **3** 2332–2352(2003).

Brill Academic Publishers
P.O. Box 9000, 2300 PA Leiden,
The Netherlands

*Lecture Series on Computer
and Computational Sciences*
Volume 4, 2005, pp. 504-508

# Theory and Applications of a Nonlinear Principal Component Analysis

Ryo Saegusa[1]and Shuji Hashimoto[2]

Department of Applied Physics,
School of Science and Engineering, Waseda University
Okubo 3-4-1 Shinjuku-ku Tokyo, 169-8555 Japan

Received 9 July, 2005; accepted in revised form 2 August, 2005

*Abstract:* Principal Component Analysis (PCA) has been applied in various areas such as pattern recognition and data compression. In some cases, however, PCA does not extract the characteristics of the data-distribution efficiently. In order to overcome this problem, we have proposed a novel method of nonlinear PCA (NLPCA) which preserves the order of the principal components and we have implemented the NLPCA with neural networks. In this paper, we discuss the property of the proposed NLPCA in regard with a curvilinear axis and a contour map with some simulated results.

*Keywords:* Nonlinear PCA, Neural Networks, Curvilinear Axis, Contour Map

*Mathematics Subject Classification:* 62H25

## 1 Introduction

In the analysis of multi-dimensional data, it is important to reduce the dimensionality of the data, because it will help to extract new knowledge from the data and to decrease the computational cost. As a method of dimensionality reduction, Principal Component Analysis (PCA) has been applied in various areas such as pattern recognition and data compression [1]-[3].

When an objective data set has nonlinear characteristics, however, a nonlinear method is considered to perform more effectively than a linear method of PCA. Recently, some methods of Nonlinear PCA (NLPCA) have been developed [4]-[7], however, these methods have some problems in efficiency to extract the principal components.

We have proposed a novel method of NLPCA that preserves the order of the principal components and examined its effectiveness [8][9]. In the proposed method, we have to construct suitable nonlinear mapping functions to carry out NLPCA. As one of the implementations of the method, we proposed the hierarchically arranged neural networks. The neural networks are trained to build a set of adequate nonlinear mapping functions to extract and reconstruct input data.

In this paper, we discuss the property of the proposed NLPCA about curvilinear axes and contour maps of principal components, and show some experimental results. In the section 2, we formulate the proposed method. In the section 3, we discuss the property of the proposed NLPCA. Finally, we present our conclusion and feature works in the section 4.

---

[1]Corresponding author. E-mail:ryos@ieee.org
[2]Corresponding author. E-mail:shuji@shalab.phys.waseda.ac.jp

## 2 Formulation of NLPCA

The aim of NLPCA is to extract low-dimensional principal components from high-dimensional data, and to reconstruct the original data from the principal components.

Let $\vec{x} \in \mathbb{R}^n$, $\vec{y} \in \mathbb{R}^m$ and $\vec{z} \in \mathbb{R}^n$ be the coordinates of an input vector in a high-dimensional space, an extracted vector (principal components) in a low-dimensional space and a reconstructed vector in the original high-dimensional space, respectively, where $n, m \in \mathbb{N}$ $(n \geq m)$ indicate the dimensionality of the input vectors and the principal component vectors, respectively. $\mathbb{R}$ and $\mathbb{N}$ represent a set of real numbers and natural numbers.

In the conventional NLPCA, the nonlinear mapping function $\vec{\phi} : \mathbb{R}^n \mapsto \mathbb{R}^m$ is defined as $\vec{y} = \vec{\phi}(\vec{x})$, while the nonlinear mapping function $\vec{\psi} : \mathbb{R}^m \mapsto \mathbb{R}^n$ is defined as $\vec{z} = \vec{\psi}(\vec{y})$. These mapping functions associate the input data with their principal components nonlinearly.

The $\vec{\phi}$ and $\vec{\psi}$ are optimized by minimizing the Mean Square Error (MSE):

$$
\begin{aligned}
E &= \langle \|\vec{x} - \vec{z}\|^2 \rangle && (1) \\
&= \langle \|\vec{x} - \vec{\psi}(\vec{\phi}(\vec{x}))\|^2 \rangle, && (2)
\end{aligned}
$$

where $\langle \cdot \rangle$ represents the expectation and $\| \cdot \|$ represents $L_2$ norm. A smaller MSE indicates the higher fidelity of the reconstruction.

In the proposed NLPCA, in order to construct principal components: $y_1, \cdots, y_m$ of $\vec{y} = (y_1, \cdots, y_m)$ in the significant order to represent an input vector $\vec{x}$, the nonlinear mapping function $\phi_i : \mathbb{R}^n \mapsto \mathbb{R}^1$ is defined as $y_i = \phi_i(\vec{x})$ for $i = 1, \cdots, m$, while the nonlinear mapping function $\vec{\psi}_i : \mathbb{R}^i \mapsto \mathbb{R}^n$ from the product space $(y_1, \cdots, y_i) \in \mathbb{R}^i$ onto the corresponding vector $\vec{z}_i \in \mathbb{R}^n$ is defined as $\vec{z}_i = \vec{\psi}_i(y_1, \cdots, y_i)$, for $i = 1, \cdots, m$.

The pairs of $\{(\phi_i, \vec{\psi}_i)\}_{i=1,\cdots,m}$ are adjusted in the increasing order of $i$. $\phi_i$ and $\vec{\psi}_i$ are optimized with the $i$th MSE:

$$
\begin{aligned}
E_i &= \langle \|\vec{x} - \vec{z}_i\|^2 \rangle && (3) \\
&= \langle \|\vec{x} - \vec{\psi}_i(y_1, \cdots, y_{i-1}, \phi_i(\vec{x}))\|^2 \rangle, && (4)
\end{aligned}
$$

where $y_1, \cdots, y_{i-1}$ are given. Consequently, the pair of $(\phi_k, \vec{\psi}_k)$ is adjusted to perform the best extractor and reconstructor combined with the previous pairs of mapping functions: $\{(\phi_i, \vec{\psi}_i)\}_{i=1,\cdots,k-1}$. Figure 1 shows the diagram of the proposed NLPCA.

The nonlinear mapping functions of the proposed NLPCA can be implemented with an ensemble of neural networks which have a hierarchical structure as shown in Figure 2. The network is composed of $m$ sub-networks. In the $i$th sub-network, the left three layers and the right three layers play the role of $\phi_i$ and $\vec{\psi}_i$, respectively. The $i$th sub-network is given the values of principal components $y_1, \cdots, y_{i-1}$ from the higher 1st, $\cdots$, $(i-1)$th sub-networks. The sub-networks are adjusted in the increasing order of $i$ with a back propagation algorithm.

## 3 Properties of NLPCA

### 3.1 Curvilinear axes and contour maps of principal components

We optimized the mapping functions of the proposed NLPCA for an artificial dataset in $\mathbb{R}^2$ to discuss the property of the functions. Let the contour surfaces of $\phi_i$ at different $y$'s be $S_i(y) = \{\vec{x} | y = \phi_i(\vec{x})\}$, and let the principal component axis of $\vec{\psi}_i$ at a constant $y_1, \cdots, y_{i-1}$ and a variable $y_i$ be $A_i(\vec{y}) = \{\vec{z} | \vec{z} = \vec{\psi}_i(\vec{y})\}$.

Figure 3 shows (a) input data $\vec{x}$, (b) the contour map $S_1$ of $\phi_1(\vec{x})$, and (c) $S_2$ of $\phi_2(\vec{x})$. Figure 4 shows (a) projection lines from an input vector $\vec{x}$ to the corresponding vector $\vec{z}_1$ reconstructed

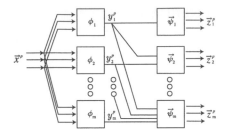

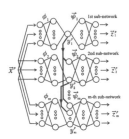

Figure 1: The diagram of the proposed NLPCA.

Figure 2: The proposed NLPCA implemented with neural networks.

by $\vec{\psi}_1(\vec{\phi}_1(\vec{x}))$, (b) the data $\vec{z}_1$ reconstructed by $\vec{\psi}_1(\vec{y})$ which is the axis $A_1$, and (c) the data $\vec{z}_2$ reconstructed by $\vec{\psi}_2(\vec{y})$. The input dataset in $\mathbb{R}^2$ were produced with $2x_1^2 + x_2 - 0.6 = 0$ for $x_1 \in [-0.8, 0.8]$ and small Gaussian noise.

In Figure 3(b) and Figure 4(b), we can find that the U-shaped axis $A_1$ is effective to represent the input data in $\mathbb{R}^2$ with one degree of freedom, and that the first contour $S_1$ which slopes along the $A_1$ is effective to map the input data onto the $A_1$. Any points on the same contour $S_1$ are mapped onto a same point on the $A_1$.

In Figure 3(c) and Figure 4(c), we can see that the second contour $S_2$ are different from $S_1$, and the data reconstruction by $\vec{\phi}_2$ is better than by $\vec{\phi}_1$, because $\vec{\phi}_2$ utilizes two principal components. In Figure 4(a), we can see that the input data tend to be projected onto the $A_1$ perpendicularly, because the perpendicular projection minimizes the distance between a point and a curvilinear axis, as well as minimizing the MSE.

In comparison to PCA, NLPCA can construct various curvilinear axes and contour surfaces, which include a closed axis and a closed contour. Actually, we can see the closed contour lines around the edge of the data-distribution in Figure 3(b). However, in the ideal case, the variety of the axis and the contour is bounded. In regard with the cases, we give propositions and their proofs in the following.

**Definition 3.1** *An identical point is defined to be $\vec{x}$ such that $\vec{x} = \vec{\psi}(\vec{\phi}(\vec{x}))$ for continuous $\vec{\phi}$ and $\vec{\psi}$.*

**Proposition 3.1** *If any point on an axis $A_i$ is an identical point, the axis does not close.*

**Proof 3.1** *Let us assume the closed axis $A_i$ such that any point on the $A_i$ is an identical point. Since $y = \phi_i(\vec{x})$ and $\vec{x} = \vec{\psi}_i(y)$ for any $\vec{x}$ on the axis $A_i$ by the assumption, $\vec{x}$ corresponds one-to-one with $y$. Consequently, $y$ varies for $\vec{x}$ monotonously because of the continuity of $\phi_i$. Let us consider a point $\vec{x}_0$ on $A_i$, which corresponds to $y_0$. When we start from $\vec{x}_0$ and follow the closed axis, corresponding $y$ increases or decreases monotonously. Then, when we return to the start point $\vec{x}_0$, the corresponding $y \neq y_0$. This goes against the one-to-one correspondence, namely against the first assumption of the closed axis. Therefore, the proposition holds.* $\square$

According to the Proposition 3.1, the training scheme is considered to have a tendency to construct an open axis, since when the MSE decreases, the number of identical points on the axis increases.

**Proposition 3.2** *If any point on an axis $A_i$ is an identical point, the axis $A_i$ and the contour surface $S_i$ intersect at one point.*

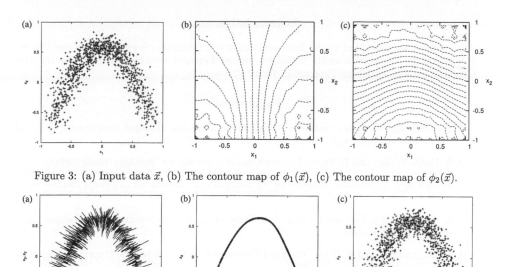

Figure 3: (a) Input data $\vec{x}$, (b) The contour map of $\phi_1(\vec{x})$, (c) The contour map of $\phi_2(\vec{x})$.

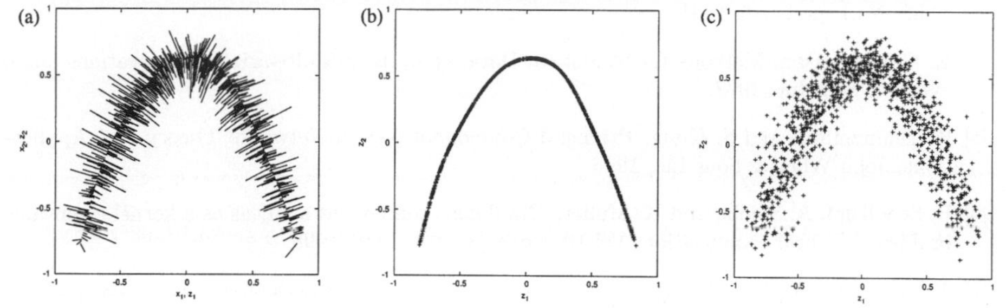

Figure 4: (a) Projection lines from an input vector $\vec{x}$ to the corresponding vector reconstructed by $\vec{\psi}_1(\phi_1(\vec{x}))$, (b) The data reconstructed by $\vec{\psi}_1(y_1)$, (c) The data reconstructed by $\vec{\psi}_2(y_1, y_2)$.

**Proof 3.2** *Let us assume that any point $\vec{x}$ on an axis $A_i$ is an identical point and that the axis $A_i$ and the contour surface $S_i$ intersect at $K$ points: $\vec{x}_1, \cdots, \vec{x}_K$. Since these points are the identical points, there exists $y_0$ such that $\vec{x}_k = \vec{\psi}_i(y_0)$ and $y_0 = \phi_i(\vec{x}_k)$ for any $k = 1, \cdots, K$. Consequently, $\vec{x}_p = \vec{\psi}_i(\phi_i(\vec{x}_p)) = \vec{\psi}_i(y_0) = \vec{\psi}_i(\phi_i(\vec{x}_q)) = \vec{x}_q$ for any $p, q = 1, \cdots, K$. Then, $\vec{x}_1 = \cdots = \vec{x}_K$, which means that the number of the intersection points is one. □*

According to the Proposition 3.2, we can consider that the closed contour surface can appear at the edge of the axis, since the closed surface around the edge can take inside the edge of the axis and the axis can intersect at one point with the surface.

# 4   Conclusion

In this paper, we formulated the NLPCA which preserves the order of principal components, and we demonstrated some simulations of the NLPCA. Moreover, we discussed the property of the NLPCA in regard with curvilinear axes and contour maps of principal components.

NLPCA has some problems originated from its nonlinearity such as local minima in optimization and its regularization. In near feature, we will approach these problems and apply the NLPCA for the pattern recognition of high-dimensional data such as images.

## Acknowledgment

This research was supported by the Japanese Ministry of Education, Science, Sports and Culture (17700247), by Waseda University Grant for Special Research Projects 2005, and by the 21st Century Center of Excellence Program in Waseda University.

## References

[1] H. Hotelling, "Analysis of complex statistical variables into principal components," Journal of Educational Psychology, Vol.24, pp.417–441 and pp.498-520, 1933.

[2] R. Duda, P. Hart and D. Stork, Pattern Classification 2nd ed., Springer-Verlag, 2000.

[3] M.Turk and A.Pentland, "Eigenfaces for recognition," Journal of Cognitive Neuroscience, Vol.3, No.1, pp.71-86, 1991.

[4] R. Gnanadesikan, Methods for Statistical Data Analysis of Multivariate Observations, John Wiley & Sons Inc, 1977.

[5] K. Diamantaras and S. Kung, Principal Component Neural Networks Theory and Applications, John Wiley & Sons Inc, 1996.

[6] B. Schölkopf, A. Smola, and K. Müller, "Nonlinear component analysis as a kernel eigenvalue problem," Neural computation, Vol.10, No.5, pp.1299-1319, 1998.

[7] T. Hastie and W. Stuetzle, "Principal curves," Journal of the American Statistical Association, Vol.84, No.406, pp.502-516, 1989.

[8] R. Seagusa, H. Sakano, S. Hashimoto, "Nonlinear principal component analysis to preserve the order of principal components," Neurocomputing, no.61, pp.57-70, 2004.

[9] Ryo Saegusa, Hitoshi Sakano, Shuji Hashimoto, "A nonlinear principal component analysis on image data," IEICE Trans. Vol.E88-D, 2005 (to appear).

Brill Academic Publishers
P.O. Box 9000, 2300 PA Leiden
The Netherlands

*Lecture Series on Computer*
*and Computational Sciences*
Volume 4, 2005, pp. 509-512

# Comparative Study on Gradient and Hessian Estimation by Kriging and Neural network approximation for Optimization

S. Sakata[1], F. Ashida[2]
Department of Electronic Control Systems Engineering,
Interdisciplinary Faculty of Science and Engineering,
Shimane University,
1060 Nishikawatsu-cho, Matsue, Shimane, Japan

M. Zako[3]
Department of Management of Industry and Technology,
Graduate School of Engineering,
Osaka University,
2-1, Yamada-Oka, Suita, Osaka, Japan

Received 10 July, 2005; accepted in revised form 4 August, 2005

*Abstract:* This paper discusses accuracy of estimated results of gradient and Hessian components of an original function. Kriging method and hierarchical neural network are used for estimation. Those methods will give a global approximated response surfaces, and gradient and Hessian components can be also estimated directly without using finite differences of estimated function values. However, those components may often include large estimation errors even if an approximated surface for objective function values can be constructed well. In this paper, therefore, accuracy of estimated results of gradient and Hessian components are investigated when objective function values are well estimated.

*Keywords:* Kriging Method, Neural Network, Gradient and Hessian Estimation, Optimization

*Mathematics Subject Classification:* 02.60.Pn, 02.60.Gf, 07.05.Mh

## 1. Introduction

Several approximate optimization methods have been studied and applied to optimization problems [1, 2, 3]. Those methods can not only reduce computational costs for iterative calculations of an objective function, but also estimate gradient and Hessian components on approximated surface directly without using finite differences. This is a great advantage for gradient-based optimization.

Neural network or Kriging method can be used for global approximation, however, enough investigations on gradient and Hessian components estimation have not been carried out yet. In this paper, therefore, accuracy of gradient and Hessian estimation by neural network or Kriging method is investigated in the case that an approximation model for function values is well constructed.

## 2. Formulation

Ordinary Kriging method using Gaussian-type semivariogram and hierarchical neural network are used for estimation. Estimation of function values, gradient and Hessian components by Kriging method at location $s_0$ are computed by Eqs. (1)-(3)[4, 5].

$$\hat{Z}(s_0) = w^T Z \tag{1}$$

$$\left. \frac{\partial \hat{Z}(s)}{\partial s} \right|_{s=s_0} = \sum_{i=1}^{n} \left. \frac{\partial w_i}{\partial s} \right|_{s=s_0} Z(s_i) \tag{2}$$

---

[1] Corresponding author. E-mail: sakata@ecs.shimane-u.ac.jp

[2] E-mail: ashida@ecs.shimane-u.ac.jp

[3] E-mail: zako@mit.eng.osaka-u.ac.jp

$$\left.\frac{\partial^2 \widehat{Z}(s)}{\partial s^2}\right|_{s=s_0} = \sum_{i=1}^{n} \left.\frac{\partial^2 w_i}{\partial s^2}\right|_{s=s_0} Z(s_i)$$

(3)

where $w$ is a weighting coefficient vector, $Z = \{Z(s_1), Z(s_2), \cdots, Z(s_n)\}^T$ is observed values at sampling locations $s_i$. Each term of differentials in Eqs. (2)-(3) can be expressed by;

$$\frac{\partial w}{\partial s} = \Gamma^{-1} \frac{\partial \gamma^*}{\partial s} + \frac{1^T \Gamma^{-1}}{1^T \Gamma^{-1} 1} \frac{\partial \gamma^*}{\partial s} \Gamma^{-1} 1$$

(4)

$$\frac{\partial^2 w}{\partial s^2} = \Gamma^{-1} \frac{\partial^2 \gamma^*}{\partial s^2} - \frac{1^T \Gamma^{-1}}{1^T \Gamma^{-1} 1} \frac{\partial^2 \gamma^*}{\partial s^2} \Gamma^{-1} 1$$

(5)

where $\Gamma$ is a constant matrix of semivariogram, $\gamma_i^*$ is a semivariogram function vector concerned with estimated location $s_0$, and

$$\frac{\partial \gamma_i^*}{\partial s} = \frac{\partial \gamma(h_i)}{\partial s} = \frac{\partial h_i}{\partial s} \times \frac{2\beta_1}{\beta_2^2} h_i \exp\left(-\frac{|h_i|^2}{\beta_2^2}\right)$$

(6)

$$\frac{\partial^2 \gamma_i^*}{\partial s^k \partial s^l} = -\frac{2\beta_1}{\beta_2^2} \exp\left(-\frac{h_i^2}{\beta_2^2}\right)\left\{\frac{2}{\beta_2^2}\left(s_i^k - s_0^k\right)\left(s_i^l - s_0^l\right) - \delta_{kl}\right\}$$

(7)

where $h_i$ shows a difference of $s_i$ and $s_0$, $\beta_i$ is a semivariogram parameter and $\delta_{kl}$ is Kronecker-delta. An approximated value $\widehat{Z}(x)$ by using hierarchical neural networks is computed by

$$\widehat{Z}(x) = z_1^N, z_i^n = f(Y_i^n), Y_i^n = \sum_j {}_j w_i^n z_j^{n-1}, f(x) = \frac{1}{1 + e^{-x}}, \quad n = 1, \cdots, N$$

(8)

where $z_i^n$ is an output from $i$th unit in $n$th layer, $Y_i^n$ is an input to $i$th unit in $n$th layer, ${}_j w_i^n$ is a weighting coefficient for output from $j$th unit in $n$-1th layer to $i$th unit in $n$th layer and $N$ is the number of total layers. $f$ is an input-output function. In this paper, $f$ is assumed as sigmoid function. Estimated gradient and Hessian components on approximated surface can be computed by;

$$\frac{\partial \widehat{Z}(x)}{\partial x} = \frac{\partial z_1^N}{\partial x}, \frac{\partial z_i^n}{\partial x} = \frac{\partial Y_i^n}{\partial x} \cdot f'(Y_i^n), \frac{\partial Y_i^n}{\partial x} = \sum_j {}_j w_i^n \frac{\partial z_j^{n-1}}{\partial x}, \quad n = 1, \cdots, N$$

(9)

$$\frac{\partial^2 \widehat{Z}(x)}{\partial x^2} = \frac{\partial^2 z_1^N}{\partial x^2}, \frac{\partial^2 z_i^n}{\partial x^2} = \frac{\partial^2 Y_i^n}{\partial x^2} \cdot f'(Y_i^n) + \left(\frac{\partial Y_i^n}{\partial x}\right)^2 \cdot f''(Y_i^n),$$

$$\frac{\partial^2 Y_i^n}{\partial x^2} = \sum_j {}_j w_i^n \frac{\partial^2 z_j^{n-1}}{\partial x^2}, \quad n = 1, \cdots, N$$

(10)

Assuming sigmoid function as input-output function, 1st and 2nd differentials of $f$ are expressed as;

$$f'(x) = \{1 - f(x)\} f(x)$$
$$f''(x) = \{1 - f(x)\} f(x) - 2f(x)\{1 - f(x)\} f(x)$$

(11)

## 3. Numerical Examples and Discussion

From equations (2)-(7) and (9)-(11), it is recognized that estimated gradient and Hessian components are computed explicitly in the case of using Kriging method or hierarchical neural network. These methods give, however, an approximated model to minimize sum or average of estimation errors at sampling locations. This fact shows that first, second or higher order differentiation cannot be adjusted at sampling locations in an approximating process.

For the purpose of investigating estimation errors for gradient and Hessian components in the case that objective function values are estimated well, a numerical example for 1-dimensional simple function is approximated, and its gradient and Hessian components are estimated.

As an example, $f_1(x) = \sin(0.2x)$ is approximated by Kriging method and hierarchical neural network. The number of hidden layers of neural network is 2, the number of units is 30. DFP method is used as update formula for weighting coefficients. In the case of using DFP updating formula, adjustment of normalizing range of input data is effective to stabilize a learning process of neural network [6]. In this paper, five different types of normalizing range are applied in learning process. 10 trials of learning for each normalizing range are carried out, and the result, which has the least estimation error, is adopted as the approximated function by neural network. The estimation region is

$0 \leq x \leq 100$, and 26, 51 and 101 sampling results with regular intervals are used for generating approximation model. Only function values are used for approximation, and gradient, Hessian or other information on an original function is not used.

Figures 1-3 shows relative estimation errors at each location. Figure 1 shows errors for function values, Fig.2 shows those for gradient, and Fig.3 shows those for Hessian. Figure (a) in each figure shows the result of Kriging estimation, Fig. (b) shows the result of neural network approximation. $n$ in Figs. 1-3 and table 1 shows the number of sampling data for approximation.

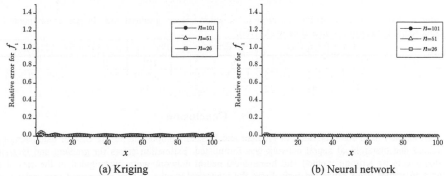

(a) Kriging          (b) Neural network

Figure 1: Relative estimation error for function values using Kriging and Neural network.

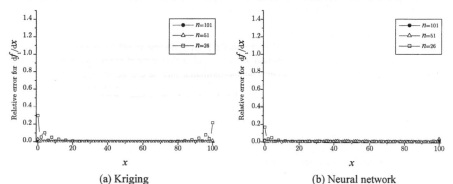

(a) Kriging          (b) Neural network

Figure 2: Relative estimation error for gradient using Kriging and Neural network.

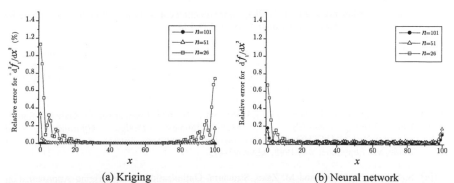

(a) Kriging          (b) Neural network

Figure 3: Relative estimation error for Hessian using Kriging and Neural network.

From Fig.1, it is recognized that function values can be estimated well by each method for all numbers of sampling results. On the other hands, comparing with the estimated results for function values, it is found from Figs. 2 and 3 that estimation errors for gradient and Hessian are large though those for objective function values are not large. Especially, estimation errors around boundaries of the estimated region tend to be larger.

Table 1 shows average estimation errors for function values, gradient and Hessian components by using Kriging method and neural network. From table 1, it can be recognized that average estimation errors increase as the number of the sampling points decreases. Estimation errors for gradient and Hessian components become large in the case of less number of sampling points, especially that of Hessian is larger than that of function values. It is also found that average estimation errors for neural network approximation are larger than that of Kriging method in the case of many sample points, but neural network approximation has the possibility of succeeding in generating a better approximating function.

Table 1: Average relative estimation error for function values, gradient and Hessian components by using Kriging method and neural network (NN).

| | Error for $f_1$ (%) | | | Error for $f_1'$ (%) | | | Error for $f_1''$ (%) | | |
|---|---|---|---|---|---|---|---|---|---|
| $n$ | 26 | 51 | 101 | 26 | 51 | 101 | 26 | 51 | 101 |
| Kriging | 1.1929 | 0.4202 | 0.0000 | 1.8239 | 0.0737 | 0.0000 | 7.9809 | 0.7465 | 0.0007 |
| NN | 0.2437 | 0.1545 | 0.2052 | 0.9771 | 0.5530 | 0.3702 | 3.7998 | 2.1526 | 1.7385 |

## 4. Conclusion

An explicit form for estimation of gradient and Hessian components in the case of using Kriging method and hierarchical neural network was formulated. Estimation errors for gradient and Hessian components by Kriging method and hierarchical neural network were investigated in the case that function values were estimated well. From the numerical results, it was recognized that estimation errors for gradient and Hessian components were larger than that of function value, especially near the boundaries of an estimated region. The number of sampling points has more influence on estimated results of gradient and Hessian components than that of function values, and estimation errors increase as the number of sample points decrease.

If enough numbers of sampling data can be used for approximation, Kriging method will give more precise estimation of gradient and Hessian. Hierarchical neural network has also the possibility of succeeding in generating a good approximation model, which can estimate gradient and Hessian components better than that of Kriging method, especially in the case of less sampling points.

## Acknowledgments

The author wishes to thank the anonymous referees for their careful reading of the manuscript and their fruitful comments and suggestions. The authors also wish to thank Mr. T. Kawaguchi for calculations of neural network approximation.

## References

[1] J. F. M. Barthelemy and R. T. Haftka, Approximation concepts for optimum structural design – a review, *Structural Optimization*, 5 (1993) 129-144.

[2] W. C. Carpenter and J.F.M. Barthelemy, A Comparison of polynomial Approximations and Artificial Neural Nets as Response Surfaces, *A collection of technical papers: the 33rd AIAA/ASME/ASCE/AHS/ASC Structures, Structural Dynamics, and Materials Conference* (AIAA, Dallas, TX, April 13-15, 1992) 2474-2482.

[3] T. W. Simpson, T. M. Mauery, J. J. Korte and F. Mistree, Comparison of Response Surface and Kriging Models for Multidisciplinary Design Optimization, *7th AIAA/USAF/NASA/ISSMO Symposium on Multidisciplinary Analysis & Optimization*, St. Louis, MI, September 2-4, 1998, AIAA-98-475.

[4] S. Sakata, F. Ashida and M. Zako, Structural Optimization Using Kriging Approximation, *Computer Methods in Applied Mechanics and Engineering*, 192-8,9 (2003), 923-939.

[5] S. Sakata, F. Ashida and M. Zako, Convexity Estimation for Global Optimization by Using Kriging Approximation, *Submitted*.

[6] F. Ashida, S. Sakata and N. Horinokuchi, Optimum Design of a Multi-Layered Composite Plate Using Neural Networks, *Journal of Thermal Stresses*, 26(2003), 1137-1150.

Brill Academic Publishers
P.O. Box 9000, 2300 PA Leiden
The Netherlands

*Lecture Series on Computer
and Computational Sciences*
Volume 4, 2005, pp. 513-517

# Energy Management System Based on Nonlinear Optimization for Direct Fuel Injection Engine and Continuous Variable Transmission

Teruji Sekozawa[1]*, Shinsuke Takahashi[2]†

[1] Dept. of Ind. Eng. & Mgmt, Kanagawa University
3-27-1 Rokkakubashi, Yokohama   221-8686, Japan
[2] Systems Development Laboratory, Hitachi Ltd.,
292 Yoshida-tyou Totuka-ku, Yokohama244-0817, Japan

Received 8 July, 2005; accepted in revised form 8 August, 2005

*Abstract:* Concern for the environment has recently led to very severe regulations on fuel consumption and exhaust emissions. We propose a scheme to minimize rate of fuel consumption for a direct fuel injection engine used by combination with a continuously variable transmission. Target values for the engine and the transmission which minimizes fuel consumption ensuring driving performance is calculated based on the nonlinear optimization method. The engine and the transmission are controlled cooperatively. The simulation in d the 10-15 mode tests showed that the proposed method is effective to achieve low fuel consumption ensuring driving performance.

*Keywords:* engine, transmission, fuel consumption, driving performance, nonlinear optimization

*PACS Numbers:* 89.40.Bb

## 1. Introduction Concept System

A direct fuel injection engine control system that   reduces fuel consumption has recently attracted attention. Because it burns fuel under very lean conditions so that the theoretical heat efficiency is improved and fuel consumption reduced.

Another way to reduce fuel consumption is to have the engine run around an optimal fuel consumption point.  This is achieved by cooperative control of the engine and a continuously variable transmission (CVT).

This paper proposes a new engine-transmission cooperative control scheme based on nonlinear optimization for DI engine and CVT. The scheme incorporates driving torque demand control in which the target driving torque is calculated based on the degree to which the accelerator is depressed.   Fuel consumption is optimized by minimizing the specific fuel consumption characteristics under various restrictions, including the achievement of target driving torque. These characteristics are previously formulated as a function of engine running conditions. As a result of optimization, target values for an engine and transmission, such as target air-fuel ratio and target gear ratio can be calculated and controlled by tracking. Under this scheme, the application of dynamic fuel consumption characteristics enables the optimization of fuel consumption in the transient condition as well as in the static condition.

## 2. A driving torque demand control scheme for DI engine and CVT

### 2.1    Concept of Low Fuel Control

Fuel consumption can be improved by using an engine with a continuously variable transmission and operating an engine around a low fuel consumption point [1]. Fuel consumption is minimized at only one point. If we operate the engine only at this point, however, the driving performance is not

---

* E-mail address: sekozawa@ie.kanagawa-u.ac.jp
† E-mail address: s-takaha@sdl.hitachi.co.jp

satisfied because the driving torque to be generated for each vehicle speed is limited. To ensure the proper driving performance, the desirable driving torque is calculated according to the degree the accelerator is depressed. The engine and transmission are controlled cooperatively to minimize fuel consumption ensuring the target driving torque. This is called driving torque demand control [2]. A 'drive by wire' structure using the electric throttle control device was adopted for this control. In DI engines, the specific fuel consumption is influenced by the air-fuel ratio and Exhaust gas recirculation (EGR) rate as well as engine torque and engine speed. Therefore target values for these variables must also be calculated to minimize fuel consumption, and the engine is controlled so as to achieve target values

## 2.2  Concept System

We consider a cooperative control system for DI engine and CVT. The drive by wire structure includes a sensor for measuring the degree the accelerator is depressed and an electronic throttle control device. An EGR system is used to purify the NOx component of exhaust emissions.

## 2.3  Concept Scheme

There are some blocks of the drive torque demand controller based on DI engine and CVT that can minimize fuel consumption. The controller consists of a target driving torque calculation block, a fuel consumption optimization block, and a dynamic control block.

(1)  Target driving torque calculation block
The target static driving torque Tdo is obtained from a two-dimensional table showing accelerator depression degree and vehicle speed. The desired driving torque data are stored in the two-dimensional table. The target transient deriving torque Td is calculated by passing the target static driving torque Tdo through a dynamic lag filter, such as

$$Td(i) = \frac{\Delta t}{T} \cdot Td0(i) + \left(1 - \frac{\Delta t}{T}\right) \cdot Td(i-1),$$ (1)

$$T = \frac{120 \cdot Vm}{N \cdot \eta \cdot Vd},$$ (2)

where, Vm is the volume of the intake manifold, Vd is total engine displacement, N is engine speed (rpm), η is volumetric efficiency, $\Delta t$ is time interval, and i is time. This procedure simulates torque response delay in the current engine control systems caused by air flow delay in the intake manifold.

(2)  Fuel consumption optimization block
This block calculates target values for an engine and transmission such as the target air-fuel ratio, target engine torque, and target gear ratio which minimize fuel

(3)  Fuel consumption optimization block
This block calculates target values for an engine and transmission such as the target air-fuel ratio, target engine torque, and target gear ratio which minimize fuel consumption. The specific fuel consumption used, Cf (g/kWh), can be expressed as

$$Cf = \frac{10^3 \times Gf}{P}$$ (3)

where Cf is specific fuel consumption per unit of time (g/kWh), Gf is fuel injection amount (kg/h), and P is engine power (kW).

Engine power P can be written as

$$P = \frac{2\pi \cdot N \cdot Te \cdot g_0}{60 \times 1000}$$ (4)

where Te is engine torque (kgfm), N is engine speed (rpm), and g0 is gravity acceleration.

There is also the following approximate relation between engine torque Te and fuel injection amount Gf:

$$Te = Tc - Tpl - Tfr - Tfp$$
$$= k \cdot \frac{Gf}{N} \cdot f(A/F, uniform \cdot stratiform, \operatorname{Re} gr) \cdot g(k\frac{Gf}{N})$$
$$- Tpl(Pm) - Tfr(N) - Tfp(N, Pf)$$ (5)

where Tc is combustion torque (kgfm), Tpl is pump loss torque (kgfm), Tfr is friction torque (kgfm), Tfp is fuel pump drive torque (kgfm), A/F is air-fuel ratio, Regr is EGR rate, Pm is intake manifold pressure, Pf is fuel pressure (atm), f is combustion efficiency correction coefficient, g is operating point correction coefficient, and k is constant.

The specific fuel consumption Cf can be written as follows by eliminating a variable Gf from expressions (3) through (5).

$$Cf = \frac{ko}{Te} \cdot G^{-1} \left[ \frac{Te + Tpl(Pm) + Tfr(N) + Tfp(N, Pf)}{f(A/F, uniform \cdot stratiform, \text{Re} gr)} \right]$$ 

(6)

$$G(x) = x \cdot g(x) \qquad \text{Here, ko is constant.}$$

Now, the specific fuel consumption function Cf is minimized under various restrictions concluding target driving torque achievement. The restrictions include   engine speed, torque, gear ratio, pulley input rotation speed, air-fuel ratio, EGR rate, ensuring a manifold pressure for the brake, achieving target drive torque, and vehicle speed. The relation between drive torque Td and engine torque Te in the table is static. On the other hand, the relation is in transition as follow,

$$Te = \ell 1 \cdot \frac{Td}{i} + \ell 2 \cdot \frac{1}{i} \cdot \frac{dv}{dt} + \ell 3 \cdot \frac{dN}{dt}$$ 

(7)

where, li ( i=1, 2, 3) is constant determined by inertia moment of rotators, final gear ratio, torque transmission efficiency, and so forth. We use expression (7) to optimize transient fuel consumption.

## (3) Dynamic control block

The dynamic control block calculates the degree to which the throttle is open, $\Theta$ th; the fuel injection amount, Gf; and the degree to which the EGR valve is open which achieve targets for engine control, that is, target engine torque Te, target air flow rate Qap, and target EGR rate Regr. There is transport delay for the air flow and EGR flow. A dynamic model is needed for controlling the air flow rate and EGR rate. The following expressions are used to derive the control scheme for achieving the target air flow rate.

$$Qap = \frac{N \cdot Vd \cdot \eta \cdot Pair \cdot Mair}{120 \cdot R \cdot Tm}$$ 

(8)

$$\frac{dPair}{dt} = \frac{R \cdot Tm}{Vm \cdot Mair} \cdot (Qat - Qap)$$ 

(9)

$$Qat = Cd \cdot \frac{Aair \cdot Pa}{\sqrt{Ta}} \cdot \sqrt{\frac{2 \cdot k \cdot Mair}{(k-1) \cdot R}} \cdot f\left(\frac{Pm}{Pa}\right) \qquad \text{and}$$ 

(10)

$$Aair = BYPATH + \frac{\pi \cdot rth^2}{4} \cdot (1 - \frac{\cos(6 + \theta th)}{\cos 6})$$ 

(11)

$$\text{When } x > \left(\frac{2.0}{k + 1.0}\right)^{\frac{k}{k-1}}$$

$$\text{When } x > \left(\frac{2.0}{k + 1.0}\right)^{\frac{k}{k-1}}, \quad f(x) = \sqrt{x^{\frac{2}{k}} - x^{\frac{k+1}{k}}}.$$ 

(12)

$$\text{When } x \leq \left(\frac{2.0}{k + 1.0}\right)^{\frac{k}{k-1}},$$

$$f(x) = \sqrt{\left(\frac{2.0}{k + 1.0}\right)^{\frac{2}{k-1}} - \left(\frac{2.0}{k + 1.0}\right)^{\frac{k+1}{k-1}}}.$$ 

(13)

Here, Qat is air mass flow rate at throttle (kg/s), Qap is air flow rate at cylinder port (kg/s), Pm is intake manifold pressure (Pa), Pair is partial pressure of air in the intake manifold (Pa), N is enigne speed (rpm), η is volumetric efficiency, Tm is temperature in the intake manifold (K), Vd is the displacement (m3), Vm is intake manifold volume (m3), R is gas constant (J/mol K), Mair is average mass of air molecular, Pa is atmospheric pressure, Ta is intake air temperature (K), Cd is the discharge coefficient, k is ratio of specific heats, BYPATH is the leak flow area (m2), rth is throttle bore radius (m2), $\Theta$ th is throttle opening degree (deg), and t is time (s).

We obtain the following expression by differentiating expression (8).

$$\Delta Pair = \frac{120 \cdot R \cdot Tm \cdot \Delta Qap - Vd \cdot \eta \cdot Pair \cdot Mair \cdot \Delta N}{N \cdot Vd \cdot \eta \cdot Mair}$$

(14)

The following expression is derived from expression (9).

$$Qat(i) = Qap(i) + \frac{Vm \cdot Mair}{\Delta t \cdot R \cdot Tm} \cdot \Delta Pair$$

,

(15)

where, i is time ( one time corresponds to $\Delta$ t), and $\Delta$ t is time interval.

When the target air flow rate at cylinder port tQap is given, the partial air pressure deviation $\Delta$ Pair is calculated based on the target air flow deviation $\Delta$ Qap and expression (14). Furthermore, the target air flow rate at throttle tQat is calculated from expression (15). If the target air flow rate at throttle is obtained, the target throttle opening degree t $\Theta$ th which achieves the target air flow rate is determined. This target signal is transmitted to the throttle control module to control the throttle.

## 3. Simulation

A simulation of the 10-15 mode tests in Japan to evaluate exhaust emissions was performed. Two schemes were evaluated for comparison. The scheme (b) is the one proposed in this paper.
(a) DI-CVT cooperative control scheme (stoichiometric setting of air-fuel ratio)
The air-fuel ratio is usually set to being stoichiometric. A value for the gear ratio which achieves minimum fuel consumption is determined and an MPI type engine and a transmission are controlled cooperatively.
(b) DI-CVT cooperative control scheme (total and dynamic optimization)
A target air-fuel ratio and target gear ratio which minimize fuel consumption are simultaneously determined by the optimization scheme discussed in Section 2.

The simulation results for the each scheme are shown in Figures 1,2. Fuel consumption is 12.13 km/l under scheme (a), 16.53 km/l under scheme (b). (b) is reduced by 26.0 % compared with scheme (a). The fuel consumption under scheme (a) is bad because the engine is not operated under lean burn conditions. The air-fuel ratio of (b) is stoichiometric in driving points in which the comparatively high engine torque is needed.

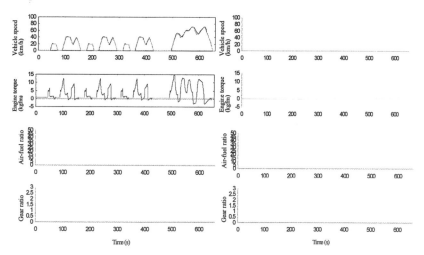

Figure 1. Simulation results of 10-15mode test
(DI-CVT stoichiometoric setting of air-fuel ratio)

Figure 2. Simulation results of 10-15mode test
(DI-CVT total and dynamic optimization)

## 4. Conclusion

(1) The proposed DI engine and CVT cooperative control scheme calculates the target values for the engine and the transmission which minimize specific fuel consumption. It ensures target drive torque and controls the engine and the transmission cooperatively. The calculation of targets is

based on the nonlinear optimization method.
(2) Control performance and fuel consumption performance are confirmed by simulations of the 10-15 mode tests.

## References

[1]H. Fujisawa and H. Kobayashi, "Electronic control gasoline injection," Sankaido publication (1987)

[2]K. Sawamura, et al., "Development of an Integrated Power Train Control System with an Electronically Controlled Throttle," Proc. of Society of Automotive Engineering of Japan 962 (1996)

[3]C. F. Aquino, " Transient A/F Control Characteristics of the 5 Liter Central FuelInjection System," SAE paper 810494 (1981)

Brill Academic Publishers
P.O. Box 9000, 2300 PA Leiden,
The Netherlands

*Lecture Series on Computer*
*and Computational Sciences*
Volume 4, 2005, pp. 518-522

# Smooth Surface and Detailed Voxel Construction with Volumetric Implicit Function for Neurosurgical Simulation

**Mayumi Shimizu and Yasuaki Nakamura[1]**

Department of Computer and Media Technologies,
Hiroshima City University,
3-4-1, Ozuka-higashi, Asa-minami-ku, Hiroshima, 731-3194, JAPAN

Received 25 July, 2005; accepted in revised form 13 August, 2005

*Abstract:* We present boundary surface construction and interactive feedback force generation for a neurosurgical simulation system. The Octree and volumetric implicit function are
utilized to manage the voxel data of a head model, and to represent piecewise continuous
volume. By sub-sampling the continuous volume, the detailed sub-voxel representation
can be obtained. To provide tactile sensation, an interactive smooth surface construction method applying a simplified Marching Cubes is proposed. Polygon surfaces of voxel
data are also generated by finding level surfaces on the volumetric function. Experimental
results reveal the effectiveness of the proposed method.

*Keywords:* Neurosurgical simulation, Smooth surface, Voxel model, Octree

*Mathematics Subject Classification:* 68U20

## 1  Introduction

Several surgical simulation systems have been proposed[1-4]. We have also been developing a neurosurgical simulation system [5] with both volume visualization and haptic presentation. Volume visualization of human organ utilizes a series of CT or MR images as voxel data. Voxel representation has advantages that any 3D shape of objects can be modeled, and that deformation can be handled by modifying values of voxels. In our simulation system, a combined method of the Octree [6] and volumetric implicit function is introduced to manage an object of voxel data. This method provides functions of smooth surface construction, comprehensive boundary surface generation, and detailed volume rendering.

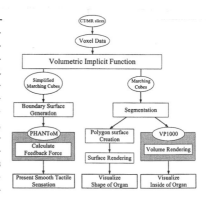

Figure 1: System configuration

[1]Corresponding author. Department of Computer and Media Technologies, Hiroshima City University. E-mail:
{m_shimizu,nakamura}@toc.cs.hiroshima-cu.ac.jp

## 2 Neurosurgical Simulation System

A surgical simulation proceeds by modifying the voxel data interactively with tactile sensation. Our neurosurgical simulation system employs the volume rendering board "VolumePro1000(VP1000)" [7] to visualize the voxel model in detail, and haptic device "PHANToM" [8] to generate tactile sensation. The Octree is used to manage the whole voxels and to perform spatial searches such as nearest neighbor and range searches efficiently. Figure 1 shows a configuration of our system.

## 3 Octree and Volumetric Implicit Function

Although the voxel representation is useful for the simulation, there are some problems in voxel representation. The first problem is in the case of enlargement. When the voxel model is zoomed in, the boundary of an object becomes more rough and jaggy. The second one is concerned with feedback force generation. In calculating the feedback force, the collision with an instrument and the voxel model must be checked. The collision check with every voxel requires a huge amount of calculation. Even if the collision is detected, the smooth and continuous reaction force cannot be obtained because the boundary of voxels is discrete. To overcome these problems, a new algorithm that approximates the voxel values using the Octree is developed.

Each voxel has density value $d$. Let the location of a voxel be $(x, y, z), (x, y, z = 0...N\text{-}1)$. Voxel data are given at discrete locations and expressed as Eq.(1). In our system, voxel value $d(x, y, z)$ is positive if the voxel is inside the object, while voxel values outside the object are assigned negative values depending on the distance from the object to the voxels. A volumetric implicit function (in short VIF), which approximates voxel data, is a tri-cubic parametric function expressed as Eq.(2); a three dimensional smooth and continuous solid can be calculated by Eq.(2).

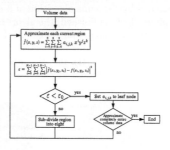

Figure 2: An Octree and VIF

$$f(x, y, z) = d_{xyz}, \quad x, y, z = 0...N-1 \qquad (1)$$

$$\hat{f}(x, y, z) = \sum_{i=0}^{3}\sum_{j=0}^{3}\sum_{k=0}^{3} a_{i,j,k} x^i y^j z^k \qquad (2)$$

$\hat{f}(x, y, z)$ is determined so as to minimize the following least square error between $\hat{f}(x, y, z)$ and $f(x, y, z)$.

$$\varepsilon = \sum_{i=0}^{N-1}\sum_{j=0}^{N-1}\sum_{k=0}^{N-1} \left| \hat{f}(x_i, y_j, z_k) - f(x_i, y_j, z_k) \right|^2 \qquad (3)$$

The procedure constructing an Octree and creating an implicit solid is shown in Figure 2. In constructing the Octree, firstly, the whole voxel data is approximated such that Eq.(3) becomes minimal. If $\varepsilon$ is greater than threshold $\varepsilon_0$, the region is divided into eight regions. Then, the sub-divided regions are approximated. If $\varepsilon$ is less than or equal to $\varepsilon_0$, the sub-division process terminates, and the parameter vector $a_{i,j,k}(i, j, k=0,...,3)$ in Eq.(2) are set to the corresponding leaf node. Until an Octree representing the collection of regions becomes to approximate the entire volume data precisely, the rest sub-regions in the Octree are sub-divided and approximated.

# 4    Surface Construction Using Volumetric Implicit Function

### 4.1    Interactive Surface Construction for Tactile Sensation

To obtain the smooth tactile sensation, the smooth boundary surfaces must be prepared for PHAN-ToM. A local smooth surface around the contact point is created interactively and used to calculate a feedback force. A simplified Marching Cubes [9] method is applied to find precise boundary points, where $\hat{f}(x, y, z) = 0$, on a voxel model. Then, mapping the boundary points to $S$-$T$ coordinate, a smooth surface is calculated.

**Simplified Marching Cubes Algorithm** Simplified Marching cubes algorithm does not seek boundary points in 3D space but 2D plane because of the processing speed. As shown in Figure 3, the boundary candidate points can be classified into five patterns with 4-neighbor points except the rotation. Symbols "+" and "−" in Figure 3 stand for the signs of voxel values. Boundary points denoted by tiny circles that satisfying equation $\hat{f}(x, y, z) = 0$ are extracted.

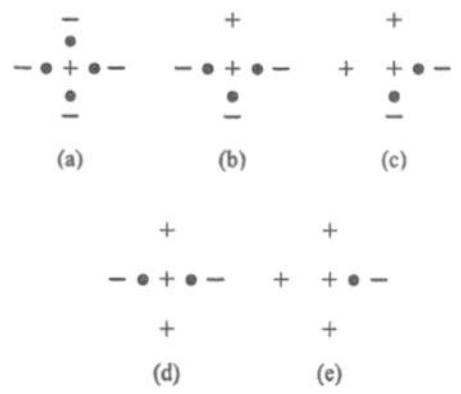

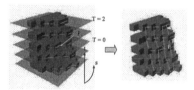

Figure 3: Patterns of boundary candidate points

The boundaries are determined by starting from the contact point and following the candidate points clockwise and counter clockwise in a plane parallel to S axis. Candidate points matching to the patterns in Figure 3 are searched by the same algorithm of 8-neighbor border following. When the boundary following for a plane, T=0, is finished, the algorithm is applied to the next plane, T=1, as shown in Figure 4. The left figure shows planes on which boundary points are followed. When the enough number of boundary points along parameter $s$ and $t$ for determining a local surface around the contact point is found, a bi-cubic parametric surface is defined by these points. The left figure in Figure 5 shows an extracted boundary and the right figure is an enlarged smooth surface and Octree subdivisions.

Figure 4:   Construction of local boundary surface

Figure 5: Constructed local surface

The boundary points are represented as $Q(s_i, t_j) = (x_{ij}, y_{ij}, z_{ij})$ $(i = -n...n, j = -m...m)$. In our experiment, $n$ and $m$ are equal to 2. Bi-cubic parametric surface expressed as Eq.(4) is computed so that the square error $\rho$ in Eq.(5) becomes minimal.

$$\hat{Q}(s, t) = \left( \sum_{k=0}^{3}\sum_{l=0}^{3} a_{k,l}s^k t^l, \sum_{k=0}^{3}\sum_{l=0}^{3} b_{k,l}s^k t^l, \sum_{k=0}^{3}\sum_{l=0}^{3} c_{k,l}s^k t^l \right) \tag{4}$$

$$\rho = \sum_{i=-n}^{m}\sum_{j=-m}^{m} |\hat{Q}(s_i, t_j) - Q(s_i, t_j)|^2 \tag{5}$$

### 4.2    Surface Rendering and Segmentation

The shapes of organs or the boundary polygons can be obtained by applying the Marching Cubes algorithm to VIF. The Marching Cubes method detects the level zero points on the continuous function (Eq.(2)) efficiently. That is, the zero points constructing the boundary surfaces are extracted, and the polygon surfaces are rendered interactively.

Figure 6 (a) and (b) are created polygon surfaces from the original voxel values and the approximated voxel values, respectively. As for the resultant image (a), the coarseness is conspicuous. On the other hand, the resultant image (b) can represent the more precise surface. Hence, a high-quality surface is extracted by finding zero points on VIF.

### 4.3  Surgical Simulation

Figure 7 (a) shows the created head model. Rotating the head model, inside structures of a head are visualized in detail interactively. Figure 7 (b) and (c) show the result of an incision. When an instrument collides with the head model, 4x4x4 voxels underlying the instrument are made transparent in this experiment.

In the interactive operation, it takes about 10 milliseconds to construct the local surface and the normal vector. The frame rate in the surgical simulation with force feedback is 7 frames per second. This frame rate is not sufficient for the interactive environment because it is well known that at least 10 frames per second are needed for the interactive simulation. To obtain more interactive frame rate, we will improve the rendering algorithm.

(a) Created from voxel values     (b) Created from values of VIF

Figure 6: Polygon surfaces

## 5   Conclusion

We introduced the method of approximating voxels with VIF. Representing voxel data by VIF, the smooth and continuous reaction force, and the visualization of smooth boundaries after the expansion, were realized.

The boundary points were extracted using the simplified Marching Cubes method. Then, by approximating these points with a bi-cubic parametric surface, smooth and continuous surfaces and normal vectors, which are necessary for calculating a feedback force, were generated interactively. As a result, the smooth and continuous reaction force was obtained.

(a) Head model

(b) Before incision     (c) After incision

Figure 7: Head model and incision operation

### References

[1] Z. Quingson, K. C. Keong, and N. W. Sing, "Interactive Surgical Planning Using Context Based Volume Visualization Techniques," Proc. MIAR 2001, pp.21-25, 2001.

[2] R. A. Kockro, et al., "Planning and Simulation of Neurosurgery in a Virtual Reality Environment," Neurosurgery, Vol. 46, No. 1, pp.118-137, 2000.

[3] J.-Y. Shi, and L.-X. Yan, "Deformation and Cutting in Virtual Surgery," Proc. MIAR 2001, pp.95-102, 2001.

[4] T. A. Galyean, and J. F. Hughes, "Sculpting: A Interactive Volumetric Modeling Technique," Computer Graphics, Vol. 25, No. 4, pp.267-274, 1991.

[5] M. Shimizu, and Y. Nakamura, "Virtual Surgical Simulation with Tactile Sensation for Neurosurgery Operation by Volume Graphics," Proc. CCCT 2004, Vol. 1, pp. 112-117, 2004.

[6] H. Samet, "The Design and Analysis of Spatial Data Structures", Addison-Wesley, 1990.

[7] VolumePro 1000 Principles of Operation, TERARECON Inc., 2001.

[8] SDK Programmer's Guide, SensAble Technologies Inc., 2000.

[9] W. E. Lorensen, and H. E. Cline, "Marching Cubes: A High Resolution 3D Surface Construction Algorithm," SIGGRAPH 87, Vol. 21, No. 4, pp. 163-169, 1987.

Brill Academic Publishers
P.O. Box 9000, 2300 PA Leiden
The Netherlands

*Lecture Series on Computer*
*and Computational Sciences*
Volume 4, 2005, pp. 523-526

# Hierarchical-Key Renewal Scheme For Multi-level Multicast Access Control

DongMyung Shin[1] , HeeUn Park , YooJae Won, ByungJin Cho

Information Security Technology Planning Team,
Information Security Technology Division,
Korea Information Security Agency,
Seoul, Korea

Received 4 August, 2005; accepted in revised form 15 August, 2005

*Abstract*: Hierarchical cryptographic system can be used to provide multi-level access control in multicast service. According to member's access level, cryptographic keys are shared in hierarchical cryptographic system. And also a member who has upper access privilege can derive keys of lower access privilege from a key of upper access privilege immediately. Research for hierarchical cryptographic system has been done steadily but it is not satisfying FS(Forward Secrecy) and BS(Backward Secrecy) on dynamic multicast environments. Thus, we propose the scheme that can update many hierarchical cryptographic keys according to member's access level. This scheme has a simple preparation step for the key update, and the distribution of public information can make both entities compute the key for the key distribution. Proposed modular operation can reduce number of exchange messages to update keys and satisfy FS and BS.

*Keywords:* multicast security, access control

## 1. Introduction

Multicast is a bandwidth-conserving technology that reduces traffic by simultaneously delivering a single stream of information to thousands of corporate recipients. Applications that take advantage of multicast include video-conferencing, corporate communications, distance learning, and distribution of software, stock quotes, and news.

Multicast delivers source traffic to multiple receivers without adding any additional burden on the source or the receivers while using the least network bandwidth of any competing technology.

Also, Multi-level access control can be applied for providing multimedia service or secret remote conferencing that have various grades.

As an example, a user who has upper access privilege can receive more messages than a user who has lower access privilege. A multicast network is composed of hierarchical virtual networks relying on access levels.

A hierarchical access control mechanism is applied to many agencies such as the ministry of national defense, departments relating government and private enterprises. For supporting this hierarchical scheme, users in higher class should be able to easily derive keys of users with lower class. However, the reverse derivation should be almost impossible. As a simple solution, a user in one class maintains all keys of lower classes. Also, overhead of key management is as deep as depth of key hierarchy in the hierarchical scheme. Thus, we need hierarchical key structure to easily get all lower keys.

There have been many researches into derivation of lower keys and creation of key hierarchy for key distribution[3,4]. However, upper and lower key has linear relationship to derive lower keys at a higher class, and any key creation results in update of total keys.

To apply the hierarchical access control for dynamic multicast environment, the key update is required whenever join and leave operation are occurred for guaranteeing FS and BS[1] in a multicast group. At this point, we have a problem of update for not one key but all hierarchical keys.

## 2. Related Works

Until now, hierarchical key is resonable to update user key in Security Class. In that case, both all

---
[1] Corresponding author.. E-mail: dmshin@kisa.or.kr, hupark@kisa.or.kr , yjwon@kisa.or.kr , bjcho@kisa.or.kr

keys of security class and all keys of users in the security class should be updated[3-4]. This does not consider the withdrawal of users belonging to security class and dynamic circumstances related to the join of new user. Nonetheless, when a joined user keeps his/her secret key in a long time, hierarchical key can be applicable to a case user's secret key needs to be updated regularly for security. This should not affect user's keys of different security class and accomplish key update efficiently.

In KSCL[2], Chang-Pan[4], when a user requests CA's secret key update for forward secrecy, CA should regenerate secret keys of all users. But in security class, a user can change his/her secret key freely and CA has only to update the public information $C$ of security class. For example, if secret key $SK_i$ of security class $SC_i$ is changed to secret key $SK_i$, public information $C_i$ is updated to $C'_i=(SK'_i \| ID_i)((HK'_i \| ID_i)+b_i) \bmod m_i$ and all other secret keys and information do not need to be changed.

But, since a leaved member can derive a member's key with lower access privilege, all the key update should be required for forward secrecy. In KSCL, Chang-Pan's scheme consider only the outside users who have no access privilege and assume that leaved members are separated. But in a hierarchical multicast access control, when a user is demoted from a higher access level to a lower access level, to block the information which can be accessed by a previous key, secrets related to all lower ones of the security class should be updated. Therefore, for forward secrecy and backward secrecy, whenever members join or left, it has limitation that CA(Central Authority)'s key update is required and all users' key should be regenerated.

## 3. Proposed Scheme

### 3.1 System Parameters

$x_k$ : secret shared key of a user x
SC : Security Class
IDsc : Security Class ID
GK : Key Encryption Key for exchange HK
$SK_n$ : Encryption key of SC
$HK_n$ : Hierarchical encryption key of SC
$U_k$ : secret key of a user in U

### 3.2 Hierarchical Encryption/Decryption over Application Layer

User information is mapped into a set of security class in the hierarchical structure. If the user is maintained in any set $U_i$, $1 \leq i \leq n$, the security class $SC_i$ assigns proper authorization to $U_i$. We assume $SC_i$ is $i$ th security class of the SC sequence $SC_1$, $SC_2$,...., $SC_n$. Also we define '$\leq$' that describes which SC is higher class of two SCs. So, $(SC, \leq)$, $SC_j \leq SC_i$ means that a user with $SC_i$ is higher or equal access authorization than one with $SC_j$. In other words, the user with $SC_i$ can read or write other users with $SC_j$. In contrast, the user with $SC_j$ cannot do it.

We calculate $HK_i$ of $SC_i$ using hash chain. A router generates a secret integer A and forms the hash chain of HK like $HK_1=H(A)$, $HK_2=H(HK_1)$, $HK_3=H(HK_2)$, ..., $HK_n=H(HK_{n-1})$. Here, H is a hash function. The recursive formation of HK has an advantage of key maintenance because of no need for store of a key list. However, overhead of continuous hash calculation is incurred. This overhead gives a system less effects due to fast and lightweight characteristics of the hash function. In addition, the number of applying hash functions is as same as total access control levels.

The user with $SC_i$ stores the cipher text C or sends C over network. The cipher text C is obtained by $E_{HKi}(M)$ for plain text M. At this time, only other users with $SC_i$ are able to the M through decrypting C, $D_{HKi}(C)$. Here, C and D is encryption and decryption function, respectively. In case of $SC_j \leq SC_i$, the user with $SC_i$ is possible to decrypt the encrypted messages with $HK_j$ because the user with $SC_i$ can obtain $HK_j$ using his own secret key $HK_i$ and public information of $SC_j$.

#### 3.2.1 *Renewal of Secret Key*

We define security class SC = $\{SC_1, SC_2, SC_3, ..., SC_n\}$ and assume $SC_1 \geq SC_2 \geq SC_3 \geq ... \geq SC_n$. Also, we set a user group U = $\{U_1, U_2, U_3, ..., U_n\}$ and suppose that SC and U is bijection.
SC=$\{SC_1, SC_2, SC_3, ..., SC_n\}$,
Group ID of SC=
$\{ID_{HK1}, ID_{HK2}, ID_{HK3}, ..., ID_{HKn}\}$
U=$\{U_1, U_2, U_3, ..., U_n\}$

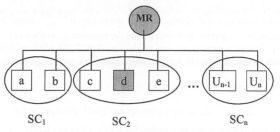

Figure 1: Multicast Security Classes and Members

In Figure 1, we assume that $U_1=\{a, b\}$ , $U_2=\{c, d, e\}$, $U_3=\{f, g\}$ and each member a, b, c, d and e have a its own secret keys $a_k$, $b_k$, $c_k$, $d_k$, $e_k$, respectively. All members are connected to a router MR. Then user groups are $U_{k1}=\{a_k, b_k\}$, $U_{k2}=\{c_k, d_k, e_k\}$ and $U_{k3}=\{f_k, g_k\}$. So members in $U_{ki}$ have a common secret key $HK_i$, $1\leq i\leq n$.

When member d leaves in its group $U_2$, all remained member in $U_2$ should update the $HK_2$. For updating key HK2, we distribute a group key $GK_i'$ to all members except for d in $U_2$. Then we redistribute a hierarchical key $HK_n'$ using $GK_i'$ and another group key.

$1^{st}$ step : A multicast router MR receives a leave message from the member d.

$2^{nd}$ step : The router broadcasts a Group-ID of $U_2$ and $H(d_k)$ to members. Through advertising $H(d_k)$, we can hide the value $d_k$ with no awareness of $d_k$ at user d.

$3^{rd}$ step : The members with $ID_{HK2}$ calculate $x_k'$ like an equation (1). α, β, ...... are hashed values of secret keys of groups with leaved members. And public integer g and n are enough large prime numbers.

$$x_{k'} = g^{(Xk-\alpha)(Xk-\beta)\cdots}\ mod\ n \tag{1}$$

In our example, $x_k$ is $H(d_k)$. Therefore,

$$c_{k'} = g^{(H(c_k) - H(d_k))}\ mod\ n \tag{2}$$

$$d_{k'} = g^{(H(d_k) - H(d_k))}\ mod\ n = 1 \tag{3}$$

$$e_{k'} = g^{(H(e_k) - H(d_k))}\ mod\ n \tag{4}$$

We can not get a new group key $GK_2'$ since eventually $d_k'$ is 1.

$4^{th}$ step : The router encrypts a new group key $GK_2'$ with $c_k'$ and $e_k'$

$$\{ID_{sk2} \| E_{ck}'[GK_2'] \| E_{ek}'[GK_2']\} \tag{5}$$

$5^{th}$ step : The router distributes new HK using a previous group and the new group key. Here, a size of HK to be distributed is equal to a number of groups regardless of a number of members.

$$\{SK\_Update\_msg \| E_{gk1}'[HK_1'] \| E_{gk2}'[HK_2'] \| \cdots \| E_{gkn}'[HK_n']\} \tag{6}$$

### 3.2.2    Efficiency Improvement of group key distribution

To improve key distribution procedure, we propose a new distribution scheme of $4^{th}$ step.

$4^{th}$ step : The router broadcasts a public integer t calculated by equation (7).

$$t = g^{ck'+ek'}\ mod\ n \tag{7}$$

The integer t only is calculated by members who know two value $ck'$ and $ek'$. Thus the member c and e derive a new group key like equation (9) and (10)

$$GK' = (t \cdot g^{-xk'})^{xk'}\ mod\ n \tag{8}$$

member c :

$$GK_2' = (t \cdot g^{-ck'})^{ck'}\ mod\ n$$
$$= (g^{ck'+ek'} \cdot g^{-ck'})^{ck'}\ mod\ n \tag{9}$$
$$= g^{ck'ek'}\ mod\ n$$

member e :

$$GK_2' = (t \cdot g^{-ek'})^{ek'}\ mod\ n$$
$$= (g^{ck'+ek'} \cdot g^{-ek'})^{ek'}\ mod\ n \tag{10}$$
$$= g^{ck'ek'}\ mod\ n$$

Thus, the two member c and e have the same group key $g^{ck'ek'}mod\ n$.

However, the equation (7) is restricted in case that the remaining member is 2.

When a member leaves a group, we can apply equation(7),(8) after splitting a group to several subgroups with only two members. If a group has n members, we distribute $\left\lfloor\frac{n}{2}\right\rfloor - t$ and total number

of groups increases.

### 3.2.3    Expansion of group key distribution procedure

To solve a restriction of 2 members in a group, we set the t like equation (11).

$$t(x) = g^{f(x)}\ mod\ n \tag{11}$$

$$f(x) = (a_k + b_k - x)x \tag{12}$$

$$f(x) = (a_k b_k + b_k c_k + c_k a_k + x^2 - a_k x - b_k x - c_k x)x \tag{13}$$

The equation(12) is case of 2 members in a group and equation (13) shows calculation procedure when the number of members is 3.

For example, , $f(a_k) = f(b_k) = a_k \cdot b_k$ when a group member is {a,b}, and $f(a_k)=f(b_k)=f(c_k)=a_k \cdot b_k \cdot c_k$ when a group member is { a, b, c }.

Besides, the number of group GN is 2Q where R=0 and GN=2*(Q-1)+3 where R=1.

$$n \geq 2, n \in \mathbf{N} \quad , \quad R : {}_{n - \left(\left\lfloor\frac{n}{2}\right\rfloor * 2\right)} \quad , \quad Q : \left\lfloor\frac{n}{2}\right\rfloor$$

## 4.   Conclusion

Proposed mechanism provides hierarchical access control service by distributing different keys with member's access privilege. Also, it is possible that the key with the upper access level derivate the one with the lower level immediately, but the reverse is difficult computationally. This research on a hierarchical encryption key has been progressed steadily, but both forward and backward secrecy are not satisfied. Therefore, In this paper, lightweight scheme that satisfy forward and backward secrecy are provided by generating hierarchical keys using hash function recursively and by update multiple encryption keys with different access privilege.

This scheme has a simple preparation step for the key update, and the distribution of public information can make both entities compute the key for the key distribution. When a member of a subnet leaves, the range of the key update is restricted within the subnet. Because of the sufficiency of the subnet bandwidth, the cost of the key update and the key exchange is very low relatively. Also, the proposed scheme is designed in the direction of the decrease of the communication frequency and quantity by the fact that CPU computing speed is faster than network's speed, and the key derivation scheme using public information is applied.

## References

[1] A. Perrig, D. Song and J. D. Tygar, "ELK, a New Protocol for Efficient Large-Group Key Distribution," IEEE Symposium on Security and Privacy, 2001

[2] F.H.Kuo, V.R.L.Shen, T.S.Chen and F.Lai, Cryptographic key assignment scheme for dynamic access control in a user hierarchy, IEE Proceedings. Computers&Digital Techniques, 1999.

[3] J.C. Birget, X. Zou, G. Noubir, B. Ramamurthy, "Secure hierarchy-based access control in distributed environments", IEEE 2002.

[4] Chang, C. C., Pan, Y. P.,"Some Flaws in a Cryptographic Key Assignment Scheme for Dynamic Access Control in a User Hierarchy, InfoSecu'2002.

[5] Qiong Zhang and Yuke Wang, "A Centralized Key Management Scheme for Hierarchical Access Control," IEEE Global Communication Conference (GLOBECOM) 2004, Dallas, TX, Nov. 2004.

Brill Academic Publishers
P.O. Box 9000, 2300 PA Leiden
The Netherlands

*Lecture Series on Computer
and Computational Sciences*
Volume 4, 2005, pp. 527-530

# Boson Peak in Amorphous Silicon

Ranber Singh* (ranber14@yahoo.com)

Department of Physics, Panjab University, Chandigarh (India)

*Present Address: Department of Physics, Doaba College, Jalandhar City (India)

Received 10 July, 2005; accepted in revised form 4 August, 2005

*Abstract:* The connection of the so-called "boson peak" to the coordination defects, strained bonds and localization of vibrational modes has been explored in amorphous silicon. It is found that the boson peak has no connection to the coordination defects but shows close connection to the strained Si-Si bonds in amorphous silicon. Further it is also found that all the vibrational modes in the vicinity of the boson peak are extended.

*Keywords:* Amorphous silicon, Boson peak, DFT
*Mathematics Subject Classification:* 61.43.Dq, 65.60.+a
*PACS:* 61.43.Dq, 65.60.+a

## 1. Introduction

The low-frequency part of the vibrational spectrum of amorphous solids is one of the major unresolved problems in condensed matter physics. One still does not understand in what way the atomic disorder in the amorphous solids modifies the low frequency vibrations of their crystalline counterparts. In amorphous solids, there exists a universally valid feature the so-called "boson peak" which is detected as an excess of low frequency modes (about 20-120 cm$^{-1}$) in the Raman and neutron scattering spectra [1] and/ or as a bump in the low-temperature specific heat [2]. The theoretical interpretation provided so far for the origin of boson peak in amorphous materials is highly controversial. There are mainly three different scenarios that have been proposed for the origin of the boson peak in amorphous materials. One scenario proposed the existence of vibrational modes in addition to the sound waves of the Debye Model [3], while another proposed that it is due to the localized vibrational modes [4]. Third scenario ascribes the origin of boson peak to the strong scattering of sound waves due to disorder in the amorphous materials [5, 6]. However, the actual origin of boson peak is still a subject of debate and provides motivations for more extensive studies of these amorphous materials. In this paper, the issues related with the correlation of boson peak to the coordination defects, strained bonds and the localization of vibrational modes have been explored in amorphous silicon.

## 2. Computational Details

The detailed description of the procedure adopted to generate the model samples of amorphous silicon by MD simulations can be found in Ref. [7]. We generated five samples of pure and hydrogenated amorphous silicon (a-Si:H) with 216 Si atoms and 24 H atoms using the density functional based tight binding (DFTB) molecular dynamics simulations [8]. Two samples of a-Si:H are generated from the liquid quench at different quenching rates while the others two are generated by the hydrogenation of pure amorphous silicon samples. The overall structural properties of these samples are in good agreement with the previous experimental and theoretical results [7]. These samples have been labeled as follows:

Pure amorphous silicon sample: **ASi**
Hydrogenated amorphous silicon samples: **ASiH, BSiH, CSiH, WSiH**

The eigenvalues (squares of vibrational frequencies) and eigenvectors (vibrational patterns) have been calculated by the direct diagonalization of the dynamical matrix. The broadening width of 20 cm$^{-1}$ is adopted to calculate the VDOS. The specific heat ($C$) per atom of simulated samples is calculated from VDOS with the help of following expression [9]:

$$C = 3k_B \int_0^\infty \left( \frac{\hbar\omega}{k_B T} \right)^2 \left( \frac{e^{\hbar\omega/k_B T}}{\left( e^{\hbar\omega/k_B T} - 1 \right)^2} \right) g(\omega) d\omega,$$

where $g(\omega)$ is VDOS normalized to unity. $k_B$ is the Boltzmann constant, $T$ is the absolute temperature, $\hbar$ is the Planck's constant and the other symbols have their usual meanings.

The localization behaviour of the vibrational modes is investigated by calculating the inverse participation ratio (*IPR*) which for a vibrational mode $j$ is defined as [10]

$$IPR \equiv P_j = \left[ \frac{\left( \sum_{i=1}^{N} \left| u_i^{\,j} \right|^2 \right)^2}{\sum_{i=1}^{N} \left| u_i^{\,j} \right|^4} \right]^{-1},$$

where $u_i^{\,j}$ is the displacement of $i$th atom from its equilibrium position in the $j$th vibrational mode and the sum is over all the $N$ atoms in the model sample. The inverse participation ratio varies from $1/N$ (all atoms have equal vibrational amplitude) to 1 (only a single atom has considerable vibrational amplitude). If all the atoms have nearly equal vibrational amplitudes in a particular vibrational mode, the mode is called an extended mode and if only a few atoms have a considerable vibrational amplitude, then the mode is said to be localized. Thus the vibrational mode, for which the participation ratio is approaching 1, will be a localized mode.

## 3. Results and Discussion

It is difficult to generate samples of amorphous materials of large size with the highly accurate methods based upon the density functional theory. But one can get some rough insight (at least) into the matter with small sized samples as well. The things which have been observed here in concern with the boson peak in amorphous silicon are quite interesting and providing motivations for further investigation of actual origin of boson peak in relation to the week strained bonds. The VDOS/$\omega^2$ plots for the model samples of pure and hydrogenated amorphous silicon are given in Fig.1 (a). There occurs a broad band peak in these plots near 100 cm$^{-1}$ in all the samples.

Table 1: Percentage of Coordination Defects (3 and 5 folded Si atoms) in different samples.

| Samples | Coordination Defects (3 and 5 folded Si atoms | | |
|---|---|---|---|
| | 3 | 5 | Total |
| ASi | 3.08 | 5.91 | 8.99 |
| ASiH | 4.47 | 7.33 | 11.50 |
| BSiH | 3.58 | 5.61 | 9.19 |
| CSiH | 2.41 | 2.07 | 4.48 |
| WSiH | 1.84 | 2.29 | 4.13 |

This peak with broader bump is identified as the boson peak. In ASi sample, this broad band peak has the broader bump at around 95 cm$^{-1}$ while in a-Si:H samples viz: ASiH, BSiH, CSiH and WSiH, it has the broader bump at around 107, 104, 105 and 105 cm$^{-1}$ respectively. The Raman scattering on bulk silicon [11] has also reported the boson peak at 114 cm$^{-1}$. In a-Si:H samples, the height of boson peak is smaller as compared to that in pure amorphous silicon sample (ASi) which means that there is a decrease in the low frequency modes in amorphous silicon upon hydrogenation. Liu et al. [12] have also reported such a reduction in the low frequency modes in amorphous silicon upon hydrogenation. Thus, the hydrogenation of pure amorphous silicon decreases the height and shifts the position of boson peak towards higher frequencies. The coordination defects in different samples are given in Table 1. The coordination defects in ASiH and BSiH samples are more, while in CSiH and WSiH samples these are less than those in the ASi sample. However, the height and position of boson peak in different samples show a different trend. While the height of boson peak is less, its position is at higher frequencies in all the a-Si:H samples as compared to the pure amorphous silicon sample (ASi). This indicates that there is no connection between the coordination defects and the boson peak in amorphous silicon. Recently, Nakhmanson et al. [13] have also reported similar conclusions.

The VDOS is further used to calculate the specific heat ($C$) of generated samples. In Fig.1 (b), we have displayed $C/T^3$ versus $T$ curves which show a clear bulge in these curves at low temperatures in all of our samples. The existence and position of this well-known bulge (the so-called boson peak) in

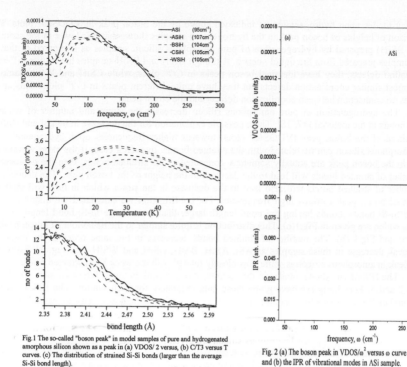

Fig.1 The so-called "boson peak" in model samples of pure and hydrogenated amorphous silicon shown as a peak in (a) VDOS/ 2 versus, (b) C/T3 versus T curves. (c) The distribution of strained Si-Si bonds (larger than the average Si-Si bond length).

Fig. 2 (a) The boson peak in VDOS/$\omega^2$ versus $\omega$ curve and (b) the IPR of vibrational modes in ASi sample.

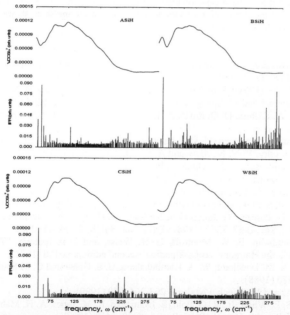

Fig. 3 The boson peak in VDOS/$\omega^2$ versus $\omega$ curves and IPR of vibrational modes in a-Si:H samples (ASiH, BSiH, CSiH and WSiH). For clarity these are displayed separately for each sample.

the specific-heat at low temperatures in our samples is in good agreement with the previous theoretical [14, 15] and experimental results [16]. The height of boson peak in the hydrogenated samples is lower as compared to that in the pure amorphous silicon sample. This can be attributed to the decrease in low frequency modes upon hydrogenation as indicated above by the boson peak in VDOS/$\omega^2$ plots. The comparison of heights of boson peak in the hydrogenated samples shows that it is lower in the samples (ASiH, BSiH) prepared by hydrogenation of pure amorphous silicon samples as compared to that in other samples prepared from the liquid quench. Although ASiH and BSiH samples have quite different coordination defects, they have identical boson peaks in $C/T^3$ plots, while CSiH and WSiH samples have almost similar coordination defects but they have quite different peaks in $C/T^3$ plots. This again indicates no connection between coordination defects and the boson peak.

The hydrogenation of pure amorphous silicon decreases the strain and number of strained (weak) bonds in the material [17, 18]. Thus, possible explanation for the decrease in height and shift in the position of the boson peak (VDOS/$\omega^2$ plots) towards higher frequencies upon hydrogenation of pure amorphous silicon can be related with the strained bonds. In such a case, if the extra vibrational modes in the boson peak are actually somehow connected to the strained bonds, then the decrease in the number of strained bonds will lead to the decrease in the height of the boson peak. The decrease in the number of strained bonds will also lead to the decrease in the strain which in turn will shift the position of boson peak towards high frequencies. For confirmation we calculated the distribution of strained Si-Si bonds (bonds having the bond length larger than the average Si-Si bond length) in our samples which are given in Fig.1(c). This distribution is quite similar to the behaviour of boson peak in Fig.1 (a) and Fig.1 (b). The number of strained bonds decreases in the same order as the height of boson peak decrease in these samples as ASi, ASiH, BSiH, CSiH and WSiH [see Fig. 1]. Thus, the boson peak in amorphous silicon is some how closely related with the week Si-Si strained bonds.

The *IPR* of vibrational modes in the vicinity of boson peak in the different samples are shown in Fig. 2 and 3. For clarity we have given these data separately for each sample. The IPR values of vibrational modes show that the vibrational modes in the vicinity of the boson peak are extended.

## 4. Conclusions

The hydrogenation of amorphous silicon reduces the height and shifts the position of boson peak towards higher frequencies. While the height and position of boson peak shows no connection to the coordination defects, but it shows a close relationship with the strained weak Si-Si bonds in amorphous silicon. We also studied the localization of vibrational modes and found that vibrational modes in the vicinity of the boson peak are extended.

## References

[1] V. K. Malinovsky, V. N. Novikov, P. P. Parshin, A. P. Sokolov and M. G. Zemlyanov, Europhys. Lett. **11**, 43 (1990); V. K. Malinovsky, V. N. Novikov and A. P. Sokolov, Phys. Lett. A **153**, 63 (1991); Seiji Kojima and Masao Kodama, Physica B **263-264**, 336 (1999); Frank Finkemeier and Wolfgang von Niessen, Phys. Rev. B **63**, 235204 (2001).
[2] P. Sokolov, A. Kisluik, D. Quitman, A. Kudlik and E. Rössler, J. Non-Cryst. Solids 172-174, 138 (1994).
[3] U. Buchenau, A. Wischnewski, D. Richter and B. Frick, Phys. Rev. Lett. 77 4035 (1996).
[4] V. K. Malinovsky, V. N. Novikov and A. P. Sokolov, J. Non-Cryst. Solids 90, 485 (1987); E. Duval, A. Boukenter and T. Achibat, J. Phys.: Condens. Matter 2, 10227 (1990).
[5] S. R. Elliot, Europhys. Lett 19, 201 (1992).
[6] W. Schirmacher, G. Diezemann and C. Ganter, Phys. Rev. Lett. 81, 136 (1998).
[7] Ranber Singh, S. Prakash, Nitya Nath Shukla and R. Prasad, Phys. Rev. B 70, 115213 (2004).
[8] M. Elstner, D. Porezag, G. Jungnickel, J. Elsner, M. Haugk, Th. Frauenheim, S. Suhai and G. Seifert, Phys. Rev. B 58, 7260 (1998) ; Phys. Stat. Sol.(b) 217/1, 41 (2000).
[9] A. A. Maradudin, E. W. Montroll, G. H. Weiss, and I. P. Ipatova, "Theory of Lattice Dyanamics in the Harmonic Approximation" second edition, p. 130 (1971).
[10] R. Biswas, A. M. Bouchard, W. A. Kamitakahara, G. S. Grest amd C. M Soukoulis, Phys. Rev. Lett. 60, 2280 (1988).
[11] C. Laermans and M. Coeck, Physica B 263-264, 280 (1999).
[12] Xiao Liu, B. E. White, Jr., R. O. Pohl, E. Iwanizcko, K. M. Jones, A. H. Mahan, B. N. Nelson, R. S. Crandall and S. Veprek, Phys. Rev. Lett. 78, 4418 (1997).
[13] S. M. Nakhmanson, D. A. Drabold and N. Mousseau Phys. Rev. B 66, 087201 (2002).
[14] J. L. Feldman, P. B. Allen, and S. R. Bickham, Phys. Rev. B 59, 3551 (1999).
[15] S. M. Nakhmanson and D. A. Drabold, Phys. Rev. B 61, 5376 (2000).
[16] M. Mertig, G. Pompe, and E. Hegenbarth, Solid State Commun. 49, 369 (1984).
[17] S. Sriraman, S. Agarwal, E. S. Aydil and D. Maroudas, Nature 418, 62 (2002).
[18] B. Tuttle and J. B. Adams, Phys. Rev. B 57, 12859 (1998).

Brill Academic Publishers
P.O. Box 9000, 2300 PA Leiden
The Netherlands

*Lecture Series on Computer*
*and Computational Sciences*
Volume 4, 2005, pp. 531-536

# The Fuzzy Neural Network:
# Application For Trends in River Pollution Prediction

K. Skowronska[1], Z. Skowronski, M. Biziuk

Department of Analytical Chemistry,
Chemistry Faculty, Gdansk University of Technology,
11/12 Narutowicza Street, 80-952 Gdansk, Poland

Received 11 July, 2005; accepted in revised form 7 August, 2005

*Abstract:* One of the main problems in environmental monitoring and management is how to control complex and nonlinear systems, like rivers. River monitoring and modeling is one of the key elements of global environmental monitoring policy. Usually mathematical models are used for this purpose, but sometimes these models require too much data and response time is too long. The behaviors of complex or nonlinear systems are difficult and sometimes impossible to describe, using numerical models. On the other hand, quantitative observations are often required to make quantitative control decisions [1]. Artificial intelligence (AI) techniques, and the integration of these methods to provide a soft computing (SC) solution, can be used to model complex problems while avoiding the disadvantages of mathematical models. AI methods, which include artificial neural networks, fuzzy logic and genetic algorithms, offer real advantages over conventional modeling, including the ability to handle large amounts of dynamic, non-linear or noisy data, and such tools can be especially useful when underlying relationships are not fully understood. Other associated benefits would include: improved performance, faster model development and calculation times and improved opportunities to provide prediction estimate confidence through comprehensive bootstrapping operations. Wherever possible, in appropriate circumstances, these systems should also utilize the knowledge of experts (so called expert systems) and attempt to incorporate the imprecision and uncertainty associated with various aspects of the modeling problem [2].

*Keywords:* Rivers pollution, environmental protection, monitoring network, fuzzy neural networks, heavy metals

*Mathematics Subject Classification:* 68T05, 68T35, 92F05, 94D05

*PACS:* 68T05, 68T35, 92F05, 94D05

Two key issues in river modeling are: precision and required amount of data. For obtain the precision, a very good understanding of the form of control model is needed and decreasing the level of the measurement noise is required. The measurement noise is a significant source of imprecision in any system. Decreasing noise means increasing sample reading numbers, but it also means that economical cost is rising. The number of replicates required to achieve a sufficient level of precision could be determine using the statistical methods. The type of the control model also determines the level of precision of obtained results. A numerical model provides high precision, but the complexity or non-linearity of a process may make this type of model unfeasible. A linguistic model (a type of fuzzy model) provides an alternative to these cases. The process is described in common language, but each particular user has a different definition of the value carried by word description and the imprecision of the model could not be measured in a statistical way [1]. Lack of success and criticism of the empirical approach based on regression techniques has arisen because the relationship between discharge and water chemistry is essentially nonlinear, time dependent and varies from site to site. At different time and space scales this relationship exhibits hysteresis with different properties [3]. Of course, as it was

---

[1] Corresponding author. E-mail: skowron@chem.pg.gda.pl

mentioned above, a lot of mathematical models exist and these models were used with good quality of results, but there is a space for another approach and AI methods could fill this gap.

The main problem of predicting chemical loads in a river system has a lot of in common with a key issue in the determination of the impact of anthropogenic impact on an aquatic ecosystem. The major problems relate to the complexity of the processes within a watershed. These problems could be defined as:

- a physical – chemical balance between atmospheric inputs,
- and/or direct anthropogenic impact,
- biogeochemical processes within the catchments, climatic factors and mechanisms of runoff generation.

Several approaches have been adopted to model the hydrochemistry. For example, some models present the stream as a mixture of soil and water from two ideally mixed reservoirs, each is described by a simple chemical equation. The key problems and perspectives within this approach are mainly related to the transition from the models which deal with average or lumped systems toward synthesized hydro chemical models which are able to incorporate the internal structure and state variables of the catchments. The quality of the different models is determined by:

- correspondence between the model and natural physical events at the time and space scales in which the model is used,
- the intrinsic nature of the model equations,
- the design of the field program used to test the model.

Usually the models contain empirical parameters which are estimated by fitting observed and simulated characteristics, and the applicability of these coefficients may be limited to the specific case [3].

There is a few popular river models existing, e.g. QUASAR [4] or QUAL2E [5]. The model QUASAR (Quality Simulation Along Rivers) has evolved over a number of years during which time there have been many applications to rivers in the UK and overseas. The model was originally developed as part of the Bedford Ouse Stud with the primary objective of simulating the dynamic behavior of flow and water quality along the river system. QUASAR should be applied for non – tidal rivers. The model combines upstream inputs from tributaries, and point and diffuse effluents, to calculate the flow and water chemistry in the river at points further downstream. Changes in river water quality may arise from natural causes or from accidental spillage and anthropogenic effluents [4]. QUAL2E is a comprehensive stream water quality model which can be operated as a steady – state or a dynamic model, and can simulate up to 15 water quality constituents. It allows multiple waste discharges and tributary flows. It assumes that the major transport mechanisms advection and dispersion are significant only along the main direction of flow (longitudinal axis of the river). The model is applicable to dendrite streams that are very mixed. Streams are divided into reaches, each reach having uniform hydraulic characteristic. The model is limited to the simulation of time periods with constant stream flow and waste loads [5].

One thing, especially worth mentioning is adaptation models dedicated for waste water treatment plants (ASM 1 model – activated sludge model no. 1) for river modeling. The adaptation of ASM 1 to river conditions consists two facts:

- the inclusion of river specific state variables and processes
- and of the modification of some of the existing processes and typical parameter and value ranges.

It needs to be stated that an ASM 1 approach for rivers involves the definition of a large number of new variables and parameters as a consequence of the fact that the methodology is more consistent and unambiguous then the traditional water quality modeling approach. For example, inclusion of the nitrogen cycle (without taking the algal growth into consideration) into an ASM 1 type river model would involve the definition of five new state variables and seven parameters [6].

This abstract presents a fuzzy neural network application in predicting water quality parameters in a river, and as the particular example, the Cu concentration prediction is shown.

Methodology described later can be used for parameter prediction for environmental analysis of possible threats and for planning of environmental management.

## 1. Study area and data collection

The object of monitoring program was the Kacza River in Gdynia (Fig. 1). The main information about this river are:

- total length is 18 km,
- the slope of terrain is about 2.5%,
- catchments area is 52.6 km$^2$, encompass an assorted mixture of small urban and rural land uses,
- the flow of the river is in the range of 0,02-3,5 m/s, depending on the level of the water [7].

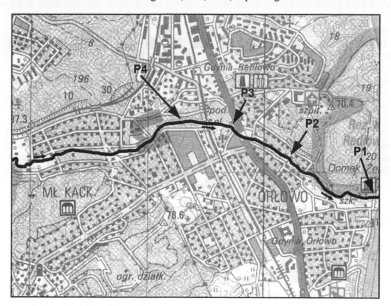

Figure 1: Sampling sites location.

Four gauging (P1-P4) stations were located throughout the catchments, along the mouth (Kacza flows into the Baltic Sea), and three points up the river. Data sets were collected for a half year period and a wide spectrum of parameter were examined, e. g. NO$_3$-, NO$_2$-, PO$_4^3$-, conductivity, COD, heavy metals: Cu, Cd, Pb, Zn. The mean sampling interval was 3 days.

The reason of choosing this particular river for investigation was that this is one of small influxes of the Baltic Sea, not covered by the main monitoring network. And the water quality in these rivers is not good enough to abandon the control, especially when the loads of individual rivers were summed up. Designing the tool for increasing the productivity of monitoring system could support the spreading the monitoring network without significant increasing the costs.

## 2. Analysis

As an example, Cu concentration was chosen as a parameter for fuzzy prediction. Cu concentration was examined by atomic spectroscopy method, with 210 VGP Buck Scientific apparatus.

The application target of FNN was checking if these methods are able to predict the value of chemical compound concentration with good accuracy. As a result of river monitoring, series of values have been collected for each analysis station. Each sample set contains measured value and time point (date) measurements. These sets can be threatened as time based series of sample. Undetermined characteristics of these series and its prediction are similar to the problem of predicting chaotic time series when FNN reaches very good results.

The general idea of prediction undetermined or chaotic time series is based on the usage of current and past values to predict further ones. As an input for FNN past values are used. The number of these values determines FNN complexity and calculation time. In the training process, past and current values are passed to inputs ($x_{t-n}, \ldots, x_{t-2}, x_{t-1}, x$) and the first further value ($x_{t+1}$) is passed to output.

The learning algorithm calculates the membership function and rules. This process is very similar to learning of ANN but in FNN, rules and membership functions are calculated instead of weights in classic ANN.

For the purpose of this paper, six levels of depth were analyzed. On the first level, only the current value was used for prediction. On the second, current and one past values were used and one more past value was taken in the third level and so on up to the sixth. To determine the degree of complexity and the time required for learning, a series of FNN was generated using from one to six input values. The learning process has been run on 500, 2500 and 5000 iterations. For FNN with more than three inputs, calculation time grows rapidly, up to few hours. For the next analysis, the first three structures were chosen. In each case, four gauss type membership functions were used in the input layer. Calculations were made using the ANFIS toolbox in Matlab.

In the next experiment, a prediction has been made to check its accuracy. 80% of the collected data sets have been used to train FNN in 1000 iterations while 20% of the data has been prepared to test results. During the test procedure, a long time prediction has been plotted and compared with test data to check what the accuracy of the prediction is and what the distance of accurate prediction is.

## 3. Results and discussions

Fuzzy neural networks and in general AI techniques were advocated as systems which could find the solution in the complex problems, like analyzing river systems. The function of pollutant (in that particular case copper) concentration versus time is strongly nonlinear, so FNN application seemed to fit perfectly.

As mentioned above, six prediction scenarios were investigated for Cu concentration. The time distance between samples was three days. The network was run over 100 iterations in the training phase with training sets with 20 data rows each. Measured and predicted Cu values are shown on Fig 2 to 4, plotted concurrently in time function.

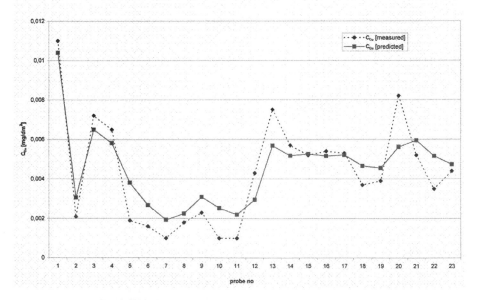

Figure 2: Results of the prediction for the first level of depth.

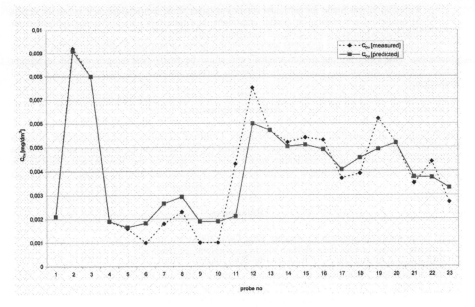

Figure 3: Results of the prediction for the second level of depth.

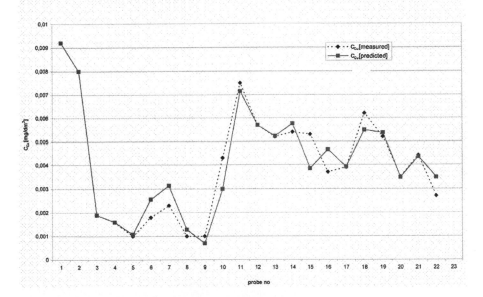

Figure 4: Results of the prediction for the third level of depth.

Based on the prediction results, the following conclusions can be drawn:
- high accuracy of the Cu concentrations predictions suggest that this approach is appropriate and properly represents water quality. For the third level, the error for the first four steps ahead is less than 0.1%. In this case, this means prediction 12 days ahead,
- all natural trends in a river's biochemical and hydrological processes are reflected in predicted results,

- a prediction was made for particular conditions (river flow, temperature, a. s. o.) so the results of prediction are reliable at similar conditions, and then this network is not applicable for modeling when extremely different conditions occur.

This experiment indicates that FNN are promising supporting tool for monitoring studies and data proceeding. This FNN is "puzzle-constructed" and could be easily redeveloped and fit to another dataset (river, or monitoring assessment). However, further analyses are indispensable for a final evaluation.

## Acknowledgments

The Department of Analytical Chemistry constitutes "Centre of Excellence in Environmental Analysis and Monitoring" which is a research project supported by the European Commission under the Fifth Framework Programme and contributing to the implementation of the Key Action "Sustainable Management and Quality of Water" within the Energy, Environment and Sustainable Development (Contract No.: EVK1-CT-2002-80010). The authors acknowledge this generous support.

## References

[1] Hayward G., Davidson V., *Analyst*, 128 , 1304 (2003)
[2] http://www.ccg.leeds.ac.uk/simon/linda/geocomp98/gc22.htm
[3] Sokolov S, Black K.P., *J. Hydrol.*, 178, 311 (1996)
[4] Ferrier R.C., Whitehead P.G., Sefton C., Edwards A.C. and Puhn K., *Wat. Res.*, 8(29), 1950 (1995)
[5] Drolc A., J. Zagorc Koncan, *Wat. Res.*, 11(30), (1996), 2587
[6] Maryns F., Bauwens W., *Wat. Sci. Tech.*, 5(36), 201 (1997)
[7] Multi-autors work, realized by: Biuro Projektów i Doradztwa Technicznego, sp. z o.o., Plan sytuacyjno-wysokościowy zlewni rzeki Kaczej, Hydrotechnika, Gdansk 1997

Brill Academic Publishers
P.O. Box 9000, 2300 PA Leiden
The Netherlands

*Lecture Series on Computer*
*and Computational Sciences*
Volume 4, 2005, pp. 537-541

# Key Concepts for Implementation a Web-Based Project Management in Worldwide Science Project

Z. Skowronski[1], K. Skowronska, M. Biziuk, J. Namieśnik

Department of Analytical Chemistry,
Chemistry Faculty, Gdansk University of Technology,
11/12 Narutowicza Street, 80-952 Gdansk, Poland

Received 11 July, 2005; accepted in revised form 8 August, 2005

*Abstract:* Evolution of the internet that can be observed last years make the web based technologies faster, more useful and easier to develop. The new architectures allows to create bigger and more complicated solutions and growing internet network makes them faster and allows transferring much more data than a few years ago.

In times of international integration of the science community, projects realized in international environment became more popular than ever. Unfortunately most of these projects didn't use potential of internet technologies and limit themselves to emails and the public web pages. The solutions that are commonly used in a software development or an electronic information management can be used in many cases.

The idea that is behind this paper grown in observing ELME project (European Lifestyle And Marine Ecosystems) that authors are involved in,. All the concepts and solutions are implemented or are currently in development phase, and are still evolving according to the requirements. The ideas presented in this paper are well known for all developers and peoples involved in a software development. Fact is that a science project is not so far from a software one. The difference is only on subject and many of the methodologies could be adapted with small modifications only.

*Keywords:* project management, web technologies, database, document management

*Mathematics Subject Classification:* 68U35, 90B05, 90B70, 90B90

*PACS:* 68U35, 90B05, 90B70, 90B90

## 1. The project workspace

A main difficulty of each project realized by the international team is a separation of his members. A distance between members and different environments of each one don't help in creation of integrated team and good team-play. Every one who has been involved in dynamic project located in one place felt this connection between peoples. Because of this, separated teams are slower and projects are less dynamic. It's just a human nature.

So what can be done to improve feeling of a team and project dynamic? Observation of internet community gives a first answer – let the people free communicate themselves; give them substitute of being together. Today's technologies and software solutions allows a full scale communication between the peoples located around the world from sending the simple text, through the voice connections up to the videoconference. The simplest way is to use instant messengers like MSN or ICQ. This software is very popular now, especially across younger generation. Most of them support file transfer, voice connections and conferences between numbers of users.

The communication is a circle that connects all members together (Fig. 1). A heart of the workspace is a project server. It's a workhorse and the centre of project development. Solutions based on that schema are very common in software development. The every well managed software project uses the CVS (Concurrent Version System), task management and others complicated functions like

---

[1] Corresponding author. E-mail: ziemowits@gmail.com

document management, analysis and project tools, requirement gathering, tests and many others. Of course many of these functions are useless for a science project. Otherwise some key functions are important for any project:

- document management,
- task management,
- global message board,
- other project specific.

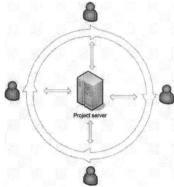

Figure 1: Communication flow in project workspace.

## 1.1. Document management

Document management is the key function realized by the project server (Fig. 2). It gives the project members ability to work together on the one document. It's a key of one of most effective agile development methodologies named extreme programming. Of course this methodology is created for the software development but is based on the human nature. Every one can imagine work over the one document with the team colleague on the one table. It's simple, fast and effective. Because of the distance between the members it's impossible to get so close relation in an international project but there is something that can be done even then. Each of the project documents can be stored in the one place where will be accessible for each of the members.

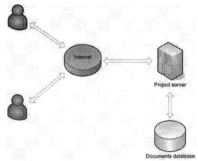

Figure 2: The structure of document management.

Document management should implement set of functions:

- document storage,
- version control,
- comments and reviews,
- security management.

Storage is a central place where documents are physically stored. In the best approach it is a database but can be also the file system. Documents are allowed to access for authorized users (project members, third party members) and author (or authors) itself. Very important function in document management is version control and change management. To explain this function, a typical document flow is presented on Figure 3. In this example, two users downloaded this same document in this same moment (approximately). In moment of downloading this document was signed with version number 1.0. Each of the users made his own modifications in the document body. These changes can be located in completely independent parts of the document, but they could also change these same parts. When the first user finishes editing he send the document back to server where it is marked as a new version of the document with number 1.1. Next the second user sends his version of the document to server. Server marks new version with number 1.2. In that moment server needs to check if both modifications are located in these same parts of the document or not. If not, then server should consolidate changes made by users into one document and mark with version number 1.3. If changes were made in these same parts, the document status should be marked as undetermined and send back to the users for review. This solution protects documents before overwriting changes without knowledge of the author. If two users made changes in this same part and in this same time, then each one hasn't opportunity to see changes made by the other.

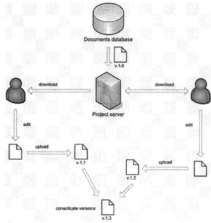

Figure 3: Document flow in document management.

This sample flow is an overall illustration of document management and detailed process is more complicated. Many implementation details depend on used technology and covers hard technical aspects.

### 1.2. Task management

Task management is direct tool for manage time and work in the project. It's designated for the project manager or task group managers. Schema of task management is simple. The authorized user can define the task and assign it to a selected user or users. After finishing the task, the user has to close it. A simple report gives the manager information about the current status of defined tasks and deadlines. Usually the task contains description, amount of work, time limit, references to documents and resources definitions. The task can be assigned to the one user or to number of them. In some task models number of assigned users and resources can affect other parameters.
There are a few common models of tasks:

- Tasks with defined time – the manager define a time required for the task completion. The user or users assigned to the task have to fulfil this time limit. This is static model and tasks almost never are changed.
- Tasks with undefined time – the manager define task and amount of work. The user defines time required for its completion. In this model time is fluid and can be changed by the user.
- Tasks with defined work – the task definition contain information about amount of work that is required to complete task. Time required for completion is calculated on the basis of assigned users. This model is the most fluid and time can change in any moment.

From other parameters defined in task, one of the most important are resources. Resources are everything but users. These are tools, materials, documents and also third party workers.
Third important parameter of the task is set of references to documents. This feature is able to connect task management with document management. Input documents for the task could be documentations, project assumptions, guidelines and results of the other tasks. An output of the task is

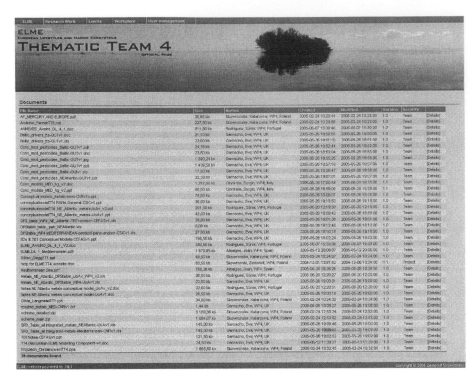

Figure 4: List of documents.

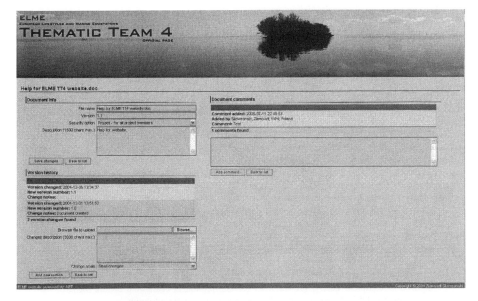

Figure 5: A document detail page.

completed work that can be also represented by document. In that moment two parts of project activity become connected.

Benefits of task management usage are serious and could have serious impact for whole project, beginning from better utilization project members and resources up to financial reports. Having well documented any work that was done in project and any resource used, manager can get clear information about costs of the project. The each resource and the each user have cost of usage. Information about its assignment to tasks allows easy calculated consumed costs. Moreover, planning further tasks allow planning costs that will be required during whole project.

Unfortunately, many projects haven't used task management even totally or in part. Full scale task management is too formal and too complicated for this type of projects but general functions of this methodology could be adapted.

### 1.3. Central communication

Direct communication circle, which was described before, allows creation of project community and workspace. Central communication is realized by more formal way. It can be compared to emails but is a bit more advanced and integrated with rest of the project management. Two different areas can be precise:

- free forum;
- comments for documents and/or tasks.

Forum is well known and very popular way of communication. Many projects use this solution to organize place for sharing ideas or just free conversation. With a little effort this well known forum can be transformed into powerful communication system. In fact all what should be done in this area is connection between forum topics and documents and tasks. In that way, efficient system of document commenting can be created.

When the new document is created the corresponding topic on forum is also created. This topic is a place where users can discuss about the associated document. This topic is also used by document management module to place system messages that inform about new versions and changes. It seems to be good idea to allow direct quotation of the document in forum messages. This requires deeper integration with document management but seems to be useful.

## 2. Summary

Key elements of the web based project management solution were presented. This description didn't fill the subject. There is many important parts to be defined like project portal which is interface for project server. Not less important are technical details of implementation.

Early version of described elements was created for the ELME project and is still in development (Fig. 4 and 5). Currently most of ideas presented above are in development phase and should be ready for use in next few months.

## Acknowledgments

This work was generously supported by ELME Project (European Lifestyle and Marine Ecosystem Project), contract number: 505576, which is lead in 6th Framework Programme, SUSTDEV-2002-3.III.2.1 - Generating models of socio-economic impacts on biodiversity and ecosystems.

## References

[1] ELME project homepage, www.ceeam.edu.pl/elme
[2] G. Booch, J. Rambaugh, I. Jacobson, *The Unified Modelling Language User Guide*, Addison Wesley Longman Inc., 1999
[3] G. R. Heerkens, *Project Management*, McGraw-Hill, 2002
[4] *Microsoft Visual C#.NET Language Reference*, Microsoft Press, 2002
[5] Microsoft Visual Studio 2005 homepage, http://lab.msdn.microsoft.com/vs2005/default.aspx
[6] Microsoft Visual Studio .NET homepage, http://msdn.microsoft.com/vstudio/

Brill Academic Publishers
P.O. Box 9000, 2300 PA Leiden
The Netherlands

*Lecture Series on Computer
and Computational Sciences*
Volume 4, 2005, pp. 542-545

# Modeling the dissociation of carbon dioxide and methane hydrate using the Phase Field Theory

A. Svandal[1], B. Kvamme[1,*], L. Grànàsy[2] and T. Pusztai[2]

[1]Department of Physics and Technology, University of Bergen, Allègaten 55, N-5007 Bergen, Norway
[2]Research Institute for Solid State Physics and Optics, H-1525 Budapest, POB 49, Hungary

Received 12 July, 2005; accepted in revised form 31 July, 2005

*Abstract:* Hydrate that is exposed to fluid phases which are undersaturated with respect to equilibrium with the hydrate will dissociate due to gradients in chemical potential. Kinetic rates of methane hydrate dissociation towards pure water and seawater is important relative to hydrate reservoirs that are partly exposed towards the ocean floor. Corresponding results for carbon dioxide hydrate is important relative to hydrate sealing effects related to storage of carbon dioxide in cold aquifers. In this work we apply a phase field theory to the prediction of carbon dioxide hydrate and methane hydrate dissociation towards pure water at various conditions both inside and at the stability region of the hydrates with respect to temperature and pressure. As expected from the differences in water solubility the methane hydrate dissolves significantly slower towards pure water than carbon dioxide hydrate.
*Keywords*: gas hydrate; phase-field theory; carbon dioxide; dissociation; methane
*PACS*-2001: 81.10.Aj, 81.10.Dn, 82.60Lf

## 1. Introduction

Gas hydrates are crystalline structures in which water forms cavities that enclathrates small non-polar molecules, so called guest molecules, like for instance $CO_2$ or $CH_4$. Macroscopically the structure looks similar to ice or snow but unlike ice these hydrates are also stable at temperatures above zero °C. The enclathrated molecules stabilize the hydrate through their volume and interactions with the water molecules which constitutes the cavity walls. Natural gas hydrate reservoirs which are partly exposed towards the seafloor will dissociate due to gradients in chemical potentials between the enclathrated guest molecules and chemical potentials of the same components in the seawater outside the hydrate structure. As a greenhouse gas methane is in the order of 25 times more aggressive than $CO_2$. It is therefore an important global challenge to be able to make acceptable predictions of the dissociation flux of methane, and the corresponding flux of methane that escapes to the atmosphere after biological consumption and conversion through inorganic and organic reactions. In a more general sense it is not likely that any hydrate reservoirs are thermodynamically stable in a rigorous thermodynamic sense. Most hydrate reservoirs are trapped by clay layers and kept in a state of extremely slow dissociation dynamics due to slow transport of dissociated molecules through clay layers above the hydrate. The stability of $CO_2$ storage in cold reservoirs may be enhanced through the formation of hydrate films. These hydrate films will form rapidly on the $CO_2$/water interface and form a more or less closed membrane which reduces the transport of water and $CO_2$ across the hydrate film. Eventually this will lead to a dynamic situation where the dissociation of hydrate towards the aqueous phase above will be the rate limiting process that determines the net flux of CO2 through the hydrate sealing.

## 2. Phase field theory

A phase field theory has previously been applied to describe the formation of $CO_2$ and $CH_4$ hydrate in aqueous solutions [1-3]. In this paper the theory is applied to model the

---

* Corresponding author. E-mail: bjorn.kvamme@ift.uib.no

dissociation of the hydrates. The solidification of hydrate is described in terms of the scalar phase field $\phi$ and the local solute concentration $c$. The field $\phi$ is a structural order parameter assuming the values $\phi=0$ in the solid and $\phi=1$ in the liquid. Intermediate values correspond to the interface between the two phases. Only a short review of the model will be given here. Full details of the derivation and numerical methods can be found elsewhere [3-6]. The starting point is a free energy functional

$$F = \int dr^3 \left[ \frac{1}{2}\varepsilon^2 T |\nabla \phi|^2 + f(\phi,c) \right], \tag{1}$$

with $\varepsilon$ a constant, T is the temperature and the integration is over the system volume. In this paper we use $c$ for concentration with units moles per volume and the mole fraction of the guest is termed $x$ and is dimensionless. Assuming equal molar volume for the two components the following relation: $c=x/v_m$ can be applied, where $v_m$ is the average molar volume. The range of the thermal fluctuations is on the order of the interfacial thickness and, accordingly, $\varepsilon$ may be fixed from knowledge of this thickness. The gradient term is a correction to the local free energy density $f(\phi,c)$. To ensure minimization of the free energy and conservation of mass, the governing equations can be written as

$$\dot{\phi} = -M_\phi \frac{\delta F}{\delta \phi} \tag{2}$$

$$\dot{c} = \nabla \cdot \left( M_c \nabla \frac{\delta F}{\delta c} \right), \tag{3}$$

where $M_c$ and $M_\phi$ are the mobilities associated with coarse-grained equation of motion, which in turn are related to their microscopic counterparts. The local free energy density is assumed to have the form

$$f(\phi,c) = wTg(\phi) + [1 - p(\phi)] f_S(c) + p(\phi) f_L(c) \tag{4}$$

where the "double well" and "interpolation" functions have the forms which emerge from the thermodynamically consistent formulation of the theory [6]. The parameter $w$ is proportional to the interfacial free energy and can be deduced from experimental measurements or predicted from molecular simulations of representative model systems. For this work the applied value used was the experimental value for water/ice reported as 29.1 mJ/m² [7]. The thermodynamic functions of the hydrate and the aqueous solutions have been determined using solubility data and molecular dynamics simulations. Here only the form of the functions is given. A full description of the derivation can be found in Ref. [2]. The free energy functions in Eq. (4), for a phase P, have the general form:

$$v_m f_P = x \cdot \mu_c + (1-x)\mu_w, \tag{5}$$

were $\mu_c$ and $\mu_w$ are the chemical potentials of the guest molecule and water in that phase. For the aqueous solution we use

$$\mu_c = \mu^\infty_c(T) + RT \ln(x\gamma_c). \tag{6}$$

Here $\mu^\infty_c(T)$ is the chemical potential at infinite dilution of component $c$ in water. $R$ is the universal gas constant and $\gamma_c$ is the activity coefficient of the guest in an aqueous solution in the asymmetric convention ($\gamma_{CO_2}$ is unity in the limit as x goes to zero). For water we have

$$\mu_w = \mu_w^{pure}(T) + RT \ln\big((1-x)\gamma_w\big). \tag{7}$$

Here $\mu_w^{pure}(T)$ is the chemical potential of pure water. The activity coefficient of water can be obtained through the Gibbs-Duhem relation. The expression for the chemical potential for water in hydrate with one type of guest molecule is

$$\mu_w^H = \mu_w^{0,H} + \sum_j RT\nu_j \ln\big(1-\theta_j\big). \tag{8}$$

Here $\mu_w^{0,H}$ is the chemical potential for water in an empty hydrate structure. The sum is over small and large cavities, where $\nu_j$ is the number of type $j$ cavities per water molecule. $\theta_j$ is the filling fraction of cavity $j$ given as $\theta_j = x_j/(\nu_L(1-x_j))$, where $x_j$ is the molar fraction with respect to cavity $j$ only. The chemical potential for the guest molecule in hydrate is

$$\mu_c^H = \Delta g_{c,j}^{inc} + RT \ln\left(\frac{\theta_j}{1-\theta_j}\right), \tag{9}$$

Here $\Delta g_{c,j}^{inc}$ is the free energy of inclusion of guest molecule $c$ in cavity $j$. The total free energy for each phase and each component are shown in Fig. 1.

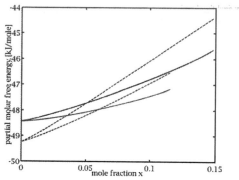

Figure 1: Free energies of hydrate (solid line) and the aqueous solution (dashed line) as a function of the mole fraction at 40 bars and 1°C. Red lines are the carbon dioxide system, and the blue lines are for the methane system.

## 3. Hydrate dissociation simulations

The model has been implemented with a narrow 2D planar geometry, simulating dissociation of a planar front. The temperature is chosen to be 1°C which is a realistic temperature for cold reservoirs and the seafloor. Simulations have been done at 10 and 40 bars representing 400 m depth which is a normal sea floor depth and inside the hydrate stability region. No flux boundary conditions at the walls were assumed and the grid resolution used was 0.4 nm. The time step was $1.6 \cdot 10^{-12}$ s. Initially a pure water solution and a hydrate film with thickness 16 nm were assumed. The movement of the front was tracked by following the $\phi=0.5$ value. In Fig. 2 simulations of carbon dioxide and methane hydrate dissociation is shown at the two different pressures. The methane dissociation rate is much slower than for carbon dioxide hydrate. This can be explained by the much lower solubility of methane in water in agreement with previous results and discussions in Ref. [1,2], that the controlling mechanism for these systems is the chemical diffusion.

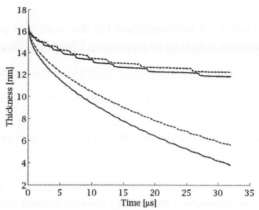

Figure 2: Thickness of the hydrate film as a function of time for the dissociation of carbon dioxide hydrate (red), and methane hydrate (blue). Dashed line is 40 bars, solid line is 10 bars.

## 4. Conclusions

Phase field theory simulations have been applied to model the dissociation of $CH_4$ and $CO_2$ exposed towards pure water. Presently there are no experimental data available for direct comparisons to the predictions presented here and the main purpose of this paper has been to demonstrate the approach and the corresponding parameterization. As expected relative to the differences in solubility of the two components in water the kinetic rates of $CO_2$ hydrate dissociation is larger than that of $CH_4$ hydrate.

## Acknowledgments

Financial support from the Norwegian Research Council through project 101204, and from Conoco-Phillips, is highly appreciated.

## References

[1] Svandal A., Kvamme B., Grànàsy L., Pusztai T.: *The Influence of Diffusion on Hydrate Growth.* J. Phase. Equilib. Diff., in press, 26(5), 2005.

[2] Svandal A., Kvamme B., Grànàsy L., Pusztai T.: *The phase field theory applied to CO₂ and CH₄ hydrate.* J. Cryst. Growth. in press, 2005.

[3] Kvamme B., Graue A., Aspenes E., Kuznetsova T., Grànàsy L., Tòth G., Pusztai T. Tegze G.: *Kinetics of solid hydrate formation by carbon dioxide; Phase field theory of hydrate nucleation and magnetic resonance imaging.* Phys. Chem. Chem. Phys., 6, 2327, 2004.

[4] Grànàsy L., Börzsönyi T., Pusztai T.: *Nucleation and Bulk Crystallization in Binary Phase Field Theory,* Phys. Rev. Lett., 88, 206105, 2002.

[5] Warren J.A., Boettinger W.J.: *Prediction of dendritic growth and microsegragation patterns in a binary alloy using the phase-field method.* Acta Metall. Mater. 43, 689-703, 1995.

[6] Wang S.L., Sekerka R.F., Wheeler A.A., Murray B.T., Coriell S.R., Braun R.J., McFadden G.B.: *Thermodynamically-consistent phase-field models for solidification.* Physica D, 69, 189-200, 1993.

[7] Hardy S.C.: *A grain boundary groove measurement of the surface tension between ice and water.* Philos. Mag., 35, 471, 1977

Brill Academic Publishers
P.O. Box 9000, 2300 PA Leiden
The Netherlands

*Lecture Series on Computer*
*and Computational Sciences*
Volume 4, 2005, pp. 546-549

# Effect of the Entropy Compensation on the Stability of Host-Guest Complexes Applied in Selective Chemical Sensors

Kornélia Szabó,[a] László Kollár,[b] Géza Nagy,[a] Sándor Kunsági-Máté [a1]

[a] Department of General and Physical Chemistry, University of Pécs, H-7624 Pécs, Hungary
[b] Department of Inorganic Chemistry, University of Pécs, H-7624 Pécs, Hungary

Received 9 July, 2005; accepted in revised form, 3 August, 2005

*Abstract:* The π-π interaction-based inclusion complexation of calix[6]arene hexasulfonate as host with neutral aromatic guest molecules was studied in aqueous media using both experimental and theoretical methods. Experimental measurements (PL and DSC) showed an increased complex stability when the electron density on the guest's aromatic rings increased, although the enthalpy change lowered. The evaluation of the thermodynamic parameters of the complex formation (enthalpy, entropy and Gibbs free energy) highlighted the excessive role of the entropy term in this particular case. To obtain atomistic view on the complex formation, DFT/B3LYP/6-31++G method and molecular dynamics simulation were performed. The stability of the complex and the complex formation were evaluated considering the Hammett parameters of the substituents of the guest. The entropy term of the complex formation was studied by simulation of the molecular rearrangement of solvent water molecules around the calixarene host when the phenol derivatives with their differently polarized aromatic rings enter into the calixarene cavity. The results show a structure of higher order of the water molecules when the aromatic ring of the entering guest molecule is electron-deficient. These results can contribute to the development of selective and sensitive chemical sensors for aromatic organic analytes.

*Keywords:* calixarenes, host-guest interaction, π-π interaction, neutral guest, molecular dynamics

*PACS:* 31.15Ar, 31.15Ew, 31.15Qg, 31.70Dk, 33.50Dq, 05.70Ce, 05.20Fw

## 1. Introduction

The recognition of neutral organic molecules by synthetic receptors is a topic of current interest in supramolecular- and also in analytical chemistry [1]. Calix[n]arenes (n=4-6,8), cyclic oligomers of phenolic units linked through the *ortho* positions, represents a fascinating class of macrocycles, because of the simplicity of their well-defined skeleton, which is associated with versatile recognition properties towards metallic or organic ions and neutral molecules.

Among several related works, in our recent papers [*e.g.* 2] the factors controlling the thermodynamic and kinetic stability or selectivity of some supramolecular calixarene complexes with neutral π-electron deficient aromatics were reported. The inclusion complexation of calix[6]arene hexasulfonate with different neutral aromatics in aqueous media have been studied recently by PL (Photoluminescence), DSC (Differential Scanning Calorimetry) and quantum-chemical methods [3].

In our recent work [4] the thermodynamic parameters were determined for the complex formation of calix[6]arene (Fig. 1, **1**) with a series of phenols (Fig. 1, **2a-d**). Comparing the thermodynamic parameters observed on the series of the guests, a decrease of the enthalpy change was observed when the electron density on the guest's aromatic ring increased. However, the Gibbs free energy and therefore the stability of the complexes increased when the enthalpy change lowered. These unexpected

---

[1] Corresponding author. E-mail: kunsagi@ttk.pte.hu

results are based on the enthalpy-entropy compensation effect and are probably due to the quite different entropy change related to the different electron density distribution on the aromatic rings of different guest molecules. Since the interactions of calixarenes with neutral species involve competition between complexation and solvation processes, it was obvious to assume that redistribution of the electron density of calixarene rings, followed by the reorganization of the solvent molecules are responsible for this unexpected entropy change at molecular level. Accordingly, in this work the evaluation of the entropy change during interaction of calix[6]arene hexasulfonate (Fig. 1, **1**) with a series of phenol derivatives (Fig. 1, **2a-2d**) will be presented.

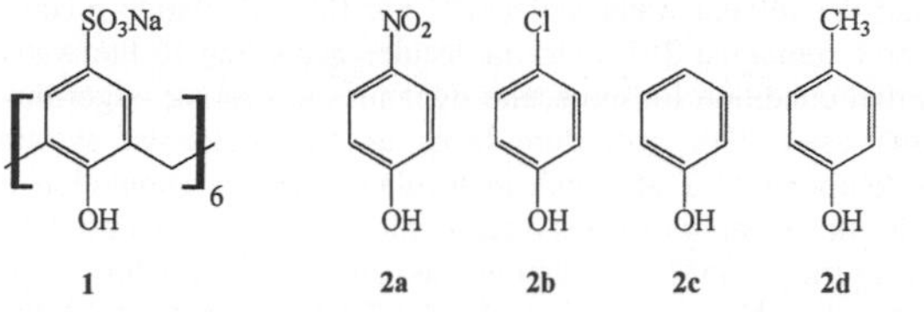

Figure 1: Calix[6]arene-hexasulfonate salt **1** and different phenol derivatives (**2a-2d**) were choosen as host and guests, respectively (**2a**: *p*-nitrophenol, **2b**: *p*-chlorophenol, **2c**: phenol and **2d** : *p*-cresol)

## 2. Methods

### 2.1 Determination of the thermodynamic parameters from the experimental measurements

Assuming a 1:1 stoichiometry, the complexation reaction can be written as follows ("H" refers to host while "G" means the guest):

$$H + G \overset{K_s}{\rightleftharpoons} HG \tag{1}$$

It is well known that in this particular case the concentration of the complex formed can be expressed as the function of the initial concentrations ([H]$_0$ and [G]$_0$):

$$[HG] = \left\{ \left([H]_0 + [G]_0 + 1/K_s\right) \pm \sqrt{\left([H]_0 + [G]_0 + 1/K_s\right)^2 - 4[H]_0[G]_0} \right\} / 2 \tag{2}$$

Assuming that the observed PL signal varies linearly with the concentration of the complex formed, $\Delta F$ is described by Eq. (3)

$$\Delta F = f_{HG} \cdot [HG] \tag{3}$$

wherein $\Delta F = F - F_0$ is a difference between the PL intensity obtained with the calixarene/phenol system and that of the free calixarene with the same concentration. The measure of the PL signals, $f_{HG}$ can be obtained for the individual HG species relative to the PL signal of pure calixarene species at the same concentrations. By definition,

$$f_{HG} = \frac{F([G]) - F([HG])}{F([H])} \Bigg|_{[HG]=[H]} \tag{4}$$

Job's method is widely used for the spectroscopic determination of complex stability constants can also be used in calixarene chemistry. The stability constant of the inclusion complex can be determined by the curve fitting of Eq. (3) to the experimental data using the expression of [HG] from Eq.(2).

However, it is known that the equilibrium in similar systems strongly depends on the temperature. The thermodynamic parameters for the individual complexes formed in the calixarene/phenol system can be determined from the van't Hoff equation:

$$\ln K = -\frac{\Delta G}{RT} = -\frac{\Delta H}{RT} + \frac{\Delta S}{R} \tag{5}$$

where $\Delta G$ is the Gibbs energy change, $\Delta S$ the entropy change and $\Delta H$ the enthalpy change associated with complex formation.

Inserting Eq. (5) for the formation constants into the Eqs. (2) and (3), the fluorescence change in Eq. (3) can be expressed as a function of the $\Delta H$, $\Delta S$ values and the $f_{HG}$ coefficient.

The thermodynamic parameters associated with the $K$ value were determined from the Job's curves by an iterative solution of Eqs. (2) and (3) using the expression of $K$ value from the van't Hoff equation (Eq. (5)).

The measurements were carried out at four different temperatures and the iterative curve-fitting procedure was done simultaneously for the experimental data. The plot of $\Delta F$ as a function of molar fraction of host gives an excellent fit, verifying the 1:1 complex stoichiometry assumed above.

### 2.2 Structure analysis and molecular modelling of the entropy term by theoretical methods

The equilibrium conformations of calixarene **1** and their complexes with phenol derivatives (series **a-d** of **2**) were studied with semiempirical AM1 (Austin Model) method, followed by *ab initio* HF/6-31G** calculations. The Fletcher-Reeves geometry optimization method was used for the investigation of the conformers. TIP3P method was applied for considering the solvent effect. The conformation of the complex in water obtained from the above calculations was compared with results derived from the geometry optimization performed at DFT/B3LYP/6-31++G level by using GAUSSIAN 03 package. The PCM (Polarizable Continuum Model) method was used to consider the solvent effect. The temperature-dependent molecular dynamic simulations were performed with AMBER forcefield. The TIP3P method is used to explicitly consider the solvent water molecules. For this calculation a cubic box with 20 Å side length is used. The box contained 265 water molecules according to the water density at 298K. To find an appropriate initial condition for molecular dynamics a 'heating' algorithm implemented in HyperChem package was used. This procedure heats up the molecular system smoothly from lower temperatures to the temperature T at which molecular dynamics simulation is desired to perform. The starting geometry for this heating phase is a static initial structure. We used the optimized geometry derived from semiempirical AM1 calculations as an initial structure. The temperature step and the time step in the heating phase were set to 2K and 0.1 fs, respectively. After equilibration at the given temperature, the MD simulations were run in 1 ps time intervals with resolution of 0.1 fs. The simulation time step was 0.1 fs. Ten thousand points were calculated in each run. Five water molecules locating closest to the calixarene's phenolic unit was chosen for data analysis. The inclination angles of the $C_2$ symmetry axis of the water molecules relating to the directions perpendicular to the planes of the appropriate phenolic units of calixarene molecules were collected during the simulation. The average values of this angles was used to represent the freedom of water molecules around the calixarene rings.

## 3. Results

### 3.1 Ab initio charge density of the guest molecules

Figure 2 shows the charge density of different phenol derivatives (**2a-2d**) as calculated by *ab-initio* DFT/6-31G** method. According to the known electron withdrawing or donating behavior of the $NO_2$, Cl, as well as H and $CH_3$ groups, the electron density on the aromatic ring of phenol derivatives increases from the *p*-nitrophenol towards *p*-cresol.

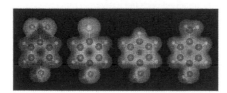

Figure 2: Charge density map of different phenol derivatives (**2a-2d**) applied as guests in this study. Calculations were performed with *ab-initio* DFT/6-31G** method.

### 3.2 Stable conformations of the calixarene-phenol host-guest complexes

Results summarized on Figures 3. a-d show that complexes formed during the interactions of the calixarene **1** and different phenol derivatives (**2a-2d**) are preferably stabilized by two attractive forces. Either by the π-π interaction between the aromatic π-electron systems of calixarene phenolic units and that of the aromatic guest molecules, or by week attractive interaction between the OH groups at the lower rim also participates in the stabilization. Thermodynamic parameters are summarized in Figure 4.

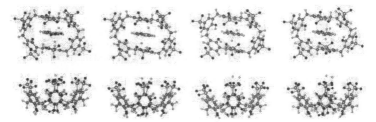

Figure 3: Top and side views of the equilibrium conformation of calix[6]arene hexasulfonate – phenol complexes. Calculations were performed at DFT/B3LYP/6-31++G level considering the solvent effect by the PCM (Polarizable Continuum Model) method.

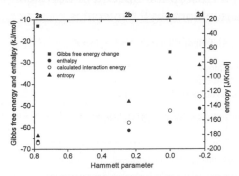

Figure 4: The experimental Gibbs free energy (squares) and enthalpy (solid circles) changes, also the calculated interaction energy (open circles), (left axis) and the measured entropy (triangles) changes (right axis) obtained for the formation of the host-guest complexes as function of the Hammett substituent constants. Hammett parameter increases from right (*p*-cresol) to left (*p*-nitrophenol) in the figure.

### 3.3 Consequences on the dynamics of water molecules at room temperature

For quantitative description of rotation of solvent water molecules, the angles of the $C_2$ symmetry axis of water molecules, relating to the directions perpendicular to the planes of the appropriate phenolic units of calixarene molecules, were collected during the simulation. The deviation from the average values of these angles of five water molecules closest to the appropriate phenolic ring of calixarene were used to represent the freedom of water molecules around the calixarene rings. It was found that the average rotation of the water molecules is nearly three times less in the case when the *p*-nitrophenol guest with its charged aromatic ring is entered into the cavity ($85^0$), while, for comparison, the average rotation value for the nearly neutral *p*-cresol was found to be $265^0$. It means, that the solvent water molecules, which lie in the outer side of the calixarene rings, has much higher ordered structure after the guest with its more electron deficient phenolic unit is entered into the cavity.

## 4. Conclusion

The interactions of calixarene with neutral species involve competition between complexation and solvation processes. Our investigation shows, that the redistribution of the electron density of calixarene rings, followed by the reordering of the solvent molecules determines the entropy change during complex formation. As a consequence, a more ordered structure is formed when the aromatic ring of the entering guest molecule is more electron-deficient. This property fits well to the experimental observation. Overall, the calculated molecular properties of the complex formation show Hammett-type correlation. Our results are hopefully applicable in the development of selective and sensitive chemical sensors for neutral organic aromatics.

## Acknowledgements

Calculations were performed on SunFire 15000 supercomputer located in the Supercomputer Center of the Hungarian National Infrastructure Development Program Office. The financial support of the Hungarian Scientific Research Fund (OTKA Grant TS044800) and that of the Hungarian PRCH Student Science Foundation (regarding participation on the ICCMSE 2005) are highly appreciated.

## References

[1] C.D. Gutsche: *Monographs in Supramolecular Chemistry, Vol.6. Calixarenes Revisited*;
The Royal Society of Chemistry: Cambridge, 1998.

[2] S. Kunsági-Máté, K. Szabó, B. Lemli, I. Bitter, G. Nagy, L. Kollár: Increased complexation ability of water-soluble calix[4]resorcinarene octacarboxylate toward phenol by assistance of Fe(II) ions, *Journal of Physical Chemistry B*, 108 (40), 15519-15522 (2004).

[3] S. Kunsági-Máté, K. Szabó, B. Lemli, I. Bitter, G. Nagy, L. Kollár: Host-guest interaction between water-soluble calix[6]arene hexasulfonate and *p*-nitrophenol, *Thermochimica Acta*, 425 (1-2), 121-126 (2005). (see also references therein to our earlier related works)

[4] S. Kunsági-Máté, K. Szabó, I. Bitter, G. Nagy, L. Kollár: Unexpected effect of charge density of the aromatic guests on the stability of calix[6]arene phenol host guests complexes, *Journal of Physical Chemistry A*, 109 (23), 5237-5242 (2005).

Brill Academic Publishers
P.O. Box 9000, 2300 PA Leiden
The Netherlands

*Lecture Series on Computer*
*and Computational Sciences*
Volume 4, 2005, pp. 550-553

# Modeling of Pressure Effects on Absorption Spectra of Solvated Chlorophyll and Bacteriochlorophyll Molecules

Tarmo Tamm[1,a], Juha Linnanto[b], Aleksandr Ellervee[c], Arvi Freiberg[a,c]

[a]Institute of Molecular and Cell Biology, University of Tartu, Estonia
[b]Department of Chemistry, University of Jyväskyla, Finand.
[c]Institute of Physics, University of Tartu, Estonia

Received 12 July, 2005; accepted in revised form 8 August, 2005

*Abstract:* Atomic level simulations of the hydrostatic pressure effects on the absorption spectra of chlorophyll *a* and bacteriochlorophyll *a* in diethyl ether solution are presented. The methodology combining CHARMm, PM3 and ZINDO/CI methods yields results in good agreement with the experimentally observed spectra, explaining the behavior of the barochromic shifts of the $Q_y$, $Q_x$ and Soret bands.

*Keywords:* absorption spectra, chlorophyll, bacteriochlorophyll, barochromic shift, modeling, ZINDO/CI

## Introduction

Chlorophyll *a* (Chl-*a*) is one of the most abundant pigment molecules on Earth. Together with the *b* modification and carotenoids it is mostly responsible for the green color of vegetation on our planet. Bacteriochlorophyll *a* (Bchl-*a*) plays similarly important role for photosynthetic bacteria. The chlorophylls and bacteriochlorophylls can be found in both light harvesting and photochemical reaction center complexes of plants and bacteria, respectively. Precise knowledge of the electronic states and spatial structure of these molecules is, therefore, crucial for the understanding of their role in the photosynthetic pathways.

In natural conditions the chlorophylls and bacteriochlorophylls readily form complexes with a variety of other molecules, including solvents. A traditional approach to study the interactions of solute electronic surfaces with the solvent is by measuring the shifts and shapes of optical spectra employing different solvents (see [1] for a review). However, this method is subject to serious shortcomings due to the discontinuous change of bulk physical properties and microscopic chemical composition of the solute environment when varying the solvent. In contrast, by applying external hydrostatic pressures, continuous tuning of solution properties over considerable range can be achieved. When combined with optical spectroscopy, this technique provides a very promising barochromic alternative for solvation studies.

The conventional solvation models operating with macroscopic experimental parameters, such as the dielectric constant of the solvent, generally fail in explaining the electronic level shifts and transition energies, least the geometry changes of the molecules, due to variations in the coordination number [2] occurring during the solvation. Atomic level approaches based on quantum chemistry are therefore necessary for calculations of these effects. Due to the size of the systems considered (140 atoms even without any solvent) the quantum chemical methods of higher accuracy (*ab initio* at Hartree Fock (HF), post-HF or DFT level with reasonably large basis sets) are unfortunately not applicable, especially for modeling solvent effects as a function of pressure. While the spectra of different chlorophylls have been calculated at various levels of sophistication for unsolvated molecules [3,4] and molecules in complex with a few solvent molecules [5], to our knowledge, no previous modeling of pressure dependent solvation effects has been done. The combination of geometry optimization at molecular mechanics and semiempirical levels followed by the estimation of the electronic spectra at the ZINDO/CI level already proved reasonably reliable for the Chl-*a* in diethyl ether solution [6].

---

[1] Corresponding author. E-mail: tarmo@chem.ut.ee

## Methodology

For the modeling of solvent effects, the structures of Chl-*a* and BChl-*a* were placed in a simulation box containing, respectively, 324 and 300 diethyl ether molecules (Figure 1). The volume of the simulation box was varied for the simulation of pressure effects while keeping the same number of molecules. The effective pressure inside the box was estimated relying on the experimental diethyl ether densities as a function of pressure [7]. The energy of the system was minimized using the CHARMm 27 [8] force field of the Tinker [9] package. The output of the force field minimization was then used as input for semi empirical quantum chemical calculations. The structure of the BChl *a* molecule surrounded by the nearest 35 solvent molecules was optimized using the PM3 method [10] from the Gaussian [11] package. The solvent molecules of the outer shells were fixed in their positions determined by the CHARMm calculations creating a cavity for the solvent molecules of the inner shell.

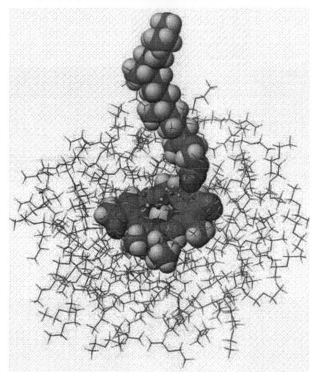

Figure 1. The simulation box containing BChl-a in diethyl ether, as optimized by CHARMm.

The electronic transition energies of the PM3 optimized supermolecule were then calculated using the semi empirical ZINDO/CIS [12] methodology as implemented in the ArgusLab [13] software. The limited CI space included 40 highest occupied and 40 lowest unoccupied MOs. The calculated transition energies were correlated with the experimental ones using the linear least squares fit, as shown previously [4]. The theoretical dipole strengths of the transitions were spread over Gaussian lineshapes having experimental bandwidths. The procedure was repeated for every step of density/pressure.

## Results and discussion

The calculated and experimental electronic absorption spectra of Chl-a in diethyl ether at atmospheric pressure are shown on Figure 2. In the 300-750 nm region of the experimental spectrum, several singlet electronic transitions are apparent: $Q_y$ at 660 nm, $Q_x$ at 578 nm, and a composite Soret band (with the first peak designated as B) below 430 nm. The main features of the simulated spectrum are in good agreement with the experimental one.

All these bands in both experimental and simulated spectra are shifted towards longer wavelengths with increased pressure. The shifts, however, are not monotonous and, what is even more interesting, different for each band. The $Q_y$ band is least sensitive to pressure change compared with the $Q_x$ and B bands for both Chl-a and BChl-a (Figure 3.). Past a certain pressure point (around 400 MPa for Chl-a) the shift rate of the $Q_y$ band is reduced while that of the $Q_x$ band accelerates. This difference is significantly more pronounced for BChl-a than for Chl-a. The simulations reveal that the central Mg atom becomes six coordinated around that pressure point from the initially 5 coordinated state, introducing a new, stronger interaction between the solute and the solvent molecule. The different courses of the band shifts can be explained by varying perturbation of the electronic orbitals relevant for specific electronic transitions and corresponding absorption bands. The more sensitive transitions/bands involve more orbitals from the central part of the porphine macrocycle, the part most influenced by the changing solvation environment and coordination of the central Mg atom.

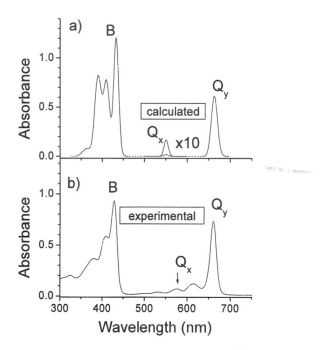

Figure 2. Calculated (a) and measured (b) absorption spectrum of Chl-a in diethyl ether at atmospheric pressure

## Conclusion

The presented atomic level simulations of the absorption spectra of chlorophyll molecules reveal important aspects of solvation effects and their pressure dependence. Close agreement with the experimental data shows the potential of the method and encourages for future studies in more complex environments, including native protein surroundings of photosynthetic complexes, our ultimate goal.

## Acknowledgments

Partial financial support from the Estonian Science Foundation (grant no. 5543) is gratefully acknowledged.

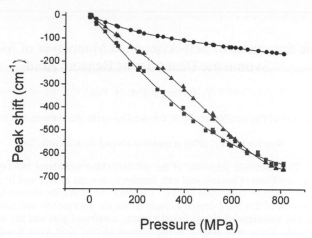

Figure 3. Relative pressure shifts of Qy (cicrcles), Qx (triangles), and Soret (squares) peaks.

## References

[1]  C. J. Cramer, D. G. Truhlar, Chem. Rev. 99 (1999) 2161.

[2]  P. M. Callahan, T. M. Cotton; J. Am. Chem. Soc. 109 (1987) 7001.

[3]  D. Sundholm, Chem. Phys. Lett. 302 (1999) 480.

[4]  J. Linnanto, J. Korppi-Tommola, Phys. Chem. Chem. Phys. (2000) 4962.

[5]  J. Linnanto, J. Korppi-Tommola, J. Comput. Chem. 25 (2004) 123.

[6]  A. Ellervee, J. Linnanto, A. Freiberg, Chem. Phys. Lett. 394 (2004) 80.

[7]  P. W. Bridgman, Proc. Am. Acad. Arts Sci. 49 (1913) 3.

[8]  N. Foloppe, A. D. MacKerell, Jr., J. Comput. Chem., 21 (2000) 86.

[9]  J. W. Ponder, Tinker Molecular Modeling Package 4.2, http://dasher.wustl.edu/tinker.

[10] J. J. P.Stewart, J. Comput. Chem. 10 (1989) 209.

[11] Gaussian 03, Revision C.02, M. J. Frisch, J. A. Pople, et. al. Gaussian, Inc., Wallingford CT, 2004.

[12] J. Ridley. M. Zerner, Theoret. Chim. Acta (Berl.), 32 (1973) 111.

[13] ArgusLab 4.0, M. A. Thompson, Planaria Software LLC, Seattle, WA. http://www.arguslab.com.

Brill Academic Publishers
P.O. Box 9000, 2300 PA Leiden
The Netherlands

*Lecture Series on Computer*
*and Computational Sciences*
Volume 4, 2005, pp. 554-557

# Electronic Structures of Self-Assembled Monolayer of Molecules of Symmetric Disulfides of Benzoic Acid

Y. -H. Tang and M. -H. Tsai [1]

Department of Physics, National Sun Yat-Sen University, Kaohsiung, 80424 Taiwan

Received 7 July, 2005; accepted in revised form 29 July, 2005

*Abstract*: The electronic properties of the self-assembled monolayer (SAM) of molecules of symmetric disulfides of benzoic acid with functional elements of H, F and Br have been studied by the first-principles calculation method. This study found that the electronic structure and the dipole moment of this SAM depend strongly on the electronegativity and size of the functional element. The variations of charge transfer among constituent ions and the dipole moment are approximately linear with respect to the external electric field even though there is a gap between the highest-occupied-molecular-orbital (HOMO) and the lowest-unoccupied-molecular-orbital (LUMO) bands. This finding suggests that the conductance of SAM may not vanish when the bias is smaller than the energy gap.

*Keywords:* self-assemble monolayer, electronic structures, dipole moment
*PACS:* 31.15.Ar, 81.16.Dn

## 1. Introduction

The self-assembled monolayer (SAM) of biomolecules has a non-vanishing dipole moment, and the absorption of SAM on semiconductors and ferroelectrics can modify their transport and dielectric properties, respectively. The former basically uses the controllable and predicable dipolar property of the organic molecules with systematically varied functional groups to tune the potential barrier at the SAM/semiconductor interface and the current in the conduction channel in the semiconductor. The latter is potentially applicable as a biosensor for detecting bacteria or viral particles. SAM's are also potentially applicable as molecular electronic [1-3] and optoelectronic devices [4,5] and the semiconductor based sensor [6].

In comparison with the energy bands of semiconductor and metal films, the energy bands of SAM's are relatively flat, so that the transport properties of SAM's are different from those of semiconductors and metals. The highest-occupied-molecular-orbital (HOMO) and the lowest-unoccupied-molecular- orbital (LUMO) electronic states contribute to the transport property of SAM. Therefore, the understanding of the electronic structures and corresponding wavefunctions, especially for LUMO and HOMO, is essential for the development of the molecular electronics. On the other hand, the understanding of the dipolar property is essential for the application of SAM/semiconductor and SAM/ferroelectrics systems.

In the conventional study of the conductivity of biomolecule systems in molecular electronics applications using the Green's function method, the electronic states of SAM are usually approximated as constant when the applied voltage is less than about 1eV [7,8]. This approximation may be inadequate because the electronic states and charge distribution in SAM response to even a small applied electric field. The neglect of the response of the electronic states to the applied field in the Green's function approach may be the source of one to two orders of magnitude of discrepancy between calculated and measured conductivities [9]. In a first step to address this problem, the dependence of the electronic property of the SAM of molecules of symmetric disulfides of benzoic acid on the applied electric field has been studied as an example.

---

[1] Corresponding author. E-mail: tsai@mail.phys.nsysu.edu.tw

## 2. The molecular structure and the calculation method

The atomic structure and the labels of atoms of a single molecule of symmetric disulfides of benzoic acid in the SAM with the functional element, X, are shown in Figs. 1(a) and (b), respectively. In this study, H, F, and Br are chosen for X. The 'ChemOffice 2002' software is used to obtain the atomic coordinates for these three cases. A square lattice is chosen for SAM with a periodicity of 11.7Å. The directions perpendicular to and parallel with the benzene-1,2-dithiolate plane are chosen as the x and y axes, and the z-axis points from the S end to the X end. Note that the planes of the two symmetric phenyls are almost perpendicular to the benzene-1,2-dithiolate plane.

The first-principles pseudofunction (PSF) [10] method implemented with the local density approximation (LDA) of Hedin-Lundqvist formulation [11] is used to obtain the electronic structures. The constant applied electric field, $E$, is perpendicular to the plane of SAM, i.e. along the z-axis. Note the PSF method uses the single-slab model, which has appropriate vacuum boundary conditions for solving the Poisson equation for the Coulomb potential. That is, the PSF method does not use the infinitely-repeated-slab super-cell model used in Pseudopotential calculations. In the single-slab model, the dipole moment density of SAM is directly related to the Coulomb potential difference across SAM.

## 3. Calculated electronic properties

The calculated energy bands of the SAM's with the three functional elements X=H, F, and Br are shown in Figs. 2(a), (b), and (c), respectively. The zero energy is chosen to be the Fermi energy ($E_F$), and the symmetry k-points, $\Gamma$, X, and M, represent the Brillouin zone center, $\left(\dfrac{\pi}{a},0,0\right)$, and $\left(\dfrac{\pi}{a},\dfrac{\pi}{a},0\right)$, respectively, where $a$ is the periodicity of the molecular lattice in the x-y plane. There are three closely spaced bands near $E_F$ for all three cases, which are composed mainly of S and C=O bond orbitals. These three bands are highlighted in Figs. 3(a)-(c). Figs. 2(a)-(c) show significant dispersions of higher-energy unoccupied energy bands, especially for X=F and Br, due to inter-molecular couplings. The dispersions of HOMO and LUMO bands are about 0.1eV for X=F and Br. There is an energy gap between the HOMO and LUMO bands of ~0.05eV for X=F and Br. In contrast, the LUMO band has a larger dispersion for X=H, which reduces the energy gap. It is expected that the dispersions will vanish when the inter-molecular separation is large enough.

The calculated dipole moments per molecule point from the X end to the S end with magnitudes of 3.52, 9.03, and 12.48 Debye, respectively, for SAM's with X=H, F, and Br. The larger dipole moments for X=F and Br are expected because they have larger electronegativity and size than H. Although F has a larger electronegativity than Br (3.98 vs. 2.96 [12]), Br has a larger size than F (1.14Å vs. 0.72Å [12]), so that the X=Br case has a larger dipole moment.

The dipole moments per molecule for X= F as a function of the applied electric field, $E$, are shown in Fig. 4. Note the positive $E$ is defined from the S end to the X end. Fig. 4 shows that the dipole moment decreases with the increase of $E$, which is because the direction of electron (negative charge) transfer in response to $E$ is the same as the dipole moment, i.e. from the X end to the S end. The variation of the dipole moment is almost linear. Effective charges of some representative ions are shown in Fig. 5, which indicate that the charge transfer in response to $E$ is also almost linear.

The molecular electronic states are usually assumed as intact unless the applied electric field is sufficiently large for the electron to overcome the energy gap between HOMO and LUMO. If this is the case, there will be a threshold of $E$ for the dipole moment and effective charges to make changes. In other words, the dipole moment and effective charges will remain constant until $E$ reaches the threshold and then rise rapidly beyond. The calculated linear response shows that this is not the case. From the calculated effective charges, one can envision a charge transfer mechanism in response to the applied electric field. The electric field not only polarizes individual atoms but also shift the bond charges. The latter causes an electronic charge transfer from one atom to a neighboring atom, which lies in the direction opposite to the electric field. In this manner, the electron charge can propagate across the molecule successively through shifted bond charges. Thus, the present result suggests that charge redistribution cannot be ignored even for a small applied electric field.

## 4. Conclusion

The first-principles calculation method has been used to study the dependence of the electronic structure of the SAM of molecules of symmetric disulfides of benzoic acid on the functional element and an applied electric field. It is found that there are three closely spaced bands near $E_F$ contributed by

S and C=O orbitals for this SAM. The molecular dipole moment depends strongly on the electronegativity and size of the functional element. The linear response of the charge redistribution in SAM to an applied electric field suggests that the use of fixed electronic states in the calculation of the conductivity of SAM may not be adequate and may be the cause of the one to two orders of magnitude of discrepancy between theory and experiment.

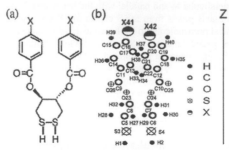

Fig. 1: The (a) atomic structure and (b) labels of atoms of a single molecule of symmetric disulfides of benzoic acid with the functional element, X.

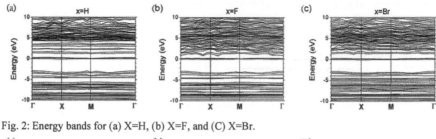

Fig. 2: Energy bands for (a) X=H, (b) X=F, and (C) X=Br.

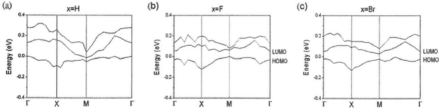

Fig. 3: Highlight of the three bands new $E_F$ for (a) X=H, (b) X=F, and (C) X=Br.

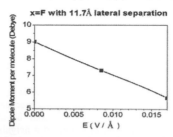

Fig. 4: The calculated dipole moments per molecule for X=F as a function of the applied electric field.

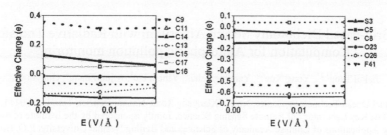

Fig. 5: The calculated effective charges for X=F as a function of the applied electric field.

## References

[1]  A. Vilan, A. Shanzer, and D. Cahen, Nature **404**, 166 (2000)

[2]  Y. Selzer and D. Cahen, Adv. Mater. **13**, 508 (2001)

[3]  D. G. Wu, J. Ghabboun, J. M. L. Martin, and D. Cahen, J. Phys. Chem. B **105**, 12011 (2001)

[4]  L. Zuppiroli, L. Si-Ahmed, K. Kamaras, F. Nuesch, M. Bussa, D. Ades, A. Siove, E. Moons, and M. Gratzel, Euro. Phys. J. B **11**, 505 (1999)

[5]  H. Sirringhaus, N. Tessler, and R. H. Friend, Science **280**, 1741 (1998)

[6]  D. G. Wu, D. Cahen, P. Graf, R. Naaman, A. Nizan, and D. Shvarts, Chem Eur. J. **7**, 1743 (2001)

[7]  J. G. Kushmerick, D. B. Holt, J. C. Yang, J. Naciri, M. H. Moore, and R. Shashidhar, Phys. Rev. Lett. **89**, 086802 (2002)

[8]  W. Tian, S. Datta, S Hong, R. Reifenberger, J. I. Henderson, and C. P. Kubiak, J. Chem. Phys. **109**, 2874 (1998)

[9]  M. Di Ventra, S. T. Panteliders, and N. D. Lang, Phys. Rev. Lett. **84**, 979 (2000)

[10] R. V. Kasowski, M. –H. Tsai, T. N. Rhodin, and D. D. Chambliss, Phys. Rev. **B34**, 2656 (1986)

[11] L. Hedin and B. I. Lundqvist, J. Phys. C **4**, 2064 (1971)

[12] *Table of periodic properties of the elements*, Sargent-Welch Scientific Company, Skokie, Illinois 1980.

Brill Academic Publishers
P.O. Box 9000, 2300 PA Leiden
The Netherlands

*Lecture Series on Computer
and Computational Sciences*
Volume 4, 2005, pp. 558-563

# Grid-based Dichotomy Search Algorithm with Radiative Transfer Computation for Atmospheric pollution monitoring

Jiakui Tang[1,2*], Yong Xue[2,3], Yanning Guan[2], Yincui Hu[2], Aijun Zhang[4], Guoyin Cai[2]

[1]National Ocean Technology Center, No.60, Xianyang Road, Nankai District, Tianjin, 300111, China
[2]State Key Laboratory of Remote Sensing Science, Jointly Sponsored by the Institute of Remote Sensing Applications of Chinese Academy of Sciences and Beijing Normal University, P.O. Box 9718, Beijing 100101, China
[3]Department of Computing, London Metropolitan University, 166-220 Holloway Road, London N7 8DB, UK
[4]School of Mechanical and Electrical Engineering, Beijing University of Chemical Technology, Beijing 100029, China
{tangjiakui@263.net}

Received 10 August, 2005; accepted in revised form 17 August, 2005

*Abstract:* Aerosol distribution retrieval for atmospheric pollution monitoring from remote sensing data involves usually in radiative transfer computation, which is very intensive and time-consumed, LUT technique is commonly used to speed up the retrieval computation, which, however, causes precision degradation because interpolation and extrapolation computation. Grid technology makes it possible to realize higher precision algorithm with higher efficiency. In this paper, we attempt to dichotomy search algorithm realization in radiative transfer computation on grid platform for atmosphere pollution monitoring using remote sensed data. Issues concerned distributed computation strategy on Grid environment are described in detail, and Experimental results obtained on HIT-SIP (High-Throughput Spatial Information Processing Prototype System) Gird consisted of commodity PCs are discussed.

*Keywords:* Aerosol, Remote Sensing, Grid, High Throughput Computing

## 1. Introduction

Grid Computing seeks to efficiently coordinate the sharing of geographically distributed computing resources, thereby bring supercomputing power to its users (Foster and Kesselman 1998). Grid computing distinguished from conventional distributed computing, cluster computing, which are both more constrained to computation on a local area networked of processors, by its focus on large-scale computing resources sharing, user's applications to utilize resources that spread across wide area networks. Grid computing can be exploited for computationally intensive tasks such as protein folding, production simulation, operation research, climate modeling, real-time distributed collaborative geo-processing (Shi et al. 2002), geo-rectification of satellite images (Teo et al. 2003), etc. There are very inspiriting progresses since the term was coined in the mid 1990s (Foster and Kesselman 1998). Globus (Foster and Kesselman 1996) is a grid computing software toolkit addressing the key technical problems in the development of grid-enabled tools, services, applications and systems. Built on top of Globus, Nimrod/G (Abramson et al. 2000) is a resource management system for scheduling of grid applications. Integrated with Globus toolkits, EuroGrid aims to create tech to for remote access to supercomputer resources and simulation codes in GRIP. Other grid systems/projects include Access Grid (http://www.fp.mcs.anl.gov/fl/access grid), Grid Physics Network (GriDPhyN) (http://www.griphyn.org/index.php), Information Power Grid (http://www.ipg.nasa.gov/), Virtual DataGrid Laboratory (iVDGL) (http://www.ivdgl.org/), TeraGrid (http://www.teragrid.org), National Grid (http://www.grid-support.ac.uk/), etc. The famous Grid focused on spatial information includes SpaceGrid, EnvirGrid and EarthObsevation Grid. ESA's SpaceGrid is an ESA funded initiative (http://sci2.esa.int/spacegrid). Grid-solvers or middlewares have also been developed on these systems

---

* Corresponding author

aforementioned. A web-based problem solving environment to simplify the submission, monitoring, and steering of Master-Worker (Goux *et al.* 2000) for Grid computing applications is introduced in by Hawick *et al.* (1998) and Good *et al.* (2000), A architecture of matching grid application requirements to a set of heterogeneous grid resources for resources allocation in computational grid is proposed by Czajkowski *et al.* (1999).

Atmosphere pollution monitoring from space observation concerns routinely Aerosol distribution retrieval, which is the procedure to separate the atmospheric molecular contribution from the TOA (Top Of Atmosphere) reflectance to abstract the aerosol information such as Aerosol Optical Thickness (AOT), size distribution using satellite remote sensing data, At the present time, the typical operational aerosol retrieval model for sea surface or land surface such as DDV (Kaufman *et al.* 1997) (Dark Dense Vegetation) pixels usually include such steps as spectrum channel (band) TOA Reflectance (Apparent Reflectance) estimation from satellite image, Ground Surface Reflectance estimation based directly on satellite image or others such in situ measurements, and computation to get the retrieval result image with radiative transfer code such as 6S (Vermote, *et al.* 1995), Lowtran, Modtran (Anderson, *et al.* 1995), etc. This procedure is heavily computationally intensive for the last step involving into large amount of RTE (Radiative Transfer Equations) calculations. LUT technique (Mobley, *et al.* 2002) is commonly used to speed up the retrieval computation, which, however, causes inevitably precision degradation because interpolation and extrapolation computation. This paper mainly discusses the application of grid computing to realize dichotomy search algorithm for the last step for atmosphere pollution monitoring using MODIS (http://modis.gsfc.nasa.gov/) data, which is expected to get higher precision with no efficiency reduced.

The rest of paper is as follows. In section 2, we introduce HIT-SIP (High-Throughput Spatial Information Processing Prototype System) Gird platform in Institute of Remote Sensing Applications, Chinese Academy of Sciences. In section 3, Dichotomy Search Algorithm is described. Section 4 describes the distributed aerosol retrieval computation on HIT-SIP Gird platform. In section 5, the experiments results and analysis are presented. Section 6 contains the concluding remarks.

## 2. HIT-SIP Grid

HIT-SIP Gird (High-Throughput Spatial Information Processing Prototype System) is constructed by Telegeoprocessing Research Team in Institute of Remote Sensing Applications, Chinese Academy of Sciences, which is a advanced high throughput Condor-based (Condor$^R$ Version 6.4.7 Manual, 2003) [which copyright is computer sciences department, university of Wisconsin-Madison] computing grid platform, Heterogeneous computing nodes including two sets of Linux computers and WINNT 2000 professional computers and one set of WINNT XP computer provide stable computing power. The grid pool uses java universe to screen heterogeneous characters. Common users can use the heterogeneous grid and share its strong computing power to achieve High-Throughput Spatial Information Processing. Condor system provide the technologies for resource discovery, dynamic task-resource matching, communication, and result collection, etc. however, it is a DOS-based command lines operation user interface and generic grid computing platform, can not directly run for Spatial Information Processing programs. HIT-SIP provides the users friendly Windows-based GUI (Graphical User Interface), computation tasks auto-partitioned, tasks auto-submission, the execution progress monitoring, the sub-results collection and the final result merging, which results in screening the complexities of grid applications programming and the Grid platform for users.

## 3. Dichotomy Search Algorithm Description

Dichotomy Search Algorithm is a powerful and commonly used technique for equation roots seeking in well-known solution space. Given a equation as follows:

$$f(x) = 0 \qquad\qquad (1)$$

In equation (1), suppose that the solution space is well-know for the root x ~[x1, x2], x1, x2 satisfies f(x1)<0, f(x2)>0 and x1<x2, the idea of Dichotomy Search Algorithm to find the root x for (1) is first to divide the original solution space [x1, x2], and to select the sub-solution space, then continue to

divide the picked sub-solution space for  further selection, the procedure of division and selection continues until that the approximate root satisfies precision condition. Dichotomy Search Algorithm for equation (1) can be described as follows:

First, we set the precision error limit $\varepsilon$, which is non-negative. One of the two instances as follows occurs.

① $\left| f(\dfrac{x_1 + x_2}{2}) \right| < \varepsilon$, (x1+x2)/2 is the root for equation (1), Dichotomy Search procedure stops.

② $\left| f(\dfrac{x_1 + x_2}{2}) \right| \geq \varepsilon$ ,we look for new sub-solution space [a1,b1] according to the value of

$f(\dfrac{x_1 + x_2}{2})$, if $f(\dfrac{x_1 + x_2}{2}) < 0$, set a1=(x1+x2)/2,b1=x2, contrarily, when $f(\dfrac{x_1 + x_2}{2}) > 0$, set a1=(x1),b1==(x1+x2)/2, substitute new sub-solution space  [a1,b1] of [x1,x2] to continue the Dichotomy Search procedure above until stop in the instance ①

Given certain precision error limit, dichotomy Search Algorithm is simple and reliable to get the root of equation to be solved, however it will be time-consumed and need powerful computing resource when high precision error limit set, which lead us to realize it on grid platform.

## 4.  Distributed Aerosol Retrieval Computation on HIT-SIP

Aerosol retrieval for atmospheric monitoring from space observation is mainly to abstract the information of aerosol distribution form satellite imagery. Remote sensing imagery from satellite visible observation mainly comes from the coupling result between surface and atmosphere information, which maybe described as follows:

$$I = f(\rho_s, atmo) \tag{2}$$

Where, I, stands for apparent reflectance at the top of atmosphere that is measured by sensor, $\rho_s$ stands for information from surface, and *atmo* for information from atmosphere, which can be further grouped into aerosol and atmosphere molecule information.

Given the parameters of  $\rho_s$ and atmosphere, Dichotomy Search Algorithm can be used for aerosol information retrieval from equation (2) just like for solving the equation (1). When the difference of apparent reflectance between measured and Dichotomy Search computation is less than Precision error limit $\varepsilon$ pre-set, Dichotomy Searching procedure stops and output the AOT value searched.

For simulating the process of the coupling  between surface and  atmosphere in equation (2), radiative transfer codes such as 6S、 Lowtran and Modtran have been developed and commonly used. In this paper, 6S code is used for radiative transfer computation. The parameters of $\rho_s$ can be estimated refer to dark target algorithm (Kaufman *et al.* 1997b). Other than aerosol optical thickness, which is the variable we are to solve for indicating atmosphere pollution, the parameters of atmosphere is all from prior knowledge according to standard American Atmospheric models.

Aerosol retrieval using Dichotomy Search Algorithm is a pixel by pixel computing procedure, which make it possible for distributed aerosol retrieval computation on grid platform. The scheme of aerosol retrieval using Dichotomy Search Algorithm HIT-SIP is as show Figure 1, which includes three phases, phase 1, Parameterization for parameters input into 6S code, phase 2, Execution of Dichotomy Searching AOT value by 6s radiative transfer computation, phase 3, Termination for merging and displaying the aerosol distribution map.

In phase 1, parameters include Geometrical Conditions, which refer to solar zenith angle, sensor zenith angle, and relative azimuth angle, Atmospheric Conditions, which refer to standard atmosphere model, mid-latitude summer atmosphere model is used in this paper, Aerosol Model (type), which refer to standard aerosol models, continental model is used in this paper, Spectral Band and Surface Reflectance. In phase 2, Dichotomy Searching solution space of AOT is set between 0.005 and 4.0, the AOT of 0.005 is the least aerosol loading and 4.0 is for the heaviest aerosol loading. In phase 3, the sub-results will be automatically merged and output onto screen and into files.

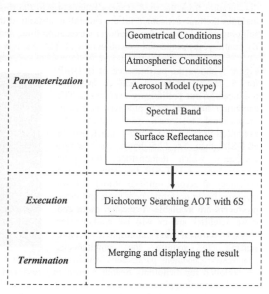

## 5. Experiments and Analysis

Our experiments were carried out on HIT-SIP which consists of 5 low commodity PCs nodes connected using 100Mbps Ethernet switch. The information of each node is shown in Table 1.

Figure 1. The scheme of aerosol retrieval using Dichotomy Search Algorithm on HIT-SIP

**Table 1.** The information of test grid computing pool

| IP | Hostname | Arch-OS | CPU (gHz) | RAM (MB) | Role |
|---|---|---|---|---|---|
| 192.168.0.5 | HU | Intel/WIN50 | 2.6 | 512 | Manager/Producer/Client |
| 192.168.0.3 | ZHONG | Intel/WIN50 | 2.6 | 512 | Producer/Client |
| 192.168.0.109 | WANG | Intel/WIN50 | 2.0 | 256 | Producer/Client |
| 192.168.0.120 | LUOY | Intel/WIN50 | 2.0 | 256 | Producer/Client |
| 192.168.0.100 | CGY | Intel/WIN50 | 2.6 | 512 | Producer/Client |

We use MODIS data to test dichotomy Search Algorithm to retrieve AOT on grid computing platform, The test area chosen is in china (about N37°~41°30", E N113°~119°30") and the data of acquisition of the MODIS data is 24 June 2003; the time is about 03:17 UTC; the image size is 512 × 512 with spatial resolution 500*m*, which covers most part of plain of north China and Bohai gulf. Figure3 shows the final AOT image of MODIS band 1 at 650 *nm* retrieved on HIT-SIP.

There are 5 producers available on HIT-SIP for our aerosol retrieval experiment, To get higher performance according to the suggestion by Tang *et al.* (2004), the whole image computation was partitioned into 5 tasks as equally as possible. To check the performance of dichotomy Search Algorithm for aerosol retrieval on HIT-SIP,

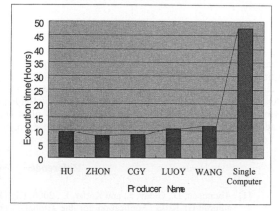

Figure 2. The performance of HIT-SIP Vs Single

contrarily, traditional computational test was carried on a single Intel CPU P4 2.0 GHz computer with 512Mb RAM, we define that the execution time Tg for aerosol retrieval on grid platform HIT-SIP was the duration from the time of tasks submission to the time of all tasks executed successfully, which includes the time consumed for the generation of tasks at the Condor manager, communication time in shipping tasks to producers, producer's execution time, and the results collection time from producers. By contrary, the sequential execution time (Ts) for a single computer.

Figure 2 shows the performance of HIT-SIP for 5 tasks partitioned and 5 available producers in our preliminary experiment. The execution time on each producer is clearly proportional to one's performance. The execution time Tg 11.8 hours is much less on HIT-SIP than Ts 47.5 hours on a single computer, which demonstrates the good performance of HIT-SIP for improving the efficiency of aerosol retrieval. Rationally, we can infer better performance if more and more producers are available. Typically, more and more satellites are or will be available for aerosol retrieval and atmosphere remote sensing to acquire the simultaneous Atmospheric Conditions parameters, which can be exploited for dichotomy Search Algorithm on grid computing platform to further improve aerosol retrieval precision with no efficiency degradation, if many producers are available, contrasting with LUT technique on a single computer.

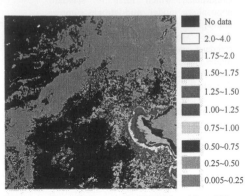

Figure 3. The AOT image of MODIS band 1 at 650 *nm* retrieved on HIT-SIP. No data pixels refer to cloudy mask and non-dark target pixels.

## 6. Conclusion and further development

The above experiments results of AOT retrieval using dichotomy Search Algorithm for atmospheric pollution monitoring on grid platform are encouraging. By means of HIT-SIP, computational efficiency of dichotomy Search Algorithm to avoid interpolation and extrapolation error of LUT method could be well improved through grid computing technique, which, by pooling and aggregating the idle CPU circle together, has high throughput computing to meet the needs of intensive computations. Our preliminary experiments results demonstrates that Grid platform make it possible to tapping time-consuming but accurate Algorithm for operational application. Ongoing work mainly focus on how to make a tradeoff due to more time needed for Condor manager to handle more available producers and more tasks and how to partition tasks un-equally according the actual performance of available producers to make producers' load balanced to get highest performance of HIT-SIP.

## Acknowledgement

This work is support by the research projects "Grid platform based aerosol fast monitoring modeling using MODIS data and middlewares development" (40471091) funded by National Natural Science Foundation of China (NSFC), "CAS Hundred Talents Program" (BRJH0101) funded by Chinese Academy of Sciences, and "Dynamic Monitoring of Beijing Olympic Environment Using Remote Sensing" (2002BA904B07-2) funded by the Ministry of Science and Technology, China. The advice and assistance of Prof. Guo Shan is gratefully acknowledged.

## Reference

[1] Abramson, D., Giddy, J., and Kotler, L., 2000, High Performance Parametric Modeling with Nimrod/G: Killer Application for the Global Grid., *Processing of the 14 international parallel and distributed processing Symposium (IPDPS)*, Mexico, IEEE CS Press, USA.

[2] Anderson, G.P., Kneizys, F.X., Chetwynd, J.H., Wang, J.,.Hoke, M.L, Rothman, L.S., Kimball, L.M., McClatchey, R.A., 1995, FASCODE/MODTRAN/LOWTRAN: Past/Present/Future, *18th Annual Review Conference on Atmopsheric Transmission Models.*

[3] Condor[R] Version 6.4.7 Manual, Condor Team, University of Wisconsin-Madison, 2003.

*[4]* Czajkowski, K., Foster, I., and Kesselman, C., 1999, Resource coallocation in computational grids, *Proceedings of the 8 IEEE International Symposium on High Performance Distributed Computing (HDPC-8).*

[5] Foster, I., and Kesselman, C., (Editors). 1998, The Grid: Blueprint for a new Computing Infrastructure. (*Morgan Kaufmann Publishers*, Inc.).

[6] Foster, I., and Kesselman, C., 1996, Globus: A Metacomputing Infrastructure Toolkits, *Proceedings of the Workshop on Environment and Tools for Parallel Scientific Computing,* SIAM Lyon, France.

[7] Good, M., and Goux, J.P., 2000, iMW: Aweb-based problem solving environment for Grid computing applications. *Technical report,* Department of Electrical and Computer Engineering, Northwestern University.

[8] Goux, J.P., Kulkani, S., Linderoth, J., and Yoder, M., 2000, An enabling framework for master-worker application on the computational grid., *Proceedings of 9 IEEE International Symposium on High Performance Distributed Computing (HDPC-9).*

[9] Hawick, K.A., and James, H.A., 1998, A web-based Interface for On-Demand Processing of Satellite Imagery Archives. *Proceedings of the Australian Computer Science Conference (ACSC) ' 98,* Perth, Australia.

[10] Kaufman, Y. J., Wald, A. E., Remer, L. A., Gao, B. C., Li, R.R, & Luke F., (1997b). The MODIS 2.1-um Channel-Correlation with Visible Reflectance for Use in Remote Sensing of Aerosol. *IEEE Transactions on Geoscience and Remote Sensing,* Vol.35, pp1286-1298

[11] Kaufman, Y.J., Tanre, D., Remer, L., Vermote, E., Chu, A., Holben, B., 1997, Operational remote sensing of tropospheric aerosol over land from EOS moderate resolution imaging spectroradiometer, *Journal of Geophisics Research,* Vol, 102, No, D14, 17051-17067.

[12] Mobley, C., Sundman, L. K., Davis, C. O., Montes, M., and Bissett, W. P., 2002, A Look-up Table Approach to Inverting Remotely Sensing Ocean Color Data, *Ocean Optics XVI,* Santa Fe, New Mexico.

[13] Shi, Y., Shortridge, A,M., and Bartholic, J., 2002, Grid Computing for Real Time Distributed Collaborative Geoprocessing, *10th International Symposium on Spatial Data Handling.* Ottawa, Canada.

[14] Tang, J.K., Xue, Y., Guan, Y.N., Liang, L.X., Hu, Y.C, Luo, Y., Cai, G.Y., Zhang, A.J., Wang, J.Q., Zhong, S.B., and Wang, Y.G., 2004, *A New Approach to Generate the Look-up Table Using Grid Computing Platform for Aerosol Remote Sensing, 2004 IEEE International Geoscience and Remote Sensing Symposium (IGARSS'04)*

[15] Teo, Y.M., Tay, S.C., and Gozali, J.P., 2003, Distributed Geo-rectification of Satellite Images Using Grid Computing, *Proceedings of the International Parallel & Distributed Processing Symposium,* IEEE Computer Society Press, Nice, France.

[16] Vermote, E., Tanre, D., Deuze, J.L., Herman, M., and Morcrette, J.J., 1995, Second Simulation of the Satellite Signal in the Solar Spectrum (6S): 6S User Guide Version 1, November 3, 1995.

Brill Academic Publishers
P.O. Box 9000, 2300 PA Leiden
The Netherlands

*Lecture Series on Computer
and Computational Sciences*
Volume 4, 2005, pp. 564-569

# Optimal Rural Water Distribution Design Using Labye's Optimization Method and Linear Programming Optimization Method

M.E. Theocharis[1*] , C.D. Tzimopoulos [2] , M. A. Sakellariou - Makrantonaki [3] , S. I. Yannopoulos [2], and  I. K. Meletiou[1]

[1] Department of Crop Production
Technical Educational Institution of Epirus, GR- 47100 Arta, Greece

[2] Department of Rural and Surveying Engineers
Aristotle University of Thessaloniki, GR- 54006 Thessaloniki, Greece

[3] Department of Agricultural Crop Production and Rural Environment
University of Thessaly, GR- 38334 Volos, Greece

Received 19 July, 2005; accepted in revised form 10 August, 2005

*Abstract:* The designating factors in the design of branched irrigation networks are the cost of pipes and the cost of pumping. They both depend directly on the hydraulic head of the pump station. It is mandatory for this reason to calculate the optimal head of the pump station as well as the corresponded optimal pipe diameters, in order to derive the minimal total cost of the irrigation network. The certain calculating methods in identified the above total cost of a network, that have been derived are: the linear programming optimization method, the non linear programming optimization method, the dynamic programming optimization method and the Labye's method. All above methods have grown independently and a comparative study between them has not yet been derived. In this paper a comparative calculation of the pump station optimal head as well as the corresponded economic pipe diameters, using the Labye's optimization method and the linear programming method, is presented. Application and comparative evaluation in a particular irrigation network is also developed. From the study it is being held that the two optimization methods in fact conclude to the same result and therefore can be applied with no distinction in the studying of the branched hydraulic networks.

*Key words:* Irrigation, head, pump station, network, cost, optimization, linear programming, Labye

## 1. Introduction

The problem of selecting the best arrangement for the pipe diameters and the optimal pumping head as like as the minimal total cost to be produced, has received considerable attention many years ago by the engineers who study hydraulic works. The knowledge of the calculating procedure in order that the least cost is obtained, is a significant factor in the design of the irrigation networks and, in general, in the management of the water resources of a region. The classical optimization techniques, which have been proposed so long, are the following: a) The linear programming optimization method [1,7,8,9,10,11,13], b) the nonlinear programming optimization method [9,10,13], c) the dynamic programming method [9,13], and d) the Labye's optimization method [2,3,4,5,6,9,12,13]. The common characteristic of all the above techniques is an objective function, which includes the total cost of the network pipes, and which is optimized according to specific constraints. The decision variables that are generally used are: the pipes diameters, the head losses, and the pipes' lengths. As constraints are used: the pipe lengths, and the available piezometric heads in order to cover the friction losses. In this study, a systematic calculation procedure of the optimal pipe diameters using the Labye's method and the linear programming method is presented. Application and comparative evaluation in an  irrigation network is also developed.

## 2. Methods

### 2.1 The Labye's optimization method

According to this method [2,3,4,5,6,9,10,11], the optimal solution of hydraulic networks is obtained considering that the pipe diameters can only be chosen in a discrete set of values corresponding to the

---

* Corresponding author: e-mail: theoxar@teiep.gr

standard ones considered. It consists of the tracing of a zigzag line in a coordinates diagram, from which the minimal cost of the network can be obtained as a function of the total piezometric losses of the network.

### 2.1.1  A network with pipes in sequence

For every pipe of the network [2,3,4,5,6,9,12] the available commercial size of diameters are selected and then calculated: **i.** the frictional head losses per meter ,$J_{ij}$ ; **ii.** the pipe cost  per meter, $c_{ij}$;

and **iii.** the  various gradients  $\varphi_{ij} = \left|\dfrac{\Delta c_{ij}}{\Delta J_{ij}}\right|$  witch are classified in decreased order. After that, the graph

P – H, (figure 1) is constructed witch is a convex zigzag line witch is called " *the characteristic*" of the network. The gradients of the zigzag line various parts, $\varphi_{ij}$, are progressively decreased from left to right. The terminal right point of the characteristic, A, corresponds to the minimal diameters for all the pipes of the network with the maximal total frictional losses, $H_A$, and the minimal total cost of the network $P_A$. Similarly the terminal left point of the characteristic, F, corresponds to the maximal diameters for all the pipes of the network with the minimal total frictional losses, $H_F$, and the maximal total cost of the network $P_F$.

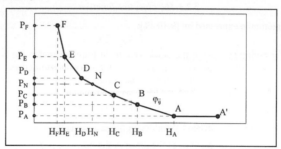

Figure 1:  A network with pipes in sequence

After that, the total cost of the network, $P_N$, corresponded to the available total frictional head loss, $H_N$, is calculated. If the point N lies in the part of the characteristic with gradient $\varphi_{ij}$, it means that only the ith pipe must be constructed with two different diameters. On the passing from the point N to the point F, linear parts with progressively increasing gradients corresponding to the various pipes of the network are detected. Each pipe is constructed with the lower diameter corresponding to the gradient $\varphi_{ij}$. Similarly any pipe, the gradients of which are detected on the right of the point N, is constructed with the higher diameter corresponding to the gradient $\varphi_{ij}$. It is concluded that only one pipe of the network is possible to be constructed with two different diameters.

### 2.1.2.  A network with two branched pipes

At first the characteristic lines of the two branches are constructed. Then the cumulative characteristic line is constructed, which is produced by adding the ordinates of the contributing branches. This characteristic line is also a convex zigzag line similar to the characteristic lines of the contributing branches [2,3,4,5,6,9,12].

### 2.1.3 Branched networks

The following  steps (figure 2) are followed [2,6,9,12]: a. The characteristic lines of the branches i.g. BC , BD , AB, AE and OA are constructed according to the paragraph 2.1.1 (case of pipes in sequence). b. The characteristic line of  the branch  BCD is constructed according to the paragraph 2.1.2.  c. The characteristic line of  the branch  BCD – AB  is constructed, adding  the characteristic lines of the branches BCD and AB according to the paragraph 2.1.1. d. The characteristic line of  the branch  (BCD – AB) – AE  is constructed, according to the paragraph 2.1.2. The procedure is continued until the head , O , of  the network. Finally  the total cost of the network, $P_N$, corresponded to the available total frictional head loss, $H_N$, is calculated  and then the economic pipe diameters are selected.

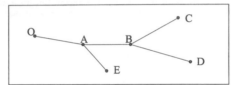

Figure 2: A branched irrigation network

## 2.2 The linear programming optimization method

According to this method [1,7,8,9,10,11,13], the search for optimal solutions of hydraulic networks is carried out considering that the pipe diameters can only be chosen in a discrete set of values corresponding to the standard ones considered. Due to that, each pipe is divided into as many sections as there are standard diameters, the length of these sections thus being adapted as decision variables. The least cost of the pipe network, $P_N$, is obtained from the minimal value of the objective function, meeting the specific functional and non negativity constraints.

### 2.2.1 The objective function

The objective function is expressed by [9,10,11]:
$$f(X) = CX \tag{1}$$
where C is the vector giving the cost of the pipes sections in Euro per meter and X is the vector giving the lengths of the pipes sections in meter. The vectors C and X are determined as:

$$C = (C_1 .... C_i ... C_n) \quad ; \quad C_i = (\delta_{i1} ...... \delta_{ij} ... \delta_{ik}) \quad ; \quad X = (X_1 .... X_i ... X_n)^T ;$$

$$X_i = (x_{i1} ...... x_{ij} ... x_{ik})^T \quad \text{for } i = 1, 2, ..., n \quad \text{and} \quad j = 1, 2, ..., k$$

where $x_{11}, x_{12}, ..., x_{nk}$, are the decision variables in meter ; $\delta_{ij}$ is the cost of jth section of ith pipe in Euro per meter ; n is the total number of the pipes in the network ; and k the total number of each pipe accepted diameters (= the number of the sections in which each pipe is divided).

### 2.2.2 The constraints of the problem

The constraints of the problem are specific functional and non negativity constraints [9,10,11]. The functional constraints are length constraints and friction loss constraints. The length constraints are expressed by $L_i = \sum_{j=1}^{k} x_{ij}$ .The friction losses constraints are expressed by: $\sum_{i=1}^{i} \Delta h_i \leq H_A - h_i$ for all the nodes ,i, where $H_A$ is the piezometric head of the water intake, $h_i$ is the minimum required piezometric head at each node i. The sum $\sum_{i=1}^{i} \Delta h_i$ is taken along the length of every route i, of the network. The non negativity constraints are expressed by: $x_{ij} \geq 0$.

### 2.2.3 The variance of the head of the pump

The calculation of the optimal pump head is achieved through the following process: a) the variance of the pump station head are determined, b) using one of the known optimisation methods, the optimal annual total cost of the project for every head of the pump station is calculated, c) The graph $P_{ET}$ - $H_A$ is constructed and its minimum point is defined, which corresponds in the value of $H_{man}$, This value constitutes the optimal pump station head. [9,10,11]

### 2.2.4 The least cost and the economic pipe diameters of the network

The least cost of the pipe network, $P_N$, is obtained from the minimal value of the objective function. The minimal value of the objective function, $\min f(X) = CX$, is obtained using the simplex method.

## 3. Application

The optimal cost of the irrigation network, which is shown in Figure 3, is calculated. The material of the pipes is PVC 10 atm and the available total head is $H_N = 60.00$ m.

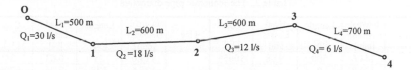

Figure 3: The under solution network

## 3.1 Selecting the acceptable commercial pipe diameters

Using the continuity equation the acceptable commercial diameters, as well as the cost per meter for every pipe of the network are selected. From the Darcy- Weisbach equation the head losses are calculated. From the values of head losses and pipe cost the various gradients $\varphi_{ij}$ are calculated. The results are presented in Table 1.

Table 1. The geometric and hydraulic characteristics off the pipes

| Pipe | Internal Diameter [mm] | Head Losses [%] | Cost [€/m] | Gradient $\varphi$ | Pipe | Internal Diameter [mm] | Head Losses [%] | Cost [€/m] | Gradient $\varphi$ |
|---|---|---|---|---|---|---|---|---|---|
| 1 | 144.6 | 1.935 | 21.31 | 0.7923 | 2 | 113.0 | 2.524 | 13.82 | 0.2809 |
|  | 180.8 | 0.640 | 31.57 | 2.5587 |  | 126.6 | 1.436 | 16.88 | 0.6418 |
|  | 203.4 | 0.359 | 38.76 | 8.3432 |  | 144.6 | 0.745 | 21.31 | 2.0681 |
|  | 253.2 | 0.123 | 58.45 |  |  | 180.8 | 0.249 | 31.57 | 6.6010 |
| 3 | 99.4 | 2.231 | 11.33 | 0.2378 |  | 203.4 | 0.140 | 38.76 |  |
|  | 113.0 | 1.184 | 13.82 | 0.6036 | 4 | 99.4 | 0.619 | 11.33 | 0.7428 |
|  | 126.6 | 0.677 | 16.88 | 1.3673 |  | 113.0 | 0.331 | 13.82 |  |
|  | 144.6 | 0.353 | 21.31 |  |  |  |  |  |  |

## 3.2. Solving the network according to the linear programming method

The objective function is expressed by :
$Z = 21.31X_1 + 31.57X_2 + 38.76X_3 + 58.45X_4 + 13.82X_5 + 16.88X_6 + 21.31X_7 + 31.57X_8 + 38.76X_9 + 11.33X_{10} + 13.82X_{11} + 16.88X_{12} + 21.31X_{13} + 11.33X_{14} + 13.82X_{15}$
Subject to:

$X_1 + X_2 + X_3 + X_4 = 500$ ; $X_5 + X_6 + X_7 + X_8 + X_9 = 600$

$X_{10} + X_{11} + X_{12} + X_{13} = 600$ ; $X_{14} + X_{15} = 700$

$1.935X_1 + 0.640X_2 + 0.359X_3 + 0.123X_4 \le H_A - h_1 = 1800$

$1.935X_1 + 0.640X_2 + 0.359X_3 + 0.123X_4 + 2.524X_5 + 1.436X_6 + 0.745 X_7 + 0.249X_8 + 0.140X_9 \le 2300$

$1.935X_1 + 0.640X_2 + 0.359X_3 + 0.123X_4 + 2.524X_5 + 1.436X_6 + 0.745 X_7 + 0.249X_8 + 0.140X_9 + 2.231X_{10} + 1.184X_{11} + 0.677X_{12} + 0.353X_{13} \le 2700$

$1.935X_1 + 0.640X_2 + 0.359X_3 + 0.123X_4 + 2.524X_5 + 1.436X_6 + 0.745X_7 + 0.249X_8 + 0.140X_9 + 2.231X_{10} + 1.184X_{11} + 0.677X_{12} + 0.353X_{13} + 0.619X_{14} + 0.331X_{15} = 3000$

The minimization of the objective function is obtained using the simplex method. The results are presented in Table 2. From the Table is produced that min Z=36929 €

Table 2. The economic pipe diameters

| Pipe | Internal Diameter [mm] | Length [m] | Head Losses [%] | Cost [€] | Pipe | Internal Diameter [mm] | Length [m] | Head Losses [%] | Cost [€] |
|---|---|---|---|---|---|---|---|---|---|
| 1 | 144.6 | 500 | 9.68 | 10655 | 2 | 113.0 | 25 | 0.63 | 345 |
| | 180.8 | 0 | | | | 126.6 | 575 | 8.26 | 9706 |
| | 203.4 | 0 | | | | 144.6 | 0 | | |
| | 253.2 | 0 | | | | 180.8 | 0 | | |
| 3 | 99.4 | 0 | | | | 203.4 | 0 | | |
| | 113.0 | 600 | 7.10 | 8292 | 4 | 99.4 | 700 | 4.33 | 7931 |
| | 126.6 | 0 | | | | 113.0 | 0 | | |
| | 144.6 | 0 | | | | | | | |

### 3.3. Solving the network according to Labye's method

For all the pipes of the network the various gradients $\varphi_{ij}$ are classified in decreased order. The results are presented in Table 2. After that, the graph P – H, (figure 4) is constructed and the total cost of the network, $P_N$ , corresponded to the available total head, $H_N = 60.00$ m, is calculated. It is resulted that $P_N = 37005$ €. According to the procedure presented in paragraph 2.1.1., the pipe diameters are calculated and the results are presented in Table 4.

Table 3. Classification of the various gradients $\varphi_{ij}$ in declining order

| $\varphi_{11}$ | $\varphi_{21}$ | $\varphi_{12}$ | $\varphi_{22}$ | $\varphi_{31}$ | $\varphi_{13}$ | $\varphi_{41}$ | $\varphi_{23}$ | $\varphi_{32}$ | $\varphi_{24}$ | $\varphi_{33}$ |
|---|---|---|---|---|---|---|---|---|---|---|
| 8.343 | 6.601 | 2.559 | 2.068 | 1.367 | 0.792 | 0.743 | 0.642 | 0.604 | 0.281 | 0.238 |

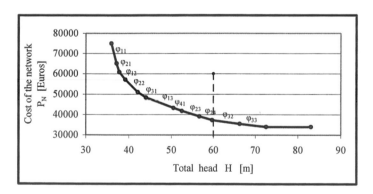

Figure 3. Variance of P per H

Table 4. The economic pipe diameters according to the Labye's optimization method

| Pipe | Internal Diameter [mm] | Length [m] | Head Losses [%] | Cost [€] | Total cost [€] |
|---|---|---|---|---|---|
| 1 | 144.6 | 500 | 9.68 | 10656 | |
| 2 | 126.6 | 600 | 8.61 | 10126 | $P_N = 37005$ € |
| 3 | 113.0 | 600 | 7.10 | 8293 | |
| 4 | 99.4 | 700 | 4.34 | 7930 | |

## 4. Conclusions

The optimal cost of the network according to the linear programming method is $P_N = 36929$ €, while according the Labye's method is $P_N = 37005$ €. These two values differ only 0.08 %. Similar differences have been detected for a great number of applications made by authors, using the above mentioned methods.

The two optimization methods in fact conclude to the same result and therefore can be applied with no distinction in the studying of the branched hydraulic networks.

The selection of optimal pipe diameters using both the optimization methods, results a vast number of complex calculations. For this reason the application as well as the supervision of these methods and the control of the calculations is time-consuming and difficult, especially in the case a network with many branches.

## References

[1] Alperovits E. and Shamir U.*Design of optimal water distribution*. Wat. Res. Res. 13 (6):885-900,1977.

[2] Labye Y., *Methodes permettant de determiner les caracteristiques optimales d 'un reseau de distribution d'eau - Methode discontinue*. Bull. Techn. du Genie Rural, No 50, Apr. 1961

[3] Labye Y., *Etude des procedés de calcul ayant pour but de rendre minimal le cout d'un reseau de distribution d'eau sous pression*. La Houille Blanche, 5: 577-583, 1966.

[4] Labye Y., *Les methodes de calcul des reseaux d'irrigation en conduites sous pression*. Colloque tenu `a Athenes de 30.3.1971 'a 1.4.1971, Irrigation Par Aspersion, Edition de la chabre Technique de Grece, pp.375-420, 1971.

[5] Labye Y., *Etude d'un probleme de dimensionnement "optimum" des reseaux d'irrigation 'a la demande,en avenir aleatoire*. Huitie mes journees Europeennes de la Comission Internationale des Irrigations et du Drainage. Colloque d'Aix en-Provence du 14 au 19 Juin 1971. France.

[6] Livaditis, E., *Méthode discontinue de U. Labye pour la recherche d'une solution de coordination économique des divers diamètres dans un d'irrigation sous pression*. Jour. Technika Chronika., Athens, 1969(10), 661 – 676, (In Greek).

[7] Shamir, U., *Optimal design and operation*. Water Res. Res, 10(1): 27-36 , 1974.

[8] Smith D. V., *Minimum cost design of linearly restrained water distribution networks*. M. Sc. Thesis, Dept. of Civil Eng., Mass. Inst. of Techncl., Cambridge , 1966.

[9] Theocharis, M., *Irrigation networks optimization. Economic diameter selection*. Ph.D. Thesis, Dep. of Rural and Surveying Engin. A.U.TH., Salonika, 2004, (In Greek with extended summary in English).

[10] Theocharis, M., et al, *Comparative Calculation of the Optimal Head of the Pump station of the Irrigation Networks using a) The Linear Programming Method and b) The Nonlinear Programming method*. Proc. 1st Int. Conf. "From Scientific Computing to Computational Engineering" (1st IC-SCCE) , Sept. 2004, CD-ROM, Athens, Greece

[11] Theocharis, M., et al, *Comparative Calculation of the Optimal Head of the Pump Station of the Irrigation Networks Using the Linear Programming Method and the Simplified Nonlinear Programming Method* , Proc. 9st Int. Conf. of the Hellenic Hydrotechnical Association, April 2003, Salonika,, pp. 249–256, (In Greek)
[10] Tzimopoulos C., Agricultural Hydraulics Vol. II, 51-94, Salonika, 1982, (In Greek).

[12] Tzimopoulos C., *Agricultural Hydraulics* Vol. II, 51-94, Salonika, 1982, (In Greek).

[13] Vamvakeridou - Liroudia L., 1990. *Pressure water supply-irrigation networks. Solution-Optimization. Hydraulics Engineering Computer Applications, (H.E.C.A.)*, Athens, 1990, (In Greek).

Brill Academic Publishers
P.O. Box 9000, 2300 PA Leiden
The Netherlands

*Lecture Series on Computer
and Computational Sciences*
Volume 4, 2005, pp. 570-571

# Resonance in an Interacting Induced-Dipole Polarization Model

Francisco Torrens
Institut Universitari de Ciència Molecular, Universitat de València,
Dr. Moliner-50, E-46100 Burjassot (València), Spain

Received 21 June, 2005; accepted in revised form 13 July, 2005

As an example of the manner in which the molecular polarizability depends on the atom polarizabilities, the isotropic molecule $CH_4$ is considered. Certain features of this dependence are illustrated. Polarizability of $CH_4$ as a function of polarizabilities of H and C is represented. Additivity of atom polarizabilities would require that the surface be a plane. It is seen that this is approximately true near the origin, where interactions are small. However, the experimental polarizability of methane is $2.62Å^3$, and this value is reached only in regions of the surface where the influence of interactions is quite marked. The most notable feature is a curve of discontinuity along which the polarizability approaches $\pm\infty$. This behaviour is seen in the polarizability surfaces of several molecules that are similarly explored. Its origin for diatomic molecules can be seen in equations $\alpha_\parallel = (\alpha_A + \alpha_B + 4\alpha_A\alpha_B/r^3)/(1 - 4\alpha_A\alpha_B/r^6)$ and $\alpha_\perp = (\alpha_A + \alpha_B - 2\alpha_A\alpha_B/r^3)/(1 - \alpha_A\alpha_B/r^6)$, where the denominators vanish when $\alpha_A\alpha_B$ approaches $r^6/4$ or $r^6$, respectively. Therefore, $\alpha_A$ and $\alpha_B$ are inversely related along the curve of discontinuity for this case. Remarkably, it is found from the computed data for $CH_4$ that the curve of discontinuity in the $\alpha_H$–$\alpha_C$ plane follows the relation $\alpha_H\alpha_C = 0.193Å^6$, which is of the form expected for diatomic molecules, though the numerical constant is not predictable from the above equations. (The general condition for infinite polarizability is $\det(\mathbf{A}) = 0$; infinite polarizability means that $\mu$ is non-vanishing when $\mathbf{E} = 0$, and this is possible only if $\det(\mathbf{A}) = 0$. Since $\det(\mathbf{A})$ is a polynomial of degree $3N$ in the atom polarizabilities, the simple inversion relation between $\alpha_H$ and $\alpha_C$ for $CH_4$ implies that this polynomial contains the factor $\alpha_H\alpha_C$—$0.193$.) The significance of a polarizability of $\pm\infty$ is that the molecule is in a state of resonance and absorbs energy from the applied field. This occurs in spite of the fact that any absorption properties of the atoms are not introduced. This behaviour of the model can be understood from its close relation to the classical system of $N$ coupled oscillators, which likewise shows resonance under conditions other than the resonance conditions of the isolated oscillators. There is evidence that the absorption properties of some types of systems can be predicted from the point dipole interaction approach used here, but it seems doubtful that this could be done reliably for molecules with our model; among other things, the neglect of electron exchange between atoms is probably serious. In the immediate vicinity of the resonance condition the polarizability is bound to be in error because the model does not take into account damping effects, which would prevent the polarizability from going to infinity. Therefore, for the present the resonance conditions are regarded wherever they appear simply as indicators that the coupling between atoms far exceeds the extent that can be treated by the model. (Indeed, it is an object of this study to learn whether coupling between atoms can be represented by the model at all.)

In the second part of the study, it is reported a set of strategies devised for applications to large systems. The following improvements are implemented in the model. (1) Damping functions are used in the calculation of the interaction tensor to prevent the polarizability from going to infinity in the models by Applequist and Birge. Böttcher proposed $T_{q,\alpha\beta}^{(2)} = 3r_{q,\alpha}r_{q,\beta}/r_q^5 - \delta_{\alpha\beta}/r_q^3$, where $\mathbf{T}^{(2)}$ is the gradient of the electric field $\mathbf{E}$ and $\delta$ represents the Kronecker delta function: $\delta(\alpha,\beta) = 1$ if $\alpha = \beta$, and $\delta(\alpha,\beta) = 0$ if $\alpha \neq \beta$. Thole proposed $T_{q,\alpha\beta}^{(2)} = 3v_q^4 r_{q,\alpha}r_{q,\beta}/r_q^5 - (4v_q^3 - 3v_q^4)\delta_{\alpha\beta}/r_q^3$, where $v_q = r_q/s_q$ if $r_q < s_q$, otherwise $v_q = 1$. The term $s$ is defined as $s = 1.662(\alpha_p\alpha_q)^{1/6}$. The values of $\alpha_p$ are parametrized by *ab initio* coupled Hartree–Fock. Miller proposed $T_{q,\alpha\beta}^{(2)} = [3r_{q,\alpha}r_{q,\beta}/r_q^5 - \delta_{\alpha\beta}/r_q^3]\{1 - \exp[-(r_q/0.7\rho_q)^{10}]\}$ with $\rho_q = \rho_p + \rho_q$ (sum of the van der Waals radii). (2) The interaction between bonded atoms and atoms with a distance lying in an interval defined by $[r^{inf}, r^{sup}]$ is neglected. The starting values for this interval are $[0, 30]$ and $r^{inf}$ is incremented if resonance conditions are detected. (3) The following tests indicating a resonance condition are implemented: (a) test whether matrix $B^{-1}$ is singular, where B is the manybody polarizability matrix; (b) test whether matrix $B^{-1}$ is not defined positive; (c) test whether matrix B is not defined positive; and (d) test whether some effective matrix $B_p$ is not defined positive.

Applications are carried out for the following systems. (1) Clusters: $Si_n$, $Ge_n$ ($1 \leq n \leq 10$), $Ga_nAs_m$ ($1 \leq n$, $m \leq 4$), $Sc_n$ ($1 \leq n \leq 74$), $C_n$-graphene ($1 \leq n \leq 96$), $C_n$-fullerene ($1 \leq n \leq 82$), endohedral metallofullerenes $Sc_n@C_m$ ($1 \leq n \leq 3$, $60 \leq m \leq 82$). (2) Single-wall carbon nanotubes: *zigzag* $(n,0)$ ($4 \leq n \leq 20$) ($16 \leq$ atoms $\leq 180$), *armchair* $(n,n)$ ($5 \leq n \leq 10$) ($90 \leq$ atoms $\leq 200$), *chiral* $(n,m)$ ($5 \leq n \leq 19$, $1 \leq m \leq 9$) ($16 \leq$ atoms $\leq 250$), as well as $n \to \infty$ and $\infty$-long extrapolations. Deformed $(8,0)$ ($28 \leq$ atoms $\leq 48$) with elliptical radial deformation. (3) Benzothiazole (A)–benzobisthiazole (B) A–$B_n$–A ($0 \leq n \leq 13$) linear oligomers and $n \to \infty$ extrapolations in three conformations: (a) fully planar (000), (b) a rotational isomer in which each unit is rotated by a fixed angle $\phi$ with respect to the previous one, all rotations performed in the same direction (+++) ($0° \leq \phi \leq 10°$), and (c) rotations performed in the alternate directions (+–+) ($0° \leq \phi \leq 10°$).

Brill Academic Publishers
P.O. Box 9000, 2300 PA Leiden
The Netherlands

*Lecture Series on Computer*
*and Computational Sciences*
Volume 4, 2005, pp. 572-575

# Effects of Two-Photon Absorption on the Propagation of Optical Solitons

G. Tsigaridas[1], I. Polyzos, V. Giannetas and P. Persephonis
Department of Physics
University of Patras
GR-26500 Patras, Greece

Received 6 July, 2005; accepted in revised form 27 July, 2005

*Abstract:* Optical solitons are a subject of intense research due to their applications in telecommunications and optical data processing. In this work, we study, both numerically and analytically, the effects of two-photon absorption on the propagation of optical solitons. Using perturbation theory, we calculated the evolution of the soliton parameters during propagation. We also studied the propagation of optical solitons under continuous and localised amplification. We have found the conditions for stable propagation in both cases, and the results were verified by numerical simulations.

*Keywords:* soliton propagation, two-photon absorption

*Mathematics Subject Classification:* 35Q51, 37K40

*PACS:* 42.65-k, 42.65.Tg, 42.81.Dp

## 1. Introduction

The stability of optical solitons makes them ideal candidates for signal transmission through optical fibers [1] as well as for applications in all-optical switching and optical data processing [2], where they are used as bits of information. In most cases, fiber losses are assumed to be linear, which is a valid approximation for ordinary silica fibers. However, two-photon absorption (TPA) effects are likely to become important for new types of fibers [3] with enhanced nonlinearity. Moreover, several applications, and especially optical switching, require low pulse energy. Therefore, the peak power required to form a fundamental soliton should be reduced. A simple way to achieve this is to use materials with higher nonlinear refractive index than that of silica. However, in many cases the enhanced nonlinear refractive index is accompanied by an enhancement of the TPA coefficient of the material. Consequently, it is important to study the effects of two-photon absorption on optical solitons.

## 2. Mathematical formulation

The propagation of a picosecond duration optical pulse in a single-mode fiber characterized by both linear and nonlinear losses is described by the equation [4]

$$\frac{\partial A}{\partial z} + \frac{1}{\upsilon_g}\frac{\partial A}{\partial t} + \frac{i}{2}\beta_2\frac{\partial^2 A}{\partial t^2} - i\gamma|A|^2 A = -\frac{\alpha_0}{2}A - \frac{\alpha_2}{2}|A|^2 A \tag{1}$$

where $A$ is the slowly varying envelope of the electric field, $\upsilon_g$ the group velocity of the pulse, $\beta_2$ the group-velocity-dispersion (GVD) coefficient, $\gamma = k_0 n_2 / A_{eff}$ a factor describing the nonlinearity of the fiber ($k_0$ is the wavenumber, $n_2$ the nonlinear refractive index and $A_{eff}$ the effective core area of the optical fiber [4]). $\alpha_0$, $\alpha_2$ are the linear and TPA coefficients respectively. If we introduce the normalized variables

---

[1] Corresponding author. E-mail: gtsig@physics.upatras.gr

$$Z = \frac{z}{L_D}, \qquad\qquad T = \frac{t - z/\upsilon_g}{T_p} \qquad\qquad (2)$$

where $L_D = T_0^2 / |\beta_2|$ is the dispersion length and $T_p$ the pulse duration, Eq. (1) becomes

$$i\frac{\partial u}{\partial Z} + \frac{1}{2}\frac{\partial^2 u}{\partial T^2} + |u|^2 u = -i\Gamma_1 u - i\Gamma_2 |u|^2 u \qquad (3)$$

where

$$u = \sqrt{\gamma L_D}\, A, \qquad \Gamma_1 = \frac{\alpha_0}{2} L_D, \qquad \Gamma_2 = \frac{\alpha_2}{2\gamma} \qquad (4)$$

In deriving Eq. (3) we have also taken into account that in order to form solitons, the group-velocity-dispersion coefficient $\beta_2$ should be negative. Eq. (3) is the well known nonlinear Srödinger (NLS) with linear and nonlinear loss terms, which can be treated as perturbations.

## 3. Calculation of the soliton parameters in the presence of linear and two-photon absorption

In the absence of perturbations Eq. (3) admits the well known one-soliton solution in the general form [1]

$$u(T, Z) = \eta \,\mathrm{sech}\left[\eta(T - T_C)\right] \exp(-i\kappa T + i\sigma) \qquad (5)$$

where

$$\frac{dT_C}{dZ} = -\kappa, \qquad \frac{d\sigma}{dZ} = \frac{1}{2}(\eta^2 - \kappa^2) \qquad (6)$$

The parameters $\kappa$, $\eta$ are related to the real and imaginary part of the soliton eigenvalue $\zeta = (\kappa + i\eta)/2$. $\tau_c$ is the location of the soliton center and $\sigma$ the time-independent phase shift. On the other hand, as it can be deduced from Eq. (6) the parameter $\kappa$ corresponds to the soliton velocity in the reference frame defined by Eq. (2). When the perturbation terms $-i\Gamma_1 u$, $-i\Gamma_2 |u|^2 u$ are present, one can assume that as far as $\Gamma_1, \Gamma_2 \ll 1$ the soliton maintain its shape, but the soliton parameters are functions of the propagation distance $Z$ (adiabatic approximation [1]). Within the limits of this approximation we find that the equations describing the evolution of the soliton parameters are the following:

$$\frac{d\eta}{dZ} = -2\Gamma_1\eta - \frac{4}{3}\Gamma_2\eta^3, \qquad\qquad \frac{d\kappa}{dZ} = 0 \qquad (7)$$

$$\frac{d\sigma}{dZ} = \frac{1}{2}(\eta^2 - \kappa^2), \qquad\qquad \frac{dT_C}{dZ} = -\kappa \qquad (8)$$

It should be noted that the soliton velocity is not altered by the loss terms. The solutions of the above equations are

$$\eta(Z) = \eta_0\left[1 + \frac{2}{3}\frac{\Gamma_2}{\Gamma_1}\eta_0^2\left[1 - \exp(-4\Gamma_1 Z)\right]\right]^{-1/2} \exp(-2\Gamma_1 Z), \qquad \kappa(Z) = \kappa_0 \qquad (9)$$

$$\sigma(Z) = \frac{3}{16\Gamma_2}\ln\left[1 + \frac{2}{3}\frac{\Gamma_2}{\Gamma_1}\eta_0^2\left[1 - \exp(-4\Gamma_1 Z)\right]\right] - \frac{1}{2}\kappa_0^2 Z + \sigma_0, \qquad T_C(Z) = -\kappa_0 Z + T_0 \qquad (10)$$

where $\eta_0$, $\kappa_0$, $T_0$, $\sigma_0$ are the initial $(Z = 0)$ values of the soliton amplitude, velocity, center location and time-independent phase respectively. It is clear that if $\kappa_0 = T_0 = \sigma_0 = 0$, which is the most common case in practice, $\kappa(Z)$, $T_C(Z)$ remain equal to zero throughout propagation.

## 4. Study of soliton propagation in the presence of continuous or localized amplification

It is evident that stable soliton propagation is not possible in the absence of amplification. There are two categories of amplification setups. The first is continuous amplification, where the gain is distributed throughout the fiber and compensates for the losses at each point. In practice this can be achieved using Raman amplification or preferably a distributed Erbium Doped Fiber Amplifier (*d*-EDFA), which provides an almost constant amplification factor [5]. The second setup is localized amplification where the pulse is amplified at certain points in the propagation line.

In the case of continuous amplification the term $gu/2$ must be added on the right hand side, where $g$ is the gain per unit length. In this case, Eq. (3)

$$i\frac{\partial u}{\partial Z}+\frac{1}{2}\frac{\partial^2 u}{\partial T^2}+|u|^2\,u = iGu-i\Gamma_1 u-i\Gamma_2|u|^2\,u \tag{11}$$

where $G = g\,L_D/2$. In this case, the evolution of the soliton amplitude $\eta$ is described by the equation

$$\frac{d\eta}{dZ}=2G\eta-2\Gamma_1\eta-\frac{4}{3}\Gamma_2\eta^3 \tag{12}$$

From the above equation it is evident that the soliton amplitude remains constant if the gain factor $G$ is equal to $\Gamma_1+(2/3)\Gamma_2\eta^2$. Consequently if the system is designed in a way that the gain factor $G$ is equal to $\Gamma_1+(2/3)\Gamma_2\eta_0^2$ where $\eta_0$ is the initial soliton amplitude it is expected that, within the limits of the adiabatic approximation, the soliton will propagate without distortion.

In the case of localized amplification, using the guiding center soliton theory [1], it can be shown that under certain conditions, solitons can propagate with negligible distortion. Specifically, setting $u(T,Z)=\tilde{\alpha}(Z)q(T,Z)$, where $\tilde{\alpha}(Z)$ is the rapidly varying pulse amplitude, Eq. (3) is equivalent to the following set of equations:

$$i\frac{\partial q}{\partial Z}+\frac{1}{2}\frac{\partial^2 q}{\partial T^2}+\tilde{\alpha}^2|q|^2\,q = i\Gamma_1^B q-i\Gamma_2\tilde{\alpha}^2|q|^2\,q \tag{14a}$$

$$\frac{d\tilde{\alpha}(Z)}{dZ}=-\Gamma_1^A\tilde{\alpha}(Z), \qquad \text{for } (n-1)Z_A < Z < nZ_A$$
$$\tilde{\alpha}(nZ_A+0)=G_0\tilde{\alpha}(nZ_A-0), \quad \text{at } Z = nZ_A \tag{14b}$$

where $G_0$ is the gain ratio at each amplifier, $Z_A$ the normalized propagation distance between two succesive amplifiers and $\delta(Z-nZ_A)$ the delta function which physically describes the localization of the gain at points $Z = nZ_A$ $(n=1,2,3,...)$. The parameters $\Gamma_1^B$, $\Gamma_1^A$ are related to the linear loss factor $\Gamma_1$ through the equation $\Gamma_1^A =\Gamma_1+\Gamma_1^B$ and the purpose of their use will be clear later. According to Eq. (14) the evolution of the pulse amplitude is given by the relation

$$\tilde{\alpha}(Z)=\begin{cases}\alpha_{n-1}\exp\{-\Gamma_1^A[Z-(n-1)Z_A]\}, & \text{for } (n-1)Z_A < Z < nZ_A \\ G_0\alpha_{n-1}\exp(-\Gamma_1^A Z_A)\equiv\alpha_n, & \text{for } Z = nZ_A\end{cases} \tag{15}$$

where $\alpha_n$ is the $\tilde{\alpha}(Z)$ exactly after the *n*-th amplifier. If $G_0 = \exp(\Gamma_1^A Z_A)$ it is clear that the pulse amplitude $\tilde{\alpha}(Z)$ will retain its initial value after each amplification stage. Further, if the average value of $\tilde{\alpha}^2(Z)$ during the propagation between two successive amplifiers is equal to unity, then Eq.(13) is equivalent to Eq. (3). Defining $\langle \tilde{\alpha}^2(Z) \rangle = \int_0^{Z_A} \tilde{\alpha}^2(Z) dZ / Z_A$, follows that $\langle \tilde{\alpha}^2(Z) \rangle = 1$ if the initial

pulse is amplified by the factor $\alpha_0 = \left\{ 2\Gamma_1^A Z_A / \left[ 1 - \exp(-2\Gamma_1^A Z_A) \right] \right\}^{1/2}$.

In this case Eq.(13) is equivalent to Eq. (3) and the amplitude evolution of the $q(T,Z)$ soliton is still given by Eq. (7). Consequently, if $\Gamma_1^B$ is set equal to $2\Gamma_2\eta_0^2/3$ or equivalently $\Gamma_1^A = \Gamma_1 + 2\Gamma_2\eta_0^2/3$, where $\eta_0$ is the initial soliton amplitude, then within the limits of the adiabatic approximation, the $q(T,Z)$ will propagate undistorted. Under these conditions the $u(T,Z)$ pulse will only undergo a periodic variation of its amplitude and its "average" form will remain unchanged. In order to verify our results we have numerically solved Eq. (3) and calculated the evolution of the pulse peak during propagation for both the cases of continuous and localized amplification. The results are shown in figure 1.

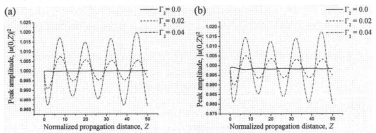

Figure 1: The evolution of the pulse peak as a function of the normalized propagation distance Z in the case of continuous (a) and localized (b) amplification. In the second case the pulse amplitude has been calculated at the half distance between two successive amplifiers.

It is clear that in both cases, for relatively small values of the nonlinear loss parameter $(\Gamma_2 \leq 0.04)$, soliton propagates with negligible distortion $(\leq 2\%)$

## 5. Conclusions

In conclusion, we have studied the propagation of solitons in the presence of two-photon absorption, both analytically and numerically. We have calculated the evolution of the soliton parameters. We have also studied soliton propagation under continuous or localized amplification and found the conditions for stable soliton propagation in each case. Our results have been verified by numerical simulations.

## Acknowledgments

We thank the European Social Fund (ESF), Operational Program for Educational and Vocational Training II (EPEAEK II), and particularly the Program PYTHAGORAS II, for funding the above work.

## References

[1] A. Hasegawa and Y Kodama: *Solitons in Optical Communications*. Clarendon Press, Oxford, 1995.

[2] A. H. Haus, Optical – fiber solitons, their properties and uses, *Proceedings of the IEEE* **81** 970-983 (1993)

[3] T. Okuno, M. Onishi, T. Kashiwada, S. Ishikawa, and M. Nishimura, "Silica-based functional fibers with enhanced nonlinearity and their applications", *IEEE J. Sel. Top. Quant.* **5**, 1385-1391 (1999)

[4] G. P. Agrawal: *Nonlinear Fiber Optics*. Academic Press, San Diego, 1989 (chap. 2)

[5] Z. M. Liao and G. P. Agrawal, Role of distributed amplification in designing high-capacity soliton systems, *Opt. Express* **9** 66-70 (2001)

Brill Academic Publishers
P.O. Box 9000, 2300 PA Leiden,
The Netherlands

*Lecture Series on Computer*
*and Computational Sciences*
Volume 4, 2005, pp. 576-581

# Explicit Eighth Order Hybrid Numerov Type Methods With Nine Stages

## Ch. Tsitouras[1]

TEI of Chalkis, Department of Applied Sciences, GR34400, Psahna, Greece

Received 12 August, 2005; accepted 12 August, 2005

*Abstract:* We present a new explicit hybrid two step method for the solution of second order initial value problem. It costs only nine function evaluations per step and attains eighth algebraic order so it is the cheapest in the literature. Its coefficients are chosen to reduce amplification and phase errors. Thus the method is well suited for facing problems with oscillatory solutions. Numerical tests justify our effort.

*Keywords:* Initial Value Problem, Second Order, Order conditions, Oscillatory solutions, Phase-Lag.

*Mathematics Subject Classification:* 65L05, 65L06

## 1 Introduction.

We consider the initial value problem of second order

$$y'' = f(y), \ y(t_0) = y^{[0]}, \ y'(t_0) = y'^{[0]}, \tag{1}$$

where $f : \Re^{N+1} \longrightarrow \Re^{N+1}$ and $y^{[0]}, \ y'^{[0]} \in \Re^{N+1}$. Observe that $y'$ is not involved in (1). The independent variable $t$ can be considered as an extra component of $y$, setting

$$y''_{N+1} = 0, \ y^{[0]}_{N+1} = t_0, \ y'^{[0]}_{N+1} = 1.$$

In this paper we investigate the class of the above problems with oscillatory solutions. Our result are methods which can be applied to many problems in celestial mechanics, quantum mechanical scattering theory, in theoretical physics and chemistry and in electronics.

Implicit hybrid two step methods were introduced by Hairer [8], Cash [2] and Chawla [3] basically for satisfying P-stability [9], a useful property for dealing periodic problems. Later Chawla [4] and Chawla and Rao [5, 6] used explicit modifications of these methods especially for reducing phase errors [1].

Their construction is usually based on interpolatory nodes. These nodes carry a lot of information which is useless even for conventional methods. So, an alternative implementation of such methods was introduced in [10, 14], and studied theoretically by Coleman [7] through B2-series. In vector notation an $s$-stage Numerov type method takes the form

$$y^{[k+1]} = 2y^{[k]} - y^{[k-1]} + h^2 \cdot (b \otimes I_s) \cdot f(Y)$$
$$Y = (e + c) \otimes y^{[k]} - c \otimes y^{[k-1]} + h^2 \cdot (A \otimes I_s) \cdot f(Y) \tag{2}$$

with $h = t_{k+1} - t_k = t_k - t_{k-1} = \cdots = t_1 - t_0$, $I_s \in \Re^s$ the identity matrix, $A \in \Re^{s \times s}, b^T \in \Re^s, c \in \Re^s$ the coefficient matrices of the method and $e = (1, 1, \cdots, 1)^T \in \Re^s$.

---

[1]E-Mail: tsitoura@teihal.gr, URL address: http://users.ntua.gr/tsitoura/

## 2  Algebraic order of the new method.

When solving (1) numerically we have to pay attention in the algebraic order of the method used, since this is the main factor of achieving higher accuracy with lower computational cost. Thus this is the main factor of increasing the efficiency of our effort. Using the notation of Nyström methods the new one can be formulated in a table like the Butcher tableau,

$$\frac{c \ \big| \ A}{\big| \ b}.$$

Following tradition we make use of known information at mesh, setting:

$$Y^{[1]} = y^{[k-1]}, \ Y^{[2]} = y^{[k]}.$$

Thus the corresponding matrices for $s = 10$ become:

$$A = \begin{bmatrix} 0 & 0 & 0 & \cdots & & 0 \\ 0 & 0 & 0 & \cdots & & 0 \\ d_{11} & d_{12} & 0 & \cdots & & 0 \\ d_{21} & d_{22} & a_{21} & 0 & \cdots & 0 \\ \vdots & & & & \ddots & \vdots \\ d_{81} & d_{82} & a_{81} & \cdots & a_{87} & 0 \end{bmatrix},$$

$$b = \begin{bmatrix} w_1 & w_2 & b_1 & b_2 & \cdots & b_8 \end{bmatrix},$$

and

$$c = \begin{bmatrix} -1 & 0 & c_1 & c_2 & \cdots & c_8 \end{bmatrix}^T.$$

The new method needs only nine function evaluations per step since $f(Y^{[1]})$ has been already evaluated in the previous step.

Under the simplifying assumptions

$$w_1 = b_1 = b_2 = 0,$$
$$(Ae)_{(3-10)} = \tfrac{1}{2} \left( c^2 + c \right)_{(3-10)}, \ (Ac)_{(3-10)} = \tfrac{1}{6} \left( c^3 - c \right)_{(3-10)}, \tag{3}$$
$$(Ac^2)_{(3-10)} = \tfrac{1}{12} \left( c^4 + c \right)_{(3-10)} \text{ and } (Ac^3)_{(5-10)} = \tfrac{1}{20} \left( c^5 - c \right)_{(5-10)}$$

with

$$c^i = \begin{bmatrix} 1 & 0 & c_1^i & c_2^i & \cdots & c_8^i \end{bmatrix},$$

and for $k_1 < k_2$

$$(v)_{(k_1 - k_2)} = (v_{k_1}, v_{k_1+1}, \cdots, v_{k_2}),$$

we get the eighth order conditions given in Table 1. In this table operation "*" may understood as component-wise multiplication.

Our methods have 62 parameters (44 entries for $A$, 10 for $b$ and 8 for $c$). Forty four equations are required assuming order conditions and satisfaction of (3). This leaves eighteen coefficients as free parameters. In order to shorten the problem we force $c_8 = 1$, $c_7 = -1$, $c_6 = \tfrac{1}{4}$, $c5 = -\tfrac{1}{4}$, $c_3 = -c_4$. Then we may express all the coefficients with respect to 13 free parameters.

Table 1: Equations of condition up to eighth order.

$$b \cdot e = 1, \qquad b \cdot c = 0,$$
$$b \cdot c^2 = \tfrac{1}{6}, \qquad b \cdot c^3 = 0,$$
$$b \cdot c^4 = \tfrac{1}{15}, \qquad b \cdot c^5 = 0,$$
$$b \cdot c^6 = \tfrac{1}{28}, \qquad b \cdot A \cdot c^4 = \tfrac{1}{840},$$
$$b \cdot c^7 = 0, \qquad b \cdot (c * (A \cdot c^4)) = \tfrac{1}{180},$$
$$b \cdot A \cdot c^5$$

## 3   Periodic problems.

Following Lambert and Watson [9] and in order to study the periodic properties of methods posed for solving (1), it is constructive to consider the scalar test problem

$$y' = -\omega^2 y, \quad \omega \in \Re. \tag{4}$$

When applying an explicit two step hybrid method of the form (2) to the problem (4) we obtain a difference equation of the form

$$y^{[k+1]} + S\left(v^2\right) y^{[k]} + P\left(v^2\right) y^{[k-1]} = 0, \tag{5}$$

where $y^{[k]} \approx y\,(nh)$ the computed approximations at $n = 1, 2, \ldots$, $v = wh$, and $S\left(v^2\right), P\left(v^2\right)$ polynomials in $v^2$.

Zero dissipation property is fulfilled by requiring $P\left(v^2\right) = 1 + v^2 b\left(I_s + v^2 A\right)^{-1} \equiv 1$, and helps a numerical method that solves (4) to stay in its cyclic orbit. The dissipation order $q$ of a method is the number satisfying $1 - P\left(v^2\right) = \mathcal{O}(v^q)$. Notice that

$$P\left(v^2\right) = 1 + \sum_{j=0}^{\infty} v^{2j+1} b \cdot A^j \cdot c.$$

A method of algebraic order $2 \cdot i$ satisfies the terms in the series above for $j = 0, 1, \cdots, i-1$. This means that for an eighth order method it is desirable to solve

$$b \cdot A^3 \cdot c = 0, \ b \cdot A^4 \cdot c = 0, \cdots \text{ etc.},$$

in order to get higher dissipation order.

The phase-lag of the method is the angle difference between numerical and theoretical cyclic solution of (4). Since the solution of (4) is

$$y(x) = e^{i\omega x},$$

we may write equation (5) as

$$e^{2iv} + S\left(v^2\right) \cdot e^{iv} + P\left(v^2\right) = \mathcal{O}(v^p), \tag{6}$$

with the number $p$ the phase-lag order of the method. Since

$$S\left(v^2\right) = 2 - v^2 b \cdot \left(I + v^2 A\right)^{-1} \cdot (e - c)$$

we observe that expression (6) is a series of the form

$$\sum_{i=2}^{\infty} v^{2i} \left( \sum_{j=1}^{i-1} \frac{1}{2(i-j)!} b \cdot A^{j-1} \cdot (e + c) + b \cdot A^{i-1} \cdot e - 2 \sum_{j=1}^{i} \frac{1}{(2j)! \cdot (2(i-j))!} \right) =$$

$$= v^2 l_2 + v^4 l_4 + v^6 l_6 + \bigcirc(v^8).$$

This series is satisfied for $v^{2 \cdot j}$, $j = 1, 2, \cdots, i$, when $2 \cdot i$ is the algebraic order of the method. Thus it is interesting to eliminate as many as possible higher order coefficients of it. We manage to get a solution with

$$z_9 = 0, \ z_{11} = 0, \ l_{10} = 0, \ l_{12} = 0, \ l_{14} = 0, \ l_{16} = 0, \ \text{and} \ l_{18} = 0.$$

The coefficients of the new method are given in Table 2.

Table 2: Coefficients of the new 8th order method

| | | |
|---|---|---|
| $c_1 = -1.618033988749895$ | $c_2 = -0.08935969452190693$ | $c_3 = -0.7180027509073757$ |
| $c_4 = 0.7180027509073757$ | $c_5 = -0.25$ | $c_6 = 0.25$ |
| $c_7 = -1$ | $c_8 = 1$ | $w_1 = 0.02267478608411768$ |
| $w_2 = b_1 = b_2 = 0$ | $b_3 = 0.1091598371161353$ | $b_4 = 0.1091598371161353$ |
| $b_5 = 0.3880338950775969$ | $b_6 = 0.3880338950775969$ | $b_7 = -0.01986851827784987$ |
| $b_8 = 0.002806267806267806$ | $d_{11} = 0.4363389981249825$ | $d_{12} = 0.06366100187501753$ |
| $d_{21} = -0.02663944838475621$ | $d_{22} = -0.02138085097354293$ | $a_{21} = 0.007333029599869930$ |
| $d_{31} = -0.05259994463359025$ | $d_{32} = 0.1179873479656171$ | $a_{31} = 0.006223764486158627$ |
| $a_{32} = -0.1728485681165938$ | $d_{41} = -0.1594931414841811$ | $d_{42} = 1.756644381705087$ |
| $a_{41} = 0.002177668974400012$ | $a_{42} = -1.462560200318788$ | $a_{43} = 0.4799966417324492$ |
| $d_{51} = -0.01315251843525407$ | $d_{52} = 0.08148753879227717$ | $a_{51} = 0.002255441346558031$ |
| $a_{52} = -0.1407999204529257$ | $a_{53} = -0.02359301393743279$ | $a_{54} = 0.00005247268677732879$ |
| $d_{61} = 0.118225140695003$ | $d_{62} = -0.2071467658425108$ | $a_{61} = -0.009902612273876664$ |
| $a_{62} = 0.2377506314405291$ | $a_{63} = -0.1720715921748083$ | $a_{64} = 0.008456715906120000$ |
| $a_{65} = 0.1809384822495436$ | $d_{71} = 0.6545342597532786$ | $d_{72} = 4.968502507588174$ |
| $a_{71} = -0.05384950599580273$ | $a_{72} = -4.016696408666935$ | $a_{73} = -1.055358930155700$ |
| $a_{74} = 0.2067362330539400$ | $a_{75} = 1.043495190976432$ | $a_{76} = -1.747363346553386$ |
| $d_{81} = -0.273125814192867$ | $d_{82} = -19.26209659195308$ | $a_{81} = 0.2868033393908071$ |
| $a_{82} = 21.50877058850632$ | $a_{83} = -1.286133152186278$ | $a_{84} = 0.7520725477949123$ |
| $a_{85} = -1.229894203564763$ | $a_{86} = 0.6765130737370460$ | $a_{87} = -0.1729097875320912$ |

## 4   Numerical Tests.

Two problems are chosen for our comparisons that are well known in the relevant literature.

### 4.1   Bessel equation

First we considered the following problem

$$y'' = \left(-100 + \frac{1}{4x^2}\right) y, \ y(1) = J_0(10), \ y'(1) = -0.5576953439142885,$$

whose theoretical solution is

$$y(x) = \sqrt{x} J_0(10x).$$

We solved the above equation in order to find the 100th root of the solution which is occurs when $x = 32.59406213134967$.

### 4.2  Inhomogeneous equation

Our second test problem was an inhomogeneous problem:

$$y'' = -100y(x) + 99\sin(x), \; y(0) = 1, \; y'(0) = 11$$

with analytical solution

$$y(t) = \cos(10x) + \sin(10x) + \sin(x).$$

We integrated that problem in the interval $x \in [0, 10\pi]$ as in [12, 15]. We recorded the end point global error achieved by our previous method of phase-lag order 22 given in [13] and the new method in Tables 3 and 4. Both problems were tested for the same computational cost.

Table 3: Accurate digits for Besell equation

| stages | 2000 | 3000 | 4000 | 5000 | 6000 | 7000 |
|--------|------|------|------|------|------|------|
| [13] | 5.3 | 7.3 | 9.3 | 9.5 | 9.9 | 10.3 |
| NEW | 6.5 | 8.0 | 9.1 | 10.0 | 10.7 | 11.4 |

Table 4: Accurate digits for Inhomogeneous equation

| stages | 1600 | 2000 | 2400 | 2800 | 3200 | 3600 |
|--------|------|------|------|------|------|------|
| [13] | 4.5 | 5.6 | 6.5 | 7.3 | 8.1 | 8.9 |
| NEW | 4.8 | 6.3 | 7.4 | 8.3 | 9.1 | 9.8 |

We observe in average an improvement of 0.75 digits which is considerable for methods of the same algebraic order. Other explicit eighth order methods that are special tuned for oscillatory problems can been found in the literature [11, 15], but it was proved that the 22-th phase-lag order method of [13] has already outperformed them.

## Acknowledgment

This research was co-founded by 75% from E.E. and 25% from Greek government under the framework of the Education and Initial Vocational Training Program - Archimedes of the TEI of Chalkis. I also thank Mrs K. Vakalopoulou (Natl. Statistical Service Greece), who was financed by this program, for valuable help during the preparation of the paper.

## References

[1] L. Brusa and L. Nigro, A one-step method for direct integration of structural dynamic equations, *Int. J. Numer. Meth. Engin.* **15** (1980) 685-699.

[2] J. R. Cash, High order P-Stable formulae for the numerical integration of periodic initial value problems, *Numer. Math.* **37** (1981) 355-370.

[3] M. M. Chawla, Two-step fourth order P-stable methods for second order differential equations, *BIT* **21** (1981) 190-193.

[4] M. M. Chawla, Numerov made explicit has better stability, *BIT* **24** (1984) 117-118.

[5] M. M. Chawla and P. S. Rao, Numerov type method with minimal phase lag for the integration of second order periodic initial value problems II. Explicit method, *J. Comput. Appl. Math.* **15** (1986) 329-337.

[6] M. M. Chawla and P. S. Rao, An explicit sixth-order method with phase-lag of order eight for $y'' = f(t, y)$, *J. Comput. Appl. Math.* **17** (1987) 365-368.

[7] J. P. Coleman, Order conditions for a class of two-step methods for $y'' = f(t, y)$, *IMA J Numer. Anal.* **23** (2004) 197-220.

[8] E. Hairer, Unconditionally stable methods for second order differential equations, *Numer. Math.* **32** (1979) 373-379.

[9] J. D. Lambert and I. A. Watson, Symmetric multistep methods for periodic initial value problems, *J. Inst. Math. Appl.* 18 (1976) 189-202.

[10] G. Papageorgiou, Ch. Tsitouras and I. Th. Famelis, Explicit Numerov type methods for second order IVPs with oscillating solutions, *Int. J. Mod. Phys. C* **12**(2001) 657-666.

[11] S. N. Papakostas and Ch. Tsitouras, High algebraic order, high phase-lag order Runge-Kutta and Nyström pairs, *SIAM J. Sci. Comput.* **21** (1999) 747-763.

[12] T. E. Simos, I. Th. Famelis and Ch. Tsitouras, Zero dissipative, explicit Numerov type methods for second order IVPs with oscillating solutions, *Numer. Algorithms* **34** (2003) 27-40.

[13] Ch. Tsitouras, Dissipative high phase-lag order methods, *Appl. Mat. Comput.* **117** (2001) 35-43.

[14] Ch. Tsitouras, Explicit Numerov type methods with reduced number of stages, *Comput. & Maths with Appl.,* **45** (2003) 37-42.

[15] Ch. Tsitouras and T. E. Simos, Explicit high order methods for the numerical integration of periodic initial value problems, *Appl. Math.& Comput.* 95 (1998) 15-26.

Brill Academic Publishers
P.O. Box 9000, 2300 PA Leiden,
The Netherlands

*Lecture Series on Computer*
*and Computational Sciences*
Volume 4, 2005, pp. 582-585

# New Approach to Monitor the Tool Condition in a CNC Machining Center

Antonio Jr Vallejo Guevara[1], Rubén Morales-Menéndez, Juan A. Nolazco Flores,
L. Enrique Sucar Succar, Ciro A. Rodriguez

Center for Innovation in Design and Technology
ITESM Monterrey campus
Monterrey NL México

Received 9 July, 2005; accepted in revised form 3 August, 2005

*Abstract:* We propose to monitor the cutting tool condition in a CNC-machining center by using continuous Hidden Markov Models (HMM). A vibration signal database was created monitoring the vibration between the cutting tool and workpiece. We trained/tested the *HMM* for 18 different operating conditions. The *HMM* were created by preprocessing the waveforms, followed by training step using the Baum-Welch algorithm. In the decoding process, the signal waveform is also preprocessed, then the trained *HMM* are used for decoding. Early experimental results validate our proposal about exploiting speech recognition frameworks in monitoring tool condition. The proposed model is capable of detecting the cutting tool condition within large variations of spindle speed and feed rate. The classifier performance was of 96%.

*Keywords:* Hidden Markov Models, Manufacturing Process, and Fault Diagnosis.

## 1 Introduction.

The successful performance of CNC machining operations involves the selection and control of a large number of parameters related to the workpiece, cutting tool and machine tool. The current trend of the development of machine tools is to increase intelligence through better process models and sensors. The capability to predict cutting tool condition can help to maintain workpiece quality and schedule tool changes [7]. We propose to implement a recognition approach for tool monitoring using continuous *HMM*. The vibration signals between the tool and the workpiece will provide the necessary data to perform this kind of tool monitoring.

## 2 State-of-the-art.

The tool failure represents $\sim 20$ % of machine tool down-time. Tool wear impacts negatively the quality in the context of dimensions, finish, and surface integrity, [8]. [3]-[4] developed an intelligent supervisory system for tool wear prediction using an Artificial Neural Network (*ANN*) output error model; also, they exploited the analysis of signals signatures in the time and frequency domains. [7] worked with multilayered *ANN* in the milling process. [5]-[1] presented an approach for feature extraction from vibrations during the drilling. They used self-organizing feature maps for extracting the features, and a discrete *HMM* classifier.

---

[1]Corresponding author, avallejo@itesm.mx

## 3   Tool-wear monitoring system

Fig. 1 shows the proposal algorithm for monitoring the tool-wear using continuous *HMM*. The

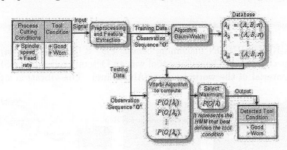

Figure 1: Flow diagram to monitor the tool condition with continuous HMM.

input signal is preprocessed and then it is separated in two branches. The training data branch leads to a *HMM*. In this training phase the system learns the patterns. The testing branch uses the preprocessed input signal and the *HMM* to compute the $P(O \mid \lambda)$ using the Viterbi algorithm for each model. The model with higher probability is the end result.

**Hidden Markov models.** *HMM*s are characterized as: $N$, number of states. We denote the states as $S = S_1, \cdots, S_N$, and the state at time $t$ as $q_t$. $M$, number of distinct observation symbols per state. State transition probability distribution $A = a_{ij}, a_{ij} = P[q_t = S_j | q_{t-1} = S_i], 1 \leq i,$ $j \leq N$. Observation symbol distribution in state $j$, $B = b_j(k)$, $b_j(k) = P[v_k | q_t = S_j], 1 \leq j \leq N, 1 \leq k \leq M$. Initial state distribution $\pi = \pi_i$, $\pi_i = P[q_1 = S_i], 1 \leq i \leq N$. $N, M$ and $\lambda$ are learned from data. Given this model and the observation we can compute $P(O | \lambda)$.

**Feature extraction step.** The vibration signals are pre-processed calculating their Mel Frequency Cesptral Coefficient (MFCC) representation [6]. First, the vibration signals are divided into short frames, and the amplitude spectrum is obtained using the Discrete Fourier Transform (DFT), and converted to the log-scale. For smoothing the scaled spectrum filter banks are used. Finally, the discrete cosine transform is applied to eliminate the correlation. We obtain a 39 dimension vector formed by 12-dimension MFCC, one energy coefficient, $13\Delta$ and $13 \Delta^2$ coefficients.

**Baum-Welch algorithm.** The well-known Baum-Welch algorithm is an iterative process for parameter estimation based on a training data set for a given model $\lambda$. The goal is to obtain a new model $\bar{\lambda}$ where the function

$$Q(\lambda, \bar{\lambda}) = \sum_Q \frac{P(O, Q \mid \lambda)}{P(O \mid \lambda)} log[P(O, Q \mid \bar{\lambda})] \tag{1}$$

is maximized. New set of parameters for the model are calculated by Baum-Welch as follow:

$$\bar{\mu}_{jk} = \frac{\sum_{t=1}^{T} \xi(j, k) o_t}{\sum_{t=1}^{T} \xi_t(t, k)} \quad \bar{U}_{jk} = \frac{\sum_{t=1}^{T} \xi_t(j, k)(o_t - \bar{\mu}_{jk})(o_t - \bar{\mu}_{jk})^t}{\sum_{t=1}^{T} \xi_t(j, k)} \quad \bar{c}_{jk} = \frac{\sum_{t=1}^{T} \xi_t(j, k)}{\sum_{t=1}^{T} \sum_k \xi_t(j, k)} \tag{2}$$

where $\xi_t(j, k)$ is the probability for changing from state $j$ to state $k$, $b_j(o)$ is a continuous output probability density function, $c_{jk}$ is the weight of the gaussian $k$ and $N(o, \mu_{jk}, U_{jk})$ is a single gaussian of mean value $\mu_{jk}$ and a covariance matrix $U_{jk}$. The term $b_{jk}$ can be written as

$$b_{jk}(o_t, \mu_{jk}, \sigma_{jk}) = \frac{1}{\prod_{i=1}^{d} \sqrt{2\pi\sigma_{jki}}} e^{-\frac{1}{2} \sum_{i=1}^{d} (\frac{o_{ti} - \mu_{jki}}{\sigma_{jki}})^2} \tag{3}$$

**Viterbi algorithm.** This algorithm [2] is used to find the single best state sequence, $Q = q_1, \cdots, q_T$, for the given observation sequence $O = O_1, \cdots, O_T$. Then, we need to compute: $\delta(i) = max_{q_1, \cdots, q_{t-1}} P[q_1, \cdots, q_t = i, O_1, \cdots, O_t \mid \lambda]$.

## 4   Experimental set up.

We worked with a KX10 HURON machining center, Fig. 2 (left), with a capacity of 20 KW, 3 axis, and a Siemens controller. The cutting tool was an Octomill face mill of SECO Carboloy, Fig. 2 (middle). The material of the workpiece was Aluminium. The vibration signals were recorded

Figure 2: Huron milling center, cutting tool, and data acquisition system

with a data acquisition system using an accelerometer installed on the flat metal support, Fig. 2 (right). We applied a full factorial design of 4 independent variables: spindle speed (1, 1.5 and 2 thousands rev/min), feed rate (600, 800, and 1,000 mm/min), depth of the tool (1 mm), and tool condition (good and worn inserts).

## 5   Results.

The database contains 153 experiments of the 18 different operating conditions. First, we selected a set of experiments $(T_r)$ for training, and other set of experiments were used for testing $(T_s)$. The utterances were processed using the Sphinx *HMM*s Toolkit developed at Carnegie Mellon University. The toolkit was configured to use several Gaussian, left-right model, and five states for the*HMM*. Fig. 3 (left) shows the performance obtained for the classifier. We observed an excellent behavior when we used enough experiments to train the *HMM*. We configured the *HMM* toolkit for recognition of two states: good and faulty(worn inserts) condition. Table 1 shows the results.

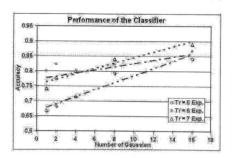 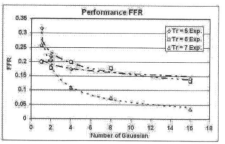

Figure 3: Performance of the Classifier with different experiments.

Table 1: Accuracy of the model

| Training Exp. | Testing Exp. | 1 Gaussian | 16 Gaussians |
|---------------|--------------|------------|--------------|
| 90 | 63 | 66.7 % | 84.0 % |
| 108 | 45 | 80.0 % | 87.0 % |
| 126 | 27 | 74.0 % | 96.0 % |

We also defined two alarms: False Alarm Rate (*FAR*) and False Fault Rate (*FFR*). FAR is the rate when decoder detects that the tool is in fault condition, but the tool is really in good condition. *FFR* is the rate when decoder detects the tool is in good condition and it is not. *FAR* condition is not a problem for the machining process. However, *FFR* could be dangerous because the poor quality of the product and the tool can be broken. We obtained a low *FFR* rate when the *HMM* are trained with several Gaussian, as we show in Fig. 3 (right).

## 6   Conclusions.

We have proposed an algorithm for monitoring the cutting tool condition in a CNC-machining center by using continuous *HMM*. A database was built with the vibration signals obtained during the machining process of an Aluminium 6061. We trained/tested the *HMM* for different operating conditions and the results were satisfactory given the limited number of experiments. Basic speech recognition framework was exploited in with successful results and great potential.

## References

[1] Les Atlas, Mari Ostendorf, and Gary D. Bernard. Hidden markov models for monitoring machining tool-wear. *IEEE*, pages 3887–3890, 2000.

[2] Jeff Bilmes. What hmms can do. Technical Report UWEETR-2002-0003, Department of Electrical Engenineering from University of Washington.

[3] Rodolfo E. Haber and A. Alique. Intelligent process supervision for predicting tool wear in machining processes. *Mechatronics*, (13):825–849, 2003.

[4] Rodolfo E. Haber, Josee E. Jiménez, C. Ronei Peres, and José R. Alique. An investigation of tool-wear monitoring in a high-spped machining process. *Sensors and Actuators A*, (116):539–545, 2004.

[5] Lane M. D. Owsley, Les E. Atlas, and Gary D. Bernard. Self-organizing feature maps and hidden markov models for machine-tool monitoring. *IEEE Transactions on Signal Processing*, 45(11):2787–2798, 1997.

[6] L.R. Rabiner and B.-H. Juang. *Fundamentals of speech recognition*. Prentice-Hall, New-Jersey, 1993.

[7] H. Saglam and A. Unuvar. Tool condition monitoring in milling based on cutting forces by a neural network. *International Journal of Production Research*, 41(7):1519–1532, 2003.

[8] Rogelio L. Hecker Steven Y. Liang and Robert G. Landers. Machining process monitoring and control: The state-of-the-art. *ASME International Mechanical Engenieering Congress and Exposition*, pages 1–12, 2002.

Brill Academic Publishers
P.O. Box 9000, 2300 PA Leiden,
The Netherlands

*Lecture Series on Computer
and Computational Sciences*
Volume 4, 2005, pp. 586-589

# Electron Energy Level Calculation for a Three Dimensional Quantum Dot

H. Voss[1]

Institute of Mathematics
Hamburg University of Technology
D-21071 Hamburg, Germany

Received 5 July, 2005; accepted in revised form 26 July, 2005

*Abstract:* In this paper we consider the rational eigenvalue problem governing the relevant energy levels and wave functions of a three dimensional quantum dot. We present iterative projection methods of Arnoldi and of Jacobi–Davidson type for computing a few eigenpairs of this system. Solving the projected nonlinear eigenvalue problems we take advantage of a minmax characterization of the eigenvalues.

*Keywords:* Quantum dot, effective mass, Rayleigh functional, Arnoldi method, Jacobi–Davidson method

*Mathematics Subject Classification:* 65F15

## 1   The governing equation

We consider the problem to compute relevant energy states and corresponding wave functions of a three dimensional semiconductor quantum dot. Let $\Omega_q \subset \mathbb{R}^3$ be a domain occupied by the quantum dot, which is embedded in a bounded matrix $\Omega_m$ of different material. A typical example is an InAs pyramidal quantum dot embedded in a cuboid GaAs matrix.

The governing equation is the Schrödinger equation

$$-\nabla \cdot \left( \frac{\hbar^2}{2m(x,\lambda)} \nabla u \right) + V(x)u = \lambda u, \ x \in \Omega_q \cup \Omega_m, \tag{1}$$

where $\hbar$ is the reduced Planck constant, $m$ is the effective electron mass depending on both, energy level $\lambda$ and position $x$, $V$ is the confinement potential depending on $x$, and $u$ is the wave function. Since the wave function decays outside the quantum dot very rapidly it is reasonable to assume homogeneous Dirichlet conditions $u = 0$ on the outer boundary of $\Omega_m$, and on the interface between the quantum dot and the matrix the Ben Daniel–Duke condition holds

$$\frac{1}{m_q} \frac{\partial u}{\partial n_q} \bigg|_{\partial \Omega_q} = \frac{1}{m_m} \frac{\partial u}{\partial n_m} \bigg|_{\partial \Omega_m}, \ x \in \partial \Omega_q \cap \partial \Omega_m. \tag{2}$$

Here $n_q$ and $n_m$ denote the outward unit normal on the boundary of $\Omega_q$ and $\Omega_m$, respectively.

Assuming non-parabolicity for the electron's dispersion relation the effective mass is given as [3]

$$\frac{1}{m_j(\lambda)} = \frac{P_j^2}{\hbar^2} \left( \frac{2}{\lambda + g_j - V_j} + \frac{1}{\lambda + g_j - V_j + \delta_j} \right) \tag{3}$$

---

[1]Corresponding author. Corresponding Member of the European Academy of Sciences. E-mail: voss@tuhh.de

where the confinement potential $V_j := V|_{\Omega_j}$ is piecewise constant, and $P_j$, $g_j$ and $\delta_j$ are the momentum matrix element, the conduction and the spin-orbit split-off band gap for the quantum dot $(j = q)$ and the matrix $(j = m)$, respectively.

Multiplying (1) by $v \in H_0^1(\Omega)$, $\Omega := \bar{\Omega}_q \cup \Omega_m$, and integrating by parts one gets the variational form of the Schrödinger equation

$$a(u, v; \lambda) := \frac{\hbar^2}{2m_q(\lambda)} \int_{\Omega_q} \nabla u \cdot \nabla v \, dx + \frac{\hbar^2}{2m_m(\lambda)} \int_{\Omega_m} \nabla u \cdot \nabla v \, dx + V_q \int_{\Omega_q} uv \, dx + V_m \int_{\Omega_m} uv \, dx$$

$$= \lambda \int_{\Omega} uv \, dx =: \lambda b(u, v) \quad \text{for every } v \in H_0^1(\Omega). \tag{4}$$

It is easily seen that for fixed $u \in H_0^1(\Omega)$, $u \neq 0$ the real equation

$$f(\lambda; u) := \lambda b(u, u) - a(u, u; \lambda) = 0 \tag{5}$$

has a unique solution $\lambda =: p(u)$. It therefore defines a functional $p : H_0^1(\Omega) \setminus \{0\} \to \mathbb{R}^+$ which is called the Rayleigh functional of the rational eigenvalue problem (4). With this functional the following variational characterization of eigenvalues of problem (4) which generalizes the well known minmax characterization of Poincaré for linear eigenproblems (cf. [7]).

**THEOREM 1**

  (i) *The Schrödinger equation (4) with effective mass given in (3), which models the energy states and wave functions of a quantum dot, has a countable set of eigenvalues $0 < \lambda_1 \leq \lambda_2 \leq \ldots$. Each of the eigenvalues has finite multiplicity, and the only cluster point is $\infty$.*

  (ii) *The $j$-th smallest eigenvalue $\lambda_j$ can be characterized as*

$$\lambda_j = \min_{\dim V = j} \max_{u \in V, u \neq 0} p(u). \tag{6}$$

Discretizing problem (4) by finite elements results in a rational matrix eigenvalue problem

$$T(\lambda)x = \lambda Mx - \frac{1}{m_1(\lambda)} A_1 x - \frac{1}{m_2(\lambda)} A_2 x - Bx = 0 \tag{7}$$

where $T(\lambda) \in \mathbb{R}^{n \times n}$ is symmetric, and also satisfies the conditions the of minmax characterization [7] for $\lambda > 0$. Hence, it has $n$ eigenvalues $0 < \tilde{\lambda}_1 \leq \ldots \tilde{\lambda}_n$, and it follows from the characterization (6) that $\lambda_j \leq \tilde{\lambda}_j$ for $j = 1, \ldots, n$.

## 2   Iterative projection methods

Typically $T(\lambda)$ is large and sparse, and its eigenvalues can be determined by iterative projection methods [4]. Here the underlying eigenproblem is projected to a sequence of subspaces of $\mathbb{R}^n$ which are expanded in the course of the algorithm by vectors with high approximation potential for the eigenvector wanted next.

Typical examples are the nonlinear Jacobi–Davidson method [1], and the nonlinear Arnoldi method [5], a template of which is listed below:

```
1: Start with an initial shift σ and initial orthonormal basis V
2: determine preconditioner M ≈ T(σ)⁻¹
3: while m ≤ number of wanted eigenvalues do
```

```
 4:    compute the m-th smallest eigenvalue μ of projected problem V^T T(λ)Vy = 0
 5:    determine Ritz vector u = Vy, ‖u‖ = 1, and residual r = T(μ)u
 6:    if ‖r‖ < ε then
 7:        accept eigenpair λ_m = μ, x_m = u
 8:        choose new shift σ and update preconditioner M if indicated
 9:        restart if necessary
10:        m=m+1
11:    end if
12:    v = Mr; v = v − VV^T v; v = v/‖v‖; V = [V, v];
13:    reorthogonalize if necessary
14: end while
```

There are many details that have to be considered when implementing the nonlinear Arnoldi method concerning the choice of the initial basis, when and how to update the preconditioner, and how to restart the method. A detailed discussion is given in [5].

A crucial point in iterative methods for general nonlinear eigenvalue problems when approximating more than one eigenvalue is to inhibit the method to converge to the same eigenvalue repeatedly. For linear eigenvalue problems this is easy to do by using Schur forms or generalized Schur forms for the projected problem and then locking or purging certain eigenvalues. For nonlinear problems, however, such Schur forms do not exist and this presents one of the most difficult tasks in achieving good convergence.

For symmetric nonlinear eigenproblems satisfying a minmax characterization (6) however, its eigenvalues can be computed safely one after the other. The proof of Theorem 1 shows that the minimum in (6) is attained by the invariant subspace of $T(\lambda_j)$ corresponding to the $j$ largest eigenvalues, and the maximum by every eigenvector corresponding to the eigenvalue 0. This suggests the safeguarded iteration for computing the $j$-th smallest eigenvalue:

```
 1: Start with an approximation μ_1 to the j-th eigenvalue of T(λ)x = 0
 2: for k = 1, 2, ... until convergence do
 3:    determine eigenvector u corresponding to the j-largest eigenvalue of T(μ_k)
 4:    evaluate μ_{k+1} = p(u)
 5:    end for
```

The safeguarded iteration has the following convergence properties [6]: It converges globally to the smallest eigenvalue $\lambda_1$. The (local) convergence to simple eigenvalues is quadratic. If $T'(\lambda)$ is positive definite, and $u$ in Step 3 of the last algorithm is replaced by an eigenvector of $T(\sigma_k)x = \mu T'(\sigma_k)x$ corresponding to the $j$-th largest eigenvalue, then the convergence is even cubic. Moreover, a variant exists which is globally convergent also for higher eigenvalues.

## 3   Numerical results

We consider a pyramidal quantum dot with width 12.4 nm and height 6.2 nm embedded in a cubic matrix of size 24.8 nm×24.8 nm×18.6 nm with the following parameters $P_q = 0.8503$, $g_q = 0.42$, $\delta_q = 0.48$, $V_q = 0$, $P_m = 0.8878$, $g_m = 1.52$, $\delta_m = 0.34$, and $V_m = 0.7$. This model was already treated by Hwang, Lin, Wang, and Wang in [2].

The authors of [2] presented a discretization of problem (4) by the finite volume method based on a uniform grid. The corresponding finite dimensional problem (7) was multiplied by the denominators, and the resulting polynomial eigenvalue problem of degree 5 was solved by the Jacobi–Davidson method.

We solved the rational eigenproblem (7) directly by the nonlinear Arnoldi method [5] (not taking advantage of the fact that the stencils are identical for all discretization points in the matrix and in the quantum dot, respectively) under MATLAB 7.0.4 on an AMD Opteron processor with 4

GByte RAM and 2.2 GHz. The following Table contains the approximations to the smallest 5 eigenvalues and the CPU times.

| dim | $\lambda_1$ | $\lambda_{2/3}$ | $\lambda_4$ | $\lambda_5$ | CPU time |
|---|---|---|---|---|---|
| 2'475 | 0.41195 | 0.58350 | 0.67945 | 0.70478 | 0.68 s |
| 22'103 | 0.40166 | 0.57668 | 0.68418 | 0.69922 | 8 s |
| 186'543 | 0.39878 | 0.57477 | 0.68516 | 0.69767 | 151 s |
| 1'532'255 | 0.39804 | 0.57427 | 0.68539 | 0.69727 | 4018 s |
| 12'419'775 | 0.39785 | 0.57415 | | | overnight |

The uniform grid does not seem to be appropriate for discretizing (4) since the wave functions corresponding to small energy levels are mainly concentrated on the quantum dot and decay rapidly outside, whereas the volume occupied by the quantum dot is only less than 3 % of $\Omega$. We discretized (4) by cubic Lagrangian elements on a tetrahedal grid with 96'640 degrees of freedom such that 43'615 DoFs where located in the quantum dot, 43'897 DoFs in the matrix, and 9'128 DoFs on the interface. We solved the rational eigenproblem by the nonlinear Arnoldi method and the Jacobi–Davidson method. The following table contains the approximations to the smallest 5 eigenvalues, the number of iterations to obtain the approximations, and the CPU times. Notice, that in this case by Theorem 1 one gets upper bounds of the corresponding eigenvalues of problem (4). Hence, the approximations to $\lambda_j$, $j = 1, 2, 3$ are definitely better than the ones obtained by the finite volume method with more than 12 million DoFs.

| dim | $\lambda_1$ | $\lambda_2$ | $\lambda_3$ | $\lambda_4$ | $\lambda_5$ | CPU time |
|---|---|---|---|---|---|---|
| 96'640 | 0.39779 | 0.57411 | 0.57411 | 0.68547 | 0.69714 | |
| Arnoldi | 44 it. | 29 it. | 29 it. | 24 it. | 21 it. | 189 s |
| JD | 9 it. | 7 it. | 9 it. | 5 it. | 6 it. | 205 s |

# References

[1] T. Betcke and H. Voss. A Jacobi–Davidson–type projection method for nonlinear eigenvalue problems. *Future Generation Computer Systems* **20** 363 – 372 (2004).

[2] T.-M. Hwang, W.-W. Lin, W.-C. Wang, and W. Wang. Numerical simulation of three dimensional quantum dot. *J. Comput.Phys.* **196** 208 – 232 (2004).

[3] Y. Li, J.-L. Liu, O. Voskoboynikov, C.P. Lee, and S.M. Sze. Electron energy level calculations for cylindrical narrow gap semiconductor quantum dot. *Comput.Phys.Comm.* **140** 399 – 404 (2001).

[4] V. Mehrmann and H. Voss. Nonlinear eigenvalue problems: A challenge for modern eigenvalue methods. *GAMM Mitteilungen* **27** 121 – 152 (2004).

[5] H. Voss. An Arnoldi method for nonlinear eigenvalue problems. *BIT Numerical Mathematics* **44** 387 – 401 (2004).

[6] H. Voss. Numerical methods for sparse nonlinear eigenproblems. (Editor: Ivo Marek), *Proceedings of the XV-th Summer School on Software and Algorithms of Numerical Mathematics, Hejnice, 2003*, pages 133 – 160, University of West Bohemia, Pilsen, Czech Republic, 2004. Available at http://www.tu-harburg.de/mat/Schriften/rep/rep70.pdf.

[7] H. Voss and B. Werner. A minimax principle for nonlinear eigenvalue problems with applications to nonoverdamped systems. *Math.Meth.Appl.Sci.* **4** 415–424 (1982).

Brill Academic Publishers
P.O. Box 9000, 2300 PA Leiden
The Netherlands

*Lecture Series on Computer*
*and Computational Sciences*
Volume 4, 2005, pp. 590-594

# A Generic Numerical Inverse Kinematic Algorithm for the Articulated Robot

Quanyu WANG[1], Zhe WANG, Jiaping HAO
Department of Computer Science and Engineering, Beijing Institute of Technology,
Beijing 100081, P.R. China

Received 9 July, 2005; accepted in revised form 17 August, 2005

*Abstract:* A fuzzy iterative method based on multi-agents with movable track equation and abstract link is introduced to solve the inverse kinematic problem, where the movable track equation is to describe the movable track of one of the link and a homogeneous vector as the abstract link stretches from one coordinate origin to another. Theoretically, this method can solve inverse kinematic problem for any serial style robots with any number of revolute links and/or the prismatic links. Practically, it is only realized to deal with those robots with any number of revolute links.

## 1 Introduction

One of the important problems in robotics is to solve the inverse kinematic model, which is full of complexities and has been the subject of numerous research works. Two main approaches, the analytical method and the numerical method are well addressed in the inverse kinematics literatures. The analytical approach is proposed to derive computationally efficient closed-form solutions, i.e., each joint variable is explicitly expressed in terms of other known quantities. However, the analytical approach lacks generality because it may be either impossible to obtain or may be only suitable to one type of the robot. Comparing with analytical approach, the numerical approach may not be so accurate, but it is quite practical in solving robot inverse kinematics, especially effective for robots with more or less than 6 degree of freedom (DOF).

In this paper a fuzzy iterative method based on multi-agents[1] with movable track equation and abstract link is introduced to solve the inverse kinematic problem. Theoretically, this method can solve inverse kinematic problem for any rigid style robots with any number of revolute links and/or the prismatic links. Practically, this method is only realized to deal with those robots with any number of revolute links [2].

## 2 Basic concepts for new method

Here below we will introduce a generic numerical inverse kinematic algorithm. Before describing the method itself, it is necessary to clarify two concepts: *Movable track equation* and *Abstract link*[3].

In fact, a link of robot, let it be $Link_i$, not only moves straightly along a line relative to $Link_{i-1}$, but also moves in other routes. The tracks can be a curve, a surface or even a single point. Let

$$O_{i+1} = f_i(\lambda_i) \quad \lambda_i \in [0,1] \tag{1}$$

be the movable track equation of $Link_i$, where $O_{i+1}$ is the position of the coordinate origin of $Link_{i-1}$ in $Link_i$'s coordinate. It is a homogeneous vector; $\lambda_i$, which ranges [0,1], is a parameter that controls $O_{i+1}$. When $\lambda_i = 0$, $O_{i+1}$ is the same with $O_n$, and when $\lambda_i = 1$, $O_{n+1}$ is at another end of the track $T_i$, as is shown in Figure 1. So just by changing the simple parameter $\lambda_i$, coordinate $O_i$ can be translated to $O_{i+1}$.

---

[1] E-mail: wangquanyu@bit.edu.cn

Abstract link is a virtual robot link, which stretches from one coordinate origin to another. Ignoring the details of complex relations, link between $O_i$ and $O_j$ is abstracted as a straight line. It is just a homogeneous vector:

$$Aa_{i-j} = O_j - O_i \qquad (2)$$

Figure 1 shows the abstract link from Link$_i$ to the end-effector.

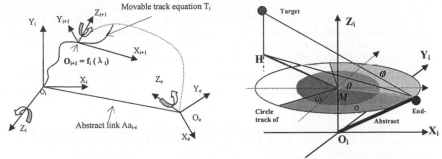

Figure 1 Movable track equation and Abstract link    Figure 2 A moving strategy for Agent$_i$

The idea of abstract link simplifies the situation and makes the calculation easy. In Link$_i$'s view Link$_j$ is a fixed part on its own link, so it's very simple for Link$_i$ to control Link$_j$'s position. Because on the one hand, it is not always necessary to calculate the position of Links between Link$_i$ and Link$_j$ when Link$_i$ is moving; on the other hand the Abstract link is just a homogeneous vector, it can be calculated from the translation matrix and revolution matrix. And when Link$_i$ uses the abstract link $Aa_{i-j}$, the movement of Links between Link$_i$ and Link$_j$ is forbidden, because if they move, the simple homogeneous vector will be no longer exist.

## 3   The Robot structure of multi-agent

Each link of the robot can treated as an agent with autonomous ability, thus a robot with multi links is an object with multi-agents. Let Agent$_i$ be the agent of Link$_i$ , which can perform two kinds of movements autonomously: translation and revolution. Translation movement is based on the movable track equation, and revolution movement is based on the abstract link which stretches from Link$_i$ to the end-effector, so the end-effector is a "fixed" part of Link$_i$, in other words, Link$_i$ has the ability to move the end-effector, and Link$_i$ can be thus treated as an end-effector. Each kind of the movements only needs one parameter, $\lambda$ for translation and $\zeta$ for revolution.

As the Agent$_i$ moves, the position and orientation of the end-effector can be described as:

$$P_e = M(Aa_{i-e}, T_i, R_i) = BaseMatrix \cdot TransMatrix(T_i) \cdot RotateMatrix(R_i) \cdot P_e' \qquad (3)$$

Where $T_i$ and $R_i$ are parameters for Agent$_i$ to autonomously control the movement of translation and rotation. $P_e$ is the position and orientation vector of the end-effector in the base frame. $P_e'$ is the previous position and orientation vector of the end-effector.

*BaseMatrix, TransMatrix(T$_i$), RotateMatrix(R$_i$)* are the base homogeneous transform matrix, local translation matrix and local revolution matrix respectively.

The existence of the abstract link Aa$_{i-e}$ simplifies the complex transform matrix from Agent$_i$ to Agent$_e$ to a transform matrix of abstract link Aa$_{i-e}$ in Agent$_i$'s coordinates. Because there is no difficulty in getting a translation matrix and a revolution matrix for a vector, after the Agent$_i$'s movement the position and orientation of the end-effector can be easily acquired. Thus,

$$P_e = \bigcup_{i=0}^{e} M(Aa_{i-e}, T_i, R_i) \qquad (4)$$

can be used to describe the consequent autonomous movement of all the link agents by the order that from base to end-effector. Let the target point be **TP**, the distance between the robot's end-effector to **TP** is $D = |TP - P_e|$.

Each agent has a desire to get a minimal **D**, but what it can do is only to perform the movement of translation or revolution of their own links. And there is only one parameter for each movement it can control. So each agent will try its best to change the parameter to get a smaller **D**. Here a better parameter means by which the parameter **D** be acquired smaller. Before the agent moves, it can try to get a better parameter than last movement several times. When the trial times are used out and it can't get a better parameter than last movement, it will give up this movement. As a consequent autonomous movement of all the link agents one by one, **D** will become smaller, at least no more than last time, and this means the robot has accomplish one time of self-adjust movement. Therefore we have:

$$\varsigma_i = D_{i-1} - D_i \tag{5}$$

Where $\varsigma_i$ is the reduction of **D** after Link$_i$'s movement. So the desire of Agent$_i$ is also equal to get the max $\varsigma_i$ at each movement.

If $D = 0$, then it means the end-effector reaches the target point. According to the translation and revolution parameters stored in each agent, the final gesture of the robot can be reached and the inverse kinematic problem is solved.

Now if it can be proved that the distance $D_m$ which is the distance after m times of robot's self-adjust movement less than the distance $D_{m-1}$, i.e., $\lim_{m \to \infty} D_m = 0$, then a conclusion that $D_m$ is convergent can be drawn. And

$$D = \bigcup_{m=1}^{K} |TP - \bigcup_{i=0}^{e} M(Aa_{i-e}, T_i, R_i)| \ (K \to \infty), \tag{6}$$

which represents the distance between end-effector to target point after K times of self-adjust movement, is also convergent.

In practice, because of the accumulative error, after several times of self-adjust movement, it is very hard for D to approach 0. But it is easy for D to reach δ--a given positive tiny constant as the motion precision. Usually this δ is precise enough for practical demand, for instance, in our experiment, δ is set to 0.1, which means 0.1mm and precise enough for industrial robot, PUMA560 only need 8 times of self-adjust movement to reach this precision. Therefore a suitable δ will make this algorithm very effective.

Let $\xi_m = D_{m-1} - D_m$, then $\sum_{m=1}^{K} \xi_m = D_0$, D$_0$ is a constant refers to the distance between end-effector and the target point at the beginning. If ξ$_m$ is given as large as possible every time, then K will become smaller and the result will come out sooner. And for a robot, the number of links' is definite, let it be **e**, so:

$$\sum_{i=1}^{e} \varsigma_i = \xi_m \tag{7}$$

From this formula we know that if each movement of the Agent$_i$ can get the max $\varsigma_i$, then the convergent speed of **D** will be dramatically accelerated. In the view of each agent, it must find out a method to get the max $\varsigma_i$ at each movement. Therefore, the moving method of agent is very important for efficiency in our algorithm. The following part is endeavored to find out an effective method of movement for articulated robot link agents.

## 4 An effective moving strategy for articulated robot link agents

For robots with only rotated links, each agent only needs to perceive the abstract link and the target, as shown in Figure 2.

Take Agent$_i$ as an example: first, the abstract link stretching from the end-effector to this agent's coordinate origin $O_i$ and the target point is transformed into this agent's local frame. Because the abstract link could only rotate around $Z_i$ axis, the moving track of the end-effector must be on a circle as shown in

Figure 2. And the plane of this circle is parallel to Agent$_i$'s XY-plane. Target point is projected to the circle-plane as point *H*. Let the point of target is *T*, the point of end-effector is *E*, and circle-plane cross $Z_i$ axis at point *M*.

As shown in Figure2, there are four kinds of rotation, $\varphi$, $\sigma$, $\theta$ and $\omega$, the direction can be determined through the right-hand rule. $\varphi$ and $\sigma$ respectively mean the positive and negative rotation range of the abstract link. $\theta$ and $\omega$ respectively mean the positive and negative rotation range from *ME* to *MH*. Let $\alpha$ be the included angle of *ME* and *MH*. Therefore the distance between target and end-effector is :

$$TE = \sqrt{|TH|^2 + |EH|^2} \tag{8}$$

and there is

$$EH = \sqrt{|MH|^2 + |ME|^2 - 2 \cdot |MH| \cdot |ME| \cdot \cos\alpha} \tag{9}$$

so

$$TE = \sqrt{|TH|^2 + |MH|^2 + |ME|^2 - 2 \cdot |MH| \cdot |ME| \cdot \cos\alpha} \tag{10}$$

For Agent$_i$ the length of *TH*, *MH*, *ME* are fixed values, and $\alpha$ have a range of 0 to 180, so the function of *TE* is a monotonic increasing function of $\alpha$. Thus the Agent$_i$ will have a minimum TE as long as $\alpha$ keeps minimum. That means Agent$_i$ need to rotate *ME* to *MH* as near as possible.

Now let $\Delta$ be the Agent$_i$'s best moving degree. As Agent has already known $\varphi$, $\sigma$, $\theta$ and $\omega$, it can easily get $\Delta$. When $\theta<\varphi$, that means MH is in ME's rotation range, so ME could superpose MH, and $\alpha$ could be zero, so $\Delta=\theta$. In the same way, if $\omega<\sigma$ then $\Delta=-\omega$. Otherwise, when $\theta>\varphi$ and $\omega>\sigma$, that means *MH* is out of *ME*'s rotation rage, so the best $\Delta$ for Agent$_i$ is $\varphi$ or $\sigma$, if $(\theta-\varphi) < (\omega-\sigma)$, that means when *ME* rotate to the end of positive range, Agent$_i$ will get the minimum $\alpha$ of this movement, so $\Delta=\varphi$. In the same way, if $(\theta-\varphi)>= (\omega-\sigma)$ then $\Delta=-\sigma$.

Therefore, when Agent$_i$ is to move, it is quite sure for it to get a best parameter, then the end-effector is sure to get close to the target.

After all the agents of the robot finished the same process described above, then a round of self-adjust movement of the robot is accomplished. After limited times of self-adjust movement, *D* will be converged to a given precision.

This inverse kinematical method is also available for space obstacle avoidance[4-6], provided each agent chooses a different abstract link and pre-approaching the obstacle, unavailable moving range is acquired, then subtract them from available moving range, after this preprocess, the method can be used in a normal way.

## 5  Results of Simulation

To validate this algorithm, we use a simulated model of PUMA560, which is a popular robot in the filed of industry. The following table gives the parameters of PUMA560's links in D-H relative coordinates:

Table 1  Parameters For Robot Puma560

| Link i | $\alpha_i$(° ) | $a_i$(mm) | $d_i$(mm) | Range of $\theta_i$(° ) | Final $\theta_i$(° ) |
|--------|--------|--------|--------|--------|--------|
| 1 | -90 | 0 | 0 | -160~+160 | -125.4 |
| 2 | 0 | 431.8 | 149.09 | -225~+45 | 4.20915 |
| 3 | 90 | -20.32 | 0 | -45~+255 | 121.699 |
| 4 | -90 | 0 | 433.07 | -110~+170 | -79.6164 |
| 5 | 90 | 0 | 0 | -100~+100 | -110 |
| 6 | 0 | 0 | 56.25 | -266~+266 | -266 |

Set each link at its negative rotation end and by the forward kinematics the end-effector can be positioned at the point (692.052, 943.604, -114.852), and the target is preset at the point (300, 400, -200), before the robot moves, the distance between the end-effector and the target is 675.618mm. By using this algorithm, it takes PUMA560 8 times of self-adjust movement and milliseconds of time to reach the target by a precision δ=0.1mm. Figure 3 shows the change of the distance between end-effector and target. And Figure 4 shows the Link's changes of rotation in degree. The final results are given in Table 1.

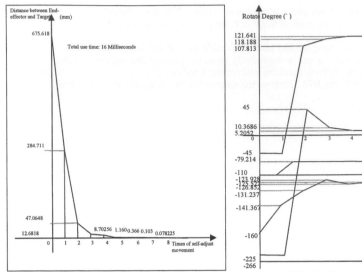

Figure 3 The change of the distance between
the end-effector and the target

Figure 4 The Link's changes of rotation in
degree

## 6  Conclusion

This paper provides a high performance general inverse kinematic algorithm for robots. Using movable
track equation and abstract link to simplify the calculation of matrix, and using agents to let links have
intelligence, so let the task to be completed autonomously. A lot of previous work has been done on
modelling and inverse kinematic problems. Further study will be focused on collision detection, speed and
acceleration level of kinematics and dynamics of robots.

## References

[1]  Hicham DJENIDI and Chakib TADJ, "Agents Paradigm to Solve a Complex Optimization Problem",
     IEEE, 2002

[2]  G. S. Chirikjian and J. W. Burdick, "A modal approach to hyper-redundant manipulator kinematics,"
     IEEE Trans. Robot. Automat., vol. 10, pp.343–354, June 1994.

[3]  Farbod Fahimi, Hashem Ashrafiuon, and C. Nataraj "An Improved Inverse Kinematic and Velocity
     Solution for Spatial Hyper-Redundant Robots" IEEE TRANSACTIONS ON ROBOTICS AND
     AUTOMATION, VOL. 18, NO. 1, FEB 2002

[4]  C. A. Klein, "Use of redundancy in the design of robotic systems," in Robotics Research: The Second
     International Symposium. Cambridge, MA: MIT Press, 1998, pp. 207–214.

[5]  A. Maciejewski and C. A. Klein, "Obstacle avoidance for kinematically redundant manipulators in
     dynamically varying environments," Int. J.Robot. Res., vol. 4, no. 3, pp. 109–117, 1985.

[6]  J. Baillieul, "Avoiding obstacles and resolving kinematic redundancy," in Proc. IEEE Conf. on
     Robotics and Automation, San Francisco, CA, Apr. 1986, pp. 1698–1704.

Brill Academic Publishers
P.O. Box 9000, 2300 PA Leiden,
The Netherlands

*Lecture Series on Computer*
*and Computational Sciences*
Volume 4, 2005, pp. 595-598

# Vector Property Generation by a Stochastic Growth Process

**Thomas Wüst[1] and Jürg Hulliger**

Department of Chemistry and Biochemistry, University of Berne,
Freiestrasse 3, CH-3012 Berne, Switzerland

Received 8 July, 2005; accepted in revised form 30 July, 2005

*Abstract:* A stochastic growth process is presented, providing a model for physical systems transforming into a macroscopic polar state upon growth, particularly molecular crystals. The process is characterized by the thermal relaxation of successively attached adlayers of polar building blocks at the interface given by a previously grown and frozen bulk. The model is applied to several real physical systems using an analytical description in terms of a mean-field approximation as well as Monte Carlo simulations for a quantitative investigation of growth-induced polarity formation.

*Keywords:* stochastic growth process, polarity formation, crystal growth, Monte Carlo simulation

## 1 Presentation of the Model

Self-assembly of functional building blocks at various length scales is of general interest for building up complex systems. Here we discuss the generation of a vector property by a stochastic growth process (Fig. 1). Building blocks are primarily molecules featuring an electrical dipole moment because of a charge distribution transforming according a polar point group. For the process of self-assembly into a particular crystallographic space group we allow just for one degree of freedom, that is, the inversion of the dipolar molecular axis during the process of molecule attachments at final surface sites. Thereafter, the single degree of freedom is kept frozen. Defined as such we reduce the real process of surface growth to a 2D Ising system, where molecules undergo thermal relaxation taking into account both, lateral interactions with nearest neighbor molecules within the same layer, as well as longitudinal bindings with molecules of the previously grown layer (substrate). Reduced as far as this we can recognize a 2D Ising system featuring anisotropic interactions placed in an external field. The system advancing layer-by-layer develops net vectorial property (polarity, $X_{net}$) because layer wise thermalization leads to a minimum free energy through orientational disorder driven by the criteria from maximum configurational entropy. The asymptotic state of the uppermost layer can be reproduced by thermalization of a two-layer system in thermodynamic equilibrium. The two-layer system shows similarity to a metamagnet.

Growth-induced polarity formation was described analytically in terms of a Markov chain like approach including a mean-field estimate in order to take into account effects of lateral cooperativity. In addition, detailed insight into growth-induced phenomena of polarity was provided by Monte Carlo simulations. Because the variety of intermolecular interactions is generally not known, input parameters were obtained either by force-field calculations or by exploring a reasonably defined space of interactions by random selections.

Conceptually, systems will be presented which were constituted using (i) a host-guest, i.e., channel

---

[1]E-mail: thomas.wuest@iac.unibe.ch

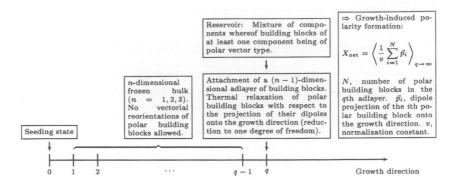

Reservoir: Mixture of components whereof building blocks of at least one component being of polar vector type.

$\Rightarrow$ Growth-induced polarity formation:

$$X_{\text{net}} = \left\langle \frac{1}{v} \sum_{i=1}^{N} \vec{p}_i \right\rangle_{q \to \infty}$$

$n$-dimensional frozen bulk ($n = 1, 2, 3$). No vectorial reorientations of polar building blocks allowed.

Attachment of a $(n-1)$-dimensional adlayer of building blocks. Thermal relaxation of polar building blocks with respect to the projection of their dipoles onto the growth direction (reduction to one degree of freedom).

$N$, number of polar building blocks in the $q$th adlayer. $\vec{p}_i$, dipole projection of the $i$th polar building block onto the growth direction. $v$, normalization constant.

Seeding state

0   1   2   · · ·   $q-1$   $q$   Growth direction

**Figure 1:** Schematic illustration of the stochastic growth process. Starting from a seeding state (with arbitrary symmetry, e.g., polar or centric), growth proceeds in discrete steps along a particular growth direction. In each growth step an adlayer of building blocks is attached at the interface between the frozen bulk consisting of previously attached adlayers and an infinite reservoir. Upon growth *orientational selectivity* of polar building blocks during their attachment at interface sites may result in a non-vanishing vector sum of dipole projections along the growth direction, and thus, to growth-induced polarity formation. $X_{\text{net}}$ denotes the dimensionless measure of polarity, $-1 \leq X_{\text{net}} \leq 1$. $\langle \rangle$ defines the average after arbitrary many growth steps.

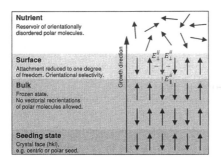

**Nutrient**
Reservoir of orientationally disordered polar molecules.

**Surface**
Attachment reduced to one degree of freedom. Orientational selectivity.

**Bulk**
Frozen state.
No vectorial reorientations of polar molecules allowed.

**Seeding state**
Crystal face (hkl), e.g. centric or polar seed.

**Figure 2:** Representation of the general growth process for the description of single-component crystals of polar molecules. In case the crystal structure induces a parallel or antiparallel alignment of all molecules, the projection of molecular dipoles relative to a particular growth direction can show solely two values: $\pm 1$ (normalized by the absolute value of the molecular dipole along this direction). Orientational selectivity occurs during thermalization of molecules at the surface, taking into account nearest neighbor interactions in longitudinal ($E_{\parallel}^{ij}$, with molecules of the previously attached adlayer) and lateral ($E_{\perp}^{ij}$, with molecules of the same adlayer) direction, $i, j \in \{$up, down$\}$.

(host)-type inclusion (guest) approach, (ii) single-component molecular crystals (Fig. 2), or (iii) solid solution crystals built up either by a polar H and a non-polar G or two polar components. Theoretical investigations have found experimental representations by the crystallization of compounds of type (i)-(iii).

## 2   Mathematical and Computational Description

The present growth model represents a stochastic process, in which each adlayer configuration (i. e., particularly the vectorial alignments of polar building blocks) depends on the configuration of the previously attached layer (substrate) as well as the orientational states of neighboring building blocks within the same adlayer. As such, the process is fully determined by the various coupling parameters and the configuration of the seeding state. However, the memory for the initial state gets lost in the course of the growth process due to ergodicity.

Based on the assumption of the thermal equilibrium formation of an adlayer, transition probabilities are calculated by normalized Boltzmann factors. The growth process has reached an asymptotic state if transition probabilities between subsequently attached adlayers become the same. For systems with a small number of different attachment possibilities for building blocks at surface sites, the asymptotic statistics ($q \rightarrow \infty$) for $X_{net}$ has been obtained analytically using this consistency condition. Thereby, the lateral neighborhood of an adlayer was taken into account within a mean-field approximation.

In case of vanishing lateral interactions between neighboring molecules, the attachment of building blocks may be represented by isolated one-dimensional chains and described by a Markov process. For such systems the asymptotic statistics ($q \rightarrow \infty$) of $X_{net}$ has an exact analytical solution. For further details, see references [1, 2, 4].

Besides, Monte Carlo simulations were carried out, describing the successive attachment and thermal equilibration of new adlayers. For the investigated physical systems, the analytical description has shown satisfactory agreement with Monte Carlo simulations.

# 3 Applications to Real Physical Systems

The general growth process presented above has found an application for the description of growth-induced polarity formation in several real physical systems:

- Mechanically driven packings of drop-shaped pearls resulting in polar arrangements [5].

- Spatially separated chains of polar guest molecules placed in channels of inclusion compounds: Here, no lateral interactions between molecules in adjacent channels are assumed [5].

- Single-component crystals of polar molecules [1, 2, 3, 5]: Statistically, most crystals crystallize in a centric space group. Because of grown-in faulted orientations of molecular dipoles, such crystals may get polar (Figs. 2, 3).

- Organic solid solutions $H_{1-X}G_X$ of polar host (H) and either polar or non-polar guest (G) molecules [4, 7]: Investigation of the interplay between polarity formation and miscibility between the two components (Fig. 4).

- One-dimensional diffusion of polar molecules into parallel channels of an open pore material (zeolite) [6].

- Biological tissue [8, 5]: Experimental work has shown that natural tissue is polar. This may be explained by a mechanism of chemical selectivity during self-assembly of short into long collagen fibrils.

# 4 Further Refinements

The present layer-by-layer growth model assumes flat crystal surfaces as well as slow growth rates, conditions which may arise only at low temperature. At higher temperature a crystal surface can become rough and thus, molecular attachments take place preferentially along step and kink sites. Therefore, we will investigate the influence of such a varying cooperativity between molecules during their attachment on surface sites on growth-induced polarity formation.

# Acknowledgment

We would like to thank C. Gervais and H. Bebie for their collaboration as well as the many fruitful discussions. Funding: NRP 47, phase I.

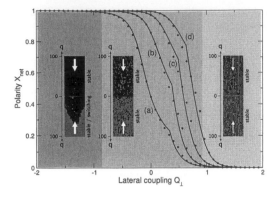

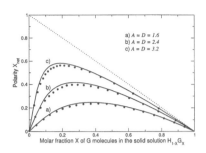

**Figure 3:** General features of polarity formation in single-component crystals. *Curves:* $X_{net}$ vs lateral coupling $Q_\perp = \Delta E_\perp/kT$ ($\Delta E_\perp$, pair energy difference between parallel and antiparallel orientations of neighboring molecules; $k$, Boltzmann constant; $T = 300$ K, temperature) for four different longitudinal couplings $(A, D)$: $(2.0,0.8)$ $(4.0,1.2)$ $(6.0,1.6)$ $(8.0,2.0)$ for (a) - (d). $A = (E_\parallel^{ud} - E_\parallel^{dd})/kT$, $D = (E_\parallel^{du} - E_\parallel^{dd})/kT$, where $d, u$ stand for *down* and *up*, respectively. Results were obtained analytically (*lines*) and by means of Monte Carlo simulations (*dots*). The cusp in the curves expresses a continuous phase transition. *Shaded backgrounds* and corresponding *inset figures* (showing typical cross sections of crystals growing along both directions upon a given seeding state; black/gray squares: down-/upwards oriented molecular dipoles) indicate regions of different phenomenological behaviors of polarity formation. *Left:* Polar ordering with little disorder. A switching from a metastable to a stable state along one growth direction may occur leading to a bi-polar state, otherwise uni-polar. *Middle:* Polar ordering with large disorder, bi-polar. *Right:* Centric ordering with disorder, bi-polar. *White arrows* indicate the orientation (and the relative number) of the majority of molecular dipoles. For more details, see reference [1].

**Figure 4:** The admixture of *non-polar* guest (G) molecules into a crystal of polar host (H) molecules upon growth forming solid solutions $H_{1-X}G_X$ ($0 < X < 1$) provides another means to the evolution of growth-induced polarity which can be described by the present growth model [4]. Even though polar (H) molecules feature no orientational selectivity ($A = D$, for definitions, see Fig. 3) among themselves ($X_{net} = 0$, for $X = 0$), the inclusion of G molecules results in nonzero vectorial alignment of H molecules. Such a *paradoxical case* is of particular interest, since polarity formation, and therefore, breaking of the centric crystal symmetry, is effected *only* by centrosymmetric molecules. Varying molar fractions of H and G molecules in the gas phase (reservoir) results in a change of the chemical potentials of the two components and thus, allows an alteration of the molar fraction $(X)$ of G molecules in the solid solution. *Lines:* Analytical solutions. *Dots:* Monte Carlo simulations.

# References

[1] J. Hulliger, H. Bebie, S. Kluge, and A. Quintel, *Chem. Mater.* **14** 1523 (2002).

[2] H. Bebie, J. Hulliger, S. Eugster, and M. Alaga-Bogdanović, *Phys. Rev. E* **66** 021605 (2002).

[3] C. Gervais, T. Wüst, N. R. Behrnd, M. Wübbenhorst, and J. Hulliger, *Chem. Mater.* **17** 85 (2005).

[4] T. Wüst and J. Hulliger, *J. Chem. Phys.* **122** 084715 (2005).

[5] J. Hulliger, *Chem. Eur. J.* **8** 4579 (2002).

[6] C. Gervais, T. Hertzsch, and J. Hulliger, *J. Phys. Chem. B* **109** 7961 (2005).

[7] C. Gervais, T. Wüst, and J. Hulliger, *J. Phys. Chem. B* **109** 12582 (2005).

[8] J. Hulliger, *Biophys. J.* **84** 3501 (2003).

Brill Academic Publishers
P.O. Box 9000, 2300 PA Leiden
The Netherlands

*Lecture Series on Computer
and Computational Sciences*
Volume 4, 2005, pp. 599-601

# Variable Population Size Made Genetic Algorithm Optimal

Xiaohua Xu[1]   Ling Chen[23]   Ping He[2]

[1]Department of Computer Science and Engineering, Nanjing University
of Aeronautics and Astronautics, Nanjing 210016, China
[2]Department of Computer Science, Yangzhou University, Yangzhou 225009, China
[3] National Key Lab of Novel Software Tech, Nanjing University, Nanjing 210093, China

Received 8 July, 2005; accepted in revised form 29 July, 2005

*Abstract:* We propose a new type of genetic algorithm called VPGA with variable population size and classical GA as its subroutine, for finding excellent solutions to difficult optimization problems. Enlightened by the law of human population evolution, we also introduce the logistic model of population size in demography into our VPGA to improve its performance. Experimental results show that VPGA gets better solutions for optimization, owns faster convergence, costs less computing time, and is more stable.
*Keywords:* Genetic algorithm, Variable population size, Logistic model

## 1. Introduction

We present a new type of genetic algorithm named VPGA (Variable Population-size GA) with variable size of population by modifying the structure of the classical GA. VPGA is a framework based on the improved GAs which are denoted as $x$GA and be used as a subroutine in VPGA. Let $s$, $t_s$, $t_e$ be the population size, the number of initial generation and the number of the last generation respectively. Denote the population of size $s$ in the $t$-th generation as Population($s$, $t$). The following algorithm $x$GA($s$, $t_s$, $t_e$) describes the procedure of evolution which starts from the $t_s$-th generation and ends at the $t_e$-th generation with population size of $s$, and the right is VPGA.

| ALGORITHM $x$GA($s$, $t_s$, $t_e$) | ALGORITHM VPGA($S$, $T$) |
|---|---|
| 1.  $t \leftarrow t_s$ | 1.  $i \leftarrow 0$ |
| 2.  initialize Population $P(s, t)$ | 2.  $x$GA($S_i$, $T_i$, $T_{i+1}$) |
| 3.  evaluate $P(s, t)$ | 3.  evaluate Population ($S_i$, $T_{i+1}$) |
| 4.  *while* ($t \leq t_e$) *do* | 4.  *while* (not terminaltion-condition) *do* |
| 5.  $t \leftarrow t + 1$ | 5.  $i \leftarrow i + 1$ |
| 6.  select $P(s, t)$ from $P(s, t\text{-}1)$ | 6.  update $x$GA's parameters $S_i \leftarrow s(t)$ |
| 7.  evolve $P(s, t)$ | 7.  $x$GA($S_i$, $T_i$, $T_{i+1}$) |
| 8.  evaluate $P(s, t)$ | 8.  evaluate Population ($S_i$, $T_{i+1}$) |
| 9.  *end while* | 9.  *end while* |

We use a logistic model of population size in demography as follows

$$s(t) = \frac{1}{\frac{1}{S_n} + (\frac{1}{S_0} - \frac{1}{S_n})e^{-\beta(t-t_0)}}$$

To illustrate that the performance of our algorithm is much higher than traditional GAs, we tested them on a set of benchmarks of non-linear programming problems F1 to F24 [9-10] which cover a wide range of non-linear functions. Let $x$GA is a traditional GA or a kind of GA with modifications on its genetic operations [6-8]. We set three types of $x$GAs with different population sizes: $x$GA1 (popSize=10), $x$GA2 (popSize=20) and $x$GA3 (popSize=30). The following table 1 shows the standard deviation of the optimal results obtained by the algorithms on the functions.

---

[1] Corresponding author. E-mail: x.xu@citiz.net.
[2] Corresponding author. E-mail: angeletx@citiz.net.

Table 1: Standard deviations of the optimal solutions of each algorithm with 100 trials

| Functions | xGA1 | xGA2 | xGA3 | VPGA |
|-----------|------|------|------|------|
| F1 | 0.0391904 | 0.0201478 | 0.0000009 | 0.0000001 |
| F2 | 14.0200884 | 3.7796196 | 1.3252E-9 | 1.293E-14 |
| F3 | 0.0060842 | 0.0035561 | 0.0042085 | 0.0042071 |
| F4 | 0.3831929 | 0.0086763 | 1.2360E-6 | 4.7716E-5 |
| F5 | 0.0770488 | 6.2449E-10 | 1.9954E-8 | 3.6548E-10 |
| F6 | 0.0624446 | 0.0305694 | 4.2655E-5 | 3.5713E-9 |
| F7 | 0.1260081 | 0.0225120 | 5.2683E-4 | 1.9730E-4 |
| F8 | 3.4882704 | 1.4545754 | 0.5080419 | 2.2602E-15 |
| F9 | 7.1564E-2 | 5.5456E-7 | 2.9685E-7 | 1.9690E-11 |
| F10 | 0.2653897 | 6.1564E-8 | 2.8513E-8 | 6.0420E-11 |
| F11 | 0.0822138 | 1.9232E-4 | 5.5660E-11 | 3.6406E-14 |
| F12 | 14.8924095 | 5.4868E-5 | 6.7413E-6 | 2.2667E-13 |
| F13 | 3.2487E-5 | 1.6047E-7 | 3.8014E-7 | 3.8801E-11 |
| F14 | 1.7945E-8 | 1.0635E-8 | 1.7066E-7 | 1.7274E-11 |
| F15 | 0.0497493 | 0 | 0.0497493 | 0 |
| F16 | 0.1399999 | 0 | 0 | 0 |
| F17 | 0.0585521 | 2.3369E-7 | 4.1895E-6 | 2.9666E-8 |
| F18 | 0.0282083 | 9.8646E-7 | 9.0300E-6 | 1.1755E-7 |
| F19 | 75.0032171 | 47.9607726 | 32.8403313 | 2.9894E-6 |
| F20 | 5.2015E-4 | 1.5482E-8 | 2.8418E-6 | 5.3692E-10 |
| F21 | 0.7643009 | 0.2699256 | 0.0989971 | 0.0000000 |
| F22 | 0.0498148 | 0.0246848 | 0.0208014 | 0.0000000 |
| F23 | 0.2550421 | 3.3956E-6 | 0.0473134 | 1.1365E-9 |
| F24 | 0.0103137 | 3.9026E-8 | 7.1072E-6 | 1.9189E-9 |

Table 2: Average probability of errors of each algorithm with 100 trails

| Functions | xGA1 | xGA2 | xGA3 | VPGA |
|-----------|------|------|------|------|
| F1 | 9% | 5% | 0% | 0% |
| F2 | 11% | 4% | 1% | 0% |
| F3 | 96% | 86% | 85% | 25% |
| F4 | 53% | 7% | 9% | 1% |
| F5 | 4% | 0% | 0% | 0% |
| F6 | 13% | 6% | 2% | 0% |
| F7 | 13% | 45% | 97% | 0% |
| F8 | 57% | 7% | 3% | 0% |
| F9 | 6% | 2% | 1% | 0% |
| F10 | 6% | 2% | 1% | 0% |
| F11 | 3% | 1% | 0% | 0% |
| F12 | 22% | 1% | 1% | 0% |
| F13 | 2% | 1% | 2% | 0% |
| F14 | 0% | 0% | 1% | 0% |
| F15 | 1% | 0% | 1% | 0% |
| F16 | 2% | 0% | 0% | 0% |
| F17 | 3% | 1% | 21% | 0% |
| F18 | 8% | 7% | 80% | 1% |
| F19 | 45% | 24% | 9% | 1% |
| F20 | 4% | 0% | 62% | 0% |
| F21 | 40% | 8% | 1% | 0% |
| F22 | 12% | 15% | 13% | 0% |
| F23 | 12% | 1% | 9% | 0% |
| F24 | 6% | 0% | 74% | 0% |

From the tables, we can see the standard deviations of the optimal results obtained by VPGA are much lower than that of the xGAs and the average probability of errors of VGPA is also much lower, which demonstrate that VPGA is superior in stability than xGAs.

Figure 1 shows the changing of the average error rate of 100 trials of VPGA and xGAs on F7 when the running time increases.

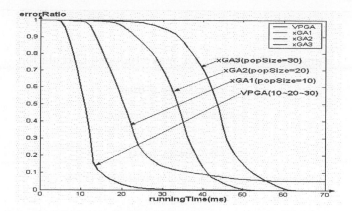

Figure 1: The changing of the error rate of VPGA and *x*GAs on F7 when the running time increases.

## 2. Conclusions

By introducing the logistic model of demography into genetic algorithm, we propose VPGA with its population size varying according to the law of human population evolution. Through a large amount of experiments, we find such improvement has higher probability to obtain optimal solutions with less computing time than the traditional GAs or other modified GAs.

## Acknowledgments

This paper is supported in part by the Chinese National Natural Science Foundation under grant No. 60473012, Chinese National Foundation for Science and Technology Development under contract 2003BA614A-14, and Natural Science Foundation of Jiangsu Province under contract BK2005047.

## References

[1] Holland, J.H., *Adaptation in Nature and Artificial Systems*, University of Michigan Press, 1975, MIT Press, 1992.

[2] Melanie, M., *An Introduction to Genetic Algorithms*, MIT Press, Cambridge, MA, 1999.

[3] Bhandarkar, S.M., Zhang, H., *Image Segment Using Evolutionary Computation*, IEEE Transactions and Evolutionary Computation, 1999, 3(1):1-21.

[4] Eiben, A.E., Hinterding, R., and Michalewicz, Z., *Parameter Controll in Evolutionary Algorithms*, IEEE Transactions on Evolutionary Computation, 1999, 3(2):124-141.

[5] Liang, K.M., Yao, X., and Newton, C.S., *Adapting Self-adaptive Paramters in Evolutionary Algorithms*, Applied Intelligence, 2001 15(3):171-180.

[6] Michalewicz, Z.,Genetic *Algorithms + Data Structures = Evolutionary Programs*, [M] 3rd ed. New York: Springer Verlag, 1996.

[7] Goldberg, D.E., Deb, K., et.al., *Rapid, Accurate Optimization of Difficult Problems Using Genetic Algorithms*, IlliGAl Technical Report No.93003, University of Illinois at Urbana-Champaign, 1993.

[8] De Jong, K.A., *An Analysis of the Behavior of Class of Genetic Adaptive Systems.(Doctoral dissertation*, University of Michigan), Dissertation Abstract International, 36(10), 5140B.(University Microfilms No 76-9381).

[9] Floudas, C.A. and Pardalos, P.M., *A Collection of Test Problems for Constrained Global Optimization Algorithms*, Springer-Verlag, Lecture Notes in Computer Science, Vol.455, 1987.

[10] Yao X., Liu Y. and Lin G., *Evolutionary programming made faster*. IEEE Transactions on Evolutionary Computation, 2004, 3(2):82-102.

Brill Academic Publishers
P.O. Box 9000, 2300 PA Leiden
The Netherlands

*Lecture Series on Computer*
*and Computational Sciences*
Volume 4, 2005, pp. 602-605

# Numerical Modelling Of Wave Induced Circulation

İ. Yıldız, A. İnan, L. Balas[1]

Department of Civil Engineering,
Faculty of Engineering and Architecture,
Gazi University,
06570 Ankara, Turkey

Received 14 June, 2005; accepted in revised form 13 July, 2005

*Abstract:* A numerical model is developed for the simulation of wave transformations and wave induced nearshore current, that is applicable to irregular bottom topographies. Combined model has two components, a wave propagation model and a wave driven current model. Wave propagation model is based on extended mild slope equation and could simulate wave shoaling, refraction, diffraction and breaking. Linear, harmonic, and irrotational waves are considered. To describe the wave motion, mild slope equation has been decomposed into three equations that are solved in terms of wave height, wave approach angle and wave phase function. Model does not have the limitation that one coordinate should follow the dominant wave direction. Different wave approach angles can be investigated on the same computational grid. Wave driven current model is based on vertically averaged non-linear shallow water equations. In the solution method, partial differential equations are replaced by a set of finite difference equations on a space staggered grid. Model has been applied to Obaköy coastline that is located at the Mediterranean Sea coast of Turkey.

*Keywords:* wave propagation, wave induced circulation, finite difference, mild slope equation

*Mathematics Subject Classification:* Basic Methods in Fluid Mechanics

*PACS:* 76M20

## 1. Introduction

Precise numerical modeling of wave propagation from the deep ocean to a shoreline and prediction of wave induced currents are quite important in coastal engineering. A combined numerical model is developed for the simulation of wave transformations and wave induced nearshore current, that is applicable to irregular bottom topographies. In the wave propagation model, mild slope equation has been decomposed into three equations related to wave phase function, wave amplitude and wave approach angle [1],[2]. Computationally, the numerical model is quite efficient for simulating wave propagation over large coastal areas subjected to varying wave conditions. To account the effect of diffraction, the wave phase function changes to consider any horizontal variation in the wave height. Three wave parameters, wave height H, local wave angle $\theta$ and phase function of the wave $|\nabla s|$ are solved by the following equations;

$$|\nabla s| = k^2 + \frac{1}{H}\left[ \frac{\partial^2 H}{\partial x^2} + \frac{\partial^2 H}{\partial y^2} + \frac{1}{CC_g}\left( \frac{\partial H}{\partial x}\frac{\partial CC_g}{\partial x} + \frac{\partial H}{\partial y}\frac{\partial CC_g}{\partial y} \right) \right] \tag{1}$$

$$\frac{\partial}{\partial x}(|\nabla s|\sin\theta) - \frac{\partial}{\partial y}(|\nabla s|\cos\theta) = 0 \tag{2}$$

$$\frac{\partial}{\partial x}\left( H^2 CC_g |\nabla s|\cos\theta \right) + \frac{\partial}{\partial y}\left( H^2 CC_g |\nabla s|\sin\theta \right) = 0 \tag{3}$$

---

[1] Corresponding author. E-mail: lalebal@gazi.edu.tr

in which, H: wave height; s: scalar phase function of the wave; $\nabla$: horizontal gradient operator; C: wave celerity; $C_g$: group velocity; k: wave number calculated by the dispersion relation; $\theta(x,y)$: angle of incidence defined as the angle made between the bottom contour normal and the wave direction.

The wave induced circulation model is based on vertically averaged nonlinear shallow water equations of constant density flows.

$$\frac{\partial u}{\partial t}+u\frac{\partial u}{\partial x}+v\frac{\partial u}{\partial y}=-g\frac{\partial \eta}{\partial x}+v\left(\frac{\partial^2 u}{\partial x^2}+\frac{\partial^2 u}{\partial y^2}\right)-\frac{gu|u|}{C^2 H}-\frac{1}{\rho H}\left(\frac{\partial S_{xx}}{\partial x}+\frac{\partial S_{xy}}{\partial y}\right) \tag{4}$$

$$\frac{\partial v}{\partial t}+u\frac{\partial v}{\partial x}+v\frac{\partial v}{\partial y}=-g\frac{\partial \eta}{\partial y}+v\left(\frac{\partial^2 v}{\partial x^2}+\frac{\partial^2 v}{\partial y^2}\right)-\frac{gv|v|}{C^2 H}-\frac{1}{\rho H}\left(\frac{\partial S_{yx}}{\partial x}+\frac{\partial S_{yy}}{\partial y}\right) \tag{5}$$

$$\frac{\partial \eta}{\partial t}+\frac{\partial (Hu)}{\partial x}+\frac{\partial (Hv)}{\partial y}=0 \tag{6}$$

where, u and v: depth averaged current velocity components in x and y directions respectively; g: gravity acceleration; $H=h+\eta$ total water depth; h: still water depth; $\eta$:water surface elevation; v: turbulent eddy viscosity coefficient; C: Chezy friction coefficient; $S_{xx}$ and $S_{yy}$ : normal radiation stress components acting on the plane perpendicular to x and y directions respectively.

## 2. Numerical Solution

In the wave propagation model, solution method is a finite difference method that uses a mesh system in Cartesian coordinates. The finite difference approximations can handle the variations in the horizontal mesh sizes. The horizontal mesh size $\Delta x$ in the x-coordinate is orthogonal to the horizontal mesh size $\Delta y$ in the y-coordinate. The horizontal mesh sizes $\Delta x$ and $\Delta y$ can be different from each other. Also, $\Delta x$ can vary along the x coordinate and $\Delta y$ can vary along the y coordinate. Input model parameters are the deep water wave parameters, wave height ($H_0$), wave approach angle ($\theta_0$) and the wave period (T). Partial derivatives in the x-direction are expressed by forward finite differences of order $O(\Delta x)$, and the partial derivatives in the y-direction are expressed by central finite differences of order $O(\Delta y^2)$ in equation (1), whereas partial derivatives in the x-direction are approximated with backward finite differences of order $O(\Delta x)$, and partial derivatives in the y-direction are expressed by central finite differences of order $O(\Delta y^2)$ in Equation (3). Wave breaking is controlled during the computations by means of a simple breaking criteria;waves are assumed to break when the ratio of wave height to water depth reaches 0.78. After waves break, model keeps the above ratio constant. A space staggered numerical grid has been used. The following numerical algorithm has been applied in the solution [3];

$$\frac{\partial u}{\partial t}=\frac{\left(u_{i,j}+|u_{i,j}|\right)\left(u_{i,j}-u_{i-1,j}\right)}{2}\frac{1}{\Delta x}-\frac{\left(u_{i,j}-|u_{i,j}|\right)\left(u_{i+1,j}-u_{i,j}\right)}{2}\frac{1}{\Delta x}-\frac{\left(v_{i,j}^u+|v_{i,j}^u|\right)\left(u_{i,j}-u_{i,j-1}\right)}{2}\frac{1}{\Delta y}$$
$$-\frac{\left(v_{i,j}^u-|v_{i,j}^u|\right)\left(u_{i,j+1}-u_{i,j}\right)}{2}\frac{1}{\Delta y}-\frac{g(\eta_{i,j}-\eta_{i-1,j})}{\Delta x}-\frac{gu_{i,j}|u_{i,j}|}{C^2 H_{i,j}^u}$$
$$-v\left(\frac{u_{i+1,j}-2u_{i,j}+u_{i-1,j}}{\Delta x^2}+\frac{u_{i,j+1}-2u_{i,j}+u_{i,j-1}}{\Delta y^2}\right)$$
$$-\frac{1}{\rho H_{i,j}^u}\left(\frac{(S_{xx})_{i,j}-(S_{xx})_{i-1,j}}{\Delta x}+\frac{1}{2}\left(\frac{(S_{xy})_{i,j+1}-(S_{xy})_{i,j-1}}{2\Delta y}+\frac{(S_{xy})_{i-1,j+1}-(S_{xy})_{i-1,j-1}}{2\Delta y}\right)\right) \tag{8}$$

$$\frac{\partial v}{\partial t}=\frac{\left(u_{i,j}^v+|u_{i,j}^v|\right)\left(v_{i,j}-v_{i-1,j}\right)}{2}\frac{1}{\Delta x}-\frac{\left(u_{i,j}^v-|u_{i,j}^v|\right)\left(v_{i+1,j}-v_{i,j}\right)}{2}\frac{1}{\Delta x}-\frac{\left(v_{i,j}+|v_{i,j}|\right)\left(v_{i,j}-v_{i,j-1}\right)}{2}\frac{1}{\Delta y}$$
$$-\frac{\left(v_{i,j}-|v_{i,j}|\right)\left(v_{i,j+1}-v_{i,j}\right)}{2}\frac{1}{\Delta y}-\frac{g(\eta_{i,j+1}-\eta_{i,j})}{\Delta y}-\frac{gv_{i,j}|v_{i,j}|}{C^2 H_{i,j}^v}$$
$$-v\left(\frac{v_{i+1,j}-2v_{i,j}+v_{i-1,j}}{\Delta x^2}+\frac{v_{i,j+1}-2v_{i,j}+v_{i,j-1}}{\Delta y^2}\right)$$
$$-\frac{1}{\rho H_{i,j}^v}\left(\frac{(S_{yy})_{i,j+1}-(S_{yy})_{i,j}}{\Delta y}+\frac{1}{2}\left(\frac{(S_{yx})_{i+1,j}-(S_{yx})_{i-1,j}}{2\Delta x}+\frac{(S_{yx})_{i+1,j-1}-(S_{yx})_{i-1,j-1}}{2\Delta x}\right)\right) \tag{9}$$

where,

$$\frac{\partial \eta}{\partial t} = \frac{u_{i+1,j} H^u_{i+1,j} - u_{i,j} H^u_{i,j}}{\Delta x} \frac{v_{i,j} H^v_{i,j} - v_{i,j-1} H^v_{i,j-1}}{\Delta y} \tag{10}$$

$$u^v_{i,j} = 0.25(u_{i,j} + u_{i,j+1} + u_{i+1,j+1} + u_{i+1,j}) \; ; \cdots v^u_{i,j} = 0.25(v_{i,j} + v_{i-1,j} + v_{i-1,j-1} + v_{i,j-1}) \tag{11}$$

$$H^u_{i,j} = 0.5(H_{i,j} + H_{i-1,j}) \; ; H^v_{i,j} = 0.5(H_{i,j} + H_{i,j+1}) \tag{12}$$

System of nonlinear equations are solved by the adaptive step size controlled Runge-Kutta Fehlberg Method. It requires six function evaluations per step, but it provides an automatic error estimate.

## 3. Model Applications

Model has been applied to Obaköy which is located at the Mediterranean Sea coast of Turkey. In the coastal waters of Obaköy a sea outfall construction has been planned by the authorities. The bathymetry for the area is shown in Figure (1). For the area, wave transformations from the dominant wave directions, which are the S and SSE directions, are simulated. The deep water wave parameters are used to specify the offshore boundary conditions and zero gradient boundary conditions are applied for wave heights and wave angles along the lateral boundaries. Deep water parameters are selected as, wave period T=6 s., wave height H=4 m. Predicted wave height distributions for waves approaching with an approach angle of $\alpha=0^0$ (from S) and of $\alpha=22.5^0$ (from SSE), are presented in Figure (2) respectively. Corresponding predicted wave induced circulations are given in Figure (3) as well. Model provides reasonable estimations for the area. Waves converge on the shoal, conveyance of energy onto shoal results in the decrease of wave heights. Model can be used successfully for the areas having complicated bathymetries.

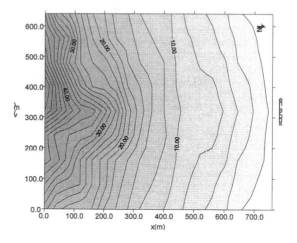

Figure 1. Water depths (m.) of Obaköy.

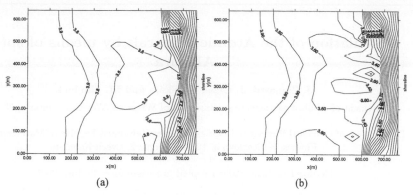

(a)                                        (b)

Figure 2. Predicted wave height distributions for waves approaching with an approach angle
of a) $\alpha=0^0$ (from S) , b)$\alpha=22.5^0$ (from SSE)

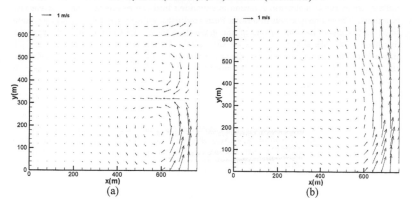

(a)                                        (b)

Figure 3. Predicted wave induced circulations for waves approaching with an approach angle
of a) $\alpha=0^0$ (from S), b) $\alpha=22.5^0$ (from SSE)

## 4.    Conclusions

A combined numerical model that has wave propagation and wave induced current sub models has been presented. Wave propagation model can simulate shoaling, refraction, diffraction and breaking effects. Wave driven current model is based on vertically averaged non-linear shallow water equations. Combined model can favourably handle complex bathymetries. Possibility of using finer resolutions in regions where the spatial gradients of the variables are sharp, is an advantage of the model. Therefore computationally, the numerical model is quite efficient for simulating wave propagation and wave induced circulation over large coastal areas subjected to varying wave conditions. Only one computational domain is enough to simulate the transformation of waves from different directions with different approach angles. Model successful application to Obaköy has been demonstrated. Developed model is a reliable tool for simulating the transformation of linear waves over complicated bathymetries.

## References

[1] L. Balas and A. İnan, A Numerical Model of Wave Propagation on Mild Slopes, *Journal of Coastal Research*, Coastal Education and Research Foundation, CERF, **SI36** 16-21 (2002).
[2] A. İnan and L. Balas, Applications of a Numerical Model to Wave Propagation on Mild Slopes, *China Ocean Engineering*, Nanjing Hydraulics Research Institute (**NHRI**), **16** 569-576 (2002).
[3] Z. Kowalik and T.S. Murty, Numerical Modeling of Ocean Dynamics, *Advanced Series on Ocean Engineering*, Vol.5, World Scientific, New Jersey, 1993.

Brill Academic Publishers
P.O. Box 9000, 2300 PA Leiden,
The Netherlands

*Lecture Series on Computer*
*and Computational Sciences*
Volume 4, 2004, pp. 606-611

# Calculation of the Atomic Integrals by means of so(2,1) Algebra

## J. Zamastil, J. Čížek, L. Skála and M. Šimánek

Department of Applied Mathematics, University of Waterloo, Waterloo, Ontario, N2L 3G1,
Canada

Department of Chemical Physics and Optics, Charles University, Faculty of Mathematics and
Physics, Ke Karlovu 3, 121 16 Prague 2, Czech Republic

Received 14 June, 2005; accepted in revised form 11 July, 2005

*Abstract:*

The use of the so(2,1) algebra for the study of the two-electron atoms is suggested. The
radial part of the two-electron function is expanded into the products of the one-electron
functions. These one-electron functions form complete, entirely discrete set and are identified as the eigenfunctions of one of the generators of the so(2,1) algebra. By applying
this algebra we are able to express all the matrix elements in analytic and numerically
stable form. For matrix elements of the two-electron interaction this is done in three steps,
all of them completely novel from the methodological point of view. First, repulsion integrals over four radial functions are written as a linear combination of the integrals over
two radial functions and the coefficients of the linear combination are given in terms of
hypergeometric functions. Second, combining algebraic technique with the integration by
parts we derive recurrence relations for the repulsion integrals over two radial functions.
Third, the derived recurrence relations are solved analytically in terms of the hypergeometric functions. Thus we succeed in expressing the repulsion integrals as rational functions of
the hypergeometric functions. In this way we resolve the problem of the numerical stability
of calculation of the repulsion integrals.

## 1  Introduction

The most accurate available method applicable for many electron problems is the configuration
interaction method (CI) (see e.g. [1, 2]).

However, the problem appears when one wants to perform very large scale CI calculation to
get very accurate results. Then one encounters what is usually refereed to as the effect of linear
dependence of the basis set. Due to numerical errors, for large basis sets the linear independence
of the basis functions is lost. Thus, from some point the inclusion of more basis functions does
not improve the variational results. To avoid this, one should keep the basis functions orthogonal.
However, this requirement causes the highly excited functions to have large number of nodes and
to change their sign frequently. Calculating matrix elements of Coulomb interaction among two
electron wave functions built from such basis functions is from numerical point of view subtracting
"two infinities". The situation about these numerical instabilities is highly unsatisfactory. People
either do not write about this at all or it seen from their results that they encountered this difficulty.

Since this effect is very general it is necessary to develop methods to deal with this. This paper
describes one such a method.

For the sake of transparency we restrict ourselves here on the so-called $s$-wave model of helium. This means that in the multipole expansion of Coulomb potential

$$r_{12}^{-1} = \frac{1}{r_>} \sum_{l=0}^{\infty} \left(\frac{r_<}{r_>}\right)^l P_l(\vec{n}_1.\vec{n}_2) \tag{1}$$

we keep just the first term $l = 0$, i.e.

$$r_{12}^{-1} \sim r_>^{-1}. \tag{2}$$

Here $r_< = r_1$, $r_> = r_2$ if $r_1 < r_2$ and $r_< = r_2$, $r_> = r_1$ if $r_1 > r_2$ and $P_l(x)$ denotes the Legendre polynomials.

This approximation is equivalent of building the configuration interaction just from the $s$-orbitals. This model gives about 96% of the correct value of the ground state energy. Also, it yields semiquantitative results when describing resonances in doubly excited states of helium. Therefore, the model attracted a great deal of attention, see e.g. [1, 3, 4]. However, this approximation is considered here from pedagogical reasons only. The extension of the method to include orbitals of higher angular momenta is not difficult [5].

## 2 so(2,1) algebra

We search for the exact wave function in the form of the linear combination of symmetrized product of one-electron basis functions

$$\psi(1, 2) = \sum_{i=0}^{\infty} f_i \psi_i(1, 2) \tag{3}$$

where

$$\psi_i(1, 2) = 2^{-(1+\delta_{n1_i, n2_i})/2} \left(R_{n1_i}(r_1) R_{n2_i}(r_2) + R_{n1_i}(r_2) R_{n2_i}(r_1)\right). \tag{4}$$

The one-electron basis functions are eigenfunctions of the operator $T_3$

$$T_3 R_n(r) = n R_n(r) \tag{5}$$

$$T_3 = \frac{1}{2} \left(r p_r^2 + r\right), \tag{6}$$

where $p_r$ is the conjugated radial momentum

$$p_r = -i \left(\frac{d}{dr} + \frac{1}{r}\right). \tag{7}$$

The explicit form of these radial functions is given as

$$R_n(r) = 2\sqrt{\frac{1}{n}} e^{-r} L_{n-1}^1(2r), \tag{8}$$

where $L_{n-1}^1(2r)$ are generalized Laguerre polynomials (see e.g. [6, 7, 8]). The use of these functions is advantageuos since they form complete, entirely discrete basis set. This should be contrasted to the eigenfunction of hydrogen where discrete part itself does not form complete basis set.

The operator $T_3$ forms together with the operators $T_1$

$$T_1 = \frac{1}{2} \left(r p_r^2 - r\right), \tag{9}$$

and $T_2$

$$T_2 = r p_r \tag{10}$$

the so-called so(2,1) algebra, i.e. they are closed under commutation

$$[T_1, T_2] = -iT_3, \ [T_2, T_3] = iT_1, \ [T_3, T_1] = iT_2 \tag{11}$$

In the similar way to angular momentum we can introduce the step-up and step-down operators

$$T_\pm = T_1 \pm iT_2 \tag{12}$$

that act on the basis functions as follows

$$T_\pm R_n(r) = \alpha^\pm(n) R_{n\pm 1}(r) \tag{13}$$

where

$$\alpha^+(n-1) = \alpha^-(n) = \sqrt{n(n-1)}. \tag{14}$$

This relation will be usefull later.

## 3   Calculation of the repulsion integrals

Using approximation (2) the matrix elements of Coulomb interaction between basis functions (4) read

$$< \psi_i | r_{12}^{-1} | \psi_j > = 2^{-(\delta_{n1_i,n2_i} + \delta_{n1_j,n2_j})/2} [X_{n1_i,n2_i,n1_j,n2_j} + X_{n1_i,n2_i,n2_j,n1_j}], \tag{15}$$

where the so-called Slater integrals are given as

$$X_{n1_i,n2_i,n1_j,n2_j} = \int_0^\infty dr_1 R_{n1_i}(r_1) R_{n1_j}(r_1) r_1^2 \int_{r_1}^\infty dr_2 R_{n2_i}(r_2) R_{n2_j}(r_2) r_2 + \tag{16}$$

$$+ \int_0^\infty dr_1 R_{n1_i}(r_1) R_{n1_j}(r_1) r_1^1 \int_0^{r_1} dr_2 R_{n2_i}(r_2) R_{n2_j}(r_2) r_2^2.$$

Numerically stable way of calculation of these integrals is achieved in three steps. First, be means of analogue of so(2,1) algebra we reduce the integrals (16) to the integrals over two radial functions. Second, combining integration by parts with algebraic technique we derive recurence relations for the reduced integrals. Third, we solve the recurence relations.

### 3.1   Wigner-Eckart theorem for so(2,1) algebra

Looking at Eq. (16) we observe that there are always the products of the wave functions of the same variable (like $R_{n1_i}(r_1) R_{n1_j}(r_1)$) that enter into the integration. Therefore, in the first step we try to write the product of two radial functions as a linear combination of the radial functions. Here, we proceed analogously to the angular integration. This trick of writing the product of two spherical harmonics as a linear combination of the spherical harmonics is known as the Wigner-Eckart theorem. Its analogue for so(2,1) algebra reads

$$r R_{n_i}(r) R_{n_j}(r) = \sum_{n=-1}^{n_i+n_j-2} c_{n_i,n_j,n} R_{n_i+n_j-n-1}(2r), \tag{17}$$

where the coefficients $c_{n_i,n_j,n}$ can be expressed via hypergeometric functions

$$c_{n_i,n_j,n} = \frac{2^{1-n_i-n_j}(n_i+n_j-2)!(n_i+n_j)!}{(n_i-1)!(n_j-1)!} \sqrt{\frac{1}{n_i}} \sqrt{\frac{1}{n_j}} \sqrt{\frac{1}{n_i+n_j-n-2}}$$
$$\left( C_{n_i,n_j,n} - \frac{n_i+n_j-2-n}{n_i+n_j-n-1} C_{n_i,n_j,n+1} \right),$$

where the coefficients $C_{n_i,n_j,n}$ are given as

$$C_{n_i,n_j,n} = \frac{F(-n_i+1,-n;-n_i-n_j+2;2)\,F(-n_i,-n;-n_i-n_j;2)}{(n_i+n_j-n-1)!n!} \tag{18}$$

for $n \geq 0$ and equal zero otherwise. Here, $F(\alpha,\beta;\gamma;z)$ denotes the hypergeometric function (see e.g. [6, 7, 8]). This result was obtained using properties of Laguerre polynomials given in [6].

Using Eq. (17) in the radial integrals (16) we rewrite them into the form

$$X_{n1_i,n2_i,n1_j,n2_j} = \sum_{n_1=-1}^{n1_i+n1_j-2} C_{n1_i,n1_j,n_1} \sum_{n_2=-1}^{n2_i+n2_j-2} C_{n2_i,n2_j,n_2}\, Q_{n1_i+n1_j-1-n_1,n2_i+n2_j-1-n_2}, \tag{19}$$

where $Q_{N_1,N_2}$ denotes the integrals over two radial functions

$$Q_{N_1,N_2} = Q^+_{N_1,N_2} + Q^-_{N_1,N_2}. \tag{20}$$

Here,

$$Q^+_{N_1,N_2} = \int_0^\infty dr_1\, R_{N_1}(2r_1)r_1 \int_{r_1}^\infty dr_2\, R_{N_2,L}(2r_2) \tag{21}$$

and

$$Q^-_{N_1,N_2} = \int_0^\infty dr_1\, R_{N_1}(2r_1) \int_0^{r_1} dr_2\, R_{N_2}(2r_2)r_2. \tag{22}$$

It turns out that it is sufficient to calculate the integrals $Q^+_{N_1,N_2}$. The integrals $Q^-_{N_1,N_2}$ are then obtained via relation

$$Q^-_{N_1,N_2} = Q^+_{N_2,N_1}. \tag{23}$$

The next task is to derive recurence relations for the integrals (21). This is done by combining integration by parts with algebraic technique.

### 3.2 Recurence relations for the integrals

Let us consider the integral

$$\int_0^{r_1} dr_2\, r_2 \left(\frac{d}{dr_2} + \frac{1}{r_2}\right)[R_{N_2}(2r_2)]. \tag{24}$$

On one hand, the integration by parts yields

$$\int_0^{r_1} dr_2\, r_2 \left(\frac{d}{dr_2} + \frac{1}{r_2}\right)[R_{N_2}(2r_2)] = r_1 R_{N_2}(2r_1). \tag{25}$$

The proof is elementary.

On the other hand, the operator $r\left(\frac{d}{dr} + \frac{1}{r}\right)$ can be expressed as the difference of the step-up and step-down operators $T_+$ and $T_-$, see Eqs. (10) and (12)

$$r\left(\frac{d}{dr} + \frac{1}{r}\right) = \frac{1}{2}(T_+ - T_-). \tag{26}$$

Thus, from the algebraic point of view

$$r\left(\frac{d}{dr} + \frac{1}{r}\right)[R_{N_2}(2r)] = \frac{1}{2}\sqrt{(N_2+1)N_2}R_{N_2+1}(2r) - \frac{1}{2}\sqrt{N_2(N_2-1)}R_{N_2-1}(2r), \tag{27}$$

Finally, combining the analytic result (25) with the algebraic one (27) we get the recurence relation for the integrals

$$\frac{1}{2}\sqrt{(N_2+1)N_2}\int_0^{r_1} dr_2\, R_{N_2+1}(2r_2) - \frac{1}{2}\sqrt{N_2(N_2-1)}\int_0^{r_1} dr_2\, R_{N_2-1}(2r_2) = r_1 R_{N_2}(2r_1). \quad (28)$$

Multiplying the last equation by $r_1 R_{N_1}(2r_1)$ and integrating it from 0 to infinity we obtain the recurence relation for the integrals (21)

$$-\frac{1}{8}\int_0^{\infty} dr\, r^2 R_{N_1}(r) R_{N_2}(r) = \frac{1}{2}\sqrt{(N_2+1)N_2}Q^+_{N_1,N_2+1} - \frac{1}{2}\sqrt{N_2(N_2-1)}Q^+_{N_1,N_2-1}. \quad (29)$$

We note that left hand side of this equation vanishes unless $|N_1 - N_2| \leq 1$.

In similar way we can derive the recurrence relations in the variable $N_1$

$$\frac{1}{8}\int_0^{\infty} dr\, r^2 R_{N_1}(r) R_{N_2}(r) = \frac{1}{2}\sqrt{(N_1+1)N_1}Q^+_{N_1+1,N_2} - \frac{1}{2}\sqrt{N_1(N_1-1)}Q^+_{N_1-1,N_2} + Q^+_{N_1,N_2}. \quad (30)$$

### 3.3 Solution of recurrence relations

The recurrence relations derived above have very simple solution

$$Q^+_{N_1,N_2} = 0 \quad (31)$$

for $N_1 > N_2$ and

$$Q^+_{N_1,N_2} = \frac{(-1)^{N_2-N_1}}{4}\sqrt{\frac{N_1}{N_2}} \quad (32)$$

for $N_2 > N_1$ and

$$Q^+_{N_1,N_1} = \frac{1}{8}. \quad (33)$$

When we include also $p$, $d$ and other orbitals, the recurrence relations (29) and (30) and their solutions are slightly more complicated, however, they can still be generally expressed in terms of the hypergeometric function [5].

## 4  Conclusions

In this paper the method for numerically stable calculation of the atomic integrals by means of so(2,1) algebra was given. The method was illustrated in detail on the calculation of the integrals within $s$-wave model of helium atom. The extension of the method to include orbitals of higher angular momenta is not difficult and it has been given in [5]. Thus, the method can be easily extended to treat all one-center integrals. Future development of the method should concetrate on two-center integrals.

## References

[1] P. O. Löwdin, Phys. Rev. **97**, 1474, 1490, 1509 (1955). See also H. Shull and P. O. Löwdin, J. Chem. Phys. **23**, 1362 (1955); **25**, 1035 (1956); **30**, 617 (1959); P. O. Löwdin and H. Shull, Phys. Rev. **101**, 1730 (1956); P. O. Löwdin and L. Rèdei, Phys. Rev. **114**, 752 (1959).

[2] O. Jitrik and C. F. Bunge, Phys. Rev. A **56**, 2614 (1997). See also C. F. Bunge, Theor. Chim. Acta (Berl.) **16**, 126 (1970).

[3] M. Draeger, G. Handke, W. Ihra, and H. Friedrich, Phys. Rev. A **50**, 3793 (1994).

[4] S. P. Goldman, Phys. Rev. Lett. **78**, 2325 (1997).

[5] J. Zamastil, J. Čížek, M. Kalhous, L. Skála and M. Šimánek, J. Math. Phys. **45**, 2674 (2004). See also J. Zamastil, M. Šimánek, J. Čížek and L. Skála, J. Math. Phys. **46**, 033504 (2005).

[6] N. Ja. Vilenkin and A. U. Klimyk, *Representation of Lie Groups and Special Functions, Vol. 1* (Kluwer Academic Publishers, Dordrecht, Boston, London, 1991).

[7] J. D. Talman, *Special Functions, A Group Theoretic Approach* (W. A. Benjamin, Inc., New York, Amsterdam 1968).

[8] *Higher Transcendental Functions, Vol. I* A. Erdélyi, Ed. (Mc Graw-Hill, New York, 1953).

Brill Academic Publishers
P.O. Box 9000, 2300 PA Leiden
The Netherlands

*Lecture Series on Computer*
*and Computational Sciences*
Volume 4, 2005, pp. 612-615

# Vector based Fault-Tolerant Routing in Hypercube Multi-computers

Qin Zheng[1];  Wang Lei[2];  Zou Jian-jun[1]

1. Department of Computer and Information Science,
Fujian University of Technology,
350014, Fuzhou, Fujian, P.R.China
2. College of Software,Dongguan University of Technology,
523106, Dongguan, Guangdong, P.R.China

Received 7 May, 2005; accepted in revised form 20 July, 2005

*Abstract:* Aiming at handling the link-faults that exist in hypercube multi-computers interconnection networks, a kind of novel fault-tolerant model named *MMSPV* (Maximum Multi-valued Safety Path Vectors) is proposed. In addition, algorithms on how to construct *MMSPV*, and fault-tolerant routing algorithms based on the new kind of fault-tolerant model are also presented. The simulation results show that the fault-tolerant routing algorithms based on *MMSPV* can record more optimal paths than the fault-tolerant routing algorithm based on *SV* respectively.

*Keywords:* ault-tolerant routing; optimal path; hypercube; Maximum Multi-valued Safety Path Vectors

## 1.  Inroduction

As the number of processors in multi-computer interconnection networks increase, the fault probability of the processors and links between them also increase. In order to realize more efficient fault-tolerant routing when large number of faults occur, designing better fault-tolerant strategies to record the information about the optimal paths existing in the networks as much as possible are of great importance.

The hypercube network [1] is of good performances in parallel and distributed processing, and is one of the most common interconnection structures used in the multi-computers system. Until now, a lot of research works have been done in the field of researches of its fault-tolerant characteristics.

In 1998, Wu [2] studied the injured hypercube interconnection networks, and proposed a fault-tolerant model *SV* (Safety Vector) and its fault-tolerant routing algorithm, which use vectors to record optimal paths. The advantage of *SV* is that it is of very low storage cost of $n$ bites for $n$ dimensional hypercube interconnection networks because of the simple form of vectors, and the drawback is that the constructing function of *SV* is too simple to be of great fault-tolerant ability.

In 2001, Al-Sadi, etc [3] proposed a new fault-tolerant model of *USV* (Unsafety Vectors) for $k$-ary $n$-cubes, and based on this fault-tolerant model, they designed a new fault-tolerant routing algorithm for the binary $n$-cube in 2002 [4]. In 2000, Narraway [5] researched the fault-tolerant hypercube networks, and proposed a new fault-tolerant routing algorithm, which can obtain alternative shortest paths in n-cubes. In 2001, Xiang [6] proposed another novel fault-tolerant model *LSI*(Local Safety Information) for hypercube multi-computers, and on the basis of which, a new fault-tolerant routing algorithm was designed simultaneously.

In this paper, a kind of novel fault-tolerant model named *MMSPV* (Maximum Multi-valued Safety Path Vectors) is proposed, which records the optimal paths of the hypercube networks by using the forms of vector. In addition, algorithms for constructing *MMSPV*, and fault-tolerant algorithms based on it are also presented simultaneously. It is proved that *MMSPV* can record the most optimal paths of the injured hypercube networks in the forms of vectors with storage cost of $n^2/2$ bites.

2. Corresponding author. Post doctor of Tsinghua university, P.R.China. E-mail: wanglei_hn@hn165.com

## 2. Preliminaries

Definition 1 (Distance): For nodes $A(w_{n-1}...w_1 \, w_0)$ and $B(w'_{n-1}...w'_1 \, w'_0)$, in an $n$-dimensional hypercube $Q_n$, where $w_j$, $w'_j \in \{0,1\}$ and $j \in [0,n-1]$, the distance between $A$ and $B$ is defined as the *Hamming* distance between them.

Definition 2 (Optimal Path with $k$-Distance): For nodes $A$ and $B$ in an $n$-dimensional hypercube $Q_n$, there exists an optimal path with $k$-distance between them if and only if the distance between $A$ and $B$ is $k$ and there exists a fault-free path with $k$-distance between them at the same time.

Table 1: The bitwise logical operations upon binary strings

| Symbols | Operation |
|---------|-----------|
| + | Bitwise OR |
| & | Bitwise AND |
| ~ | Bitwise NOT |
| $\otimes$ | Bitwise XOR |
| $2^i$ | Binary string, of which the $i$th bit is 1 and all the other bits are 0 |
| *sour* | Source node of the information |
| *dest* | Destination node of the information |
| *cur* | Current node |
| *REL* $(A, B)$ | Relative address of $A$ and $B$ |
| $REL_j (A, B)$ | The $j$th bit of $REL$ $(A, B)$ |
| *nei(A, i)* | The $i$ th neighboring node of $A$ |
| *Dist* $(A, B)$ | Distance between $A$ and $B$ |

## 3. Maximum Multi-valued Safety Path Vector

Definition 3 (Maximum Multi-valued Safety Path vector): The maximum multi-valued safety path vector $MMSPV_A$ of node $A$ in an $n$-dimensional hypercube interconnection network $Q_n$ has $n$ bits. The values of $MMSPV_A [k]$, where $1 \le k \le n$, are calculated by the following rules:

$$MMSPV_A [k] = \sum_{j=0}^{n-1} f_j$$

Where $f_j = \begin{cases} 2^j : if \ M_A^k[j] = 1; \\ 0 : Otherwise. \end{cases}$

Where $M_A^k$, which is the $n$ bit link state vector obtained after $k$-1 rounds of information exchanges among the node $A$ and all of its neighboring nodes. The values of $M_A^k$ can be obtained through the following rules:

$$M_A^1 [j] = \begin{cases} 1 : \text{If the } j\text{th link of node } A \text{ is fault - free;} \\ 0 : \text{Otherwise} \end{cases}$$

When $k \ge 2$, $M_A^k$ can be obtained from the $M^{k-1}$ of the neighboring nodes of node $A$ as follows:

$$M_A^k = (a_{n-1}, ..., a_1, a_0)$$

Where the values of $a_i (0 \le i \le n-1)$ can be obtained through the following rules:

$$a_i = \begin{cases} 1 : \lambda^i + \sum_{j=0}^{n-1} (b_j^i + c_j^i) > n - k; \\ 0 : \text{Otherwise} \end{cases}$$

Where $b_j^i$, $c_j^i$, $\lambda^i$ can be obtained through the following rules:

1) $b_j^i = \begin{cases} 1 : f_j^i + g_j^i = 2; \\ 0 : \text{Otherwise} \end{cases}$

Where:

$$f^i_j = \begin{cases} 1: \text{If the } i\text{th link of node } A \text{ is fault;} \\ \text{the } j\text{th bit of } (\sim M^{k-1}_{nei(A,i)}) \, \& \, (\sim 2^i): \text{Otherwise} \end{cases}$$

$$g^i_j = \begin{cases} 0: \text{If the } i\text{th link of node } A \text{ is fault or } j = i; \\ M^{k-1}_{nei(A,j)}[i]: \text{Otherwise} \end{cases}$$

2) $c^i_j = \begin{cases} 0: \text{If the } i\text{th link of node } A \text{ is fault} \\ \text{the } j\text{th bit of } M^{k-1}_{nei(A,i)} \, \& \, (\sim 2^i): \text{Otherwise} \end{cases}$

3) $\lambda^i = \begin{cases} 1: \sum\limits_{j=0}^{n-1} d^i_j \geq 1; \\ 0: \text{Otherwise} \end{cases}$

Where: $d^i_j = $ the $j$th bit of

$$\{ \overset{r}{\underset{t=1}{+}} \{ M^{k-1}_{nei(A,m_t)} \, \& [\sim(\overset{p}{\underset{w=1}{+}} 2^{s_w})] \& [\sim(\overset{q}{\underset{w=1}{+}} 2^{l_w})] \& (\sim 2^i) \} \};$$

Where:

$m_1,\ldots,m_r\{m_t \notin \{s_1,\ldots,s_p, l_1,\ldots, l_q, i\}, t \in [1,r], r \leq n-p-q-1\}$ represent all of the fault-free links of node $A$, which satisfy $M^{k-1}_{nei(A,m_t)}[i]=0$;

$s_1,\ldots,s_p$ are the bits in $M^{k-1}_{nei(A,i)}$, which equal to 1;

$l_1,\ldots,l_q\{l_t \notin \{s_1,\ldots,s_p\}, t \in [1,q]$ represent all of the fault-free links of node $A$ that satisfy $M^{k-1}_{nei(A,l_t)}[i]=1$.

According to the definition 5, the construction algorithm of *MMSPV* can be described as follows:

Algorithm *MMSPVFilling*(){
    Collect link fault information to construct $M^1$;
    Determine *MMSPV*[1] using $M^1$;
    Send $M^1$ via non-faulty link;
    For($k$=2;$k \leq n$;$k$++){Receive all neighbors' $M^{k-1}$;
    Calculate $M^k$ with all neighbors' $M^{k-1}$ according to definition 5;
    Determine *MMSPV*[$k$] using $M^k$;
        IF ($k$!=$n$) send $M^k$ out via non-faulty links;}
}

Theorem 1: For any node $A$ in an $n$-dimensional hypercube interconnection network $Q_n$, $MMSPV_A[k] = \chi = 2^{t_1} +\ldots+ 2^{t_m}$ $(1 \leq k \leq n) \Leftrightarrow$ there exists an $k$-distance optimal path between node $A$ and any node, which is of $k$-distance to node $A$ and there is one bit among its $t_1$ th,$\ldots,t_m$ th bits that is different from the same bit of node $A$.

## 4. Fault-tolerant routing algorithm based on *MMSPV*

The maximal multi-valued safety path vector *MMSPV* has recorded the information of optimal paths in the hypercube network, so we can design a fault-tolerant routing algorithm based on it, and use it to send messages to destinations along the optimal paths. The detailed algorithm can be described as follows:

Algorithm *MMSPV_Route* {
    If ($cur == dest$) Return (SUCCESS);
    For($link$=0;$link < n$;$link$++) //to find optimal paths
    {If ($link \in \{j \mid REL_j (cur,dest)=1\}$&& $MMSPV_{nei(A,link)}[Dist(cur,dest)-1] \, \& \, 2^{link} =$
        $= 2^{link}$ ){Send message with this link; Return(*SUCCESS*);}
    }
    For($link$=0;$link < n$;$link$++) //to find sub-optimal paths

$$\{\text{If } (link \in \{j \mid REL_j \ (cur, \ dest)=0\} \ \& \ MMSPV_{nei(A,link)}[Dist(cur,dest)+1] \ \& \ 2^{link} =$$
$$=2^{link} )\{\text{Send message with this link; Return } (SUCCESS);\}$$
$$\}$$
Return $(FALSE)$;
$$\}$$

In order to compare the fault-tolerant performances of the routing algorithms based on *SV, MMSPV*,we simulate their ability to send messages along optimal paths in the 8 dimensional and 10 dimensional hypercube interconnection networks separately. For a given fault links number, we select 100 kinds of faulty distributed models, and send messages from source node to the destination node for one time. By considering all of the messages sent between fault-free nodes as 100%, we obtain the percentages of messages sent along optimal paths according to *SV* and *MMSPV* as following table2.

Table 2 Comparative results of *SV* and *MMSPV* (8-D hypercube with fault links)

| Total numbers of fault links | OP in *SV* | OP in *MMSPV* | *OP Exist* |
|---|---|---|---|
| 15 | 99.9774 | 99.9962 | 99.9969 |
| 30 | 99.5550 | 99.9931 | 99.9938 |
| 40 | 97.2506 | 99.9907 | 99.9914 |
| 50 | 86.4614 | 99.9879 | 99.9884 |
| 60 | 62.3946 | 99.9853 | 99.9861 |

From the simulation results of table 2, it is easy to know that the fault-tolerant ability of algorithm based on *MMSPV* is better than *SV*.

## 5. Conclusions

Aiming at handling the link-faults that exist in hypercube multi-computers networks, this paper proposes the concept of *MMSPV*. And in addition, algorithm on how to construct *MMSPV*, and fault-tolerant algorithm based on *MMSPV* are also presented. Due to that excellent characteristics of *MMSPV*, we solve the problem arisen during the research of *n*-dimensional hypercube interconnection networks: for any node *A* in an *n*-dimensional hypercube network, with the storage cost of $n^2/2$ bits, how to record the optimal paths information as much as possible by the *n*-1 rounds of information exchanges among neighboring nodes only.

## References

[1] Wang Lei, Lin Ya-ping, Chen Zhi-ping, *Fault-tolerant Routing for Hypercube Multicomputers Based on Maximum Safety-Path Matrix*. Journal of Software 994-1004(2004).

[2] Wu J. *Adaptive fault-tolerant routing in cube-based multicomputers using safety vectors*. IEEE Trans Parallel and Distributed Systems 321-334(1998).

[3] J. Al-Sadi, K. Day, M. Ould-Khaoua. *Unsafety vectors: A new fault-tolerant routing for k-ary n-cubes*. Microprocessors and Microsystems 239-246(2001).

[4] J. Al-Sadi, K. Day, and M. Ould-Khaoua. *Unsafety vectors: A new fault-tolerant routing for the binary n-cube*. Journal of Systems Architecture 783-793(2002).

[5] JJ.Narraway. *Alternative shortest paths in n-cubes*. IEEE Transactions on Electronics Letters, 1916-1918 (2000).

[6] D. Xiang. *Fault-Tolerant Routing in Hypercube Multicomputers Using Local Safety Information*. IEEE Transactions on parallel and Distributed Systems, 942-951(2001).

Brill Academic Publishers
P.O. Box 9000, 2300 PA Leiden
The Netherlands

*Lecture Series on Computer*
*and Computational Sciences*
Volume 4, 2005, pp. 616-618

# Automatic Reconstruction in Single Photon Emission Computed Tomography Using Inhomogeneous Markov Random Fields Models

S. Zimeras[1]

Department of Statistics and Actuarial-Financial Mathematics,
University of the Aegean,
83200 Karlovassi, Samos, Greece

Received 18 April, 2005; accepted in revised form 12 June, 2005

*Abstract:* In this work a complete approach for simultaneous and automatic parameter estimation and image reconstruction is presented, which allows spatially varying amounts of local smoothing. An inhomogeneous Markov random field (M.r.f.) model is described which allows spatially varying degrees of smoothing in the reconstructions and a re-parameterization is proposed which implicitly introduces a local correlation structure in the smoothing parameters. A MCMC/EM procedure is described having first derived a calibration curve for prior parameter estimation. The automatic procedure is applied using the OSL and Metropolis-Hastings algorithms. The analysis of SPECT data of the brain is used to illustrate the procedures and synthetic data to demonstrate that the proposed procedures lead to successful reconstruction. The inhomogeneous model allows greater flexibility; small features are not masked by the smoothing and constant regions obtain sufficient smoothing to remove the effects of noise.

*Keywords:* Markov random fields, Inhomogeneous models, image reconstructions, Single photon emission computed tomography.

## 1. Introduction

Procedures using Bayesian approaches incorporate information regarding the nature of the true image in terms of prior models. Previously, various authors (Shepp and Vardi, 1982; Vardi, Shepp and Kauffman, 1985) had adopted maximum likelihood approaches with only moderate success. The prior model and data likelihood are combined to give the posterior distributions on which estimation is based. The most common choice of prior distribution promotes local smoothness, using random field models defined as Gibbs distributions. The incorporation of prior information into the reconstruction was first suggested by Besag (1983), Geman and Geman (1984) and Besag (1986). This approach was subsequently used by many authors to produce much improved reconstructions. Procedures including prior models usually assume that any prior parameters are known. In practice the choice of these parameters are often determined by ``trial and error''. When the prior parameters are constant across the image, a homogeneous Markov random field model is defined (Cross and Jain; 1983). A combined MCMC/EM approach was proposed by Geman and McClure (1987) and has been used by Chalmond (1989) for homogeneous prior models. Also fully Bayesian approaches have been used by Weir (1997), Aykroyd (1998) and Higdon et. al. (1997).

In many applications the form of the likelihood is determined by physical considerations, though there may be unknown parameters. For example in medical imaging involving radioactive emissions a Poisson model is generally accepted and in other applications a Gaussian model may be appropriate. The next choice is the form of the prior distribution. In many cases there is no single obvious choice of the form of this distribution. If a vague description of the prior information is given, such as ``the truth should be smooth'', then there are many apparently appropriate choices. The exact choice, however, may change the final detail of the image reconstructions. A common choice in image applications is a homogeneous Gibbs distribution. This, however, is made with little comment and model checking is rarely considered. A more realistic model may say that there are homogenous regions. This may suggest an edge-site model such as that suggested by Geman and Geman (1984). A more usual

---

[1] Corresponding author. Lecturer- Grad. Stat. E-mail:zimste@aegean.gr

approach is to select a prior which does not overly bias true discontinuities inhomogeneous image, for example using edge-preserving or implicit-discontinuity priors (Geman and McClure, 1987), Green P. J., 1990, Geman and Reynolds, 1992)

The use of sub-regions suggests that different smoothing parameters are needed. An obvious generalization is to allow separate parameters for each pixel, which defines an inhomogeneous model (Aykroyd and Zimeras; 1999, Zimeras, 1997). This extension allows variable amounts of spatial smoothing across the image. Controlling the local spatial variation, the features appear much clearer especially in cases where jumps and constant regions occur. The procedure is based on calibration equations describing the relationship between prior and posterior energies and the parameter. Initial experiments show that, when the energy of the whole image is considered, the estimate smoothing parameter for the image is large compared to the values that were found by trial and error. Avoiding the background and using only a region of interest, the estimate is the same as the trial and error values. The need for automatic estimation and spatially varying smoothing has been discussed by many researchers. The need for such models was noted in Besag (1986) and by Titterington, Clifford, Kent, Coleman, and Ford and McCall in the discussion. Green (1990) suggested that the choice of prior would depend on the typical pattern expected and that there was ``potential for a richer class of functions allowing prior parameters to vary stochastically across the scene". Weir (1997) restated possible benefits of allowing the smoothing to vary across the scene. In this work, automatic estimation procedures for inhomogeneous models have been studied. The procedure for automatic reconstruction is applied using the OSL and Metropolis-Hastings algorithms and uses real SPECT data to demonstrate that the procedures work well and synthetic data to demonstrate that the proposed procedures lead to successful reconstruction (Zimeras, 1997). The procedures represent a substantial advance for general image and spatial analysis.

## Conclusions

Over the years inhomogeneous prior models in image processing have been widely accepted. For most applications the simplest Gaussian model tends to over-smooth reconstructions, indicating the unsatisfactory results. This work is directed to the assumption of homogeneity where a easy to use inhomogeneous Markov random field model is applied. The simplicity of the model is allows rapid calculation, and the flexibility of the spatially variation of the prior parameters, allowing different degree of spatial smoothness. The analysis of real single photon emission computed tomography data is used to illustrate the method, leading to more accurate reconstructions than the edge-preserving homogeneous alternative. The inhomogeneous model allows greater flexibility; more features are not masked by the smoothing and constant regions obtain sufficient smoothing to remove the effects of the noise (Figures 1, 2).

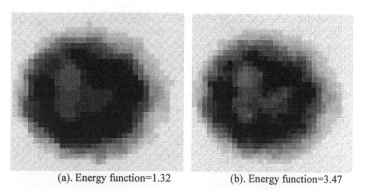

(a). Energy function=1.32          (b). Energy function=3.47

Figure 1: Reconstructions using the homogeneous model for slice 35: a. OSL algorithm for slice 35; b. Metropolis-Hastings algorithm

(a). Energy function=1.345        (b). Energy function=1.48

Figure 2: Reconstructions using the inhomogeneous model for slice 35: a. OSL algorithm; b. Metropolis-Hastings algorithm

## Acknowledgments

The author would like to thank Professor Peter Green University of Bristol, and Dr. Ian Weir for providing a very efficient computer program for evaluating the tomographic weight and providing the SPECT data.

## References

[1] Aykroyd R. G. (1998): Bayesian estimation of Homogeneous and Inhomogeneous Gaussian Random Fields, IEEE Trans. PAMI, 20, 533-539.

[2] Aykroyd R. G., Zimeras S. (1999): Inhomogeneous prior models for image reconstruction, Journal of American Statistical Association (JASA), Vol 94, No 447, 934-946.

[3] Besag J. (1974): Spatial interaction and the statistical analysis of lattice systems, J. Royal Statistical Society, Series B, 36, 192-236.

[4] Besag J. (1986): On the statistical analysis of dirty pictures, J. Royal Statistical Society, Series B, 48, 259-302.

[5] Besag J. (1986): Discussion of paper by P. Switzer, Bulletin of the International Statistical Institute, Ser A 332, 323-342.

[6] Chandler D. (1978): Introduction to Modern Statistical Mechanics, Oxford University Press, N. York.

[7] Chalmond B. (1989): An Iterative Gibbsian Technique for Simultaneous Structure Estimation and Reconstruction of M-ary Images, Pattern Recognition, 22 (6), 747-761.

[8] Cross G. R. and Jain A. K. (1983): Markov random field texture models, IEEE Trans. PAMI, 5(1), 25-39.

[9] Geman S. and Geman D. (1984): Stochastic relaxation, Gibbs distribution and the Bayesian restoration of images, IEEE Trans. Pattern Anal. Mach. Intell, 6, 721-741.

[10] Geman D. and Reynolds G. (1992): Constrained restoration and the recovery of discontinuities, IEEE Trans. PAMI, 9, 39-55.

[11] Green P. J. (1990): On use of the EM algorithm for penalized likelihood estimation. J. R. Stat. Soc. B., 52, 443-452.

[12] Geman S. and McClure D. E. (1987): Statistical methods for tomographic image reconstruction, Bull. Int. Statist. Inst., 52, No 4, 5-21.

[13] Higdon D. M., Bowsher J. E., Johnson V. E., Turkington T. G., Gilland D. R. and Jaszczak R. J. (1997): Fully Bayesian Estimation of Gibbs Hyperparameters for Emission Computed Tomography Data, IEEE Trans. on Medical Imaging, 16, 516-526.

[14] Shepp L. A. and Vardi Y. (1982): Maximum likelihood reconstruction in positron emission tomography, IEEE Trans Medical Imaging, Vol 1, 113-122.

[15] Vardi Y., Shepp L. A. and Kauffman L. (1985): A statistical model for positron emission tomography, J. Amer. Stat. Assoc, No 80, 8-37.

[16] Weir I. S. (1997): Fully Bayesian SPECT Reconstructions, JASA, 92, 49-60.

[17] Zimeras S. (1997): Statistical models in medical image analysis, Ph.D. Thesis, Leeds University, Department of Statistics.

Brill Academic Publishers
P.O. Box 9000, 2300 PA Leiden,
The Netherlands

*Lecture Series on Computer
and Computational Sciences*
Volume 4, 2005, pp. 619-622

# Convergence of a Colocated Finite Volume Scheme for the Incompressible Navier–Stokes Equations

**S. Zimmermann**[1]

Department of Applied Mathematics,
Centrale Lyon University,
36 avenue de Collongue
69134 Ecully, France

Received 5 August, 2005; accepted 16 August, 2005

*Abstract:* We study a finite volume scheme for the approximation of the incompressible Navier-Stokes equations in two dimensions. It uses a projection (fractional step) method to ensure the incompressibility constraint. The unknowns for the velocity and the pressure are colocated at the center of the cells of a triangular mesh. The discrete operators are consistent and satisfy a discrete inf-sup (Babŭska-Brezzi) condition. We state the stability and the convergence of the scheme, and give some numerical results.

*Keywords:* Navier-Stokes equations, colocated discretization, projection method, finite volume methods.

*Mathematics Subject Classification:* 65N12, 65N15, 76D05, 76M12.

*PACS:* 02.60.-x, 02.70.Fj, 47.11+j.

## 1   Introduction

Let $\Omega$ be a bounded polygonal set of $\mathbb{R}^2$. We consider the flow of an incompressible fluid in $\Omega$. We assume that the adimensional velocity field $\mathbf{u} : [0,T] \times \Omega \to \mathbb{R}^2$ and pressure $p : [0,T] \times \Omega \to \mathbb{R}$ are solution of the following incompressible Navier-Stokes equations

$$\mathbf{u}_t - \frac{1}{\mathrm{Re}}\Delta\mathbf{u} + (\mathbf{u}\cdot\boldsymbol{\nabla})\mathbf{u} + \nabla p = \mathbf{f}, \tag{1}$$

$$\mathrm{div}\,\mathbf{u} = 0, \tag{2}$$

$$\mathbf{u}|_{\partial\Omega} = 0, \tag{3}$$

$$\mathbf{u}|_{t=0} = \mathbf{u}_0. \tag{4}$$

In the above equations, $\mathbf{f}$ is an external force applied to the fluid, and $\mathbf{u}_0$ is the initial velocity. The Reynolds number is defined as $\mathrm{Re} = \frac{UL}{\nu}$, where $U$ and $L$ are characteristic scales for velocity and length, and $\nu$ is the viscosity of the fluid. There exists numerous computational methods for (1)-(4) using finite difference, finite element and finite volume schemes; see for example [2] and [3]. In particular, colocated finite volume schemes are useful because they easily take into account the coupling with other equations. Unfortunately, few mathematical results are available in that case. Let us cite [4] where a penalty method is used to ensure the incompressibility constraint (2), and [5] which uses a projection method to that end. We study in this paper a colocated finite volume scheme for (1)-(4) using a projection method as in [5]. However, we consider a triangular mesh, whereas in [5] the mesh is built up from squares.

---

[1]E-mail: Sebastien.Zimmermann@ec-lyon.fr

## 2 Discrete setting

Let $\mathcal{T}_h$ be a triangular mesh for $\Omega$, $\mathcal{E}_h$ be the set of faces. For each face $\sigma \in \mathcal{E}_h$, $K_\sigma$ and $L_\sigma$ are the neighbouring cells. For each cell $K \in \mathcal{T}_h$, $\mathbf{x}_K$ is the orthocenter of $K$. The parameter $h$ is the maximum diameter of the cells of the mesh. Let $L^2 = L^2(\Omega)$ be the usual Lebesgue space. We define the following spaces

$$P_0 = \{ v \in L^2 \; ; \; \forall K \in \mathcal{T}_h, \; v|_K \text{ is a constant} \}, \qquad P_0^0 = \left\{ v \in P_0 \; ; \; \int_\Omega v(\mathbf{x}) \, d\mathbf{x} = 0 \right\},$$

$$P_1^{nc} = \{ v \in L^2 \; ; \; \forall K \in \mathcal{T}_h, \; v|_K \text{ is affine} \; ; \; \forall \sigma \in \mathcal{E}_h, \; v \text{ is continous at the middle of } \sigma \},$$

and $\mathbf{P}_0 = (P_0)^2$. We note $\Pi_{\mathbf{P}_0} : L^2 \rightarrow \mathbf{P}_0$ and $\Pi_{P_1^{nc}} : L^2 \rightarrow P_1^{nc}$ the projection operators for the $L^2$ norm on the spaces $\mathbf{P}_0$ and $P_1^{nc}$ respectively.

The Navier-Stokes equations (1)-(4) use the following operators: gradient, divergence, laplacian. Their discrete counterpart for the spaces $P_0$ and $\mathbf{P}_0$ are defined as follows. The building of the discrete gradient operator $\nabla_h : P_0 \rightarrow \mathbf{P}_0$ uses an affine interpolation. For each face $\sigma$, a value is deduced using the values at $\mathbf{x}_{K_\sigma}$ and $\mathbf{x}_{L_\sigma}$ (figure 1). The discrete divergence operator $\text{div}_h : \mathbf{P}_0 \rightarrow P_0$ is the adjoint of the gradient operator. Lastly, let $D_\sigma$ be the diamond cell defined for each $\sigma$ with $\mathbf{x}_{K_\sigma}$ and $\mathbf{x}_{L_\sigma}$ (figure 1). The definition of the discrete laplacian operator $\tilde{\Delta}_h : P_0 \rightarrow P_0$ uses another gradient operator defined on $D_\sigma$. We note $\tilde{\Delta}_h = (\tilde{\Delta}_h)^2$. The second discrete laplacian operator is $\Delta_h = \text{div}_h(\nabla_h)$. The discrete operators $\nabla_h$ and $\tilde{\Delta}_h$ are consistent.

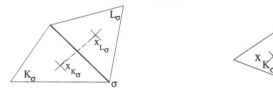

Figure 1: *Definition of the discrete gradient and laplacian.*

The Navier Stokes equations (1)-(4) use also the advective term $(\mathbf{u} \cdot \nabla)\mathbf{u}$. Its discrete counterpart is defined as follows. For each $\sigma \in \mathcal{E}_h$, let $\mathbf{n}$ be the normal of $\sigma$ pointing opposite to $K_\sigma$ (figure 4). We set

$$\mathbf{v}_\sigma = \begin{cases} \mathbf{v}_{K_\sigma} \text{ if } \mathbf{u}_{K_\sigma} \cdot \mathbf{n} \geq 0 \\ \mathbf{v}_{L_\sigma} \text{ if } \mathbf{u}_{K_\sigma} \cdot \mathbf{n} < 0 \end{cases} \qquad \text{or} \qquad \mathbf{v}_\sigma = \frac{1}{2}(\mathbf{v}_{K_\sigma} + \mathbf{v}_{L_\sigma}).$$

For the first choice, we use the cell ($K_\sigma$ or $L_\sigma$) located *upstream* of the local direction of the flow. For the second choice, we compute a mean value (*centered* scheme). The centered scheme is less stable than the upstream one, but it gives more accurate results. We then use the values $\mathbf{v}_\sigma$ to de-

Figure 2: *Definition of the discrete advection operator.*

fine a discrete advection operator $\mathbf{b}_h : \mathbf{P}_0 \times \mathbf{P}_0 \rightarrow \mathbb{R}^2$. It is consistent.

## 3   The scheme

The scheme uses a projection (fractional-step) method. At each time step we compute a first velocity field, without taking into account the incompressibility constraint (2). We then compute the pressure and recover a second velocity field that is divergence-free. We use a second-order BDF scheme for the time discretization.

The algorithm goes as follows. Let $N \in \mathbb{N}^*$. The time step is $k = \frac{T}{N}$. For all $m \in \{0, \ldots, N\}$, let $t_m = mk$. We start with $\mathbf{u}_h^0 \in \mathbf{P}_0$, $\mathbf{u}_h^1 \in \mathbf{P}_0$ and $p_h^0 \in P_0^0$, $p_h^1 \in P_0^0$. Then for all $n \in \{1, \ldots, N-1\}$, $(\tilde{\mathbf{u}}_h^{n+1}, \mathbf{u}_h^{n+1}, p_h^{n+1})$ is deduced from $(\tilde{\mathbf{u}}_h^n, \mathbf{u}_h^n, p_h^n)$ as follows.

- **First step:** $\tilde{\mathbf{u}}_h^{n+1} \in \mathbf{P}_0$ satisfies

$$\frac{3\,\tilde{\mathbf{u}}_h^{n+1} - 4\,\mathbf{u}_h^n + \mathbf{u}_h^{n-1}}{2\,k} - \tilde{\boldsymbol{\Delta}}_h \tilde{\mathbf{u}}_h^{n+1} + \mathbf{b}_h(2\,\mathbf{u}_h^n - \mathbf{u}_h^{n-1}, \tilde{\mathbf{u}}_h^{n+1}) + \nabla_h p_h^n = \Pi_{\mathbf{P}_0} \mathbf{f}(t_{n+1}). \tag{5}$$

- **Second step (projection):** $p_h^{n+1} \in P_0^0$ is given by

$$\Delta_h(p_h^{n+1} - p_h^n) = \frac{3}{2\,k} \operatorname{div}_h \tilde{\mathbf{u}}_h^{n+1} \tag{6}$$

and $\mathbf{u}_h^{n+1} \in \mathbf{P}_0$ is given by

$$\mathbf{u}_h^{n+1} = \tilde{\mathbf{u}}_h^{n+1} - \frac{2\,k}{3} \nabla_h(p_h^{n+1} - p_h^n). \tag{7}$$

The boundary conditions for equations (5)–(7) are included in the definition of the discrete operators. Note that the discrete velocity and pressure are both piecewise constant. The discrete unknowns are thus both located at the center of the cells of the mesh (figure 3).

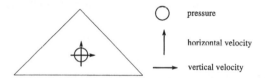

Figure 3: *Location of the discrete unknowns.*

## 4   Mathematical analysis

We have the following stability result

**Theorem 1** *Let us assume that there exists $K > 0$ such that $k \leq Kh^2$. Under suitable hypothesis on the mesh, the data $\mathbf{u}_0$, $\mathbf{f}$, and the initial values of the scheme $\mathbf{u}_h^0$, $\mathbf{u}_h^1$, $p_h^0$, $p_h^1$, there exists $C > 0$ such that for all $m \in \{2, \ldots, N\}$ we have*

$$|\mathbf{u}_h^m|^2 + k \sum_{n=0}^{m} |\Pi_{P_1^{nc}} p_h^n|^2 \leq C.$$

We now state the convergence result. Let $\varepsilon = \max(h, k)$. We define $\mathbf{u}_\varepsilon : \mathbb{R} \to \mathbf{P}_0$ as follows. For all $n \in \{0, \ldots, N-1\}$ and all $t \in [t_n, t_{n+1}]$ we set $\mathbf{u}_\varepsilon(t) = \mathbf{u}_h^n$; if $t \in ]-\infty, 0[\cup]T, \infty[$ we set $\mathbf{u}_\varepsilon(t) = 0$.

We then have

**Theorem 2** *Under the hypothesis of theorem 1, there exists a weak solution* $\mathbf{u}$ *of* (1)–(4) *such that*

$$\mathbf{u}_\varepsilon \to \mathbf{u} \quad in \ L^2(0,T;\mathbf{L}^2)$$

*as* $\varepsilon \to 0$.

## 5   Numerical Results

The scheme has been implemented for the classical lid-driven cavity problem. The fluid flow tends to a stationary state. We show below (figure 4) the level sets of the streamline function for Re = 1000 and Re = 5000. Here, the centered scheme has been used for the discrete advection term. The results are similar to those given by other methods (finite difference, finite element).

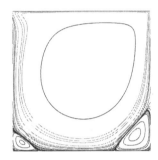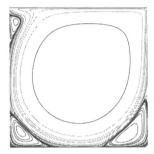

Figure 4: *Level sets of the streamline function for Re=1000 (left) and Re=5000 (right).*

## 6   Conclusion

We have introduced a new colocated finite volume method for the incompressible Navier-Stokes equations in two dimensions. We have proved its convergence, and given some numerical results. We can now generalize the scheme to incompressible fluids obeying to more complex equations, *e.g.* quasi-newtonian ones. Also, the colocated discretization makes it easy to build multilevel methods for the numerical simulation of turbulence.

## References

[1] J. Chorin, Numerical solution of the Navier Stokes equations, *Math. Comp.* **22** 745-762(1968).

[2] D. Kim and H. Choi, A second-order time-accurate finit volume method for unsteady incompressible flow on hybrid unstructured grids, *J. Comp. Phys.* **162** 411-428(2000)

[3] R. Temam: *Navier-Stokes equations: theory and numerical analysis.* North-Holland, 1984.

[4] R. Eymard, J. C. Latché and R. Herbin, Convergence analysis of a colocated finite volume scheme for the incompressible Navier Stokes equations on general 2D or 3D meshes, Preprint.

[5] S. Faure, Stability of a colocated finite volume scheme for the Navier Stokes equations, *Num. Methods P.D.E.* **21** 242-271(2005).

Brill Academic Publishers
P.O. Box 9000, 2300 PA Leiden
The Netherlands

*Lecture Series on Computer
and Computational Sciences*
Volume 4, 2005, pp. 623-624

# Electric (hyper)polarizability: From atoms and molecules to the nonlinear optics of materials.
## *A celebration of David Bishop's contribution*

George Maroulis

Department of Chemistry,
University of Patras,
University of Patras,
GR-26500 Patras, Greece

Electric polarizability is a universal characteristic of atomic and molecular systems. It is present as key element of the rational approach to wide spectrum of phenomena. From the analysis of fundamental spectroscopic observations to pharmacological studies and the nonlinear optics of materials, the reliable knowledge of electric polarizabilities is a necessary element of progress. The importance of these properties has prompted extensive theoretical investigations focusing on their nature and determination. In recent years, theory is not only in a position to support experimental investigations but has clearly shown that it can provide guidance to purpose-oriented experimental research. In this Symposium we honour a scientist that has worked consistently on the great project of an effective rapprochement between theory and experiment in the field of electric polarizability.

Eminent people have accepted to contribute the Symposium. An eloquent and moving presentation of the man and his work has been penned by **A.J.Thakkar**. This will be followed by a short talk by **R.J.Bartlett**, a top scientist and eminent specialist in the field. The Symposium also includes invited lectures, contributed papers and a Special Session on *Vibrational Polarizability* organiged by the invited speakers **M.G.Papadopoulos** and **B.Kirtman**.

A. Invited Lectures.
    **Benoit Champagne** will talk on the nonlinear optical properties of chiral species.
    **Bernard Kirtman** will present a paper on the polarization of quasilinear polymers in electric fields.
    **Manthos Papadopoulos** will read a paper on the electron correlation, relativistic and vibrational effects on electric polarizabilities and hyperpolarizabilities.
    **Mark Pederson** will about DFT calculations of molecular polarizabilities
    **Claude Pouchan**, in collaboration with D.Y.Zhang, D. Jacquemin and E. A. Perpète is presenting a paper on the hyperpolarizability of phosphorous and silicon containing oligomers.
    **Peter Schwerdtfeger** will talk on the polarizability on Ne clusters.
    **Gustavo E. Scuseria** is presenting a paper on the polarizability of single-walled carbon nanotubes.
    **Patrick Sennet** will talk on the polarizability of water clusters.

B. Contributed Papers
    This part of the Symposium includes papers by **Marina Balakina, Feng-Long Gu, Uwe Hohm, Ji-Kang Feng, George Maroulis, Akihiro Morita, Sourav Pal, Rafaele Resta, Hideo Sekino** and **Demetrios Xenides**.

C. Special Session on *Vibrational Polarizability*
    This part includes the papers of the organizers and contributions by **Benoit Champagne** (with **Milena Spassova** and **Bernard Kirtman**), **M. del Zoppo, Josep Luis, Olivier Quinet, Antonio Rizzo, Kenneth Ruud** and **Hajime Torii**.

## George Maroulis

**Greek**, born in 1953. His intellectual activities include History, Hittite and Mycenaean studies and Forgotten Religions. His passions include Arabic, Persian, Chinese and Japanese poetry. He has a certain taste for Epic poetry. He studied Chemistry in the National and Capodistriac University of Athens and obtained a PhD on Quantum Chemistry from the Catholic University of Louvain in 1981. In the early stage of his career he was fortunate enough to meet and work with people like David Bishop and Ajit Thakkar. His research field is Computational Quantum Chemistry. He teaches Quantum and Computational Chemistry in the Department of Chemistry of the University of Patras.

Brill Academic Publishers
P.O. Box 9000, 2300 PA Leiden,
The Netherlands

*Lecture Series on Computer*
*and Computational Sciences*
Volume 4, 2005, pp. 625-629

# David M. Bishop, the quintessential gentleman-scientist

**Ajit J. Thakkar**[1]

Department of Chemistry,
University of New Brunswick,
Fredericton, New Brunswick E3B 6E2,
Canada

Received 10 August, 2005; accepted 10 August, 2005

*Abstract:* A tribute to Professor David M. Bishop is presented together with a brief biographical sketch, and some personal reminiscences.

*Keywords:* biography, tribute

*PACS:* 01.60.+q

## 1   Biographical sketch

David Bishop was born in London during the short-lived reign of King Edward VIII. He was educated at various schools in the United Kingdom and graduated from Caterham School in Surrey. He obtained his B.Sc. in 1957 from University College London and stayed there under the tutelage of David Craig to earn a Ph.D. in 1960.

He then crossed the Atlantic Ocean to begin a post-doctoral position with Robert Parr at the Carnegie Institute of Technology, as it was known in those days. Bishop stayed on for a year after Parr left for Johns Hopkins University. In 1963, Bishop accepted an appointment at the University of Ottawa where he worked his way through the ranks, picking up a D.Sc. (1972) from the University of London along the way. He remains in Ottawa to this day, having reached the status of Emeritus Professor in 2000.

He has served on the editorial boards of several journals. In 1992 he was elected a Fellow of the Royal Society of Canada. He received the 2003 John C. Polanyi Lecture Award for excellence in research in theoretical and physical chemistry. He was honored by the publication of a special issue of the *Journal of Molecular Structure* (THEOCHEM) [Volume 633, Issues 2–3, 2003]. In 2005, we are gathered in Loutraki, Greece at this symposium to celebrate his scientific work.

## 2   Contributions to theoretical chemistry

His interest in theoretical chemistry has centered on its relationship to experimental work. His Ph.D. thesis, entitled "A Theory of Inductive Substitution in Some Aromatic Molecules", dealt with the anilinium ion represented as a benzene molecule perturbed by the $NH_3^+$ group. He got involved with the construction of single-center wave functions for molecules during his post-doctoral stint and pursued this, and his interest in group theory, during the 1960's.

In the 1970's he was absorbed with very accurate calculations on $H_2^+$ and $H_2$, many of them made in collaboration with Lap Cheung. The accuracy of the spectroscopic properties of $H_2^+$

---

[1] E-mail: ajit@unb.ca, Web page: http://www.unb.ca/chem/ajit/

Figure 1: David M. Bishop

came close to refining some of the fundamental constants. His $H_2$ calculations led, in part, to a re-assessment of the experimental dissociation energy of hydrogen.

In the 1980's his focus switched to electric response properties. Benchmark computations for $H_2^+$, $H_2$, He and Li followed. His highly accurate determination of the second hyperpolarizability of helium has frequently been used in standardizing gas-phase nonlinear optical experiments. He did pioneering work on vibronic effects on polarizabilities and hyperpolarizabilities.

In 1990, Bishop collected his thoughts on vibrational, rotational and vibronic effects on response properties in a paper in Reviews of Modern Physics that has since been quoted widely. He established the theoretical basis for a common empirical dispersion formula for electronic hyperpolarizabilities thereby allowing one to relate the results of different nonlinear optical processes. In collaboration with Bernard Kirtman, he developed efficient schemes for the computation of vibrational hyperpolarizabilities for larger molecules. Bishop then turned his attention to similar properties where a magnetic field is involved, in particular the Cotton-Mouton Effect and the Faraday Effect. Calculations were made of magnetizabilities, rotational g-factors, electric field gradients, Sternheimer shielding constants, and nuclear shielding constants, notably in association with Mark Cybulski and Janusz Pipin. Kirtman sparked Bishop's interest in long-chain polymers and infinite periodic systems.

David Bishop's numerous review articles [1, 3, 6, 13–18] give a more complete overview of his contributions to theoretical chemistry made in more than 230 publications. Many of these topics will be discussed at this symposium. A measure of the impact of Bishop's work is summarized in Table 1 which lists current citation counts for his most influential publications [1–12].

Table 1: David Bishop's most cited publications as of August 10, 2005

| Reference | Publication year | Citations |
|-----------|------------------|-----------|
| [1] | 1990 | 282 |
| [2] | 1991 | 180 |
| [3] | 1994 | 121 |
| [4] | 1992 | 121 |
| [5] | 1979 | 116 |
| [6] | 1998 | 115 |
| [7] | 1982 | 109 |
| [8] | 1973 | 105 |
| [9] | 1977 | 100 |
| [10] | 1965 | 98 |
| [11] | 1985 | 91 |
| [12] | 1980 | 82 |

David did not have many graduate students, preferring instead to work with post-doctoral fellows and young scientists starting their independent careers. He wrote a textbook on group theory [19] in an attempt to make the subject understandable without oversimplifying it. The success of his endeavor is reflected by the reissue of this book as a Dover paperback.

## 3 Reminiscences

This would be a sterilized tribute if I did not make an effort to give you an idea of David, the person. I think I first met David in 1974 at the 5th Canadian Symposium on Theoretical Chemistry in Ottawa. He was a cochairman of that meeting and I was a lowly first-year graduate student. Thus it is not surprising that I remember thinking of him as somewhat aloof. More interestingly, I also recall the banquet at which he seemed to have been charmed by the lovely classical guitarist, Liona Boyd, who was seated beside him at the head table.

I got to know him better over the next few years as I met him repeatedly at conferences and especially several regional meetings held at Queen's University, Kingston. He encouraged me to have patience and persist with theoretical chemistry during a period when faculty positions were rather difficult to obtain in Canada. Our academic paths crossed frequently in the late 1980s and early 1990s, and we saw more of each other. We shared an interest in highly accurate calculations on few-body systems, in polarizabilities, and in travel. Dinner with David was something I looked forward to at many conferences. In the late 1980s, David flirted with the idea of taking early retirement and devoting his time to an organization like Oxfam. However, he gave up this notion as his work on vibrational polarizabilities began to attract attention around the world.

David, the quintessential gentleman-scientist, is a private man who has his telephone number listed under a pseudonym. It has been my privilege to know him well enough to have seen the mischievous little boy behind the sartorial elegance and immaculate manners. We have shared in the scribbling of graffiti, the drinking of fine wine, gossiping up about the goings-on in academia, and the exchange of confidences about our private lives. It gives me great pleasure to wish David many more pleasant and productive years.

# References

[1] D. M. Bishop, Molecular vibrational and rotational motion in static and dynamic electric fields, Rev. Mod. Phys. 62 (1990) 343–374.

[2] D. M. Bishop, B. Kirtman, A perturbation method for calculating vibrational dynamic dipole polarizabilities and hyperpolarizabilities, J. Chem. Phys. 95 (1991) 2646–2658.

[3] D. M. Bishop, Aspects of non-linear-optical calculations, Adv. Quantum Chem. 25 (1994) 1–45.

[4] D. M. Bishop, B. Kirtman, Compact formulas for vibrational dynamic dipole polarizabilities and hyperpolarizabilities, J. Chem. Phys. 97 (1992) 5255–5256.

[5] D. M. Bishop, L. M. Cheung, Quadrupole moment of the deuteron from a precise calculation of the electric field gradient in $D_2$, Phys. Rev. A 20 (1979) 381–384.

[6] D. M. Bishop, Molecular vibration and nonlinear optics, Adv. Chem. Phys. 104 (1998) 1–40.

[7] D. M. Bishop, L. M. Cheung, Vibrational contributions to molecular dipole polarizabilities, J. Phys. Chem. Ref. Data 11 (1982) 119–133.

[8] D. M. Bishop, R. W. Wetmore, Vibrational spacings for $H_2^+$, $D_2^+$ and $H_2$, Mol. Phys. 26 (1973) 145–157.

[9] D. M. Bishop, L. M. Cheung, Calculation of transition frequencies for $H_2^+$ and its isotopes to spectroscopic accuracy, Phys. Rev. A 16 (1977) 640–645.

[10] D. M. Bishop, K. J. Laidler, Symmetry numbers and statistical factors in rate theory, J. Chem. Phys. 42 (1965) 1688–1691.

[11] D. M. Bishop, G. Maroulis, Accurate prediction of static polarizabilities and hyperpolarizabilities. A study on FH ($X^1\Sigma^+$), J. Chem. Phys. 82 (1985) 2380–2391.

[12] D. M. Bishop, L. M. Cheung, Dynamic dipole polarizability of $H_2$ and $HeH^+$, J. Chem. Phys. 72 (1980) 5125–5132.

[13] D. M. Bishop, Single-center molecular wavefunctions, Adv. Quantum Chem. 3 (1967) 25–59.

[14] D. M. Bishop, L. M. Cheung, Accurate one- and two-electron diatomic molecular calculations, Adv. Quantum Chem. 12 (1980) 1–42.

[15] D. M. Bishop, Effect of the surroundings on atomic and molecular properties, Int. Rev. Phys. Chem. 13 (1994) 21–39.

[16] C. Rizzo, A. Rizzo, D. M. Bishop, The Cotton-Mouton effect in gases: Experiment and theory, Int. Rev. Phys. Chem. 16 (1997) 81–111.

[17] D. M. Bishop, P. Norman, Calculations of dynamic hyperpolarizabilities for small and medium-sized molecules, in: H. S. Nalwa (Ed.) Handbook Adv. Electron. Photonic Mater. Devices 9 (2001) 1–62.

[18] B. Champagne, D. M. Bishop, Calculations of nonlinear optical properties for the solid state, Adv. Chem. Phys. 126 (2003) 41–92.

[19] D. M. Bishop, Group Theory and Chemistry, Oxford University Press, Oxford, 1973.

Brill Academic Publishers
P.O. Box 9000, 2300 PA Leiden
The Netherlands

*Lecture Series on Computer*
*and Computational Sciences*
Volume 4, 2005, pp. 630-632

# Anisotropic Continuum Model for the description of the Effect of Deformable Polarizable Polymer Matrix on the Nonlinear Optical Response of Incorporated Chromophores

M.Yu. Balakina.[1]

A.E. Arbuzov Institute of Organic and Physical Chemistry of Kazan Scientific Center,
Russian Academy of Sciences, Arbuzov str., 8, 420088, Kazan, Russia

Received 12 June, 2005; accepted in revised form 6 August, 2005

*Abstract:* In the present work the modeling of the effect of polymer matrix on the Nonlinear optical response of the incorporated chromophore is performed in the framework of the continuum approach, the anisotropy and deformability of the polarizable polymer matrix being taken into account. The chromophore is considered as embedded in the cavity of the ellipsoidal shape, the main axes of the cavity ellipsoid being oriented along the main axes of the characteristic ellipsoid of the generalized permittivity tensor depending on the applied electric field as well as on temperature and strain tensor of the medium; the ellipsoids are chosen to be conformal. Thus, the cavity inherits the symmetry of the permittivity tensor of the anisotropic medium in the external field. The electrostatic treatment of this system with the account of standard boundary conditions is performed. The analytical expression for the local electric field experienced by the chromophore in the anisotropic medium is appropriate for modeling the macroscopic response of the polymer.

*Keywords:* polymer materials, nonlinear optical response, chromophores, anisotropic medium, cavity, generalized dielectric permittivity

*PACS:* 31.70.Dk, 33.15.Kr, 77.84.J, 71.20.R, 82.35.E, 82.35.L

## 1. Introduction

New polymer materials with nonlinear optical (NLO) response to external electric field are of great interest due to their possible applications in photonics and optoelectronics [1]. For quadratic NLO response noncentrosymmetric materials are required, and polymers with incorporated NLO chromophores – conjugated molecules with large dipole moment and (hyper)polarizability, which are the microscopic origins of the NLO response of the material, - are poled by applying an electric field above the glass transition temperature, then appropriate measures are taken to preserve the orientational order of the chromophores. For anisotropic poled polymers the NLO response of the chromophores depends on the properties of the polymer matrix in the applied fields of various nature (electromagnetic, temperature field etc.) One of the principal requirements for the optimization of the NLO response of the material is the account of the effect of environment in the molecular design of the organic NLO chromophores. Thus, establishing the relationship between *micro*scopic hyperpolarizabilities of chromophores and *macro*scopic properties of the material as a whole is a principal problem in this field of research.

The effect of the environment is assumed to result in two manifestations: first, the medium changes the electronic distribution of the incorporated chromophore comparing to that in gas phase (the effect of the so-called static Reaction field); second, the dielectric medium changes the external electromagnetic fields which are experienced by the chromophore [2, 3]. The latter effect is conventionally described through the introduction of the Local Field factors (LFF). The expressions for LFF were obtained in classical electrostatics for pure liquid, whose molecules possess equal molecular volumes and have a spherical shape; besides, this pure liquid is considered as homogeneously polarizable [4]. The usual situation encountered in NLO experiments in condensed media, however, is

---

[1] Corresponding author. E-mail: marina@iopc.knc.ru

a binary system consisting in general of a solute with large and anisotropic polarizability immersed in a medium with considerably smaller polarizability. Besides, conventional approaches elaborated up to now are meant for the liquid phase system – a chromophore placed in a solvent, and are hardly appropriate for the account of the effect of polymer matrix on NLO properties of the chromophore. The need to overcome this difficulty resulted in the search for novel approaches, in particular a new model has been suggested to calculate optical and dielectric properties of polymer material from the microscopic structure and molecular response [5, 6]. The progress achieved in the investigation of NLO properties in the solid phase is reviewed in [7]. Here the new treatment including the locally anisotropic environment concept and its mathematical description is presented.

## 2. General Description of the Approach

The model suggested here considers polymeric matrix as homogeneous and intrinsically anisotropic medium, and is based on *the continuum approach*, considering the interaction between chromophore and the averaged field of the molecules composing the medium [4]. Continuum approach involves the concept of a *cavity* – a certain volume, surrounding the incorporated chromophore and inaccessible for other molecules. For the purpose of modeling, the most natural is the use of *ellipsoidal* cavity [8, 9] as more adequate than the simplified spherical one for the allocation of the elongated chromophore. However, according to our opinion, the most significant advantage of ellipsoidal form of the cavity is conditioned by the fact that the characteristic surface of symmetrical second rank tensor (strain tensor, permittivity and magnetic permeability tensors, etc.) is also represented by the elliptic surface [8, 9]:

$$\tilde{x}^i \; \tilde{x}^k \varepsilon_{ik}^{-1} = \frac{R_0^2}{\varepsilon_0}, \qquad \rightarrow \qquad \frac{x^{1\,2}}{\varepsilon_1} + \frac{x^{2\,2}}{\varepsilon_2} + \frac{x^{3\,2}}{\varepsilon_3} = \frac{R_0^2}{\varepsilon_0}$$

here the right formula is written in the coordinates $x^i$, in which $\varepsilon_{ik}$ becomes diagonal; $\varepsilon_i$ - eigenvalues of the permittivity tensor; $R_0$ is the effective size of the cavity.

The approach suggested here is based on the assumption that the cavity used for the modeling may be chosen conforming to one of such characteristic surfaces. This gives reason for the special *concept of the cavity*, inheriting the symmetry of the generalized permittivity tensor of anisotropic media in the external fields. According to this concept, we choose for a cavity an ellipsoid with main axes aligned to those of characteristic ellipsoid of the generalized permittivity tensor. The corresponding semi-axes lengths of the ellipsoids are required to be proportional to each other, the cavity size being appropriate for the incorporation of the chromophore molecule.

Dielectric properties of the medium are described by the generalized permittivity tensor $\varepsilon_{ik}(T,\sigma,E)$, defined according to [10] from constitutive equation coupling induction vector $D_i$ with electric field strength $E^k$ in the medium with temperature $T$ and strain $\sigma^{mn}$:

$$D_i = P_i^0(T,\sigma) + \varepsilon_{ik}(T,\sigma,E) \cdot E^k;$$
$$P_i^0(T,\sigma) \equiv p_i \cdot (T - T_0) + d_{imn} \cdot \sigma^{mn} + \dots$$
$$\varepsilon_{ik}(T,\sigma,E) \equiv \varepsilon_{ik}^0(T) + \chi_{ikl}^{(2)} \cdot E^l + \chi_{iklm}^{(3)} \cdot E^l E^m + \dots + Q_{ikmn} \cdot \sigma^{mn} + \tilde{Q}_{iklmn} \cdot E^l \cdot \sigma^{mn} + \dots$$

Here $P_i^0(T,\sigma)$ is the so-called spontaneous polarization, which is independent of the applied electric field [9, 10]; $p_i$ and $d_{imn}$ are the pyroelectric and the piezoelectric coefficients, respectively [9, 10]; $\chi_{ikl}^{(2)}$ and $\chi_{iklm}^{(3)}$ are nonlinear susceptibilities of the 2$^{nd}$ and 3$^{rd}$ orders [11]; the coefficients $Q_{ikmn}$ and $\tilde{Q}_{iklmn}$ describe the electrostriction effect [9, 10].

The solution of the corresponding electrostatic problem (for the ellipsoidal cavity inside the anisotropic medium in the framework of the above-mentioned *concept*, accounting the standard boundary conditions [4, 8]), gives the following expression for the electric field inside the cavity – the local field affecting the chromophore:

$$\vec{E}_{in} = 3 \cdot \left\{ \vec{e}_1 \frac{\varepsilon_1 E_\infty^1}{2\varepsilon_1 + \varepsilon_0} + \vec{e}_2 \frac{\varepsilon_2 E_\infty^2}{2\varepsilon_2 + \varepsilon_0} + \vec{e}_3 \frac{\varepsilon_3 E_\infty^3}{2\varepsilon_3 + \varepsilon_0} \right\}.$$

Here $\vec{e}_i$ are basic vectors of the coordinate system associated with the cavity (eigenvectors of the $\varepsilon_{ik}(T, \sigma, E_\infty)$ tensor); $\varepsilon_i$ - are eigenvalues of this tensor; $\varepsilon_0$ - dielectric constant of model isotropic medium inside the cavity; $E_\infty^i$ - are the components of the applied macroscopic electric field, having the magnitude $E_\infty$ at infinity. It is worth pointing that at $\varepsilon_1 = \varepsilon_2 = \varepsilon_3 = \varepsilon$ the expression is identical to the conventional one for the field inside the spherical cavity [4, 8].

Thus, when the macroscopic polymer sample is affected by the external fields of various nature (temperature gradients, strains etc.), this results in the change of the generalized permittivity tensor $\varepsilon_i(T, \sigma, E_\infty)$, being in fact the functions of temperature, strain and applied electric field, and since its eigenvalues enters into the equation for $\vec{E}_{in}$ one gets the interconnection between the external fields and the local field experienced by the chromophore.

## 3. Conclusions

The eigenvalues of the generalized permittivity tensor of anisotropic medium in the external fields, become the main set of guiding parameters of the model, linking micro- and macro- levels of the description. Other guiding parameters are the directional cosines of the angles between the laboratory ("macroscopic") frame of reference and eigenvectors of the $\varepsilon_{ik}(T, \sigma, E_\infty)$ tensor, as well as the orientation parameters of the chromophore with respect to the principal axes of the cavity ellipsoid. The dielectric constant $\varepsilon_0$ may be considered as an additional modeling parameter.

The detailed analysis and applications of the suggested concept to specific molecular systems are planned in the near future.

## References

[1]. Nonlinear Optical Effects and Materials, P. Guenter (Ed.), Springer, 2002.

[2]. D.M. Bishop, Effect of surroundings on atomic and molecular properties, *Int. Rev. Phys. Chem.* **13** (1994) 21-39.

[3]. D.M. Bishop, R. Wortman. Effective polarizabilities and local field corrections for nonlinear optical experiments in condensed media. *Journal of Chemical.Physics* **108** 1001-1007 (1998)

[4 ]. C.J.F. Boettcher. Theory of Electric Polarization, Elsevier, 1973, v.1, 375 p

[5]. R. W. Munn. Microscopic theory of linear and nonlinear optical and dielectric response in polymer materials. *Journal of Chemical.Physics* **114** 5404-5414 (2001)

[6]. A. Eilmes, R. W. Munn, V. G. Mavrantzas, D. N. Theodorou, A. Go´ra. Microscopic calculation of the static electric susceptibility of polyethylene. *Journal of Chemical.Physics* **119** 11458-11466 (2003)

[7]. B. Champagne, D.M. Bishop, Calculations of Nonlinear optical properties for the solid state. *Advances in Chemical Physics* **126** 41-92 (2003)

[8]. J.D. Jackson. Classical Electrodynamics. N.Y. Weinheim: Wiley, 1999, 808 p.

[9]. L. D. Landau, E. M. Lifchitz, L. P. Pitaevskii. Electrodynamics of Continuous Media.Butterworth Heinemann, Oxford, 1996.

[10]. J.F. Nye, Physical properties of Crystals, Oxford, 1957, 385 p.

[11]. A.D. Buckingham. Permanent and Induced Molecular Moments and Long Range Intermolecular Forces. *Advances in Chemical Physics* **12** 107-142 (1967)

Brill Academic Publishers
P.O. Box 9000, 2300 PA Leiden
The Netherlands

*Lecture Series on Computer
and Computational Sciences*
Volume 4, 2005, pp. 633-636

# *Ab initio* Investigation of Doping-effects on the Electronic and Vibrational Second Hyperpolarizabilities of Polyacetylene Chains

Benoît CHAMPAGNE [1], Milena SPASSOVA [1,2], and Bernard KIRTMAN [3]

[1] Laboratoire de Chimie Théorique Appliquée, Facultés Universitaires Notre-Dame de la Paix,
Rue de Bruxelles, 61, B-5000 Namur (BELGIUM)

[2] Institute of Organic Chemistry, Bulgarian Academy of Sciences, 1113 Sofia (BULGARIA)

[3] Department of Chemistry and Biochemistry, University of California, Santa Barbara, CA. 93106 (USA).

Received 11 July, 2005; accepted in revised form 17 August, 2005

*Abstract:* The effect of charging on the longitudinal electronic and vibrational second hyperpolarizabilities of PA chains is investigated *ab initio* by characterizing chains with and without an explicit alkali atom (Li, Na, K) as dopant.

*Keywords: nonlinear optical properties, ab initio, polyacetylene chains, doping effects*

*PACS: 33.15.Kr, 42.65.An*

## 1. Introduction

Applications in optoelectronics and photonics require materials exhibiting large second- and third-order nonlinear optical (NLO) responses. The third-order NLO responses are described at the macroscopic level by $\chi^{(3)}$, the third-order nonlinear susceptibility, and at the microscopic level by $\gamma$, the second hyperpolarizability. Several strategies based on a combination of experimental and theoretical investigations have been employed to design compounds presenting large $\chi^{(3)}$ and $\gamma$ values. Since large $\gamma$ values are found in organic $\pi$-conjugated compounds, these studies have addressed the effects of the nature of the conjugated segment (polyacetylene (PA) versus polypyrrole, polythiophene) and the presence of donor/acceptor groups at specific positions along the $\pi$-conjugated segment [1]. In the latter instance, one-dimensional (1-D) quadrupolar compounds (D-$\pi$-D and A-$\pi$-A and also D-$\pi$-A-$\pi$-D and A-$\pi$-D-$\pi$-A) have been compared to dipolar compounds (D-$\pi$-A) [2]. As in the case of second-order NLO responses [3], $\Lambda$-shaped two-dimensional chromophores could also constitute efficient means for enhancing the third-order NLO responses. Another approach consists in charging the chromophores by appropriate chemical or electrochemical dopings, which are known to modify the structures and properties of conjugated polymers [4]. The present paper summarizes some studies that have recently addressed the effects of doping on the second hyperpolarizability of PA chains by employing *ab initio* approaches. The calculations have been performed at the coupled-perturbed Hartree-Fock (CPHF) level of approximation [5], which is equivalent to a finite field approach based on Hartree-Fock energies/dipole moments. Most of the results were obtained using the 6-31G basis set. However, a basis set investigation, not reported here, has been carried out for a selection of small systems at both the Hartree-Fock and Møller-Plesset second-order (MP2) levels of approximation to assess the impact of including electron correlation effects.

## 2. Results and discussion

The formation of a soliton defect is accompanied by the separation of the PA chain into two phases having positive and negative bond length alternation (BLA), separated by a region where the BLA is zero. In the absence of a counterion, the soliton width attains seven CC double bonds but upon adding a counterion, the geometrical defect is delocalized over 6±1 CH units (HF/6-31G* results). In addition, in the presence of a counterion, the chain is bent (Fig. 1). When taking into account electron correlation at the MP2 level, the soliton width is slightly larger [6]. The soliton defect can also be characterized by the charge distribution. Although different charge definitions and levels of treatment provide different charge amplitudes and delocalizations [7], it is important to note they all show that the counterion pins the defect to a small region around the cation position and an excess charge on the chain of less than one (per alkali atom) [8].

Figure 1: Schematic representation of the $C_{23}H_{25}M$ chain doped by one alkali atom.

The static longitudinal electronic second hyperpolarizability [$\gamma_L^e(0)$] of PA chains doped with a single alkali atom (Li, Na, K) was investigated first. Fig. 2 displays the evolution of $\gamma_L^e(0)/N_C$ as a function of the PA chain length in comparison with the undoped case and with positively and negatively charged chains without a counterion. Whereas charging dramatically enhances $\gamma_L^e(0)$ of an isolated chain at intermediate chain lengths, the presence of an alkali atom counterion substantially reduces this effect. As the size of the alkali atom increases the hyperpolarizabilities approach those of the isolated chain [8]. At the same time, there is an increase in the equilibrium distance between the counterion and the chain; the chain deviates less from planarity; and the amount of charge transfer to the chain becomes closer to one electron. The behavior of $\gamma_L^e(0)$ is most simply explained in terms of a reduced pinning potential at longer counterion-chain distances. This was substantiated by performing calculations on chains bearing a negative soliton defect interacting with a partial positive charge of the same magnitude as the corresponding Mulliken charge [8]. At all chain lengths in this study [8] $\gamma_L^e(0)$ of PA is enhanced by alkali-doping. For $N_C = 50$, K-doping leads to an increase of about $9 \times 10^7$ a.u. in $\gamma_L^e(0)$, which more than doubles the undoped value for a similar chain length. This would be the effect of 2.0 mole % uniform doping assuming negligible soliton-soliton interaction. Unlike $\gamma_L^e(0)$ itself, $\gamma_L^e(0)/N_C$ can exhibit a maximum for sufficiently long chains. Such a maximum is seen in our calculations at about $N_C = 61$ for the isolated charged soliton, and the value of the hyperpolarizability per carbon atom at that point is over four times that of an infinite undoped (and unbent) PA chain. When the alkali dopant is taken into account the maximum is considerably reduced in height and shifts to larger $N_C$ than we have considered. In comparison with the maximum for the undoped species (which occurs at $N_C = \infty$) there is a small enhancement of $\gamma_L^e(0)/N_C$ for K-doping, but none for either Li- or Na-doping at the CPHF/6-31G level of theory.

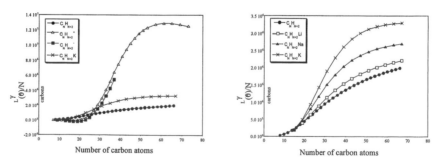

Figure 2: Evolution with chain length of the CPHF/6-31G $\gamma_L^e(0)/N_C$ (in a.u.) of neutral, charged (positive or negative), and alkali-doped polyacetylene chains.

The longitudinal vibrational second hyperpolarizabilities ($\gamma_L^v$) were estimated for non-resonant processes by combining the perturbation approach [9] – which provides a sum-over-modes expression for the $[\alpha^2]_L^0$ (Raman) term – and the finite field/nuclear relaxation (FF/NR) approach [10] for the $[\mu\beta]_L^0$ (hyperRaman) and $[\mu^2\alpha]_L^1$ (first-order anharmonic) terms. Our estimates are based on including only the lowest order (in anharmonicity) contribution of each type and using the infinite optical frequency approximation [10]. This treatment leads to expressions where static values of the different terms mentioned above are multiplied by fractional coefficients, which depend on the NLO process but not on the frequency:

DFWM: $\qquad \gamma_L^v(-\omega;\omega,-\omega,\omega)_{\omega\to\infty} = \frac{2}{3}[\alpha^2]_{L,\omega=0}^0$

dc-SHG: $\qquad \gamma_L^v(-2\omega;\omega,\omega,0)_{\omega\to\infty} = \frac{1}{4}[\mu\beta]_{L,\omega=0}^0$

dc-Kerr: $\qquad \gamma_L^v(-\omega;\omega,0,0)_{\omega\to\infty} = \frac{1}{3}[\alpha^2]_{L,\omega=0}^0 + \frac{1}{2}[\mu\beta]_{L,\omega=0}^0 + \frac{1}{6}[\mu^2\alpha]_{L,\omega=0}^1$

For DFWM and, especially, dc-SHG, calculations on intermediate length chains show a major increase in the vibrational hyperpolarizability of the isolated charged soliton as compared to undoped PA [8]. In fact, vibrational contributions for the charged soliton can be 10-100 times larger than the corresponding static electronic hyperpolarizability. Once again, however, the presence of an alkali counterion strongly reduces these effects although there remains a substantial enhancement due to doping for the DFWM. For dc-SHG, the vibrational contribution is of opposite sign to the electronic counterpart, which results in a decrease of the second hyperpolarizability amplitude (see Table 1 in the case of Na-doped chains). The dc-Kerr effect is different from DFWM and dc-SHG in that anharmonicity can make a significant contribution in the infinite optical frequency nuclear relaxation approximation. Indeed, for the Na-doped systems presented in Table 1, neglecting anharmonicity contributions leads to $\gamma_L^v(-\omega;\omega,0,0)_{\omega\to\infty}$ values of −1.58, -0.09, and 5.22 × $10^6$ a.u. for $C_{15}H_{17}Na$, $C_{19}H_{21}Na$, and $C_{23}H_{25}Na$, respectively. Because of the three separate contributions, that may partially cancel one another, it is more difficult to extrapolate the chain length dependence than for the other two processes we have examined. Further studies at longer chain lengths are thus needed to determine the enhancement possibilities for the dc-Kerr effect and to verify our conclusions for the DFWM and dc-SHG vibrational hyperpolarizabilities.

Table 1: CPHF/6-31G electronic and vibrational second hyperpolarizability terms (in $10^6$ a.u.) for Na-doped PA chains of different intermediate lengths. (columns 2-4) static quantities; (columns 5-8) dynamic vibrational second hyperpolarizability for the most common NLO processes determined within the infinite optical frequency approximation.

| $N_C$ | $\gamma_L^e(0)$ | $[\alpha^2]_{L,\omega=0}^0$ | $[\mu\beta]_{L,\omega=0}^0$ | $[\mu^2\alpha]_{L,\omega=0}^1$ | $\gamma_L^v(-\omega;\omega,-\omega,\omega)$ | $\gamma_L^v(-2\omega;\omega,\omega,0)$ | $\gamma_L^v(-\omega;\omega,0,0)$ |
|---|---|---|---|---|---|---|---|
| 15 | 3.03 | 8.38 | -8.74 | -17.1 | 5.58 | -2.18 | -4.45 |
| 19 | 8.88 | 21.07 | -14.18 | -27.6 | 14.03 | -3.55 | -4.71 |
| 23 | 18.66 | 40.09 | -16.35 | -26.3 | 26.68 | -4.11 | 0.75 |

The role of dopant concentration was subsequently studied, although only for the static electronic second hyperpolarizability [11]. Consistent with experimental work [12], uniform distributions of counterions were considered. For the dopant concentrations y = 1/5, 1/7, and 1/9, we examined the structures H-[$C_3H_3M$-$C_2H_2$]$_N$-$C_3H_3M$-H, H-[$C_3H_3M$-$C_4H_4$]$_N$-$C_3H_3M$-H, and H-[$C_3H_3M$-$C_6H_6$]$_N$-$C_3H_3M$-H, where N is a positive integer and M = Li, Na, K. The alkali atoms alternate above and below the plane of the PA backbone although, except for chain bending when all alkali atoms are on the same side, the conclusions change little.

For Li-doping there is a small increase in $\gamma_L^e(0)/N_C$ with respect to undoped PA (~ 10%) regardless of doping level and chain length. In the long chain limit the largest enhancement is attained for the lowest doping level (1/5). Fig. 3 shows a comparison between Li, Na, and K for y = 1/9. There is evidently a substantial increase in $\gamma_L^e(0)$ due to replacing Li by Na and, particularly, by K . For the longest chains (93 carbon atoms) the values of $\gamma_L^e(0)$ are in the ratio 1.00 : 1.99 : 4.83. Furthermore, at concentrations in the range 1/5 > y > 1/9 the value of $\gamma_L^e(0)$ increases for Na and K (as opposed to Li) as the doping level increases. A rather large maximum theoretical enhancement factor of ~ 60 occurs for K-doped chains at a length of 24 carbon atoms and doping level y = 1/7 (Fig. 3). In general, this factor diminishes with chain length until reaching its asymptotic value [11]. In the case of K-doping at a level y = 1/9, for example, it is approximately halved in going from 24 carbons to an (extrapolated) infinite chain value of 5.3. Since the charge per CH unit increases with dopant level it appears that the best enhancement strategy, besides optimizing the chain length, is to maximize the charge transferred to the chain and minimize the electrostatic interaction between dopant and chain. Further investigations should also consider the dopant concentration effects on the vibrational second hyperpolarizability.

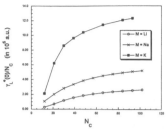 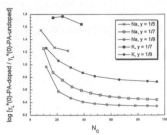

Figure 3: (left) CPHF/6-31G $\gamma_L^e(0)/N_C$ for different dopants at $y = 1/9$; (right) enhancement factor for CPHF/6-31G $\gamma$ of Na- and K-doped vs. undoped PA chains as a function of chain length.

## 3.    Outlook

Potentially important effects not considered thus far for doped PA chains include the interchain interactions [13] and disorder effects.   Then, a significant further enhancement of the second hyperpolarizabilities might be achieved by using dopants for which the hole created by electron charge transfer is more delocalized than it is for an alkali atom; this could lead to reduced pinning of the charge.   Another direction of investigation is to consider other polymers like polythiophene and polyparaphenylenevinylene, more stable and already used in different conducting polymer devices [4].

## Acknowledgments

This study results from a scientific cooperation established and supported by the Bulgarian Academy of Sciences, the Belgian National Fund for Scientific Research (FNRS) and the Commissariat Général aux Relations Internationales de la Communauté française Wallonie-Bruxelles.  This work has been financially supported by the Bulgarian Fund for Scientific Research under the project X-827. B.C. thanks the FNRS for his Research Director position.  The calculations have been performed on PCs installed in the Institute of Organic Chemistry of the Bulgarian Academy of Sciences as well as on the Interuniversity Scientific Computing Facility (ISCF), installed at the Facultés Universitaires Notre-Dame de la Paix (Namur, Belgium), for which the authors gratefully acknowledge the financial support of the FNRS-FRFC and the "Loterie Nationale" for the convention n° 2.4578.02, and of the FUNDP.

## References

[1]  See for instance B. Champagne and B. Kirtman, in *Handbook of Advanced Electronic and Photonic Materials and Devices*, edited by H.S. Nalwa, Vol. 9, *Nonlinear Optical Materials*, (Academic Press, San Diego, 2001), Chap. 2, p. 63.

[2]  E. Zojer, D. Beljonne, P. Pacher, and J.L. Brédas, Chem. Eur. J. **10**, 2668 (2004)

[3]  M. Yang and B. Champagne, J. Phys. Chem. A **107**, 3942 (2003).

[4]  See for instance A.J. Heeger, J. Phys. Chem. B **105**, 8475 (2001).

[5]  G.J.B. Hurst, M. Dupuis, and E. Clementi, J. Chem. Phys. **89**, 385 (1988).

[6]  B. Champagne and M. Spassova, Phys. Chem. Chem. Phys. **6**, 3167 (2004).

[7]  V. Monev, M. Spassova, and B. Champagne, Int. J. Quantum Chem. **104**, 354 (2005).

[8]  B. Champagne, M. Spassova, J.B. Jadin, and B. Birtman, J. Chem. Phys. **116**, 3935 (2002).

[9]  D.M. Bishop and B. Kirtman, J. Chem. Phys. **95**, 2646 (1991).

[10] D.M. Bishop, M. Hasan and B. Kirtman, J. Chem. Phys. **103**, 4157 (1995).

[11] M. Spassova, B. Champagne, and B. Kirtman, Chem. Phys. Lett., in press.

[12] N.S. Murthy, L.W. Shacklette, and R.H. Baughman, Phys. Rev. B **40**, 12550 (1989).

[13] B. Champagne and D.M. Bishop, Adv. Chem. Phys. **126**, 41 (2003).

Brill Academic Publishers
P.O. Box 9000, 2300 PA Leiden
The Netherlands

*Lecture Series on Computer*
*and Computational Sciences*
Volume 4, 2005, pp. 637-638

# New Conjugated Organic Systems With Large Vibrational Polarizabilities

*M. Del Zoppo, A. Lucotti, G. Zerbi*

Dept. Chimica, Materiali, Ing. Chimica; Politecnico di Milano, Piazza Leonardo da Vinci, 32 Milano
(ITALY)

Received 7 July, 2005; accepted in revised form 1 August, 2005

## 1. Introduction

The calculation of vibrational polarizabilities has been, in the last few years, the subject of active research among several groups.

Our approach is based on a very simple semiclassical model [1] which relates directly vibrational polarizabilities to spectroscopic observables, namely frequencies and intensities, which can be either calculated with quantum chemical programs or experimentally measured [2].

Thus, for example, it is possible to experimentally evaluate the vector component of the tensor, according to:

$$\beta_\mu^v \approx \beta_{xxx}^v \approx \frac{1}{4c} \sqrt{\frac{3}{\pi^5 N}} \sum_i \frac{\sqrt{A_i I_{i//,//}}}{\nu_i^2}$$

where it is supposed that only the $_{xxx}$ component is relevant (which is indeed the case in several classes of molecules with large $^v$ values); $A_i$ is the integrated absorbance of band i, $I_{i//,//}$ is the absolute Raman intensity obtained in an experiment where the incident and scattered light are polarized with mutually parallel electric field vectors, $_i$ are the vibrational frequencies and c is the sample concentration.

Or we can evaluate the average of the tensor $< ^v>$ according to:

$$< \gamma^v > = \frac{1}{15} \frac{1}{4\pi^2 c^2} \sum_i \frac{I_i}{\nu_i^2}$$

Although our method is approximated, for example it neglects anharmonicities which in some cases may be relevant, it has the great advantage to highlight in a simple and direct way which are the molecules able to yield large vibrational contributions.

The analysis carried out on a large variety of molecules has enabled the identification of the necessary requirements in order to have large vibrational polarizabilities. One of the most important factors is the existence of delocalized electrons whose presence guarantees the existence of large transition moments and low transition energies. Both these factors contribute to enhance Raman cross sections and infrared intensities. The latter benefits to a large extent also from the existence of intramolecular charge transfer. This is particularly important in order to have large vibrational second order polarizabilities ( ) as in the case of linear push-pull conjugated molecules [3].

Relevant examples of how this method has been used in order to "design" and synthetize new molecules that fulfil the requirements to obtain a large vibrational contribution to molecular polarizabilities, will be discussed. In particular, we will focus on photochromic molecules and polymers which can exist in two stable states (one much more conjugated than the other). As predicted, these two forms show large differences in their third order polarizability.

Moreover it will be shown how suitable chemical functonalization can be used in order to optimize the infrared intensity of these molecules and hence enhance the first and second order vibrational polarizabilities. Here quantum chemical calculations have proven to be of fundamental importance in suggesting the "route" to be followed by the chemists. The experimental results nicely confirmed the predicted conclusions.

Another interesting class of molecules is that of thiophene derivatives which have a "quinoid" character in the ground state and which have been studied together with their isoelectronic "aromatic" precursors.

A further important consequence of our approach is that it allowed us to point out that in some cases the vibrational and electronic polarizabilities have a similar behaviour. Namely, whenever there exists a molecular vibration which induces an oscillation of the molecular polarizability analogous to the charge redistribution induced by the transition to the most relevant excited states, the two contributions are somehow related. This means that similar values and similar trends in series of molecules in which a structural parameter (e.g. delocalization length, donor/acceptor strength, etc.) has been changed, are found. When this is not the case the vibrational contribution may be extremely large (e.g. cyanine-like molecules [4]) but completely unrelated to the electronic counterpart.

The idea of using vibrational spectroscopy as a guiding line in order to obtain large vibrational polarizabilities has recently suggested to exploit the intensity enhancement induced by the presence of metal nanoclusters. This approach lead to promising results which are presently under study and which certainly deserve a careful modelling [5].

## References

[1] C. Castiglioni, M. Gussoni, M. Del Zoppo, G. Zerbi, Solid State Comm., **82**, 13 (1992)
[2] C. Castiglioni, M. Tommasini, M. Del Zoppo, J. Mol. Struct., **521**, 137 (2000)
[3] M. Del Zoppo, C. Castiglioni, P. Zuliani, G. Zerbi, in "Handbook of Conducting Polymers", II Edition, (T. Skotheim Ed.), Dekker, New York (1998)
[4] A. Bianco, M. Del Zoppo, G. Zerbi, Synth. Met. **125**, 81 (2002)
[5] A. Lucotti, M. Del Zoppo, G. Zerbi, J. Raman Spectr. in press

Vi pe-VSP International
Science Publishers
P.O. Box 346, 3700 AH Zeist
The Netherlands

*Lecture Series on Computer
and Computational Sciences*
Volume 4, 2005, pp. 639-642

# Localization Scheme for the Elongation Method in the Presence of an Electric Field

Feng Long Gu [a,*], Bernard Kirtman [b], and Yuriko Aoki [a,c]

a) Department of Molecular and Material Sciences, Interdisciplinary Graduate School of
   Engineering Sciences, Kyushu University, 6-1 Kasuga-Park, Fukuoka, 816-8580, Japan
b) Department of Chemistry and Biochemistry, University of California, Santa Barbara,
   California 93106, USA
c) Group, PRESTO, Japan Science and Technology Agency (JST), Kawaguchi Center Building,
   Honcho 4-1-8, Kawaguchi, Saitama, 332-0012, Japan

Received 15 July, 2005; accepted in revised form 12 August, 2005

*Abstract:* The elongation method works in a localized molecular orbital basis and the localization quality is essential to the accuracy of the calculations. Recently, a new localization scheme has been developed based on regional molecular orbitals. This scheme is efficient and accurate even for covalently bonded and strongly delocalized systems. In the present study, our procedure is extended to a system in the presence of an external electric field in order to obtain static and dynamic (hyper)polarizabilities.

*Keywords:* elongation method, localization, perturbation theory

## 1. Introduction

Some time ago Imamura *et al* [1] proposed the Hartree-Fock (or DFT) elongation method for calculating electronic states of large quasi-one-dimensional aperiodic polymers. This method mimics experimental polymerization and copolymerization procedures. It builds up polymers by stepwise adding monomer units to a starting cluster.

In the elongation procedure, the first step is to localize the canonical molecular orbitals (CMOs) of a starting cluster into frozen and active regions. Next, a monomer is added to the active region of the cluster. The SCF problem is solved in this interactive region by disregarding the LMOs that have weak interaction with the attacking monomer. By repeating this procedure, a polymer is elongated step by step until it reaches the desired length. The advantage of this approach lies in the fact that one can avoid solving very large secular equations for large aperiodic systems. By working in the LMO representation one can also reduce the number of two-electron integrals that have to be evaluated. This results in a major reduction in computation time.

It is essential to have a localization scheme for the elongation method that is both reliable and efficient. A new scheme has been developed, [2] based on a density matrix approach, which yields regional localized molecular orbitals (RLMOs) that satisfy the above criteria. It has been demonstrated that the RLMO scheme works well even for highly delocalized systems. For the (hyper)polarizability calculations, one needs a localization scheme for the perturbed MOs. In section 2, the procedures to obtain unperturbed RLMOs are presented. In section 3, the localization scheme is described for a system perturbed by an external electric field.

## 2. Localization scheme for the elongation method

For a starting cluster containing $N$ monomer units, the canonical molecular orbitals (CMOs), $\varphi^{(0)CMO}$, are delocalized over the whole space, i.e.

---

* Email: gu@cube.kyushu-u.ac.jp

$$\varphi^{(0)\mathrm{CMO}} = C_{\mathrm{AO}}^{(0)\mathrm{CMO}} \chi^{\mathrm{AO}}$$

(1)

where vector $\chi^{\mathrm{AO}}$ is the entire set of atomic orbitals (AOs) and $C_{\mathrm{AO}}^{(0)\mathrm{CMO}}$ is the matrix of transformation coefficients from AOs to CMOs obtained by solving the Hartree-Fock-Roothaan equations in the conventional quantum chemistry approach. The superscript (0) is for the unperturbed quantities.

In the elongation procedure, the starting cluster is first divided into frozen and active regions. These regions are chosen bearing in mind that an attacking monomer is going to interact with the active region. Then, the CMOs of the cluster are split into two sets. One set is localized onto the frozen region while the other set is localized onto the active region. The elongation eigenvalue problem is solved on the basis of the active LMOs and the CMOs of the attacking monomer instead of the entire set of AOs as in conventional quantum chemistry calculations.

The density matrix in an atomic orbital (AO) basis is defined as:

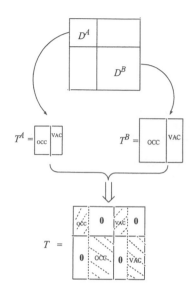

Figure 1. Construction of the transfor,mation matrix from OAO basis to RO basis.

$$D^{(0)\mathrm{AO}} = C_{\mathrm{AO}}^{(0)\mathrm{CMO}} d\, C_{\mathrm{AO}}^{(0)\mathrm{CMO}\dagger}$$

(2)

where $d$ is the diagonal occupation number matrix. For a restricted Hartree-Fock wavefunction the occupation number is 2 for doubly-occupied spatial orbitals or 0 for unoccupied orbitals. The orthonormal condition is given as,

$$C_{\mathrm{AO}}^{(0)\mathrm{CMO}\dagger} S^{\mathrm{AO}} C_{\mathrm{AO}}^{(0)\mathrm{CMO}} = 1$$

(3)

where $S^{\mathrm{AO}}$ is the overlap matrix in the AO basis and $1$ is the identity matrix. From Eqs. (2) and (3), one obtains the well-known idempotency relation in the non-orthogonal AO basis:

$$D^{(0)\mathrm{AO}} S^{\mathrm{AO}} D^{(0)\mathrm{AO}} = 2 D^{(0)\mathrm{AO}}.$$

(4)

This density matrix can be transformed to an orthogonal AO (OAO) basis by adopting Löwdin's symmetric orthogonalization procedure,[3] which gives an orthogonal basis that is least distorted from the original AOs. The transformation matrix $X$, obtained by diagonalizing the AO overlap matrix, gives

$$D^{(0)\mathrm{OAO}} = X D^{(0)\mathrm{AO}} X^{\dagger}$$

(5)

Using Eqs.(2), (5), and the fact that $X^{\dagger} X = X X^{\dagger} = S^{\mathrm{AO}}$ one can easily verify that

$$D^{(0)\mathrm{OAO}} D^{(0)\mathrm{OAO}} = 2 D^{(0)\mathrm{OAO}}$$

(6)

which is the idempotency relation in the OAO basis.

The desired RLMOs are obtained in two steps. In the first step, a regional orbital (RO) space is constructed by separately diagonalizing $D^{(0)\mathrm{OAO}}$ (A) and $D^{(0)\mathrm{OAO}}$ (B), which are the sub-blocks of $D^{(0)\mathrm{OAO}}$ with AOs corresponding to the A and B regions, respectively. Thus,

$$T^{(0)A^{\dagger}} D^{(0)OAO}(A) T^{(0)\mathrm{A}} = \lambda^{(0)A}$$

(7a)

and

$$T^{(0)B^{\dagger}} D^{(0)OAO}(B) T^{(0)\mathrm{B}} = \lambda^{(0)B}.$$

(7b)

The transformation from OAOs to ROs is given by the direct sum of $T^{(0)\mathrm{A}}$ and $T^{(0)\mathrm{B}}$.

$$T^{(0)} = T^{(0)\mathrm{A}} \oplus T^{(0)\mathrm{B}}$$

(8)

Figure 1 shows the matrices $D^{(0)\mathrm{OAO}}$ (A) and $D^{(0)\mathrm{OAO}}$ (B) and the schematic construction of the $T$ matrix. The corresponding eigenvalues are divided into three sets corresponding to ROs that are approximately doubly-occupied, singly-occupied and empty. The singly-occupied orbitals occur in pairs which are combined to form new ROs so that one RO is approximately doubly-occupied and the other is approximately empty. For a non-covalently bonded system there are only two sets of ROs, either doubly-occupied or empty. The resulting RO density matrix is

$$D^{(0)RO} = T^{(0)^\dagger} D^{(0)OAO} T^{(0)} \tag{9}$$

and, using Eqs. (2) and (5), the transformation coefficients from ROs to CMOs may be written as

$$C_{RO}^{(0)CMO} = T^{(0)^\dagger} X C_{AO}^{(0)CMO}. \tag{10}$$

From Eq.(6) and the unitary condition $T^{(0)} T^{(0)^\dagger} = T^{(0)^\dagger} T^{(0)} = 1$, one can confirm that

$$D^{(0)RO} D^{(0)RO} = 2 D^{(0)RO} \tag{11}$$

The second, and final, step is to carry out a unitary transformation between the occupied and unoccupied blocks of $D^{(0)RO}$ in such a way as to preserve the localization as much as possible. This is done using the same Jacobi procedure that is employed in ref. 4 to convert natural bond orbitals into localized molecular orbitals. If the resulting transformation is denoted by $\zeta^{(0)}$, then $D^{(0)RLMO}$ is diagonal and the only non-zero elements are equal to 2 (*cf.* Eq.(11)). The unitary transformation from CMO to RLMO orbitals is given as:

$$C_{RLMO}^{(0)CMO} = \zeta^{(0)^\dagger} T^{(0)^\dagger} X C_{AO}^{(0)CMO} \tag{12}$$

Finally, the transformation from the original AOs to RLMOs is given by,

$$C_{AO}^{(0)RLMO} = X^{-1} T^{(0)} \zeta^{(0)} \tag{13}$$

## 3. Localization scheme for the perturbed elongation method

For a system perturbed by a time-dependent electric field, $E(\omega,t)$, the same overall procedure can be followed as in the absence of a field except that each field-dependent quantity must now be expanded as a Taylor series. In solving the resulting time-dependent Hartree-Fock equations we utilize the non-canonical formulation of Karna and Dupuis (KD).[5] For the purposes of illustration, one may consider the case of a static field and the expansions:

$$D^{AO} = D^{(0)AO} + E D^{(1)AO} + E^2 D^{(2)AO} + \dots \tag{14a}$$

$$T^A = T^{(0)A} + E T^{(1)A} + E^2 T^{(2)A} + \dots \tag{14b}$$

$$\lambda^A = \lambda^{(0)A} + E \lambda^{(1)A} + E^2 \lambda^{(2)A} + \dots \tag{14c}$$

$$\zeta = \zeta^{(0)} + E \zeta^{(1)} + E^2 \zeta^{(2)} + \dots \tag{14d}$$

etc.

Inserting the expansions of Eqs.(14a) and (14b) into the analog of Eq.(7a) for the perturbed system gives the first order expression -

$$T^{(0)A^\dagger} D^{(0)OAO}(A) T^{(1)A} + T^{(0)A^\dagger} D^{(1)OAO}(A) T^{(0)A} + T^{(1)A^\dagger} D^{(0)OAO}(A) T^{(0)A} = \lambda^{(1)A} \tag{15}$$

$T^{(1)A}$ may be written as -

$$T^{(1)A} = T^{(0)A} \eta^{(1)A} \tag{16}$$

which yields

$$\lambda^{(0)A} \eta^{(1)A} + g^{(1)A} + \eta^{(1)A^\dagger} \lambda^{(0)A} = \lambda^{(1)A} \tag{17}$$

where $g^{(1)A} = T^{(0)A^\dagger} D^{(1)OAO}(A) T^{(0)A}$. Following the treatment in KD, the block-diagonal elements of $\eta^{(1)A}$ are set to zero while the non-diagonal blocks are obtained as

$$\eta_{ij}^{(1)A} = \frac{g_{ij}^{(1)A}}{\lambda_j^{(0)A} - \lambda_i^{(0)A}} \tag{18}$$

The $T^{(1)B}$ matrix can be obtained analogously and the total $T^{(1)}$ is the direct sum of $T^{(1)A}$ and $T^{(1)B}$,

$$T^{(1)} = T^{(1)A} \oplus T^{(1)B} \tag{19}$$

The first-order density matrix in the RO basis is then given by

$$D^{(1)RO} = T^{(0)^\dagger} D^{(0)OAO} T^{(1)} + T^{(0)^\dagger} D^{(1)OAO} T^{(0)} + T^{(1)^\dagger} D^{(0)OAO} T^{(0)} \tag{20}$$

In diagonalizing the field-dependent RO density matrix perturbatively we follow exactly the same procedure as in Eqs.(15)-(18). Hence, the first-order transformation matrix is expressed as

$$\zeta^{(1)} = \zeta^{(0)} \vartheta^{(1)} \tag{21}$$

and if $q^{(1)} = \zeta^{(0)^\dagger} D^{(1)RO} \zeta^{(0)}$, then

$$\vartheta_{ij}^{(1)} = \frac{q_{ij}^{(1)}}{\phi_j^{(0)} - \phi_i^{(0)}} \tag{22}$$

where the difference in eigenvalues $\phi_j^{(0)} - \phi_i^{(0)}$ is + or − 2. Finally, the first-order matrix for transformation from AOs to perturbed RLMOs is

$$C_{AO}^{(1)RLMO} = X^{-1} T^{(1)} \zeta^{(0)} + X^{-1} T^{(0)} \zeta^{(1)} \tag{23}$$

The treatment presented here may be readily generalized to higher-order and to frequency-dependent fields. Illustrative computations will be discussed.

## Acknowledgments

Financial support from the Research and Development Applying Advanced Computational Science and Technology of the Japan Science and Technology Agency (ACT-JST) and the Japanese Ministry of Education, Culture, Sports, Science and Technology (MEXT) are acknowledged.

## References

[1]  A. Imamura, Y. Aoki, and K. Maekawa, J. Chem. Phys. 95, 5419 (1991).
[2]  F. L. Gu, Y. Aoki, J. Korchowiec, A. Imamura, and B. Kirtman, J. Chem. Phys. 121, 10385 (2004).
[3]  P-O Löwdin, J. Chem. Phys. 18, 365 (1949).
[4]  A. E. Reed and F. Weinhold, J. Chem. Phys. 83, 1736 (1985).
[5]  S. P. Karna and M. Dupuis, J. Comput. Chem. 12, 487 (1991).

Brill Academic Publishers
P.O. Box 9000, 2300 PA Leiden,
The Netherlands

*Lecture Series on Computer
and Computational Sciences*
Volume 4, 2005, pp. 643-646

# Fuzzy electro-optics:
# Approximations of linear and non-linear electro-optical properties

Uwe Hohm[1]

Institut für Phyikalische und Theoretische Chemie
der TU Braunschweig, Hans-Sommer-Str. 10
D-38106 Braunschweig, Germany

Received 2 July, 2004; accepted in revised form 30 July, 2004

*Abstract:* In this contribution some relationships between linear, non-linear, and higher-order polarizabilities of atoms, molecules, and clusters are discussed. It is shown that some electro-optical properties can be estimated with reasonable accuracy from fairly simple models.

*Keywords:* Polarizability, Hyperpolarizability, Cluster

*PACS:* 32.10.Dk, 33.15.Kr, 42.65.An

## 1 Introduction

The precise experimental and theoretical determination of electro-optical properties of atoms, molecules, and clusters is a very active field of basic research. Despite the need for precise values heuristic considerations of relationships between various optical and other physico-chemical properties have a long history (see e.g. [1]). Interestingly this kind of research is still an active field where especially conceptual density-functional theory has shed a new light on these qualitative and semi-quantitative relations [2]. Here we only mention concepts interrelating polarizability, (critical) volume and ionization energy of atoms and molecules [3, 4], questions connected with the minimum polarizability principle in chemical reactions [5], or even the apparent mutual dependence of physical and chemical hardness [6, 7]. In this contribution we only want to stress on two points:

a) The relationships between the linear dipole-polarizability $\alpha$ and the second hyperpolarizability $\gamma$ of atoms and small molecules and

b) energetics of cluster formation and the intrinsic change of the linear dipole-polarizability $\alpha$ .

## 2 Non-linear optics

The dipole-polarizabilities are defined via a series expansion of the induced dipole moment $\vec{\mu}$ in terms of the electric field $\vec{E}$

$$\vec{\mu} = \hat{\alpha} \cdot \vec{E} + \frac{1}{2}\hat{\beta} : \vec{E}\vec{E} + \frac{1}{6}\hat{\gamma} \vdots \vec{E}\vec{E}\vec{E} + \dots \tag{1}$$

---

[1] E-mail: u.hohm@tu-bs.de

Although the expansion coefficients do not depend on each other it would be attractive to have a connection between the linear and non-linear dipole-polarizabilities. We concentrate on the second hyperpolarizability $\hat{\gamma}$ which is the first non-vanishing non-linear electro-optical property for all systems. Some approximations of $\gamma$ exist in the literature. However the validation of these estimates requires some care. Due to its strong frequency dependence only the zero-frequency limit of the electronic part $\gamma = \gamma^e(0; 0, 0, 0)$ and a suitable invariant has to be taken into account which in the case of atoms is given by $\gamma = \gamma_\parallel$ [9]. Some semi-empirical considerations [8] have led to the proportionality $\gamma \sim S(-3)S(-4)/S(-2)$ where e.g. the even Cauchy moments $S(-2k)$ are defined via a series expansion of the linear dipole-polarizability in terms of the squared frequency $\alpha(\omega) = \sum_k S(-2k)\omega^{2k}$ [10]. During the last decade more precise data for the second hyperpolarizability became available and our findings [8] require an update. It is interesting to note that the inclusion of more recent data results in a nearly linear relationship between $\gamma$ and the Cauchy moment $S(-4)$, see Fig. 1. In this plot results for 27 atoms and small molecules are included. $S(-4)$ are taken from experiment whereas $\gamma$ is obtained from both, experiment and theory. The observed proportionality

$$\gamma = \phi S(-4) \tag{2}$$

yields $\phi = 51.76(90)$ from a linear least-squares fit. This result might also be useful in considerations of the invariants of the incremental pair-polarizability where contributions from hyperpolarizabilities are discussed to give a small but essential contribution to e.g. collision-induced light scattering and the non-linear density dependence of the refractive index in the vapor phase [11].

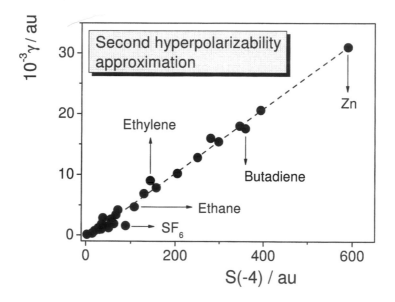

Figure 1: Second hyperpolarizability $\gamma$ as a function of the Cauchy moment $S(-4)$

# 3  Polarizabilities $\alpha$ of homonuclear clusters

For reactions of the type (atomization)

$$\Lambda_m B_n \ldots \;\longrightarrow\; mA + nB + \ldots \tag{3}$$

it was found that there is a linear relation between the change of the polarizability $\Delta\alpha = \sum_n \nu_i \alpha_i$, $\nu_i$ being the stoichiometric coefficient of the reactant, and the observed atomization energy $D$ [5]. A similar relation holds for the change of the cube-root of the polarizability, $\Delta\alpha_{CR} = \sum_n \nu_i \alpha_i^{1/3}$ where about 80 molecules were included in our earlier study. We also found that $\Delta\alpha_{CR}$ is positive irrespective of the molecule which lead to the formulation of a minimum polarizability principle in chemical reactions. Due to the vast amount of experimental and theoretical data on the polarizabilities of homonuclear clusters of type $X_n$ we now extend our studies to the atomization of clusters. In this case $\Delta\alpha_{CR}$ is given by

$$\Delta\alpha_{CR} = n \times \alpha(X)^{1/3} - \alpha(X_n)^{1/3}. \tag{4}$$

We observe two groups with similar behaviour: one is given by the clusters made of Na and Hg, the second group contains Si, S, Ge, Ni, Nb, and Al, see Fig. 2. As in the case of molecules for all clusters the important relation $\Delta\alpha_{CR} > 0$ holds. It is easy to show that molecules and clusters which show this linear dependence of $D$ on $\Delta\alpha_{CR}$ follow a polarizability minimum principle: The most stable isomer should be the least polarizable.

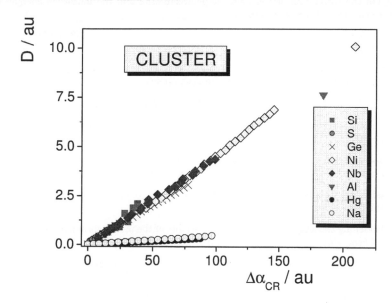

Figure 2: Atomization energy $D$ of different homonuclear clusters as a function of $\Delta\alpha_{CR}$, Eq.(4).

## 4  Final Remarks

Although concentrating on linear and non-linear dipole polarizabilities there is also the need for a precise determination of higher-order polarizabilities like dipole-quadrupole or dipole-octopole polarizability which might be of considerable importance in surface enhanced Raman spectroscopy [12]. Calculation and especially the experimental determination of these properties are extremely difficult and approximate methods are still in use. In this context we mention especially the geometric rules given by Luzanov [13]. The next step is connecting property derivatives like Raman scattering intensity with molecular softness or hardness as it was done only very recently by Torrent-Sucarrat et al. [14].

## References

[1] J. R. Partington, *An Advanced Treatise on Physical Chemistry, Vol. 4*. Longmans, London 1960.

[2] P. Geerlings, F. De Proft, and W. Langenaecker, Conceptual Density Functional Theory, *Chem. Rev.* **103** 1793-1874 (2003).

[3] P. Politzer, P. Jin, and J. S. Murray, Atomic polarizability, volume and ionization energy, *J. Chem. Phys.* **117** 8197-8202 (2002).

[4] B. Fricke, On the correlation between electric polarizabilities and the ionization energies of atoms, *J. Chem. Phys.* **84** 862-866 (1986).

[5] U. Hohm, Is there a minimum polarizability principle in chemical reactions?, *J. Phys. Chem. A* **1004** 8418-8423 (2000).

[6] W. Yang, R. G. Parr, and L. Uytterhoeven, New Relations between Hardness and Compressibility of Minerals, *Phys. Chem. Minerals* **15** 191-195 (1987).

[7] J. J. Gilman, Chemical and physical hardness, *Mat. Res. Innovat.* **1** 71-76 (1997).

[8] U. Hohm Empirical relation between linear and nonlinear dipole polarizabilities, *Chem. Phys. Lett.* **183** 304-308 (1991).

[9] D. P. Shelton and J. E. Rice, Measurements and calculations of the hyperpolarizabilities of atoms and small molecules in the gas phase, *Chem. Rev.* **94** 3-29 (1994).

[10] U. Fano and J. W. Cooper, Spectral Distribution of Atomic Oscillator Strengths, *Rev. Mod. Phys.* **40** 441-507 (1968).

[11] U. Hohm Modelling the long-range part of the invariants of the incremental pair polarizability tensor $\Delta\hat{\alpha}$, *Chem. Phys. Lett.* **211** 498-502 (1993).

[12] G. S. Kedziora and G. C.Schatz, Calculating dipole and quadrupole polarizabilities relevant to surface enhanced Raman spectroscopy, *Spectrochim. Acta A* **55** 625-638 (1999).

[13] A. V. Lutzanov and L.N. Lisetskii, Topological Model for Dipole-Quadrupole Polarizability and its Application to Analysis of Discriminating Intermolecular Interactions, *J. Struct. Chem.* **42** 544-549 (2001).

[14] M. Torrent-Sucarrat, F. De Proft, and P. Geerlings, Stiffness and Raman Intensity: a Conceptual and Computational DFT Study, *J. Phys. Chem. A* **109** 6071-6076 (2005).

Brill Academic Publishers
P.O. Box 9000, 2300 PA Leiden
The Netherlands

*Lecture Series on Computer
and Computational Sciences*
Volume 4, 2005, pp. 647-650

# Theoretical Investigation on Two-photon Absorption of $C_{60}$ and $C_{70}$

Ji-Kang Feng *[a,b], Xin Zhou[a]

a State Key Laboratory of Theoretical and Computational Chemistry, Institute of Theoretical
Chemistry, Jilin University, Changchun 130023, China
b College of Chemistry, Jilin University, Changchun 130023, China

Received 13 July, 2005; accepted in revised form 15 August, 2005

*Abstract:* We present a theoretical study on the two-photon absorption (TPA) properties of $C_{60}$ and $C_{70}$. The results show that the calculated TPA cross section peak of $C_{70}$ ($2757.5 \times 10^{-50}$ $cm^4 \cdot s/$ photon) is about 2.7 times as that of $C_{60}$ ($995.7 \times 10^{-50}$ $cm^4 \cdot s/photon$) Another interest phenomenon is that $C_{60}$ possesses the distinct TPA process in contrast to other conjugated molecules and $C_{70}$ in terms of three-state approximation.

*Keywords:* two-photon absorption, nonlinear optics, ZINDO-SOS, $C_{60}$, $C_{70}$

## 1. Introduction

Cage-like carbon clusters, fullerenes, are the subject of wide interdisciplinary interest in materials science. Amongst the various kinds of fullerenes, $C_{60}$ and $C_{70}$ are the most abundant and have been extensively studied [1, 2]. Many researchers have examined theoretically and observed experimentally the superconductivity, magnetic ordering, photophysical and excited-state kinetic properties and optical limiting properties [3-5]. However, less attention has been paid to the two-photon absorption (TPA) properties of $C_{60}$ and $C_{70}$. The TPA process considering here involves the simultaneous absorption of two photons, and the transition probability for absorption of two identical photons is proportional to $I^2$, where $I$ is the intensity of the laser pulse. The TPA phenomena of materials have now received more and more consideration for the potentially practical applications [6]. Most of the studied molecules possess one-dimensional structure or two-dimensional structure, but the researches on the TPA properties of three-dimensional fullerenes such as $C_{60}$ and $C_{70}$ are lacked. It is known that $C_{60}$ and $C_{70}$ exhibit large third-order nonlinear optics (NLO) response and the TPA cross section $\delta(\omega)$ of materials can be directly related to the imaginary part of the third-order optical susceptibility. According to this relationship, one can anticipate that $C_{60}$ and $C_{70}$ should have large TPA cross sections.

## 2. Theoretical Methods

TPA is one of NLO phenomena, and the process of it corresponds to simultaneous absorption of two photons. TPA cross section can be directly related to the imaginary part of the second hyperpolarizability $\gamma(-\omega; \omega, -\omega, \omega)$ by [6, 7]:

---

* Corresponding author. E-mail: jikangf@yahoo.com

$$\delta(\omega) = \frac{8\pi^2 \hbar \omega^2}{n^2 c^2} L^4 \, \text{Im} \, \gamma(-\omega; \omega, -\omega, \omega) \qquad (1)$$

where $\hbar$ is Planck's constant divided by $2\pi$, $n$ is the refractive index of medium, $c$ is the speed of light, $L$ is a local field factor (equal to 1 for vacuum).

The SOS expression to evaluate the components of the second hyperpolarizability $\gamma_{ijkl}$ can be induced out using perturbation theory and density matrix method. By considering a power expansion of the energy with respect to the applied field, the $\gamma_{ijkl}$ Cartesian components are given by [8, 9]:

$$\gamma_{ijkl}(-\omega_\sigma;\omega_1,\omega_2,\omega_3) = \frac{4\pi^3}{3h^3} P(i,j,k,l;-\omega_\sigma;\omega_1,\omega_2,\omega_3)$$

$$\left[ \sum_{m\neq o}\sum_{n\neq o}\sum_{p\neq o} \frac{\langle o|\mu_i|m\rangle\langle m|\bar{\mu}_j|n\rangle\langle n|\bar{\mu}_k|p\rangle\langle p|\mu_l|o\rangle}{(\omega_{mo}-\omega_\sigma-i\Gamma_{mo})(\omega_{no}-\omega_2-\omega_3-i\Gamma_{no})(\omega_{po}-\omega_3-i\Gamma_{po})} \right. \tag{2}$$

$$\left. - \sum_{m\neq o}\sum_{n\neq o} \frac{\langle o|\mu_i|m\rangle\langle m|\mu_j|o\rangle\langle o|\mu_k|n\rangle\langle n|\mu_l|o\rangle}{(\omega_{mo}-\omega_\sigma-i\Gamma_{mo})(\omega_{no}-\omega_3-i\Gamma_{no})(\omega_{no}+\omega_2-i\Gamma_{no})} \right]$$

$\Gamma_{mo}$ is the damping factor of excited state $m$. Considering the higher the excited state is and the shorter its lifetime is, the $\Gamma_{mo}$ is expressed as follows:[9]

$$\Gamma_{mo} = 0.08 \times \frac{\omega_{mo}}{\omega_{1o}} \tag{3}$$

To compare the calculated $\delta$ value with the experimental value measured in solution, the orientationally averaged (isotropic) value of $\gamma$ is evaluated, which is defined as

$$\langle\gamma\rangle = \frac{1}{15}\sum_{i,j}\left(\gamma_{iijj}+\gamma_{ijij}+\gamma_{ijji}\right) \quad i,j = x,y,z \tag{4}$$

Whereafter <$\gamma$> is taken into the Eq.(1), and then the TPA cross section $\delta(\omega)$ is obtained.

## 3. Results and Discussion

We have obtained the equilibrium geometry of $C_{60}$ and $C_{70}$ using B3LYP/6-31G method. For $C_{60}$, Sixty carbon atoms occupy 60 vertices of a truncated icosahedron, $R_{6-6}$=1.397Å, $R_{5-6}$=1.4601Å, in excellent agreement with the neutron powder diffraction experiment [10]. As regard to $C_{70}$, its optimized geometric parameters are in accord with the values in literature [11].

According to expressions (1)-(4), we compiled a program to calculate the third-order optical susceptibility $\gamma$ and the TPA cross section $\delta(\omega)$. The absorption of light by matter is a consequence of the interaction of an electromagnetic field with optically induced electric dipoles in a molecule. For the molecule like $C_{60}$ that has central symmetry, a change in the parity between the initial and final states (wave functions) is required for every photon involved in the transition for electric dipole transitions. Thus the selection rule for TPA is different from that of OPA. One change of parity is required for a one–photon transition, while two–photon transitions must have initial and final states with the same parity. However, for the molecule without a center of inversion symmetry such as $C_{70}$, every state is of mixed parity and hence all electronic states involving any number of photons are allowed. For $C_{60}$, at 518 nm, the TPA cross section gives a maximum, 995.7 $\times 10^{-50}$ cm$^4$·s/photon (as shown in Table 1). The result is in good agreement with the observed result of literatures [12] and [13]. The states corresponding to the peak TPA cross section have $^1H_g$ symmetry, which is in line with the standpoint mentioned in literature [13]. With regard to $C_{70}$, the TPA spectrum is much more complicated than that of $C_{60}$. There are four TPA peaks compared with only one TPA peak in the TPA spectrum of $C_{60}$. In the four peaks of TPA spectrum of $C_{70}$, the maximum TPA cross section value is $2757.5\times10^{-50}$ cm$^4$·s/photon at 408.6 nm and the corresponding excited state is $1A_2''$ symmetry. This maximum TPA cross section of $C_{70}$ is about 2.7 times larger than that of $C_{60}$.

It is interesting to point out another difference between $C_{60}$ and $C_{70}$ on the TPA process. For ordinary conjugated molecules, the position and relative strength of the two-photon resonance are to be predicted using the following simplified form —three-state approximation of the SOS expression [6]. This formula can considerably facilitate a search and design of new molecules with strongly enhanced TPA, because it is a simple relationship between the TPA cross section value and some usual linear absorption parameters. For ordinary conjugated molecules with large TPA cross sections, the intermediate state of TPA (namely the final state of OPA) should locate lower than the final state of TPA. However, for $C_{60}$, the intermediate state of TPA corresponding to the first strong OPA peak appears to be higher state than the final state of TPA (displayed in Figure 1 (a)). As to $C_{70}$, the similar TPA process to the ordinary conjugated molecules with large TPA cross sections is exhibited and shown in Figure 1 (b), that is, the TPA intermediate state (the first strong OPA peak) lies lower than the final state of TPA.

## 4. Conclusion

In this paper, we have performed the theoretical research on the TPA properties of $C_{60}$ and $C_{70}$. The calculated TPA cross section peak of $C_{70}$ is about 2.7 times as that of $C_{60}$, which is in line with results in literatures. It is notable that, the TPA processes of both are entirely different on the basis of the concept of three-state approximation of the SOS expression.

## Acknowledgments

This work is supported by the National Nature Science Foundation of China (20273023) and the Key Laboratory for Supramolecular Structure and Material of Jilin University.

Table 1: Two-photon absorption wavelengths, second hyperpolarizabilities γand TPA cross–sections δ(ω).

| final states | Sym. | $\lambda^{(2)}_{max}$ (nm) | $\gamma \times 10^{-34}$ (esu) | $\delta(\omega) \times 10^{-50}$ (cm$^4$·s/photon) |
|---|---|---|---|---|
| $S_1$-$S_3$ | $^1T_{1g}$ | 1077.6 | -10.62-1.181i | -20.99 |
| $S_4$-$S_6$ | $^1T_{2g}$ | 1068.0 | -10.64-1.179i | -21.36 |
| $S_7$-$S_{10}$ | $^1G_g$ | 1044.2 | -10.69-1.172i | -22.19 |
| $S_{11}$-$S_{15}$ | $^1H_g$ | 939.0 | -11.28-0.7089i | -16.60 |
| $S_{35}$-$S_{39}$ | $^1H_g$ | 660.2 | -13.41-0.1562i | -7.403 |
| $S_{40}$-$S_{42}$ | $^1T_{1g}$ | 659.0 | -13.49-0.1631i | -7.759 |
| $S_{43}$-$S_{45}$ | $^1T_{1g}$ | 658.0 | -13.56-0.1749i | -8.345 |
| $S_{51}$-$S_{54}$ | $^1G_g$ | 641.6 | -14.19-0.7382i | -37.03 |
| $S_{95}$-$S_{97}$ | $^1T_{2g}$ | 534.0 | -12.53+6.664i | 482.6 |
| $S_{98}$-$S_{101}$ | $1G_g$ | 533.6 | -12.58+6.845i | 496.5 |
| **$S_{102}$-$S_{106}$** | **$^1H_g$** | **518.0** | **-19.97+12.94i** | **995.7** |
| $S_{107}$-$S_{109}$ | $^1T_{1g}$ | 515.6 | -23.04+12.44i | 966.4 |
| $S_{110}$-$S_{113}$ | $^1G_g$ | 515.0 | -23.52+12.28i | 956.2 |
| $S_{114}$-$S_{116}$ | $^1T_{2g}$ | 514.4 | -24.06+12.01i | 948.3 |
| $S_{117}$-$S_{121}$ | $^1H_g$ | 495.2 | -28.98+3.998i | 336.7 |
| $S_{131}$-$S_{135}$ | $^1H_g$ | 467.8 | -32.14+2.339i | 220.7 |
| $S_{136}$-$S_{139}$ | $^1G_g$ | 466.6 | -32.54+2.131i | 202.1 |
| $S_{147}$-$S_{149}$ | $^1T_{2g}$ | 459.6 | -34.32+0.276i | 26.99 |
| $S_{153}$-$S_{155}$ | $^1T_{2g}$ | 453.8 | -34.95-1.323i | -132.7 |
| $S_{156}$-$S_{159}$ | $^1G_g$ | 451.6 | -35.08-1.823i | -184.2 |
| $S_{169}$-$S_{173}$ | $^1H_g$ | 446.8 | -35.29-2.684i | -277.7 |
| $S_{174}$-$S_{176}$ | $^1T_{2g}$ | 446.0 | -35.34-2.802i | -290.9 |
| $S_{177}$-$S_{179}$ | $^1T_{2g}$ | 442.4 | -35.54-3.252i | -343.1 |
| $S_{180}$-$S_{183}$ | $^1G_g$ | 440.4 | -35.70-3.462i | -368.7 |
| $S_{184}$-$S_{188}$ | $^1H_g$ | 435.2 | -36.24-3.942i | -429.8 |
| $S_{192}$-$S_{196}$ | $^1H_g$ | 413.8 | -40.58-5.615i | -677.2 |

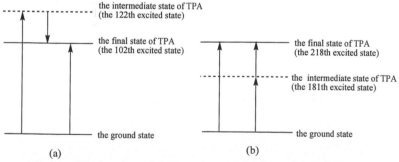

Figure 2: The sketch map of the two photon absorption of (a) $C_{60}$ and (b) $C_{70}$

## References

[1] H.W.Kroto, J.R.Heath, S.C.O'Brien, R.F.Curl, R.E.Smalley, Nature 318 (1985) 162.

[2] W.Krätchmer, L.D.Lambd, K.Fostiropoulos, D.R.Huffman,Nature 347 (1990) 354.

[3] J.N.Arvogast, A.P.Darmanyan, C.S.Foote, Y.Rubin, F.N.Diedelich, M.M.Albarez, S.J.Anz, R.L.Whetten, J.Phys.Chem.95 (1991) 11.

[4] S.D.Sibley, S.M.Argentine, A.H.Francis, Chem.Phys.Lett.188 ( 1992) 187.

[5] C.V.Bindhu, S.S.Harilal, V.P.N.Nampoori, C.P.G.Vallabhan, Appl.Phys.B 70 (2000) 429.

[6] M.Albota, D.Beljonne, J.L.Brédas, J.E.Ehrlich, J.Fu, A.A.Heikal, E.Hess, T.Kogej, M.D.Levin, S.R.Marder, D.McCord-Maughon, J.W.Perry, H.RÖeckel, M.Rumi, G.Subramaniam, W.W.Webb, X.Wu, C.Xu, Science 281 (1998) 1653.

[7] C.L.Caylor, I.Dobrianow, C. Kimmr, R.E. Thome, W. Zipfel, W.W. Webb, Phys. Rev. E. 59(4) (1999) R3831.

[8] B.J. Orr,; J.F.Ward, Mol.Phys. 20 (1971) 513.

[9] D. Beljonne, J. Cornil, Z. Shuai, J.L. Bredas, F. Rohlfing, D.D.C. Bradlley, W.E. Torruellas, V. Ricci, G.I. Stegeman, Phys.Rev.B 55 (1997) 1505.

[10] W.I.F.David, R.M.Ibberson, J.C.Mattewman, K.Prassides, T.J.S.Dennis, J.P.Hare, H.W.Kroto, R.Taylor, D.R.M.Walton, Nature 12 (1991) 147.

[11] J.Baker, P.W.Fowler, P.Lazzeretti, M.Malagoli, R.Zanasi, Chem.Phys.Lett. 184 (1991) 182.

[12] S.Couris, E.Koudoumas. A.A.Ruth, S.Leach, J.Phys.B:At. Mol. Opt. Phys. 28 (1995) 4537.

[13] Y.Wang, L.Cheng, J.Phys.Chem. 96 (1992) 1530.

Brill Academic Publishers
P.O. Box 9000, 2300 PA Leiden
The Netherlands

*Lecture Series on Computer
and Computational Sciences*
Volume 4, 2005, pp. 651-654

# Treatment of Vibrational Contribution to Resonant and Non-Resonant Optical Properties for Molecules with Large Anharmonicity

Josep M. LUIS[1], Miquel TORRENT-SUCARRAT[1], Nicolas S. BONNESS[3],
David M. BISHOP[2], and Bernard KIRTMAN[3]

[1] Institute of Computational Chemistry and Department of Chemistry, University of Girona, Campus de Montilivi,
17071 Girona, Catalonia, (SPAIN)

[2] Department of Chemistry, University of Ottawa, Ottawa, K1N 6N5 (CANADA)

[3] Department of Chemistry and Biochemistry, University of California, Santa Barbara, CA. 93106 (USA).

Received 11 July, 2005; accepted in revised form 8 August, 2005

*Abstract:* A variational approach for calculating non-resonant vibrational linear and nonlinear optical properties of molecules with large electrical and/or mechanical anharmonicity is presented. This approach utilizes the 'exact' numerical field-dependent potential energy surface and is based on a self-consistent solution of the vibrational Schrödinger equation. Vibrational correlation is introduced through second-order Møller-Plesset perturbation theory. In addition, we discuss a simple new method for evaluating Franck-Condon factors, which govern the intensity of vibronic transitions in linear and nonlinear optical spectra. Both intra-mode and inter-mode anharmonic interaction terms are taken into account by means of a second-order perturbation treatment analogous to the Bishop-Kirtman method used for non-resonant properties. Our procedure is used to simulate photoelectron spectra and it is shown that molecules with more than 20 normal modes are easily handled due to an efficient vibrational basis set truncation algorithm and the deletion of unimportant modes.

*Keywords:* linear and nonlinear optical properties, vibrational Schrödinger equation, Franck-Condon factors, mechanical anharmonicity.

*PACS:* 33.15.Kr, 42.65.An

## 1.    Introduction

The theoretical prediction of nonlinear optical (NLO) properties could have a key role in the design of new materials with applications in communications, medicine, optical computers, and holography. At the microscopic level the low-order NLO properties are governed by the first and second hyperpolarizabilities ($\beta$ and $\gamma$). Within the Born-Oppenheimer approximation the hyperpolarizability can be split into electronic and vibrational contributions, with the latter often being as large as, or even larger than, the former [1].

Bishop and Kirtman (BK) have developed a perturbation treatment of vibrational hyperpolarizabilities at non-resonant frequencies [2]. In zeroth-order the BK perturbation treatment utilizes the harmonic approximation for the field-free potential energy surface (PES) and assumes that the electrical properties depend linearly on the normal modes. Thus, in the case of highly anharmonic systems, or even some ordinary $\pi$-conjugated NLO molecules, it is not entirely unexpected that the double perturbation series in mechanical and electrical anharmonicity may be divergent either initially [3] or in higher-order. There exists an alternative to the perturbation approach for the static and most important dynamic nonlinear processes, which is based on finite field (FF) geometry relaxation [4]. Because of computational difficulties previous FF calculations have relied on approximations that are valid only when the anharmonicity is not too large. In the present paper we summarize a recent study that represents the first attempt to extend this procedure to the highly anharmonic case using modern variational treatments of the vibrational Schrödinger equation which parallel those of electronic structure theory [5].

Within the Born-Oppenheimer approximation, the leading term in the expression for the spectral intensity is proportional to the square of the vibrational overlap integral, also known as the Franck-

Condon factor (FCF), between the initial and final states. There are, of course, smaller contributions to the intensity due to higher-order terms governed by Herzberg-Teller factors, but we focus here on FCFs. Since the equilibrium geometry and potential energy surface (PES) of the initial and final electronic states are different the 3N-6 (or 3N-5 for linear molecules) dimensional FCF overlap integrals cannot be factored into a simple product of one-dimensional normal mode integrals. A wide variety of methods, based on the harmonic approximation for the PES of both the initial and final states, have been presented to solve this problem. Most of them are based on the generating function approach of Sharp and Rosenstock [6], or on the recursion relations of Doctorov *et al* [7]. Including vibrational anharmonicity always implies a strong increase in computational cost and, for most methods, the formulation becomes much more complex as well. This explains why studies in the literature that include anharmonicity are limited in practice to triatomic molecules [8], or assume separability of the normal modes [9]. We summarize our recent work on a new method for evaluating FCFs [10],[11] which is more broadly applicable. Results will be shown that illustrate the feasibility of this procedure for evaluating FCFs of a polyatomic molecule including the effect of non-diagonal (i.e. mode-mode) anharmonic couplings and of Duschinsky rotations.

## 2. Variational vibrational hyperpolarizabilities

The vibrational self-consistent field treatment (VSCF) [12] is a variational method to solve the vibrational Schrödinger equation which accounts for intra-modal anharmonicity 'exactly' and treats anharmonic mode-mode coupling within a mean field approximation. Thus, the vibrational wavefunction is written as a simple product of single-mode anharmonic vibrational wavefunctions, which are obtained by solving vibrational Hartree-type equations:

$$\left(-\frac{1}{2}\frac{\partial^2}{\partial Q_i^2} + \overline{V_i^0}(Q_i)\right)\phi_i^0(Q_i) = \varepsilon_i^0\phi_i^0(Q_i) \qquad \forall i \qquad (1)$$

Each mode vibrates in the average potential generated by all other modes. By definition there is no correlation between modes. The simplest method for introducing such correlation is vibrational Møller-Plesset perturbation theory (VMP) [13]. The VMP1 energy, which is the sum of the zeroth- and first-order terms, is just the VSCF energy. Thus, the first correction to the VSCF energy is given by a second-order Møller-Plesset perturbation treatment, i.e VMP2.

The vibrational contribution to the static hyperpolarizability may be calculated by evaluating the change in the field-dependent zero-point vibrational averaged energy ($E^{zpva}$) associated with the shift from the field-free to the field-dependent equilibrium geometry. In order to calculate the field-dependent $E^{zpva}$ using the VSCF and VMP2 procedures we linked the VSCF code implemented in GAMESS to the GAMESS subroutines that compute the field-dependent electronic energy. The VSCF and VMP2 calculations were performed an 'exact' numerical *ab initio* PES. In order to provide a reference point for the analysis of our results we also performed full vibrational CI (FVCI) and BK (PT2) calculations. However, in our FVCI and BK calculations the ab initio PES was truncated at the quartic terms in a normal coordinate expansion.

Our new approach has been applied to three molecules: $H_2O$, HOOH, and HSSH [5]. As expected beforehand anharmonic effects are small for $H_2O$ and larger for the other two molecules. $H_2O$ is nearly harmonic for all properties; HOOH is quite anharmonic, particularly for (hyper)polarizabilities; and HSSH lies in between. Our results indicate when the commonly used Bishop-Kirtman perturbation procedure can be expected to break down and variational methods are likely to be required. BK perturbation method is a reliable procedure for treating the effect of the anharmonicity for all properties of $H_2O$ as well as for $\mu$ and $\alpha$ of HSSH. However, the remaining cases involve large anharmonicity and the BK perturbation method breaks down. In highly anharmonic situations the new variational approach is required to ensure reliable values.

VSCF hyperpolarizabilities turn out to be a reasonable starting point even for large anharmonicity [5]. Nonetheless, for quantitative results in the latter case vibrational correlation corrections are necessary. VMP2 works well as long as the anharmonic coupling terms are not too large as they are for the vibrational hyperpolarizabilities of HOOH. In such highly anharmonic situations a higher level method like vibrational coupled clusters (VCC) or VCI is required. One of our future goals is to develop and implement the necessary methodology to compute vibrational contributions to NLO properties using these higher level methods.

The effect of terms not included in the quartic PES is roughly comparable to, or smaller than, the effect of vibrational correlation. These terms are automatically taken into account by using an 'exact' numerical *ab initio* PES as implemented in GAMESS. We also investigated the importance of 3-mode coupling. It turns out to be small, though non-negligible, when anharmonicity is very important as in the vibrational second hyperpolarizability of HOOH.

## 3. New approach for evaluating Franck-Condon factors

The starting point to derive our approach to calculating FCFs is just the Schrödinger equation for nuclear motion in the ground and excited electronic states [10]. Taking into the fact that: (i) the vibrational Hamiltonian is Hermitian, (ii) the vibrational eigenfunctions for the excited electronic state form a complete set, and (iii) the nuclear kinetic energy operator is the same for both the ground and excited electronic states, one can obtain the following set of homogenous linear simultaneous equations [10]:

$$\sum_{\mu_e} S_{v_g \mu_e} \left[ \left\langle \psi_{\mu_e}^e \middle| \hat{V}^g - \hat{V}^e \middle| \psi_{v_e}^e \right\rangle + \left( E_{v_e}^e - E_{v_g}^g \right) \delta_{\mu_e v_e} \right] = 0, \; \forall v_e \quad (2)$$

In Eq. (2), $\hat{V}^g$, $\psi_{v_g}^g$ and $E_{v_g}^g$ are the PES, vibrational wavefunction and energy of the ground electronic state; $\hat{V}^e$, $\psi_{v_e}^e$ and $E_{v_e}^e$ are their counterparts for the electronic excited state; and $S_{v_g v_e} = \left\langle \psi_{v_g}^g \middle| \psi_{v_e}^e \right\rangle$ is an FC overlap integral. In practice this infinite set of equations must be truncated and, then, the FCFs are determined by solving Eq. (2) subject to the normalization condition.

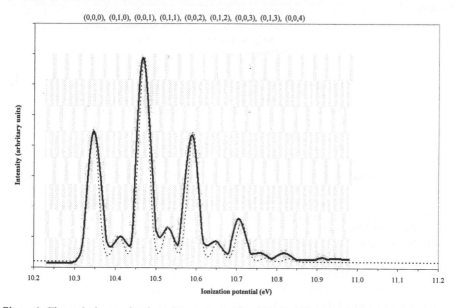

Figure 1: Theoretical second-order anharmonic simulated (dashed line) and experimental (solid line) first band of the ClO2 He I PE spectrum. See reference [10] for computational details.

Duschinsky rotations are incorporated by expanding $\hat{V}^g - \hat{V}^e$ as a power series in the excited electronic state normal coordinates, $\mathbf{Q}^e$. This is done using the relation $\mathbf{Q}^g = \mathbf{J}\mathbf{Q}^e + \mathbf{K}$ where $\mathbf{J}$ is the Duschinsky rotation matrix given by $\mathbf{J} = \mathbf{L}^{g\dagger}\mathbf{L}^e$ and $\mathbf{K}$ is the difference in equilibrium geometry between the ground and excited electronic states expressed in terms of ground state normal coordinates.

The effect of anharmonicity can be accounted for by perturbation theory. In first-order we have:

$$\sum_{\mu_e} S_{v_g \mu_e}^{(1)} \left[ \left\langle \psi_{\mu_e}^e \middle| \hat{V}^g - \hat{V}^e \middle| \psi_{v_e}^e \right\rangle + \left( E_{v_e}^e - E_{v_g}^g \right) \delta_{\mu_e v_e} \right]^{(0)}$$

$$+ \sum_{\mu_e} S_{v_g \mu_e}^{(0)} \left[ \left\langle \psi_{\mu_e}^e \middle| \hat{V}^g - \hat{V}^e \middle| \psi_{v_e}^e \right\rangle + \left( E_{v_e}^e - E_{v_g}^g \right) \delta_{\mu_e v_e} \right]^{(1)} = 0 \quad (3)$$

In this equation the first-order corrections to the vibrational wavefunctions and energies are determined by the terms in $\hat{V}^g$ and $\hat{V}^e$ that are cubic in the normal coordinates of the appropriate electronic state.

The solution is obtained by imposing the first-order normalization condition (i.e. $S^{(1)\dagger} S^{(0)} = 0$ ). A similar procedure may be followed for the second-order corrections [10].

One of the most critical steps in our new procedure is the truncation of the vibrational basis set for the excited electronic state. The algorithm used involves an iterative build-up procedure whereby the range of vibrational quantum numbers is increased while, simultaneously, removing unnecessary functions [11]. Our procedure for excluding wavefunctions drastically reduces the growth of the basis set thereby leading to a major improvement in efficiency.

As a test of the method we have applied it to simulate the first band of the $ClO_2$ He I PE spectrum [10]. Our theoretical second-order anharmonic results are in excellent agreement with previous theoretical simulations and with experimental spectra (see Figure 1). This simple procedure was also used to characterize the first band of the $C_2H_4$ and $C_2D_4$ photoelectron spectra [11]. In that case the computer time was reduced from more than one hour to less than one second, without loss of accuracy, by eliminating all states from the vibrational basis other than those corresponding to excitations of the C=C stretch, the $CH_2$ scissors bend, and the torsion. The torsional motion leads to a twisted $D_2$ geometry in the cationic state and must be treated in a unique manner because it is so highly anharmonic. This motion is assumed to be separable from the other modes and the 1D anharmonic vibrational equation is solved exactly. Good agreement with experiment is, then, obtained using second-order perturbation theory for the remaining modes. Finally, we discuss the successful application to the photoelectron spectrum of furan, a molecule that contains 21 vibrational degrees of freedom

## Acknowledgments

Support for this work under Grants No. BQU2002-04112-C02-02 and BQU2002-03334 from the Dirección General de Enseñanza Superior e Investigación Científica y Técnica (MEC-Spain) is acknowledged. J.M.L. acknowledges financial support from the Generalitat de Catalunya through the Gaspar de Portolà and BE2003 programs. D.M.B and J.M.L. thank the Natural Sciences and Engineering Research Council of Canada for funding. M.T. thanks the Generalitat of Catalunya for financial help through CIRIT Project No. FI/01-00699.

## References

[1]  See for instance B. Champagne and B. Kirtman, in _Handbook of Advanced Electronic and Photonic Materials and Devices_, edited by H.S. Nalwa, Vol. 9, _Nonlinear Optical Materials_, (Academic Press, San Diego CA, 2001), Chap. 2, p. 63.

[2]  D.M. Bishop, J.M. Luis, and B. Kirtman, J. Chem. Phys. **108**, 10013 (1998).

[3]  See for instance M. Torrent-Sucarrat, M. Solà, M. Duran, J.M. Luis, and B. Kirtman, J. Chem. Phys. **120**, 6346 (2004).

[4]  See for instance B. Kirtman, J.M. Luis, and D.M. Bishop, J. Chem. Phys. **108**, 10008 (1998).

[5]  M. Torrent-Sucarrat, J.M. Luis, and B. Kirtman, J. Chem. Phys. **122**, 204108 (2005).

[6]  T.E. Sharp and H.M. Rosenstock, J. Chem. Phys. **41**, 3453 (1964).

[7]  E.V. Doctorov, I.A. Malkin, and V.I. Man'ko, J. Mol. Spectrosc. **64**, 302 (1977).

[8]  See for instance E.P.F. Lee, D.K.W. Mok, F.-T. Chau, and J.M. Dyke, J. Chem. Phys. **121**, 2962 (2004).

[9]  A. Hazra, H.H. Chang, and M. Nooijen, J. Chem. Phys. **121**, 2125 (2004).

[10] J.M. Luis, D.M. Bishop, and B. Kirtman, J. Chem. Phys. **120**, 813 (2004).

[11] J.M. Luis, M. Torrent-Sucarrat, M. Solà, D.M. Bishop, B. Kirtman, J. Chem. Phys. **122** 184104 (2005).

[12] See for instance R.B. Gerber and M.A. Ratner, Adv. Chem. Phys. **70**, 97 (1988).

[13] See for instance J.-Q. Jung and R.B. Gerber, J. Chem. Phys. **105**, 10332 (1996).

Brill Academic Publishers
P.O. Box 9000, 2300 PA Leiden
The Netherlands

*Lecture Series on Computer
and Computational Sciences*
Volume 4, 2005, pp. 655-658

# Polarizability in Small Cadmium Selenide Clusters
# (CdSe)n n=1, 2, 3 and 4

P. KARAMANIS

*Department of Chemistry University of Patras, GR-26500 Patras, Greece*

C. POUCHAN[1]

*Laboratoire de Chimie Structurale, UMR 5624, UniVersite' de Pau et des Pays de l'Adour,
F-64000 Pau, France*

G. MAROULIS[1]

*Department of Chemistry University of Patras, GR-26500 Patras, Greece*

Received 4 July, 2005; accepted in revised form 25 July, 2005

*Abstract*: Conventional ab initio and DFT/B3LYP-B3PW91 values for the static polarizability of small CdSe clusters are reported. Our study extends our previous work on small CdSe clusters [J. Phys. Chem. B, 107, 39 (2003)] relying on second order Møller-Plesset perturbation theory (MP2), coupled-cluster (CC) techniques and specially designed basis sets. B3LYP, B3PW91 and conventional ab initio values are not in good agreement for the monomer but the gap is drastically reduced for the dimer and the tetramer.

*Keywords*: polarizability, CdSe clusters, ab initio, DFT.

## 1. Introduction

Research on clusters and nanostructures has attracted a lot of attention due to their unique size-tuned physical and chemical properties [1, 2]. Beyond the consideration that they represent a transition between molecules and solids, clusters and nanostructures are studied as a special class of molecular systems. Size dependence of electronic properties of clusters, such as the HOMO-LUMO gaps, binding energies and polarizability, which in its dynamic form is directly linked to the absorption spectra, turned out to be of great importance in cluster research. [3]. In particular, II-VI semiconductor clusters and nanostuctures of CdS and CdSe have been extensively studied due to their non linear optical properties, their unusual fluorescence behaviour, their catalytic properties or their structure and phase transitions [4].

Pseudopotential calculations on (CdS)n and (CdSe)n clusters by Troparevsky and Chelicowsky [5], showed a strong reduction of the (per atom) mean dipole polarizability with size. This was supported from our previous ab initio study on small (CdS)n clusters [6] with conventional ab initio methods. In the present study we extend this work by presenting conventional ab initio and DFT results for the static dipole polarizability of (CdSe)n n= 1, 2, 3 and 4. Especially, the planar (CdSe)2 and hexagonal planar (CdSe)3 are also of some importance as they are basic building units of larger three dimensional spheroid structures of (CdSe)n, as proposed very recently by Matxain et al [7]. Our goal is to present accurate ab initio values of static dipole polarizabilities of these four systems in order to explore trends in the evolution of polarizability with size. For this purpose, we have designed basis of Gaussian type functions especially for these molecular systems. We rely on Møller-Plesset perturbation theory (MP) and coupled-cluster (CC) techniques to study electron correlation effects. The DFT methods chosen are the widely used B3LYP and B3PW91.

## 2. Computational details

All calculations were performed at MP2(Full)/3-21G* geometries which are stationary points characterized by computations of the fundamental vibrational frequencies at the same level of theory

---

[1] Corresponding authors: E-mail maroulis@upatras.gr and Claude.Pouchan@univ-pau.fr.

(Figure 1). Two basis set substrates, relative small but flexible, were chosen for this study. The first was the spit valence 3-21G basis (as implemented in the GAUSSIAN 98 set of programs) and the second Cd : $(17s11p8d)$, Se: $(12s9p3d)$ from Koga et *al.* [8] contracted to [6s4p2d /5s4p1d]. We obtained equivalent basis sets for each cluster of the following types: SV1≡[7s6p4d/6s5p3d] SV2≡[7s6p5d/6s5p4d] SV3≡[7s6p5d1f/6s5p4d1f] (from 3-21G) and T1≡ [7s6p3d/6s5p2d], T2≡ [7s6p5d/6s5p4d] and T3≡ [7s6p5d1f /6s5p4d1f]. Basis set construction in this work follows a well-tested computational philosophy [9]. The finite-field method used to obtain the polarizabilities has

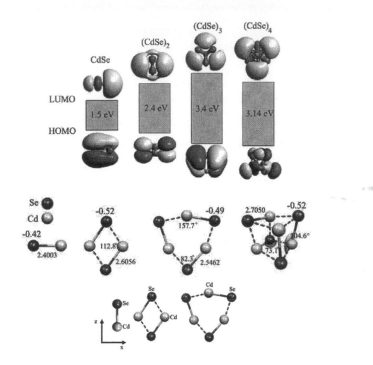

Figure 1: MP2(full)/3-21G* optimized geometries ) and MP2 Mulliken atomic charges, HOMO-LUMO gaps at the equilibrium geometries, visual representation of HOMO and LUMO orbitals and molecular orientation used in all calculations. Bond lengths are in Å, angles in degrees, atomic charges in au and orbital energies in eV.

been presented in sufficient detail elsewhere [10]. In all post Hartree-Fock calculations the innermost MOs were kept frozen (27 for CdSe, 54 for (CdSe)$_2$, 82 for (CdSe)$_3$ and 108 for (CdSe)$_4$). The GAUSSIAN 98 set of programs were used in all calculations. Atomic units are used for the molecular properties throughout this work.

## 3. Results

In table 1 we show SCF, MP2, B3LYP, B3PW91 values of the dipole polarizability tensors, of dipole mean polarizability and the polarizability anisotropy of all systems calculated with basis set SV3. CCSD(T) values are presented only for the monomer and dimer with SV3 and SV1 basis sets respectively. In both cases $\overline{\alpha}_{CCSD(T)} > \overline{\alpha}_{MP2} > \overline{\alpha}_{B3LYP} > \overline{\alpha}_{B3PW91} > \overline{\alpha}_{SCF}$. As size increases from CdSe to (CdSe)$_2$ the distance between $\overline{\alpha}_{MP2}$ and $\overline{\alpha}_{CCD(T)}$ is reduced. This is verified from T1 basis set which yields $\overline{\alpha}_{SCF} = 127.99$ au $\overline{\alpha}_{MP2} = 148.53$ au and $\overline{\alpha}_{CCSD(T)} = 150.75$ for the dimer while in the monomer case, the corresponding values with basis set T3 are $\overline{\alpha}_{SCF} = 93.62$ au $\overline{\alpha}_{mp2} = 92.62$ au and $\overline{\alpha}_{CCSD(T)} = 102.15$ au.

Figure 2 shows the (per atom) polarizability evolution in the series with (SCF, MP2, B3LYP, B3PW91) /SV3 methods and the comparison with (CdSe)$_{n=1,2,4}$ from our previous work at SCF level of theory. All methods predict strong reduction of the previous quantity, and the most interesting feature is that B3LYP, B3PW91 and MP2 values come to an agreement as cluster size increases.

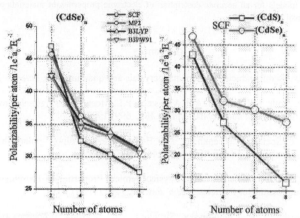

Figure. 2 Mean dipole polarizabilities per atom of (CdSe)$_n$ and (CdS)$_n$ vs the number of atoms in the cluster

Table 1: Dipole moment ($\mu$), dipole polarizability tensors ($\alpha_{xx}$, $\alpha_{yy}$, $\alpha_{zz}$), mean values of the dipole polarizability ($\overline{\alpha}$) and the polarizability anisotropy ($\Delta\alpha$) of (CdSe)$_n$, n=1,2,3 and 4.

| | Method | $\mu$ | $\alpha_{zz}$ | $\alpha_{xx}$ | $\alpha_{yy}$ | $\overline{\alpha}$ | $\Delta\alpha$ |
|---|---|---|---|---|---|---|---|
| Monomer | SCF/SV3 | -3.2269 | 132.33 | 74.56 | | 93.82 | 57.77 |
| | MP2/SV3 | -2.8337 | 115.92 | 78.95 | | 91.27 | 36.98 |
| | CCSD(T)/SV3 | -2.5569 | 129.91 | 85.44 | | 100.26 | 44.47 |
| | B3LYP/SV3 | -2.4139 | 114.33 | 70.17 | | 84.89 | 44.16 |
| | B3PW91/SV3 | -2.5281 | 114.64 | 69.96 | | 84.85 | 44.68 |
| Dimer | SCF/SV3 | | 150.86 | 140.38 | 97.21 | 129.48 | 69.66 |
| | MP2/SV3 | | 169.81 | 157.27 | 107.58 | 144.89 | 80.62 |
| | CCSD(T)/SV1 | | 169.16 | 159.73 | 108.99 | 145.96 | 79.27 |
| | B3LYP/SV3 | | 175.28 | 149.54 | 98.34 | 141.05 | 95.94 |
| | B3PW91/SV3 | | 167.95 | 148.73 | 97.28 | 137.99 | 89.50 |
| Trimer | SCF/SV3 | | 207.71 | 130.35 | | 181.93 | 109.40 |
| | MP2/SV3 | | 230.95 | 141.62 | | 201.17 | 126.33 |
| | B3LYP/SV3 | | 235.53 | 135.13 | | 202.06 | 141.98 |
| | B3PW91/SV3 | | 231.32 | 133.85 | | 198.83 | 137.84 |
| Tetramer | SCF/SV3 | | 220.51 | | | 220.51 | 0 |
| | MP2/SV3 | | 247.17 | | | 247.17 | 0 |
| | B3LYP/SV3 | | 248.65 | | | 248.65 | 0 |
| | B3PW91/SV3 | | 244.70 | | | 244.70 | 0 |

## Acknowledgments

Panos Karamanis is happy to acknowledge the warm hospitality of the Laboratoire de Chimie Structurale during his stay in Pau. This work is part of the COST action **D26/0013/02** project of the EEC "Development of Density Functional Theory models for an accurate description of electronic properties of materials possessing potential high non-linear optical properties".

## References

[1] A. P. Alivisatos, Semiconductor clusters, nanocrystals and quantum dots, *Science* **271**, 933 (1996).

[2] D.L. Klein, R. Roth, A.K.L. Leim, A.P. Alivisatos, P.L. Mceuen, A single-electron transistor made from a cadmium selenide nanocrystal, *Nature* **389,** 699 (1997).

[3] S. Schlecht, R. Scafer, J. Woenkhaus, J.A. Becher, Electric dipole polarizabilities of isolated gallium arsenide clusters, *Chem. Phys. Lett.* **246,** 315-320 (1996).

[4] A. Hanglein, Small-Particle Research: Physicochemical Properties of Extremely Small colloidal Metals and Semiconductor Particles, *Chem. Rev.* **89**, 1861 (1989).

[5] M.C Troparevsky, J. R. Chelikowsky, Structural and electronic properties of CdS and CdSe clusters, *J. Chem. Phys.* **114**, 943 (2001)

[6] G. Maroulis, C. Pouchan, Size and Electric Dipole (Hyper)polarizability in Small Cadmium Sulfide Clusters: An ab Initio Study on $(CdS)n$, $n = 1, 2$, and 4, *J. Phys. Chem. B* **107**, 10683 (2003).

[7] J. Matxain, J. M. Mercero, J.Fowler and J.M.Ugalde,Cluster II-VI Materials: $Cd_iX_i$, X= S, Se, Te, $i \le 16$, *J. Phys. Chem. A* **108**, 10502 (2004).

[8] T. Koga, H. Tatewaki, H. Matsuyama, Y. Satoh, Atoms K through Kr, *Theor. Chem. Acc.,* **102,** 105 (1999). T. Koga, S. Yamamoto, T. Shimazaki, H. Tatewaki, Contracted Gaussian-type basis functions revisited IV. Atoms Rb to Xe, *Theoret. Chem. Acc.* **108,** 41 (2002).

[9] P. Karamanis, G. Maroulis, Finite-field Møller-Plesset perturbation theory and Coupled Cluster calculations of the electric multipole moments and the dipole polarizability of $As_2$, *Chem. Phys.* **269**, 137 (2001).

[10]G. Maroulis, A systematic study of basis set, electron correlation, and geometry effects on the electric multipole moments, polarizability, and hyperpolarizability of HCl, *J. Chem. Phys.* **108**, 5432 (1998).

Brill Academic Publishers
P.O. Box 9000, 2300 PA Leiden
The Netherlands

*Lecture Series on Computer
and Computational Sciences*
Volume 4, 2005, pp. 659-662

# (hyper)polarizability Evolution in the Series  H-C≡C-C≡C-CX₃,
# X = H, F, Cl, Br and I

P. KARAMANIS

*Department of Chemistry University of Patras,
GR-26500 Patras, Greece*

G. MAROULIS[1]

*Department of Chemistry University of Patras,
GR-26500 Patras, Greece*

Received 4 July, 2005; accepted in revised form 25 July, 2005

*Abstract*: We studied the evolution of dipole moment ($\mu_\alpha$), dipole polarizability ($\alpha_{\alpha\beta}$), first ($\beta_{\alpha\beta\gamma}$) and second ($\gamma_{\alpha\beta\gamma\delta}$) (hyper)polarizabilities in the series H-C≡C-C≡C-CX₃, X = H, F, Cl, Br and I, relying on second order Møller Plesset perturbation theory (MP2) and specially designed basis sets. Studies of the hyperpolarizability evolution in such systems are of grate importance for the non-linear optical behaviour of molecules and material science.

*Keywords*: polarizability, hyperpolarizability, substituted diacetylenes, conjugated molecules.

## 1. Introduction

Macroscopic linear and nonlinear susceptibilities of organic compounds are strongly related with the microscopic (hyper)polarizabilities [1] thus a systematic study of small model-systems with high accuracy *ab initio* methods can give reliable information for the non-linear optical behavior of molecules and material science [2]. In previous works [3, 4] we studied the (hyper)polarizability evolution of mono-substituted diacetylene with halogens. We showed that for molecules of that magnitude, basis set effect is dominant and correlated methods (Möller-Plesset perturbation theory, coupled cluster techniques) are required in order to achieve realistic predictions. The present effort extends that work by studying the dipole moment and dipole (hyper)polarizability evolution in the series: H-C≡C-C≡C-CX₃, X = H, F, Cl, Br and I, relying on second order Møller Plesset perturbation theory (MP2) and specially designed basis sets.

## 2. Theory and Computational details

The energy of an uncharged molecule interacting with a weak homogeneous electric field can be written as:

$$E(F_\alpha) = E^0 - \mu_\alpha F_\alpha - (1/2)\alpha_{\alpha\beta}F_\alpha F_\beta - (1/6)\beta_{\alpha\beta\gamma}F_\alpha F_\beta F_\gamma$$
$$- (1/24)\gamma_{\alpha\beta\gamma\delta}F_\alpha F_\beta F_\gamma F_\delta + ... \qquad (1)$$

where $F_\alpha$ is the field $E^0$ is the energy of the free molecule and $\mu_\alpha$, $\alpha_{\alpha\beta}$, $\beta_{\alpha\beta\gamma}$, $\gamma_{\alpha\beta\gamma\delta}$ are the dipole moment and dipole (hyper)polarizabilities. The subscripts denote Cartesian components and the repeated subscript implies summation over $x$, $y$ and $z$.

All calculations were performed at MP2(Full) using the basis sets of Table 1 geometries for each molecule and the produced geometries (Fig. 1) are stationary points characterized by computations of the fundamental vibrational frequencies at the same level of theory. Basis set construction was based on the D95 substrate for C, H, F and Cl atoms, while for Br and I we used substrates from J. Andzelm and M. Klobukowski [5] consisting of (13*s*10p4*d*) contracted to [4*s*3p1*d* ] and I: (16*s*13p7*d*) contracted to [5*s*4p2*d*], respectively. Construction details and the composition of the obtained basis

---

[1] Corresponding author: E-mail address maroulis@upatras.gr

sets is given systematically elsewhere [6]. GAUSSIAN 94 and GAUSSIAN 98 programs were used in all calculations and all electric property values are in atomic units.

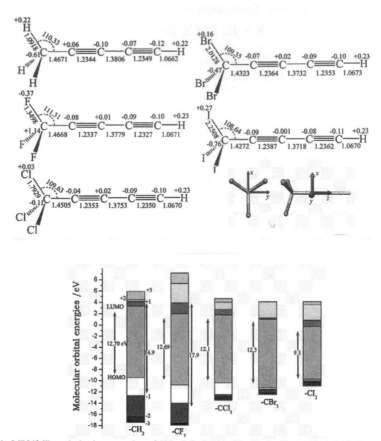

Figure 1: MP2(full) optimized geometries of HC≡C-C≡C-CX₃ (X= H, F, Cl, Br, I), MP2(full) NBO atomic charges HOMO, LUMO, HOMO±(1,2,3) LUMO±(1,2,3) energies and HOMO LUMO gaps at the eqilibrium geometries. Bond lengths are in Å, angles in degrees, atomic charges in au and orbital energies in eV. Atomic charge calculations were performed using optimized basis set for each molecule [6] and orbital energies were calculated using the MP2(full) /6-311G* method.

Table 1: Basis sets used in geometry optimization for each molecule.

| Molecule | Atomic basis sets |
|---|---|
| H-C≡C-C≡C-CH₃ | D95** (5d) |
| H-C≡C-C≡C-CF₃ | D95** (5d) |
| H-C≡C-C≡C-CCl₃ | D95* * (5d) |
| H-C≡C-C≡C-CBr₃ | C, H: D95** (5d), Br(13s10p4d) contracted to [4s3p1d] (5d) [5] |
| H-C≡C-C≡C-CI₃ | C, H: D95** (5d) I: (16s13p7d) contracted to [5s4p2d] (5d) [5] |

## 3. Results

In Table 2 one we can observe the evolution of dipole moment, the mean values of the first and second (hyper)polarizability and the polarizability anisotropy in the series H-C≡C-C≡C-CX₃ X=H, F, Cl, Br, and I, at Hartree-Fock and MP2 levels of theory. Fig. 2 shows the atomic charge transfer on H-

C≡C-C≡C-CH₃ H-C≡C-C≡C-CF₃ and H-C≡C-C≡C-CCl₃ in the presence of static homogenous electric field.

Table 1: Dipole moment ($\mu$), the mean value of the dipole polarizability ($\bar{\alpha}$), the polarizability anisotropy ($\Delta\alpha$), and the mean values of first and second hyperpolarizabilities ($\bar{\alpha}$, $\bar{\beta}$, $\bar{\gamma}$) at SCF and MP2 levels of theory using equivalent basis sets for each molecule.

| Method | $\mu$ | $\bar{\alpha}$ | $\Delta\alpha$ | $\bar{\beta}$ | $\bar{\gamma}$ |
|---|---|---|---|---|---|
| H₃C—(C≡C)₂—H | | | | | |
| SCF/ [5s3p3d/3s3p] | -0.5191 | 64.68 | 73.00 | -151.93 | 14997 |
| MP2/[5s3p3d/3s3p] | -0.4657 | 63.97 | 71.43 | -204.38 | 20405 |
| F₃C—(C≡C)₂—H | | | | | |
| SCF/[5s3p3d/3s3p1d/5s3p3d] | 0.7854 | 136.51 | 79.97 | -85.18 | 9557 |
| MP2/[5s3p3d/3s3p1d/5s3p3d] | 0.7924 | 138.90 | 82.20 | -55.65 | 13469 |
| Cl₃C—(C≡C)₂—H | | | | | |
| SCF/[5s3p3d/3s3p1d/7s5p3d] | 0.8253 | 109.75 | 79.02 | 28.99 | 20971 |
| MP2/[5s3p3d/3s3p1d/7s5p3d] | 0.8159 | 107.93 | 79.09 | 120.75 | 32708 |
| Br₃C—(C≡C)₂—H | | | | | |
| SCF/[5s3p3d/3s3p/5s4p4d] | 0.7854 | 136.51 | 79.97 | 250.38 | 35685 |
| MP2/[5s3p3d/3s3p/5s4p4d] | 0.7924 | 138.90 | 82.20 | 443.83 | 57599 |
| I₃C—(C≡C)₂—H | | | | | |
| SCF/[5s3p3d/3s3p/7s5p4d] | 0.5268 | 182.77 | 71.86 | 527.65 | 65165 |
| MP2/[5s3p3d/3s3p/7s5p4d] | 0.5825 | 184.98 | 76.53 | 725.40 | 105251 |

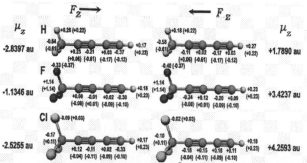

Fig. 2. MP2(full) NBO atomic charges, and dipole moments in the presence of static homogenous electric field in the direction of the dipole moment (C₃ axis). Values in parenthesis are atomic charges of the free molecules at the same level of theory. In both cases, atomic charge calculations were performed using optimized basis sets [6] and the chosen field strength is of 0.02 e⁻¹a₀⁻¹Eₕ.

## Acknowledgments

The authors gratefully acknowledge support of this work by the University of Patras through a grant from the CARATHEODORY project (Grant 02449).

## References

[1] R. A. Hann, D. Bloor, *Organic Materials for nonlinear optics II*, Royal society of Chemistry, Cambridge, 1993.

[2] S. R. Marder, J. E. Sohn (Eds), Materials for Nonlinear Optics: Chemical Perspectives, ACS Symposium Series 455 (ACS, Washington DC, 1991).

[3] P. Karamanis, G. Maroulis, Electric properties of substituted diacetylenes September 12-16, Kastoria, Greece. Computational Methods in Sciences and Engineering 2003, pp 285-288 World Scientific, Singapore (2003). International Conference on Computational Methods in Sciences and Engineering (ICCMSE 2003).

[4] P. Karamanis, G.Maroulis, Static electric dipole polarizability and hyperpolarizability of fluorodiacetylene, *J. Mol. Struct.* (THEOCHEM) **621**, 157-162 (2003).

[5] J. Andzelm, M. Klobukowski, E. Radzio-Andzelm, *J. Comput. Chem.* **5** 146 (1984).

[6] P. Karamanis: *Contribution to the systematic study of substitution effects on the linear and nonlinear polarizability of hydrocarbons with conjugated bonds. Phd Thesis* (University of Patras 2004).

Brill Academic Publishers
P.O. Box 9000, 2300 PA Leiden,
The Netherlands

*Lecture Series on Computer
and Computational Sciences*
Volume 4, 2005, pp. 663-666

# Charge Response Kernel Theory based on Ab Initio and Density Functional Calculations

**Akihiro Morita[1] and Tateki Ishida**

Department of Computational Molecular Science
Institute for Molecular Science
Myodaiji, Okazaki 444-8585, Japan

Received 9 July, 2005; accepted in revised form 14 August, 2005

*Abstract:* The charge response kernel (CRK) of a molecule, $\partial Q_a/\partial V_b$, is formulated and calculated via *ab initio* molecular orbital and density functional theory, where $Q_a$ is the partial charge of the site $a$ and $V_b$ is the electrostatic potential at the site $b$. The non-empirical definition of CRK provides a general and rigorous scheme of polarizable molecular modeling.

*Keywords:* charge response kernel, polarizability, density functional theory

## 1 Introduction

The electronic polarization is of vital importance in intermolecular interactions in condensed phases, in particular, to correctly describe solvation energetics and structures. Therefore, most molecular models in common use incorporate the electronic polarization effects in either implicit or explicit way. The former, implicit way employs a nonpolarizable molecular model with augmented polarization, such as the SPC [1] or TIP4P [2] water model, while the latter uses a polarizable molecular model. Although the latter is computationally more demanding than the former for use of molecular simulation, it is capable of describing both spatial and temporal fluctuation of molecular polarization, which often plays crucial roles in local solvation or interface structures [3, 4].

The charge response kernel (CRK) is considered as a general model to explicitly represent the molecular polarization [5]. It is defined as $(\partial Q_a/\partial V_b)_N$ on the basis of the interaction site representation, where $Q_a$ is the partial charge at the site $a$ and $V_b$ is the electrostatic potential at the site $b$. This property is readily incorporated into the molecular dynamics simulation based on the interaction site molecular models, introducing fluctuating partial charges on the interaction sites.

We propose the *ab initio* formulation of the CRK, due to the following advantages. First, the *ab initio* theory obviates empirical parameterization to determine the CRK, and therefore it is generally applicable to any molecule by *ab initio* calculations. Second, the accuracy and reliability of the CRK are warranted by those of the underlying electronic structure theory at any level of accuracy. Third, these general features of the *ab initio* CRK provide an equal footing to treat the electronic polarization of solute and solvent molecules, including novel solute species that are often difficult to deal with by an empirical model. As one example, the CRK model in combination

---

[1]Corresponding author. E-mail: amorita@ims.ac.jp

of molecular dynamics simulation was successfully utilized to elucidate the anomalous diffusion of some aromatic transient radicals in solutions [6].

In the present paper, first we briefly summarize the theoretical framework of the CRK model in the *ab initio* molecular orbital theory. Then we extend the formulation to the density functional theory, and demonstrate the accuracy of the calculated results.

## 2 Ab Initio Theory

In our theoretical treatment, the partial charge $Q_a$ is defined as a electrostatic potential charge, which is optimized to reproduce the surrounding electric field generated by the molecule. The partial charge $Q_a$ is given by an one-electron operator $\hat{Q}_a$ as $Q_a = \langle \Psi | \hat{Q}_a | \Psi \rangle$, where $\Psi$ denotes the wavefunction of the many-electron system. The detailed form of $\hat{Q}_a$ is presented in our previous works [5].

The molecular Hamiltonian in the condensed phase, $\hat{H}$, is given by $\hat{H} = \hat{H}_0 + \hat{H}'$, where $\hat{H}_0$ is the molecular Hamiltonian in the gas-phase and $\hat{H}'$ is the perturbation Hamiltonian $\hat{H}'$ by the solvation environment. The latter, $\hat{H}'$, is given in the interaction site representation as

$$\hat{H}' = \sum_a^{\text{sites}} \hat{Q}_a V_a, \tag{1}$$

where $V_a$ is the electrostatic potential at the site $a$ generated by the ambient solvent.

The CRK $K_{ab} = (\partial Q_a / \partial V_b)$, by definition, is represented via the first derivative of the wavefunction $\Psi^b$ with respect to the external electrostatic potential $V_b$ at site $b$,

$$K_{ab} \equiv \frac{\partial Q_a}{\partial V_b} = \frac{\partial^2 E}{\partial V_a \partial V_b} = \langle \Psi^b | \hat{Q}_a | \Psi \rangle + \langle \Psi | \hat{Q}_a | \Psi^b \rangle. \tag{2}$$

Note that the CRK is equivalent to the second derivative of the total energy $E$. The derivative of the wavefunction is generally formulated by the coupled-perturbed Hartree-Fock equation in the *ab initio* theory [7]. We first illustrate this scheme in a simple case of the closed-shell Hartree-Fock theory.

Suppose the molecular orbital and its derivative by $V_a$ are denoted by $C_{pi}$ and $C_{pi}^a$. In the followings, suffixes $p$, $q$, ... refer to atomic orbital (AO), and $i$, $j$, ... to molecular orbital (MO). The derivative coefficient $C_{pi}^a$ is expressed in the MO representation using the transformation matrix $U$,

$$C_{pi}^a = \sum_j^{\text{MO}} C_{pj} U_{ji}^a, \tag{3}$$

and $U$ is formulated by the following coupled-perturbed Hartree-Fock equation,

$$(\epsilon_l - \epsilon_i) U_{li}^a + \sum_j^{\text{occ}} \sum_k^{\text{vir}} H_{likj}^{(MO)} U_{kj}^a = -Q_{a,li}^{(MO)}, \tag{4}$$

where $j$ and $k$ indicate the occupied and virtual MO's, respectively. $\epsilon_i$ is the orbital energy of the $i$-th canonical MO, and $H$ is given with the two-electron integrals (in the MO representation) as

$$H_{likj} = 4(li|kj) - (lk|ij) - (lj|ik). \tag{5}$$

Eq. (4) is a set of linear equations for $U$, which derives the CRK as follows,

$$K_{ab} = \sum_i^{\text{occ}} \sum_j^{\text{vir}} 4 Q_{a,ij}^{(MO)} U_{ji}^b. \tag{6}$$

## 3  Density Functional Version

The density function theory (DFT) is in widespread use in the last decade in the field of quantum chemistry [8]. Its amazing success hinges on the exchange-correlation functionals developed to date, such as B3LYP [9, 10], which can take account of the electron correlation effects fairly accurately at a modest computational cost. Since the Kohn-Sham formalism is quite analogous to the Hartree-Fock formalism, the extension of the CRK model to the density functional theory is straightforward, as shown in the followings.

The Kohn-Sham matrix is represented as

$$F_{pq}^{KS} = h_{pq} + \sum_{r,s} (pq|rs)D_{rs} + V_{pq}^{xc}[\rho], \tag{7}$$

where $V_{pq}^{xc}[\rho] = \delta E^{xc}[\rho]/\delta D_{pq}$ is the exchange-correlation potential, which is the derivative of the exchange-correlation energy functional $E^{xc}[\rho]$ with respect to the density matrix $D_{pq}$.

The formulation of the Kohn-Sham matrix, (7), coincides with that of the Fock matrix by replacing the exchange-correlation potential $V_{pq}^{xc}$ with the exchange term, $-\sum_{r,s}(pr|qs)D_{rs}$. Accordingly, eq. (4) is modified in the DFT case by replacing $H$ with the Kohn-Sham equivalent as,

$$H_{pqrs} \rightarrow 4(pq|rs) + \delta V_{pq}^{xc}/\delta D_{rs}. \tag{8}$$

The DFT version of eq. (4) is often called coupled-perturbed Kohn-Sham equation. Details for the second term are given in the general analytical second derivative of the density functional theory [11].

We calculated the CRK via the DFT and compared them to those by the Hartree-Fock case. Table 3 illustrates the results of $NH_3$. The program code for the CRK calculation was implemented into the GAMESS-UK package [12]. To evaluate the reliability of the calculated CRK, the electronic polarizability derived from the CRK is compared to the experimental value. The polarizability $\alpha_{ij}$ is derived from the CRK as

$$\alpha_{ij} = -\sum_{a,b}^{\text{sites}} \frac{\partial Q_a}{\partial V_b} r_i(a) r_j(b), \qquad i,j = x,y,z, \tag{9}$$

where $r_i(a)$ is the $i$-th Cartesian coordinate of the site $a$. The calculated CRK derives an isotropic polarizability of $\bar{\alpha} = 12.13$ a.u. (HF) and 13.59 a.u. (B3LYP), while the experimental value is 15.25 a.u. While both results are fairly consistent to the experiment, the density functional calculation shows some improvement in evaluating the electronic polarization. Computational cost of the CRK is almost equivalent to that of the analytical calculation of the conventional dipole polarizability.

Table 1: Charge Response Kernel of $NH_3$ molecule, calculated by (a) HF and (b) B3LYP. The basis set is d-aug-cc-pVTZ for N and aug-cc-pVTZ for H, with $s$ and $p$ diffusion functions only, (20s7p2d1f/6s3d1d) / [6s5p2d1f/4s3p1d]. Unit: a.u.

| | (a) HF | | | | | (b) B3LYP | | | |
|---|---|---|---|---|---|---|---|---|---|
| | 1 | 2 | 3 | 4 | | 1 | 2 | 3 | 4 |
| 1(N) | -23.98 | | | | 1(N) | -28.39 | | | |
| 2(H) | 7.99 | -4.38 | | | 2(H) | 9.46 | -5.03 | | |
| 3(H) | 7.99 | -1.81 | -4.38 | | 3(H) | 9.46 | -2.22 | -5.03 | |
| 4(H) | 7.99 | -1.81 | -1.81 | -4.38 | 4(H) | 9.46 | -2.22 | -2.22 | -5.03 |

## 4  Summary

In this work, the Charge Response Kernel $(\partial Q_a/\partial V_b)$ formulated via the ab initio molecular orbital theory by us is extended to the density functional theory. These definitions of CRK rigorously based on the quantum chemical theories can provide a general scheme to construct polarizable molecular models. We plan to adopt this CRK model to molecular simulation of the interfacial nonlinear spectroscopy. Such work is now in progress in our group.

## Acknowledgment

This work was supported by the Grant-in-Aid (no. 15550012) and National Research Grid Initiative (NAREGI) of the Ministry of Education, Japan.

## References

[1] H. J. C. Berendsen, J. P. M. Postma, M. F. van Gunsteren, and J. Hermans. *Intermolecular Forces*, p. 331. Reidel, Dordrecht, 1981.

[2] W. L. Jorgensen, J. Chandrasekhar, J. D. Madura, R. W. Impey, and M. L. Klein. *J. Chem. Phys.*, **79**, 926 (1983).

[3] J. Caldwell, L. X. Dang, and P. A. Kollman. *J. Am. Chem. Soc.*, **112**, 9144 (1990).

[4] P. Jungwirth and D. J. Tobias. *J. Phys. Chem. B*, **106**, 6361 (2002).

[5] A. Morita and S. Kato. *J. Am. Chem. Soc.*, **119**, 4021 (1997).

[6] A. Morita and S. Kato. *J. Chem. Phys.*, **108**, 6809 (1998).

[7] P. Pulay. *Adv. Chem. Phys.*, **69**, 241 (1987).

[8] R. G. Parr and W. Yang. *Density-Functional Theory of Atoms and Molecules*. Oxford Univ., New York, 1989.

[9] A. D. Becke. *J. Chem. Phys.*, **98**, 5468 (1993).

[10] C. Lee, W. Yang, and R. G. Parr. *Phys. Rev. B*, **37**, 785 (1988).

[11] N. C. Handy, D. J. Tozer, G. J. Laming, C. W. Murray, and R. D. Amos. *Israel J. Chem.*, **33**, 331 (1993).

[12] GAMESS-UK is a package of ab initio programs written by M.F. Guest, J.H. van Lenthe, J. Kendrick, K. Schoffel, and P. Sherwood, with contributions from R.D. Amos, R.J. Buenker, H.J.J. van Dam, M. Dupuis, N.C. Handy, I.H. Hillier, P.J. Knowles, V. Bonacic-Koutecky, W. von Niessen, R.J. Harrison, A.P. Rendell, V.R. Saunders, A.J. Stone, D.J. Tozer, and A.H. de Vries. The package is derived from the original GAMESS code due to M. Dupuis, D. Spangler and J. Wendoloski, NRCC Software Catalog, Vol. 1, Program No. QG01 (GAMESS), 1980.

Brill Academic Publishers
P.O. Box 9000, 2300 PA Leiden
The Netherlands

*Lecture Series on Computer*
*and Computational Sciences*
Volume 4, 2005, pp. 667-673

# Ab Initio Studies of Electronic First Hyperpolarizability of AB Systems Containing Phosphorus and Silicon Atoms

D. Y. Zhang[1], C. Pouchan[1], D. Jacquemin[2], E. A. Perpète[2]

[1]Laboratoire de Chimie Structurale, UMR 5624
Université de Pau et des Pays de l'Adour, IFR-Rue Jules Ferry
BP 27540, 64075 Pau Cedex, France

[2]Laboratoire de Chimie Théorique Appliquée
Facultés Universitaires Notre-Dame de la Paix
Rue de Bruxelles, 61,
B-5000, Namur, Belguim

Received 16 July, 2005; accepted in revised form 10 August, 2005

*Abstract* Longitudinal first hyperpolarizability ( ) calculations were carried out in three phosphorus- and one silicon-containing oligomers among which three are isovalent to polymethineimine (PMI). Extremely large values have been predicted in two of the phosphorus-containing oligomers, namely, polyphosphaacetylene (PPA), -(P=CH)$_n$- and polyphosphasilyne (PPS), -(P-SiH)$_n$-, i.e. 212777 and 90811 a.u. respectively, at n=16. These values are 5.3 and 2.3 times that of polymethineimine (PMI), which is well known for its higher nonlinear response. In striking contrast, extremely small values are predicted in another phosphorus-containing polymer, namely, polyborophosphene (PBP), -(PH=BH)$_n$-, with a /n approaching zero at large n. Furthermore, consistently negative values are predicted for the polysilanitrile (PSA) oligomers, which are opposite to all the phosphorus-containing polymers as well as PMI. The causes for the relative values in all these series of oligomers are discussed, in light of the bonding structure, the bond strength, the size of the nuclei, and the degree of delocalization.

*Keywords:* Ab initio, hyperpolarizability, phosphorus, silicon, -bond strength, zwitterionic structures

## 1. Introduction

Conjugated systems as potential materials with nonlinear optical properties have been widely studied since the late 80's, as nonlinear optical properties play a vital role in phononic and electro-optic devices [1]. The most well-known strategy is the so-called push-pull system, where a conjugated, polyacetylene-like segment, is capped by an electron-donating group on one end of the chain, and an electron-accepting group on the other end [1]. Recently, we have been exploring another system as potential NLO materials, which is referred to as the AB system, where two different nuclei are involved in the conjugated chain, with no extra electron-donating or electron-accepting group at either end [2-8].

Such systems carry two key qualities that contribute to large NLO responses, namely, the presence of delocalized electrons and the asymmetry of the system. What is not so trivial is the fact that the increase of delocalization of the electrons inevitably leads to a decreased degree of asymmetry; and when asymmetry disappears, decreases to 0. Therefore, the challenge of obtaining large values in an NLO material becomes the problem of designing a molecular system which contains both a highly conjugated system, and also the asymmetry of the molecular frame. Unlike the push-pull chains whose polymeric values disappear to zero due to the lack of asymmetry, the AB system may present a non-zero /n value, because in such a system, the asymmetry arises from not only the chain-ends (CE), but also the unit cell (UC). The evolution of /n value with respect to the number of unit cells in the oligomer, n, can be categorized into three cases:

Case 1. /n value increases initially (in its amplitude) and then saturates to the polymeric limit. Polymethineimine (PMI) provides an example of such behavior, with a positive polymeric /n, value and the *cis-transoids* conformation of polysilaacytelene (PSA) is another example of such a case but with the  /n values to be negative at all chain length, as well as at polymeric limit.

Case 2.  /n value increases (in its magnitude) initially, then turns around and decreases. Eventually it reaches zero. This is the typical behavior of the push-pull system [7], where asymmetry disappears at long chain length. However, some AB systems, such as PSN, also belong to such category [8].

Case 3.  /n value is initially small and negative; and continues to be negative and increases in its amplitude and reaches a minimum. Then it turns around and eventually converges to the polymeric limit. An example for this type of polymers is given by polyphosphinoborane (PPB). Such behavior is owing to the fact that the CE and UC contributions to  having opposite signs, with the former dominating the total response at short chain length, while the latter at long chain length; and the polymeric  /n value is eventually determined by the nature of UC of the system.

In this paper, we report theoretical studies of the first hyperpolarizabilities of four AB type of oligomers with the size ranging from 2 to 16. The structures for these polymers are displayed in Figure 1.

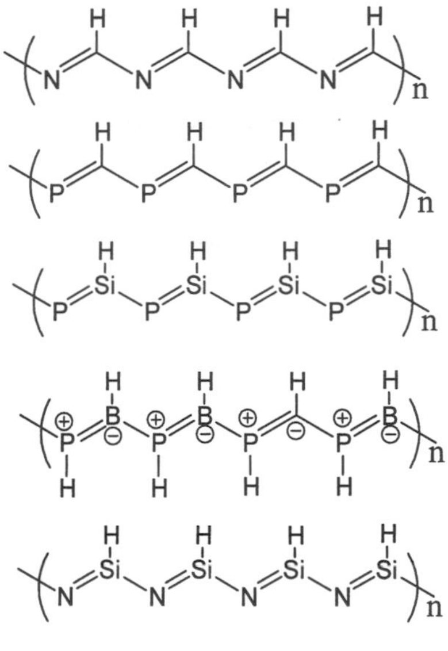

Figure 1: Chemical structures (from top to bottom) of PMI, PPA, PPS, PBP, and PSN.

As their chemical structures show, three of the four series are isovalent to each other, and the series containing boron has a different bonding structure, which involves charge separation. Structural data, natural bond analysis results, molecular orbital energy values for all four series are presented, and comparisons are made between the series. The causes for the large  values in PPA and PPS, the small  values in PBP, and the negative  values in PSN are discussed.

## 2. Computational Methodology

Calculations were carried out with the Gaussian03 suite of program [10]. Geometry optimizations were carried out using oligomers kept planar in their trans-transoid (TT) conformation, and the TT chains were fixed in a linearized form in which all skeleton bond angles were set equal in order for a meaningful definition of the longitudinal direction. The choice of the HF/6-31(d) level of theory for ground-state optimization has been dictated by the accuracy/time balance. Indeed MP2 calculations are too cpu-intensive for PPA and PSN, and B3LYP provides very poor geometries for the PMI oligomers. Based on the calculations on PMI and PSN, it is expected that HF predict sufficiently good geometries for obtaining accurate $\beta$ values at by the MP2, even though it is known that HF predicts too long bond lengths for such molecules, as it is the bond length alternation ($\Delta r$) that most significantly affects the $\beta$ values rather than the absolute bond lengths.

Static electronic first hyperpolarizability ($\beta$) were computed at MP2/6-31G(d) level of theory at each optimized geometry by using the finite-field procedure [11]. As hybrid DFT functional fail completely to reproduce the $\beta$ of PMI, and as the incorporation of electron correlation effects are absolutely required for such calculations, perturbational method is mandatory. The use of 6-31(d) has been found to be sufficient in our recent study of PMI, PSN, and polysilaacetylene (PSA), and further extension of the basis set are shown to be modify the $\beta$ values by less than 30% [8]. Only the longitudinal component of ($\beta$) are reported here. The notations for $\beta$ is such that positive values correspond to the orientation along with the dipole moment.

## 3. Results and Discussion

Table 1 lists the static longitudinal first hyperpolarizability ($\beta$) values for all five AB type of oligomers, namely, NC, PC, PSi, PB, and NSi, with n=2 – 16. In comparison with NC, the $\beta$ values of PC and PSi increase much faster down the column with the increase of the chain length, and reach values of 212777 and 90811 a.u., respectively, at n=16. These $\beta$ values are 5.3 and 2.3 times the value for NC at the same chain length. In striking contrast, the PB oligomers display a completely different behavior, such that the $\beta$ value starts to be the highest among the three series, but essentially stops increasing at around n=12, and ends up to be 4956 a.u. at n=16, an extremely small value of only 12% of NC at the same chain length. Similar behavior to PB is observed in NSi, expect that all $\beta$ values are predicted to be negative.

Table 1: Calculated first hyperpolarizability, $\beta$, (in a.u.) for the five of AB-type oligomers, with n = 2 – 16

| N | –(N=CH)$_n$- | –(P=CH)$_n$- | –(P=SiH)$_n$- | –(PH=BH)$_n$- | –(N=SiH)$_n$- |
|----|-----|-----|-----|-----|-----|
| 2  | 64    | -26    | 367   | 421  | -455  |
| 4  | 653   | 1104   | 793   | 1665 | -2233 |
| 6  | 2390  | 6146   | -829  | 2997 | -2830 |
| 8  | 5890  | 19623  | 1635  | 3886 | -3131 |
| 10 | 11494 | 45651  | 14171 | 4403 | -3353 |
| 12 | 19192 | 86367  | 35896 | 4698 | -3524 |
| 14 | 28764 | 142212 | 62672 | 4868 | -3638 |
| 16 | 39917 | 212777 | 90811 | 4968 | -3717 |

The $\beta$/n values are listed in Table 2. The trends in $\beta$/n versus n are similar trend for the three isovalent series, which suggest that both PPA and PPS belong to the same category of the AB system as PMI, namely, case 1. As expected, the $\beta$/n curves for both PB and NSi are completely different from the other three series. The trends, with a projected zero $\beta$/n value at the polymeric limit, categorize them both as the case 2 AB systems.

Table 2: The ̄n values (in a.u.) for the five AB-type oligomers, with n = 2 – 16

| N | –(N=CH)$_n$- | –(P=CH)$_n$- | –(P=SiH)$_n$- | –(PH=BH)$_n$- | –(N=SiH)$_n$- |
|---|---|---|---|---|---|
| 2 | 32 | -13 | 184 | 211 | -228 |
| 4 | 163 | 276 | 198 | 416 | -558 |
| 6 | 398 | 1024 | -138 | 500 | -472 |
| 8 | 736 | 2453 | 204 | 486 | -391 |
| 10 | 1149 | 4565 | 1417 | 440 | -335 |
| 12 | 1599 | 7197 | 2991 | 2392 | -294 |
| 14 | 2055 | 10168 | 4477 | 348 | -260 |
| 16 | 2495 | 13299 | 5676 | 311 | -232 |

What is responsible for the much larger    values in PPA and PPS relative to PMI? As is known, the P=C and P=Si    bonds are much weaker than the N=C    bond, due to the poor    overlap involving second-row elements. Calculations by Schleyer et al [13], for instance, predicted the P=C and P=Si bond energy to be 49.4 and 29.7 kcal/mol, which are both much smaller than N=C    bond strength of 80.8 kcal/mol. Also is known that weaker    bonds often result in smaller HOMO-LUMO gap in a conjugated system, as the    orbitals are of higher energy and the    * orbitals, lower energy. Indeed, the HOMO and LUMO energy values reflect perfectly this correlation to the    bond strength.

Table 3 lists the HOMO and LUMO energy values computed at B3LYP/6-31G(d) for the four oligomers at n=12. In all cases, the HOMO are of    symmetry, with the orbital containing the lone pairs on N or P lying lower than the highest    symmetry orbital. As expected, the HOMO in PC (-0.198437 hartrees) is much higher in energy than NC (-0.249014 hartrees), by as much as 0.05 Hartree, or, 31 kcal/mol. Remarkable, this values matches exactly with the    bond energies calculated by Schleyer et al as 80.0 – 49.4 = 31.4 kcal/mol. Also as expected, the HOMO in PSi (-0.183423) is of even higher energy than PC, due to the even weaker P=Si bond. The difference between P=C and P=Si is 0.015 Hartree, or, 9.4 kcal/mol. This difference is smaller than what is predicted by Schleyer et al of 19.7 kcal/mol. Slightly higher P=Si bond strength of between 28 and 37.9 kcal/mol has been reported by Schlegel and coworkers [14]; and with these data, the difference between the P=Si and P=C    bond strength would be between 11.5 and 21.4 kcal/mol. The HOMO in PB is about the same energy as PC; however, the LUMO in PB is much higher in energy than the LUMO in PC. In fact PB has the highest LUMO in all four series This is consistent with the fact that the P=B bond in PBP is a dative bond .

The HOMO-LUMO gaps are the smallest in the two oligomers with the highest    values, as the gap values are 0.05 and 0.02 hartrees for PC and PSi, but as large as 0.11 and 0.10 for NC and PB. These correlations are also consistent with the understanding that systems with smaller HOMO-LUMO gap usually give higher    values. However, as the HOMO-LUMO gap is not the only factor that contributes to the size of    , correlation does not always exist. Between PC and PSi, the reversed correlation is predicted, as PC has a much higher    value as PSi, but also a slightly larger HOMO-LUMO gap.

Table 3: The ⁻n values (in a.u.) for the five series of AB-type oligomers, with n = 2 – 16

| Oligomer | $r_1, r_2$ (Å) | ∟ABA (°) | . (a.u.) | LUMO HOMO (Hartree) |
|---|---|---|---|---|
| –(N=CH)$_{12}$– | 1.376, 1.260 | 119.0 | 19192 | -0.139451 -0.249014 |
| –(P=CH)$_{12}$– | 1.779, 1.672 | 110.9 | 86367 | -0.149445 -0.198437 |
| –(P=SiH)$_{12}$– | 2.174, 2.104 | 110.0 | 35896 | -0.161525 -0.183423 |
| –(PH=BH)$_{12}$– | 1.853, 1.852 | 124.5 | 4698 | -0.096325 -0.197751 |

Besides the HOMO-LUMO gap, another factor that significantly contributes to the size of    is the degree of conjugation of the    system. The bond length alternation ( r) values serve as a good indication of such character. The    r values are summarized in Table 4.

Table 4: Calculated bond length alternation,    r (in Å), for the four series of AB-type oligomers.

| n | –(N=CH)$_n$– | –(P=CH)$_n$– | –(P=SiH)$_n$– | –(PH=BH)$_n$– | –(N=SiH)$_n$– |
|---|---|---|---|---|---|
| 2 | 0.156 | 0.159 | 0.152 | 0.054 | 0.093 |
| 4 | 0.139 | 0.146 | 0.133 | 0.015 | 0.066 |
| 6 | 0.129 | 0.133 | 0.112 | 0.006 | 0.043 |
| 8 | 0.124 | 0.123 | 0.096 | 0.003 | 0.030 |
| 10 | 0.120 | 0.115 | 0.082 | 0.002 | 0.020 |
| 12 | 0.116 | 0.107 | 0.070 | 0.001 | 0.014 |
| 14 | 0.114 | 0.101 | 0.060 | 0.001 | 0.010 |
| 16 | 0.112 | 0.090 | 0.051 | 0.001 | 0.007 |

As can be seen,    r decrease by only 28% in NC between n=2 and n=16, but 43% and 66% in PC and PSi. The greater changes in the    r values for PC and PSi indicate larger conjugation in these two oligomers, which further contributes positively to the    values. Chemically, the larger delocalization can be expected when N is replaced by P, as phosphorus is able to form "hypervalent" structures involving more than 4 covalent bonds due to its size, which effectively increases conjugation and therefore delocalization of the electrons. However, as in the same case of correlations between    and the HOMO-LUMO gap, reversed correlation is again observed between    and  r in the series of PC and PSi series - the change in the r values seem to suggest that PSi should have a larger than PC, however, the calculation predicts a much higher value for PC. What lowers the value in PSi can be due to the larger volume of the PSi polymer, as it consists of both nuclei of the second-row elements. The bond lengths in PSi with n=12 is between 2.104 and 2.174 Å, which are on average 25% larger than the PC bonds of 1.672 and 1.779 Å. The larger volume in PSi decreases the "density" of the electrons which are the means for polarization.

That the r values decreases to essentially zero at n=12 in PB is in complete agreement with the small values at long chain length. The lack of asymmetry in PB can be explained by the nature of bonding in PB. Three major resonance structures (Figure 2) can be expected to contribute to the overall structure of PBP. Natural bond order analysis results show that I and II are by far the most important resonance structures, in spite of the charge separation, and the dative bonds are preferably formed. The combination of both I and II would diminish the asymmetry of the oligomer, so is the contribution of III, if any, as it is completely symmetric.

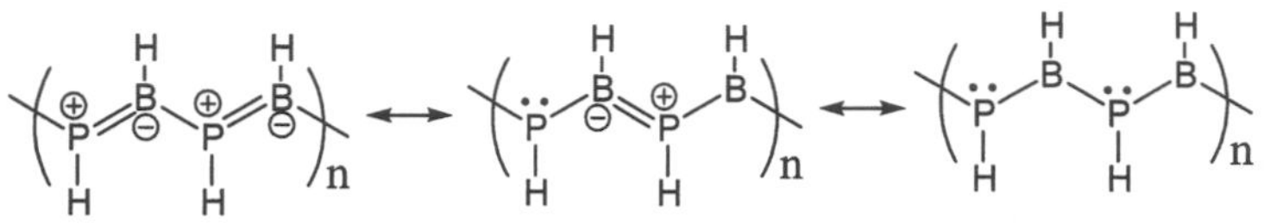

I                     II                     III

Figure 2: Resonance structures of PBP

The negative values in PSA oligomers indicates that the dipole moment in the ground state is larger than in the excited state, which, further suggest that there is a large amount of charge separation in the ground sate of PSN. The sign difference of the values between PSA and PMI, as well as the different behavior of the correlation between versus r in these two series can furthermore be rationalized chemically by the resonance structures.

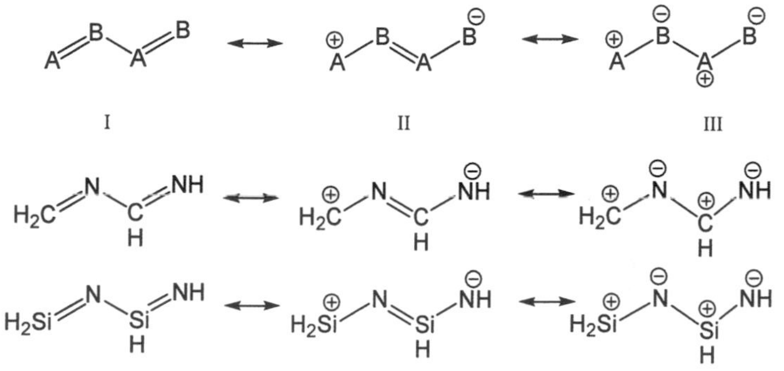

I                     II                     III

Figure 3: Resonance structures of PMI and PNS oligomers with n=2

As shown in Fig. 3, the weight of contribution of each resonance structure is mainly determined by the electronegativities of A and B, as well as the bond strength between A and B. Due to the low electronegativity of Si, structures II and III are expected to have a much larger weight to the overall structures of PSN than in PMI. This explains why the dipole moment is large in PSN and why values in PSN is negative, while for PMI is consistently positive. An increase of the contribution of either II or III would effectively results in a reduction of r, as both II and III add single bond character into the terminal double bonds in I, and structure II also introduces double bond character into the internal single bond in I. However, even though the contribution of II and III have the same effect on r, they have, in contrast, opposite effects on the size of . Specifically, the contribution of II increases conjugation, whereas contribution of III decreases conjugation by localizing the electrons as lone pairs on the more electron negative atom. This interplay of the two resonance structures on r and $_L$ is manifested in the calculation results for PSN. In spite of the smaller    r values in PSN at any chain length, the magnitude of    for PSN is consistently small. Therefore, while reducing the bond lengths alternation value, the contribution of III reduces    conjugation as well, which reduces the first hyperpolarizability. In striking contrast, the contribution of III is expected to be minimal in PMI, due to the high    bond energy of C=N. This is consistent with the large positive    values predicted by calculations.

## 4. Conclusions

The first hyperpolarizability (   ) values of five AB type of oligomer series, with A=N and P, and B=C, Si and B, show the order of PC > PSi >NC >> PB ≈ NSi. The larger    values in the two phosphorus-containing polymers in comparison to PMI are due to, (A) the weaker    bonds of P=C in PPA and P=Si in PPS, relative to both N=C in PMI; and (B) the larger delocalization of the    system in PPA and PPS due to the size of the phosphorus atom that allows for "hypervalent" type of structures and therefore increases the conjugation of the    electrons. Between PPA and PPS, the even larger    values for the former, in spite of the fact that the    bonds are stronger in the former and the delocalization is less, is probably due to the smaller size of PPA relative to PPS, where the "    electron density" is higher in the former. The large difference in the behavior of the    values in PBP and the other three polymers is due to a completely different bonding structure, where charge transferred resonance structures involving dative    bonds are the main contributors to the overall structure of the molecule, and the small    values are due to the lack of asymmetry in the molecular framework.

The small and negative    values for the PSN oligomers show that for certain (AB)$_n$ type of conjugated systems, where A and B have very different electronegativity, a smaller    r does not always correspond to larger    values, as one needs to consider the contribution of zwitterionic resonance structures, which can simultaneously reduce both    r and    .

## Acknowledgments

D. J. and E. A. P thank the Belgian National fund for their research associate positions. Authors gratefully acknowledge the financial support from COST D-26 for STSM, as well as the FNRS-FRFC and the "Loterie Nationale" (convention number 2.4578.02) and the FUNDP for the CPU time at ISCF. D. Y. Zhang would like to thank CNRS for the "poste rouge, 2004 – 2005" during which this work was carried out.

## References

[1]  D. R. Kanis, M. A. Ratner, and T. J. Marks, *Chem. Rev.* **94**, 195 (1994).
[2]  D. Jacquemin, *J. Phys. Chem. A* **108**, 500 (2004).
[3]  D. Jacquemin, E. A. Perpère, and J. M. André, *J. Chem. Phys.* **120**, 4389 (1004).
[4]  D. Jacquemin, *J. Phys. Chem. A* **108**, 9260 (2004).
[5]  D. Jacquemin, J. M. André, and E ; A ; Perpère, *J. Chem. Phys.* **121**, 4389 (2004).
[6]  D. Jacquemin, *J. Chem. Theory Comput.* **1**, 307 (2005).
[7]  D. Jacquemin, M. Medved, and E. A. Perpète, *Int. J. Quantum Chem.* **103**, 226 (2005).
[8]  D. Y. Zhang, C. Pouchan, D. Jacquemin, and E. Q. Perpère, *Chem. Phys. Lett.* **408**, 226 (2005).
[9]  D. Jacquemin, B. Champagne, E. A. Perpète, J. Luis, and B. Kirtman, *J. Phys. Chem. A* **105**, 9748 (2001).
[10] M. J. Frisch, et al. Gaussian 03, Revision B.04, Gaussian, Inc., Wallingford, CT, 2004.
[11] D. Jacquemin, E. A. Perpète, and B. Champagne, *Phys. Chem. Chem. Phys.* **4**, 432 (2002).
[12] D. Jacquemin, B. Champagne, and J. M. André, *Chem. Phys. Lett.* **284**, 24 (1998).
[13] P. von R. Schleyer and D. Kost, *J. Am. Chem. Soc.* **110**, 2105 (1988).
[14] A. G. Baboul and H. B. Schlegel, J. Am. Chem. Soc. 118, 8444 (1996).

Brill Academic Publishers
P.O. Box 9000, 2300 PA Leiden,
The Netherlands

*Lecture Series on Computer
and Computational Sciences*
Volume 4, 2005, pp. 674-677

# Fully Analytical TDHF Schemes for the Determination of Mixed Derivatives of the Energy : Application to the ZPVA Corrections of $\beta(-\omega_\sigma; \omega_1, \omega_2)$

O. Quinet[1]

Laboratoire de Chimie Théorique Appliquée
Facultés Universitaires Notre-Dame de la Paix,
B-5000 Namur, Belgium

Received 10 July, 2005; accepted in revised form 7 August, 2005

*Abstract:* Two different methods are briefly introduced: the first consists into evaluating iteratively the (mixed) derivatives whereas the second is based on the $2n + 1$ rule and the interchange theorem for computing the first and second order derivative of the dynamic polarizability and first hyperpolarizability. The later permits to reduce the complexity of the computations by avoiding the explicit determination of the most time-consuming derivatives or by replacing them by simpler ones. Those schemes have been applied to compute the first-order Zero-Point Vibrational Average (ZPVA) correction to the dynamic electronic first hyperpolarizability $\beta^{\text{elec}}$ for five mono-substituted benzene molecules.

*Keywords:* Time-Dependent Hartree-Fock, analytical derivatives, Zero-Point Vibrational Average, Dynamic First Hyperpolarizability

## 1   Introduction

The Zero-Point Vibrational Average (ZPVA) correction is defined as the difference between the average electronic property in the vibrational ground state and the property evaluated at the equilibrium geometry.

$$\Delta\beta^{\text{ZPVA}} \quad = \quad \langle \Psi^{\text{v}} | \beta(R) | \Psi^{\text{v}} \rangle - \beta(R_0) \tag{1}$$

In the above studies [1–4] the ZPVA correction was evaluated, as in most investigations on polyatomic molecules, to first-order in electrical and mechanical anharmonicity following the general variation-perturbation treatments of Kern, Raynes, Spackman and co-workers : [5–7], i.e.

$$\Delta P^{\text{ZPVA}} \quad \equiv \quad [P]^{0,1} + [P]^{1,0} \tag{2}$$

Here the exponents $n, m$ in $[P]^{n,m}$ refer to the order of perturbation theory in electrical and mechanical anharmonicity, respectively and $P = \alpha, \beta, \gamma$. This approach requires the computationally intensive evaluation of first and, especially, second derivatives of the hyperpolarizability with respect to each normal coordinate $Q_a$ as well as the determination of the cubic force contants, $F_{abb}$.

---

[1]Corresponding author. E-mail: olivier.quinet@fundp.ac.be

For the first hyperpolarizability, the terms in Eq. 2 are

$$[\beta]^{1,0} = \frac{1}{4} \sum_a \frac{\partial^2 \beta^e(-\omega_\sigma; \omega_1, \omega_2)/\partial Q_a^2}{\omega_a} \tag{3}$$

$$[\beta]^{0,1} = -\frac{1}{4} \sum_a \left( \sum_b \frac{F_{abb}}{\omega_b} \right) \frac{\partial \beta^e(-\omega_\sigma; \omega_1, \omega_2)/\partial Q_a}{\omega_a^2} \tag{4}$$

## 2   Determining the mixed derivatives of the energy

The Time-Dependent Hartree-Fock (TDHF) expression for the electronic contribution to the dynamic first hyperpolarizabilty ($\beta^e_{\xi\zeta\eta}$) may be written in matrix form as

$$\beta^e_{\xi\zeta\eta}(-\omega_\sigma; \omega_1, \omega_2) = -\mathrm{Tr}\left[H^\xi D^{\zeta\eta}(\omega_1, \omega_2)\right] \tag{5}$$

where $H^\xi$ is the dipole moment matrix and $D^{\zeta\eta}(\omega_1, \omega_2)$ is the second-order perturbed density matrix associated with the dynamic electric fields $F_\zeta(\omega_1)$ and $F_\eta(\omega_2)$. For the sake of clarity, and to avoid confusion with geometrical indices, the exponent 'e' is eliminated in the following formalism. The first and second derivatives of $\beta_{\xi\zeta\eta}$ with respect to the $a$ and/or $b$ atomic Cartesian coordinates, which are needed to evaluate Eqs 3-4, require, in addition to the second-order field-perturbed density matrix, the determination of third- and fourth-order perturbed density matrices or, equivalently, third- and fourth-order LCAO coefficient matrices.

Various schemes to determine the desired quantities are based on a multiple perturbation theory treatment of the TDHF equation ($FC - i\partial(SC)/\partial t = SC\epsilon$) subject to the normalization condition ($C^\dagger SC = 1$) and using the density matrix ($D = CnC^\dagger$). The matrices involved are expanded as a power series in the simultaneous perturbations due to the dynamic electric fields $\lambda^\zeta = E_\zeta e^{+i\omega_\zeta t}$ and the atomic Cartesian coordinates $\lambda_a = x_a$. This leads to a set of TDHF perturbation equations obtained by equating terms of the same order and time-dependence.

*Iterative schemes.*   One may solve the perturbation equations iteratively for the mixed (i.e. different fields and/or geometrical coordinates) derivatives of the density matrix. This is the procedure followed in the CPHF [8], TDHF [9–11] and extended TDHF [12, 13] methodologies. The main idea in implementing the iterative schemes is to express the perturbed LCAO coefficients (i.e. the coefficient derivatives) in terms of the unperturbed ones.

*Non-iterative schemes.*   The iterative determination of derivatives of the $U$ matrix can be time-consuming especially second derivatives w.r.t. atomic Cartesian coordinates. Fortunately, this step can be avoided by using the $2n + 1$ rule [14]. However, the resulting expression for the second derivatives of $\beta$ w.r.t. atomic Cartesian coordinates then involves corresponding second derivatives of the coefficient matrix. In order to avoid the time-consuming computations that would be entailed in evaluating the latter we use interchange relations [15, 16], which introduce third derivatives of the coefficient matrix w.r.t. electric fields. Although the $2n + 1$ rule is abandoned in this regard the computational effort is considerably reduced.

## 3   Applications

Our computational method to obtain the TDHF ZPVA correction for $\beta$ has been implemented in the GAMESS quantum chemistry package[17]. In the present application to a set of mono-substituted benzenes (Ph-R with R = $CH_3$, CHO, OH, $NH_2$, and $NO_2$), this fully analytical, partially non-iterative, procedure was used to evaluate the electronic hyperpolarizabilities as well as their first and second derivatives with respect to geometrical coordinates. On the other hand, the anharmonic force constants ($F_{abb}$) required for Eq. 4 were obtained by numerical differentiation of the Hessian. For that purpose we computed the Hessian at geometries displaced from equilibrium along the different normal coordinates. The displacement amplitudes were $\pm 0.050$ and $\pm 0.100$ a.u. and a Romberg procedure[18] was employed for increased numerical accuracy. In order to study basis set effects both the 6-31G and 6-31G* basis sets were utilized. The results are summarized in Table 1.

Table 1: Theoretical Hartree-Fock conformation of the optimized ground state and TDHF hyperpolarizabilities of Ph-CH$_3$, Ph-CHO, Ph-OH, Ph-NH$_2$ and Ph-NO$_2$. The optical frequency is $\hbar\omega$ = 0.042823 au = 1.16527 eV or, equivalently, $\lambda$ = 1064 nm.

| Molecule | Basis set | Conformation[a] | $\beta_{//}(0;0,0)$ | | $\beta_K(-\omega;\omega,0)$ | | $\beta_{//}(-2\omega;\omega,\omega)$ | |
|---|---|---|---|---|---|---|---|---|
| | | | $[\beta]^I$ | $\beta^e$ | $[\beta]^I$ | $\beta^e$ | $[\beta]^I$ | $\beta^e$ |
| Ph-CH$_3$ | 6-31G | NP (C$_s$) | 2.742 | 2.303 | 2.772 | 2.245 | 2.880 | 2.610 |
| | 6-31G* | NP (C$_s$) | 3.208 | 3.084 | 3.257 | 3.013 | 3.392 | 3.290 |
| Ph-CHO | 6-31G | P (C$_s$) | -4.918 | 56.487 | -4.746 | 62.979 | -4.057 | 80.385 |
| | 6-31G* | P (C$_s$) | -4.881 | 47.027 | -4.772 | 52.335 | -4.280 | 66.608 |
| Ph-OH | 6-31G | P (C$_s$) | -1.089 | -38.031 | -1.132 | -38.605 | -1.288 | -40.393 |
| | 6-31G* | P (C$_s$) | -1.369 | -20.469 | -1.419 | -20.690 | -1.586 | -21.508 |
| Ph-NH$_2$ | 6-31G | P (C$_{2v}$) | -5.349 | 131.204 | -5.333 | 136.134 | -5.210 | 149.957 |
| | 6-31G* | NP (C$_s$) | 0.708 | 37.385 | 0.779 | 39.021 | 1.049 | 43.539 |
| Ph-NO$_2$ | 6-31G | P (C$_{2v}$) | -2.622 | 98.193 | -1.694 | 110.331 | 1.549 | 142.440 |
| | 6-31G* | P (C$_{2v}$) | -3.774 | 46.682 | -3.329 | 53.489 | -1.705 | 71.982 |

[a] P: planar, NP: non-planar

The ZPVA correction to the static and dynamic $\beta$ is always less than about 10% of the electronic value except in the case of toluene where the electronic and ZPVA terms are of similar magnitude. Even when the magnitude of the ratio is small, however, there is considerable variation depending upon the particular substituent.

Symmetry, when present, plays a key role in the distribution of the ZPVA corrections between electrical and mechanical anharmonicity. In the absence of degenerate vibrations the anharmonic force constants $F_{abb}$ vanish unless $Q_a$ is totally symmetric. Thus, the ZPVA correction associated with non-totally symmetric modes is due entirely to electrical anharmonicity. For that reason the electrical anharmonicity often, though not always, is the major contributor to the total $[\beta]^I$. The situation is complicated by the fact that mechanical anharmonicity can be more important than electrical anharmonicity for the totally symmetric vibrations and, in general, the sign of each individual term may be either positive or negative. It is noteworthy that the particular vibration(s) making the largest contribution to the ZPVA correction is(are) almost always associated with the substituent group. The predominant role of the substituent group may be explained by the fact that its presence destroys the center of symmetry and allows for beta to be non-zero.

## Acknowledgment

O.Q. and B.C. thank the Belgian National Fund for Scientific Research for their Postdoctoral Researcher and Senior Research Associate positions, respectively. The calculations were performed thanks to the Interuniversity Scientific Computing Facility (ISCF), installed at the Facults Universitaires Notre-Dame de la Paix (Namur, Belgium), for which the authors gratefully acknowledge the financial support of the FNRS-FRFC and the "Loterie Nationale" for the convention n 2.4578.02, and of the FUNDP.

# References

[1] D. M. Bishop and S. P. A. Sauer, J. Chem. Phys. **107**, 8502 (1997).

[2] D. M. Bishop and P. Norman, J. Chem. Phys. **111**, 3042 (1999).

[3] P. Norman, Y. Luo, and H. Ågren, J. Chem. Phys. **109**, 3580 (1998).

[4] V. E. Ingamells, M. G. Papadopoulos, and S. G. Raptis, Chem. Phys. Lett. **307**, 484 (1999).

[5] W. C. Ermler and C. W. Kern, J. Chem. Phys. **55**, 4851 (1971).

[6] W. T. Raynes, P. Lazzaretti, and R. Zanasi, Mol. Phys. **64**, 1061 (1988).

[7] A. J. Russell and M. A. Spackman, Mol. Phys. **84**, 1239 (1995).

[8] Y. Yamaguchi, Y. Osamura, J. D. Goddard, and H. F. Schaefer III, *A New Dimension to Quantum Chemistry: Analytic Derivative Methods in Ab Initio Molecular Electronic Struture Theory* (Oxford University Press, Oxford, 1994).

[9] H. Sekino and R. J. Bartlett, J. Chem. Phys. **85**, 976 (1986).

[10] S. P. Karna, M. Dupuis, E. Perrin, and P. N. Prasad, J. Chem. Phys. **92**, 7418 (1990).

[11] S. P. Karna and M. Dupuis, J. Comp. Chem. **12**, 487 (1991).

[12] O. Quinet and B. Champagne, J. Chem. Phys. **115**, 6293 (2001).

[13] O. Quinet and B. Champagne, J. Chem. Phys. **117**, 2481 (2002), publisher's note : **118**, 5692 (2003).

[14] T. S. Nee, R. G. Parr, and R. J. Bartlett, J. Chcm. Phys. **64**, 2216 (1976).

[15] A. Dalgarno and A. L. Stewart, Proc. Roy. Soc. (London) **A242**, 245 (1958).

[16] B. Kirtman, J. Chem. Phys. **49**, 3895 (1968).

[17] M. W. Schmidt, K. K. Baldridge, J. A. Boatz, S. T. Elbert, M. S. Gordon, J. H. Jensen, S. Koseki, N. Matsunaga, K. A. Nguyen, S. J. Su, T. L. Windus, M. Dupuis, and J. A. Montgomery, J. Comp. Chem. **14**, 1347 (1993).

[18] P. J. Davis and P. Rabinowitz, in *Numerical Integration* (Blaisdell Publishing Company, London, 1967), p. 166.

Brill Academic Publishers
P.O. Box 9000, 2300 PA Leiden,
The Netherlands

*Lecture Series on Computer
and Computational Sciences*
Volume 4, 2004, pp. 678-678

# About the Longitudinal Polarizability of Quasilinear Systems

R. Resta [1], K.N. Kudin [2], and R. Car[2]

[1] INFM-Democritos National Simulation Center, Trieste, and University of Trieste, Italy
[2] Department of Chemistry and Princeton Institute for Science and Technology of Materials (PRISM),
Princeton University, USA

Received 11 July, 2005; accepted in revised form 10 August, 2005

Whenever a quasilinear system is nonconducting and macroscopically homogeneous, its longitudinal linear polarizability is an extensive quantity: the dipole linearly induced along $z$ by an applied unit field parallel to $z$ is asymptotically proportional to the length $L$ of the system. Ergo, the polarizability per unit length $\alpha(L)/L$ is an intensive quantity and goes to the finite limit $\alpha_\infty$ . The main issue addressed here is how and why the asymptotic value is approached. Our finding is that the asymptotic value $\alpha_\infty$ is approached with a universal power law in $L$. We emphasize that, instead, the polarizability of a dielectric *slab* of thickness $L$ converges *exponentially* with $L$ to the corresponding asymptotic (bulk) value.

The dominant role of electrostatics in such properties is demonstrated by means of a one-dimensional Clausius-Mossotti-like classical model. The main conclusions derived from the model are applicable to the realistic polarizability of many-electron systems, owing to the modern theory of polarization [1], according to which the dipolar polarizations of the Wannier-function charge distributions are enough to determine the macroscopic (electronic) polarization. Ergo, the many-electron system can be *exactly* mapped into a Clausius-Mossotti-like one, where the localized polarizable charges are identified with Wannier-function charge distributions.

We demonstrate our findings for stereoregular polymers, by means of ab-initio calculations (both Hartree-Fock and density-functional) on linear stacks of $H_2$ molecules, and on polyacetylene chains, going up to very high $L$ values [2].

[1] R. Resta, *Macroscopic polarization in crystalline dielectrics: The geometric phase approach*, Rev. Mod. Phys. **66**, 899 (1994).

[2] K.N. Kudin, R. Car, and R. Resta, *Longitudinal polarizability of long polymeric chains: Quasi-one-dimensional electrostatics as the origin of slow convergence*, J. Chem. Phys. **122**, 134907 (2005).

Brill Academic Publishers
P.O. Box 9000, 2300 PA Leiden
The Netherlands

*Lecture Series on Computer
and Computational Sciences*
Volume 4, 2005, pp. 679-681

# The Infrared and Vibrational Circular Dichroism Spectra of (S)-Proline in Water: an ab initio study

Chiara Cappelli,[¥] Susanna Monti, Antonio Rizzo

Istituto per i Processi Chimico Fisici del Consiglio Nazionale delle Ricerche, Area della Ricerca di Pisa,
loc. San Cataldo, Via Moruzzi 1, I-56124, Pisa, Italy

Received 19 July, 2005; accepted in revised form 11 August, 2005

## 1. Introduction

As main subject of the talk, the infrared and vibrational circular dichroism of proline in water solution are investigated *ab initio* employing density functional theory and the integral equation formalism (IEF) version of the polarizable continuum model (PCM). Three solvent models are exploited to evaluate solvent effects in the 1000-2000 cm$^{-1}$ frequency range: a pure implicit continuum approach, a pure explicit model (limited to three solvent molecules) and a combined specific/continuum approach. Effects on spectra arising from different protonation states (neutral, zwitterionic, cationic and anionic are analyzed.[1]

The effect of specific (hydrogen bond) solute-solvent interactions, arising due to the unique features of the aqueous environment, and which can only partially be estimated by a pure continuum approach is accounted for by introducing cluster structures including explicitly three water molecules arranged around the proline moiety.

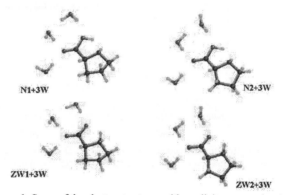

Figure 1. Some of the cluster structures with explicit water molecules

The conformational analysis, yielding information on free energy and Boltzmann populations for each individual conformer, gives insights on the distribution among different conformers of proline in the gas phase and in water. Although the effect of environment on the geometry of a given conformer is relatively small, the ordering as determined by the Boltzmann population may change dramatically, and, with the stabilization provided by the solvent to H-bonded and charged dipolar structures, it leads in some cases to the disappearance in the aqueous environment of some of the low lying gas-phase favored conformational structures.

---

¥ Current affiliation: PolyLab - INFM, Pisa

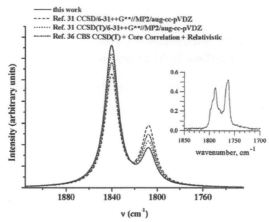

Figure2. The calculated IR spectrum of neutral non-zwitterionic proline in the gas phase in the 1725–1925 cm_1 region (solid curve). The experimental spectrum is in the inset. Also shown are simulated spectra obtained by using our B3LYP/6-31G(d) frequencies and intensities together with Boltzmann populations taken from the literature. The temperature is 435 K.

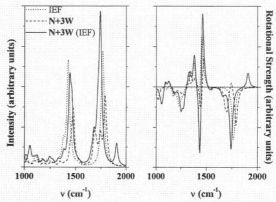

Figure3. The simulated Boltzmann averaged IR (left) and VCD (right) spectra in the 1000–2000 cm$^{-1}$ frequency range for the neutral non-zwitterionic praline in water obtained by using the three different salvation models

The effect of the environment on IR and VCD spectra, analyzed in the 1000-2000 cm$^{-1}$ frequency range, is remarkable, causing shifts to lower frequencies and changes in intensities and, in the case of VCD, also changes of band sign. The structural model employed (proline moiety, or proline with three water molecules in the cluster, with or without inclusion of the IEF continuum) influences remarkably in some cases the outcome. Also, protonation has a noticeable effect on the predicted spectra, as does the balance of zwitterionic and non-zwitterionic neutral structures in solution.

The speaker plans also to present and discuss very briefly the highlights of his studies of the effects of molecular vibrations on high order mixed electric and magnetic properties, with emphasis on those involved in the rationalization of birefringences. These are often essential for a correct interpretation of experiment. Examples are given for the electric (Kerr), magnetic (Cotton-Mouton) and electric-field-gradient (Buckingham) birefringences.

The results of a combined experimental/theoretical study of the Kerr birefringence in molecular oxygen in its ground $^3\Sigma_g^-$ electronic state will be shown.[2] These include the determination of the effect of molecular vibrations (both zero point vibrational average and pure vibrational contributions) carried out exploiting a multiconfigurational self-consistent field wave function model for the analytical response calculation of the molecular frequency dependent properties.

The effect of molecular vibrations on the molecular magnetizability anisotropies, electric dipole polarizabilities and molecular quadrupole moments which are often measured in experiments combining Kerr, Cotton-Mouton and Buckingham birefringences will be also discussed, if time allows. Examples from a now rather extensive list of systems, of varying size and complexity, will be presented.

## References

[1] C. Cappelli, S. Monti and A. Rizzo, *"Effect of the Environment on Vibrational Infrared and Circular Dichroism Spectra of (S)-Proline."*, Int. J. Quantum Chem., in press,. On-line since March 22nd, 2005, http://www3.interscience.wiley.com/cgi-bin/fulltext/110432109/HTMLSTART)

[2] F. Bielsa, R. Battesti, C. Robilliard, G. Bialolenker, G. Bailly, G. Trénec, A. Rizzo and C. Rizzo, *"Kerr effect of molecular oxygen at λ=1064 nm: experiment and theory."*, Eur. Phys. J. D , in press.

Brill Academic Publishers
P.O. Box 9000, 2300 PA Leiden,
The Netherlands

*Lecture Series on Computer*
*and Computational Sciences*
Volume 4, 2005, pp. 682-685

# Computational Challenges in Vibrational Contributions to (non)linear Optical Properties in the Gas and Liquid Phase

**Kenneth Ruud.**[1]

Department of Chemistry, University of Tromsø, N-9037 Tromsø, Norway

Received July 10, 2005; accepted in revised form 16 August, 2005

*Abstract:* In this paper I will discuss some of the challenges that face the *ab initio* calculation of vibrational contributions to linear and nonlinear optical properties, with particular emphasis on the computational costs of such calculations. A few of the most recent advances made in my research group to overcome these obstacles will be presented, such as parallel calculations with superlinear scaling, a fully parallel polarizable continuum model, and the use of a optimized, partial force field for calculating the most dominating vibrational contributions through mode selection.

*Keywords:* Parallel calculations, Polarizable Continuum Model, Vibrational contributions, Mode selection

## 1   Introduction

Since attention was drawn to the possible importance of vibrational contributions to linear and nonlinear optical properties [1], the field has been growing steadily. This growth has to a large extent been fueled by the observation that the vibrational contributions can in many case be as large as, or in some cases also be larger than, the electronic contributions, in particular for static (hyper)polarizabilities, see for instance Ref. [2]. For an account of recent work in the field of *ab initio* calculations of vibrational contributions to (non)linear optical properties I refer to recent reviews [3, 4].

One of the common ways of calculating vibrational contributions to (non)linear optical properties is based on the perturbation analysis first introduced by Bishop and Kirtman [5, 6]. In this approach, the vibrational contributions to the (non)linear optical properties are calculated as a linear combination of products of geometrical derivatives of lower-order (non)linear optical properties and the harmonic and/or anharmonic force fields. Although the optical properties involved are of lower order than the electronic (non)linear optical property of interest, the need for, in most cases, performing numerical differentiation with respect to nuclear displacements increases the scaling of the calculations by a factor of at least $6N$, where $N$ is the number of nuclei in the molecule. As such, the cost of calculating vibrational corrections to nonlinear optical properties increases steeply with the size of the molecule. Any means for reducing the time required for a single-point calculation of the molecular properties, or the number of geometrical derivatives, is therefore of great importance. Note that recent work have enable analytical calculations of first-order geometrical derivatives of both the linear polarizability and the first hyperpolarizability [5, 6].

In the following I will briefly outline some of the most recent efforts in my research group to reduce the computational time needed for single-point property calculations, and of the number of geometrical derivatives needed for calculating vibrational corrects to these properties.

---

[1] E-mail: ruud@chem.uit.no

## 2  Superlinear scaling in parallel calculations

Traditionally, in Hartree–Fock and Density Functional Theory calculations, the time-consuming two-electron integrals were written to disk, and then reread in all subsequent evaluations of Fock or generalized Fock matrices. This put limits to the size of the systems that could be effectively calculated, due to the need for very large two-electron integral files. A major improvement came with the introduction of the direct Hartree–Fock method [7], in which the two-electron integrals were not stored on disk, but rather recalculated when needed, at the price of significantly increased computational time. To reduce the computational time, most Hartree–Fock and DFT codes are today parallelized, by distributing the calculation of two-electron integrals to multiple processors. Assuming perfect linear scaling, such a parallelization scheme will reduce the wall time needed for a calculation by a factor $1/K$, where $K$ is the number of processors. Using Global Array tools to perform parallel I/O [8], even larger reductions in computational time can be achieved, so-called superlinear scaling.

We have extended our master-slave parallel calculation of two-electron integrals to take advantage of the increasing amount of memory available when running a calculation on several processors. In the first evaluation of the two-electron integrals, batches of two-electron integrals are distributed to the different slaves. The slave then stores as many of these two-electron integrals in memory as possible, reporting back to the master whether a batch of integrals have been stored or not. In all subsequent Fock matrix calculations, the batches of integrals stored in memory are first assigned to the slaves having stored them, whereas any remaining integrals are then distributed to the slaves for recomputation in a load-balanced manner—that is, the master assigns a computational task to a slave as soon as it finishes a batch of two-electron integrals. Each slave constructs it's own partial (generalized) Fock matrix, which are added together at the end of the two-electron integral evaluation on the master node, which then performs the necessary matrix operations.

The calculation of response functions can be formulated using only generalized Fock matrices [9], and the two-electron integrals stored in memory can therefore be used both in the wave function optimization step as well as in the calculation of the linear, quadratic of cubic response functions [10]. As less computing time is needed as more processors are added (until all two-electron integrals can be stored in memory), superlinear scalings can be expected. Indeed, in calculations of the second hyperpolarizability, we have seen a speed-up by a factor 55 compared to the time used on a single node when calculating the second hyperpolarizability for a molecule with 350 basis function on a total of 16 slave nodes [10].

## 3  A parallel Polarizable Continuum Model

There are only a very limited number of studies of solvent effects on vibrational contributions to nonlinear optical properties (see for instance Ref. [11]). However, since most experiments take place in a liquid solution, or in some kind of condensed phase, being able to take into account medium effects is important in order to give a reliable prediction of the importance of vibrational contributions to nonlinear optical properties. Furthermore, since a solvent may often reduce the vibrational frequencies of the solute, vibrational effects may turn out to be even more important in the liquid or condensed phase [11].

The Polarizable Continuum Model [12, 13, 14] is a popular approach to account for solvent effects on molecular properties using a polarizable, dielectric continuum as a model for the collective effect of the solvent molecules on the solute of interest. PCM distinguishes itself from other, similar approaches in that the cavity in which the solute is places is shaped after the molecule, the cavity surface being discretized into small elements (tesserae) on which the induced surface charge and the potential due to the solute is assumed to be constant. The electrostatic problem can in this way be discretized and the interaction energy be written as a sum over all tesserae. The free-energy functional of the solute is then

$$\mathcal{G} = \left\langle \Psi \left| \hat{H}_0 \right| \Psi \right\rangle + \frac{1}{2} \sum_{pq} D_{pq} \sum_{\tau} \left( \frac{1}{2} \left( V_{pq}^{\tau} \cdot q_N^{\tau} + V_N^{\tau} \cdot q_{pq}^{\tau} \right) + V_{pq}^{\tau} \cdot q_e^{\tau}(\rho) \right), \tag{1}$$

where $\hat{H}_0$ is the vacuum Hamiltonian, $D_{pq}$ the one-electron density matrix, $V_{pq}^{\tau}$ and $V_N^{\tau}$ the electrostatic potential at the tessera due to the electron density and the nuclear framework, respectively, and $q_{pq}$ and $q_N$ the corresponding charges induced on the cavity surface.

The free energy functional only depend on one-electron quantities (the potential at the tessera) and thus formally scales as $N^2$ with respect to the number of basis functions $N$, but due to the large number of tesserae needed to enclose the solute, the effective scaling may be closer to $N^3$. In parallel calculations this may lead to significant computational bottlenecks, as these one-electron terms may become time-consuming compared to the two-electron integrals that are calculated in parallel.

To overcome these potential problems, we have now parallelized all summations over tesserae in a load-balanced manner, both in the calculation of the molecular energy, as well as the one- and two-index transformed potentials and charges needed for the calculation of linear and quadratic response functions. Initial studies show the scaling of this approach to be good (close to linear scaling), and we expect to present more detailed scaling results in the near future [15].

## 4  Mode selection

One might expect certain vibrational modes to be more important than others in determining a vibrational response of a molecule. In such cases, it would simplify our calculations if we could calculate only these modes and their contribution to, for instance, the vibrational contribution to a nonlinear optical property without determining the full vibrational force field.

It was recently suggested that selective calculations of only a few normal modes could be be achieved using a Davidson-like approach for the vibrational modes of interest [16]. Thus, starting from one or more vibrational modes of interest, using as start vectors the normal coordinates obtained in a low-level calculation, such as for instance a Hartree–Fock calculation using an STO-3G basis, the normal modes can be iteratively improved

$$[\mathbf{H} - \lambda_{\mu}^{(i)}]\mathbf{Q}_{\mu}^{(i)} = \mathbf{r}_{\mu}^{(i)}, \tag{2}$$

where $\mathbf{H}$ is the vibrational Hessian, $\mathbf{Q}_{\mu}^{(i)}$ the $i$th iterative solution to normal coordinate $\mu$ and $\lambda_{\mu}^{(i)}$ the corresponding eigenvalue, and the iterative procedure is stopped when the residual $\mathbf{r}_{\mu}^{(i)}$ becomes less than a predefined threshold.

Since the normal coordinates are limited, we only need to perform a numerical differentiation along the normal coordinates of interest until convergence is achieved. Such a scheme is thus actually more adapted to numerical differentiation than for analytical evaluation of the vibrational modes, since a similar decoupling of the differentiated quantities cannot be achieved in analytical calculations, since the normal coordinates themselves are highly delocalized over the atomic distortions (which are the building blocks of the analytical derivative calculations).

This selective mode optimization scheme is directly parallelizable over the normal mode distortions, and we have combined this functionality [16] with the Dalton quantum chemistry program [17]. In the case where only a limited number of normal coordinates are of interest, the parallelization is very coarse-grained. In order to improve parallel efficiency, as well as take advantage of the parallel improvements described in the two preceding sections, we have enabled a parallel distribution of the distorted geometries, and at each geometry calculated simultaneously we perform a parallel Dalton calculation [18]. In this way, we ensure the most efficient use of the computational resources.

Let us finally note that the use of mode selection allows us to perform, in a computationally efficient manner, another important set of calculations: In cases where direct interactions between solute and solvent molecules are important, the use of mode selection allows us to consider only the vibrational modes of the solute without having to explicitly determine the force field for the intermolecular solute-solvent and solvent-solvent vibrations, nor of the intramolecular solvent vibrations, and this may lead to significant computational savings when considering large molecules in highly coordinating solvents.

## Acknowledgment

This work has received support from the Norwegian Research Council through a Strategic University Program in Quantum Chemistry (Grant No 154011/420), a YFF grant to KR (Grant No 162746/V00), from the Nanomat program (Grant No 158538/431), as well as through a grant of computer time from the Norwegian Supercomputing Program.

## References

[1] Bishop, D. M. *Rev. Mod. Phys.* **1990**, *62*, 343.

[2] Quinet, O.; Champagne, B. *J. Chem. Phys.* **1998**, *109*, 10594.

[3] Bishop, D. M.; Norman, P. . In *Handbook of Advanced Electronic and Photonic Materials*; Nalwa, H. S., Ed.; Academic Press: San Diego, 2000.

[4] Bishop, D. M. *Adv. Chem. Phys.* **1998**, *104*, 1.

[5] Bishop, D. M.; Kirtman, B. *J. Chem. Phys.* **1991**, *95*, 2646.

[6] Bishop, D. M.; Kirtman, B. *J. Chem. Phys.* **1992**, *97*, 5255.

[7] Almlöf, J.; Fægri Jr., K.; Korsell, K. *J. Comput. Chem.* **1982**, *3*, 385.

[8] Lindh, R.; Krogh, J. W.; Schütz, M.; Hirao, K. *Theor. Chem. Acc.* **2003**, *110*, 156.

[9] Norman, P.; Jonsson, D.; Ågren, H.; Dahle, P.; Ruud, K.; Helgaker, T.; Koch, H. *Chem. Phys. Lett.* **1996**, *253*, 1.

[10] Fossgård, E.; Ruud, K. In preparation.

[11] Norman, P.; Luo, Y.; Ågren, H. *J. Chem. Phys.* **1998**, *109*, 3580.

[12] Miertuš, S.; Scrocco, E.; Tomasi, J. *J. Chem. Phys.* **1981**, *55*, 117.

[13] Cammi, R.; Tomasi, J. *J. Comput. Chem.* **1995**, *16*, 1449.

[14] Tomasi, J.; Persico, M. *Chem. Rev.* **1994**, *94*, 2027.

[15] Ruud, K.; Ferrighi, L.; Frediani, L. In preparation.

[16] Reiher, M.; Neugebauer, J. *J. Chem. Phys.* **2003**, *118*, 1634.

[17] "Dalton, an ab initio electronic structure program, Release 2.0. See http://www.kjemi.uio.no/software/dalton/dalton.html", 2005.

[18] Herrmann, C.; Reiher, M.; Ruud, K. Work in progress.

Brill Academic Publishers
P.O. Box 9000, 2300 PA Leiden
The Netherlands

*Lecture Series on Computer
and Computational Sciences*
Volume 4, 2005, pp. 686-688

# Dynamic Response Property Evaluation by Multi-resolution Multi-wavelet Basis Set

H. Sekino[1], Y. Maeda and S. Hamada

Department of Knowledge Based Information Engineering,
Toyohashi University of Technology,
1-1 Tenpaku-cho Toyohashi, Japan

Received 22 July, 2005; accepted in revised form 18 August, 2005

*Abstract:* An Exponential Linear Part (ELP) algorithm for solving a differential equation in time domain is implemented for evaluation of dynamic response properties which may reveal an instability. For evaluation of less problematic problems, frequency domain algorithm is developed. The method based on multi-resolution multi-wavelet basis is implemented and applied for high-precision evaluation of dynamic response properties such as frequency dependent polarizabilities. Results prove the guaranteed convergence property within a given precision.

*Keywords:* Dynamic Response Property, Polarizability, Multi-resolution Multi-wavelet basis, ELP Time-domain equation solver

## 1. Multi-resolution representation of the response equation

The response equation is formulated in non-standard form where the operators are represented on scaling function in hierarchical space $V_n$ and on wavelet in difference space $W_n$.

$$\begin{bmatrix} T & C \\ B & A \end{bmatrix} \begin{bmatrix} x \\ y \end{bmatrix} = \begin{bmatrix} S \\ d \end{bmatrix}$$

where $x$, $S \in V_{n-1}$ and $y$, $d \in W_n = V_n - V_{n-1}$.
The solution $x$ at resolution level $n-1$ is corrected by the contribution from the difference space between the resolution of $n-1$ and higher resolution $n$.

$$(T - CA^{-1}B)x = s - A^{-1}d$$

through

$$y = A^{-1}(d - Bx).$$

We used the scaling function and wavelet created with Legendre polynomial assigned in the 3-dimensional boxes for each level of the. hierarchical space in order to represent physical operators. Gauss-Legendre quadrature formula is used for the evaluation of the operators.

## 2. Time Domain Equation Solver

The time domain equation solver is formulated for the problems where time progression of wavefunction involves high-order differentiation in the space coordinate. The situation may be encountered in transportation problems where the wavefunction representing the transport can be largely deviated from the one at the stationary bound state.
The time dependent equation can be formulated as

$$\frac{d\Psi}{dt} = L\Psi + N[\Psi]$$

[1] Corresponding author. E-mail: sekino@tutkie.tut.ac.jp

where $L$ and $N$ indicate linear and nonlinear operators respectively. We employed Exponential Linear Part (ELP) algorithm [1] where linear part is treated exactly on the exponential anzaz whereas nonlinear part is computed in a regular implicit predictor-corrector scheme. The dynamic response of the system can be computed from the time progression of the wavefunction.

## 3. Dynamic Property Evaluation in Frequency Domain

For less problematic problems where electron dynamics does not involve drastic reconstruction of the electronic state, we can formulate a regular perturbative form of the response equation for the corresponding frequency. However, since Multiresolution representation employed focuses on the convergent property towards completeness, the zero-th order equation does not provide redundant quantity such as virtual space. The projection operator is not defined on the virtual-occupied space but uniformly defined using the zero-th order density matrix only. The first-order density matrix for N electron systems is characterized by 2N parameters.

$$u_p^{(+)}(\omega) = (1 - \rho^0)\rho(\omega)\rho^0$$

and

$$u_p^{(-)}(\omega) = \rho^0 \rho(\omega)(1 - \rho^0)$$

which are determined by the first-order equation of motion for the first-order density matrix $\rho(\omega)$ in frequency domain.

$$(1 - \rho^0)[(\hat{F}^0 - \varepsilon_p^0)u_p^{(+)}(\omega) + \{r_\zeta + \tfrac{\partial \tilde{g}}{\partial \rho}[\rho^0] * (\sum_i^{occ} u_p^{(+)}\phi_i^+(r) + \sum_i^{OCC} u_p^{(-)}(\omega)\phi_i(r)\}\phi_p(r)]\rho^0$$

$$= \omega u_p^{(+)}(\omega)$$

and

$$\rho^0[(\hat{F}^0 - \varepsilon_p^0)u_p^{(-)}(\omega) + \{r_\zeta + \tfrac{\partial \tilde{g}}{\partial \rho}[\rho^0] * (\sum_i^{occ} u_p^{(+)}\phi_i^+(r) + \sum_i^{OCC} u_p^{(-)}(\omega)\phi_i(r)\}\phi_p(r)](1 - \rho^0)$$

$$= \omega u_p^{(-)}(\omega)$$

Here we applied the formalism for evaluation of optical properties such as dynamic polarizabilities. The operators for nuclear attraction, electron repulsion, dipole moment as well as density functional are evaluated using the wavelet basis created by the Legendre polynomial of k-th order. The kinetic energy part is evaluated using BSH kernel for the Green's function. Detailed algorithm is described in Ref [2-4]. All the operators including kinetic part are evaluated by the program system MADNESS [5] The dynamic polarizability is defined as linear coefficient of the polarization response with respect of external perturbation $E(\omega)$.

$$\mu_\xi(\omega) = \mu_0 + \alpha_{\xi\zeta}(\omega) \bullet E_\zeta(\omega)$$

and can be evaluated straightforwardly.

$$\alpha_{\xi\zeta}(\omega) = -Tr\langle r_\xi \bullet \rho_\zeta(\omega)\rangle.$$

The obtained polarizabilities for typical small molecules are compared with ones by different size of Gaussian basis sets. The results for typical small molecules indicate the guaranteed convergence property in the multi-resolution multi-wavelet basis set method and insufficiency of Gaussian basis sets for property evaluation. The results for larger molecules and other properties will be presented in the conference. The algorithm taken here for evaluation of response properties is quite similar to the one developed for evaluation of excitation energy of the molecules [6]

H₂O

| | static(ω=0.0) | | dynamic(ω=0.0656) | |
|---|---|---|---|---|
| | HF | Svwn5 | HF | Svwn5 |
| Gaussian function basis | | | | |
| cc-pvdz(24) | 5.0087 | 5.2918 | 5.0598 | 5.3670 |
| cc-pv5z(201) | 7.8416 | 9.0932 | 7.9264 | 9.2436 |
| aug-cc-pvdz(41) | 8.1319 | 9.8633 | 8.2257 | 10.0555 |
| aug-cc-pv5z(287) | 8.5189 | 10.5244 | 8.6208 | 10.7314 |
| d-aug-cc-pvdz(58) | 8.5123 | 10.6071 | 8.6158 | 10.8255 |
| d-aug-cc-pv5z(373) | 8.5324 | 10.5870 | 8.6359 | 10.8042 |
| | | | | |
| MRMW basis | | | | |
| k=5 | 8.5178 | 10.5734 | 8.6243 | 10.7825 |
| k=7 | 8.5323 | 10.5872 | 8.6359 | 10.8051 |
| k=9 | 8.5323 | 10.5869 | 8.6359 | 10.8047 |

## Acknowledgement

We thank Prof. R.J. Harrison at Oakridge National Laboratory to provide us MADNESS program. Special thanks go to Dr.T. Yanai for providing TDKS code based on MR algorithm. This research is partly supported by Grant-in-Aid for Science from the Ministry of Education, Culture, Sports, Science and Technology (No.17300094) and CREST-JST.

## References

[1] G..Betlkin, J. M. Keiser, *J.l of Comptational Physics*, **147**, 362-387 (1998). .
[2] B. Alpert, G. Beylkin, D. Gines, and L. Vozovoi, *J. Comput. Phys.* **182**, 149 (2002); University of Colorado,.APPM preprint #409, June 1999
[3] R. J. Harrison, G.I. Fann, T. Yana, Z. Gan and G. Beylkin, *J. Chem. Phys.*, **121**, 11537-11597 (2004)
[4] T. Yanai, G. I. Fann, Z. Gan, R. J. Harrison and G. Beylkin, *J. Chem. Phys.*, **121**, 6680-6688
[5] Gneral quantum chemical program system based on Multi-resolution Multi-wavelet basis set, created by Robert Harrison (Oak Ridge National Laboratory)
[6] T. Yanai, G.I.Fann,R.J.Harrison and G. Beylkin, to appear in *J. Chem. Phys.*

Brill Academic Publishers
P.O. Box 9000, 2300 PA Leiden
The Netherlands

*Lecture Series on Computer*
*and Computational Sciences*
Volume 4, 2005, pp. 689-692

# Responses of Molecular Vibrations to Electric Fields and Their Effects on Vibrational Spectra

Hajime Torii[1]

Department of Chemistry, School of Education, Shizuoka University,
836 Ohya, Shizuoka 422-8529, Japan

Received 21 July, 2005; accepted in revised form 15 August, 2005

*Abstract:* Various responses of molecular vibrations to electric fields (including both radiation fields and intermolecular electric fields) are examined. The role of the electrical property derivatives (dipole derivatives, etc.) in describing those responses is illustrated on the basis of the expansion formula of the potential energy with respect to the electric field and vibrational coordinates. The effects of those responses on the linear and nonlinear vibrational spectroscopic features are discussed.
*Keywords:* molecular vibration, electric field, dipole derivative

*PACS:* 33.15.Kr, 33.20.Ea, 33.20.Fb, 33.20.Tp, 33.70.Fd, 33.70.Jg, 78.30.Cp

## 1. Introduction

Molecular vibrations respond to electric fields in various ways, depending on the nature and the time scales of both the molecular vibrations and the electric fields. As a result, we can see various vibrational spectroscopic features arising from those responses, including the vibrational frequency shifts, vibrational transition intensity modulations, and vibrational band shapes.

Important quantities for describing those responses are the derivatives of the electrical properties (dipoles $\mu$, polarizabilities $\alpha$, etc.) with respect to the molecular vibrational coordinates, which are generally called electrical property derivatives [1]. This is because these derivatives appear in the expansion of the potential energy ($V$) with respect to the electric field ($E$) and vibrational coordinates ($q$). The expansion formula is given as

$$V = V_0 + \frac{1}{2}\sum_p k_p q_p^2 + \dots$$

$$- \sum_j \left[ \mu_j + \sum_p \frac{\partial \mu_j}{\partial q_p} q_p + \frac{1}{2}\sum_{p,r} \frac{\partial^2 \mu_j}{\partial q_p \partial q_r} q_p q_r + \dots \right] E_j$$

$$- \frac{1}{2}\sum_{i,j} \left[ \alpha_{ij} + \sum_p \frac{\partial \alpha_{ij}}{\partial q_p} q_p + \frac{1}{2}\sum_{p,r} \frac{\partial^2 \alpha_{ij}}{\partial q_p \partial q_r} q_p q_r + \dots \right] E_i E_j + \dots \quad (1)$$

where $i$ and $j$ (=1, 2, 3) stand for the $x$, $y$ and $z$ directions, $q_p$ is the $p$th (mass-weighted) normal mode of a given molecule, and $k_p$ is its quadratic force constant. Here, $E_j$ may be the electric field of radiation or the external electric field operating between molecules in condensed phases. As a result, the electrical property derivatives are important for describing the responses of molecular vibrations to both of these types of electric fields.

---

[1] Corresponding author. E-mail: torii@ed.shizuoka.ac.jp, hajime.torii@nifty.com

In the present study, we show a few cases where linear and nonlinear vibrational spectroscopic features are affected by those responses, and discuss the role of the electrical property derivatives.

## 2. Vibrational Transition Intensity Modulations

In condensed phases, the molecular dipole moments and polarizabilities are modulated by intermolecular electrostatic interactions. According to the dipole-induced dipole (DID) mechanism [2,3], the dipole moment $\mu_m$ of the $m$th molecule enhanced by intermolecular electrostatic interactions is given as

$$\mu_m = \mu_m^{\text{M}} + \alpha_m^{\text{M}} \sum_{\substack{n=1 \\ (n \neq m)}}^{N} T_{mn} \mu_n \tag{2}$$

where the superscript M stands for the quantities intrinsic to the molecule, i.e., without perturbation by intermolecular electrostatic interactions, and $T_{mn}$ is the dipole interaction tensor between the $m$th and $n$th molecules, which is given as

$$T_{mn} = \frac{3 r_{mn} r_{mn} - r_{mn}^2 I}{r_{mn}^5} \tag{3}$$

where $r_{mn} = r_m - r_n$ is the distance vector between the two molecules, $r_{mn}$ is the length of this vector, and $I$ is a $3 \times 3$ unit tensor. The sum of $T_{mn}\mu_n$ over $n$ in the second term of Eq. (2) represents the electric field coming from the dipole moments of the surrounding molecules. The polarizability $\alpha_m$ of the $m$th molecule is obtained as the derivative of $\mu_m$ with respect to the electric field, and is expressed as [4]

$$\alpha_m = \alpha_m^{\text{M}} + \sum_{\substack{n=1 \\ (n \neq m)}}^{N} \left( \alpha_m^{\text{M}} T_{mn} \alpha_n + \beta_m^{\text{M}} T_{mn} \mu_n \right) \tag{4}$$

where $\beta_m$ is the first hyperpolarizability of the $m$th molecule. The enhancement mechanism including the second term of the summation in Eq. (4) is called extended DID (XDID) mechanism [4].

As is well known, the infrared (IR) and Raman intensities are proportional to the square of the dipole derivatives and polarizability derivatives, respectively. This is the case where $E_j$ in Eq. (1) is the electric field of radiation, which is called *case 1* [1]. The derivatives of $\mu_m$ and $\alpha_m$ with respect to the librational coordinates $R_{l,n}$ ($4 \leq l \leq 6$; $1 \leq n \leq N$) are given as

$$\frac{\partial \mu_m}{\partial R_{l,n}} = \delta_{mn} \left[ \frac{\partial \mu_m^{\text{M}}}{\partial R_{l,m}} + \frac{\partial \alpha_m^{\text{M}}}{\partial R_{l,m}} \sum_{\substack{k=1 \\ (k \neq m)}}^{N} T_{mk} \mu_k \right] + \alpha_m^{\text{M}} \sum_{\substack{k=1 \\ (k \neq m)}}^{N} T_{mk} \frac{\partial \mu_k}{\partial R_{l,n}} \tag{5}$$

$$\frac{\partial \alpha_m}{\partial R_{l,n}} = \delta_{mn} \left[ \frac{\partial \alpha_m^{\text{M}}}{\partial R_{l,m}} \left( 1 + \sum_{\substack{k=1 \\ (k \neq m)}}^{N} T_{mk} \alpha_k \right) + \frac{\partial \beta_m^{\text{M}}}{\partial R_{l,m}} \sum_{\substack{k=1 \\ (k \neq m)}}^{N} T_{mk} \mu_k \right]$$

$$+ \alpha_m^{\text{M}} \sum_{\substack{k=1 \\ (k \neq m)}}^{N} T_{mk} \frac{\partial \alpha_k}{\partial R_{l,n}} + \beta_m^{\text{M}} \sum_{\substack{k=1 \\ (k \neq m)}}^{N} T_{mk} \frac{\partial \mu_k}{\partial R_{l,n}} \tag{6}$$

These formulas indicate that, in some cases, the first hyperpolarizability derivatives may also be important for calculating the Raman intensities of the intermolecular modes.

The effects of the modulations of $\mu_m$ and $\alpha_m$ by intermolecular electrostatic interactions shown above are seen in the profiles of the far-IR and low-frequency Raman spectra as well as in the time dependence of the optical Kerr effect (OKE) signals. The existence of those effects has been discussed for the spectra of many liquids [2–6]. It has been shown that the intensity modulations due to the XDID term are sufficiently recognizable in the cases of liquid formamide and $N$-methylformamide [4].

### 3. Structural Changes Induced by Electric Field and Their Consequences

When $E_j$ is static or slowly varying as compared with the time scale of $q_p$ in Eq. (1), the position of the potential energy minimum is displaced from the original one at $q_p = 0$. This is called *case 2* [1]. To the first order in $E_j$ and neglecting higher-order derivatives of dipole moment, the displacement is expressed as

$$\delta q_p = \frac{1}{k_p} \sum_j \frac{\partial \mu_j}{\partial q_p} E_j \tag{7}$$

This displacement $\delta q_p$ gives rise to an induced dipole moment $\delta \mu_i$ as

$$\delta \mu_i = \sum_p \frac{\partial \mu_i}{\partial q_p} \delta q_p = \sum_j \sum_p \frac{1}{k_p} \frac{\partial \mu_i}{\partial q_p} \frac{\partial \mu_j}{\partial q_p} E_j \tag{8}$$

As a result, it contributes to the molecular polarizability. The formula for this contribution (called vibrational polarizability [7–9]) in the static limit is easily derived from Eq. (8), and is given as

$$\alpha_{ij}^{(v)} = \sum_p \frac{1}{k_p} \frac{\partial \mu_i}{\partial q_p} \frac{\partial \mu_j}{\partial q_p} \tag{9}$$

Another effect of the displacement $\delta q_p$ is a shift of the quadratic force constant of a mechanically anharmonic vibrational mode. This effect is expressed as $\delta k_p = f_p \, \delta q_p$, where $f_p$ is the cubic force constant of the mode $q_p$, neglecting the cross terms of the mechanical anharmonicity. Substituting Eq. (7) into this equation, and adding the effect of the electrical anharmonicity (the dipole second derivative), we have

$$\delta k_p = \sum_j \left( \frac{f_p}{k_p} \frac{\partial \mu_j}{\partial q_p} - \frac{\partial^2 \mu_j}{\partial q_p^2} \right) E_j \tag{10}$$

Because of this effect, the vibrational frequencies of the modes with large cubic (relative to quadratic) force constants and large dipole first and second derivatives are likely to be affected significantly by the intermolecular electrostatic interactions.

As an example, the result obtained for the oligomers of $N$-methylacetamide (NMA) [10] is shown in Figure 1. It is seen that, because of the effect of mechanical anharmonicity, $\delta k_p$ is linearly related with $\delta q_p$. Since $\delta q_p$ is controlled by the electric field as shown in Eq. (7), the sum of the effects of the mechanical and electrical anharmonicities [Eq. (10)] is also linearly related.

### 4. Transition Dipole Coupling between Molecules

When the electric field on the $m$th molecule, denoted as $E_m$, is modulated by *intramolecular* vibrations of different molecules ($n$), the vibrations of those molecules are directly coupled with each other (called *case 3* [1]). Because $E_m$ is expressed within the dipole approximation as

$$E_m = \sum_{n(\neq m)} T_{mn} \left( \mu_n + \sum_r \frac{\partial \mu_n}{\partial q_{r,n}} q_{r,n} + \dots \right) \tag{11}$$

where $\mu_n$ is the dipole moment of the $n$th molecule and $q_{r,n}$ is its $r$th normal mode, we obtain the term bilinear in $q_{p,m}$ and $q_{r,n}$, given as

$$V = V_0 + \dots - \sum_{m,n(m>n)} \sum_p \sum_r \frac{\partial \mu_m}{\partial q_{p,m}} T_{mn} \frac{\partial \mu_n}{\partial q_{r,n}} q_{p,m} q_{r,n} + \dots \tag{12}$$

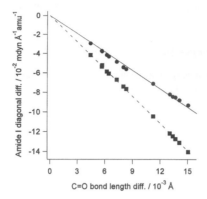

Figure 1: Relation between the shifts in the (diagonal) force constants of the amide I mode and the C=O bond length of the NMA oligomers from the values of an isolated NMA molecule calculated at the B3LYP/6-31+G(2df,p) level [10]. ■: Contribution of the mechanical anharmonicity; ●: total value; solid and broken lines: the least-squares fitted lines (constrained to pass the origin) for these points.

This intermolecular bilinear coupling between $q_{p,m}$ and $q_{r,n}$ is called transition dipole coupling (TDC). The effect of this coupling is most clearly seen as the noncoincidence effect [11] (the phenomenon that the frequency positions of the IR, isotropic Raman, and anisotropic Raman components of a vibrational band do not coincide) for some modes of polar liquids, such as the C=O stretching mode of liquid acetone, and the decay of the transient IR absorption anisotropy [12]. In the case of the amide I band of peptides, the effect is seen as the sensitivity of the band profiles to the secondary structures [13].

## References

[1]   H. Torii, *Vib. Spectrosc.* **29**, 205–209 (2002).
[2]   T. Keyes, D. Kivelson and J. P. McTague, *J. Chem. Phys.* **55**, 4096– 4100 (1971).
[3]   B. M. Ladanyi and T. Keyes, *Mol. Phys.* **33**, 1063–1097 (1977).
[4]   H. Torii, *Chem. Phys. Lett.* **353**, 431–438 (2002).
[5]   D. Frenkel and J. P. McTague, *J. Chem. Phys.* **72**, 2801–2818 (1980).
[6]   P.A. Madden and D.J. Tildesley, *Mol. Phys.* **55**, 969–998 (1985).
[7]   D. M. Bishop, *Rev. Mod. Phys.* **62**, 343–374 (1990).
[8]   D. M. Bishop, *Adv. Chem. Phys.* **104**, 1–40 (1998).
[9]   B. Kirtman, B. Champagne and J. M. Luis, *J. Comput. Chem.* **21**, 1572–1588 (2000).
[10]   H. Torii, *J. Mol. Struct.* **735/736**, 21–26 (2005).
[11]   H. Torii, Computational methods for analyzing the intermolecular resonant vibrational interactions in liquids and the noncoincidence effect of vibrational spectra, *Novel Approaches to the Structure and Dynamics of Liquids: Experiments, Theories and Simulations*, J. Samios and V. A. Durov, Eds, Kluwer, pp. 343–360 (2004).
[12]   S. Woutersen and H. J. Bakker, *Nature* **402**, 507–509 (1999).
[13]   H. Torii and M. Tasumi, Theoretical analyses of the amide I infrared bands of globular proteins, *Infrared Spectroscopy of Biomolecules*, H. H. Mantsch and D. Chapman, Eds, Wiley-Liss, pp. 1–18 (1996).

Brill Academic Publishers
P.O. Box 9000, 2300 PA Leiden,
The Netherlands

*Lecture Series on Computer*
*and Computational Sciences*
Volume 4, 2005, pp. 693-696

# On the (hyper)polarizability of $N_4$ ($T_d$)

## Demetrios Xenides[1]

Theoretical Chemistry Division, Insitute of General, Inorganic and Theoretical Chemistry
University of Innsbruck, Innrain 52a, A-6020 Innsbruck, Austria

Received 14 July, 2005; accepted in revised form 4 August, 2005

*Abstract:* We present an extensive *ab initio* and DFT study on the electric properties of the $N_4$ tetrahedron. At the CCSD(T)/[7s5p4d1f] 196 Gaussian type functions (GTFs) basis set level of theory we calculated $\bar{\alpha}$=24.96 $e^2 a_0^2 E_h^{-1}$, $\beta_{xyz}$=15.1 $e^3 a_0^3 E_h^{-2}$, and $\bar{\gamma}$=22.76 x $10^{-2} e^4 a_0^4 E_h^{-3}$ at the equilibrium distance $R_e$=1.4516 Å. We found that the $m$PW1PW91 DFT method behaves similar to the CCSD(T).

*Keywords:* polarizability, hyperpolarizability, ab initio, density functional theory.

## 1  Introduction

Enviromentaly friendly energy resources are always a demand. In this course High Energy Density Materials (HEDMs), meaning compounds that can store large amounts of energy, are seem to be good alternatives to the traditional energy resources. One of the simplest such materials the tetranitrogen ($N_4$) molecule. The singlet tetrahedral ($^1A_1$, $T_d$) structure of the molecule is preffered over the linear triplet ($^3A_1$, $C_s$), to most *ab initio* studies [1] when this is still an open question to experiment [2].

In present study we calculated the electric (hyper)polarizabilities of ($N_4$). We relied on *ab initio* quantum mechanical methods such as Hartree-Fock (HF), Møller−Plesset Perturbation Theory (MPn) [3] (n=1,2,3,4), and the more sophisticated coupled cluster techniques (CC) [4] (CCSD and CCSD(T)). In addition we have employed the B3LYP [5a,b], B3P86 [5a,c], B3PW91 [5a,d], MPW1PW91 [5e,d] and O3LYP [5f,b] density functional methods in order to examine their performance in electric properties calculations. Four independent basis set sequences were constructed upon initial substrates by Dunning and Hay [6a], by Schäfer, Huber and Ahlrichs [6b], by Thakkar et al. [6c], and by Partridge [6d]. All calculations have been performed at the theoretically predicted geometry $R_e$=1.4516 Å (Ref.1i).

## 2  Theory

*Methodology:* The energy of molecule in weak, homogeneous electric field can be written as [8]:

$$E = E^0 - \mu_\alpha F_\alpha - \frac{1}{2!}\alpha_{\alpha\beta}F_\alpha F_\beta - \frac{1}{3!}\beta_{\alpha\beta\gamma}F_\alpha F_\beta F_\gamma - \frac{1}{4!}\gamma_{\alpha\beta\gamma\delta}F_\alpha F_\beta F_\gamma F_\delta + \cdots \quad (1)$$

$E^0$ is the energy of the unperturbed molecule, $F_\alpha \ldots$ is the strength of the applied electric field, $\mu_\alpha$ is the dipole moment, $\alpha_{\alpha\beta}$ is the dipole polarizability, and $\beta_{\alpha\beta\gamma}$ and $\gamma_{\alpha\beta\gamma\delta}$ are the first and second hyperpolarizability tensors, respectively. The independent components of the the $N_4$ $T_d$

---

[1]Corresponding author. E-mail: Demetrios.Xenides@uibk.ac.at

molecule for dipole polarizability, first hyperpolarizability and second hyperpolarizability are one, $\alpha_{zz} = \bar{\alpha}$, one, $\beta_{xyz}$, and two, $\gamma_{zzzz}$ and $\gamma_{xxzz}$, respectively [9]. In addition we calculated the mean value of second hyperpolarizability as $\bar{\gamma} = \frac{3}{5}(\gamma_{zzzz} + 2\gamma_{xxzz})$. The molecule's center of mass has been chosen as the center of the Cartesian axis. Throughout this paper properties are given as plain numbers that as: $\bar{\alpha}/e^2 a_0^2 E_h^{-1}$, $\beta_{\alpha\beta\gamma}/e^3 a_0^3 E_h^{-2}$, and $\gamma_{\alpha\beta\gamma\delta}/e^4 a_0^4 E_h^{-3}$.

All properties were calculated from finite differences between the perturbed energy (perturbation was introduced by fields scaled as F, 2F, 4F with F=0.0050 $e^{-1}a_0^{-1}E_h$) and the energy of the unperturbed system. All calculations were performed with Gaussian 98 [7a] and Gaussian 03 [7b] programs.

*Basis set*: A judicious choice can yield to reliable results [10, 11, 12, 13]. The followed strategy has been established by Werner and Meyer [11], and elucidated paradigms can be found in works by Maroulis and his collaborators [14]. The following basis sets have been used in present study (further details are available upon request): A0 $\equiv$ [7s5p2d] 128 CGTF, A1 $\equiv$ [7s5p3d1f] 176 CGTF, A2 $\equiv$ [7s5p3d1f] 176 CGTF, A3 $\equiv$ [7s5p4d1f] 196 CGTF, A4 $\equiv$ [7s5p4d2f] 224 CGTF, A5 $\equiv$ [7s5p5d3f] 272 CGTF, Dhf $\equiv$ [6s4p4d2f] 208 CGTF, Khf $\equiv$ [9s6p5d3f] 292 CGTF, Phf $\equiv$ [16s11p5d2f] 352 GTF. For all the sequences 5d and 7f GTF were used.

# 3  Results

*Hartree-Fock calculations*: The results from HF calculations are collected in Table 1. The striking

Table 1: Basis set convergence at the Hartree-Fock level.

| Property | A0 | A1 | A2 | A3 | A4 | A5 | Dhf | Khf | Phf |
|---|---|---|---|---|---|---|---|---|---|
| $\bar{\alpha}$ | 24.19 | 24.25 | 24.24 | 24.27 | 24.30 | 24.30 | 24.30 | 24.29 | 24.30 |
| $\beta_{xyz}$ | 20.0 | 20.6 | 20.4 | 20.4 | 20.5 | 20.4 | 20.8 | 20.4 | 20.3 |
| $\gamma_{zzzz}$ | 1509 | 1566 | 1660 | 1664 | 1666 | 1673 | 1674 | 1674 | 1670 |
| $\gamma_{zzxx}$ | 535 | 550 | 565 | 565 | 566 | 567 | 574 | 567 | 566 |
| $\bar{\gamma}$ | 1548 | 1599 | 1674 | 1676 | 1679 | 1685 | 1693 | 1685 | 1681 |

feature of Table 1 is the fast convergence of independent values, i.e. calculated with different basis sets, to those obtained with the very large (Phf) basis set. Thus, the Phf results can be taken as near-HF (nHF) values. Remarkably good agreement has been observed between A3 and Phf properties (columns 5 and 10 in Table 1. The largest members of the produced sequences (A5, Dhf, Khf and Phf) produced indistinguishable results (columns 7, 8, 9, and 10 in Table 1). Therefore, the constructed sequences are free of systematical errors.

*post-Hartree-Fock Calculations*: Electron correlation effects were estimated with the MPn (n=1, 2, 3, 4), and CC (CCSD, CCSD(T)) methods using the A3 basis set. DFT results are included in the present section, as well, since they introduce some electron correlation not definable though. Our best CCSD(T) estimates are 24.96, 15.1, 2271, 761 for $\bar{\alpha}$, $\beta_{xyz}$, $\gamma_{zzzz}$, and $\gamma_{zzxx}$, respectively. The $m$PW1PW91 produces results in agreement with the CCSD(T) ones.

# 4  Conclusion

We have implemented a large basis set (A3$\equiv$[7s5p4d1f], 196 GTFs) and a high level of theory [CCSD(T)] in the post-HF calculation of electric (hyper)polarizabilities of $N_4$, thus we obtained the following estimates: $\bar{\alpha}$=24.96, $\beta_{xyz}$=15.1 and $\bar{\gamma}$=2276. The DFT method with the most acceptable performance is the $m$PW1PW91.

Table 2: *ab initio* and DFT results A3≡[7s5p4d1f] 196 GTFs.

| Method | $\bar{\alpha}$ | $\beta_{xyz}$ | $\gamma_{zzzz}$ | $\gamma_{zzxx}$ | $\bar{\gamma}$ | Method | $\bar{\alpha}$ | $\beta_{xyz}$ | $\gamma_{zzzz}$ | $\gamma_{zzxx}$ | $\bar{\gamma}$ |
|---|---|---|---|---|---|---|---|---|---|---|---|
| HF | 24.27 | 20.4 | 1664 | 565 | 1676 | B3LYP | 24.95 | 17.4 | 2495 | 850 | 2517 |
| MP2 | 24.81 | 13.4 | 2364 | 790 | 2367 | B3P86 | 24.70 | 16.4 | 2285 | 778 | 2304 |
| MP4 | 25.38 | 15.9 | 2467 | 828 | 2473 | B3PW91 | 24.68 | 16.6 | 2315 | 787 | 2333 |
| CCSD | 24.67 | 15.8 | 2104 | 724 | 2132 | O3LYP | 24.90 | 16.4 | 2656 | 901 | 2675 |
| CCSD(T) | 24.96 | 15.1 | 2271 | 761 | 2276 | *mPW1PW91* | 24.61 | 16.5 | 2272 | 770 | 2287 |

# 5 Acknowledgement

The author gratefully acknowledges the warm and generous hospitality of the Department of Theoretical Chemistry of the University of Innsbruck and wishes to express his intebteness to Dr. Michael Fink, system administrator at the Zentral Informatik Dienst (ZID) of the University of Innsbruck, for specific computer resources allocations.

# References

[1] a)T. J. Lee and J. E. Rice, *J. Chem. Phys.* **94**, 1215-1221 (1991);b)M. M. Francl and J. P. Chesick, *J. Phys. Chem.* **94**, 526-528 (1990);c)W. J. Lauderdale, J. F. Stanton, and R. J. Bartlett, *J. Phys. Chem.* **96**, 1173-1178 (1992);d)D. R. Yarkony, *J. Am. Chem. Soc.* **114**, 5406-5411 (1992);e)K. M. Dunn and K. Morokuma, *J. Chem. Phys.* **102**, 4904-4908 (1995);f)A. A. Korkin, A. Balkova, R. J. Bartlett, R. J. Boyd, and P. v. R. Schleyer, *J. Phys. Chem.* **100**, 5702-5714 (1996);g)S. A. Perera and R. J. Bartlett, *Chem. Phys. Lett.* **314**, 381-387 (1999);h)H. Östmark, O. Launila, S. Wallin, and R. Tryman, *J. Raman Spectrosc.* **32**, 195-199 (2001);i)T. J. Lee and J. M. Martin, *Chem. Phys. Lett.* **357**, 319-325 (2002).

[2] F. Cacace, G. de Petris, and A. Troiani, *Science* **295**, 480-481 (2002); M. T. Nguyen, *Coord. Chem. Rev.* **244**, 93-113 (2003).

[3] C. Møller and M. S. Plesset, *Phys. Rev.* **46**, 618-622 (1934).

[4] J. Čížek, *Adv Chem Phys* **14**, 35-89 (1969);J. A. Pople, R. Krishnan, H. B. Schlegel, and J. S. Binkley, *Int. J. Quant. Chem.* **14**, 545-560 (1978); R. J. Bartlett, *Annu. Rev. Phys. Chem.***32**, 359-401 (1981); G. D. Purvis and R. J. Bartlett, *J. Chem. Phys.* **76**, 1910-1918 (1982); A. Szabo and N. S. Ostlund, Modern Quantum Chemistry, Mac Millan, New York, (1982).S. Wilson, Electron Correlation in Molecules, Clarendon, Oxford, (1984); M. Urban, I. Cernusak, V. Kellö, and J. Noga, Electron correlation in molecules. In Methods in Computational Chemistry Vol. I. Electron Correlation in Atoms and Molecules, Ed. S. Wilson, Plenum Press, New York, 117-250, (1987); G. E. Scuseria, C. L. Janssen, and H. F. Schaefer III, *J. Chem. Phys.* **89**, 7382- 7387 (1988);G. E. Scuseria and H. F. Schaefer III, *J. Chem. Phys.* **90**, 3700-3703 (1989);J. A. Pople, M. Head-Gordon, and K Raghavachari, *J. Chem. Phys.* **87**, 5968-5975 (1987).

[5] a)A. D. Becke, *Phys. Rev. A* **98**, 5648-5652 (1993);b)C. Lee, W. Yang, and R. G. Parr, *Phys. Rev. B* **37**, 785-789 (1988);c)J. P. Perdew, *Phys. Rev. B* **33**, 8822-8824 (1986);d)J. P. Perdew, K. Burke and Y. Wang, *Phys. Rev. B* **54**, 16533-16539 (1996);e)C. Adamo and V. Barone, *J. Chem. Phys.* **108**, 664-675 (1998);f)N. C. Handy and A. J. Cohen, *Mol. Phys.* **99**, 403-412 (2001).

[6] a)T. H. Dunning and P. J. Hay, Gaussian basis sets for molecular calculations in Modern Theoretical Chemistry, Ed. H. F. Schaefer III, Plenum Press, New York, 1-28, (1977);b)A. Schäfer, C. Huber, and R. Ahlrichs, *J. Chem. Phys.* **100**, 5829-5835 (1994);c)T. Koga, M. Sato, R. E. Hoffmeyer, and A. J. Thakkar, *J. Mol. Struct. THEOCHEM* **306**, 249-260 (1994); d)H. Partridge, Near Hartree Fock quality gaussian type orbital basis sets for the first- and third-row atoms, NASA Technical Memorandum 101044, Jan 1989.

[7] a)M. J. Frisch et al, Gaussian 98 Revision A.11.3, Gaussian Inc. Pittsburgh PA, 2002;b)M. J. Frisch et al, Gaussian 03 Revision C.03, Gaussian Inc. Wallingford CT, 2004.

[8] J. Stiehler and J. Hinze, *J. Phys. B: At. Mol. Opt. Phys.* **28**, 4055-4071 (1995).

[9] A. D. Buckingham, Basic theory of intermolecular forces: Applications to small molecules in Intermolecular Interactions: From Diatomics to Biopolymers, Ed. B. Pullman, Wiley, New York, 1-67, (1978).

[10] I. Shavitt, Methods in Electronic Structure Theory, Ed. H. F. Schaefer III, Plenum Press, New York, 243, (1977).

[11] H. J. Werner and W. Meyer, *Mol. Phys.* **31**, 855-872 (1976).

[12] S. Wilson, Advances in Chemical Physics (Ab Initio Methods in Quantum Chemistry), Ed. K. P. Lawley, Wiley, New York, p. 439, (1987).

[13] E. R. Davidson and D. Feller, *Chem. Rev.* **86**, 681-696 (1986).

[14] G. Maroulis and A. J. Thakkar, *Chem. Phys. Lett.* **156**, 87-90 (1989);G. Maroulis and P. Karamanis, *Chem. Phys.* **269**, 137-146 (2001);G. Maroulis and D. Xenides, *J. Phys. Chem. A* **107**, 712-719 (2003).

Brill Academic Publishers
P.O. Box 9000, 2300 PA Leiden
The Netherlands

*Lecture Series on Computer
and Computational Sciences*
Volume 4, 2005, pp. 697-698

# Trends and perspectives in Computational Chemistry

George Maroulis

Department of Chemistry,
University of Patras,
University of Patras,
GR-26500 Patras, Greece

This general symposium comprises papers in the wide area of Computational Chemistry. Most of them focus on subjects of central importance to Computational Chemistry. Some of the contributions are clearly close to the interface of Chemistry with Biology and Medicine. Thus, they are expected to reach a very wide audience. In addition to the highlighted and invited lectures the Symposium includes important contributions by **Yuriko Aoki, Hamid Berriche, Styliani Consta, Valerij Gurin, Kikuo Harigaya, Agnes Mylona-Kosma, Rudolph Magyar, George Maroulis, Heinz Oberhammer, Ponadurai Ramasami, Georg Rollmann, V.F.Sokolov, Ioannis Thanopoulos** and **Tomasz Wesolowski**.

## Highlighted Lectures

1. R.J.Bartlett, University of Florida, USA.
   *Electronic Structure of Molecules and Materials: From Coupled-Cluster Theory to DFT to the Transfer Hamiltonian.*
2. D.Clary, University of Oxford, England.
   *Quantum theory of chemical reaction dynamics*
3. E.R.Davidson, University of Washington, USA.
   *Can Heisenberg spin coupling parameters be reliably extracted from unrestricted calculations?*

## Invited Lectures

4. X.Assfeld, Université Henri Poincaré, France.
   *Recent improvements in the LSCF treatment of large molecular systems*
5. K.Balasubramanian, University of California (Davis), USA.
   *Relativistic Effects in very heavy and superheavy molecules*
6. S.Farantos, University of Crete, Greece.
   *Grid Computing Algorithms for Classical and Quantum Dynamics*
7. W.A.Goddard III, California Institute of Technology, USA.
   *First Principles Simulations of materials using Multi-paradigm Multi-scale methods*
8. M.Kosmas, University of Ioannina, Greece.
   *Conformational properties of dendritic block copolymers*
9. J.Leszczynski, Jackson State University, USA.
   *From the Gas Phase to Crystals: Comprehensive Investigations on Formation of the Clusters*
10. M.Meuwly, University of Basel, Switzerland.
    *Computer Simulations of Dynamical Processes in Proteins*
11. A.Painelli, University of Parma, Italy.
    *Charge instabilities in molecular materials: cooperative behavior from electrostatic interactions*
12. C.Pouchan, Université de Pau, France.
    *Vibrational spectra of medium size molecular systems from DFT an ab initio quartic force field calculations*
13. V.Renugopalakrishnan, University of Florida, USA.

*Protein Engineering : Role of Computational Methods in Design of Thermally Stable Proteins.*

14. B.M.Rode, University of Innsbruck, Austria
*Recent Developments and Challenges in Chemical Simulations.*

15. P.Schwerdtfeger, Massey University, New Zealand.
*The Quest for Absolute Chirality*

16. A.J.Thakkar, University of New Brunswick, Canada.
*Structure and energetics of hydrogen-bonded clusters.*

17. A.van der Avoird, University of Nijmegen, The Netherlands.
*Potential surfaces and non-adiabatic dynamics of hydrogen-bonded complexes containing radicals.*

18. P.Weinberger, Technical University of Vienna, Austria.
*Multi-scale approaches in computational materials science*

19. S. S. Xantheas, Chemical Sciences Division, Pacific Northwest National Laboratory, Richland, WA, USA.
*'Strong' vs. 'weak' nearest-neighbor hydrogen bonds in water clathrates: Interplay between relative stability and IR spectral features*

The texts of the invited lectures will published in a separate volume.

VSP International
Science Publishers
P.O. Box 346, 3700 AH Zeist
The Netherlands

*Lecture Series on Computer
and Computational Sciences*
Volume 4, 2005, pp. 699-702

# Performance of the Elongation Method for Large Systems

Y. Aoki[1,2], M. Makowski[1], and F.L. Gu[1]

[1] Department of Material Science, Faculty of Engineering Sciences, Kyushu University, Fukuoka, 816-8580, Japan
[2] Group, *PRESTO*, Japan Science and Technology Corporation (JST), Kawaguchi Center Building, Honcho 4-1-8, Kawaguchi, Saitama 332-0012, Japan

Received 7 July, 2005; accepted in revised form 7 August, 2005

*Abstract:* The elongation method has been developed for efficiently determining the electronic structure of large systems, such as biopolymers. The electronic states of large periodic/aperiodic polymers can be obtained by the elongation method, treating only a few units at a time of the total system using localized molecular orbital technique. The performance of the elongation method including the accuracy and efficiency is presented in this paper for polyglycine model system. It is demonstrated that the elongation method is so efficient compared to the conventional method with high accuracy in the total energy.

*Keywords:* Elongation method, Polyglycine

## 1. Introduction

The elongation method was first proposed by Imamura in 1990s for efficiently calculating the electronic structure of periodic/aperiodic polymers. [1] The elongation method is based on O($N$) quantum chemical approach since a constant number of localized molecular orbitals (LMOs) is involved in the diagonalization procedure throughout the elongation processes. The elongation method is much more appropriate to treat huge systems like proteins or DNA even at high accuracy than the conventional MO treatment with large basis sets.

The idea of the elongation method is to mimic the polymerization procedures in experiment. By the elongation method, the electronic structure of a polymer is synthesized step by step only for the interaction site between oligomer and an attacking fragment similar to the polymerization reaction in experiment. As a start point, the MOs of a suitable size of a cluster are localized into two regions, frozen and active regions. The active region is close to the attacking fragment and the frozen region is remote from the interaction site.

The elongation method is implemented into *ab initio* program package GAMESS for the restricted closed-shell Hartree-Fock (RHF), restricted open-shell Hartree-Fock (ROHF) and unrestricted Hartree-Fock (UHF) methods. The results for polyglycines by the elongation and conventional methods will be presented in the Section 2 and the outlooks for future development are given in section 3.

## 2. Review of the elongation method

In this section, the elongation method is briefly reviewed. Figure 1 schematically illustrates the elongation processes. A random polymer is built up by adding a monomer unit to a starting cluster step by step. It is general in the sense that the adding monomers during the elongation steps are random, which means any random polymer can be theoretically synthesized. However, If all the adding monomers are the same, a periodic polymer is constructed.

First, the canonical molecular orbitals (CMOs) of a starting cluster are localized into frozen and active regions by a unitary transformation. Next, a monomer is attacking to the starting cluster from the active region. The elongation SCF is solved by disregarding the LMOs which have very weak interaction with the attacking monomer. The reason for doing so is that the elongation method is

---

[1] E-mail: aoki@cube.kyushu-u.ac.jp.

working in the LMO representation, from which the interaction between the frozen and attacking monomer is minimized. Thus, the frozen part can be disregarded in the elongation SCF procedure. This reduces the number of variational degrees of freedom. And more importantly, the number of two-electron integrals that have to be evaluated in the system can be drastically reduced. As the system is elongated, the most remote frozen part can be physically cut off without significant loss of accuracy and by this way it achieves major saving in the computational timings.

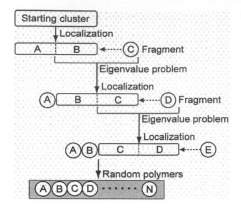

Figure 1. Schematic illustration for the elongation method.

## 3. Performance of the elongation method

We choose polyglycine as an example to test the performance of the elongation calculations. There are four different conformations in polyglycine, e.g. C5, C7, $3_{10}$-helix, and $\alpha$-helix. In Fig. 2, the energy differences for these four conformations between the elongation calculation and the conventional one are depicted. From Fig. 3, one can see that the energy differences for C5 and C7 are very small compared to the other two conformations, $3_{10}$-helix and $\alpha$-helix. The plots for $3_{10}$-helix and $\alpha$-helix are almost linear against the number of unit cells. This means the error by the elongation method is not accumulated. Although the energy difference is not so small due to the strong interactions between neighboring unit cells, it can be improved by enlarge the size of the starting cluster.

The total CPU time of the elongation calculations is plotted in Fig.3 together with the conventional CPU time. Approximately, a factor of four is obtained in the CPU timing compared to the conventional calculations. It is, however, the elongation CPU time is not linear scaling against the size of the system. The reason is that for each elongation step we have to re-calculate the two-electron integrals for the direct SCF. If the elongation is performed in a disk-base calculation, the SCF CPU time will be linearly scaling. The elongation SCF CPU time of the C5 polyglycine is plotted in Fig. 4. It can be seen that the CPU time is linear scaling to the size of the system.

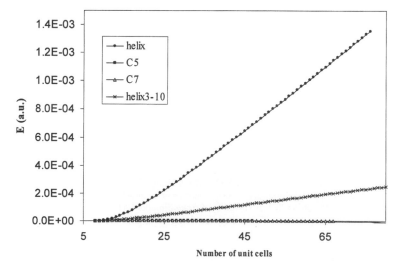

Fig. 2. RHF/STO-3G energy difference between the elongation and conventional calculations for four polyglycince conformations.

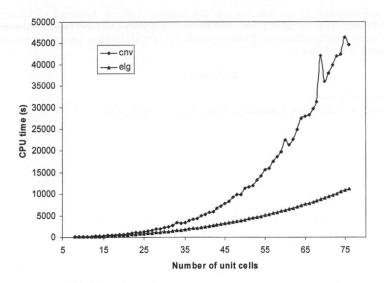

Fig. 3. CPU times for the conventional and elongation calculations for helix polyglycine at RHF/STO-3G.

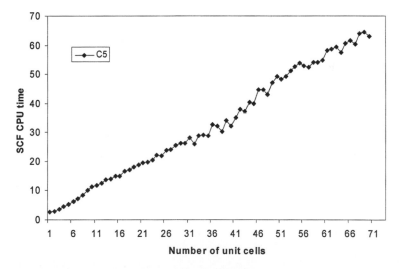

Figure 4. SCF CPU time for C5 polyglycine at RHF/STO-3G.

## 4. Outlook and further development

The two important techniques of the elongation method are the localization procedure and cut-off technique. The results here clearly show the efficiency as well as the accuracy of the elongation method. The elongation calculations can be performed at *ab initio* level for any extended basis-sets. The application of the elongation method to huge biopolymers, such as collagen, is very much promising. Another ways we are planning to go are including the electron correlation and implementing the elongation method in the presence of external fields.

## Acknowledgments

This work was supported by the Research and Development Applying Advanced Computational Science and Technology of the Japan Science and Technology Agency (ACT-JST) and the Ministry of Education, Culture, Sports, Science and Technology (MEXT). The calculations were performed on the Linux PC cluster in our laboratory.

## References

[1]  A. Imamura, Y. Aoki, and K. Maekawa, J. Chem. Phys., **95**, 5419 (1991) .

Brill Academic Publishers
P.O. Box 9000, 2300 PA Leiden
The Netherlands

*Lecture Series on Computer
and Computational Sciences*
Volume 4, 2005, pp. 703-708

# Theoretical investigation of the KRb$^+$ ionic molecule: potential energy, K and Rb polarisabilities and long-range behaviour

C. Ghanmi, H. Berriche$^*$ and H. Ben Ouada

*Laboratoire de Physique et Chimie des Interfaces, Département de Physique,
Faculté des Sciences de Monastir, Avenue de l'Environnement, 5019 Monastir, Tunisia*

Received 5 August, 2005; accepted in revised form 12 August, 2005

*Abstract:* The potential energy of the electronic states of KRb$^+$ ionic molecule dissociating up to K(6s) + Rb$^+$ and K$^+$ + Rb(7s) are calculated using an ab-initio approach involving a non-empirical pseudopotentials for K and Rb atoms. The spectroscopic constants were determined and compared with the available works showing the high accuracy of our calculation. Such accuracy was exploited to realize a long range vibrational level spacing analysis using the WKB semi-classical approximation in order to extract the K and Rb atomic polarisabilites.

*Keywords:* Pseudopotential, Spectroscopic constants, Polarisability, WKB approximation, vibrational analysis

## 1. Introduction

Intermolecular potential and interactions at long range play an important role for the calculation of the properties of atoms, molecules and clusters as well as the electric polarisability and hyperpolarisability. These properties present an interesting role in the treatment of core-polarisability and core-valence correlation effects as well as in the evaluation of core-core interaction in ab initio calculation based on a pseudo-potential techniques [1]. In addition, the electric polarisabilities are important to evaluate the weak coulombic interaction between an alkali atom and alkali cation at long range, which governs the potential energy curves of ionic molecules at large interaction distances [2, 3].

The XY$^+$ alkali ionic systems present an interesting long-range behaviour dominated by charge-dipole interactions. In a previous work, using the long range LiK$^+$ electronic states behavior, we have determined the atomic polarisabilities of Li and K [4] atoms. In this study, in addition to the potential energy curves and the spectroscopic constants of KRb$^+$ cation, we determined the atomic polarisabilities of K and Rb atomic levels. We have exploited the accuracy of the *ab initio* calculation of the potential energy at long-range distances, to perform a vibrational spacing analysis for the ground and numerous excited states of $^2\Sigma$, *and* $^2\Pi$ symmetries.

The paper will be structured as follows. In section 2, a summary of the ab inition calculation and the numerical method based on the WKB approximation is presented. Section 3 is devoted to the presentation and discussion of the potential energy curves, the spectroscopic constants and the atomic polarisbilities. Finally, we summarize our conclusions in section 4.

## 2. Summary of the method

Basically we use the same pseudopotential method as in our previous works on LiNa$^+$ and LiK$^+$ ionic molecules [5-6]. The use of pseudopotentials technique [7] for KRb$^+$ cores has reduced the number of active electrons to only one electron, where the SCF calculation has produced the exact energy in the basis set. Furthermore, the SCF energy is corrected by taking into account the core-core and core-valence electron correlation following the formalism of Faucroult et al [8]. The nonemprical pseudopotentials permit the use of very large basis set for the valence and Rydberg states and allow accurate description for the highest excited states. The K and Rb atoms are treated through the one-electron pseudopotential proposed by Barthelat and Durand [7]. For potassium, we used the Gaussian basis set (7s,6p,5d,3f/6s,5p,4d,2f) of Ref. 6, while for rubidium we used the Gaussian basis set (7s,4p,5d,1f/6s,4p,4d,1f) of Ref . 9. The core dipole polarisability of the ionic atoms K$^+$ and Rb$^+$ have been taken to the experimental ones, respectively, 5.354 $a_0^3$ and 9.245 $a_0^3$ .

For the vibrational spacing analysis we have used the WKB semi-classical method. An excellent review of the semi-classical approximation applied to the vibrational diatomic molecules can be found in Ref. 10. Here we present briefly the used formalism to perform a long-range potential analysis.

The WKB semi-classical approximation applied to a potential of the following expression:

$$V(R) = D - \frac{C_n}{R^n} \tag{1}$$

where D is the asymptotic limit, leads to an analytical expression for the vibrational energy levels:

$$E(v) = D - \left( (v_D - v) \frac{n-2}{2n} K_n \right)^{2n/(n-2)} \tag{2}$$

$v_D$ is the real (non-integer) number corresponding to the last vibrational level near the dissociation energy limit and $K_n$ is given by:

$$K_n = \hbar \sqrt{\frac{2\Pi}{\mu}} \frac{\Gamma(1+1/n)}{\Gamma(1/2+1/n)} \frac{n}{(C_n)^{1/n}} \tag{3}$$

In our case, at long-range distance, the potential energy behaves as a charge-dipole interaction and n=4.

Using the last expression for vibrational energy levels, the $(E_v - D)^{1/4}$ can be written as following.

$$(E_v - D)^{1/4} = -(v_D - v) \frac{K_n}{4} \tag{4}$$

The plot of $(E_v-D)^{1/4}$ should be purely linear. A least square fit of this linear law allows us to determine rather accurately the $v_D$ and $C_4$ constants from which all vibrational levels near the dissociation limit and also the related classical turning points ( $R_c(v)$) can be easily determined according to

$$E(v) = D - 11.817045(v_D - v)^4 C_4^{-1} \mu^{-2} \tag{5}$$

$$R_c(v) = C_4^{1/4} (D - E(v))^{-1/4} \tag{6}$$

## 3. Results and discussion

### 3. 1. Potential Energy Curves and spectroscopic constants

In our recent theoretical investigations on $XY^+$ alkali cation systems, we have determined the structure and spectroscopic properties of $LiNa^+$ [5], $LiK^+$ [6] and $NaK^+$ [11] ionic molecules. In this paper, we present our results for numerous molecular states of $^2\Sigma$, $^2\Pi$ and $^2\Delta$ symmetries of the $KRb^+$ ionic molecule dissociating into K(4s, 4p, 5s, 3d, 5p, 4d,6s) + $Rb^+$ and $K^+$ + K (5s, 5p, 4d, 6s, 6p, 5d, 7s). The potential energy curves are displayed in figure 1. The ground state has the deepest well ($D_e$ =5521cm$^{-1}$) compared to the excited states or other symmetries showing the electron delocalization and the formation of a real chemical bond.

The spectroscopic constants ($R_e$, $D_e$, $\omega_e$, $\omega_e\chi_e$, $B_e$) of the ground and the first excited states are presented in table 1 and compared with other theoretical works [12-15]. A very good agreement between our spectroscopic constants and those of the most recent work of Korek et al [15] is observed for the ground and the first excited states. Our equilibrium distance ($R_e$) and well depth ($D_e$) for the ground state are, respectively, $R_e$=8.90 a. u and $D_e$=5521 cm$^{-1}$ to be compared with Korek et al values $R_e$=8.92 a. u and $D_e$=5739 cm$^{-1}$. Such good agreement is not surprising as we used similar methods and it's expected to obtain similar agreement for the excited states. A detailed comparison for these states will be done in a future paper. Such good agreement shows the high accuracy of our ab initio calculation, which will be used in the vibrational spacing analysis to extract the atomic polarisabilities.

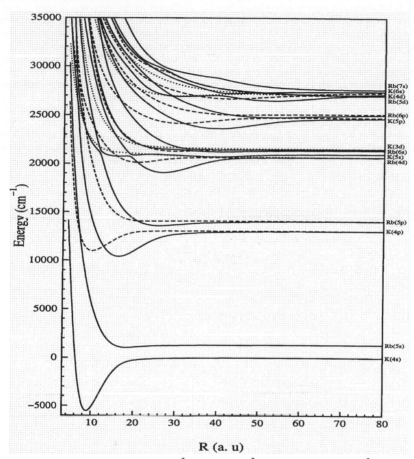

**Figure 1**: Potential energy curves for the $^2\Sigma$ (solid line), $^2\Pi$ (dashed line) and the $^2\Delta$ (dotted line) electronic states of KRb$^+$.

**Table 1:** The spectroscopic constants of the X$^2\Sigma$ and A$^2\Sigma$ electronic states of KRb$^+$.

| State | $R_e$(u. a) | $D_e$(cm$^{-1}$) | $\omega_e$ (cm$^{-1}$) | $\omega_e\chi_e$ (cm$^{-1}$) | $B_e$(cm$^{-1}$) | References |
|-------|-------------|------------------|------------------------|------------------------------|------------------|------------|
| X$^2\Sigma$ | 8.90 | 5521 | 59.45 | 0.20 | 0.028316 | this work |
|       | 8.74 | 5751 |        |        |          | [12] |
|       | 8.70 | 5896 |        |        |          | [13] |
|       | 8.10 | 6025 |        |        |          | [14] |
|       | 8.92 |      | 67.50  |        | 0.0282   | [15] |
| A$^2\Sigma$ | 18.27 | 261 | 11.64 | 3.36 | 0.006745 | this work |
|       | 18.17 | 273 | 11.59  |        | 0.0068   | [15] |
|       | 17.00 | 807 |        |        |          | [14] |

### 3. 2. Atomic polarisabilities

The use of the WKB semi classical approximation give us additional information on the long range potential through an analysis of the preliminary vibrational level progression and based on the remarkable adequacy of the linear fit. According to the relation (4), a plot of $(E_v - D)^{\frac{1}{4}}$ versus v for several $N^2\Sigma^+$ (N=1-4, 6-7, 9-10) electronic states is purely linear (See Fig. 2). Except for the last points of some curves, these plots are remarkably linear. In fact the deviation from the linear low for the last points can be taken as a measure of the numerical limitations of the Numerov propagation method near the dissociation limit. This assumption is particularly clear for $1^2\Sigma^+$ state. Furthermore, using this fit we find the total number of vibrational level trapped by each state and specially the last excited vibrational levels near the dissociation limit. We reported in table 2 our extracted static polarisabilities compared with other theoretical and experimental works for various atomic states ns $^2$S of K and Rb atoms. Our ground state static dipole polarisabilities for K and Rb atoms are in good agreement with Magnier et al [16] calculated values as well as the experimental values [17-18] . We find for K and Rb, respectively 295.40 and 323.60 a. u to be compared with the experimental values 305 and 329 a. u. Our static dipole polarisabilities for K (4s) and Rb (6s) atoms are also in good agreement with the calculated values of Magnier et al [16] given by, respectively, 302 and 335 a. u. We report in tables 3, the static polarisabilities of the two lowest $^2$P excited levels of K and Rb atoms. A good agreement is also observed between our values and those of Magnier et al [16].

**Table 2a**: The static dipole polarisabilities of various atomic ns $^2$S levels for K and Rb atoms

| | K | $v_D$ | | Rb | $v_D$ | References |
|---|---|---|---|---|---|---|
| 4s | 295.40 | 198.4 | 5s | 323.60 | 77.1 | this work |
| | 302 | | | 335 | | [16] |
| | 290 | | | 320 | | [19] |
| | 263±6 | | | 319±6 | | [17] |
| | 305±2 | | | 329±23 | | [18] |
| 5s | 4861.7 | 148.8 | 6s | 5116.3 | 140.8 | this work |
| | 4940 | | | 5270 | | [16] |
| | 4990 | | | | | [20] |

**Table 2b**: The static dipole polarisabilities for various atomic np$_m$ (m=±1) $^2$P of K and Rb atoms

| | K | $v_D$ | | Rb | $v_D$ | References |
|---|---|---|---|---|---|---|
| $4p_0$ | 514.5 | 301.4 | $5p_0$ | 1082.7 | 200.6 | this work |
| | 509 | | | 700 | | [16] |
| $4p_{\pm1}$ | 666 | | $5p_{\pm1}$ | 786.9 | | this work |
| | 662 | | | 623 | | [16] |
| $5p_0$ | 6485.3 | 402.5 | $6p_0$ | 14858.6 | 314.1 | this work |
| | 6038 | | | 10913 | | [16] |
| $5p_{\pm1}$ | 7570.3 | | $6p_{\pm1}$ | 13567.1 | | this work |

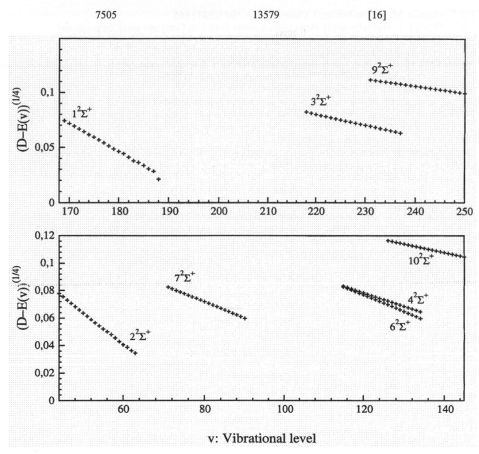

**Figure 2:** Top: plot of $(E_v - D)^{1/4}$ versus v for 1, 3 and 9 $^2\Sigma^+$ electronic states of the KRb$^+$ molecule.

Bottom: plot of $(E_v - D)^{1/4}$ versus v for 2, 4, 6, 7 and 10 $^2\Sigma^+$ electronic states of the KRb$^+$ molecule.

## 4. Conclusion

In this paper, we have determined the potential energy curves and spectroscopic constants of the KRb$^+$ electronic states dissociating into K(4s, 4p, 5s, 3d, 5p, 4d, 6s) + Rb$^+$ and K$^+$ + Rb(5s, 5p, 4d, 6s, 6p, 5d, 7s). The spectroscopic constants of the ground and the first excited states were compared with the work of Korek et al [15] showing a very good agreement and a high accuracy of our calculation. Such accuracy has been exploited to perform a long range vibrational spacing analysis using the WKB approximation. The plot of $(E_v - D)^{1/4}$ versus v has shown a pure linear behavior. A least squares fit of this linear law has allowed us to extract rather accurately the $v_D$ and the static polarisabilities from which all vibrational levels near the dissociation limit and also the related turning points have been calculated. Despite the simplicity of the used method, the extracted dipole polarisabilities are in good agreement with the available theoretical and experimental works.

## Acknowledgments

The authors wish to thank the anonymous referees for their fruitful comments and suggestions.

## References

[1] M. Foucrault, Ph Millié, JP. Daudey, J. Chem. Phys. 96 (1992) 1257.

[2] S. Magnier, S Rousseau, AR. Allouche, G. Hadinger, M. Aubert-Frécon, Chem. Phys. 64 (199) 57.

[3] S. Magnier, M. Aubert-Frécon, J. Chem. Phys. A 105 (2001) 165.
[4] C. Ghanmi, H. Berriche and H. Ben Ouada, Lecture Series on Computer and Computational Sciences 1 (2004) 1061.
[5] H. Berriche. J. Molecular Structure, 663 (2003) 101.
[6] H. Berriche, C. Ghanmi and H. Ben Ouada, J. Mol. Spec. 230 (2005) 161.
[7] Ph. Durand and J. C. Barthelat, Theoret. Chim. Acta 38 (1975) 283;
    J. C. Barthelat and Ph. Durand, Gazz. Chim. Ital. 108 (1978) 225.
[8] M. Foucrault, Ph. Millié and J. P. Daudey, J. Chem. Phys. 96 (1992) 1257.
[9] D. Pavolini, T. Gustavsson, F. Spiegelmann and J- P Daudey, J. Phys. B:
    At. Mol. Opt. Phys. 22 (1989) 1721.
[10] J. Vigué, Ann. Phys. (Paris) 3 (1982) 155.
[11] C. Ghanmi, H. Berriche and H. Ben Ouada, (submitted to the Journal of Molecular Spectroscopy).
[12] L. Von Szentplay, P. Fuentealba, H. Preuss and H. Stoll, Chem. Phys. Lett. 93 (1982) 555.
[13] W. Schwartmann, Ph. D. Thesis, Bochum, FRG (1979).
[14] A. Valance, A. Bernier and M. El Maddarsi, Chem. Phys. 103 (1986) 151.
[15] M. Korek, G. Younes, A. R. Allouche, Int. J. Q. Chem. 92 (2003) 376.
[16] S. Magnier, M. Aubert-Reécon, J. Quant. Spect. Rad. Trans. 75 (2002) 121.
[17] R. W. Molof, H. L. Schwartz, Th. M. Miller, and B. Bederson, Phys. Rev. A 10 (1974) 1131.
[18] W. D. Hall, J. C. Zorn, Phys. Rev. A 10 (1974) 1141.
[19] P. Fuentealba, O. Reyes, J. Phys. B: At. Mol. Opt. Phys. 26 (1993) 2245.
[20] M. Réat, M. Mérawa, B. Honvault-Bussery, J. Chem. Phys. 109 (1998) 7246.

Brill Academic Publishers
P.O. Box 9000, 2300 PA Leiden,
The Netherlands

*Lecture Series on Computer*
*and Computational Sciences*
Volume 4, 2005, pp. 709-712

# Detecting Reaction Pathways and Computing Reaction rates in Condensed Phase

S. Consta[1]

Department of Chemistry,
The University of Western Ontario,
London, Ontario, Canada N6A 5B7

Received 18 July 2005; accepted 11 August, 2005

*Abstract:* In this contribution, methods for the computation of rate constants that characterize classical reactions occurring in the condensed phase are discussed. Methods of transition path sampling and minimal action are reviewed and their adaption in the study on physical fragmentation of charged aqueous clusters is discussed.

*Keywords:* reaction coordinate, chemical reaction pathways, physical fragmentation of clusters, condensed phase, molecular dynamics, Monte Carlo

*PACS:* 34.10.+x, 36.40.Wa, 36.40.Qv

## 1   Introduction

Theoretical and experimental research in the study of reactions in the condensed phase has made remarkable progress in the understanding of the role of solvent on the mechanisms of a wide array of chemical reactions. While microscopic expressions for these transport properties are well known [1, 2], their computation presents challenges for simulation since reactive events often occur rarely, and the long time scales that are typical for reactive processes are not accessible using simple molecular dynamics methods. Furthermore, the underlying free energy surface is very complex with many saddle points that prevent sampling of possible reaction pathways. A conventional approach for the study of reaction mechanisms and rates is to define a reaction coordinate and use the reactive-flux method to compute the rate coefficient. The saddle point defined by the free energy profile computed along the reaction coordinate facilitates the computation of the rate. Because of the role of both solvent and reacting system in the reaction, the reaction coordinate may be a complex many-body function of the system's degrees of freedom. Since there is no an *a priori* way to define a "good" reaction coordinate, methods are being developed to assist in a systematic construction of a reaction coordinate. Such methods are the transition path sampling (TPS) [3], the steepest descent [4] and minimum action methods [5].

## 2   Systematic ways of finding RCs

*Transition path sampling*   Transition path sampling (TPS) developed by Bolhuis, Chandler, Dellago, Geissler [3] is an alternative technique to find possible reaction paths. TPS has two parts: the sampling of reactive trajectories and the computation of the reaction rate using a type of

---

[1]Corresponding author. E-mail: sconstas@uwo.ca

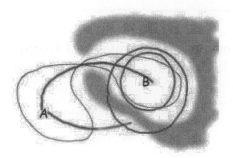

Figure 1: Illustration of problems arising in refining the phase space in the course of thermodynamic integration. Dashed blue lines show the most probable paths corresponding to sets $B_k$ and $B_{k+1}$. Dashed area signifies high values of the free-energy barrier. One observes singular changes in most probable path connecting $A$ and $B_{k,k+1}$ when the difference between $B_k$ and $B_{k+1}$ is small.

thermodynamic integration. For the sampling of the dynamic trajectories an initial reactive path is generated using constant energy deterministic or stochastic dynamics. Then, a state that corresponds to a randomly chosen time slice in the trajectory is modified by specific moves. The new state(s) is accepted by an MC scheme that satisfies detailed balance provided that the new trajectory starts in the reactant and finishes in the product states. Otherwise, the trajectory is immediately rejected. By this scheme the reactive trajectories are sampled and their analysis can provide knowledge of the reaction mechanism and give insight into ways of defining reaction coordinates. The computation of the reaction rate is slightly more complicated and is based on the use of an order parameter. One may express the rate in a direct way as the number of successful trajectories $n_{AB}$ from A to B within time $t$ over the total number of trajectories, $n_A$, that start from A

$$k_{AB}(t) = \frac{n_{AB}(t)}{n_A} = \frac{\langle I_A(\xi(0))I_B(\xi(t))\rangle}{\langle I_A(\xi(0))\rangle}. \tag{1}$$

$I_{A(B)}(\xi)$ is an indicator function defined as:

$$I_A(\xi) = \begin{cases} 1 & \text{if} \quad \xi \in A \\ 0 & \text{else} \end{cases} \tag{2}$$

In the TPS method the rate (1) can be computed using thermodynamic integration as follows: An order parameter ($\lambda$) is introduced such that the coordinate space is foliated into subsets

$$B(s) = \{q|\lambda(\xi) > s\}. \tag{3}$$

One considers an ordered sequence of values of the order parameter $\{s_0, s_1, \ldots, s_n\}$ and the corresponding subsets:

$$R^{3N} = B(s_0) \supset B(s_1) \supset \cdots \supset B(s_n) = B \tag{4}$$

One may verify by direct substitution that rate expressed by eq. (1) can be factorized as:

$$k_{AB}(t) = \prod_{i=1}^{n} \frac{\langle I_A(\xi(0))I_{B_i}(\xi(t))\rangle}{\langle I_A(\xi(0))I_{B_{i-1}}(\xi(t))\rangle}. \tag{5}$$

Each factor in the above product is the conditional probability for a trajectory to finish in subset $B_k$ provided that the trajectory reached subset $B_{k-1}$. By refining the sequence $B_k$ one may expect to achieve high degree of accuracy in the computation of the rate. In Fig. 1 we illustrate importance of adequate description of the free energy surface by the order parameter. When this is not the case ensemble of paths connecting $A$ (reactants) and $B_i$ may undergo drastic changes for small changes in the value of the order parameter. This may affect directly the efficiency of the MC scheme. The fraction of paths contributing to the factor $k_{B_k B_{k+1}}$ is very small even for a small difference between sets posing a challenge for accurate sampling. The above example emphasizes the role of study of chemical mechanisms for accurate computations of the rates.

*Minimal action path* A scheme proposed by Elber, Olender, Cárdenas, Ghosh and Shalloway [5] finds the most probable reaction path with fixed ends in the reactant and product states from minimization of an action:

$$S = \inf \int_{t_1}^{t_2} d\tau \left\| \varsigma \frac{d\xi}{dt} + \nabla W(\xi) \right\|^2 . \tag{6}$$

Motivation for considering the above form of an action functional comes from Langevin equation in the large friction limit

$$\varsigma \frac{d\xi}{dt} + \nabla W(\xi) = \mathbf{f}(t). \tag{7}$$

The most probable path (6) may reproduce the reaction mechanism when a reaction is determined by energetic factors and the majority of trajectories are found close to the minimum energy path. In high temperature where entropy effects are important, several trajectories that pass from different locations on the underlying free energy surface may be important for the reaction mechanism. A single trajectory that is found by this technique may not be representative of the reaction mechanism.

Alternatively, instead of minimizing the action on a fixed interval, we may look for the steepest descent reaction paths $\xi(\lambda)$ such that

$$\frac{d\xi(\lambda)}{d\lambda} = \frac{\nabla W(\xi)}{\|\nabla W(\xi\|} . \tag{8}$$

In the above equation we changed parametrization of curve $\xi$ from time $t$ to a natural parameter $\lambda$. Differentiating the identity

$$\left\| \frac{d\xi(\lambda)}{d\lambda} \right\|^2 = 1 \tag{9}$$

and using equation (8) we deduce that for the steepest descent path the component of $\nabla W(\xi)$ perpendicular to curve $\xi$ is zero:

$$\nabla W(\xi) \cdot \frac{d\xi(\lambda)}{d\lambda} = 0. \tag{10}$$

This condition

$$\nabla W(\xi)^\perp = 0 \tag{11}$$

is taken as the basis of minimal energy path approach. For a smooth energy landscape a solution of eq. (11) connecting two local minima consists of a connected path between extrema of the energy profile $\nabla W(\xi) = 0$.

Ulitsky and Elber[5] proposed a parametrized path evolution technique to find piecewise steepest descent path by integrating the following equation:

$$\frac{d\xi(\alpha, s)}{ds} = -\nabla W(\xi)^\perp(\alpha, s) \tag{12}$$

In the evolution of eq. (12) neither natural parametrization nor natural length of $\xi$ is preserved.

For a complex energy surface steepest descent method produces many possible paths connecting the product and reactant states. The direct enumeration of the paths is impossible except for the simplest cases. Direct integration of eq. (12) will produce only one solution for a given initial condition. To overcome this problem the finite temperature string method [4] was proposed where eq. (12) is replaced with a stochastic differential equation

$$\frac{d\xi(\alpha, s)}{ds} = -\nabla W(\xi)^{\perp}(\alpha, s) + \mathbf{f}^{\perp}(\alpha) \tag{13}$$

where $\mathbf{f}^{\perp}(\alpha)$ is gaussian noise with zero component in the direction $d\xi(\alpha, s)/d\alpha$. This approach, while having no physical foundation, provides an efficient way for sampling multiple steepest descent pathways.

## 3   Applications in classical systems

Physical fragmentation [6] of liquid clusters composed of water and ions of the same sign is presented as an example of process where TPS and minimization of action techniques cannot be applied. Physical fragmentation of charged clusters is a multi-channel processes without an a priori knowledge of the products. The study of fragmentation of charged liquid clusters is important for understanding fundamental questions of the physical and chemical processes in electrospray techniques [7] used in experiments. Fragmentation is controlled by two competing factors: the repulsive Coulomb interactions of the ions with similar charge that tend to fragment the droplets and the opposing effect of the surface tension and the polarization of the solvent. For certain ratio of number of ions to solvent molecules, these effects give rise to a free energy barrier between the compact structure of the system that corresponds to the reactant state and the fragmented states that are the products. The products are not well defined since the ions may escape from the droplet with variable number of solvent molecules. The reaction is studied by using a new "non-conventional" reaction coordinate that we called "transfer RC" (TRC). The TRC contains both ion and solvent spacial coordinates and it associates the motion of the ions, for example the escape of the ion, with the location of the solvent molecules in the cluster. Using the TRC, cluster configurations with bottlenecks that separate ions are distinguished from spherical or elliptical configurations.

## References

[1] P. Hänggi, P. Talkner and M. Borkovec, Rev. Mod. Phys. **62**, 251 (1990).

[2] R. Kapral. S. Consta and L. McWhirter, in *Classical and Quantum Dynamics in Condensed Phase Simulations*, pp. 587, eds. B. J. Berne, G. Ciccotti, D. F. Coker (1998).

[3] C. Dellago, P. Bolhuis and R. L. Geissler, Adv. in Chem. Phys. **123**, 1 (2002).

[4] E. Weinan, W. Ren and E. Vanden-Eijnden, Phys. Rev. B **66**, 05230 (2002).

[5] R. Elber, J. Meller, R. Olender, J. Phys. Chem. **103**, 899 (1999); A. Ulitsky and R. Elber, J. Chem. Phys. **92**, 1510 (1990).

[6] S. Consta, J. of Mol. Struct.-Theochem **591**, 131 (2002); S. Consta, K. .R. Mainer and W. Novak, J. Chem. Phys. **119**, 10125 (2003).

[7] P. Kebarle, J. Mass Spectrom. **35**, 804 (2000); J. B. Fenn, J-. Rosell and C. K. Meng, J. Am. Soc. Mass Spectrom. **8**, 1147 (1997).

Brill Academic Publishers
P.O. Box 9000, 2300 PA Leiden
The Netherlands

*Lecture Series on Computer*
*and Computational Sciences*
Volume 4, 2005, pp. 713-716

# SAC-CI study of a series of the lowest states of HCl+

V. S. Gurin[1], Vitaly E. Matulis
Physico-Chemical Research Institute,
Belarusian State University,
Leningradskaya str., 14, 220050, Minsk, Belarus

M. V. Korolkov
Institute of Physics,
National Academy of Sciences of Belarus,
Independence Av., 220072, Minsk, Belarus

S. K. Rakhmanov
Belarusian State University,
Independence Av., 4, 220050, Minsk, Belarus

Received 3 July, 2005; accepted in revised form 28 July, 2005

*Abstract:* The symmetry-adapted-cluster configuration interaction (SAC-CI) method used for calculation of electronic structure of HCl+ molecular ion. Potential energy curves and dipole moments are obtained for a series of low-lying states and transition dipole moments are evaluated from the ground $X^2\Pi$ state. The various dependencies of dipole moments on distance reveal the corresponding asymptotics and interpreted as the two pathways of HCl+ dissociation. The behavior of transition dipole moments shows the possibility of selective excitation of the different states leading to control of the dissociation pathways.

*Keywords:* diatomic molecule, potential curves, SAC-CI, transition dipole moments
*PACS:* 31.10.+z; 31.15.Ar; 31.25.Nj; 33.15-e

## 1. Introduction

Under application of different calculation methods in quantum chemistry diatomic molecules are good tools for testing as far as they are examples of simplest systems in which a chemical bonding appears. On the other hand, experimental studies for interaction with an intense optical field are strongly developed last years, and diatomics are under first attention since the fields ($10^{14}$ W/cm$^2$) change remarkably electronic structure of molecules resulting to very complicated tasks in the case of polyatomic molecules. HCl and HCl+ are the examples of rather popular system in molecular spectroscopy [1,2], however, no complete theoretical analysis has been performed to date at the modern ab initio level. In the present paper, we consider application of the symmetry-adapted clusters – configuration interaction (SAC-CI) method to the calculation of a series of electronic states of HCl+ analyzing the full potential curves together with transitions between states. This method has been developed for calculations of ground and excited states of molecules and was successfully adopted for molecular spectroscopy purposes with molecules of different level of complicacy [3]. In particular, HCl ionization spectrum has been reproduced quite successful [4]. The SAC-CI approach includes an advanced cluster expansion to take into account CI due to a number of excited states, and selection of expansion terms based on symmetry provides both more physical insight and better convergence [5]. As compared with MCSCF approaches like CASSCF, CCSD, MRSDCI, etc., SAC-CI uses an optimized selection of the reference states to account CI of one- and multi-electron excitations.

## 2. Details of calculation technique

In the SAC-CI calculations (Gaussian 03) we use the all-electron basis set of the 6-311G type with addition of two diffuse s- and p-functions with exponents $\xi_s$=0.06, 0.015 and $\xi_p$=0.04, 0.01 and four d-

---

[1] Corresponding author. E-mail: gurin@bsu.by; gurinvs@lycos.com

functions with $\xi_d$=0.7, 0.25, 0.08 and 0.02 for Cl. For hydrogen atom two p-functions with (p=1.0 and 0.3 were added to the 6-311G basis set. This choice of basis was adopted from [4], it gives good consistence with experiment for ionization spectrum of HCl, and the modifications of this basis increasing its complicacy did not result in better consistence with known spectroscopic constants of HCl+. CI accounting was done selecting 1s, 2s and 2p atomic orbitals of Cl as frozen. Active space of orbitals included the valence orbitals (beginning from 3s- for Cl and 1s- for H): 4 occupied and 58 virtual. This choice of active orbitals is typical also for different molecules with 3p-elements.

Potential curves of a series of the lowest states assigned to the C∞h point group was calculated in the range of distances 1-20 ao. The distance-dependent dipole moments (DM) and transition dipole moments (TDM) from the ground X(2( are collected below and analysed to reveal possibility of selective excitation of the molecule under light absorption.

## 3. Results and discussion

The potential curves (Fig. 1) show that the bound states appear for A 2(+ and X(2( states and there exist it also for 4 2(. These results are in agreement with experimental data for the number of bound states of HCl+ and the calculations in [6]. The bound states correspond to different dissociation limits, evidently, H+ + Cl and H + Cl+, respectively. Our calculations for 4 2( indicate the maximum, which was not found in [6]. Starting from the distance 8-9 a0 all curves demonstrate the asymptotic behavior, and variations of energy are negligible. A state of full dissociation may be taken already about 10 a0. As the whole, good consistence of the present results and the calculations with different approach, CASSCF, in [6] evidences that the set of states under study is properly determined and the corresponding wave functions may be used for analysis of electronic transitions.

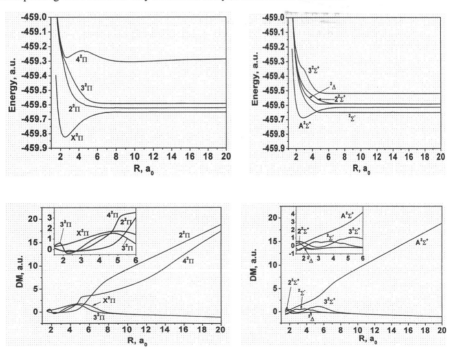

Figure 1: Potential curves and DM for a series of 2(, 2(+, 2(- and 2( states.

DM reveal rather different behavior associated with the corresponding pathways of dissociation. The large values of DM at long distance are related to the states which dissociate onto H$^+$ and Cl and the asymptotic behavior occurs as ~$(35/36)eR$, while the dissociation pathway to H and $Cl^+$ results in the asymptote ~$(1/36)eR$ [7]. DM are increasing with distance for the bound state of $\Sigma^+$ symmetry, A $^2\Sigma^+$, and for the second and fourth state of $\Pi$ symmetry, $2\,^2\Pi$ and $4\,^2\Pi$. Dependencies with maxima at the distances more than the equilibrium ones and approaching some small negative value at large distances occur for the bound $\Pi$-state, X $^2\Pi$, the third $\Pi$-state, $3\,^2\Pi$, and the couple of $\Sigma^+$-states, $2\,^2\Sigma^+$ and $3\,^2\Sigma^+$. There are several points of sign inversion of DM for these states that show that the dissociation is accompanied by strong changes in bonding of the atoms. The latter couple of states indicate different positions for maxima, while the maximal values of DM are close. DM for $\Sigma$ and $\Delta$ states have the maxima of the lower heights only for small distances, and at $R>2.5a_0$ they are very low and vanish at $R>6a_0$ as well the values of DM for other states with the extreme R-dependence. $\Sigma^+$ and $\Pi$- states reveal also complicated behavior at small distances. Thus, the strong rebuilding of the molecule at distances less than the equilibrium takes place for all states under calculation.

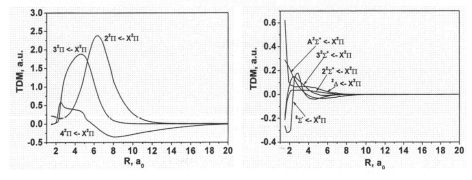

Figure 2: Calculated TDM from X $^2\Pi$ state to a series of the higher $^2\Pi$-states and to some low-lying $^2\Sigma^+$, $^2\Sigma^-$ and $^2\Delta$ states.

The calculation results for TDM from the ground state, $X\,^2\Pi$, are summarized in Fig. 2. The transitions to the two nearest $^2\Pi$-states are consistent with the data of Ref. 6 quite well besides some deviation at very small distances $R<2a_0$. Qualitatively, the behavior of TDM for these transitions appears with pronounced maxima, and the maximum for the lower final state, $2\,^2\Pi$, is located at the larger distances than for the $3\,^2\Pi$ final one. This can mean the possibility of selective excitation of this molecule from $X\,^2\Pi$ to different final states in explicit dependence on the dissociation degree. This supposition is more interesting since $2\,^2\Pi$ and $3\,^2\Pi$ states correspond to different pathways of $HCl^+$ dissociation. Both maxima of TDM occurs at the distances essentially more than the equilibrium, $R_0$, for the ground $X\,^2\Pi$ state, while at this $R_0$ the values of TDM are close to zero. Thus, most intensive transitions proceed at the beginning of predissociation. In contrast, at the large distances, $R>12a_0$, TDM for both transitions become practically zero. Meanwhile, it should be noted that the corresponding potential curves attain a plateau (see above) at the less distances, $R>8a_0$., i.e. intensive transitions (in particular, for the case $X\,^2\Pi\text{->}2\,^2\Pi$) can be even for weakly interacting atoms (ions) at considerable late dissociation steps.

TDM from ground to $4\,^2\Pi$ state also have a maximum but less pronounced comparatively to the transition to the above lower states, and the whole dependence of this TDM on distance is more complicated. In particular, TDM change the sign at $R\approx5a_0$.

TDM for transitions from the ground state $X\,^2\Pi$ to the states of other symmetries ($\Sigma^+$, $\Sigma^-$ and $\Delta$) reveal very different behavior (Fig. 2). The maximal values are much less than those for $\Pi$-$\Pi$ transitions. Also, they invert the sign (but, absolute value is less than 0.1 a.u.) in some range of R more than the equilibrium $R_0$. At the larger distances these TDM are approaching zero being negative. The transition between bound states X $^2\Pi$ and A $^2\Sigma^+$ decreases from the value about 0.3 a.u. crossing the zero line at about R=4 a.u. The transitions to the two next states, $2\,^2\Sigma^+$ and $3\,^2\Sigma^+$, show the R-dependence with low maxima in the range of R around the values $R_0$ for X $^2\Pi$ state (but not exactly at $R_0$), and these TDM are negative for small distances. These dependencies are similar to the corresponding data in Ref. 6 in qualitative behavior, however, numerically they slightly deviate. The transitions from X $^2\Pi$ to $^2\Delta$ and $^2\Sigma^-$ states discover another features, however, they both are very small and increase only in the range of shortest distance (Fig. 2).

## 4. Conclusions

Using the SAC-CI approach we have calculated a series of electronic states of $HCl^+$ and summarized potential curves and dipole moments for them and transition dipole moments from the ground state. The potential curves reproduce well both experimental data available and another calculations at the compatible theory level. DM demonstrate correct asymptotic behavior for the different dissociation pathways and some states reveal the pronounced maxima at the distances approximately twice larger than the equilibrium. TDM of $\Pi$-$\Pi$ type attain considerable large values also at the distances more than $R_o$ and vanish at separation of the atoms. The transitions to the states of other symmetries ($\Sigma^+$, $\Sigma^-$ and $\Delta$) appear to be of much lower intensity with maxima either close to $R_o$ ($2\,^2\Sigma^+$, $3\,^2\Sigma^+$, and $\Delta$) or at the lower distances (A $^2\Sigma^+$, $\Sigma^-$). This behavior of TDM in dependence on R and type of the states means the possibility of selective excitation of $HCl^+$ and proper control of its dissociation.

## Acknowledgments

The work is supported by the project INTAS-Belarus 2003, N 5765.

## References

[1] W.J. Stevens and M. Krauss, The electronic structure and photodissociation of HCl, *J. Chem. Phys.* **77** 1368-1372 (1982)..

[2] M. Michel, M.V. Korolkov, and K.-M. Weitzel, State-selective photodissociation spectroscopy of $HCl^+$ and $DCl^+$ ions, *J. Phys. Chem.* **A108** 9924-9930 (2004).

[3] H. Nakatsuji, SAC-CI method: theoretical aspects and some recent topics, *Computational Chemistry: Reviews of Current Trends - V. 2* (Editor: J. Leszczynski), World Scientific Publ. Co., Singapore, 1997.

[4] M. Ehara, P. Tomasello, J. Hasegawa, H. Nakatsuji, SAC-CI general-R study of the ionization spectrum of HCl, *Theor. Chem. Acc.* **102** 161-164 (1999).

[5] H. Nakatsuji and K. Hirao, Cluster expansion of the wavefunction. Symmetry-adapted-cluster expansion, its variational determination, and extension of open-shell orbital theory, *J. Chem. Phys.* **68** 2053-2065 (1978).

[6] A. D. Pradhan, K. P. Kirby, and A. Dalgarno, Theoretical study of $HCl^+$ : Potential curves, radiative lifetimes, and photodissociation cross sections, *J. Chem. Phys.* **95** 9009-9023 (1991).

[7] A. Conjusteau, A. D. Bandrauk, and P. B. Corkum, Barrier suppression in high intensity photodissociation of diatomics: Electronic and permanent dipole moment effects, *J. Chem. Phys.* **106** 9095-9104 (1997).

Brill Academic Publishers
P.O. Box 9000, 2300 PA Leiden,
The Netherlands

*Lecture Series on Computer
and Computational Sciences*
Volume 4, 2005, pp. 717-720

# Exciton Effects in Möbius Conjugated Polymers

## K. Harigaya[1]

Nanotechnology Research Institute, AIST, Tsukuba 305-8568, Japan
Synthetic Nano-Function Materials Project, AIST, Tsukuba 305-8568, Japan

Received 16 July, 2005; accepted in revised form 12 August, 2005

*Abstract:* Optical excitations in Möbius conjugated polymers are studied theoretically by the tight binding model of poly(*para*-phenylene), taking exciton effects into account. Calculated results of optical absorption spectra are reported. Certain components of optical absorption for an electric field perpendicular to the polymer axis mix with absorption spectra for an electric field parallel to the polymer axis. Therefore, the polarization dependences of an electric field of light enable us to detect whether conjugated polymers have the Möbius boundary.

*Keywords:* Möbius conjugated polymers, optical excitations, theory

*PACS:* 78.66.Qn, 73.61.Ph, 71.35.Cc

Recently, low-dimensional materials with peculiar boundary conditions, *i.e.*, Möbius boundaries have been synthesized: they are NbSe$_3$ [1] and aromatic hydrocarbons [2], for example. A Möbius strip consists of one surface that does not have the difference between the outer and inner surfaces. The orbitals of electrons are twisted while traveling along the strip axis, and electronic states can be treated with antiperiodic boundary conditions mathematically. Even though the presence of twisted $\pi$-electron systems has been predicted theoretically about forty years ago [3], structural perturbations due to topological characters might result in optical properties.

We have studied boundary condition effects in nanographite systems where carbon atoms are arrayed in a one-dimensional shape with zigzag edges [4-6]. Due to the presence of a Möbius boundary, magnetic domain wall states and helical magnetic orders are realized in spin alignments, and a domain wall also appears in charge density wave states. Such abundant properties have been experimentally observed by their unique magnetic properties [7,8].

In this paper, we investigate another candidate for twisted $\pi$-electron systems: conjugated polymers. We choose poly(*para*-phenylene) (PPP) as a model material of conjugated polymers. The structure of PPP is shown in Fig. 1(a). Optical excitations in periodic systems have been previously studied using a tight binding model with long-range Coulomb interactions for poly(*para*-phenylenevinylene) (PPV) [9], PPP, and so forth [10]. Exciton effects have been taken into account by the configuration interaction method. The model has three parameters: the neighboring hopping integral $t$, the onsite $U$ and long-range $V$ repulsions. As in the paper [9], we use the followings: $U = 2.5t$, $V = 1.3t$, and the Lorentzian broadening $\gamma = 0.15t$ for the optical spectra.

---

[1]Corresponding author. E-mail: k.harigaya@aist.go.jp

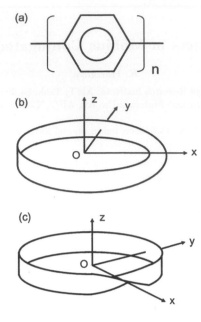

Fig. 1. (a) Polymer structures of poly(*para*-phenylene) (PPP). Geometries of PPP strips used for the calculations: (b) spatially uniform Möbius polymer, and (c) the Möbius polymer with spatially localized twists. The PPP chain is represented schematically by the strip. The $\pi$-orbitals of carbons extend perpendicularly with respect to the strip.

We consider two cases of Möbius boundaries in order to look at optical excitation spectra. In the first case (A) [Fig. 1(b)], the ring torsions occur uniformly over the polymer with the torsion angle $\Phi = 180°/M$, $M$ being the number of phenyl rings. The phenyl rings rotate helically along with the polymer. The polymer axis is represented by a circle in the $x$-$y$ plane in Fig. 1(b). In the second case (B) [Fig. 1(c)], the twist due to the Möbius boundary is localized among five phenyls where the polymer circle crosses the $x$-axis. The twist angle from the torsion is taken as $\Phi = 30°$. Hereafter, the calculated absorption spectra are shown for $M = 20$.

In the case (A), the molecular plane becomes almost parallel to the $x$-$y$ plane in the part where the polymer axis crosses the $x$-axis ($x > 0$). Among this part, almost perpendicular polarization is realized, and therefore mixing of the absorption coming from perpendicular polarization is expected. Figure 2(a) shows the actual calculated spectrum. There are two weak features in the energy region between $2.2t$ and $3.0t$ owing to the mixing. There is no difference between the $x$- and $y$-polarizations, because the system is uniform. On the other hand, the absorption of the perpendicular polarization does not show any polarization dependence [Fig. 2(b)].

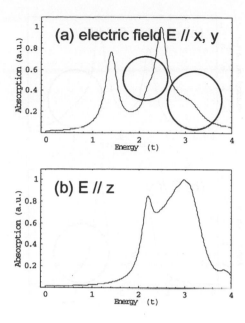

Fig. 2. Optical absorption spectra for the Möbius boundary case (A): (a) electric field of light parallel with the polymer axis, (b) electric field perpendicular to the polymer axis. The circles show features coming from the mix of perpendicular polarization. The number of phenyl rings is $M = 20$. The vertical axis is shown in the arbitrary units, and the Lorentzian broadening $\gamma = 0.15t$ is used in the calculations.

In the case (B), the polymer plane can become parallel with the $x$-$y$ plane at the region where the twist is present near the $x$-axis ($x > 0$). There is nearly perpendicular polarization among this region. The mixing of the perpendicular polarization occurs around the energy of $3.0t$ as shown in Fig. 3(a), where the electric field of light is parallel with the $x$-axis. For the case that the field is along the $y$-axis, the mixing of the perpendicular polarization is very weak because of the localization of the twist [Fig. 3(b)].

The difference between the periodic and Möbius boundary conditions is regarded as due to the addition of antiperiodicity, and the allowed wave numbers are different. In oligomers with a few phenyl rings, the allowed states are so sparsely distributed that the boundary condition effects could be measured. The recent synthesis of Möbius aromatic systems [2] will promote the synthesis of Möbius polymers in view of the presence of many types of ring polymers [11,12].

In summary, electronic structures and optical excitations in PPP with Möbius boundaries have been studied by taking exciton effects into account. In the calculated optical absorption spectra, certain components of optical absorption for an electric field perpendicular to the polymer axis mix with absorption spectra for an electric field parallel to the polymer axis. The polarization dependences of an electric field of light enables us to detect whether conjugated polymers have the Möbius boundary. Therefore, the experimental synthesis of materials with unique geometries is expected.

We note that detailed discussion has been published as a letter paper recently [13].

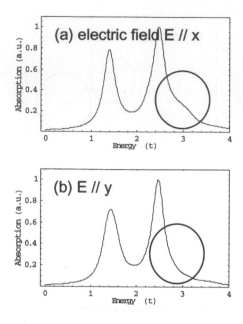

Fig. 3. Optical absorption spectra for the Möbius boundary case (B): (a) electric field of light parallel with the $x$-axis, and (b) $y$-axis. The circles show features coming from the mix of perpendicular polarization. The number of phenyl rings is $M = 20$. The vertical axis is shown in the arbitrary units, and the Lorentzian broadening $\gamma = 0.15t$ is used in the calculations.

## Acknowledgments

This work has been supported partly by Special Coordination Funds for Promoting Science and Technology, and by NEDO under the Nanotechnology Program.

## References

[1] S. Tanda et al.: Nature **417** (2002) 397.
[2] D. Ajami et al.: Nature **426** (2003) 819.
[3] E. Heilbronner: Tetrahedron Lett. **5** (1964) 1923.
[4] K. Wakabayashi and K. Harigaya: J. Phys. Soc. Jpn. **72** (2003) 998.
[5] A. Yamashiro, Y. Shimoi, K. Harigaya and K. Wakabayashi: Phys. Rev. B **68** (2003) 193410.
[6] A. Yamashiro, Y. Shimoi, K. Harigaya and K. Wakabayashi: Physica E **22** (2004) 688.
[7] Y. Shibayama et al.: Phys. Rev. Lett. **84** (2000) 1744.
[8] H. Sato et al.: Solid State Commun. **125** (2003) 641.
[9] K. Harigaya: J. Phys. Soc. Jpn. **66** (1997) 1272.
[10] K. Harigaya: J. Phys.: Condens. Matter **10** (1998) 7679.
[11] E. Mena-Osteritz: Adv. Mater. **14** (2002) 609.
[12] M. Mayor and C. Didschies: Angew. Chem. Int. Ed. **42** (2003) 3176.
[13] K. Harigaya: J. Phys. Soc. Jpn. **74** (2005) 523 (Letters section).

Brill Academic Publishers
P.O. Box 9000, 2300 PA Leiden
The Netherlands

*Lecture Series on Computer
and Computational Sciences*
Volume 4, 2005, pp. 721-723

# Complex Orbital Calculation of Singlet-Triplet Molecular Oxygen Spacing and its Effect on the Theoretical Investigation of a Large Number of Atmospheric Reactions

Agnie M. Kosmas[1] and Evangelos Drougas

Department of Chemistry, Physical Chemistry Sector,
University of Ioannina, GR-451 10 Ioannina, Greece

Received 5 June, 2005; accepted in revised form 13 July, 2005

*Abstract:* A large number of gas-phase reactions of atmospheric significance involve molecular oxygen production pathways either through the singlet or the triplet potential energy surfaces. In the investigation of such systems, it has been repeatedly observed that very sophisticated quantum mechanical electronic structure methods fail to produce the correct energy level of singlet $^1\Delta$ $O_2$. In the present work a series of quantum mechanical calculations of the singlet $O_2$ ($^1\Delta$) electronic energy are carried out using the complex orbitals methodology at the HF and MP2 levels of theory. It is found that although the Hartree-Fock method considerably underestimates the energy splitting, the MP2 level produces the correct electronic energy and the derived singlet-triplet energy spacing is found to be 22.4 kcal mol$^{-1}$, in excellent agreement with the experimental findings. The consequences of this result in the lowering of the $O_2$ ($^1\Delta$) production pathways in a large number of atmospheric reactions is discussed in detail.

*Keywords:* Singlet oxygen, triplet oxygen, complex orbitals, atmospheric reactions

*PACS:* 31.15.-p

## 1. Introduction

A large number of gas-phase reactions that play a significant role in the stratospheric ozone depletion cycle [1], have been studied computationally in the literature over the last years. A lot of them involve halogen oxides and release molecular oxygen among other products, through either the singlet or the triplet potential energy surfaces. An example of such systems are the reactions of hydroperoxy, $HO_2$, and methylperoxy, $CH_3O_2$, radicals with halogen monoxides [2] which proceed on the singlet potential energy surface according to the following scheme :

$$
\begin{aligned}
YO_2 + XO &\rightarrow XOO + YO \rightarrow X + {}^1O_2 + YO & (1)\\
&\rightarrow YOX + {}^1O_2 & (2)\\
&\rightarrow OXO + YO & (3)\\
&\rightarrow HCHO + HX + {}^1O_2 \quad \text{(present only for Y=CH}_3\text{)} & (4)\\
&\rightarrow YX + O_3 & (5)
\end{aligned}
$$

where Y= H, $CH_3$ and X=Cl, Br, I. We readily see that reaction channels (1), (2) and (4) which are found experimentally to be the most probable reaction pathways involve the formation of singlet $O_2$ that relaxes to the triplet ground state. Consequently, the computation of the correct energy level of ($^1\Delta$) $O_2$ is vital for the proper construction of the dynamics of the reaction in order to determine the correct reaction probability of each channel. Repeatedly in the literature, the electronic energy of singlet oxygen in a large number of relevant studies has been miscalculated and the singlet-triplet energy gap has been overestimated compared to the experimental value of 22.5 kcal mol$^{-1}$ , found to range from ~4.5 to ~8.5 kcal mol$^{-1}$ higher, depending on the methodology used. The result is the underestimation of the role of the molecular oxygen production channels in the mechanism of the

---

[1] Corresponding author. E-mail: amylona@cc.uoi.gr

reaction, since these pathways have been erroneously placed higher in the potential energy surface compared to where they are really located. The above remarks demonstrate clearly that the singlet-triplet oxygen energy spacing is a very essential quantity that determines the correct reactive surface and the correct reactive probability of each specific pathway that involves the production of molecular oxygen.

## 2. Computational details

In the present work we have carried out a series of calculations of the $O_2$ ($^1\Delta$) electronic energy using the ab initio RHF, MP2, CCSD(T), QCISD(T) and G2MP2 methodologies combined with the 6-31G* real basis set and density functional theory models (B3LYP) also combined with the 6-31G* real basis set. In the second series of calculations the restricted HF and MP2 methods in combination with the complex 6-31G* basis set were employed. All calculations have been performed using the Gaussian 98 series of programs [3].

## 3. Results and Discussion

Table 1 contains the results for the singlet-triplet molecular oxygen splitting based on the calculations for the electronic energy of $O_2$ ($^1\Delta$) at various levels of theory using both real and complex orbital methodologies. We readily see that all theory levels when used in combination with real orbitals fail to produce the correct singlet-triplet splitting and locate the singlet molecular

Table 1: Calculated energy spacing of $O_2$ ($^1\Delta$) – $O_2$ ($^3\Sigma$) in kcal mol$^{-1}$ at various levels of theory

|  | B3LYP | RHF | MP2 | CCSD(T) | QCISD(T) | G2MP2 |
|---|---|---|---|---|---|---|
| Real 6/31G* | 39.0 | 29.1 | 33.7 | 30.3 | 30.1 | 26.2 |
| Complex 6/31G* |  | 15.1 | 22.4 |  |  |  |
| Experimental | 22.5 |  |  |  |  |  |

oxygen higher than where it is really found. In addition, the restricted Hartree-Fock method using complex orbitals underestimates the energy splitting [4]. The MP2 method however, which includes electron correlation effects produces the correct energy spacing, in excellent agreement with the experimental result [5].

The serious consequence of the incorrect quantum mechanical calculation of $O_2$ ($^1\Delta$) electronic energy has been the underestimation of the theoretical reaction probability of molecular oxygen production pathways on the singlet potential energy surface. Thus, the theoretical description of various reactions such as the $HO_2$, $CH_3O_2$ +ClO,[6,7] $HO_2$, $CH_3O_2$ +BrO [8,9] and OH+OClO [10] systems, has repeatedly failed to agree with the experimental observations, as it has been pointed out in the study of $HO_2$+IO reaction [11].

Indeed, for the system $CH_3O_2$+ClO for example, the various experimental measurements have produced large values of the overall rate constant [12-19], ranging from $4.0 \times 10^{-12}$ cm$^3$ molecule$^{-1}$ s$^{-1}$ at 200 K and $3.1 \times 10^{-12}$ cm$^3$ molecule$^{-1}$ s$^{-1}$ at 300 K at constant pressure. The product analysis in the earlier studies have supported the conclusion that the reaction leads primarily to $CH_3O$+ClOO formation with ClOO readily dissociating into molecular oxygen and active atomic chlorine. On the absence of detection of any other products such as $CH_3OCl$, OClO and $O_3$ at that time, it was concluded that channel (1) is the most important reaction pathway. Based on this conclusion, Crutzen et al. [20] have reported a very interesting modeling study regarding the role of reaction (1) in the stratospheric ozone depletion cycle. They have shown that a mechanism in which reaction (1) is coupled with the heterogeneous reaction HCl+HOCl → $Cl_2$+$H_2O$ (taking place on pure and nitric acid doped ice-surfaces) may explain the accelerated rates of ozone depletion caused by active chlorine. Later however, Helleis et al. [14, 17] have measured a non-negligible value for the rate coefficient of methyl hypochlorite, $CH_3OCl$, formation pathway and they have thus, shown that channel (4) does occur to a considerable extent, lowering the percentage ratio of channel (1) in the overall rate of the reaction. These experimental observations are in excellent agreement with the present calculations that place the singlet molecular oxygen ~4.5 to ~8.5 kcal mol$^{-1}$ lower than it was previously assumed, indicating the higher exothermicity and the higher reaction probability of this channel. The important atmospheric implications of both the more recent experimental findings and the present theoretical results is that the severe effect of reaction (1) on the catalytic ozone depletion cycle is weakened to a considerable degree compared to what it was previously assumed, due to the non-negligible probability of channel (4) that involves the significant trapping of Cl into the $CH_3OCl$ species.

## Acknowledgments

The authors wish to thank the University of Ioannina Computer Center for the Computer services provided.

## References

[1]   M.J. Molina and F.S. Rowland, *Nature* **249** 810-815 (1974).

[2]   Z. Alfassi (Ed.), *Peroxyl Radicals* , Wiley, N. York (1997).

[3]   M.J. Frisch et al. GAUSSIAN 98, Gaussian Inc. Pittsburgh, PA (1998).

[4]   N. Kaltsoyannis and D.m. Rowley, *Phys Chem Chem Phys* **4** 419 (2002).

[5]   D. Buttar and D.M. Hirst, *J. Chem. Soc. Faraday Trans.* **90** 1811 (1994).

[6]   S.L. Nickolaisen, C.M. Roehl, l.K. Blakeley, R.R. Friedl, J.S. Francisco, R. Liu and S.P. Sander, *J. Phys. Chem. A* **104** 308 (2000).

[7]   E. Drougas, A.F. Jalbout and A.M. Kosmas, *J. Phys. Chem. A* **107** 11386 (2003).

[8]   S. Guha and J.S. Francisco, *J. Phys. Chem. A* **103** 8000 (1999).

[9]   S. Guha and J.S. Francisco, *J. Chem. Phys.* **118** 1779 (2003).

[10]   Z-F Xu, R. Zhu and M.C. Lin, *J. Phys. Chem. A* **107** 1040 (2003).

[11]   E. Drougas and A.M. Kosmas, *J. Phys. Chem. A* **109** 3887 (2005).

[12]   F.G. Simon, J.P. Burrows, W. Schneider, G.K. Moortgat and P.J. Crutzen, J. Phys. Chem. 93 7807 (1989).

[13]   W.b. DeMore, *J. Geophys. Res.* **96** 4995 (1991).

[14]   F. Helleis, J.N. Crowley and G.K. Moortgat, *J. Phys. Chem.* **97** 11464 (1993).

[15]   R.D. Kenner, K.R. Ryan and J.C. Plumb, *Geophys. Res. Lett.* **20** 1571 (1993).

[16]   A.S. Kukui, T.P.W. Jungkamp and R.N. Schneider, Ber. Bunseges. Phys. Chem. 98 1298 (1994).

[17]   F. Helleis, J.N. Crowley and G.K. Moortgat, *Geophys. Res. Lett.* **21** 1795 (1994).

[18]   P. Biggs, C.E. Canosa-Mass, J.M. Frachbound, D.E. Shallcross and R.P. Wayne, *Geophys. Res. Lett.* **22** 1221 (1995).

[19]   V. Daele and G. Poulet, *J. Chim. Phys.* **93** 1081 (1996).

[20]   P.J. Crutzen, R. Muller, C. Bruhl and T. Peter, *Geophys. Res. Lett.* **19** 1113 (1992).

Brill Academic Publishers
P.O. Box 9000, 2300 PA Leiden,
The Netherlands

*Lecture Series on Computer*
*and Computational Sciences*
Volume 4, 2005, pp. 724-726

# Chemically Induced Magnetism in Nano-clusters

R.J. Magyar[†] [1,2], V. Mujica[1,2], C. Gonzalez[2], and M. Marquez[3]

[1] INEST * Group,
Philip Morris USA Postgraduate Research Program,
4201 Commerce Road,
Richmond, VA 23234

[2] National Institute of Standards and Technology,
100 Bureau Dr. MS 8380,
Gaithersburg, MD 20899

[3] Philip Morris USA Research Center,
4201 Commerce Road,
Richmond, VA 23234

Received 5 August, 2005; accepted in revised form 21 August, 2005

*Abstract:* Intrinsic magnetism in thiolated gold nano-clusters is explored from a first-principles perspective. The computed electronic structure may elucidate the relevant mechanism whereby these nano-clusters develop a permanent magnetic moment.

*Keywords:* nano-clusters, magnetism, density functional theory

*PACS:* 73.22.-f, 75.75.+a, 71.15.Mb

## 1 An First-Principles Investigation of Nano-scale Magnetism

Modern synthesis and characterization techniques have enabled the controlled production of nano-sized metal and semi-conductor clusters. Because of the small and tunable size, these nano-clusters exhibit adjustable optical and other properties that may differ drastically from their bulk properties. Since these properties often depend critically upon the size and shape of the nano-cluster, it is possible that nano-clusters can be precisely tuned to be suitable for precise technological applications.

Recent experimental evidence demonstrates that nano-clusters of non-magnetic bulk materials may undergo a transition to a magnetic state as the nano-cluster size is reduced. In particular, thiolated gold nano-clusters of about 100-300 atoms retain a magnetic moment under zero field bias, and their magnetization exhibits a hysteresis curve typical of ferromagnetic systems. This is quite surprising as bulk gold is diamagnetic. The onset of tunable magnetic properties on a nano-scale could be quite useful. For example, magnetic clusters could be used in nano-scale magnetic field sensors to probe properties difficult to study using traditional optical sensors. Gold is enticing because it is both easy to manipulate and chemically inert. These magnetic nano-particles might also find applications in nano-scale magneto-resistance and could lead to nano-scale computer memory devices.

---

[†]Corresponding author. E-mail:rjmagyar@nist.gov
*Interdisciplinary Network of Emerging Science and Technology

Evidence for magnetism in gold nano-clusters has been accumulating. Zhang and Sham used X-ray spectrography techniques on alkane-thiolated gold nano-clusters to explore how lattice contraction and surface ligands affect the magnetic behavior in these clusters [1]. A subsequent experiment by Crepo and coworkers used X-ray absorption near-edge structure (XANES) to show that in thiolated clusters, the gold atoms have a magnetic moment of $\mu = 0.036\mu_B$ per atom [2]. They found that the presence of magnetism depends strongly on the nature of the surfactant. When Sulfur-based capping ligands are used, the magnetism is detected. However, when Nitrogen-based ligands are used, the magnetism is quenched. This ligand dependence indicates that ligand-cluster interactions might play a vital role. This dependence is seemingly in contradiction with results seen by Yamamoto obtained using XMCD [3]. Yamamoto and coworkers have observed intrinsic spin polarization of gold nano-clusters even without thiol ligands. The exact mechanism for magnetic behavior is still unknown. Our calculations aim to understand this magnetization and whether it is predominantly a size effect or ligand effect.

There have already been several attempts to model gold nano-clusters theoretically [4, 5, 6, 7]. Traditional tight-binding models used to describe magnetism for bulk systems will not work as accurately since the surface of these nano-clusters play a vital role that can not simply be added ad hoc into the models. On the other hand, the subtle spin correlations responsible for magnetism are often hard to capture within a first-principle framework. In order to determine the role of the surface, density functional theory combined with perturbative methods will be used. Our main goal is to characterize the mechanism which causes the onset of ferromagnetism in these systems. If possible, we would like to show explicitly how the mechanism depends on the cluster size and geometry, and we will address whether surface effects alone in bare nano-clusters can induce magnetism.

Density functional theory, a first principles electronic structure method, is well-suited to describe for clusters of 10-100s of atoms. Nevertheless, in order to model these properties, many assumptions must be made. In our work, we consider only single isolated nano-clusters. Density functional theory is based on the principle that type exact energy and physical properties of an interacting quantum solution can be found by the analysis of a corresponding non-interacting analogue with effective potential. In practice, this potential must be approximated. We will employ several approximations starting with simple generalized gradient approximations (PBE,BP86), then to hybrids, (B3LYP), and then the sophisticated exact-exchange (within the KLI approximation). These methods are available is several popular quantum chemical codes. Primarily, we will use Octopus [9] and Siesta [8]. Both codes use the pseudo-potential formulation in which the core electrons are approximated by an effective potential. The capping ligands on the gold nano clusters have long carbon chains which we replace by single thiols or otherwise truncated ligands. One criterion for a magnetic ground-state is that the unrestricted ground-state is significantly spin-polarized. If it is found that this criterion is inaccurate, we will consider using a perturbative determination of the tendency towards a spin-polarized state through a Fermi-contact interaction term.

The effects of thiolation should be apparent even at the two gold atom limit. In this limit, we will investigate how thiolation breaks the spin symmetry and induces a magnetic moment on the system. Size and shape will play critical roles in determining the magnitude of the moment. We will investigate how these effects come into play by considering larger and more complex clusters. We have already developed geometries for highly symmetric 24 and 38 atom gold clusters. One model attributes the magnetization of these clusters to the decrease of $d$ holes due to the thiol ligands [2]. These first principle calculation will directly show if this is the case. By replacing the thiols with other ligands, we can systematically see how the $d$ hole density changes. For a certain critical size, the formation of domain walls will be energetically favored, and it is unclear whether our first principles methods will be able to capture this.

I will talk in detail about the results of our ongoing calculations and consider the implications for understanding magnetism in larger nano-systems.

## References

[1] P. Zhang and T.K. Sham, Physics Review Letters, 245502, 2003.

[2] P. Crespo, R. Litran, T.C. Rojas, M. Multigner, J.M. de la Fuenta, J.C. Sanchez-Lopez, M.A. Garcia, A. Hernando, S. Penades, and A. Fernandez, Physics Review Letters, 087204, 2004.

[3] Y.Yamamoto, T. Miura, M. Suzuki, N. Kawamura, H. Miyagawa, T. Nakamura, K. Kobayashi, T. Teranishi, and H. Hori, Physics Review Letters, 116801, 2004.

[4] C. Gonzalez, Y. Simon-Manso, M. Marquez-Sanchez, and V. Mujica, In Preparation (2005).

[5] H. Haekkinem, and U. Landman, Physical Review B, **62**, R2287 (2000).

[6] H. Haekkinem, B. Yoon, U. Landman, X. Li, H. Zhai, and L. Wang, Journal of Physical Chemistry A, **107**, 6168 (2003).

[7] D. van Leeuwen, J.M. van Ruitenbeek, L.J. de Jongh, A. Ceriotti, G. Pacchioni, O.D. Haeberlen, and N. Roesch, Physics Review Letters, **73**, 1432 (1994).

[8] P. Ordejan, E. Artacho, and J.M. Soler. Physical Review B **53** (1996), p. R10441. J.M. Soler, E. Artacho, J.D. Gale, A. Garcia, J. Junquera, P. Ordejan, and D. Sanchez-Portal. Journal of Physics: Condensed Matter **14** (2002), p. 2745. D. Sanchez-Portal, P. Ordejan, E. Artacho, and J.M. Soler. International Journal Quantum Chemistry **65** (1997), p. 453.

[9] M.A.L. Marques, A Castr.o, G. F. Bertsch, and A. Rubio Computational Physics Communications 151, **60** (2003).

Brill Academic Publishers
P.O. Box 9000, 2300 PA Leiden,
The Netherlands

*Lecture Series on Computer
and Computational Sciences*
Volume 4, 2005, pp. 727-730

# A Quantum Chemical Study of Doped Aragonite and Comparison with Calcite

M. Menadakis and G. Maroulis[1]

Department of Chemistry,
University of Patras,
GR-26500 Patras, Greece

P. G. Koutsoukos[2]

Department of Chemical Engineering,
University of Patras and FORTH-ICEHT,
PO Box 1414 GR 26504 Patras, Greece

Received 7 March, 2004; accepted in revised form 10 March, 2004

*Abstract:* We have investigated the presence of foreign ions into the bulk structure and the [011] surface of aragonite using periodic ab-initio methods. Four cations isovalent to $Ca^{2+}$ were studied: $Mg^{2+}$, $Sr^{2+}$, $Ba^{2+}$ and $Zn^{2+}$. The calculations were performed at structures (bulk, surface) that contain four and eight $CaCO_3$ units. Our results, at the Hartree-Fock level, show that the incorporation of those ions into aragonite depends strongly on their size. $Mg^{2+}$ and $Zn^{2+}$, due to their smaller size, can substitute $Ca^{2+}$ ions in the crystal lattice while the incorporation of $Sr^{2+}$ and $Ba^{2+}$ into aragonite is energetically less favored. Examination of the [011] surface of aragonite revealed that the incorporation is more pronounced than in bulk structure. The incorporation of $Mg^{2+}$ in the bulk structure of aragonite has almost the same energetic content with the corresponding one on the [104] surface of calcite. Bulk and surface incorporation exhibit different trends in aragonite and calcite. The surface incorporation is more favored in aragonite while in calcite ions prefer to be incorporated in the bulk structure.

*Keywords:* Aragonite, Calcite, crystal impurities

*PACS:* 61.50.Ah, 61.72.-y

## 1 Introduction

Calcium carbonate is an exceptional mineral. The chemical formula $CaCO_3$ covers a raw material, which is widespread throughout nature, whether dissolved in rivers and oceans, or solid as mineral in the form of stalactites, stalagmites or as the major constituent of whole mountain ranges. Plants and animals need calcium carbonate to form their skeletons and shells and almost every product in daily life either contains calcium carbonate or has some associaion with the mineral during its production. Calcium carbonate, one of the most abundant natural inorganic minerals, exists as three polymorphs namely calcite, aragonite and vaterite, two hydrated forms monohydrocalcite and calcium carbonate hexahydrate and an unstable amorphous one.

---

[1]Corresponding author. E-mail:maroulis@upatras.gr
[2]Corresponding author. E-mail:pgk@chemeng.upatras.gr

Although calcite is the thermodynamically most stable polymorph, various polymorphs and morphologies of $CaCO_3$ are formed depending on the experimental conditions. Calcium carbonate nucleation and growth can be manipulated through the addition of crystal modifiers to the crystal growing solution. Examples of additives include simple organic molecules[1], inorganic salts[2], and proteins[3, 4]. Other factors that control the polymorphs selectivity and the morphology of $CaCO_3$ are pH, the temperature of the solution and the extent of supersaturation[5]. One frequently suggested mechanism for controlling $CaCO_3$ polymorph formation involves the introduction of a specific inhibitor of calcite which favors the aragonite polymorph [6]. Magnesium is known to induce aragonite formation from sea water, and in vitro at ratios of Mg/Ca>4. Motivated by this behaviour in this paper we examine the way of incorporation of Mg ions into bulk and [011] surface of aragonite. Along with $Mg^{2+}$, three other isovalent to $Ca^{2+}$ ions were studied ($Sr^{2+}$, $Ba^{2+}$ and $Zn^{2+}$) regarding their ability to substitute the calcium ions. These results can be compared with our previous work on calcite[8] to assess the effect on the structural and energetic characteristics of the incorporation on the two most stable phases of calcium carbonate.

## 2   Computational details

The CRYSTAL98[7] program permits the calculation of the wavefunction and properties of crystalline materials within the Hartree-Fock Linear-Combination-of-Atomic-Orbitals (HF-LCAO). We have utilised Gaussian basis functions for the quantum chemical description of the Ca, C and O centers present in the unit cell of aragonite. The corresponding basis sets have been previously tested in calcite and other crystalline compounds with the outer d orbitals for each atom reoptimized in the environment of aragonite with respect to the total energy. Calcium carbonate in the form of aragonite crystallize in the orthorhombic crystal system with four $CaCO_3$ units in the unit cell. Building structures (bulk or surface) with multiple $CaCO_3$ units and subsequent substitution of the Ca centers could provide an estimate of the effect of the degree of contamination in the defect formation energy. The substitutional defect formation energy is given by

$$\Delta E_X = E_{CaCO_3-X} - N E_{CaCO_3} + E_{Ca^{2+}} - E_{X^{2+}} \qquad (1)$$

where $E_{CaCO_3-X}$ (X=$Mg^{2+}$, $Sr^{2+}$, $Ba^{2+}$ and $Zn^{2+}$) is the total energy of the defective structure, N is the number of the $CaCO_3$ units in the corresponding perfect lattice supercell, $E_{CaCO_3}$ is the bulk energy per formula unit of aragonite and $E_{Ca^{2+}}$, $E_{X^{2+}}$ are the total energies of the isolated ions. The construction of the [011] surface of aragonite, used for the examination of surface incorporation, was based on the slab model (two sided surface). The incorporation of $Mg^{2+}$ ion into bulk aragonite was further analysed by relaxing the environment aroud the defect. $Mg^{2+}$ is surrounded by nine O atoms that are organized in four groups. The first neighbour of the $Mg^{2+}$ defect consist of one O atom and the relaxation was performed by changing its cartesian coordinates. The defect formation energy was calculated both on the relaxed and unrelaxed structure yielding the response of the bulk structure to the formation of the defect.

## 3   Results

Table 1 shows the defect formation energy for the incorporation of $Mg^{2+}$, $Sr^{2+}$, $Ba^{2+}$ and $Zn^{2+}$ ions into aragonite bulk structure and into the [011] surface. The calculations were performed at two different cells (bulk or surface) with four and eight structural units. The degree of contamination ranged from 13 up to 100% meaning that all the $Ca^{2+}$ are substituted with the foreign ions. A negative value of the defect formation energy indicates that the incorporation is energetically favoured while its absolute value measures the ability of the foreign ions for the incorporation. In bulk aragonite, ions that are smaller than $Ca^{2+}$ ($Mg^{2+}$, $Zn^{2+}$) can enter the host lattice substituting Ca centers while the incorporation of $Sr^{2+}$ and $Ba^{2+}$ is less favoured.

Table 1: Defect formation energy [eV] of $X^{2+}$ doped aragonite.

| % | $Mg^{2+}$ | | $Sr^{2+}$ | | $Ba^{2+}$ | | $Zn^{2+}$ | |
|---|---|---|---|---|---|---|---|---|
| | | | | 4 CaCO$_3$ units | | | | |
| | Bulk | [011] | Bulk | [011] | Bulk | [011] | Bulk | [011] |
| 25 | -2.2580 | -1.8636 | 1.8559 | 1.3592 | 5.4579 | 3.8608 | -2.5317 | -2.3140 |
| 50 | -4.5082 | -3.8069 | 3.7141 | 2.8284 | 10.9123 | 8.0808 | -5.0503 | -4.6204 |
| 75 | -6.7465 | -5.7462 | 5.5854 | 4.3086 | 16.4033 | 12.2926 | -7.5587 | -6.9291 |
| 100 | -8.9772 | -7.5884 | 7.4594 | 5.6799 | 21.8911 | 16.1885 | -10.0537 | -9.2200 |
| | | | | 8 CaCO$_3$ units | | | | |
| | Bulk | [011] | Bulk | [011] | Bulk | [011] | Bulk | [011] |
| 13 | -2.2601 | -1.8638 | 1.8550 | 1.3592 | 5.4517 | 3.8603 | -2.5351 | -2.3149 |
| 25 | -4.5162 | -3.7275 | 3.7117 | 2.7184 | 10.9155 | 7.7218 | -5.0633 | -4.6283 |
| 38 | -6.7682 | -5.6708 | 5.5691 | 4.1875 | 16.3640 | 11.9415 | -7.5864 | -6.9348 |
| 50 | -9.0164 | -7.6140 | 7.4283 | 5.6566 | 21.8248 | 16.1614 | -10.1014 | -9.2408 |
| 63 | -11.2569 | -9.5534 | 9.2988 | 7.1368 | 27.3104 | 20.3731 | -12.6148 | -11.5506 |
| 75 | -13.4931 | -11.4927 | 11.1710 | 8.6171 | 32.8062 | 24.5852 | -15.1177 | -13.8591 |
| 88 | -15.7258 | -13.3350 | 13.0441 | 9.9883 | 38.2893 | 28.4805 | -17.6188 | -16.1510 |
| 100 | -17.9544 | -15.1772 | 14.9189 | 11.3596 | 43.7823 | 32.3767 | -20.1088 | -18.4408 |

The relaxation procedure was performed only for $Mg^{2+}$ due to the high computational cost. The results are presented in Table 2 where comparison is made with the analogous results for the incorporation of $Mg^{2+}$ into calcite bulk structure. In the relaxed structure of the $Mg^{2+}$ doped aragonite there is an energy gain of approximately 0.06eV while the first neighbouring O atom exhibits an inwards relaxation (the Mg-O distance decreases). Fig. 1 summarizes the incorporation of $Mg^{2+}$ into aragonite and calcite both on bulk and surface structures. The incorporation of $Mg^{2+}$ seems to be energetically more favored in calcite bulk structure. Moreover, $Mg^{2+}$ in aragonite prefers to be incorporated into the [011] surface rather than the crystal lattice while in calcite this trend is reversed.

Table 2: Defect formation energy [eV] of $Mg^{2+}$ doped aragonite and calcite in the unrelaxed and relaxed structures. $\Delta R$ is the difference in Mg-O distance upon relaxation.

| | Aragonite | Calcite | |
|---|---|---|---|
| | 25% | 50% | 25% |
| Unrelaxed | -2.2580 | -2.8744 | -2.8740 |
| Relaxed | -2.3102 | -3.3507 | -3.4278 |
| Mg-O [Å] | 2.3262 | 2.2467 | 2.2457 |
| $\Delta R$ [Å] | -0.0895 | -0.1252 | -0.1262 |

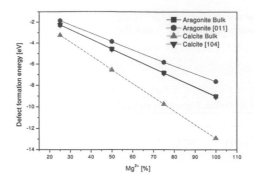

Figure 1: Incorporation of $Mg^{2+}$ into aragonite and calcite.

## Acknowledgment

M. Menadakis gratefully acknowledges a scholarship form the Institute of Chemical Engineering and High Temperature Chemical Processes of the Foundation of Research and Technology-Hellas (FORTH/ICE-HT).

## References

[1] F. Manoli and E. Dalas, Spontaneous precipitation of calcium carbonate in the presence of ethanol, isopropanol and diethylene glycol, *J. Cryst. Growth* **218** 359-364(2000).

[2] E. Loste, R. M. Wilson, R. Seshadri and F. C. Meldrum, The role of magnesium in stabilising amorphous calcium carbonate and controlling calcite morphologies, *J. Cryst. Growth* **254** 206-218(2003).

[3] S. Albeck, I. Addadi and S. Weiner, Regulation of calcite crystal morphology by intracrystalline acidic proteins and glycoproteins, *Connect. Tissue Res.* **35** 365-370(1996).

[4] A. M. Belcher, X. H. Wu, R. J. Christensen and P. Hansma, Control of crystal phase switching and orientation by soluble mollusc-shell proteins, *Nature* **381** 56-58(1996).

[5] T. Sabbides, E. K. Giannimaras and P. G. Koutsoukos, The Precipitation of Calcium Carbonate in Artificial Seawater at Sustained Supersaturation, *Environ. Technol.* **13** 73-80(1992).

[6] R. Giles, S. Manne, S. Mann, D. E. Morse, G. D. Stucky and P. K. Hansma, Inorganic Overgrowth of Aragonite on Molluscan Nacre Examined by Atomic Force Microscopy, *Biol. Bull.* **188** 8-15(1995).

[7] V. R. Saunders, R. Dovesi, C. Roetti, C. Causa, N. M. Harrison, R. Orlando and C. M. Zicovich-Wilson, *CRYSTAL98 User's Manual*, Torino: Universita di Torino 1999.

[8] M. Menadakis, G. Maroulis and P. G. Koutsoukos, A Quantum Chemical Study of Doped $CaCO_3$ (calcite), *Lecture Series on Computer and Computational Sciences* **1** 1029-1032(2004).

Brill Academic Publishers
P.O. Box 9000, 2300 PA Leiden
The Netherlands

*Lecture Series on Computer
and Computational Sciences*
Volume 4, 2005, pp. 731-731

# Molecular structure and conformations of some thioperoxides RS-OR': experiment and theory

Areti Kosma,[a] Sonia E. Ulic,[b] Carlos O. Della Vedova,[b] and Heinz Oberhammer.[a1]

[a] Institut für Physikalische und Theoretische Chemie, Universität Tübingen, D-72076, Tübingen, Germany

[b] Departamento de Química. Facultad de Ciencias Exactas. Universidad Nacional de La Plata, 47 esq. 115 (1900) La Plata, República Argentina

Received 5 August, 2005; accepted in revised form 12 August, 2005

All peroxides RO-OR' and disulfanes RS-SR', whose gas phase structures are known, possess gauche structures and most of them possess dihedral angles between 80° and 120°. Thioperoxides of the type RS-OR' are much less stable than peroxides and disulfanes and very little is known about their structural properties. Sulfenic acid, HSOH, dimethoxysulfane, $CH_3OSOCH_3$ and dimethoxydisulfane, $CH_3OSSOCH_3$, possess gauche structures with dihedral angles between 75° and 90°. A combination of vibrational spectroscopy, gas electron diffraction and quantum chemical calculations results for trifluoromethanesulfenyl acetate, $CF_3S-OC(O)CH_3$, and for trifluoromethanesulfenyl trifluoroacetate, $CF_3S-OC(O)CF_3$, in a mixture of two conformers with the prevailing component possessing gauche structure. The minor form (18(5)% and 11(5)%, respectively) possesses an unexpected trans structure around the S-O bond. The C=O bond of the acetyl group is oriented syn with respect to the S-O bond in both conformers. A similar experimental and theoretical study of S-(fluoroformyl)O-(trifluoroacetyl)thioperoxide, $FC(O)S-OC(O)CF_3$, results also in a mixture of two conformers. In this thioperoxide, however, both conformers possess gauche structure around the S-O bond and differ only by the orientation of the FC(O) group (C=O bond syn or anti with respect to the S-O bond).

---

[1] Corresponding author. E-mail address heinz.oberhammer@uni-tuebingen.de

Brill Academic Publishers
P.O. Box 9000, 2300 PA Leiden
The Netherlands

*Lecture Series on Computer
and Computational Sciences*
Volume 4, 2005, pp. 732-734

# Gauche and Trans Conformers of 1,2-Dihaloethanes: A Study by *ab initio* and Density Functional Theory Methods

P. Ramasami[1]

Department of Chemistry,
Faculty of Science,
University of Mauritius,
Réduit, Mauritius

Received: April 29, 2005; accepted in revised form: May 25, 2005

*Abstract:* This work deals with a systematic study of the gauche and trans conformers of 1,2-dihaloethanes ($XCH_2CH_2X$, X=F, Cl, Br and I). The methods used are Hartree-Fock (HF), second order Møller-Plesset theory (MP2) and density functional theory (DFT). The basis sets used are 6-311+(d,p) and 6-311++(d,p) except for 1,2-diiodoethane where 3-21G is used. The functional used for DFT method is B3LYP. G2/MP2 calculations have also been done. The results reveal that there is a preference for the trans conformer except for 1,2-difluoroethane. The energy difference between the gauche and trans forms generally increases with an increase in the size of the halogen. Vibrational analysis has also been carried out.

*Keywords:* 1,2-Dihaloethanes, Ab initio, HF, MP2, DFT, Conformer, Energy, Vibrational analysis

## 1. Introduction

Conformers of small molecules [1] are suitable prototypes for more complex conformational processes and molecular modelling. 1,2-Disubstituted ethane molecules are important in the study of internal rotation leading to conformational analysis. One of the interesting features of these molecules is that among their rotamers, there are two stable conformations namely the trans and gauche forms. The difference in energy between these two conformations is termed as the conformational energy difference ($\Delta E$). The latter decides the amount of each of the two conformers.

There have been studies involving 1,2-disubstituted ethanes [1-7] but there have not been any systematic theoretical studies of the conformers of 1,2-dihaloethanes ($XCH_2CH_2X$, X=F, Cl, Br and I). The small energy difference between the trans and gauche conformers and the reasonably large barrier due to eclipsed halogens make these molecules suitable for computational investigations. These investigations are becoming more important with the availability of small but fast computers and quantum mechanical programs such as Gaussian 03W [8].

This work aims to obtain the energy difference between the gauche and trans conformers of the series of 1,2-dihaloethanes molecules in the gas phase. The conformers have also been subjected to vibrational analysis.

---

[1] Corresponding author. E-mail: p.ramasami@uom.ac.mu

## 2. Methods

Calculations have been done using Hartree-Fock (HF), second order Møller-Plesset perturbation theory (MP2) and density functional theory (DFT). The basis sets used are 6-311+G(d,p) and 6-311++(d,p) except for 1,2-diiodoethane where the basis set 3-21G is used. The functional used for DFT is B3LYP. A conformer has first been optimized and the optimized structure has been used for frequency calculations using the same method and basis set as for optimization. G2/MP2 calculations have also been done except for 1,2-diiodoethane where the basis set used in the method is not available for iodine.

All calculations have been done using Gaussian 03W running on a Pentium IV computer having 1Gb RAM. GaussView 3.0 has been used for visualizing the molecules.

## 3. Results and discussion

The optimised parameters for 1,2-difluoroethane using the HF, MP2 and B3LYP methods and 6-311+G(d,p) as the basis set are as given in Table 1.

Table1: Optimised parameters for 1,2-difluoroethane

| Parameter | HF | | MP2 | | B3LYP | |
|---|---|---|---|---|---|---|
| | Trans | Gauche | Trans | Gauche | Trans | Gauche |
| r (C-C) | 1.513 Å | 1.503 Å | 1.516 Å | 1.504 Å | 1.519 Å | 1.505 Å |
| r (C-F) | 1.368 Å | 1.365 Å | 1.392 Å | 1.389 Å | 1.399 Å | 1.397 Å |
| ∠ (C-C-F) | 108.0° | 110.5° | 107.9° | 110.3° | 108.0° | 110.8° |
| τ (F-C-C-F) | 180.0° | 70.3° | 180.0° | 69.9° | 180.0° | 72.2° |

The energies of the trans and gauche conformers of 1,2-dihaloethanes and $\Delta E$ are reported in Table 2. These energies were calculated at MP2 level after full geometry optimizations. G2/P2 calculations are reported in Table 3. The results indicate that except for 1,2-difluoroethane, the trans conformer is more stable then the gauche one and $\Delta E$ increases with the size of the halogens. These results are in agreement with literature [4,5] and the abnormal behaviour of 1,2-difluoroethane has been reported [5].

Table 2: MP2 calculated energies of 1,2-dihaloethanes

| Compound | Basis set | Trans (Hartrees) | Gauche (Hartrees) | $\Delta E$ (kJ/mol) |
|---|---|---|---|---|
| 1,2-difluoroethane | 6-311++G(d,p) | -277.719167 | -277.720358 | -3.13 |
| 1,2-dichloroethane | 6-311++G(d,p) | -997.684535 | -997.682220 | 6.08 |
| 1,2-dibromoethane | 6-311++G(d,p) | -5223.381418 | -5223.378069 | 8.79 |
| 1,2-diiodoethane | 3-21G | -13853.565670 | -13853.561124 | 11.94 |

Table 2: G2/MP2 calculated energies of 1,2-dihaloethanes

| Compound | Trans (Hartrees) | Gauche (Hartrees) | $\Delta H$ at 0 K (kJ/mol) |
|---|---|---|---|
| 1,2-difluoroethane | -277.925725 | -277.926917 | -0.75 |
| 1,2-dichloroethane | -997.913779 | -997.911872 | 1.20 |
| 1,2-dibromoethane | -5223.589499 | -5223.586297 | 2.01 |

The vibrational frequencies for the trans and gauche conformers have been calculated. As an example, the symmetric stretching mode of $CH_2$ in 1,2-difluoroethane using the DFT method, B3LYP as the functional and 6-311+G(d,p) as the basis set is found to be at 3047.78 cm$^{-1}$ (uncorrected) with an intensity of 39.08. A reasonably good agreement is obtained between the experimental and calculated wavenumbers for both conformers, after appropriate scaling. All the 18 fundamentals of the trans and gauche conformers have been assigned.

A critical assessment of the results indicate the MP2 and DFT methods are better than the HF method and this can be due to correlation energy being neglected in the HF method.

## Acknowledgments

This work was supported by the University of Mauritius Research Fund. The author acknowledges contribution of M Ramkalam. The author wishes to thank anonymous referees for their careful reading of the manuscript and their fruitful comments and suggestions.

## References

[1] B. L. Mc Clain and D. Ben-Amotz, Global Quantitation of Solvent Effects on the Isomerization Thermodynamics of 1,2-Dichloroethane and *trans*-1,2-Dichlorocyclohexane, *J. Phys. Chem. B.* **106**, 7882-7888 (2002).

[2] D. E. Brown and B. Beagley, The Gas-phase Rotamers of 1,1,2,2-Tetrafluoroethane - Force field, Vibrational amplitudes and Geometry – A Joint Electron-diffraction and Spectroscopic Study, *J. Mol. Struct.* **38**, 167-176 (1977).

[3] K. B. Wiberg and M. A. Murcko, Rotational Barriers. 2. Energies of Alkane Rotamers. An Examination of Gauche Interactions, *J. Am. Chem. Soc.* **110**, 8029-8038 (1988).

[4] D. A. Dixon, N. Matsuzawa and S. C. Walker, Conformational Analysis of 1,2-Dihaloethanes: A Comparison of Theoretical Methods, *J. Phys. Chem.* **96**, 10740-10746 (1992).

[5] K. B. Wiberg, T. A Keith, M. J. Frisch and M. A. Murcko, Solvent Effects on 1,2-Dihaloethane Gauche/Trans Ratios, *J. Phys. Chem.* **99**, 9072-9079 (1995).

[6] X. Zheng and D. L. Phillips, Vibrational Spectra and Assignments of 1-bromo-2-iodoethane, *Vib. Spectrosc.* **17**, 73-81 (1998).

[7] A. Horn, P. Klaeboe, B. Jordanov, C. J. Nielsen and V. Aleksa, Vibrational Spectra, Conformational Equilibrium and *ab initio* Calculations of 1,2-Diphenylethane, *J. Mol. Struct.* **695-696**, 77-94 (2004).

[8] Gaussian 03, Revision A.1, M. J. Frisch, G. W. Trucks, H. B. Schlegel, G. E. Scuseria, M. A. Robb, J. R. Cheeseman, J. A. Montgomery, Jr., T. Vreven, K. N. Kudin, J. C. Burant, J. M. Millam, S. S. Iyengar, J. Tomasi, V. Barone, B. Mennucci, M. Cossi, G. Scalmani, N. Rega, G. A. Petersson, H. Nakatsuji, M. Hada, M. Ehara, K. Toyota, R. Fukuda, J. Hasegawa, M. Ishida, T. Nakajima, Y. Honda, O. Kitao, H. Nakai, M. Klene, X. Li, J. E. Knox, H. P. Hratchian, J. B. Cross, C. Adamo, J. Jaramillo, R. Gomperts, R. E. Stratmann, O. Yazyev, A. J. Austin, R. Cammi, C. Pomelli, J. W. Ochterski, P. Y. Ayala, K. Morokuma, G. A. Voth, P. Salvador, J. J. Dannenberg, V. G. Zakrzewski, S. Dapprich, A. D. Daniels, M. C. Strain, O. Farkas, D. K. Malick, A. D. Rabuck, K. Raghavachari, J. B. Foresman, J. V. Ortiz, Q. Cui, A. G. Baboul, S. Clifford, J. Cioslowski, B. B. Stefanov, G. Liu, A. Liashenko, P. Piskorz, I. Komaromi, R. L. Martin, D. J. Fox, T. Keith, M. A. Al-Laham, C. Y. Peng, A. Nanayakkara, M. Challacombe, P. M. W. Gill, B. Johnson, W. Chen, M. W. Wong, C. Gonzalez, and J. A. Pople, Gaussian, Inc., Pittsburgh PA, 2003.

Brill Academic Publishers
P.O. Box 9000, 2300 PA Leiden,
The Netherlands

*Lecture Series on Computer
and Computational Sciences*
Volume 4, 2005, pp. 735-738

# Electron Correlation Effects in Small Iron Clusters

## G. Rollmann[1] and P. Entel

Theoretical Low-Temperature Physics,
Physics Department, University of Duisburg–Essen, Campus 47048 Duisburg, Germany

Received 7 March, 2004; accepted in revised form 10 March, 2004

*Abstract:* We present results of first-principles calculations of structural, magnetic, and electronic properties of small Fe clusters. It is shown that, while the lowest-energy isomers of $Fe_3$ and $Fe_4$ obtained in the framework of density functional theory within the generalized gradient approximation (GGA) are characterized by Jahn-Teller-like distortions away from the most regular shapes (which is in agreement with other works), these distortions are reduced when electron correlation effects are considered explicitly by using the GGA+$U$ approach. At the same time, the magnetic moments of the clusters are enhanced with respect to the pure GGA case, resulting in maximal moments (in the sense of Hund's rules) of 4 $\mu_B$ per atom for the ground state structures of $Fe_3$ and $Fe_4$ already for moderate values of the Coulomb repulsion parameter $U \sim 2.0$ eV. We explain this by changes in the electronic structures of the clusters.

*Keywords:* Density functional theory, GGA + U, Fe clusters, Magnetic moments

*PACS:* 31.15.Ew, 36.40.Cg, 36.40.Mr

## 1   Introduction

Magnetic transition metal (TM) particles are an important ingredient in many state-of-the-art technological applications like, e.g., ultra-high density magnetic storage devices. During the fabrication of these structures, miniaturization has already reached a point where particle sizes are so small that quantum effects start to play a role in determining their magnetic properties, which, in turn, show a pronounced size dependence. One example for this are the magnetic moments of free Fe, Co, and Ni clusters, which are not just given by simple interpolations between the limiting values of the isolated atoms and the corresponding bulk materials, but rather change non-monotonically with cluster size in an oscillatory fashion [1]. Therefore, a detailed understanding of the electronic structure of small TM clusters is of fundamental interest, especially with respect to the design of new materials with unique properties.

But although it is already possible to examine the bahavior of small TM clusters and even single atoms deposited on surfaces in detail, a systematic investigation of geometries and magnetic properties of their free counterparts is still not feasible in experiments. In the case of Fe clusters, which are the subject of the present work, this leads to the fact that the major part of information about their structural and magnetic properties has been derived from *ab initio* simulations, mainly based on density functional theory (DFT). For an overview, we refer to [2] and references therein. It has turned out that it is important to go beyond the local density approximation (LDA) and include gradient corrections (GGA) when calculating the exchange and correlation energy, as well as to

---

[1]Corresponding author. E-mail: georg@thp.uni-duisburg.de

allow for free relaxation of the atoms without imposing symmetry constraints, because the presence of degenerate electronic states in highly symmetric clusters originating from the interaction of the $d$-manifold leads to Jahn-Teller distorted ground-state geometries with lower symmetry [3, 4]. Resulting from different computational schemes, proposed magnetic moments vary from 8 to 10 $\mu_B$ for $Fe_3$ and from 10 to 14 $\mu_B$ for $Fe_4$. The true complexity of the problem has been revealed in two recent comprehensive investigations where a high number of local minima close in energy have been found for $Fe_3$ and $Fe_4$ [2]. Although a definite assignment of the ground states has up to day neither been achieved for $Fe_3$ nor $Fe_4$, the spin multiplicities were the same for all low-lying states, giving rise to the assumption that within DFT/GGA, the lowest-energy isomers of $Fe_3$ and $Fe_4$ possess magnetic moments of 10 and 14 $\mu_B$, respectively.

But the fact, that electron correlation due to intra-atomic Coulomb repulsion of localized $d$ or $f$ electrons is not described very well within conventional LDA (and also GGA), remains as a major drawback in all these calculations. Different methods have been proposed to overcome this limitation. Of these, the LDA + $U$ method where a Hubbard-like term $U$ is incorporated into the density functional, has been applied successfully to a variety of problems in strongly correlated systems where conventional LDA gives qualitatively wrong results, see, e.g., [6]. In this article, we discuss the influence of electronic correlation on geometric and magnetic properties of small Fe clusters, by explicitly investigating the effect of the size of the parameter $U$ on various properties of selected clusters.

## 2 Computational method

Starting from different lowest-energy states found in earlier calculations, we have optimized the geometries of $Fe_3$ and $Fe_4$ without imposing any symmetry constraint, using a convergence criterion of 1 meV/Å for the forces. The calculation of the electronic structures of the clusters was performed in the framework of DFT in combination with the GGA for the description of exchange and correlation [7], periodic boundary conditions, and a plane wave basis set with a cutoff energy of 335 eV. A number of 8 valence electrons was taken into account for each Fe atom, the remaining core electrons together with the nuclei were described by following the projector augmented wave method [8] as implemented in the Vienna *ab initio* simulation package VASP [9].

For the GGA + $U$ calculations, we have adopted a version proposed by Dudarev *et al.* [5]. In this implementation, the total energy depends on the difference $U - J$, where $U$ and $J$ represent the spherically averaged on-site Coulomb interaction and screened exchange integrals, respectively. It can be shown that, within this formalism, unoccupied $d$ states are shifted towards higher energies by $(U - J)/2$, while the opposite is true for occupied $d$ states. As the value of $J$ was kept constant at $J = 1$ eV in our calculations, the case $U = 1$ eV corresponds here to the pure GGA limit, because in this case $U - J = 0$ eV.

## 3 Results and discussion

The energetic relationships for the $Fe_3$ cluster as obtained with different values for the parameter $U$ and fixed spin multiplicities of 9, 11, and 13 are shown in Fig. 1. In the case of conventional GGA ($U = 1.0$ eV), we find that, within $D_{3h}$ symmetry constraints, an equilateral triangle with a magnetic moment of 10 $\mu_B$ is lowest in energy. But, as will be discussed below, this state is only metastable with respect to a Jahn-Teller distortion, leading to the $^{11}A_1$ state which we obtain as ground state when the atoms are allowed to relax freely. An $^{11}A'$ state with $C_s$ symmetry is located some 12 meV/atom higher in energy. In contrast to the case of $M = 10\,\mu_B$, equilateral triangles with spin multiplicities of 9 and 13 are stable with respect to structural deformations. But in both states, as well as in the $^{13}A_1$ state, the aufbau principle is violated (see below), and they do not

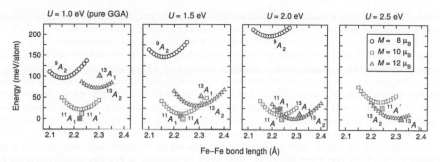

Figure 1: Relative total energy of different isomers of the Fe₃ cluster as a function of the average bond length, total magnetic moment $M$, and symmetry, for different values of $U$. Open symbols refer to equilateral triangles, filled symbols denote energies of relaxed clusters.

play a role in the search for the ground state. A detailed discussion of the nature of these states has been given elsewhere [2] and is beyond the scope of the present work. We ascribe the small differences between the results presented here and those obtained in [2] to the use of plane wave basis sets – which are in the sense superior to localized orbitals that calculated quantities converge smoothly with the cutoff energy – here as compared to localized basis functions in the latter.

When we now increase the value of the Hubbard $U$ parameter, we observe that the relative positions of the states are shifted. While the $^9A_2$ state becomes energetically unfavorable, states with a total moment of 12 $\mu_B$ are lowered. At the same time, the order of $^{11}A_1$ and $^{11}A'$ is reversed, so that the latter becomes the lowest-energy isomer already for $U = 1.5$ eV. For $U = 2.0$ eV, isomers with multiplicities of 11 and 13 are degenerate, before we finally encounter the $^{13}A_1$ state, with a magnetic moment of 12 $\mu_B$ and a distortion of 3.4 %, being lowest in energy for $U = 2.5$ eV.

In order to obtain a better understanding of the reasons behind the relative stability of the different states, we take a closer look at the electronic structures of selected isomers, but not without bearing in mind that, in principle, there is no direct physical meaning associated with the calculated one-electron orbitals in DFT. From Fig. 2, the origin of the Jahn-Teller distortion of the $^{11}A_1$ as well as the instabilities of the states with a total moment of 13 $\mu_B$ within GGA due to the violation of the aufbau principle can be understood. But when $U$ is increased, we observe, that

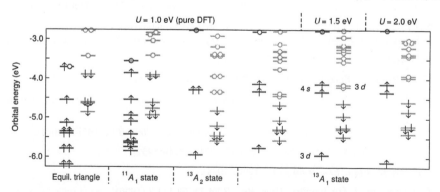

Figure 2: Kohn-Sham eigenvalues (horizontal bars) of selected states of the Fe₃ cluster for GGA and GGA + $U$. Arrows represent electrons (of either spin), circles denote holes.

occupied majority spin $d$ states are shifted towards lower energies, while the empty minority spin $d$ states are moved upwards. This leads to a state crossing at the Fermi energy for $U \sim 1.5$ eV and to the fact that, from around $U = 2.0$ eV, the $^{13}A_1$ state becomes the ground state.

For Fe$_4$, we find a similar situation. When tetrahedral symmetry constraints are imposed, the energy is minimal for a cluster with a magnetic moment of 12 $\mu_B$ and interatomic distances of 2.28 Å. But this state is also unstable with respect to a Jahn-Teller distortion, resulting in a state with $D_{2d}$ symmetry with 2 short (2.20 Å) and 4 longer bonds (2.32 Å). Although the lowest-energy isomer of Fe$_4$ found within DFT/GGA belongs to the same point group, it is characterized by a magnetic moment of 14 $\mu_B$, 2 long (2.54 Å) and 4 shorter bonds (2.23 Å), and a distortion of 6.2 %.

When we increase the value for the Hubbard $U$, we observe similar effects as in the case of the Fe trimer. Low-moment states are shifted towards higher energies, and isomers with a spin multiplicity of 17 are becoming favorable. But, while in the case of Fe$_3$ the overall situation had not yet changed for $U = 1.5$ eV, we here find the $M = 16$ $\mu_B$ state already lowest in energy among all perfect tetrahedra for this value of $U$. Finally, for $U = 2.0$ eV, also the lowest-energy isomer of Fe$_4$ possesses a moment of 16 $\mu_B$. Its structural distortion is calculated to 0.4 %.

To summarize our findings, we have shown that the order of states of small Fe clusters changes drastically compared to conventional GGA when electronic correlation effects are considered explicitly. For values of $U \sim 2.0$ eV, total magnetic moments of the lowest-energy isomers are by 2 $\mu_B$ higher, yielding values of 12 and 16 $\mu_B$ for Fe$_3$ and Fe$_4$, respectively. We expect this trend observed for the small clusters to continue as the number of atoms is increased, and therefore predict the calculated moments of larger Fe clusters to be higher (by at least 2 $\mu_B$) when electron correlation is explicitly taken into account. In order to justify reasonable choices of $U$, we emphasize that it is inevitable to perform further investigations of other properties related to the electronic structure of these clusters, like ionization potentials, electron affinities, and binding energies, as these have been obtained to good accuracy in experiments.

## Acknowledgment

This work has been supported by the German Science Foundation through the SFB 445 *"Nano-Particles from the Gasphase: Formation, Structure, Properties"*.

## References

[1] I.M.L. Billas, A. Châtelain, and W.A. de Heer, *J. Magn. Magn. Mater.* **168** 64-84(1997).

[2] S. Chrétien and D.R. Salahub, *Phys. Rev. B* **66** 155425(2003).
G.L. Gutsev and C.W. Bauschlicher, Jr., *J. Phys. Chem. A* **107** 7013-7023(2003).

[3] M. Castro, *Int. J. Quant. Chem.* **64** 223-230(1997).

[4] G. Rollmann, S. Sahoo, and P. Entel, *Phys. Stat. Solidi (a)* **201** 3263-3270(2004).

[5] S.L. Dudarev, G.A. Botton, and S.Y. Savrasov, C.J. Humphreys, and A.P. Sutton, *Phys. Rev. B* **57** 1505-1509(1998).

[6] G. Rollmann, A. Rohrbach, P. Entel, and J. Hafner, *Phys. Rev. B* **69** 165107(2004).

[7] J.P. Perdew and Y. Wang, *Phys. Rev. B* **45** 13244-13249(1992).

[8] P.E. Blöchl, *Phys. Rev. B* **50** 17953-17979(1994).

[9] G. Kresse and J. Furthmüller, *Phys. Rev. B* **54** 11169-11186(1996).
G. Kresse and D. Joubert, *Phys. Rev. B* **59** 1758-1775(1999).

Brill Academic Publishers
P.O. Box 9000, 2300 PA Leiden
The Netherlands

*Lecture Series on Computer*
*and Computational Sciences*
Volume 4, 2005, pp. 739-739

# The Quest for Absolute Chirality

Peter Schwerdtfeger

*Theoretical Chemistry, Bldg.44, Institute of Fundamental Sciences, Massey University*
*(Albany Campus), Private Bag 102904, North Shore MSC, Auckland, New Zealand*

Received 2 August, 2005; accepted in revised form 11 August, 2005

*Abstract:* Emil Fischer's studies of peptides and sugars lead to chiral molecules (D-sugars and L-amino acids) and to the confirmation of Pasteur's conjecture that the universe is disymmetric. Friedrich Hund regarded the existence of stable chiral molecules as paradoxial (the Hund Dilemma). A fundamental discovery of this century is that our Universe is left handed (Weinberg-Salam-Glashow theory), the electro-weak interaction (parity-odd) gives rise to primarily left-spinning electrons during beta decay. In 1957 Lee and Yang discovered parity violation (PV) in the $K^+$ decay, which was confirmed shortly after by Wu *et al.* for the -decay. In the last decade PV effects in atomic transitions have been measured and calculated very accurately confirming the so-called standard model in particle physics. It is now well accepted that PV can lead to a small energy difference between enantiomers of chiral molecules ($V_n$ -$A_e$ coupling for the Z-exchange), although there is no experimental verification yet of this symmetry breaking effect. Current high resolution optical spectroscopy carried out in the $CO_2$ laser frequency range (878-1108 cm$^{-1}$) currently achieves resolutions of about $ = 1$ Hz. Recent calculations show that PV effects in vibrational transitions of chiral methane derivates CFXYZ (X,Y,Z= H, Cl, Br, I) are in the mHz range and below the detection limit. Our research group is currently searching for molecules including heavy elements ($Z^5$-scaling) to achieve PV effects in the Hz range. Two new compounds are promising candidates, Cl-Hg-CHFCl and PR$_3$-Au-CHFCl. A short discussion on the Yamagata hypothesis is provided, which states that one chiral form is stabilized by PV giving rise to biomolecular homochirality in nature.

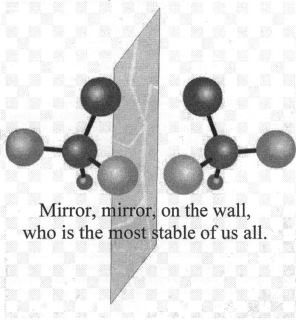

Mirror, mirror, on the wall,
who is the most stable of us all.

Brill Academic Publishers
P.O. Box 9000, 2300 PA Leiden
The Netherlands

*Lecture Series on Computer*
*and Computational Sciences*
Volume 4, 2005, pp. 740-743

# Efficient Computation of Hydrophobic Interactions Based on Fundamental Measure Treatment

V.F. Sokolov[1], G.N. Chuev

Laboratory of Biophysics of excitable mediums,
Institute of Theoretical and Experimental Biophysics,
Russian Academy of Sciences,
142290 Pushchino, Russia

Received 14 June, 2005; accepted in revised form 20 June, 2005

*Abstract:* The main goal of our investigations is to apply fundamental measure theory to a variety of problems in chemistry and biology. For this purpose, we have developed FMT codes to treat both atomic fluid models with hydrocarbon solutes. We have developed as well as three-dimensional real space and Fourier space algorithms. The former rely on a matrix-based Newton's method while the latter couple fast Fourier transforms with a matrix-free Newton's method. In the paper, we describe our algorithm development work as well as briefly discuss a few applications including hydrophobic hydration of linear, branched and cyclic hydrocarbons.

*Keywords:* Fundamental measure theory, fast Fourier transform, Newton method of iterations, hydrophobic hydration.

*Mathematics Subject Classification: PACS:* 05.10.-a, 31.15.Ew, 45.10.-b
*PACS:* 61.20.Gy, 61.20.Ne, 82.70.Uv.v

## 1. Introduction

Hydrophobic interactions play an important role in stabilization of various biomacromolecular complexes including nucleic acids, proteins, and lipids, since these complexes contain a large number of nonpolar groups (aminoacid and aliphatic sites of a peptide chain, hydrocarbonic tails of lipids)[1]-[3]. Despite of long history of research of hydrophobic interactions, at present there is no complete understanding of nature of hydrophobic interactions because of difficult multiscale character of these effects, which can reveal as microscopic changes of water structure near small hydrophobic groups, as well as conformations and aggregation of biomacromolecules at mesoscopic scales up to several tens of angstrom [4].

The theory of hydrophobic interaction in water solvent is traditionally connected to the idealized concept of cavitation, i.e. formations of a cavity of the excluded volume in pure solvent in which the particle of a substratum is located [5]. As against usual two-partial intermolecular interactions, free energy of cavitation represents complex many-particle effect corresponding, basically, to entropy changes that accompany with process of solvation. Because of nonlocal character, arising structural changes of solvent, the physical understanding of this phenomenon is not full, and its calculation represents a serious problem.

The methods of the molecular dynamics, Monte Carlo [6], [7] in many cases allow to feature adequately properties of various fluids with the specified interaction potentials and to obtain the data on structure of substance. However, numerical evaluations of hydrophobic solvation by Monte Carlo or molecular dynamics methods are rather difficult, especially for nano-sized particles. Based on the fundamental measure theory [8] the calculation of correlation functions of the solutes considerably becomes simpler, and not only in the symmetric one-dimensional case, but also three-dimensional too. In addition, it is not necessary for these purposes of major expenses of computing efforts likes as in MC or MD methods. Moreover, the FMT is more preferable at evaluation of correlation functions because of it does not impose restriction on solute and solvent sizes and geometries of them.

---

[1] Corresponding author. E-mail: vicvor@mail.ru.

## 2. Fundamental Measure Theory

Density functional theory is based on searching the extreme of the grand free energy functional, $\Omega$, which we write for hard spheres system as

$$\Omega = F_{id} + F_{HS} - \int \rho(\vec{r})(V(\vec{r}) - \mu)d\vec{r},\tag{1}$$

where $F_{id}$ is the ideal gas contribution to the Helmholtz free energy, $F_{HS}$ is excess free energy of the hard sphere, $V(\vec{r})$ is the external field acting on solute and $\mu$ is the chemical potential. Present the terms as

$$F_{id} = k_B T \int \rho(\vec{r})(\ln(\Lambda^3 \rho(\vec{r})) - 1)d\vec{r},\tag{2}$$

$$F_{HS} = k_B T \int \Phi[n_\alpha(\vec{r})]d\vec{r},\tag{3}$$

where $k_B$ is the Boltzmann constant, $T$ is temperature, $\Lambda$ is the thermal de Broglie wavelength, $\Phi$ is a nonlocal free energy density, $n_\alpha(\vec{r})$ are the nonlocal densities. We have applied the hard sphere free energy densities [9]. In this approach of the fundamental measure theory (FMT), the nonlocal densities are

$$n_\alpha(\vec{r}) = \int \rho(\vec{r}')w_\alpha(\vec{r} - \vec{r}')d\vec{r}',\tag{4}$$

where $w_\alpha(\vec{r})$ are weight functions (4 scalars and 2 vectors) based on the fundamental measures (volume, surface, radius, etc) of the fluid particles. The weight functions are

$$w_3(r) = \theta(r - R),$$

$$w_2(r) = 4\pi R w_1(r) = 4\pi R^2 w_0(r) = \delta(r - R),\tag{5}$$

$$\vec{w}_{v2}(\vec{r}) = 4\pi R \vec{w}_{v1}(\vec{r}) = \delta(r - R)\frac{\vec{r}}{r},$$

where $R$ is the radius of solvent particle. The free density of hard sphere $\Phi$ defines as

$$\Phi[n_\alpha] = -n_0 \ln(1 - n_3) + \frac{n_1 n_2 - \vec{n}_{v1}\vec{n}_{v2}}{1 - n_3} + \frac{n_2^3 - 3n_2(\vec{n}_{v2}\vec{n}_{v2})}{24\pi(1 - n_3)^2}.\tag{6}$$

The density distribution $\rho(\vec{r})$ can be calculated from the solution of the integral equation obtained

from the variational principle $\left.\dfrac{\partial \Omega[\rho(\vec{r})]}{\partial \rho(\vec{r})}\right|_\mu = 0$ for the grand canonical potential (1) $\Omega$

$$\rho(\vec{r}) = \rho_b \exp\left((k_B T)^{-1}(\mu - V(\vec{r}) - \frac{\partial F_{HS}[\rho(\vec{r})]}{\partial \rho(\vec{r})}\right),\tag{7}$$

where $\rho_b$ is the bulk density.

## 3. Numerical approaches

The main goal of our investigation is to solve the equation (7) and to find the excess chemical potential (3) for hard spheres solute. We use a fully coupled Newton iteration to locate equilibrium solutions. The unknowns in the problem are represented by $\vec{X}$ (e.g. $\rho(\vec{r})$) and $\vec{f}(\vec{X}) = 0$ is the system of equations to be solved. While we eventually discretized the unknown functions into a vector of unknown coefficients, and the functional equations into a set of algebraic equations, the equations will be presented in functional form. Newton's method is posed as the solution to

$$J\Delta\vec{X} = -\vec{f},\tag{8}$$

where $\Delta\vec{X} = \vec{X}^{(k+1)} - \vec{X}^{(k)}$ is the difference between successive iterations of the unknowns and the Jacobian matrix is given by

$$J_{ij}(\vec{r}, \vec{r}') = \frac{\partial f_i(\vec{r})}{\partial X_j(\vec{r}')}.\tag{9}$$

Subscripts $i$ and $j$ represent different functional equations and unknowns, such as when solving for multiple components. We note that, while we are seeking solutions to the nonlinear equations

$\vec{f}(\vec{X}) = 0$, multiple solutions may exist. Away from a phase transition, only one of the solutions will be thermodynamically stable; other solutions will be either unstable or metastable.

One approach we have implemented is based on forming integration stencils for the nonlocal densities defined above. The computational strategy is to form the Jacobian matrix based on the real space integration stencils [10]. The precise nature of the Jacobian matrix depends on how one chooses to formulate the systems of equations. The residual equation for hard sphere fluids was given in equation (7). From a numerical point of view, the critical question is how to formulate the Jacobian. Taking the functional derivative in equation (9) yields

$$J_{ij}(\vec{r},\vec{r}') = \frac{\delta_{ij}\delta(\vec{r}-\vec{r}')}{\rho(\vec{r}')} + \frac{1}{k_B T}\int\sum_{\alpha}\sum_{\beta}\frac{\partial^2\Phi}{\partial n_\alpha \partial n_\beta}(\vec{r}'')w_\alpha(\vec{r}_i-\vec{r}'')w_\beta(\vec{r}''-\vec{r}_j')d\vec{r}'', \qquad (10)$$

where $\delta_{ij}$ is Kronecker delta. The physical meaning of the product of weight functions in the above equation is clear. This product will be nonzero only if two weight function stencils (centered at $\vec{r}$ and $\vec{r}'$) are overlapping. Each entry in the Jacobian contains an integral of the intersecting volumes of all possible products of weight functions. Performing these integrals is time consuming because it is an operation that is $o(N_s)$, where $N_s$ is the number of points in a given integration stencil. In addition to computational complexity, the memory required to store the Jacobian is another important factor in the feasibility of large three-dimensional calculations using these accurate FMTs. In this case, there are nonzero Jacobian entries for every node within two stencil lengths (this is the particle radius, $R$, in the Rosenfeld functional) of the row position, $\vec{r}$. Thus the required memory is proportional to the number of nonzeros in the matrix, $M$, where

$$M = \frac{4\pi}{3}\left(\frac{2R}{h}\right)^3 N_c^2 N, \qquad (11)$$

$N_c$ is number of components in the system and $N$ is the number of mesh points, $h = 0.1\sigma$ is step.

## 4. Application FMT for hydrocarbons

At first, based on the equation (7) and the method described above, we have calculated the radial distribution functions for different Lennard-Jones solutes, which have been described in paper [11]. The results are presented on Figure 1.

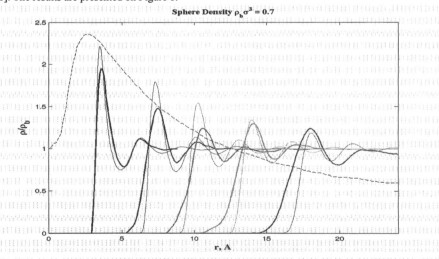

Figure 1: The distribution functions for hard LJ solutes: narrow lines are plotted for FMT calculations, bold lines are for MD simulation and black dashed line is plotted for scaled particle theory.

The bulk density and diameter of water molecules are 0.0333 Å⁻³ and $\sigma_{water} = 2.77$ Å, correspondingly. Discrepancy between maxima of obtained from molecular dynamic simulation and by fundamental measure treatment are in limits 5-15%. Zeros of radial distribution functions are almost coinciding but the widths of the peaks received by FMT are a bit narrow then from MD simulation [11]. Supposing, that such difference will incidentally affect integrated thermodynamic magnitudes we have found the radial distribution functions for different (linear, branched, cyclic) hydrocarbons at temperature 298 K. In this case, we modeled solvent particles as hard spheres and interaction between solute and solvent was modeled as Lennard-Jones potential with corresponding parameters solute-oxygen diameter $\sigma_{so}$ and $\varepsilon_{so}$ [12].

Using the radial distribution functions, we calculated the excess chemical potentials for various hydrocarbons. The calculated and experimental data in Table 1 are presented with respect to the solvent accessible surface area (SASA). The SASA is a commonly used predictor of the hydration properties of nonpolar compounds [13] it is used here as a reference to discuss the observed trends of the experimental and calculated data.

Table 1. The experimental, calculated by SASA and by FMT excess chemical potentials

| Molecule | $\Delta G_{FMT}$, kcal/mol | $\Delta G_{SASA}$, kcal/mol | $\Delta G_{exp}$, kcal/mol |
|---|---|---|---|
| 1. methane | 2.1226 | 2.404 | 2.01 |
| 2. ethane | 2.2078 | 2.630 | 1.84 |
| 3. propane | 1.9464 | 2.890 | 1.96 |
| 4. butane | 1.9770 | 3.210 | 2.08 |
| 5. pentane | 2.0296 | 3.447 | 2.33 |
| 6. hexane | 2.1005 | 3.781 | 2.49 |
| 7. isobutane | 1.9733 | 3.029 | 2.24 |
| 8. 2-methylbutane | 2.0156 | 3.510 | 2.44 |
| 9. neopentane | 2.0113 | 3.234 | 2.51 |
| 10. cyclopentane | 1.9909 | 2.802 | 1.21 |
| 11. cyclohexane | 2.0246 | 2.338 | 2.25 |

## 5. Results and discussion

In this work, we used the fundamental measure theory to the quantitative description of the hydrophobic phenomena based on the density functional theory. For achievement of this purpose, we used Newton iterative algorithm for three-dimensional molecules, which has high efficiency and best stability in convergence of radial distribution function. Based on the FMT and a matrix-free Newton iterative method we have calculated radial distribution function for linear, branched and cyclic hydrocarbons in liquid water. Using found radial distribution functions, we have calculated the excess chemical potential for them. These values received in frameworks FMT are rather well corresponded with the experimental results. It is surprising, that for such relatively simple solvent model (hard spheres) the experimental and the calculated values of the excess chemical potential pretty close coincide.

## References

[1] A.L. Fink , *Folding Design* 3 R9(1998).
[2] W. Kauzmann, *Adv. Protein Chem.* 14 1(1959).
[3] C. Tanford. *The Hydrophobic Effect: Formation of Micelles and Biological Membranes*; John Wiley & Sons: New York, 1973.
[4] K. Lum, D. Chandler, J. D. Weeks, *J. Phys. Chem. B* 103 4570(1999).
[5] B. Guillot, Y. Guissani, and S. Bratos, *J. Phys. Condensed Matter* 2 165(1990).
[6] W. L. Jorgensen, J. F. Blake, and J. K. Buckner, *Chem. Phys.* 129 193(1989).
[7] K. Watanabe and H. C. Andersen, *J. Phys. Chem.* 90 795(1986).
[8] Y. Rosenfeld, M. Schmidt, H. Lowen and P. Tarazona, *Phys. Rev. E* 55 4245(1997).
[9] Y. Rosenfeld, *Phys. Rev. Lett.* 63 980(1989).
[10] L. J. D. Frink and A. G. Salinger, *J. Comput. Phys.* 159 407(2000).
[11] Henry S. Ashbaugh, and Michael E. Paulaitis, *J. Am. Chem. Soc.*, **123**, 10721(2001).
[12] E. Gallicchio, M. M. Kubo, and R. M. Levy, *J. Phys. Chem. B* **104**, 6271(2000).
[13] D. Sitkoff, K. Sharp, B. Honig, *Biophys. Chem.* **51**, 397(1994).

Brill Academic Publishers
P.O. Box 9000, 2300 PA Leiden,
The Netherlands

Lecture Series on Computer
and Computational Sciences
Volume 4, 2005, pp. 744-747

# Switching Nucleotide Base Pairs by Coherent Light

## Ioannis Thanopulos* [1] and Moshe Shapiro*†

*Department of Chemistry, The University of British Columbia, Vancouver V6T1Z1, Canada
† Department of Chemical Physics, The Weizmann Institute, Rehovot 76100, Israel

Received 19 July, 2005; accepted in revised form 13 August, 2005

*Abstract:* We show that phase-coherent optical techniques allow for the detection and
automatic repair of mutations in nucleotide pairs. We demonstrate computationally that
there is a laser pulse sequence that can detect the occurrence of a mutation caused by a
double proton transfer between hydrogen-bonded nucleotide pairs and automatically repair
it by converting the mutated nucleotide-pair to the non-mutated one. The specific system
chosen for this demonstration is the hydrogen-bonded 2-pyridone·2-hydroxypyridine dimer
at typical inter-nucleotide distances, a well-established model for tautomeric acid base
pairs.

*Keywords:* nucleotide base pairs; tautomerization; detection of a mutation; automatic
repair; laser fields; coherent control

*PACS:* 32.80.Qk, 33.80.-b

## Introduction

Hydrogen bonding between nucleotide pairs is a major factor in the observed stability and fidelity
of replication of DNA[1]. It has been recognized for quite some time that double proton transfer
between hydrogen-bonded di-nucleotides can transform one di-nucleotide pair to another, leading
to loss of recognition of the correct base pair in DNA and RNA. For example, proton transfer in
the Guanine-Cytosine (G·C) pair is considered a crucial part of the radiation-induced damage to
DNA[2].

The mechanism of the tautomerization of hydrogen-bonded base-pairs, induced by double proton transfer (DPT) from one stable configuration $A$ to another (almost energetically equivalent)
stable configuration $B$, in either a sequential or concerted fashion, has been extensively discussed
in the literature[3, 4, 5, 6]. Control over the DPT process is a major objective, as well as techniques
that would enable the detection and repair of an undesired tautomerization once it has occurred[7].

In this work[8] we show that control of such a process can be achieved with optical means,
using *Coherent Control* (CC)[9] techniques. CC is based on guiding a material system (employing
"tailor-made" external laser fields) to arrive at a given final state via a number of indistinguishable quantum pathways. It has been amply documented, both experimentally[10, 11, 12] and
theoretically[9], that *selectivity* in a host of physical and chemical processes can be achieved by
controlling the interference between such quantum pathways.

A complementary process, in which the material system is guided in an adiabatic fashion along
a single pathway (termed "Adiabatic Passage" (AP)[13]) enables one to transfer populations in a
*complete* way from one state to another. The merging of the two techniques, appropriately named

---

[1] Corresponding author. E-mail: ioannis@chem.ubc.ca

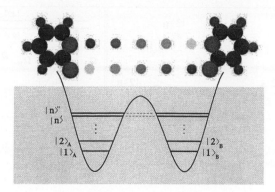

Figure 1: (Upper panel) The di-nucleotide dimer: The double proton (cyan) transfer takes place between the two nitrogen atoms (green) and the two oxygen atoms (red). (Lower panel) One dimensional potential energy cut along the DPT coordinate $s$. Each of the localized states $|i\rangle_k = (|i'\rangle \pm |i''\rangle)/\sqrt{2}$, $k = A, B$ is a superposition of an $|i'\rangle$ and an $|i''\rangle$ vibrational eigenstate, where $|i'\rangle$ denotes the $s$−symmetric $i$-th vibrational eigenstate, and $|i''\rangle$ - the $s$−antisymmetric $i$-th eigenstate.

"Coherently Controlled Adiabatic Passage" (CCAP), achieving both *selectivity* and *completeness* has been recently accomplished[14, 15, 16].

Here, we demonstrate by computational means that the CCAP method enables one to both distinguish between base-pairs residing in two energetically equivalent $A$ and $B$ minima, as well as induce and control the inter-conversion of the system between these two configurations. This demonstration gives rise to the hope that we would in the future be able to detect, as well as correct by purely optical means, undesired mutagenesis and manipulate genomes in a sequence-specific manner.

## The base-pair model system, mutation detection and repair

As a vehicle for this demonstration we choose the well studied[4, 5] nucleic acid base-pair model - the 2-pyridone (2PY) and 2-hydroxypiridine (2HP) dimer (shown in Fig. 1) embedded in a "pocket" of a nucleic acid. We keep the dimer at a separation similar to that found in nucleic acids, where the base-pair members are unable to approach each other too closely due to their attachments to the nucleotide backbone[6]. In this way we slow down the natural rate of DPT tunneling in isolated dimers, in keeping with environments typical of base-pairs embedded in nucleic acids.

The concerted motion of the two protons during the 2PY·2HP tautomerization process may be described as a motion along a *linear reaction path*[17] connecting the two equilibrium configurations $A$ and $B$. This type of path is particularly suitable for studying the transfer of light particles between two heavier moieties[17], which are thus naturally kept immobile. In addition to the in-plane motion we also consider the out-of-plane vibration that involves out-of-phase motions of the hydrogen atoms as the system propagates along the reaction path. This is the minimal space needed to properly describe the dynamics of our control scheme using the energetically lowest states of the dimer.

The potential energy and the molecular structures are obtained in an *ab initio* way using the

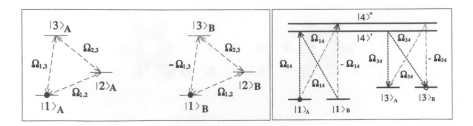

Figure 2: (Left panel) The detection scheme: The coupling scheme for the discriminator. (Right panel) The repair scheme: The levels considered and their couplings.

hybrid B3LYP density functional (DFT) method and the valence triple-zeta 6-311++G(d,p) basis set of the *Gaussian98* electronic structure calculation package. The two equilibrium configurations of the dimer are planar at this level of *ab initio* theory and consequently the linear reaction path from A to B is planar as well. Our calculations yield a linear reaction path potential-energy barrier height of $\approx 8550$ cm$^{-1}$. Our two-dimensional configuration space model predicts tunneling times from A to B of $\approx 0.3$ $\mu$s. Although this is still much shorter than the estimated tunneling times in nucleic acids[18], which may be also due to the fact that B3LYP in general underestimates energy barriers for proton transfer, it is by far longer than the tunneling times in the isolated unconstrained dimers where the monomers are allowed to get much closer to one another[5]. However, this tunneling time is sufficiently long to allow us to consider the localized states at the A or B minima as "legitimate" initial or final states.

The fact that there exist degenerate A and B tautomers makes the separate addressing of these tautomers by optical means difficult. We have overcome this difficulty using a two step approach: In the first step we affect the *discrimination* between the tautomers. In the second step we use the discrimination afforded by the first step to control the *interconversion* between the tautomers. These two steps[14, 15] makes use of the different symmetries of the $x$, $y$ and $z$ components of the dipole moment to manipulate molecules in configuration A and B differently.

The cyclic laser coupling scheme, which is at the heart of the first "discrimination" step, is depicted in the left part of Fig. 2. As shown in the figure, the laser fields inter-couple three vibrational states of the A tautomer, as well as inter-couple separately three vibrational states of the B tautomer. We can, by controlling one overall phase of the laser fields[14, 15], excite only the $|1\rangle_B$ ground state to a higher $|i\rangle_B$ vibrational state of the B tautomer, while leaving all the A-tautomer states untouched.

Following this selective excitation, it is now possible to transfer in the second "converter" step depicted in the right part of Fig. 2 the $|1\rangle_A$ (ground state of the A tautomer) population, if such exists, to a $|j\rangle_B \neq |i\rangle_B$ B-tautomer excited state. In this way, any population residing initially in $|1\rangle_A$, is automatically converted to the B tautomer. Thus, we are able both to detect the existence of a mutation *and* to repair it automatically. Moreover, if it is the A tautomer that we desire, we can, by tuning the overall laser phase, interchange the process, with the excitation of the B states switched over to the excitation of the A states and the transfer occurring from B to A.

We find that after the "discriminator" and "converter" steps, all the population in tautomer A has been converted to tautomer B. In fact, our method is also capable of purifying a mixture in which there is an initial population in both tautomers. In a complementary way, our approach can be also used on the *single molecule* level to identify, for a given dimer, if the ground state configuration is that of A or B.

# References

[1] G.A. Jeffrey and W. Saenger. *Hydrogen Bonding in Biological Structures*. Springer, Berlin, 1991.

[2] E. Nir, K. Kleinermanns, and M.S. de Vries. Pairing of isolated nucleic-acid bases in the absence of the DNA backbone. *Nature*, 408:949, 2000.

[3] J. Catalán, P. Pérez, J.C. del Valle, J.L.G. de Paz, and M. Kasha. H-Bonded N-heterocyclic base-pair phototautomerizational potential barrier and mechanism: The 7-azaindole dimer. *Proc. Natl. Acad. Sci. USA*, 101:419, 2004.

[4] J.A. Frey, A. Müller, H.-M. Frey, and S. Leutwyler. Infrared depletion spectra of 2-aminopyridine·2-pyridone, a Watson-Crick mimic of adenine·uracil. *J. Chem. Phys.*, 121:8237, 2004.

[5] J.R. Roscioli, D.W. Pratt, Z. Smedarchina, W. Siebrand, and A. Fernandez-Ramos. Proton transfer dynamics *via* high resolution spectroscopy in the gas phase and instanton calculations. *J. Chem. Phys.*, 120:11351, 2004.

[6] V. Zoete and M. Meuwly. Double proton transfer in the isolated and DNA-embedded guanine-cytosin base pair. *J. Chem. Phys.*, 121:4377, 2004.

[7] T. Schultz, E. Samoylova, W. Radloff, I.V. Hertel, A.L. Sobolewski, and W. Domcke. Efficient deactivation of a model base pair via excited-state hydrogen transfer. *Science*, 306:1765, 2004.

[8] I. Thanopulos, and M. Shapiro. Detection and Automatic Repair of Nucleotide Base-Pair Mutations by Coherent Light. *J. Am. Chem. Soc.*, in press.

[9] Moshe Shapiro and Paul Brumer. *Principles of the Quantum Control of Molecular Processes*. John Wiley & Sons, Inc., New Jersey, 2003.

[10] T. Brixner, N.H. Bamrauer, P. Nicklaus, and G. Gerber. Photoselective adaptive femtosecond quantum control in the liquid phase. *Nature*, 414:57, 2001.

[11] H.A. Rabitz, M.H. Hsieh, and C.M. Rosenthal. Quantum optimally controlled transition landscapes. *Science*, 303:1998, 2004.

[12] J.L. Herek, W. Wohlleben, R.J. Cogdell, D. Zeidler, and M. Motzkus. Quantum control of energy flow in light harvesting. *Nature*, 417:533, 2002.

[13] N.V. Vitanov, M. Fleischhauer, B.W. Shore, and K. Bergmann. Coherent manipulation of atoms and molecules by sequential laser pulses. *Adv. At. Mol. Opt. Phys.*, 46:55, 2001.

[14] P. Kral, I. Thanopulos, M. Shapiro, and D. Cohen. Two-step enantio-selective optical switch. *Phys. Rev. Lett.*, 90:033001, 2003.

[15] I. Thanopulos, P. Kral, and M. Shapiro. Theory of two-step enantiomeric purification of racemic mixtures by optical means: The $D_2S_2$ molecule. *J. Phys. Chem.*, 119:5105, 2003.

[16] I. Thanopulos, P. Kral, and M. Shapiro. Complete Control of Population Transfer between Clusters of Degenerate States. *Phys. Rev. Lett.*, 92: 113003 (2004).

[17] W.H. Miller, B.A. Ruf, and Y.-T. Chang. A diabatic reaction path hamiltonian. *J. Phys. Chem.*, 89:6298, 1988.

[18] Per-Olov Löwdin. Proton Tunneling in DNA and its Biological Implications. *Rev. Mod. Phys.*, 35:724, 1963.

Brill Academic Publishers
P.O. Box 9000, 2300 PA Leiden,
The Netherlands

*Lecture Series on Computer
and Computational Sciences*
Volume 4, 2005, pp. 748-750

# Multi-level Computer Simulations of Condensed Matter Based on Subsystem Formulation of Density Functional Theory

**T.A. Wesołowski[1]**

Department of Physical Chemistry,
30, quai Ernest-Ansermet, CH-1211 Genève 11, Switzerland

Received 5 July, 2005; accepted in revised form 10 Aug 2005.

*Keywords:* orbital-free embedding, non-additive kinetic energy functional, one-electron equations, multi-scale simulations, electron density partitioning

*PACS:* 31.15.Ew, 31.15.Bs, 71.10.Ca, 71.70.-d

In conventional Kohn-Sham formulation of density functional theory (DFT) [1], the total energy is expressed as an explicit functional of one-electron functions (Kohn-Sham orbitals). Euler-Lagrange minimization of this functional leads to the celebrated Kohn-Sham equations. In 1991, Cortona introduced an alternative formulation of DFT which uses similar concepts. The total energy, however, is expressed as an explicit functional of several sets of one-electron functions ($\{\phi_i^j\}$). For the case of two closed-shell subsystems, this functional reads in atomic units:

$$
\Xi^E[\{\phi_i^A\}; \{\phi_i^B\}] = 2\sum_{i=1}^{N^A} < \phi_i^A| - \frac{1}{2}\nabla^2|\phi_i^A > +2\sum_{i=1}^{N^A} < \phi_i^B| - \frac{1}{2}\nabla^2|\phi_i^B > +T_s^{nad}[\rho_A, \rho_B]
$$
$$
+ \quad V[\rho] + J[\rho] + E_{xc}[\rho] \tag{1}
$$

where

$$
\rho = 2\sum_{i-1}^{N^A} |\phi_i^A|^2 + 2\sum_{i=1}^{N^B} |\phi_i^B|^2 \tag{2}
$$

and where $T_s^{nad}[\rho_A, \rho_B] \equiv T_s[\rho_A + \rho_B] - T_s[\rho_A] - T_s[\rho_B]$ is the bi-functional of the non-additive kinetic energy $V[\rho]$. The symbols $V[\rho]$, $J[\rho]$, $E_{xc}[\rho]$, and $T_s[\rho]$ denote the components of the total energy functional defined in the conventional Kohn-Sham formalism - energy of the interaction with the external field, classical electron-electron repulsion energy, the exchange-correlation energy, and the kinetic energy in the reference system of non-interacting electrons, respectively. The embedded orbitals ($\{\phi_i^j\}$) will be referred here as *embedded orbitals*. This distinction important because the embedded orbitals are not related to the Kohn-Sham orbitals in a straightforward manner. Euler-

---

[1]Corresponding author. E-mail: tomasz.wesolowski@chiphy.unige.ch

Lagrange minimization leads to two coupled sets of one-electron equations[2]:

$$\left[ -\frac{1}{2}\nabla^2 + V_{eff}^{KSCED}\left[\rho_A, \rho_B; \vec{r}\right] \right] \phi_i^A = \epsilon_i^A \phi_i^A \quad i = 1, N^A \tag{3}$$

$$\left[ -\frac{1}{2}\nabla^2 + V_{eff}^{KSCED}\left[\rho_B, \rho_A; \vec{r}\right] \right] \phi_i^B = \epsilon_i^B \phi_i^B \quad i = 1, N^B \tag{4}$$

We will refer to each of these equations as Kohn-Sham Equations with Constrained Electron Density (KSCED).

The effective potential expressed by means of universal density functionals: $E_{xc}[\rho]$ (exchange-correlation energy) and $T_s^{nad}[\rho_A, \rho_B]$ reads:

$$V_{eff}^{KSCED}\left[\rho_A, \rho_B; \vec{r}\right] = V_{eff}^{KS}\left[\rho_A; \vec{r}\right] + V_{eff}^{emb}\left[\rho_A, \rho_B; \vec{r}\right], \tag{5}$$

where

$$V_{eff}^{KS}\left[\rho_A; \vec{r}\right] = \sum_{i_A}^{N_{nuc}^A} -\frac{Z_{i_A}}{|\vec{r} - \vec{R}_{i_A}|} + \int \frac{\rho_A(\vec{r}')}{|\vec{r}' - \vec{r}|} d\vec{r}' + \left.\frac{\delta E_{xc}[\rho]}{\delta\rho}\right|_{\rho=\rho_A} \tag{6}$$

where $V_{eff}^{KS}\left[\rho_A; \vec{r}\right]$ is the Kohn-Sham effective potential [1] for the *isolated* subsystem A, and where the part representing the environment is expressed by means of universal density functionals[3]:

$$
\begin{aligned}
V_{eff}^{emb}\left[\rho_A, \rho_B; \vec{r}\right] = {} & \sum_{i_B}^{N_{nuc}^B} -\frac{Z_{i_B}}{|\vec{r} - \vec{R}_{i_B}|} + \int \frac{\rho_B(\vec{r}')}{|\vec{r}' - \vec{r}|} d\vec{r}' \\
& + \left.\frac{\delta E_{xc}[\rho]}{\delta\rho}\right|_{\rho=\rho_A+\rho_B} - \left.\frac{\delta E_{xc}[\rho]}{\delta\rho}\right|_{\rho=\rho_A} + \frac{\delta T_s^{nad}[\rho_A, \rho_B]}{\delta\rho_A},
\end{aligned}
\tag{7}
$$

A more complete description of the subsystem formulation of density functional theory and its use in multi-level computer simulations can be found in our recent review[4]. Practical applications of this formalism hinge on approximations to $E_{xc}[\rho]$ and to $T_s^{nad}[\rho_A, \rho_B]$. Currently, we use the gradient-dependent approximation to $T_s^{nad}[\rho_A, \rho_B]$ which emerged from our dedicated studies of numerical accuracy of different approximations in the case of weakly overlapping $\rho_A$ and $\rho_B$[5].

One of possible applications of the Cortona's formalism is its use in multi-level computer modelling studies in which different types of theoretical descriptions are used for different subsystem. In particular, if the subsystem of primary interest (say $A$) is described using Eqn. 3 whereas the other subsystem is described using a simpler method yielding $\rho_B$ we have the case of *orbital-free embedding*.

In the first part, we review our recent developments concerning the applied approximations to the relevant functionals:

*i)* The new tests of the accuracy of the gradient-free and gradient-dependent approximations based on comparisons with the benchmark *ab initio* results for the potential energy surfaces for an extended set of intermolecular complexes including hydrogen-bonded systems[6, 7].

*ii)* A model study concerning exploration of a new possibility offered by Eqns. 1, 3, and 4 to choose the most appropriate approximation for the relevant functional for different subsystems[8].

In the second part, we will review our recent applications of the orbital-free embedding effective potential in the computer simulation studies of various complex materials. The applications are selected in a way to illustrate possible types of application of the embedding potential given in Eqn. 7. They include:

*iii)* Multi-level computer simulation studies of electronic excitations in solvated acetone[9] and aminocoumarin-155[10]. In these studies, the electron density of the environment is obtained using Car-Parrinello or classical molecular dynamics simulations and the adequacy of various simplified methods to generate the electron density $\rho_B$ needed in the evaluation of the embedding potential given in Eqn. 3 is analyzed in detail.

*iv)* Multi-level computer simulation studies of a solvated model radical molecule[11]. Similar computer simulation strategies to the ones in the studies of solvated chromophores were applied and analyzed. These studies focus, however, not on electronic excitations but on the hyperfine structure.

*v)* Studies of the electronic structure of lanthanide centers in crystal environment applying the embedding potential given in Eqn. 7 to represent the ligands[12].

In the final part, our recent efforts in development of non-empirical approximations to the non-additive kinetic energy functional $T_s^{nad}[\rho_A, \rho_B]$ will be reviewed.

*Acknowledgement:* This work is supported by the Swiss National Science Foundation.

## References

[1] W. Kohn, L.J. Sham, *Phys. Rev.*, **140** (1965) A1133.

[2] P. Cortona, *Phys. Rev. B*, **44** (1991) 8454.

[3] T.A. Wesolowski, A. Warshel, *J. Phys. Chem.*, **97** (1993) 8050.

[4] T. A. Wesolowski, *One-electron equations for embedded electron density: challenge for theory and practical payoffs in multi-level modelling of soft condensed matter.* In: *Computational Chemistry: Reviews of Current Trends*, Vol. XI, World Scientific, 2005 *in press*.

[5] T.A. Wesolowski, *J. Chem. Phys.*, **106** (1997) 8516.

[6] M. Dulak, T.A. Wesolowski, *in preparation*.

[7] R. Kevorkiants, M. Dulak, T.A. Wesolowski, *in preparation*.

[8] C. Jacobs, T.A. Wesolowski, L. Visscher, *in preparation*.

[9] J. Neugebauer, M.J. Louwerse, E.J. Baerends, T.A. Wesolowski, *J. Chem. Phys.*, **122** (2005) 094115.

[10] J. Neugebauer, C.R. Jacob, T.A. Wesolowski, E.J. Baerends, *J. Phys. Chem. A, in press* (2005)

[11] J. Neugebauer, M.J. Louwerse, P. Belanzoni, T.A. Wesolowski, E.J. Baerends, *J.Chem. Phys., in press* (2005)

[12] M. Zbiri, M. Atanasov, C. Daul, J.-M. Garcia Lastra, T.A. Wesolowski, *Chem. Phys. Lett.*, **397**, (2004) 441; Zbiri *et al.*, (2005) *in preparation*.

Brill Academic Publishers
P.O. Box 9000, 2300 PA Leiden,
The Netherlands

*Lecture Series on Computer*
*and Computational Sciences*
Volume 4, 2005, pp. 751-754

# On the Structure of Liquid Water: An ab initio QM/MM MD Simulation Study

## D. Xenides, B. R. Randolf and B. M. Rode [1]

Theoretical Chemistry Division, Insitute of General, Inorganic and Theoretical Chemistry
University of Innsbruck, Innrain 52a, A-6020 Innsbruck, Austria

Received 5 July, 2005; accepted in revised form 10 August, 2005

*Abstract:* Pattern and dynamics of hydrogen bonds in liquid water were investigated by a quantum mechanical/molecular mechanical molecular dynamics (QM/MM MD) simulation at Hartree-Fock (HF) level of theory. A large subregion of the whole system comprising two complete coordination shells was treated quantum mechanically in order to include all polarization and charge transfer effects and to obtain accurate data about structure and dynamics of the intermolecular bonds. The results of this investigation are in agreement with recent experimental findings and suggest that in liquid water every molecule forms in average 2.8, but almost as a rule less than four intermolecular hydrogen bonds.

*Keywords:* QM/MM MD; Liquid water; ab initio HF; Hydrogen bonds

## 1 Introduction

The microstructure of liquid water has been and still is the subject of many experimental and theoretical studies. Its elucidation along with the knowledge of the ultrafast dynamics that govern the phenomena occuring in liquid water could provide sufficient information to understand the peculiar behaviour of this liquid. Nilsson et al. in a recent [1] refinement of their previous study [2] performed a combined X–ray absorption spectroscopy (XARS) and X–ray Raman scattering (XRS) study to investigate the local structure and more specifically the intermolecular hydrogen bonds (HBs) of liquid water. Their results have indicated that the number of four–fold bound water molecules in liquid water is rather small in comparison to two– and three–fold hydrogen bonded units. This observation confirmed some previous conclusions [2, 3, 4, 5, 6, 7, 8, 9, 10] and provokes new investigations and challenges for both theory and experiment.

## 2 Theory

*QM/MM Molecular Dynamics Simulation*: A detailed description of the rationale behind the definition of the size of the QM subregion is given in previous publications [22, 23, 24]. The procedure starts with a classical simulation and from the graphical representation of the radial distribution functions (RDF) the sizes of the coordination shells are determined. The radius of the subregion included in the present quantum mechanical simulation was set to the minimum after the second O–O peak, i.e. 5.6 Å. Thus, the simulation included a central water molecule plus 2 full coordination shells and was carried out for a total of 40 ps. All quantum mechanical force calculations were performed with a DZP basis set [18] using our TCI QM/MM MD code implementing TURBOMOLE [19, 20, 21].

---

[1] Corresponding author. E-mail: Bernd.M.Rode@uibk.ac.at

## 3   Results - Discussion

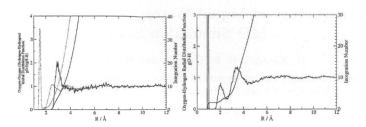

Figure 1: **left** Oxygen–Oxygen (full line) and Hydrogen-Hydrogen (dotted line), **right** Oxygen–Hydrogen RDF from a HF simulation.

After evaluation of the gO–H graph the average number of intermolecular HBs has been found close to 4. However, how many of these bonds are formed simultaneously, can only be revealed after a detailed analysis of the MD trajectory files. This evaluation was based on the following geometrical criteria for the formation of a hydrogen bond: a) The O–O distance ($R_{O-O}$) must take values in between the borders of the first peak in Fig. 1 (left graph, full line), i.e. $2.5 \leq R_{O-O} \leq 3.4$ Å, b) The intermolecular hydrogen bond distance ($R_{O\cdots H}$) must not exceed the borders of the second peak in (right) Fig. 1, thus $1.5 \leq R_{O-H} \leq 2.5$ Å, and c) The hydrogen bond angle ($\vartheta = \angle O \cdots H–O$) should be $\geq 100°$ accounting for the loss of hydrogen bond stabilization at lower angles. After application of the above mentioned criteria in the analysis of the MD trajectories the total average value of simultaneously formed HBs was found to be 2.8. It can be seen from

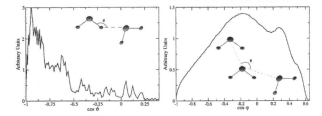

Figure 2: **left** Cosine distribution of $O\cdots H–O$ ($\vartheta$) and **right** O–O–O ($\varphi$) angles from a HF simulation.

the wide distribution of both angles (Fig. 2) that the structure of the first coordination shell is far from being regular. Snapshots characterising some structural patterns of liquid water have been produced by the MOLVISION 4D [17] programm and are depicted in (right picture) Fig. 3 along with the distribution of the coordination numbers (left picture).

## 4   Conclusion

In accordance with recent experimental data, our ab initio QM/MM simulation shows an enormous flexibility of the hydrogen bond network in liquid water at ambient conditions. This flexibility and the associated high mobility provide a good explanation for the extremely fast adaptability of the

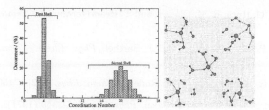

Figure 3: **left** Distribution of coordination numbers within the QM region of the simulation. Limits for $1^{st}$ and $2^{nd}$ shell were set to 3.4 and 5.6 Å.**right** Different structural entities in liquid water (The large circle denotes the oxygen atom of the central water molecule).

solvent to changes in temperature and/or pressure and to any solute. The simulation suggests that any accurate water model used for the interpretation of spectroscopic and other experimental data has to include other than 4–coordinated entities, in particular 3– and 5–coordinated ones, which are formed in considerable amounts and will be present simultaneously with the distorted tetrahedral structures, forming the main basis of the liquid at room temperature. However, even in these four–coordinated clusters, one cannot expect that all water molecules form four hydrogen bonds at the same time.

## 5 Acknowledgement

Financial support from Austrian Science Foundation (FWF) (Project 16221) and an Ernst Mach grant from the Austrian Ministry for Education, Science and Culture for D.X. are gratefully acknowledged.

## References

[1] P. Wernet, D. Nordlund, U. Bergmann, M. Cavalleri, M. Odelius, H. Ogasawara, Å. Näslund, T. K. Hirsch, L. Ojamäe, P. Glatzel, L. G. M. Petterson, A. Nilsson, *Science* **304**, 995-999 (2004).

[2] S. Myneni, Y. Luo, L. Å. Näslund, M. Cavalleri, L. Ojamäe, H. Ogasawara, A. Pelmenschikov, P. Wernet, P. Väterlein, C. Heske, Z. Hussain, L. G. M. Petterson, and A. Nilsson, *J Phys: Condens Matter* **14**, L213-L219 (2002).

[3] A. Rahman and F. H. Stillinger, *J. Am. Chem. Soc.* **55**, 7943-7948 (1973).

[4] A. V. Okhulkov, Y. N. Demianets, and Y. E. Gorbaty, *J. Chem. Phys.* **100**, 1578-1588 (1994).

[5] A. K. Soper, *J. Chem. Phys.* **101**, 6888-6901 (1994).

[6] S. Woutersen, U. Emmerichs, and H. J. Bakker, *Science* **278**, 658-660 (1997).

[7] P. Jedlovsky, J. P. Brodholt, F. Bruni, M. A. Ricci, A. K. Soper, and R. Vallauri, *J. Chem. Phys.* **108**, 8528-8540 (1998).

[8] R. Bucher, J. Barthel, and J. Stauber, *Chem. Phys. Lett.* **306**, 57-63 (1999).

[9] J. M. Sorenson, G. Hura, R. M. Glaeser, and T. Head-Gordon, *J. Chem. Phys.* **113**, 9149-9161 (2000).

[10] R. Bucher, C. Hölzl, J. Stauber, and J. Barthel, *Phys. Chem. Chem. Phys.* **4**, 2169-2179 (2002).

[11] D. Xenides, B. R. Randolf, and B. M. Rode, *J. Chem. Phys.* **122**, 174506 (2005).

[12] A. Rahman and F. H. Stillinger, *J. Chem. Phys.* **55**, 3336-3359 (1971).

[13] F. H. Stillinger and A. Rahman, *J. Chem. Phys.* **68**, 666-670 (1978).

[14] K. Laasonen, M. Sprik, M. Parrinello, and R. Car, *J. Chem. Phys.* **99**, 9080-9089 (1993).

[15] M. Sprik, J. Hutter, and M. Parrinello, *J. Chem. Phys.* **105**, 1142-1152 (1996).

[16] P. Silvestrelli and M. Parrinello, *J. Chem. Phys.* **111**, 3572-3580 (1999).

[17] H. T. Tran and B. M. Rode, *MOLVISION 4D programm: visualization of chemical systems*, www.molvision.com, 2002.

[18] T. H. Dunning Jr. and P. J. Hay, *in Modern Theoretical Chemistry*, Ed. H. F. Schaefer III, Plenum Press, New York, p. 1-28, (1976).

[19] R. Ahlrichs, M. Bär, and M. Häser, and C. Kölmel, *Chem. Phys. Lett.* **162**, 165-169 (1989).

[20] R. Ahlrichs and M. von Arnim, *in Methods and Techniques in Computational Chemistry: METECC-95 STEF*, Cagliari, vol. 3, 1995).

[21] M. von Arnim and R. Ahlrichs, *J. Comput. Chem.* **19**, 1746-1757 (1998).

[22] C. F. Schwenk, H. H. Loeffler, B. M. Rode, *J. Chem. Phys.* **115**, 10808-10813 (2001).

[23] R. Armunanto, C. F. Schwenck, and B. M. Rode, *Chem. Phys.* **295**, 63-70 (2003).

[24] C. F. Schwenk, T. H. Hofer, and B. M. Rode, *J. Phys. Chem. A* **108**, 1509-1514 (2004).

[25] P. Bopp, G. Jancsó, and K. Heinzinger, *Chem. Phys. Lett.* **98**, 129-133 (1983).

[26] G. Jancsó, P. Bopp, and K Heinzinger, *Chem. Phys.* **85**, 377-387 (1984).

[27] M. P. Allen and D. J. Tildesley, *in Computer Simulation of Liquids*, University Press, Oxford,(1987).

[28] H. J. Berendsen,J. R. Grigera, and T. P. Straatsma, *J. Phys. Chem.* **91**, 6269-6271 (1983).

[29] P. Bopp, *Chem. Phys.* **106**, 205-212 (1986).

[30] E. Spohr, G. Pálinkás, K. Heinzinger, P. Bopp, and M. M. Probst, *J. Phys. Chem.* **92**, 6754-6761 (1988).

Brill Academic Publishers
P.O. Box 9000, 2300 PA Leiden
The Netherlands

*Lecture Series on Computer*
*and Computational Sciences*
Volume 4, 2005, pp. 755-756

# Preface to the Minisymposium on Mathematical and Computational Approaches to Structure, Dynamics and Biology.

K. Balasubramanian[1]

Chemistry & Material Science Directorate
Lawrence Livermore National Laboratory, PO Box 808, L-268
Livermore, CA 94550
And
California State University, Eastbay,
Hayward CA 94542

This minisymposium deals with an exciting interdisciplinary topic that is emerging to be of great importance in many scientific disciplines from chemistry to biology. The spirit of this minisymposium is to highlight such computational and mathematical techniques that have made significant impact, especially in chemistry and biology. I have put together distinguished scientists who are working on the cutting edges of computational and mathematical approaches to chemistry and biology. As can be seen from the ensuing extended abstracts, the minisymposium covers a variety of topics dealing with mathematical and computational techniques to chemistry and biology. The presentations by Drs. Balasubramanian, Carbó-Dorca, Hosoya, King, Paldus, Silagi, and Takahashi deal with both novel mathematical and computational approaches to chemical problems of structural origin. The presentations by Drs. Fried and Gee deal with computational applications concerning dynamics and simulations pertinent to highly reactive fluids, spinodal-assisted polymer crystallization. The present author, Dr. Basak, Dr. Bonchev, Dr. Lionelleo, Dr. Moudgal and Dr. Sumners will present talks concerning both mathematical and computational applications to predictive toxicology, biology and the use of DNA knot theory to assay packing of DNA. In many of these complex biological and chemical applications experimental data are very hard to find and numerous practical applications ranging from the environment to drug industry demand for powerful predictions. Experimental studies on a large number of such chemicals and biological agents are also impractical and one must rely on such powerful predictive computational tools. This is the area were both mathematical and computational techniques can converge to make powerful impact and perhaps ahead of experimental science. The cost of making a new drug, for example, has dramatically been reduced due to the advent of such mathematical and computational techniques.

The symposium was put together with the novel mathematical and computational threads as the weaving materials for this novel interdisciplinary fabric. The two methods are intertwined as advances in mathematical methods could result in novel algorithms for new computations. Together the mathematical and computational methods can make powerful impact in many practical areas of science as seen from the ensuing abstracts. These topics are quite varied and may appear to be too diverse yet the common link is the spirit of the symposium. It is hoped that by bringing together such diverse computational and mathematical scientific topics one can create fusion of ideals and emerge into new scientific frontiers that have not been visited before.

---

[1]. E-mail: balu@llnl.gov

## Krishnan Balasubramanian

Krishnan Balasubramanian (Balu) received his master's degree from Birla's Institute of Technology and Science (BITS), Pilani. Where he worked with a mathematician and deputy director, Prof. V. Krishnamurthy on "Combinatorial Enumeration of Chemical Isomers" Balu received his PhD from Johns Hopkins University in about 2 years in 1980 for his work on "Chemical Applications of Discrete Mathematics" under the sponsorship of Professor Walter S. Koski. Balu then moved to UC Berkeley in 1980 to work with Professor Kenneth S. Pitzer. At Berkeley, Balu focused on the art and science of relativistic quantum chemistry, an area that he has gotten addicted to since that time. After being in the faculty of Arizona State University, Tempe, AZ for over 18 years, Balu became the youngest Professor Emeritus at ASU at 43, and returned to the bay area to take up a senior position jointly at Lawrence Livermore National Lab, University of California Davis and Glenn T. Seaborg Center at Berkeley where is he continuing his active research now. Prof Balu is also an adjunct professor in the dept of Mathematics & computer science, College of Science at California State University, east bay, Hayward CA.

Balu has published nearly 500 research papers, two Wiley books on "Relativistic Effects in Chemistry". Balu has received an Alfred P. Sloan Fellowship (1984), Camille Henry Dreyfus Teacher-Scholar award (1985), Fulbright Distinguished professorships (1997), one of the most cited chemists in 1987-1997, Joint Prize of Ministry of Education, Poland (1998), LLNL associate director's distinguished service award (2003), Robert S. Mulliken Lecturer (2003), election to International Academy of Mathematical Chemistry (2005) and an award for distinction in graduate teaching (1991). Balu is on the editorial board of Journal of Mathematical Chemistry. Balu has served on the National Academy of Science panel for the Air force office of scientific research and US department of Energy panels. Balu has mentored over 50 graduate students, post-docs and visiting researchers. Balu's research is in the general area of theoretical, computational and mathematical chemistry.

Brill Academic Publishers
P.O. Box 9000, 2300 PA Leiden
The Netherlands

*Lecture Series on Computer
and Computational Sciences*
Volume 4, 2005, pp. 757-758

# Using DNA Knots To Assay Packing of DNA in Phage Capsids

J. Arsuaga[1], M. Vazquez[1], J. Roca[2], D.W. Sumners[3]

[1]Department of Mathematics,
San Francisco State University
San Francisco, CA USA 94132

[2]Department of Molecular Biology
Institut de Biologia Molecular de Barcelona
Consejo Superior Investigaciones Cientificas
08034 Barcelona, Spain

[3]Department of Mathematics
Florida State University,
Tallahassee, FL USA 32306

Received 6 July, 2005; accepted in revised form 3 August, 2005

*Abstract:* Bacteriophages are viruses that infect bacteria. They pack their double-stranded DNA genomes to near-crystalline density in viral capsids and achieve one of the highest levels of DNA condensation found in nature. Despite numerous studies some essential properties of the packaging geometry of the DNA inside the phage capsid are still unknown. Although viral DNA is linear double-stranded with sticky ends, the linear viral DNA quickly becomes cyclic when removed from the capsid, and for some viral DNA the observed knot probability is an astounding 95%. By comparing the observed viral knot spectrum with the simulated knot spectrum, we conclude that the packing geometry of the DNA inside the capsid is non-random and writhe-directed.

*Keywords:* Bacteriophage, Capsid, Chiral, DNA, Knots, Simulation, Writhe

*Mathematics Subject Classification:* 92C40, 57M25
*PACS:* 82.39.Pj

## 1. DNA Knots in Phage P4

Icosahedral bacteriophages pack their double-stranded DNA genomes to near-crystalline density and achieve one of the highest levels of DNA condensation found in nature. Despite numerous studies some essential properties of the packaging geometry of the DNA inside the phage capsid are still unknown. We present a new approach [1,2,3] to the problems of randomness and chirality of the packed DNA. We recently showed [2] that most DNA molecules extracted from bacteriophage P4 are highly knotted due to the cyclization of the linear DNA molecule confined in the phage capsid. Here we show [4] that these knots provide information about the global arrangement of the DNA inside the capsid. First, we analyze the distribution of the viral DNA knots by high-resolution gel electrophoresis. Next, we perform Monte-Carlo computer simulations of random knotting for freely jointed polygons confined to spherical volumes. Comparison of the knot distributions obtained by both techniques produces the first topological proof of non-random packaging of the viral DNA. Moreover, our simulations show that the scarcity of the achiral knot $4_1$ and the predominance of the torus knot $5_1$ over the twist knot $5_2$, observed in the viral distribution of DNA knots, cannot be obtained by confinement only, but must include writhe bias in the conformation sampling. These results indicate that the packaging geometry of the DNA inside the viral capsid is writhe directed.

## References

[1] S. Trigueros, J. Arsuaga, M.E. Vazquez, D.W. Sumners and J. Roca, *Novel display of knotted DNA molecules by two-dimensional gel electrophoresis*, Nucleic Acids Research **29**(2001), 67-71.

---

[3]Corresponding Author, summers@math.fsu.edu

[2] J. Arsuaga, M. Vazquez, S. Trigueros, D.W. Sumners and J. Roca, *Knotting probability of DNA molecules confined in restricted volumes: DNA knotting in phage capsids*, Proc. National Academy of Sciences US*A* **99**(2002), 5373-5377.

[3] J. Arsuaga, R.K-Z Tan, M. Vazquez, D.W. Sumners, S.C. Harvey, *Investigation of viral DNA packing using molecular mechanics models*, Biophysical Chemistry **101-102** (2002), 475-484.

[4] J. Arsuaga, M. Vazquez, P. McGuirk, D. W. Sumners, J. Roca, *DNA Knots Reveal Chiral Organization of DNA in Phage Capsids*, Proc. National Academy of Sciences USA **102**(2005), 9165-9169.

Brill Academic Publishers
P.O. Box 9000, 2300 PA Leiden
The Netherlands

*Lecture Series on Computer
and Computational Sciences*
Volume 4, 2005, pp. 759-764

# Relativistic Effects in the Chemistry of very Heavy and super heavy Molecules

K. Balasubramanian[1]

Chemistry & Material Science Directorate
Lawrence Livermore National Laboratory, PO Box 808, L-268
Livermore, CA 94550
And
California State University, East bay,
Hayward CA 94542

Received 15 July, 2005; accepted in revised form 7 August, 2005

*Abstract:* Relativistic effects are very significant for molecules and clusters containing very heavy and super heavy elements. We demonstrate this further with our recent results of relativistic computations that included complete active space multi-configuration interaction (CAS-MCSCF) followed by multi-reference configuration interaction (MRSDCI) computations with up to 50 million configurations of transition metal and main group clusters. We shall also be presenting our recent works on substituted fullerenes and actinide complexes of environmental concern. My talk will emphasize these unusual features and trends concerning structure and spectroscopic properties of these very heavy species. We compare the properties of not only the ground electronic states, but also several excited electronic states. We have also carried out extensive computations on very heavy clusters such as gold clusters; ruthenium clusters and assignment of the observed spectra have been suggested. It is shown that the gold clusters exhibit anomalous trends compared to copper or silver clusters. For example, Jahn-Teller distortion is quenched in the case of $Au_3$ by spin-orbit coupling, and for the first time the spin-orbit component of the $Au_3$ ground state has been observed experimentally. We have also carried out relativistic computations for the electronic states of the newly discovered super heavy elements and yet to be discovered elements such as 113 (eka-thallium) 114 (eka-lead) and $114^+$. Many unusual periodic trends in the energy separations of the electronic states of the elements 114, 113, $114^+$, (113)H, 114H, and Lawrencium and Nobelium compounds. We will be presenting our results on uranyl, plutonyl and meputunyl complexes in aqeous solution using a combined quantum chemical and PCM models for solvation. We have employed coupled cluster levels of theory to obtain the frequencies and equilibrium geometries of these complexes.

*Keywords:* Relativistic Effects, Very heavy Clusters, super heavy clusters

## 1. Relativistic Effects

This is the year of centennial celebration of publication of Einstein's famous 1905 paper on special theory of relativity. Indeed it is an exciting and very fitting that the celebrated work of Einstein penetrated many branches of science and changed our thinking in many ways. In atomic theory and more recently in chemical context, it is now well recognized that Einstein's special theory of relativity becomes important particularly in the properties of very heavy and superheavy elements and their molecules, as inner electrons in these species travel with speeds comparable to the speed of light. This is a consequence of a large nuclear charge, which is the driving force of the inner electrons, which speed up in order to keep balance with the increased electrostatic attraction. For example, the speed of the 1s electron of gold, which has 79 protons is estimated to be ~60% of the speed of light. Thus to treat the properties of such species one needs to embrace both relativity and quantum mechanics, the latter was viewed with skepticism by Einstein. Nevertheless the combination of two theories provides

[1]. E-mail: balu@llnl.gov

the founding blocks of modern relativistic quantum mechanics, which is most relevant in dealing with the properties of molecules containing very heavy and superheavy elements.

Relativistic effects are defined as differences in the observable properties of electrons as consequence of the correct speed of light, as opposed to infinite speed. Many relativistic quantum approaches start with the relativistic analog of the Schrödinger equation well known as the Dirac equation. Then electron correlation effects are introduced in some combination with relativistic techniques since relativistic and electron correlation effects can be coupled in such species with very heavy elements. Although relativistic effects are more important for the core electrons, the valence electrons too experience such relativistic effects as to cause substantial differences in the chemical and spectroscopic properties of atoms and molecules containing heavy atoms such as third row transition metal species, sixth-row main group elements, actinides and superheavy elements.

Relativistic effects can cause substantial changes to te properties of molecules containing very heavy atoms[1,2]. For example, the contraction of the 6s valence orbital of the gold atom leads to shorter bonds in gold clusters and gold compounds the ionization potentials of the copper, silver and gold atoms. As one goes down a given group in the periodic table, one expects a monotonic decrease in the ionization energy, since the outermost electron is farther from the nucleus. However, gold is an anomaly in that its ionization potential is not only higher than silver but also copper. This anomaly of the gold atom is due to the relativistic stabilization of the outermost 6s electron of the atom, thus, making it strongly bound, leading to higher ionization energy. Yellow color of gold is attributed to relativity.

The inertness of the $6p^2$ shell of the lead atom and even more so the 7p2 shell of element 114 are both due to both mass-velocity and spin-orbit relativistic effects. The relativistic stabilization of the 6s orbital of mercury leads to a larger 6s-6p promotion energy in Hg compared to Cd and Zn. The increase for mercury is a consequence of relativistic stabilization of the $6s^2$ shell, which leads to a larger $^1S$-$^1P$ separation for Hg compared to Cd. The $6s^2$ shell cannot, thus, form a very strong bond, unless promotion into 6p is achieved. This is the reason for the fact that $Hg_2$, in its ground state, forms only a van der Waals complex, while in its excited state, it is bound. Mercury is thus a liquid at room temperature due to the formation of weak clusters of mercury atoms, which undergo metal-nonmetal transition by hybridization with 6p as the cluster size increases.

Spin-orbit coupling is another relativistic effect, which can alter the reactivity and spectroscopic properties of very heavy and superheavy species. The spin-orbit coupling can increase or decrease the bond lengths depending on, which states mix. Likewise it can destabilize the chemical bonding as in the case of $Bi_2$, which forms a less stable bonding than a triple bond. On the other hand, the PtH molecule is considerably more stable compared to PdH. The ground state of the platinum atom is a triplet arising from $5d^96s^1$, while it is singlet for Pd arising from the $4d^{10}$ configuration. This is primarily due to the relativistic stabilization of the 6s orbital of Pt, which overcomes the enhanced stability attributed to the closed-shell $d^{10}$ configuration. Consequently, the differences in the chemistry of Pt and Pd containing systems arise from this and the larger spin-orbit splitting in Pt compared to Pd. The entire third row transition metal atoms react more compared to the second row atoms due to relativity. We will demonstrate analogy between mercury and element 114.

In mathematical terms relativity results in a "four-vector" formalism of angular momentum that couples both the spin and spatial angular momenta via spin-orbit coupling and thus a relativistic electron has memory of only total angular momentum symmetry. This results in novel double group symmetry due to the spin-orbit coupling **L.S** operator. Since this operator changes sign upon rotation by 360°, the periodicity of the identity operation is broken and is thus no longer the identity operation of the group, as exemplified in Figure 1 with a Mobius strip, as one completes a 360° rotation along the Mobius surface there is a sign change since one goes from inside of the surface to the outside. This requires the introduction of a new operation R in the normal point group of a molecule that corresponds to the rotation by 360° which is not equal to E, the identity operation. Hence we have the double group and double-valued representations in relativistic quantum chemistry of molecules with heavy atoms.

Fig 1 A Mobius strip depiction for the double group symmetry of a relativistic hamiltonian. The introduction of spin-orbit coupling into the relativistic Hamiltonian changes the periodicity of the normal point group symmetry into a double group symmetry, as rotation by 360 deg is not the identity operation.

The author[3] has recently formulated the symmetry double groups of non-rigid molecules recently. The double group character tables of such species have been derived using wreath product groups and their double groups. These character tables grow astronomically in sizes due to large number of operations in the double group. It may also be noted that the double group is not a direct product of the normal single group and the two-valued operational groups. It is this aspect hat introduces even-dimensional two-valued representations into the double group. Thus the relativistic spinor representations of molecules containing heavy atoms do not conform ot normal spin and spatial symmetries. Two electronic states that have the same symmetry or transform as the same double group representation can mix regardless of their spin multiplicities and spatial symmetries.

We provide here two different set of species to demonstrate the importance of relativity in the chemistry and bonding of very heavy and superheavy species. The first set of species is from the compounds of late actinides namely, compounds of Nobelium and Lawrencium[4]. The compounds of Nobelium and lawrencium surprisingly exhibit unusual non-actinide properties in that the chemistry of these species is principally determined by the 7s and 7p orbitals rather than the 5f or 6d shells. Since hydrides are the simplest of all species, we have considered high-level relativistic computations for the lawrencium and nobelium dihydrides. The ground and first excited states of lawrencium and nobelium arise from the 7s and 7p shells, and thus the potential energy surfaces of these species are unusual in having 7p characteristics. Both molecules form stable bent ground states reminiscent of a $sp^2$ hybridization with equilibrium bond angles near 120°. The lawrencium compounds exhibit unusual characteristics due to avoided crossings of the potential energy surfaces. As a result of spin-orbit coupling, the $^2B_2$ state of $LrH_2$ undergoes avoided crossing with the $^2A_1$ state in the spin double group, which reduces the barrier for insertion of Lr into $H_2$. The Nobelium compounds are considerably less stable compared to the lawrencium compounds due to the relativistic stabilization of the 7s shell of the nobelium atom. The barrier for the insertion of Lr into $H_2$ is lowered by relativity (spin-orbit coupling) while No has to surpass a larger barrier due to the relativistic stabilization of the $7s^2$ shell. Lawrencium is the only element in the actinide series with unusually low ionization potential, and $NoH_2$ has an unusually large dipole moment of 5.9 Debye. We have found that the lawrencium and nobelium compounds have periodic similarities to the thallium and radium compounds, respectively.

The ground state of Lr is not $7s^26d$ as listed in many websites, for example, the Los Alamos periodic table, but $7s^27p^1$. One could have expected Lr to be similar to Ac or La but in fact the change of this configuration makes it different. The $^2D_{3/2}$ state of Lr is however close to the $^2P_{1/2}$ state. That is, the $^2D_{3/2}$-$^2P_{1/2}$ splitting seems quite sensitive to the level of theory. In any case we find that the ground state of Lr is the $^2P_{1/2}$ state arising from the 7s7p configuration and the spin-orbit splitting of 7p as measured by the $^2P_{3/2}$-$^2P_{1/2}$ energy separation is 7790 cm$^{-1}$. The ionization potential of Lr is 4.87 eV, substantially smaller than No or any other actinide. Thus unusually low ionization energy of Lr would result in the ionic character of the Lr compounds. The large drop in the IP from No to Lr is a consequence of relativistic stabilization of the 7s orbital of Lr which leads to very stable Lr$^+$ with $7s^2$

closed-shell configuration, while in the case of No the ground state of the neutral No is stabilized due to the $7s^2$ configuration of No relative to $No^+$. Thus the IP of No is substantially larger than Lr. Our computations on the energy levels and atomic states of Lr and $Lr^+$ are consistent with the previous relativistic Fock space coupled cluster study of Eliav et al[5]

Figure 2 shows our computed potential energy surfaces of the electronic states of $LrH_2$ in the

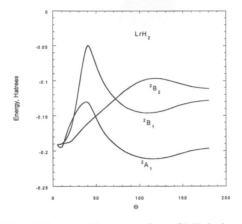

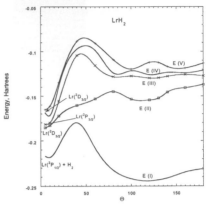

Figure 2 The potential energy surfaces of $LrH_2$ in the absence of spin-orbit coupling

Figure 3 The computed potential energy surfaces of $LrH_2$ including spin-orbit effects

absence of spin-orbit coupling. All three lowest double electronic states considered here correlate into the $Lr(^2P) + H_2$ dissociation limit in the absence of spin-orbit coupling. Among these $^2A_1$ state is the lowest, which forms a stable ground state near a bond angle of 120° and this minimum is more stable than the $Lr(^2P) + H_2$ dissociation limit. The ground state of $LrH_2$ is 12 kcal/mole more stable than $Lr(^2P) + H_2$ at the CASSCF level and 16 cal/mole more stable at the second-order CI level. Figure 3 shows the computed potential energy surfaces of $LrH_2$ including spin-orbit effects. Comparison of thse potential energy surfaces reveals dramatic differences. These differences arise from avoided crossings introduced by spin-orbit coupling. As seen from Fig.2 the $^2B_2$ curve crosses the $^2A_1$ curve. Both states correlate into the same E representation in the double group and thus the $^2A_1$ state with open-shell spin can strongly couple with the $^2B_2$ state with spin in the region of curve crossing. The near proximity of the $^2B_1$ state near the dissociation limit could also introduce strong mixing due to spin-orbit coupling. Thus the $^2B_2$ state and $^2A_1$ state undergo avoided crossing, which leads to the lowering of the barrier for insertion of Lr into $H_2$.

The potential energy surfaces dissociate quite differently when spin-orbit effects are included, as can be seen from Figure 2. The ground state of $LrH_2$ including spin-orbit effects dissociates into $Lr(^2P_{1/2}) + H_2$. The first excited state arises from the $Lr(^2D_{3/2}) + H_2$ dissociation, and the potential energy surface looks very different from the corresponding $^2B_1$ state in the absence of spin-orbit effects due to avoided crossings. The curve has a substantially smaller barrier followed by a shallow minimum. The potential energy curves arising from $Lr(^2D_{5/2})$ and $Lr(^2P_{3/2})$ exhibit substantially larger barriers, and all of these states form obtuse minima. The substantial differences and the shapes of the potential energy surfaces can be rationalized by consideration of the composition of the electronic states including spin-orbit effects as a function of the H-Lr-H bond angle. At = 20°, the E(I) state is composed of 61% $^2B_2$, 6.5% $^2B_1$, 9% $^4B_2 + ^2B_2$ ($1a_1^2 2a_1 3a_1 1b_2$). At = 40° this state becomes 80% $^2B_2$, 4% $^4A_1$, and 3% $^2B_2$ (II), but at = 60° this state is 92% $^2A_1$ 0.8% $^2B_2$, and 0.5% $^2B_1$. Thus the $^2A_1$ and $^2B_2$ states undergo avoided crossing near 50° regions. Near the minimum bond angle of 118°, the E(I) state becomes 94% $^2A_1$ ($1a_1^2 2a_1 1b_2^2$), 0.5% $^2B_1$ The dominant contribution remains the same at longer bond distances but the second important contributor changes to $^4A_2$ near linear geometries. It is also interesting to note that at the linear limit the state attains substantial Lr (6d) character as the g orbital arises from the interaction of the Lr(6d ) with the hydrogen 1s orbitals all with same signs. This contrasts with the bent geometry for which the Lr (7p) makes a substantial contribution. The E(II) state is 59% $^2B_1$, 10% $^2B_2$ (II), 6% $^2B_2$(III), 5% $^2A_1$ at = 20° but at = 40° this state becomes 77% $^2A_1$, 9% $^2B_2$, and 3% $^2B_2$. At = 60° this state becomes 80% $^2B_2$, 3% $^2B_2$ (II), 1% $^2A_1$. Near the minimum ( = 120° ) this state becomes 34% $^2B_1$, 32% $^2B_1$ (II), 7% $^2A_1(6a_1)$, an d3% $^2A_1(5a_1)$. Near the linear geometry this state becomes 43% $^2B_1$, 37% $^2A_1$ (II) so that the state would correlate into a state of 3/2. The analysis of the selected states in the double group reveals the complexity of the electronic states and how the states vary as a function of the bond angle and geometry. The number of avoided crossings exhibited by these electronic states including spin-orbit effects leads to the shapes of the potential energy curves in Figure 2.

Figure 4 shows our computed potential energy surfaces of $NoH_2$ in the absence of spin-orbit coupling while Fig 5 shows the corresponding curves with spin-orbit effects. Since the No atom has a closed-shell $5f^{14}7s^2$ configuration in contrast to the open-shell configuration of Lr, we expect No to be less reactive with $H_2$. The $No(^1S_0)$ state does not insert into $H_2$ as it has to surpass a large barrier in the absence of spin-orbit coupling. On the other hand, the $^3B_2$ state crosses the $^1A_1$ state before the barrier is reached, and thus spin-orbit effect may have substantial impact in reducing the barrier height. The $^3A_1$ and $^3B_1$ states are considerably higher in energy and they arise from the excited $No(^3P) + H_2$ species. Among these the $^3A_1$ state forms a shallow minimum in the obtuse bond angle. Comparison of Figs4 and 5 exemplify the differences due to spin-orbit coupling in these electronic states.

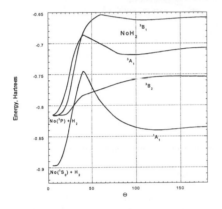

Figure 4 The potential energy surfaces of NoH2 in the absence of spin-orbit coupling

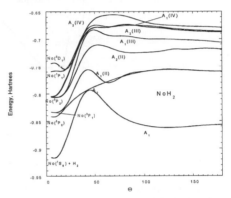

Figure 5 The potential energy surfaces of NoH2 in the absence of spin-orbit coupling

We have been interested in relativistic computations of the properties of molecules containing superheavy elements such as 113[6] and 114. We have computed the electronic states of (113)H, the eka-thallium hydride and 114H$_2$ the eka-lead dihydride. We have demonstrated that the 6d-electron correlation-spin-orbit effects are so large for (113)H that they lead to significant shortening of the (113)-H bond and stabilization of the bond. It is shown that the periodic trends of (113)H are such that (113)H has a knight's move relation to AuH in exhibiting unusual stability and d correlation-spin-orbit effects. We have also shown that (114)H$_2$ exhibits similarity to HgH$_2$ and also considerable spin-orbit coupling that mixes the singlet ($^1A_1$) state of (114)H$_2$ with the triplet state ($^3B_1$). The computed potential energy curves of (113)H are shown below on Fig. 6

An unusual feature of the ground state of (113)H is dramatic shortening of the bond length by valence spin-orbit coupling and core-valence spin-orbit coupling. That is, the $r_e$ value of the $0^+$ ground state is 1.782 Å when no excitations from the 6d shells of 113 are allowed. The corresponding dissociation energy is 1.45 eV. However when excitations are allowed from the 6d orbitals from a multi-reference set of configurations, there is significant bond contraction and stabilization. The origin of the contraction can be understood through analysis of the composition of the $0^+$ state. The $^3$ $(0^+)$ state originating from the excitation of the 6d ( ) shell to 7p ( ) orbital is the most important contributor to the contraction of the (113)-H bond. This leads to 4 real configurations, viz., 6d( ) $1_x$..6d( ) $1_x$..6d( ) $1_y$..and6d( ) $1_y$. in addition to other reference configurations discussed already. These reference configurations couple with other valence multi-configurations causing the contraction of the (113)-H bond. Consequently this is a core-valence multi-reference spin-orbit and correlation effect. We have also considered the importance of relativity in bonding and spectroscopic properties of actinide complexes of relevance to the environment.[7-10] Since these topics have been considered in depth elsewhere, we refer to these works for further details concerning such actinide complexes.

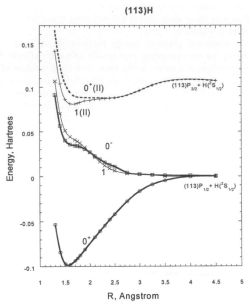

Figure 6 The potential energy surfaces (113)H with of spin-orbit coupling.

## Acknowledgments

This research was performed under the auspices of the US department of Energy by the University of California, LLNL under contract number W-7405-Eng-48 while the work at California State university was supported by National Science Foundation and basic energy sciences division of US department of energy.

## References

[1] K. Balasubramanian, "Relativistic Double Group Spinor Representations of Non-rigid Molecules", *Journal of Chemical Physics*, **120**, 5524-5535(2004)

[2] K. Balasubramanian, *Relativistic Effects in Chemistry, Part A: Theory and Techniques*, Wiley-interscience, New York, NY p301 1997.

[3] K. Balasubramanian, *Relativistic Effects in Chemistry, Part B: Applications*, Wiley-interscience, New York, NY p527 1997.

[4] K. Balasubramanian, *J. Chem. Phys.*, **116**, 3568 (2002).

[5] E. Eliav, U. Kaldor and Y. Ishiwata, *Phys., Rev. A.*, **52**, 291 (1995).

[6] K. Balasubramanian, *Chem. Phys. Lett.*, **361**, 297 (2002).

[7] Z. Cao and K. Balasubramanian, "Theoretical Studies of hydrated complexes of uranyl, neptunyl and plutonyl in aqueous solution: $UO_2^{2+}(H_2O)_n$, $NpO_2^{2+}(H_2O)_n$, and $PuO_2^{2+}(H_2O)_n$", J. Chem. Phys. (2005)

[8] D. Chaudhuri and K. Balasubramanian, "Electronic Structure and Spectra of Plutonyl Complexes and their hydrated forms: $PuO_2CO_3$ and $PuO_2CO_3.nH2O$ (n=1,2)", *Chemical Physics Letters*, 399, 67-72 (2004)

[9] D. Majumdar and K. Balasubramanian, "Theoretical studies on the nature of uranyl–silicate, uranyl–phosphate and uranyl–arsenate interactions in the model $H_2UO_2SiO4 \cdot 3H_2O$, $HUO_2PO_4 \cdot 3H_2O$, and $HUO_2AsO_4 \cdot 3H_2O$ molecules", *Chemical Physics Letters, 397*, 26-33 (2004).

[10] K. Balasubramanian, "Relativity and the Periodic Table", (D. Rouvary & R. B. King, editors), Periodic Table into the 21st Century, 2004

Brill Academic Publishers
P.O. Box 9000, 2300 PA Leiden
The Netherlands

*Lecture Series on Computer*
*and Computational Sciences*
Volume 4, 2005, pp. 765-767

# Computational and Mathematical Approaches to Clusters, Fullerenes and Proteomes

K. Balasubramanian[1]

Chemistry & Material Science Directorate
Lawrence Livermore National Laboratory, PO Box 808, L-268
Livermore, CA 94550
And
California State University, Eastbay,
Hayward CA 94542

Received 15 July, 2005; accepted 28 July, 2005

*Abstract:* We propose to consider several exciting applications of computational and mathematical approaches to cluster with special emphasis on combinatorics, graph theory based algorithms to a number of scientific disciplines, particularly in fullerene chemistry and computational biology. We shall demonstrate that enumerative combinatorics and algorithms find numerous applications to enumeration of structure sand three dimensional fullerene cages and their nuclear spin statistics and spectroscopy. We propose to consider cross fertilization of graph theory with chemistry and biology which seems to provide a fertile ground for interdisciplinary research in bioinformatics, predictive toxicology, molecular and drug design. Examples of applications from each of these fields will be provided and would include DNA algorithm and graph theoretical characterization of proteomics and genomics

*Keywords:* Combinatorics of fullerenes, spectroscopy fullerenes, proteomics, graph theory

*Mathematics Subject Classification:* 92B99, 92E10

*PACS:* 31.15.Ar, 87.10.+e

## 1. Introduction

Computational and mathematical techniques have made substantial impact in all branches of science, particularly in chemistry and biology in recent years. With the advent of supercomputers and very powerful desktop workstations, it is becoming increasingly feasible to carry out sophisticated computations concerning electronic structure, spectroscopy and dynamics of many large molecules and chemical systems. It is also noteworthy that these computers have made it possible to solve some of the complex and demanding algorithms and applications concerning combinatorics and graph theory which could not be done otherwise. In fact some unsolved mathematical problems in the field of graph theory and combinatorics could be addressed with such powerful computers. We show here that a combination of graph theory, combinatorics and computational algorithms result in exciting solutions to a number of problems concerning fullerene cages, their spectroscopy, nuclear spin statistics and so on.

Applications of quantum chemistry, combinatorics and graph theory to biological systems and predictive toxicology[1-2] are emerging in recent years. We show that the application of graph theory combined with complex algebra can provide powerful tools for the characterization of proteome patterns and response of proteomes induced by externally applied chemicals. Graph theory which is basically algebra of connectivity can provide powerful tools and algorithms for complex patens such as the proteome of a living cell. Typically the proteome is characterized by a two-d gel electrophoresis technique which involves separation of a complex set of proteins in the proteome into individual

---

[1]. E-mail: balu@llnl.gov

proteins on the basis of their mass and charge. Graph theory provides a powerful method to characterize the proteome in terms of mass and charge and relative abundances.

We have applied mathematical techniques to main-group molecular structures such as water clusters, clusters of TATB, etc[3-6]. We have also carried out methods developments for these species. We provide here only a brief of summary of highlights. Doped fullerene cages are of recent interest as novel materials for faster devices, lasers and high-energy materials. Doping fullerenes with Group 13 and 15 elements have received particular attention. We have considered the vibrational and rovibronic spectra of $C_{48}N_{12}$ cages and the group theoretical analysis of the nuclear spin statistics of these cages as $^{14}N$ nuclei are spin-1 particles.[4] A recent experimental work indicated the possibility of 3 stable isomers of $C_{60}H_{36}$ shown on Fig. 1.

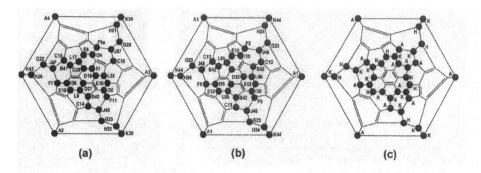

(a)                              (b)                              (c)

Fig 1 Three Isomers of $C_{60}H_{36}$.

We obtained the computed ESR spectra of these isomers to contrast them as shown in Fig 2 for 2 of the isomers.[5]

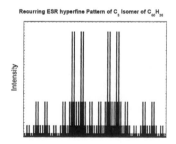
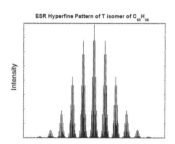

Relativistic double group spinor representations were developed for non-rigid molecules containing very heavy atoms.[6] The representations facilitate spinor wavefunctions for non-rigid molecules containing very heavy atoms. Moreover the rovibronic levels of such species would be split by both torsional tunneling and spin-orbit coupling. Thus it is necessary to have such a formalism. We have developed these techniques for the first time. Several group theoretical developments have also been carried for the development of group theory of water pentamer.[3]

We have combined mathematical and computational quantum chemical computations to predict biological phenomena. In particular we have been focusing our efforts in two areas, one concerns with toxicity prediction[1,2] and the other deals with the characterization of proteome and its response to applied chemicals. The toxicity of chemicals that have received considerable attention in the field of environmental chemistry. Both halocarbons and dioxins are known to be harsh pollutants of the environment and it is important to understand the mechanisms of their toxicity and which ones are most harmful. Thus our quantum chemical computations of the main group species had environment and atmospheric chemistry as their central theme and thus we focused our work on halocarbons[1], polychlorinated Dibenzo-*p*-dioxins, and dibenzo-p-dioxins.[2]. In the case of halocarbons we considered 55 species with known toxicity and we used quantum chemical and other statistical methods to predict and correlate the toxicity of these species using electron attachment as the key mechanism governing heptotoxicity of these species.

In collaboration with Basak and coworkers, We have developed a complex graph matrix representation to characterize proteomics maps obtained from 2d-gel electrophoresis. In this method each bubble in a 2d-gel proteomics map is represented by a complex number with components being charge and mass. Then a graph with complex weights is constructed by connecting the vertices in the relative order of abundance. This yields adjacency matrices and distance matrices of the proteomics graph with complex weights. We have computed the spectra, eigenvectors and other properties of complex graphs and the Euclidian/graph distance obtained from the complex graphs. The leading eigenvalues, eigenvectors and likewise the smallest eigenvalues and eigenvectors and the entire graph spectral patterns of the complex matrices derived from them yield novel weighted biodescriptors that characterize proteomics maps with information of charge and masses of proteins. We have also applied these eigenvector and eigenvalue maps to contrast the normal cells and cells exposed to four peroxisome proliferators, namely, clofibrate, DEHP, PFDA and PFOA.

## Acknowledgments

This research was performed under the auspices of the US department of Energy by the University of California, LLNL under contract number W-7405-Eng-48 while the work at California State university was supported by National Science Foundation and basic energy sciences division of US department of energy.

## References

[1] S. C. Basak, K.Balasubramanian, B.D. Gute, D. Mills, A. Gorczynska, and S. Roszak, "Prediction of cellular toxicity of halocarbons from computed chemodescriptors: A hierarchical QSAR approach", *J. Chem. Inf. Comput. Sci.* **43**, 1103-1109 (2003)

[2] J. Lee, W. Choi, B. J. Mhin, K. Balasubramanian, "Theoretical Study on the Reaction of OH Radicals with Polychlorinated Dibenzo-*p*-dioxins", J. Phys. Chem A., **108**, 607-614 (2004)

[3] K. Balasubramanian, "Non-rigid Group Theory, Tunneling Splittng and Nuclear Spin Statistics of Water Pentamer: $(H_2O)_5$", *J. Physical Chemistry. A*, **108**, 5527-5536 (2004)

[4] K. Balasubramanian, "Group Theoretical Analysis of Vibrational Modes and Rovibronic Levels of extended aromatic $C_{48}N_{12}$ Azafullerene", *Chemical Physics Letters*, **391**, 64-68 (2004)

[5] K. Balasubramanian, "Combinatorics of NMR and ESR Spectroscopy of $C_{60}H_{36}$ Isomers", *Chemical Physics Letters,* **400**, 78-85 (2004)

[6] K. Balasubramanian, "Relativistic Double Group Spinor Representations of Non-rigid Molecules", *Journal of Chemical Physics*, **120**, 5524-5535(2004)

Brill Academic Publishers
P.O. Box 9000, 2300 PA Leiden
The Netherlands

*Lecture Series on Computer
and Computational Sciences*
Volume 4, 2005, pp. 768-771

# Use of Proteomics Based Biodescriptors Versus Chemodescriptors in Predicting Halocarbon Toxicity: An Integrated Approach

S.C. Basak[1]

Center for Water and the Environment,
Natural Resources Research Institute,
University of Minnesota Duluth,
Duluth, MN 55811, USA

Received 9 July, 2005; accepted in revised form 24 July, 2005

*Abstract:* Halocarbons constitute an important group of pollutants. Therefore, prediction of potential toxicity of this group of chemicals is important in the management of human health and environment. Attempts have been made by various researchers to predict toxicity of halocarbons from their computed chemodescriptors. On the other hand, experimental proteomics data on the effects of these chemicals gives us information about their mode of action at the level of genomic transcription, translation, and post-translational processes. Our research team at NRRI has developed a number of schemes for the quantification of proteomics patterns in terms of a novel class of descriptors called biodescriptors by us. This presentation will deal with the relative importance of chemodescriptors and biodescriptors in the predictive toxicology of halocarbons.

*Keywords:* 2-DE gel electrophoresis, biodescriptor, chemodescriptor, proteomics, toxicoproteomics

*Mathematics Subject Classification:* 92B99, 92E10

*PACS:* 31.15.Ar, 87.10.+e

## 1. Use of Proteomics Based Biodescriptors Versus Chemodescriptors

An important aspect of contemporary computational toxicology is the prediction of toxicity of chemicals from properties which can be calculated directly from their structure without the input of any other experimental data. While traditional experts over the decades have relied on the use of bioassay data and a suite of experimentally determined relevant physicochemical properties for hazard assessment of chemicals, this process is becoming more and more difficult with time. One of the reason is that there are very large number of chemicals that need to be evaluated for their toxic effects on human and environmental health and most of them have very little or no experimental data. For example, the Toxic Substances Control Act (TSCA) Inventory, the data base of industrial chemicals in USA, has over 81,000 entries and the number is increasing by 2,000 to 3,000 per year. More than 50% of TSCA chemicals have no data at all. A very small minority, ~15%, have mutagenicity, genotoxicity or chronic toxicity data.

One viable approach to address this quagmire has been to develop models for the estimation of toxicity and toxicologically relevant properties from calculated molecular descriptors. A molecular descriptor is a mapping of the molecular structure to the set of real numbers. In other words, one represents the molecule using a chosen model, *e.g.*, graph theoretic, 3-D or quantum chemical, and a mathematical model associates a number or a set of numbers to the molecular structure. Such descriptors are based on size, shape, complexity, symmetry, and stereoelectronic aspects of molecular architecture.

It has to be understood that any representation or modeling of a chemical species is an exercise in abstraction. This necessarily emphasizes certain aspects of the molecular architecture, ignoring, at the same time, other features based on the experience of the individual scientist. As Hans Primas put it: "In order to describe an aspect of holistic reality, we have to ignore certain factors such that the remainder separates into facts. Inevitably, such a description is true within the adopted partition of the world, that

---

[1] Corresponding author. E-mail: sbasak@nrri.umn.edu

is, within the chosen context." The different models of molecular structure, *e.g.*, topostructural, topochemical, geometrical, and quantum chemical, quantify different aspects of the reality represented by the molecule.

It is possible that the various molecular descriptors developed by different formalisms quantify non-overlapping, some time complementary, aspects of molecular reality. So, it makes sense that we begin the modeling process using the simplest descriptors and add more complicated/ resource intensive ones if the simpler descriptors do not provide reasonable models. Our research team has taken such an approach, called the hierarchical quantitative structure-activity relationship (HiQSAR) approach [1–10]. Sometimes it is possible that predictive models based only on theoretical descriptors cannot capture very complex aspects of toxicity and toxic modes of action. In the post-genomic era, catapulted by the Human Genome Project, the sciences of genomics, proteomics, and metabolomics can provide a lot of information relevant to biomedicinal and toxic effects of chemicals. Because these "omics" techniques can produce a lot of information, Basak and collaborators have embarked on the development of various approaches in the compact quantification of relevant information from toxicoproteomics data: 1) graphs/ matrices associated with proteomics maps [11–20]; 2) information theoretic biodescriptors [21, 22]; 3) biodescriptors from spectrum-like representations of proteomics maps [23]; and 4) statistical approaches to discover critical protein biomarkers [24].

Taking chemodescriptors and biodescriptors together, the hierarchical QSAR approach in the post-genomic era may be summarized by the following figure:

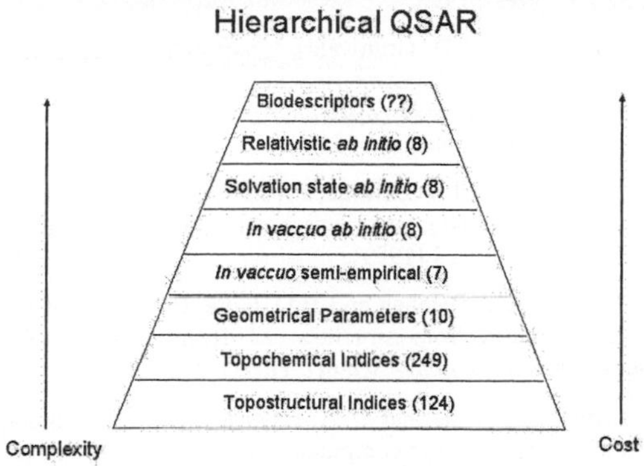

## Hierarchical QSAR

Figure 1: Hierarchical approach to modeling with chemodescriptors and biodescriptors.

Halocarbons are an important class of industrial chemicals used in large quantities both in USA and Europe. So, it is important to carry out hazard assessment of these chemicals. These chemicals have been tested using various unicellular organisms and hepatocytes. The hepatocytes have been analyzed for the effects of these chemicals on the proteomics patterns using 2-D gel electrophoresis.

This presentation will discuss the utility of calculated chemodescriptors versus proteomics biodescritors in predicting toxicity of halocarbons. It will also explore how far an integrated QSAR (I-QSAR) involving both chemo and biodescriptors can be more powerful as compared with QSARs using either of the two classes of descriptors.

## Acknowledgments

This manuscript is contribution number 385 from the Center for Water and the Environment of the Natural Resources Research Institute. This material is based on research sponsored by the Air Force Research Laboratory, under agreement number F49620-02-1-0138. The U.S. Government is authorized to reproduce and distribute reprints for Governmental purposes notwithstanding any copyright notation thereon.

The views and conclusions contained herein are those of the authors and should not be interpreted as necessarily representing the official policies or endorsements, either expressed or implied, of the Air Force Research Laboratory or the U.S. Government.

## References

[1] S.C. Basak, B.D. Gute and G.D. Grunwald, A comparative study of topological and geometrical parameters in estimating normal boiling point and octanol–water partition coefficient, *Journal of Chemical Information and Computer Science* **36** 1054–1060 (1996).

[2] B.D. Gute and S.C. Basak, Predicting acute toxicity of benzene derivatives using theoretical molecular descriptors: a hierarchical QSAR approach, *SAR and QSAR in Environmental Research* **7** 117–131 (1997).

[3] S.C. Basak, B.D. Gute and G.D. Grunwald, Use of topostructural, topochemical, and geometric parameters in the prediction of vapor pressure: a hierarchical approach, *Journal of Chemical Information and Computer Science* **37** 651–655 (1997).

[4] S.C. Basak, B.D. Gute and G.D. Grunwald, Relative effectiveness of topological, geometrical, and quantum chemical parameters in estimating mutagenicity of chemicals, *Quantitative Structure–Activity Relationships in Environmental Sciences VII*, (Editors: F. Chen, G. Schuurmann), SETAC Press, 1998.

[5] S.C. Basak, B.D. Gute and G.D. Grunwald, A hierarchical approach to the development of QSAR models using topological, geometrical and quantum chemical parameters, *Topological Indices and Related Descriptors in QSAR and QSPR*, (Editors: J. Devillers, A.T. Balaban), Gordon and Breach Science Publishers, 1999.

[6] B.D. Gute, G.D. Grunwald and S.C. Basak, Prediction of the dermal penetration of polycyclic aromatic hydrocarbons (PAHs): a hierarchical QSAR approach, *SAR and QSAR in Environmental Research* **10** 1–15 (1999).

[7] S.C. Basak, D.R. Mills, A.T. Balaban and B.D. Gute, Prediction of mutagenicity of aromatic and heteroaromatic amines from structure: a hierarchical QSAR approach, *Journal of Chemical Information and Computer Science* **41** 671–678 (2001).

[8] S.C. Basak, D. Mills, B.D. Gute and D.M. Hawkins, Predicting mutagenicity of congeneric and diverse sets of chemicals using computed molecular descriptors: a hierarchical approach, *Quantitative Structure–Activity Relationship (QSAR) Models of Mutagens and Carcinogens*, (Editor: R. Benigni), CRC Press, 2003.

[9] S.C. Basak, K. Balasubramanian, B.D. Gute, D. Mills, A. Gorczynska and S. Roszak, Prediction of cellular toxicity of halocarbons from computed chemodescriptors: a hierarchical QSAR approach, *Journal of Chemical Information and Computer Science* **43** 1103–1109 (2003).

[10] B.D. Gute, K. Balasubramanian, K. Geiss and S.C. Basak, Prediction of halocarbon toxicity from structure: a hierarchical QSAR approach, *Environmental Toxicology and Pharmacology* **16** 121–129 (2004).

[11] M. Randić, F. Witzmann, M. Vračko and S.C. Basak, On characterization of proteomics maps and chemically induced changes in proteomes using matrix invariants: Application to peroxisome proliferators, *Medicinal Chemistry Research* **10** 456–479 (2001).

[12] M. Randić and S. C. Basak, A comparative study of proteomics maps using graph theoretical biodescriptors, *Journal of Chemical Information and Computer Science* **42** 983–992 (2002).

[13] M. Randić, J. Zupan, M. Novič, B.D. Gute and S.C. Basak, Novel matrix invariants for characterization of changes of proteomics maps, *SAR and QSAR in Environmental Research* **13** 689–703 (2002).

[14] Ž. Bajzer, M. Randić, D. Plavšić and S.C. Basak, Novel map descriptors for characterization of toxic effects in proteomics maps, *Journal of Molecular Graphics and Modelling* 22 1–9 (2003).

[15] M. Randić, N. Lerš, D. Plavšić and S.C. Basak, On invariants of a 2-D proteome map derived from neighborhood graphs, *Journal of Proteome Research* 3 778–785 (2004).

[16] M. Randić, N. Lerš, D. Plavšić and S.C. Basak, Characterization of 2-D proteomics maps based on spots nearest neighborhoods, *Croatica Chemica Acta* 77 345–351 (2004).

[17] Ž. Bajzer, S.C. Basak, M. Vračko Grobelsek and M. Randić, Use of Proteomics Based Biodescriptors in the Characterization of Chemical Toxicity, *Genomic and Proteomic Applications of Toxicity Testing*, (Editor: M.J. Cunningham), Humana Press, in press.

[18] M. Randić and S. C. Basak, On similarity of proteome maps, *Medicinal Chemistry Research*, in press.

[19] M. Randić, N. Lerš, D. Vukičević, D. Plavšić, B.D. Gute and S.C. Basak, Canonical labeling of proteome maps, *Journal of Proteome Research*, in press.

[20] M. Randić, F.A. Witzmann, V. Kodali and S.C. Basak, On the dependence of a characterization of proteomics maps on the number of protein spots considered, *Journal of Chemical Information and Modeling*, submitted.

[21] S.C. Basak, B.D. Gute and F.A. Witzmann, Application of information-theoretic biodescriptors in toxicity prediction, *Journal of Chemical Information and Modeling*, submitted.

[22] S.C. Basak, B.D. Gute and F.A. Witzmann, Information-theoretic Biodescriptors for Proteomics Maps: Development and Applications in Predictive Toxicology, *Proceedings of the 9th WSEAS International Conference on Computers*, in press.

[23] M. Vračko and S.C. Basak, Similarity study of proteomic maps, *Chemometrics and Intelligent Laboratory Systems* 70 33–38 (2004).

[24] D.M. Hawkins, S.C. Basak, J. Kraker, K. Geiss and F.A. Witzmann, Combining chemodescriptors and biodescriptors in quantitative structure-activity relationship modeling, *Journal of Chemical Information and Modeling*, submitted.

Brill Academic Publishers
P.O. Box 9000, 2300 PA Leiden
The Netherlands

*Lecture Series on Computer*
*and Computational Sciences*
Volume 4, 2005, pp. 772-775

# Networks – The Universal Language of the Complex World

D. G. Bonchev[1]

Department of Mathematics, and
Center for the Study of Biological Complexity,
Virginia Commonwealth University,
Richmond, VA 23284, USA

Received 4 July, 2005; accepted in revised form 4 August, 2005

*Abstract:* The universality of network approach to systems in nature and technology is discussed, alongwith a review of the most common quantitative network descriptors. Special attention is devoted to the quantitative measures of network complexity: substructure count, overall connectivity, total walk count, and the information index for vertex degree distribution. The complexity measures are adapted for use in largebiological networks. A new definition for the average distance in directed networks is presented, which restores the correct "small-world ness" ratio of the directed and the parent undirected networks.

*Keywords:* networks, complexity, graph theory, topology, network descriptors

*Mathematics Subject Classification:* AMS-MOS 05C40 and 05C12

## 1. Introduction

Networks have long been used as a tool for describing complex system in nature and technology. However, the progress in network theory has been slow, due to the limitations of the random network theory used as a basis. In the year 1999, a short paper published in Science[1] initiated a revolution in network theory. It was shown that dynamic systems are essentially nonrandom, and have common features, independent of their nature. Thus, nodes in dynamic networks have only few degrees of separation (the "small-world" property)[2]. There are always some highly connected nodes (hubs), whereas the majority of nodes is weekly connected (the "scale-free" property, expressed mathematically as a power law of node connectivity).[1] Network architecture always includes some characteristic motifs, which may serve as topological fingerprints.[3] Important element is the network modular structure, the grouping of nodes into clusters.[4] Confirmations of these views came from wide variety of areas of science and technology spanning from quantum gravity (space-time networks) to biology (gene-, metabolic-, and protein-protein interaction networks), ecology (food webs), sociology (social networks), economics (market networks), computer science (electronic networks), electrical engineering (power grids), and communication (world-wide-web, telephone nets, transportation nets), etc. Thus, dynamic evolutionary networks became the universal language describing the complex world we live in. Simptomatically, studies on networks recently accounted for at least 50% of the Santa Fe Institute of Complexity research publications.[5] Scientists of very diverse background joined their efforts to develop network theory.[6,7] Chemical graph theory and, more generally, mathematical chemistry also contributes to these developments proceeding from the accumulated experience of at least 30 years studies of topology of molecules and chemical reaction networks. This report summarizes some of these contributions related to quantifying network complexity.[8-10]

---

[1] Elected corresponding member of the Bulgarian Academy of Sciences. E-mail: dgbonchev@vcu.edu

## 2. Quantitative Characterization of Networks

Topological description of networks is usually limited to few indices characterizing local and global connectivity, clustering, and distances. Local connectivity is presented by node degree distribution (node degree $k$ = number of links). It is described by a power law, showing the probability $p(k)$ of a node to have $k$ links to decrease with $k$:

$$p(k) \sim k^{-\gamma} \tag{1}$$

The exponent $\gamma$ is usually within the 2 to 3 range, except for food webs for which $\gamma \approx 1$. The global connectivity of a network with $V$ vertices is quantified either by the *average node degree*, $<k>$, or by *connectedness, Conn*:

$$<k> = \frac{1}{V}\sum_{i=1}^{V} k_i \ ; \quad Conn = \frac{2}{V(V-1)}\sum_{i=1}^{V} k_i \tag{2a,b}$$

The property of dynamic network nodes to group around a local center $i$ depends on the number of links $E_i$ between them. It is expressed by the *cluster coefficient* $c_i$, which is averaged over all nodes to produce the network *average cluster coefficient C*:

$$c_i = \frac{2E_i}{V_i(V_i-1)} \ ; \quad C = \frac{1}{V}\sum_{i=1}^{V} c_i \tag{3a,b}$$

The "small-world" property is verified by calculating the network *average distance*, $<d>$, and maximum distance, $d_{max}$, called *network radius* and *network diameter*, respectively:

$$<d> = \frac{2}{V(V-1)}\sum_{i,j=1}^{V} d_{ij} \tag{4}$$

## 3. Network Complexity

Complexity of molecules and chemical reaction networks has been studied intensively since the beginning of the 1980s.[6,11] Hierarchical concepts of complexity have been devised, and criteria for a complexity measure have been discussed. A consensus emerged that a complexity measure should increase with the number of links and that of cycles, as well as with the patterns of increasing branching, cyclicity, centrality, and clustering, and to decrease with the degree of symmetry. A number of reliable, second generation complexity descriptors appeared at the end of the 1990s. [12-14] Simultaneously and independently, Bertz[12] and Bonchev[13,14] proposed to use the subgraph count, *SC*, a descriptor that varies monotonically with the major complexity factors. The idea of a complete structural characterization by accounting for all subgraphs was developed further[13,14] by weighting each subgraph with the value of certain graph invariant, and calculating the total weight for the entire structure. Several invariants were examined, such as the Wiener number, the first and second Zagreb index, and the Hosoya index, however, the best fit to the variety of complexity factors was shown to be produced by the overall connectivity index, *OC* (termed initially too broadly "topological complexity", and denoted by *TC*). This index is based on the simplest graph invariant, that of the total adjacency $A$, which is the sum of all vertex degrees $a_i$. Thus, this complexity measure is expressed as the sum of total adjacencies of all subgraphs. Both, the subgraph count and the overall connectivity were presented as series of terms of null-, first-, second-, etc. order, the numeral meaning the number of edges $e$ = 0, 1, 2, ... in the subgraphs of the graph $G$ having a total of $E$ edges:

$$SC(G) = {}^0SC + {}^1SC + {}^2SC + ... + {}^ESC; \quad \{SC\} = \{{}^0SC, {}^1SC, {}^2SC,..., {}^ESC\} \tag{5a,b}$$

$$OC(G) = {}^0OC + {}^1OC + {}^2OC + ... + {}^EOC; \quad \{OC\} = \{{}^0OC, {}^1OC, {}^2OC,..., {}^EOC\} \tag{6a,b}$$

$$OC(G) = \sum_{e=1}^{E} {}^{e}OC({}^{e}G) = \sum_{e=1}^{E} \sum_{k=1}^{K(e)} {}^{e}A_{k}({}^{e}G_{k}) = \sum_{e=1}^{E} \sum_{k=1}^{K(e)} \sum_{i=1}^{N(i)} a_{i}({}^{e}G_{i}) \qquad (7)$$

In eq. (7), the overall connectivity index $OC$ $(G)$ of graph $G$ is more rigorously defined as the sum of the total adjacencies ${}^{e}A_{k}$ $({}^{e}G_{k})$ of all $K(e)$ subgraphs ${}^{e}G_{k}$ of $G$ having $e$ edges and $N(i)$ vertices.

Another reliable complexity measure was proposed by Rücker and Rücker[15] proceeding from summing-up all walks $w_{i}$ within the graph. The total walk count $TWC$ thus introduced can also be presented as a series of terms ${}^{l}TWC$ each one incorporating all walks ${}^{l}w_{i}$ of the same length $l$:

$$TWC(G) = \sum_{l=1}^{V-1} {}^{l}TWC = \sum_{l=1}^{V-1} \sum_{i=1}^{V} {}^{l}w_{i} \; ; \; \{TWC\} = \{{}^{1}TWC, {}^{2}TWC, ..., {}^{V-1}TWC\} \qquad (8a,b)$$

When dealing with large networks, such us cellular networks having thousands of nodes, the calculation of the topological complexity measures (5-8) would lead to combinatorial explosion. Our studies[8-10] have shown that the first several orders of the subgraph count (${}^{1}SC$, ${}^{2}SC$, ${}^{3}SC$), overall connectivity (${}^{1}OC$, ${}^{2}OC$, ${}^{3}OC$), and total walk count (${}^{1}TWC$, ${}^{2}TWC$, ${}^{3}TWC$) suffice to discern among networks of close size having different topological structure. Additional convenience for comparing networks complexity provide the averaged per vertex descriptors ${}^{k}X_{a}$, and the normalized complexity descriptors ${}^{k}X_{n} \equiv {}^{k}SC_{n}, {}^{k}OC_{n}, {}^{k}TWC_{n}$ of order $k$, the normalizing factor being the value the descriptor has for the complete graph $K_{V}$ having the same number of vertices $V$:

$$ {}^{k}X_{a} = \frac{{}^{k}X}{V} \; ; \qquad {}^{k}X_{n} = \frac{{}^{k}X}{{}^{k}X(K_{V})} \qquad (9a,b)$$

Equations were derived[10] for the ${}^{k}X(K_{V})$ terms to facilitate their use in the characterization of biological networks.

The specificity of node degree distribution in networks according to the power law (1) was also used[10] as a basis for another sensitive measure of network complexity: the information-theoretic index of vertex degree distribution, $I_{vd}$:

$$I_{vd} = \sum_{i=1}^{V} a_{i} \log_{2} a_{i} \qquad (10)$$

Eq. (10) directly demonstrates that dynamic networks are much more complex than random networks, due to the presence of hubs having high vertex degree ($a_{i} \gg 1$). More generally, the four complexity measures defined by eqs. (5-10) can be used for evolutionary studies and comparative analyses of organisms and dynamic processes, stability/fragility analysis, classification purposes, and others.

## 4. The "Small World" of Directed Networks

The interactions that are encoded by network links are often directed. Typical examples are the metabolic and signaling interactions in the living cell, and the "who-eats-whom" interaction in food webs. Not all node pairs in directed networks are connected by a path. Such distances are usually assumed as zeros, whereas they are infinity. This leads to the absurd conclusion for a smaller degree of separation in a directed network relative to the parent undirected one. A more realistic approach[16] defined the *adjusted average distance AD* of directed graph DG by accounting for the limited accessibility *Acc* of network nodes, which in turn is defined as the ratio between the number of finite distances $N_{d}$ in the directed graph *DG* and that in the parent undirected graph *G*:

$$AD(DG) = \frac{<d(DG)>}{Acc(DG)} \; ; \; Acc(DG) = \frac{N_{d}(DG)}{N_{d}(G)} \qquad (11a,b)$$

Eqs. (11a,b) restore the correct picture of a "smaller world" of the parent undirected network and the "larger world" of the derived directed network.

# References

[1]  A.-L Barabási and R. Albert, Emergence of Scaling in Random Networks. *Science* **286** 509-512 (1999).

[2]  D. J. Watts and S. H. Strogatz, Collective Dynamics of "Small-World" Networks. *Nature* **393** 440-442(1998).

[3]  R. Milo, S. Shen-Orr, S. Itzkovitz, N. Kashtan, D. Chklovskii and U. Alon, Network Motifs: Simple Building Blocks of Complex Networks, *Science* **298** 824-827(2002).

[4]  H. Jeong, B. Tombor, Z. Albert, and A.-L. Barabási, The Large-Scale Organization of Metabolic Networks, *Nature* **407** 651-654(2000).

[5]  http://www.santafe.edu/sfi/publications/working-papers.html.

[6]  D. Bonchev and D. H. Rouvray (Eds): Mathematical Chemistry, Vol. VII, *Complexity in Chemistry*. Taylor and Francis, London, U. K., 2003.

[7]  D. Bonchev and D. H. Rouvray (Eds) *Complexity in Chemistry, Biology, and Ecology*, Springer, New York, 2005.

[8]  D. Bonchev. Complexity of Protein-Protein Interaction Networks, Complexes and Pathways, *Handbook of Proteomics Methods*, (Editor: M. Conn) Humana, New York (2003) 451-462.

[9]  D. Bonchev, Complexity Analysis of Yeast Proteome Network, Chem. & Biodiversity, **1** 312-332(2004).

[10] D. Bonchev and G. A. Buck, Quantitative Measures of Network Complexity. *Complexity in Chemistry, Biology and Ecology* (Editors: D. Bonchev and D. H. Rouvray), Springer, New York, 191-235(2005).

[11] O. N. Temkin, A. V. Zeigarnik, and D. Bonchev, *Chemical Reaction Networks. A Graph Theoretical Approach*. CRC Press, Boca Raton, FL, 1996.

[12] S. H. Bertz and T. J. Sommer, Rigorous Mathematical Approaches to Strategic Bonds and Synthetic Analysis Based on Conceptually Simple New Complexity Indices. *Chem. Commun.* 2409-2410(1997).

[13] D. Bonchev, Novel Indices for the Topological Complexity of Molecules, *SAR QSAR Environ. Res.* **7** 23-43(1997).

[14] D. Bonchev, Overall Connectivities /Topological Complexities: A New Powerful Tool for QSPR/QSAR, *J. Chem. Inf. Comput. OCi.* **40** 934-941(2000).

[15] G. Rücker and C. Rücker, Walk Count, Labyrinthicity and Complexity of Acyclic and Cyclic Graphs and Molecules, *J. Chem. Inf. Comput. Sci.* **40** 99-106(2000).

[16] D. Bonchev, On the Complexity of Directed Biological Networks. *SAR QSAR Envir. Sci.* **14** 199-214(2003).

Brill Academic Publishers
P.O. Box 9000, 2300 PA Leiden
The Netherlands

*Lecture Series on Computer*
*and Computational Sciences*
Volume 4, 2005, pp. 776-780

# Molecular Quantum Similarity Measures in Minkowski Metric Vector Semispaces

Ramon Carbó-Dorca

Department of Inorganic and Physical Chemistry
Ghent University ; Krijgslaan 281, B-9000 Gent, Belgium
and
Institut de Química Computacional
Universitat de Girona, Girona 17071 Catalonia (Spain)

Received 12 July, 2005; accepted in revised form 8 August, 2005

*Abstract:* Minkowski metric vector semispaces can be chosen as the natural mathematical framework, where quantum similarity measures are described and evaluated. The obtained results in this study show that the Minkowski metric option is easily feasible, providing a new set of computationally simpler expressions, computationally faster when compared with Euclidian based quantum similarity measures.

*Keywords:* Quantum Similarity Measures. Vector Semispaces. Unit Shell. Minkowski Metric. General Weighted Minkowski Metric. Shape Functions.

## 1. Introduction

Quantum molecular similarity measures have been applied in many chemical fields [[1] ,[2] ], since its first appearance in the literature [[3] ] they have been mainly employed in a computational structure provided with a metric based on Euclidian norms and scalar products. The development of the quantum similarity theoretical background on the other hand, has brought essentially to the definition and properties of semispaces [[4] ] within their natural Minkowski metric [[5] ]. These findings had raised the expectative to use such a metric in the quantum similarity field; thus some of the related questions will be tentatively answered in a first instance within the present work.

## 2. Minkowski metric vector semispaces

A Minkowski norm is defined within a vector semispace (VSS) in the following way:
a) In infinite dimensional functional VSS, Minkowski norms are described by an integral:

$$\forall f(t) \in V_\infty(\mathbf{R}^+) \to \langle f \rangle = \int_D f(t)dt ,$$

b) In finite dimensional column or row VSS, the Minkowski norm is simply evaluated by summing up their elements, for example:

$$\forall \mathbf{x} = \{x_i\} \in V_n(\mathbf{R}^+) \to \langle \mathbf{x} \rangle = \sum_i x_i .$$

Shells $S(\mu)$ in VSS are constructed taking into account the Minkowski norms of their elements as follows:

$$\forall x \in V(\mathbf{R}^+) : \langle x \rangle = \mu \in \mathbf{R}^+ \to x \in S(\mu) \subset V(\mathbf{R}^+).$$

Among the infinite set of shells in a VSS, the unit shell $S(1)$ can be considered the VSS subset made of all probability distributions; also, an element of any shell can be transformed into a unit shell one by a simple homothecy. This property has been discussed recently in order to set the relationship and associated properties between first order density functions and shape functions [[7] ].

Scalar products in VSS shall be performed taking into account the square roots of the involved vectors' elements, in order they yield the Minkowski norm, when both vectors appearing into the scalar product are the same. That is, it can be written:

$$\langle f|g\rangle = \int_D \left(f(t)g(t)\right)^{\frac{1}{2}} dt \wedge \langle \mathbf{x}|\mathbf{y}\rangle = \sum_i \left(x_i y_i\right)^{\frac{1}{2}}$$

for the two kinds of possible Minkowski norms in VSS as commented before. To distinguish the Minkowski metric from the Euclidian one, the symbol $\left(\mathbf{a}|\mathbf{b}\right)$ will be used to write an Euclidian scalar product; for instance, if: $\mathbf{a} = \{a_i\} \wedge \mathbf{b} = \{b_i\}$, are two n-tuple arrays, then the Euclidian scalar product symbol is computed as: $\left(\mathbf{a}|\mathbf{b}\right) = \sum_i a_i b_i$.

### 3. Cosine of the subtended angle between two vectors in a Minkowski metric VSS

The cosine of the subtended angle between two vectors in a Minkowski metric VSS can, thus, be computed as in the usual vector algebra:

$$\cos(\alpha) = \left(\langle \mathbf{x}\rangle\langle \mathbf{y}\rangle\right)^{-\frac{1}{2}} \langle \mathbf{x}|\mathbf{y}\rangle \wedge \mathbf{x} = \mathbf{y} \to \cos(\alpha) = 1,$$

because employing the previous Minkowski metric definitions: $\langle \mathbf{x}|\mathbf{x}\rangle = \langle \mathbf{x}\rangle$.

Suppose now that the following generating rules [[4]] are set for two vectors in a known Minkowski metric VSS: $R(\mathbf{a} \to \mathbf{x}) \wedge R(\mathbf{b} \to \mathbf{y})$. This construction, if both generated VSS vectors are elements of the unit shell, implies that the generating vectors are normalized in the Euclidean norm sense, and vice versa: whenever the generating vectors are normalized in the Euclidian sense, the generated vectors belong to the unit shell. In this circumstance the cosine of the VSS vectors and the one of their generators is the same, that is:

$$\cos(\alpha) = \langle \mathbf{x}|\mathbf{y}\rangle = \sum_i \left(x_i y_i\right)^{\frac{1}{2}} = \sum_i a_i b_i = \left(\mathbf{a}|\mathbf{b}\right),$$

moreover, the cosine is coincident with the Minkowski product of the generated vectors and the Euclidian product of the generating vectors.

### 4. Overlap quantum similarity measures

The equivalence between unit shell Minkowski metric and generating Euclidian scalar products permits the evaluation of the, so-called, overlap similarity measures (OSM) [[10]] and, thus, by extension the so-called Carbó index [[11]] is obtained, involving infinite dimensional VSS made of quantum mechanical density functions as elements.

Indeed, suppose two density functions, scaled by the respective number of electrons, that is, the so-called shape functions attached to a pair of molecular structures: $\{\rho_A, \rho_B\}$, the OSM according to a Minkowski metric can be written:

$$\langle \rho_A | \rho_B \rangle = \int_D \left(\rho_A \rho_B\right)^{\frac{1}{2}} dV.$$

On the other hand, it has been described that to any density function, $\rho$, a pseudo-wave function, $\psi$, which generates it in the previous sense:

$$R(\psi \to \rho) \equiv \rho(\mathbf{r}) = \psi^*(\mathbf{r}) * \psi(\mathbf{r})$$

can be easily described [[6], [12]] under some specific circumstances. To gain insight into this generating process, suppose that for the molecule $A$, a MO description for the density function is known and that a pseudo-wave function is constructed, then the density function can be described as the inward product [[8], [9]]:

$$\rho_A = \sum_i \omega_i^A \left| \phi_i^A \right|^2 \wedge \psi_A = \sum_i \left( \omega_i^A \right)^{\frac{1}{2}} \phi_i^A \rightarrow \rho_A = \psi_A^* * \psi_A, \tag{1}$$

also the same holds for molecule $B$. The signs of the square roots of the scaled MO occupation numbers $\left\{ \omega_i^A = N_A^{-1} v_i^A \right\}$ are taken to be positive, in order to obviate the undetermined signature of the pseudo-wave function [[6], [12]], which acts as an undetermined phase factor. It must be now commented that this set up is a generalized one in MO theory, as both the density expression and the attached pseudo-wave function, conveniently written in a similar manner as in equation (1), can be taken as feasible formal expressions valid up to any density function order [[13]].

If the densities in systems $A$ and $B$ are normalized in the Minkowski sense, then due to the property already studied, the cosine of the subtended angle between both functions is written:

$$\langle \rho_A | \rho_B \rangle = \left( \psi_A | \psi_B \right) = \sum_i \sum_j \left( \omega_i^A \omega_j^B \right)^{\frac{1}{2}} \int_D \left( \phi_i^A \right)^* \phi_j^B dV, \tag{2}$$

but, as in molecule $A$, for instance, the MO set $\left\{ \phi_i^A \right\}$ within the LCAO theory can be expressed as a superposition of atomic centered functions $\left\{ \chi_\mu^A \right\}$, and the same holds for molecule $B$. Therefore, one can write the overlap between two MO belonging to systems $A$ and $B$ as:

$$\int_D \left( \phi_i^A \right)^* \phi_j^B dV = \sum_\mu \sum_v \left( c_{\mu i}^A \right)^* c_{vj}^B \int_D \left( \chi_\mu^A \right)^* \chi_v^B dV = \mathbf{c}_{A;i}^+ \mathbf{S}_{AB} \mathbf{c}_{B;j},$$

with the overlap matrix between basis functions of systems $A$ and $B$ defined by:

$$\mathbf{S}_{AB} = \left\{ S_{AB;\mu v} = \int_D \left( \chi_\mu^A \right)^* \chi_v^B dV = \left( \chi_\mu^A | \chi_v^B \right) \right\}$$

and the MO coordinate coefficients are collected into the column vector: $\mathbf{c}_{A;i} = \left\{ c_{\mu i}^A \right\}$ with a similar form for molecule $B$. The OSM conforming to a VSS Minkowski metric can be written as:

$$\langle \rho_A | \rho_B \rangle = \sum_i \sum_j \left( \omega_i^A \omega_j^B \right)^{\frac{1}{2}} \mathbf{c}_{A;i}^+ \mathbf{S}_{AB} \mathbf{c}_{B;j}.$$

In order to evaluate the OSM above described in an efficient computational manner, one can define the auxiliary column vectors, which are nothing else than the MO coordinates scaled by the square root of the occupation numbers:

$$\mathbf{d}_{A;i} = \left( \omega_i^A \right)^{\frac{1}{2}} \mathbf{c}_{A;i},$$

and the same for molecule $B$. With the aid of the scaled MO coordinate vectors, the following matrix:

$$\mathbf{T}_{AB} = \left\{ T_{AB;ij} = \mathbf{d}_{A;i}^+ \mathbf{S}_{AB} \mathbf{d}_{B;j} \right\}$$

can be easily constructed. Finally, the Minkowski OSM is expressible as a complete sum [[13]] of the elements of matrix $\mathbf{T}_{AB}$:

$$\langle \rho_A | \rho_B \rangle = \langle \mathbf{T}_{AB} \rangle = \sum_i \sum_j T_{AB;ij}. \tag{3}$$

The OSM obtained in this way still will depend on the relative positions of both molecules, as the associated Euclidean expression is, see for example [[2]], but the Minkowski MO algorithm is connected to integrals with a pair of orbital basis functions only, instead of a four-function dependence as encountered in the usual Euclidean representation [[10]].

## 5. General operator weighted Minkowski metric OSM

For other kind of weighting operators than the unity, the intermolecular overlap matrix shall be substituted by the corresponding matrix representation. It is only necessary to consider a positive definite operator, which can be written in a module like form: $\Omega\left(\mathbf{r}_1;\mathbf{r}_2\right) = \left|\Theta\left(\mathbf{r}_1;\mathbf{r}_2\right)\right|$, then the weighted OSM can be immediately evaluated as:

$$\left\langle\rho_A\left|\Omega\right|\rho_B\right\rangle = \int_{D_1}\int_{D_2}\left(\rho_A\left(\mathbf{r}_1\right)\left|\Theta\left(\mathbf{r}_1;\mathbf{r}_2\right)\right|^2\rho_B\left(\mathbf{r}_2\right)\right)^{\frac{1}{2}}d\mathbf{r}_1 d\mathbf{r}_2$$

$$= \left(\psi_A\left|\Omega\right|\psi_B\right) = \int_{D_1}\int_{D_2}\psi_A^*\left(\mathbf{r}_1\right)\Omega\left(\mathbf{r}_1;\mathbf{r}_2\right)\psi_B\left(\mathbf{r}_2\right)d\mathbf{r}_1 d\mathbf{r}_2$$

so, a matrix representation over the basis functions shall be considered to be computed. Therefore employing the following matrix expression:

$$\mathbf{Z}_{AB} =$$

$$\left\{Z_{AB;\mu\nu} = \int_{D_1}\int_{D_2}\left(\chi_\mu^A\left(\mathbf{r}_1\right)\right)^*\Omega\left(\mathbf{r}_1;\mathbf{r}_2\right)\chi_\nu^B\left(\mathbf{r}_2\right)d\mathbf{r}_1 d\mathbf{r}_2 = \left(\chi_\mu^A\left|\Omega\right|\chi_\nu^B\right)\right\}$$ (4)

a new matrix can be constructed by using the scaled MO coordinate vectors:

$$\mathbf{A}_{AB} = \left\{A_{AB;ij} = \mathbf{d}_{A;i}^+\mathbf{Z}_{AB}\mathbf{d}_{B;j}\right\},$$

in this way, the weighted Minkowski OSM can be written as the total sum:

$$\left\langle\rho_A\left|\Omega\right|\rho_B\right\rangle = \left\langle\mathbf{A}_{AB}\right\rangle = \sum_i\sum_j A_{AB;ij}.$$

The weighted Minkowski metric OSM in this easy manner present a generalized structure associated to positive definite operators. The matrix (4) will behave as a positive definite metric matrix.

## 6. Conclusions

The basic procedure described here constitutes a correct and generalized process to compute quantum similarity measures within the LCAO MO theory, conforming to a VSS Minkowski metric. The deducible algorithms in the present theory appear to be, in general, computationally faster than the usual ones, appeared so far in the literature, based on Euclidian metric.

## Acknowledgements

The author expresses his acknowledgement to the Ministerio de Ciencia y Tecnología for the grant: BQU2003-07420-C05-01; which has partially sponsored this work and for a Salvador de Madariaga fellowship reference: PR2004-0547, which has made possible his stage at the Ghent University. The author also wishes to thank Professor Patrick Bultinck of the Ghent University for the deep discussions on the problems met in an early stage of the construction of the present work, his warm hospitality is also heartily acknowledged.

## Bibliography

[1]    R. Carbó-Dorca, D. Robert, L. Amat, X. Gironés, E. Besalú; *Molecular Quantum Similarity in QSAR and Drug Design* Lecture Notes in Chemistry . Vol 73. Editorial Springer Verlag, Berlin (2000).

[2]    P. Bultinck, X. Girones, R. Carbó-Dorca; *Molecular Quantum Similarity: Theory and Applications*; Rev. Comput. Chem. Vol. 21, K.B. Lipkowitz, R. Larter and T. Cundari (Eds.), John Wiley & Sons, Inc., Hoboken (USA) (2005) pag. 127.

[3]    R. Carbó, L. Leyda, M. Arnau; Intl. J. Quantum Chem., 17 (1980) 1185.

[4]    R. Carbó-Dorca; *Fuzzy sets and Boolean tagged sets, vector semispaces and convex sets, QSM and ASA density functions, diagonal vector spaces and quantum chemistry*, Adv. Molec. Simil. Vol. 2, R. Carbó-Dorca, P. G. Mezey (Eds.), JAI Press (London) (1998) pag. 43.

[5]     I. M. Vinogradov (Editor); *Encyclopaedia of Mathematics*, Reidel-Kluwer Academic Publishers, Dordrecht (1987).

[6]     R. Carbó-Dorca; J. Math. Chem., 32 (2002) 201.

[7]     R. Carbó-Dorca, P. Bultinck; J. Math. Chem., 36 (2004) 191.

[8]     R. Carbó-Dorca; J. Mol. Struct. (Theochem), 537 (2001) 41.

[9]     R. Carbó-Dorca; *Quantum Quantitative Structure-Activity Relationships (QQSAR): A Comprehensive Discussion Based On Inward Matrix Products, Employed As A Tool To Find Approximate Solutions Of Strictly Positive Linear Systems And Providing A QSAR-Quantum Similarity Measures Connection.* Proceedings of European Congress on Computational Methods in Applied Sciences and Engineering. ECCOMAS 2000. Barcelona,11-14 September 2000; ISBN-84-89925-70-4, (2000) pags. 1-31.

[10]    R. Carbó-Dorca, E. Besalú; J. Molec. Struct. (Theochem), 451 (1998) 11.

[11]    R. Carbó, E. Besalú, Ll. Amat, X. Fradera; J. Math. Chemistry, 19 (1996) 47.

[12]    R. Carbó-Dorca; *Density Functions and Generating Wave Functions*, in: "Reviews of Modern Quantum Chemistry". A Celebration of the Contributions of Robert G. Parr, Vol I, K. D. Sen (Editor), World Scientific (Singapore) (2002) pag. 401.

[13]    E. Besalú, R. Carbó; *Applications of Nested Summation Symbols to Quantum Chemistry:Formalism and Programming Techniques*, in: "Strategies and Applications in Quantum Chemistry: from Astrophysics to Molecular Engineering" An Hommage to Prof. G. Berthier, M. Defranceschi, Y. Ellinger (Editors), Kluwer Academic Pub. (Amsterdam) (1996) pag. 229.

Brill Academic Publishers
P.O. Box 9000, 2300 PA Leiden
The Netherlands

*Lecture Series on Computer
and Computational Sciences*
Volume 4, 2005, pp. 781-784

# Simulations of Highly Reactive Fluids

Laurence E. Fried[1], M. Riad Manaa, and Evan J. Reed
Chemistry and Materials Scince Directorate.
Lawrence Livermore National Laboratory,
L-282, 7000 East Avenue,
Livermore, CA, 94550, USA

Received 20 July, 2005; accepted in revised form 12 August, 2005

*Abstract:* We report density functional molecular dynamics simulations to determine the early chemical events of hot (T = 3000 K) and dense (1.97 g/cm$^3$, V/V$_0$ = 0.68) nitromethane (CH$_3$NO$_2$). The first step in the decomposition process is an intermolecular proton abstraction mechanism that leads to the formation of CH$_3$NO$_2$H and the aci ion H$_2$CNO$_2^-$, in support of evidence from static high-pressure and shock experiments. An intramolecular hydrogen transfer that transforms nitromethane into the aci acid form, CH$_2$NO$_2$H, accompanies this event. This is the first confirmation of chemical reactivity with bond selectivity for an energetic material near the condition of fully reacted specimen. We also report the decomposition mechanism followed up to the formation of H$_2$O as the first stable product.

*Keywords:* Density Functional Theory, Molecular Dynamics, Energetic Materials, Nitromethane, Reaction mechanisms

*PACS:* 71.10, 71.15, 71.20, 61.20

## 1. Introduction

The reaction chemistry of energetic materials at high pressure and temperature is of considerable importance in understanding processes that these materials experience under impact and detonation conditions. Basic questions such as: (a) which bond in a given energetic molecule breaks first, and (b) what type of chemical reactions (unimolecular versus bimolecular, etc.) that dominate early in the decomposition process, are still largely unknown. The most widely studied, and archetypical example of such materials is nitromethane (CH$_3$NO$_2$), a clear liquid with mass density 1.13 g/cm$^3$ at 298°K. Static high-pressure experiments[1] showed that the time of explosion for deuterated nitromethane is approximately ten times longer than that for protonated materials, suggesting that a proton or hydrogen atom abstraction is involved in the rate determining step. Isotope-exchange experiments, using diamond cells methods, also gave evidence [2] that the aci ion concentration (H$_2$CNO$_2^-$) increases with increased pressure. Other studies[3] also suggested that reactions occur more rapidly and are pressure enhanced when small amount of bases are present, giving further support to the aci ion production. Shock wave studies of the reaction chemistry are still inconclusive and at odds: mass spectroscopic studies suggesting condensation reactions[4], time-resolved Raman spectroscopy suggesting a bimolecular mechanism[5], UV-visible absorption spectroscopy indicating no sign of chemical reaction[6], or the production of H$_3$CNO$_2^-$ intermediate for amine-sensitized nitromethane[7]. It was noted, however, that part of the discrepancy is due to the fact that the ring-up experiments are mapping lower temperature regimes (≈1000 °K) than experienced under detonation conditions (T≈2500-5000 °K)[4].

In this work, we use spin-polarized, gradient corrected density-functional calculations to determine the interatomic forces, and simulate the initial decomposition steps of hot (T = 3000K) dense (1.97 g/cm$^3$, V/V$_0$ = 0.68) nitromethane at constant-volume and temperature conditions. The studied state is in the neighborhood of the Chapman-Jouget state, which is achieved behind a steady detonation front when the material has fully reacted. This state could be achieved through a sudden heating of nitromethane in a diamond anvil cell under constant volume conditions. Our results emphatically show

---

[1] Corresponding author.. E-mail: fried1@llnl.gov

that the first chemical event is a proton extraction to form $CH_3NO_2H$, the aci ion $H_2CNO_2^-$, and the aci acid $H_2CNO_2$ H. These results are uniquely associated with the condensed-phase rather than the energetically favored C-N decomposition expected in the gas-phase.

## 2. Computational Approach

The electronic structure calculations of the molecular forces were performed using density functional theory (DFT).[8] For the exchange-correlation potential, we used the spin-polarized generalized gradient corrected approximation of Perdew -Wang (PW91)[9]. Electron-ion interactions were described by Vanderbilt-type ultrasoft pseudopotentials,[10] and orbitals were expanded in a plane wave basis set with kinetic energy cutoff of 340 eV. We used two k-point spacing in the Brillouin zone, each with a total number of 2921 plane waves. Minimization of the total density functional from DFT utilized the charge density mixing scheme[11]. Calculations on a single unit cell were performed using the CASTEP program [12], while those on larger cells employed the VASP program.[11,13]

Molecular dynamics simulations were carried out under constant volume and temperature using a Nose thermostat. For each MD run, random initial velocities were chosen, and a first-order Verlet extrapolation of the wave functions was used. Periodic boundary conditions, whereby a particle exiting the cell on one side is reintroduced on the opposing side with the same velocity were imposed. A dynamical time step of 0.25 fs was employed for all runs, the longest of which was 4.5 ps. Simulations were performed at a constant temperature of 3000 K using either one unit cell of nitromethane crystal (4 molecules, 28 atoms), a supercell with 8 molecules, and a supercell with 16 molecules. The unit cell was fully optimized at the reduced (compressed) volume, $V = 205.36 \text{ Å}^3$.

## 3. Results and Discussion

The initial configuration at density 1.974 g/cm³ was determined by compressing the simulation cell and performing full relaxation of all atomic coordinates. From this initial structure, molecular dynamics were performed using the Nose-Hoover thermostat at 3000K.

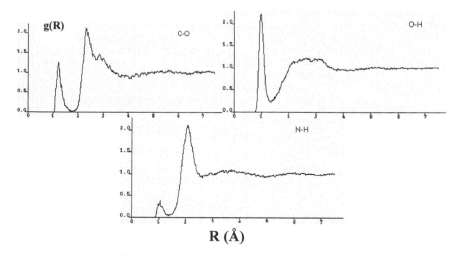

**R (Å)**

Figure 1. Calculated radial distribution functions g(R) for various intra and intermolecular bonds.

Here, we report the results obtained from the simulation using the largest supercell, consisting of four unit cells of nitromethane molecules with a repetition of 2x2x1 of the unit cell and corresponding to a volume of 821.5 Å³. The total time of this simulation was 1.16 ps.

Figure 1 shows the C-O, O-H, and N-H radial distribution functions obtained over the total time of the simulation. It is evident from the distributions for the C-O, O-H and N-H that significant rearrangement of the bonds have occurred and chemistry has ensued. For C-O, the dominant population around 1.2 Å is due mainly to the formation of $CO_2$, while for N-H, the small population around 1.0 Å is due to the formation of radical intermediates of $CH_2NHO$. Most interestingly is the significant population growth of O-H at 1.0 Å, which encompasses the formation of $H_2O$, and in the early stages, to inter and intramolecular hydrogen bonding that leads to proton transfer.

A snapshot of the MD simulation at 59 fs where the formation of $CH_3NO_2H$ and $CH_2NO_2$ takes place is shown in Figure 2. This process of proton transfer is initially facilitated by enhancement in the C-N double bond character, and an accelerated rotation of the methyl groups ($CH_3$), rotations that are omnipresent even at ambient temperatures. [14]

The proton transfer process described above is uniquely associated with the condensed fluid phase of nitromethane. This bond specificity is remarkable, since in the gas phase the C-N bond is the weakest in the molecule ($D_0$ = 60.1 kcal/mol) [15], and is therefore expected to be the dominant dissociation channel and the initial step in the decomposition of nitromethane even at high temperature. In contrast, the C-H bond is the strongest in the nitromethane molecule. In the condensed phase, however, vibrational energy is the highest in the C-H mode. Due in part to a caging effect, this vibrational motion eventually leads to a proton extraction.

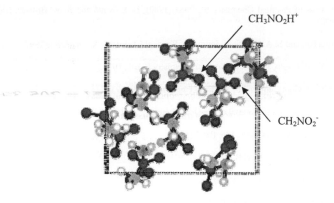

Figure 2. A snapshot of the MD simulation at 59 fs. The formation of $CH_3NO_2H$ and $CH_2NO_2$ due to intermolecular hydrogen abstraction is shown.

The formation of $CH_3NO_2H$ and $CH_2NO_2$ via proton extraction was observed in all three simulations of different supercell sizes. In the simulation on a single unit cell, the event occurs at 785 fs of the simulation time. We performed Mulliken charge analysis and determined the net charges on the two moieties $CH_3NO_2H$ and $CH_2NO_2$. We notice that the negative charge on the carbon atom of $CH_2NO_2$ is larger than in $CH_3NO_2H$, while the opposite trend is exhibited for the positive charge on nitrogen. This is a manifestation of electronic charge redistribution in the region between the C and N atoms.

It is noteworthy that all three simulations (one, two and four unit cells) have yielded the same results in the formation of $CH_3NO_2H$, $H_2CNO_2^-$, and $CH_2NO_2H$. Experimental concurrence for the production of the aci ion in highly pressurized and detonating nitromethane abound. Shaw et al.[1] observed that the time to explosion for deuterated nitromethane is about ten times longer than that for the protonated materials, suggesting that a proton or hydrogen atom) abstraction is the rate-determining step. Isotope-exchange experiments provided evidence that the aci ion concentration is increased upon increasing pressure[2], and UV sensitization of nitromethane to detonation was shown to correlate with the aci ion presence. [16] Finally, we note that a recent electronic structure study of solid nitromethane determined a significant C-H stretch upon compression, which eventually lead to proton dissociation[17].

## 4. Conclusion

We studied the early chemical events of hot (T = 3000 K) and dense (1.97 g/cm$^3$, V/V$_0$ = 0.68) nitromethane using density functional molecular dynamic simulations. Three simulations on one, two , and four unit cells of crystal nitromethane have shown that the first step event in the decomposition process is an intermolecular proton abstraction mechanism that leads to the formation of $CH_3NO_2H$ and the aci ion $H_2CNO_2^-$, which lends support to experimental results from static high-pressure and shock experiments. An intramolecular hydrogen transfer that transforms nitromethane into the aci acid form, $CH_2NO_2H$, accompanies this event. This is the first confirmation of chemical reactivity with bond selectivity for an energetic material near the condition of fully reacted specimen.

## Acknowledgments

This work was performed under the auspices of the U.S. Department of Energy by the Lawrence Livermore National Laboratory under contract number W-7405-Eng-48.

## References

[1] R. Shaw, P. S. Decarli, D. S. Ross, E. L. Lee, and H. D. Stromberg, Combust. Flame 35, 237 (1979); R. Shaw, P. S. Decarli, D. S. Ross, E. L. Lee, and H. D. Stromberg, Combust. Flame 50, 123 (1983).

[2] R. Engelke, D. Schiferl, C. B. Storm, and W. L. Earl, J. Phys. Chem. 92, 6815 (1988).

[3] J. W. Brasch, Journal of Physical Chemistry 84, 2084 (1980); D. L. Naud and K. R. Brower, High-Pressure Research 11, 65 (1992).

[4] N. C. Blais, R. Engelke, and S. A. Sheffield, Journal of Physical Chemistry A 101, 8285 (1997).

[5] J. M. Winey and Y. M. Gupta, Journal of Physical Chemistry B 101, 10733 (1997).

[6] J. M. Winey and Y. M. Gupta, Journal of Physicall Chemistry A 101, 9333 (1997).

[7] Y. A. Gruzdkov and Y. M. Gupta, Journal of Physical Chemistry A 102, 2322 (1998).

[8] P. Hohenberg and W. Kohn, Phys. Rev. 136, B864 (1964).

[9] J. P. Perdew and Y. Wang, Phys. Rev. B 46, 6671 (1992).

[10] D. Vanderbilt, Phys. Rev. B 41, 7892 (1990).

[11] G. Kresse and J. Furthmuller, Phys. Rev. B 54, 11169 (1996).

[12] A. Inc., Cerius 2 Modling Environment (Accelrys Inc., San Diego, 1999).

[13] G. Kresse and J. Hafner, Phys. Rev. B 47, RC558 (1993).

[14] M. E. Tuckerman and M. L. Klein, Chem. Phys. Lett. 283, 147 (1998); D. C. Sorescu, B. M. Rice, and D. L. Thompson, J. Phys. Chem. B 104, 8406 (2001).

[15] J. B. Pedley, R. D. Naylor, and S. P. Kirby, *Thermochemical Data of Organic Compounds*, 2nd ed. ed. (Chapman, New York, 1986).

[16] R. Engelke, W. L. Earl, and C. M. Rohlfing, J. Phys. Chem. 90, 545 (1986).

[17] D. Margetis, E. Kaxiras, M. Elstner, T. Frauenheim, and M. R. Manaa, J. Chem. Phys. 117, 788 (2002).

Brill Academic Publishers
P.O. Box 9000, 2300 PA Leiden
The Netherlands

*Lecture Series on Computer
and Computational Sciences*
Volume 4, 2005, pp. 785-789

# Molecular Dynamics Simulations of Spinodal-Assisted Polymer Crystallization

Richard H. Gee[1], Naida M. Lacevic, and Larry E. Fried

Chemistry and Materials Science Directorate
Laurence Livermore National Laboratory
P.O. Box 808, L-268, Livermore, CA 94550

Received 8 July, 2005; accepted in revised form 28 July, 2005

*Abstract:* Large scale molecular dynamics simulations of bulk melts of polar (poly(vinylidene fluoride) (pVDF)) polymers are utilized to study chain conformation and ordering prior to crystallization under cooling. While the late stages of polymer crystallization have been studied in great detail, recent theoretical and experimental evidence indicates that there are important phenomena occurring in the early stages of polymer crystallization that are not understood to the same degree. When the polymer melt is quenched from a temperature above the melting temperature to the crystallization temperature, crystallization does not occur instantaneously. This initial interval without crystalline order is characterized as an induction period. It has been thought of as a nucleation period in the classical theories of polymer crystallization, but recent experiments,[1] computer simulations,[2] and theoretical work[3] suggest that the initial period in polymer crystallization is assisted by a spinodal decomposition type mechanism. In this study we have achieved physically realistic length scales to study early stages of polymer ordering, and show that spinodal-assisted ordering prior to crystallization is operative in polar polymers suggesting general applicability of this process.

*Keywords:* poly(vinylidene fluoride), polymer crystallization, early stages, molecular dynamics

*Subject Classification:* Polymer Physics

*PACS:* 61.41.+e, 81.05.Kf

## 1. Introduction

One of the most challenging problems in polymer physics is understanding the dynamics and thermodynamics of polymer crystallization. When the polymer melt is quenched from a temperature above the melting temperature to the crystallization temperature, crystallization does not occur instantaneously. This interval without crystalline order is characterized as an induction period. Imai and coworkers performed a detailed investigation of this induction period via small angle x-ray scattering (SAXS) experiments of poly(ethylene terephthalate) (PET). They have shown that density fluctuations occur in the induction period. Inspired by the work of Imai,[4] Olmstead and coworkers[3] proposed a generalized phase diagram that describes the polymer crystallization from the melt. This phase diagram suggests that there are three types of primary nucleation: direct, binonal and spinodal.[1]

In this study we focus on the spinodal-type primary nucleation. The observation of an induction period in experiments is highly dependent on the sensitivity of the detector,[5] therefore molecular dynamics (MD) simulations are an excellent tool to probe the length scales of this process. Increased computational power has helped to achieve realistic length scales in molecular dynamics simulations that are necessary to probe the early stages of primary nucleation[6] and test theoretical predictions.[3]

The purpose of this study is to demonstrate general applicability of spinodal-type primary nucleation prior to crystallization via simulations and to obtain physical insight in polymer ordering prior to crystallization upon deep quench.

## 2. Computational Methods

### 2.1 pVDF Melt Representation and Force Fields

MD simulations of the polymer melts were carried out on bulk entangled amorphous ensembles of both explicit atom (EA) and united-atom (UA) linear polymers. The simplified UA model affords the investigation of a much larger polymer ensemble as compared to explicit atom representations, thus allowing one to resolve the important SAXS peaks identifiable experimentally. The polymer melt is composed of either ten 120-mer polymer (240 backbone carbons) EA chains (7,220 explicit-atoms; $M_w$=7,682 g/mol), or 8,640 240-bead polymer chains (2,073,600 united-atoms; $M_w$=7,682 g/mol). The polymer models are representative of poly(vinylidene fluoride) (pVDF). In the UA polymer models, the hydrogen or fluorine atoms are lumped onto the carbon backbone atoms to which they are attached. The EA simulations employed the COMPASS[19] force field parameter set, while the UA simulations employed the force field parameter set of Goddard *et al.*[8, 9] The force field parameters consist of both valence (stretch, bending, and torsion terms) and non-bonded potential terms (vdW and Coulomb). All valence degrees of freedom were explicitly treated and unimpeded.

### 2.2 Molecular Dynamics Method

Molecular dynamics is a widely used method in the investigation of various physical processes related to the dynamics of macromolecules (e.g. deformation of solids, polymer crystallization, aging in supercooled liquids, motion of biomolecules). MD provides static and dynamics properties for a collection of particles, which allow atomic scale insights that are difficult to gain otherwise.

We used three-dimensional cubic periodic boundary conditions. The simulations were carried out using constant particle number, pressure, and temperature (*NPT*) dynamics at zero pressure. All computations were carried out using the LAMMPS code.[10] The particle-particle particle-mesh Ewald (PPPM) method[14] was used for the treatment of long-range electrostatic interactions.

The polymer melts were initially equilibrated at a temperature well above the temperature at which ordering occurs, $T_c$. The periodic box was allowed to relax under *NPT* conditions for a minimum of 5 ns. Following this equilibration step, the melts were cooled in *NPT* runs in increments of 50 K and equilibrated for a minimum of 5 ns at temperatures above $T_c$. For the temperature at which the polymer melts order, the simulations were carried out for more than 30 ns.

## 3. Results and Discussions

Figure 1 demonstrates that a microphase separation from the amorphous melt into many ordered liquid domains occurs very early in the simulations for both EA and UA polymer melt models.

We next turn our attention to the process of nucleation and growth, and the question of why the pVDF forms three-dimensional ordered structures. The overall evolution of the orientationally ordered structure of pVDF is easily identified by comparing the initial starting configuration (Figure 2*a*) of the polymer melt to the final orientationally ordered structure (Figure 2*e*). The nucleation and growth process is illustrated for the ten chain 120-mer ensemble in Figure 4*a-e*, where snapshots of the chain conformation at various times throughout the simulation (*t* = *5, 7.5, 10, and 15 ns*) at a temperature above $T_c$ (panel *a-c*) and at a temperature at $T_c$ (panel *d-e*). The melt-to-crystal phase transition occurs via an effective three stage process, consisting of first, local chain ordering, in which random sequences of the polymer backbone form small all *trans* segments on the order of two or three *trans* conformations (*"extended trans sequence formation"*), second, cluster formation, in which the randomly oriented small *trans* segments aggregate (*"cluster formation"*), and third, a stage which primarily consists of stem growth (gauche defect removal), polymer chain registration, and "lamella" thickening (*"growth"*). It is during this stage that the *trans* states become increasingly populated, and the polymer disentangles itself.

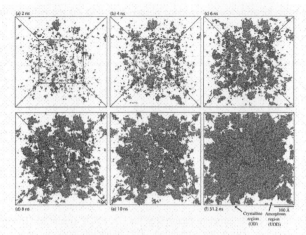

Figure 1. Snapshots from MD simulations showing the evolution of ordered domains of UA pVDF (at times of 2, 4, 6, 10, and 31.2 ns, panels (a)-(f), respectively. Only oriented domains (OD) are shown (the unordered domains (UOD) (amorphous regions) are shown in white.

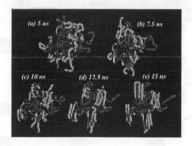

Figure 2. Illustration of the time evolution of EA pVDF ordering. (*a*) initial starting configuration at 50 K above $T_c$, (*b*) after 2.5 ns at 50 K above $T_c$, (*c*) after 5 ns 50 K above $T_c$, (*d*) after 2.5 ns at $T_c$, (*e*) after 5 ns at $T_c$. "Loops" into and out of the ordered domains and "amorphous" regions are colored purpule. Only the carbon backbone atoms are shown for clarity. Chain segments colored read and yellow illustrate the polycrystalline nature of the ensemble. Time labels are for the total simulation time.

Figure 3*a* shows the evolution of the structure factor *S(q,t)* for UA pVDF. *S(q,t)* shows three distinct peaks that increase in magnitude during the simulations. Those peaks are: the induction peak, long peak and Bragg peak. Experimentally, the induction and long peak are observed via small angle x-ray scattering and the second-order Bragg peak is observed in wide angle x-ray scattering (WAXS). The induction peak is related to density fluctuations of the rigid segments. The long peak is related to alternating unordered and ordered domains. The second-order Bragg peak, which corresponds to the nearest-neighbor packing distance between polymer chains, is related to densification of the system and transition from unoriented to oriented domains.

In order to demonstrate that we observe a spinodal-assisted ordering process we calculate the growth rate $R(q)$ from $S(q,t) \propto S(q,0)\exp[2R(q)t]$ for UA pVDF melts. $R(q)$ is a central quantity in the Cahn-Hilliard theory of spinodal decomposition,[15] and according to this theory, density fluctuations grow exponentially for a liquid undergoing a demixing instability. Figure 2*b* shows a Cahn plot ($R(q)/q^2$ versus $q^2$) for UA pVDF. The linear behavior with a negative slope observed in $R(q)$ for the polymer melts suggests that spinodal-assisted ordering prior to crystallization.

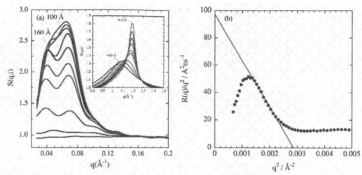

Figure 3. Evolution of the structure factor S(q) for UA pVDF (panel *a*). The induction, long and Bragg peaks increase during the simulation. The induction peak merges with the long peak suggesting that spinodal-assisted ordering is completed. The inset shown the time evolution of the second-order Bragg peak. Panel *b* shows the Cahn-Hilliard plot derived from the early stages of the nucleation process in the polymer crystallization simulations for pVDF. The solid lines are a fit to the data. The linear nature of the plot demonstrates the presence of a spinodal-assisted crystallization process in undercooled polymer melts.

In order to characterize structural evolution of $S(q,t)$ , we calculated the integral of $S(q,t)$ in the $q$-range (0.027,0.05) (integrated induction peak) and (0.05-0.1) (integrated long peak), and time evolution of the second-order Bragg peak location. Figure 4*a* shows the time dependence of the integrated induction peak and the integrated long peak for UA pVDF. The integrated induction peak grows in time until approximately 15 ns and saturates after this time, while the integrated long peak grows in time during the entire simulation. This suggests that spinodal-assisted ordering prior crystallization is completed after approximately 15 ns, and ordered domains continue to grow throughout the simulation. Figure 4*b* shows the time evolution of the second-order Bragg peak location, converted to an effective interchain distance, *d*. The abrupt change in *d* at approximately 10 ns for UA pVDF indicates a phase transition between unoriented and oriented domains.

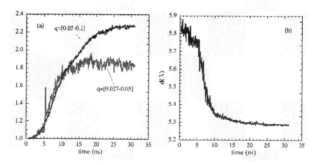

Figure 4. The time evolution of integrated intensity of the small-angle scattering peaks [$S(q,t)$ integrated for $q$ values as labeled] are shown in for UA pVDF (panel *a*). The time evolution of the second-order Bragg peak location, converted to an effective interchain distance, *d*, for UA pVDF (panel *b*).

## 4. Conclusions

In this paper we presented a study of polar pVDF polymer melt properties upon undercooling. We find that polymer chains cooled from the melt attain critical rigid stem lengths, and illustrate the mechanism by which ordering occurs. Further, we provide evidence that ordering is a spinodal-assisted nucleation and is driven by the growth of rigid segments. The spinodal-assisted nucleation is demonstrated showing the linearity of the growth rates in the Cahn-Hiliard plots. The detailed comparison of spinodal-assisted nucleation of both polar and non-polar polymer melts will be presented elsewhere.

## Acknowledgments

The work was performed under the auspices of the U.S. Department of Energy by the University of California Lawrence Livermore National Laboratory under Contract W-7405-Eng-48. We thank the Livermore Computing for generous amounts of CPU time on Thunder and MCR clusters.

## References

1.  Kaji, K., Nishida, K., Matsuba, G., Kanaya, T. & Imai, M. Details of Structure Formation During the Induction Period of Spinodal-Type Polymer Crystallization. Journal of Macromolecular Science-Physics B42, 709-715 (2003), and references therein.
2.  Gee, R. H. & Fried, L. E. Ultrafast crystallization of polar polymer melts. J. Chem. Phys. 118, 3827 (2003), and references therein.
3.  Olmsted, P. D., W.C.K.Poon, McLeish, T. C. B., Terrill, N. J. & A.J.Ryan. Spinodal-Assisted Crystallization in Polymer Melts. Physical Review Letters 81, 373 (1998), and references therein.
4.  Imai, M., Kaji, K., Kanaya, T. & Sakai, Y. Ordering Process in the induction period of crystallization of poly(ethylene terephthalate). Physical Review B 52, 12696 (1995).
5.  Mandelkern, L., Martin, G. M. & F.A. Juinn, J. J. Res. Nat. Bur. Stand 58, 2745 (1957).
6.  Gee, R. H., Lacevic, N. M. & Fried, L. E. Atomistic Simulation of Spinodal-Assisted Polymer Crystallization. Nature Materials (in review) (2005).
7.  Paul, W., Yoon, D. Y. & Smith, G. D. An optimized united atom model for simulations of polymethylene melts. J. Chem. Phys. 103, 1702-1709 (1995).
8.  Mayo, S. L., Olafson, B. D. & Goddard, W. A. J.Phys. Chem. 94, 8897 (1990).
9.  Gao, G. (California Institute of Technology, 1998).
10. Plimpton, S. J. Fast Parallel Algorithms for Short-Range Molecular Dynamics. J. Comp. Phys. 117, 1-19 (1995).
11. Verlet, L. Computer "Experiments" on Classical Fluids. I. Thermodynamical Properties of Lennard-Jones Molecules. Phys. Rev. 159, 98 (1967).
12. Nose, S. A unified formulation of the constant temperature molecular dynamics methods. J. Chem. Phys. 81, 511 (1984), and references therein.
13. Melchionna, S., Ciccotti, G. & Holian, B. L. Hoover's style Molecular Dynamics for systems varying in shape and size. Mol. Phys. 78, 533 (1993).
14. Hockney, R. W. & Eastwood, J. W. Computer Simulation using Particles (McGraw-Hill, New York, 1981).
15. Cahn, J. W. & Hilliard, J. E. Free energy of a nonuniform system I. Interfacial free energy. J. Chem. Phys. 28, 25 (1958).

Brill Academic Publishers
P.O. Box 9000, 2300 PA Leiden
The Netherlands

*Lecture Series on Computer*
*and Computational Sciences*
Volume 4, 2005, pp. 790-791

# Quantum Chemistry Literature Data Base

Haruo Hosoya[1]

Emeritus Professor
Ochanomizu University,
Bunkyo-ku, Tokyo 112-8610, Japan

Received 8 July, 2005; accepted in revised form 2 August, 2005

*Abstract:* Quantum Chemistry Literature Data Base (QCLDB) is a database of papers dealing with the results of *ab initio* calculations of atomic and molecular electronic structure. History, purpose, function, and on-line usage of QCLDB are described.

*Keywords:* ab inito calculation, electronic states, literature data base, on-line version, internet.
*Mathematics Subject Classification:* 81-00

*PACS:* 81-08, 92E99

## 1. What is QCLDB?

Quantum Chemistry Literature Data Base (QCLDB) is a database of papers published after 1978 which deal with the results of *ab initio* calculations of atomic, molecular, and some crystal electronic structure. From about thirty of internationally available core journals those papers are collected, surveyed, and given proper set of tags representing the content and essence therein by the big group (more than a hundred members in number) of young Japanese quantum chemists. Those theoretical works even without reporting any computational results are also included if they are judged to have significant relevance to *ab initio* calculations, while semi-empirical calculations are excluded. After a few steps of data check QCLDB is finally edited and copyrighted by Quantum Chemistry Data Base Group (QCDBG) of which the present author is working as chairperson.

For more than a quarter century QCLDB has compiled almost 75,000 original papers which can be retrieved and searched through internet as will be explained later in this paper. Although several changes have occurred during this long period of time, now each paper is tagged with (1) Author names, (2) Journal names, (3) Volume, pages, and year, (4) Substance formula, (5) IUPAC names or common names of the substances, (6) Computational methods, (7) Basis sets, (8) Physical properties, and (9) Comment. The user can search all the stored documents by combining the keywords selected freely from the above categories. The user can get perspective view of each paper just by looking through the Comment column, where important keywords other than what are already included in the former eight categories are selected in free format.

In this manner QCLDB provides quite useful information not only to chemists but also to both theoretical and experimental scientists, in physics, astronomy, life science, and material sciences etc., who want to know the sate of the art theoretical aspects of chemical substances and materials.

## 2. History of QCLDB

Every year a special issue "Quantum Chemistry Literature Data Base. Supplement. xx. Bibliography of Ab initio Calculations for yyyy", (xx=1–23, yyyy=1981–2003) has been published in Journal of Molecular Structure (THEOCHEM) by Elsevier carrying the latest data annually edited [1]. These printed versions comprise a series of supplement to the hard covered book of QCLDB published in 1982 [2,3], which contains only 2465 papers. Presently more than 7000 papers including DFT (Density-Functional Theory) papers are collected and stored in QCLDB annually.

Before the QCLDB project begins, Richards in Oxford University has published similar but smaller data base of ab initio calculations for the period from 1969 to 1980 [4]. However, in 1980 he declared

---

[1] E-mail: hosoya@is.ocha.ac.jp

to quit from his project and asked Ohno to continue QCLDB. Since then QCLDB has been the only internationally appreciated data base in this area of science.

The idea of QCLDB was first hinted in as early as 1976 by the late Professor Takehiko Simanouti of University of Tokyo to Kimio Ohno, the first chairperson of QCDBG, of Hokkaido University, who was the leader of "ab initioists" in Japan at that time. After a few years of preparatory period, in 1979 the first version of QCLDB was installed in the supercomputer in the Institute for Molecular Science (IMS) at Okazaki, Aich, Japan. This data base was freely provided for quantum chemists working in universities or in governmental and non-profit institutions. For other people working in and outside of Japan cassette-tape version was leased through Japan Association for International Chemical Information (JAICI). Although the number of Institutions exceeded one hundred and fifty at the peak time, those users are decreasing with the popularization of internet. Now QCLDB is open to the public through internet as will be explained below.

## 3. QCLDB-on-Line

Everybody who wants to know and use the QCLDB can get into QCLDB II by opening the website
http://qcldb2.ims.ac.jp/cqi-bin/qcldb.cqi,
where he/she will be asked to input his/her password. Even if he/she is unfamiliar to this system and has no idea about his/her own password to QCLDB, instantly his/her temporal password will be sent to his/her machine upon his/her request by opening "Registration."

Once entering QCLDB II system, one can freely execute bibliographical search through the pull-down menu boxes by composing questions using his/her own keywords with "and", "or", and "not". Then the computer returns the number of hit papers, which can be displayed on the screen. The number of hit papers can be adjusted to a manageable size by modifying the question. As already explained, each document carries nine categories and its ID number. The obtained information will help the user decide in which direction his/her research or survey should go.

Finally the active members of QCDBG will be introduced: Masahiko Hada and Kenro Hashimoto (Tokyo Metropolitan Univ.), Haruo Hosoya (Ochanomizu Univ.), Nobuaki Koga (Nagoya Univ.), Toshio Matsushita (Osaka City Univ.), Hidenori Matsuzawa (Chiba Inst. of Technology), Shigeru Nagase (IMS), Umpei Nagashima (Natl. Inst. of Advanced Industrial Science and Technology), Shinko Nambu (IMS), Keiko Takano (Ochanomizu Univ.), Hiroaki Wasada (Gifu Univ.), Shin'ichi Yamabe (Cochair,Nara Univ. of Education).

## Acknowledgments

The author and the active members of QCDBG are grateful to all the colleagues who have been engaged in the everlasting project of editing and compiling QCLDB for more than a quarter century. Special thanks go to the former chairperson Kimio Ohno and the retired QCDBG members, Hiroshi Kashiwagi, Kiyoshi Tanaka, Yoshihiro Osamura, Masatomo Togasi, Shigeru Obara, Kazuhiko Honda, Shigeyoshi Yamamoto, Satoshi Yabushita, Satoshi Minamino, and Keiji Morokuma. Ministry of Education, Culture, Sports, Science, and Technology has supported this project as the main sponsor through IMS. Cooperation by JAICI is highly acknowledged.

## References

[1] Suuplement 22. Bibliography of ab initio Calculations for 1981, *Journal of Molecular Structure (THEOCHEM)* **91** 1–250 (1982).

[2] K. Ohno and K. Morokuma: *Quantum Chemistry Literature Data Base. Bibliography of ab initio Calculations for 1978-1980.* Elsevier, Amsterdam, 1982.

[3] Y. Osamura et al.,Quantum Chemistry Literature Data Base, *Journal of Chemical Information and Computer Sciences* 21 86–90 (1981).

[4] W. G. Richards et al.: *A Bibliography of ab initio Molecular Wave Functions.* Clarendon, Oxford, 1971.

Brill Academic Publishers
P.O. Box 9000, 2300 PA Leiden
The Netherlands

*Lecture Series on Computer
and Computational Sciences*
Volume 4, 2005, pp. 792-795

# The Mathematics of Zeolite-like Possible Carbon and Boron Nitride Allotropes

R. Bruce King[*]

Department of Chemistry
University of Georgia
Athens, Georgia 30602, USA

Received 27 June, 2005; accepted in revised form 21 July, 2005

*Abstract:* Structures for low-density zeolite-like possible carbon and boron nitride allotropes can be constructed from networks of trigonal ($sp^2$) carbon atoms containing hexagons, heptagons, and/or octagons. The most symmetrical such structures are obtained by leapfrog transformations of either the Dyck tessellation of 12 octagons or the Klein tessellation of 24 heptagons. The Dyck and Klein tessellations can be obtained by subjecting a cube to quadrupling (chamfering) and septupling (capra) transformations, respectively. Both the Dyck and Klein tessellations can be embedded into genus 3 infinite periodic minimal surfaces of negative curvature. The automorphism group, $^4O$, of the Dyck tessellation is a solvable group of order 96, the structure of which relates to the dual of the Dyck tessellation as the regular tripartite graph $K_{4,4,4}$. However, the automorphism group, $^7O$, of the Klein tessellation is a simple group of order 168, which can be generated from the prime number 7 in a way analogous to the generation of the icosahedral rotation group $I$ ($\approx$ the alternating group $A_5$) from the prime number 5.

*Keywords:* Zeolite structures, carbon allotropes, Dyck tessellation, Klein tessellation, infinite periodic minimal surface

*Mathematics Subject Classification:* Riemann surfaces, negative curvature manifolds, minimal surfaces, finite automorphism groups

*PACS:* 14H55, 32Q05, 53A10, 20B25

## 1. Introduction

The experimentally known allotropes of carbon include diamond, graphite, the fullerenes, and carbon nanotubes. Among these carbon allotropes diamond and graphite have been known since antiquity, the fullerenes were discovered in the 1980s, and the nanotubes in the 1990s. The building blocks of all of the known carbon allotropes other than diamond are trigonal $sp^2$ carbon atoms, which are favored to form hexagons. Introduction of pentagons of $sp^2$ carbon atoms as in the fullerenes and nanotubes leads to sites of positive curvature.

A question of interest is whether carbon allotropes can be obtained having structures based on networks of trigonal carbon vertices with heptagons and/or octagons as well as hexagonal carbon rings. The presence of heptagons or octagons leads to sites of negative curvature and precludes the formation of closed polyhedra in the absence of compensating pentagons or squares. The resulting infinite structures resemble those of zeolites and provide the possibility of a new porous low-density form of carbon having cavities suitable for trapping small molecules. Such carbon allotropes would be non-oxidic non-metallic analogues of zeolites.

The possibility of negative curvature allotropes of carbon was apparently first recognized by Mackay and Terrones [1] in 1991.In 1992 Vanderbilt and Tersoff [2] first postulated the so-called D168 carbon structure with a unit cell of genus three containing 24 heptagons and 80 hexagons and a total of 168 carbon atoms. A major obstacle in the synthesis of such carbon allotropes is the selective formation of carbon rings of the desired sizes (e.g., hexagons and heptagons) without forming carbon rings of undesired sizes (e.g., pentagons).

---

[*] Corresponding author. E-mail: rbking@sunchem.chem.uga.edu.

Boron nitride analogues of diamond and graphite are known with the stoichiometry $(BN)_x$ in which pairs of adjacent carbon atoms are replaced by isoelectronic BN pairs thereby suggesting the possible existence of negative curvature boron nitrides. In this connection, boron nitride structures $(BN)_x$ are most favorable energetically if they contain only B–N bonds without any B–B or N–N bonds. However, the presence of carbon rings with an odd number of atoms makes impossible analogous $(BN)_x$ structures if only the most energetically favorable B–N bonds are allowed. Thus in order to obtain $(BN)_x$ structures having only B–N bonds, only $(BN)_r$ rings with even numbers of atoms having alternating boron and nitrogen atoms are allowed. Therefore, the most favorable positive curvature BN cage analogous to the fullerenes contains only $(BN)_2$ squares and $(BN)_3$ hexagons and the most favorable negative curvature IPMS boron nitride structure contains only $(BN)_4$ octagons and $(BN)_3$ hexagons. In this connection the truncated octahedral $B_{12}H_{12}$ structure has been shown computationally [3, 4] to be a favorable structure for a cage $(BN)_{x_x}$ boron nitride.

The renaissance of elemental carbon chemistry arising from the discoveries first of fullerenes and then of carbon nanotubes has raised new and interesting mathematical questions relating to their topologies and symmetries. The possibility of creating even more complicated zeolite-like carbon or boron nitride allotrope structures based on negative curvature networks containing hexagons, heptagons, and/or octagons opens up even richer areas of mathematics. This presentation provides an overview of some of the areas of mathematics relevant to understanding these structures including pertinent aspects of geometry, topology, and group theory.

## 2. Geometry:  The Dyck and Klein Surfaces

The ancient Greeks already knew about the five regular polyhedra, namely the tetrahedron of $T_d$ symmetry, the cube and octahedron of $O_h$ symmetry (a dual pair), and the dodecahedron and icosahedron of $I_h$ symmetry (another dual pair). These regular or Platonic polyhedra have all faces, vertices, and edges equivalent. Furthermore, all faces of these regular polyhedra are regular polygons. These regular polyhedra are homeomorphic to the sphere and are therefore of genus zero. Extension of the concept of Platonic polyhedra to Platonic tessellations on surfaces of non-zero genus [5] leads to the genus 3 Dyck tessellation with 32 vertices and 12 regular octagons and the likewise genus 3 Klein tessellation with 56 vertices and 24 regular heptagons. These tessellations are of particular interest as templates for the construction of zeolite-like carbon allotropes containing heptagons and octagons as well as hexagons.

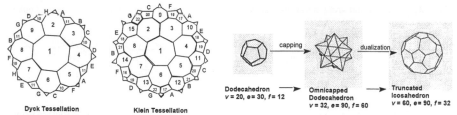

Dyck Tessellation    Klein Tessellation

Dodecahedron → Omnicapped Dodecahedron → Truncated Icosahedron
$v = 20, e = 30, f = 12$    $v = 32, e = 90, f = 60$    $v = 60, e = 90, f = 32$
capping    dualization

**Figure 1:** (a) The Dyck and Klein tessellations. In both cases the edges with identical letters (A→H or G) are identified to embed the tessellation in a genus 3 surface; (b) The leapfrog transformation of the regular dodecahedron to the truncated icosahedron of $C_{60}$.

A hypothetical carbon allotrope $C_{20}$ with 12 interlocking five-membered rings and icosahedral symmetry based on the regular dodecahedron is not stable because of the adjacent pentagons in this structure leading to antiaromatic pentalene units. In order to obtain a stable polyhedral allotrope of carbon, the network of pentagons must be diluted with enough hexagons so that no pair of pentagons in the final structure shares any edges. Introducing the minimum number of hexagons to achieve this objective triples the number of vertices in $C_{20}$ while preserving its icosahedral symmetry to give the famous truncated icosahedral $C_{60}$ structure in a so-called leapfrog transformation. Similar tripling of the Dyck and Klein tessellations leads to the unit cells of possible low-density carbon allotropes containing 96 or 168 carbon atoms, respectively; the latter is the structure proposed by Vanderbilt and Tersoff [2].

The leapfrog transformation is the simplest of a series of processes multiplying the number of vertices of a map by a small integer. Also of interest are the quadrupling or chamfering transformation and the septupling or capra transformation which multiply the number of vertices by four and seven, respectively [6]. Applying the quadrupling and septupling processes to the cube lead to the Dyck and Klein tessellations, respectively (Figure 2).

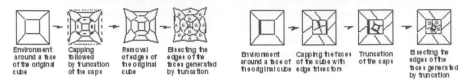

Figure 2: (a) The local environment of a face of a cube during the quadrupling (chamfering) transformation; (b) The local environment of a face of a cube during the septupling (capra) transformation.

## 3. Topology: Curvature and Infinite Periodic Minimal Surfaces

Consider surfaces of constant curvature that can be embedded in three-dimensional space. If the constant curvature is positive, the resulting surface is a sphere. Analogous surfaces with constant negative curvature do not include any shape as simple as a sphere. Useful surfaces with negative curvature for the description of chemical structures are known as infinite periodic minimal surfaces (IPMSs) [7]. In this context minimal surfaces are surfaces where the mean curvature at each point is zero so that such surfaces are negative curvature surfaces constructed from hyperbolic planes. Minimal surfaces are saddle-shaped everywhere except at certain "flat points" that are higher order saddles. IPMSs are obtained by smoothly joining patches of varying negative curvature and zero mean curvature to give an infinite surface with zero mean curvature that is periodic in all three directions. Such IPMSs cannot be defined by analytic functions in the usual Cartesian space of three dimensions but instead require elliptic or hyperelliptic integrals, which must be solved numerically. The finite surface element building block that is repeated periodically throughout space in an IPMS plays a role analogous to the unit cell in a crystal structure. Of particular interest in constructing the zeolite-like low-density carbon allotropes are the P and D IPMSs with cubic and tetrahedral unit cells, respectively (Figure 3a). Such unit cells are naturally divisible into eight equivalent octants, represented by the "plumber's nightmare" consisting of an octahedral junction of six pipes through the faces of a cube (Figure 3b). For the P surface all octants must be equivalent (Figure 3a) and contain the symmetry elements of the $O_h$ point group describing the symmetry of the underlying octahedron or its dual cube.

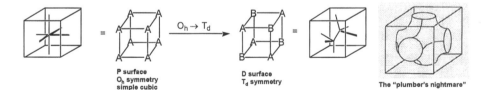

Figure 3: (a) The relationship between the P and D surfaces; (b) The "plumbers nightmare."

## 4. Symmetry and Group Theory: Regular Tripartite Graphs and Hidden Symmetry

The permutational symmetry group of the Dyck tessellation, $^4O$, is a solvable group with 96 operations that break down into the conjugacy classes $E + 24C_8 + 6C_4 + 3C_2 + 32C_3 + 12C_2' + 12C_4' + 6\ C_4''$. This is the symmetry of the $K_{4,4,4}$ tripartite graph, which is the dual of the Dyck tessellation [8]. In a similar sense the octahedron, which is the dual of the cube, can be considered to be a $K_{2,2,2}$ tripartite graph. The group $^4O$, although four times the size of the octahedral rotation group $O$, does not contain $O$ as a normal subgroup. Furthermore, $^4O$ is not a normal subgroup of the four-dimensional analogue of the octahedron.

The permutational symmetry group of the Klein tessellation, $^7O$, is a simple group with 168 operations that break down into the conjugacy classes $E + 24C_7 + 24C_7{}^3 + 56C_3 + 21C_2 + 42C_4$ [9]. This group can be generated from the prime number 7 in a way analogous to the generation of the icosahedral group $I$ (or the isomorphic alternating group $A_5$) from the prime number 5. Embedding the Klein tessellation into a cubic lattice based on the P surface (Figure 3a) destroys its seven-fold permutational symmetry so that this seven-fold symmetry is an example of its "hidden" symmetry.

## Acknowledgments

The author is indebted to the National Science Foundation for partial support of this work under Grant CHE-0209857.

## References

[1] A. L. Mackay and H. Terrones, Diamond from Graphite, *Nature*, **1991**, *352*, 762–762.

[2] D. Vanderbilt and J. Tersoff, Negative Curvature Fullerene Analogues of $C_{60}$, *Phys. Rev. Lett.*, **1992**, *68*, 511–513.

[3] I. Silaghi-Dumitrescu, F. Lara-Ochoa, and I. Haiduc, $A_{12}B_{12}$ (A = B, Al; B = N, P) 4/6 Fullerene-Like Cages and their Hydrogenated Forms Stabilized by Exohedral Bonds. An AM1 Molecular Orbital Study, *J. Mol. Struct. (Theochem)*, **1996**, *370*, 17–23.

[4] G. Seifert, P. W. Fowler, D. Mitchell, D. Porezag, and T. Frauenheim, Boron-Nitrogen Analogues of the Fullerenes: Electronic and Structural Properties, *Chem. Phys. Lett.*, **1997**, *268*, 352–358.

[5] R. B. King, Platonic Tessellations of Riemann Surfaces as Models in Chemistry: Non-zero Genus Analogues of Regular Polyhedra, *J. Mol. Struct.*, **2003**, *656*, 119–133.

[6] R. B. King and M. V. Diudea, From the Cube to the Dyck and Klein Tessellations: Implications for the Structures of Zeolite-like Carbon and Boron Nitride Allotropes, *J. Math. Chem.*, **2005**, *00*, 000.

[7] S. Andersson, S. T. Hyde, K. Larsson, and S. Lidin, Minimal Surfaces and Structures: From Inorganic and Metal Crystals to Cell Membranes and Biopolymers, *Chem. Rev.*, **1988**, *88*, 221–242.

[8] R. B. King, The Dual of the Dyck Graph as a Regular Tripartite Graph: Relevance to Hypothetical Zeolite-like Boron Nitride Allotropes, *MATCH—Communications in Mathematical Chemistry*, **2003**, *48*, 155–162.

[9] A. Ceulemans, R. B. King, S. A. Bovin, K. M. Rogers, A. Troisi, and P. W. Fowler, The Heptakisoctahedral Group and its Relevance to Carbon Allotropes with Negative Curvature, *J. Math. Chem.*, **1999**, *26*, 101–123.

Brill Academic Publishers
P.O. Box 9000, 2300 PA Leiden
The Netherlands

*Lecture Series on Computer*
*and Computational Sciences*
Volume 4, 2005, pp. 796-799

# STATE-UNIVERSAL MULTIREFERENCE COUPLED-CLUSTER METHOD USING A GENERAL MODEL SPACE

Xiangzhu Li and Josef Paldus[1]

Department of Applied Mathematics,
University of Waterloo,
Waterloo, Ontario, Canada N2L 3G1

Received 23 July, 2005; accepted in revised form 12 August

*Abstract:* A recently developed version of the multireference (MR) state-universal (SU) coupled-cluster (CC) method that employs a general model space (GMS), while preserving the intermediate normalization and size consistency, will be described. A number of examples of the performance of this technique will be presented.
*Keywords:* Multireference, Coupled Cluster, Electron Correlation, Excited States

*Mathematics SubjectClassification:* 81Q05, 81-08

*PACS:* 31.15.Dv, 31.15.Ar, 31.25.Qm

## 1. Basic Formalism

For a chosen set of references $\{\Phi_i\}$, the target wave functions $\Psi_i$ are expressed in terms of the SU CC Ansatz [1],

$$\Psi_i = \Sigma_j c_{ij} e^{T(j)} \Phi_j \ . \tag{1}$$

The cluster amplitudes $t_k(i)$ defining the cluster operators $T(i)$, $T(i) = \Sigma_k t_k(i) G_k(i)$, are obtained by solving SU CC equations

$$\langle G_l(i) \Phi_i | e^{-T(i)} H e^{T(i)} | \Phi_i \rangle = \Sigma_{j(\neq i)} \Gamma^{ij}(l) H_{ji}^{(\text{eff})} , \tag{2}$$

where

$$H_{ji}^{(\text{eff})} = \langle \Phi_j | H e^{T(i)} | \Phi_i \rangle \tag{3}$$

and

$$\Gamma^{ij}(l) = \langle G_l(i) \Phi_i | e^{-T(i)} e^{T(j)} | \Phi_j \rangle \tag{4}$$

are the effective Hamiltonian matrix elements and coupling coefficients, respectively. Further, $G_k(i)$ designate the pertinent excitation operators. The energies and the $c_{ij}$ coefficients are obtained by diagonalizing the effective Hamiltonian matrix $\|H_{ji}^{(\text{eff})}\|$.

## 2. C-conditions for General-Model-Space

Comparing with the standard SU CC method that uses a complete model space, the key idea that allows the use of a general model space (GMS), while preserving the size consistency, are the constraint or connectivity conditions (C-conditions) that are imposed on the internal cluster amplitudes (i.e., those associated with excitations transforming one reference configuration into another one) [2-13]. The C-conditions simply require that the amplitude of a higher-order connected internal excitation be fixed to cancel out all the possible products of its disconnected counterparts[2]. The C-conditions can be

---

[1] Also at Department of Chemistry and Guelph-Waterloo Center for Graduate Work in Chemistry, University of Waterloo, Waterloo, Ontario, Canada N2L 3G1. E-mail: paldus@scienide.uwaterloo.ca

derived from the fact that for an arbitrarily chosen reference set, the exact (i.e., the full configuration interaction) wave function must be expressible in terms of the SU CC Ansatz (1) [2]. The resulting C-conditions are generally applicable and can be employed in other MR CC methods [12] that are based on the cluster Ansatz (1). It is very easy to implement the C-conditions in the SU CC theory: We solve the above SU CC equations (2) for the external clusters, while at the same time enforce the C-conditions for the internal clusters.

## 3.   General Implementation

The general applicability of the code is achieved by relying on the ``local active orbital'' concept [2, 4] that facilitates the handling of required coupling terms between different reference configurations. The flexibility provided by the arbitrariness of the chosen GMS is of a great help in avoiding the well-known intruder state problems that often plague the standard MR CC approaches.

## 4.   Size-consistency

The C-conditions can be shown to guarantee the connectivity of both the effective Hamiltonian matrix elements and of coupling coefficients of the SU MR CC formalism [8, 9]. Yet, when we wish to describe a given dissociation channel, the chosen GMS must be size-consistent. For a separating super system A+B, we require that the reference space be a direct product of the reference spaces for A and B. Then the energies for A+B are given by the (composite) sum of the corresponding energies for A and B, i.e., $E_{ij}(A+B)=E_i(A)+E_j(B)$. In other words, the size-consistency is achieved not only for the ground state, but for the excited states as well [9].

## 5.   GMS SU CCSD and CCSD(T) Methods

The GMS SU CC methods have been implemented at the GMS SU CCSD [2, 4] and CCSD(T) [14] levels. The SD model approximates $T(i)$ by one- and two-body operators, $T_1(i)+T_2(i)$, relative to a reference $\Phi_i$ as a vacuum. In the SD(T) case, non-iterative or perturbative triple corrections are added after MR CCSD has converged. The computational cost of GMS SU CCSD scales as $Kn^6$, where $K$ is the number of references, while $n$ defines the size of the system considered. The cost of GMS SU CCSD(T) is one additional step of $Kn^7$. The GMS SU CCSD(T) is particularly useful when the MR CC approach is required to properly describe the spin coupling in those excited states, whose zero-order wave function requires multi-determinants.

## 6.   $(N,M)$-CCSD Method

Beside the perturbative approach correcting for triples, the effect of higher-than-pair clusters can also be accounted for through an externally corrected version of the GMS SU CCSD formalism, namely the $(N,M)$-CCSD method [6], representing the MR generalization of single-reference RMR CCSD. The method exploits $N$-reference MR CISD wave functions as a source of higher-order cluster amplitudes for the $M$-reference, $(M{\leq}N)$, GMS SU CCSD. In this method, $T(i)$ is given by $T_1(i)+T_2(i)+T_3^0(i)+T_4^0(i)$, where $T_3^0(i)$ and $T_4^0(i)$ are obtained from $N$-reference MR CISD [7] and fixed when solving $M$-reference CCSD. This kind of external corrections is important and useful for the handling of intruders.

The $(N,M)$-CCSD method can be exploited for a systematic improvement of computations. In the $(N,M)$-CCSD hierarchy, the standard SR CCSD method corresponds to (0,1)-CCSD. Also $(N,1)$-CCSD is a RMR CCSD corrected by $N$-reference MR CISD (i.e. $N$R-RMR CCSD) and $(0,M)$-CCSD is equivalent to $M$R SU CCSD. When $N$ is so large that the resulting $N$-reference MR CISD becomes FCI, we recover the exact FCI energies for $M$ states.

## 7.   Example: Excited States of $H_2O$

A detailed study of the low-lying excited states of water as obtained with various GMS SU CCSD methods was given in Ref.[11]. Most recently, we also examined the role of the triple corrections [14]. A selection of the resulting computed and experimental excitation energies is given in Table 1. Note that the triple corrections generally improve the computed energies, particularly for the low-lying states, where there is less uncertainty about the experimental values.

Table 1: $N$-reference CCSD and CCSD(T) ($N$=8 or 9) singlet and triplet vertical excitation energies (in eV) of $H_2O$ obtained with cc-pVTZ-plus-diffuse-function basis set. The values are related to the ground state energies of 9R CCSD (-76.32900 a.u.) or 9R CCSD(T) (-76.33661 a.u.). The experimental energies (Exp.) are given for the sake of a comparison.

| State | 9R | 9R(T) | Exp. | State | 8R | 8R(T) | Exp. |
|---|---|---|---|---|---|---|---|
| $^3A_1$ | 9.49 | 9.40 | 9.3 | $^3B_1$ | 7.18 | 7.07 | 7.0 |
| $^1A_1$ | 9.84 | 9.76 | 9.67 | $^1B_1$ | 7.51 | 7.41 | 7.42 |
| $^3A_1$ | 9.87 | 9.80 | 9.81 | $^3B_1$ | 10.03 | 9.96 | 9.98 |
| $^1A_1$ | 10.19 | 10.12 | 10.17 | $^1B_1$ | 10.06 | 9.99 | 10.01 |
| $^3A_1$ | 12.11 | 12.04 | | $^3B_1$ | 10.70 | 10.61 | 11.01 |
| $^1A_1$ | 12.34 | 12.27 | | $^1B_1$ | 10.80 | 10.70 | 11.11 |
| $^3A_1$ | 12.98 | 12.89 | | $^3B_1$ | 12.23 | 12.17 | |
| $^1A_1$ | 13.12 | 13.03 | | $^1B_1$ | 12.27 | 12.20 | |
| State | 8R | 8R(T) | Exp. | State | 8R | 8R(T) | Exp. |
| $^3A_2$ | 9.15 | 9.08 | 8.9 | $^3B_2$ | 11.39 | 11.30 | 11.11 |
| $^1A_2$ | 9.30 | 9.24 | 9.1 | $^1B_2$ | 11.65 | 11.57 | |
| $^3A_2$ | 11.01 | 10.94 | 10.68 | $^3B_2$ | 13.05 | 12.94 | |
| $^1A_2$ | 11.06 | 11.00 | | $^3B_2$ | 13.44 | 13.31 | |
| $^1A_2$ | 13.52 | 13.44 | | $^1B_2$ | 13.42 | 13.34 | |
| $^3A_2$ | 13.52 | 13.44 | | $^1B_2$ | 13.82 | 13.67 | |
| $^3A_2$ | 14.42 | 14.32 | | $^3B_2$ | 15.61 | 15.50 | |
| $^1A_2$ | 14.62 | 14.52 | | $^1B_2$ | 15.93 | 15.82 | |

## Acknowledgments

The continued support by NSERC (Canada) is greatly acknowledged (.J.P).

## References

[1] B. Jeziorski and H. J. Monkhorst, Coupled Cluster Method for Multideterminantal Reference States, *Physical Review A: 24*, 1668 (1981).

[2] X. Li and J. Paldus, General-Model-Space State-Universal Coupled-Cluster Theory: Connectivity Conditions and Explicit Equation, *Journal of Chemical Physics*, **119**, 5320 (2003).

[3] X. Li and J. Paldus, The General-Model-Space State-Universal Coupled-Cluster Method Exemplified by the LiH Molecule, *Journal of Chemical Physics*, **119**, 5346 (2003).

[4] X. Li and J. Paldus, Performance of the General-Model-Space State-Universal Coupled-Cluster Method, *Journal of Chemical Physics*, **120**, 5890 (2004).

[5] J. Paldus, X. Li, and N. D. K. Petraco, General-Model-Space State-Universal Coupled-Cluster Method: Diagrammatic Approach, *Journal of Mathematical Chemistry*, **35**, 213 (2004).

[6] X. Li and J. Paldus, *N*-Reference, *M*-State Coupled-Cluster Method: Merging the State-Universal and Reduced Multireference Coupled-Cluster Theories, *Journal of Chemical Physics*, **119**, 5334 (2003).

[7] J. Paldus and X. Li, Analysis of the Multireference State-Universal Coupled-Cluster Ansatz, *Journal of Chemical Physics*, **118**, 6769 (2003).

[8] J. Paldus and X. Li, Can We Avoid the Intruder-State Problem in the State-Universal Coupled-Cluster Approaches While Preserving Size Extensivity, *Collection of Czechoslovak Chemical Communications*, **69**, 90 (2004).

[9] X. Li and J. Paldus, Size Extensivity of a General-Model-Space State-Universal Coupled-Cluster Method, *International Journal of Quantum Chemistry*, **99**, 914(2004)

[10] J. Paldus and X. Li, CC and CI Approaches to Quasidegeneracy, *Progress in Theoretical Chemistry and Physics* (Editor: S. Wilson) (Springer, Berlin, 2005), in press.

[11] X. Li and J. Paldus, General-Model-Space State-Universal Coupled-Cluster Method: Excitation Energies of Water, *Molecular Physics*, in press.

[12] J. Pittner, X. Li and J. Paldus, Multi-reference Brillouin-Wigner Coupled-Cluster Method with a General Model Space, *Molecular Physics*, **103**, 2239 (2005).

[13] X. Li, General-Model-Space State-Universal Coupled-Cluster Method: Excited States of Ozone, *Collection of Czechoslovak Chemical Communications*, in press.

[14] X. Li and J. Paldus, unpublished results.

Brill Academic Publishers
P.O. Box 9000, 2300 PA Leiden
The Netherlands

*Lecture Series on Computer
and Computational Sciences*
Volume 4, 2005, pp. 800-801

# Applying Computational Methods to Assess the Toxicity of Chemical and Biological Agents that Lack Empirical Data

*Chandrika J. Moudgal*[1]
*Threat and Consequence Assessment Division, National Homeland Security Research Center,
Office of Research and Development, U.S.EPA
22350 SW 111th Avenue, Tualatin, OR 97062*

Received 12 July, 2005; accepted in revised form 16 August, 2005

*Abstract:* As terrorism-related activity increases in the US and other countries, it is important that the scientific community lends assistance to the affected population regarding potential health effects of the threat agents used in terrorism. Since traditional toxicity data are often absent for most chemical and biological agents, alternative methods such as QSARs and VFARs have to be applied to estimate the heath effects of the aforementioned threat agents. This presentation will summarize the risk assessment process within U.S.EPA and will outline QSAR methods that have been applied to date to fill in the experimental data gaps. In addition, the presentation will also cover some future research plans to estimate hazards from chemical and biological agents using computational methods.

*Keywords:* Threat agents, toxicity, QSAR, VFAR, risk assessment
*Mathematics Subject Classification:* 92B99, 92E10
*PACS:* 31.15.Ar, 87.10.+e

With an increase in terrorism-related activity in the US and other countries in the world today, and the ready availability of chemical and biological agents/weapons of mass destruction to terrorists, it is important for the scientific community to lend assistance to the affected areas/populations regarding the potential health effects of the aforementioned agents on a timely basis. More often than not, the potential health effects of the various agents, revealed by conducting a risk assessment, are either not available in the literature, or are not elucidated using standardized experimental protocols. A lack of experimental data and/or poorly generated data which do not meet recommended guidelines, such as the 1983 National Academy of Sciences (NAS)[1] risk assessment paradigm pose a problem in assessing the acute and long-term health risks. The NAS paradigm and the EPA risk assessment/risk management paradigm are depicted in Figure 1.

Figure 1 outlines the data requirements to conduct a risk assessment and also outlines the processes involved in conducting a risk assessment and risk management.

Recent high-profile events in highly secure areas such as US Senate offices have pushed for a greater understanding of the health effects of threat agents, especially since exposure to these agents may lead to severe disability, including death. Since empirical data on many chemical and biological agents are unavailable in the literature, and standard empirical testing methods are time consuming and resource intensive, it is important that innovative and resource saving methods be utilized to assess the short and long-term effects posed by these agents. One such innovative method that is commonly used with great success is Quantitative Structure Activity Relationship (QSAR) models. QSAR models can help estimate health effects of chemicals that lack empirical data by solely using the physical and chemical properties of a given chemical. Since the models are dependent on physicochemical properties alone, QSARs are in a position to estimate potential health effects of current, as well as new/modified chemical or biological agents. In addition, the time required to generate an estimate using QSARs is a fraction of the time required to assess the health effects using conventional empirical testing. Another upcoming method that can be applied to assess the health effects of biological agents that lack traditional experimental data is Virulence Factor Activity Relationships (VFARs). Virulence-factor

---

[1] *Corresponding Author E-mail:* <u>moudgal.chandrika@epa.gov</u>

activity relationship (VFAR) is a concept that was developed as a way to relate the architectural and biochemical components of a microorganism to its potential to cause human disease [2].

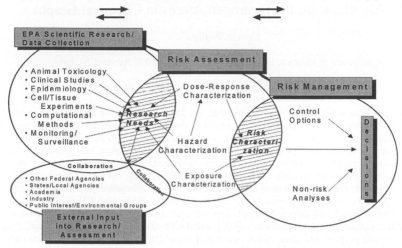

Figure 1: NAS Risk Assessment Paradigm and U.S EPA's Risk Assessment/Risk Management Paradigm
*Source: Courtesy, Dr. Hugh McKinnon, U.S. EPA, National Risk Management Research Laboratory*

The Threat and Consequence Assessment Division (TCAD) at the National Homeland Security Research Center (NHSRC), U.S. EPA intends to modify existing risk assessment methods [3] and apply them to selected priority chemical and biological agents. Since traditional experimental data for agents on this list are incomplete, TCAD scientists are currently utilizing existing QSAR methods, while also developing *de novo* QSARs to fill the data gaps. In addition to QSARs, the scientists are also experimenting with applying VFARs to fill the data gaps for biological agents.

In the case of QSAR application, commercial QSAR models (TOPKAT, DEREK, METEOR) will be used to estimate both the short- and long-term risks of chemical agents. These models will enable the NHSRC to set research goals and develop a strategic plan for an effective decontamination process.

## Acknowledgments

The author wishes to thank Dr. Andrew P. Avel, Director National Homeland Security Research Center, Dr. Jonathan Herrmann, Deputy Director, National Homeland Security Research Center, Dr. Peter Jutro, Associate Director for Science, National Homeland Security Research Center, and Dr. Cynthia Sonich-Mullin, Director, Threat and Consequence Assessment Division for their support in the QSAR and VFAR related projects. U.S.EPA, Office of Research and Development, National Homeland Security Research Center provides the funding for the projects.

## References

[1] *Risk Assessment in the Federal Government: Managing the Process*. National Research Council; Academic Press: Washington, DC, 1983.

[2] T.M. Jenkins, T.M. Scott, J.R. Cole, S.A. Hashsham, and J.B. Rose; *Assessment of virulence-factor activity relationships (VFARs) for waterborne diseases*; Water Science & Technology Vol 50 No 1 pp 309–314 © IWA Publishing 2004.

[3] U.S.EPA, 2005, Integrated Risk Information System; http://www.epa.gov/iris/backgr-d.htm.

Brill Academic Publishers
P.O. Box 9000, 2300 PA Leiden
The Netherlands

*Lecture Series on Computer*
*and Computational Sciences*
Volume 4, 2005, pp. 802-803

# An Algorithm for the Hydrogen Atoms in Chemical Graphs

Lionello Pogliani[*]

Dipartimento di Chimica, Università della Calabria, 87030 Rende (CS), Italy.

Received July 19, 2005; accepted in revised form 17 August, 2005

*Abstract:* A new algorithm for the $\delta^v$ number, the basic parameter of molecular connectivity indices, is proposed. The new algorithm, which is also based on complete graphs for the core electrons, is able to encode the bonded hydrogen, without the need to introduce any new graph concept. The contribution for the bonded hydrogen is introduced taken into account through a perturbation parameter whose power can be optimized. This new algorithm is able to differentiate among compounds which have atoms with differing number of bonded hydrogen atoms but with the same number of heteroatoms i.e., among hydrogen depleted chemical graphs with the same number of connections, like, for example, $BHF_2$ and $CH_2F_2$. The proposed algorithm is tested with different properties of different classes of compounds and the resulting model is compared with the previous QSPR (quantitative structure-property relationship) model for the same property and class, which was obtained with the former $\delta^v$ algorithm, which did not include any hydrogen perturbation term.

*Keywords*: chemical graph theory, molecular connectivity, hydrogen contribution, model study.
*Mathematical subject Classification*: computer aided design, computer modeling and simulation.
*PACS*: 07.05.Tp

## 1. Methods and Results

Complete graphs, [1] have been used to encode the core electrons of atoms and to define a $\delta^v$ algorithm which takes into account their contribution, i.e., [2-5]

$$\delta^v = q \cdot \delta^v(ps) / (p \cdot r + 1) \tag{1}$$

The parameter $p \cdot r$ equals the sum of all vertex degrees in complete graphs, and this sum equals twice the number of connections. A $\delta^v$ algorithm which introduces the hydrogen atom contribution as a kind of perturbation has, recently, been proposed, which is not solely based on graph concepts but which has as a bonus, the possibility to 'graph'-differentiate compounds like $BHF_2$ and $CH_2F_2$, whose either simple or general graphs are identical. The algorithm is the following,

$$\delta^v = \frac{(q + f_\delta^n)\delta^v(ps)}{(pr+1)} \tag{2}$$

The hydrogen atom perturbation, $f_\delta$, is defined in the following way,

$$f_\delta = [\delta^v_m(ps) - \delta^v(ps)] / \delta^v_m(ps) = n_H / \delta^v_m(ps) \tag{3}$$

Here $\delta^v_m(ps)$ is the maximal $\delta^v(ps)$ value an heteroatom can have in a chemical HS general graph when all bonded hydrogens are substituted by heteroatoms, and $n_H$ is the number of bonded hydrogen atoms. We can now define the following molecular connectivity indices, which make up the best descriptor of the analyzed property (where: $I = (\delta^v_i + 1) / \delta_i$),

$$^0\chi = \Sigma_i(\delta_i)^{-0.5}, \quad ^0\psi_I = \Sigma_i(I_i)^{-0.5}, \quad D_F = \Sigma_F \delta_F, \quad D_{Cl} = \Sigma_{Cl} \delta_{Cl} \tag{4}$$

[*] E-mail address: lion@unical.it

The first results obtained with the new $\delta^v$ algorithm, including the hydrogen perturbation, are encouraging, as can be seen from the following model (Table 1) of the partition coefficient, Log$P$, of halogenated organic compounds in different media, where the power of the hydrogen perturbation is $n = 2$,

Table 1. The $F$ - $r$ - $s$ values due to $\{^0\chi, {}^0\psi_i, D_F, D_{Cl}\}$ for $f_\delta^2 = 0$ and $\neq 0$, and $N = 25$.

| $f_\delta^2$ | Saline | Blood | Liver | Oil | Muscle | Fat |
|---|---|---|---|---|---|---|
| 0 | 26-0.917-0.2$_5$ | 38-0.941-0.2$_5$ | 52-0.955-0.2 | 99-0.976-0.2 | 41-0.944-0.2 | 150-0.984-0.1 |
| $\neq 0$ | 19-0.888-0.3 | 53-0.956-0.2 | 81-0.970-0.1$_5$ | 219-0.989-0.1 | 48-0.951-0.2 | 283-0.991-0.1 |

Let us now leave out, first, five compounds and then ten compounds, but keeping $n_H/n_O =$ constant, in this case we obtain,

Table 2. The $F$ - $r$ - $s$ values due to $\{^0\chi, {}^0\psi_i, D_F, D_{Cl}\}$ for $f_\delta^2 = 0$ and $\neq 0$, and $N = 20$ and 10.

| $f_\delta^2$ | $N$ | Saline | Blood | Liver | Oil | Muscle | Fat |
|---|---|---|---|---|---|---|---|
| 0 | 20 | 21-0.920-0.3 | 32-0.946-0.2 | 36-0.952-0.2 | 75-0.976-0.2 | 29-0.941-0.2 | 138-0.987-0.2 |
| $\neq 0$ | 20 | 15-0.894-0.3 | 44-0.960-0.2 | 57-0.969-0.2 | 164-0.989-0.1$_5$ | 36-0.951-0.2 | 257-0.993-0.1 |
| 0 | 15 | 11-0.905-0.3 | 19-0.941-0.3 | 27-0.975-0.2 | 47-0.975-0.2 | 21-0.944-0.2 | 88-0.986-0.2 |
| $\neq 0$ | 15 | 9.1-0.886-0.4 | 25-0.954-0.2 | 48-0.975-0.2 | 96-0.987-0.1$_5$ | 29-0.960-0.2 | 281-0.996-0.1 |

Excluding the *Fat* case where the $f_\delta^2 = 0$ and $\neq 0$ description diverge as $N$ gets smaller favoring the $f_\delta^2 \neq 0$ case, practically the other Log$P$ cases show no consistent changes with $N$, an interesting result, actually an expected result with

## References

1. N. Trinajstić, *Chemical Graph Theory*, CRC: Boca Raton, 1992.
2. L. Pogliani, Encoding the Core Electrons with Graph Concepts. *J.Chem.Inf.Comput.Sci.* **44**, 42-49 (2004).
3. L. Pogliani, Modeling Molecular Polarizabilities with Graph-Theoretical Concepts. *J.Comput.Meth.Sci.&Eng.* **4**, 737-751 (2004).
4. L. Pogliani, Model of Physical Properties of Halides with Complete Graph based Indices. *Int.J.Quant.Chem.* **102**, 38-52 (2005).
5. L. Pogliani, A Natural Graph-Theoretical Model for the Partition and Kinetic Coefficients. *New J. Chem.* **29**, in press, (2005).

Brill Academic Publishers
P.O. Box 9000, 2300 PA Leiden
The Netherlands

*Lecture Series on Computer
and Computational Sciences*
Volume 4, 2005, pp. 804-806

# Germanium Cluster Polyhedra: A Density Functional Theory Study

Ioan Silaghi-Dumitrescu,[*] Attila Kun, Alex Lupan, and R. Bruce King

Department of Chemistry, Babe„s-Bolyai University, Cluj-Napoca, Roumania, and Department of
Chemistry, University of Georgia, Athens, Georgia 30602, USA

Received 7 March, 2004; accepted in revised form 10 March, 2004

*Abstract:* Hybrid DFT B3LYP geometry optimizations on $Ge_n^z$ (n=5-11, z=-6, -4, -2, 0, +2, +4, +6) have been undertaken in order to understand the structure of some post transition metal cluster anions experimentally found by X-ray crystallography. Distortions from the expected deltahedral geometries are interpreted in terms of the hypo- or hyperelectron counting of the skeletal electrons.

*Keywords:* Germanium clusters, polyhedra, density functional theory

*Mathematics Subject Classification:* Molecular structure

*PACS:* 92E10

## 1. Introduction

In the 1930s Zintl and co-workers discovered that a number of post-transition elements can be reduced with alkali metals in liquid ammonia to give anionic species. Work by Corbett and co-workers a few decades later led to the elucidation of the structures of several types of metal cluster anions found in these systems using X-ray crystallographic methods. Many of the polyhedra in these structures were found to be similar to the polyhedra in borane structures. However, in some cases, particularly with the Group 13 metals In and Tl, polyhedra cluster polyhedra were found that are unknown in borane chemistry.

Since the mid-1970s one of us (R. B. K.) has been studying three-dimensional aromaticity in boranes using models based on topology and graph theory. Shortly after the initial discovery of unusual polyhedra in anionic Group 13 metal clusters in the mid 1990s work was initiated on similar methods to understand the chemical bonding in these apparently anomalous metal cluster structures. However, it became apparent that it would be desirable to support these ideas by theoretical calculations using density functional methods. For this reason the collaboration between the University of Georgia and Babe„s-Bolyai University was initiated to study these post-transition metal clusters using density functional (DFT) methods. The initial computational work [1] was done on the six-vertex Group 13 clusters $M_6^z$ (M = In, Tl; $z = -8, -6, -4$) since $Tl_8^{8-}$ and $Tl_6^{6-}$ were known experimentally to have regular and oblate (compressed) octahedral structures, respectively.

A general difficulty with these initial calculations on anionic Group 13 clusters was the relatively high charges on the species of interest. These high charges led in some cases to convergence to the individual atoms beyond bonding distances. In order to avoid this difficulty analogous computations were performed on clusters isoelectronic with the clusters of interest but with lower overall charges. Germanium proved to be a good model post-transition element for this purpose since the magnitudes of the charges on bare germanium clusters were minimized for the skeletal electron counts of interest for comparison with experimental data. This paper summarizes the status of our work on $Ge_n^z$ clusters $Ge_n^z$ containing five, six, seven, eight, nine, ten and 11 germanium atoms with charges ranging from –6 to +6 in some cases.

## 2. Computational Methods

Geometry optimizations were carried out at the hybrid DFT B3LYP level[2] with the 6-311G(d) (valence) triple zeta quality basis functions extended by adding one set of polarization (d) functions. The Gaussian 94 package of programs[3] was used in which the fine grid (75302) is the default for

---

[*] Corresponding authors. E-mail:isi@chem.ubbcluj.ro

numerically evaluating the integrals and the tight ($10^{-8}$) hartree stands as default for the self-consistent field convergence. Computations were carried out using a variety of initial geometries for each cluster size including examples of possible symmetries for the final configurations. The symmetries were initially maintained during the geometry optimization processes. In addition, symmetry breaking using modes defined by imaginary frequencies was used to determine optimized structures with minimum energies. Vibrational analyses show that all of the final optimized structures are genuine minima at the B3LYP//6-311G(d) level without any significant imaginary frequencies ($N_{imag} = 0$). In a few cases the calculations ended with acceptable small imaginary frequencies[4] and based on the recent discussions in the literature [4] the corresponding structures are interpreted also as genuine minima.

## 3. Results and Discussion

### 3.1 Structures derived from bipyramids for $Ge_5{}^Z$, $Ge_6{}^Z$, and $Ge_7{}^Z$

Bipyramidal global minima were found for $Ge_5{}^{2-}$ and $Ge_7{}^{2-}$. These bipyramids were computed to become oblate (i.e., compressed along a major axis) without reduction in symmetry upon loss of two electrons to form the corresponding neutral species. Our computations on the six-vertex $Ge_6{}^{2-}/Ge_6$ system indicate that the oblate tetragonal bipyramid structure previously found for the 12-skeletal electron $In_6{}^{6-}$ undergoes further distortion to give a less symmetrical $C_s$ structure best regarded as an edge-capped trigonal bipyramid. Further removal of electrons from the neutral $Ge_n$ clusters to give the dications $Ge_n{}^{2+}$ having $2n - 2$ skeletal electrons was found to lead to more complicated structural changes. Thus for $Ge_5{}^{2+}$ the lowest energy structure is a completely unsymmetrical ($C_1$) array of three fused triangles whereas for $Ge_6{}^{2+}$ the edge-capped trigonal bipyramid found for $Ge_6$ undergoes further distortion to give a somewhat more symmetrical looking structure best regarded as an edge-bicapped butterfly. Only for the lowest energy computed structure of $Ge_6{}^{2+}$ does the oblate pentagonal bipyramid found for $Ge_7$ remain recognizable although it undergoes further distortion to an unsymmetrical prolate (elongated) bipyramid related to the prolate trigonal antiprisms previously computed for $In_6{}^{4-}$ and $Tl_6{}^{4-}$.

### 3.2 Eight-vertex structures

For $Ge_8{}^{2-}$ the $D_{2d}$ bisdisphenoid structure predicted by the Wade-Mingos rules is not computed to be the global minimum but instead lays 3.9 kcal/mole above the $T_d$ tetracapped tetrahedron global minimum predicted to exhibit spherical aromaticity. The hyperelectronic clusters $Ge_8{}^{4-}$ and $Ge_8{}^{6-}$ have *nido* $B_8H_{12}$ and square antiprism structures, respectively, as global minima in accord with the Wade-Mingos rules and experimental data on $E_8{}^{2+}$ (E = Sb, Bi) cations. Hypoelectronic eight vertex clusters isoelectronic and isolobal with $Ge_8$, $Ge_8{}^{2+}$, and $Ge_8{}^{4+}$ are not known experimentally. Their computed structures include smaller polyhedra having one or more capped triangular faces as well as more open non-polyhedral structures.

### 3.3 Nine-vertex structures

The global minimum for $Ge_9{}^{2-}$ is the tricapped trigonal prism expected by Wade's rules for a $2n + 2$ skeletal electron structure. An elongated tricapped trigonal prism is the global minimum for $Ge_9{}^{4-}$ (Figure 1) similar to the experimentally found structure for the isoelectronic $Bi_9{}^{5+}$. However, the capped square antiprism predicted by Wade's rules for a $2n + 4$ skeletal electron structure is only 0.21 kcal/mole above this global minimum indicating that these two nine-vertex polyhedra have very similar energies in this system. Tricapped trigonal prismatic structures are found for both singlet and triplet $Ge_9{}^{6-}$, with the latter being lower in energy by 3.66 kcal/mole and far less distorted. The global minimum for the hypoelectronic $Ge_9$ is a bicapped pentagonal bipyramid. However, a second structure for $Ge_9$ only 4.54 kcal/mole above this global minimum is the $C_{2v}$ flattened tricapped trigonal prism structure found experimentally for the isoelectronic $Tl_9{}^{9-}$. For the even more hypoelectronic $Ge_9{}^{2+}$ the lowest energy structure consists of an octahedron fused to two trigonal bipyramids. For $Ge_9{}^{4+}$ the global minimum is an oblate (squashed) pentagonal bipyramid with two pendant Ge vertices.

### 3.4 Ten-vertex structures

The global minimum for $Ge_{10}{}^{2-}$ is the $D_{4h}$ 4,4-bicapped square antiprism found experimentally in $B_{10}H_{10}{}^{2-}$ and other 10-vertex clusters with 22 skeletal electrons.

$Ge_9^{4-}$       $Ge_9^{3-}$       $Ge_9^{2-}$       $Ge_9 - Tl_9^{9-}$

Figure 1: Some optimized nine-vertex germanium cluster structures.

The global minima found for the electron-rich clusters $Ge_{10}^{4-}$ and $Ge_{10}^{6-}$ are not those known experimentally. However, experimentally known structures for *nido*-$B_{10}H_{14}$ and the pentagonal antiprism of *arachno*-$Pd@Bi_{10}^{4+}$ are found at higher but potentially accessible energies for $Ge_{10}^{4-}$ and $Ge_{10}^{6-}$. The global minimum for the neutral $Ge_{10}$ is the $C_{3v}$ 3,4,4,4-tetracapped trigonal prism predicted by the Wade-Mingos rules and found experimentally in the isoelectronic $Ni@Ga_{10}^{10-}$. However, only slightly above this global minimum for $Ge_{10}$ (+3.3 kcal/mole) is the likewise $C_{3v}$ *isocloso* 10-vertex deltahedron found in metallaboranes such as $(\eta^6\text{-arene})RuB_9H_9$ derivatives. Structures found for the more electron-poor clusters $Ge_{10}^{2+}$ and $Ge_{10}^{4+}$ include various capped octahedra and pentagonal bipyramids. Several 10-vertex cluster structures that have not yet been realized experimentally but would be interesting targets for future synthetic 10-vertex cluster chemistry using vertex units isolobal with the germanium vertices are predicted on the basis of the present calculations.

### 3.5 Eleven-vertex structures

The global minimum within the $Ge_{11}^{2-}$ set is an elongated pentacapped trigonal prism distorted from $D_{3h}$ to $C_{2v}$ symmetry. However, the much more spherical edge-coalesced icosahedron, also of $C_{2v}$ symmetry, expected by the Wade-Mingos rules for a $2n + 2$ skeletal electron system and found $B_{11}H_{11}^{2-}$ and isoelectronic carboranes, is of only slightly higher energy (+5.2 kcal/mole). Even more elongated $D_{3h}$ pentacapped trigonal prisms are the global minima for the electron-rich structures $Ge_{11}^{4-}$ and $Ge_{11}^{6-}$. For $Ge_{11}^{4-}$ the $C_{5v}$ 5-capped pentagonal antiprism analogous to the dicarbollide ligand $C_2B_9H_{11}^{2-}$ is of significantly higher energy (~28 kcal/mole) than the $D_{3h}$ global minimum. The $C_{2v}$ edge-coalesced icosahedron is also the global minimum for the electron-poor $Ge_{11}$ similar to its occurrence in experimentally known 11-vertex "isocloso" metallaboranes of the type $(\eta^6\text{-arene})RuB_{10}H_{10}$. The lowest energy polyhedral structures computed for the more hypoelectronic $Ge_{11}^{4+}$ and $Ge_{11}^{6+}$ clusters are very similar to those found experimentally for the isoelectronic ions $E_{11}$ ($E_{11}^{7-}$ = Ga, In, Tl) and $Tl_9Au_2^{9-}$ in intermetallics in the case of $Ge_{11}^{4+}$ and $Ge_{11}^{6+}$, respectively. These DFT studies predict an interesting $D_{5h}$ centered pentagonal prismatic structure for $Ge_{11}^{2+}$ and isoelectronic metal clusters.

### Acknowledgments

The authors are indebted to the National Science Foundation (Grant CHE-0209857) and Babes-Bolyai University Cluj-Napoca for partial support of this work.

### References

[1]  R. B. King, I. Silaghi-Dumitrescu, and A. Kun, *Inorg. Chem.*, **2001**, *40*, 2450.

[2]  Becke, A. D., *J. Chem. Phys.*, **1993**, *98*, 5648.

[3]  Frisch, M. J.; Trucks, G. W.; Schlegel, H. B.; Gill, P. M. W.; Johnson, B. G.; Robb, M. A.; Cheeseman, J. R.; Keith, T.; Petersson, G. A.; Montgomery, J. A.; Raghavachari, K.; Al-Laham, M. A.; Zakrzewski, V. G.; Ortiz, J. V.; Foresman, J. B.; Peng, C. Y.; Ayala, P. Y.; Chen, W.; Wong, M. W.; Andres, J. L.; Replogle, E. S.; Gomperts, R.; Martin, R. L.; Fox, D. J.; Binkley, J. S.; Defrees, D. J.; Baker, J.; Stewart, J. J. P.; Head-Gordon, M.; Gonzalez, C.; Pople, J. A., Gaussian 94, Revision C.3; Gaussian, Inc.: Pittsburgh PA, 1995.

[4]  Xie, Y.; Schaefer, H. F., III; King, R. B. *J. Am. Chem. Soc.*, **2000**, *122*, 8746.

Brill Academic Publishers
P.O. Box 9000, 2300 PA Leiden
The Netherlands

*Lecture Series on Computer*
*and Computational Sciences*
Volume 4, 2005, pp. 807-810

# Prediction of Reaction Pathways and New Properties in Organosilicon Compounds Based on *ab initio* MO and Density Functional Theory Calculations

Masae Takahashi[1]

Institute for Materials Research,
Tohoku University,
Sendai 980-8577, Japan

Received 2 July, 2005; accepted in revised form 29 July, 2005

*Abstract:* The paper consists of two topics. In the first part, mechanisms of 1,2-addition reactions to Si=C, Si=Si, and C=C are presented. All reactions start from electrophilic and nucleophilic initial complexes, $C_E$ and $C_N$, those are related to the frontier molecular orbital theory. In the reaction of $H_2Si=SiH_2 + H_2O$, $C_E$ and $C_N$ determine the whole reaction path and finally lead to different products. In the addition reaction of acidic alcohol or hydrogen halides, only electrophilic channels are found. For the reactions of methyldisilene with water and of ethynyldisilene with water, four initial complexes and therefore four reaction channels were found. Only two complexes were found, however, in the reactions of fluorodisilene and aminodisilene with water/ hydrogen fluoride. The reaction of polar substrate, aminodisilene, shows the similar reaction profiles to that of polar substrate, silene. In addition, product switching of the stereochemistry, depending on the acidity of the reagent was found for the reaction of aminodisilene. In the second part, we describe the design of planar anionic polysilicon chains and cyclic $Si_6$ anions with $D_{6h}$ Symmetry. Among 18 calculated molecules, planar equilibrium structures are obtained for seven anionic chains ($Si_2H_2^{2-}$, $Si_4H_2^{2-}$, $Si_4H_2^{4-}$, $Si_6H_2^{4-}$, $Si_6H_2^{6-}$, $Si_8H_2^{4-}$, $Si_8H_2^{6-}$) and three anionic six-membered rings ($c$-$Si_6^{2-}$, $c$-$Si_6^{4-}$, $c$-$Si_6^{6-}$). The number of π electrons formally accommodated in the out-of-plane π orbitals is the same as the number of silicon atoms and is independent of the number of doped electrons. An anionic six-membered ring, $c$-$Si_6^{2-}$, shows a large negative value of nucleus-independent chemical shifts (NICS), which indicates that $c$-$Si_6^{2-}$ is an aromatic molecule.

*Keywords:* Organosilicon, ab initio MO, DFT, Reaction Pathway, Unsaturated Silicon

## 1. Introduction

Two topics are introduced here: how to predict reaction pathways and how to design new molecules by computer simulations based on the author's recent publications. Target molecules are organosilicon compounds. The prediction of reaction pathways or new properties of materials with reliable accuracy by ab initio molecular orbital (MO) and density functional theory calculations is an attractive subject for theoretical chemists.

## 2. Reaction Pathways

The frontier molecular orbital (FMO) theory works well at the early stage of chemical reactions [1]. On the other hand, modern theoretical approach to reaction mechanisms is based on the determination of the transition states and on finding the pathways from the transition states followed by the intrinsic reaction coordinates (IRCs). The target reactions here are 1,2-addition to Si=C, Si=Si, and C=C.

### 2.1. Initial Steps of Water Addition Reaction to Disilene

In the reaction mechanisms of disilenes with alcohol, a four-centered cyclic mechanism including a nucleophilic attack of the alcoholic oxygen to the unsaturated silicon has been first proposed based on the theoretical and experimental studies [2,3]. However, Apeloig and Nakash experimentally found

---

[1] E-mail: masae@imr.edu

that an electrophilic attack prompts the reaction of disilene with phenol having electron-withdrawing substituents [4]. We have reinvestigated the water addition reaction to disilene by _ab initio_ quantum mechanical calculations and have found both an electrophilic and a nucleophilic channel [5,6]. The first step of both channels is the appearance of the initial complexes, $C_E$ and $C_N$. The structure of these complexes can be understood with the help of FMO theory [1] as they are formed by an electrophilic interaction between the LUMO of water and the HOMO of disilene (electrophilic channel), or by a nucleophilic interaction between the HOMO of water and the LUMO of disilene (nucleophilic channel). $C_E$ and $C_N$ determine the whole reaction path and finally lead to different products, _syn_- and _anti_-adducts, respectively, _via_ two different transition states, $T_E$ and $T_N$. The path from the Lewis adduct $C_L$ is common in the two reaction channels and similar to that proposed by Nagase, Kudo, and Ito [2]. The rate determining steps are before $C_L$. The two pathways well explain the experimentally observed solvent-dependent _anti_-product formation [4c] and also the kinetic isotope effects of the electrophilic channel [4a].

## 2.2. Sensitivity of the Reaction Mechanisms to the Acidity of Reagents and the Polarity of Substrates

In case of acidic alcohol ($CF_3OH$) [5-7] or hydrogen halides [8] only electrophilic channels could be found. This mechanism is, however, different from that of the usual addition reactions to C=C bond in a few steps. The reaction profile of water addition to silene having a polar double bond is also different from the case of disilene [7]. Initial nucleophilic and electrophilic complexes, $C'_E$ and $C'_N$, were also found to lead to two reaction channels. The reaction from $C'_E$ proceeds in one step giving silylmethanol _via_ a four centered, energy rich transition state, $T'_L$. On the other hand, the reaction from $C'_N$ occurs in two steps with intervening a Lewis adduct, $C'_L$. The latter reaction channel is a facile process with low activation energy. This result immediately explains the experimental fact: the product of alcohol addition to silene is always an alkoxysilane [9]. The regioselectivity of these reactions is caused by an inductive effect (I effect) due to the difference of electronegativity between silicon and carbon.

## 2.3. Possible Variation of the Reaction Mechanisms of 1,2-Additions to Substituted Disilenes

In these cases, the two silicon atoms are different which indicates four theoretically possible products with different product ratio depending on the effects of the substituent. The study of the substituent effects may explain and predict the variation in the regioselectivity and stereoselectivity and may help in the synthesis of the required derivatives [10]. To study the 1,2-addition reaction of polar substrate with homogeneous double bond, the water additions to the following disilenes are investigated: 1-methyldisilene, 1-fluorodisilene, 1-ethynyldisilene, and 1-aminodisilene. The expected effects of these substituents are the following. Methyl group: weak inductive (-I), fluorine: strong inductive (-I) and weak conjugative (+M), ethynyl group: inductive (-I) and π-conjugative (-M) and amino group: strong inductive (-I) and strong conjugative (+M) effects.

Here the initial complexes of those reactions play an important role [11]. All reactions start from the electrophilic and nucleophilic initial complexes, $C_E$ and $C_N$. Four initial complexes and therefore four reaction channels were found for the reactions of methyldisilene and of ethynyldisilene with water. Only two complexes were found, however, in the reactions of fluorodisilene and aminodisilene with water/ hydrogen fluoride. The reaction of polar substrate, aminodisilene, shows the similar reaction profiles to that of polar substrate, silene. In addition, product switching of the stereochemistry, depending on the acidity of the reagent was found for the reaction of aminodisilene.

## 3.  Design of Planar Silicon  -Electron Systems

Silicon π-electron systems remarkably differ in structure and stability from carbon π-electron systems, although formally the same hybridization occurs. The carbon π system is usually stable, while enormous efforts have been required for the isolation of stable unsaturated silicon compounds. The isolation of a stable disilene in 1981 by West _et al._ was a breakthrough in unsaturated silicon chemistry [12]. Recently, the synthesis and X-ray crystal analysis of the formally _sp_-hybridized trisilaallene have been reported by Kira _et al._ [13] and the Si=Si=Si skeleton of trisilaallene has been revealed to be significantly bent with a bond angle of 136.5°. It differs from the linear C=C=C skeleton of allene ($H_2C$=C=$CH_2$). The bonding at the central silicon atom of trisilaallene cannot be described as a simple _sp_-hybridization. Very recently, linear trisilaallenes have been designed theoretically by Apeloig _et al._

[14], where σ-donor, π-acceptor $R_2B$ substituents are used. A recent highlight in unsaturated silicon chemistry was the synthesis of a silicon-silicon triple-bonded compound reported by Sekiguchi *et al.* [15], which has been desired ever since the first synthesis of the double-bonded compound by West *et al.* The revelations of the *trans* bending of disilene and disilyne and of the chair form of a silicon analogue of benzene all go back over twenty years. A donor-acceptor bonding model [16] has well explained the *trans* bending of disilene and disilyne. The planar tendency in ethylene and benzene is essential for the π conjugation and $sp^2$-hybridization. However, planar polysilicon chains and planar silicon six-membered rings have not been obtained for neutral molecules, neither theoretically nor experimentally.

Planar acyclic polysilicon and aromatic six-membered rings of silicon with $D_{6h}$ symmetry is theoretically designed as follows [17]. Chains and rings consisting of disilyne units are considered, where each silicon atom has no substituent. In addition, electrons are doped forming anion, to stabilize the planar structures.

The structures of neutral and anionic hydrogen-terminated oligomers (H–$Si_n$–H: n = 2, 4, 6, 8) and silicon six-membered rings ($c$-$Si_6$) have been investigated using quantum-chemical methods to realize stable planar polysilicon chains and aromatic silicon six-membered rings. Among 18 calculated molecules, planar equilibrium structures are obtained for seven anionic chains ($Si_2H_2^{2-}$, $Si_4H_2^{2-}$, $Si_4H_2^{4-}$, $Si_6H_2^{4-}$, $Si_6H_2^{6-}$, $Si_8H_2^{2-}$, $Si_8H_2^{6-}$) and three anionic six-membered rings ($c$-$Si_6^{2-}$, $c$-$Si_6^{4-}$, $c$-$Si_6^{6-}$). The number of π electrons formally accommodated in the out-of-plane π orbitals is the same as the number of silicon atoms and is independent of the number of doped electrons. An anionic six-membered ring, $c$-$Si_6^{2-}$, shows a large negative value of nucleus-independent chemical shifts (NICS), which indicates that $c$-$Si_6^{2-}$ is an aromatic molecule.

## Acknowledgments

The author wishes to thank Professor Y. Kawazoe (Tohoku University), Professor M. Kira (Tohoku University), and Professor T. Veszprémi (Technical University of Budapest), for their fruitful discussion. The author appreciates the financial supports from the Ministry of Education, Science, Sports, and Culture of Japan (Grant No. 16310080), the Hayashi Memorial foundation, and the interdisciplinary research program of Tohoku University.

## References

[1] K. Fukui, Recognition of Stereochemical Paths by Orbital Interaction, *Acc. Chem. Res.* **4** 57-64(1971).

[2] S. Nagase, T. Kudo, K. Ito, Structures, Stability, and Reactivity of Doubly Bonded Compounds Containing Silicon or Germanium, *Applied Quantum Chemistry* (Editor: V. H. Smith Jr., H. F. Schaefer, K. Morokuma), Reidel: Dordrecht 249-267(1986).

[3] A. Sekiguchi, I. Maruki, H. Sakurai, Regio- and Stereochemistry and Kinetics in the Addition Reactions of Alcohols to Phenyl-Substituted Disilenes, *J. Am. Chem. Soc.* **115** 11460-11466(1993).

[4] (a) Y. Apeloig, M. Nakash, The Mechanism of Addition of Phenols to Tetramesityldisilene. Evidence for Both Nucleophilic and Electrophilic Rate-Determining Steps, *J. Am. Chem. Soc.* **118** 9798-9799(1996). (b) Y. Apeloig, M. Nakash, Arrhenius Parameters for the Addition of Phenols to the Silicon-Silicon Double Bond of Tetramesityldisilene, *Organometallics* **17** 2307-2312(1998). (c) Y. Apeloig, M. Nakash, Solvent-Dependent Stereoselectivity in the Addition of $p$-$CH_3OC_6H_4OH$ to *(E)*-1,2-Di-*tert*-butyl-1,2-dimesityldisilene. Evidence for Rotation around the Si-Si Bond in the Zwitterionic Intermediate, *Organometallics* **17** 1260-1265(1998).

[5] M. Takahashi, T. Veszprémi, B. Hajgató, M. Kira, Theoretical Study on Stereochemical Diversity in the Addition of Water to Disilene, *Organometallics* **19** 4660-4662(2000).

[6] M. Takahashi, T. Veszprémi, M. Kira, Importance of Frontier Orbital Interactions in Addition Reaction of Water to Disilene, *I. J. Quantum Chem.*, **84** 192-197(2001).

[7] T. Veszprémi, M. Takahashi, B. Hajgató, M. Kira, The Mechanism of 1,2-Addition of Disilene and Silene. 1. Water and Alcohol Addition, *J. Am. Chem. Soc.* **123** 6629-6638(2001).

[8] B. Hajgató, M. Takahashi, M. Kira, T. Veszprémi, The Mechanism of 1,2-Addition of Disilene and Silene: Hydrogen Halide Addition, *Chem. Eur. J.8* 2126-2133(2002).

[9] (a) M. Kira, T. Maruyama, H. Sakurai, Stereochemistry and Mechanism for the Addition of Alcohols to a Cyclic Silene, *J. Am. Chem. Soc.* **113** 3986-3987(1991). (b) G. W. Sluggett, W. J. Leigh, Reactive Intermediates from the Photolysis of Methylpentaphenyl- and Pentamethylphenyldisilane, *J. Am. Chem. Soc.* **114** 1195-1201(1992). (c) W. J. Leigh, G. W. Sluggett, Aryldisilane Photochemistry. A Kinetic and Product Study of the Mechanism of Alcohol Additions to Transient Silenes, *J. Am. Chem. Soc.* **116** 10468-10476(1994). (d) C. J. Bradaric, W. J. Leigh, Arrhenius Parameters for the Addition of Nucleophiles to the Silicon-Carbon Double Bond of 1,1-Dipheylsilene, *J. Am. Chem. Soc.* **118** 8971-8972(1996).

[10] M. Takahashi, T. Veszprémi, M. Kira, 1,2-Addition Reaction of Monosubstituted Disilenes: An *ab initio* Study, *Organometallics* **23** 5768-5778(2004).

[11] M. Takahashi, T. Veszprémi, K. Sakamoto, M. Kira, Role of Initial Complexes in 1,2-Addition Reactions of Disilene Derivatives, *Mol. Phys.* **100** 1703-1712(2002).

[12] R. West, M. J. Fink, J. Michl, Tetramesityldisilene, a Stable Compound Containing a Silicon-Silicon Double Bond, *Science* **214** 1343-1344(1981).

[13] S. Ishida, T. Iwamoto, C. Kabuto, M. Kira, A Stable Silicon-Based Allene Analogue with a Formally *sp*-Hybridized Silicon Atom, *Nature* **421** 725-727(2003).

[14] M. Kosa, M. Karni, Y. Apeloig, How to Design Linear Allenic-Type Trisilaallenes and Trigermaallenes, *J. Am. Chem. Soc.* **126** 10544-10545(2004).

[15] A. Sekiguchi, R. Kinjo, M. Ichinohe, A Stable Compound Containing a Silicon-Silicon Triple Bond, *Science* **305** 1755-1757(2004).

[16] (a) G. Trinquier, J.-P. Malrieu, Nonclassical Distortions at Multiple Bonds, *J. Am. Chem. Soc.* **109** 5303-5315(1987). (b) J.-P. Malrieu, G. Trinquier, Trans Bending at Double Bonds. Occurrence and Extent, *J. Am. Chem. Soc.* **111** 5916-5921(1989). (c) G. Trinquier, Double Bonds and Bridged Structures in the Heavier Analogues of Ethylene, *J. Am. Chem. Soc.* **112** 2130-2137(1990). (d) G. Trinquier, J.-P. Malrieu, Trans Bending at Double Bonds. Scrutiny of Various Rationales through Valence-Bond Analysis, *J. Phys. Chem.* **94** 6184-6196(1990).

[17] M. Takahashi, Y. Kawazoe, Theoretical Study on Planar Anionic Polysilicon Chains and Cyclic $Si_6$ Anions with $D_{6h}$ Symmetry, *Organometallics* **24** 2433-2440(2005).

Brill Academic Publishers
P.O. Box 9000, 2300 PA Leiden
The Netherlands

*Lecture Series on Computer
and Computational Sciences*
Volume 4, 2005, pp. 811-814

# 2$^{nd}$ Symposium on
# Industrial and Environmental Case Studies

## Preface

The Symposium on 'Industrial and Environmental Case Studies' includes research works pertaining to at least one of the following domains: (I) Industrial Operation / Process, (II) Environmental Impact, (III) Modelling/Simulation, (IV) Data / Information Processing – AI System. The kind of colligation is quoted in the simplified ontological diagram shown below. As the first two domains refer to content, the semantic links are mainly of the *is-a* or *has-a* type; as the last two domains refer to methodology, the semantic link *is-examined-by* (*means of a*) dominates.

I would like to thank Prof. Theodore Simos, who very perceptively and masterly embedded the Symposium in the Conference and Dr. Christina Siontorou for her contribution in the interactive communication with the participating research team leaders, as well as with the teams that, although strongly interested in, were unable to attend this year but they are going to send a research work well in advance for the next year's Symposium.

The current Symposium is hopefully anticipated to establish the conditions and requisites for both (a) a general and wide multi-lateral forum for exchanging ideas and experience in the field of Computer Aided Industrial / Environmental Applications, and (b) certain partial cooperative schemes in specialized topics where the participating teams have similar or complementary interests.

Fragiskos Batzias
Symposium Organizer

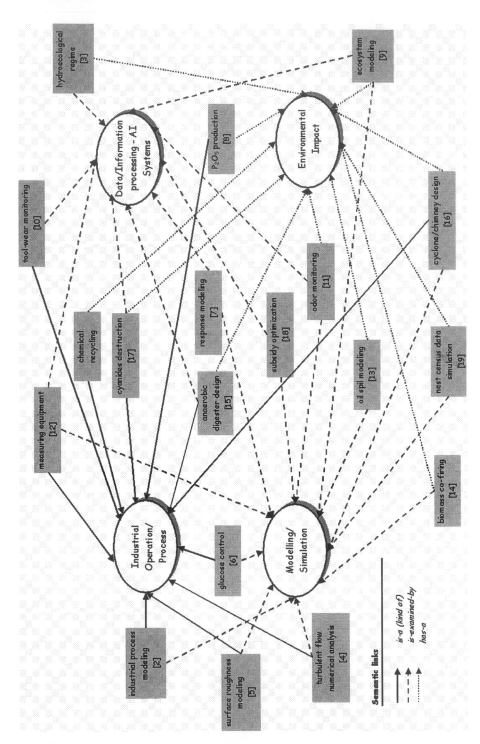

Simplified ontology designed *ad hoc* by induction at a surface phenomenological level on the basis of the contributed works; only certain semantic links are depicted.

**Symposium Papers as Components of the Ontological Network**

[1]  Preliminary Engineering Study for Evaluating the Industrial Feasibility of the Chemical Recycling of PET through Glycolysis

[2]  Micro-Climatic and Anthropogenic Impacts on the Hydroecological Regime of a Large Freshwater Body

[3]  Surface Roughness Modeling in High Speed Machining

[4]  Glucose Optimal Control System in Diabetes Treatment

[5]  Applying Response Modeling Methodology to Model Temperature-Dependency of Vapor Pressure

[6]  Industrial Case Study of Reaction Kinetics Applied for Continuous Dihydrate – Hemihydrate Method of $P_2O_5$ Production for Low Grade Phosphorites

[7]  Simultaneous Assimilation of Physical and Biological Observations into a Coupled Ecosystem Model using Joint and Dual Kalman Filtering

[8]  New Approach to Monitor the Tool Condition in a CNC Machining Center

[9]  Odor Fingerprinting/Monitoring within a Processing Industry Environment by Means of Distributed Biosensors – The Case of Oil Refineries

[10] Introducing Chemical Engineering Processes into Optimal Design of Measuring Systems Equipped with Biosensors

[11] Modeling Oil Spills in Areas of Complex Topography

[12] Co-Firing Different Types of Biomass as a Contribution to Households' Energy Independence and Environment Protection at Local Level

[13] Computer Aided Design of an Anaerobic Digester with Low Biogas Production for Decision Making Under Uncertainty

[14] Optimal Design of Cyclone/Chimney System – Implementation within an Olive Pomace Oil Mill

[15] Contribution to Pre-Selection and Re-Design of Cyanides Destruction Method at Industrial Level for Environmental Protection

[16] Determination of Optimal Subsidy for Renewable Energy Sources Exploitation by means of Interval Analysis

[17] Simulating Nest Census Data for the Endangered Species of Marine Turtle Caretta-Caretta

## Fragiskos Batzias

Fragiskos Batzias is Associate Professor and Vice Head of the Department of Industrial Management and Technology at the University of Piraeus, Greece. He is also teaching at the interdepartmental postgraduate courses (i) Systems of Energy Management and Protection of the Environment, running by the University of Piraeus in cooperation with the Chem. Eng. Dept. of the Nat. Tech. Univ. of Athens, and (ii) Techno-Economic Systems, running by the Electr. & Comp. Eng. Dept. of the Nat. Tech. Univ. of Athens in cooperation with the University of Athens and the University of Piraeus. His research interests are in Chem. Eng. Systems Analysis and Computer Aided Decision Making with techno-economic criteria. He holds a 5years Diploma and a PhD degree in Chem. Eng., and a BSc in Economics. He has also studied Mathematics and Philosophy.

Brill Academic Publishers
P.O. Box 9000, 2300 PA Leiden
The Netherlands

*Lecture Series on Computer
and Computational Sciences*
Volume 4, 2005, pp. 815-821

# Preliminary Engineering Study for Evaluating the Industrial Feasibility of the Chemical Recycling of PET through Glycolysis

Nicolas Abatzoglou and Maher Boulos

Department of Chemical Engineering,
Université de Sherbrooke, Sherbrooke, Quebec, Canada

Received 3 July, 2005; accepted in revised form 28 July, 2005

*Abstract:* The department of chemical engineering of the University of Sherbrooke has established since more than 12 years a course of integration (Design of Industrial Chemical Processes) at final year of its programme; during this last year the students form working groups like in Engineering companies and, under the scientific and technical 'coaching' of their professors-engineers and their assistants proceed to the preliminary engineering of a chosen industrial process. This activity includes an intense computational effort to simulate the operation of all modules of the industrial production facility and leads to results which have a considerable research value and it is considered by interested investors for industrial applications. This work presents the results of this effort applied in the case of the chemical recycling of a plastic residue, the PET. Many programs of paper, metal, glass and plastic recycling have been implemented during the last 20 years. Plastic recycling is by far the most complex and it is still under heavy R&D. This is mainly due to the complexity of the formulations and the low raw materials cost in industry. Indeed, in the majority of the cases, the monomers of synthesis are less expensive than the recycled plastic which results mainly from mechanical recycling (shredding and extrusion). This work focuses on the chemical recycling by means of Glycolysis of one of the principal plastics used in the industry of food packing: the Poly-(Ethylene Terephthalate) (PET). The product obtained by glycolysis is the Bis-(Hydroxyl-Ethyl) Terephthalate (BHET) which is the building unit of the PET. The chemically recycled PET, said "food rank", has a commercial value much higher than that of the mechanically recycled PET because of its higher purity, higher specifications and low risk of cross-contamination. In the United States, according to FDA's rules, the chemically recycled PET can be used in concentrations up to 25% in the manufacture of food packing. The results of this work show that the ideal capacity of production for a chemically recycled PET production facility is of 25 000 ton/y and the most favourable site, according to the market research and techno-economic evaluation, is the state of South Carolina, USA. The factory obtains a ROI of 11.41%. The sensitivity analysis shows that the parameters influencing more the profitability of the factory are: the selling price of the BHET, the production volume (output), as well as the cost of the raw material (used PET).

*Keywords:* Chemical Recycling, Glycolysis, Design, Case Study, Industrial, PET, Analysis, Engineering

## 1. Market Study

The market research was focused on a new prospect: the production of BHET obtained by glycolysis. Indeed, new studies confirm the interest of certain major companies in this field (Wellman, Coke, Eastman, KoSA, DuPont) on purchasing pure BHET (1). To reach these companies, however, it is necessary that the quality of the BHET respects the requirements and specifications of their products, and this, at a cost lower than the production cost of the virgin resin. The site of production was thus selected consequently, and is at 12Km of Columbia, capital of the State of South Carolina; this insures the necessary proximity with prospective customers and raw material sources. For reasons related to the availability of the raw material and the capacity of selling out the entire production the output of the factory was fixed at 25 000 tones of BHET per year, and the industrial unit was designed to be operated continuously.

## 2. Choice of Technology

The chemical recycling of the PET is based on the depolymerization which can be done by various technologies, including glycolysis (2), methanolysis (3), a combination of methanolysis/glycolysis, acid hydrolysis (4, 5), alkaline hydrolysis (6) and simple hydrolysis at neutral pH (7). A thorough analysis of the advantages and disadvantages of the above processes is available in the literature (8). Based on this analysis it has been concluded that the glycolysis is, at the present time, the most appropriate technology to evaluate due to its technical, economical and environmental superiority.

The glycolysis is already used on an industrial scale and can proceed in a Batch or a Continuous Stirred Tank Reactor (CSTR), at reaction times typically lower than 1h. As the reaction proceeds at pressures (P≤0.3MPa) and temperatures (T≤280°C), which are lower than required in the case of methanolysis, the glycolysis requires conditions of standard safety, less expensive than those of their competitors. Moreover, it is even possible to catalyze this reaction by adding zinc acetate, which will increase the reaction rate. The reagent is the post-consumption PET. The solvent used during the reaction, which is also a reagent, is the ethylene glycol (EG); it is less toxic and less flammable than methanol required in Methanolysis. The various products obtained are the (a) (2-hydroxy-ethyl) terephthalate (BHET) (the targeted product), EG and some short BHET oligomers (2 or 3 BHET units). These products have a good resale value on the market. The conversion obtained is close to 100%. Since Glycolysis is already used worldwide at much greater extend than the acid and alkaline hydrolysis, it's scaling-up and industrial operation are easier and less troublesome. Moreover, the fact of not using acids and strong bases strongly reduces corrosion on equipment, and there are less Health and Safety concerns. The FDA accredited this process for use in food packing. Besides, the BHET has a better resale value on the market in comparison with the other products generated by the alternative depolymerization processes. The BHET is the intermediate polymer in the production of the PET; thus, a process using directly BHET has a stage less than other processes starting with Terephtalic acid. The

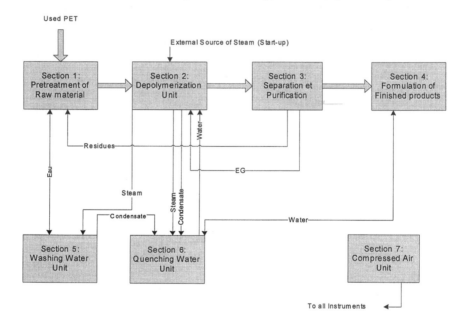

Figure 1: Process block diagram

block diagram of the process is presented in Figure 1. Each section is detailed in the flow diagrams presented in Figures 2-7.

## 3. Process Description

The process comprises four major sections (1-4): Treatment of the Raw Material, Depolymerization, Separation and Purification and Finished Products. A description is provided below and Design details are shown in Figures 2-5. The other 3 sections are "Utilities" and details are provided for Sections 5 and 6 in Figures 6-7.

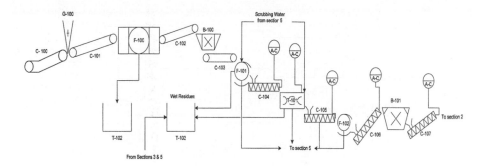

Figure 2: Flow Diagram of Section 1

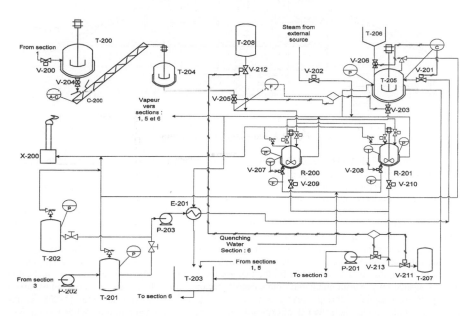

Figure 3: Flow Diagram of Section 2

### 3.1 Raw Material Reception and Pre-treatment

The bundles of post-consumer PET of footprint dimensions of 0,8m*0,8m arriving in this section are cut in a coarse way by a guillotine (knife) cutter (G-100). Once their initial size reduced, they are sent into a drum screen (F-100) to be disencumbered (separated) of their impurities (sand, rock, metals). Thereafter, they will go straight towards the primary crusher (B-100), where they will be crushed in flakes of approximately 2-3cm. After the crusher the flakes are pre-washed in a centrifuge (F-101), and then washed in a washer/liquid-solid separator (T-100). All the water used in this part is recycled and treated in section 5. Once washed the flakes of PET are dried, then crushed in a second crusher (B-101) down to an average size of 4mm (optimal size for depolymerization). Finally, the flakes are stored and sent to the depolymerization (Section 2). The transport of the PET flakes from one module to another is insured by a number of appropriately designed and operated conveyors (belt, screw). Depolymerization: The PET flakes, coming from section 1, arrive in a Chute equipped with a gate valve, isolating section 1 from section 2, and they fall into the tank T-200. A stirring mechanism insures that the flakes do not agglomerate prohibitively. They are, then, sent in the basin (T-204) feeding the depolymerization

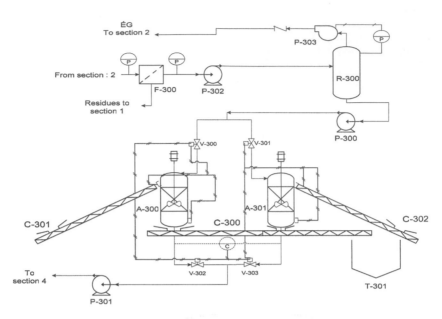

Figure 4: Flow Diagram of Section 3

reactors (twin vessels R-200 or R-201) by gravity passing through a rotary valve (V-205). Then, a heated and perfectly agitated flow of premixed EG and catalyst is pumped into the reactors. An analyzer reads continuously the mixture composition in order to insure that the ratio EG/Catalyst is at the set point. The reaction time (residence time of the reagents) in the CSTR is one hour. The liquid mixture leaving the reactors moves towards section 3. Thermocouples are incorporated in the reactors controlling the temperature set point. The control is achieved by adjusting the cooling water flow rate and the catalyst percentage. Relief valves as well as relief reception tanks are installed in the reactors in order to avoid overpressure and explosion. These valves redirect gases towards the flare (X-200). The cooling water coming from section 6, used to remove the heat produced by the exothermic reaction taking place, is turned to steam; the latter is used to fulfill the majority of the process energy needs, namely preheating of the reagents flows. Steam purchased from a near-by factory fulfills the remainder of the industrial unit energy needs, especially during start-up of production. A tank of reaction inhibitor (T-208) is also designed in order to minimize the risk of explosion in the event of reaction runaway. When this inhibitor is used, the content of the reactor (un-reacted mixture) is sent in an electrically heated tank (T-207); this mixture has to be disposed of according to the prevailing Health & Safety rules. All the tanks are equipped with level controller. At the exit of the reactors, the transport of the fluid is done by pumping and gravity.

### 3.2 Separation and Purification

The fluid exiting the depolymerization reactors of section 2 is sent to the filter (F-300), where the solid part is collected as a cake. The filter is automated and pressure drop is monitored. The liquid part passes through the filter and moves towards the BHET distillation column (R-300). This column separates the EG from the BHET. The EG vapour is aspired by the vacuum pump (P-303), and it is, afterwards, condensed. Temperature and pressure gauges collect data along the distillation column. Thereafter, the condensed EG is sent to its recovery and recycle tank (T-201). The BHET, the catalyst and the impurities, which are coming from the PET additives, dyes and the reaction by-products, are collected at the bottom of the column and sent to the adsorption columns (P-300). The so purified BHET is forwarded to the section 4. The columns of adsorption are packed with activated carbon which must be regenerated or simply replaced after saturation. These two columns are equipped with a stirring mechanism, whose function is to facilitate the purge of the used activated carbon. Since the amounts of activated carbon used are relatively small it is preferable to avoid regeneration.

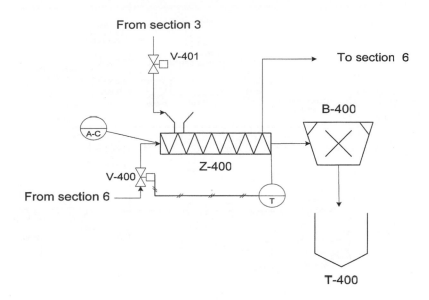

Figure 5: Flow Diagram of Section 4

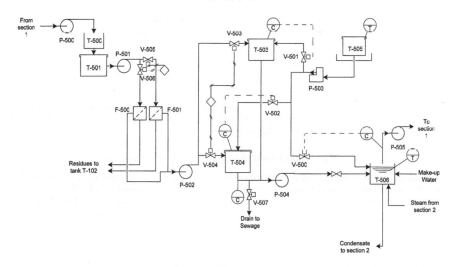

Figure 6: Flow Diagram of section 5

### 3.3 Finished products

The BHET arrives from section 3 and moves towards the extrusion module (Z-400), at the exit of which it is cooled under controlled temperature (V-400), and then cut in rollers of 2 feet length. Thermocouples are installed at the entrance and the outlet side of the extrusion machine in order to make sure that the viscosity of the BHET is adequate thus avoiding premature wearing of the equipment parts and the other components. Moreover, an Amber-meter indicates any loss of power and dictates preventive stopping of the extrusion machine. The so formulated BHET passes in a crusher (B-400), where it is cut out in customer-ready pellets.

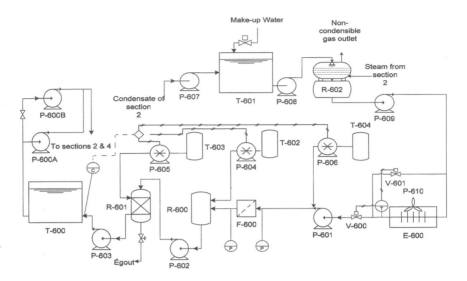

Figure 7: Flow Diagram of Section 6

### 3.4 The Other sections

In the washing water section there is conductivity controllers placed in the water storage basins to detect the accumulation of electrolytes (salts) of salts which are detrimental to the equipment. In the cooling water section, pH-meters are installed in order to control the water alkalinity in the softener. Moreover, water samples are analyzed periodically in the laboratory to detect the presence of harmful micro-organisms. When necessary, water chlorination is an added option. Finally, in order to reduce the corrosion of the equipment, a vapour injection is used to remove 90-95% of the dissolved oxidative gases ($CO_2$ and $O_2$).

The Design methodology used follows the principles and methods available in the literature (9-13), adapted to the needs of the studied project particularities. A schematic view of the methodology, the steps and the control loops used are presented in (14).

## 4. Economic Evaluation and Sensitivity Analysis

The process described above has been simulated using both commercial Simulation Software (HYSYS and ASPEN) and Custom-made Mathematical Modeling according to the needs.

The simulation produced full mass and Energy balances and allowed to proceed to the Specifications of the major and minor equipments, choice of modules technology, cost estimation and budget estimation through requests of quotations.

In the actual economic context characterized by fluctuations difficult to predict, the price of polymers is also subject to this influence. However, under the pressure of the Kyoto Protocol, the continuously increasing Energy cost and the growing Environmental concerns, the PET recycling industry has a high thriving potential. The Industry can use an established technology, as Glycolysis, and foment a new market to be established and compete with the producers of PET and recycled PET. The project can benefit from the reorganization of the recycling markets and the sustainable development in North America.

Based on the assumption that the raw material purity is only 80% (pessimistic scenario), the results of the economic evaluation of this project are as follows:

The benefit from this investment was estimated at 19 800 310 $ over a 10 years period. This represents a return on investment (ROI) which varies from 9.5% the first year up to 18.6% in the tenth year. This rise is explained by a better control of the costs over the 10 years production period but it has to be considered as probably optimistic because some obsolete equipment will have to be replaced after 10 years. Regarding the evaluation of the project by the method of the net value present (NPV) and a Minimum Acceptable Return on Investment (MAROI) fixed at 10%, this project proves profitable before (7 286 995$) and after tax (6 731 821$).

The undertaken sensitivity analysis has shown that the parameters influencing more the profitability of the factory are: the selling price of the BHET, the product output and the purchase price of the FART.

## Acknowledgments

The contribution of all Design group members is gratefully acknowledged. These members are: Mr. Mathieu Bégin, Mr. Pascal Charbonneau, Mrs. Kathleen Gordian, Mr. Jean-Charles Mankeyi, Mr. Marc-Antoine Michel, Mr. Philippe Pettigrew and Mr. Patrick Villemure. Special thanks to Mr. François Hudon and Mrs. Heïdi Brochu for their helpful technical assistance.

## References

[1] NEXANT, ChemSystems Rep., http://www.chemsystems.com/newsletters/perp/Oct200099s4-abs.cfm.

[2] Carta, D., Cao, G. and D'Angeli, C., Chemical Recycling of Poly(ethylene terephthalate) (PET) by Hydrolysis and Glycolysis, ESPR - Environmental Science and Pollution Research – International, 2003 , v. 10 , n. 6 , p. 390 , 5 p.

[3] Genta M., Yano F., Kondo Y., Matsubara W. and Oomote S., Development of Chemical Recycling Process for Post-Consumer PET Bottles by Methanolysis in Supercritical Methanol, Mitsubishi Heavy Industries Ltd., Technical Review Vol. 40, Extra No 1, Jan. 2003.

[4] Mishra, S., Goje, A. and Zope, V., Chemical Recycling, Kinetics, and Thermodynamics of Poly (Ethylene Terephthalate) (PET) Waste Powder by Nitric Acid Hydrolysis, Polymer Reaction Engineering, 2003 , v. 11 , n. 1 , p. 79.

[5] Yoshioka, T., Sato, T. and Okuwaki, A., Hydrolysis of Waste PET by Sulfuric Acid at 150°C for a Chemical Recycling, Journal of Applied Polymer Science, 1994 , v. 52 , n. 9 , p. 1353.

[6] Mishra, S. and Goje, A., Chemical Recycling, Kinetics, and Thermodynamics of Alkaline Depolymerization of Waste Poly (Ethylene Terephthalate) (PET), Polymer Reaction Engineering, 2003, v. 11, n. 4 , p. 963.

[7] Mancici S. D., and Zanin M., Optimization of Neutral Hydrolysis Reaction of Post-consumer PET for Chemical Recycling, Progress in Rubber Plastics and Recycling Technology, 2004, v.20 , n.2, p. 117.

[8] Bégin M, Charbonneau P., Gordian K., Mankeyi J.-C., Michel M.-A., Pettigrew P. and Villemure P., Le recyclage chimique du PET, Rapport GCH-425, Design des procédés chimiques II, Université de Sherbrooke, Dpt. Of Chemical Engineering, Sherbrooke (Quebec), Canada, Dec. 10, 2004.

[9] Peters, M.S., K.D. Timmerhaus and R.W. West, Plant Design and Economics for Chemical Engineers, Editor McGraw Hill, 2003, 5th Edition, ISBN 0-07-239266-5.

[10] Sinnott, R.K., Coulson & Richardson's Chemical Engineering, Editor Butterworth-Heinemann, Vol.6, 3rd Edition, 2003, ISBN 0 7506 4142 8.

[11] Seider, W.D., J.D. Seader and R.L. Lewin, Product & Process Design Principles: Synthesis, Analysis and Evaluation, Editor Wiley, 2004, ISBN 0-471-21663-1.

[12] Ulrich, G.D. and P.T. Vasudevan, Chemical Engineering Process Design and Economics: A Practical Guide, Editor Process (Ulrich) Publishing, 2nd Edition, 2004, ISBN 0-9708768-2-3.

[13] Ray, M.S. and M.G. Sneesby, Chemical Engineering Design Project: A Case Study Approach, Gordon and Breach Science Publishers, 2nd Edition, 1998, ISBN 90-5699-137-X.

[14] Abatzoglou, N. and Boulos M., Curriculum Integration in Chemical Engineering Education at the Université de Sherbrooke, Proceedings of 2nd CDEN International Conference, Kananaskis, Alberta, Canada, July 18-20, 2005.

Brill Academic Publishers
P.O. Box 9000, 2300 PA Leiden
The Netherlands

*Lecture Series on Computer*
*and Computational Sciences*
Volume 4, 2005, pp. 822-826

# Micro-climatic and Anthropogenic Impacts on the Hydroecological Regime of a Large Freshwater Body

E. Dimitriou[1] and I. Zacharias[2]

[1]Institute of Inland Waters, Hellenic Centre for Marine Research,
46,7 km of Athens- Sounio Ave., 19013, GREECE,
2 Department of Environmental and Natural Resources Management, University of Ioannina,
2 Seferi Str., 30100 Agrinio, GREECE,

Received 8 July, 2005; accepted in revised form 9 August, 2005

*Abstract:* In the past four decades, natural wetlands have undergone devastating stresses as a result of significant hydrologic changes. The latter are primarily caused by the increase in water irrigation demands and the climate change phenomenon. This study attempted to analyze the past climatic and water management alterations in Trichonis Lake, located in W. Greece, to quantify the exact impacts on the surrounding wetlands that belong to the NATURA 2000 protection network. Statistical elaboration of rainfall and water level time series have been applied, including regression analysis and Cumulative Sum method, to identify relevant past trends while remote sensing and GIS techniques have been used to map and illustrate the past and present morphological conditions of the riparian area. The temporal changes in the regional water management scheme such as the alterations in irrigation demands during the last 40 years have been also taken into consideration. Thus, a comparative assessment performed between the estimated alterations in rainfall and anthropogenic water abstractions to identify the contribution of each one of these factors on the measured water level fluctuations in the lake and the associated wetlands extent changes during the study period. The results indicated that the observed decrease in the wetland area depends mainly on human activities such as the expansion of agricultural land in the catchment, the development of a large scale irrigational network to meet the increasing water demands and the unsustainable water management strategies. The climate change patterns that have been identified in the area concern a slight decrease in rainfall, which indicated relatively lower impacts on the wetlands since the associated hydrologic transitions are relatively limited on temporal and spatial basis.

*Keywords:* Wetlands, Land Use, Climate change, Water resources management, hydrology

## 1. Introduction

Hydrologic impacts can affect many aspects of the environment including river channels, riparian and wetland habitats, aquatic communities, ecological diversity and human health and welfare [1]. Several studies have been conducted in the last two decades targeting to estimate past water levels of freshwater bodies and to relate the recorded changes with specific human interventions and micro-climatic trends [2, 3]. In this study, the past and present water level fluctuations in Trichonis lake have been investigated and the associated impacts on the environment have been identified and analytically quantified. The land cover changes in the riparian zone during the last 40 years have been mapped, using remote sensing and GIS techniques and the effects of all the aforementioned environmental alterations on the regional wetlands have been illustrated.

Trichonis lake, located in Western Greece, is the largest freshwater body in the country regarding water volume ($2.8 \times 10^9$ m$^3$) and presents significant annual and monthly water level fluctuations (~1m and 0.5m respectively).

---

[1] Corresponding author: Dr. Elias Dimitriou, email: elias@ath.hcmr.gr

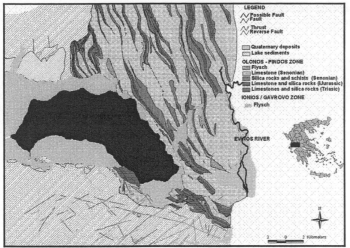

Fig. 1. Geological map of the Trichonis lake catchment and its location in Greece

## 2. Methodology

To identify the past fluctuations of rainfall and water level elevation in Trichonis lake daily respective measurements have been used for the period 1951-1997. At the first stage of this study, descriptive statistic analysis has been carried out, the extreme events have been identified and cross-correlation elaboration on the data series has been applied, so as to quantify the degree of linearity existing between these parameters.

The Cumulative Sum (CUSUM) technique has been implemented in the second stage of the study to reveal the fluctuation in the past trends regarding rainfall and water level. This method is widely used to present medium and long-term alterations in the average value of a parameter [4]. Particularly, the cumulative value of a parameter is given by the following equation:

$$s_i = \sum_{j=1}^{i} (x_j - \bar{x}), \text{ for } i<n \qquad (1)$$

where $x$ is the mean of $n$ successive measurements and $x_i$ is a single measurement $i \in [1, 2, ..., (n-1)]$.

The impacts from the past alterations in the water level of Trichonis lake have been quantified by using the Digital Bathymetric Model (DBM) of the lake, which has been developed by elaborating the lake's bathymetric values, obtained by topographic maps (1:5.000, designed by the Greek Military Geographical Service) with 3D Analyst software (GIS component).

## 3. Results

The monthly rainfall records for the period 1951-2002 are presented in figure 2 and a seasonality analysis of the data indicates a long-term pattern of seasonality cycles that last approximately 10 years each. The trendline of this diagram, obtained through the linear correlation of the monthly rainfall values, indicates a decreasing trend in the rainfall throughout the entire study period, which has been further investigated by using the CUSUM approach. The average rainfall value for each decade during the period 1951-2001, accredits the aforementioned decreasing trend of the rainfall. The total change rate observed during the period 1951-2001 for the area of Trichonis Lake is -6.53% and the greatest decrease occurred in the decade 1981-1991, when extensive dry conditions had caused significant water shortage problems in the entire country. During the last decade, rainfall trend presents the highest increase but the long-term trend for the whole study period (1951-2001) remains negative.

The average water level elevation in Trichonis lake is +15.96m for the period 1951-1997, while the minimum recorded value is +14.83m (November 1991) and the maximum value is +17,84m (April 1957).

The correlation coefficient estimated for the daily values of the parameters rainfall – water level elevation for the period 1951-1997 is very low (0,25), which means that the degree of linearity between these specific parameters is relatively low. The Cumulative Sum diagram of the annual water level elevations (figure 3) indicates 5 different alteration stages in the parameter's trends for the period 1951-1997. The most significant stage is observed during the period 1967 and 1980 with a substantial decrease in water level caused mainly by the extension of the irrigation abstraction schemes in the area as well as by the expansion of agricultural land.

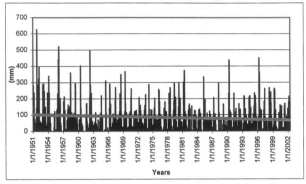

Fig. 2. Monthly rainfall measurements for the period 1951-2002 and the associated trend for Trichonis lake catchment

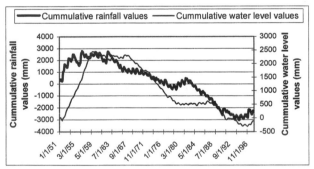

Fig. 3. Cumulative Sum diagram of monthly rainfall and annual water level elevation values for the period 1951-2002

The development of the lake's Bathymetric Model in combination with the specialized algorithm of the 3D Analyst and water level measurements led to the creation of figure 4 which illustrates the relationship between the lake's surface extent, volume and water level elevation. This is a significant output that can facilitate regional water management, since it can contibute to the quantification of the potential impact on the lakeshore drawback by various water abstractions scenarios that affect directly the water level elevation.

The average annual water level elevation was approximately +18m prior to the construction of the main controlled flow canal that connects Trichonis and Lysimachia lakes and outflows significant water volumes on an annual basis ($273 \times 10^6 m^3$), in order to cover irrigation needs of remote areas [5]. After the construction and operation of this canal the average annual water level elevation fell more than 2m in Trichonis lake, which had significant impacts. Particularly, the surface area of the lake decreased by $2.89 \times 10^6$ m² (3% decrease in relation to the total area of the Lake), which means that the riparian zone that hosts significant wetlands have been drought up to an extent of approximately equal to the above. Furthermore, the water storages in the lake have also been decreased by 188 $\times 10^6$ m$^3$ (6.6% of the total water volume), thus imposing constraints on the long-term sustainable water management through potential destabilization for the local hydrologic regime. The wetlands around the lake belong to European Protection Network 'NATURA 2000' and a significant part of them have been drought up, as a result of the aforementioned practice leading to direct environmental loss (figure 5). This has increased the agrochemical inflows to the lake and has therefore caused deterioration of the water quality, which imposes certain health impacts regarding local biota and people who cover their water needs from this freshwater body. The wetland zone that was dried up in the period 1951-1997 has been estimated to be over 80% of its natural extent

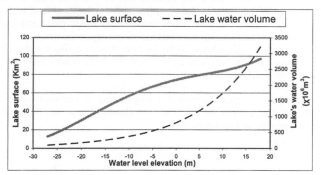

Fig. 4. Trichonis lake surface area-volume-water level relationship.

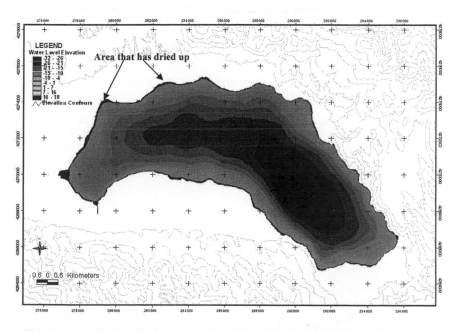

Fig 5. Lake area that has dried up as a result of the water level decline in the past few decades.

## 4. Concluding Remarks

In Trichonis lake catchment the climate change during the last 50 years, which is summarized by a decrease in the average annual rainfall, had caused a respective decrease in the average water level elevation. However, this decrease in the water level has significantly been enhanced by human activities and mainly by the construction and operation of the controlled flow canal that connects Trichonis and Lysimachia lakes. The very high annual outflow from Trichonis lake to cover irrigation needs of remote areas has led to the substantial fall in the average annual water level (approximately 2m), which has drought up a large extent of ecologically important riparian zone (3%) and has permanently decreased the water storages in the lake over 6.6%. These alterations in the hydrologic regime have imposed specific impacts and constraints on the area's environment, on the sustainable water management potential and on the future water quality status of the lake.

The observed land use alterations in the period 1945-1986 indicated that the significant increase of cropland eliminated the wetland areas in the southern part of Trichonis Lake and caused negative effects in the forested areas. Furthermore, the expansion of the agricultural land leads also to adverse hydrologic conditions since increases water losses through evapotranspiration and lowers the amount of water stored in the soil, which is a very important hydrologic aspect for wetlands.
The riparian zone has been strongly affected by the land use changes during the above period since the water level decline revealed new terrestrial areas around the lake, which enhanced the agricultural intensification and the wetlands degradation.

The methodology used in this study comprised a combination of statistical elaborations and GIS techniques that managed to quantify the effect of both climate change and human activities on the lake's water level in temporal and spatial basis.

## References

[1] W. Mitsch and J. Gosselink, *Wetlands, 2nd edition*, John Wiley & Sons, 1993, New York

[2] I. Jones and J. Banner, Estimating recharge thresholds in tropical karst island aquifers: Barbados, Puerto Rico and Guam, *Journal of Hydrology,* Vol. 278, 2003, pp. 131-143

[3] F. P.Crapper, M. P. Fleming, J. D. Kalma, Prediction of Lake levels using water balance models, *Environmental Software,* Vol.11, No 4, 1996, pp.:251-258

[4] A.C. McGilchrist and D. K. Woodyer, Note on a distribution-free CUSUM technique, *Technometrics,* Vol. 17, 1975, pp. 321-325

[5] I. Zacharias, E. Dimitriou, Th. Koussouris, Estimating groundwater discharge into a lake through underwater springs by using GIS technologies, *Environmental Geology Journal*, Vol.44, 2003, pp.843-851

Brill Academic Publishers
P.O. Box 9000, 2300 PA Leiden
The Netherlands

*Lecture Series on Computer
and Computational Sciences*
Volume 4, 2005, pp. 827-830

# Surface Roughness Modelling in High Speed Machining

Rubén Morales-Menéndez[1], Sheyla Aguilar[2], Antonio Jr Vallejo[3],
Miguel Ramírez Cadena[4], Francisco J Cantú Ortiz[5]

ITESM Monterrey campus
64,849 Monterrey NL México

Received 13 July, 2005; accepted in revised form 21 August, 2005

*Abstract:* Surface roughness of machined parts is a key attribute in determining product quality. Functional features of parts such as fatigue life, friction, wear, etc. are closely related with surface roughness. Online control systems of surface roughness are not an industrial solution for several reasons. The absence of sensors that provide measurements reliably and effectively in a hostile machining environment is one of the main reasons. Surface roughness is difficult to measure online; however, combining variables from several kinds of sensors allow us to cope with this problem. We propose a predictive model based on the sensor-fusion concept for on-line estimation of the surface roughness. Statistical multiple regression and artificial neural nets frameworks are exploited. Early experimental results with aluminium show high accuracy rate for predicting surface roughness.

*Keywords:* Modelling, Artificial Neural Nets, Surface Roughness

## 1.  Introducction

Milling is a very important operation among CNC industrial machining processes. Surface roughness plays a crucial role in the performance of end milling, and fatigue life, friction, wear, etc. are features closely related with surface finish. Surface roughness is influenced by several factors including cutting parameters; these cutting parameters such as spindle speed, feed rate, and depth of cut can be defined in advance.

Intelligent control techniques are theoretical solutions for online milling process control. However, no industrial applications in surface roughness exist because there are no sensors that provide valuable measurements in this harsh environment. Sensor fusion is a mathematical method that integrates several sensor signals into one fused measurement. These integrated measurements can predict states such as surface roughness, more accurately than single sensor measurement.

## 2.  Background

Measuring surface roughness in High Speed Machining (*HSM*) has traditionally received considerable attention [1]. [2] have classified modelling techniques in: *machining theory* based approach, *experimental investigation* approach, *designed experiments* approach, and *Artificial Intelligent (AI)* approach [3,4,5,6]. We will assume an *AI* approach for this paper.

Sensor fusion is an indirect measuring method which extracts key unmeasurable information from several readily available measurements. Few sensors are used in machining's harsh environment making sensor fusion a research area for unmeasurable states. Selection of key sensors and effective models are crucial factors in sensor fusion, and no specific methodologies exist for choosing these factors in the field of machining.

---

[1] R. Morales-Menéndez is with Center for Innovation in Design and Technology at ITESM, rmm@itesm.mx
[2] S. Aguilar is with the Graduate Program at ITESM, A00792459@itesm.mx
[3] A. Vallejo is with the Mechatronic Dept at ITESM Laguna campus, avallejo@itesm.mx
[4] M Ramírez is with the Instituto de Automática Industrial, CSIC, Spain, mramirez@iai.csic.es
[5] F. Cantú is with the Research and Graduate Studies Office, ITESM, fcantu@itesm.mx

We choose independent variables (i.e. s,d,f,Po) which have the highest sensitivity to the dependent variable (i.e. Ra). Two modelling approaches will be exploited in this paper: statistical multiple regression and artificial neural nets.

### 2.1 Statistical Multiple Regression.

A model was developed by using spindle speed (s) , feed rate (f), depth of cut (d) and vibration signal (Po) as the independent variables and the surface roughness as the dependent variable. We are proposing 2 structures: **Structure I** given by eqn (1):

$$Ra_i = b_0 + b_1 s_i + b_2 f_i + b_3 d_i + b_4 Po_i + b_5 s_i f_i + \cdots + b_{11} s_i f_i d_i + \cdots + b_{15} s_i f_i d_i Po_i + \varepsilon_i \quad (1)$$

where $\varepsilon_i \sim \mathcal{N}(0,\sigma^2)$ , $\{b_j\}_{j=0}^{15}$ are unknown parameters. We assume that *Ra* follows a normal distribution for every combination of the independent variables. This structure contains the main effects and interactions of the independent variables. **Structure II** given by eqn (2) has a similar definitions as *structure I*:

$$Ra_i = \alpha (S_i)^{\vartheta} (f_i)^{\varsigma} (d_i)^{\tau} (Po_i)^{\gamma} + \varepsilon_i \quad (2)$$

### 2.2 Artificial Neural Nets (ANN)

*ANN*s are a form of multiprocessor computer systems based on the parallel architecture of the human brain. Multi-layer feed-forward networks are the most common architecture. These networks can consist of any number of neurons arranged in a series of layers. Layer *0* of the network is the input data, and the outputs of the last layer *n* of artificial neurons is the output of the network. Layers between *0* and *n* are named hidden layers. At the moment, no clear guidelines exist for determining the number or neurons required at each level for a given application.

There are several learning algorithms used to *train* neural networks, i.e. Backpropagation and Radial Basis Function training. Both of these training methods were developed off the well-known Delta rule for single layer networks. Backpropagation has been proved successful in most industrial applications and it is easily implemented in practice.

## 3. Experiments

Numerous factors influence the surface roughness, [2]. We focus on 3 preliminary cutting parameters: **f**eed rate (6 values), **s**pindle speed (5 values), **d**epth of cut (3 values), and we also recorded one cutting phenomena parameter: the vibration signal[6].

### 3.1 CNC Machining Center.

The experimental tests were conducted in a KX10 HURON machining center, with a capacity of 20 KW, 3 axis, and equipped with a SIEMENS open-Sinumerik 840D controller, Fig.1 (left pic). The cutting tool was a Octomill face mill of SECO Carboloy, with a diameter of 80 mm, depth of cut 3.5 mm, and 6 inserts of the SECO Carboloy, Fig. 1 (middle pic).

Vibration signals were measured online with an accelerometer installed on the flat metal support, Fig.1 (right pic). We analyze each vibration signal as a Power Spectral Density (*PSD*), where the spectral estimation describes the power density. We compute the (*PSD*), using the periodogram nonparametric method where the power spectrum of a signal is obtained by the squared magnitude of a discrete Fourier transform. Based on this result, we compute the average power (Po). We exploit Po as a vibration signal feature.

---

[6] All the experiments were realized without chatter phenomena

## 4. Results

We trained the models' parameters using the full dataset. Also, the dataset was randomly divided into two sets as training (75 %) and testing (25 %) sets in order to measure its generalization capacity. An uniform sampling process was conducted for splitting the data.

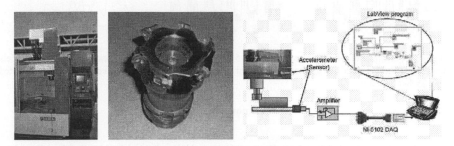

Figure 1: Huron machining center, cutting tool and data acquisition system

Several models were trained with both general frameworks. However, only the best models are shown in Tables 1. In Table 1, the LSE are shown as $\frac{1}{n}\sum e^2$ (in $\mu m^2$ per sample). We show different models and their variables (left columns), then LSE for both structures. The LSE for 100% indicates that we used 100% of the dataset for training and testing. However, 25% indicates that we learned the models' parameters with 75% of the dataset and we evaluated the model with the 25% of the remaining dataset. Then similar information is presented for the ANN approach with varying numbers of neurons in the hidden layer. We validate the 25 % LSE by recomputing it with 10 different random dataset for the full model # 5 (structure I); we got 0.71 as average LSE with 0.21 of variance.

Table 1: Experimental models using Statistical Multiple Regression and Artificial Neural Nets.

| Models | | | | | Least Square Error | | | | | | | |
|---|---|---|---|---|---|---|---|---|---|---|---|---|
| # | Variables | | | | Structure I | | Structure II | | # neurons in hidden layers | | | | |
| | s | d | F | Po | 100% | 25% | 100% | 25% | 4 | 5 | 6 | 7 | 8 |
| 1 | 0 | 1 | 1 | 1 | 1.39 | 2.56 | 16.94 | 5.67 | 0.872 | 0.652 | 0.559 | | |
| 2 | 1 | 0 | 1 | 1 | 2.14 | 2.46 | 4.58 | 2.48 | 0.993 | 1.180 | 0.773 | | |
| 3 | 1 | 1 | 0 | 1 | 0.83 | 1.42 | 0.75 | 0.87 | 0.400 | 0.659 | 0.361 | | |
| 4 | 1 | 1 | 1 | 0 | 0.59 | 1.33 | 0.67 | 0.90 | 0.223 | 0.088 | 0.078 | | |
| 5 | 1 | 1 | 1 | 1 | 0.40 | 0.71 | 0.68 | 0.88 | 0.079 | 0.077 | 0.088 | 0.062 | 0.042 |

## 5. Conclusions

The surface roughness could be effectively predicted with both approaches. As we can see in Table 1 and Figure 1 the ANN approach gives much better results. The characteristics of an *ANN* allow them to model more complex relationships, i.e. nonlinearities. For multiple regression models, *structure I* shows better performance than *structure II*. Also, we noticed Po is not an important vibration signal feature due to an insignificant reduction in the LS estimate for both models. Therefore, we have to look for another vibration signal feature. Although, this cannot be a general conclusion because this is an early work with few data and with only one material (Aluminum 6061).

We want to control the surface roughness in an online fashion. Considering the process has many stochastic variables, probabilistic models could be an excellent approach for combining this model with automatic decision systems based on evidence and probabilistic information.

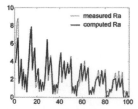 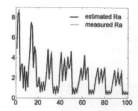 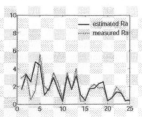

Figure 2: Performance of structure I (model # 5) and ANN (4-8-1). In left and middle plots, models were trained/tested with full dataset. In the right plot, ANN (4-8-1) was trained with 75 % and tested with 25 % of dataset.

## Acknowledgment

The authors wish to thank Adam Terry Simmons (Auburn University AL) for his careful reading of the manuscript and its suggestions.

## References

[1] M. Correa, M. Ramírez, J. R. Alique, and C. A. Rodríguez. Factores que afectan el acabado superficial en los procesos de mecanizado: Técnicas de análisis y modelos. In *XXV Jornadas de Automática Industrial CEA-IFAC*, pages 75-82, 2004.

[2] P. G. Benardos and G. C. Vosniakos. Predicting surface roughness in machining: a review. *Int. J. of Machine Tools and Manufacture*, pages 833-844, 2003.

[3] S. S. Lee and J. C. Chen. On-line surface roughness recognition system using artificial neural networks system in turning operations. *Int. J. of Advanced Manufacturing Technology*, 22:498-509, 2003.

[4] F. Dweiri, M. Al-Jarrah, and H. Al-Wedyan. Fuzzy surface roughness modeling of CNC down milling of alumic-79. *J. of Materials Processing Technology*,(133) : 266-275, 2003.

[5] T. Ozel and Y. Karpat. Predictive modeling of surface roughness and tool wear in hard turning using regression and neural networks. *Machine tools and Manufacture*, (45):467-479, 2005.

[6] P. V. S. Suresh, P. Venkateswara Rao, and S. G. Deshmukh. A genetic algorithmic approach for optimization of surface roughness prediction model. *Int. J. of Machine Tools and Manufacture*, 42:675-680, 2002.

Brill Academic Publishers
P.O. Box 9000, 2300 PA Leiden
The Netherlands

*Lecture Series on Computer*
*and Computational Sciences*
Volume 4, 2005, pp. 831-834

# Glucose Optimal Control System in Diabetes Treatment

Irma Yolanda Sánchez Chávez[a][1], Rubén Morales-Menéndez[b], Sergio Omar Martínez Chapa[c]

[a]Mechatronics and Automation Department, [b]Center for Innovation in Design and Technology and
[c]Electrical Engineering Department, ITESM Campus Monterrey
Eugenio Garza Sada 2501 Sur, 64849 Monterrey NL, México

Received 27 June, 2005; accepted in revised form 19 July, 2005

*Abstract:* A control system for optimal insulin delivery in a type I diabetic patient is presented based on the Linear Quadratic Regulatory Problem (LQRP) theory. The glucose-insulin dynamics is simulated with the Sorensen's high order nonlinear model. The process is linearized and reduced to a second order model whose state variables are the glucose and the insulin concentrations in the blood. The meaningful definition of the state variables allows the formulation of a practical cost function for a diabetes treatment in terms of the deviation from the normal glucose level and the dosage of exogenous insulin. The optimal control law is computed from this cost function. The regulatory approach is applied and the weighting factors of the optimization problem are adjusted to improve closed loop performance. The velocity of the correction of a glucose level deviation can be increased with a higher value for its weighting factor $\eta$, while an augment of the weighting factor $\rho$ for insulin supply may be necessary to prevent saturation of the controller output and oscillation of the blood glucose concentration. Simulation results show the flexibility and practicability of the LQRP approach in biomedicine.

*Keywords:* biomedical engineering, glucose optimal control, closed loop drug delivery

## 1. Introduction

Computational tools have enabled the simulation of complex processes for analysis and control. Automatic control techniques can improve the functionality of biomedical systems, including disease treatments. Diabetes mellitus has been declared the first cause of death in Mexico in 2003, and this fact inserts into a worldwide reality of an increasing number of cases. Juvenile-onset diabetes is characterized by the defective or null capacity to produce insulin, necessary for the absorption of glucose in the cells. This problem itself determines a medical treatment based on the dosing of exogenous insulin, which is done frequently in a fixed prescribed amount without continuous monitoring or feedback information.

Feedback control systems for precise blood glucose regulation in a diabetic patient have been proposed based on decision support systems [1], PID type controllers and adaptive and predictive control algorithms [2]. Advanced control strategies involve optimization techniques to achieve a normal blood glucose level with minimum supply of insulin. This work explores the design of an optimal controller using the Linear Quadratic Regulatory Problem (LQRP) formulation. The Sorensen's model of the physiological process of glucose-insulin as well as its simplification for design purposes is presented in section 2. Then, in section 3, the controller derivation using LQRP principles under different design considerations is explained. Simulation results are shown in section 4, and the conclusions are given in section 5.

---

[1] Corresponding author. E-mail: isanchez@itesm.mx

## 2. Mathematical model of glucose metabolism

The glucose-insulin dynamics is represented by the Sorensen's model [3]. This model is a nonlinear 19th order equation system based on mass balances between different organs or compartments (figure 1). The present work incorporates the meal disturbance modeling to the Sorensen model and uses the corresponding physiologic parameters available in the literature [4].

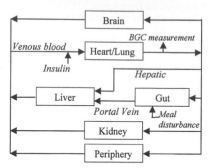

Figure 1: The Sorensen's glucose-insulin physiologic model (BGC: blood glucose concentration).

In order to derive a *linear* state space representation of the Sorensen's model, only the arterial glucose and insulin concentrations are defined as state variables $(g(t), h(t))$ with deviation values from initial steady state conditions $(G_0, H_0)$, the insulin infusion rate $u(t)$ is taken as the input of the system and the whole human body is considered as a single compartment. These simplifications lead to the Ackerman model [5] given by $\dot{g}(t) = -m_1 g(t) - m_2 h(t)$ and $\dot{h}(t) = m_4 g(t) - m_3 h(t) + u(t)$. The terms containing $m_1$ and $m_3$ account for self-removal of glucose and insulin $(m_1 > 0, m_3 > 0)$. The terms with $m_2$ and $m_4$ represent the reduction of the glucose level with insulin and the increment of blood insulin with glucose, respectively $(m_2 > 0$ and $m_4 \geq 0$ ). For a type I diabetic patient, $m_4=0$. The final state-space model is:

$$\dot{x}(t) = Ax(t) + Bu(t), \quad x(0)=0, \ u(0)=0 \tag{1}$$
$$y(t)=Cx(t)+Du(t)$$

where $x(t)=[x_1(t) \ x_2(t)]^T =[g(t) \ h(t)]^T$, $A=[-m_1 \ -m_2; \ 0 \ -m_3]$, $B=[0 \ 1]^T$, $C=[1 \ 0]$ and $D=0$.
The transition matrix A is obtained by the evaluation of the Jacobian:

$$A = \begin{bmatrix} \partial f_1/\partial x_1 & \partial f_1/\partial x_2 \\ 0 & \partial f_2/\partial x_2 \end{bmatrix} \tag{2}$$

where $f_1$ and $f_2$ are the first-order time derivatives of glucose $(x_1)$ and insulin $(x_2)$ concentrations, respectively, in the heart and lungs compartment. This linearization procedure preserves meaningful parameters and states and mathematical precision, since the controller is derived from a practical criteria and the model considers explicitly the main physical variables in hyperglycemic episodes in diabetes.

## 3. Design of optimal controller

The diabetes treatments aim at keeping the patient's glucose level normal by minimum infusion of insulin. Therefore, a performance criteria $J$ can be defined as:

$$J(u) = \int_0^\infty \left[ \eta (x_1(t) - x_d)^2 + \rho u(t)^2 \right] dt \tag{3}$$

where $x_d$ represents the desired glucose concentration deviation with respect to the same steady state condition used for $x_1$, and $\eta$ and $\rho$ are weighting factors.
The solution of the optimization problem, according to the LQCP technique, is the following:

$$u(t) = -\rho^{-1} B^T \lambda(t) \text{ with } \lambda(t) = Px(t) + \mu \tag{4}$$

where $\lambda(t)$ is the costate vector, $P$ is a symmetric positive semidefinite matrix and $\mu$ is a column vector. The dynamic behavior of the costate vector is given by:

$$\dot{\lambda}(t) = -Q(x(t) - x_D(t))Px(t) + \mu \tag{5}$$

with $x_D(1) = x_d$ and $Q = [2\eta \quad 0; \quad 0 \quad 0]$.

The control law based on the linear model (1) and the objective function (3) is expressed as:

$$u(t) = -K_c x + K \tag{6}$$

where $K_c = [K_{c1} \quad K_{c2}]$, $K_{c2} = \sqrt{m_1^2 + m_3^2 + 2\sqrt{m_1^2 m_3^2 + m_2^2 \rho^{-1} \eta}} - (m_1 + m_3)$,

$K_{c1} = -\left(K_{c2}^2 + 2m_3 K_{c2}\right)/2m_2$ and $K = -m_2 x_d \rho^{-1} \left(m_1^2 m_3^2 + m_2^2 \rho^{-1} \eta\right)^{-0.5}$.

The existence of an independent term in the control law depends on the exploited design approach in the formulation of the linear quadratic problem [6]. The servocontrol approach considers the initial condition of the patient as a reference for the definition of deviation values and specifies a desired change in blood glucose concentration ($x_d = G_d - G_0$) to produce a control action. On the other hand, the regulatory approach avoids a constant term in the control law by considering a normal glucose level as a reference state ($G_0 = G_d$; $x_d = 0$; $x_1(t) = G(t) - G_d$) and manages any deviation from that reference as a disturbance.

The closed loop performance can be affected by modifying the weighting factors $\eta$ for the glucose concentration deviation and $\rho$ (min$^{-2}$) for the insulin delivery or controller action. Since the product $\rho^{-1}\eta$ is used in the calculation of the elements of the controller gain vector Kc and the constant term K, a single tuning parameter can be managed as either $\rho$ or $\eta$ or the product $\rho^{-1}\eta$ itself.

In order to evaluate the control law both state variables have to be quantified. The blood glucose level is considered the output of the process or measured variable. The blood insulin concentration is estimated using an extended Kalman filter to account for process noise due to the linearization and for noisy glucose measurements.

## 4. Closed loop simulation

The proposed glucose feedback control system is compared with previous works [7] in which the control objective is the correction of an initial hyperglycemic condition of a type I diabetic patient with 300 mg/dl once its cause has ceased. This abnormal condition is produced by the simulation of a glucose meal input to the gut compartment of the Sorensen's model. The glucose meal input is increased exponentially from zero and kept at 408.7 mg/min to stabilize the arterial glucose concentration in 299.65 mg/dl. Then, the glucose meal input is suspended and the metabolism of the diabetic patient takes about 8 hours to reach a steady glucose concentration of 119.5 mg/dl.

The time for recovery of the normal glucose concentration of 119.5 mg/dl can be reduced by the continuous feedback control system of figure 2, where the optimal controller is used to determine the insulin infusion rate for the patient. The controller output is limited to 133 mU/min considering the capacity of insulin pumps and the risk of hypoglycemia [4].

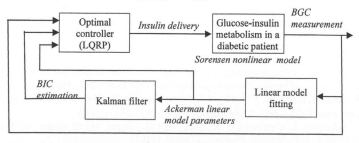

Figure 2: Glucose feedback control system (BIC: blood insulin concentration).

The regulatory control approach is used to avoid an offset error. The normal glucose level is restored after 250 min, according to the simulation results shown in figure 3 (top plots). The magnitude of the initial slope and the undershoot of the closed loop response increase with higher $\eta$ and lower $\rho$. Figure 3 also shows the corresponding effect of the weighting factors on insulin delivery (bottom plots). The order of magnitude of $\rho$ needs to be very high in order to prevent saturation and an ON-OFF controller behavior.

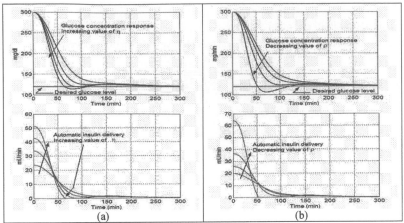

Figure 3: Effect of weighting factors on closed loop response. (a) The values for $\eta$ are 1, 1.5, 2 and 2.5, with $\rho=3.3 \times 10^{11}$ min$^2$. (b) The factor $\rho$ is varied as 1, 2, 3 and 4 times $10^{11}$ min$^2$ with $\eta=1$.

## 5. Conclusions

Previous reported results [7] show a change in blood glucose concentration of 58.5% and 90.5% at 60 and 120 minutes. The closed loop implemented with the process dynamics given by the Sorensen's model under a regulatory approach (with $\eta=1$ and $\rho=3.3 \times 10^{11}$ min$^2$) achieve 60.87% and 93.06% of the total desired change at the same times, ensuring no steady state deviation. Thus, the comparison shows agreement with prior researches and validates the present approach for optimal glucose control.

The results based on the interaction of a controller designed with LQRP principles with a highly nonlinear plant show the robustness of the controller and the adequacy of the design approach for a particular biomedical problem. Unavailable measurements demand reliable state estimation which needs noise management and on line identification. Further investigation could be oriented to the use of unscented particle filters for nonlinear processes to improve state estimation. The tests for the closed control system should include the regulation during prolonged disturbances such as exercising or feeding which may require the extension of the physiological model for simulation analysis.

Continuous feedback control systems for disease treatment are potentially bloodless, painless and more precise therapies. Their successful application in biomedicine requires the development of microsystems with reliable sensors and actuators and embedded control algorithms.

## References

[1] R. Bellazi, G. Nucci and C. Cobelli, The Subcutaneous Route to Insulin-Dependent Diabetes Therapy, *IEEE Eng. in Medicine and Biology*, 20, (1) 54—64 (2001).

[2] R. Parker, F. Doyle III and N. Peppas, The Intravenous Route to Blood Glucose Control, *IEEE Eng. in Medicine and Biology*, 20, 1 65-73 (2001).

[3] J. Sorensen: *A Physiologic Model of Glucose Metabolism in Man and its Use to Design and Assess Improved Insulin Therapies for Diabetes*. PhD Thesis, MIT,1985.

[4] R. Parker, F. Doyle III, J. Ward and N. Peppas, Robust $H_\infty$ Glucose Control in Diabetes Using a Physiological Model, *AIChE Journal*, 46, 12 2537-2549 (2000).

[5] E. Ackerman, L. Gatewood, J. Rosevear, and G. Molnar, Model Studies of Blood-Glucose Regulation, *Bull. Mathem. Biophys.*, 27 suppl. (1965).

[6] I.Y.Sánchez Chávez, R.Morales-Menéndez and S.Martínez Chapa, Linear Quadratic Control Problem in Biomedical Engineering, L. Puigjaner and A. Espuña (Eds): Computer Aided Chemical Engineering Vol. 20: European Symposium on Computer Aided Process Engineering-15, Elsevier, 1195-1200 (2005).

[7] M. Kikuchi, N. Machiyama, N. Kabei, A. Yamada and Y. Sakurai, Homeostat to Control Blood Glucose Level. *Int. Symp. Med. Inf. Syst.*, 541-545 (1978).

Brill Academic Publishers
P.O. Box 9000, 2300 PA Leiden
The Netherlands

*Lecture Series on Computer*
*and Computational Sciences*
Volume 4, 2005, pp. 835-838

# Applying Response Modeling Methodology to Model Temperature-Dependency of Vapor Pressure

D. Benson-Karhi[a][1], H. Shore[a], M. Shacham[b]

[a]Department of Industrial Engineering and Management,
[b]Department of Chemical Engineering,
Faculty of Engineering Sciences,
Ben-Gurion University of the Negev,
Beer-Sheva 84105, Israel

Received 19 July, 2005; accepted in revised form 21 August, 2005

*Abstract:* In modeling chemical properties, theory-based relationships often represent available data accurately. However, when goodness-of-fit is not satisfactory, empirical modeling is called for. Recently, a new empirical modeling methodology has been developed, denoted Response Modeling Methodology (RMM). The new approach is intended to model monotone convex relationships. In this paper we apply RMM to model the temperature dependence of vapor pressure. The resulting models are compared to well-known and widely used property correlation equations. These include the "Acceptable Models", currently recommended by DIPPR[2] and models recommended by "Table Curve", a dedicated empirical modeling software. Recently developed methodologies for data-based comparison of models are employed to select the best model. Results show that RMM can represent satisfactorily curves of different shapes. Further research would extend the results introduced to other combinations of chemical properties and substances.

*Keywords:* AIC, Curve Fitting, Empirical Modeling, Physical and Thermodynamic Properties, Response Modeling Methodology (RMM)

Mathematical modeling of chemical properties aims to provide prediction with satisfactory accuracy. When theory-based models do not satisfy this requirement, empirical modeling is called for. A new empirical modeling methodology is applied to model vapor pressure in order to obtain more accurate empirical models than currently exist. A second objective is to evaluate the new methodology as a uniform platform for modeling chemical properties. The advantage of the latter becomes evident for property prediction based on co-linearity between the molecular descriptors of different compounds (Structure-Structure Correlation, Shacham *et al., 2004*).

The RMM approach models monotone convex/concave relations between a response and the affecting factor. When more than a single factor is known to transmit variation to the response, it is assumed that the affecting agent is the "linear predictor" (henceforth LP), a linear combination of effects that are related to the response. In that case, it is assumed that a monotone convex/concave relationship exists between the LP ( plus an additive random error) and the response. This characterization of the multi-factor relationship is characteristic to all current empirical modeling approaches, like linear regression and generalized linear modeling (henceforth GLM).

The RMM model is developed axiomatically from five basic assumptions (refer to Shore, 2003, 2004). A dual-error structure is assumed. The two error terms, $\varepsilon_1$ and $\varepsilon_2$, are assumed to be correlated, and to derive from a bi-variate normal distribution with correlation $\rho$.

The final basic RMM model (one of eight possible variations) is

$$Y = \exp\{\frac{\alpha}{\lambda}[(\eta + \varepsilon_1)^\lambda - 1] + \mu_2 + \varepsilon_2\} = \exp\{\frac{\alpha}{\lambda}[(\eta + \sigma_{\varepsilon 1} Z_1)^\lambda - 1] + \mu_2 + \sigma_\varepsilon Z_2\} , \qquad (1)$$

where $\eta$ is the LP, $Z_1$ and $Z_2$ are standard normal variables derived form a bi-variate standard normal distribution with correlation $\rho$, and

---

[1] Corresponding author. E-mail: karhi@bgu.ac.il
[2] Design Institute for Physical Properties

$$\theta_1 = \{\alpha, \lambda, \sigma_{\varepsilon 1}\}, \; \theta_2 = \{\mu_2, \sigma_{\varepsilon 2}\} \tag{2}$$

are real-valued parameters. Introducing $Z_2$ as a linear combination of $Z_1$ and another *independent* (uncorrelated) standard normal variable, $Z$,

$$Z_2 = \rho Z_1 + (1 + \rho^2)^{\frac{1}{2}} Z, \tag{3}$$

the RMM model may be re-written in terms of two independent standard normal variables, $Z_1$ and $Z_2$:

$$W = \log(Y) = \frac{\alpha}{\lambda} [(\eta + \sigma_{\varepsilon 1} Z_1)^\lambda - 1] + \mu_2 + \sigma_{\varepsilon 2} [\rho Z_1 + (1 - \rho^2)^{\frac{1}{2}} Z_2] \tag{4}$$

Find details about the other variations of the RMM model, for example, in Shore (2005a,b).

Wagner et al. (1973, 1976) have published high-precision data for vapor pressure for Argon, Nitrogen and Oxygen. The ranges of the data were from the triple point to the critical point for Argon and Nitrogen and from the normal boiling point to the critical point for Oxygen. Some additional information for characterization of these data is given in Table 1.

The experimental vapor pressure data and the calculated RMM curve for Oxygen are plotted versus temperature in Figure 1. The latter shows that all of the experimental data points are virtually located on the calculated curve (very small errors). Similar results were obtained for Argon and Nitrogen.

The performance of RMM for a wide range of vapor-pressure data was compared to that of the following existent models (relate to Daubert, 1998, for details):

### 1. Antoine equation:

$$\log(P) = A + \frac{B}{T + C} \tag{5}$$

where A, B and C are constants determined by the regression of experimental data. This equation is widely used for vapor-pressure correlation.

### 2. Wagner's equation:

$$\log(P_R) = \frac{a\tau + b\tau^{1.5} + c\tau^3 + d\tau^6 + e\tau^9}{T_R} \tag{6}$$

where a, b, c, d and e are adjustable parameters, $T_R$ is the reduced temperature ($T_R = T/T_c$, where $T_c$ is the critical temperature of the particular substance), $P_R$ is the reduced pressure ($P_R = P/P_c$, where $P_c$ is the critical pressure of the particular substance) and $\tau = (1 - T_R)$. This specific equation form was developed for Oxygen data only. (Relate to Wagner, 1973, 1976, for details).

### 3. The extended Riedel equation:

$$\log(P) = A - \frac{B}{T} + C \log(T) + DT^2 + \frac{E}{T^2} \tag{7}$$

where A, B, C, D and E are 5 adjustable parameters.

### 4. Selected model(s) by "Table Curve" (TC):

This software selects the best model/models out of about 4,000 different equations stored in its data-bank. Using "Model Selection" procedures surveyed in Burnham and Anderson (2004), where the AIC statistic is used for the final selection of the model, running TC with Oxygen data results in the following "best" model:

$$\log(P) = a + \frac{b}{T} + \frac{c}{T^2} + \frac{d}{T^3} + \frac{e}{T^4} + \frac{f}{T^5} + \frac{g}{T^6} \tag{8}$$

### 5. "Acceptable Model" recommended by DIPPR:

The model recommended by DIPPR for vapor pressure is:

$$\log(P) = A + \frac{B}{T} + C\log(T) + DT^E \tag{9}$$

## 6. RMM

leads to the following two-parameter version :

$$\log(P) = \log(M_P) + \frac{a}{b}\{\exp[\frac{b(T - M_T)}{Std_T}] - 1\} + \frac{a(T - M_T)}{Std_T} \tag{10}$$

where $M_P$ is the median of the measured vapor pressure data, $M_T$ is the temperature median, $Std_T$ is the standard deviation of the temperature data, and "a" and "b" are 2 adjustable parameters.

The calculations reported in this study were carried out with MATHEMATICA (a trademark of Wolfram Research, Inc.[3]) and Table Curve (by Systat software Inc.[4]).

The results of this comparison for Oxygen are shown in Table 2. AIC is also given. AIC is a statistic, calculated for given data set and a specified model, that estimates the relative loss of information due to the model used (without specifying the true model). It is commonly used to compare alternative models. Lower AIC indicates a better fit. (Burnham *et al., 2002).*

The three-parameter Antoine equation could not be fitted to data covering this wide temperature range of vapor pressure. It can be seen that the accuracy of the two-parameter RMM equation is significantly higher than that of the five-parameter DIPPR equation. Wagner equation provides the most accurate correlations for the vapor pressure of Oxygen, Argon and Nitrogen, but it should be noted that this equation was actually optimized (by stepwise regression) for the very same substances. From the results of this comparison, it can be concluded that the general purpose RMM competes favorably with the most accurate specific vapor-pressure equations for representing data covering a wide temperature range.

Applying RMM to model available data sets relating to temperature-dependent physical properties has proved to result in improved accuracy relative to existing comparable models (Shore, Brauner and Shacham, 2002, Shore, 2003). The same experience has been repeated more recently, when five substances and thirteen properties, provided by DIPPR, were analyzed via the RMM approach[5].

Based on the comparison of the resulting goodness-of-fit relative to that obtained with the respective "Acceptable Models", as currently provided by DIPPR, it can be concluded, at least for this limited-scope preliminary study, that modeling via RMM most commonly provides better goodness-of-fit statistics than the existing models. Furthermore, no "guess-work" is required to find the structure of the best-fitting model since the RMM estimation procedure determines not only the parameters' values but also the very structure of the estimated model.

Further research is planned to extend the use of RMM to represent physical and thermodynamic properties as a function of state variables such as temperature, for different substances. The advantage of RMM being a single uniform model can be used for property prediction based on co-linearity between the molecular descriptors of different compounds. The prediction approach of Shacham *et al.* (2004) requires the use of one model form for all compounds/properties relevant to a specified prediction equation. Thus, models are obtained that can be used for prediction of properties for a wide range of compounds, without the need to change a model's structure and/or its number of parameters as we move from one compound/property to the next.

Table 1: Data Characterization: Vapor Pressure Data

|  | Oxygen | Argon | Nitrogen |
|---|---|---|---|
| **Sample size (n)** | 177[6] | 57 | 68 |
| **Temp units** | K | K | K |
| **Temp range** | 90.188-154.581 | 83.804-150.651 | 63.148-126.2 |
| **Pressure units** | MPa | bar | bar |
| **Pressure range** | 0.10128-5.04337 | 0.6895-48.578 | 0.1252-34.002 |

---

[3] http://www.wolfram.com

[4] http://www.systat.com

[5] Fall Technical Conference (Roanoke, Virginia, October, 2004), a joint conference of the Chemical and Process Industry Division and the Statistics Division of the ASQ and of the Section on Physical and Engineering Sciences and of the Quality & Productivity Section of ASA.

[6] The original data set contains 183 observations. Measurements 178-183 close to the critical point were found outliers with respect to most of the thousands of models been checked.

**Table 2**: Comparison of Vapor-Pressure Correlation Equations for Oxygen Data

| MODEL | No. of parameters | MSE | AICc |
|---|---|---|---|
| Antoine | 3 | ------ | ------ |
| Wagner | 5 | $6.08*10^{-9}$ | -3341 |
| Riedel | 5 | $3.94*10^{-8}$ | -3011 |
| TC | 7 | $1.04*10^{-8}$ | -3244 |
| DIPPR | 5 | $2.98*10^{-6}$ | -2245 |
| RMM | 2 | $1.49*10^{-7}$ | -2779 |

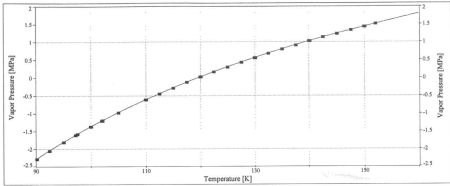

Figure 1: Experimental data and calculated curve using RMM for Oxygen's vapor pressure.

## References

[1] Burnham, K.P., Anderson, D.R. (2002). *Model Selection and Multimodel Inference*, Springer-Verlag, New York.

[2] Daubert, T. E. (1998) Evaluated Equation Forms for Correlating Thermodynamic and Transport Properties with Temperature. *Ind. Eng. Chem. Res.,* 37, 3260-3276.

[3] Shacham, M., Brauner, N., Cholakov, G. St., Stateva R. P. (2004). " Property Prediction by Correlations Based on Similarity of Molecular Structures", *AIChE J.* 50(10), 2481-2492.

[4] Shacham, M., H. Shore, N. Brauner, "A General Procedure for Linear and Quadratic Regression Model Identification", pp. 674-676 in T. Simos and G. Maroulis (Eds.), Lecture Series on Computer and Computational Sciences, VSP International Science Publishers, 2004.

[5] Shore, H., Brauner, N., Shacham, M. (2002). Modeling physical and thermodynamic properties via inverse normalizing transformations. *Industrial and Engineering Chemistry Research*, 41, 651-656.

[6] Shore, H. (2003). Response Modeling Methodology (RMM)- A new approach to model a chemo-response for a monotone convex/concave relationship. *Computers and Chemical Engineering*, 27(5), 715-726.

[7] Shore, H. (2004). Response Modeling Methodology (RMM)- Validating evidence from engineering and the sciences. *Quality and Reliability Engineering International*, 20, 61-79.

[8] Shore, H. (2005a). Response Modeling Methodology (RMM)- Maximum likelihood estimation procedures. *Computational Statistics and Data Analysis*, 49, 1148-1172.

[9] Shore, H. (2005b). *Response Modeling Methodology (RMM): Empirical Modeling for Engineering and Sciences*. World Scientific Publishing Co. Ltd., Singapore.

[10] Wagner, W. (1973). New Vapor Pressure Measurements for Argon and Nitrogen and a New Method for Establishing Rational Vapor Pressure Equations. *Cryogenics,* 13, 470.

[11] Wagner, W., Ewers, J. and Pentermann, W. (1976). New Vapor Pressure Measurements and New Rational Vapour-Pressure Equation. *J. Chem. Thermodyn,* 8, 1049-1060.

Brill Academic Publishers
P.O. Box 9000, 2300 PA Leiden
The Netherlands

*Lecture Series on Computer*
*and Computational Sciences*
Volume 4, 2005, pp. 839-842

# Industrial Case Study of Reaction Kinetics Applied for Continuous Dihydrate – Hemihydrate Method of P2O5 Production for Low Grade Phosphorites

I.V. Soboleva[1], E.M. Koltsova, A.V. Jensa

Department of Cybernetics of Chemical Engineering,
Mendeleyev Chemical-Technological University of Russia (MUCTR)
Miusskaya sq. 9, Moscow, Russia, 125047

Received 1 July, 2005; accepted in revised form 2 August, 2005

*Abstract:* More than 60 % of phosphorus containing fertilizers are being produced on the basis of extraction phosphoric acid, the main methods of its production being dihydrate and dihidrate-hemihydrate methods. Kinetics of calcium sulfate crystallization and recrystallization during the process of phosphoric acid extraction was studied in this work. Mechanisms of calcium sulfate crystal nucleation, growth and dissolution were found out. In this work we found out dihydrate-hemihydrate process optimum conditions. This provides more concentrated phosphoric acid yield with comparison of widely spread industrial dihydrate method. The common industrial dihydrate method requires significant energy supply on 22% $P_2O_5$ weak acid concentration by evaporation. The proposed method also provides energy saving on evaporation and intermediate filtration (according to the method involved more concentrated phosphoric acid – 34 – 35% $P_2O_5$, is obtained without intermediate filtration between stages) and it is friendly to environment due to wastes decrease because of pure calcium sulphate $\alpha$-hemihydrate production which can be used in construction industry as a binding material.

*Keywords:* modelling, phosphoric acid, dehydrate - hemihydrate method

## 1. Introduction

Acidic methods of phosphorus-containing raw material treatment are the main for phosphoric fertilizers production. More than 60 % of phosphorus containing fertilizers all over the world is produced on the basis of extraction phosphoric acid.

The process of phosphoric acid extraction is a concurrent process including both natural phosphates dissolution in sulfuric acid solutions and crystallization of calcium sulfate modifications (calcium sulfate dihydrate or calcium sulfate hemihydrate) depending on dihydrate, hemihydrate or dihydrate - hemihydrate methods carried out under different temperatures conditions.

Main purpose of this paper was to investigate experimentally and by means of mathematical modelling kinetics of main processes occurring during dehydrate – hemihydrate method of phosphoric acid extraction, namely: kinetics of phosphates dissolution, of passivation film formation on phosphates grains, of calcium sulphate dihydrate nucleation and growth in solution at the first stage and at the second stage, with the rise of temperature, calcium sulphate dihydrate dissolution and calcium sulphate hemihydrate nucleation and growth; to find out optimal conditions for industrial dehydrate – hemihydrate process and, on the basis of experimental and computer investigations of kinetics, to propose a new method of extraction phosphoric acid production by dihydrate – hemihydrate method.

## 2. Objectives and techniques of mathematical modelling of reaction kinetics applied for dehydrate- hemihydrate method of phosphoric acid extraction

On the base of experimental kinetics investigations the mathematical model of dihydrate stage was developed. The model takes into account the main physical-chemical regularities both for poly dispersed phosphorites decomposition with sulphate filming on them and dehydrate calcium sulphate(DH) nucleation and growth.. For the mathematical model developing for the second stage (recrystallisation of dihydrate calcium sulphate into hemihydrate calcium sulphate under periodic

---

[1] Corresponding author. E-mail: isoboleva@freemail.ru

conditions) an ideal mixing model was used. For the description of recrystallisation dihydrate calcium sulphate into hemihydrate calcium sulphate process we took an assumption that was checked during the experimental tests, that recrystallisation process mainly took place in liquid phase and crystal growth of hemihydrate calcium sulphate occurs in the kinetic area. Mathematical model of the second stage includes mathematical description of dihydrate calcium sulphate crystals dissolution process under simultaneous hemihydrate calcium sulphate crystals formation from solution. Model includes the equations of dihydrate calcium sulphate particles amount balance and hemihydrate calcium sulphate particles amount balance that are expressed in first three moments of their distribution function density for the quickness of computer calculations.

Let us consider dehydrate stage more detailed. Laboratory and industrial experimental accumulated materials allowed us to carry out analysis of phosphates dissolution and calcium sulphate crystallisation elementary acts and to define a type of processes rate dependencies on their moving forces. For the calcium sulphate crystallisation description the mechanism of a homogeneous nucleation and crystals growth in the kinetic area was accepted. Scheme of the processes interaction, described by the mathematical model, is shown in the fig. 1.

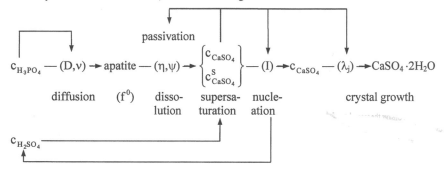

Figure 1: Scheme of physical-chemical phenomena interaction in the combined mathematical model of phosphates dissolution and calcium sulphate crystallisation processes.

Moving forces type of phosphates dissolution and crystallisation processes are based on the equation of universal mass transfer moving force, including enthalpy component, velocity component of non-equilibrium phases and on differences in chemical potentials of dissolved substance in the flow core and near the phases' interface surface.

Formatting alongside with a phosphoric acid calcium sulfate dihydrate has a shape of stretched parallelepiped. Therefore when modeling calcium sulfate dihydrate production we should consider growth of two facets.

Taking into account that both processes (phosphorite dissolution and calcium sulphate crystallization) are described within the framework of ideal mixing isothermal reactor models the following equations for velocities of the main processes were developed:

for phosphorite particles dissolution rate $\eta$ and passivation filming rate $\psi$:

$$\eta = \frac{K_2}{\rho_a}\left(\frac{\varepsilon D^4}{v l^2}\right)^{1/6}\frac{c_{H_3PO_4}e^{-K_1 h}}{\rho_\sigma} \;;\; \psi = R(a + bl)^q\left(c_{CaSO_4} - c^S_{CaSO_4}\right), \qquad (1)$$

for calcium sulphate nucleation rate I, for average rate of calcium sulphate crystal growth $\lambda_j$ on linear parameters $L_j$:

$$I = K_3\left(\frac{c_{CaSO_4} - c^S_{CaSO_4}}{c^S_{CaSO_4}}\right)^n \;;\; \lambda_j = K_{j+3}\left(\frac{c_{CaSO_4} - c^S_{CaSO_4}}{c^S_{CaSO_4}}\right)^{m_j}, \qquad (2)$$

where $\rho_a$ and $\rho_\sigma$ - the densities of phosphorite and passivation films with the thickness h; D - molecular diffusion factor, $\varepsilon = dl/dt$ - a linear rate of phosphate particle sizes transformations at the dissolution; e - specific capacity on mixing; l - size of phosphate particles ; h - a thickness of sulfuric film on phosphate surfaces; n - viscosity; $\nu$ - rate of dissolving, $c_{CaSO_4}$ and $c_{CaSO_4}^{S}$ - actual and equilibrium concentrations of $CaSO_4 \cdot 2H_2O$ in phosphoric acid solutions, accordingly. K, R, a, b, q - constants.

For the description of sulphate calcium mass crystallisation and polydispersed phosphorite dissolution the equations for the particles size distribution functions were used. The particles ensemble evolution of phosphorite was examined in two-dimensional phase space with co-ordinates: l - size of phosphorite non-dissolved grain, h - thickness of the passivation film. For the crystallisation process two characteristic linear sizes of crystal L1 and L2 were selected as phase space co-ordinates. For description of dissolution process experimental data, received by the method of radioactive isotopes on laboratory installation were used.

The mathematical model of extraction phosphoric acid production under the periodic conditions for dihydrate process was developed on the basis of mentioned above equations. As result of mathematical modelling the dissolution rate of polydispersed composition phosphorite, the rate of nucleation and the crystal growth rate were calculated, kinetic constants were found.

Comparing kinetic parameters of dihydrate process for phosphorites (Karatau, Kazakhstan) and Hibin apatites we can see that decomposition process for phosphorites is much more efficient for phosphorites, i.e. decomposition process for phosphorites goes much more rapidly. This is due to difference in crystal structure of phosphorites and apatites. By mathematical modelling it was also found out that crystallization rate of sulphate films does not depend on the size of initial particles that results from small parameter *b* in comparison with the same parameter for Hibin apatites.

With the help of computing experiment it was established that when concentration $P_2O_5$ are low than 26.3%, passivation films don't appear on phosphate grains. With the help of computer calculation according to mathematical model a number of dependences for dihydrate stage of the process were obtained which measurement with experiment are very difficult.

Let us consider the second stage. For the mathematical model developing for the second stage (recrystallisation of dihydrate calcium sulphate into hemihydrate calcium sulphate under periodic conditions) an ideal mixing model was used. For the description of re-crystallisation dihydrate calcium sulphate into hemihydrate calcium sulphate process we took an assumption that was checked during the experimental tests, that recrystallisation process mainly took place in liquid phase and crystal growth of hemihydrate calcium sulphate occurs in the kinetic area. Mathematical model of the second stage includes the mathematical description of dihydrate calcium sulphate crystals dissolution process under simultaneous hemihydrate calcium sulphate crystals formation from solution. Model includes the equations of dihydrate calcium sulphate particles amount balance and hemihydrate calcium sulphate particles amount balance that are expressed in first three moments of their distribution function density for the quickness of computer calculations.

In this work it was set up the problem of dihydrate-hemihydrate process optimum conditions determination. In a given work mathematical model of dihydrate-hemihydrate process in continuous conditions are also developed. Technological scheme of phosphoric acid production for industrial installation is offered.

According to the scheme, developed by us, the phosphorous containing raw material is getting mixed with the reversible phosphoric acid, entered from filtration hemihydrate pulp stage. Forming pulp goes into the reactor of the dihydrate stage, where a concentrated sulfiric acid 93-95% is also loaded (see fig. 3). At a temperature of 75-80°C the phosphate decomposition on this stage ends in 0,8 - 1,2 hour with forming of a calcium sulphate dihydrate. After decomposition is over the pulp is loaded into reactor of recrystallization, where a certain amount of sulphiric acid is added to create its excess in a liquid phase. The second stage is accompanied by the crystallization of calcium sulphate hemihydrate and by recrystallization of the calcium sulphate dihydrate, which formed on the first stage, to a calcium sulphate hemihydrate. The second stage duration is about 3,5 - 4 hours. After that the pulp is filtered and washed. Dihydrate stage is carried out in a one-section reactor of "Rhythm" type, and hemihydrate stage in an industrial 10-sectioned reactor with the capacity of 740 м$^3$.

The rectangular ten-sectional reactor with the working capacity of 740 m$^3$ was chosen as the most widely spread type of reactors used in phosphoric acid production. Taking into consideration a number of assumptions, this type of reactor can be described as the model consisting of five units of ideal mixing with the following process flow diagram: where $\upsilon_{H_2SO_4}, \upsilon_{H_3PO_4}$ - volume flow rates of the sulphuric and phosphoric acids; $\upsilon_a, \upsilon_f, \upsilon_p$ - volume flow rates of apatite, pulp on stage of filtration and circulation, accordingly; $V_{1-5}$ - capacities of units, corresponding to the capacity of two adjacent sections of simulated reactor; $\upsilon_i$ - pulp flow into i unit of reactor model that is equal to the amount of loaded reagents.

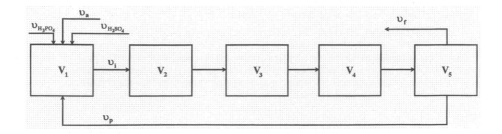

Figure 2: Process flow diagram for continuous industrial process integration.

At the development of the mathematical model of continuous process, taking into account the thermodynamic conditions in the reactor, the following assumptions are accepted: the extractor of a dihydrate stage presents by itself one unit of an ideal mixing, and the extractor of a hemihydrate stage consists of five cells of ideal mixing (one cell consists of two sections).

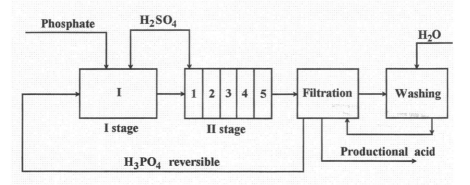

Figure 3: Designed technological scheme of production of extractional phosphoric acid for the industrial production.

By means of mathematical modelling the optimum sulphiric acid flow rates on first and second stage of dihydrate-hemihydrate process, the reversible phosphoric acid flow rate and the phosphorite rate on the first stage and volume of first stage reactor were found. This allows to support the optimum components concentrations in liquid phase for the achievement of maximum extraction coefficient on the first stage and maximum degrees of re-crystallization on the second stage. Under these optimum conditions the extraction coefficient of $P_2O_5$ to solutions is 98%, and the degree of calcium sulphate dihydrate re-crystallization to a calcium sulphate hemihydrate is 99%.

## 3. Conclusion

On the base of experimental investigations and mathematical modelling the principle new technological scheme of extraction phosphoric acid production (from phosphates with high concentrations of undesirable for acid technologies impurities) by dihydrate-hemihydrate method was developed.
Such technological scheme allows to obtain clean calcium sulphate α-hemihydrate, which further can be used in construction as a binding material; it results in reduction of calcium sulphate wastes. This method also allows to produce phosphoric acid with higher concentrations of $P_2O_5$, what results in power consumption reduction on the stage of this acid.

## Acknowledgments

The work was carried out under support of grants RFBR № 03-01-00567, FP6 programme INCO-Russia PL 0113359 "ECOPHOS".

Brill Academic Publishers
P.O. Box 9000, 2300 PA Leiden,
The Netherlands

*Lecture Series on Computer*
*and Computational Sciences*
Volume 4, 2005, pp. 843-847

# Simultaneous assimilation of physical and biological observations into a coupled ecosystem model using Joint and Dual Kalman filtering

G. Triantafyllou[a1], G. Korres[a], I. Hoteit[b], A. Pollani[a], and G. Petihakis[a]

[a] HCMR, Institute of Oceanography, PO BOX 712, 19013, Anavyssos, Greece
[b] Scripps Institution of Oceanography, La Jolla 92093-0224, USA

Received 7 July, 2005; accepted in revised form 11 August, 2005

*Abstract:* In this study, two assimilation systems based on a suboptimal extended Kalman filter have been developed to simultaneously assimilate physical and biochemical data into an ecosystem model of the Eastern Mediterranean. The ecosystem model is composed of two on-line coupled sub-models: the three-dimensional Princeton Ocean Model (POM) and the European Regional Seas Ecosystem Model (ERSEM). The filter is a variant of the extended Kalman filter which makes use of low-rank error covariance matrices to reduce computational burden. Two different approaches have been considered: the "joint approach" and the "dual approach". In the first approach, a unique filter is used in which the state vector of the filter is composed of the prognostic variables of both POM and ERSEM models. Basically, the numerical models are integrated forward in time to produce the (physical and biochemical) forecasts. The observations are then assimilated simultaneously to correct all forecast variables using the cross-correlations between all physical and biochemical forecast errors, providing the analyses for the physics and for the ecology. The dual approach consists of two filters, operating separately on the physics and on the ecology. The two assimilation systems were implemented and numerical experiments were performed to evaluate their performances.

*Keywords:* Data Assimilation, Joint Kalman filter, Dual Kalman filter, Ecological Modeling

## 1 Introduction

Data assimilation have known tremendous progress in physical oceanography during the last few years and advanced assimilation techniques have been developed to tackle the specific problems. However, this area is still in its infancy in marine ecology, due to the high complexity of the ocean ecosystem, to its dependency on the hydrodynamical forcing (advection/diffusion of ecological quantities), and to its significant spatial and temporal variability. Furthermore, the ecosystem modeling requires coupling of two complex models: the physical model that describes the currents of the area, and the biochemical model that describes the interactions between the different ecological species, making the assimilation into these systems very challenging. Up today, most of the studies consider only the assimilation into one of the two models, assuming that the other model is perfect. However, data assimilation into one system may result in misalignments of the physical and biological fronts giving rise to spurious cross-frontal fluxes of biological quantities. For instance, assimilation of only biological data often lead to spurious ecological responses (e.g.

---

[1]Corresponding author. E-mail: *gt@ath.hcmr.gr*, Phone: +30 22910 76402, Fax: +30 22910 76323

enhanced productivity). Moreover, a perfect model assumption is far too optimistic and one may not obtain reliable estimates of the ecology if the ocean circulations are not well simulated by the physical model. It is therefore necessary to constrain both models simultaneously with physical and biological observations in order to improve their behaviors and to assure a consistency between their respective analyses. In other words, a successful ecosystem assimilation system requires the coupling of a biological assimilation system with a hydrodynamic one capable of producing relevant physical fields, supportive of the newly updated biology. This problem is addressed in this paper with a complex 3D ecosystem model of the eastern Mediterranean, while focusing on: the effectiveness of a suboptimal Kalman filter in simultaneously assimilating physical and biological data into a coupled ecosystem model; the comparison between the joint and the dual approach; the capacity of an observing system based on one observed biological variable only (usually the ocean colour related to the model through chlorophyll); the control of numerous non-observed biological variables through the multivariate character of the assimilation scheme; and the effect of the physical data on the ecology.

## 2  The Eastern Mediterranean Ecosystem Model

The coupled physical-biogeochemical model consists of two, on-line coupled, sub-models: the Princeton Ocean Model (POM) [1], which describes the hydrodynamics of the area, and provides the physical forcing to the second sub-model, the European Regional Seas Ecosystem Model (ERSEM) [2]. POM is a three dimensional, time dependent, primitive equations ocean model; the equations are solved over an Arakawa-C differencing scheme and a sigma-coordinates discretization in the vertical. Time integration is achieved through an explicit scheme with the barotropic and baroclinic modes being integrated separately using a leap frog scheme with different time steps. The model state variables are the sea surface elevation, the zonal and meridional components of velocity $U, V$ and their depth-averaged counterparts $(\overline{U}, \overline{V})$, potential temperature $T$, salinity $S$, the turbulent kinetic energy $q^2$ and the turbulent kinetic energy times the turbulent length scale $q^2 l$. ERSEM describes the biogeochemical cycles and has been successfully applied in a wide variety of regimes from coastal eutrophic to open sea oligotrophic systems and on a variety of spatial scales [2, 3]. The use of a functional group idea where organisms with similar processes are grouped together, rather than that of a species, increases ERSEM portability to any area. The biotic system encompasses the three major types (producers, consumers and decomposers) with each type being further subdivided, increasing thus the required complexity into 88 state variables, which include oxygen, carbon dioxide, phosphate, nitrate, silicate, ammonia, all the phytoplankton and zooplankton groups, particulate organic matter, dissolved organic matter, bacteria and benthic particulate organic matter. The modeled area covers the Eastern Mediterranean from $20°E$ to $36.4°E$ and $30.7°N$ to $41.2°N$. A horizontal resolution of 6 minutes is chosen producing $165 \times 106$ elements. To better represent the increased velocity gradients near the surface, a logarithmic vertical distribution is used with a total of 24 $\sigma$-layers. The model is initialized with climatological objectively analyzed temperature and salinity profiles from the Mediterranean Ocean DataBase (MODB-MED4). Initial velocities are set to zero. Wind stress fields derived from the ECMWF 6-hour reanalysis data. The initialization of the biogeochemical model variables is described in [4].

## 3  The assimilation scheme

The assimilation scheme is a simplified square-root extended Kalman (EK) filter, called the Singular Evolutive Extended Kalman (SEEK) filter. It has been developed by [5] for application with highly dimensional $(n)$ systems. It basically reduces the prohibitive computational burden of the EK filter by using low-rank matrices approximations of the filter's error covariance matrices. The main idea behind this filter is to initialize the EK filter by a singular low-rank ($r \ll n$) matrix of the form $P_0 = L_0 U_0 L_0^T$, where $L_0$ and $U_0$ are of dimension $n \times r$ and $r \times r$, respectively. It can then be shown that the filter's error covariance matrices $P_k$ always remain of low-rank $r$ of the same form as $P_0$, i.e. $P_k = L_k U_k L_k^T$, if the model error is projected onto the subspace generated by $L_k$. One can then avoid the manipulation of the matrices $P_k$ while only iteratively updating $L_k$ and $U_k$ as described in the filter's algorithm presented hereafter.

Consider the dynamical system $X_k = M_{k,k-1}(X_{k-1}) + \eta_k$ and $Y_k = H_k(X_k) + \varepsilon_k$ where $X_k$ and $Y_k$ denote the system state and the associated observational vector at time $t_k$, $M_{k,k-1}$ is the transition operator from time $t_{k-1}$ to time $t_k$ and $H_k$ is the observational operator, $\eta_k$ and $\varepsilon_k$ represent the dynamical and observational noise. They are assumed to be independent Gaussian vectors of mean zero and covariance matrices $Q_k$ and $R_k$, respectively.

The SEEK filter proceeds in two steps apart from an initialization step. The initial analysis state $X_0^a$ and the corresponding error covariance matrix $P_0^a$ are generally estimated from a long sequence of model outputs $H_S$, as the mean and the sample covariance matrix of $H_S$, respectively. A low-rank $r$ approximation of $P_0^a$ is then obtained via the application of a principal component (PC) analysis on $H_S$.

1- **Forecast step:** Assuming that an estimate $X_{k-1}^a$ of the system state and the corresponding rank $r$ error covariance matrix $P_{k-1}^a = L_{k-1} U_{k-1} L_{k-1}^T$ are available at time $t_{k-1}$. The model of the dynamical system is integrated forward in time to compute forecast starting from the analysis state $X_k^f = M_{k,k-1}(X_{k-1}^a)$. The forecast error covariance matrix is then approximated by $P_k^f = L_k U_{k-1} L_k^T + Q_k$, where $L_k = \mathbf{M}_{k,k-1} L_{k-1}$ and $\mathbf{M}_{k,k-1}$ is the gradient of $M_{k,k-1}$ evaluated at $X_{k-1}^a$.

2- **Correction step:** The new observation $Y_k$ at time $t_k$ is used to correct the forecast with the formula $X_k^a = X_k^f + L_k U_k L_k^T \mathbf{H}_k^T R_k^{-1}[Y_k - H_k(X_k^f)]$, where $\mathbf{H}_k$ is the gradient of $H_k$ evaluated at $X_k^f$, and $U_k$ is updated with $U_k^{-1} = \left[U_{k-1} + (L_k^T L_k)^{-1} L_k^T Q_k L_k (L_k^T L_k)^{-1}\right]^{-1} + L_k^T \mathbf{H}_k^T R_k^{-1} \mathbf{H}_k L_k$. The corresponding filter error covariance matrix is then equal to $P_k^a = L_k U_k L_k^T$.

The simultaneous assimilation into two coupled models is based on *joint* and *dual* filtering approaches, which originally designed to estimate the model state concurrently with the model parameters using an analogous filter [6]. Here we generalize them estimating the state of the ecological model concurrently with the physical forcings.

We begin with the joint approach which is simpler to conceptualize: the physical state vector is simply appended into the ecological state vector, to form one state vector for the coupled model. The physical and ecological observations are also appended together into one observational vector. The ecological and the physical part are updated by each model, but the entire augmented covariance matrix is propagated as one. The dual filtering basically intertwines a pair of distinct Kalman filters; one estimating the ecology and the other estimating the physics. This approach respects more the one-way coupling nature of our ecological model, allowing to correct the ecology independently from the physics. In its general form, biological data can be also assimilated with the "physical filter". However this was not applied in the present work to reduce heavy computational burden required for the linearization of the biological model with respect to the physics.

## 4    Results and Discussion

The effectiveness of the joint and the dual filtering was evaluated following the "twin-experiments" approach. In these experiments, the "truth" is assumed to be provided by the model itself. This allows to assess the filter's behavior on non-observed variables as all uncertain parameters are known by design. The model statistics were also used for the initialization of the filter, which requires an initial state estimate and a corresponding low-rank (to be chosen) error covariance matrix.

After a four years integration period of the coupled ecosystem model to achieve a quasi adjustment of the model climatological dynamics, another integration of two years was carried out to generate a historical sequence $H_S$ of 365 model state vectors which were sampled every two days. The filter was then initialized by the mean of $H_S$ and a low-rank approximation of the initial error covariance matrix was obtained by applying a Principal Component (PC) analysis on $H_S$. Prior to the analysis, the model variables were normalized by the inverse of their variance averaged over all the model domain to make the distance between model state vectors independent from unit of measure. In the joint approach, the rank of the filter was set to 40 as the first 40 (physical-ecological) PCs were found to explain more than 96% of the total variance of $H_S$. In the dual approach, the "physical filter" and the "ecological filter" were respectively initialized by error covariance matrices of ranks 40 and 10, as the first 40 physical PCs and 10 ecological PCs approximately resume 90% and 91% of the system total variance.

For the assimilation experiments, pseudo-observations of sea surface height (SSH) and chlorophyll (CHL) were assumed to be available every two days over the whole surface of the model domain. These observations were extracted from a set of 45 reference states $X^t$ simulated by the model over a three months period (from March $5^{th}$ to June $5^{th}$). Random Gaussian noise of zero mean and 5% of $H_S$ mean as standard deviation were added to the pseudo-observations in order to build a more realistic framework. The $X^t$ are also later used to evaluate the performance of the filters by comparing them to the filters estimates, relative to the model free-run results. The free-run consists of integrating the model without any assimilation starting from the filter initial conditions. The relative error (RRMS) for each model variable was defined as the ratio between assimilation error and forecast error. Relative errors smaller than unity indicate that the assimilation estimates are in closer to the true run than free-run estimates. Assimilation

Table 1: Summary of the filters runs

| Experiment | Filter type | Assimilated data |
|:----------:|:-----------:|:----------------:|
| Exp.1 | DUAL | SSH |
| Exp.2 | DUAL | SSH & CHL |
| Exp.3 | JOINT | SSH |
| Exp.4 | JOINT | SSH & CHL |

runs were preformed to study the filter(s) behavior as summarized in the Table 1. Figure 1 plots the evolution of the averaged RRMS of the filters runs for the physical variables (left panel) and ecological variables (right panel) and compare them to the one obtained from the model free-run. In all cases, the assimilation schemes significantly improve the model's behavior with respect to the free-run. The assimilation of SSH improves the performance of the hydrodynamic model by reducing the error about 70%, with the best performance being obtained in Exp.1 and Exp2. The corrected physics were also shown to have an important impact on the evolution of the ecology and this is depicted in Exp.1, in which SSH data were used to constrain the physical sub-system only, without any assimilation of ecological data. Additional assimilation of surface CHL data with the dual filter further improves the estimation of the ecology. The best performance for the ecological assimilation sub-system was achieved in Exp.3 and Exp.4, where the joint approach was used to constrain the state vector of all physical-ecological variables. This approach highly improves the convergence to the ecological estimates when only physical data are assimilated by efficiently using the cross-correlations between the physical and ecological variables, allowing the filter to control

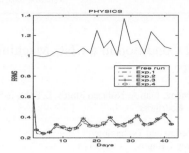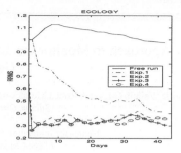

Figure 1: Evolution in time of the average RRMS for the physical variables (left panel) and the ecological variables (right panel) for the different filters runs.

the behavior of both models. As it can be expected, additional assimilation of surface CHL data with SSH (Exp.4) further improves the estimation of the ecological variables, but has very little impact on the physics which is consistent with the one-way coupling of our model. Overall, both approaches were capable of controlling the behavior of the ecosystem, with a slightly better performance obtained by the joint approach particularly in the case of assimilation of physical data only. This, however, can be explained by the absence of linearization in the ecology with respect to the physics in our dual system (section 3). Additionally, the idealized twin-experiments approach was favored by the joint system since physical-ecological cross-correlations were a priori very well estimated, which is often not true in realistic applications.

## Acknowledgment

This work has been supported by Mediterranean ocean Forecasting System: Toward Environmental Predictions (MFSTEP) project.

## References

[1] A. F. Blumberg and G. L. Mellor. A description of a three-dimensional coastal ocean circulation model. In N.S. Heaps, editor, *Three-Dimensional Coastal Ocean Circulation Models*, Coastal Estuarine Science, pages 1–16. American Geophysical Union, Washington, D.C., 4 edition, 1987.

[2] J.W. Baretta and H. Baretta-Bekker. Observational requirements for validation of marine ecosystem models in the mediterranean. *Rapp. Comm. int. Mer Medit.*, 35:18–20, 1998.

[3] G. Triantafyllou, I. Hoteit, G. Korres, and G. Petihakis. Ecosystem Modeling and Data Assimilation of Physical-Biogeochemical processes in Shelf and Regional Areas of the Mediterranean Sea. *Applied Numerical Analysis & Computational Mathematics*, (in print), 2005.

[4] G. Triantafyllou, I. Hoteit, and A. Pollani. Toward a Pre-Operational Data Assimilation System for the E. Mediterranean using Kalman Filtering Techniques. *Lecture Series on Computer and Computational Sciences*, 1:500–505, 2004.

[5] D.T. Pham, J. Verron, and M.C. Roubaud. Singular evolutive kalman filter with eof initialization for data assimilation in oceanography. *Journal of Marine Systems*, 16:323–340, 1997.

[6] A. T. Nelson. *Nonlinear estimation and modeling of noisy time series by dual Kalman filtering methods*. PhD thesis, Oregon Graduate Institute, 1999.

Brill Academic Publishers
P.O. Box 9000, 2300 PA Leiden
The Netherlands

*Lecture Series on Computer
and Computational Sciences*
Volume 4, 2005, pp. 848-851

# New Approach to Monitor the Tool Condition in a CNC Machining Center

Antonio Jr. Vallejo Guevara[1], Rubén Morales-menéndez, Juan A. Nolazco Flores, L. Enrique Sucar Succar, Ciro A. Rodríguez

Center for Innovation in Design and Technology,
ITESM Monterrey Campus,
Monterrey, N.L. México

Received 14 July, 2005; accepted in revised form 8 August, 2005

*Abstract:* We propose to monitor the cutting tool condition in a CNC-machining center by using continuous Hidden Markov Models (HMM). A vibration signal database was created monitoring the vibration between the cutting tool and workpiece. We trained/tested the *HMM* for 18 different operating conditions. The *HMM* were created by preprocessing the waveforms, followed by training step using the Baum-Welch algorithm. In the decoding process, the signal waveform is also preprocessed, then, the trained *HMM* are used for decoding. Early experimental results validate our proposal about exploiting speech recognition frameworks in monitoring tool condition. The proposed model is capable of detecting the cutting tool condition within large variations of spindle speed and feed rate. The classifier performance was of 96%.

*Keywords:* Hidden Markov Models, Manufacturing Process, and Fault Diagnosis.

## 1.  Introduction.

The successful performance of CNC machining operations involves the selection and control of a large number of parameters related to the workpiece, cutting tool and machine tool. The current trend of the development of machine tools is to increase intelligence through better process models and sensors. The capability to predict cutting tool condition can help to maintain workpiece quality and schedule tool changes [7]. We propose to implement a recognition approach for tool monitoring using continuous *HMM*. The vibration signals between the tool and the workpiece will provide the necessary data to perform this kind of tool monitoring.

## 2.  State-of-the-art.

The tool failure represents ~20% of machine tool down-time. Tool wear impacts negatively the quality in the context of dimensions, finish, and surface integrity, [8]. [3]-[4] developed an intelligent supervisory system for tool wear prediction using an Artificial Neural Network (*ANN*) output error model; also, they exploited the analysis of signals signatures in the time and frequency domains. [7] worked with multilayered *ANN* in the milling process. [5]-[1] presented an approach for feature extraction from vibrations during the drilling. They used self-organizing feature maps for extracting the features, and a discrete *HMM* classifier.

## 3.  Tool-wear monitoring system.

Fig. 1 shows the proposal algorithm for monitoring the tool-wear using continuous *HMM*. The input signal is preprocessed and then it is separated in two branches. The training data branch leads to a

---

[1] Corresponding author. avallejo@itesm.mx

*HMM*. In this training phase the system learns the patterns. The testing branch uses the preprocessed input signal and the *HMM* to compute the *P(O\λ)* using the Viterbi algorithm for each model. The model with higher probability is the end result.

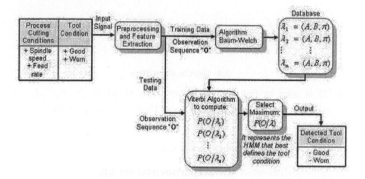

Figure 1: Flow diagram to monitor the tool condition with continuous HMM.

### 3.1 Hidden Markov models

*HMM*s are characterized as: $N$, number of states. We denote the states as $S = S_1,...,S_N$, and the state at time $t$ as $q_t$. $M$, number of distinct observation symbols per state. State transition probability distribution $A = a_{ij}, a_{ij} = P[q_t = S_j | q_{t-1} = S_i]$, $1 \le i, j \le N$. Observation symbol distribution in state $j$, $B = b_j(k), b_j(k) = P[v_k | q_t = S_j]$, $1 \le j \le N$, $1 \le k \le M$. Initial state distribution $\pi = \pi_i, \pi_i = P[q_1 = S_i], 1 \le i \le N$. $N, M$, and λ are learned from data. Given this model and the observation we can compute $P(O|\lambda)$.

### 3.2 Feature extraction step

The vibration signals are pre-processed calculating their Mel Frequency Cesptral Coefficient (MFCC) representation [6]. First, the vibration signals are divided into short frames, and the amplitude spectrum is obtained using the Discrete Fourier Transform (DFT), and converted to the log-scale. For smoothing the scaled spectrum filter banks are used. Finally, the discrete cosine transform is applied to eliminate the correlation. We obtain a 39 dimension vector formed by 12-dimension MFCC, one energy coefficient, 13Δ and 13Δ² coefficients.

### 3.3 Baum-Welch algorithm

The well-known Baum-Welch algorithm is an iterative process for parameter estimation based on a training data set for a given model λ. The goal is to obtain a new model $\overline{\lambda}$ where the function

$$Q(\lambda, \overline{\lambda}) = \sum_Q \frac{P(O, Q|\lambda)}{P(O|\lambda)} \log[P(O, Q|\overline{\lambda})] \tag{1}$$

is maximized. New set of parameters for the model are calculated by Baum-Welch as follow:

$$\overline{\mu}_{jk} = \frac{\sum_{t=1}^T \xi(j,k)o_t}{\sum_{t=1}^T \xi_t(t,k)} \quad \overline{U}_{jk} = \frac{\sum_{t=1}^T \xi_t(j,k)(o_t - \overline{\mu}_{jk})(o_t - \overline{\mu}_{jk})'}{\sum_{t=1}^T \xi_t(j,k)} \quad \overline{c}_{jk} = \frac{\sum_{t=1}^T \xi_t(j,k)}{\sum_{t=1}^T \sum_k \xi_t(j,k)} \tag{2}$$

Where $\xi_t(j,k)$ is the probability for changing from state $j$ to state $k$, $b_j(o)$ is a continuous output probability density function, $c_{jk}$ is the weight of the Gaussian $k$ and $N(o,\mu_{jk},U_{jk})$ is a single Gaussian of mean value $\mu_{jk}$ and a covariance matrix $U_{jk}$. The term $b_{jk}$ can be written as

$$b_{jk}(o_t,\mu_{jk},\sigma_{jk}) = \frac{1}{\prod_{i=1}^{d}\sqrt{2\pi\sigma_{jki}}} e^{-\frac{1}{2}\sum_{i=1}^{d}\left(\frac{o_{ti}-\mu_{jki}}{\sigma_{jki}}\right)^2} \tag{3}$$

### 3.4 Viterbi algorithm

This algorithm [2] is used to find the single best state sequence, $Q = q_1,...,q_T$, for the given observation sequence $O = O_1,...,O_T$ . Then, we need to compute: $\delta(i) = \max_{q_1,...,q_{t-1}} P[q_1,...,q_t = i, O_1,...,O_t|\lambda]$ .

## 4. Experimental set up.

We worked with a KX10 HURON machining center, Fig. 2 (left), with a capacity of 20 KW, 3 axes, and a Siemens controller. The cutting tool was an Octomill face mill of SECO Carboloy, Fig. 2 (middle). The material of the workpiece was Aluminum. The vibration signals were recorded with a data acquisition system using an accelerometer installed on the flat metal support, Fig. 2 (right). We applied a full factorial design of 4 independent variables: spindle speed (1, 1.5 and 2 Thousands rev/min), feed rate (600, 800, and 1,000 mm/min), depth of the tool (1 mm), and tool condition (good and worn inserts).

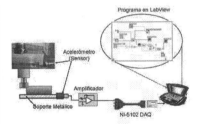

Figure 2: Huron milling center, cutting tool, and data acquisition system.

## 5. Results.

The database contains 153 experiments of the 18 different operating conditions. First, we selected a set of experiments $(T_r)$ for training, and other set of experiments were used for testing $(T_s)$ . The utterances were processed using the Sphinx *HMM*s Toolkit developed at Carnegie Mellon University. The toolkit was configured to use several Gaussian, left-right model, and five states for the *HMM*. Fig. 3 (left) shows the performance obtained for the classifier. We observed an excellent behavior when we used enough experiments to train the *HMM*. We configured the *HMM* toolkit for recognition of two states: good and faulty (worn inserts) condition. Table 1 shows the results.

We also defined two alarms: False Alarm Rate (*FAR*) and False Fault Rate (*FFR*). *FAR* is the rate when decoder detects that the tool is in fault condition, but the tool is really in good condition. *FFR* is the rate when decoder detects the tool is in good condition and it is not. *FAR* condition is not a problem for the machining process. However, *FFR* could be dangerous because the poor quality of the product and the tool can be broken. We obtained a low *FFR* rate when the *HMM* are trained with several Gaussian, as we show in Fig. 3 (right).

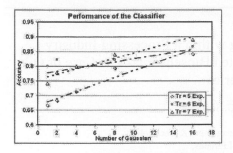
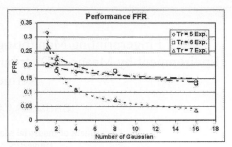

Figure 3: Performance of the Classifier with different experiments.
Table 1: Accuracy of the model

| Training Exp. | Testing Exp. | 1 Gaussian | 16 Gaussians |
|---|---|---|---|
| 90 | 63 | 66.7% | 84.0% |
| 108 | 45 | 80.0% | 87.0% |
| 126 | 27 | 74.0% | 96.0% |

## 6. Conclusions.

We have proposed an algorithm for monitoring the cutting tool condition in a CNC-machining center by using continuous *HMM*. A database was built with the vibration signals obtained during the machining process of Aluminum 6061. We trained/tested the *HMM* for different operating conditions and the results were satisfactory given the limited number of experiments. Basic speech recognition framework was exploited in with successful results and great potential.

## References

[1] Les Atlas, Mari Ostendorf, and Gary D. Bernard. Hidden markov models for monitoring machining tool-wear. *IEEE*, pages 3887-3890, 2000.

[2] J. Bilmes. What HMMs can do. Technical Report UWEETR-2002-0003, Department of Electrical Engineering from University of Washington.

[3] R. E. Haber and A. Alique. Intelligent process supervision for predicting tool wear in machining processes. *Mechatronics*, (13): 825-849, 2003.

[4] R. E. Haber, J. E. Jiménez, C. R. Peres, and J. R. Alique. An investigation of tool-wear monitoring in a high-speed machining process. *Sensors and Actuators,* (39):583-605, 2004.

[5] L. M. D. Owsley, L. E. Atlas, and G. D. Bernard. Self-organizing feature maps and hidden markov models for machine-tool monitoring. *IEEE transactions on Signal Processing*, 45(11):2787-2798, 1997.

[6] L. R. Rabinier and B. H. Juang. *Fundamentals of speech recognition*. Prentice-Hall, New Jersey, 1993.

[7] H. Saglam and U. Unuvar. Tool condition monitoring in milling based on cutting forces by a neural network. *Int. Journal of Production Research*, 41(7):1519-1532, 2003.

[8] S.Y. Liang, R. L. Hecker, and R. G. Landers. Machining process monitoring and control: The state-of-art. *ASME Int. Mechanical Engineering Congress and Exposition*, pages 1-12, 2002.

Brill Academic Publishers
P.O. Box 9000, 2300 PA Leiden
The Netherlands

*Lecture Series on Computer
and Computational Sciences*
Volume 4, 2005, pp. 832-838

# Odor Fingerprinting/Monitoring within a Processing Industry Environment by Means of Distributed Biosensors – The Case of Oil Refineries

F.A. Batzias, C.G. Siontorou[1]

Department of Industrial Management & Technology,
University of Piraeus,
GR-185 34 Piraeus, Greece

Received 17 July, 2005; accepted in revised form 16 August, 2005

*Abstract:* The escalating public concern about odor pollution prompts the efforts to establish control programmes in many countries. In practice, controlling odorants is a very complex problem: sources and emissions have to be identified, analytical methods have to be evaluated, risks have to be assessed, critical emissions have to be controlled, and economical aspects have to be integrated. This paper addresses odor monitoring through a holistic approach, involving odor fingerprinting, source identification and remedial activities to reduce emissions in facility design and operation, through a biosensor network and an effective communications between the offending facility and those groups that may be negatively impacted by odors. Biomonitoring is realized through a suitable developed computer-aided selection program, that, considering the odorous mixture and its source, is almost tailor-made for the specific needs at hand in order to minimize uncertainty and monitoring costs and increase reliability of odor control and abatement. The response of a range or array of sensors can be monitored through separate temporal and/or spatial Geographical Information System (GIS) layers and subsequently correlated by means of a relational database, so that odor levels can be recorded either systematically or as function of time and/or space. Significantly, a novel system in the form of a rational framework at the conceptual design level has been developed, that actually contributes towards achieving a cost-effective long-term monitoring program, with the flexibility to counter on-course any (anticipated or not) variations/modifications of the surveillance environment. The advantages of this novel and pioneering approach will further offer a dynamic system utilized in (a) odor impact studies and risk assessment (positive/analytic approach), (b) decision-making in the short-run (normative/tactic approach), and (c) policymaking in the long-run (normative/strategic approach).
*Keywords:* biosensors, decision support system, odor monitoring, ontology

## 1. Introduction

Of the various categories of air pollutants, odors have been ranked as the major generators of public complaints to regulatory agencies worldwide. These complaints are addressed to a wide variety of industries and operations including agriculture, sewage treatment works, paint, plastics, resin and chemical manufacturers, refining operations, rendering plants, pulp mills and landfills, among others.

Odors are defined as sensations that occur when chemical substances (odorants) stimulate receptors in the nasal cavity [1]. Most odors perceived in the environment are made up of a multifaceted mixture of odorants. The compounds that make up particular odors are often present in small concentrations and can act in the human nose in a complex effect, making their regulation by the setting of emission limits, similar to other ambient air pollutants, complicated [2]. The effects of odors are equally complicated and range from the associative and the psychological, such as unease, discomfort, irritation or depression, to the measurable and the physiological, including sensory irritations, headaches, respiratory problems, nausea, or vomiting [3]. Sub-irritant levels of odorants may trigger acute symptoms through non-toxicological, odor-related mechanisms that have been postulated to include innate odor aversions, odor exacerbation of underlying conditions, odor-related aversive conditioning, stress-induced illness, and mass psychogenic illness. The associative and emotional characteristics of the sense of smell may be important in the field of odor regulation because negative associations to odors, once formed, seem to be difficult to change. For instance, a neighborhood may develop a

---

[1] Corresponding author. E-mail: csiontor@unipi.gr

negative association to a particular odor during a period of intense odorous emissions, that it is usually maintained even after odors are substantially and measurably reduced.

Odor problems occur with sufficient frequency and sufficient impact to warrant intervention. Although odor is attributed to certain chemical compounds, these are not regulated solely for their odor impacts [2]. Compounds such as hydrogen sulfide and volatile organic compounds are regulated for their impacts on human health and the environment, and while the standards do recognize the concepts of annoyance and nuisance, the numbers do not necessarily relate well to odor thresholds. Interesting enough, although the detection threshold concentrations of substances that evoke a smell are slight, a concentration only 10 to 50 times above this threshold value is often the maximum intensity that can be detected by humans [1]. In this sense, using currently established federal and state standards for chemical emissions is not a good proxy for odor, as odors can be detected well before the emission levels reach a regulated threshold. In addition, the offending industries lack the methodology to aid them in assessing odor levels and testing odor reduction technologies prior to and during full-scale implementation. Failure to acknowledge the potential for odors and to work to prevent and monitor odor emissions can result in complaints, shutdowns, expensive retrofits, and non-acceptance of the finished product.

The measurement of odors is performed with scentometric techniques based on the perception of a human panel [4]. The disadvantages of human sensory panels include subjectivity, poor reproducibility (i.e., results fluctuate depending on time of day, health of the panel members, prior odors analyzed, fatigue, etc.), time consumption, and large labor expense. Also, human panels cannot be used to assess hazardous odors, work in continuous production, or remote operation. Instrumental odor monitoring is currently performed only on some registered toxic compounds, quantitated, however, at the toxic limits and not the much lower odor thresholds. The discrimination of organic compounds in the gas phase at such low levels in complex mixture can be achieved *ex situ* by the cumbersome, expensive and time-consuming GC/MS, HPLC and NMR, largely unsuitable for long-term field surveillance around the odor source or impractical for monitoring unknown odorous compounds, further to the lack of correlation between the instrumental results and the human perception.

Biosensors, utilizing natural biorecognition mechanisms coupled to chemical transducers for signal detection [5,6], have been long recognized as nature-mimicking, low-cost, real-time, and in-situ monitoring devices able to reliably and reproducibly measure qualitatively/quantitatively target analyte(s) at ppt detection limits [6,7]. For instance, the electronic nose, in accordance with the human olfaction mechanism, contains a series of receptors, each specific to an odorant, the complexation of which produces a physicochemical change (mass change, thermal change, altered surface charge, flow of current or change in optical properties), detected by the suitable transducer [8]. The raw signal generated by the array of odor sensors is typically a collection of different electrical measures, which need to be processed in order to allow the recognition of a particular odor.

This paper addresses odor monitoring through a holistic approach, involving odor fingerprinting, source identification and remedial activities to reduce emissions in facility design and operation, through a biosensor network and an effective communications between the offending facility and those groups that may be negatively impacted by odors. Biomonitoring is realized through a suitable developed computer-aided selection program, that, considering the odorous mixture and its source, is almost tailor-made for the specific needs at hand in order to minimize uncertainty and monitoring costs and increase reliability of odor control and abatement. The response of a range or array of sensors can be monitored through separate temporal and/or spatial Geographical Information System (GIS) layers and subsequently correlated by means of a relational database, so that odor levels can be recorded either systematically or as function of time and/or space. Significantly, a novel system in the form of a rational framework at the conceptual design level has been developed, that actually contributes towards achieving a cost-effective long-term monitoring program, with the flexibility to counter on-course any (anticipated or not) variations/modifications of the surveillance environment. The advantages of this novel and pioneering approach will further offer a dynamic system utilized in (a) odor impact studies and risk assessment (positive/analytic approach), (b) decision-making in the short-run (normative/tactic approach), and (c) policymaking in the long-run (normative/strategic approach).

## 2. Methodology

### 2.1 The development of the Knowledge Base

Biosensors for industrial and environmental uses have been mostly studied on a bench scale and very few have reached field-testing and commercialization. This status provides for both major disadvantages and advantages; the former lie mainly on the need for R&D prior to large-scale application, whereas the latter offer many possibilities as per the costume-tailoring of the device to a

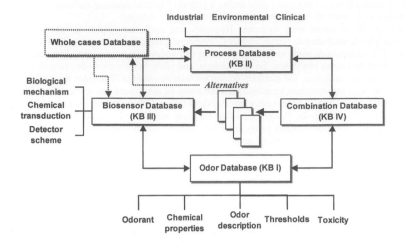

Figure 1. Infrastructure of the relational database, designed for optimal decision making in odor fingerprinting.

specific application. For the benefits of this technology to be realized, much depends upon the proper selection of the biochemical system, the transducing principle, the sensor preparation and the sample presentation; the latter is very important as it sets the specifications and the requirements, derived from the specific process that emits the target analyte. The successful implementation of sensors to odor fingerprinting, presupposes the association between biosensors and odor inventories through studied/applied combinations that would also allow for the presentation of potentially suitable physicochemical mechanisms, to account for the lack of field exploitation. For example $\alpha_2$-adrenoreceptors are quite useful in detecting dimethylamine emitting from animal rendering sources but markedly ineffective in the food processing industry, where the emissions of putrescine, used in fruits for accelerating ripening, compete with the target analyte for the receptor binding sites.

The authors have constructed a KB using a relational database management system (RDBMS) for mining the data gathered and generate the relations. The domains of investigation (primary domains or independent databases) – KB I: odor; KB II: process; KB III: biosensor (Figure 1) - have been linked to all levels of constraints through rules-based reasoning in order to provide secondary domains (KB IV: combinations database), which by means of compatibility indices they provide the alternatives. As the primary domains are mostly taxonomic, the semantic links are mainly of the *is-a* or *has-a* type, while in the secondary domains the semantic links *is-part-of-a* and *means-of-a* (*is-expressed/examined-by*) dominate.

Data can be derived either from (a) whole cases, introduced in the whole cases database and decomposed to parts in the databases of the primary domains, or (b) by information related solely to the primary domains. The data input and the intermediate/final results output are stored into the RDBMS, an adaptation of the ABEPE Database [9]. The RDBMS model provides a user friendly and efficient tool to handle large quantities of data; moreover, since SQL is an ANSI standard, widely accepted by the computer industry, a large amount of other commercial software can cooperate with SQL applications and increase/expand functionality. Since there are natural language interpreters translating voice or keyboard entered commands into SQL statements (e.g. Microsoft English Query), the potential program user can submit queries in a natural language. The broad availability of MS Access renders this RDB the most suitable development environment. Not to mention that such an implementation of the database could be easily upsized to a multiprocessor system running SQL Server RDBMS because of the transparent migration of an Access application to SQL Server.

Knowledge Base I, contains an odorant inventory, derived from the substances released from industrial processes, also providing information on health impacts, such as toxicity, medical cases and available treatments on inhalation. Knowledge Base II has been produced form the decomposition of industrial units to individual processes, serving as the main link to KB I. Knowledge Base III provides the biosensors which have developed/implemented for the detection of various compounds in a variety of matrices; given a specific odorant, all available biosensing mechanisms are retrieved and forwarded to KB IV where through a computer-aided multicriteria selection process [10], prescreening occurs for the identification of those biosensors that could potential be suitable for the specific application. The selection of the optimal (most suitable) alternative is performed by Fuzzy Multicriteria Analysis, through a methodology developed by the authors to deal with uncertainty, especially in the case where no data is available on the efficiency and reliability of a proposed process in conjunction with a specific biosensor [10].

## 2.2 Odor fingerprinting and management

The algorithmic procedure especially designed and developed by the authors for odor monitoring and management, includes the stages shown below. Figure 2 illustrates the connection of the stages, represented by the corresponding number or letter, in case of activity or decision node, respectively.

1. Construction of geomorphology and population density layers of the region under consideration in a GIS.
2. Collection of meteorological/microclimatic/pollution data and construction of the corresponding GIS layers.
3. Decomposition of the industrial plant into unit processes via KB II.
4. Determination (by means of technical literature survey) of odorants involved in each unit process and their characteristics via KB I.
5. Collection/storing/processing of information concerning usual and incidental emissions of odorants from each process units of the industrial plant.
6. Odor impact assessment on the human factor, by (a) circulating a properly designed questionnaire among the employees, (b) interviewing the unit medical inspector.
7. Odor impact assessment on productivity, using a quantifier (like time, value, resources), as estimated by the administration of the industrial plant.
8. Odor impact assessment as estimated by (a) common people leaving and/or working in the vicinity, (b) the Local Authorities of nearby communities, (c) the Health Care personnel/professionals Units working in these communities, (d) the representatives of local non-governmental organizations having an interest on environment and/or health.
9. Estimation of capital available for the investment required for the project of odor fingerprinting/monitoring.
10. Design of the biosensors network under the constraint of the budget formed in stage (9).
11. Selection of the proper biosensors (via KB III), to form the fingerprinting/monitoring network, using a multicriteria method for optimal choice [10].
12. Design of a composite measuring system, equipped with the proper biosensor, according to [10].
13. Real time operation of the biosensors network.
14. Creation/enrichment of the KB IV, especially designed for the project at hand.
15. Signal transmission indicating that at least one odorant exceeded the desired upper value set a priori on a long-term basis.
16. Design of such a control process and documented proposal for its installation by the industrial company operating the plant (subsidize investment).
17. Checking for low efficiency.
18. FTA performance in the control process to find out the cause and (after confirmation) the corresponding remedial proposal.
19. Search for locating the fault in the unit process within the production line.
20. FTA performance in the unit process to find out the cause and (after confirmation) the corresponding remedial proposal.
21. Quality assessment of raw/auxiliary/intermediate materials used.
22. Revision/supplement of/to specification referring to quality of these materials.
23. Relaxation of the odor threshold set a priori by allowing the required minimal increase of the desired upper value on a short-term basis.
24. Preparation of report and information processing for the enrichment/updating of KB IV.
P. Is there at least one required odorant measurement not covered by a corresponding biosensor?
Q. Is there an odor control process for this odorant within the industrial plant?
R. Is the efficiency low indeed?
S. Is there such a fault?
T. Is there any evidence for contribution of certain characteristics of those materials to the fault under investigation?
U. Is the new threshold accepted on a short-term basis?

The procedure starts with the determination of the boundaries and parameters of the area under surveillance (stages 1 and 2), which will provide for the network architecture, signal transmission method, required ruggdness of biosensors, and odor distribution patterns. The offending industrial plant is decomposes to processes and sub-processes (stage 3), establishing an inventory of sources of odor emissions (stage 4). The health impact and/or level of nuisance and annoyance to the workers (stage 6) and the area communities (stage 8) set the extent of the severity of the situation, thereby indicating the

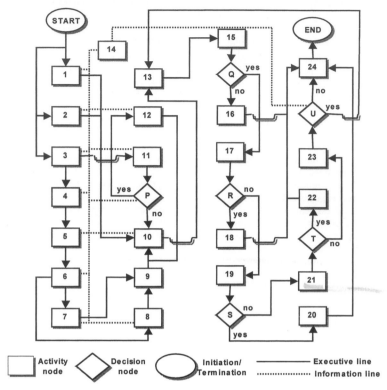

Figure 2. The algorithmic procedure designed/developed for odor pollution control, management and abatement.

intensity of monitoring, as well as the required concentration range for detection (i.e., based on people's response, threshold values could be less than those registered [2]), serving as a guide for proper biosensor selection. Data gathered is stored/processed in KB IV (stage 14), which is created ad hoc, for the needs of the specific project. Following the establishment of the necessary budget (stage 9), the biosensor selection process is initiated (stage 11) and the best available alternative (stage 12) is field-tested (stage 13). Stages 16 onwards, are activated when one at least odorant exceeds upper threshold values (stage 15). The identification of the offending source, either in operation (stages 19 and 20) or in the inefficiency of the control measures (stage 18) leads to the determination of the recommended remedial activity. At this point an effective communication is initiated with the near-by community and based on the level of community co-operation and tolerance, the offending facility proceeds with establishing its odor reduction plan.

## 3. Implementation

The methodology described above, has been implemented in the case of oil-refinery surveillance. The facility selected, is located in the outskirts of Korinthos (Greece), an area with moderate inhabitancy, mainly from the employers and a limited local population. Among the various odorous emissions from the installation, the monitoring of ammonia has been selected in order to test suitability of the proposed program for odor control and abatement.

KB II provided all refinery processes (stage 3), from which sulfur recovery has been selected. The facility uses the hydrosulfreen method in tail gas cleanup, i.e., in removing sulfur from the $H_2S$ gas streams. The process is a dry-bed, sub-dew point absorption, based on the extension of the Claus reaction (the catalytic oxidation of $H_2S$ to S), preceded by $TiO_2$-catalyzed hydrolysis of COS and $CS_2$

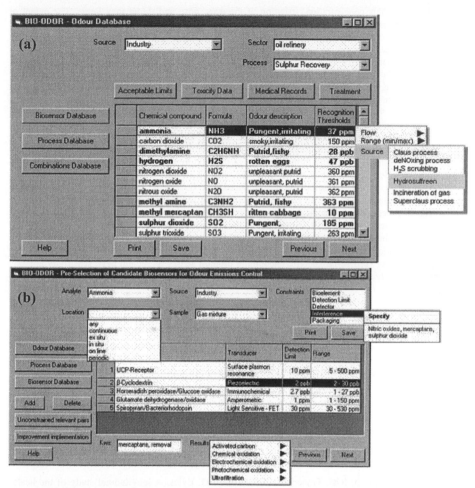

Figure 3. Sample screenshots of the developed program for odor monitoring with biosensors. (a) The odorous substances emitted form the hydrosulfreen process in the tail gas clean-up have been retrieved by KB I. (b) the available biosensors for the detection of ammonia in the gas mixture (also containing NOx, mercaptans and $SO_2$ are pre-screened through a multicriteria process, to provide the best alternatives.

to $H_2S$. The 'odor database' button, linking KB II to KB I, retrieved all odorants potentially emitted, by design (and accounted for) or accidentally, from the sulphur recovery unit (Figure 3a), highlighting the emissions that derive from the selected process (stage 4). Ammonia has been chosen for implementation, owing to (a) the increased odor intensity, (b) the high adaptation threshold (odor fatigue) at moderate concentrations (persistent odor), (c) the high water solubility (indicating the ability and strong potency for the molecule to be engaged in biological and chemical reactions), and (d) the great difficulty in its quantitation due to the complexity of the sample.

The solubility of ammonia, and hence its odor, is markedly influenced by the solution pH. The $NH_3$ in an acid medium accepts $H_3O^+$ to produce ammonium ion ($NH_4^+$) which stays in solution and does not volatilize up to a pH value of 8; above a pH of 9, however, ammonia is rapidly volatilized. The methods for ammonia odor reduction include flaring (burning of ammonia to nitrogen and water), (bio)filtration, acidification or physical methods, such as reducing downwind dust particle concentrations and odor concentration via windbreak walls or air dams. Ammonia gas is very corrosive to body tissues, reacting with body moisture on contact. The odor threshold for ammonia is on average 0.040 ppm although the range of sensitivity ranges from 0.017 ppm to 50 ppm for acclimatized individuals [3]. Generally, concentrations of up to 25 ppm are tolerated although unpleasant and pungent. Above this concentration, irritation of the eyes, nose and throat may begin. The extent of irritation increases with increasing ammonia concentration.

The biosensors that have been developed for the detection of ammonia have been retrieved from KB III, as per the constraint of industrial gas mixture sample only, i.e.,.no devices are shown that are suitable for water, soil or clinical samples (stage 11). The available transduction mechanisms are mostly electrochemical, although some optical and piezoelectric have been developed. The smaller detection value is given by the gramicidin biosensor, which however is inhibited by $H_2S$, certain to be found in the ammonia sample of interest.

Combining the process, the odorant and the biosensors, the computer program pre-selected, through multi-criteria process (stage 12) [10], the five sensor devices shown in Figure 3b (KB IV). From the seven potential sensors deriving after setting the interference constraint, available for assessment through the 'Unconstrained relevant pairs' button, the proposed devices are offer the highest possible signal reliability in the presence of the mixture compounds (nitric oxides, mercaptans and sulfur dioxide). Human expertise can intervene with the pre-selection procedure, via the 'Add' and 'Delete' buttons, while the information introduced can be stored to the database. Furthermore, improvements can be proposed on request ('Improvement implementation' button), that can provide pretreatment process for the blocking of particular interefrents [10]; for instance, the query regarding the removal of mercaptans (Figure 3b) resulted in the retrieval of available trapping processes.

The best solution as per the device type suitable for ammonia monitoring, produced through the fuzzy multicriteria run of stage 12 [10], is the glutamate dehydrogenase/oxidase, amperometric sensor with UCP-receptor (brown-fat specific uncoupling protein) sensor as second best; sensitivity analysis proved the stability of the best solution.

## Acknowledgments

This work was performed within the framework of Pythagoras II EU-GR Research Programme (Section: Environment) for the design/development/implementation of biosensors/ bioindicators (Co-financed by the European Social Fund and the Greek Ministry of Education within the framework of the 2nd Operational Program of Education and Professional Training). The authors also acknowledge the contribution of the Research Center of the University of Piraeus.

## References

[1] I.L. Kratskin, O. Belluzzi, Anatomy and neurochemistry of the olfactory bulb. In *Handbook of Olfaction and Gustation,* 2nd edition, (Ed.) R.L. Doty, Marcel Dekker, NY, 2003.

[2] Integrated Pollution Prevention and Control (IPPC): Draft Horizontal Guidance for Odour Part 1 —Regulation and Permitting. Scottish Environment Protection Agency, 2002.

[3] I. N. Luginaah, S.M. Taylor, S.J. Elliott and J.D. Eyles. A longitudinal study of the health impacts of a petroleum refinery, *Soc. Sci. & Med.*, 50, pp. 1155-1166, 2000.

[4] Integrated Pollution Prevention and Control (IPPC): Draft Horizontal Guidance for Odour Part Part 2 —Assessment and Control. Scottish Environment Protection Agency, 2002.

[5] D.R. Thévenot, K. Toth, R.A. Durst, G. S. Wilson, Electrochemical Biosensors: Recommended Definition and Classification, *Pure App. Chem.*, 71, pp. 2333-2348, 1999.

[6] A.P.F. Turner, I. Karube, G. Wilson, Biosensors, Fundamentals and Applications, NY, Oxford University Press, 1987.

[7] A. Mulchandani, K.R. Rogers (Eds): Enzyme and Microbial Biosensors. Techniques and Protocols. Methods in Biotechnology Vol. 6, Humana Press, Totowa, N. Jersey, 1998.

[8] D. Gao, C. Mingming, Y. Ji, Simultaneous estimation of classes and concentrations of odors by an electronic nose using combinative and modular multilayer perceptrons, *Sens. Act. B: Chem.l*, 107, pp. 773-781, 2005.

[9] F. A. Batzias, D. K. Sidiras, and E. K. Spyrou, Evaluating livestock manures for biogas production: a GIS based method, *Ren. Energy*, 30, pp. 1161-1176, 2005.

[10] F.A. Batzias and C.G. Siontorou, Introducing Chemical Engineering Processes into Optimal Design of Measuring Systems Equipped with Biosensors, *Proceedings of ICCMSE 2005*, 21-26 October, Greece.

Brill Academic Publishers
P.O. Box 9000, 2300 PA Leiden
The Netherlands

*Lecture Series on Computer*
*and Computational Sciences*
Volume 4, 2005, pp. 859-865

# Introducing Chemical Engineering Processes into Optimal Design of Measuring Systems Equipped with Biosensors

F.A. Batzias, C.G. Siontorou[1]

Department of Industrial Management & Technology,
University of Piraeus,
GR-185 34 Piraeus, Greece

Received 15 July, 2005; accepted in revised form 5 August, 2005

*Abstract:* Biosensors as 'dip in and register' devices, generically offer simplified reagentless analyses for a range of biomedical and industrial applications and for this reason this area has continued to develop into an ever-expanding and multidisciplinary field during the last decades. Just moving from the proof-of-concept stage to field testing and commercialization, they require the establishment of chemical engineering processes for their advancement to measuring equipments. This work presents a decision support system for the selection of sample treatment to precede biosensing, in order to minimize interference. The proposed system is comprised of (a) a Knowledge Base developed through rules based reasoning from four separate databases relating to biosensors, treatment processes, samples and application environment, used to provide the alternatives that are re-evaluated through case-based reasoning by a separate whole cases database and (b) an integrated fuzzy multicriteria analysis method (FMCA) to provide the optimal solutions, tailor-made by the authors to help researchers engaged in biosensor development to select the best biosensor-pretreatment method suited for a specific application while fulfilling a set of technoeconomic criteria. The algorithmic procedure developed by the authors for the realization of the selection procedure has been implemented in the case of reduced glutathione measurement in serum.

*Keywords:* biosensors, fuzzy multicriteria analysis, rules-based reasoning, case-based reasoning, knowledge base

## 1. Introduction

Biosensors offer the prospects of simplified measurement of complex biochemical parameters that are usually destined for cumbersome laboratory-based measurement. They comprise a natural or mimic biological recognition element (enzyme, antibody, cell receptor, DNA) juxtaposed to a chemical or physical transducer to create a single unit (Figure 1) [1-3]; the latter is used to register some physicochemical change resulting from the binding reaction between the recognition element and the target compound (mass change, thermal change, altered surface charge, flow of current or change in optical properties). Biosensors, as 'dip in and register' devices, generically offer simplified reagentless analyses for a range of biomedical and industrial applications and for this reason this area has continued to develop into an ever-expanding and multidisciplinary field during the last decades.

Their current attractiveness lies in the fact that they are amenable to microfabrication further to their ability to provide a direct readout in optically opaque clinical, environmental or industrial samples. In principle, the high selectivity of the device, presented inherently by the bioelement that is tailored-made to meet the specific analyte(s), excludes interference problems and matrix effects. In practice, both constitute a major problem mainly due to the inability to precisely control the transducer and other materials used in the construction of the biosensors. The main contributing factors to biosensor interference derive from: (a) the transduction mechanism: e.g. in electrochemical devices detecting $H_3O^+$, other electroactive species in the sample ($NH_4^+$, ascorbic acid, etc.) interfere with measurement [2-4], (b) the mimicking nature of the device, accomplished with the immobilization of biological moieties on the transducer surface: e.g. in the case of proreinaceous molecules, sample proteins may adsorb on the biorecognition surface providing false signals, induce fouling or poison the transducer; furthermore, the desorption/denaturation/saturation of the bioelement is common, especially in FIA systems or after several cycles of operation.

---

[1] Corresponding author. E-mail: csiontor@unipi.gr

Bench-scale research addressed these problems by using chemical cross-linkers to stabilize the bioelement and protective outer membranes to exclude interferents. Field-testing, however, proved the

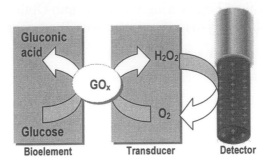

Figure 1. Simplified representation of a biosensor detection scheme for glucose [3]. The target analyte is converted to gluconic acid by glucose oxidase (Gox), immobilized on the electrode surface. Secondary to this reaction is the production of the hydrogen peroxide 'transducer', voltammetrically detected when it is oxidized at the electrode.

inadequacy of these practices, since real sample matrix and field conditions are vastly unknown, variable and unpredicted. For the benefits of this technology to be realized, much depends upon sensor preparation and sample presentation. The term 'pretreatment' started to appear in the relevant scientific literature almost at the same time that these devices got through the physicochemical study into equipment development. Although much attention has been given to bioelement immobilization, introducing costume-designed innovative techniques as sol-gel polymerization, sample pretreatment has been a much neglected area of research.

Pretreatment is by its nature an iterative process. The initial requirements lead to a first selection that suggests modifications to the requirements. If they are insufficiently restricted, too many solutions are obtained; if over-restrictive, none may remain requiring a relaxation of constraints or the need to develop a completely new process or hybrid of two or more processes. This puts strict requirements on selection tools: they must allow the iterative procedure and accommodate a range of databases appropriate to the various types of sensing devices.

The aim of this work is to present a decision support system, comprised of (a) a Knowledge Base (KB) developed through rules based reasoning from four separate databases relating to biosensors, treatment processes, samples and application environment, used to provide the alternatives, which are re-evaluated through case-based reasoning by a separate whole cases database and (b) an integrated fuzzy multicriteria analysis method (FMCA) to provide the optimal solutions, tailor-made by the authors to help researchers engaged in biosensor development to select the best sample pretreatment method suited for the biosensor type and application while fulfilling a set of technoeconomic criteria.

## 2. Methodology

*The development of the Knowledge Base.* The authors have constructed a KB using a relational database management system (RDBMS) for mining the data gathered and generate the relations. The domains of investigation (primary domains or independent databases) - biosensor, application area, sample properties, process (Figure 2) - have been linked to all levels of constraints through rules-based reasoning in order to provide secondary domains, which by means of process compatibility indices they provide the alternatives. For successfully coupled/combined with biosensors, the pretreatment process should eliminate interference contained in the sample while not altering sample conditioning or producing new interferents. The latter is depended mainly upon the biosensor type as per chemical transduction and biochemical mechanism: in electrochemical biosensors all electroactive species should be excluded [3, 4], whereas in immunobiosensors, harsh chemicals or conditions should be eliminated [2]. As the primary domains are mostly taxonomic, the semantic links are mainly of the *is-a* or *has-a* type, while in the secondary domains the semantic links *is-part-of-a* and *means-of-a* (*is-expressed/examined-by*) dominate. The database can be used to retrieve available processes that fit the pre-set constraints of 'biosensor', 'sample' and 'application', ranked by their compatibility indices. It also provides information on process details (procedure, materials, requirements, etc.), efficiency, possible physicochemical mechanism, means of control, etc.

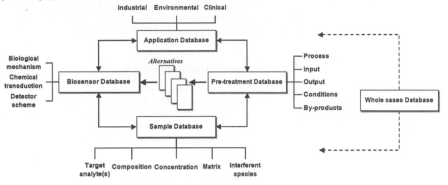

Figure 2. Infrastructure of the relational database, designed for optimal decision making in pre-treatment process selection.

Data can be derived either from (a) whole cases, introduced in the whole cases database and decomposed to parts in the databases of the primary domains, or (b) by information related solely to the primary domains. The whole cases database is used in a later stage for selection verification and/or modification of the suggested process through case-based reasoning (vide infra). For example, the measurement of BOD (sample domain) can be performed using a microbial biosensor that measures oxygen consumption (biosensor domain), is heavily interfered by non-easily biodegradable organic compounds as phenols [3, 4]. The removal of organic compounds from the sample prior to biosensing has been studied in the case of the photocatalytic BOD biosensor [5], stored in the whole cases database, and after decomposition, the parts have been stored in the independent databases. The degradation of organic compounds, considered separately and not necessarily associated with any biosensor, adapt various techniques, such as chemical, electrochemical, or photochemical oxidation, all stored as parts in the pretreatment domain.

The data input and the intermediate/final results output are stored into the RDBMS, an adaptation of the ABEPE Database [6]. The RDBMS model provides a user friendly and efficient tool to handle large quantities of data; moreover, since SQL is an ANSI standard, widely accepted by the computer industry, a large amount of other commercial software can cooperate with SQL applications and increase/expand functionality. Since there are natural language interpreters translating voice or keyboard entered commands into SQL statements (e.g. Microsoft English Query), the potential program user can submit queries in a natural language. The broad availability of MS Access renders this RDB the most suitable development environment. Not to mention that such an implementation of the database could be easily upsized to a multiprocessor system running SQL Server RDBMS because of the transparent migration of an Access application to SQL Server. An SQL Select query was created as an ADD-ON software device to combine and appropriately relate the eco-regional parameters. In fact, the use of a RDBMS has been proven effective, since the formation of complex queries comes easily through the exploitation of SQL.

*Fuzzy multicriteria selection.* The selection of the optimal (most suitable) alternative is performed by Fuzzy Multicriteria Analysis (FMCA), through a methodology developed by the authors to deal with uncertainty, especially in the case where no data is available on the efficiency and reliability of a proposed process in conjunction with a specific biosensor. The methodology is a combination of the deterministic PROMETHEE method [7], the Tseng and Klein method that compares fuzzy sets of alternatives in pairs [8], and a modified 4-stage Delphi, method especially tailored for the needs of the present work. The multicriteria problem that must be solved is: $\text{Max}\{f_1(a),\ldots,f_K(a) \mid a \in A\}$ where A is the set of $T$ alternative instrumental methods and $f_i$, $i=1,\ldots,K$, are the $K$ criteria under which the alternative measurement are evaluated. The FMCA adopted herein consists basically of two steps: (i) the formulation of the decision matrix ($K \times T$), where each element $x_{kt}$ is the evaluation of alternative $a_t$ with respect to criterion $f_k$ (ii) the ranking of the alternatives, according to the rules of the method developed herein. The preference measurement is the simple sum of the scores $x_{kt}$ multiplied by the weight of the particular criterion $w_k$, over all criteria $K$. To overcome the basic assumption that complete compensation between attributes takes place, several outranking methods are considered that allow partial compensation. Also, they allow for incomparability ($aRb$) and weak preference ($aQb$) between the alternatives $a$, $b$, in addition to the strict preference ($aPb$) and indifference ($aIb$) that the 'classical' methods are based on. The evaluation of the alternatives with respect to each criterion in the decision matrix is given by the experts, who participated within the DELPHI procedure.

## 3. Optimal Design of Measuring System

The algorithmic procedure especially designed and developed by the authors for the development/ selection of biosensor-pretreatment combination, includes the stages shown below. Figure 3 illustrates the connection of the stages, represented by the corresponding number or letter, in case of activity or decision node, respectively.

1. Search for creating an initial set of candidate biosensors suitable for the specific measurement, from the biosensor knowledge base via stages 10, 11.
2. Search for creating an initial set of candidate pretreatment processes suitable for the specific measurement from the processes knowledge base via stages 10, 11.
3. Determination of the combination constraints for the specific measurement from the sample and application databases via stages 10, 11.
4. Formation of possible pairwise combinations between the sets produced at stages 1 and 2.
5. Pre-selection by screening among the combinations produced at stage 4 for compatibility.
6. Assignment of values to (i) the weights' vector and (ii) the multicriteria matrix under the form of fuzzy numbers to account for uncertainty.
7. Run of the fuzzy multicriteria ranking (FMR) algorithmic procedure.
8. Performance of sensitivity analysis of the proposed (ranked first) combination.
9. Revision of the multicriteria matrix by score reduction, as indicated by the sensitivity analysis, and re-run the FMR algorithmic procedure.
10. Link to/enrichment of the Knowledge Base (KB).
11. Information retrieval via external knowledge/information bases by means of an intelligent agent [9].
12. Choice of the best as a result of comparison among all examined combinations.
13. Data mining within the whole cases knowledge base.
14. Pre-selection by screening among alternative matching solutions.
15. Assignment of values to (i) the weights' vector and (ii) the multicriteria matrix under the form of fuzzy numbers to account for uncertainty.
16. Run of the fuzzy multicriteria ranking (FMR) algorithmic procedure.
17. Performance of sensitivity analysis of the proposed (ranked first) matching solution.
18. Revision of the multicriteria matrix by score reduction, as indicated by the sensitivity analysis, and re-run the FMR algorithmic procedure.
19. Choice of the best as a result of comparison among all examined matching solutions.
20. Integration of the selected biosensor with the pretreatment process into a measuring system.
21. Report preparation for scale-up.
P. Is this combination sensitive?
Q. Has this combination been re-examined in the previous run?
R. Is there any combination left unexamined?
S. Is the proposed matching solution sensitive?
T. Has this matching solution been re-examined in the previous run?
U. Is there any matching solution left unexamined?

The procedure starts with the retrieval of biosensors (stage 1) and processes (stage 2) that could potentially fit the desired analyte measurement (stage 4), though a mechanical procedure that considers only incompatibilities of the sensing *per se*, i.e., posed by the physicochemical properties of the biosensor and/or the sample matrix properties (stage 3). In stage 5, however, the pre-selection/screening is based on a multicriteria process, according to the criteria $f_i$ ($i$=1, ..., 10):

$f_1$ – High efficiency as regards purification of the product (sample) to be subsequently measured for the target analyte(s).

$f_2$ – Capability of correspondence as regards the desired degree of purification to the one accomplished, i.e., in case total elimination of interference (100% efficiency of the pre-treatment process) cannot be achieved, a separation stage should precede biosensing; it is thus imperative to be able to quantify the pre-treatment output.

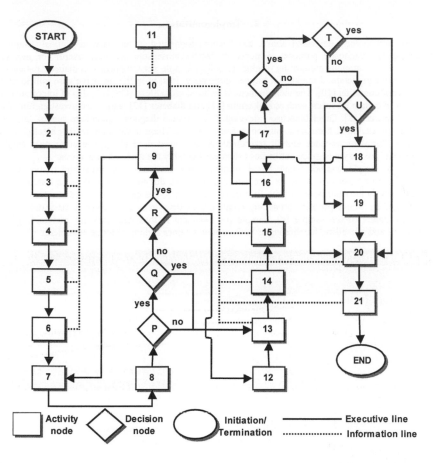

Figure 3. The algorithmic procedure designed/developed for introducing chemical enginccring processes in the optimal design of measuring systems based on biosensors

$f_3$ – Reliability, defined as the stability of the output product (i.e., of the efficiency of the method) regardless of the variation in input composition.

$f_4$ – No co-production of by-products that could themselves interfere with measurement, mask biosensor response, poison the device or react with the target analyte(s).

$f_5$ – Long shelf-life (especially in the case of catalytic reactors that commonly suffer from raw material impurities, leading eventually to poisoning or ageing of the reactor).

$f_6$ - Easy and inexpensive replacement/maintenance/regeneration of the pretreatment assembly.

$f_7$ – Capability of continuous operation and input rate adjustment (and thus adjustment of the sample retention time)

$f_8$ – Size compatible with the geometry of the accompanying system.

$f_9$ – Capability of independent efficiency monitoring/control

$f_{10}$ – Known physicochemical/biochemical background

The pre-selected combinations are processed through FMR (stage 7) with metrological criteria for the determination of the best alternative, as this results from the loop mechanism of stages 8, 9 and 12. The whole cases database is used for the retrieval of cases that 'contain' the selected combination, and an FMR is performed with Case Base Reasoning criteria for matching (stage 16). If no whole case is available at a sufficient degree of matching, the combination resulted from stage 12 is forwarded to development, testing, scale-up and implementation.

## 4. Implementation

The methodology described above, has been implemented in the case of clinical sample pre-treatment for diagnostic purposes. Glucose or lactate biosensors are very useful for monitoring metabolism. Continuous monitoring of glucose is for example very important in diabetic patients. The measurement of lactate, a marker for oxygen deficiency, is used in the intensive care unit to monitor the patients' condition [10]. The measurement of reduced glutathione (GSH) is also of importance, been a marker in certain disorders such as leukaemia [11] and diabetes [12], and in the investigation of some types of cancer [13]. Other biosensors developed for clinical diagnosis quantitate sugars, fatty acids, blood gases, enzymes, hormones, antibodies and microbes. These devices are mainly electrochemical, although chemiluminescence and IR transductions have been used [1-3]. The commonly employed bioelements are enzymes, since antibodies extent considerably the duration of analysis.

Clinical samples include whole blood, plasma, serum, saliva, sweat, urine and tissue. Although biosensors are by means suitable for biomedical monitoring, due to fouling and blood clotting, *in vivo* performance is far from optimal [10]. Protective membranes, designed to exclude interferent species, are blocked by sample proteins, producing eventually a barrier to analyte transport, or destroyed by the corrosive nature of the sample [4]. Sample pre-treatment is thus necessary for maintaining device operability and reliability, functioning as both, protein retainer and interference neutralizer.

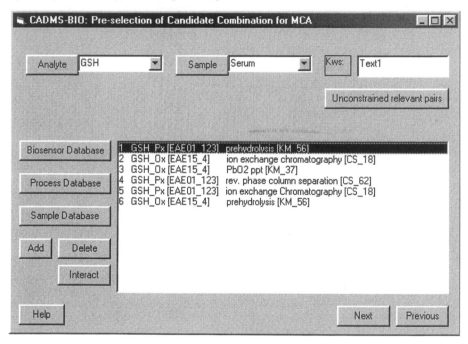

Figure 4. Sample of the developed computer program for selecting biosensor-pretreatment combinations for the development of a measuring system for reduced glutathione in serum.

The electrochemical enzymatic biosensors that measure blood-related analytes in whole blood, plasma or serum, are hindered by ascorbic acid, oxones, ketones, and a variety of ions, further to the non-selective protein deposition, mainly of albumin [4]. The pretreatments suitable for the specific sample types include chromatographic separation, filtration, transformation, chemical precipitation and denaturation.

The case examined was the development of a measuring system for GSH in serum. The computer program pre-selected combinations shown in Figure 4, that corresponds to stage 5 of the algorithmic procedure. The available biosensors are two, based on glutathione oxidase or glutathione peroxidase (stage 1), whereas the alternative pre-treatment processes proposed by the program at stage 2, are pre-hydrolysis, $PbO_2$ precipitation, ion exchange chromatography, and reversed phase column separation.

From the eight potential combinations deriving from the coupling of sensors to treatments, available for assessment through the 'Unconstrained relevant pairs' button, the pre-screening process of stage 5 produced six (Figure 4). The user can retrieve relevant information on the biosensor device, the process or the sample properties through the respective buttons. Human expertise can intervene with the pre-selection procedure, via the 'Add' and 'Delete' buttons, while the information introduced can be stored to the database by pressing 'interact'.

The FMR run produced as best alternative glutathione oxidase with $PbO_2$ precipitation with second best the glutathione peroxidase with ion exchange chromatography; sensitivity analysis proved the stability of the solution (stage 8). Consulting the whole cases database (stage 13), three cases have been retrieved (stage 14), none of which showed sufficient matching (stage 19).

## Acknowledgments

This work was performed within the framework of Pythagoras II EU-GR Research Programme (Section: Environment) for the design/development/implementation of biosensors/ bioindicators (Co-financed by the European Social Fund and the Greek Ministry of Education within the framework of the 2nd Operational Program of Education and Professional Training). The authors also acknowledge the contribution of the Research Center of the University of Piraeus.

## References

[1] D.R. Thévenot, K. Toth, R.A. Durst and G. S. Wilson, Electrochemical Biosensors: Recommended Definition and Classification, *Pure and Applied Chemistry*, 71, pp. 2333-2348, 1999.

[2] A.P.F. Turner, I. Karube and G. Wilson, Biosensors, Fundamentals and Applications, NY, Oxford University Press, 1987.

[3] A. Mulchandani and K.R. Rogers (Eds): Enzyme and Microbial Biosensors. Techniques and Protocols. Methods in Biotechnology Vol. 6, Humana Press, Totowa, N. Jersey, 1998.

[4] F.A. Batzias and C.G. Siontorou, Investigating the causes of biosensor SNR decrease by means of FTA, *IEEE Journal of Instrumentation and Measurement*, August 2005, in press.

[5] G.-J. Chee, Y. Nomura, I. Ikebukuro, I. Karube, Development of photocatalytic biosensor for the evaluation of biochemical oxygen demand, *Biosensors Bioelectronics*, 21, pp. 67-73, 2005.

[6] F. A. Batzias, D. K. Sidiras, and E. K. Spyrou, Evaluating livestock manures for biogas production: a GIS based method, *Renewable Energy*, 30, pp. 1161-1176, 2005.

[7] J.P. Brans, J.P., Ph. Vincke, and B. Mareschal, How to select and how to rank projects: The PROMETHEE method, *European Journal of Operational Research*, 24, pp. 228 – 238, 1986.

[8] T.Y. Tseng, and C.M. Klein, New algorithm for the ranking procedure in fuzzy decisionmaking, *IEEE Transactions On Systems, Man and Cybernetics*, 19(5), pp. 1289 – 1296, 1989.

[9] F. A. Batzias and E. C. Markoulaki, Restructuring the keywords interface to enhance CAPE knowledge via an intelligent agent, *Computer-Aided Chemical Engineering*, 10, pp. 829-834, 2002.

[10] P.H. Wang, Tight glucose control and diabetic complications, *Lancet*, 342, pp. 129-135, 1993.

[11] G.S. devi, M.H. Prasad, I. Saraswathi, D. Raghu, D.N. rao, P.P. Reddy, Free radicals antioxidant enzymes and lipid peroxidation in different types of leukaemias, *Clinica Chimica Acta*, 293, 53-62, 2000.

[12] M. McDonagh, L. Ali, A. Kahn, P.R. Flatt, Y.A. Barnett, C.R. Barbett, Antioxidant status, oxidative stress and DNA damage in the etiology of malnutrition related diabetes mellitus, *Transactions of Biochemical Society*, 25, 146S, 1997.

[13] T.D. Oberly, L.W. Oberly, Antioxidant enzyme levels in cancer, *Histology and Histopathology*, 12, 525-535, 1997.

Brill Academic Publishers
P.O. Box 9000, 2300 PA Leiden,
The Netherlands

*Lecture Series on Computer*
*and Computational Sciences*
Volume 4, 2005, pp. 866-871

# Modeling Oil Spills in Areas of Complex Topography

## A. I. Pollani[1]

Hellenic Centre for Marine Research, Institute of Oceanography,
PO BOX 712, GR-19013, Anavyssos, Greece

Received 6 August, 2005; accepted in revised form 18 August, 2005

*Abstract:* This work triggered from the recently signed memorandum of collaboration among Greece, Russia and Bulgaria for the construction of the Burgas - Alexandroupolis oil pipeline, and deals with the development and the application of a pre-operational management tool for the Thracian Sea. The tool is a part of the Poseidon system which has been designed to provide real-time data and forecasts for marine environmental conditions in the Greek Seas.

*Keywords:* oil spills, nested models, operational management tools

## 1 Introduction

Although advances in computer technology are very high, when dealing with sensitive nearshore areas and estuaries or complex islander topography, details of hydrodynamic phenomena and local interaction waves may be of great importance for the planning of the protection and spill response operations and therefore for the sustainability of these areas. Nesting a limited-area high-resolution model within a coarser-grid model of a wider area is the only viable approach to capture all these details making at the same time affordable the required computational effort.

After the agreement among Greece, Russia and Bulgaria for the construction of the Burgas-Alexandroupolis oil pipeline, the need for a better forecast and prevention of oil spill dispersion in the area is rising. The Burgas - Alexandroupolis pipeline will constitute an additional to the Bosporus way of oils transportation from the Black Sea to the Mediterranean. It will have length roughly 280 kilometers and includes two terminal installations,one in the Burgas of Bulgaria and the other in Alexandroupolis in Greece. The installations in Alexandroupolis will include reservoirs of capacity of about 940.000 $m^3$ and navigable points of oil loading (SBM) of about 8-9 kilometers in the sea. The oil will be transported with tankers from the Novorossiysk and other harbors of Black Sea in the Burgas and from there via the pipeline will be transported in Alexandroupolis.

Considering the large amount of crude oil that will be loaded in the tankers through the terminals in Alexandroupolis, the Oil Spill Model (OSM), on-line coupled with the Princeton Ocean Model (POM) was implemented. The developed system of models is a part of the Poseidon system. The Poseidon system is an operational monitoring, forecasting and information system for the marine environmental conditions of the Greek Seas.

---

[1]Corresponding author. E-mail: ank@ath.hcmr.gr

# 2 Materials and Methods

## 2.1 The area

The operational POSEIDON system is a forecast system for the Greek Seas, covering the area of the Aegean with a resolution of 1/20 x 1/20 degrees. As it is obvious the system is not capable to describe the detailed hydrodynamic field in the nearshore areas of Alexandroupolis and the Dardanelles. However, as an operational tool is ideal for contributing by giving initial and boundary conditions to a nested high-resolution model chosen to study in more details this area. The computational domain of the high-resolution model covers the area from 24.9E to 26.85E and 38.85N to 41N, with a resolution of 1/60 x 1/60 degrees.

## 2.2 Basic assumptions and initial spreading

For the present application the Poseidon's oil spill model (OSM) was used. Following, a short description of its major processes is given. As is described by [3] and [2] the oil might be represented by a large number of material particles or parcels, each of which represents in turn a group of oil droplets of like size and composition. For modeling purposes, the whole mass of oil slick is simulated with 'parcels' characterized by evolving physicochemical properties and may represent, in reality, many $m^3$ of oil, occupying a considerable space on the sea surface. The parameters used to attribute the physicochemical properties for each parcel and to initialize the OSM are the $x, y, z$ coordinates, the initial volume, the density, and the droplet diameter. It has to be mentioned that both models (OSM and POM), use the same orthogonal lattice of points in the horizontal dimension. In the vertical, the OSM uses a Cartesian system of reference, while POM uses sigma coordinates.

The process of initial spreading, dominated by the inertial gravity, viscosity and surface tension forces, has described satisfactorily by Fay's law [7].

## 2.3 Oil transport

The hydrodynamic processes are described by two modules, the circulation module and the wind generated waves module. The circulation module used is the Princeton Ocean Model (POM) [1], where both formulation and numerical solutions are described in detail.

The three-dimensional velocity field is calculated from the vector summation at each particle position, by supplying the advective forces caused by the water currents and the wave drift.

$$\vec{v}_{part} = \vec{v}_{current} + \vec{v}_{wave} \tag{1}$$

The oil parcels are transported as passive tracers, and the horizontal displacement of each particle, due to advection, is $L_{(x,y)} = (u_{part}, v_{part})dt$, where $u_{part}, v_{part}$ are the local velocities in the x and y direction in $ms^{-1}$, and $dt$ is the time step $(s)$.

The vertical transport of the oil particles is described by the hydrodynamic vertical velocity field and by the buoyancy velocity which is function of the density and the diameter of the droplet. The size distribution of the particles influence the proportion of those particles which are submerged, and those which resurface. The distinction between small and large oil droplets is based on the concept of critical diameter [2].

$$d_c = \frac{9.5\nu^{2/3}}{g^{1/3}(1 - \frac{\rho_o}{\rho_w})^{1/3}} \tag{2}$$

where $d_c$ is the critical diameter, $\nu$ is the kinematic viscosity $(m^2 s^{-1})$, $g$ is the acceleration, due to gravity $(ms^{-2})$, $\rho_o$ and $\rho_w$ are the oil density and the water density in $(kg/m^3)$. Small particles

$(d < d_c)$ in a high viscosity environment rise with velocity:

$$w_b = \frac{gd^2(1 - \frac{\rho_o}{\rho_w})}{18\nu} \tag{3}$$

while large particles $(d > d_c)$ in a low viscosity environment rise with velocity:

$$w_b = \frac{8}{3}gd(1 - \frac{\rho_o}{\rho_w})^{1/2} \tag{4}$$

Finally, the vertical movement of particles is calculated by $L_z = (w_b + w_{water})dt$, where $L_z$ is the rise distance (m), $w_b, w_{water}$ are the rise velocities due to buoyancy and due to water currents, and $dt$ is the time step.

The wave drift (Stokes drift) is the net current speed caused by linear waves and has the same direction as the wind [2]:

$$u_{wave} = \frac{\omega k a^2 cosh(2k(H - z))}{2sinh^2(kH)} + C \tag{5}$$

where $\omega = 2\pi/T$ is the wave frequency $(s^{-1})$, $k = 2\pi/L$ is the wave number $(m^{-1})$, $T$ is the wave period (s), $L$ is the wave length (m), $H$ is the wave depth (m), $z$ is the particle depth (m), $H_s$ is the significant wave height (m), $a = H_s/2$ (m), and $C$ is a depth independent term. The depth independent term might be neglected because is likely to be small [2]. Typically, Stokes drift has much smaller values than the wind current drift.

### 2.4 Horizontal and vertical diffusion

For the current component, the Mellor-Yamada 2.5 turbulence closure model [6] is used to calculate the vertical diffusion of the oil particles in the water column. This model characterizes the turbulence using equations for the turbulent kinetic energy $q^2/2$ and a turbulent macro scale $l$. The vertical mixing coefficient is defined as $K_{vel} = lqS_{vel}$, where $S_{vel}$ is a stability factor which is a function of the Richardson number and depends on stratification. A complete description of these equations can be found in [6]. The wave component is varying exponentially from the free surface down to a substantial limit of $z = -L/2$, where $L$ is the wave length [3]:

$$K_w = 0.028\frac{H_s^2}{T_s}e^{(4\pi/L)z} \tag{6}$$

$K_w$ is the vertical diffusion coefficient due to waves $(m^2s^{-1})$, $H_s$ is the significant wave height (m), $T_s$ is the significant wave period (s) and z is the depth (m). The horizontal diffusion coefficient $A_s$ is calculated according to Smagorinsky and is also provided by the hydrodynamic model. Then, for each particle, an excursion is computed [4]:

$$d = \sqrt{6Ddt} \ (2[R]_0^1 - 1) \tag{7}$$

where $[R]_0^1$ represents a random number in the range between 0 and 1, and $dt$ is the time step. This excursion can be positive or negative. $D$ could be either $K_{vel} + K_w$ or $A_s$ to describe the vertical or the horizontal diffusion.

The combined effect of depth varying vertical and horizontal diffusion describe the entrainment and the dispersion of the oil slick. These processes form the specific oil slick elongation with thick surface head and a tail of particles distributed vertically along a certain depth.

## 2.5 Evaporation

The rate of evaporation depends on surface area, thickness, vapor pressure and mass transfer coefficient, which, in turn, are functions of the composition of the oil, wind speed and temperature. The method used to characterize evaporation of oil has suggested by [10, 9]. In the model the oil is assumed to consist of a mixture of hydrocarbons of various densities, and is predetermined the heavier fraction that can undergo evaporation. Next, is checked if the parcel under examination is on the surface and the slick thickness in the adjacent area of the parcel is computed. The difference of the evaporated volume during the present time step, from the total evaporated fraction, is computed according to:

$$dF_v = -U_c \frac{P_i}{RT} d\theta \tag{8}$$

and

$$d\theta = K\alpha dt / V_o \tag{9}$$

where $V_o$ is the initial volume ($m^3$), $\alpha$ is the spill area ($m^2$), $K$ is the mass transfer coefficient ($m/s$), $R$ is the gas law constant, $T$ is the absolute temperature in $°K$, $U_c$ is the mean molar volume for all components ($4. \times 10^4 m^3 / mol$), and $P_i$ is the equilibrium vapor pressure ($Pa$).

## 2.6 Emulsification

Emulsification is a process commencing after a predefined time since the generation of the spill. Water droplet become mixed in oil up to concentrations of 80 % water and 20 % oil. Factors that influence the emulsification process are the wind speed, the wave characteristics, the make-up of the oil, the degree of weathering of the oil, the environmental temperature, the local thickness, and the time [7]. In the model the emulsification is influencing the hydrocarbon fractions with specific gravity bigger than a predetermined critical value, after a time longer than four hours since the spill generation, for wave steepness greater than a critical one $H_s/T_s > 0.028mm$, and for local oil slick thickness less than a critical value of 0.75 mm [8]. Following the model suggested by Mackay and colleagues [5]:

$$\frac{dy}{dt} = K_A(c_u + u_w)^2(1 - K_b Y_w) \tag{10}$$

where $Y_w$ is the water content of the oil, $K_B$ is a rate constant depending on the content of surfactants $\approx 1.25(m^3/m^3)$, $u_w$ is the wind speed (knots), $c_u$ is a background velocity constant $= 1m/s$, and $K_A$ is a rate constant depending on the sea state $\approx 4.5 \times 10^{-6}$. The solution of the differential equation is:

$$Y_w = \frac{1}{K_B}(1 - e^{-K_A K_B (1+u_w)^2 t}) \tag{11}$$

where $t$ is the time (s).

## 3   Results and Discussion

Two experiments have taken place. In the first one a ship accident was assumed giving an instant spill of 50000 tonnes of crude oil. This amount of oil has been represented by 10000 parcels. The accident is supposed to happen on January 2, 2004 at midnight and at Lon 25.94E and Lat 40.75N. The second experiment was supposed to be a slick accidentally happened at the time of loading on the same day and time, at Lon 25.88E and Lat 40.74N, represented by 3000 parcels. Following the downscaled simulation results of the operational Poseidon system, the model run for 48 hours providing results every 6 hours. Figure 1 shows in the left panel the wider area with the mean circulation for the period January 2 to January 4, and in the right panel depicts a schematic representation of the oil loading navigable points and the two trajectories of of the oil's statistical center of gravity. It is worth noting that none of the oil spills hit on the coasts for the period of the run. Instead the parcels remain quite grouped and move slowly, something that might be attributed to the low wind speeds of the simulation period. Since the surrounding area is a very sensitive one, with high environmental importance, the results show that if a contingency plan could be activated quickly then the consequences at all levels (environmental, social, economic) should be eliminated. Another point that is also very clear from the runs, is that in such complicated areas the only viable approach is a high-resolution nested model, and of course a permanent monitoring system that will allow the management authorities to respond rapidly and to minimize the risk.

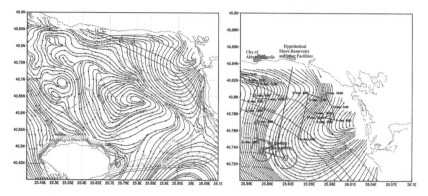

Figure 1: Mean circulation field and bathymetry of the wider area (left panel), Schematic representation of the shore and offshore facilities and oil loading navigable points over a mean circulation field of the area. Simulation results of two hypothetical oil spills: trajectory of the center of gravity every 6 hours (right panel)

# References

[1] A. F. Blumberg and G. L. Mellor. A description of a three-dimensional coastal ocean circulation model. In N.S. Heaps, editor, *Three-Dimensional Coastal Ocean Circulation Models*, Coastal Estuarine Science, pages 1–16. American Geophysical Union, Washington, D.C., 4 edition, 1987.

[2] A. Elliott. Shear diffusion and the spread of oil in the surface layers of the North Sea. *Dt. Hydrogr.*, Z.39:113–137, 1986.

[3] Ø. Johansen. Particle in fluid model for simulation of oil drift and spread. Oceanographic Center, SINTEF Group, Trondheim, Norway, 40 pp, 1985.

[4] C.G. Koutitas. Mathematical models in coastal engineering. *Pentech Press*, 156 pp, 1988.

[5] D. Mackay, I. Buist, R. Mascarenhas, and S. Paterson. Oil spill processes and models. Environment Canada, Ottawa. Report EE-8, 1980.

[6] G.L. Mellor and T. Yamada. Development of a turbulence closure model for geophysical fluid problems. *Review of Geophysics and Space Physics*, 20:851–875, 1982.

[7] D. Rasmussen. Oil spill modeling - A tool for cleanup operations. *Proc. 1985 Oil Spill conference*, pp. 243–249.

[8] J. Riemsdijk van Eldik, R. J. Ogilvie, W. W. Massie MS4: Marine spill simulation software set. Process descriptions. Dept. Civil Engineering, Delft Univ. of Technology, Delft, The Netherlands, 74 p., 1986.

[9] W. Stiver, W. Shiu, D. Mackay. Evaporation times and rates of specific hydrocarbons in oil spills. *Envir. Sci. Technology*, 23:101–105, 1989.

[10] W. Stiver and D. Mackay. Evaporation rate of spills of hydrocarbons and petroleum mixtures. *Envir. Sci. Technology*, 18(11), 1984.

Brill Academic Publishers
P.O. Box 9000, 2300 PA Leiden
The Netherlands

*Lecture Series on Computer
and Computational Sciences*
Volume 4, 2005, pp. 872-877

# Co-firing Different Types of Biomass as a Contribution to Households' Energy Independence and Environment Protection at Local Level

P. Stathatos, D.K. Sidiras[1] and F.A. Batzias

Department of Industrial Management and Technology,
University of Piraeus,
GR-134 85 Piraeus, Greece

Received 11 July, 2005; accepted in revised form 28 July, 2005

*Abstract:* This paper deals mainly with (i) establishing a network of processing/storing/transshipment (PST) points/nodes receiving raw biomass from concentration points and dispatching processed biomass to consumers or downstream stores and (ii) appling a GIS-based technique for locating the PST points that may help potential supply of biomass to meet the potential demand of households, under the least total cost assumption. A bioenergy sustainability index (BSI) is also proposed, based on thermochemical calculations, for providing a measure of substituting processed biomass for conventional fuels that are regularly used for heating in household activities. An implementation is presented for the district of East Macedonia and Thrace in North Greece, including five prefectures. We have proved that a GIS properly applied can correlate biomass exploitation policy with the particular characteristics of each region, leading to optimal location of biomass processing units and generally to better distribution/allocation of the available resources with technoeconomic and environmental criteria.

*Keywords:* biomass, agricultural residues, co-firing, energy, local level, GIS

## 1. Introduction

Biomass is a renewable source of primary energy and its sustainable burning does not emit carbon dioxide, as a net product, according to the following photosynthesis-combustion cycles.

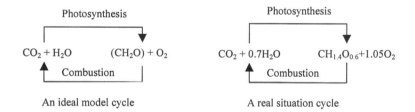

An ideal model cycle      A real situation cycle

Therefore burning of biomass does not contribute to the greenhouse effect, *ceteris paribus* (that is neglecting culture/collection/transportation). The European Union targets for renewable energy sources (RES) and the changes in national socio-economic conditions (public awareness for environmental protection, the change of regulations in the energy market, etc.) are expected to enforce the evolution of biomass applications [1, 2]. Particularly, bioenergy is expected to supply 15% of global primary energy by 2050 [3].

The aim of the present work is (a) to investigate the economic incentives for establishing a network of processing/storing/transshipment (PST) points/nodes (see Fig. 1) receiving raw biomass from concentration points (where farmers leave their agricultural residues, collecting/transferring them from the field after harvesting by their own technical means) and dispatching processed biomass to consumers or downstream stores, (b) to determine the econmo-technical criteria/specifications of desired biomass mixtures prepared for co-firing under the form of briquettes/pellets, taking into

---
[1] Corresponding author. E-mail: sidiras@unipi.gr

account the corresponding thermochemical properties, (c) to apply a GIS-based technique for locating the PST points that may help potential supply of biomass to meet the potential demand of households,

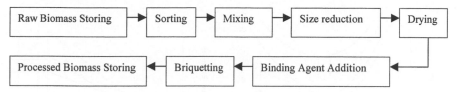

Figure 1: Operations taking place within a PST unit or node of the network

under the least total cost assumption, (d) to propose a bioenergy sustainability index (BSI), based on thermochemical calculations, for providing a measure of substituting processed biomass for conventional fuels that are regularly used for heating in household activities, possibly extending to include space cooling for complete air conditioning, especially in the case of bigger installations/ buildings in the nearby communities.

## 2. Analysis – Proposed Methodology

To investigate the economic incentives for establishing a biomass exploitation network, we examine / depict the household energy economics, initially in absence of this network, by means of the tradeoff shown in Fig. 2; assuming that the biomass inventory of each household increases in batches, which is the result of raw biomass collecting/transporting by the farmer himself, holding cost $C_1$ increases with batch size $L$, according to the standard inventory control theory (i.e., due to mean stock level increase); on the other hand, collection/transportation cost $C_2$ is a decreasing function of $L$ (i.e., $dC_2/dL < 0$), due to lower number of batches (resulting to the same cumulative raw biomass consumption per time period), implying scale economies, and lower fixed cost. The abscissa at minimum value of total cost $C_T = C_1 + C_2$, measured on the axis of ordinates, gives the optimal L-value $L_{opt}$. In the presence of the biomass exploitation network, the holding cost function $C_1$ corresponding to each household inventory is moved to $C'_1$, because a significant part of the required biomass (in processed form) is stored in the respective PST installation, where scale economies are at a high level; when biomass stock level over a household inventory is lower that a predetermined value, an order is placed to the PST warehouse and the quantity demanded is transferred to the household 'just in time' (possibly the farmer himself may execute his order directly, to save time and money). In the region of high $L$-values, the difference $C_1 - C'_1$ increases as L increases (i.e., $d(C_1 - C'_1)/dL > 0$), because room saving in the household ware-place becomes more valuable in this region due to lack of space (in the low $L$-region, this problem is not significant as farmers usually maintain some space available). On the other hand, the cost function $C_2$ is moved to $C'_2$ as the equivalent to collection/transportation cost paid by the farmer for the network operation is lower at each $L$-level (parallel shifting of the $C_2$-curve in respect to its original placement is assumed for simplicity, as no other tendency seems to prevail). As a result, the new total cost curve $C_T'$ shifts downwards to the right giving a minimum $L'_{opt} > L_{opt}$. This means that the farmer has a higher stock level of (processed) biomass and consequently can keep a high energy independence level all year around and over all his house, which nowadays is quite larger.

As regards the PST installations themselves, it is interesting to note that when the crude oil price (and consequently the prices of all its downstream products and their substitutes within the market) increases the break even point increases and profit increases, under the *ceteris paribus* clause (see Fig. 3, where F, V, R are depreciation, variable expenses, revenue, respectively, before and, primed, after the oil price increase); this means that even in case that such a network does not operate economically nowadays, it is likely to do so in the near feature, under the assumption that the oil price, as a function of time, is increasing in the medium/long run.

According to the research we have performed, the economo-technical criteria for the successful design of biomass mixtures for efficient co-firing are:
1. Easy/economic size reduction and briquettes forming.
2. High specific heat of combustion, expressed as formal energy released per unit mass burned.
3. High burning time on a briquette basis, achieved by high density (compactness) and proper composition of the mixture.
4. Low concentration of constituents (natural and artificial, including binding agents) producing smog during combustion or releasing volatile compounds during storage.

5.  Minimal presence or absolute absence of insects or microbial population that might deteriorate lignocellulosics.

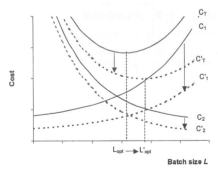

Figure 2. Optimization and sensitivity analysis for determining the shift of Lopt when setting up the network.

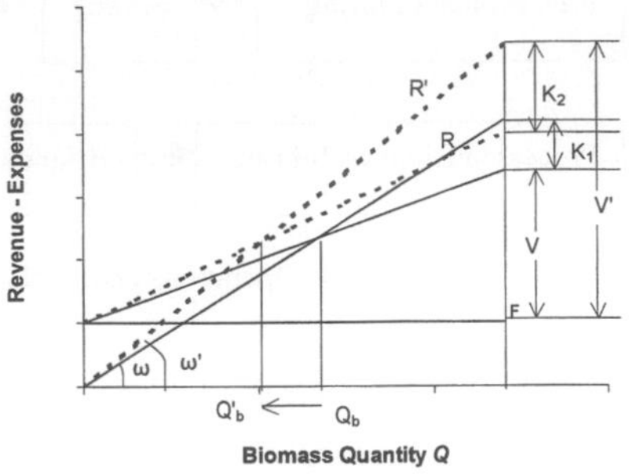

Figure 3. Break even point analysis for a PST unit, indicating the profit shift from K1 to K2 , when oil price increases.

6.  Minimal humidity within the briquettes.
7.  High auto-ignition point, even under extreme conditions that might prevail either in the PST installation or tin the household-storing place where no standards are followed by the farmer.

The location of the PST units has been made by means of the following non-linear mixed integer program, where the processing/storing cost is a non-linear function of capacity of each PST installation due to scale economies.

$$\text{minimize} \sum_{i=1}^{m}\sum_{j=1}^{n} C_{ij}X_{ij} + \sum_{j=1}^{n} P_{j} \,,\; X_{ij} \geq 0 \,,\, (i=1, 2, ..., m;\; j=1, 2, ..., n)$$

$$\text{subject to} \sum_{j=1}^{n} X_{ij} = S_{i} \;\text{(supply constraint)},\; \sum_{i=1}^{m} X_{ij} = D_{j} \;\text{(demand constraint)},\; \sum_{j=1}^{n} D_{j} = D$$

for different *n*-values taken from the GIS PST-units positioning layer, ($P_j$ and $D_j$ are depended on the n-value), where

$m$ = number of raw biomass concentration points (pre-selected)
$n$ = number of biomass PST points (to be determined)
$X_{ij}$ = number of biomass units to be transported from concentration point i to PST point j
$C_{ij}$ = cost of transporting one raw biomass unit from concentration point i to PST point j
$S_i$ = supply at raw biomass concentration point i
$D_j$ = demand at PST point j
$P_j$ = processing/storing cost at PST point j
$D$ = total demand

Potential raw biomass supply as a total as well as at pre-selected (according to spatial/temporal biomass distribution) concentration points is estimated by means of a modified version of BBEPE (agricultural By-products dataBase for Energy Potential Estimation), a resource management tool described in [4, 5]. This tool emerged from the experience gained by the development of 'BIOBASE', a spreadsheet-using application [6], and 'BIOBASE I+', a computer-aided biomass resource management tool, incorporating 'MapInfo' as a GIS [7]. This tool can cooperate with an other tool, short-named by initials ABEPE (Animals dataBase for Energy Potential Estimation), also developed by our research group, as a GIS based biomass resource assessment application using a relational database management system to estimate biogas production from livestock manures [8]. A tailor-made computer software in VB was developed to estimate a satisfactory solution using heuristics and myopic rules as the heterogeneity of space (depicted on the various GIS layers) does not permit application of a pure mathematical procedure aiming at estimating global optimal solutions of general validity.

As a measure of substituting processed biomass or conventional fuels, which are regularly used for heating in households, we propose the following BSI, based on thermochemical calculations: BSI = raw biomass required to release the same amount of thermal energy with a mass unit of the conventional fuel being used until replacement (w/w on dry basis). Bearing in mind that in co-firing several kinds of lignocellulosic materials are burned we conclude that a general expression for mean biomass composition might simplify the calculations; such an expression, based on data from over 200 species of biomass [9], is the following.

$$HHV = 0.3491C + 1.1783 H - 0.1034 O - 0.0211 A + 0.1005 S - 0.0151 N$$

where HHV is the high heating value (in MJ/kg) based on the complete combustion of the sample to carbon dioxide and liquid water, C is the weight fraction of carbon; H of hydrogen; O of oxygen; A of ash; S of sulfur and N of nitrogen appearing in the ultimate analysis. The BSI serves also as a link between biomass potential/availability and households requirements (supply and demand side analysis, respectively); it is also useful for the estimation of subsidy from the State, which is expressed as a percentage of the required capital to be spent for the investment to set up the biomass network and especially the PST units.

## 3. Implementation

The statistical data Input and the intermediate/final results Output, referring to present and future estimations, are stored into a Relational Database Management System (RDBMS). The broad availability of MS Access and its friendly interface with the GIS used renders this RDB the most suitable development environment; not to mention that such an implementation of 'BBEPE' could be easily upsized to a multiprocessor system running SQL. The GIS software used was ArcView version 8.2, which can extract data from an RDBMS through an OLE DB provider interface. Finally, a customised forecasting routine was built to predict biomass production for a given year in the future; automatic best-fit model selection is also possible on the basis of the minimal standard error of estimate.

The district of East Macedonia and Thrace in North Greece, including five prefectures (Drama, Kavala, Rodhopi, Xanthi, Evros), was chosen for implementation. The raw biomass is coming from lignocellulosic residues obtained after harvesting / cutting, by taking into account availability for each species separately. In Fig. 4 a screenshot of the connection between BBEPE and the corresponding GIS layers for the district of East Macedonia and Thrace in North Greece is shown. The pentagons in the map indicate the locations of the PST units, as they are estimated by means of the non-linear mixed integer program presented in the previous chapter. In Fig. 5, two primary GIS layers (a,b) referring to residues coming from straw and wood are presented which (together with other layers referring to tops/leaves, cobs, hull/pod, cake/pulp and process wastes) are the biomass constituents used for briquettes preparation (used also for the thermochemical calculation of BSI).

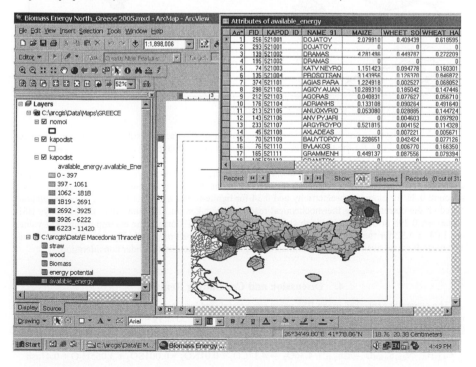

Figure 4: Screenshot of the connection between BBEPE and the corresponding GIS layers for the district of E. Macedonia and Thrace in N. Greece; the pentagons indicate the locations of the PST units.

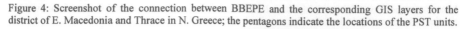

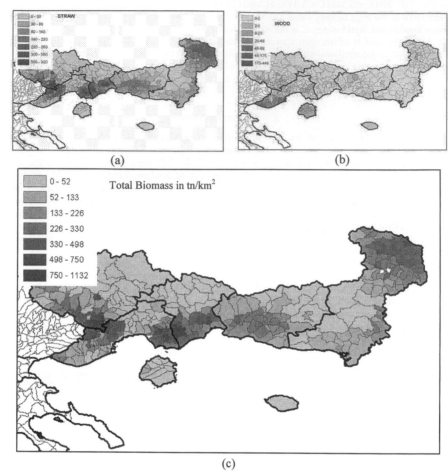

(c)

Figure 5: GIS layers presenting the spatial distribution of biomass constituents used for briquettes preparation (used also for the thermochemical calculation of BSI); layer (c) depicts a total result.

The BSI, estimated on the basis of available dry biomass is 0.409 tn/tn lignite as extracted (i.e. including moisture) or 0.908 tn/tn dry lignite, when the following efficiency values are taken into account: 0.25 for biomass-to-electricity and 0.34 for lignite-to-electricity. If only enthalpy calculations based on the corresponding thermochemical reactions are taken into account, then the respective BSI-values are 0.298 or 0.663. It is worthwhile noting that for total biomass the corresponding BSI-values are 1.04 or 2.32 for the same efficiencies and 0.761 or 1.69 when only enthalpy calculations are used. The mean biomass availability was estimated to be 39.2% for the district under consideration.

## 4. Discussion and Concluding Remarks

The generally expressed mean BSI-value can be replaced by a more specific index to obtain precise results. For example, if wood with a typical composition of C = 50%, H = 6.1%, N = 0.20%, S = 0.03%, O = 43.2%, ash 0.41% w/w on dry basis, then the following chemical reaction can be used:

$$C_{4.17}H_{6.13}N_{0.01}S_{0.001}O_{2.70}ash_{0.41} + 4.36O_2 \rightarrow 4.17CO_2 + 0.01NO_2 + 0.001SO_2 + 3.065H_2O + 0.41ash$$

Special kinds of biomass should also be considered when they form a significant constituent in the mixture prepared for co-firing. Such a constituent might be the kelp family, which has the following empirical formula on dry basis and gives the corresponding combustion equation:

$$C_{2.61}H_{4.63}N_{0.10}S_{0.01}O_{2.23}ash_{26.7} + 2.7625O_2 \rightarrow 2.61CO_2 + 0.10NO_2 + 0.01SO_2 + 2.315H_2O + 26.7ash$$

A representative value of heat evolved by combustion of kelp is 12.39 MJ/kg with product water in the liquid state [9]. For computerized applications, the following parametric equation can be adopted, witch covers the stoichiometry needs of almost all combustion reactions that may be met in co-firing:

$$C_xH_yN_zS_wO_aash_b + (2x+2z+2w+y/2-a)/2O_2 \rightarrow xCO_2 + zNO_2 + wSO_2 + y/2H_2O + b(ash)$$

In conclusion, by combining thermochemical equations either for each lignocellulosic species separately and then applying the superposition principle to obtain total thermal energy or for the mixture, taking into account empirical formulae, we can obtain both, the required lignocellulosics supply for co-firing and the corresponding sustainability index. Thus, the policy for RES promoting at the macro-economic level is related to the RES usage at household level, contributing to the energy independence at local level, which, in its turn, affects regional development, especially of the most remote/isolated communities where the electric energy supply is inadequate and/or its price is high for the end user. We have proved that a GIS properly applied can correlate this policy with the particular characteristics of each region, leading to optimal location of biomass processing units and generally to better distribution/allocation of the available resources with technoeconomic and environmental criteria.

## References

[1] E.G. Koukios, D.K. Sidiras, I. Daouti-Koukios, C.Rakos, G. Kunze, O. Danielsen B. Halkier, and E. Danielsen: An 'Express Path' to Biomass use for Sustainable Regional Development, Driving Forces and Barriers for Technology Diffusion in Denmark, Austria and Greece, 8th European Biomass Conf., Vienna, 1994.

[2] O. Danielsen, , R. Hackstock, , E. Koukios, G. Kunze, C. Rakos, and D. Sidiras: What makes renewable energy work. E.V.A - Energieverwertungsagentur - the Austrian Energy Agency, http://www.eva.wsr.ac.at/publ/pdf/ren_work.pdf 2001.

[3] G. Fischer, L. Schrattenholzer: Global bioenergy potentials through 2050. *Biomass and Bioenergy* **20** 151-59 (2001)

[4] F.A.Batzias, D.K. Sidiras: GIS-assisted planning of a multi-plant agro-industrial scheme for converting crop residues to bioethanol. Proc. 2nd World Conference on Biomass for Energy and Industry and Climate Protection, Rome, Italy, 577-580, 2004.

[5] F.A. Batzias, N. P. Nikolaou, A. S. Kakos and D.K. Sidiras: Computer Aided Optimal Design of a Biomass Processing Unit Based on Economic Criteria, Proc. 2nd World Conference on Biomass for Energy and Industry and Climate Protection, Rome, Italy, 2182-2185, 2004.

[6] D.K. Sidiras, E.G. Koukios: BIOBASE, A Database for Accessing the Biomass Potential in National and Regional Level, Proc. 9th European Bioenergy Conf., Biomass for Energy and the Environment, Copenhagen, Denmark, 787-792, 1996.

[7] D.K.Sidiras: 'BIOBASE I+' an agricultural solid residues management system. Proc. 2nd Int. Conf. on Ecological Protection of the Planet Earth, Bio-Environment and Bio-Culture. Sofia, Bulgaria, 2003

[8] F.A. Batzias, D.K. Sidiras, E.K. Spyrou: Evaluating livestock manures for biogas production: a GIS based method, *Renewable Energy* **30**(8) 1161-1176 (2005).

[9] D.L. Klass: *Biomass for Renewable Energy, Fuels, and Chemicals*. Academic Press, San Diego, 1998.

Brill Academic Publishers
P.O. Box 9000, 2300 PA Leiden
The Netherlands

*Lecture Series on Computer*
*and Computational Sciences*
Volume 4, 2005, pp. 878-886

# Computer Aided Design of an Anaerobic Digester with Low Biogas Production for Decision Making Under Uncertainty

F.A. Batzias[1]

Department of Industrial Management & Technology,
University of Piraeus,
Karaoli & Dimitriou 80,
GR-185 34 Piraeus, Greece

Received 3 July, 2005; accepted in revised form 4 July, 2005

*Abstract:* The main advantage of anaerobic digestion of organic waste is biogas production bearing, however, the disadvantage of the need for high level monitoring of process parameters. In this work, an algorithmic procedure is presented for computer aided design of an anaerobic digester to cope with problems arising from (i) low biogas production and (ii) high heterogeneity of the raw waste material, leading to difficulties in achieving/maintaining steady state conditions at optimal level. An implementation of this algorithmic procedure is also presented in the case of the anaerobic contact process (system with a cycle via a thickener/clarifier set in series with the bioreactor); the same software offers the possibility of running either complete models to obtain precise results for the sake of accuracy or simplified models to obtain approximate results for the sake of comparability with data produced under similar conditions and cited in technical literature.

*Keywords:* anaerobic digester, biogas production, computer aided design, decision making, uncertainty

## 1. Introduction

Digestion is a complex biochemical process in which anaerobic/facultative microorganisms break down organic substrates. The major advantages of anaerobic over aerobic treatment of organic wastes are (i) biogas production, containing about 60% $CH_4$, with significant economic value, which can provide thermal energy to the digester for keeping temperature at 35 °C and/or be used for electric/thermal energy co-production for clients in the vicinity or the electric grid [1] or the transportation network [2], provided that high organic load is processed, (ii) less biomass produced per unit of organic substrate utilized, which also means a decrease in the requirement for nitrogen and phosphorous, (iii) higher organic loading potential as the process is not limited by the oxygen transfer capability at high oxygen utilization rates. Disadvantages of the anaerobic treatment are (i) the relatively slow bioreaction rate for the methane production step, which is rate-limiting in the overall process, possibly resulting to incomplete stabilization of organic waste at economical residence time, and (ii) the need for continuous monitoring, because if the rate of methane production does not keep up with the rate of preceding acid formation, the pH will decrease below 6.5 and the methane formation will stop. Consequently, the main anticipated problems of anaerobic digestion are (i) inadequate biogas production and (ii) uncertainty that makes monitoring tedious.

The aim of the present work is to develop an algorithmic procedure for computer aided design of an anaerobic digester producing low amounts of biogas and operating under real conditions implying uneven probability density function for the input variables and uncertainty as regards certain parameter-values since waste water is raw material of high heterogeneity.

## 2. Methodological Framework

We have collected 17 different anaerobic processes from technical literature and stored properly in a Knowledge Base (KB) together with relevant information referring to variables/parameters values, conditions of operation, models performance, and simulated/real results. The algorithmic procedure we

---

[1] Corresponding author. E-mail: fbatzi@unipi.gr

have developed as a methodological framework for the computer aided design of any anaerobic digester stored in the KB is described by the following stages (for their interconnection, see Fig. 1).

1. Selection of representative values for maximum specific substrate utilization rate $k$, saturation constant $K_s$, digester temperature for maximum yield $T$, safety factor $SF$, active biomass concentration $X$, specific heat coefficient $c$, air mean temperature $T_0$.
2. Input of microbial decay coefficient $k_d$ and biomass synthesis constant $Y$ as fuzzy numbers for a linguistic variable taking lipid content values according to a Linkert scale.
3. Defuzzification to obtain the corresponding crisp numbers.
4. Collection of data for (i) fresh feed substrate concentration, $S_0$, measured as COD, and (ii) volumetric flow rate $Q$.
5. Determination of probability density functions corresponding to these data.
6. Estimation of $S_0$- and $Q$- values to enter the subsequent algorithmic procedure, based on an initial preset significant level $\alpha$.
7. Selection of models for minimum biological solids retention time $\Theta_m$, substrate concentration in the effluent $S_e$, digester volume $V$, observed yield coefficient $Y_p$, biomass production $QDX$, biogas production $G$, thermal energy surplus $H$, thermal energy demand $H_0$.
8. Model ranking for each dependent variable, according to preset criteria, and proposition of the first of each category for running the simulation.
9. Calculation of biological solids retention time $\Theta_c$ as a product of the input $SF$-value and the estimated (through the corresponding bioreactor kinetics model) $\Theta_m$-value.
10. Estimation of the $S_e$-value and the respective efficiency $E=1-S_e/S_0$.
11. Comparison with the results obtained in similar installations.
12. Reactor scale down for experimentation at laboratory level to confirm/identify the parameter values used so far.
13. Experimental design and evaluation of measurements.
14. Knowledge Base (KB) enrichment and support/consultation activity.
15. Intelligent Agent functions for data mining from external information sources [3].
16. Estimation of bioreactor volume $V$.
17. Comparison with operating volumes of corresponding bioreactors.
18. Estimation of biomass production $QDX$ and its thermal content $H$.
19. Estimation of biogas production $G$.
20. Comparison with the $G$-values obtained from similar bioreactors operating under similar conditions.
21. Estimation of thermal energy demand $H_0$ for the bioreactor operation.
22. Calculation of energy balance.
23. Determination of kind and amount of renewable energy required to cover the deficit (thermal losses of the whole system included).
24. Determination of kind and amount of conventional energy required to cover the deficit (thermal losses of the whole system included).
25. Preparation of final report.

P. Is the comparison of the $S_e$-values in absolute and relative terms (i.e., as concentration and efficiency, respectively) satisfactory?
Q. Is there at least one ranked $S_e$-model not yet used in simulation?
R. Are these new parameter-values significantly different in comparison with the old ones?
S. Is the comparison satisfactory?
T. Is there at least one ranked $V$-model not yet used in simulation?
U. Is the comparison satisfactory?
V. Is there at least one ranked $G$-model not yet used in simulation?
W. Is there a deficit?
X. Can the deficit be covered by a renewable energy source available at local level?

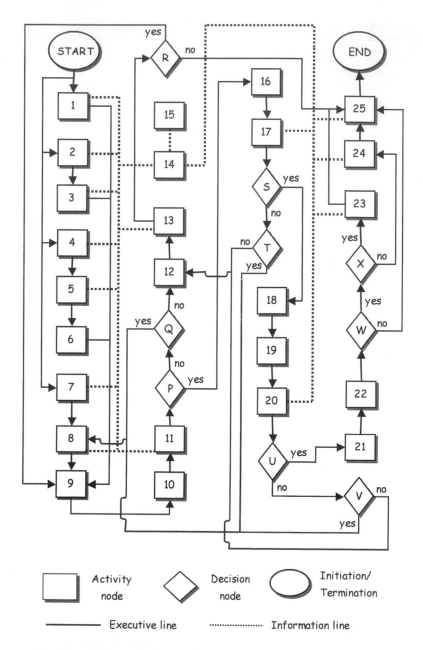

Figure 1: Flow chart of the knowledge-based algorithmic procedure developed as a methodological framework for the computer aided design of an anaerobic digester.

To account for uncertainty $k_d$ and $Y$ are entered as fuzzy numbers in stage 2, for low, moderate, and high lipid content of wastewater; for the same reason, $S_0$ and $Q$ are entered as stochastic variables and the corresponding probability density functions are chosen from the KB (stage 14) by best fitting to data. Stage 22 can be used for sensitivity analysis of the energy break-even-point, where $DH=H-H_0=0$, to find out (i) the minimum $Q$-value required for the installation to be energy–independent, and (ii) the optimal recycle ratio $R=Q_r/Q$ used as a control variable for both, the $X$-

value and the real substrate concentration value $S_i$ of the stream entering the bioreactor, calculated by applying a substrate mass balance at the point where fresh feed is mixed with the recycle stream. It is worthwhile noting that $Q$ can be increased in case of co-treatment, i.e., by transporting wastewater from nearby industrial producers to take advantage of the scale economies, provided that the transportation cost permits such a cooperative scheme [4].

## 3. Implementation

The anaerobic contact process was chosen for implementation because it is rather simple, with known functional parameters, and can be applied for both, pure industrial and mixed industrial/domestic wastewater treatment, when BOD in the influent stream exceeds a threshold depended on the spatial characteristics of each installation. In this process, untreated wastes are mixed with recycled sludge solids and then digested in a reactor sealed to the entry of air. The mixture is completely mixed and, after digestion, is separated in a thickener/clarifier or a vacuum flotation unit. The supernatant is discharged as effluent, usually for tertiary treatment, while the sludge is partly recycled to seed the raw influent and partly further treated either for ultimate disposal or as a soil conditioner within proper mixtures. Because of the low synthesis rate of anaerobic microorganisms, the excess sludge that must be disposed of is minimal [5]. This process has been used successfully for the stabilization of food industries wastewater (e.g., meat processing/packing).

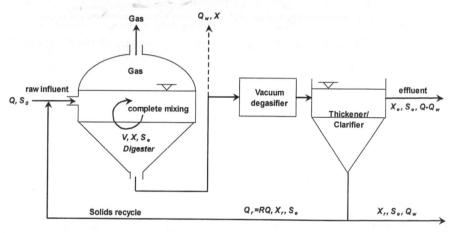

Figure 2: Flow chart of the anaerobic contact process used for implementation, showing also the two alternatives for sludge wasting.

A typical flow chart of the anaerobic contact process is depicted in Fig. 2, where two alternatives for sludge wasting are shown. Sludge wasting, with volumetric flow rate $Q_w$, can be performed either after the thickener or just after the digester. Although the former is more common, the latter contributes to better plant control and is also beneficial to subsequent sludge thickening, as it has been shown that higher solids concentration can be achieved when dilute mixed liquor rather than concentrated return sludge is thickened (a view supported also for aerobic treatment in [6]).

The procedure of calculations is shown in the Appendix and some arithmetic results (for $T = 35\ °C$, $T_0 = 20\ °C$, $SF = 2$) are shown in Figs. 3 and 4 in the form of diagrams. The former presents a kind of 'intensity effect', i.e., the influence of more intense/concentrated microbial culture on design capacity, while the latter shows a kind of sensitivity analysis of the break-even-point from the energy point of view.

The algorithmic procedure we have developed allows the run of the complete expressions for the estimation of the intermediate variables without the simplifications usually adopted for convenience; furthermore, by combining stages 7, 8, and 14, we can solve the 'inverse problem', i.e., to produce comparable to the technical literature results by allowing the usual simplifications to be made. Thus, we can assess the observed deviations and whether they are justified or not; these deviations are actually significant, as shown in Figs. 5 and 6. The former illustrates the deviation $DS_e = g(S_0)$ in the estimation of the $S_e$-value, and consequently in the efficiency that is by definition $E=1-S_e/S_0$, when $k_d$ is not used (as proposed by [7] and adopted by [8]) in the estimation of (I) only $\Theta_m=f_1$ $(S_0, K_s, Y, k, k_d)$, and (II) both $\Theta_m$ and $S_e=f_2(\Theta_m, SF, K_s, Y, k, k_d)$. Fig. 6 presents the deviation $DV=h(S_0)$ in the estimation of the capacity, expressed by the bioreactor volume, when $k_d$ is not used in the estimation of (I) $\Theta_m$, (II) $\Theta_m$ and $S_e$, (III) all values of the dependent variables $\Theta_m$, $S_e$, $V=f_3$ $(\Theta_m, SF, S_0, S_e, Q, X, K_s, Y, k, k_d)$.

Larger deviations have been observed when $K_s$ is also eliminated in the estimation of $\Theta_m$ (as proposed by [9] and partially adopted as an alternative in [8]), as shown in Figs. 7 and 8. The former illustrates the deviation $DS_e=g'(S_0)$ in the estimation of the $S_e$-value, when (I) $k_d$ is used in $f_2$ and (II) $k_d$ is not used in $f_2$, while the latter shows the deviation of $DV=h'(S_0)$ in the estimation of the capacity when $k_d$ is (I) used in $f_2, f_3$, (II) used only in $f_3$, (III) totally eliminated.

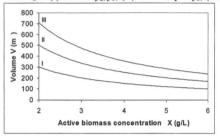

Figure 3. Dependence of bioreactor design capacity $V$ (m³) on biomass concentration $X$ (g/L) for different values of $S_0$ (g/L): (I) 4; (II) 7; (III) 10.

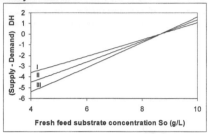

Figure 4. Dependence of (Supply-Demand) of thermal energy DH ($10^9$ cal/day) on $S_0$ (g/L) for different values of $Q$ (m³/day): (I) 400; (II) 500; (III) 600.

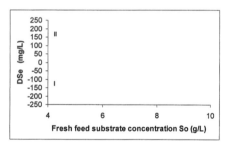

Figure 5. Observed deviation $DS_e$ (mg/L) as a function of $S_0$ (g/L), when $k_d$ is eliminated from (I) $f1$, (II) both $f_1$ and $f_2$.

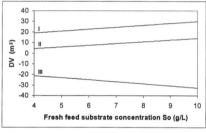

Figure 6. Observed deviation $DV$ (m³) as a function of $S_0$ (g/L), when $k_d$ is eliminated from (I) $f1$, (II) both $f_1$ and $f_2$, (III) $f_1, f_2, f_3$.

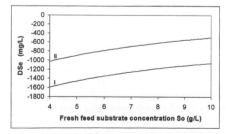

Figure 7. Observed deviation $DS_e$ (mg/L) as a function of $S_0$ (g/L), when $k_d$ and $K_s$ are eliminated from $f1$ and $k_d$ is (I) used in $f_2$, (II) not used in $f_2$.

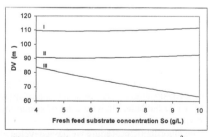

Figure 8. Observed deviation $DV$ (m³) as a function of $S_0$ (g/L), when $k_d$ and $K_s$ are eliminated from $f1$ and $k_d$ is used (I) in both $f_2$ and $f_3$, (II) only in $f_3$, (III) not at all.

## 4.     Discussion and Concluding Remarks

The stages 11, 17, 20, we have developed for comparing intermediate/final results obtained by applying in each case the algorithmic procedure shown in Fig. 1, permit also revision of aggregate/empirical models towards individualistic/mechanismic ones, since separate estimation for each substrate species is feasible and evaluation of simultaneous interaction is possible, provided that either the biochemical mechanism has been revealed or adequate numerical data for simulation are available. It is worthwhile noting that several aggregate models have been derived by (mis)using the superposition principle within a dynamic system, which seems to be (incorrectly) a static one, simply because it is (assumed to be) in steady state conditions, implying disappearance of certain variables from the final quantitative expression, although they are present in differential/dynamic form.   Due to lack of space we shall analyze only the aggregate model proposed by O'Rourke [10] and adopted by technical literature (e.g., see [8]), according to which $S_e$ for a mixture of $n$ fatty acids found or produced in the waste to be treated can be given by the expression

$$(S_e)_{overall} = \frac{1+k_d\Theta_c}{\Theta_c(Yk-k_d)-1}K_c, \qquad K_c = \sum_{i=1}^{n}K_{s,i} \qquad (1)$$

This expression, which have been derived by assuming that the values of $Y$, $k_d$ and $k$ are equal for all fatty acid fermentations, cannot be used as an aggregate model, because the corresponding relation

$$S_e = \frac{(1+k_d\Theta_c)K_s}{\Theta_c(Yk-k_d)-1} \qquad (2)$$

supposed to be valid for each constituent fatty acid separately, comes from a steady state condition as follows.

Biomass balance as a word statement: Accumulation = Generation − Removal

Since biomass appears as a result of growth and disappears as a result of hydraulic action in both the sludge waste line and the effluent, the word statement of the biomass balance quoted above is expressed as

$$(\frac{dX}{dt})V = (\frac{dX}{dt})_g V - [Q_wX +(Q-Q_w)X_e] \qquad (3)$$

where net growth $(dX/dt)_g$ equals total growth $(dX/dt)_T$ minus biomass lost to endogenous respiration $(dX/dt)_e$; since the rate at which biomass is lost to endogenous respiration is proportional to the biomass present and the total growth is proportional to the organic substrate utilized (assuming that all substrate is channeled into the growth operation, whereas the energy for maintenance comes from the oxidation of cellular components), i.e., $(dX/dt)_e = k_dX$ and $(dX/dt)_T = Y(dS/dt)$, we obtain the expression $(dX/dt)_g = Y(dS/dt) - k_dX$. By introducing $\Theta_c \equiv X_T/(\Delta X/\Delta t)_T$, the biomass balance shown in Fig. 2 gives $\Theta c = XV/[Q_wX+(Q-Q_w)X_e]$ and finally equation (3) becomes

$$(\frac{dX}{dt})V = [Y\frac{dS}{dt} - k_dX]V - \frac{XV}{\Theta_c} \qquad (4)$$

According to Lawrence and McCarty [11], $dS/dt = kXS/(K_s+S)$, therefore the last equation for steady state conditions as regards biomass (i.e., $Vdx/dt=0$) gives the equation (2) for minimum $\Theta_c$, named $\Theta_m$ (obtained for $S_e=S_0$, when wash-out conditions prevail). Therefore, equation (1) cannot be valid, because steady state conditions are achieved/maintained (via the control variable recycle ratio $R=Q_r/Q$) only for constant $X$-value, which has not been included in the assumptions.

In conclusion, it is proved that the knowledge-based algorithmic procedure we have developed provides a methodological framework for the design of an anaerobic digester as it uses as input stochastic and fuzzy variables, to account for uncertainty, together with conventional variables, given as crisp numbers when availability/precision/reliability of data is satisfactory. The same procedure can (i) perform sensitivity analysis of the energy break-even-point to help in finding out a solution for the energy dependence problem of the plant when the biogas produced is low, (ii) run of the complete expressions for the estimation of the intermediate variables without the simplifications usually adopted for convenience, (iii) solve the 'inverse problem', i.e., to produce comparable to the technical literature results by allowing the usual simplifications to be made. Therefore, we can assess the observed deviations and whether they are justified or not; in the case example of the anaerobic contact process we examined for a wide range of values of system variables and representative parameter-values we proved that these deviations are mostly significant and thus the corresponding simplifications should be avoided.

## 5. Pseudocode

The following brief pseudocode includes the most characteristic steps for covering the calculation requirements of the implementation of the algorithmic procedure for optimal design of the anaerobic contact process (digester with recycle).

For high, medium, low lipid content, define Kd, Y values as triangular LR fuzzy numbers, let them be Kd_fuzzy, Y_fuzzy.

Input lipid content as low, medium, high in percentage

Transform combined fuzzy to crisp numbers by defuzzification

```
// Definitions of function routine:
// LingVar: Linguistic variable, (e.g. Kd_fuzzy, Y_fuzzy), which contains fuzzy sets
// m, alpha, beta: center value, left and right interval of LR fuzzy number
// a, b, c: actual abscissas of the triangle (projections of the triangle vertices on the
//       universe of discourse)
// h: height at which the triangle is truncated – in our case lipid content, 0 to 1
// E: Area of the truncated triangle
// x: center of gravity of the truncated triangle

Function DefuzzCOG(LingVar)
    S1 = 0: S2 = 0
    For each FuzzySet of LingVar
        If h > 0 then
            a = m – alpha
            b = m
            c = m + beta
            E = 0.5 * (2 – h) * (c – a) * h
            d1 = (b – a) * h + a
            d2 = c – (c – b) * h
            x = (1/3) * (Sqr(c) + d2 * c + Sqr(d2) – Sqr(a) – d1*a – Sqr(d1)) / (d2 – d1 + c – a)
            S1 = S1 + x * E
            S2 = S2 + S2 * E
        End If
    Next FuzzySet
    y = S1 / S2
    DefuzzCOG = y
End Function

Y = DefuzzCOG(Y_fuzzy)
Kd = DefuzzCOG(Kd_fuzzy)
```

Input So, Q data

Construction of the corresponding histograms

Determination of best fit probability density functions

Input significance level a
Estimation of So, Q input values at significance level a (one tail)

Example: IF the best-fit x squared test suggests the normal distribution
THEN the following subroutine is automatically retrieved from KB

```
Function z(a)
    p = 1 – a
    x = -5
```

```
    y = 5
    Do
        z = (x + y) / 2
        t = z
        s = z
        r = 1
        Do
            t = (-t * z * z * (2 * r - 1))
            t = t / ((2 * r + 1) * 2 * r)
            s = s + t
            r = r + 1
        Loop while Abs(t) > 0.00001
        a = 0.5 - (s / sqr(8*atn(1)))
        if abs(a - p) < 0.0000001 then exit function
        if a < p then y = z
        if a > p then x = z
    Loop
End Function
```

So = mean(So()) – StDev(So()) * z(a)
Q = mean(Q()) – StDev(Q()) * z(a)

Input k, ks, T, Sf, X, c, T0 as crisp values

Thm = 1 / ((Y * k * So / (ks + So)) - kd)

Thc = Sf * Thm

Se = ks * (1 + kd * Thc) / (Thc * (Y * k - kd) - 1)

E = 1 - Se / So

V = Thc * Y * Q * (So - Se) / (X * (1 + kd * Thc))

Yp = Y / (1 + kd * Thc)

QDX = Yp * Q * (So - Se)

N = 0.122 * QDX

P = 0.023 * QDX

Go = (273 + T) * 22.4 / (273 * 64)

G = Go * ((So - Se) * Q - 1.42 * QDX)

h = 8560 * 273 * G / (273 + T)

ho = Q * (10 ^ 6) * c * (T – T0)

DH = 0.75 * h – ho

## References

[1] F.A. Batzias, D.K. Sidiras and E.K. Spyrou, Evaluating livestock manures for biogas production: a GIS based method, *Renewable Energy* **30** 1161-1176 (2005).

[2] J.D. Murphy and K. McCarthy, The optimal production of biogas for use as a transport fuel in Ireland, *Renewable Energy* **30** 2111-2127 (2005).

[3] F.A. Batzias and E.C. Markoulaki, Restructuring the keywords interface to enhance CAPE knowledge via an intelligent aagent, *Computer-Aided Chemical Engineering* **10** 829-834 (2002).

[4] F.A. Batzias, Contribution to optimal design of bioreactors for cooperative wastewater treatment, *Environmental Science and Pollution Research* **9** 170 (2002).

[5] Matcalf and Eddy, *Wastewater Engineering – Treatment, Disposal, Reuse*. 3$^{rd}$ Edition, McGraw-Hill International Editions, 1991.

[6] P.L. McCarty, Sludge concentration – Needs, accomplishments and future goals, *Journal of the Water Pollution Control Federation* **38** 493 (1966).

[7] P.L. McCarty, Anaerobic treatment of soluble wastes, *Advances in Water Quality Improvement* (Editors: E.F. Gloyna and W.W. Eckenfelder), University of Texas Press, 1968.

[8] L.D. Benefield and C.W. Randall, *Biological Process Design for Wastewater Treatment*. Prentice-Hall, NJ, 1980.

[9] A.W. Lawrence and P.L. McCarty, Kinetics of methane fermentation in anaerobic treatment, *Journal of the Water Pollution Control Federation* **41** R1 (1969).

[10] J.T. O'Rourke: *Kinetics of Anaerobic Treatment at Reduced Temperatures*. Doctoral thesis, Stanford University, Stanford, Kalifornia, 1968.

[11] A.W. Lawrence and P.L. McCarty, Unified basis for biological treatment design and operation, *Journal of the Sanitary Engineering Division, ASCE* **96** 757 (1970).

Brill Academic Publishers
P.O. Box 9000, 2300 PA Leiden
The Netherlands

*Lecture Series on Computer
and Computational Sciences*
Volume 4, 2005, pp. 887-893

# Optimal Design of Cyclone/Chimney System – Implementation within an Olive Pomace Oil Mill

F.A. Batzias and D.K. Sidiras[1]

Department of Industrial Management and Technology,
University of Piraeus,
GR-134 85 Piraeus, Greece

Received 5 July, 2005; accepted in revised form 17 July, 2005

*Abstract:* Atmospheric pollution modeling is widely used in maintaining environmental standards within preset limits applicable to a wide national or international/transboundary area. At local level, however, additional parameters should be considered, such as the distribution and attitude of the population and the sensitivity of ecosystems. In the present work, we have developed a hierarchical model under the form of a GIS-based algorithmic procedure where local environmental parameters are taken under consideration; following a sub-procedure of successive numerical filtering, the most strict parameter defines the specifications that the agro-industrial polluting source should fulfill. The suggested procedure enables an iterative process for the design parameter estimation to achieve a balanced tradeoff between sustaining the environmental standards and keeping construction/operating cost of a cyclone/chimney system within acceptable limits, defined by the competition among the companies of the same sector. The implementation presented on the optimal cyclone/chimney design for an olive pomace oil unit that will operate within a NATURA region of Greece, indicates that these targets can be met through sensitivity analysis of the positive models used, which are transformed to normative ones by selecting those parameter values that produce the desired results.

*Keywords:* air pollution modeling, chimney, cyclone, GIS, olive pomace oil mill, algorithmic procedure

## 1. Introduction

Atmospheric pollution modeling is a useful means to predict incidence of poor local air quality and monitor pollutant releases in order to maintain environmental standards within preset limits applicable to a wide national or international/transboundary area. At local level, nevertheless, additional parameters should be considered, such as the distribution and attitude of the population and the sensitivity of ecosystems. To prevent the deterioration of the environment at local level, EU has issued several Directives and Regulations, the application of which demands a detailed mapping of the region under consideration, so that the specific critical environmental parameters can be checked just before and after a potentially polluting activity, especially under the requirements posed within the NATURA network [1]. In the case of remote agro-industrial units, there is a difficulty in mapping the effects of pollution as these units operate usually a few months per year and release their gaseous or liquid pollutants without a steady flow and permanent bed.

In the present work, we have developed a hierarchical model under the form of an algorithmic procedure where local environmental parameters are taken under consideration; following a sub-procedure of successive numerical filtering, the most strict parameter defines the specifications that the agro-industrial polluting source should fulfill all year around. Incorporated with this procedure is a BATNEEC (Best Available Technique Not Entailing Excessive Cost) scheme for pollution mapping, based on an ad hoc modified Geographical Information System (GIS) coupled with a Global Positioning System (GPS) receiver for determining the environmental impact (e.g., accumulated particulate matter via a dry deposition natural process) of a source of pollution during the period of the year that this source does not operate. The BATNEEC scheme we developed is especially designed/adapted for use by communities which are responsible for maintaining/preserving a protected area; it concerns not a production but rather a service activity, namely the process of

---

[1] Corresponding author. E-mail: sidiras@unipi.gr

observing/measuring/monitoring the impact of air pollution over the district under the jurisdiction of a municipality which usually has limited economic and technical resources. An illustration of the algorithmic procedure is presented, concerning the pollution (and its consequences) caused by an olive pomace oil mill operating in a NATURA 2000 protected area in the Greek Island of Crete.

## 2. Methodology

There are three constraints which determine the GIS-based design of a cyclone/chimney system within a given environment: (i) the expected maximum atmospheric pollution should conform with the corresponding environmental standards, (ii) pollution level at pre-selected representative point(s) should be sustained under the limits set after agreement between local and central authorities, and (iii) the maximum permitted values of pollution in ecological sensitive regions, set by concerned organizations on the basis of scientific/technical literature, according to the principle of BATNEEC. Such reasoning can be realized through the following algorithmic procedure developed by the authors as a methodological framework for the design of a cyclone/chimney system suitable for preventing particle pollution; the interconnection of the activity and decision nodes is shown in Fig. 1.

1. Data collection and GIS layer formation for the geomorphology of the region round the point source of atmospheric pollution.
2. Data collection and GIS layers formation for population spatial distribution; pre-selection of representative point(s) for setting upper limits for maximum and mean atmospheric pollution.
3. Location of ecologically sensitive sub-regions (ESS).
4. Determination of maximum permitted values of gas pollutants in these ESS's (including accumulation on the ground due to dry deposition) and strict location of the most sensitive point(s) for setting upper limits for ground/water pollution, by taking into account accumulation too.
5. Olive pomace drying in similar small-scale dryer under conditions expected to prevail in real/large scale.
6. Design of the minimum efficiency cyclone system, capable to remove particles from released gases to prevent atmospheric pollution.
7. Calculation of total and marginal cost for the cyclone system.
8. Design of minimum height chimney, capable to disperse adequately the remaining particles in order to meet the environmental standards.
9. Calculation of total and marginal cost of the chimney.
10. Estimation of particle distribution in the chimney output.
11. Pre-selection of point source diffusion models based on the Gaussian plume type.
12. *Apriori* ranking (based on technical literature survey) of pre-selected models according to the criteria of adaptability, reliability and simplicity (or maximum falsifiability, after K. Popper).
13. Numerical scanning application of the model which is ranked first among the remaining untested models.
14. Testing of the diffusion model *per se*, according to criteria stated in stage 12.
15. Choice of the best diffusion model among the ones examined so far.
16. Simulation of drying and gas releasing for testing of cyclone performance by comparing estimated values of particle distribution at the expected maximum atmospheric pollution point with the corresponding environmental standards.
17. Estimation of particle distribution at the pre-selected representative point(s) by Monte Carlo simulation.
18. Estimation of particle distribution at the most sensitive point(s) located in stage 4.
19. Creation/development/enrichment of Knowledge Base (KB).
20. Scale-up of the cyclon-chimney system.
21. Comparison of marginal costs corresponding to the cyclone system and the chimney, separately, in respect to pollutant concentration decrease at the critical measuring point (estimated by means of worst-case analysis).

P. Is this testing satisfactory?
Q. Are there models that have not been examined yet?
R. Are the environmental standards satisfied?
S. Are the environmental standards for the representative point(s) satisfied?
T. Are the set upper limits for ground/water pollution satisfied?
U. Is the chimney differential cost (required for technical improvement capable to satisfy the corresponding environmental constraint) less than the respective cost for the cyclone system?

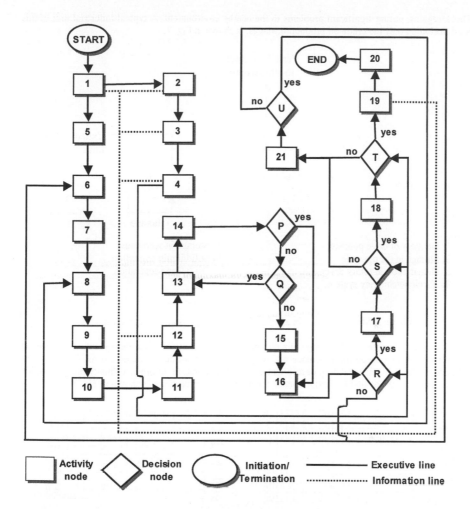

Figure 1. Flowchart of the algorithmic procedure developed for the optimal design of the cyclone-chimney system, based on economo-technical and environmental criteria.

The procedure quoted above adopts a positive/analytic point of view in contrast with the normative/managerial aspect adopted by other authors in corresponding procedures (see e.g. [2-4]). The three critical constraints mentioned in the beginning are indicated in stages 16, 17, and 18 of the algorithmic procedure; as they are connected in series, the most strict of them will determine the technical characteristics of the cyclone/chimney system.

Moreover, special emphasis is put on the creation/enrichment/usage of the local Knowledge Base (stage 19), which is expected to offer expertise at local level, whereas the data collection/processing stages 1-3, 17, 18, can be also used for revealing other possible sources of pollution to be located either by the local authorities or by the environmental experts who use a Central Knowledge Base, where tools for simulating pollution episodes are available.

## 3. Implementation

An olive pomace oil mill has been selected for the implementation of the proposed methodological framework, owing to the large number of such industrial units operating in several olive oil producing

Greek regions, posing significant problems to the nearby environment. A typical industrial unit of this kind, including only the crude oil production stages, is shown in Fig. 2.

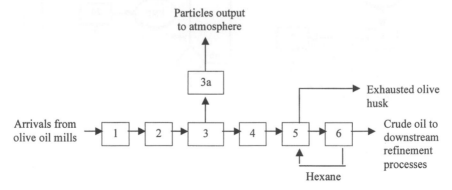

1. Delivery of olive pomace
2. Storage in the open air
3. Dehydration in rotary dryer
3a. Cyclone/chimney system

4. Storage in warehouse
5. Solid/liquid extraction
6. Oil-hexane separation

Figure 2. Main processes in a typical crude olive pomace oil production unit.

The case study refers to the design of a unit near the community of Alkalohorio in the Greek island of Crete, producing crude olive pomace oil, which will be forwarded to other downstream units for refinement, through neutralization, decolorization and deodorization, in order to finally produce edible oil. According to the available measurements, the distribution of particles is shown in Fig. 3.

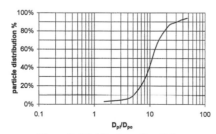

Figure 3: Distribution of particles

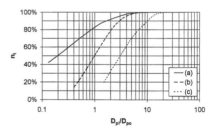

Figure 4: Cyclone efficiency *vs* particle distribution according to three different models: (a) ref. [5], (b) ref. [6] and (c) ref. [7].

Following the model proposed by Leith and Licht [5], cyclone efficiency is shown in Fig. 4a, whereas the total cyclone efficiency is $n = 99.5\%$. According to the methodology proposed/ adopted by McCabe and Smith [6], cyclone efficiency is given by

$$n_i = 1 - \exp\left[-k(D_{pi}/D_{pc})^v + m\right] \qquad (1)$$

where $D_{pi}$ is the particle diameter, $D_{pc}$ the cut diameter, whereas the empirical parameter-values for $k$, $m$ and $v$ have been estimated to 0.991, 0.287 and 0.894, respectively (Fig. 4b); the total cyclone efficiency is thus estimated to $n = 98.8\%$. This value is further reduced to $n = 91.2\%$, if the model quoted in Coulson & Richardson [7] is used. Adopting the value obtained by applying the methodology quoted in McCabe and Smith [6], i.e., 98.8%, as the most moderate estimation, the distribution of the particles released from the chimney can be calculated as follows.

The distribution of a pollutant released from a point source is simulated by means of a bidimensional stochastic model (e.g. see [8]).

$$c = \frac{Q}{2\pi u \sigma_y \sigma_z} \exp\left(-\frac{y^2}{2\sigma_y^2}\right) \left\{ \exp\left[-\frac{(z-H)^2}{2\sigma_z^2}\right] + \exp\left[-\frac{(z+H)^2}{2\sigma_z^2}\right] \right\} \qquad (2)$$

where

$c$     concentration of particles, kg/m$^3$
$Q$     continuous release rate of the particles from the chimney, kg/s
$u$     wind speed, m/s
$\sigma_x, \sigma_z$ dispersion coefficients, m
$x$     horizontal distance from the chimney base parallel to wind direction, m
$y$     horizontal distance from the chimney base vertical to wind direction, m
$z$     vertical distance from the chimney base, m
$H$     height of mixing, m

Also,

$z = Y - Y_0$, where $Y$ is the altitude from the sea level, m, and $Y_0$ is the altitude at the chimney base, m
$H = H_r + \Delta h$, where $H_r$ is the chimney height and $\Delta h$ is height of release above the outlet, and

$$\Delta h = \frac{V_s D}{u}\left[1.5 + 2.68 \times 10^{-3} PD\left(\frac{T_s - T_a}{T_s}\right)\right] \qquad (3)$$

where

$V_s$     speed of release of the carrier gas from the chimney, m/s
$D$     chimney diameter, m
$P$     pressure, millibars
$T_s$     temperature of the carrier gas in the chimney, K
$T_a$     atmospheric temperature, K

According to [9], for an unstable atmosphere (stability class A):
$\sigma_y = 0.493x^{0.88}$
$\sigma_z = 0.087x^{1.10}$ for $100 \leq x \leq 300$ and
$\log \sigma_z = -1.67 + 0.902 \log x + 0.181(\log x)^2$ for $300 \leq x \leq 3000$

For a neutral atmosphere (stability class C):
$\sigma_y = 0.195x^{0.90}$
$\sigma_z = 0.112x^{0.91}$ for $100 \leq x \leq 10000$

For a stable atmosphere (stability class F):
$\sigma_y = 0.067x^{0.90}$
$\sigma_z = 0.057x^{0.80}$ για $100 \leq x \leq 500$ and
$\log \sigma_z = -1,91 + 1,37 \log x - 0,119(\log x)^2$ για $500 \leq x \leq 10000$

Also, for continuous releases above the ground level:

$$C_{max} = \frac{2Q}{e\pi u H_r^2}\left(\frac{\sigma_z}{\sigma_y}\right) \qquad \text{with } \sigma_z = \frac{H_r}{\sqrt{2}} \qquad (4)$$

where $C_{max}$ is the maximum concentration of particles (kg/m$^3$) downwind along the central axis of the plume [9].

Subsequently we apply the simulation model shown in (2) in order to optimize the chimney height of the specific unit by taking into account the following arithmetic data, included in a preliminary study of Environmental Impact Assessment.

Height $Y_0 = 329$ m
continuous release rate of the particles from the chimney $Q = 0.00185$ kg/s
wind speed $u = 2$ m/s

speed of release of the carrier gas from the chimney $V_s = 9.5$ m/s.
chimney diameter, $D = 2.2$ m (initially suggested to start the algorithmic procedure)
pressure, $P = 1000$ millibars
temperature of the carrier gas in the chimney, $T_s = 273 + 90 \,°C = 363$ K
atmospheric temperature, $T_a = 273 + 20 \,°C = 293$ K
chimney height $H_r = 28$ m

From the expression (3), we obtain: $\Delta h = 27{,}6$ m

Fig. 5 presents the dependence of maximum particle concentration on the chimney height while Fig. 6 presents the dependence of particle concentration at ground level on the distance, for the olive pomace oil mill studied, along with the effect of climatic conditions and geomorphology. At 'stable atmosphere' conditions, a particle pollution level more than twice the alarm threshold (50 $\mu g/m^3$) was estimated.

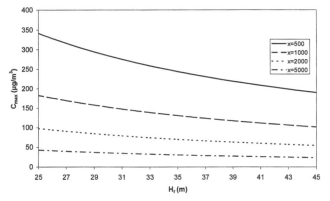

Figure 5: The dependence of the maximum particle concentration on the chimney height for various distance values (stable atmosphere; wind speed $u = 2$ m/s).

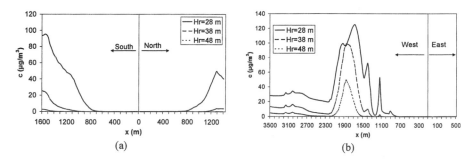

Figure 6: The dependence of the particle concentration at the ground level on the distance for various chimney height values (stable atmosphere; olive pomace oil mill location at $x = 0$; wind speed $u = 2$ m/s); (a) North-South, (b) East-West

## 4. Discussion and Concluding Remarks

The analysis presented above exploits only part of the proposed algorithmic procedure shown in Fig. 1, due to the limited information available form the preliminary study of Environmental Impact Assessment, submitted to the authorities with a view for the granting of license for the operation of an olive pomace oil mill. In practice, it is uncommon to obtain a feedback from the authorities containing optimal construction design based on the same data in order to provide the applicant company with appropriate technical specification to revise the initial study.

The herein suggested procedure enables an iterative process for the design parameter estimation to achieve a balanced tradeoff between sustaining the environmental standards and keeping construction/operating cost of the cyclone/chimney system within acceptable limits, defined by the competition among the companies of the same sector. The implementation presented indicates that these targets can be met through sensitivity analysis of the positive models used, which are transformed to normative ones by selecting those parameter values that produce the desired results. A similar procedure can be followed in the case of water pollution monitoring, as we have proved elsewhere [10].

## References

[1] European Commission 'Managing NATURA 2000 sites – The provisions of article 6 of the 'Habitats' Directive 92/43/EEC', Official Publ. EC, Luxembourg, 2000.

[2] A.C. Caputo, P. M. Pelagagge and F. Scacchia, 'GIS-assisted waste management in a protected area', *Environmental Management and Health* **13**, pp. 71-79 (2002).

[3] A. Marino, M. Pecci and F. Silvestri, 'Remote sensing and GIS techniques for the environmental management of areas exposed to industrial pollution events', *Proc. SPIE – The International Society for Optical Engineering*, pp. 87-93(1999).

[4] G. Senes, and A. Toccolini, 'Sustainable land use planning in protected rural areas in Italy' *Landscape and Urban Planning*, **June**, pp. 104-117 (1998).

[5] *Separation Techniques 2: Gas /Liquid/Solid Systems*. Edited by Larry Ricci and the Staff of Chemical Engineering. Chemical Engineering. McGraw – Hill Publications Co, New York, 1980.

[6] W.L. McCabe and J.C. Smith, *Unit Operations of Chemical Engineering* (6$^{td}$ Edition), McGraw-Hill, 2001.

[7] Coulson & Richardson, *Chemical Engineering*, Volume 6. Second Edition. Chemical Engineering Design. By R.K. Sinnott. Butterworth – Heinemann, Oxford, 1993.

[8] Noel de Nevers, *Air Pollution Control Engineering*, McGraw-Hill, New York, 1995.

[9] *Perry's Chemical Engineering Handbook*, 7$^{th}$ edition, (eds) R.H. Perry, D.W. Green, J.O. Maloney. McGraw-Hill, New York, 1997.

[10] F.A. Batzias, D.K. Sidiras and D. Marinakis, 'A GIS - Based Mapping of Pollution Caused by an Olive Pomace Oil Mill Operating in a NATURA 2000 Protected Area', *Proc. 2nd International Conference on Ecological Protection of the Planet Earth, Bio-Environment and Bio-Culture*, Sofia, Bulgaria, June 2003.

Brill Academic Publishers
P.O. Box 9000, 2300 PA Leiden
The Netherlands

*Lecture Series on Computer*
*and Computational Sciences*
Volume 4, 2005, pp. 894-901

# Contribution to Pre-Selection and Re-Design of Cyanides Destruction Method at Industrial Level for Environmental Protection

F.A. Batzias [1]

Department of Industrial Management and Technology,
University of Piraeus,
GR-134 85 Piraeus, Greece

Received 22 June, 2005; accepted in revised form 23 June, 2005

*Abstract:* Fuzzy multicriteria ranking of pre-selected cyanide destruction methods gave the following order of preference: $SO_2$/Air Oxidation, $H_2O_2$ Oxidation, Alkaline Chlorination, Ozone Treatment, Combined Oxidation, Acidification-Volatilization, Biological Oxidation; the criteria used were capital and operating costs, friendliness to environment, reliability, maturity of technology, and perspective for improvement. Although sensitivity analysis indicates that the proposed method as first in order of preference provides a robust solution, incomparability is observed between certain alternatives when parameter values giving low preferability/resolution/ discrimination are used. On the other hand, experimental evidence is in accordance with the view that Biological Oxidation is a quite promising method although still with low reliability/efficiency at least in the short run.

*Keywords:* Cyanides, environmental protection, biochemical kinetics, fuzzy multicriteria ranking

## 1. Introduction

Although waste reduction and recycling are preferable to the treatment and disposal of wastes, they are not always possible. This is particularly true for many wastes contaminated by cyanide compounds which will, for the foreseeable future, continue to be land disposed [1]. These waste streams arise from different process industries such as those wastes from manufacturing synthetic fiber (acrylonitrile), coal conversion wastes or coking effluents (from the iron and steel industries), electroplating waste, petrochemical wastes, automobile industry waste (from plating shops) and wastes from the processing of precious metals resources by cyanidation. In view of toxicity of cyanides, EU and subsequently national/regional/local authorities have been forced to impose stricter environmental regulations. Without a doubt the best way to get rid of this environmental problem is to replace cyanides with a substitute. Such alternative lixiviants are the halogens and among them the iodine seems to be the most promising, as reported by certain studies [2, 3].

Cyanides are usually characterized/measured as free, weak acid dissociable (WAD), and total. The first category includes hydrogen cyanide (HCN) and cyanide ion in solution ($CN^-$); the proportions of HCN and $CN^-$ in solution are determined by their equilibrium equation, which is highly influenced by pH. The second category includes free cyanides and either weak or moderately strong cyanide complexes of Ag, Cd, Cu, Hg, Ni, Zn and some other metals with similar low dissociations constants. The third category includes all the above mentioned species as well as all strong complexes of Fe, Co, Pt, Au; actually, only cyanate ($CNO^-$) and thiocyanate ($SCN^-$) are excluded.

The most important source of these important species is gold ore leaching by cyanidation to form stable complexes, according to the Elsner equation

$$2\ Au + 4\ NaCN_{(s)} + H_2O + \tfrac{1}{2}\ O_{2\,(aq)} \rightarrow 2\ Na[Au(CN)_2]_{(aq)} + 2\ NaOH_{(aq)}$$

This leaching process generates about 620 million tons of hazardous solid and liquid waste worldwide per year. Industrially generated waste waters are usually stored in unsealed tailing ponds. The result is uncontrolled release of cyanide by volatilization, photodegradation, chemical oxidation and, to a lesser extent, by microbial oxidation. The waste can contain up to 120 ppm free cyanide, up to 400 ppm total cyanide and various cyanide complexed heavy metals (esp. copper). This represents a potential risk for

---

[1] Corresponding author. fbatzi@unipi.gr

the environment, because of leakages, dam failures and discharges into adjacent rivers. Another huge problem in developing countries is caused by so called "small scale mining", which summarizes all mining activities, which do not underlie public authority control. Waste waters generated in this way are discharged in the environment without further clarification processes. Especially for these unregulated informal mining activities a need exists for a low tech and economical waste treatment procedure, like wetland phytoremediation (see Fig. 1).

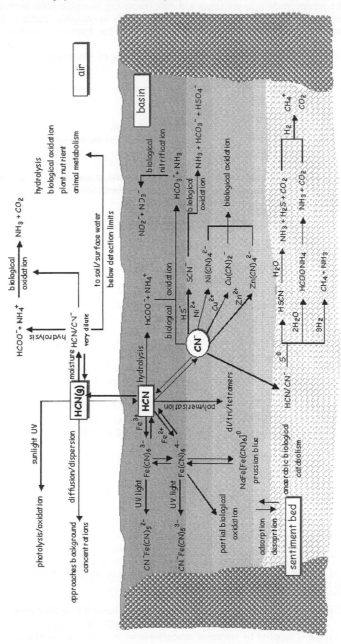

Figure 1: The main physical and chemical processes responsible for the transformation/destruction of cyanides in the natural environment; the same processes are met in both, extensive and intensive treatment (e.g. unsupervised open wetlands and supervised closed tanks, respectively).

In countries, where land availability is very low, due to intensive industrial activities and/or high density of population and/or development related to tourism (like the region of Chalkidiki in North Greece, where there are mines of auriferous pyrites), the possibility of using a nature-imitating method as the main treatment process for cyanides destruction is negligible. On the other hand, the industrial-scale methods currently in operation are met in various installations over all the world in quite different versions and combinations, which are not transparent as they incorporate proprietary sub-processes. The aim of the present study is to pre-select and rank (according to multicriteria analysis, based on fuzzy reasoning/algebra to count for uncertainty) these methods as categories rather than as specific processes usually recognized under the name of the company that has developed each one of them.

## 2. Pre-selection and Final Setting of Alternatives

The criteria fi ($i = 1, ..., 6$) used for pre-selection and final ranking of cyanides destruction methods are the following:

f1. Capital or fixed cost                 f4. Reliability
f2. Operating or variable cost            f5. Maturity of technology
f3. Friendliness to environment           f6. Perspective of improvement

The methods pre-selected according to these criteria, on the additional condition that (a) they have been used in large scale for industrial applications and (b) relevant information is available to make comparative evaluation feasible, are mentioned below.

*Biological Oxidation* (BO); this method can be described by the following simplified biochemical reactions:
$$CN^- + \tfrac{1}{2}O_2 \text{ [enzymatic catalysis]} \rightarrow CNO^- \quad CNO^- + 3H_2O \text{ [enzymes]} \rightarrow NH_4^+ + HCO_3^- + OH^-$$
Further microbial action converts ammonium nitrate:
$$NH_4^+ + \tfrac{3}{2}O_2 \rightarrow NO_2^- + 2H^+ + H_2O \qquad NO_2^- + \tfrac{1}{2}O_2 \rightarrow NO_3^-$$
This treatment is achieved by feeding wastewater to a system of rotating biological contactors (RBCs), which consists of a shaft of circular plastic elements revolving partly submerged in a contour-bottomed tank; when the discs, which are spaced such that wastewater can enter between them, are rotating out of the tank, air enters the spaces while the liquid trickles out over films of biological growth attached to the media; the active biomass consumes nutrients in solution and converts cyanide to ammonium while a second RBCs system in series is used for final conversion to nitrate. Heavy metals and excess biomass is precipitated by the aid of ferric chloride enriched with a polymeric agent; the effluent is polished by sand filtering before its final discharge to the environment.

*Alkaline Chlorination* (AC); this method can be described by the following two-step chemical reaction:
$$Cl_2 + CN^- \rightarrow CNCl + Cl^- \qquad CNCl + H_2O \rightarrow CNO^- + Cl^- + 2H^+$$
If there is only a slight excess of chlorine, cyanate is hydrolyzed to yield ammonium, according to the following catalytic reaction:
$$CNO^- + 3H_2O \text{ [Cl}_2 \text{ catalyst]} \rightarrow NH_4^+ + HCO_3^- + OH^-$$
In the presence of excess chlorine, further chlorination takes place until complete oxidation is achieved:
$$2NH_4^+ + 3Cl_2 \rightarrow N_2 + 8H^+ + 6Cl^-$$
The AC method contributes also to the oxidation of thiocyanates, which usually coexist in the same solution, leading to increase of operating cost due to excessively high consumption of chlorine:
$$SCN^- + 4Cl_2 + 5H_2O \rightarrow CNO^- + 8Cl^- + SO_4^{2-} + 10H^+$$
The reactions are carried out at pH>10.0 to ensure complete hydrolysis of cyanogen chloride (CNCl) to cyanate ($CNO^-$); with the advancement of the cyanide destruction, metals previously complexed with cyanide (e.g., Cu, Zn, Ni) are precipitated as metal-hydroxide compounds; last, as the above reaction generate an acid solution, due to presence of $H^+$, alkalinization is necessary by means of lime or caustic soda to prevent HCN release to atmosphere.

*SO₂ / Air Oxidation* (SO); this method can be described by the following simplified reaction:
$$CN^- + SO_2 + O_2 + H_2O \text{ [Cu}^{2+} \text{ catalyst]} \rightarrow CNO^- + SO_4^{2-} + 2H^+$$
The required oxygen is supplied by scattering atmospheric air into the reaction vessels. The required copper is supplied as $CuSO_4 \cdot 5H_2O$ after preparing an aquatic solution of about 10 to 50 mg/L; the required $SO_2$ is supplied either as liquid sulphur dioxide or as sodium metabisulphite ($Na_2S_2O_5$).

*H₂O₂ Oxidation* (HO); this method can be described by the following simplified reactions:

$$CN^- + H_2O_2 \text{ [Cu}^{2+}\text{ catalyst]} \rightarrow CNO^- + H_2O \quad CNO^- + 2H_2O \rightarrow NH_4^+ + CO_3^{2-}$$

The first of them takes place in high pH, which is already present in gold processing effluents, while the second occurs at slightly acidic aquatic environment, without the need for cupric ions that catalyze only the cyanate formation.

*Combined Oxidation* (CO); which is a combination of the SO and HO methods, described by the following reactions:

$$H_2O_2 + H_2SO_4 \rightarrow H_2SO_5 + H_2O \text{ (destructive agent preparation)}$$
$$CN^- + H_2SO_5 \rightarrow CNO^- + H_2SO_4 \text{ (destruction of cyanides)}$$
$$CNO^- + 2H_2O \rightarrow NH_3 + CO_2 + OH^- \text{ (cyanate hydrolysis)}$$

*Ozone Treatment* (OT); this method can be described by the following chemical reactions:

$$CN^- + O_3 \rightarrow CNO^- + O_2 \quad 3CN^- + O_3 \rightarrow 3CNO^- \quad 2CNO^- + 3O_3 + H_2O \rightarrow N_2 + 2HCO_3^- + 3O_2$$

If the waste contains sulphur compounds, then SCN⁻ are most likely to result as intermediates which subsequently may give HCN; pH in the vicinity of slightly less than 11.0 is needed; as regards the capital cost, it is rather increased for large capacity installations in comparison with other methods where the scale economies are higher.

*Acidification – Volatilization* (AV); this method runs directly counter to the principle of keeping pH of cyanide solutions in the alkaline range so that toxic hydrogen cyanide (HCN) gas will not be released: initially, pH is lowered by the addition of sulphuric acid so that HCN gas is formed within a sealed mixing vessel; the liquid stream leaving the reactor is stripped with a current of air in a packed column; the HCN- laden air is absorbed in a second column containing a downward-flowing stream of caustic soda, forming sodium cyanide, which returns to the leaching process:

$$HCN_{(g)} + NaOH_{(aq)} \rightarrow NaCN_{(aq)} \quad CN^-_{(aq)} + H^+_{(aq)} \rightarrow HCN_{(g)}$$

One potential drawback relates to the safety of operators of such plants because HCN gas, which is highly toxic, is present in concentrated form in the plant pipeline system, though no incident has been recorded in some 8 plants that have operated using this technology.

## 3. Fuzzy Multicriteria Ranking of Alternatives

The multicriteria problem that must be solved for ranking the cyanides destruction alternatives is: $\text{Max}\{f_1(a),\ldots, f_K(a) \mid a \in A\}$ where A is the set of $T$ alternatives and $f_i$, $i=1,\ldots,K$, are the $K$ criteria under which the alternatives are evaluated. Multi Attribute Decision Making (MADM), a special division of Multi Criteria Decision Making (MCDM), is mostly used in the comparison of several distinct alternatives, as it examines a finite set of alternatives. It consists basically of two steps: (i) the formulation of the decision matrix ($K$ x $T$), where each element $x_{kt}$ is the evaluation of alternative $a_t$ with respect to criterion $f_k$ (ii) the ranking of the alternatives, according to the rules of the selected MADM method. The preference measurement in these methods is the simple sum of the scores $x_{kt}$ multiplied by the weight of the particular criterion $w_k$, over all criteria $K$. To overcome the basic assumption that complete compensation between attributes takes place, several outranking methods have been developed that allow partial compensation. Also, they allow for incomparability ($aRb$) and weak preference ($aQb$) between the alternatives $a$, $b$, in addition to the strict preference ($aPb$) and indifference ($aIb$) that the 'classical' methods are based on. PROMETHEE [4] is such an outranking method, generally preferred for its simplicity, clearness and stability.

The notion of a generalized criterion is used to construct an outranking relation by defining the preference index $\Pi(a,b) = \Sigma w_i P(a,b)/ \Sigma w_i$ as the weighted average of the preference functions $P_i$, that quantifies the preference of the decision maker of alternative $a$ over $b$, taking into consideration all the criteria. In terms of topology, the preference index values can be represented as a valued outranking graph, the nodes of which are the alternatives. By summing the column elements in each row of the outranking relation matrix, the flow leaving each node is obtained, which shows its outranking character, while by summing the row elements in each column, the entering flow is obtained for each alternative, which shows its outranked character. By considering the leaving and entering flows, as well as the fact that the higher the leaving flow and the lower the entering flow the better the alternative, the partial preorder (PROMETHEE I) is obtained. Although the partial preorder carries more realistic information, sometimes the total preorder (PROMETHEE II) is requested to avoid any incomparabilities; this preorder is induced by the net flows, i.e. the difference between the leaving and

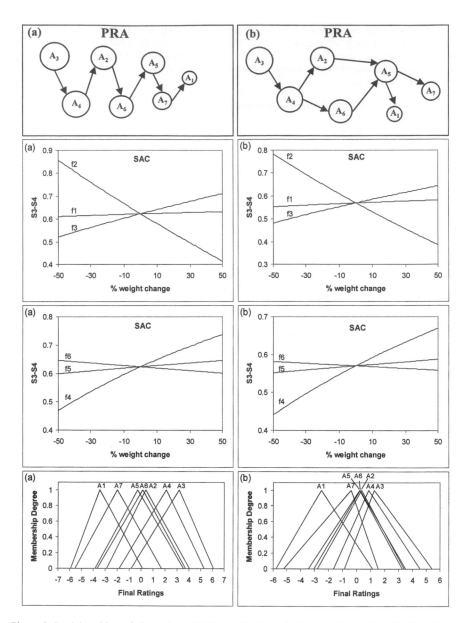

Figure 2: Partial ranking of alternatives (PRA), sensitivity analysis for each criterion (SAC) and total ranking of alternatives (TRA), at (a) high preferability level with low q, p values and (b) low preferability level with medium/high q, p values. The arrow → means 'better than'.

the entering flows. The generalized criterion used is a piecewise linear preference function $P = H(d) \in [0, 1]$, where $d$ is the difference of the evaluation of two alternatives $a$, $b$. The parameters of $H(d)$ are an indifference threshold $q$, the greatest value of $d$ below which there is indifference, and a preference threshold $p$, the lowest value of $d$ above which there is strict preference – the interval between $q$ and $p$ can be considered as the weak preference region.

The evaluations of the alternatives w.r.t. each criterion in the decision matrix are given by the $n$ experts, who participated within an *ad hoc* modified DELPHI procedure. There are two modes of performing fuzzy multicriteria analysis, as regards the synthesis of experts' evaluations: (a) a consensus is formed form all experts' fuzzy data to enter as input, and (b) a consensus is formed based on each expert's fuzzy output; in the first case, the algorithmic procedure runs once while in the second (followed herein) it runs $n$ times.

As fuzzy sets deal realistically with unsharp figures, fuzzy algebra was integrated into the PROMETHEE algorithm. In particular, L-R fuzzy triangular fuzzy numbers (with their arithmetic operations) established by Dubois and Prade [5] were used for the decision matrix and criteria weights. There are many ways to conclude a partial or complete preorder from the resulting fuzzy sets, and these usually refer to defuzzification methods. Two of the most often used ones are: (a) the center of gravity of the fuzzy set of the alternative, (b) the Tseng and Klein [6] method (used herein) that compares the fuzzy sets of the alternatives in pairs by calculating the (crisp) dominating areas of them (partial preorder); then a summation of the elements of each row (alternative) of the domination matrix gives a measure of the strength of each alternative that leads to a complete preorder.

The results shown in Fig. 2 are at (a) high preferability level, with low parameter q, p values (q = 0.5, p = 1.0), and (b) low preferability level, with medium/high parameter q, p values (q = 1.5, p = 3.0). The partial ranking of alternatives (PRA), according to fuzzy PROMETHEE I, reveals that incomparability is observed only at low preferability level (b) where resolution is low as it is seen in total ranking of alternatives (TRA), according to fuzzy PROMETHEE II. In the PRA diagrams, the area of each circle is proportional to the crisp number (named ranking quantifier – S) which indicates the corresponding relative value in the ranking vector. The TRA output is shown as a set of triadic fuzzy numbers in the usual form of triangles to reveal the common parts which constitute a measure of overlapping along the relative scale of preorder (i.e. the horizontal axis), especially for lower membership function values.

## 4. Discussion of Results and Concluding Remarks

At high preferability level, the range of ranking quantifiers is 4.48 - 1.30 = 4.18, i.e. broader in comparison with the range corresponding to low preferability level 5.13 - 1.57 = 3.56, giving grounds for higher resolution. In Table 1, the vertical distance SVD between successive S-values gives a measure of local robustness at each level while the horizontal distance SHD gives a measure of difference of discrimination ability between the two levels. On the other hand Table 2 contains numerical information about the sensitivity analysis diagrams of Fig. 2, where the difference S3 - S4 remains positive, implying the absolute robustness of the solution, at least as regards the proposed alternative as most preferable.

Table 1: Sensitivity analysis of ranking quantifiers at intra-preferability and inter-preferability levels (SVD and SHD, respectively)

| Rank | Alternative | High Preferability | | Low Preferability | | |
|------|-------------|-----|------|-----|------|------|
| | | S | SVD | S | SVD | SHD |
| 1 | A3 (SO) | 5.48 | | 5.13 | | |
| | | | 0.62 | | 0.57 | 0.05 |
| 2 | A4 (HO) | 4.86 | | 4.56 | | |
| | | | 0.99 | | 0.81 | 0.18 |
| 3 | A2 (AC) | 3.87 | | 3.75 | | |
| | | | 0.33 | | 0.09 | 0.24 |
| 4 | A6 (OT) | 3.54 | | 3.66 | | |
| | | | 0.19 | | 0.18 | 0.01 |
| 5 | A5 (CO) | 3.35 | | 3.48 | | |
| | | | 1.26 | | 1.13 | 0.13 |
| 6 | A7 (AV) | 2.09 | | 2.35 | | |
| | | | 0.79 | | 0.78 | 0.01 |
| 7 | A1 (BO) | 1.30 | | 1.57 | | |

Table2: Absolute and relative sensitivity according to each criterion, at both preferability levels
examined (see also Fig. 2 for the respective graphs)

| Criteria | High Preferability | | Low Preferability | |
|---|---|---|---|---|
| | Absolute sensitivity | Relative sensitivity | Absolute sensitivity | Relative sensitivity |
| f1 | 0.020 | 0.045 | 0.029 | 0.073 |
| f2 | 0.443 | 1.000 | 0.396 | 1.000 |
| f3 | 0.190 | 0.429 | 0.162 | 0.409 |
| f4 | 0.269 | 0.607 | 0.230 | 0.581 |
| f5 | 0.047 | 0.106 | 0.035 | 0.088 |
| f6 | 0.044 | 0.099 | 0.022 | 0.056 |

The grades each expert assigned to the multicriteria matrix were statistically tested, after defuzzification, for agreement (according to each criterion, taken separately) by means of the Kendall's coefficient of concordance, applied in both versions, simple and top-down. Probably the most surprising finding is that BO was ranked first when only criteria f6 and f2 were applied separately; this means that Biological Oxidation, although last according to multicriteria analysis is considered first as regards 'perspective for improvement' and 'operating cost'. For testing experimentally the view that this method is promising as a cheap alternative, possibly to be used in developing countries instead of wetland phytoremediation (see Introduction), $CN^-_{WAD}$ solution was treated within a 2 L batch reactor by using waste activated sludge; both, $CN^-_{WAD}$ decrease, due to biological destruction, a ammonia increase as a reaction product were measured by means of a Hack spectrophotometer, according to techniques described in [7, 8, 9]. The results, depicted as points in Fig. 3, seem to follow a first order kinetic equation with constant term which is the asymptote:

$$\frac{dC}{dt} = k(C_\infty - C) \text{ or } C = C_\infty - (C_\infty - C_0)e^{-kt} \text{ , where } C_0, C, C_\infty \text{ is the cyanide concentration (in mg/L)}$$

at time 0, t, $\infty$ (in hours), respectively.

In the case of cyanides destruction, when all measurements are taken into account, the parameter values and the standard error of estimates SEE estimated through non-linear regression are $C_0 = 19.49$, $C_\infty = -0.558$, k = 0.0847, SEE = 0.893; while if the $C_\infty$-value is preset at zero (forced/constrained regression) then $C_0 = 19.57$, k = 0.0903, SEE = 0.868. When the first-stage measurements (that correspond to acclimation/establishment/development of active biomass) are excluded, $C_0 = 21.74$, $C_\infty = 0.40$, k = 0.112, SEE = 0.408; if a priori $C_\infty = 0$, then $C_0 = 21.15$, k = 0.1035, SEE = 0.406.

In the case of ammonia production, when all measurements are taken into account, the parameter values and the SEE estimated through non-linear regression are $C_0 = 0.59$, $C_\infty = 15.10$, k = 0.0694, SEE = 0.805; while if the $C_0$-value is preset at zero then $C_\infty = 14.63$, k = 0.0796, SEE = 0.842. When the first-stage measurements are excluded, $C_0 = -3.85$, $C_\infty = 13.66$, k = 0.122, SEE = 0.459; if a priori $C_0 = 0$, then $C_\infty = 14.66$, k = 0.0792, SEE = 0.815.

These results show that there is satisfactory fitting of the kinetic model to experimental results, especially when the medium/long run measurements (which determine reaction's efficiency) are taken into account, at lab-scale level; this means that the problem of low reliability (BO is ranked last according to this criterion) is actually a scale-up problem, which is expected to be solved with the advancement of bioengineering.

In conclusion, it seems that the $SO_2$/Air Oxidation method is ranked first, due to its dominance mainly in capital cost, reliability and maturity of technology, and secondarily in friendliness to environment; the $H_2O_2$ Oxidation Method is ranked second as it has been assigned very good (but not maximal) grades according to all criteria, except for the last one, which does not influence the result significantly, because of its minimal weight. The Biological Oxidation method, although last in total ranking, seems to be the most promising for the near future because its field (Bioengineering/Biotechnology) is rapidly growing up and out experimental results are in agreement with the view that at lab-scale level (and, according to available information, at semi-pilot plant level too) the important criteria of reliability and technical efficiency are satisfied.

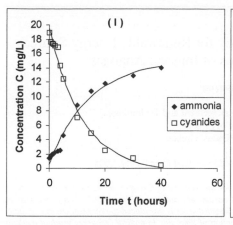 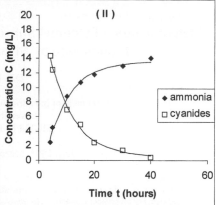

Figure 3: Kinetics of biological destruction of cyanides in a lab-scale batch reactor within a 40-hours period, when (I) all measurements are taken into account, (II) first-stage measurements (that correspond to acclimation/establishment/development of active biomass) are excluded.

## References

[1] J.R. Parga, S.S. Shukla, F.R. and Carrillo-Pedroza, Destruction of cyanide waste solutions using chlorine dioxide, ozone and titania sol, *Waste Management* **23** 183-191 (2003).

[2] F.A. Batzias and D.K. Sidiras, Wastewater Treatment with Gold Recovery through Adsorption by Activated Carbon, *Water Pollution IV: Modelling, Measuring and Prediction*, Ed. Brebbia C.A., Series: Progress in Water Resources, Vol 3, WIT Press, Southampton, 143-152 (2001).

[3] P.A.M. Teirlinck, F.W. Petersen, Factors influencing the adsorption of gold-iodide on to activated carbon, *Separation Science & Technology* **30** 3129-3142 (1995).

[4] J.P. Brans, Ph. Vincke, B. Mareschal, How to select and how to rank projects : The PROMETHEE method. *European Journal of Operational Research* **24** 228-238 (1986).

[5] D. Dubois, H. Prade, Operations on fuzzy nymbers, Innternational Journal on Systems Science, **9(6)** 613-626 (1978).

[6] T.Y. Tseng, C.M. Klein, New algorithm for the ranking procedure in fuzzy decisionmaking, *IEEE Transactions on Systems, Man & Cybernetics* **19(5)** 1289-1296 (1989).

[7] D.M. White, W. Schnabel, Treatment of cyanide waste in a sequencing batch biofilm reactor, *Water Research* **32(1)** 254-257 (1998).

[8] D.J. Barkley, J.C. Ingles, A rapid colorimetric, solvent extraction procedure: the determination of cyanide in gold mine effluents and receiving waters. Research Report R-221. Canadian Department of Energy and Mines, Mines Branch, Ottawa, Canada, 1970.

[9] *Hack Spectrophotometer Handbook DR /4000*, 255-264 & 497-545 (1997).

Brill Academic Publishers
P.O. Box 9000, 2300 PA Leiden
The Netherlands

*Lecture Series on Computer
and Computational Sciences*
Volume 4, 2005, pp. 902-909

# Determination of Optimal Subsidy for Renewable Energy Sources Exploitation by Means of Interval Analysis

D.F. Batzias[1]

Department of Industrial Management and Technology,
University of Piraeus,
GR-134 85 Piraeus, Greece

Received 8 July, 2005; accepted in revised form 9 July, 2005

*Abstract: In this work, the optimal subsidy $I_{opt}$% for the Renewable Energy Sources Exploitation (RESE) is determined as a function of economo-technical parameters and control variables, when a conventional energy saving investment is planned. An innovative parameter is introduced to represent change of State's participation to investment as an increasing function of time, due to natural resources gradual depletion implying excessive mining cost. An alternative model for estimating $I_{opt}$ is also presented in the discussion section taking into consideration the part of subsidy the users/customers return directly or indirectly to the state. Interval analysis used to count for uncertainty seems to be a valuable technique, provided that a correction is performed within each arithmetic stage to avoid overestimation, otherwise this technique may lead to a rather conservative policy. Sensitivity analysis is performed based on input data representative for the Greek market and the relevant legislation that sets the rules for subsidizing within the EU official policy for supporting RESE.*

*Keywords:* subsidy, optimization, renewable energy sources, interval analysis, interval algebra, sensitivity.

## 1. Introduction

Subsidy is an economic benefit granted by the government or a public sector agency to the producer or the consumer of a product or service intended to make its price lower than it otherwise would be; this benefit will also in general have the effect of raising the expected profit of the recipient above the level it would otherwise have reached. The subsidy can be direct or indirect (e.g. a cash grant or low-interest export credits guaranteed by a governmental agency, respectively); as it may somehow distort free competition at national and international level, certain agreements have been made to discriminate between permissible and non-permissible subsidies.

There is no agreed definition of energy subsidies among European Union (EU) Member States. The term may include cash transfers paid directly to producers, consumers and related bodies, as well as less transparent support mechanisms, such as tax exceptions and rebates, price controls, trade restrictions, planning consent and limits on market access. It may also cover government failure to correct market imperfections, such as external costs arising from energy production or consumption. This results in a wide range of economic estimates and confusing policy arguments.

Beyond the annual report of direct state aid for the coal industry, there is no harmonized reporting mechanism for energy subsidies. In two reports, the European Parliament and the European Commission [1,2] have attempted to provide a full audit of energy support in the EU 15 (corresponding to the pre-May 1 2004 EU Member States). Both reports are snapshots based upon best available data, rather than structured ongoing reviews, and have not been updated since. A recent report [3] has synthesized data from a range of sources to estimate the size of support to the energy sector within the EU 15. Total subsidies (excluding external costs) were estimated to be in the order of EUR 29 billion in 2001.

Directive 2001/77/EC of the European Parliament and the Council 'on the promotion of electricity produced from renewable energy sources in the internal electricity market' recognizes that (a) the potential for the renewable energy sources exploitation (RESE) is underused in the Community, and (b) the promotion of RESE contributes to environmental protection and sustainable development, while can also (i) create local employment, (ii) have a positive impact on social cohesion, (iii) contribute to

---

[1] Corresponding author. E-mail: fbatzi@unipi.gr

security of supply, and (iv) make it possible to meet Kyoto targets more quickly. As a result, it accepts the operation of different mechanisms of support for RESE at national level by the member states,

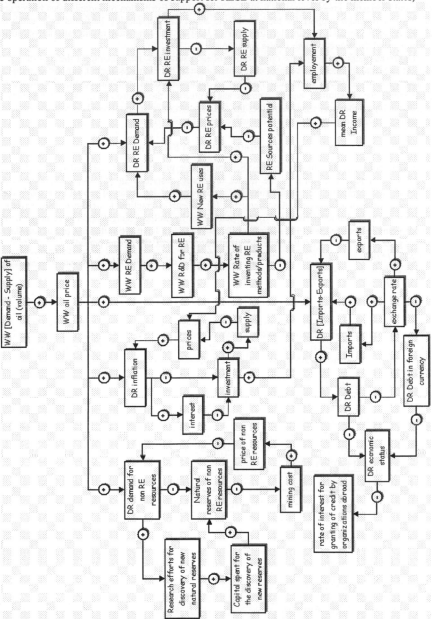

Figure 1: Part of an ontology built on the kernel of the 'world wide (WW) oil price' under the form of a digraph to show the influence on Renewable Energy (RE) issues (DR: Domestic/Regional).

including green certificates, investment aid, tax exemptions or reductions, tax refunds and direct price support schemes; moreover it states that an important means to achieve the aim of the Directive is to guarantee the roper functioning of these mechanisms until a Community framework is put into operation, in order to maintain investor confidence.

Certain cases where a subsidy may apply to support activities for the promotion of RESE can be found in Fig.1; e.g. the increase of demand for renewable energy (RE) may lead to corresponding R&D effort and productive investment with better spatio-temporal distribution of capital/expenditures under the oriented financial support for the part of the State, i.e. the subsidy performs as a catalyst to accelerate an action that is expected to occur at least in the long run, if the rational of positive economics and the ceteris *paribus clause* are adopted.

The present work deals with the optimization of such subsidies, introducing for the first time in the relevant technical literature the concept of differentiation of the State's economic support as a function of time, due to depletion of natural reserves of non-renewables and the continually increase of requirements for substituting energy sources and protecting the environment; since the parameter-values which play the most significant roles for optimization refer to future trends within the respective forecasting procedure, interval analysis is used to count for uncertainty.

## 2. Methodological Framework

To determine optimal subsidy for RESE, a normative model is used, derived from a positive / mechanismic model, according to the model taxonomy described in [4]. We can illustrate this derivation by determining the optimal subsidy $I_{opt}$, expressed as a fraction of capital invested in RESE equipment within a factory [5]. Assuming that (a) at least a fraction $K$ of the energy saving (in monetary units) is deducted annually by the State from its welfare budget and (b) re-investment of annual 'gains' is realized with the common interest rate $i$, we have the following periodical (herein, annual) gains $U_i$ ($i = 1, 2, ..., t$) for the State, estimated at the end of a $t$ years period:

$U_1 = K \cdot F \cdot (1+i)^{t-1}$

$U_2 = K \cdot (1+b) \cdot F \cdot (1+i)^{t-2} \cdot (1+f)$

$U_3 = K \cdot (1+b)^2 \cdot F \cdot (1+i)^{t-3} \cdot (1+f)^2$

$: : : : : : : : : : : :$

$U_t = K \cdot (1+b)^{t-1} \cdot F \cdot (1+i)^{t-t} \cdot (1+f)^{t-1}$

where $F$ is the first year energy cost saving, $f$ is the rate of $F$-icrease, $b$ is the rate of $K$-increase (i.e. a proposed measure of differentiation of the State's economic support, as a function of time), and $t$ the time periods (dimensionless) taken into account. By transforming the right hand part of the above expressions to show that they are terms of a geometric series, we obtain:

$U_1 = K \cdot F \cdot (1+i)^{t-1}$

$U_2 = K \cdot F \cdot (1+i)^{t-1} \cdot \dfrac{(1+b) \cdot (1+f)}{1+i}$

$U_3 = K \cdot F \cdot (1+i)^{t-1} \cdot \left( \dfrac{(1+b) \cdot (1+f)}{1+i} \right)^2$

$: : : : : : : : : : : : :$

$U_t = K \cdot F \cdot (1+i)^{t-1} \left( \dfrac{(1+b) \cdot (1+f)}{1+i} \right)^{t-1}$

$$U = \sum_{i=1}^{t} U_i = K \cdot F \cdot (1+i)^{t-1} \cdot \left[ 1 + \frac{(1+b)(1+f)}{1+i} + \left( \frac{(1+b)(1+f)}{1+i} \right)^2 + ... + \left( \frac{(1+b)(1+f)}{1+i} \right)^{t-1} \right] \tag{1}$$

The expression within the brackets is a geometric series of the type $1 + x + x^2 + ... + x^{t-1}$, the sum of which is given by $(x^{t-1}x - a)/(x - 1)$, where $a = 1$ and $x = (1+b)(1+f)/(1+i)$, i.e., the first term and the constant multiplier (or ratio), respectively. Thus, equation (1) is rewritten as

$$U = K \cdot F \cdot (1+i)^{t-1} \cdot \frac{\left( \dfrac{(1+b)(1+f)}{1+i} \right)^t - 1}{\dfrac{(1+b)(1+f)}{1+i} - 1} \tag{2}$$

On the other hand, we can estimate the opportunity cost as the potential loss $Y$ that is the value of the alternatives or other opportunities which have to be foregone in order to subsidize an energy saving investment of initial capital $S$, with an amount of money $IS$. If $r$ is the return on the best alternative investment (called 'the second best' in comparison with the first best for the State, which is the subsidized fraction $I$), then $Y$, expressed at the end of a $t$ years period, is given by the following relation:

$Y = I \cdot S \cdot (1+r)^t \tag{3}$

Obviously, the optimal value of $I$ is obtained for $U = Y$, so neither the State nor the investor make a surplus profit causing a corresponding loss to the other part (condition of equilibrium). From the equations (2) and (3), we obtain

$$I_{opt} = \frac{K \cdot F \cdot (1+i)^{t-1}}{S \cdot (1+r)^t} \cdot \frac{\left(\frac{(1+b)(1+f)}{1+i}\right)^t - 1}{\frac{(1+b)(1+f)}{1+i} - 1} \quad (4)$$

To count for uncertainty, interval numbers are used, which have the general form of an ordered pair of real numbers $[a, b] = \{x | a \leq x \leq b\}$. The basic arithmetic operations are:

$$[a, b] + [c, d] = [a + c, b + d], \ [a, b] \cdot [c, d] = [min(ac, bd, bc, bd), max(ac, ad, bc, bd)]$$
$$[a, b] - [c, d] = [a - d, b - c], \quad [a, b] / [c, d] = [a, b] \cdot [1/d, 1/c] \mid 0 \notin [c, d]$$

As interval arithmetic is a generalization or an extension of real arithmetic, we can write rounded interval arithmetic subroutines for a given computer system based on these formulae. Such subroutines produce intervals which contain not only the infinite precision results but also the unrounded machine-architect results. However, these bounds computed by interval arithmetic become overestimated, especially when the number of intermediate arithmetic calculations carried out to reach the final result is big. Several algorithms including nested form, centered form, and subdivision have been developed to prevent the growth of interval width [6]. Interval analysis has been used, *inter alia* , in (a) the application of intelligent methods to the design/testing of measurement models [7], (b) worst- and best-case (WBC) analysis [8], (c) constrained optimization [9], and (d) multi-criteria optimal design [10]. A neural network approach has been also proposed for improving interval algebra [11].

## 3. Implementation

For the implementation of the methodological framework described above, the Greek scheme for supporting/promoting investment in RESE is adopted. According to this scheme, (i) a cash grant, (ii) a loan interest rate subsidy, (iii) a leazing subsidy of up to 40% is offered by the State to new investors; for the prefectures of Xanthi, Rodopi, Evros, Kavala, and Drama (all in Northestern Greece), (iv) additional (percentage) units are applicable for investment proposals submitted within a relatively sort period after the Investment Law was published in order to create economic 'takeoff' conditions in this region which has high unemployment and consequent low economy is characterized by a relatively high marginal propensity to consume (MPC); this means that the multiplier, given as 1/[1-(MPC)], is also high contributing to significant increase of regional income/demand.

The parameter values (for $i, f, r, t, b$) used for determining $I_{opt}$ were estimated as intervals after interviewing experts in welfare economics and RESE. A subjective semi-quantitative method was followed by adopting initial values from [4] to form a reference set and subsequently letting each expert to change initial values towards a new dynamic equilibrium, giving lower and upper bound for each parameter; this technique was followed in three steps according to a modified Delphi method. The results were rounded properly (e.g. the rates-parameters $i, f, r, b$ to half percentage unit), within the third step of the Delphi method to achieve maximum consensus and increase comprehension of the corresponding dependence/sensitivity diagrams. The parameter values estimated in this way are $i = (0.025 - 0.045)$, $f = (0.050 - 0.080)$, $r = (0.040-0.070)$, $b = (0.010-0.030)$, $t = (15-25)$. Fig 2 contains the diagrams of dependence of $I_{opt}$ on i, r, f, t, when the ratio F/S is given as an interval (0.110-0.140), dimensionless, implying a payback period of 8 years) estimated to be valid for most medium size RESE investments in industrial applications. In each diagram, the parameter central values are used, except for the parameter represented on the horizontal axis and the ratio $F/S$ which is always taken as an interval. From the algebraic point of view, the resulting diagram in interval form (i.e. the area in grey) coincides with the diagram obtained by carrying out simple arithmetic calculus twice, i.e. for the upper and the lower bound of the ratio $F/S$, separately, forming the upper and the lower curve, respectively. From these four diagrams, it is shown that $0.15 < I_{opt} < 0.40$ which means that the results are in accordance with the Greek legislation and practice. This is not the case for the upper bound of the resulting interval when both F/S and another parameter are taken as input intervals to carry out the calculation by means of interval algebra; the upper curve always exceeds not only the common upper limit of 40% set by the Greek legislation for most prefectures but also the limit of 45%, applying for the 5 prefectures in northestern Greece, already mentioned above. This is a problem of overestimation and a simple solution is to take as a more reliable result the centered curve, which is actually very close to the accepted range (see Fig. 3).

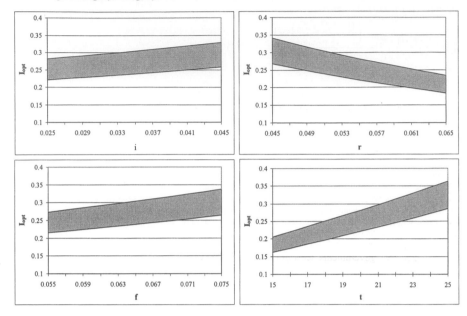

Figure 2. Dependence of optimum subsidy on *i, r, f, t,* successively, when *F/S* is given as an interval (0.11-0.14, dimensionless).

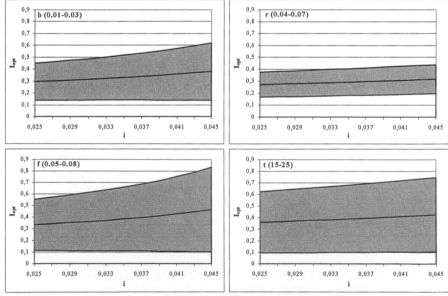

Figure 3. Dependence of optimum subsidy on rate of interest *I*, when both *F/S* and the parameter shown in the upper left corner of each diagram are given as intervals.

## 4. Discussion and Concluding Remarks

By performing sensitivity analysis as regards the influence of b-value on $I_{opt}$. A peculiar behavior of the resulting upper curve is shown (see Fig. 4): although it is within the accepted range when only F/S is set in interval form, it is extremely high and concave when f is also taken as an interval; in the latter case, the centered curve remains nearly within the accepted interval but if a sudden increase is given to the upper bound input the model does not function as there is not enough time to counterbalance this increase by changing the ratio *F/S*, that might be achieved through reengineering, leading to new investment.

The ranking of influence each parameter *P* exerts on $I_{opt}$ is shown by performing first and second order sensitivity analysis in the region $P_o \pm 0.25P_o$, where $P_o$ the centered or mean value of each parameter; to achieve comparability, the dimensionless ratio $p = P / P_o$ is used, in which case, all the first order sensitivity curves pass through the same point $p = 1$ or $P = P_o$; the mean absolute slope in the first order or the corresponding derivative in the second order sensitivity analysis gives this ranking, which is $t > r > f > i > b$, where the symbol '>' means 'more influential than' (the corresponding weighted order is $1.00 > 0.90 > 0.58 > 0.23 > 0.18$, only as an approximation, because the corresponding partial derivatives are not linear functions in respect to each parameter).

Further to the capital subsidies and the fiscal mechanisms which can be both characterized as direct methods (DM) of subsidizing, there are the following methods.

(a) Fixed feed-in tariffs (FT), which have been widely and successfully deployed throughout Europe to support renewable technologies, most notably in Denmark, Germany and Spain. According to this also DM, governments set a price at which the country's electricity supply companies must purchase all renewable energy delivered to the distribution grid; price premiums are passed on to consumers under the form of higher electricity prices.

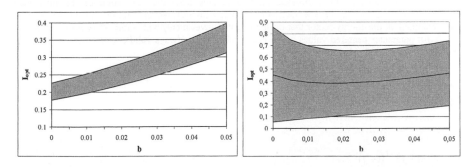

Figure 4. Monoparametric sensitivity analysis indicating the dependence of optimum subsidy $I_{opt}$ on the rate of increase *b* of *K* when (I) only *F/S* and (II) both *F/S* and *f* are given as intervals; the upper bound of the interval in (II) is concave.

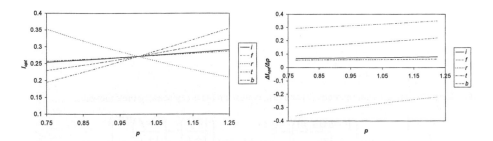

Figure 5. First (I) and second (II) order sensitivity analysis of optimal subsidy $I_{opt}$ in the region $P_0 \pm 0.25P_0$, where $P_0$ is the mean value of parameter *P* (that is *i, f, r, t, b*); on the horizontal axis, the dimensionless values $p = P/P_0$ are quoted.

(b) Purchase obligations (PO), which set targets for consumption of electricity (usually percentage based) that should be sourced from a certain fuel source. This mechanism has been deployed for renewables and combined heat and power in several EU Member States. Energy distribution companies must prove the origin of purchase, pay a penalty or produce the required amount themselves, creating an artificial demand and price premium for renewable generation. If the overall system target cannot be met, prices rise until new market entrants and inventors are attracted. Tradable certificates often accompany such schemes. The cost of this subsidy is borne by consumers.

(c) Competitive tender (CT), which invites producers to bid to provide specific amounts and types of renewable energy from the market at cap (maximum price) or below cap (lower) prices. Contracts are then signed with the lowest cost bid to deliver output over a number of years. End consumers pay the premium on wholesale price through a levy.

PO and CT can be characterized as indirect methods and a system of subsidies containing both direct and indirect methods can be named mixed system (MS), otherwise it is a direct system (DS). Since the EU member countries have all capital subsidies and relevant fiscal mechanisms we can classify them as follows mentioning also which of the rest three methods (FT, PO, CT) is/are included in their policy making.

| Austria | FT.PO-MS | Germany | FT-DS | Netherlands FT.PO-MS | |
|---------|----------|---------|-------|-----------|--|
| Belgium | FT.PO-MS | Greece | FT-DS | Portugal | FT-DS |
| Denmark | FT-DS | Ireland | CT-MS | Spain | FT-DS |
| Finland | DS | Italy | PO-MS | Sweden | PO-MS |
| France | FT.CT-MS | Luxemburg | FT-DS | UK | PO.CT-MS |

Expression 4 should be transformed as follows, if at least on of the methods FT, PO, CT has been included in the subsidy policy because after $m$ periods from the realization of the investment consumers/users start to pay back a fraction $N$ of energy savings, which is reinvested (possibly by the same company of the energy sector that made the initial investment) with a rate of return $j$ ($b < i < j < r < f$), for parameter centered values, as an empirical rule not always valid); all estimated refer at the end of a t years period, where $U - V = Y$:

$$V_m = NF(1+f)^{m-1}(1+j)^{t-m}$$

$$V_{m+1} = NF(1+f)^m(1+j)^{t-m-1}$$

$$\cdots\cdots\cdots\cdots\cdots\cdots\cdots\cdots\cdots$$

$$\cdots\cdots\cdots\cdots\cdots\cdots\cdots\cdots\cdots$$

$$V_t = NF(1+f)^{t-1}(1+j)^{t-t}$$

$$V = \sum_{k=m}^{t} V_k = NF(1+j)^{t-1}\left[\left(\frac{1+f}{1+j}\right)^{m-1} + \left(\frac{1+f}{1+j}\right)^m + \ldots + \left(\frac{1+f}{1+j}\right)^{t-1}\right]$$

$$V = NF(1+j)^{t-1}\frac{\left(\frac{1+f}{1+j}\right)^t - \left(\frac{1+f}{1+j}\right)^{m-1}}{\frac{1+f}{1+j} - 1}$$

By taking also into account relations (2) and (3), expression (4) is replaced by the following:

$$I_{opt} = \frac{F}{S(1+r)^t}\left[K(1+i)^{t-1}\frac{\left(\frac{(1+b)(1+f)}{1+i}\right)^t - 1}{\frac{(1+b)(1+f)}{1+i} - 1} - N(1+j)^{t-1}\frac{\left(\frac{1+f}{1+j}\right)^t - \left(\frac{1+f}{1+j}\right)^{m-1}}{\frac{1+f}{1+j} - 1}\right] \tag{5}$$

In conclusion, the expressions derived can be effectively used when steady state economics apply or smooth development paths are followed. Interval analysis used to count for uncertainty seems to be a valuable technique, provided that a correction is performed within each arithmetic stage to avoid overestimation, otherwise this technique may lead to a rather conservative policy. The rate of increase b and the decrease of U, when the users/consumers return (indirectly) part of the subsidy, both concepts introduced herein for the first time, can be used not only for policy making but also for clarifying the mechanism of the optimal subsidy determination under uncertainty.

## References

[1] European Commission (2003a), Inventory of public aid granted to different energy sources, Staff Working paper SEC (2002) 1275.

[2] F. Oosterhuis: *Energy Subsidies in the European Union*, Final Report, European Parliament, July 2001.

[3] European Environment Agency, Energy subsidies in the European Union: A brief overview, EEA Technical report Jan 2004.

[4] F. Batzias and Z. Res: *Decision Making*, HOU Press, Patras, Greece, 2005.

[5] A.P. Oliveira Francisco, F.A. Batzias and H.A. Matos, Computer aided determination of optimal subsidy for installing energy saving equipments within an industrial complex, *European Symposium on Computer Aided Process Engineering – 14* **18** 409-414 (2004).

[6] R.E. Moore, *Interval Analysis*, Prentice Hall, New Jersey, 1966

[7] L. Reznik and K.P. Dabke, Measurement models: application of intelligent methods, *Measurement* **35** 47-58 (2004).

[8] H.S. Jacobsen, A new method for evaluating worst- and best- case (WBC) economic consequences of technological development, *Int. J. Production Economics* **46-47** 241-250 (1996).

[9] K. Ichida, Constrained optimization using interval analysis, *Computers ind.. Engng* **31** 933-937 (1996).

[10] F. Hao and J.P. Merlet, Multi-criteria optimal design of parallel manipulations based on interval analysis, *Mechanism and Machine Theory* **40** 157-171 (2005).

[11] J. Dinerstein and P. Egbert, A neural net approach to improving interval arithmetic, *Neural Information Processing/Letters and Reviews* **1**(1) 2003.

Brill Academic Publishers
P.O. Box 9000, 2300 PA Leiden
The Netherlands

*Lecture Series on Computer and Computational Sciences*
Volume 4, 2005, pp. 910-914

# Simulating Nest Census Data for the Endangered Species of Marine Turtle Caretta-caretta

Y.G. Matsinos[1], A.D. Mazaris

Biodiversity Conservation Laboratory,
Department of Environmental Studies,
University of the Aegean,
Mytilene 81100, Greece

Received 17 July, 2005; accepted in revised form 12 August, 2005

*Abstract:* In sea turtle studies, population trend assessments are primarily based on the number of recorded nesting females. However, it is often difficult to make a direct count of all nesting females, hence their abundance is often estimated by dividing the total number of nests laid during a season by the mean number of clutches deposited annually per individual, termed clutch frequency. In the present study, the potential effect of individual variability in clutch frequency has been investigated with respect to population trend predictions. We used two simulation models to perform our analysis: a stochastic model that simulates breeding performance as an individual based process, and a stochastic exponential growth model (Wiener – drift). Our analysis indicates that highly variable nesting performance strongly affected the model results regarding both population trends. Therefore, we conclude that population assessments based on annual censuses produced with respect to mean clutch frequencies should be viewed and used with increased caution.

*Keywords:* sea turtles, monitoring, simulation, conservation, nesting frequency, uncertainty

## 1. Introduction

It has been suggested that extensive data series on the numbers of nesting marine turtle females provide sufficient information to objectively assess sea turtle population trends[1-3]. Due to the practical difficulties that arise when monitoring large areas across long periods of time, the annual numbers of nesting individuals is often estimated using secondary field data, rather than directly counting individuals. Such data include the number of nests laid per year, the number of eggs and hatchlings produced and counts of the annual egg harvest, from which the numbers of nesting females are estimated [4].

However, assessing population dynamics, by using nesting emergences data or relative female turtle abundance collected from available data sets such as nest count census data, could result in highly biased estimates [5-7]. Aside from inaccuracies that can arise from (i) sampling errors due to possible inaccurate observations [8] and (ii) measurement errors due to changes in sampling methods, sampling protocols or personnel, this approach is subject to further bias due to erroneous assumptions and conclusions made when extrapolating the available data. Clutch frequency, defined as the number of clutches laid in a season, is not taken into consideration as a stochastic variable that significantly varies between individuals during the same nesting season but also for the same turtle across successive nesting seasons.

The effect of different parameters to calculate the number of nesting individuals and then use the produced censuses to assess population dynamics has been well documented at a theoretical level [9]. However, so far, little attention has been given towards quantitatively investigating their effect.

---

[1] Corresponding author. E-mail: matsinos@aegean.gr

## 2. Materials and Methods

In the present study we have attempted to analyse the importance of individual variation in clutch frequency, by applying two stochastic modelling approaches. First we developed a stochastic model that simulates the breeding performance as an individual based process. The second model is based on the approach developed by [10] and represents a stochastic formulation for estimating population growth that also measures the probability of extinction under a series of optimality assumptions. An evaluation of our results and a short discussion is provided.

### 2.1 Simulation Model of Nesting Behaviour

We developed a stochastic simulation model that can assimilate the behaviour of individuals and the mechanisms associated with breeding performance, in this instance for sea turtles. Since our aim was to explicitly model nesting behaviour, only mature nesting females are simulated. Mortality risk or other life history processes were not taken into consideration. Each simulation represents a single breeding season. In each simulation, nesting behaviour and reproductive output of each individual were sequentially modelled.

  In order to evaluate the possible variation in the number of nesting females as a result of the individual variability in the number of nests deposited in a season, we established a threshold value for the maximum numbers of nests laid annually. The value was set to 2018 nests, which is the maximum number of nests that have ever been recorded in the nesting rockery of loggerhead sea turtles in Zakynthos Island, Greece [11]. Nesting individuals of the population were modelled sequentially until this maximum value was reached. On completion of each individual's breeding performance total nest laid was estimated. If the number of nests constructed until this time step has exceeded the given maximum value, the simulation was terminated and the total number of turtles that participated in reproductive activities in the current simulation was calculated.

  To incorporate nesting activities we developed two simple sub-models: the first one with respect to clutch frequency, and the second with respect to nesting success (successful and abandoned nesting attempts). In the first sub-model (i), the clutch frequency of each nesting female was individually determined by drawing random numbers within an observed range of values (1-5, derived by Broderick et al. 2003). Nesting activity of an individual can either result in the successful construction of nest or to an abandoned nesting attempt. The proportion of the total attempts that successfully result in a nest is defined as 'nesting success' ($Ns$). Obviously, the exclusion of abandoned attempts could significantly alter the estimated number of nesting females present; therefore the second sub-model (ii) incorporated this additional feature. At each simulation, each individual ($i$) was characterised by a maximum number of nesting attempts ($at_{max,i}$) and a maximum number of nests ($cl_i$) that could constructed. For each turtle, the fate of each nesting attempt (abandoned or successful) was determined by comparing a random deviate $\in [0, 1]$, with a given value of nesting success. The attempt was assumed to be successful whenever the random number exceeded that given for nesting success. The same process was repeated until each individual had finally constructed the maximum assigned number of nests given or had reached its maximum assigned number of attempts. The mean value of nesting success was derived by [11], while the number of clutches deposited by each individual was estimated using a mean value and deviation term. The number of nesting attempts of each individual was also determined by drawing values within the range $\in [0, at_{max}]$ with $at_{max}$, the maximum number of nesting attempts. However, since no field data was available on the number of individual nesting attempts, we assumed that it could be estimated as the ratio of the total emergences divided by the number of nesting females. A mean value was obtained over the 19 year of nesting data for Zakynthos nesting rockery [11].

$$at_{mean} = \sum_{t=1}^{19} \frac{E_t}{Nl_t / cl_t}$$

with $at_{mean}$ the mean number of nesting attempts, $E_t$ the total number of emergences at year $t$, $Nl_t$ the total number of nests constructed at year $t$, and $cl_t$ the mean number of clutches deposited in the same year ($t$) calculated by using a mean and deviation term derived by [12]. However, there is an underlying source of error to this approach, in that the number of individuals is calculated with respect to a mean annual value of clutch frequency. To minimise this fact from producing biased estimates, $at_{mean}$ was calculated 5000 times by varying the annual mean number of clutches laid per individual. The maximum calculated value of the produced $at_{mean}$ was then determined and defined as $at_{max}$. Each sub-model was ran 10,000 times, and prediction intervals for the estimated number of nesting individuals were produced.

## 2.2. Stochastic growth model

The importance of individual variability in the number of nests laid on population growth was also examined by using a stochastic approach, with some basic optimality assumptions. Our analysis was built on the model designed by [10], which provides the basic statistics for determining maximum likelihood estimates of extinction by treating population data as a simple diffusion process (Wiener with drift). The Dennis model uses log-transformed census data as: $y_i = [ln(N_i / N_{i-1})]/^{\tau_i}$, with $N_i$ population size at time $i$ ($i=0,1,2...q$) and $^{\tau_i}$ the time interval between two successive observations. By using $yi$ as the dependent and $\sqrt{\tau_i}$ as the independent variable, a linear regression without interception is applied, where the slope represents the unbiased maximum likelihood estimator of $\mu$, while the variance of the residuals can be taken as $\sigma^2$. Following Dennis et al. (1991), after estimating $\mu$ and $\sigma^2$, the continuous rate of increase ($r$) and its variance $Var(r)$ were given as: $r = \mu + (\sigma^2 / 2)$ and $Var(r) = (\sigma^2/t_q) + \sigma^4/[2(q-1)]$, with $t_q$ the difference between the first and last census and $q$ the total number of population censuses. The probability of extinction was defined as the probability of attaining a lower threshold $N_{low}$ before a given time $T$, was obtained as a continuous probability density distribution (Eq. 88, 16, in [10]). The prediction interval of the extinction probability, was estimated by randomly sampling pairs of $\mu$ and $\sigma^2$ within the 95% confidence limits (given as $\mu \pm t_{\alpha/2,q-1}\sqrt{\sigma^2/t_q}$ for $\mu$ and as $q\sigma^2 / \chi^2_{\alpha_1,q-1}, q\sigma^2 / \chi^2_{1-a_2,q-1}$ for $\sigma^2$) and repeating the process 1000 times. For a full mathematical description, detailed discussion of the method and examples, see Dennis et al. 1991. In order to avoid wide prediction intervals of the estimated extinction probability we used a short time frame. Thus, the maximum given time ($T_{max}$) before attaining the lower threshold, was set to $T_{max} \leq n/5$, with $n$ being the number of observations. Extinction probabilities were then examined by gradually increasing the given time starting from year 1 to $T_{max}$. For this analysis we used only nesting females, assuming that the natural logarithm of annual censuses was approximated by the Wiener diffusion process with drift (means that $\mu \sim normal~(\mu,~\sigma^2 / t_q)$. We also assumed a closed population, where the numbers of nesting females reflect the population structure, with the census data free of observation error. We assumed that the annual size of the nesting population was obtained by dividing the total annual number of nests laid by the mean annual clutch frequency. The annual nesting data were derived from a 19-year study on loggerhead turtles nesting on Zakynthos, Greece [11], while the mean value and s.d. of the mean annual number of nests laid were derived from the loggerhead population nesting on the island of Cyprus [12]. The analysis was initially performed by assuming a stable annual number of nests laid by individuals that was equal to the mean value (1.9). The effect of varying the mean number of nests laid on model predictions was then examined by performing the same analysis for 100 new population censuses. For each census, the annual number of individuals was determined by drawing values for clutch frequency from a normal distribution, with mean and variance derived from field data (mean value as before: 1.9, and s.d. equal to 1.2).

Threshold population sizes were set to 100, 50 and 10 individuals. A maximum time ($T_{max}$) of 5 years was assumed to be required to reach the threshold value. The sensitivity of the model predictions to the changes in the population census as a result of the fluctuating mean annual number of nests laid per female was investigated by comparing prediction intervals of population growth and extinction probabilities. Simply put, the results produced by simulating the additional 100 stochastic population sizes were displayed as a density distribution.

## 3. Results and Discussion

In Fig. 1. the simulation results produced by running sub-model (ii) for 20 individual animals are illustrated. It is apparent that the inclusion of individual stochastic variation with respect to the maximum number of nests and the maximum number of nesting attempts made clearly affects model output (number of nests actually laid).

When combining various population sizes, the estimated extinction probability varied throughout the entire interval from 0 to 1. On the other hand, when analysing a constant population size, the probability of extinction was found to be highly sensitive to both the threshold value and the time required to reach the threshold value.

These results indicate that to reach extinction there is an association between time and larger confidence intervals, however the broad span of the intervals produced by the combined population data invalidated the use of this method for modelling such population dynamics, which suggests that unreliable population data sets such as those used in the present analysis could lead to highly uncertain estimates of extinction risk. Our analysis produced some essential information highlighting the degree of error obtained when assessing sea turtle population size by using counts of nesting females produced by the mean values of clutch frequency. Overall, the results of the stochastic models used in this paper suggest that individual variability in the number of clutches deposited could significantly impact population size assessments. In the stochastic model simulating nesting activity, the estimated number of nesting individuals varied significantly. This great difference in the number of nesting individuals illustrates the caution that should be used when interpreting natural processes and associated behaviours. When values for nesting attempts (sub-model (ii)) were incorporated into our model and thus a higher degree of stochasticity in the nesting process was included, highly upward results were obtained. Although, these estimates may not be precise, since we used a rather arbitrary measure for modelling nesting attempts, the results are striking.

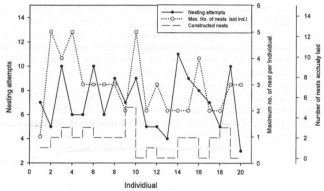

Fig. 1. Simulation results of the nesting activity of 20 individuals produced when using sub-model (ii). The solid and dotted lines indicate the maximum number of failed nesting attempts and successful nests assigned for each individual at the beginning of the simulation, the edges of the dashed lines indicate the number of nests that were actually laid by each individual at the end of the simulation.

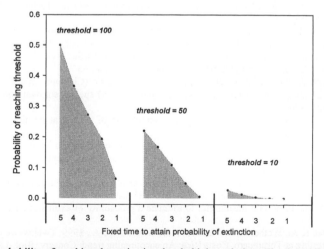

Fig. 2. The probability of reaching the extinction threshold through time. The dotted lines represent the upper limits of the predicted probabilities as produced by using a threshold of 100, 50 and 10 individuals. For each threshold the probabilities of extinction were estimated through a short time scale of 1-5 years.

Therefore a more detailed interpretation of nesting behaviour could significantly affect nesting outputs and hence population size predictions. As would be expected, population growth and probability of extinction was found to be highly sensitive to changes in population size. Considering that significant fluctuations in annual population sizes could be produced by the erroneous interpretation of the mean number of nests laid, we believe that this approach impoverishes the precision and accuracy of population assessments. Hence, the number of nests per individual as the source of data variance could in turn lead to highly biased estimates. The diffusion process has been found to produce a good approximation of population changes and has been used in several real data studies (Dennis et al. 1991). On the other hand, imitations of Dennis stochastic model estimators', which might lead to poor precision, have been well documented. Despite this criticism the Dennis model has been applied to numerous studies, since it provides a maximum likelihood estimator, which in some cases could have relatively high precision and could provide significant biological information. For this reason the relative measurement of growth rate and extinction probability, as provided by the Dennis model algorithms, were considered as appropriate estimators for our study proposes. Furthermore, in order to further reduce the erroneous interpretation of model results, we performed our analysis under a short time scale, while in our analysis we considered the confidence intervals of the produced measures rather than mean values. Our analysis shows that an additional source of bias arises when assessing population size and trends when using annual censuses of nesting individuals as a function of the total number of nests laid and the mean nest production. Thus, we conclude that sea turtle population assessments based on counts of nesting females should be viewed with caution, while strong attention should be given when these counts are produced with respect to the mean values of the number of nests laid. Considering the fact that such erroneous results could underestimate extinction risk of a population, we identify the need for better understanding of species biology and behaviour. Since collection of reliable data on clutch frequency is relatively impractical [12] we suggest that the operation of quality monitoring programs should be maintained, while future research should also focus on the sources leading to variability in natural processes. We believe that the factors that lead to strong variation in breeding performance (i.e. number of nests laid, interannual variation in breeding season) could contribute towards better management options and accurate conservation actions.

## References

[1] Meylan, A.B., 1982. Estimation of population size in sea turtles, in Bjorndal, K.A., (Ed.), Biology and Conservation of Sea Turtles, Smithsonian Institutional Press, Washington DC, USA., pp. 135-138.

[2] Limpus, C.J., 1996. Myths, reality and limitations of green turtle census data, in: Keinath, J.A., Barnard, D.E. J., Musick, A. & Bell, B.A.(comp), Proceedings of the 15th Annual Symposium on Sea Turtle Biology and Conservation: NOAA Technical Memorandum NMFS-SEFSC-387, pp. 170-173.

[3] Schroeder, B., Murphy, S., 1999. Population surveys (ground and aerial) on nesting beaches, in Eckert, K.L., Bjorndal, K.A., Abrue-Grobois, F.A., Donnelly, M. (Eds), Research and Management Techniques for the Conservation of Sea Turtles. IUCN/SSC Marine Turtle Specialist Group, Washington, DC. Publication No 4, pp. 45-55

[4] Seminoff, J. 2004. Global status assessment, Green turtle (Chelonia mydas). Marine turtle specialist group. The World Conservation Union (IUCN) species survival commission. Red list programme.

[5] Ross, J.P., 1996. Caution Urged in the Interpretation of Trends at Nesting Beaches. Mar. Turtle Newsletter 74, 9-10.

[6] Bjorndal, K. A. Bolten, A. B. (Eds)., 2000. Proceedings of a workshop on assessing abundance and trends for in-water sea turtle populations. U.S. Dept. of Commerce, NOAA Tech. Memo. NMFS-SEFSC-445.

[7] Chaloupka, M., Limpus, C., 2001. Trends in the abundance of sea turtles resident in southern Great Barrier Reef waters. Biol. Conserv. 102, 235-249.

[8] Ludwig, D., 1999. Is it meaningful to estimate a probability of extinction? Ecology 80, 298-310.

[9] Bjorndal, K.A., Wetherall, J.A. Bolten, A.B., Mortimer, J.A., 1999. Twenty-six years of green turtle nesting at Tortuguero, Costa Rica: An encouraging trend. Conserv. Biol. 13, 126-134.

[10] Dennis B., Munholland, P.L., Scott, J.M., 1991. Estimation of growth and extinction parameters for endangered species. Ecol. Monogr. 61, 115–143.

[11] Margaritoulis, D., 2005. Nesting Activity and Reproductive Output of Loggerhead Sea Turtles (Caretta Caretta) Over 19 Seasons (1984-2002) at Laganas Bay, Zakynthos, Greece: The Largest Rookery in the Mediterranean. Chelonian Conserv. Biol. 4, 916-930.

[12] Broderick, A.C., Glen, F., Godley, B.J., Hays, G.C., 2002. Estimating the number of green and loggerhead turtles nesting annually in the Mediterranean. Oryx 36, 227-235.

Brill Academic Publishers
P.O. Box 9000, 2300 PA Leiden,
The Netherlands

*Lecture Series on Computer
and Computational Sciences*
Volume 4, 2005, pp. 915-915

# Modelling and Analysis of Nonlinear Physical Problems

## A. Bratsos

This session will focus on computational and theoretical aspects concerning the numerical modelling of differential equations, which are mainly appear in applied physical problems as propagation of waves, reaction-diffusion equations etc. The methods will be presented with their relevant numerical algorithms and results.

## Athanassios G. Bratsos

Department of Mathematics,, Technological Educational Institution (T.E.I.) of Athens,, 122 10 Egaleo, Athens, Greece.,
E-mail: bratsos@teiath.gr,
URL: http://www.math.teiath.gr/~bratsos/

A. Bratsos has B.Sc. (1972) from the Department of Mathematics, University of Athens, Greece, M.Sc. (1987) and Ph.D (1993) from the Department of Mathematics and Statistics, Brunel University, England. He taught in high schools in Greece from 1977 to 1980. He was Associate Professor from 1980 to 1984 in the Centre of Higher Technical Education of Kozani, Greece and from 1984 to 1987 in the Technological Educational Institution (T.E.I.) of Athens, Greece. Since 1987 he is Professor in T.E.I. Athens, Greece. He is member of IMACS and ESCMSE.

Professor Bratsos research interests are in the numerical solution of partial differential equations with applications to applied sciences.

Brill Academic Publishers
P.O. Box 9000, 2300 PA Leiden,
The Netherlands

*Lecture Series on Computer*
*and Computational Sciences*
Volume 4, 2005, pp. 916-919

# On the Numerical Solution of the Sine-Gordon Equation in 2+1 Dimensions

## A. G. Bratsos[1]

Department of Mathematics,
Technological Educational Institution (T.E.I.) of Athens,
122 10 Egaleo, Athens, Greece

Received 7 August, 2005; accepted in revised form 8 August, 2005

*Abstract:* The method of lines is used to transform the initial/boundary-value problem associated with the Sine-Gordon equation in two space variables, into a first-order, initial-value problem. The finite-difference methods are developed by replacing the matrix-exponential term in a recurrence relation by rational approximants. The resulting finite-difference methods are analyzed for local truncation error, stability and convergence. Numerical solutions for cases involving the most known from the bibliography ring and line solitons are given.

*Keywords:* Soliton; Sine-Gordon equation; Finite-difference methods; Method of lines

*Mathematics Subject Classification:* 35Q51, 35Q53, 65M06, 65M20, 65N40

*PACS:* 02.60.Lj, 05.45.Yv, 03.75.Lm, 02.70.Bf

## 1 Introduction

Solitons represent essentially special wave-like solutions to nonlinear dynamic equations. These waves have been found to a variety of nonlinear differential equations such as the Korteweg & de Vries equation, the Schrödinger equation, the Sine-Gordon equation etc. Physical applications of solitons have been found among others to shallow-water waves, optical fibres, Josephson-junction oscillators etc. The development of analytical solutions of soliton type equations, especially the inverse scattering transform (see Ablowitz and Segur [1] etc.) are also known long time ago.

In higher dimensions models which possess soliton-like solutions have attracted great interest for numerical investigation. The two-dimensional Sine-Gordon (SG) equation in two space variables or as it is also known in $2+1$ dimensions, which belongs in this category, is given by

$$\frac{\partial^2 u}{\partial t^2} + \rho \frac{\partial u}{\partial t} = \frac{\partial^2 u}{\partial x^2} + \frac{\partial^2 u}{\partial y^2} - \phi(x,y)\sin u \tag{1.1}$$

with $u = u(x,y,t)$ in the region $\Omega = \left\{(x,y),\ L_x^0 \le x \le L_x^1,\ L_y^0 \le y \le L_y^1\right\}$ for $t > t_0$, where the parameter $\rho$ is the so-called *dissipative* term, which is assumed to be a real number with $\rho \ge 0$.

The two-dimensional SG equation arises in extended rectangular Josephson junctions, which consists of two layers of super conducting materials separating by an isolating barrier. A typical arrangement is a layer of lead and a layer of niobium separated by a layer of niobium oxide. A

---

[1]Corresponding author. E-mail: bratsos@teiath.gr

quantum particle has a nonzero significant probability of being able to penetrate to the other side of a potential barrier that would be in penetrable to the corresponding classical particle. This phenomenon is usually referred to as quantum tunneling. When $\rho = 0$, Eq. (1.1) reduces to the undamped SG equation in two space variables, while, when $\rho > 0$ to the damped one.

For the undamped SG equation in higher dimensions exact solutions have been obtained in [11], [9] and [13] using Hirota's method, in [10], [5] and [16] using Lamb's method, in [6] and [14] by Bläcklund transformation and in [12] by Painlevè transcendents. A numerical solution for the undamped SG equation has been given by Christiansen and Lomdahl [4], who used a generalized leapfrog method and by Argyris et *al* [2] using finite-elements. Both methods using appropriate initial conditions have been applied successfully with the latter one giving slightly better results. A numerical approach to the undamped SG equation appears in Djidjeli et *al* [7], where the method arises from a two-step, one-parameter leapfrog scheme, which is a generalization to that used by Christiansen and Lomdahl [4] and by Bratsos [3] who used a predictor-corrector scheme. In the case of the damped SG equation the presence of the dissipative term corresponds to a physically relevant effect in real Josephson junctions (see Nakajima et *al* [15]). This problem has been studied numerically by Djidjeli et *al* [7] and Bratsos [3].

Initial conditions associated with Eq. (1.1) will be assumed to be of the form

$$u(x, y, t_0) = g(x, y) \; ; \; L_x^0 \le x \le L_x^1 \, , \; L_y^0 \le y \le L_y^1 \tag{1.2}$$

with initial velocity

$$\left. \frac{\partial u(x, y, t)}{\partial t} \right|_{t=t_0} = \hat{g}(x, y) \; ; \; L_x^0 \le x \le L_x^1 \, , \; L_y^0 \le y \le L_y^1 \, . \tag{1.3}$$

In Eq. (1.1) the function $\phi(x, y)$ may be interpreted as the Josephson current density, while in Eqs. (1.2)-(1.3) the functions $g(x, y)$ and $\hat{g}(x, y)$ represent wave modes or kinks and velocity respectively.

Boundary conditions will be assumed to be of the form

$$\frac{\partial u(x, y, t)}{\partial x} = 0 \text{ for } x = L_x^0 \text{ and } x = L_x^1 \, , \; L_y^0 < y < L_y^1 \tag{1.4}$$

and

$$\frac{\partial u(x, y, t)}{\partial y} = 0 \text{ for } y = L_y^0 \text{ and } y = L_y^1 \, , \; L_x^0 < x < L_x^1 \, , \tag{1.5}$$

when $t > t_0$.

## 2 The finite-difference method

To obtain a numerical solution the region $R = \Omega \times [t > t_0]$ with its boundary $\partial R$ consisting of the lines $x = L_x^0$, $L_x^1$, $y = L_y^0$, $L_y^1$ and $t = t_0$ is covered with a rectangular mesh, $G$, of points with coordinates $(x, y, t) = (x_k, y_m, t_n) = \left( L_x^0 + kh_x, L_y^0 + mh_y, t_0 + n\ell \right)$ with $k, m = 0, 1, ..., N+1$ and $n = 0, 1, ...$, in which $h_x = \left( L_x^1 - L_x^0 \right) / (N+1)$ and $h_y = \left( L_y^1 - L_y^0 \right) / (N+1)$ represent the discretization into $N + 1$ subintervals of the space variables, while $\ell$ represents the discretization of the time variable. The solution of an approximating finite-difference scheme at the same point will be denoted by $u_{k,m}^n$.

Let the solution vector be

$$\mathbf{u^n} = \mathbf{u}(t_n) = \mathbf{u}(t) = \left[ u_{0,0}^n, u_{1,0}^n, ..., u_{N+1,0}^n; \; u_{0,1}^n, u_{1,1}^n, ..., u_{N+1,1}^n; \right.$$

$$\left. ...; \; u_{0,N+1}^n, u_{1,N+1}^n, ..., u_{N+1,N+1}^n \right]^T \tag{2.1}$$

Replacing the space derivative in Eq. (1.1) by the familiar second-order central-difference approximant and applying Eq. (1.1) to all $(N+2)^2$ mesh points of the grid $G$ at time level $t = n\ell$ with $n = 0, 1, \ldots$ subject to the boundary conditions given by Eq. (1.4) it leads to an initial-value problem of the form

$$D^2 \mathbf{u}(t) + \rho D \mathbf{u}(t) = A \mathbf{u}(t) - \mathbf{G}(\mathbf{u}(t)) \ ; \ t > t_0,$$

$$\mathbf{u}(t_0) = \mathbf{g} \ , \ D\mathbf{u}(t_0) = \hat{\mathbf{g}}, \tag{2.2}$$

where $D^2 = \text{diag}\left\{d^2/dt^2\right\}$, $D = \text{diag}\left\{d/dt\right\}$ both diagonal matrices of order $(N+2)^2$, $\mathbf{G}(\mathbf{U}(t))$ is a vector of order $(N+2)^2$ arising from the nonlinear term $\sin u$ and $A = h_x^{-2} B + h_y^{-2} C$ is a matrix of order $(N+2)^2$ with $B$ a block diagonal matrix with tridiagonal blocks $B_1$ of order $N+2$ given by

$$B = \begin{bmatrix} B_1 & & & \\ & B_1 & & \\ & & \cdots & \\ & & & B_1 \end{bmatrix} \text{ with } B_1 = \begin{bmatrix} -2 & 2 & & & \\ 1 & -2 & 1 & & \\ & \cdot & \cdot & \cdot & \\ & & 1 & -2 & 1 \\ & & & 2 & -2 \end{bmatrix} \tag{2.3}$$

and $C$ a block tridiagonal matrix of order $(N+2)^2$ with diagonal blocks given by

$$C = \begin{bmatrix} -2I & 2I & & & \\ I & -2I & I & & \\ & \cdot & \cdot & \cdot & \\ & & I & -2I & I \\ & & & 2I & -2I \end{bmatrix} \tag{2.4}$$

with $I$ the identity matrix of order $N+2$.

Using the relations

$$\mathbf{u}(t \pm \ell) = \exp(\pm \ell D)\, \mathbf{u}(t) \tag{2.5}$$

it leads to the following three-time level recurrence relation for solving Eq. (1.1)

$$\mathbf{u}(t + \ell) = [\exp(\ell D) + \exp(-\ell D)]\, \mathbf{u}(t) - \mathbf{u}(t - \ell) \ ; \ t = \ell, 2\ell, \ldots \tag{2.6}$$

Numerical methods will be developed by replacing the matrix-exponential term in the recurrence relation (2.6) by rational replacements, which are also known as the $(\mu, \nu)$ Padé approximants, of the form

$$\exp(\ell D) \approx \left(I + a_1 \ell D + b_1 \ell^2 D^2\right)^{-1} \left(I + c_1 \ell D + d_1 \ell^2 D^2\right) \tag{2.7}$$

with $a_1$, $b_1$, $c_1$ and $d_1$ parameters, which are real numbers having appropriate values for each type of approximants.

To avoid solving the nonlinear system arising from Eq. (2.6) a *Predictor-Corrector* scheme analogous to that introduced by Bratsos [3] was used.

The behavior of the numerical method arising from Eq. (2.6) was tested to selected examples used by Christiansen and Lombahl [4], Argyris et al [2], Djidjeli et al [7] and Bratsos [3].

## Acknowledgements

This research was co-funded by 75% from E.E. and 25% from the Greek Government under the framework of the Education and Initial Vocational Training Program - Archimedes, Technological Educational Institution (T.E.I.) Athens project *"Computational Methods for Applied Technological Problems"*.

# References

[1] M.J. Ablowitz and H. Segur, *Solitons and the Inverse Scattering Transform*, SIAM Studies in Applied Mathematics 4, Society for Industrial and Applied Mathematics, Philadelphia, 1981.

[2] J. Argyris, M. Haase and J.C. Heinrich, Finite element approximation to two-dimensional Sine-Gordon solitons, *Computer Methods in Applied Mechanics and Engineering*, **86** 1-26(1991).

[3] A.G. Bratsos, An explicit numerical scheme for the Sine-Gordon equation in 2+1 dimensions, *Appl. Num. Anal. Comp. Math.* **2** No. 2, 189-211(2005).

[4] P.L. Christiansen and P.S. Lomdahl, Numerical solutions of 2+1 dimensional Sine-Gordon solitons, *Physica 2D* 482-494(1981).

[5] P.L. Christiansen and O.H. Olsen, On Dynamical Two-Dimensional Solutions to the Sine-Gordon Equation, *Z. Angew. Math. Mech.* **59** T30(1979).

[6] P.L. Christiansen and O.H. Olsen, Ring-shaped quasi-soliton solutions to the two and three-dimensional Sine-Gordon equations, *Physica Scripta 20* 531-538(1979).

[7] K. Djidjeli, W.G. Price and E.H. Twizell, Numerical solutions of a damped Sine-Gordon equation in two space variables, *Journal of Engineering Mathematics* **29** 347-369(1995).

[8] R.K. Dodd, J.C. Eilbeck, J.D. Gibbons and H.C. Morris, *Solitons and Nonlinear Wave Equations*, London Academic Press, 1982.

[9] J.D. Gibbon and G. Zambotti, The Interaction of $n$-Dimensional Soliton Wave Fronts, *Nuovo Cimento* **25B**, 1(1976).

[10] G. Grella and M. Marinaro, Special Solution of the Sine-Gordon Equation in 2+1 Dimensions, *Lett. Nuovo Cimento* **23** 459(1978).

[11] R. Hirota, Exact three-soliton solution of the two-dimensional Sine-Gordon equation, *J. Phys. Soc. Japan* **35** 1566(1973).

[12] P. Kaliappan and M. Lakshmanan, Kadomtsev-Petviashvili and two-dimensional sine-Gordon equations: reduction to Painlevè transcendents, *J. Phys. A: Math. Gen.* **12** L249(1979).

[13] K.K. Kobayashi and M. Izutsu, Exact Solution of the $n$-Dimensional Sine-Gordon Equation, *J. Phys. Soc. Japan* **41** 1091(1976).

[14] G. Leibbrandt, New exact solutions of the classical Sine-Gordon equation in $2+1$ and $3+1$ dimensions, *Phys. Rev. Lett.* **41** 435-438(1978).

[15] K. Nakajima, Y. Onodera, T. Nakamura and R. Sato, Numerical analysis of vortex motion on Josephson structures, *Journal of Applied Physics* **45**(9) 4095-4099(1974).

[16] J. Zagrodzinsky, Particular solutions of the Sine-Gordon equation in $2+1$ dimensions, *Phys. Lett.* **72**A 284-286(1979).

Brill Academic Publishers
P.O. Box 9000, 2300 PA Leiden,
The Netherlands

*Lecture Series on Computer
and Computational Sciences*
Volume 4, 2005, pp. 920-923

# On the Numerical Solution of the Kadomtsev-Petviadhvilli Equation

**A. G. Bratsos**[1] and **I. Th. Famelis**

Department of Mathematics,
Technological Educational Institution (T.E.I.) of Athens,
122 10 Egaleo, Athens, Greece

Received 7 August, 2005; accepted in revised form 8 August, 2005

*Abstract:* A finite-difference method is used to transform the initial/boundary-value problem associated with the Kadomtsev-Petviashvilli equation into an algebraic system. For this purpose the space and the time partial derivatives are replaced with familiar finite-difference replacements. To avoid solving the resulting nonlinear system an appropriate linearized scheme is proposed.

*Keywords:* Soliton; Korteweg & de Vries equation; Kadomtsev-Petviashvilli equation; Finite-difference method.

*Mathematics Subject Classification:* 35Q51, 35Q53, 65M06

*PACS:* 02.60.Lj, 05.45.Yv, 02.70.Bf

## 1  Introduction

In their original paper Korteweg & de Vries [9] derived an equation, referred from now on as the KdV equation, equivalent to

$$u_t + \left(3u^2\right)_x + u_{xxx} = 0 \; ; \; t > 0 \, , \; x \in \Re \tag{1}$$

in order to describe approximately the slow evolution of long water waves of moderate amplitude, as they propagate under the influence of gravity in one direction in shallow water of uniform depth. It is known that the KdV equation describes approximately the solution of long, one-dimensional waves in many physical settings, including long internal waves in a density-stratified ocean, ion-acoustic waves in a plasma, acoustic waves on a crystal lattice and many more.

The KdV equation's two-dimensional form, the Kadomtsev-Petviashvili (KP) equation,

$$u_{xt} + \left(3u^2\right)_{xx} + u_{xxxx} + 3u_{yy} = 0 \; ; \; t > 0 \, , \; x \in \Re \tag{2}$$

was first introduced by Kadomtsev and Petviashvili [7] in order to discuss the stability of one-dimensional solitons against transerve perturbations. It was shown by Hirota [4] that Eq. (2) has a soliton solution, which can be expressed in the form

$$u\left(x, y, t\right) = 2\frac{\partial^2 \log_e f}{\partial x^2}, \tag{3}$$

---

[1]Corresponding author. E-mail: bratsos@teiath.gr

where the function $f = f(x, y, t)$ satisfies the bilinear form

$$\left(D_x^4 + D_x D_t + 3D_y^2\right) f \cdot f = 0 \tag{4}$$

in which

$$D_x^n D_t^m f \cdot f = \left[\left(\frac{\partial}{\partial x} - \frac{\partial}{\partial x'}\right)^n \left(\frac{\partial}{\partial t} - \frac{\partial}{\partial t'}\right)^m f(x,t) f(x',t')\right]_{\substack{x = x' \\ t = t'}} \tag{5}$$

Lamb [10] obtained an $N$-soliton solution of the Eq. (2) using inverse scattering theory, which it can be written in the form

$$f_N(x, y, t) = |\delta_{ij} + p_{ij} \exp(\Theta_i + \Psi_j)|, \tag{6}$$

where $\Theta_i = l_i \chi_i - l_i^2 \psi_i - 4l_i^3 t$, $\Psi_i = n_i \chi_i + n_i^2 \psi_i - 4n_i^3 t$, with $\chi_i = x_i - x_{0,i}$, $\psi_i = y_i - y_{0,i}$; $i = 1, 2, ..., N$ and $x_{0,i}$, $y_{0,i}$ are the initial positions of the soliton relative to the $x$, $y$ axes with $l_i$, $n_i$, $a_i$ real constants, $p_{ij} = a_i / (l_i + n_j)$; $i, j = 1, 2, ..., N$, where $\delta_{ij}$ is the Kronecker symbol. Determinant (6) is a Wronskian (see also Wahlquist [11], Freeman and Nimmo [5]). If $x_{0,1} = y_{0,1} = 0$, the solution $u$ for the single-soliton can be written as

$$u_1(x, y, t) = 2a_1(l_1 + n_1) \frac{\exp\left[(l_1 + n_1)x + k_1\right]}{\{1 + p_{11} \exp\left[(l_1 + n_1)x + k_1\right]\}^2}, \tag{7}$$

where $k_1 = -\left(l_1^2 - n_1^2\right)y - 4\left(l_1^3 + n_1^3\right)t$ (see also Bratsos and Twizell [2]).

Initial conditions will be assumed to be of the form

$$u(x, y, t_0) = g(x, y), \quad L_0 \le x, y \le L_1. \tag{8}$$

## 2   The numerical method

To obtain a numerical solution the time and space partial derivatives are replaced by central-difference replacements. To this effect the region $R = \Omega \times [t > t_0]$, where $\Omega$ is a square defined by the lines $x$, $y = L_i$; $i = 0$, 1 with boundary $\partial\Omega$ and initial axis $t = t_0$, is covered with a rectangular mesh, $G$, of points with coordinates $(x, y, t) = (x_k, y_m, t_n) = (L_0 + kh, L_0 + mh, t_0 + nl)$ with $k$, $m = 0, 1, ..., N+1$ and $n = 0, 1, ...$ The theoretical solution of Eq. (2) at the typical mesh point $(x_k, y_m, t_n)$ is $u(x_k, y_m, t_n)$ which may be denoted, when convenient, by $u_{k,m}^n$. The solution of an approximating difference scheme at the same point will be denoted by $U_{k,m}^n$.

Let the solution vector at time $t = n\ell$ be

$$\mathbf{U}^n = \left[U_{1,1}^n, U_{1,2}^n, ..., U_{1,N}^n; U_{2,1}^n, U_{2,2}^n, ..., U_{2,N}^n; ...; U_{N,1}^n, U_{N,2}^n, ..., U_{N,N}^n\right]^T. \tag{9}$$

Consider the following boundary conditions

$$\frac{\partial u(x, y, t)}{\partial x} = 0 \text{ for } x = L_0, L_1; \ L_0 < y < L_1, \tag{10}$$

$$\frac{\partial u(x, y, t)}{\partial y} = 0 \text{ for } y = L_0, L_1; \ L_0 < x < L_1, \tag{11}$$

when $t > t_0$ and

$$u(x, y, t) = \tilde{b}(x, y, t), \tag{12}$$

when $x = y = L_0$ or $x = y = L_1$ with $\tilde{b}$ an appropriate function describing the boundary values.

Using forward-difference approximants for the time derivative and central-difference approximants for the space derivative it can be easily shown that

$$u_{xt}|_{(x_k, y_m, t_n)} = \frac{1}{2h\ell}\left(u_{k+1,m}^{n+1} - u_{k-1,m}^{n+1} - u_{k+1,m}^{n} + u_{k-1,m}^{n}\right). \tag{13}$$

Applying Eq. (13) to Eq. (2) at each interior point of the grid $G$ and using the Crank-Nickolson method, it leads to the following two-time level, finite-difference scheme

$$U_{k+1,m}^{n+1} - U_{k-1,m}^{n+1} - U_{k+1,m}^{n} + U_{k-1,m}^{n} + 3r\delta_x^2\left(q_{k,m}^{n+1} + q_{k,m}^{n}\right)$$

$$+p\delta_x^4\left(U_{k,m}^{n+1} + U_{k,m}^{n}\right) + 3r\delta_y^2\left(U_{k,m}^{n+1} + U_{k,m}^{n}\right) = 0 \tag{14}$$

for $k, m = 1, 2, ..., N$, where $q(U) = U^2$, $r = \ell/h$ and $p = \ell/h^3$. Finally, Eq. (14) leads to a nonlinear system for the unknown vector $\mathbf{U}^{n+1}$.

The stability analysis of the method arising from Eq. (14) with a linearizing technique applied to the nonlinear term $q(U) = U^2$ has proved that the method is unconditionally stable.

## 2.1 The linearized scheme

Let

$$q_{k,m}^{n+1} = U_{k,m}^{n+1}U_{k,m}^{n} \tag{15}$$

for $k, m = 1, 2, ..., N$ and $n = 0, 1, ....$ Then Eq. (14) is written as

$$U_{k+1,m}^{n+1} - U_{k-1,m}^{n+1} - U_{k+1,m}^{n} + U_{k-1,m}^{n} + 3r\delta_x^2\left(U_{k,m}^{n+1}U_{k,m}^{n} + q_{k,m}^{n}\right)$$

$$+p\delta_x^4(U_{k,m}^{n+1} + U_{k,m}^{n}) + 3r\delta_y^2\left(U_{k,m}^{n+1} + U_{k,m}^{n}\right) = 0 \tag{16}$$

for $k, m = 1, 2, ..., N$. Eq. (16) using Eq. (15) leads to a two-time level *linear* finite-difference scheme, which can be written in matrix-vector form as

$$A\,\mathbf{U}^{n+1} = \mathbf{F}\left(\mathbf{U}^n\right) + \tilde{\mathbf{b}} \tag{17}$$

with

$$A = \begin{bmatrix} A_1 & B_1 & P & & & \\ C_2 & A_2 & B_2 & P & & \\ P & C_3 & A_3 & B_3 & P & \\ & \cdot & \cdot & \cdot & & \\ & & P & C_{N-1} & A_{N-1} & B_{N-1} \\ & & & P & C_N & A_N \end{bmatrix}, \quad A_k = \begin{bmatrix} a_{k,1} & 3r & & & \\ 3r & a_{k,2} & 3r & & \\ & \cdot & \cdot & \cdot & \\ & & 3r & a_{k,N-1} & 3r \\ & & & 3r & a_{k,N} \end{bmatrix}, \tag{18}$$

where $A$ a block quindiagonal matrix of order $N^2$, $A_k$; $k = 1, 2, ..., N$ tridiagonal matrices of order $N$ in which $a_{1,m} = 7p - 6r\left(1 + U_{1,m}^n\right)$ for $m = 1, ..., N$, $a_{k,m} = 6\left[p - r\left(1 + U_{k,m}^n\right)\right]$ for $k = 2, 3, ..., N-1$ and $m = 1, ..., N$, $a_{N,m} = 7p - 6r\left(1 + U_{N,m}^n\right)$ for $m = 1, ..., N$. Also $B_k = \text{diag}\{b_{k,m}\}$ with $b_{k+1,m} = -\left(4p - 1 - 3rU_{k+1,m}^n\right)$ for $k = 1, 2, ..., N-1$ and $m = 1, ..., N$, $C_k = \text{diag}\{c_{k-1,m}\}$ with $c_{k-1,m} = -\left(4p + 1 - 3rU_{k-1,m}^n\right)$ for $k = 2, 3, ..., N$ and $m = 1, ..., N$ and $P = \text{diag}\{p\}$ all diagonal matrices of order $N$. Finally $\tilde{\mathbf{b}} = \tilde{\mathbf{b}}\left(\tilde{\mathbf{b}}^n, \tilde{\mathbf{b}}^{n+1}\right)$ is the vector of the boundaries conditions of order $N^2$.

The behavior of the method arising from Eq. (14) was applied successfully to the known from the bibliography numerical schemes by Bratsos and Twizell [2], Feng and Mitsui [3].

## Acknowledgement

This research was co-funded by 75% from E.E. and 25% from the Greek Government under the framework of the Education and Initial Vocational Training Program - Archimedes, Technological Educational Institution (T.E.I.) Athens project *"Computational Methods for Applied Technological Problems"*.

## References

[1] J.M. Ablowitz and J. Villarroel, Solutions of the Time Dependent Schrödinger and the Kadomtsev-Petviashvili Equations, *Phys. Rev. Let.* **78** No. 4 (1997).

[2] A.G. Bratsos, E.H. Twizell, An Explicit Finite-difference scheme for the solution of the Kadomtsev-Petviashvili equation, *Intern. J. Computer Math.*, **68** 157-187(1998).

[3] B.F. Feng, T. Mitsui, A finite difference method for the Korteweg-de Vries and the Kadomtsev-Petviashvili equations, *Comput. Appl. Math.*, **90** No. 1, 95-116(1998).

[4] R. Hirota, *Direct methods in soliton theory*, In Solitons (Bullough, R.K. and Caudrey, P.J., eds). Berlin: Springer-Verlang, 1980.

[5] N.C. Freeman, Soliton solutions of Non-linear Evolution Equations, *IMA Journal of Applied Mathematics*, **32** 125-145(1984).

[6] N.C. Freeman, J.J.C. Nimmo, Soliton solutions of the Korteweg-de Vries and the Kadomtsev-Petviashvili equations: the Wronskian technique, *Proc. R. Soc.* A 389(1983).

[7] B.B. Kadomtsev and V.I. Petviashvili, On the stability of solitary waves in weakly dispersive media, *Sov. Phys. Dokl.*, **15** 539-541(1970).

[8] C.S. Gardner, J.M. Greene, M.D. Kruskal, R.M. Miura, Methods for solving the Kortweg-de Vries equation, *Phys. Rev. Lett.*, **19** 1095-1097(1967).

[9] D.J. Korteweg & G. de Vries (1895), On the change of form of long waves advancing in a rectangular channel, and on a new type of long stationary wave, *Phil. Mag. Ser.* 5, **39** pp. 422-433(1895).

[10] G.L.JR. Lamb, *Elements on Soliton Theory*, New York, Chichester, Brisbane, Toronto: John Wiley, 1980.

[11] H.D. Wahlquist (1974), *Bäcklund transformation of potentials of the Korteweg-de Vries equation and the interaction of solitons with cnoidal waves*, In Bäcklund Transformations, Berlin: Springer-Verlang, 1974.

[12] X.P. Wang, J.M. Ablowitz and H. Segur (1994), Wave collapse and instability of solitary waves of a generalized Kadomtsev-Petviashvili equation, *Physica D*, **78** 241-265(1994).

[13] N.J. Zabusky, M.D. Kruskal (1965), Interaction of solitons in a collisionless plasma and the recurrence of initial states, *Phys. Rev. Lett.*, **15** 240-243(1965).

Brill Academic Publishers
P.O. Box 9000, 2300 PA Leiden
The Netherlands

*Lecture Series on Computer
and Computational Sciences*
Volume 4, 2005, pp. 924-927

# Numerical Analysis of Turbulent Fluid Flow in Rotor Stator Gap with Injection from Rotor Surface

J. A. Esfahani[1] , Y. Daghighi , A. Nabovati

Department of Mechanical Engineering,
Faculty of Engineering,
Ferdowsi University of Mashhad,

Received 15 July, 2005; accepted in revised form 11 August, 2005

*Abstract:* Developing turbulent flow of cooling fluid in the rotor-stator gap of electrical machines with axial flow (Taylor Couette flow with axial flow) has been numerically studied, with and without injection from rotor surface. Results of two turbulence model, Standard k-ε and RNG k-ε, have been compared with published experimental data. It has been shown that RNG k-ε has better agreement with experimental data and needs less CPU time. Effect of inlet flow velocity, rotor rotational velocity and rate of injection from rotor surface on the flow pattern, average tangential and axial velocity have been studied.

*Keywords:* Taylor Couette Flow-Numerical Modeling- Turbulent Flow- Rotor Stator Gap

*Mathematics Subject Classification:* Navier-Stokes equations – k-ε modeling – finite volume methods – equations and system with constant coefficients.

*PACS:* 76D05- 76F60 - 76M12 - 35E99

## 1. Introduction

Flow between two coaxial cylinders, when one or both of them are rotating with constant velocity is called Taylor Couette flow. Velocity field and stability boundaries [1,2], effect of buoyancy [3,4], heat transfer and it's effect on stability [5,6] are widely studied in the literature. Despite of wide experimental studies, a few researchers have paid attention to numerical modeling of Taylor Couette flow [7,8]. Taylor Couette flow with axial flow has also been studied due to its importance on cooling of rotary electrical machines. The main concern of designers of such machines is the heat transfer rate and pressure drop of working fluid in the rotor stator gap and temperature distribution. In the some industrial applications such as cooling large electrical machine, blowing and suction of cooling fluid from rotor and stator surface is used to enhance the heat transfer. In this case, flow in the rotor stator gap will never be fully developed.

In the present work, developing turbulent flow of cooling fluid in the rotor-stator gap of electrical machines with axial flow (Taylor Couette flow with axial flow) has been numerically studied, with and without injection from rotor surface to present a details of flow field. In the analysis, results of two different turbulence model, Standard k-ε and RNG k-ε model, have been compared with the experimental results of Mayle et al. [9] to provide validation of the work. Effect of inlet flow velocity, rotor rotational velocity and rate of injection from rotor surface on the flow pattern, average tangential and axial velocity have been studied.

## 2. Physical Model

Figure (1) shows the problem geometry. The geometry used in the present study is the same as the one used by Mayle et al. [9]. Rotor stator radius ratio is η=0.87, Taylor number is $3.8 \times 10^5$ and the rotor length to gap distance ratio is equal to 40. Inlet axial flow Reynolds number and injection mass flow rate are the varying parameters. Cooling fluid is injected into the gap through 48 circular holes on the rotor surface distributed from z/h=14 to z/h=20. Main inlet and outlet of cooling fluid are perpendicular to the rotor axis of rotation. Points A, B and C are the points on which experimental results are available. Cooling fluid is air with constant properties.

---
[1] Corresponding author. . E-mail: abolfazl@ferdowsi.um.ac.ir

## 3. Mathematical Model

The standard equations presented by Versteeg and Malalasekera [10] have been used as well as continuity and turbulent momentum equations, in order to describe the three dimensional turbulent flow of cooling fluid in rotor-stator gap. In post processing of the out put average tangential velocity and average axial velocity are defined as bellow and have been calculated numerically:

$$\overline{V}_{axial} = \int_{R}^{R+h} V_{axial}\, r dr \left/ \int_{R}^{R+h} r dr \right. = 2\pi \int_{R}^{R+h} V_{axial}\, r dr \left/ A \right. \tag{1}$$

$$\overline{V}_{tan} = \int_{R}^{R+h} r^2 V_{axial} V_{tan}\, dr \left/ (R+h/2) \int_{R}^{R+h} r V_{axial}\, dr \right. = 2\pi \int_{R}^{R+h} r^2 V_{axial} V_{tan}\, dr \left/ (R+h/2)\overline{V}_{axial} A \right. \tag{2}$$

Rotor velocity and rotor length have been selected as references velocity and length. Thus dimensionless axial position z/h, normalized injection mass flow rate $\dot{m}_{injection} / (2\pi\rho \times RV_{rotor}\Delta Z)$, normalized average tangential velocity, $\overline{V}_{tan}/V_{rotor}$ normalized average axial velocity (inlet flow) $\overline{V}_{axi}/V_{rotor}$ have been defined. Dimensionless governing parameters are radius ratio $\eta = R/(R+h)$, Taylor number[2] (when the outer cylinder is stationary and the inner one is rotating), $Ta = R\omega_i h / \upsilon$, and axial flow Reynolds number $Re_a = \overline{V}_{axi} h / v$ in which R, R+h, $\omega_i$, h, $\overline{V}_{axi}$ and v are inner cylinder radius, outer cylinder radius, inner cylinder rotational speed, gap distance, average velocity of axial flow, and fluid viscosity, respectively. In above integrations A, $\dot{m}$, $V_{rotor}$, $V_{tan}$, $\Delta$ z, z and $\rho$ mean cross section of Rotor Stator gap, injection mass flow, tangential velocity of rotor at outer surface ($R\omega_{rotor}$), fluid tangential velocity, injection zone length, axial coordinate and cooling fluid density.

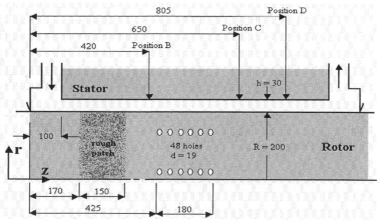

Figure1- Geometry and dimensions of model for numerical modeling
(all the dimensions are in millimeter)

## 4. Results

Taylor number for all numerical simulations and experimental tests was equal to $3.8\times10^5$.
Figure (2) represents numerical results using standard k-ε and RNG k-ε turbulence models and the experimental results of Mayle et al. [9] for three different cases without injection from rotor surface. Experimental results have been presented for the normalized average axial velocity (inlet flow), $\overline{V}_{axi}/V_{rotor}$, from 0.09 to 0.2.The used Taylor number is in the range of large air-cooled turbo generators. Numerical results have been presented for three inlet flow, normalized average axial velocity of 0.09, 0.14 and 0.2, indicated by M1, M2, M3, respectively applying standard k-ε and RNG k-ε turbulence models. Results of standard k-ε and RNG k-ε turbulence models are nearly the same

---

[2] In some literature, it has been referred as rotor Reynolds number.

accuracy of prediction; more deviation of standard k-ε at M1, of RNG k-ε at M2 than the other one, and same deviation at M3, but the RNG k-ε needs a little less iteration for convergence.

Figure (3) shows the comparison between numerical and experimental results of Mayle et al. [9] for normalized average tangential velocity as a function of dimensionless axial position for three different injection mass fluxes and three different normalized average inlet axial velocities. Table 1 contains the

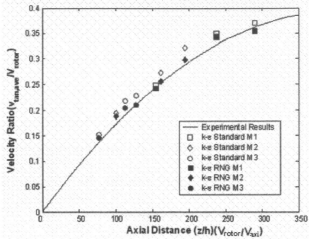

Figure 2: Comparison of results of numerical modeling and experimental results of Mayle et al. [9]

non-dimensional inlet axial velocity, normalized injection mass flux which had been used for three injection tests(mentioned as T1, T2and T3). It can be seen that by increasing the mass flow of injected fluid between z/h=14 to z/h=20 from rotor surface, the more increase in the average tangential velocity of moving fluid is observed. This could be the effect of tangential momentum of injected fluid. Also increasing the average inlet velocity will decrease the average tangential velocity of moving fluid. For the cases 1 and 2 (low injection mass flows), the average

Table 1: Conditions for experimental tests and numerical simulation presented in figure (3)

| Test number | $\dfrac{\overline{V}_{axi}}{V_{rotor}}$ | $\dfrac{\dot{m}_{injection}}{2\pi\rho RV_{rotor}\Delta z}$ |
|---|---|---|
| T1 | 0.179 | 0.0057 |
| T2 | 0.088 | 0.0106 |
| T3 | 0.026 | 0.0187 |

tangential velocity of moving fluid is increased even after the end of the injection. But for the case 3, where the injection mass flow rate is relatively high , ψ =0.0187, average tangential velocity of moving fluid at the z/h=18 is about 75% of the rotor tangential velocity. In this case, after z/h=20 the average tangential velocity of moving fluid is decreased due to the friction. It is observed that results of RNG k-ε have better agreement with available experimental data.

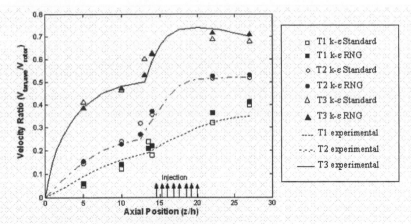

Figure 3: Comparison of numerical modeling results to Mayle et al. experimental results [9]at Ta=$3.8\times10^5$ for normalized average tangential velocity as a function of dimensionless axial position for three different injection mass flow inlet flow, presented in table 1.

## 5. Conclusion

Developing turbulent flow of cooling fluid in the rotor-stator gap of electrical machines with axial flow (Taylor Couette flow with axial flow) has been numerically studied, with and without injection from rotor surface. Two different turbulence model, Standard k-ε and RNG k-ε, have been used.

- In the case of no injection, fluid average tangential momentum is increased due to the friction with rotor surface.
- Decreasing the normalized average axial velocity will increase the average tangential velocity of cooling fluid along the rotor length.
- In the case of injection of cooling fluid from rotor surface, increase of fluid average tangential velocity is mainly because of injected fluid.
- Increasing the average inlet velocity of fluid into the rotor stator gap from main inlet, will decrease the change of the average tangential velocity due to the injection.
- Results of RNG k-ε have better agreement with available experimental data and saving CPU time.

## References

[1] A. Akonur, R. M. Lueptow, "Three-dimensional velocity field for wavy Taylor–Couette flow", Physics of Fluids, Vol. 15, No. 4, pp. 1-13, 2003.
[2] S. T. Wereley, R. M. Lueptow, "Spatio-temporal character of non-wavy and wavy Taylor Couette Flow", Journal of Fluids Mechanics, Vol. 364, pp. 59-80, 1998.
[3] D. C. Kuo, K. S. Ball, "Taylor–Couette flow with buoyancy: Onset of spiral flow", Physics of Fluids, Vol. 9, No. 10, pp 2872-2884, 1997.
[4] K. S. Ball and B. Farouk, "A flow visualization study of the effects of buoyancy on Taylor vortices", Physics of Fluids A, Vol. 1, pp. 1502-1513, 1989.
[5] K. S. Ball and B. Farouk, "An experimental study of heat transfer in a vertical annulus with a rotating inner cylinder", International Journal of Heat and Mass Transfer, Vol. 32, pp. 1517-1526, 1989.
[6] M. Ali, P.D. Weidman, "On the stability of circular Couette flow with radial heating", Journal of Fluid Mechanics, Vol. 220, pp. 53-84, 1990.
[7] M. J. Braun, V. Kudriavtsev, R. K. Corder, "Flow visualization of the evolution of Taylor instabilities and comparison with numerical simulations", Computational Technologies for Fluid/Thermal/Structural/Chemical Systems with Industrial Applications, Vol. 1, ASME 2002.
[8] J. A. Esfahani, A. Nabovati, "Numerical Analysis of Taylor Couette Flow", Proceeding of the 8th conference of Fluid Dynamics, Tabriz, Iran, 2003, (In Farsi).
[9] R. E. Mayle, S. Hess, C. Hirsch, J. von Wolfersdorf, "Rotor-Stator gap flow analysis and experiments", IEEE Transaction on Energy Conversion, Vol. 13, No. 2, 1998.
[10] H. K. Versteeeg, W. Malalasekera, An Introduction to Computational Fluid Dynamics-The finite volume method, 2nd Edition, Longman Group, 1996.

Brill Academic Publishers
P.O. Box 9000, 2300 PA Leiden
The Netherlands

*Lecture Series on Computer
and Computational Sciences*
Volume 4, 2005, pp. 928-932

# Numerical Analysis of Taylor-Couette Flows in Wavy and Non-Wavy Regimes

J. A. Esfahani[1] , A. Mokhtari, Y. Daghighi

Department of Mechanical Engineering,
Faculty of Engineering,
Ferdowsi University of Mashhad,

Received 15 July, 2005; accepted in revised form 11 August, 2005

*Abstract:* Taylor-Couette flow is a classic case of an unstable flow; in this case, the centrifugal instability manifests itself in the form of Taylor vortices stacked like donuts in the annulus between a rotating inner cylinder and a fixed outer one. The objectives of this investigation are to numerically verify the experimental results which have been reported in different publishes and to further study detailed flow fields and bifurcations related to different Taylor numbers. Grids effect on result accuracy, wavy and non-wavy flow for both laminar and turbulent Taylor-Couette flow, and the range of Taylor number for different regimes have been studied. It is demonstrated that 'shift-and-reflect' symmetry holes in Taylor-Couette flow without an imposed axial flow.

*Keywords:* Rotating Flows, Wavy Taylor-Couette Flow, Non Wavy Taylor-Couette Flow, Instability, Transition

*Mathematics Subject Classification:* Navier-Stokes equations – Finite volume methods – Equations and system with constant coefficients.

*PACS:* 76D05- 76M12 - 35E99

## 1. Introduction

The flow between two concentric cylinders with inner one rotating and outer one stationary, called Taylor-Couette flow, has been studied by many researchers for decades [1-4]. Taylor-Couette flow is often observed in various types of engineering application, for example, journal bearing lubrication and cooling of rotating machinery among others. If the Taylor number (Ta) based on $\Omega i$ goes over a critical one ($Ta_{c1}$), the flow instability caused by the curved streamlines of the main flow produces axisymmetric Taylor vortices. This fact was first noticed by Taylor in an analytical study of the related flow instability [5]. Since then, many researchers have studied the instability causing Taylor vortices [6-10]. In the early days of studying Taylor vortices, researchers' attention was mainly focused on determining $Ta_{c1}$ by experimental or analytical methods. As $Ta$ further increases over a higher threshold value ($Ta_{c2}$), the Taylor vortices become unsteady and non-axisymmetric, called wavy vortices [11]. Coles was the first who show the final state of the flow strongly depends on the initial conditions [12]. Davey et al.[11] analytically determined the value of $Ta_{c2}$; that was subsequently confirmed by Eagles' experiment[13]. Seven different states can be identified for a given Reynolds number. As the speed changes these states replace each other. When the inner cylinder is accelerated from rest a series of flow transitions with the following modes occur: Circular Couette Flow (CCF) → Laminar Vortex Flow (LVF) → Wavy Vortex Flow (WVF) → Modulated Wavy Vortex Flow (MWVF) → Turbulent Modulated Wavy Vortex Flow (TMWVF) →Turbulent Wavy Vortex Flow (TWVF) → Taylor Vortex Flow (TTVF)[14].

In this study, the characteristics of the vortices in Taylor-Couette flow is investigated which have the identical cross-sectional geometry as in the experiments of Wereley and Lueptow [10 and 14]. Analysis of wavy and non-wavy laminar and turbulent Taylor-Couette flow, Range of Taylor number for different regimes and the effect of grid on result accuracy are studied.

---

[1] Corresponding author. . E-mail: abolfazl@ferdowsi.um.ac.ir

## 2. Geometric Model

The problem geometry is shown in figure 1. Ratio of inner to outer radius is η=0.8. Γ is the aspect ratio between the height of the working fluid, $L$, and the clearance of the annulus, $d = R_o - R_i$, and denoted as $\Gamma = L \times (R_o - R_i)^{-1} = 50$. In this study Γ was constant while Taylor number is defined by $Ta = \Omega R i d \times (v)^{-1}$, where Ω is a constant angular velocity. Density and kinematic viscosity of working fluid, oil, are equal to $1700\, Kg/m^{-3}$ and $3.1 \times 10^{-6}\, m^2/s$, respectively. The upper and bottom ends of the annulus are solid.

## 3. Mathematical Model

The fluid motion is governed by three-dimensional incompressible Navier-Stokes and continuity equations in the cylindrical coordinate system $(r, \theta, z)$:

$$\nabla.U = 0 \qquad (1)$$

$$\frac{\partial U}{\partial t} + \nabla.(UU) = -\nabla P + \frac{1}{Re}\Delta U \qquad (2)$$

Where U=(u, v, w). The velocity components on the cylinder walls and both end walls were non-slip conditions. Over the whole flow field, initial velocity was zero also $v_r = v_\theta = v_z = 0$ at $r = R_o$ and $v_r = v_z = 0, v_\theta = R_i\Omega$ at $r = R_i$.

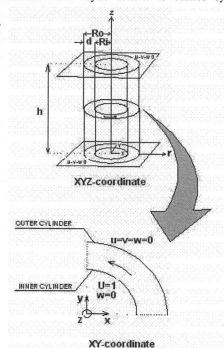

XYZ-coordinate

XY-coordinate

**Figure 1**- Sketch of cylindrical system

## 4. Solution Methodology

Finite volume method and SIMPLE algorithm[15] have been used in order to solve the mentioned equations. The inner cylinder is rotating and the outer is at rest. Classical two-equations models of turbulence (e.g. k-$\varepsilon$ and Reynolds stress model) which set for case of flow with high Reynolds number were confirmed as well as Spalart-Allmaras and k-$\omega$ models, which are used to solve turbulent flows with low Reynolds numbers. The numbers of the grid in a respective analysis is an important factor in forming a flow field and accuracy of the results. In our preliminary analysis, the optimum grid was investigated as 4, 102,and 110 in radial, azimuthal, and axial direction, respectively.

## 5. Results and Discussion

The most significant numerical modeling is achievement of the wavy vortices, which are form in the middle part of the cylinder and close to the end walls of cylinder the wavy vortices are not reached because the influence of boundary conditions are evident. Obviously, the much longer the cylinder is, the better assumptions for wavy vortices formation exist. Whenever, acceptable laminar model and k-$\omega$ model for transient flow which are confirmed, give similar results.

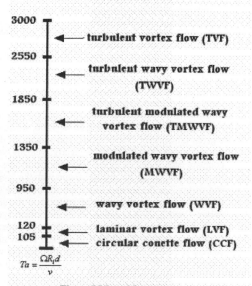

**Figure 2**-Map of flow regime

## 6. Map of flow regimes

Determining the different flow regimes and related Taylor numbers for each area is the most principal result of this study which has been show in figure 2. The axis indicates the Taylor number. This figure should not be considered as sharp boundaries between flow regimes, but simply as a visual guide. The figure describes that at low Taylor numbers, circular Couette flow (CCF) exists in the annulus and transition to laminar vortex flow (LVF) occurs by increasing the Taylor number. After that, Laminar vortex flow becomes unstable and as the result, the vortices develop an azimuthally waviness so the flow is similar in appearance to wavy vortex flow (WVF). Wereley and Lueptow[12] reported that the value of $Ta_{c1}$ over which Taylor vortices are formed is 102 and that the value of $Ta_{c2}$ is in the range of 124< $Ta_{c2}$ <131.Modulated wavy vortices (MWVF) appeared at higher Taylor numbers. The observation of waviness faces to difficulty when the vortex become turbulent at slightly higher Taylor numbers, although the modulated waviness was remained (TMWVF) as shown in Figure2. Regime of turbulent Taylor vortices (TVF), occurs at very high Taylor number. The compression between experimental[12] and numerical results is given at following table. By a flash look at figure 2 and table 1, one can says that there is a good agreement in numerical and experimental results.

Table 1-Comparison between numerical and experimental results of transition between flow regimes at different Taylor numbers.

| REGIMES | CC | LV | WV | MWV | TMWV | TWV | TV |
|---------|----|----|----|-----|------|-----|-----|
| NUM | 0 | 105 | 120 | 950 | 1350 | 1850 | 2550 |
| EXP[12] | 0 | 103 | 125 | 945 | 1400 | 1900 | 2500 |

## 7. Non-Wavy Taylor Couette

Figure 3(a) shows the velocity vector field on an radial-axial plane at Ta=110 that is just above the transition to first critical flow; figure 3(b) shows the result at Ta= 120 which slightly below the transition to wavy vortex flow. The solid curves donated the magnitude contours of azimuthal velocity, while the lower and upper lines represent the inner and the outer cylinder surfaces, respectively. One can notice in both figures that are on the boundaries of Taylor vortices the magnitude of the outer directional velocity is larger than that of the inner directional velocity, and that the two adjacent vortices which have outer directional velocity at their boundary (here after called "vortex pair" ) get closer to each other.  Obviously, the outflow and also the distance between the two vortex centers of a vortex pair are shorter than the inflow and distance between two adjacent vortex pairs.

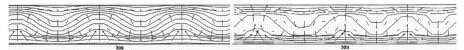

**Figure 3:** Radial-axial velocity vectors in a radial-axial plane for non-wavy Taylor vortex flow.(a) Just above transition to laminar vortex flow, Ta=110 (b) below the transition to wavy vortex flow ,Ta=120. the upper and lower lines represent the inner and the outer cylinder surfaces, respectively.

## 8. Wavy Taylor-Couette

Wavy vortices exhibit oscillations in the azimuthal direction as well as the axial direction. This fact can be explained by a traveling wave. On a given *r-z* plane, oscillation is notice as a traveling wave passes through that plane. Therefore, time-dependent characteristics on that plane must be consistent with those along the azimuthal direction. Figure 4 shows time-dependent result of present study on a typical r-z plane at Ta=130, Ta=250, and Ta=1000, where each frame is taken along the azimuthal direction at a given time (every five second). The symbol (•) denotes the centers of the vortices. In order to show the computed flow field more clearly, velocity vectors are slightly magnified in figure 4(a). If we compare each frame of figure 4(a) with that of five second later, a symmetry rule can be identified. For instance, comparing the second frame with the fourth one in same figure, one can notice that vortex 1 and vortex 2 in the second frame are shifted to the counterparts in the fourth frame and that vortex 2 in the fourth frame is reflection of vortex 1 in the second frame with respect to the boundary between vortex 1 and 2. This is called 'shift-and-reflect' symmetry. (This is a repetitive rule at figure 4(b) and 4(c)). In addition, the effect of the Taylor number on the flow is evident from comparing figure 4. Clearly, increasing Taylor number leads to having stronger vortex. The axial motion of the vortex centers is evident in figure-4(a) and 4(b), but the vortex center motion is diminished at higher Taylor number in figure 4(c). Our numerical results are in a good agreement with experimental results [12].

## 9. Conclusion

In this study, numerical simulations of Taylor-Couette flow are carried out for the identical flow geometry and flow parameters as in the experimental researches [12 and 16]. The most principal result of this study is determining the different flow regimes and also related Taylor numbers for each area. The vortexes in various flow regimes are all consistently reproduced with their experiments. It is confirmed that the 'shift-and-reflect' symmetry is valid for Taylor-Couette flow without an imposed axial flow. The compute speed of vortex translation and the normalized phase speed of traveling waves are in consistent with the experimental and theoretical values. Growth rate of the flow instability is defined and used in predicting the type of the vortices. It is seen that increasing the Taylor number above the second critical Taylor number, makes the fluid velocity much complex, so, one can say that the numerical results obtained are in excellent agreement with their experimental results both qualitatively and quantitatively; therefore, the numerical simulation employed for the current investigation is suitable and economical.

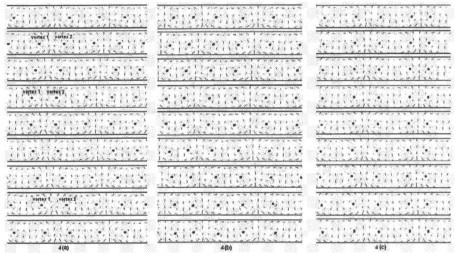

Figure 4: Velocity vector in a radial-axial plane for wavy vortex flow. (a) Just above transition to wavy vortex flow, Ta=130. (b) Ta=250. (c) Ta=1000(MWVF). The rotating inner cylinder is the upper boundary and the fixed outer cylinder is the lower boundary of each frame.

## References

[1] Roesner, K.G., "Hydrodynamic stability of cylindrical Couette flow," Arch. of Mech., 1978, Vol. 30, pp. 619-627.

[2] Brandstater, A. and Swinney, H.L., 1987, "Strange attractors in weakly turbulent Taylor-Couette flow," Phys. Rev. A, Vol.35, No. 5, pp. 2207-2220.

[3] Canuto, C., Hussaini, M.Y., Quarteroni, A., and Zang, T.A., 1988, Spectral Methods in Fluid Dynamics, 3rd edn. Springer-Verlag, Berlin.

[4] Ali, M. and Weidman, P.D., 1990, "On the stability of circular Couette flow with radial heating," J. Fluid Mech. Vol.220, pp.53-84.

[5] Taylor, GI., "Stability of a viscous liquid contained between two rotating cylinders," Phil Trans R Soc 1923; A223:289-343.

[6] Stuart, JT., "ON the non-linear mechanics of hydrodynamic stability," J. Fluid Mech. 1958; 4:1-21.

[7] Davey, A., "The growth of vortices in flow between rotating cylinders," J. Fluid Mech. 1962;14:336-68.

[8] Esfahani, J. A., Nabovati, A., "Numerical Analysis of Taylor Couette Flow", Proceeding of the 8th conference of Fluid Dynamics, Tabriz, Iran, 2003, (In Persian).

[9] Akonur, A., Lueptow, RM., "Three-dimensional velocity field for wavy Taylor–Couette flow", Physics of Fluids, Vol. 15, No. 4, pp. 1-13, 2003.

[10] Wereley, ST., Lueptow, RM., "Spatio-temporal character of non-wavy and wavy Taylor Couette Flow", Journal of Fluid Mechanic, Vol. 364, 1998. pp. 59-80.

[11] Davey, A., Diprima, RC., Stuart, JT., "On the instability of Taylor vortices," J Fluid Mech. 1968;31:17-52.

[12] Coles, D., "Transition in circular Couette flow," J. Fluid Mech., 1965, Vol.21, pp. 385-425.

[13] Eagles, PM., "On the stability of Taylor vortices by fifth-order amplitude expansions," J. Fluid Mech. 1971;49:529-50.

[14] Wereley, ST., Lueptow, RM., "Velocity field for Taylor-Couette flow with an axial flow," Phys. Fluids, 1999; 11(12):3637-49.

[15] Versteeeg, H.K., Malalasekera, W., An Introduction to Computational Fluid Dynamics-The finite volume method, $2^{nd}$ Edition, Longman Group, 1996.

Brill Academic Publishers
P.O. Box 9000, 2300 PA Leiden
The Netherlands

*Lecture Series on Computer
and Computational Sciences*
Volume 4, 2005, pp. 933-936

# Application of Nonlinear Regression Method in the Evaluation of Piezoelectric Materials

Z.X. Yu[1], Y.R Shi, H.Y Hsu, L.X Kong

Centre for Advanced Manufacturing and Research,
University of South Australia,
Mawson Lakes, Adelaide

Received 2 July, 2005; accepted in revised form 21 July, 2005

*Abstract:* This article applies nonlinear regression method in characterizing a lossy piezoelectric material, i.e. PVDF (polyvinylidene fluoride), which has been widely used for ultrasonic transducer design. The characterization is based on the measurement of electrical impedance. A computer based data acquisition system is built to collect the data. The fitting process is then applied to the collected data. The best fit parameters are achieved at the selected frequency range, which is in good agreement between the modeled impedance curve and experimental data points at resonance region. The whole data collection and evaluation process can be easily and fast implemented within a few minutes. The design and simulation of piezoelectric materials based MEMS ultrasonic transducer is one of the motivations for developing this method. The evaluated piezoelectric material parameters will be imported to the later transducer modeling to improve the design accuracy of MEMS devices with a reduced design period.

*Keywords:* PVDF, Piezoelectric, Characterization, Nonlinear Regression Method

*PACS:* 77.8.-s; 43.38.+n

## 1. Introduction

As a lossy piezoelectric material, PVDF exhibits higher mechanical loss and dielectric loss, and lower electromechanical coupling factor than piezoceramics. Due to its special features in mechanical abilities and acoustic performance [1], it has found many successful applications in ultrasound transducers. However the high dielectric and elastic losses of polymers present special challenges for characterizing their material properties and modeling their acoustic performance, since the conventional piezoceramics characterization method is not any more applicable to piezopolymers. This paper will investigate application procedures using nonlinear regression method to characterize piezoelectric polymers, meanwhile, the methods also apply to other piezoelectric materials. It will then help us to improve our ability in the design of MEMS transducers by improving the accuracy of material modeling.

Losses in piezoelectric transducers can be divided into three types: mechanical, dielectric and piezoelectric losses. They are introduced by considering the corresponding parameters (elastic, dielectric and piezoelectric coefficients— $C^D$, $\varepsilon^s$, $k$ ) as imaginary magnitudes that are frequency dependent . To account for the losses, all these three material parameters are treated as complex quantities. An asterisk can be added to each parameter as the superscript to represent the complex value. Since the transducer to be designed is basically working in the thickness mode, all the parameters and relationships will be considered in the thickness direction only. Therefore the general complex expressions for these three parameters can be written as :

$$C_{33}^{D*} = C_{33}^{D}(1 + i \tan \delta_M) \tag{1}$$

$$\varepsilon_{33}^{S*} = \varepsilon_{33}^{S}(1 - i \tan \delta_E) \tag{2}$$

$$k_t^* = k_t(1 + i \tan \delta_K) \tag{3}$$

where $\tan \delta_M$ , $\tan \delta_E$, and $\tan \delta_K$ are the so called loss factors in relation to mechanical, dielectric and piezoelectric losses, respectively.

---

[1] Corresponding author. E-mail: zhenxian.yu@postgrads.unisa.edu.au

## 2. Impedance Measurement

The electrical impedance is measured by HP8752A Network Analyzer, which has a wide sweeping frequency ranging from 3000 KHz to 3 GHz. The measurement principle can be referred to the impedance measurement handbook by Agilent [2]. To reduce the loss in the transmission line, the connection cable is calibrated before the measurement. The communication between the instrument and computer is realized by a National Instrument GPIB card, and an operation program written in LabVIEW is run to instruct the operator to finish the whole setup, calibration, and data collection process. The collected data is stored in a prescribed folder for the later use in fitting process.

The samples used for the evaluation are purchased from MSI, which are double side metallized PVDF thin films. Figure 1 shows the schematics of the film element and its measurement model. The lateral dimensions of the elements are large comparing with their thicknesses. Therefore the electrical impedance in thickness mode can be expressed as [3],

$$Z(f) = \frac{t}{i2\pi f \varepsilon_{33}^S A}\left[1 - k_t^2 \frac{\tan\left(\pi f t\sqrt{\frac{\rho}{C_{33}^D}}\right)}{\pi f t\sqrt{\frac{\rho}{C_{33}^D}}}\right] \tag{4}$$

where $t$, $A$ and $\rho$ are corresponding to the material physical parameters: thickness, effective area and density, respectively. The detailed values of the sample are listed in Table 1.

Table 1: Sample properties

| Sample | Thickness | Density | Dimension [mm] | Electrode | Electrode Thickness |
|---|---|---|---|---|---|
| PVDF thin film | 70 um | 1780 kg/m³ | 16(12)×41(30) | Ag ink | <0.1um |

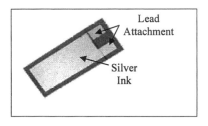
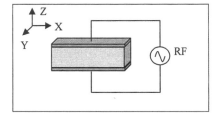

Figure 1: PVDF thin film element and its measurement model.

## 3. Fitting Process

Nonlinear regression is a very popular tool to analyze experimental data and determine the model parameters. The least squares principle is used to estimate the parameters in nonlinear models. A detailed introduction of the application of nonlinear regression method can be found in reference [4]. Here we will follow the seven basic steps summarized in [5] to implement the fitting process:

Step 1:   Clarify the goal,
Step 2:   Prepare the data
Step 3:   Choose your model
Step 4:   Decide which model parameters to fit and which to constrain
Step 5:   Choose a weighting scheme
Step 6:   Choose your initial values
Step 7:   Perform the curve fit and interpret the best-fit parameter values

The goal of this fitting process is to determine the three complex material parameters given in Equation (4). The data collection has been described in section 2. Due to the complexity of the impedance values, the experimental measurement results usually include two components—amplitude and phase angle. According to Equations (1) ~ (4), there are three complex parameters to be determined, which can be separated into six real numbers for these two fitting models.

To choose the right parameters for each fitting curve, the sensitivity of each parameter to both fitting curves are studied. The results indicate that the phase curve is very sensitive to the change of $C_{33}^D, \tan\delta_M, k_t, \tan\delta_K$, and $\tan\delta_E$, while $\varepsilon_{33}^S$ has a strong impact on the amplitude curve. Therefore $C_{33}^D, \tan\delta_M, k_t, \tan\delta_K, \tan\delta_E$ are chosen as the fitting parameters in phase model, and $\varepsilon_{33}^S$ as the amplitude model. With the determination of the parameters for each fitting model, an iterative fitting process composed of three major steps is developed (Figure 2). In the first two steps, the material parameters are extracted by nonlinear regression method, where all the parameters are considered as frequency independent. In the third step, the dielectric relaxation phenomenon is considered in the model, and the corresponding Cole-Cole distribution of dielectric constant is extracted to explain the discrepancy between the calculated curve and experimental curve.

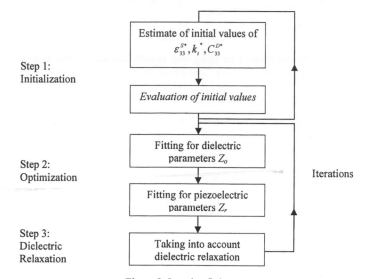

Figure 2: Iterative fitting process.

The initial values are calculated with the procedures developed above, which are also compared with Sherrit's method [6]. Although the starting values are quite different with these two methods, the fitting process converged to the same result. It is obvious that by using nonlinear fitting regression method the influences of initial values will be reduced, which will then simplify the piezoelectric material characterization process. At the resonance region, the piezo material will have least loss, hence a corresponding frequency range is selected for the fitting process. Figure 3 shows the fitting result without considering dielectric relaxation, while the discrepancies are observed outside the resonant frequency region, which is considered to be caused by the frequency dependencies of those fitting parameters. According to previous sensitivity analysis, it is found that the impedance is mostly contributed by the dielectric properties of the material outside the resonance region. Therefore the frequency dependency of complex dielectric constant can be modeled to compensate the discrepancy. The frequency dispersion of the complex dielectric permittivities can be modeled by Cole-Cole distribution of relaxation times. With the consideration of dielectric relaxation, an improved fitting result can be achieved.

## 4. Conclusion

This research work aims to improve the piezoelectric material modeling in the simulation of a MEMS transducer. The evaluated material parameters will then be imported to our transducer model to improve the simulation results. As we expected, the material parameters evaluated by the nonlinear regression fitting process improved the agreement between the simulation and experimental results after being imported to our transducer model built in ANSYS, see Figure 4.

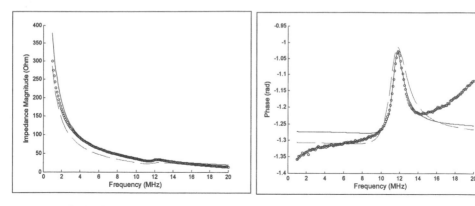

Figure 3: Calculated results (solid line) after fitting with initially calculated curve from initial values (dashed line) and experimental results ('o' line), fitted frequency ranges from 8.5 MHz to 15MHz.

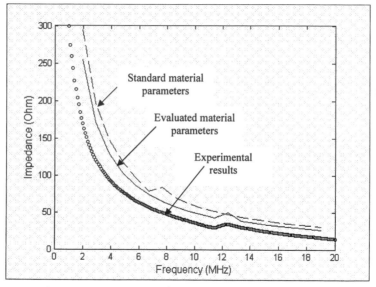

Figure 4: Verification of evaluated material parameters in ANSYS.

## Acknowledgments

The authors wish to thank the support of the Institute for Telecommunications Research at the University of South Australia for providing the measurement equipment and generous help.

## References

1. Brown, L.F., *Design considerations for piezoelectric polymer ultrasound transducers.* IEEE transactions on ultrasonics, ferroelectrics, and frequency control, 2000. **47**(6): p. 1377-1396.
2. *Impedance measurement handbook.* 2003, Agilent Technologies.
3. *IEEE standards on piezoelectricity.* 1987.
4. Rawlings, J.O., *Applied regression analysis: a research tool.* 2rd ed. 1998, New York: Springer.
5. Motulsky, H. and A. Christopoulos, *Fitting models to biological data using linear and nonlinear regression method.* 2003, GraphPad Software., Inc.
6. Sherrit, S., et al., *0-3 Piezoelectric-glass composites.* Ferroelectrics, 1992. **134**: p. 65-69.

Brill Academic Publishers
P.O. Box 9000, 2300 PA Leiden,
The Netherlands

*Lecture Series on Computer
and Computational Sciences*
Volume 4, 2005, pp. 937-941

# Computational Molecular Science: Hybrid Classical-Quantum and Force Field Methods

## Per-Olof Åstrand[1]

Department of Chemistry,
Norwegian University of Science and Engineering (NTNU),
7491 Trondheim, Norway

Received 14 August, 2004; accepted 14 August, 2004

*Abstract:* In theoretical chemistry, different levels of sophistication are used for molecular models. In this mini-symposium, different "classical" models are discussed that are used either as an alternative to or combined with molecular quantum mechanics. The models include force field methods for intermolecular interactions to calculate properties of liquids and solutions by molecular dynamics simulations, force field methods to obtain atomic properties such as atomic charges and polarizabilities, and combined quantum mechanical and classical models to study solvation of molecules. Here, these methods are briefly introduced.

*Keywords:* Solvation, Intermolecular interactions, Polarization, Molecular mechanics, Force fields, Dielectric models, QM/MM

*PACS:* 31.15.Gy, 31.15.Qg, 31.50.Bc, 31.70.Dk, 31.90.+s, 33.15.Kr, 34.20.Gj

## 1 Introduction

Molecular modelling has today a central role in chemistry and includes a large variety of methods such as molecular quantum mechanics, molecular mechanics (force field) methods, statistical mechanical simulations of macromolecules, liquids and solutions [1]. In molecular quantum mechanics (quantum chemistry), the Schrödinger equation is solved for the molecular Hamiltonian, which results in an *ab initio* representation of the electronic charge distribution in terms of molecular orbitals [2]. Alternatively, density functional theory (DFT) methods may be adopted to obtain the electronic charge distribution [3]. However, quantum mechanical methods are still relatively demanding from a computational point of view, and their usage in the study of macromolecules and in simulations of liquids and solutions are limited.

Alternative, less sophisticated, methods exist that allow for studies of large systems and that can be used to obtain statistical mechanical averages. In molecular mechanics, intra- and intermolecular interaction energies are obtained from a simple analytical function and a set of atom-type parameters [4], which results in a very rapid calculation of interaction energies. These force fields are central in molecular dynamics and Monte Carlo simulations of liquids and solutions [5, 6]. Force fields, however, rely on a set of atom-type parameters that has to be parametrized either from experimental data or quantum chemical calculations.

Solvent effects on molecular properties are modelled with a large variety of methods. In the study of molecular clusters, the entire cluster may be treated with quantum chemical methods.

---

[1] E-mail: per-olof.aastrand@chem.ntnu.no

For macroscopic solvation, however, hybrid quantum-classical methods are often employed [7]. The molecule of interest is treated with *ab initio* methods, whereas the surroundings is modelled with less sophisticated. In continuum models, the molecule is enclosed in a cavity surrounded by a dielectric medium [8, 9]. Alternatively, the surroundings is described with molecules represented by a force field in combined quantum mechanical and molecular mechanics (QM/MM) models [10].

## 2 Force fields and simulations

In force fields, the potential energy, $V_{\text{pot}}$ is often divided into different energy contributions,

$$V_{\text{pot}} = V_{\text{intra}} + V_{\text{ele}} + V_{\text{rep}} + V_{\text{ind}} + V_{\text{disp}} , \tag{1}$$

where $V_{\text{intra}}$ contains intramolecular contributions, $V_{\text{ele}}$ is the electrostatic energy, $V_{\text{rep}}$ is the short-range repulsion energy, $V_{\text{ind}}$ is the induction energy, and $V_{\text{disp}}$ is the dispersion energy. The intramolecular energy term normally consists of bond stretch terms, angle bend terms, and torsional contributions [4]. The functional form of the intermolecular terms, $V_{\text{ele}}$, $V_{\text{rep}}$, $V_{\text{ind}}$ and $V_{\text{disp}}$, may be investigated by using perturbation theory [11, 12, 13, 14].

The electrostatic energy is in most cases calculated by a Coulomb term, where the charge distribution is represented by atomic charges and sometimes atomic dipole and quadrupole moments [14]. Atomic charges may be obtained by several different methods, as for example the partitioning of the molecular wave function in terms of Mulliken charges or the fitting of the electrostatic potential around the molecule. One method, the electronegativity equalization method (EEM), is based on the assignment of an atomic electronegativity and chemical hardness to each atom and that the molecular electronegativity (chemical potential) is equalized by a charge redistribution from among the atoms in a molecule [15].

The induction (electronic polarization) energy is in most cases calculated from an atomic induced dipole moment, $\mu_{I,\alpha}^{\text{ind}}$, as

$$\mu_{I,\alpha}^{\text{ind}} = \alpha_{I,\alpha\beta} E_{I,\beta}^{\text{tot}} , \tag{2}$$

where $\alpha_{I,\alpha\beta}$ is an atomic polarizability, and $E_{I,\beta}^{\text{tot}}$ is the total electric field at atom $I$. The total electric field consists of contributions from both the permanent electric moments and the atomic induced dipole moments of all surrounding molecules. Thus, the induction energy is a true many-body energy and the calculation of the induction energy involves the solution of a set of coupled equations to obtain the atomic induced dipole moments. The molecular polarizability may be divided into atomic contributions by many different methods [13, 14]. One method is based on the classical point-dipole interaction (PDI) model, where a set of atomic polarizabilities is studied in an external electric field and parametrized to reproduce known molecular polarizabilities [16].

The main application of force fields is in molecular dynamics or Monte Carlo simulations of liquids and solutions. Both simulation methods rely on a sampling of the phase space which requires that the energy of the system and in molecular dynamics, the force on each particle can be calculated repeatedly (perhaps $10^9 - 10^{12}$ times) [5, 6]. Thus, the calculation of energies and forces have to be efficient. In principle all structural, dynamical and thermodynamical properties of a solution can be obtained from simulations. It is noticed, that the calculation of free energies are particularly cumbersome and that specific methods have been developed to obtain free energy differences [17].

## 3 Solvent models

Implicit solvent models are based on an *ab initio* model of a molecule surrounded by a solvent represented by a dielectric constant. If the classical Onsager model of a point dipole moment, $\mu_\alpha$, placed in the centre of a spherical cavity of radius $R$ within a dielectric continuum with the dielectric constant $\varepsilon$, the reaction field, $E_\alpha^{RF}$, at the point dipole moment arising from the polarization of the dielectric medium is given as [8]

$$E_\alpha^{RF} = \frac{2\,(\varepsilon - 1)}{2\varepsilon + 1}\frac{\mu_\alpha}{R^3}\ .\tag{3}$$

Similarly, the reaction field may be obtained from a molecular charge distribution, and thus it may be included in the Hamiltonian, $\hat{H}$, in addition to the molecular Hamiltonian,

$$\hat{H} = \hat{H}_{mol} + \hat{H}_{RF}\ .\tag{4}$$

A quantum chemical model including a dielectric medium requires a specification of the dielectric constant and the shape of the cavity. A commonly adopted model is the polarizable continuum model (PCM) which relies on a molecule-shaped cavity in the sense that the cavity is determined by a van der Waals radius for each atom of the system [9].

Alternatively, the surrounding medium may be modelled with molecules represented by a force field in combined quantum mechanical and molecular mechanics (QM/MM) models [10]. Normally, the classical electrostatics and sometimes the polarization terms of the force field are included in the Hamiltonian,

$$\hat{H} = \hat{H}_{mol} + \hat{H}_{QM/MM} + \hat{H}_{MM}\ ,\tag{5}$$

that gives the perturbed molecular wave function. In Eq. (5), $\hat{H}_{mol}$ is the regular quantum mechanical molecular Hamiltonian, $\hat{H}_{QM/MM}$, is the interaction between the core molecules described by quantum mechanics and the surrounding molecules modelled by molecular mechanics, and $\hat{H}_{MM}$ contains the interactions between the molecules described by molecular mechanics.

An implicit dielectric model is an average over the solvent and thus only one quantum mechanical calculation is required to obtain a solvent contribution to a molecular property. On the other hand, a configuration of molecular mechanics molecules describes only one single point in the phase space, and a statistical mechanical averaging of the positions of the surrounding molecules is required. Normally, only a polarization term is included in dielectric models, whereas both electrostatic and polarization terms are in most cases included in QM/MM models. Therefore, QM/MM models are more suitable to study for example systems where hydrogen bonding is important since the hydrogen bonding is in principal governed by an orientationally dependent electrostatic energy term. In very few cases, dispersion and exchange repulsion has been included in the Hamiltonian coupling the quantum mechanical system and the surroundings.

## Acknowledgment

The author has received support from the Norwegian Research Council (NFR) through a Strategic University Program (Grant no 154011/420) and a NANOMAT program (Grant no 158538/431).

## References

[1] A. R. Leach, Molecular modelling: Principles and applications, 2nd Edition, Prentice Hall, 2001.

[2] T. Helgaker, P. Jørgensen, J. Olsen, Molecular electronic-structure theory, Wiley, Chichester, 2000.

[3] R. G. Parr, W. Yang, Density-Functional Theory of Atoms and Molecules, Oxford Science, Oxford, 1989.

[4] A. K. Rappé, C. J. Casewit, Molecular mechanics across chemistry, University Science Books, Sausalito, 1997.

[5] M. P. Allen, D. S. Tildesley, Computer Simulations of Liquids, Clarendon, Oxford, 1987.

[6] D. Frenkel, B. Smit, Understanding Molecular Simulation: From algorithms to applications, 2nd Edition, Academic Press, San Diego, 2002.

[7] C. J. Cramer, Essentials of Computationl Chemistry: Theories and Models, Wiley, Chichester, 2002.

[8] C. J. F. Böttcher, Theory of Electric Polarization, 2nd Edition, Vol. 1, Elsevier, Amsterdam, 1973.

[9] J. Tomasi, M. Persico, Molecular interactions in solution: an overview of methods based on continuous distributions of the solvent, Chem. Rev. 94 (1994) 2027–2094.

[10] A. Warshel, S. T. Russell, Calculations of electrostatic interactions in biological systems and in solutions, Quart. Rev. Biophys. 17 (1984) 283–422.

[11] A. D. Buckingham, Permanent and induced molecular moments and long-range intermolecular forces, Adv. Chem. Phys. 12 (1967) 107–142.

[12] M. Margenau, N. R. Kestner, Theory of Intermolecular Forces, Pergamon, Oxford, 1969.

[13] A. J. Stone, The Theory of Intermolecular Forces, Clarendon Press, Oxford, 1996.

[14] O. Engkvist, P.-O. Åstrand, G. Karlström, Accurate intermolecular potentials obtained from molecular wave functions: Bridging the gap between quantum chemistry and molecular simulations, Chem. Rev. 100 (2000) 4087–4108.

[15] P. Geerlings, F. De Proft, W. Langenaeker, Conceptual density functional theory, Chem. Rev. 103 (2003) 1793–1873.

[16] J. Applequist, An atom dipole interaction model for molecular optical properties, Acc. Chem. Res. 10 (1977) 79–85.

[17] P. Kollman, Free energy calculations: applications to chemical and biochemical phenomena, Chem. Rev. 93 (1993) 2395–2417.

## Per-Olof Åstrand

Prof. Per-Olof Åstrand was borned in 1965 in Sätila, Sweden. He received his Ph.D. in 1995 in Theoretical Chemistry at the University of Lund with Prof. Gunnar Karlström as his supervisor. In the period 1995-2002, he worked as an assistant professor in Denmark at the University of Aarhus aand at the University of Copenhagen (with Prof. Kurt V. Mikkelsen), and at Research Centre Risø. Since 2002, he is a professor in theoretical chemistry at the Norwegian University of Science and Technology (NTNU) in Trondheim, Norway. He has around 60 publications in international journals, and his research interests include intermolecular forces, force field methods for electric properties, and molecular simulations.

Brill Academic Publishers
P.O. Box 9000, 2300 PA Leiden,
The Netherlands

*Lecture Series on Computer*
*and Computational Sciences*
Volume 4, 2005, pp. 942-944

# Proton Binding to Proteins: Insights from Continuum Models with a Linear Response Approximation

## G. Archontis[a1] and T.Simonson[b]

[a]Department of Physics, University of Cyprus, PO20537, CY1678, Cyprus [b]Laboratoire de Biochimie (UMR7654 du CNRS), Department of Biology, Ecole Polytechnique, 91128 Palaiseau, France.

Received 14 August, 2005; accepted 14 August, 2005

*Abstract:* We describe $pK_a$ results for two buried and one solvent-exposed aspartate in two proteins, using a Linear Response approach, as well as a 'standard' Poisson-Boltzmann (PB) approach. The methods account explicitly for protein relaxation by averaging over conformations from the two endpoints of the proton binding reaction. For two buried carboxylates, the model yields $pK_a$ shifts in better agreement with experiment, than a molecular dynamics free energy calculation. For a solvent-exposed carboxylate, desolvation by the protein cavity is overestimated by 2.9 $pK_a$ units; several effects could explain this. 'Standard' PB methods downscale incorrectly charge-charge interactions by using a large, *ad hoc* protein dielectric; as a result, they give an unbalanced description of the reaction and a large error for the shifted $pK_a$'s of Asp26 and Asp14.

*Keywords:* Poisson-Boltzmann, molecular dynamics, free energy calculations, proton binding

*PACS:* 87.15Aa, 87.15.He, 87.15.Nn, 87.15.Kg

## 1 Introduction

The accurate determination of aminoacid $pK_a$'s in proteins is of fundamental interest in biophysical chemistry. The ionization state of titratable aminoacids in solution is known from experiment. However, in a folded protein, the $pK_a$'s can be shifted with respect to the solution values. Theoretical methods to calculate protein $pK_a$'s have been the focus of considerable efforts in the past two decades [1, 2, 3, 4]. A widely-used "standard" approach treats the protein as a low-dielectric cavity immersed in a high-dielectric solvent, and determines the electrostatic free energy at various ionization states by solving the Poisson-Boltzmann (PB) equation. In the standard approach it is customary to treat the protein dielectric constant as an adjustable parameter, which accounts for protein structural reorganization upon a change in the ionization state (proton binding or release), as well as for nonelectrostatic effects. [1, 5].

In this work, we analyze three carboxylate proton-binding reactions with a linear response approximation, combined with a Poisson-Boltzman model (referred to as the PB/LRA method [6]). Two of the carboxylates correspond to interior Asp residues, with high $pK_a$ shifts; Asp26 in thioredoxin, with a large $pK_a$ of 7.5 [7] and Asp14 in ribonuclease A, with a low $pK_a$ of 2.0 [8]. Such interior residues represent the proper benchmarks to test the accuracy of a $pK_a$ calculation [1]. The third carboxylate is the solvent-exposed Asp20 in thioredoxin, with an unshifted $pK_a$ of

---

[1]Corresponding author. E-mail: archonti@ucy.ac.cy

4.0 [8]. The results are compared to experiment and to a recent calculation of the same $pK_a$ shifts by MDFE simulations in explicit solvent [2]. Semi-quantitative agreement with MDFE and with the experimental $pK_a$ shifts is obtained using a low, physically reasonable value of the protein dielectric constant (one or two).

## 2 Methods

**The PB/LRA expression:** The aspartate proton binding is treated by inserting a +1 charge on sidechain atoms of the titratable aspartate. The total free energy change due to the insertion of a single point charge at position "0" is in the linear response approximation [5, 6]

$$\Delta G = \frac{1}{2}q(V_0^{\text{prod}} + V_0^{\text{reac}}).$$ (1)

where $V^{\text{reac/prod}}$ is the equilibrium electrostatic potential at the insertion site "0", in the reactant or product state. The above equation shows that the total electrostatic free energy change can be calculated by averaging the interaction energies between the inserted charge and the permanent and induced charges of the reactant and product states. Generalizing these results to the insertion of several charges is straightforward (see, eg, [5] for details).

**Numerical methods:** The electrostatic potentials for a particular protein structure and protonation state are evaluated by solving the Poisson equation numerically with the UHBD and MEAD programs [9, 10] for several hundred protein structures, generated by MD simulations. The protein and solvent are treated as two homogeneous dielectric media, with dielectric constants $\epsilon = 1$–4 and $\epsilon^{\text{w}} = 80$, respectively. The reacting sidechain is treated as a cavity of $\epsilon = 1$, unless otherwise noted. More details are supplied in [6].

## 3 Results

We have applied the PB/LRA method to three aspartate sidechains in two proteins; the buried aspartates Asp26 (in thioredoxin) and Asp14 (in RNAseA), with significant $pK_a$ shifts, and solvent-exposed Asp20 in thioredoxin, with an unshifted $pK_a$. The results are listed in Table 1, along with predictions from molecular dynamics free energy simulations (MDFE), experimental $pK_a$ measurements, and computations with a standard PB protocol.

For the buried aspartates (Asp26 in thioredoxin and Asp14 in RNAseA), the method yields better agreement with experiment than do the MDFE simulations [2]. For these aspartates, the method captures correctly the balance between protein reorganization, unfavorable desolvation, and favorable interactions of the Asp carboxylate with proximal protein residues. Importantly it also reproduces semi-quantitatively the electrostatic potentials, the protonation free energies and the Marcus reorganization free energies (see ref. [6]), essentially without adjustable parameters. Agreement with experiment is rather poor for the unshifted Asp20. In this case, desolvation is overestimated, possibly due to charge rearrangement on the Asp sidechain, bias in the MD conformations, or underestimated local solvent ordering.

The $pK_a$s are also calculated by a "standard" PB protocol [3]. This protocol uses a single protein endpoint structure. Because it does not include the protein relaxation explicitly, the "standard" method yields $pK_a$ shifts that depend strongly on the structural model assumed for the protein (see Table 1). When a low dielectric value is used for the protein ($\epsilon = 4$, table 1 or 2 [6]), the dependence on the structural model is more pronounced; the method fails in all three carboxylate calculations. When the protein dielectric is increased to $\epsilon \approx 20$, the "standard" method yields $pK_a$ shifts in reasonable agreement with experiment for two cases, thanks to a fortunate compensation of errors.

Table 1: Summary of pK$_a$ shifts, calculated with PB/LRA, the "standard" PB protocol and MDFE.

| residue | PB/LRA[†] | Standard | | | | MDFE | Exp. |
|---|---|---|---|---|---|---|---|
| | | Cavity[‡] | No Cavity[‡♯] | Cavity[*] | No Cavity[*♯] | | |
| **Asp26** | 5.8(1.5) | 1.4/-0.7[‖] | 0.0/-1.4 | 5.8/-2.8 | 3.5/-3.7 | 7.8 | 3.5 |
| **Asp14** | -1.4(1.5) | 2.2/0.0 | -0.7/-2.0 | 5.0/-3.6 | 0.4/-5.6 | 0.0 | <-2.0 |
| **Asp20** | 3.7(0.4) | 2.2/1.4 | 0.8/0.7 | 2.9/2.2 | 1.7/1.2 | 0.6 | 0. |

[†] The PB/LRA values correspond to averages of the results with a protein $\epsilon = 1$ and $\epsilon = 2$. [‡] Using a protein dielectric of twenty. [♯] The sidechain is embedded directly in the protein medium. [‖] Throughout the table, $x/y$ denote results using structures from the protonated/ionized state. [*] Using a protein dielectric of four.

However, this high dielectric value is unphysical for the present systems, because it is inconsistent with the atomic charge set and the actual magnitude of the protein dielectric relaxation [6].

# 4 Conclusions

The ionization free energy of a titrating sidechain depends on the desolvation penalty to transfer a net charge from solution into the protein cavity, and on the interactions between the transferred charge and polar residues in the protein. The relative magnitude of these two factors depends on the geometry of the particular protein, the location of the aminoacid with respect to the protein-solvent interface, and the protein structural reorganization upon ionization. By accounting explicitly for the protein reorganization and using a physically consistent, low dielectric constant, the PB/LRA method yields better agreement with experiment for the two buried aspartates studied here than a molecular dynamics free energy approach or a standard PB protocol.

# References

[1] C. N. Schutz and A. Warshel. *Proteins:Structure, Function and Genetics* **44** (2001). 400-417.

[2] J. Carlsson T. Simonson and D.A. Case. *JACS* **126**, 4167–4180 (2004).

[3] D. Bashford and M. Karplus. *Biochemistry* **29**, 10219–10225 (1990).

[4] J. Warwicker. *Protein Science* **8**, 418–425 (1999).

[5] T. Simonson, G. Archontis, and M. Karplus. *J. Phys. Chem. B* **103** (1999). 6142-6156.

[6] G. Archontis and T. Simonson. *Biophys. J.* **88**, 3888–3904 (2005).

[7] K. Langstemo, J. Fuchs, and C. Woodward. *Biochemistry* **30**, 7603–7609 (1991).

[8] W. Forsyth, J. Antonisiewicz, and A. Robertson. *Proteins* **48**, 388–403 (2002).

[9] J. Madura et. al. *Comp. Phys. Comm.* **91**, 57–95 (1995).

[10] D Bashford. In Vol. 1343 of Lecture Notes in Computer Science Berlin, 1997. Y. Ishikawa, R.R. Oldehoeft, J.V.W. Reynders, and M. Tholburn, editors. ISCOPE97, Springer, (1997).

Brill Academic Publishers
P.O. Box 9000, 2300 PA Leiden,
The Netherlands

*Lecture Series on Computer*
*and Computational Sciences*
Volume 4, 2005, pp. 945-948

# Multicenter Multipolar Representation of Electrostatic Potential for Flexible Molecules

## M. L. Chodkiewicz[1]

Chemistry Department
University of Warsaw,
Pasteura 1, 02-093 Warsaw, Poland

Received 8 July, 2005; accepted in revised form 17 July, 2005

*Abstract:* A method for generating a compact multicenter multipolar representation of the electrostatic potential (EP) for flexible molecules is presented. The fitting procedure used adopts the Least Absolute Shrinkage and Selection Operator (LASSO) technique (R. Tibshirani, J. Roy. Stat. Soc. B 58, 267 (1996)) which can be seen as penalized ordinary least squares (OLS). The constrains optimized for the particular molecule of interest effectively removes redundant multipoles. It is shown that the use of multiple conformations is crucial for the predictive ability of the EP model for flexible molecules. The multipole local coordinate systems are chosen in a way that best reflects the key conformational changes.

*Keywords:* electrostatic potential, distributed multipoles

*Mathematics Subject Classification:* 62H99

*PACS:* 34.20.Gj, 31.15.-p, 33.15.Bh, 02.30.Sa, 02.60.-x

## 1 Introduction

An accurate representation of the electrostatic part of the intermolecular interaction potential is crucial for molecular simulations. Multipole expansion is one of the most convenient approaches. The proposed method of generating an EP representation is based on fitting multipole components to the EP. The approach is designed to produce an EP model fulfilling the following requirements: (1) an ability to represent the EP over a range of possible molecular conformations; (2) elimination of redundant multipoles which provide only minor improvement of the EP representation; (3) optimizing the multipole description with respect to predictive power, rather than its ability to describing just a training set of conformations; (4) to accurately reflect the chemical structure of the molecule (e.g. bond directionality) in the definition of multipoles.

## 2 The method

The method minimizes an error expressed as a difference between the quantum mechanical electrostatic potential $\phi_{QM}$ and model-based $\phi_M$ EPs. The error takes the form of a weighted sum over individual conformations of a given molecule:

---

[1] Corresponding author. E-mail: mchod@alfa.chem.uw.edu.pl

$$Q = \sum_{i=1}^{N} w_i \sum_{k=1}^{k} \left( \phi_{QM}^i(\mathbf{r}_k) - \phi_M^i(\mathbf{r}_k) \right)^2 = \sum_{i=1}^{N} w_i \sum_{k=1}^{k} \left( \phi_{QM}^i(\mathbf{r}_k) - \sum_{j=1}^{M} b_j m_j^i(\mathbf{r}_k) \right)^2 \qquad (1)$$

subject to:

$$\sum_{j=1}^{M} |b_j| = \|\mathbf{b}\|_1 \leq t \qquad (2)$$

Here $t$ is a tuning parameter; $m_{ji}(\mathbf{r})$ is the EP generated by j-th normalized multipole in the i-th molecule; and the $b_j$ are multipole components.

Set of conformations is divided into training and testing sets. Training set is used to obtain multipole components from eq.(1) and testing set is used for estimation of predictive power of the obtained model. Final model is generated using the tuning parameter which maximize predictive power of the model.

In ordinary least squares (OLS) approach (which is commonly used for fitting multipoles), an error is reduced with respect to training instead of testing set. OLS is known to be reliable only when the correlation matrix is close to the identity matrix [1]. Otherwise (if some variable are near correlated) changes in some variable can be compensated by an adjustment of another, it allows to describe some details of EP of the training set but fails with the testing set. This is known as an "overfitting" problem.

We applied the LASSO (the least absolute shrinkage and selection operator) method [3] to overcome "overfitting" problem. This method both selects variables and penalizes values of predictors. The set of constraints implied by eq.(2) is equivalent to adding a penalty term $\lambda \|\mathbf{b}\|_1$ to the OLS error expression. The LASSO method can produce some coefficients that are exactly equal to zero, therefore eliminating the corresponding variable from the model. Another feature of the penalty function used is invariance of the resulting model with respect to variable rescaling - if one would like to fit "half charges" insted of charges to the electrostatic potential, the final result would be the same as for charges (this property is unique for LASSO penalty function). Moreover, the penalties imposed on variables are independent of their magnitudes.

In order to preserve a correspondence between the EP generated by the model and the varying molecular geometry, the definition of multipoles are based on the molecular structure (specifically bond directions). EP corresponding to n-th order multipole component be expressed as follows:

$$\phi = \frac{1}{r^{2n+1}} \hat{R} p(\mathbf{r}, G), \qquad (3)$$

$$p(\mathbf{r}, G) = (\mathbf{r}^T \mathbf{r}_1)(\mathbf{r}^T \mathbf{r}_2)...(\mathbf{r}^T \mathbf{r}_n) \qquad (4)$$

where $\hat{R}$ projects $p(\mathbf{r}, G)$ on the space of regular solid harmonics, $G$ stands for a molecular geometry and $\mathbf{r}_i$ is a versor in i-th direction. When more than three directions are assigned to a center, multipole components can not be fitted unambiguously by the EP fit for single molecule, but linear dependence can be removed in the case of the fit based on multiple conformations.

## 3 Calculations

Five molecules were chosen for testing purposes: ethanol; $NH_2CH_2CH_2OH$ ; n-butane; vigabatrin and alanine dipeptide. For vigabatrin, we used geometries corresponding to set of conformers. Alanine dipeptide conformations where taken from NVE MD. For other molecules conformations were generated by rotating the main chain dihedral angles in steps of 30° and optimizing with frozen corresponding dihedral angles. Atom-centered multipoles were used. Multipoles sets are described using the notation $\{x_{1,,}x_n\}\{y_{1,,}y_n\}$ where $x_i$'s stand for the ranks of multipoles centered on the non-hydrogen atoms, and $y_i$'s for the ranks of multipoles centered on the hydrogen atoms. (For example, $\{012\}\{01\}$ places monopoles+dipoles+quadrupoles on non-hydrogens, and monopoles+dipoles on hydrogen atoms ). Quality of fit is assessed by relative root mean square (RRMS) calculated using conformations not included in the fitting procedure.

## 4   Results and discussion

For accurate conformational transferability multiconformational fitting procedure is needed to be performed. It is especially evident when higher multipoles are used, since an enlargement of the variable set results in a less precise estimation of the model parameters [5]. The general predictive ability of the model for vigabatrin converges slowly with the number of conformations when a traditional (OLS) method is applied (Fig. 1). LASSO solutions are optimized with respect to the prediction error, and as a result the prediction error converges much faster than in case of OLS.

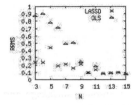

Figure 1: RRMS for LASSO and OLS model as a function of training set size (vigabatrin $\{012\}\{01\}$).

It was shown [6] that RRMS of the monopole-only EP expansion for polar molecules can be reduced by 2-3 orders of magnitude by addition of the higher multipoles (up to quadrupoles). Unfortunately, it holds only for a single molecule fit. Typically, the RRMS of the training set is much lower than RRMS of the testing set, but this situation changes with increasing number of conformations used in the fit (Fig. 2). The use of the multipole set $\{012\}\{01\}$ set for the polar molecules decreases the RRMS only 1.5-3 times in comparison to the use of monopoles only, but increases the number of multipoles about ten times.

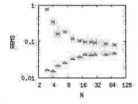

Figure 2: RRMS of training and testing set (calculated using OLS) versus size of t raining set(alanine dipeptide $\{012\}\{01\}$).

There is a wide range of tuning parameter values which cause only slightly higher than the minimal prediction error, and as a result number of the multipoles can be reduced dramatically in comparison to the size of the initial set at a cost of a low increase of RRMS, as illustrated in Fig. 3. The choice of the tuning parameter is associated with a trade-off between loosing predictive ability of a model, and reducing the number of multipoles.

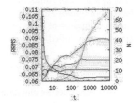

Figure 3: Fig. 5. RRMS (red) and number of multipoles (charges(green), dipoles (blue), quadrupoles(pink) and octupoles(cyan)) vs. tuning parameter for $NH_2CH_2CH_2OH$.

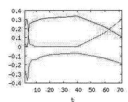

Figure 4: Some coefficients of multipoles centered on vigabatrin COOH group vs. tuning parameter({012}{01} multipoles set) H centered charge (red), H centered dipole directed to O (green), O centered dipole directed to H (blue).

It is known [1] that when variables used in a least-squares model are correlated, then their values may be devoid of interpretation. It is intrinsic feature of this type of fit. As an example, the values of some vigabatrin multipoles as a function of t are presented in Fig. 4. It is clear that they are mutually dependent, so individual values do not make sense when considered separately.

The conformational dependence of multipoles i sespecially important in the case of high variability of angles between bonds. For example RRMS for alanine dipeptide using {012}{01} set of multipoles is considerably higher (0.117) when multipoles are defined using orthogonal local frame (OLFM) of coordinates (0.917 otherwise).

## References

[1] A. E. Hoerl, R. W. Kennard, *Technometrics* **12**, 55 (1970).

[2] L. Breiman, *Technometrics* **37**, 373 (1995).

[3] R. Tibshirani *J. Roy. Stat. Soc. B* **58**, 267 (1996).

[4] J. Fan and R. Li, *J. Am. Stat. Assoc.* **96**, 1348 (2001).

[5] R. H. Myers Classical and modern Regression Analysis with Application (PWS-K ENT, Boston, MA, 1990).

[6] D. E. Williams, *J. Comp. Chem.* **9**, 745 (1988).

Brill Academic Publishers
P.O. Box 9000, 2300 PA Leiden
The Netherlands

*Lecture Series on Computer
and Computational Sciences*
Volume 4, 2005, pp. 949-953

# Development and Validation of a Force-Field for Biologically Important Sterols in Biomembranes

Z. Cournia[1,*], J.C. Smith[1] and G.M. Ullmann[2]

[1] Computational Molecular Biophysics, Interdisciplinary Center for Scientific Computing (IWR),
Im Neuenheimer Feld 368, Universität Heidelberg, 69120 Heidelberg, Germany
[2] Structural Biology/Bioinformatics, Bayreuth Center of Molecular Biosciences, Universität Bayreuth
Universitätsstr. 30, BGI, 95447 Bayreuth, Germany

Received 11 July, 2005; accepted in revised form 19 July, 2005

*Abstract: We have performed a molecular mechanics force field parametrization for the biologically important sterols cholesterol, ergosterol and lanosterol using an automated refinement method. The parameters derived have been incorporated in the CHARMM force field. Parameter sets are designed to fit results of quantum chemical calculations. The force field is validated on 2ns MD simulations of hydrated sterol:lipid systems (cholesterol, ergosterol and lanosterol, respectively). The results show different ordering effects of the hydrocarbon chains upon addition of the different sterols and with respect to their orientation in the pure bilayer. Different sterol hydration levels in the membrane are also presented.*

*Keywords: cholesterol, ergosterol, lanosterol, force-field parametrization, MD simulations, biomembranes*

*PACS:* 87.15.Aa, 87.14.Cc, 02.70.Ns, 33.20.Tp

## 1. Introduction

In higher vertebrate cells, cholesterol has been found to account for up to 50% of the lipid concentration in the plasma membrane. Cholesterol has been shown to influence the physical properties of membranes, such as regulating their fluidity [1,2], the formation of lipid-raft domains [3] and increase the membrane's mechanical strength [4]. Ergosterol (provitamin $D_2$) can be found in the membranes of fungi, yeasts and protozoans. Lanosterol, the evolutionary and biosynthetic precursor of cholesterol [5], is the major constituent of prokaryotic cell membranes. Cholesterol, ergosterol and lanosterol have very similar chemical structures. However, they exert different influences on membrane properties [1,6]. Using Molecular Dynamics (MD) Simulations, it is possible to interpret experimental results on complex membrane systems and to gain insight into the interactions at the atomic level. In MD, the potential energy of the system is described by an empirical energy function or "force field". The functional form of the force field includes a set of empirical parameters that are molecule dependent and must be optimized prior to performing simulations. This optimization step is generally referred to as parametrization of the force field. Therefore, the first necessary step towards a reliable MD simulation is the parametrization procedure. In this paper, the development of a force field for cholesterol, ergosterol and lanosterol is discussed as well as its validation through MD simulations for a cholesterol/DPPC, ergosterol/DPPC and lanosterol/DPPC membrane.

## 2. Computational Details

The parameters for the three sterols were derived using an automated frequency matching method (AFMM) [7,8]. AFMM provides an efficient automated way to generate intramolecular force field parameters using normal modes. The method involves careful choice of an initial parameter set, which is then fitted to match vibrational eigenvector and eigenvalue sets derived from quantum chemical calculations. The resulting parameter set is thus able to reproduce both the intramolecular energetics (resulting from the fitted eigenvalues) and the dynamics (resulting from the fitted eigenvectors) of the system. The method is suitable for modeling physical properties of rigid molecules which do not undergo major conformational changes, such as steroids. AFMM projects each of the Molecular Mechanics (MM) eigenvectors $\left[ \begin{array}{c} MM \\ i \end{array} \right]$ (the subscript i indicates the normal mode number and the superscript MM that the modes are calculated with MM) onto a reference set of eigenvectors

---

[1*] Corresponding author. E-mail: zoe.cournia@iwr.uni-heidelberg.de

$\left\{ \chi_i^{QM} \right\}$ (the superscript QM indicates that the modes are calculated with QM) to find the frequency $v_j^{max}$ corresponding to the to the highest projection ( $j : \chi_i^{MM} \cdot \chi_i^{QM} = \max$ ) and compares this frequency with the corresponding frequency, $v_i$. In the ideal case $v_i = v_j^{max}$ and $\chi_i^{MM} \cdot \chi_i^{QM} = \delta_{ij}$, where $\delta_{ij}$ is the Kronecker delta. AFMM is based on minimizing a penalty function, $Y^2$, which is the deviation from the ideal situation:

$$Y^2 = \sum_{3N-6} (v_i - v_j^{max})^2 \tag{Eq.1}$$

where N is the number of atoms in the molecule.

For the automatic optimization of the parameter set, a standard Monte Carlo (MC) scheme is used to minimize $Y^2$. At each step n, all chosen parameters are iteratively varied in the MC algorithm and $Y_n^2$, and if $Y_n^2 < Y_{n-1}^2$, the new parameter set is used in the next step, n+1.

The quantum mechanical calculation of normal modes for the three sterols was performed using the DFT/B3LYP level of theory with the basis set SBJKC [9] within the NWChem 4.5 package [10]. Partial atomic charges were calculated using the 6-31G(d) baseis set level and the CHELPG analysis.

All molecular mechanics calculations were performed with the CHARMM27 package [11]. We have used CHARMM27 force field for DPPC and our derived force field for cholesterol, ergosterol and lanosterol. In CHARMM the empirical potential energy function is given by Eq.2:

$$E_{bonded} = \sum_{bonds} k_b (b - b_0)^2 + \sum_{angles} k_\theta (\theta - \theta_0)^2 + \sum_{dihedrals} k_\varphi (1 + \cos[n\varphi - \delta]) + \sum_{impropers} k_\omega (\omega - \omega_0)^2 +$$

$$\sum_{ub} k_{ub} (s - s_0)^2 + \sum_{vdW} \varepsilon_{ij} \left[ \left( \frac{R_{ij}^{min}}{r_{ij}} \right)^{12} - \left( \frac{R_{ij}^{min}}{r_{ij}} \right)^6 \right] + \frac{q_i q_j}{D r_{ij}} \tag{Eq.2}$$

where $k_b$, $k_\theta$, $k_\varphi$, $k_\omega$, $k_{ub}$ are, respectively, bond lengths, angle, dihedral, improper torsion and Urey-Bradley constant, and $b$, $\theta$, $\varphi$, $\omega$, and s represent, respectively bond lengths, bond angles, dihedral angles, improper torsion angles and Urey-Bradley 1-3 distances (the subscript zero corresponds to equilibrium values). Nonbonded interactions are described by the Lennard-Jones 6-12 (LJ) term for the van der Waals interactions and the Coulomb interaction term for electrostatics.

To construct the sterol:DPPC (sterol: cholesterol, ergosterol and lanosterol) membrane for the MD simulations we used coordinates for DPPC (1,2-dipalmitoyl phosphatidylcholine) molecules determine by Sundaralingam [12]. Coordinates for the cholesterol molecule were taken from the crystal structure by Shieh et al. [13]. The ergosterol and lanosterol structures were created from the cholesterol crystal structure using the program Insight II (Accelrys). Three MD simulations were performed on systems of 120 DPPC, 80 sterol molecules (cholesterol, ergosterol and lanosterol respectively) and 1600 TIP3P [14] waters. The production rum of the simulation was performed at constant pressure (1 atm) and temperature (Nose-Hoover thermostat at 309 K) with periodic boundary conditions for 2 ns.

## 3. Results

### 3.1 Parametrization of the sterols

As a first step, the partial atomic charges for each sterol were determined based on the QM-optimized structure with the CHELPG method. Then, an initial parameter set based on analogy with already existing CHARMM parameters and chemical intuition was used for minimization and calculation of normal modes with CHARMM. AFMM was then employed to match the normal modes (both eigenvalues and eigenvectors) obtained with the reference normal modes obtained from quantum chemical calculations. Special care was taken to reproduce correctly the torsional potential of the hydroxyl group region. This rotational barrier along the $H-O-C_3-C_2$ dihedral of each sterol is important because it will influence the residence time and stability of the hydrogen bonds between each sterol and water and lipid head groups in a membrane simulation. The torsional force constants for this dihedral were derived from the energy barrier for rotation at the DFT/SBJKC level of theory using single-point calculations of the geometry-optimized structures to scan the potential energy landscape [15].

### 3.2 Deuterium Order Parameters from the MD Simulation

The most popular quantity to characterize the order of the hydrocarbon chains in lipid bilayers is the deuterium NMR order parameters. The order parameter may be defined for every $CH_2$ group in the lipid hydrocarbon chains as:

$$S^i_{CD} = \frac{1}{2}\left(3\left\langle \cos^2\theta^i_{CD}\right\rangle - 1\right)$$  (Eq.3)

where $^i_{CD}$ is the angle between a C-D bond (D: deuterium in the experiment) or a C-H bond (in the simulation) of the *ith* carbon on the alkyl chain and the membrane normal (z-axis). The brackets indicate averaging over the two bonds in each $CH_2$ group, all the lipids and time. Our results indicate that all three sterols induce order to the lipid chains with respect to the pure bilayer system. Ergosterol is shown to induce more order than cholesterol. Lanosterol shows the smallest ordering effect on the DPPC membrane. Our results show good overall agreement with those obtained in the studies of Urbina et al. [16] and Faure et al. [17]. Our order parameter profile (see Table 1) is also consistent with all simulation results obtained for similar conditions [6,16-21].

### 3.3 Solvation of the sterol hydroxyl in the membrane

The water in the polar lipid region solvates the polar lipid head groups. These interactions can be best described by radial distribution functions for the water oxygen surrounding the phosphate atom, choline nitrogen, carbonyl carbon, and sterol hydroxyl hydrogen atoms. The radial distribution function $g(r)$ describes the probability of finding a particle *y* at a distance between *r* and *dr* away from particle *x* in a simulation box of volume V containing N particles. The distances between the atoms on different molecules were binned and the resulting $g(r)$ was normalized by dividing by $4\pi r^2 dr$ in which *r* is the distance in the middle of the bin and *dr* is the bin width, set at 0.07Å. The volume around each particle is divided into concentric spherical shells, and the number of particles in each shell is counted and divided by the shell volume (given by the difference between two spherical volumes), to obtain the local density. In Figure 1, we plot the radial distribution functions for each sterol hydroxyl hydrogen and the water oxygen taken from the membrane simulations. We note that all three sterol hydroxyls show a sharp first peak at 2 Å, indicating tight solvation shell around the hydroxyl group. By integrating the corresponding $g(r)$ functions to the first minimum we obtain the first water hydration shell around the hydroxyl groups, which corresponds to 3 waters/cholesterol, 2.5 waters/ergosterol and 2 waters/lanosterol.

## 4. Conclusions

We have presented the development of a force field for three biologically-important sterols for the CHARMM simulation package. Three 2ns MD simulations of hydrated cholesterol:DPPC, ergosterol:DPPC and lanosterol:DPPC systems were performed to validate the developed force field and show the different effect of the three sterols on the lipid bilayer. The results show an ordering of the hydrocarbon chains to the membrane normal, with respect to their ordering in the pure bilayer. More specifically, the ordering effect follows the order ergosterol>cholesterol>lanosterol, in agreement with experimental and simulation results. The first hydration shell of the hydroxyl of each sterol was determined by calculating the time-averaged radial distribution function from the membrane simulations. Cholesterol has the tightest hydration shell with three water molecules on average.

Although structural aspects of cholesterol in membranes have been investigated so far, our knowledge for cholesterol dynamics in biomembranes is quite limited. A combination of quasielastic neutron scattering experiments and MD simulations will help us investigate the dynamical effects of sterols in membranes and examine biologically-relevant structure-function relationships from a dynamical point of view.

| Carbon -S_CD | 2 | 3 | 4 | 5 | 6 | 7-9 | 10 | 11 | 12 | 13 | 14 | 15 |
|---|---|---|---|---|---|---|---|---|---|---|---|---|
| Cholesterol | 0.26 | 0.38 | 0.41 | 0.43 | 0.44 | 0.45 | 0.44 | 0.43 | 0.42 | 0.40 | 0.36 | 0.30 |
| Ergosterol | 0.24 | 0.40 | 0.42 | 0.44 | 0.45 | 0.46 | 0.46 | 0.45 | 0.45 | 0.43 | 0.40 | 0.34 |
| Lanosterol | 0.24 | 0.36 | 0.39 | 0.42 | 0.44 | 0.44 | 0.43 | 0.42 | 0.40 | 0.37 | 0.33 | 0.26 |

Table 1 Deuterium order parameters for cholesterol, ergosterol and lanosterol averaged over the 2 ns trajectory for the *sn2* chain of DPPC. The ordering effect on the DPPC alkyl chain follows the order cholesterol > cholesterol > lanosterol.

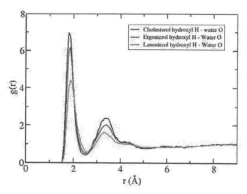

Fig.1 Radial distribution functions for the sterol hydroxyl hydroxyl and the water oxygen. By integrating the corresponding *g(r)* functions to the first minimum we obtain the first water hydration shell around the hydroxyl groups, which corresponds to 3 waters/cholesterol, 2.5 waters/ergosterol and 2 waters/lanosterol.

## Acknowledgments

The authors would like to thank Dr.A.C.Vaiana for fruitful discussions. The work was supported by BMBF project 03SHE2HD. QM and MD calculations have been performed on HELICS, IWR - Universität Heidelberg (HBFG funds, hww cooperation).

## References

[1] K. Bloch, Cholesterol, evolution of structure and function, pp. 1-24, in *Biochemistry of Lipids and Membranes*, Eds. J.E. Vance and D.E. Vance, Benjamin/Cummins Pub. Co. Inc., New York, 1985

[2] A. Kusumi, M. Tsuda, T. Akino, O. Ohnishi, an Y. Terayama, Protein-phospholipid-cholesterol interaction in the photolysis of invertebrate rhodopsin, Biochemistry 22:1165-1170, 1983

[3] K. Simons, and E. Ikonen, Functional rafts in cell membranes, Nature, 387:569-572, 1997

[4] M. Bloom and O.G. Mouritsen, The evolution of membrane, in *Structure and Dynamics of Membrane*, Eds. R. Lipowski and E.Sackmann, Elsevier, Amserdam, pp.69-95, 1995

[5] J.M. Risley. Cholesterol Biosynthesis: Lanosterol to Cholesterol. J. Chem Ed., 79:377-384, 2002

[6] A. Smondyrev and M.L. Berkowitz, MD simulation of the structure of DMPC bilayers with cholesterol, ergosterol and lanesterol, Biophys. J., 80:1649-1658, 2001

[7] A.C. Vaiana, A. Schulz, J. Worfrum, M. Sauer, and J.C. Smith, Molecular mechanics force field parametrization of the fluorescent probe rhodamine 6G using automated frequency matching, J. Comp. Chem., 24:632, 2002

[8] A.C. Vaiana, Z. Cournia, I.B. Castescu, and J.C. Smith, AFMM: A Molecular Mechanics Force Field Parametrization Program. Comp Phys Commun, 167:34-42, 2005

[9] W.J. Stevens, H. Basch and M. Krauss, Compact effective potentials and efficient shared-exponent basis sets for the first- and second-row atoms. J Comp. Phys., 81:6026-6033, 1984

[10] NWChem, A Computational Chemistry Package for Parallel Computers, Version 4.5, Pacific Northwest National Laboratory, Richland, Washington, USA, 2003.

[11] B.R. Brooks, R. Bruccoleri, B.D. Olafson, D.J. States, S. Swaminathan, and M. Karplus, CHARMM: A program fo Makromolecular Energy, Minimization and Dynamics Calculations, J. Comp. Chem. 4:187, 1983

[12] M. Sundaralingam, Molecular structures and conformations of the phospholipids and shingomyelins, Ann. N.Y. Acad. Sci., 195:324, 1972

[13] H. Shich, L.G. Hoard, and C.E. Nordman, The structure of cholesterol, Acta Cryst. B37:1538, 1981

[14] W.L. Jorgensen, J. Chrandasekhar, J.D. Madura, R.W. Imprey, and M.L. Klein, Comparison of simple potential functions simulating liquid water, J. Chem. Phys., 79:926, 1983

[15] Z. Cournia, J.C. Smith and G.M. Ullmann, A Molecular Mechanics Force Field for Biologically-Important Sterols, in press J.Comp.Chem., 2005

[16] J.A. Urbina, S. Pekerar, H. Le, J. Patterson, B. Montez, E. Oldfield, Molecular order and dynamics of phosphatidylcholine bilayer membranes in the presence of cholesterol, ergosterol and lanesterol: a comparative study using 2H-, 13C- and 31P-NMR spectroscopy, Biochim. Biophys. Acta, 1283:163, 1995

[17] C. Faure, J. Transchant, E.J. Dufourc, Comparative effects of cholesterol and cholesterol sulfate on hydration and ordering of DMPC membranes, Biophys. J. 70:1380, 1996

[18] K. Tu, M. Klein, and D. Tobias, Constant-Pressure MD investigation of Cholesterol effects in a DPPC bilayer, Biophys. J., 75:2147, 1998

[19] A.M. Smondyrev and M.L. Berkowitz, Structure of DPPC/Cholesterol bilayer at low and high cholesterol concentrations: molecular dynamics simulation, Biophys. J., 77:2075, 1999

[20] C. Hofsaess, E. Lindahl, and O. Edholm, Molecular dynamics simulations of phospholipid bilayers with cholesterol, Biophys. J., 84:2192, 2003

[21] R. Murari, M. Murari, and W.J. Baumann, Sterol orientations in phosphatidylcholine liposomes as determined by deuterium NMR, Biochemistry, 25:1062, 1986

Brill Academic Publishers
P.O. Box 9000, 2300 PA Leiden,
The Netherlands

*Lecture Series on Computer
and Computational Sciences*
Volume 4, 2005, pp. 954-957

# Third-Order Optical Properties of Molecular Systems in Solution by Means of Response Theory

Luca Frediani[1]

Department of Chemistry,
University of Tromsø,
9037 Tromsø, Norway

Received 11 July, 2005; accepted in revised form 28 July, 2005

*Abstract:* The Polarizable Continuum Model (PCM) which allows a faithful representation of electrostatic solvation effects onto molecular solute described at the *ab initio* or DFT level, has recently been developed to calculate third order optical properties by means of quadratic response theory both for singlet and for triplet excitations. The current implementation allows to calculate solvent effects on, e.g. hyperpolarizabilities, two-photon absorption cross-section and phosphorescence lifetimes. Here the theory development will be presented and a few selected applications will briefly be described..

*Keywords:* Response, *ab initio*, solvent effect, hyperpolarizability, quantum mechanics.

## 1    Introduction

The field of nonlinear optical properties of molecules has known a continued interest in the scientific community in view of its numerous and wide applications.[1,2] From the point of view of theoretical and computational modelling a wide variety of approaches have been developed to reproduce and rationalize the behavior of molecules displaying nonlinear optical characteristics. One very successful approach is to employ Response Theory:[3] a formalism allowing to obtain information about linear and nonlinear optical properties of molecules in an analytical fashion, once the wavefunction (for Hartree-Fock (HF) and Multiconfiguration Self-Consistent Field (MCSCF) approaches) or the electronic density (for the Density Functional Theory (DFT) approach) is known.

Since the experimental determination of nonlinear optical properties is carried out in environmental conditions which are generally far away from the idealized picture of an isolated molecule, it is important to be able to reproduce effectively and efficiently how such properties are affected by the surrounding environment. Within continuum models[4] the environment is represented by a homogeneous dielectric medium surrounding the molecule. In this way the solvated system is represented in the same way as the isolated molecule, therefore the environmental effect can conveniently be estimated by comparing the results for the isolated molecule to the ones obtained for the solvated molecule. One such model is called Polarizable Continuum Model[4] (PCM) which employs a molecular-shaped cavity where the molecule is inserted.

---

[1]Corresponding author. E-mail: luca@james.chem.uit.no

# 2  Theory

## 2.1  Response formalism

We assume to have a molecule which is determined by the coordinates and charges of its nuclei and by the total number of electrons. We also assume that the electronic wavefunction $|0\rangle$ in the clamped nuclei approximation has been obtained through a variational SCF or MCSCF procedure. When an external time-dependent perturbation $V(t)$ is added to the Hamiltonian $H_0$ of the isolated molecule the wavefunction is no longer stationary: $|0\rangle = |0\rangle(t)$. The time evolution of expectation values of operators $A(t) = \left\langle 0(t)\left|\hat{A}\right|0(t)\right\rangle$ can conveniently be obtained by an expansion on the linear, quadratic and higher order response functions:

$$
\langle 0(t)|A|0(t)\rangle = \langle 0|A|0\rangle|_{t=0} + \int e^{-(i\omega+\epsilon)t}\langle\langle A; V^\omega\rangle\rangle\,d\omega +
$$
$$
\frac{1}{2}\iint e^{-(i(\omega_1+\omega_2)+2\epsilon)t}\langle\langle A; V^{\omega_1}, V^{\omega_2}\rangle\rangle\,d\omega_1\,d\omega_2 + \dots
\tag{1}
$$

In equation (1) $\langle\langle A; V^\omega\rangle\rangle$ and $\langle\langle A; V^{\omega_1}, V^{\omega_2}\rangle\rangle$ are the linear and quadratic response functions respectively. In particular, by choosing the frequencies $\omega_1$ and $\omega_2$, the hyperpolarizability $\beta(\omega; \omega_1, \omega_2)$ (with $\omega = -\omega_1 - \omega_2$) can be obtained.[3]

## 2.2  The Polarizable Continuum Model

In this section we outline how the solvent effects is described by the Polarizable Continuum Model[4]. The solute molecule is treated *ab initio* whereas the solvent is modeled as a macroscopic medium. The solute-solvent interaction is obtained by placing the solute in a molecular-shaped cavity $C$ inside the dielectric environment defined by its macroscopic properties. For a homogeneous and isotropic solvent only the static ($\epsilon_{stat.}$) and optical ($\epsilon_{opt.}$) dielectric constants are needed in order to calculate the solvent reaction field. The solute-solvent interaction energy is then given by:

$$
E_{int} = \int_\Gamma V(s)\sigma(s)\,ds,
\tag{2}
$$

where $V(s)$ is the electrostatic potential produced by the molecule and $\sigma(s)$ is the apparent surface charge (ASC) representing the solvent polarization. In practice the problem is formulated in such a way that the solute-solvent interaction is added to the isolated-molecule Hamiltonian:

$$
H = H_0 + J + X(0)
\tag{3}
$$

where $H_0$ is the *in-vacuo* Hamiltonian, $J$ represents the interaction of the electrons with the nuclear apparent charges and $X(0)$ is the interaction of the electrons with the electronic apparent charges. The wavefunction is then obtained by minimizing the Free Energy Functional

$$
\mathcal{G} = \langle 0|G|0\rangle = \left\langle 0\left|H_0 + J + \frac{1}{2}X(0)\right|0\right\rangle,
\tag{4}
$$

with respect to the parameters defining the wavefunction. The details of the PCM-MCSCF wavefunction calculation have been given in a previous paper.[5]

## 2.3  Solvation terms in the quadratic response expression

In response theory, the expectation value $\left\langle 0(t)\left|\hat{A}\right|0(t)\right\rangle$ is obtained by writing the time dependent Schrödinger equation and solving it at the successive orders of the perturbations in a matrix form.

The solvent terms arise when one is concerned with the Hamiltonian operator. Without giving the details of the derivation which can be found elsewhere,[6,7] it is possible to write:

$$G^{[n+1]}_{j,l_1,l_2,\ldots,l_n} = E^{[n+1]}_{j,l_1,l_2,\ldots,l_n} + V^{[n+1]}_{j,l_1,l_2,\ldots,l_n} \cdot [q^N + q^i(\rho^i)] + \sum_{k=0}^{n} V^{[n-k+1]}_{j,l_{k+1},l_{k+2},\ldots,l_n} \cdot q^{d[k]}_{l_1,l_2,\ldots,l_k} \quad (5)$$

where $E^{[n+1]}_{j,l_1,l_2,\ldots,l_n}$ is the vacuum expression and the other terms represent contributions due to the solute-solvent interactions. For quadratic response $n = 2$, the corresponding solvent terms ($V^{[3]}_{jlm}$, $V^{[2]}_{jl}$, $V^{[1]}_j$, $q^{d[2]}_{l_1,l_2,\ldots,l_2}$, $q^{d[1]}_{l_1,l_2,\ldots,l_1}$) have been obtained in Ref. 7.

# 3 Applications

The implementation described in the previous section has been employed on a few applications. Here we will briefly present two such applications.

## 3.1 SCF and MCSCF hyperpolarizabilities of push-pull systems

The formalism described in section 2 has first been employed to study the solvent effect on the static and frequency-dependent hyperpolarizabilities of $NO_2(CH{=}CH)_nNH_2$ ($n = 1, 2, 3$) and para-nitroaniline (see structures in Fig. 1(a)) has been investigated at the HF and MCSCF levels of theory by comparing the results in gas phase with the corresponding values in cyclohexane ($\epsilon_{stat.} = \epsilon_{opt.} = 2.023$) and water ($\epsilon_{stat.} = 78.39$ and $\epsilon_{opt.} = 1.776$). The hyperpolarizability tensor $\beta_{ijk}$ has been reduced to a scalar quantity by calculating the parallel component of the averaged hyperpolarizability $\beta_\parallel = \sum_i \beta_i \cdot \frac{\mu_i}{||\boldsymbol{\mu}||}$ where the averaged hyperpolarizability $\beta_i = 1/5 \sum_j \beta_{ijj} + \beta_{jij} + \beta jji$ and $\hat{\boldsymbol{\mu}}$ is the dipole moment. The obtained results have been analyzed razionalizing the observed trends upon change of environment, chain length, inclusion of correlation and change of basis set.[7]

## 3.2 Two-Photon absorption of trans-stilbene and derivatives

The experimental TPA cross section for an excitation from the ground state $|0\rangle$ to a final state $|f\rangle$ is defined in terms of the normalized line shape function, $g(\omega_\nu + \omega_\mu)$, and the two-photon absorption transition amplitude tensor, $\mathbf{T}^{\omega_\nu \omega_\mu, f}$:[8]

$$\delta_{TPA} = \frac{(2\pi e)^4 \omega_\nu \omega_\mu}{c^2} g(\omega_\nu + \omega_\mu)|\mathbf{T}^{\omega_\nu \omega_\mu, f}|^2 . \quad (6)$$

In response theory, the TPA transition amplitude $|\mathbf{T}^{\omega_\nu \omega_\mu, f}|^2$ can be extracted from a single residue of the quadratic response function $\langle\langle \hat{A}; \hat{B}, \hat{C} \rangle\rangle_{\omega_B, \omega_C}$

$$\lim_{\omega_C \to \omega_f} (\omega_C - \omega_f)\langle\langle \hat{A}; \hat{B}, \hat{C} \rangle\rangle_{\omega_B, \omega_C} = -\mathbf{T}^{AB,f}\mathbf{T}^{C,f} . \quad (7)$$

where $(T^{C,f})$ is the first order transition amplitude between ground state $|0\rangle$ and excited state $|f\rangle$. A detailed description of the QR-DFT equations for evaluation of TPA tensors as well as numerical recipes for their solution, can be found in Ref.9, whereas the details concerning the solvent implementation have been given in Ref. 10 This methodology has been applied to obtain the TPA cross sections in gas-phase, water and cyclohexane for trans-stilbene and three disubstituted derivatives (structures reported in Fig. 1(b)).

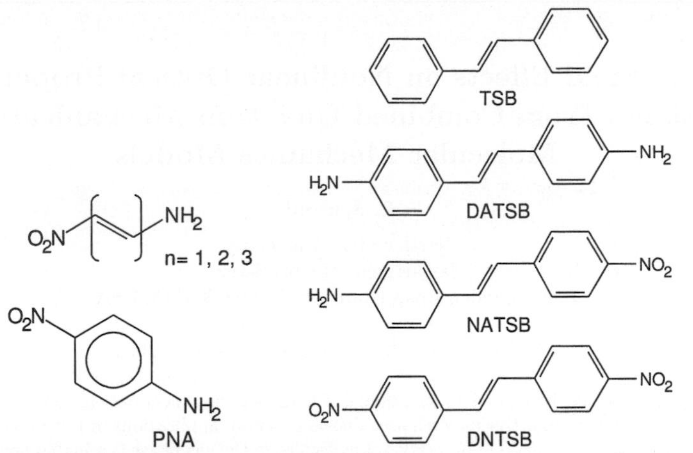

(a) Push-pull systems employed in hyperpolarizability calculations
(b) trans-stilbene and disubstituted derivatives employed in TPA calculations

Figure 1: Molecular structures employed in the present study

# 4 Acknowledgments

The financial support from the Norwegian Research Council and the Carl Trygger foundation through post.Doc grants to the author is greatly acknowledged.

# References

[1] P. N. PRASAD and D. J. WILLIAMS, *Introduction to nonlinear optical effects in molecules and polymers*, Wiley, New York, 1991.

[2] D. R. KANIS, M. A. RATNER, and T. J. MARKS, *Chem. Rev.* **94**, 195 (1994).

[3] J. OLSEN and P. JØRGENSEN, in *Modern Electronic Structure Theory*, edited by D. R. YARKONY, p. 857, World Scientific, New York, 1995.

[4] J. TOMASI and M. PERSICO, *Chem. Rev.* **94**, 2027 (1994).

[5] R. CAMMI, L. FREDIANI, B. MENNUCCI, J. TOMASI, K. RUUD, , and K. V. MIKKELSEN, *J. Chem. Phys.* **117**, 13 (2002).

[6] R. CAMMI, L. FREDIANI, B. MENNUCCI, and K. RUUD, *J. Chem. Phys.* **119**, 5818 (2003).

[7] L. FREDIANI, L. FERRIGHI, H. ÅGREN, and K. RUUD, submitted, 2005.

[8] W. M. MCCLAIN, *J. Chem. Phys.* **55**, 2789 (1971).

[9] P. SALEK, O. VAHTRAS, J. D. GUO, Y. LUO, T. HELGAKER, and H. ÅGREN, *Chem. Phys. Lett.* **374**, 446 (2003).

[10] L. FREIDANI, Z. RINKEVICIUS, and H. ÅGREN, *J. Chem. Phys.* **122**, 244104 (2005).

Brill Academic Publishers
P.O. Box 9000, 2300 PA Leiden,
The Netherlands

*Lecture Series on Computer*
*and Computational Sciences*
Volume 4, 2005, pp. 958-960

# Local Field Effects on Nonlinear Optical Properties: Insights From Combined Quantum Mechanical and Molecular Mechanics Models

L. Jensen[1]

Northwestern University
Department of Chemistry
2145 Sheridan Road,Evanston, IL 60208-3113, USA

Received 9 July, 2005; accepted 4 August, 2005

*Abstract:* The calculation and understanding of linear and nonlinear optical (NLO) properties of materials is an active research area due to potential applications in future technologies. The calculation of NLO properties of molecules in the gas phase is a highly nontrivial problem and even more so for molecules in the condensed phase. Comparing computed results with experimental results obtained in the condensed phase requires that solvent and local field effects are included directly in the calculations. We will in this work review some recent applications of the discrete solvent reaction field (DRF) model for calculating nonlinear optical (NLO) properties of molecules in the condensed phase with emphasize on the effects of including local fields in the calculations.

*Keywords:* NLO properties; solvent effects; QM/MM; TDDFT; fullerenes; push-pull molecules

## 1  Introduction

Understanding the linear and nonlinear optical (NLO) properties of materials is important since materials exhibiting NLO effects are of great technological importance for use in furture application within electronics and photonics. [1] Therefore, accurate predictions of molecular response properties in both gas and condensed phases are of particular interest both from a theoretical and a technological point of view. The use of quantum chemical methods enables accurate calculations of molecular NLO response properties. However, the accurate prediction of these properties for molecules requires large basis sets, high-level electron correlation, frequency-dependence and both electronic and vibrational contributions to be included. Even in the gas phase this is only feasible for small molecules due to a very high computational burden.

The calculation of response properties of molecules in the condensed phase is developed to a much lesser degree than in the gas phase. Although several methods exist for which solvent effects can be considered, the comparison with experiments are still often done at the microscopic level. The fact that microscopic experimental values are extracted from macroscopic quantities, usually using Lorentz local field factors, further complicates any comparison. In recent years it has been realized that it is better to calculate the macroscopic quantitites, i.e., the (non-)linear susceptibilites, which can be related to experimental results without the need for extracting microscopic properties.

[1]Corresponding author. E-mail: l.jensen@chem.northwestern.edu

## 2   The Discrete Solvent Reaction Field model

We have developed a polarizable quantum mechanical / molecular mechanics (QM/MM) model within time-dependent density functional theory (TDDFT) to study the molecular response properties in the condensed phase. [2, 3, 4] This Discrete Solvent Reaction Field model (DRF) combines a DFT description of the solute molecule with a classical description of the discrete solvent molecules. The latter are represented using distributed atomic charges and polarizabilities. The combination of DRF with TDDFT allows for a computational attractive solution when the explicit solvent structure is of interest.

The advantages of including polarizabilities in the MM part is that all parameters can be obtained from gas phase properties. The atomic charges are straightforward obtained using Multipole Derived Charges (MDC) and the distributed polarizabilities by adopting standard parameters or refitting them to match the calculated polarizability tensor. This allows for a simple procedure to obtained the solvent model parameters which subsequently can be used in the molecular dynamics (MD) and QM/MM simulations.

The DRF model has been extended to also include the so-called local field factors, i.e. the difference between the macroscopic electric field and the actual electric field felt by the solute. [5, 6, 7] This enables the calculation of effective microscopic properties which can be related to the macroscopic susceptibilities. The macroscopic susceptibilities can then be compared directly with experimental results.

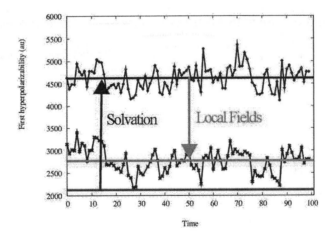

Figure 1: The first hyperpolarizability of pNA in gas phase and in 1,4-dioxane solution. Top line is for $\beta^{sol}$, the middle line for $\beta^{eff}$ and the bottom line is the gas phase value for $\beta$. Results are presented for 100 QM/MM configurations as a function of time steps in the MD simulations and are taken from Ref. [8]. Solid lines corresponds to average values.

## 3   Results

We will review some applications of the DRF model for calculating NLO properties of molecules in the condensed phase with emphasize on the effects of including local fields in the calculations. Examples will include pure solutions, a push-pull molecule in solution and fullerene clusters. Results for both microscopic properties ,like polarizabilitites and hyperpolarizabilities, and for macroscopic properties, like refractive index and NLO susceptibilities, will be presented. We wil show that it is important to include both the solvent and the local field effects in the calculations in order to predict correct solvation shifts. This is illustrated in Fig. 1 where the first hyperpolarizability ($\beta$) of p-nitroaniline (pNA) in 1,4-dioxane solution is prestented. From the figure it is clear that the solvent effects increases $\beta$ significantly whereas the local field effects decreases $\beta$ giving an overall small increase of $\beta$ in solution.

# References

[1] P. N. Prasad and D. J. Williams. *Introduction to nonlinear optical effects in molecules and polymers.* Wiley, New York, 1991.

[2] L. Jensen, P. Th. van Duijnen, and J. G. Snijders. *J. Chem. Phys.*, **118**, 514, 2003.

[3] L. Jensen, P. Th van Duijnen, and J. G. Snijders. *J. Chem. Phys.*, **119**, 3800, 2003.

[4] L. Jensen, P. Th. van Duijnen, and J. G. Snijders. *J. Chem. Phys.*, **119**, 12998, 2003.

[5] P. Th. van Duijnen, A. H. de Vries, M. Swart, and F. Grozema. *J. Chem. Phys.*, **117**, 8442, 2002.

[6] L. Jensen, M. Swart, and P. Th. van Duijnen. *J. Chem. Phys.*, **122**, 034103, 2005.

[7] L. Jensen and P. Th. van Duijnen. *Int. J. Quant. Chem.*, **102**, 612, 2005.

[8] L. Jensen and P. Th. van Duijnen, 2005. J. Chem. Phys. *Accepted.*

Brill Academic Publishers
P.O. Box 9000, 2300 PA Leiden,
The Netherlands

*Lecture Series on Computer*
*and Computational Sciences*
Volume 4, 2005, pp. 961-965

# New frontiers for Continuum Solvation Models

## B. Mennucci[1]

Department of Chemistry,
University of Pisa,
Via Risorgimento 35, 56-126 Pisa, Italy

Received 4 July, 2005; accepted in revised form 17 July, 2005

*Abstract:* In the last years solvation continuum models have been coupled with modern quantum chemical calculations; this has permitted a detailed description of the electronic structure of a huge variety of solvated species. The aim of this paper is to present a brief overview on the most recent developments of these models and on possible new frontiers for their applications.

*Keywords:* Continuum models, nonequilibrium, solvent relaxation, response properties, intermolecular interactions.

## 1 Introduction

Continuum models have a long history (going back to Born, Onsager and Kirkwood to cite the authors of the oldest models[1]), however only recently they have been coupled to quantum-mechanical (QM) techniques in order to get an accurate description of the solute structure and charge distribution and of the solvent dielectric polarization.[2, 3, 4] Thanks to this coupling, in the last decade continuum solvation models have become the default approach to study energies and geometries of solvated systems. In parallel, new frontiers have appeared: in the following sections three of these frontiers will be considered, namely (i) time dependent phenomena and solvent relaxations, (ii) molecular properties and spectroscopies, and (iii) solute-solvent specific effects.

Before passing to describe the status of the art in these new fields and possible future developments, let us recall the main elements of the basic theory here expressed within the Polarizable Continnum Model (PCM)[5, 6].

The solute, represented as a charge density (either classical or quantum-mechanical) is inside a cavity of proper shape and dimension surrounded by an infinite dielectric with permittivity $\epsilon$. The polarization of the solvent due to the presence of this charge density is represented by a set of "apparent surface charges" placed on the cavity surface. When a QM description of the solute is used, an Effective Hamiltonian is introduced:

$$\hat{H}_{eff}|\Psi\rangle = \left[\hat{H}^0 + \hat{V}^R\right]|\Psi\rangle = E|\Psi\rangle \tag{1}$$

where $\hat{H}^0$ is the Hamiltonian of the isolated molecule, and $V^R$ is the solvent *reaction potential*. The treatment of the operator $\hat{V}^R$ is delicate, as this term depends on the solute wavefunction (through the apparent charges) and thus it induces a nonlinear character to the solute Schrödinger equation. Equation (1) can be solved with all the techniques developed for isolated systems. It is,

---

[1]Corresponding author. E-mail: bene@dcci.unipi.it

however, important to note that the correct energetic quantity to consider is now the free energy functional,

$$\mathcal{G} = E + \frac{1}{2}\langle \Psi \left| V^R \right| \Psi \rangle \tag{2}$$

and not the eigenvalue of the nonlinear Hamiltonian, here indicated as $E$. The difference between $E$ and $\mathcal{G}$ has, however, a clear physical meaning; it represents the polarization work which the solute does to create the charge density inside the solvent. It is worth remarking that this interpretation is equally valid for zero-temperature models and for those in which the thermal agitation is implicitly or explicitly taken into account.

## 2    Time dependent solvent polarization

In the modeling of time-dependent (TD) phenomena in solution, one of the open questions is how to take into account the evolution of the solvent polarization and of the related solute-solvent interaction. In fast processes, such as electronic excitations, electron transfers or ionizations, the time-scale of the change in the charge density of the solute is usually much smaller than the time-scale in which a polar solvent fully relaxes to reach a new equilibrium state. During this relaxation, the solvent nuclear and molecular motions act as inertia on the solvation response and a nonequilibrium regime is established.

This nonequilibrium regime may appear at different timescales due to the the different contributions which enter in the solvent polarization. In particular, the orientational contribution (characterized by response longer times) will first start to lag behind, followed by the atomic term (with times around $10^{-14}$ sec), and finally, for very fast changes, by the electronic term (with times around $10^{-16}$ sec). Within the continuum framework the different response times of the various terms constituting the solvent polarization $P(t)$ can be taken into account.

If we limit the analysis to the initial step of the TD process, i.e. a sudden change in the solute state such a vertical electronic transition, a Franck-Condon response of the solvent is assumed exactly as for the solute molecule. The solvent molecules will not be able to immediately re-orient according to the new solute state and the corresponding part of the response will remain frozen in the state immediately previous to the change. The actual polarization will thus be the sum of an inertial (or fixed) term and a dynamic (or instantaneous) term. In the PCM framework, this will lead to a partition of the apparent charges in two separate sets:[4] the dynamic set of charges will depend on the instantaneous charge distribution of the solute and on the optical dielectric constant while the inertial charges will still depend on the solute charge distribution of the initial state.

This nonequilibrium regime, however, is just an instantaneous situation that will relax towards a new equilibrium state corresponding to the final solute state. This relaxation of the solvent which is generally indicated as solvation dynamics occurs in a timescale and through relaxation modes which are typical of the structure of the solvent molecules and of the solute-solvent interactions. In order to study the effects of the solvent relaxation we have to generalize the PCM model to a real time dependence. This is possible by assuming that the change in the solute occurs instantaneously (at time $t = 0$) and the corresponding variation of the electrostatic potential on the cavity surface between the initial (0) and the final ($fin$) state is a step change $V(t) = V^0 + \theta(t)\Delta V$ where $\theta(t)$ is the usual step function and $(\Delta V) = V^{fin} - V^0$ The TD solvent charges at a generic time $t$ can thus be written as:

$$\mathbf{q}(t) = \mathbf{q}^0 + \delta\mathbf{q}\left(\Delta \mathbf{V}, t\right) \tag{3}$$

with the proper boundary conditions $\mathbf{q}(t \to -\infty) = \mathbf{q}^0$ and $\mathbf{q}(t \to \infty) = \mathbf{q}^{fin}$ where $\mathbf{q}^0$ and $\mathbf{q}^{fin}$ are the polarization charges when the initial and the final solute-solvent equilibrium is valid, respectively.

By applying the results of the linear response we obtain[7]

$$\delta \mathbf{q} \left( \Delta \mathbf{V}, t \right) = \int_{-\infty}^{t} dt' \mathbf{R} \left( t - t' \right) \theta \left( t' \right) \Delta \mathbf{V} \tag{4}$$

This expression is transformed in a numerical procedure by passing from the time domain to the frequency domain. This change is required as the dielectric response of the solvent is described in terms of its complex dielectric permittivity as a function of a frequency $\omega$. The $\omega$ dependence of $\hat{\varepsilon}$ can either be modeled using pure diffusive expressions (as in the Debye relaxation expression [1]), or calculated on the basis of experimental measurements of the absorption.

# 3  Response properties

Most properties can be defined as the response of a wave function, an energy or an expectation value of an operator to one (or more) perturbations. Here in particular we will consider three types of perturbations: external electric ($\mathbf{F}$) or magnetic $\mathbf{B}$ and a change in the nuclear geometry ($\mathbf{R}$). Following the general derivative formulation, the basic idea is to expand the energy in a Taylor series in the perturbation(s) strength and define the *nth*-order property as the *nth*-order derivative of the energy.

Translating this analysis to molecular solutes in the presence of solvent interactions, requires to introduces specificities which are not present in isolated systems.

First of all the energy functional used in the expansion has to be substituted by the the the free energy analog we have defined in eq.(2). In this framework a general property can be defined as

$$property \propto \frac{\partial^{n_F + n_B + n_R} G}{\partial \mathbf{F}^{n_F} \partial \mathbf{B}^{n_B} \partial \mathbf{R}^{n_R}} \tag{5}$$

where $n_F$, $n_B$ and $n_R$ are the order corresponding to the perturbation represented by external electric ($\mathbf{F}$) or magnetic $\mathbf{B}$ and a change in the nuclear geometry ($\mathbf{R}$). Within a QM description these expressions are solved by introducing analytical derivative approaches. In this framework the property will be determined both by the derivatives of the Hamiltonian but also by the derivatives of the density matrix with respect to the perturbation(s). This means that even when the derivatives of the Hamiltonian with respect to a given perturbation do not contain explicit solvent contributions (as the latter do not depend on such a perturbation and thus the corresponding derivatives are zero), the solvent still affects the derivatives of the density matrix. In such a way the final property will be always modified by the solvent.

Besides the basic differences in the definition of the property expressed in eq.(5), the solvent induces also other specificities.

For example, to compute derivatives of the free energy with respect to nuclear coordinates in continuum solvation models we have to take into account the presence of the cavity. The infinitesimal displacement of a nucleus, considered in a given partial derivative, entails an analogous infinitesimal motion in the corresponding part of the cavity. The neglect of this infinitesimal motion of the cavity gives origin to apparent forces acting on the surface which can have an important effect, as it results from the comparison of the derivatives computed with or without this approximation.

In addition, when the perturbation is an external electric field, the presence of the solvent modifies not only the solute charge distribution but also the probing electric field acting on the molecule. This is a problem of general occurrence when an external field interacts with a molecule in a condensed phase (historically it is known as 'local field effect').

Finally, when the external field has an oscillatory behavior, we have to account for solvent relaxation effects as those described in the previous section.

What it is important to stress here is that all these specificities can be included in the PCM framework so to give a complete and coherent model. In addition, the great advantage of the PCM approaches to reduce solvent effects to a set of operators which can be cast in a physically and formally simple form, allows one to extend the study of the properties of molecular systems in contact with more and more complex environments such as liquid crystalline phases or symmetric crystalline frames, charged solutions like those representing the natural neighborhood of proteins and other biological molecules and membranes, immiscible solvents with a contact surface, metal bodies inside a dielectric, etc.[8]

# 4   Specific effects

Until now we have described the potentialities of polarizable continuum models; these, however, present also important weaknesses. Among them, we have to cite specific interactions between solute and solvent molecules, such as hydrogen-bonds. The electronic aspects of these interactions in fact cannot be properly accounted for by continuum models. It becomes thus necessary to complement the model with different descriptions and investigate the possibility of combining different techniques. This has indeed been the topic of a number of recent studies in which the analysis is focused on the combination of complementary representations of the solvent, namely, explicitly by a small number of solvent molecules (microsolvation) and implicitly by a continuum.[4] This kind of generalized continuum models in which the solute includes also a few explicit solvent molecules (namely those belonging to the first solvation shell/s) is generaly indicated as "solvated supermolecule".

The coupling of the supermolecule approach with a continuum is indeed an interesting technique which has proved to be extremely useful in studying solvent effects on electronic and vibrational spectra and response molecular properties as well in studying reaction mechanisms and energetics in solution. One of the main reasons to explicitly treat some solvent molecules is to get an accurate description of strong specific interactions (e.g., hydrogen bonding), however such an approach is not straightforward and its correct realization requires a detailed analysis on the nature and the strength of such interactions. In the case of very strong solute-solvent interactions such as solutes with strong H-bond acceptor centers in water or in alcohols, a sufficient description is often obtained by including all the solvent molecules necessary to saturate such strong acceptor centers (in general few solvent molecules are enough) in the QM optimization procedure. The resulting structure is then used to compute the corresponding vibrational spectrum or the properties of interest or just to compute the thermal free energy necessary to get the energetics of a reaction. In all the steps of this strategy an external continuum is used in order to account for long-range (or bulk) effects. When we are in the presence of weaker interactions, however, a representation of the supermolecule in terms of a single rigid structure obtained as the minimum of the potential energy surface of the supermolecule cannot be adopted. The real situation is in fact dynamic and a variety of different representative structures can and do occur. A possible way to get such a picture is to consider structures derived from either classical or ab initio MD shots taken at different simulation times: for each of these selected structures the number of solvent molecules to be used in the supermolecule calculation is determined by a threshold imposed in the distance between the solute H-bond acceptor/donor center and the corresponding donor/acceptor center in the solvent molecules. Each of the resulting clusters (generally involving more than the solvent molecules really needed to saturate H-bonding centers) is then embedded in the external continuum and studied at the proper QM level.

# References

[1] (a) C.J.F. Böttcher: *Theory of electric polarization, vol.I.* Elsevier, Amsterdam, 1973; (b) C.J.F. Böttcher and P. Bordewijk: *Theory of electric polarization, vol.II.* Elsevier, Amsterdam

1979.

[2] J. Tomasi and M. Persico, Molecular Interactions in solution: an overview of methods based on continuous distributions of the solvent, *Chem. Rev.* **94** 2027 (1994).

[3] C.J. Cramer and D.G. Truhlar, Implicit solvation models: Equilibria, structure, spectra, and dynamics, *Chem. Rev.* **99**, 2161 (1999).

[4] J. Tomasi, B. Mennucci and R. Cammi, Quantum mechanical continuum solvation models, *Chem. Rev.*, in press (2005).

[5] (a) S. Miertus, E. Scrocco and J. Tomasi, Electrostatic interaction of a solute with a continuum. A direct utilizaion of ab-initio molecular potentials for the prevision of solvent effects, *Chem. Phys.* **55** 117 (1981); (b) R. Cammi and J. Tomasi, Remarks on the Use of the Apparent Surface-Charges (Asc) Methods in Solvation Problems - Iterative Versus Matrix- Inversion Procedures and the Renormalization of the Apparent Charges, *J. Comput. Chem.* **16** 1449 (1995).

[6] (a) E. Cancès, B. Mennucci and J. Tomasi, A new integral equation formalism for the polarizable continuum model: Theoretical background and applications to isotropic and anisotropic dielectrics, J Chem Phys 107 3032 (1997); (b) B. Mennucci, E. Cancès and J. Tomasi, Evaluation of solvent effects in isotropic and anisotropic dielectrics and in ionic solutions with a unified integral equation method: Theoretical bases, computational implementation, and numerical applications, *J. Phys. Chem. B*, **101** 10506 (1997).

[7] F. Ingrosso, B. Mennucci and J. Tomasi, Quantum mechanical calculations coupled with a dynamical continuum model for the description of dielectric relaxation: Time dependent Stokes shift of coumarin C153 in polar solvents, *J. Mol. Liq.* **108**, 21 (2003); (b) M. Caricato, F. Ingrosso, B. Mennucci, and J. Tomasi, A time-dependent Polarizable Continuum Model: theory and application, *J. Chem. Phys.* **122**, 154501 (2005).

[8] (a) S. Corni and J. Tomasi, Surface enhanced Raman scattering from a single molecule adsorbed on a metal particle aggregate: A theoretical study, *J. Chem. Phys.* **116**, 1156 (2002); (b) L. Frediani, R. Cammi, S. Corni, and J. Tomasi, A polarizable continuum model for molecules at diffuse interfaces, *J. Chem. Phys.* **120** 3893 (2004); (c) B. Mennucci, J. Tomasi and R. Cammi, Excitonic splitting in conjugated molecular materials: A quantum mechanical model including interchain interactions and dielectric effects, *Phys. Rev. B* **70** 205212 (2004); (d) O. Andreussi, S. Corni, B. Mennucci and J. Tomasi, Radiative and non radiative decay rates of a molecule close to a metal particle of complex shape, *J. Chem. Phys.* **121** 10190 (2004); (e) M. Pavanello, B. Mennucci and A. Ferrarini, Quantum-mechanical studies of NMR properties of solutes in Liquid Crystals: a new strategy to determine orientational order parameters, *J. Chem. Phys.* **122**, 064906 (2005);

[9] B. Mennucci, J. M. Martínez and J. Tomasi, Solvent effects on nuclear shieldings: Continuum or discrete solvation models to treat hydrogen bond and polarity effects?, *J. Phys. Chem. A* **105** 7287 (2001); (b) B. Mennucci, Hydrogen bond versus polar effects: An ab initio analysis on n-$\pi^*$ absorption spectra and n nuclear shieldings of diazines in solution, *J. Am. Chem. Soc.* **124** 1506 (2002); (c) B. Mennucci and J. M. Martinez, How to model solvation of peptides? Insights from a quantum mechanical and molecular dynamics study of N-methylacetamide. I. Geometries, Infrared and Ultraviolet spectra in water, *J. Phys. Chem. B* **109** 9818 (2005); (d) How to model solvation of peptides? Insights from a quantum mechanical and molecular dynamics study of N-methylacetamide. II. $^{15}N$ and $^{17}O$ nuclear shielding in water and in acetone, *ibidem*, 9830.

Brill Academic Publishers
P.O. Box 9000, 2300 PA Leiden,
The Netherlands

*Lecture Series on Computer*
*and Computational Sciences*
Volume 4, 2005, pp. 966-971

# Correlated Electronic Structure Nonlinear Response Methods for Structured Environments

Kurt V. Mikkelsen [1]

Department of Chemistry,
H. C. Ørsted Institute,
University of Copenhagen,
DK-2100 Copenhagen Ø, Denmark

Received 9 July, 2005; accepted 4 August, 2005

*Abstract:* We consider a brief outline of structural environment models where correlated electronic structure response methods are utilized for the determination of nonlinear optical properties of molecules. The presentation provides theory and applications of a heterogeneous dielectric media model and a quantum mechanical-classical mechanical model at the level of correlated electronic structure response methods. The correlated electronic structure response methods include (i) the multiconfigurational self-consistent field (MCSCF) method and (ii) the coupled cluster (CC) method.

*Keywords:* Heterogeneous dielectric media, quantum mechanical-classical mechanical, correlated electronic structure response methods, multiconfigurational self-consistent field, coupled cluster

## 1   Introduction

The development of quantum chemical methods investigating molecules embedded in a dielectric medium is an increasingly active research area[1, 2, 3, 4, 5, 6, 7, 8, 9, 10, 11, 12, 13, 14, 15, 16, 17, 18, 19, 20, 21, 22, 23, 24, 25, 26, 27, 28, 29, 30, 31, 32, 33, 34, 35] but the the electronic structure methods have generally been uncorrelated[1, 2, 3, 4, 5, 6, 7, 8, 9, 10, 11, 14, 15, 16, 17, 18]. Few research groups have undertaken the task of improving the standards by developing correlated electronic structure methods coupled to a classical system and the efforts include methods at the second order Møller-Plesset (MP2) level[27, 29], the multiconfigurational self-consistent reaction field (MCSCRF) level[12, 19] and the coupled-cluster self-consistent reaction field (CCSCRF) level[35].

These methods do not provide realisitic investigations of molecules that are exposed to structured environments. Molecules in structured environments could be a molecule in a dielectric film being absorbed on a metallic surface, a heterogeneously solvated molecule or within a biological system. For molecules in structured environments, it is in many cases crucial to be able to compute energy terms along with linear and nonlinear electromagnetic properties. This is clearly seen for many areas of biochemistry and chemistry in relation to electrochemistry, photo-catalysis, surface photo-chemistry and surface-enhanced two-photon transitions or hyperpolarizabilities.

This presentation will focus on the development of highly accurate methods for investigating nonlinear electromagnetic properties of molecules that are situated in a heterogeneous environment. We consider two different methods for the determination of molecular properties of molecules in

---

[1]E-mail: kmi@theory.ki.ku.dk

structured environments, (i) a heterogeneous dielectric media method by Jørgensen et al.[36, 37, 38] and (ii) a quantum mechanical-classical system method by Poulsen et al. and Osted et al. and Kongsted et al.[39, 40, 41, 42, 43, 44, 45, 46, 47, 48, 49, 50].

The methods provide calculation procedures for molecular properties such as:

- frequency-dependent second hyperpolarizabilities ($\gamma$),

- three-photon absorptions,

- two-photon absorption between excited states,

- frequency-dependent polarizabilities of excited states,

- frequency dependent first hyperpolarizability tensors,

- two-photon matrix elements.

- frequency dependent polarizabilities

- excitation and deexcitation energies along with their corresponding transition moments.

## 2  Energy functional

The Hamiltonians and the energy functionals for the two different structural environment methods are presented. The representations of these are crucial for determining the molecular properties of molecules situated in environments. Approaches where one part of the total system is described by quantum mechanics whereas the another part is pictured by a much coarser method have successfully been applied within quantum chemistry[1, 2, 3, 4, 5, 6, 7, 8, 9, 10, 11, 12, 13, 14, 15, 16, 17, 18, 19, 20, 21, 22, 23, 24, 25, 26, 27, 28, 29, 30, 31, 32] and molecular reaction dynamics[51, 52, 53, 18, 54]. The quantum subsystem and the classical subsystem interact through classical interactions between charges and/or induced charges and a van der Waals term[1, 2, 3, 4, 5, 6, 7, 8, 9, 10, 11, 12, 13, 14, 15, 16, 17] and this is achieved by an effective interaction operator. The solution of the quantum mechanical equations including the effective interaction operator provides a solution of the quantum mechanical problem for the prinicipal system interacting with the classical subsystem[1, 2, 3, 4, 5, 6, 7, 8, 9, 10, 11, 12, 14, 15, 16, 18, 23, 25, 27, 28, 29, 30, 31, 32].

### 2.1  The heterogeneous dielectric media method

Within this model, the total system is divided into two subsystems where the solvated molecule is encapsulated in a cavity C that has the surfaces $\Sigma_m$ and $\Sigma_l$. The cavity is surrounded by two dielectric media seperated into two parts $S_m$ and $S_l$ and the they are in contact with the cavity at $\Sigma_m$ and $\Sigma_l$. The dielectric media are given as linear, homogeneous and isotropic media. We have constructed the two dielectric media with the following spatial positions:

$$V_{S_m} = \left\{ \vec{r} = (r, \theta, \phi) | \theta \notin \left[ -\frac{\pi}{2}; \frac{\pi}{2} \right] \right\}$$

$$V_{S_l} = \left\{ \vec{r} = (r, \theta, \phi) | r \geq R, \theta \in \left[ -\frac{\pi}{2}; \frac{\pi}{2} \right] \right\} \tag{1}$$

The cavity, $C$, is a half sphere with a radius of $R$ and is placed on the surface of $S_m$ and embedded in $S_l$. The volume of cavity is

$$V_C = \left\{ \vec{r} = (r, \theta, \phi) | r \leq R, \theta \in \left[ -\frac{\pi}{2}; \frac{\pi}{2} \right] \right\}. \tag{2}$$

For nonlinear molecular properties of molecules in structured environments, it is crucial to solve simultaneously the quantum mechanical problem and the classical electrostatic problem with the appropriate the boundary conditions. The Hamiltonian consists of two terms:

- a part that gives the Hamiltonian, $H_M^0$, for the isolated molecular system in vacuum.

- Another part gives the interaction operator, $W_{pol}$, including the interactions between the molecular system and the structured environment.

The interaction operator depends on two items:

- the induced potential, $U_{pol}$, in the two dielectric media and

- the molecular charge distribution $\rho_m(\vec{r})$.

### 2.2 The quantum mechanical-classical system method

The Hamiltonian is given by three parts:

- the Hamiltonian of the molecular system in vacuum ($\hat{H}_{QM}$),

- the Hamiltonian for the classical system ($\hat{H}_{CM}$)

- the interactions between the quantum mechanical and the classical system ($\hat{H}_{QM/CM}$).

The Hamiltonian is written as

$$\hat{H} = \hat{H}_{QM} + \hat{H}_{QM/CM} + \hat{H}_{CM} \tag{3}$$

The second term $\hat{H}_{QM/CM}$ contains contributions from

- the electrostatic interactions denoted by $\hat{H}^{el}$,

- the polarization interactions given by $\hat{H}^{pol}$

- along with the van der Waals interactions represented by $\hat{H}^{vdw}$

and therefore the interaction operator is defined as

$$\hat{H}_{QM/CM} = \hat{H}^{el} + \hat{H}^{vdw} + \hat{H}^{pol} \tag{4}$$

## 3   Response Equations

Response theory methods provide a solid starting point for determining molecular properties of a quantum mechanical subsystem coupled to a structured environment and we consider the response of a given reference state of the molecular system to an externally applied time-dependent electromagnetic field produced by a laser. The solution to the time-dependent Schrödinger equation is found be utilizing Frenkel's variation principle in the form of the Ehrenfest's equation. We evaluate the time evolution of the expectation value of any operator $A$ and we have

$$\frac{d\langle A\rangle}{dt} = \left\langle \frac{\partial A}{\partial t} \right\rangle - i\langle [A, H]\rangle \tag{5}$$

where we have the following for the the the total Hamiltonian $H$

$$H = H_0 + W_{se} + V(t). \tag{6}$$

Here we utilize the following definitions:

- the part of the total Hamiltonian represented by $(H_0 + W_{se})$ forms the time independent Hamiltonian of the quantum subsystem

  - including the coupling to the structured environment by the interaction operator $W_{se}$
  - and $H_0$ denotes the Hamiltonian of the isolated quantum subsystem.

- The operator $V(t)$ represents the interactions between the externally applied time-dependent electromagnetic field and the quantum subsystem.

- Time-dependent expectation values are determined using the time dependent wave function, $|0^t >$

$$\langle \cdots \rangle = \langle {}^t 0 \mid \cdots \mid 0^t \rangle \tag{7}$$

We find for an arbitrary time-independent operator A

$$\langle 0^t \mid A \mid 0^t \rangle = \langle 0 \mid A \mid 0 \rangle + \int_{-\infty}^{\infty} d\omega_1 exp[(-i\omega_1 + \epsilon)t] \langle\langle A; V^{\omega_1} \rangle\rangle_{\omega_1}$$

$$+\frac{1}{2} \int_{-\infty}^{\infty} d\omega_1 \int_{-\infty}^{\infty} d\omega_2 exp[(-i(\omega_1 + \omega_2) + 2\epsilon)t] \langle\langle A; V^{\omega_1}, V^{\omega_2} \rangle\rangle_{\omega_1,\omega_2} + ... \tag{8}$$

and we denote[55]:

- the function $\langle\langle A; V^{\omega_1} \rangle\rangle_{\omega_1}$ is the linear response function and
- the function $\langle\langle A; V^{\omega_1}, V^{\omega_2} \rangle\rangle_{\omega_1,\omega_2}$ is the quadratic response function.

## Acknowledgment

K.V.M. thanks Statens Naturvidenskabelige Forskningsråd, Statens Tekniske Videnskabelige Forskningsråd, the Danish Center for Scientific Computing and the EU-network NANOQUANT for support.

## References

[1] M. D. Newton. *J. Phys. Chem.*, 79:2795, 1975.

[2] J.O. Noell and K. Morokuma. *Chem. Phys. Lett.*, 36:465, 1975.

[3] D. L. Beveridge and G. W. Schnuelle. *J. Phys. Chem.*, 79:2562, 1975.

[4] J. Hylton, R. E. Christoffersen, and G. G. Hall. *Chem. Phys. Lett.*, 24:501, 1974.

[5] R. Contreras and A. Aizman. *Int. J. Quant. Chem.*, 27:193, 1985.

[6] H. Hoshi, M. Sakurai, Y. Inone, and R. Chujo. *J. Chem. Phys.*, 87:1107, 1987.

[7] O. Tapia. *Molecular Interactions.* Wiley, New York, 1980. by H. Ratajczak and W.J. Orville-Thomas.

[8] A. Warshel. *Chem. Phys. Lett.*, 55:454, 1978.

[9] E. Sanchez-Marcos, B. Terryn, and J. L Rivail. *J. Phys. Chem.*, 87:4695, 1985.

[10] D. Rinaldi. *Comput. Chem.*, 6:155, 1982.

[11]  O. Tapia. *Quantum Theory of Chemical Reactions*. Wiley, Dordrecht, 1980. by R. Daudel and A. Pullman and L. Salem and A. Veillard, vol. 3, 25.

[12]  G. Karlström. *J. Phys. Chem.*, 93:4952, 1989.

[13]  M. Karelson and M. Zerner. *J. Am. Chem. Soc.*, 112:9405, 1990.

[14]  J.H. Jensen, M.S. Gordon, S.P. Webb, W.J. Stevens, M. Krauss, D. Garmer, H. Basch, and D. Cohen. *J. Chem. Phys.*, 105:1968, 1996.

[15]  W. Chen and M.S. Gordon. *J. Chem. Phys.*, 105:11081, 1996.

[16]  C.J. Cramer and D.G. Truhlar. *J. Am. Chem. Soc.*, 113:8305, 1991.

[17]  C.J. Cramer and D.G. Truhlar. *Science*, 256:213, 1992.

[18]  K. V. Mikkelsen, E. Dalgaard, and P. Svanstøm. *J. Phys. Chem.*, 91:3081, 1987.

[19]  K. V. Mikkelsen, H. Ågren, H. J. Aa. Jensen, and T. Helgaker. *J. Chem. Phys.*, 89:3086, 1988.

[20]  K. V. Mikkelsen, P. Jørgensen, and H. J. Aa. Jensen. *J. Chem. Phys.*, 100:6597–6607, 1994.

[21]  K. V. Mikkelsen, Y. Luo, H. Ågren, and P. Jørgensen. *J. Chem. Phys.*, 100:8240, 1994.

[22]  K. V. Mikkelsen, Y. Luo, H. Ågren, and P. Jørgensen. *J. Chem. Phys.*, 102:9362, 1995.

[23]  K. V. Mikkelsen and K. O. Sylvester-Hvid. *J. Phys. Chem.*, 100:9116, 1996.

[24]  S. Di Bella, T. J. Marks, and M. A. Ratner. *J. Am. Chem. Soc.*, 116:4440, 1994.

[25]  J. Yu and M. C. Zerner. *J. Chem. Phys.*, 100:7487, 1994.

[26]  R. Cammi, M. Cossi, B. Mennucci, and J. Tomasi. *J. Chem. Phys.*, 105:10556, 1996.

[27]  A. Willetts and J. E. Rice. *J. Chem. Phys.*, 99:426, 1993.

[28]  R. Cammi, M. Cossi, and J. Tomasi. *J. Chem. Phys.*, 104:4611, 1996.

[29]  M.W. Wong, M.J. Frisch, and K.B. Wiberg. *J. Am. Chem. Soc.*, 113:4776, 1991.

[30]  J.G. Angyan. *Chem. Phys. Lett.*, 241:51, 1995.

[31]  F.J. Olivares del Valle and J. Tomassi. *Chem. Phys.*, 150:139, 1991.

[32]  C. Chipot, D. Rinaldi, and J.L. Rivail. *Chem. Phys. Lett.*, 191:287, 1991.

[33]  C. J. Cramer and D. G. Truhlar. *Chem. Rev.*, 99:2161, 1999.

[34]  J. Tomasi, R. Cammi, and B. Mennucci. *Int. J. Quant. Chem.*, 75:783, 1999.

[35]  O. Christiansen and K. V. Mikkelsen. *J. Chem. Phys.*, 110:1365, 1999.

[36]  S. Jørgensen, M. A. Ratner, and K. V. Mikkelsen. *J. Chem. Phys.*, 115:3792, 2001.

[37]  S. Jørgensen, M. A. Ratner, and K. V. Mikkelsen. *J. Chem. Phys.*, 2001.

[38]  S. Jørgensen, M. A. Ratner, and K. V. Mikkelsen. *J. Chem. Phys.*, 2002.

[39]  T. D. Poulsen, J. Kongsted, A. Osted, P. R. Ogilby, and K. V. Mikkelsen. *J. Chem. Phys.*, 115:2393–2400, 2001.

[40] T. D. Poulsen, P. R. Ogilby, and K. V. Mikkelsen. *J. Chem. Phys.*, 116:3730–3738, 2002.

[41] T. D. Poulsen, P. R. Ogilby, and K. V. Mikkelsen. *J. Chem. Phys.*, 115:7843–7851, 2001.

[42] J. Kongsted, A. Osted, K.V. Mikkelsen, P-.O. Åstrand and O. Christiansen, J. Chem. Phys. **121**, 8435 (2004).

[43] K. Aidas, J. Kongsted, A. Osted, K.V. Mikkelsen and O. Christiansen, J. Phys. Chem. A. (2005). Accepted for publication in J. Phys. Chem. A.

[44] J. Kongsted, A. Osted, K.V. Mikkelsen and O. Christiansen, J. Phys. Chem. A. **107**, 2578 (2003).

[45] J. Kongsted, A. Osted, K.V. Mikkelsen and O. Christiansen, J. Chem. Phys. **118**, 1620 (2003).

[46] J. Kongsted, A. Osted, K.V. Mikkelsen and O. Christiansen, J. Chem. Phys. **119**, 10519 (2003).

[47] J. Kongsted, A. Osted, K.V. Mikkelsen and O. Christiansen, J. Chem. Phys. **120**, 3787 (2004).

[48] J. Kongsted, A. Osted, K.V. Mikkelsen and O. Christiansen, Mol.Phys. **100**, 1813 (2002).

[49] A. Osted, J. Kongsted, K.V. Mikkelsen and O. Christiansen, J. Phys. Chem. A. **108**, 8646 (2005).

[50] A. Osted, J. Kongsted, K.V. Mikkelsen and O. Christiansen, Mol. Phys. **101**, 2055 (2003).

[51] D. Borgis, S. Lee, and J. T. Hynes. *Chem. Phys. Lett.*, 162:19, 1989.

[52] D. Borgis and J. T. Hynes. *J. Chem. Phys.*, 94:3619, 1991.

[53] D. Borgis and J. T. Hynes. *Chem. Phys.*, 170:315, 1993.

[54] K. V. Mikkelsen. *Z. Phys. Chem.*, 170:129–142, 1991.

[55] J. Olsen and P. Jørgensen. *J. Chem. Phys.*, 82:3235, 1985.

Brill Academic Publishers
P.O. Box 9000, 2300 PA Leiden,
The Netherlands

*Lecture Series on Computer
and Computational Sciences*
Volume 4, 2005, pp. 972-975

# Aromatic Association in Aqueous Solution: a Potential of Mean Force Study

## S. Polydoridis and G. Archontis[1]

Department of Physics, University of Cyprus, PO20537, CY1678, Cyprus

Received 4 August, 2005; accepted 14 August, 2005

*Abstract:* Interactions between aromatic sidechains affect the stability of protein structures. In the present work we calculate the free-energy profile for a pair of aromatic Phenylalanine (Phe) aminoacids in aqueous solution, as a function of the distance between the $C_\alpha$ atoms of the peptide main chains. As a comparison system, we also study the Alanine (Ala) pair, in which the side chains (methyl groups) have hydrophobic but not aromatic character. In contrast to other studies of aromatic association, we include the peptide main chains in the simulation system. The free-energy profiles show that the Phe association is weakly favored in water, whereas in the Ala pair the separated conformation is the most stable. An energetic interaction analysis is employed to interpret these differences.

*Keywords:* molecular dynamics, free energy calculations, aromatic interactions, peptide association

*PACS:* 87.15Aa, 87.15.He, 87.15.Nn, 87.15.Kg

## 1 Introduction

Aromatic interactions are important for the stabilization of protein structures. Recent experiments [1, 2] and modeling studies [3] have suggested that these interactions play an important role in the formation of amyloid fibers, multipeptide or protein complexes implicated in a variety of diseases [4]. To gain some understanding of the stabilization of peptide complexes due to aromatic interactions, we compute and compare the association free energy for the peptide pairs Phe-Phe and Ala-Ala in aqueous solution. In contrast to other aromatic association studies with simpler systems [5], we include the peptide main chain in the simulation (the structure of the Phe peptide model is shown in the left panel of fig. 1); this increases considerably the number of degrees of freedom, but enables us to assess the contribution due to main chain and sidechain interactions in the stabilization of the peptide complexes.

## 2 Methods

To ensure that the simulations obtain a representative sampling of the peptide conformations, we employed long runs ($\approx$ 300 ns for each pair), combined with umbrella-sampling techniques [6]. **MD Simulations:** The MD simulations were performed with the Molecular Dynamics program CHARMM, version c30b1 [7]. Force-field parameters were taken from the CHARMM22 all-atom force field [8]. The water was reproduced by a modified TIP3P water model [9]. Each peptide pair was solvated by a 16 Å-radius sphere, containing 625 water molecules. The van der Waals

---

[1]Corresponding author. E-mail: archonti@ucy.ac.cy

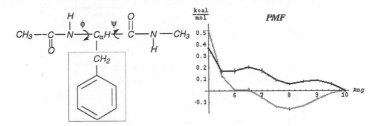

Figure 1: **Left:** Structure of the model for the Phe peptide. The main chain contains two peptide groups with orientations defined by two flexible dihedral angles ($\phi$, $\psi$), and is terminated by two methyl groups. The Phe sidechain is enclosed in the rectangle. **Right:** Potential of mean force curves for the Ala-Ala pair (black) and Phe-Phe pair (gray) in aqueous solution. Both curves are shifted to zero at a $C_\alpha - C_\alpha$ distance of 10 Å.

interactions were switched off at interatomic distances beyond 14 Å. Electrostatic interactions were calculated exactly below 14 Å and by use of a multipole expansion approximation (extended electrostatics) beyond 14 Å [10]. The electrostatic and van der Waals effects due to the bulk solvent beyond the simulation system were modeled by the Spherical Solvent Boundary Potential (SSBP) method [11]. Constant-temperature simulations were conducted by subjecting the water oxygens in the spherical layer between 9–16 Å from the simulation center to random and frictional Langevin forces mimicking a 300 $K$ thermal bath. The length of all bonds involving hydrogen atoms, and the internal geometry of the water molecules was fixed with the SHAKE algorithm [12]. The peptide pair was kept at the center of the solution sphere by a harmonic restraint of 2.0 kcal/mol/Å$^2$, applied to the pair center of mass.

**Potential of Mean Force Calculations:** For each system, umbrella-sampling simulations were performed, with the peptide $C_\alpha - C_\alpha$ distance restrained successively at values $r_i \equiv 4.5$, 5.0, 5.5, 6.0, 6.5, 7.0, 7.5, 8.0, 8.5, 9.0, 9.5 and 10.0 Å by a harmonic energy term. The biased probability distribution of the $C_\alpha - C_\alpha$ distance $p_i^b(r)$ was obtained directly from each simulation $i$. The corresponding unbiased probability distribution $p_i^u(r)$ was obtained by the equation [6]

$$p_i^u(r) = p_i^b(r) \exp[\beta U_i(r)] \exp[-\beta F_i(r)] \tag{1}$$

where $U_i(r) \equiv k_b(r - r_i)^2$ is the restraining potential, $\beta = 1/k_B T$, with $k_B$ Boltzmann's constant and $T$ the temperature, and $F_i(r)$ are free-energies related to the average of the restraining potential [6]. The quantities $F_i(r)$ were calculated self-consistently and the probabilities $p_i^u(r)$ were merged to a global probability function $p(r)$ by the weighted histogram analysis method (WHAM) [6, 13]. The free-energy profile is linked to $p(r)$ by the relation $F(r) = -k_B T \ln[p(r)] + C$; $C$ is a constant.

The $C_\alpha$–$C_\alpha$ restraining force constant was set to 3.0 kcal/mol; with this magnitude, there was good overlap between the $C_\alpha$–$C_\alpha$ distance distributions of adjacent simulations. The simulations employed a 0.002 ps timestep. For each peptide pair, 18–24 independent runs of 1.2-ns length were conducted for each restraining distance. The first 0.2 ns of a run were treated as equilibration stage; in the last 1 ns, structures were stored at every 1 ps and used later for analysis. The total simulation length for each restraining distance was 21.6–28.8 ns; for the entire profile it was ≈ 300 ns.

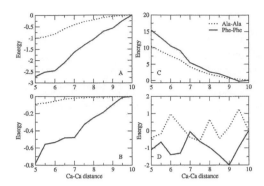

Figure 2: Average energy component profiles (in kcal/mol). **A.** Intermolecular main chain-sidechain energy; **B.** Sidechain-sidechain energy. **C.** Peptide-water energy. **D.** Total energy.

## 3  Results

The potentials of mean force (pmf) for the Ala-Ala and Phe-Phe pair are shown in fig. 1 (right panel); both curves have been shifted to a 0.0 value at the maximum restrained $C_\alpha - C_\alpha$ distance (10.0 Å). For the Ala-Ala pair in solution, the total free energy increases as the peptides approach; the two local minima at distances of 8 Å and 5.5–6 Å are positive. Thus, the most stable configuration corresponds to separated aminoacids. For the Phe-Phe pair, the free-energy curve has a stable well in the range 6.5–10 Å, with a negative (-0.15 kcal/mol) minimum centered at 8 Å; in this region, the peptide dimer is stabilized compared to the infinitely separated dimer.

To understand better the differences betwee the two curves, we calculated the energy of the two systems as a function of the $C_\alpha - C_\alpha$ distance. Fig. 2 includes profiles for selected energy components. All curves have been shifted to zero at a $C_\alpha - C_\alpha$ distance of 10.0 Å.

Figs. 2A and 2B show, respectively, the average intermolecular main chain-sidechain and sidechain-sidechain interaction energies. As expected, the energies become more negative at smaller ·distances, as the peptides approach and interact with each other. This trend is much more pronounced for the Phe pair. Thus, each aromatic sidechain interacts much more strongly with the sidechain or main chain of the second peptide, compared to the methyl sidechains in the Ala pair. The energetic stabilization of the Phe pair at small distances is partly cancelled by an increase in the peptide-water energy (fig. 2C) and the water-water energy (not shown). Due to these cancellations, the *total* energy profiles of the two systems (shown in fig. 2D) have a much smaller dependence on the $C_\alpha - C_\alpha$ distance. Nevertheless, the total energy curve of the Phe pair has a broad minimum in the region 7–10 Å, indicating that the Phe pair configuration at the $C_\alpha - C_\alpha$ distance of 7.5 Å is stabilized energetically.

At $C_\alpha - C_\alpha$ distances beyond 8 Å the peptides are separated and interact with water. At smaller distances they form hydrophobic interactions (via the main chain methyl blocking groups, and/or the sidechain groups). In the Phe pair, interactions between the aromatic rings also observed, with stacked (parallel) orientations being the most probable. At distances closer than 8.0 Å, the main-chain peptide groups form intermolecular hydrogen bonds (hb); at 7.5 Å the probability to observe at least one intermolecular hb is 8.7% (Ala pair ) and 6.3% (Phe pair). At smaller distances, the hb frequency is consistently more pronounced in the Phe pair. At 5.0 Å it is 63.1%, compared to 49.1% for the Ala pair. Thus, the increased sidechain-sidechain and sidechain-main chain interactions in the Phe pair seem to facilitate intermolecular main chain hb formation.

# 4  Conclusions

The two free-energy profiles have a different dependence on the $C_\alpha - C_\alpha$ distance. In the Phe pair, the aromatic sidechains stabilize the peptide association. Analysis of the energy components suggests that this stabilization is due to direct interactions involving the aromatic sidechains, as well as due to enhanced interactions between the main chains. The screening of interactions by the aqueous solvent and the entropic contributions due to the large number of accessible molecular conformations at each $C_\alpha - C_\alpha$ distance result in shallow free-energy profiles. Indeed, the Phe-pair minimum corresponds to a free energy of only -0.15 kcal/mol, compared to the value at 10 Å. In larger peptides the stabilization should be more pronounced, due to the increase in the number of intermolecular interactions and the higher exclusion of solvent.

**Acknowledgement:** This research has been funded by a RPF/PENEK grant (to GA and SP).

# References

[1] Y. Porat et. al. Completely different amyloidogenic potential of nearly identical peptide fragments. *BIOP* **69**, 161–164 (2003).

[2] R. Nelson et al. Structure of the cross-$\beta$ spine of amyloid fibers. *NAT* **435**, 773–778 (2005).

[3] G.G. Tartaglia et. al. Prediction of aggregation rate and aggregation prone segments in polypeptide sequences. *PSCI* (2005). To appear.

[4] C. Dobson. Protein misfolding, evolution and disease. *TIBS* **24** (1999). 329-332.

[5] R. Chelli et. al. Stacking and T-shape competition in aromatic-aromatic aminoacid interactions. *J. Am. Chem. Soc.* **124**, 6133–6143 (2001).

[6] M. Souaille and B. Roux. Hydrophilicity of polar amino acid side-chains is markedly reduced by flanking peptide bonds. *Comput. Phys. Commun.* **135**, 40–57 (2001).

[7] B. R. Brooks et. al. CHARMM: A program for macromolecular energy, minimization, and dynamics calculations. *J. Comput. Chem.* **4**, 187–217 (1983).

[8] A. D.Mackerell et. al. An all-atom empirical potential for molecular modelling and dynamics study of proteins. *J. Phys. Chem. B* **102**, 3586–3616 (1998).

[9] W. L. Jorgensen et. al. Comparison of simple potential functions for simulating liquid water. *J. Chem. Phys.* **79**, 926–35 (1983).

[10] R. Stote, D. States, and M. Karplus. On the treatment of electrostatic interactions in biomolecular simulation. *J. Chim. Phys.* **88**, 2419–2433 (1991).

[11] D. Beglov and B. Roux. Finite representation of an infinite bul system: Solvent Boundary Potential for computer simulations. *JCP* **100**, 9050–9063 (1994).

[12] J. P. Ryckaert et. al. Numerical integration of the cartesian equations of motion of a system with constraints: molecular dynamics of n-alkanes. *J. Comput. Phys.* **23**, 327–41 (1977).

[13] A. M. Ferrenberg and R. H. Swendsen. Optimized monte carlo data analysis. *Phys. Rev. Lett.* **63**, 1195–1198 (1989).

Brill Academic Publishers
P.O. Box 9000, 2300 PA Leiden,
The Netherlands

*Lecture Series on Computer*
*and Computational Sciences*
Volume 4, 2005, pp. 976-979

# Molecular Dynamics Simulations of Conformational Changes in 7-helix Trans-membrane Receptors

**C.F. Sanz-Navarro**

Department of Chemistry, Norwegian University of Science and Technology (NTNU).
N-7491 Trondheim, Norway.

**C. Dezi, J. Brea, M. Pastor[1] and F. Sanz**

Research Group on Biomedical Informatics (GRIB)-IMIM/UPF,
E-08003 Barcelona Spain

Received 9 July, 2005; accepted in revised form 17 July, 2005

*Abstract:* We sketch a complete methodology to perform molecular dynamics simulations (MD) of both inactive and partially activated states of 7-helix trans-membrane (7TM) receptors. In particular, we give details of a protocol for setting up the atomistic system and posterior MD simulations. The methodology is then applied to the case of bovine rhodopsin (bRho) and 5-HT$_{2A}$ (serotonin) receptors, showing how the models can provide some insights into the activation process of G-protein coupled receptors (GPCRs).

*Keywords:* molecular dynamics simulation, conformational changes, 7TM receptor, GPCR.

## 1  Introduction

Communication with the inner part of a living cell is often mediated by 7TM receptors, which explains why these trans-membrane proteins are among the most valuable drug-targets in pharmaceutical research. Bacteriorhodopsin, channelrhodopsin and the superfamily of GPCRs are three typical examples of this wide group. Although numerous scientific papers have been devoted to the subject, still many questions on their activation-dependent conformational changes, and so their functional activity, remain completely unanswered. The main reason is the technical hindrances facing the extraction of intact 7TM receptors from their natural membrane environment. Even if we could soon visualize new 7TM receptors with sufficiently high resolution, they will most surely be in an inactive conformation. Thus, so far, information on activation mainly comes from indirect experimental observations (e.g. mutagenesis data), low-resolution images (e.g. AFM data) and atomistic simulations. Regarding the latter, MD simulations of conformational changes during activation of 7TM receptors entail several challenges in terms of computational effort. The overall receptor activation typically lasts for around a millisecond and this time span is currently unreachable by MD approaches. Alternatively, a possible strategy consists of the modification of the position of some of the side chains of an inactive receptor structure in order to drag the system towards a partially activated state according to results suggested by indirect experimental observations. Then the resulting unstable conformation is let evolve by an MD simulation, expecting to discover additional conformational changes to accommodate the artificially introduced modifications. Here we illustrate this computational methodology by a couple of representative examples.

---

[1]Corresponding author. E-mail: mpastor@imim.es

Figure 1: Typical image of an MD simulation of a system comprising receptor (5-HT$_{2A}$), ligand (ketanserin), bilayer membrane (POPC) and involving layers of water.

## 2  MD Simulation Methodology

Fig. 1 shows the initial image of one of our MD simulations, which were performed with version 3.1 of the GROMACS software package [1]. This widely employed set of MD tools contains a database with force-field parameters for many building blocks of amino acids and a few more common molecules such as water. Nonetheless, most receptor ligands and even the bilayer membrane require additional force-field information. Templates and tools for both force-field parameters and atom positions were found on the web. For instance, an initial template for a cubic box containing a bilayer membrane and water was downloaded from the database website *http://www.ucalgary.ca/~tieleman/download.html*. Regarding the bRho receptor, atom coordinates of the dark-adapted structure [2] and a partially activated state were taken from the website *http://mosberglab.phar.umich.edu/resources/*. The actived conformation considered several steric modifications on the inactive receptor; the retinal was isomerized from the 11-cis to the all-trans conformation and the receptor helix VI was slightly rotated [3]. GROMACS format topology for the retinal was built from the PDB coordinates by using the web tool PRODRG [4] and atomic charges were taken from *ab-initio* quatum mechanics calculations [3].

To date, bRho is the only GPCR whose 3D structure has been solved by high-resolution X-ray diffraction. For the rest of GPCRs, initial atom positions must be obtained by computational approaches. An inactive conformation of the 5-HT$_{2A}$ receptor was derived by homology modelling. The resulting structure was transformed into a partially activated receptor conformation by changing two specific residues from *gauche+* to *trans* conformation following ref. [5]. Finally, a ligand was inserted into the receptor-binding pocket by docking techniques. Further details about our procedure for setting the initial positions of the receptor-ligand complexes can be read somewhere else (C. Dezi et al., *to be published*).

Next step was the insertion of the receptor-ligand complex into the bilayer membrane. Since 7TM receptors are not cylindrically symmetric, a simple cylindrical hole made in the membrane would leave large vacuum regions between the receptor and the membrane. In order to tailor a hole in the membrane similar in shape to that of the receptor, we employed a modified version of *mdrun* (core GROMACS tool for MD simulations) which was downloaded from the web address *http://www.gromacs.org/contributions/uploaded_contributions/mdrun_make_hole.tar.gz*. Posteriorly the ligand-receptor complex was inserted and the whole system was allowed to relax through a

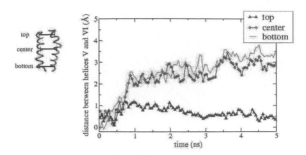

Figure 2: Variation in the interaxial distance between helices V and VI computed by taking three centers of mass (namely, *top, center and bottom*) of the backbone of three different sets of 9 residues along each helix.

4-stage minimization protocol. During the first minimization, the protein was frozen, but the bi-layer membrane and water atoms were left completely free. Then a second minimization with the hydrogen atoms defrost run in order to stabilize the hydrogen bonds before the protein scaffold was left free. The receptor side chains were freed during the third minimization and finally only the carbon-$\alpha$ atoms were fixed during the fourth minimization. Thus the system was ready for heating, which was achieved by a prior 1-ns equilibration stage with the positions of the carbon-$\alpha$ atoms restrained by a harmonic potential. System temperature and pressure were maintained at 310K and 1 bar respectively by the Berendsen algorithm. Finally, the system was left completely unrestrained and a proper NPT MD simulation was carried out.

## 3  Results

Recently several research groups have suggested that the conformational changes upon activation of GPCRs finally lead to a distancing of helices V and VI [6]. However, still little is known about where exactly the activation starts and how the conformational changes are propagated. In order to give some insights into the hypothesis of interhelical distancing, we applied the aforementioned methodology to the cases of both active and partially activated structures of bRho and 5-HT$_{2A}$ receptors. In the 5-HT$_{2A}$ case, both serotonin (agonist) and ketanserin (inverse agonist) were considered as ligands.

As expected, the inactive state of bRho remained oscillating around the initial positions during a 5-ns MD simulation and so no major changes in the separation between helices V and VI were observed, whereas the distance between such helices increased considerably in the partially acti-vated case as shown in Fig. 2. Moreover, as displayed in Fig. 3(a), it was found that for the case of the inactive state, Trp-265 in helix VI established a hydrogen bond with Tyr-301 in helix VII. This is consistent with experimental evidences [7] of hydrogen bond established by Trp-265, although experiments did not reveal with which second residue the hydrogen bond is formed. Moreover, the hydrogen bond contributes to the immobilization of helix VI and so helps to maintain the initial separation between helices V and VI. In contrast, the rotation of helix VI for the partial activation of bRho reallocated Trp-265 to a new position where it was in close contact with Phe-212. In order to relax the tension between side chains, Trp-265 and Phe-212 moved in such a way that their aromatic rings faced each other as shown in Fig. 3(b). The simulation showed that this rearrangement of side chains can only be achieved by a separation between the centers of helices V and VI. Subsequently this distancing was propagated down to IL3.

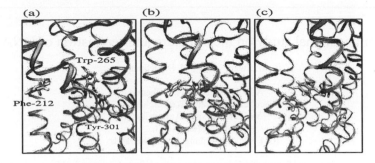

Figure 3: MD images revealing the cause of the distancing between helices V and VI during activation. (a) Inactive bRho case: Trp-265 establishes a H-bond with helix VII. (b) Partially activated bRho case: aromatic rings of Trp-265 and Phe-212 take a face-to-face stacked conformation and as a result the distance between helices V and VI increases considerably. (c) Partially activated 5-HT$_{2A}$ receptor case: face-to-face stacked conformation between Phe-243 and Trp-336 side-chains.

Simulations considering the 5-HT$_{2A}$ receptor confirmed the above mentioned observations. The center of the helices V and VI distanced from each other in the partially activated case for a similar reason to that commented for the bRho case (c.f. Figs. 3(b) and (c)). However, contrarily to the case of bRho, the interhelical distance between the bottom sides decreased over the course of the MD simulation. Further inspection revealed that Tyr-254 established a bond between helices V and VI, decreasing the interhelical distance. It could be a spurious effect due to a wrong initial orientation of the Tyr-254 side chain. On the other hand, it could also happen that Tyr-254 is actually bound to a specific residue of IL3, which was not considered in our simulations. New experiments are currently being performed to discern between these proposed explanations.

## Acknowledgment

The present work has been supported in part by the CICYT (SAF2002-04195-C03-03) and "Fundació La Marató de TV3" grants. C. Dezi wishes to thank a PhD fellowship (FI) awarded by Generalitat de Catalunya. We are also very grateful to B. Oliva for useful discussion about the usage of several MD tools.

## References

[1] H.J.C. Berendsen, D. van der Spoel and R. van Drunen, *Comp. Phys. Comm.* **95**, 43 (1995).

[2] K. Palczewski et al., *Science*, **289**, 739 (2000).

[3] J. Saam, E. Tajkhorshid, S. Hayashi, and K. Schulten, *Biophys. J.*, **83**(6), 3097 (2002).

[4] A.W. Schuettelkopf and D.M.F. van Aalten, *Acta Crystallographica*, **D60**, 1355 (2004).

[5] S.L.G. Lei et al., *J. Biol. Chem.* **277**(43), 40989 (2002).

[6] R. Arimoto, O.G. Kisselev, G.M. Makara, and G.R. Marshall, *Biophys. J.*, **81**(6), 3285 (2002).

[7] S.W. Lin and T.P. Sakmar, *Biochemistry*, **35**(34), 11149 (1996).

Brill Academic Publishers
P.O. Box 9000, 2300 PA Leiden
The Netherlands

*Lecture Series on Computer
and Computational Sciences*
Volume 4, 2005, pp. 980-982

# Computational study on the structure and replication of DNA

M. Swart[1] and F. M. Bickelhaupt

Afdeling Theoretische Chemie,
Faculteit der Exacte Wetenschappen,
Vrije Universiteit Amsterdam,
De Boelelaan 1083,
1081 HV Amsterdam, The Netherlands

Received 8 July, 2005; accepted in revised form 4 August, 2005

*Abstract:* We present here our first attempts to use QM/MM and QM/QM methods to study the structure of DNA oligomers in solution and the process of replication of DNA, as well as the influence of the molecular environment on them. For the hydrogen bonds between DNA base pairs, the latter influence was previously shown to be of utmost importance for the comparison of experimental and theoretical results. The basic reaction of the replication process is a nucleophilic substitution ($S_N2$) reaction, for which we recently performed a benchmark study to assess the performance of various Density Functional Theory (DFT) approaches with respect to high-level reference data.

*Keywords:* DNA, Density Functional Theory, QM/MM method, QM/QM method

*Mathematics Subject Classification:*

*PACS:* 03.65.-w, 05.20.-y, 31.25.Qm, 71.15.Mb, 82.39.Pj, 87.14.Gg

The discovery in 1953 of the double stranded helical structure of DNA has led to several new insights on the origin and continuation of life.[1] One of the key features of the DNA structure comprises the hydrogen bonds between the different DNA base pairs, that are strong and specific and well understood by high-level theoretical calculations.[2-4] The process of DNA replication, e.g. the creation of an exact copy of one of the strands by fusing together DNA nucleotides, occurs with high fidelity, the origin of which is still under debate.[5] The basic reaction taking place in the DNA replication process is a ($S_N2$) nucleophilic substitution of the sugar moiety of the partly made strand (primer) on the phosphate group of the to-be-added nucleotide (see Scheme 1).

Scheme 1. Nucleophilic substitution ($S_N2$) reaction taking place in DNA replication process

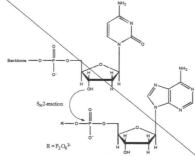

Nucleophilic substitution reactions are usually associated with organic compounds, in many cases with halide anions as leaving group or nucleophile. Because of its importance for organic synthesis and the relative simplicity of intermediates, often involving highly symmetric structures, the $S_N2$ reaction has

---

[1] Corresponding author. E-mail: m.swart@few.vu.nl

been the subject of a large number of theoretical investigations,[6] ranging from low-level semi-empirical to high-level coupled cluster studies with basis sets close to the basis set limit. We have recently performed a benchmark study for the performance of various density functionals[7] with respect to high-level reference data, i.e. CCSD(T) with basis sets close to the basis set limit.[8] The best performing functional(s) from that study will be used in subsequent investigations of the replication process of DNA.

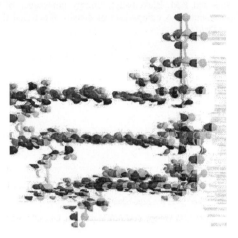

Figure 1: Schematical representation of the $S_N2$ reaction taking place in the replication of DNA.

In the present contribution, we report our first attempts to use QM/MM and QM/QM methods to study the structure of DNA oligomers in solution and the $S_N2$ reaction taking place in the DNA replication process. In QM/MM methods, only the most relevant parts of the system are described by DFT, while the others are treated at a lower, yet sufficiently accurate, MM level; the QM/QM method treats the complete system by QM methods, but different QM methods are used for different regions of the system. For the study on the replication process of DNA, we take the template (shown on the left in Figure 1), the primer (bottom right) and the to-be-added nucleotide (top right) into account explicitly.

## Acknowledgments

The authors wish to thank the Netherlands organization for Scientific Research (NWO-CW) for financial support.

## References

[1] James Watson, DNA: The secret of life, 2004.

[2] C. Fonseca Guerra and F.M. Bickelhaupt, Charge transfer and environment effects responsible for characteristics of DNA base pairing, *Angewandte Chemie-International Edition* **38** 2942-2945(1999).

[3] C. Fonseca Guerra, F.M. Bickelhaupt, J.G. Snijders and E.J. Baerends, The nature of the hydrogen bond in DNA base pairs: The role of charge transfer and resonance assistance, *Chemistry-a European Journal* **5** 3581-3594(1999).

[4] C. Fonseca Guerra, F.M. Bickelhaupt, J.G. Snijders and E.J. Baerends, Hydrogen bonding in DNA base pairs: Reconciliation of theory and experiment, *Journal of the American Chemical Society* **122** 4117-4128(2000).

[5] C. Fonseca Guerra and F.M. Bickelhaupt, Orbital interactions in strong and weak hydrogen bonds are essential for DNA replication, *Angewandte Chemie-International Edition* **41** 2092-2095(2002).

[6] J.K. Laerdahl and E. Uggerud, Gas phase nucleophilic substitution, *International Journal of Mass Spectrometry* **214** 277-314(2002).

[7] W. Koch and M.C. Holthausen, *A Chemist's Guide to Density Functional Theory*, Wiley-VCH, Weinheim, 2000.

[8] M. Swart, M. Solà and F.M. Bickelhaupt, Energy landscapes of bimolecular nucleophilic substitution ($S_N2$) reactions: A comparison of density functional theory and coupled cluster methods, *(to be) submitted*.

Brill Academic Publishers
P.O. Box 9000, 2300 PA Leiden,
The Netherlands

*Lecture Series on Computer*
*and Computational Sciences*
Volume 4, 2005, pp. 983-986

# Interaction and Equalization Models to Calculate Molecular Dipole Moments and Polarizabilities

Gaétan Weck[1] and Per-Olof Åstrand

Department of Chemistry,
Norwegian University of Science and Technology,
7491 Trondheim, Norway

Received 10 August, 2005; accepted in revised form 12 August, 2005

*Abstract:* Many classical approaches give good values of molecular dipole moments and polarizabilities but conjugated systems are usually a problem. In this study, the point dipole method and the electronegativity equalization method are coupled in one single interaction method. Variations around this method are presented and discussed. An exchange term proportional to the overlap of the atomic electronic clouds was investigated.

*Keywords:* polarizability, dipole moment, point-dipole interaction, electronegativity equalization, exchange

*PACS:* 33.15.Kr

## 1 Introduction

Quantum mechanical calculations give accurate values of the polarizability and the dipole moment of isolated molecules if the method has been chosen carefully. However, very large systems (proteins, solvated molecules) are still in most cases out of the scale of quantum mechanical calculations. If we can build a classical model that can reproduce the properties of small systems correctly, we can hope that the same model will give good estimations of the properties of large molecules or molecules surrounded by other molecules at a very low computational cost.

Classical methods use atom-type parameters. Usually atomic parameters are attractive because of their physical meaning. They can be optimized by adopting experimental or quantum mechanical values. We have chosen to fit our parameters using quantum mechanical values of the dipole moments and polarizabilities of isolated molecules, avoiding intermolecular interaction effects which appear in experimental measurements.

**PDI.** Applequist exploited a point-dipole interaction (PDI) model[1, 2] where a molecule is described by a set of atomic polarizabilities. Each atom interacts with an electrostatic

---

[1]Corresponding author. E-mail: weck@phys.chem.ntnu.no

field and polarizes itself. Induced atomic dipole moments interact with the other polarizable atoms. In the PDI model the molecule has no dipole moment if no external field is applied. This is a major drawback of this model since no charge transfer is allowed in the molecule.

**EEM.** In the electronegativity equalization model (EEM, see in [3] for a review), each atom has an electronegativity. An electron redistribution in the molecule will minimize the energy resulting in that atoms with high electronegativity will gain electrons and atoms with low electronegativity will lose electrons, thereby stabilizing the molecule. An additional term prevents electrons to leave the atoms. Like the potential of a spring, a quadratic term in the atomic charge (times a chemical hardness) is included in the Hamiltonian. In this model each atom will be represented as a point charge. The total molecular charge is conserved.

This model gives a permanent dipole moment. If a small external field is applied to the molecule, a new distribution of the electrons will be more favorable, leading to a new dipole moment. This model can be used to reproduce molecular dipole moments and polarizabilities of molecules. However, the calculated out-of-the-plane polarizability of for example benzene is zero. This model certainly lacks atomic polarizabilities.

## 2 Improvements through a EEM+PDI model

A suggestion is to couple the two previous models in a single Hamiltonian[4]. With given atomic parameters, a minimization of the energy will lead to a redistribution of the electrons and the creation of atomic induced dipole moments.

Atomic charges and dipole moments will give the molecular dipole moment. An applied field will change the atomic charge distribution and the atomic induced dipole moments. These new atomic variables will induce a new molecular dipole moment. Some improvements have been considered in this study.

**Gaussian distribution.** A damping is required in the PDI model in order to reduce the very strong interaction between two neighboring atoms[5, 6]. Because the interatomic distance can be very small, a small induced dipole moment can create a large field on the closest atoms. This large field will create a large induced dipole moment, which leads to a polarization catastrophe.

We have considered a Gaussian distribution of electrons around the nucleus of each atom[7]. The electrostatic interaction energy between two Gaussian distributions is lower than the interaction between two point charges (or point dipoles).

Each atom type will have an extra parameter characterizing the width of the electronic distribution. A Slater distribution could have been taken into account at the cost of a higher complexity in the equations and higher computational time.

**Electrons and nuclei.** In the electronegativity equalization model the atoms have a positive or negative charge. The electrons and nucleus of a given atom are one single charge located at a point in space. However, it can be interesting to consider each of them separately. By doing this we can put different widths for the electrons and the nucleus distributions of a same atom type. The width of the positive charge distribution can be 0 (point charge), identical to the width of the electronic distribution or something else.

Of course, all electrons can be included in the electronic Gaussian distribution. We can also put some of them in the positive charge, along with the nucleus. If some electrons are included in the positive charge (core electrons), the width of the positive charge distribution should be investigated.

**Exchange term**. An extra term trying to mimic the Pauli repulsion can be added to our Hamiltonian. This term should be described with our classical variables. We have chosen to consider the exchange term as proportional to the overlap of the electronic distributions[8]. This model gives an equation linear or quadratic with respect to the variables, resulting in a minor modification of the PDI and EEM equations.

This extra term should be sensitive to the number of electrons in the negative charge distributions of the neighboring atoms. If we want to include this "pseudo-exchange term", a distinction should then be made between valence and core electrons.

**Charge flow**. When a small uniform electrostatic field is applied to the system, each atom will polarize and a charge flow in the molecule is possible. Three different charge flow models have been investigated: a total redistribution of the electrons or a partial redistribution between only sp and sp2 atoms or no redistribution at all.

In the first case, it is known that the polarizability of large molecules will grow faster than their size[9]. In the last case, the applied field is considered to be too small to overcome the band gap and no conduction electrons are created. Allowing a charge transfer between sp or sp2 atoms (once the charges of sp3 atoms have been calculated), should reproduce the easily polarizable bonds.

## 3  Results

In this study, the different variations on the initial model are discussed. The obtained dipole moments and polarizabilities (for a chosen set of molecules) are plotted as a function of the expected quantum mechanical values (see for example Figure 1). These plots give almost straight lines (good correlation between classical and quantum mechanical values). However, the way used to optimize the parameters will be critical in the final discussion.

## Acknowledgment

The authors have received support from the Norwegian Research Council (NFR) through a Strategic University Program (Grant no 154011/420), a NANOMAT program (Grant no 158538/431), and a grant of computer time from the Norwegian High Performance Computer Consortium (NOTUR).

## References

[1] J. Applequist, J. R. Carl and K.-F. Fung, An atom dipole interaction model for molecular polarizability. Application to polyatomic molecules and determination of atom polarizabilities, *J. Am. Chem. Soc.* **94** 2952–2960(1972).

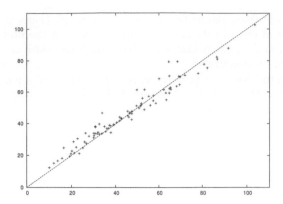

Figure 1: Obtained polarizabilities (a.u.) of 31 molecules (with hydrogen, carbon and oxygen atoms) as a function of the quantum mechanical reference values, for one set of atomic parameters. Here the widths of positive and negative atomic charges are identical, exchange has been taken into account, number of valence electrons is 4 for C and 2 for O. No charge flow allowed when polarizing the molecules.

[2] J. Applequist, An atom dipole interaction model for molecular optical properties, *Acc. Chem. Res.* **10** 79–85(1977).

[3] P. Geerlings, F. De Proft and W. Langenaeker, Conceptual Density Functional Theory, *Chem. Rev.* **103** 1793-1873(2003).

[4] H. A. Stern, G. A. Kaminski, J. L. Banks, R. Zhou, B. J. Berne and R. A. Friesner, Fluctuating charge, polarizable dipole, and combined models: parametrization from ab initio quantum chemistry, *J. Phys. Chem. B* **103** 4730–4737(1999).

[5] R. R. Birge, Calculation of molecular polarizabilities using an anisotropic atom point dipole interaction model which includes the effect of electron repulsion, *J. Chem. Phys.* **72** 5312–5319(1980).

[6] B. T. Thole, Molecular polarizabilities calculated with a modified dipole interaction, *Chem. Phys.* **59** 341–350(1981).

[7] L. Jensen, P.-O. Åstrand, A. Osted, J. Kongsted and K. V. Mikkelsen, Polarizability of molecular clusters as calculated by a dipole interaction model, *J. Chem. Phys.* **116** 4001-4010(2002).

[8] A. J. Stone: *The theory of intermolecular forces.* Clarendon Press, Oxford, 1996.

[9] L. Jensen, P.-O. Åstrand and K. V. Mikkelsen, An atomic capacitance-polarizability model for the calculation of molecular dipole moments and polarizabilities , *Int. J. Quant. Chem.* **84** 513–522(2001).

Brill Academic Publishers
P.O. Box 9000, 2300 PA Leiden,
The Netherlands

*Lecture Series on Computer*
*and Computational Sciences*
Volume 4, 2005, pp. 987-989

# Symposium on

# Computational Methods for Atomic and Molecular Clusters

René Fournier and Michael Springborg

Clusters form an intermediate state of matter between atoms and molecules, and the condensed phase. Researchers have adapted ideas and computational methods from quantum chemistry (the "molecular approach") and from solid state physics to study clusters. But, in the intermediate size regime, clusters are really different from molecules and the bulk, and neither quantum chemistry nor solid state physics offers an entirely adequate theoretical framework. Likewise, algorithms and computing techniques developed for molecules or solids often fall short of expectations when studying clusters. Therefore, much time and effort has been (and still is) spent on devising new models, algorithms, computational techniques, and simulation protocols to overcome computational problems that, while not unique to clusters, are particularly severe with clusters. Here is a partial list of those problems.

**(1)** Unlike molecules of main group elements, the lowest energy structures of most clusters can not be predicted by simple rules. There is a huge number of local minima on the potential surface, conservatively on the order of $2^{N-3}$ where $N$ is the number of atoms. Therefore, a determination of geometric structure necessitates an efficient global optimization technique.

**(2)** Many clusters have very low-lying excited electronic states (ie, a very small HOMO-LUMO gap). This is certainly the case for metal clusters. But it also occurs when some atoms have a nominal coordination or oxidation state very different from their usual values, giving an open-shell character to the cluster. Small gaps make it very difficult, and sometimes impossible, to reach a converged set of orbitals and electron density by the usual first-principles self-consistent field (SCF) methods.

**(3)** Another problem with small HOMO-LUMO gaps is the need to consider several possible spin states. This can be tedious, and the choice of theoretical model can be critical.

**(4)** Some properties that depend on dispersion $D$ (the ratio of surface atoms to total number of atoms) should ideally be studied over a size range from a few atoms ($D \approx 1$) up to roughly 1000 atoms ($D \approx 1/3$) in order to obtain meaningful trends. But a full-fledged study of clusters with hundreds of atoms is too demanding for first-principles quantum chemistry methods; and because they lack symmetry, these systems are also very demanding for solid state techniques.

**(5)** In modelling surface sites with clusters, one normally has a two-part model — local (cluster) and global (solid). Some properties converge slowly with size, for example surface electronic states, so the global part of the model is sometimes necessary for an overall qualitatively correct description. Some approaches use clusters embedded in a semi-infinite solid, others use a hierarchy of methods with only the smallest piece (cluster) being treated with atomic resolution and a high level of theory. Dividing the system into two or more parts brings fundamental and practical difficulties.

**(6)** Making a connection to experimental work is essential but very difficult. Most clusters are inherently unstable and can be made in laboratory only in very small quantities or with very short lifetimes. Special experimental setups and spectroscopic techniques are used to generate, detect, and characterize clusters, and one needs careful modelling and interpretation of calculations to make a meaningful comparison of theory to experiment. Moreover, in some cases, kinetic and not energetic effects determine which clusters are observed in experiment, making the connection to theory very difficult.

(7) Many interesting experiments are done on clusters that are far from equilibrium conditions or undergo dynamical processes, and modelling the dynamics — always a difficult task — becomes essential. Some clusters are fluxional at room temperature, others display quantum dynamics at the relevant temperatures ($He_n$). Some experiments involve high energy processes, such as Coulomb explosions, or accelerated cluster ions impacting on surfaces. Modelling the dynamics in all those cases is challenging as each experiment is different.

The contributions assembled for this symposium will give, we hope, a good idea of recent efforts in overcoming these kinds of problems. They illustrate that, over the years, every type of problem has prompted many possible solutions. For example, there are many different ways of modelling the potential energy surfaces of clusters depending on elemental composition and size: first-principles, semi-empirical, analytical functions, physical models such as "jellium" and "embedded atom", with many variations within each of those categories. Another example is the variety of strategies for discovering the lowest-energy geometric structures of clusters: genetic algorithms, cluster growth algorithms, simulated annealing and related methods, tabu search, and many others. In modelling clusters, various approaches can be useful within their range of applicability. Results obtained with different approaches must be combined and summarized to arrive at a coherent picture with concepts that have general validity. The variety of computational methods that are required makes the study of clusters challenging for theoreticians. But it also makes cluster research a fertile ground for innovations in computing methods and for the expression of researchers' creativity.

René Fournier
Michael Springborg

### René Fournier

Born in 1962.
PhD Chemistry, Universite de Montreal (1989) under the supervision of Dennis Salahub.
NSERC Postdoctoral Fellow, Iowa State University (Ames), 1989-91 in the group of Andrew DePristo.
Postdoctoral Researcher, Steacie Institute for the Molecular Sciences, National Research Council (Ottawa), 1991-94.
Postdoctoral Researcher, University of Nevada (Las Vegas), 1995-96, in the group of Changfeng Chen and Tao Pang.
Assistant Professor in the Department of Chemistry of York University 1996-2001; Associate Professor since 2001.

### Michael Springborg

Michael Springborg, born 1956 in Denmark. After studies and PhD at the Technical University of Denmark he has been working both at chemistry and at physics institutions in Denmark and Germany. Currently he is the holder of a professorship at the University of Saarland, Germany, in physical chemistry. His work concentrates on the development and application of theoretical methods, including accompanying computer programs for the calculation of electronic and structural properties of specific materials, most notably polymers and chain compounds as well as clusters and colloids.

Brill Academic Publishers
P.O. Box 9000, 2300 PA Leiden,
The Netherlands

*Lecture Series on Computer*
*and Computational Sciences*
Volume 4, 2005, pp. 990-993

# Computational Spectroscopy of Supported Cluster nanocatalysts

## Hannu Häkkinen[1] and Michael Walter

Department of Physics
Nanoscience Center
University of Jyväskylä
FIN-40014 Jyväskylä, Finland

Received 28 June, accepted in revised form 26 July

*Abstract:* Spectroscopic properties of magnesia-supported, chemically active gold clusters are investigated using density functional theory (DFT). We discuss features in simulated STM images, generated through simple theory by Tersoff and Hamann, and theoretical optical spectra, calculated from the linear response formulation of time-dependent DFT.

*Keywords:* Time-dependent density functional theory, optical absorption, gold, electronic structure, nanocatalysis, scanning tunneling microscopy

*PACS:* 36.40.Vz, 68.47.Jn, 31.10.+z,31.15.Ew

## 1 Introduction

Traditionally, a working view of a catalyst has been the one where the surface of the active system has a number of well-defined adsorption sites for the reactant molecules and possibly for the reaction products. Maximising the surface area, e.g., by using supported nanoparticles or porous materials, leads then to an enhanced reaction output. Recently, chemical activity of metal particles, clusters and nanostructures has attracted increasing interest, motivated by experiments that have shown that it is possible to fine-tune the sensitivity and selectivity of important chemical reactions by controlling the *size* of nanometer-scale metal clusters. A prominent example is that of gold, which as bulk material is chemically most uninteresting, yet in nanoscale surprisingly active as shown, e.g., in recent reports (and in references therein) of catalytic activity of magnesia-supported gold clusters towards low-temperature CO oxidation [1]. It has been found that in those cases the *finite-size quantum effects on the electronic structure of the catalytic centre as a whole* have an important role in binding of the reactant and product molecules as well as in modifying the reaction barriers. Here detailed computer simulations with an advanced level of theory are indispensable, and combined with high-resolution experiments, they can lead to a novel understanding of the functionality of model "nanocatalysts".

The atomic structure, binding modes and electronic structure of magnesia-supported gold clusters $Au_n/F_s@MgO(100)$ with $n \leq 8$, bound at a surface color center (oxygen vacancy) $F_s$ of the $MgO(100)$ face, have been discussed in detail previously[1]. Very recently, their optical spectra were calculated from the linear-response time-dependent density functional theory[2]. Here we present complementary data on the spatial features of the electron density of these clusters at Fermi level, $\rho_F(\mathbf{r})$. Following the theory of Tersoff and Hamann [3] this data also serves as a simple prediction for STM imaging. We also discuss optical absorption of $Au_8/F_s/MgO$ and $Au_{20}/F_s/MgO$.

[1]Corresponding author. E-mail: hannu.hakkinen@phys.jyu.fi

## 2 Methods

The atomic and electronic structure of the supported $Au_n/MgO$ system (comprising the vicinity of the color center of the MgO(100) surface, the $Au_n$ cluster, and the adsorbed molecules) were calculated within the DFT in combination with Born-Oppenheimer (BO) Molecular Dynamics (MD)[4] including self-consistent gradient corrections via the so-called PBE-GGA functional. $Au(5d^{10}6s^1)$, $Mg(3s^2)$ and $O(2s^22p^4)$ electrons were included in the valence, and the interaction to the ion cores was described by scalar-relativistic non-local norm conserving pseudopotentials devised by Troullier and Martins. The Kohn-Sham states are expanded in a plane wave basis with 62 Ry cutoff.

The MgO surface is modelled by a two-layer ab initio cluster $Mg_mO_m$ or $F_s@Mg_mO_{m-1}$, embedded in an extended point-charge lattice to include effects of the long-rage Madelung potential [5]. For the embedding lattice, around 2100 alternating charges of +2 and -2 representing Mg and O ions, respectively, were used. In addition, those positive point charges that would be nearest neighbors to the peripheral O atoms of the central ab initio $Mg_mO_n$ cluster have been replaced by "empty" Mg pseudopotentials (MgPP) in order to prevent unphysical polarization effects to O ions [5]. The lattice parameter of the embedding part is fixed to the experimental lattice constant (4.21 Å) of bulk MgO. The cluster, molecules and nearest-neighbor Mg ions to the $F_s$ are treated dynamically in structural optimizations that included both steepest-descent and quenched molecular dynamics runs.

The optical spectra were calculated from the linear response time dependent DFT formulated by Casida, [6] as implemented in ref. [7]. Briefly, in order to get the weights $F_I$ and energies $\hbar\omega_I$ of optical transitions $\{I\}$, one solves an eigenvalue problem $\Omega F_I = \omega_I^2 F_I$ where the $\Omega$ matrix elements are given by

$$\Omega_{ij,kl} = \delta_{ik}\delta_{jl}\varepsilon_{ij}^2 + 2\sqrt{n_{ij}\varepsilon_{ij}n_{kl}\varepsilon_{kl}}K_{ij,kl}. \tag{1}$$

Here $\varepsilon_{ij} - \varepsilon_j$ $\varepsilon_i$ and $n_{ij} - n_i - n_j$ are the difference of the KS particle-hole eigenvalues and occupation numbers, respectively. $K_{ij,kl}$ is a coupling matrix that describes the linear response of the electron density $\rho$ to the single-particle – single-hole excitations in the basis spanned by the ground state KS orbitals $|i\rangle$. The transition matrix element in polarisation direction $\hat{e}_\nu$ is

$$(M_I)_\nu = \sum_{ij} \sqrt{\varepsilon_{ij}n_{ij}}\langle j|r_\nu|i\rangle(F_I)_{ij} \tag{2}$$

and the corresponding polarization-dependent and polarization-averaged oscillator strengths are $(f_I)_\nu = 2|(M_I)_\nu|^2$ and $\bar{f}_I = (1/3)\sum_{\nu=1}^3(f_I)_\nu$, respectively.

## 3 Results

Figure 1 shows the lowest energy structures of $Au_8/F_s@MgO$, $O_2/Au_8/F_s@MgO$ and the corresponding images of $\rho_F$. The center images show densities at a constant height $z = 7.4$Å above the MgO(100) plane whereas the images on the right show the height information for constant-density contour of $10^{-5}$ a.u. These would then simulate STM images at constant-height and constant-current modes, respectively [3]. The images predict clear changes in the spatial distribution of the Fermi density upon adsorption of the oxygen molecule. A significant portion of the $\rho_F$ is localised to oxygen, since the initially empy $2\pi^*$ antibonding orbitals are pulled down below Fermi level and get occupied via charge-transfer from the supported gold cluster, which originally became negatively charged upon adsorption to the $F_s$ site. In this case, we calculated local charges for the gold cluster to be $-0.37e$ and $+0.73e$ before and after oxygen adsorption, respectively, and the oxygen molecule gains a charge of $-1.26e$. This charge-transfer-induced activation of $O_2$ was shown to be the key to the catalytic activity[1].

Figure 2 shows comparison of optical spectra of MgO-supported $Au_8$ and $Au_{20}$ clusters[8]. The tetrahedral structure of $Au_{20}$ was previously shown to be the ground state structure in gas phase[9]. Figure 2 shows that while optical absorption of the gold octamer is still rather "molecule-like", i.e., consisting of prominent peaks in the visible and near-UV range, the $Au_{20}$ has a continuous and monotonous spectrum with a smooth absorption edge around 2 eV. The characteristic spectra of small gold clusters and modification of their optical absorption by adsorbed molecules[2] give confidence that structures and functions of chemically active supported metal clusters could in many cases be resolved by applications of surface-sensitive optical spectroscopic tools, i.e., the cavity ringdown spectroscopy[10].

## Acknowledgment

This work is partially supported by the Academy of Finland. We acknowledge CSC - the Finnish IT Center for Science in Espoo for providing the computational resources.

## References

[1] B. Yoon, H. Häkkinen, U. Landman, A. Wörz, J.-M. Antonietti, S. Abbet, K. Judai, U. Heiz, Charging Effects on Bonding and Catalyzed Oxidation of CO on $Au_8$ clusters on MgO, *Science* **307** 403 (2005); and references therein.

[2] M. Walter and H. Häkkinen, Optical absorption by magnesia-supported gold clusters and nanocatalysts: effects from the support, cluster and adsorbants, submitted, preprint available at arXiv:physics/0506121 v1.

[3] J. Tersoff and D.R. Hamann, Theory and application for the scanning tunneling microscope, *Physical Review Letters*, **50** 1998 (1983).

[4] R.N. Barnett and U. Landman, Born-Oppenheimer molecular-dynamics simulations of finite systems: Structure and dynamics of $(H_2O)_2$, *Physical Review B* **48** 2081 (1993).

[5] G. Pacchioni, Cluster modelling of oxide surfaces: Structure, adsorption and reactivity, in *Chemisorption and Reactivity on supported Clusters and Thin Films*, edited by R.M. Lambert and G. Pacchioni, Kluwer, Dortdrecht, 1997).

[6] M.E. Casida, in *Recent Developments and Applications of Modern Density Functional Theory*, edited by J.M. Seminario (Elsevier, Amsterdam 1996).

[7] M. Moseler, H. Häkkinen and U. Landman, Photoabsorption Spectra of $Na_n^+$ clusters: Thermal Line-Broadening Mechanisms, *Physical Review Letters* **87** 053401 (2001).

[8] Details of structures, binding and optical absorption of magnesia-supported $Au_{20}$ clusters will be discussed elsewhere.

[9] J. Li, X. Li, H.-J. Zhai and L.-S. Wang, $Au_{20}$: A tetrahedral cluster, *Science* **299** 864 (2003).

[10] J.M. Antonietti, M. Michalski, U. Heiz, H. Jones, K.H. Lim, N. Rösch, A.D. Vitto and G. Pacchioni, Optical absorption spectrum of gold atoms deposited on $SiO_2$ from cavity ringdown spectroscopy, *Physical Review Letters* **94** 213402 (2005).

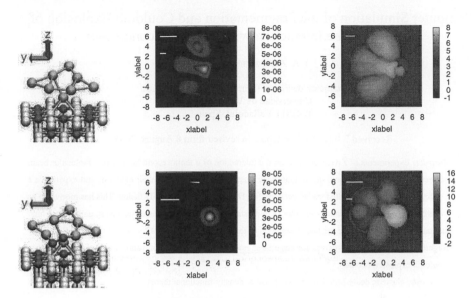

Figure 1: Left: an atomic view of a supported $Au_8$ (top) and $O_2/Au_8$ (bottom) clusters adsorbed at $F_s/MgO$. Center and right: corresponding images of local electron densities at the Fermi level. Center: $\rho_F(z)$ with $z = 7.4$ Å, Right: $z$-scale of $\rho_F = 10^{-5}$ a.u.

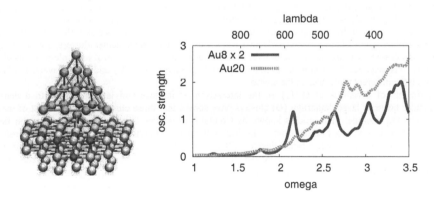

Figure 2: Left: The structure of tetrahedral $Au_{20}$ adsorbed at $F_s/MgO$. Right: Comparison of polarization-averaged optical spectra of supported $Au_8$ and $Au_{20}$ clusters.

Brill Academic Publishers
P.O. Box 9000, 2300 PA Leiden
The Netherlands

*Lecture Series on Computer
and Computational Sciences*
Volume 4, 2005, pp. 994-997

# Computer Simulation of the Fragmentation and Coulomb Explosion of Clusters by the Interaction with a Femtosecond Laser

J. A. Alonso[1] and M. Isla

Departamento de Física Teórica, Atómica y Optica,
Universidad de Valladolid,
E 47011 Valladolid, Spain

Received 7 July, 2005; accepted in revised form 8 August, 2005

*Abstract:* Experiments of Zweiback et al. on the interaction of a femtosecond laser with a molecular beam of deuterium (D) clusters have shown that those clusters can lose most of their electrons and experience a Coulomb explosion. The collisions between the fast D nuclei give rise to D-D fusion. This has motivated us to carry out computer simulations in order to understand the ultrafast processes occurring under these high excitations. We have studied the laser irradiation of the $D_{13}^+$ cluster. For not too large intensities the cluster fragments in a slow way, whereas for large laser intensities substantial ionization takes place and then a Coulomb explosion occurs.

*Keywords:* clusters; deuterium; Coulomb explosion; density functional theory

*PACS:* 36.40.-c; 36.40.Qv; 36.40.Wa

## 1. Introduction

The study of the interaction between matter and intense laser fields (intensities higher than $10^{14}\,\text{Wcm}^{-2}$) with ultra-short pulse duration ($<10^{-13}$ s) has been an area of growing activity over the last two decades. The technology of ultra-short pulsed lasers has progressed to such an extent that femtosecond lasers are now widely available. Interesting new developments have occurred for atoms and molecules. But, as a form of matter intermediate between molecules and solids, atomic and molecular clusters have attracted attention because they exhibit unique properties. Hydrogen clusters are a typical example of molecular clusters, characterized by a high intra-molecular binding energy (4.8 eV) and a very weak binding between the different molecules in the cluster.
Experiments of Zweiback et al. [1] on the interaction of femtosecond laser pulses with a dense molecular beam of large deuterium (D) clusters have shown that those clusters can lose most of their electrons and explode, in a process known as Coulomb explosion. The collisions between the fast deuterium nuclei give rise to D-D fusion. This has motivated us to perform computer simulations using the time-dependent density functional theory (TDDFT) in order to understand the processes occurring under these high excitations. In particular we study the irradiation of the single-charged cluster $D_{13}^+$, through the simulation of the first stages in the experiments of Zweiback.

## 2. Method

The computational scheme employed in the computer simulations is a real-space, real-time implementation of the TDDFT, where the time-dependent Kohn-Sham equations giving the evolution of the one-electron orbitals are explicitly integrated. A full description of the state of the art in this theory can be found elsewhere [2]. The ground state of the initial field-free cluster is calculated using

---

[1] Corresponding author. E-mail: jaalonso@fta.uva.es

the DFT formalism; then this ground state is perturbed by a laser pulse, treated classically, and the time-dependent Kohn-Sham equations are propagated in time. For the exchange-correlation energy functional we have used the adiabatic Local Density Approximation. The calculations have been performed with the OCTOPUS code [3]. The simulation cell is a sphere with a radius of 15 or 17 a.u. In this volume, a Cartesian grid with a constant spacing of 0.4 a.u. is used. A time step of 0.1 a.u. has been chosen in the time propagation of the wave functions, with a total integration time between 1000 and 2000 a.u. (1 a.u is equal to 0.0242 femtoseconds). The algorithms employed in the time propagation of the Kohn-Sham equations are the Approximated Enforced Time-Reversal Symmetry algorithm (AETRS) to approximate the evolution operator $U(t+\Delta t,t)$, and the Lanczos Subspace Approximation for the exponential of a matrix operator [4].

## 3. Results

The sizes of the deuterium clusters in the experiment are too large, so for the calculations we have selected the small cationic cluster $D_{13}^{+}$, since we expect that the qualitative picture of the fragmentation may not be too sensitive to the cluster size. The optimized ground state geometry is shown in the leftmost upper and lower panels of Fig. 1. It is formed by a triangular charged trimer, $D_3^{+}$, surrounded by a first solvation shell formed by three parallel $D_2$ molecules. The axes of those molecules are oriented nearly perpendicular to the plane of the trimer and their centers of mass are in the plane of the trimer. The remaining two molecules form a second shell at a larger distance.

Our objective is the theoretical study of the fragmentation of $D_{13}^{+}$ induced by the interaction with an intense laser. We performed a calculation of the optical excitation spectrum of the cluster in order to identify large absorption frequencies. The first large peak in the absorption spectrum was found at a frequency $\omega = 0.352$ a.u. (energy $\hbar \omega = 9.58$ eV). We now present the results of a simulation in which a 9.6 fs laser pulse is applied to the cluster and the dynamical evolution is monitored by propagating the system for 20000 time steps according to the time-dependent Kohn-Sham equations; this gives a total simulation time of 2000 a.u. (about 48 fs). The shape of the laser pulse is given by a cosinoidal envelope with an amplitude of 0.02 a.u., giving a total pulse intensity of $1.40 \times 10^{13}$ W cm$^{-2}$. The results of the evolution of the cluster structure are shown in Figure 1, where a few snapshots have been selected of the instantaneous structure for increasing time.

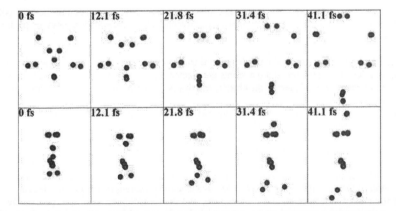

Figure 1: Snapshots of the cluster structure at different times after the application of the laser pulse, for laser intensities leading to slow fragmentation. Two mutually perpendicular views are shown for each snapshot. The two leftmost panels (upper and lower) give the initial structure of the $D_{13}^{+}$ cluster.

The main features of this process are clearly seen. The central trimer absorbs energy from the laser pulse and splits into two fragments: a $D_2$ molecule is emitted upwards (in the orientation used in the Figure) and the third atom moves opposite to that molecule. As the molecule flies it passes near two molecules of the first solvation shell and the intermolecular interaction potential sets those two molecules in motion. On the other hand, the third atom moves down towards the other molecule of the first shell and an exchange collision occurs: a new molecule is formed and one of the atoms of the original molecule is ejected. The two molecules of the second solvation shell appear not to be seriously perturbed in this process.

The laser energy is first absorbed in the form of electronic excitations of the cluster. A part of that excitation energy is employed in dissociating the central trimer and in breaking the weak bonds between the trimer and the surrounding molecules, and another part is transformed into kinetic energy of the molecules; but this does not exhaust the absorbed energy and some energy still remains as electronic excitations during the simulation. The negative potential energy remains very substantial after the dissociation because the system still maintains most of its initial binding energy in the intramolecular bonds. In summary, the fragmentation of the cluster begins with the dissociation of the central trimer, which is the most important step in the process. In a second step the dissociated central trimer causes the breaking of the weak intermolecular bonds which maintained the cluster stable. The process can be qualified as a slow dissociation.

When the amplitude of the laser pulse is substantially increased the outcome of the simulation changes drastically. We have performed simulations with pulses of frequency $\omega = 0.352$ a.u. and a shape given by a cosinoidal envelope with an amplitude of 0.1 a.u., five times larger than the amplitude used in the slow fragmentation case. The duration of the pulse is again 9.6 fs, giving rise to a total intensity of $3.51 \times 10^{14}$ W cm$^{-2}$. The time-step used in the simulation is 0.1 a.u., the same as before, but since the fragmentation is faster in this case, it is enough to propagate the electronic wavefunctions for 10000 time steps; that is, the total integration time is 1000 a.u. The results of the evolution of the cluster structure are presented in Figure 2, where a few snapshots have been selected.

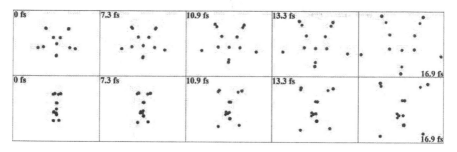

Figure 2: Snapshots of the cluster structure at different times after the application of the laser pulse, for laser intensities leading to the Coulomb explosion. Two mutually perpendicular views are presented for each snapshot.

The main feature is now that not only the central trimer dissociates; the intramolecular bonds of the other molecules are broken too and the atoms are ejected like in an explosion. This sudden explosion occurs simultaneously in all parts of the cluster, and this is due to the substantial ionization experienced by the cluster.

The negative potential energy of the cluster quickly changes (decreasing in absolute value) when the cluster interacts with the laser pulse. The potential energy even becomes positive (repulsive interaction), reaching a maximum value at 7 - 8 fs. This is a manifestation of the charging of the cluster due to the ionization of the atoms. This becomes clear by counting the number of electrons in the cluster as the simulation progresses in time. The number of electrons in the cluster is identified with the total electronic charge inside the simulation cell. In a short interval, the first 9 fs, roughly corresponding to the duration of the laser pulse, the cluster loses five electrons (moreover, the loss of electrons does not stop at the end of that interval, and it continues), and then a Coulomb explosion occurs. During the explosion the repulsive potential energy accumulated at the top of the Coulomb barrier is transformed into kinetic energy of the ionized atoms. When the atoms have lost their electrons, two kinds of Coulomb repulsion occur: intramolecular Coulomb repulsion and repulsion between nuclei of different molecules. The first one is stronger because the two atoms in the molecule are at a distance of 1.44-1.46 a.u., substantially shorter than the intermolecular distances.

The velocity of the fastest nuclei resulting from this Coulomb explosion is still very small compared to the 1-10 keV range of the colliding particles in typical fusion processes, from controlled fusion reactors to solar reactions [5]. This indicates that the Coulomb explosion process simulated in this work delivers kinetic energies that are still far from those required to produce D-D fusion. This is an expected result since the radius of the $D_{13}^+$ cluster is 5 a.u., whereas the radii of the clusters in the experiments of Zweiback [1] are ten times bigger, so the number of atoms is one thousand times larger.

Nevertheless we expect that the mechanisms of fragmentation for larger clusters are the same analyzed here for the smaller clusters.

## Acknowledgments

Work supported by MCYT (Grant MAT2002-04499). M. I. is greatful to the University of Valladolid for a postgraduate fellowship.

## References

[1] J. Zweiback, R. A. Smith, T. E. Cowan, G. Hays, K. B. Wharton, V. P. Yanovsky and T. Ditmire, *Physical Review Letters* **84**, 2634 (2000).

[2] M. A. L. Marques and E. K. U. Gross, in: *A Primer in Density Functional Theory* (Editors: C. Fiolhais, F. Nogueira and M. A. L. Marques), Lecture Notes in Physics, Springer Verlag, Berlin, **620**, 144 (2003).

[3] M. A. L. Marques, A. Castro, G. F. Bertsch and A. Rubio, *Computer Physics Communications* **151**, 60 (2003). See also http://www.tddft.org/programs/octopus.

[4] A. Castro, M. A. L. Marques and A. Rubio, *Journal of Chemical Physics* **121**, 3425 (2004).

[5] K. S. Krane, *Introductory Nuclear Physics*, John Wiley & Sons, New York (1988). See Chapter 14.

Brill Academic Publishers
P.O. Box 9000, 2300 PA Leiden
The Netherlands

*Lecture Series on Computer*
*and Computational Sciences*
Volume 4, 2005, pp. 998-1001

# First-Principles Molecular and Reaction Dynamics Simulations: Application to the Structure, Thermodynamics and Photochemistry of Ionic Aqueous Clusters

Gilles H. Peslherbe, Qadir K. Timerghazin and Denise M. Koch

Centre for Research in Molecular Modeling
and Department of Chemistry & Biochemistry
Concordia University
7141 Sherbrooke St. West
Montréal, Québec, Canada H4B 1R6

Received 15 July, 2005; accepted in revised form 12 August, 2005

*Abstract:* Realistic computer simulations of the structure and thermodynamics of complex chemical systems governed by subtle intermolecular interactions, as well as the real time-evolution of such systems in their electronically excited states, resulting for example from photoexcitation, represent many challenges for modern computational chemistry. In this contribution, we discuss the application of first-principles molecular dynamics simulations to the ground-state structure and thermodynamics and the excited-state dynamics of iodide-water clusters. One of the fascinating features of these clusters lies in the possibility of photochemical transfer of an electron from the ion to the solvent, giving rise to so-called charge-transfer-to-solvent excited states, which arises from surface solvation structures in small to medium-sized clusters and eventually lead to precursor states of the solvated electron. We will thus discuss a rigorous investigation of surface vs. interior solvation thermodynamics for these species, we will show how reliable calculations of energetics, vertical excitation energies and ionization potentials can be combined with experiment to probe cluster solvation structures, and we will highlight the importance of solvation dynamics in the relaxation of photoexcited ionic clusters, in connection with recent femtosecond photoelectron spectroscopy experimental results.

*Keywords:* First-principles simulations, molecular dynamics, reaction dynamics, ab initio calculations, electronically excited states, ionic aqueous clusters, solvation structure, thermodynamics, photochemistry, charge transfer to solvent, photoelectron spectra

*PACS:* 31.15.Qg, 31.70.Dk, 34.70.+e, 36.40.-c, 36.40.Wa, 31.15.Ar, 31.50.Df, 31.70.Hq, 33.60.-q

The role of solvation on chemical species or processes, and the hydration of simple ions in particular, has been the focus of numerous studies over the years, due to its fundamental importance in chemistry, biology and environmental science. Since the development of supersonic jets, clusters – an intermediate form of matter between the gas and bulk phases – have been used to probe the evolution of various properties from the gas to bulk phases. For instance, cluster studies provide means for detailed investigations of the role of individual solvent molecules in the solvation of ionic species. A very fundamental and important issue in ionic cluster studies is the hydration structure of halide ions. Numerous experimental [1] and theoretical [2] studies suggest, but not unequivocally, that large halide ions such as iodide preferentially sit at the surface of small-to-medium sized aqueous clusters, rather than being fully surrounded by water molecules, as they would be in the liquid phase. A fascinating feature of the resulting surface-solvation cluster structures lies in the possibility of photochemical transfer of an electron from the ion to the solvent network, giving rise to so-called charge-transfer-to-solvent (CTTS) excited states, which eventually lead to precursor states of the solvated electron, even in very small halide-solvent clusters [3]. The hydrated electron itself is extremely important in the context of a number of physical, chemical, and biological processes, including radiation damage to biologically important molecules, electron transfer, and charge-induced reactions. Since studies of solvent cluster anions may provide further insight into electron-solvent interactions at the microscopic level, significant experimental and theoretical efforts have been devoted to detailed studies of these species, and up till now, the ultrafast dynamics of the solvated electron remains a key fundamental problem [4, 5]. Our goal is thus to perform rigorous, realistic simulations of 1) the ground-state structure and thermodynamics of iodide-water clusters and 2) their excited-state dynamics, especially in connection with recent femtosecond experiments.

As mentioned above, large halides tend to adopt surface structures in aqueous clusters. Since the ion-water binding energy is of a similar magnitude as the water-water binding energy, the water molecules tend to bind to each other rather than to the ion, and the ion-water interaction is not strong enough for the ion to disrupt the relatively stable water network. A key question that arises from the existence of surface solvation states for large halide ions in aqueous clusters concerns the cluster size at which the solvation behavior of these ions converge to bulk, *i.e.* at what cluster size does the ion adopt an interior, bulk-like, solvation structure? Markovich *et al.* [6] analyzed the $I^-(H_2O)_n$ (n=1-60) vertical detachment energies and solvent electrostatic stabilization energies resulting from photoelectron spectra, and could not conclude with certainty whether the ions adopted an interior or a surface solvation structure. Coe [7] then combined small cluster experimental solvation data, the results of simulations in model polar solvent clusters of intermediate size, and limiting continuum dielectric trends for large clusters to determine the evolution of the ion solvation free energy from the smallest cluster size to bulk. This analysis suggests that a gradual transition from surface to interior solvation occurs over the range n=1–60, and the transition to bulk behavior occurs around n ~ 60.

In previous work [8], we performed a rigorous investigation of the structural properties and thermodynamics of surface *vs.* interior solvation in iodide-water clusters in order to identify the possible transition from surface to bulk behavior. Two convenient coordinates were chosen in order to discriminate between surface and interior solvation. The first coordinate is the distance between the ion and the solvent center of mass $r_{cm}$, which would be close to zero for an ideally spherically solvated interior ion, and obviously deviates significantly from zero for a surface solvation state. The other, equally important, coordinate is the angle $\theta$ between individual solvent molecules, the ion and the aqueous cluster center of mass. In this case, large deviations from an isotropic angular probability distribution $P(\theta)=sin\theta$ are an indication of surface solvation. We evaluated the potential of mean force $W(r_{cm})$, *i.e.* the free energy change as the ion is forced to move towards or away from the solvent cluster center of mass, by means of constrained Monte Carlo simulations with both non-polarizable (OPLS) [9] and polarizable (OPCS) [10] model potentials. The $r_{cm}$ probability distributions $P(r_{cm})$ for $I^-(H_2O)_{32}$ and $I^-(H_2O)_{64}$ resulting from the simulated potentials of mean force and calculated as $P(r_{cm}) = 4\pi r_{cm}^2 e^{-W(r_{cm})/kT}$ are shown in Figs. 1a and 1b, while the OPLS angular probability distributions for the most probable $r_{cm}$ values are shown in Figs. 1c and 1d. These results suggest that the iodide ion tends to reside at the surface of a water cluster of size 32, whereas both interior and surface solvation states coexist at cluster size 64. We also calculated the equilibrium constants for interior and surface solvation states for $I^-(H_2O)_{64}$ and found that interior solvation is favored over surface solvation by cluster size 64. However, an order of magnitude difference between the OPLS and OPCS equilibrium constants was found, suggesting that polarization favors interior solvation. Our simulation results also indicate that entropy favors interior solvation for cluster size 64, since the cluster enthalpies are very similar for the interior and surface solvation states [8]. Moreover, in a more recent study of the thermodynamics and structural properties of halide-water clusters, we found a serious discrepancy in the results predicted by the OPLS and OPCS model potentials for chloride and bromide-water clusters, such that it is not possible to determine at which cluster size interior solvation is favored over surface solvation. In

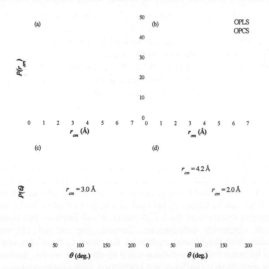

Figure 1: Probability distributions of the distance between the ion and the solvent center of mass $P(r_{cm})$ for (a) $I^-(H_2O)_{32}$, and (b) $I^-(H_2O)_{64}$ with both the OPLS (solid line) and OPCS (dashed line) model potentials, and angular probability distributions $P(\theta)$ for (c) $I^-(H_2O)_{32}$ and (d) $I^-(H_2O)_{64}$ predicted by the OPLS model evaluated at the peaks in the $P(r_{cm})$ distributions.

order to obtain more reliable equilibrium constants and potentials of mean force for halide-water clusters, we perform and report the results of first-principles constrained molecular dynamics simulations, employing semiempirical quantum chemistry reparameterized on the basis of small cluster geometries and binding energies.

Furthermore, quantum chemical refinements [11] of the semiempirical potentials of mean force are performed with density functional theory to ensure a fully quantitative description of surface *vs.* interior solvation in halide-water clusters. These quantum chemical refinements are based on single-point quantum chemical energies obtained for a properly weighted subset of configurations [11] generated from the semiempirical molecular dynamics simulations. Finally, quantum effects on nuclear motion are also investigated by means of path integral molecular dynamics with reparameterized semiempirical quantum chemistry, in order to assess the importance of zero-point energy and low-temperature effects on the structural properties and thermodynamics of halide-water clusters.

As mentioned earlier, the surface structure of iodide-water clusters gives rise to charge transfer to solvent (CTTS) upon photoexcitation. The nature of the CTTS states in clusters is similar to that in the bulk: upon photoexcitation an electron is transferred from the halide anion to a diffuse orbital supported by the solvent network [12]. The excess electron–solvent interactions are essentially the same as in the case of the dipole-bound solvent cluster anions and the bulk solvated electron. Neumark and co-workers [13] employed femtosecond photoelectron spectroscopy to follow the relaxation dynamics of small $I^-(H_2O)_n$ (n=4–6) clusters after photoexcitation. For $I^-(H_2O)_4$, the electron vertical binding energy (VBE) was found to remain constant throughout the measurement time-window, and the population of photoexcited clusters decayed exponentially due to spontaneous ejection of the dipole-bound electron from the vibrationally excited water cluster. On the other hand, clusters with five and six water molecules exhibited much more interesting dynamics: an induction period of ca. 500 fs was observed, during which the VBE is constant, and after which it increases by ca. 0.3 eV, suggesting that the cluster undergoes some reorganization that leads to the stabilization of the excited electron.

Neumark and co-workers [13] proposed that the rise in electron binding energy after the first 500 fs is due to a reorganization of the water molecules in $I^-(H_2O)_5$ and $I^-(H_2O)_6$ clusters, i.e. the water molecules in $I^-(H_2O)_5$ and $I^-(H_2O)_6$ gain configurations which can support the excess electron much more efficiently than in the initial cluster configuration. This model is based solely on the consideration of solvent dynamics, neglecting the possible role of the neutral iodine atom formed upon photoexcitation, and is hereafter referred as the "solvent-driven" relaxation model. This model also accounts for the pronounced isotope effects observed, i.e. a longer induction period for $I^-(D_2O)_n$ than for $I^-(H_2O)_n$ [13]. An alternative interpretation of the femtosecond experiments was proposed by Chen and Sheu [14], in which the stabilization of the excited electron is rationalized by the ejection of the neutral iodine atom from the water cluster. Static quantum-chemical calculations (with the water cluster moiety frozen in the initial cluster geometry) show that the presence of the neutral iodine atom can considerably destabilize the excited electron in excited $I^-(H_2O)_n$, and thus as the iodine atom departs from the water cluster, the electron binding energy must increase significantly. The main caveat with this "iodine-driven" relaxation model is the complete neglect of solvent motion.

Our preliminary classical trajectory study [15] of the model $I^-(H_2O)_3$ excited-state dynamics showed that the relaxation of excited iodide-water clusters is characterized by a combination of the solvent and iodine-driven mechanisms, i.e. both solvent reorganization and iodine atom motion appear to be important for interpreting the existing femtosecond photoelectron spectroscopy results. Accordingly, in this work, we present the first realistic first-principles classical trajectory study of the CTTS relaxation dynamics of iodide-water clusters with four and five water molecules, in order to make a definite connection with experiment.

First-principles excited-state molecular dynamics simulations require a fast and robust model chemistry, which should also be able to reproduce quantitatively the potential energy surface of the system under investigation. The diffuse nature of weakly-bound excited/excess electron in CTTS excited states of halide-solvent clusters and solvent cluster anions presents a number of challenges for quantum chemistry. First, because of the large spatial extent of the excited electron cloud (>30 Å) one needs to use highly diffuse basis sets for proper treatment of the CTTS states, which leads to significant convergence difficulties in the quantum chemistry calculations. Second, the excited electron stabilization due to its dispersion interaction with the solvent molecules is known to be of the same magnitude as the electrostatic interactions between solvent and the excited electron. However, through cancellation of errors, the computationally inexpensive configuration interaction with single excitations (CIS) method, along with an economic double-zeta basis set augmented by a set of diffuse functions (denoted DZ+), was found to provide an adequate description of the potential energy surface of the ground and excited-state iodide-solvent clusters, when compared to very high-level complete active space second-order perturbation theory (CASPT2) with large basis sets. The floppy nature of iodide-water clusters poses yet another challenge for realistic simulations of CTTS relaxation dynamics. Halide-water clusters possess a number of almost iso-energetic isomers which can easily interconvert and proper sampling of initial cluster configurations should take into account all these isomers. Proper sampling of initial configurations should also take into account nuclear quantum effects, since the zero-point energy differences between various $\Gamma(H_2O)_n$ isomers can affect their relative population. In this work, we use the first-principles path-integral molecular dynamics simulations described above to produce a proper set of initial conditions for the trajectories. The classical equations of motion are then propagated with the standard constant-energy velocity Verlet algorithm, with energy and gradient calculated at each time step with CIS/DZ+. Finally, since the CIS method reproduces fairly well the cluster geometries during relaxation but greatly underestimates the VBEs, the photoelectron spectra are evaluated from high-level single-point CASPT2 calculations for all cluster configurations along the trajectories. The resulting spectra are compared with their experimental counterpart, providing insight into the CTTS relaxation dynamics.

In this contribution, we demonstrated how first-principles computer simulations can paint a realistic picture of the ground-state thermodynamics and structural properties of systems involving subtle intermolecular interactions such as ionic aqueous clusters, as well as of their charge-transfer-to-solvent excited-state dynamics, which can provide insight into experiment.

## Acknowledgments

This research was supported by Natural Sciences and Engineering Research Canada (NSERC). Some calculations were performed at the Centre for Research in Molecular Modeling (CERMM), which was established with the financial support of the Concordia University Faculty of Arts & Science, the Ministère de l'Education du Québec (MEQ) and the Canada Foundation for Innovation (CFI). Additional computing resources were provided by the CFI-funded Réseau Québecois de Calcul de Haute Performance (RQCHP) and Westgrid consortia. GHP holds a Concordia University Research Chair, QKT is the recipient of a Concordia Graduate Fellowship, and DMK the recipient of Concordia and Canada Graduate Scholarships.

## References

[1] See e.g. J.H. Choi, K.T. Kuwata, Y.B. Cao, M. Okumura, Vibrational spectroscopy of the $Cl^-(H_2O)_n$ anionic clusters, n – 1-5, *J. Phys. Chem. A.* **102** 503 (1998), and references therein.

[2] See e.g. E. Knipping, M.J. Lakin, K.L. Foster, P. Jungwirth, D.J. Tobias, R.B. Gerber, D. Dabdub, G.J. Finlayson-Pitts, Experiments and simulations of ion-enhanced interfacial chemistry on aqueous NaCl aerosols, *Science* **228** 301 (2000), and references therein.

[3] D. Serxner, C.E.H. Dessent, M.A. Johnson, Precursor of the I-aq CTTS band in $\Gamma(H_2O)_n$ clusters, *J. Chem. Phys.* **105** 7231-7234 (1996).

[4] D.H. Paik, I.R. Lee, D.-S. Yang, J.S. Baskin, A.H. Zewail, Electrons in finite-sized water cavities: hydration dynamics observed in real time, *Science* **306** 672-675 (2004).

[5] A.E. Bragg, J.R.R. Verlet, A. Kammrath, O. Cheshnovsky, D.M. Neumark, Hydrated electron dynamics: From clusters to bulk, *Science* **306** 669-671 (2004).

[6] G. Markovich, S. Pollack, R. Giniger, O. Cheshnovsky, Photoelectron spectroscopy of $Cl^-$, $Br^-$, and $I^-$ solvated in water clusters, *J. Chem. Phys.* **101** 9344 (1994).

[7] J.V. Coe, Connecting cluster anion properties to bulk: Ion solvation free energy trends with cluster size and the surface vs internal nature of iodide in water clusters, *J. Phys. Chem. A* **101** 2055 (1997).

[8] D.M. Koch, G.H. Peslherbe, On the transition from surface to interior solvation in iodide-water clusters, *Chem. Phys. Lett.* **359** 381-389 (2002).

[9] J. Chandrasekhar, D.C. Spellmeyer, W.L. Jorgensen, Energy component analysis for dilute aqueous solutions of $Li^+$, $Na^+$, $F^-$, and $Cl^-$ ions, *J. Am. Chem. Soc.* **106** 903 (1984).

[10] G.H. Peslherbe, B.M. Ladanyi, J.T. Hynes, Free energetics of NaI contact and solvent-separated ion pairs in water clusters, *J. Phys. Chem. A* **104** 4533 (2000).

[11] R. Iftimie, D. Salahub, J. Schofield, An efficient Monte Carlo method for calculating ab initio transition state theory reaction rates in solution, *J. Chem. Phys.* **119** 11285 (2003).

[12] Q.K. Timerghazin, G.H. Peslherbe, Theoretical investigation of charge transfer to solvent in photoexcited iodide-acetonitrile clusters, *Chem. Phys. Lett.* **354** 31-37 (2002).

[13] L. Lehr, M.T. Zanni, C. Frischkorn, R. Weinkauf, D.M. Neumark, Electron solvation in finite systems: femtosecond dynamics of iodide (water)_n anion clusters, *Science* **284** 635-638 (1999).

[14] H.-Y. Chen, W.-S. Sheu, Iodine effect on relaxation pathway of photoexcited $\Gamma(H_2O)_n$ clusters, *Chem. Phys. Lett.* **335** 475-480 (2001).

[15] Q.K. Timerghazin, G.H. Peslherbe, Further insight into the relaxation dynamics of photoexcited $\Gamma(H_2O)_n$ clusters, *J. Am. Chem. Soc.* **125** 9904-9905 (2003).

Brill Academic Publishers
P.O. Box 9000, 2300 PA Leiden
The Netherlands

*Lecture Series on Computer
and Computational Sciences*
Volume 4, 2005, pp. 1002-1005

# Insight into Energetic Cluster Impacts on Solid Surfaces Provided by Computer Simulation

Roger P. Webb[*]

Ion Beam Centre, Advanced Technology Institute,
School of Electronics and Physical Sciences,
University of Surrey,
Guildford, GU2 7XH, UK

Received 5 July, 2005; accepted in revised form 3 August, 2005

*Abstract:* For the past 30 years the effect of the interaction of energetic clusters with solid surfaces has been extensively studied both computationally and experimentally. Here we will give examples of the role that computer simulation is playing in developing an understanding of the processes involved in this topic. In particular we will look at examples from surface scattering and cluster impact induced molecular desorption.

## 1. Introduction

The computational modeling of the interaction of energetic clusters with surfaces is bounded on the one hand by the collision of large blocks of material at acoustic velocities and on the other by single ion bombardment. In the former case this can be modeled very successfully using continuum methods and in the latter by atomistic simulations.

In the case of a single atomic ion impact the damage created and sputtering of the surface is largely determined by the energy density deposited in nuclear collisions. This is determined largely by the ion and target masses and the target density. A light ion does not deposit as high an energy density as a heavier ion. Light ions lose a larger portion of their energy through collision with electrons and also penetrate deeper into the target than heavier ions. The consequence of this is that lighter ions have a lower deposited energy density than heavier ions. If the ion energy is increased then the cross-section for nuclear collisions decreases with the result that an increasing proportion of energy is lost in electronic collisions and the ions penetrate deeper and the deposited nuclear energy density decreases at high implantation energy. This is shown in figure 1 for a selection of light and heavy ions and for different density targets. A wide range of energy densities can be produced for single atomic ion impacts provided that the choice of atomic species is not an issue. But even in this case there is a maximum energy density that can be obtained for a particular target matrix. Using clusters and molecular species for impacts multiplies the implanted energy density. Shown in the figure is the deposited energy density for a fullerene $C_{60}$ implantation. It can be seen that it is possible, by using a molecular cluster, to increase the deposited energy density. In this case the deposited energy density of the molecule can be increased to be similar to that of a heavy ion, whilst at the same time the penetration depth and stopping characteristics are similar to that of a light ion. Hence, by using an energetic cluster instead of a single atomic ion, we can vary the energy density that the cluster liberates in the target surface by varying the number of atoms in the cluster. By varying the energy density in the vicinity of the impacting cluster we can potentially allow new and different reactions to take place both chemically and physically.

The effects of the impact of an energetic cluster on a solid surface and the fate of the cluster itself depend on many factors. These will include: the velocity of the cluster; the mass of the constituent atoms in the cluster relative to the mass of the target atoms; the number of atoms in the cluster; and the binding of the atoms to each other in the cluster.

---

[*] Corresponding author. E-mail: R.Webb@surrey.ac.uk

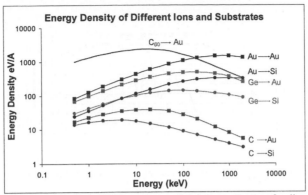

Figure 1: Energy Density of Different Ions and Substrates. See text for discussion

The velocity of the cluster will determine whether the cluster will embed itself in the target or scatter from the surface. At low impacting velocities: the binding energy of the cluster to the solid surface will determine if the cluster succeeds in bouncing free of the surface or remains attached to the surface after impact; and the binding energy of the cluster atoms to each other will determine if the cluster breaks-up and shatters on impact or remains intact. If the cluster shatters on impact then the scattering of the resulting constituent pieces of the cluster across the surface can result in the removal of loosely bound debris and particles on the surface as well as the general smoothing of an initially uneven surface.

The relative mass of the constituent cluster atoms and the target will determine at high impact velocities the penetration depth and the probability of the atoms of the cluster penetrating through the solid in close proximity to each other. The more they remain close to each other the more likely that the cluster impact will behave differently from the impact of the same number of single atom ions one after the other and their combined deposited energy will increase the deposited energy density to a level far higher than is achievable through single ion impact. This will often results in an effect which is "non-linear".

The number of atoms in the cluster will determine the total amount of kinetic energy delivered to the surface. The mass ratio of the cluster atoms to target atoms and the initial velocity determines the depth over which this energy is distributed. Hence, the combination of these two factors, allow the energy density to be changed for the same constituent atoms. This can not be done in conventional single atom ion irradiation as the amount of energy being deposited can not be decoupled from the depth over which the energy is deposited, without changing the constituent materials.

The use of computer simulation is invaluable in determining when these effects become important.

## 2. Fullerene Scattering from Graphite Surfaces

Ion scattering spectroscopy has been used over a number of years, normally to investigate the properties of surfaces. Typically mono-energetic ions are scattered from a surface and the reflected ions are detected and their angular and energy distribution measured. Scattering of molecular species from solids can also give some information about the strength and mechanical properties of the scattered molecule.

A series of experiments were performed[1] to verify the understanding of molecular scattering. In these experiments He, Xe and $C_{60}$ ions were scattered from a graphite surface. Very good agreement between the experiment and simulations were found[1]. However a number of interesting "side-effects" were identified. In particular it was noted that Xe scattered from the surface of the graphite as a result of a combination of up to 7 collisions with 14 surface atoms as shown in figure 2 above. Initially it was expected that as Xe was heavier than carbon it would not scatter from the graphite at all. It was also noted that unexpectedly the $C_{60}$ did not scatter in a specular direction as might be expected for a spherical molecule. Instead it was found to bounce from the surface substantially below the specular direction.

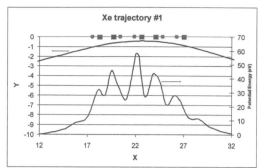

Figure 2. Potential Energy of a Xe atom scattering from a graphite surface showing 7 collisions

Figure 3 shows a visualization of the scattering events for both $C_{60}$ and Xe derived from the simulations. From the figure it can be seen that the $C_{60}$ causes substantial deformation of the surface and the whole scattering process is severely in-elastic. The $C_{60}$ itself picks up a substantial amount of internal energy – often leading to fragmentation in the gas phase above the target. But it is the energy lost to deforming the target which inevitably leads to the non-specular scattering. A large part of the problem here is that graphite is made up from a set of almost uncoupled planes. These planes easily deform like an elastic membrane, as the scattering process is over quickly the surface is still distorting after the molecule has scattered. This means that energy is lost to the surface in deforming the layer on the inward path and it does not get returned to the molecule on the outward path as the deformation is still continuing. In the case of the Xe the surface deformation is relatively minor so that any loss of energy from the ion to the surface in creating the deformation is small in comparison to the overall scattering energy. Scattering simulations of $C_{60}$ performed on diamond substrates show a much more specular reflection. The problem with the diamond scattering is that the fullerene molecule invariably shatters on impact leaving only scattered fragments.

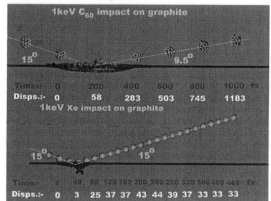

Figure 3. Visualization of the scattering event for $C_{60}$ and Xe as derived from the simulation results. Overlaid snapshots of the projectile and first surface layer atoms of the graphite surface are shown every 200 and 20 fs for $C_{60}$ and Xe respectively. The lower row in each picture lists the number of surface atoms displaced from their equilibrium positions at each time.

## 3. Impact Induced Desorption

Simulations have also been made in which a layer of benzene molecules has been placed on top of a solid substrate, either carbon[2], diamond[3] or silver[4]. In these simulations the mechanisms for desorbing molecules from surfaces by cluster impact have been investigated. Simulations of both a carbon and diamond substrate with a benzene monolayer adsorbed onto it were run for 3-4ps after impact by a $C_{60}$ fullerene molecule[2,3]. The benzene adsorbed layer was then analysed to see how many of the benzene molecules had been ejected intact from the surface and how many were broken by the interaction. If this technique is to be used as a probe for large (and fragile) organic molecules on surfaces then information of this type will be needed to be able to pick the best impact cluster and energy combination to provide a high yield of intact desorbed/sputtered molecules and at the same time create as little surface disruption as possible to both the substrate surface and the adsorbate layer. Fragmentation of the organic over-layer will give a confused signal if the same region is sampled by the analysing cluster beam.

It has been shown in computer simulations that fullerene impacts on graphite can cause strong acoustic waves that propagate from the impact site[5-8]. These acoustic waves are not seen to the same magnitude on a diamond surface. It has been shown that this transverse wave can be used to encourage the desorption of both physisorbed and chemisorbed molecules from a graphite surface[3]. The yield of desorbed benzene molecules is enhanced by this effect for low collision energies (<1keV) for a graphite surface when compared with a diamond surface. The yield of desorbed benzene molecules is enhanced by a factor of 3 by this effect[3].

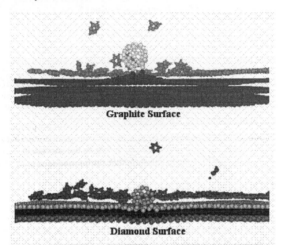

Figure 4: Cross sectional view of 250eVfullerene impact on benzene over-layer on a graphite and diamond surface.

Figure 4 shows how the acoustic wave created by the fullerene impact on the graphite surface lifts the surface up sufficiently to cause the surface to assist in the desorption process of the benzene molecules. By contrast the diamond surface response is insufficient to encourage desorption this way

## 4. Conclusions

In conclusion it has been shown that atomistic computer simulation is capable of providing great insight into the physical processes that occur in complex and often chaotic systems which result from the impact of energetic clusters and molecules with solid surfaces.

## References

[1]  M Hillenkamp, J Pfister, M Kappes, RP Webb, *J. Chem. Phys.* **111(22)**, (1999), 10303

[2]  M Kerford, R Webb, *Nuclear Instruments & Methods*, **B180**, (2001), 44

[3]  RP Webb, M.Kerford, E.Ali, M.Dunn, L.Knowles, K.Lee, J.Mistry, F.Whitefoot, *Surface and Interface Analysis*, **31**, (2001), 297

[4]  Z.Postawa, *Applied Surface Science*, **231-232**, (2004), 22

[5]  R.Smith R.Webb, *Proc R. Soc. Lond.* **A 441**, (1993), 495

[6]  M Kerford, RP Webb, *Nuclear Instruments & Methods*, **B153**, (1999), 270

[7]  M Kerford, RP Webb, *Carbon*, **37**, (1999), 859

[8]  RP Webb, *Applied Surface Science*, **231-2**, (2004), 59

Brill Academic Publishers
P.O. Box 9000, 2300 PA Leiden,
The Netherlands

*Lecture Series on Computer
and Computational Sciences*
Volume 4, 2005, pp. 1006-1009

# Rotations in Doped Quantum Clusters

**Stephanie Wong, Nicholas Blinov, and Pierre-Nicholas Roy**[1]

Department of Chemistry
University of Alberta
Edmonton, Alberta
T6G 2G2, Canada

Received 18 July, 2005; accepted in revised form 17 August, 2005

*Abstract:* We discuss computational methods for the study of rotational dynamics of doped quantum clusters. Exact basis set methods are amenable to the study of clusters containing a linear molecule and one or two solvent atoms. For larger clusters, path integral Monte Carlo (PIMC) techniques have been used in order to extract various properties such as the importance of quantum exchange. An important building block of a PIMC simulation is the high temperature density matrix. We propose here an approach for the calculation of this density matrix based on a recursive Lanczos method. Calculation results are presented for the Helium dimer to illustrate the usefulness of the method.

*Keywords:* quantum clusters, quantum dynamics, path integral simulations, spectroscopy

*PACS:* PACS numbers: 36.40.-c,36.40.Ei,36.40.Mr,61.46.+w,67.90.+z

## 1 Introduction

Doped quantum clusters offer a unique environment to study a variety of phenomena such as quantum solvation and rotational dynamics. Spectroscopic studies on OCS [9] and $N_2O$ [10] doped helium clusters have in particular revealed an unexpected size evolution of the rotational constants of the complex. Theoretical studies of doped quantum clusters are used to rationalize the observed experimental observations and also to make predictions.

In Section 2 we discuss some computational approaches used to study the rotational dynamics of doped clusters. Exact basis set and Path Integral Monte Carlo (PIMC) methods are covered. We propose in Section 3 a convenient iterative approach for the calculation of the pair density matrix used in PIMC simulations. We provide concluding remarks in Section 4.

## 2 Simulation approaches for doped clusters

When dealing with one or two solvent atoms, it is possible to directly calculate the energy levels and wavefunctions of doped clusters using finite basis set methods. We have for example calculated microwave transition frequencies for the He-$N_2O$ system and obtained excellent agreement with experiment [8]. This kind of study allows the assessment of the quality of the solvent-solute potential energy surface. Calculations of this type are also possible for two solvent atoms.

When dealing with larger systems, quantum Monte Carlo techniques are usually used. We have developed a PIMC approach for the study of rotational dynamics in doped helium clusters [3]. The

---

[1]Corresponding author. E-mail: pn.roy@ualberta.ca

technique has been used investigate the importance of quantum exchanges in $N_2O$ doped helium clusters [7]. It was observed that for smaller clusters containing 8 helium atoms, exchange effects are quenched due to localization. For larger clusters, exchange becomes important and can be used to explain the decoupling between a solute and its cluster environment. Our approach has also been used in the study of exchange effects in $CO_2$ doped helium clusters [2]. An important quantity used in PIMC simulations is the high temperature density matrix and we present an approach for its calculation in the following section.

## 3 Iterative calculation of the pair density matrix: Helium dimer example

We propose here a method for the calculation of the high temperature pair density matrix. The approach is based on the iterative Lanczos algorithm for the solution of sparse eigenvalue problems [5]. The helium dimer will be used for illustration. The pair density matrix for two atoms is defined as,

$$\rho(r) = \langle r|e^{-\beta \hat{H}}|r \rangle , \tag{1}$$

where $\beta = 1/k_B T$ is the reciprocal temperature. For total angular momentum $J$, the Hamiltonian, $\hat{H}$ of the above equation is,

$$\hat{H} = \frac{-\hbar^2}{2\mu}\frac{\partial^2}{\partial r^2} + \frac{\hbar^2}{2\mu r^2}J(J+1) + V(r) , \tag{2}$$

where $\mu$ is the reduced mass and $r$ is the relative distance. The quantity $V(r)$ corresponds to the interaction potential between two helium atoms for which we use the Aziz model[1]. The Hamiltonian operator is represented as matrix, $H$, in the discrete variable representation (DVR) with an equally spaced grid [6]. To calculate the elements of the diagonal density matrix given in Eq. 5, we use the iterative Lanczos approach. The Lanczos recursion is given by,

$$\hat{H}|v_j\rangle = \beta_{j-1}|v_{j-1}\rangle + \alpha_j|v_j\rangle + \beta_j|v_{j+1}\rangle \tag{3}$$

where $\alpha_j = \langle v_j|H|v_j \rangle$ and $\beta_{j-1} = \langle v_{j-1}|H|v_j \rangle$. After a number $M$ of Lanczos iterations, we have a symmetric tridiagonal representation, $H_T$, of the Hamiltonian (the diagonal entries are the $\alpha_j$'s, and the upper and lower diagonal entries are the $\beta_j$'s) and a series of Lanczos vector, $\{v_0, v_1, ..., v_{M-1}\}$. The Lanczos vectors are orthogonal and span a Krylov subspace. The initial Lanczos vector is now chosen to be a localized position vector, $\langle v_0|j \rangle = \delta_{ij}$, where $|i\rangle$ is localized on grid point $r_i$. The expression for the density matrix can now be written as,

$$\rho(r_i) = |<i|r_i>|^2 \sum_{ll'}\langle i|v_l\rangle\langle v_l|e^{-\beta \hat{H}}|v_{l'}\rangle\langle v_{l'}|i\rangle , \tag{4}$$

where the first factor on the R.H.S. is a weight to convert from the DVR to the position representation. The above can be simplified by using the orthogonality of the Lanczos vectors and the fact that $v_0$ is localized onto point $r_i$,

$$\rho(r_i) = |<i|r_i>|^2\langle v_0|e^{-\beta \hat{H}}|v_0\rangle , \tag{5}$$

Diagonalization of $H_T$ yields some of the eigenvalues, $E_n$, of $H$ and the above equation becomes,

$$\rho(r_i) = |<i|r_i>|^2 \sum_{n}^{M}|\langle v_0|n\rangle|^2 e^{-\beta E_n} , \tag{6}$$

where $\langle v_0|n \rangle$ corresponds to the first elements of the eigenvectors of $H_T$.

In summary, one selects a position-localized initial Lanczos vector, performs a number $M$ of Lanczos iterations, diagonalizes the resulting $H_T$ matrix to obtain eigenvalues $E_n$, along with only the first elements $\langle v_0|n \rangle$ of the corresponding eigenvectors, and applies Eq. 6 to compute the density matrix. We present in Figure 1 the diagonal pair density matrix for the helium dimer at 94.72 K. A grid size of 20 bohrs and $N = 50$ grid points were sufficient at that temperature. The

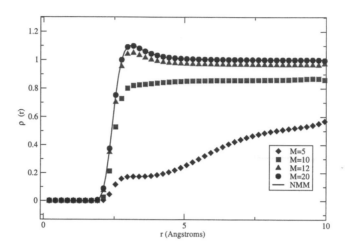

Figure 1: Calculated pair density matrix at $T = 94.72$ K for different number, $M$, of Lanczos iterations.

matrix size is therefore $50 \times 50$ for this simple model problem. Since $J$ is a good quantum number, calculations were performed one $J$ at a time and then combined. The maximum value of $J$ was set to 100. Results are given for increasing numbers of Lanczos iterations, $M$, to illustrate the convergence of the method. A value of $M - 20$ is sufficient in the present case (more iterations are required at lower temperature). The calculations took less than a minute of CPU time on a 2.0 GHz Pentium 4 processor. Results obtained using the numerical matrix multiplication method (NMM) [4] are shown for comparison. The present approach scales as $M \times N^2$ for each $J$ whereas the NMM calculation required 8 matrix squaring steps, each step scaling as $N^3$. For larger $N$ and sparse matrices, our method offers a definite advantage over the NMM method. Calculations of the off-diagonal density matrix elements are more costly to obtain but still accessible.

## 4  Concluding Remarks

We have briefly discussed methods for the simulation of doped quantum clusters. Exact basis set techniques are presently only amenable to the study of systems with one or two solvent atoms while PIMC techniques can be used to study the finite temperature quantum properties of larger clusters. The high temperature density matrix is an important building block of PIMC simulations and we have suggested a Lanczos-type approach for its calculation. The proposed method is fast and accurate as revealed by our test calculations on the helium dimer.

## Acknowledgment

This work was supported by the Natural Sciences and Engineering Research Council of Canada and The Research Corporation.

## References

[1] R. A. Aziz, A. R. Janzen, and M.R. Moldover. *Ab Initio* calculations for helium: A standard for transport property measurements. *Phys. Rev. Lett.*, 74(9):1586–1589, 1995.

[2] N. Blinov and P.-N. Roy. Effect of exchange on the rotational dynamics of doped helium cluster. *J. Low Temp. Phys*, 140, 2005.

[3] N. Blinov, X. Song, and P.-N. Roy. Path integral Monte Carlo approach for weakly bound van der Waals complexes with rotations: Algorithm and benchmark calculations. *J. Chem. Phys.*, 120(13):5916–5931, 2004.

[4] D. M. Ceperley. Path integrals in the theory of condensed helium. *Rev. Mod. Phys.*, 67:279–355, 1995.

[5] G. H. Golub and C. F. van Loan. *Matrix Computations*. Johns Hopkins University Press, Baltimore, 1989.

[6] J.C. Light and T. Carrington Jr. Discrete-variable representations and their utilization. *Adv. Chem. Phys.*, 114:263–310, 2000.

[7] S. Moroni, N. Blinov, and P.-N. Roy. Quantum monte carlo study of helium clusters doped with nitrous oxide: Quantum solvation and rotational dynamics. *J. Chem. Phys.*, 121(8):3577–3581, 2004.

[8] X. Song, Y. Xu, P.-N. Roy, and W. Jäger. Rotational spectrum, potential energy surface, and bound states of the weakly bound complex $hen_2o$. *J. Chem. Phys.*, 121(24):12308–12314, 2004.

[9] J. Tang, Y. Xu, A. R. W. McKellar, and Wolfgang Jäger. Quantum solvation of carbonyl sulfide with helium atoms. *Science*, 297:2030–2033, 2002.

[10] Y. Xu, W. Jäger, J. Tang, and A. R. W. McKellar. Spectroscopic studies of quantum solvation in $^4he_n - N_2O$ clusters. *Phys. Rev. Lett.*, 91:163401, 2003.

Brill Academic Publishers
P.O. Box 9000, 2300 PA Leiden,
The Netherlands

*Lecture Series on Computer*
*and Computational Sciences*
Volume 4, 2005, pp. 1010-1013

# Structural and Electronic Properties of Nanostructured HAlO and AlO

Yi Dong,[a1] Michael Springborg,[a2] Markus Burkhart,[b3] and Michael Veith[b4]

[a]Physical and Theoretical Chemistry
and [b]Inorganic Chemistry
University of Saarland,
66123 Saarbrücken,
Germany

Received 10 July, 2005; accepted in revised form 10 July, 2005

*Abstract:* The results of theoretical studies of nanostructured HAlO and AlO are presented. We have considered isolated clusters, the interactions between two clusters, and two-dimensional layers. In most of the calculations we used a parameterized density-functional tight-binding method in the calculation of the electronic properties for a given structure, combined with two different unbiased approaches, i.e., an 'Aufbau' and a genetic-algorithm method, for optimizing the structure for clusters. The results for the isolated clusters are analyzed by means of similarity, stability, and shape parameters. Smaller structures were also studied with parameter-free DFT methods.

*Keywords:* Clusters, structure, stability, density-functional calculations

*PACS:* 36.40.-c, 36.90.+f, 61.46.+w, 73.22.-f

## 1   Introduction

HAlO is an interesting material that can be used as a substrate for organized structures of organic materials, but only little is known about its precise structure. Here, we shall show that theoretical studies can give useful information that ultimately turns out to extend and support the experimental information about it.

HAlO can be prepared either by CVD (chemical vapor deposition) at low temperatures as a thin glassy layer using the precursor bis-(*tert*-butoxyalane) $[AlH_2(OtBu)]_2$ and various metals as target substrates [1, 2], or as an amorphous powdered nanostructured material by the reaction of different methylsiloxanes with the alane $H_3Al \cdot NMe_3$ in either ether or aromatic solvents under mild conditions [3, 4].

The purpose of the present work is to obtain further information on HAlO by considering both finite clusters and infinite, periodic layers and ultimately present a proposal for the structure of the HAlO compounds. In agreement with experimental indications, we find that the HAlO clusters consist of an AlO core with H atoms on the surface. We shall therefore also consider isolated AlO clusters, here.

---

[1]e-mail: y.dong@mx.uni-saarland.de
[2]Corresponding author. e-mail: m.springborg@mx.uni-saarland.de
[3]e-mail: m.burkhart@mx.uni-saarland.de
[4]e-mail: veith@mx.uni-saarland.de

We studied theoretically the structural properties of the $(HAlO)_n$ clusters with $n$ up to 26 using a density-functional-theory tight-binding method. This method can give the electronic properties for a given structure as well as determine a structure of a local total-energy minimum once an initial structure has been chosen. However, the method is not directly able to determine the structure of the global total-energy minimum. In order to search for that for the $(HAlO)_n$ clusters, we have used two unbiased approaches, i.e., a method we have called the 'Aufbau' method as well as genetic algorithms.

To verify the results of these methods we also performed parameter-free DFT calculations on smaller $(HAlO)_n$ units with $n = 1, 2, 3, 4, 6$. For each $n$ we considered several different isomers, as well as calculated their vibrational spectra.

Since the experimentally produced nanostructured material is extended, it must contain nanostructures in close contact. Therefore, we also studied the interactions between pairs of optimized $(HAlO)_{n_1}$ and $(HAlO)_{n_2}$ clusters for different values of $n_1$ and $n_2$. Finally, the fact that HAlO can be synthesized as a layer compound made us study infinite, periodic, two-dimensional layers of HAlO, too.

Finally, we also used the parameterized density-functional method together with the genetic algorithms in studying stoichiometric AlO clusters.

## 2 Theoretical Methods

For most of the calculations, we used the density-functional tight-binding method (DFTB) of Seifert and coworkers [5, 6]. With this method, the binding energy is written as the difference in the orbital energies of the compound minus those of the isolated atoms, i.e., as $\sum_i \epsilon_i - \sum_m \sum_i \epsilon_{mi}$ (with $m$ being an atom index and $i$ an orbital index), augmented with pair potentials, $\sum_{m_1 \neq m_2} U_{m_1,m_2}(|\vec{R}_{m_1} - \vec{R}_{m_2}|)$ (with $\vec{R}_m$ being the position of the $m$th atom). In calculating the orbital energies we need the Hamilton matrix elements $\langle \chi_{m_1 n_1} | \hat{H} | \chi_{m_2 n_2} \rangle$ and the overlap matrix elements $\langle \chi_{m_1 n_1} | \chi_{m_2 n_2} \rangle$. Here, $\chi_{mn}$ is the $n$th atomic orbital of the $m$th atom. The Hamilton operator contains the kinetic-energy operator as well as the potential. The latter is approximated as a superposition of the potentials of the isolated atoms, $V(\vec{r}) = \sum_m V_m(|\vec{r} - \vec{R}_m|)$, and subsequently we assume that the matrix element $\langle \chi_{m_1 n_1} | V_m | \chi_{m_2 n_2} \rangle$ vanishes unless at least one of the atoms $m_1$ and $m_2$ equals $m$. Finally, the pair potentials $U_{m_1,m_2}$ are obtained by requiring that the total-energy curves from parameter-free density-functional calculations on the diatomics are accurately reproduced.

Finally, we used the parameter-free, density-functional program package TURBOMOLE [7] for the smallest $(HAlO)_n$ clusters.

We used two different methods in determining the structures of the clusters.

In some of the calculations we used our own *Aufbau* method [8]. The method is based on simulating experimental conditions, where clusters grow by adding atom by atom to a core. By repeating this process **very** many times, we can identify the structures of the lowest total energy. Alternatively, we optimized the structures using the so-called genetic algorithms [8, 9, 10]. Here, from a set of structures we generate new ones through cutting and pasting the original ones. Out of the total set of old and new clusters those with the lowest total energies are kept, and this process is repeated until the lowest total energy is unchanged for a large number of generations.

## 3 Results for the $(HAlO)_n$ Clusters

First, we optimized the structure of $(HAlO)_n$ clusters with $n$ up to 26 using our 'Aufbau' and with $n$ up to 18 using the genetic-algorithms approaches, respectively. We define a radial distance $\vec{r}_i$ for each of the $3n$ atoms as $r_i = |\vec{R}_i - \vec{R}_0|$, where $\vec{R}_0$ is the center of the cluster, $\vec{R}_0 = \frac{1}{3n} \sum_{i=1}^{3n} \vec{R}_i$. Fig. 1 shows the radial distance for the different clusters and atoms. The figure shows, for each

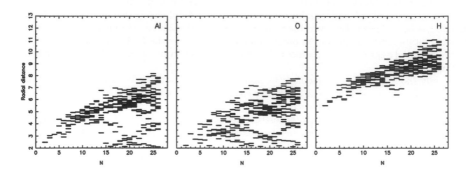

Figure 1: The radial distances (in a.u.) for Al, O, and H atoms, separately, as a function of the size of the cluster $n$ for $(HAlO)_n$ clusters. In each panel, a small horizontal line shows that at least one atom of the corresponding type has that distance to the center of the cluster for a given value of $n$.

type of atoms separately, the radial distances for the different values of $n$. It is very clear that the largest radial distances are found for H atoms and, in addition, that essentially all H atoms always have larger radial distances than the Al and O atoms. On the other hand, the central part of the clusters are clearly formed by both Al and O atoms.

Also the parameter-free DFT calculations on various (in total, 28) isomers of $(HAlO)_6$ supports this consensus. For each of the 28 $(HAlO)_6$ clusters we calculated the radial distances of the different atoms. Subsequently we plotted this as a function of the total energy per unit and show in Fig. 2 the results. When comparing the clusters of roughly the lowest total energies we see that Al and O are those atoms with smaller radial distances, whereas those of hydrogen are larger. For the higher total energies, the radial distances of Al and O show a weak tendency to increase and simultaneously those of H are slightly decreasing. In total this analysis confirms the tendency for the $(HAlO)_n$ clusters to possess a Al-O core covered with H atoms.

By analysing the overall shape of the clusters we also found that there are particularly stable clusters (most pronounced for $n = 4$) for which the structure is roughly spherical.

## 4 Conclusions

The results of this study are unique as they have been obtained by using several different experimental and theoretical approaches independently of each other. By combining the results from all approaches, we have arrived at an unusually detailed picture of the structural properties of nanostructures HAlO despite the complications determining this directly in experimental studies.

All approaches indicate that nanostructured HAlO consists of subsystems with an AlO core covered by H atoms. This structure can, e.g., be obtained by keeping the $(HAlO)_n$ clusters not too large, but, alternatively, also two-dimensional sheets HAlO can also satisfy these constraints. The calculations (not shown here) give indeed that also such sheets are stable with a binding energy comparable with that of the largest clusters considered here.

Finally, we stress that some of the conclusions could only be obtained after a careful analysis of the results of the calculations by means of special descriptors.

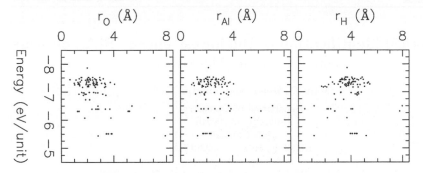

Figure 2: The radial distances for the different smaller clusters as function of the total energy per unit. The different panels show the different types of atoms. The results are from the DFT calculations for the $(HAlO)_n$ clusters with $n = 1, 2, 3, 4, 6$.

## Acknowledgment

We gratefully acknowledge *Fonds der Chemischen Industrie* for very generous support. This work was supported by the SFB 277 of the University of Saarland.

## References

[1] M. Veith, K. Andres, S. Faber, J. Blin, M. Zimmer, Y. Wolf, H. Schnöckel, R. Köppe, R. Masi, S. Hüfner, Eur. J. Inorg. Chem. **2003**, 4387 (2003).

[2] M. Veith, K. Andres, J. Metastable Nanocryst. Mater. **15-16**, 279 (2003).

[3] S. W. Liu, R. J. Wehmschulte, C. M. Burba, J. Mater. Chem. **13**, 3107 (2003).

[4] S. W. Liu, J. Fooken, C. M. Burba, M. A. Eastman, R. J. Wehmschulte, Chem. Mater. **15**, 2803 (2003).

[5] G. Seifert and R. Schmidt, New J. Chem. **16**, 1145 (1992).

[6] G. Seifert, D. Porezag, and Th. Frauenheim, Int. J. Quant. Chem **58**, 185 (1996).

[7] R. Ahlrichs, M. Bär, M. Häser, H. Horn, C. Kölmel, Chem. Phys. Lett. **162**, 165 (1989).

[8] Y. Dong and M. Springborg, in *Proceedings of 3rd International Conference "Computational Modeling and Simulation of Materials"*, Ed. P. Vincenzini *et al.*, Techna Group Publishers, p. 167 (2004).

[9] J.-O. Joswig, M. Springborg, and G. Seifert, Phys. Chem. Chem. Phys. **3**, 5130 (2001).

[10] J.-O. Joswig and M. Springborg, Phys. Rev. B **68**, 085408 (2003).

Brill Academic Publishers
P.O. Box 9000, 2300 PA Leiden,
The Netherlands

*Lecture Series on Computer
and Computational Sciences*
Volume 4, 2005, pp. 1014-1017

# Theoretical identification of the lowest energy $Si_xL_y$ clusters using Genetic Algorithms

## John D. Head[1] and Yingbin Ge[2]

Department of Chemistry,
University of Hawaii,
2545 The Mall,
Honolulu, Hawaii 96822, USA

Received 8 July, 2005; accepted in revised form 20 July, 2005

*Abstract:* An overview of a strategy for theoretically finding the lowest energy structure of $Si_xL_y$ nanoclusters using a genetic algorithm will be presented. The ab initio global minima identified for various $Si_xH_y$ and $Si_xF_y$ clusters will be discussed.

*Keywords:* Global optimization, genetic algorithms, $Si_xH_y$ clusters, $Si_xF_y$ clusters, ab initio calculations, semiempirical calculations, electronic structure of silicon nanoclusters.

## 1  Introduction

Nanometer-sized silicon clusters are of considerable interest because they exhibit a novel bright photoluminescence which might eventually be developed into a practical optoelectronic device. Several groups have performed quantum mechanical calculations on different $Si_xL_y$ clusters with assumed structures where L serves as a passivating ligand. One potential problem with the calculations is that experimentally it is very difficult to determine the geometry of an individual cluster or even if the cluster has the anticipated stoichiometry. In this presentation we will describe the strategy we have been developing using genetic algorithms (GA) to globally optimize $Si_xL_y$ clusters [1, 2, 3, 4, 5] so that cluster structure responsible for the bright photoluminescence might be identified. The presentation will consist of three main parts. An overview of the computational strategy will be given first. This will be followed by a discussion of the structures we have obtained for $Si_xH_y$ clusters. In a practical optoelectronic device, H is not going to be a very stable passivating ligands. The last part of the talk will describe our work on $Si_xF_y$ clusters and how we plan to extend our calculations to clusters with more inert passivating ligands.

## 2  Computational Strategy

The cluster global minimum search (CGA) is performed using a genetic algorithm which works directly on the cartesian coordinates of the atoms composing the clusters. An initial population of typically 100 clusters is randomly generated and locally optimized. The fitter, lower energy, clusters are then mated to produce an offspring population of locally optimized structures. The CGA effectively lets good structural features present in a cluster to be passed onto succeeding generations until the global minimum is found. In our initial approach we used the Deaven and

---

[1]Corresponding author. E-mail: johnh@hawaii.edu
[2]Current address: Ames Lab, Iowa State University, Ames, Iowa 50011, USA.

Ho cut-and-paste mating method [6] along with some other relatively simple coordinate averaging mating operations [1, 2, 3] The Deaven and Ho operator is a crossover method which cuts two parent clusters into halves along randomly oriented planes and recombines two halves from different parents into an offspring structure. The Deaven and Ho operator facilitates selecting low energy features present in an existing cluster structure whereas our coordinate averaging mating operations were intended to produce new structural features and enhance the diversity of the overall population.

While this initial CGA could find low energy and presumably the ab initio global minimum for various $Si_xH_y$ clusters it was not able to find the $Si_{14}H_{20}$ global minimum [3]. Eventually we realized this was because our initial mating operators did not readily produce structures where a $SiH_2$ unit might be shifted between different parts of a cluster. This lead to the development of a second CGA which in addition to Deaven and Ho cut-and-paste operator uses the new mating operators shown in figure 1 [4].

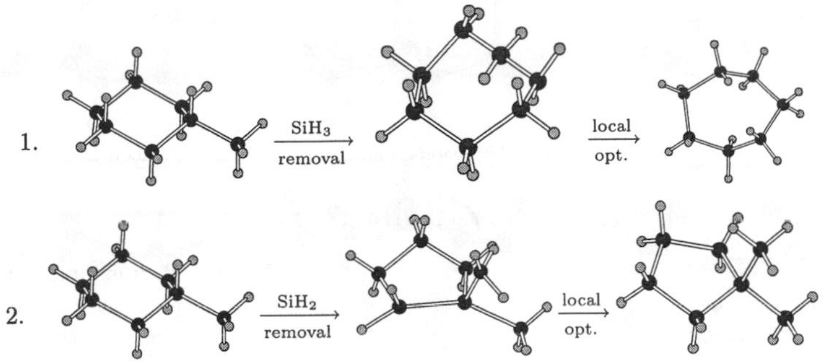

Figure 1: Examples illustrating two of the new mutation operators used in the cluster GA which act on a parent cluster to produce an offspring which is then locally optimized [4].

The CGA clearly requires a very large number of cluster energy calculations. After consideration of a number of fast energy methods we selected the AM1 semiempirical procedure for prescreening the relative energies of the $Si_xH_y$ clusters prior to performing any ab initio calculations [1]. The AM1 methods has a Hartree-Fock basis and has the capability to track the different type of bonding situations which might occur as the $Si_x$ cluster core is surrounded by greater numbers of ligands L. For instance, since it was not clear how many H atoms might be needed to passivate the Si core, when there are excess H atoms the AM1 calculation automatically allows $H_2$ loss and the resulting $Si_xH_{y-2}$ cluster is globally optimized.

Although the AM1 method proved to be useful fast energy method we soon found that the AM1 ranking of the various possible cluster structures differed significantly from the ab initio ranking. This lead us to the idea of reoptimizing the AM1 parameters using data obtained from ab initio calculations on several low energy $Si_xH_y$ clusters [2, 3]. As described below, the resulting GAM1 parameters produced cluster energy rankings more consistent with the ab initio calculations thereby facilitating a more reliable search for the $Si_xH_y$ global minimum.

## 3   Global minima for $Si_xH_y$ clusters

The initial CGA strategy enabled us to find low energy cluster structures at the ab initio level for both low and fully H passivated silicon clusters [1]. The CGA successfully found cluster global minima for the AM1 parameters but these were significantly different to the ab initio global minima which we expect to agree better with chemical reality. For instance, the ab initio $Si_{10}H_{16}$ global minimum was the tenth lowest AM1 energy structure. Using the GAM1 method, where the the AM1 parameters were reoptimized using a $Si_7H_{14}$ training set, greatly improved the cluster structure energy prescreening with semiempirical calculations [3]. The same $Si_{10}H_{16}$ global minimum structure was obtained by both the GAM1 and ab initio calculations. Coupling GAM1 energy prescreening with the revised CGA operators in figure 1 lead to finding the global minima shown in figure 2.

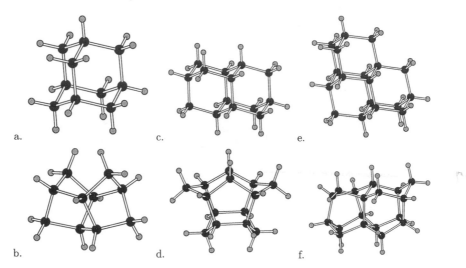

Figure 2: *Ab initio* global minima determined for: (a) $Si_{10}H_{16}$, (b) $Si_{10}H_{14}$, (c) $Si_{14}H_{20}$, (d) $Si_{14}H_{18}$, (e) $Si_{18}H_{24}$ and (f) $Si_{18}H_{22}$ [4].

The $Si_xH_y$ global minima we have obtained so far suggest that the low H passivated clusters, such as $Si_6H_2$, $Si_6H_6$ and $Si_{10}H_y$, $y = 4, 8$ favor forming structures with compact Si cores somewhat analogous to elemental Si clusters where many of the Si atoms cannot achieve tetrahedral coordination [1]. For the higher H passivated we obtain global minimum of the type shown in figure 2. When there are enough H atoms, as with the $Si_{10}H_{16}$, $Si_{14}H_{20}$ and $Si_{18}H_{24}$ stoichiometries the Si core does favor forming a fragment of the bulk Si diamond lattice. However, for the slightly under passivated clusters $Si_{10}H_{14}$, $Si_{14}H_{18}$ and $Si_{18}H_{22}$ the global minima are more difficult to predict by using chemical intuition alone. Figure 2 shows that the $Si_{10}H_{14}$, $Si_{14}H_{18}$ and $Si_{18}H_{22}$ global minima do not have any obvious structural similarities. The $Si_{18}H_{22}$ global minimum does appear to retain more of the diamond-lattice like structure than in the smaller $Si_{14}H_{18}$ and $Si_{10}H_{14}$ clusters. One important conclusion from these studies is that the lowest energy distorted lattice structure in $Si_xH_{y-2}$ cannot be simply computed by removing two H atoms from the diamond-lattice like $Si_xH_y$ global minimum and then performing a local geometry optimization.

## 4 Extending the $Si_xL_y$ GM search to other ligands

Air inert Si nanoclusters for use in a real world optoelectronic device will require some other passivating ligand besides H. One of difficulties of going beyond H as the ligand L is that the semiempirical energy prescreening calculation becomes much slower. In the presentation we will describe our progress at developing faster prescreening methods for the heavier and more chemically inert ligands. Improved prescreening is clearly important as indicated by our preliminary calculations, illustrated in figure 3, which show that the global minima for $Si_x F_y$ have structures very different to the corresponding $Si_x H_y$ global minimum [5]. It appears the H atoms in low-energy well H-passivated Si clusters prefer to be fairly evenly distributed over the $Si_x$ core enabling five- and six-membered Si rings to be formed. In contrast, the low-energy well F-passivated clusters like to form structures containing one or more trifluorosilyl groups at the expense of forming highly strained four- and even three-membered Si rings.

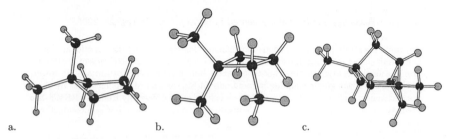

a.                              b.                              c.

Figure 3: Example B3LYP $Si_x F_y$ global minima: (a) $Si_7 H_{14}$, (b) $Si_7 F_{14}$ and (c) $Si_{10} F_{16}$ [5].

## Acknowledgment

We are grateful for the generous supply of computer time provided by the Maui High Performance Computing Center.

## References

[1] Global Optimization of H-Passivated Si Clusters with a Genetic Algorithm. Ge, Y., and Head, J. D., *J. Phys. Chem. B*, **106**, 6997 (2002).

[2] Global optimization of $Si_x H_y$ at the Ab Initio Level via an Iteratively Parametrized Semiempirical Method. Ge, Y., and Head, J. D., 2003, *Inter. J. Quant. Chem.*, **95**, 617 (2003).

[3] Global Optimization of H-Passivated Si Clusters at the Ab Initio Level via the GAM1 Semiempirical Method. Ge, Y., and Head, J. D., *J. Phys. Chem. B*, **108**, 6025 (2004).

[4] Fast global optimization of $Si_x H_y$ clusters: new mutation operators in the cluster genetic algorithm. Ge, Y., and Head, J. D., *Chem. Phys. Lett.*, **398**, 107 (2004).

[5] Ligand effects on $Si_x L_y$ cluster structures with L = H and F. Ge, Y., and Head, J. D., *Mol. Phys.*, **103**, 1035 (2005).

[6] Molecular Geometry Optimization with a Genetic Algorithm, Deaven, D. M., and Ho, K. M., *Phys. Rev. Lett.*, **75**, 288 (1995).

Brill Academic Publishers
P.O. Box 9000, 2300 PA Leiden,
The Netherlands

*Lecture Series on Computer*
*and Computational Sciences*
Volume 4, 2005, pp. 1018-1021

# Magnetism in Pure and Doped Manganese clusters

## Mukul Kabir[†1], Abhijit Mookerjee[†] and D. G. Kanhere[‡]

[†] S. N. Bose National Centre for Basic Sciences, JD Block, Sector III, Salt Lake City,
Kolkata 700 098, India

[‡] Department of Physics and Centre for Modeling and Simulation, University of Pune,
Pune - 411 007, India

Received 1 July, 2005; accepted in revised form 5 August, 2005

*Abstract:* We study pure and As-doped manganese clusters from density functional theory. Our results show excellent agreement with the SG molecular beam experiment. Introduction of a single As atom in $Mn_x$ clusters enhances binding energies substantially. However, the Mn-Mn ferromagnetic coupling is found only in $Mn_2As$ and $Mn_4As$ clusters. We show the exchange coupling in these doped clusters behave quite differently from the RKKY-like predictions as cluster size increases. It's possible relevance to the III-V semiconductor is also discussed.

*Keywords:* cluster, magnetism, semiconductor ferromagnetism

*PACS:* 61.46.+w, 36.40.Cg, 75.50.Pp

## 1 Introduction

Magnetism in transition metal clusters is largely motivated by both technological and fundamental point of view: How magnetic properties behave in reduced dimension and how does it evolve up to the bulk? Manganese clusters are particularly more interesting because isolated Mn atom has $4s^2 3d^5$ electronic structure, which gives a large magnetic moment of 5 $\mu_B$, according to Hund's rule. Through Stern-Gerlach (SG) molecular experiment [1, 2], Knickelbein found that $Mn_{5-99}$ posses finite (0.4-1.7 $\mu_B$/atom) despite of the fact that bulk manganese is antiferromagnetic. This observed finite magnetic moment of $Mn_x$ clusters could arise either one of the reasons that the individual atomic moments are small and aligned ferromagnetically or they still have large moment but their orientation flips. Recently we studied free $Mn_x$ clusters and our results show excellent agreement with SG experiment and we show that the experimentally observed sudden drop in the magnetic moment is due to their particular geometric structure and we argue the possible presence of several closely (to the ground state) lying isomers in the SG beam [3].

Now it is interesting to see that how magnetic behaviour changes when a single impurity is added to the $Mn_x$ clusters? Moreover, it is interesting to see the magnetism in $Mn_xAs$ clusters as (Ga,Mn)As and (In,Mn)As show ferromagnetic behaviour and promises to be a good candidate for the spintronic materials[4, 5]. Due to the low temperature molecular-beam-epitaxy growth of these samples, the presence of several kinds of defects are possible and are, indeed, responsible for the reported wide variation of Curie temperature. Recently we discussed the effect of one such defects, clustering of Mn around As, through studying free $Mn_xAs$ ($x =$2-10) clusters[6].

---

[1]Corresponding and presenting author. E-mail: mukul@bose.res.in

## 2 Results and Discussions

We perform density functional theory calculations within the pseudopotential method. We employ projector-augmented-wave method [7, 8] and Perdew-Bruke-Ernzerhof exchange-correlation functional [9] for spin-polarized generalized gradient approximation as implemented in the VASP package[10]. The $3d$ and $4s$ electrons of Mn and $4s$, $4p$ electrons of As are treated as valence electrons and simple cubic supercells are used with at least 12 Å of vacuum region between two neighbouring clusters. The atomic moments are treated collinearly. To confirm the ground state geometry as well as the ground state magnetic configuration, we considered several geometrical and all possible spin multiplicities for a particular sized cluster.

*Pure manganese clusters:* Due to the lack of hybridization between the $3d$ and $4s$ electrons and due to the high enough (2.14 eV) $4s^2 3d^5 \rightarrow 4s^1 3d^6$ promotion energy, the binding energy of $Mn_x$ ($x$=2-20) clusters are small. Although it increases with the cluster size as $s - d$ hybridization increases with co-ordination number. But the improvement is not much and remains lowest compared to other $3d$ transition metal clusters. Figure 1 (left panel) shows kinks at $x = 7$, 13 and 19 due to their relatively large stability compared to their respective neighbours. This effect is even more prominent from the second difference in energy, $\triangle_2 E(x) = E(x+1) + E(x-1) - 2E(x)$ plots, where $E(x)$ is the total energy of the x-atom cluster (Figure 1, inset): peaks represent greater stability. This observed large stability is due to their "closed" geometric structure: $Mn_7$ is a pentagonal bipyramid which is basically the building blocks of larger sized clusters, $Mn_{13}$ is the first closed icosahedron and $Mn_{19}$ has double icosahedral structure[3].

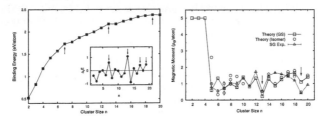

Figure 1: (Left panel) Plot of binding energy per atom as a function of cluster size. Inset represents the variation of the second difference in energy. (Right panel) Size dependent variation of magnetic moment and comparison is made with SG experiment. Magnetic moments of the isomers which lie close to the ground state are also shown.

Very small clusters up to $x = 4$ are found to be ferromagnetic with $5\mu_B$/atom magnetic moment and a transition to the ferrimagnetic ground state take place at $x = 5$ and remains the same up to clusters with 20 atoms. Calculated magnetic moments per atom is depicted in the Figure 1 (right panel) which shows an excellent agreement with the recent SG molecular beam experiment. For example, the predicted total magnetic moment of $5\mu_B$ is exactly the same with the SG experimental value, 0.72 ($\pm$ 0.42) $\mu_B$/atom[2]. Here it is important to note the large (58% of the measured value) uncertainty in the SG measurement[3]. We find two other isomers with different magnetic states, 7 and 3 $\mu_B$ of total moment, lie very close to the predicted ground state. Therefore, the possible presence of these two isomers along with the ground state in the SG molecular beam explains the observed large uncertainty in the measurement. One other interesting experimental feature is the "sudden" drop in the magnetic moment at $x = 13$ and 19. It is generally true that for almost all the clusters there exists isomers with different magnetic structure, which lie close to the ground state[3]. We find this is due to their closed geometric structure. The local magnetic moment can

Table 1: Total cluster magnetic moments $\mu_x$ of pure $\mathrm{Mn}_x$ and $\mathrm{Mn}_x\mathrm{As}$ clusters, corresponding to the ground state, for cluster size $x \leq 10$.

| $x$ | $\mu_x(\mu_B)$ | | $x$ | $\mu_x(\mu_B)$ | |
|---|---|---|---|---|---|
| | $\mathrm{Mn}_x$ | $\mathrm{Mn}_x\mathrm{As}$ | | $\mathrm{Mn}_x$ | $\mathrm{Mn}_x\mathrm{As}$ |
| 1 | 5 | 4 | 6 | 8 | 9 |
| 2 | 10 | 9 | 7 | 5 | 6 |
| 3 | 15 | 4 | 8 | 8 | 7 |
| 4 | 20 | 17 | 9 | 7 | 10 |
| 5 | 3 | 2 | 10 | 14 | 13 |

be defined as $\mathcal{M} = \int_0^R [\rho^\uparrow(\mathbf{r}) - \rho^\downarrow(\mathbf{r})]d\mathbf{r}$, where $\rho^\uparrow(\mathbf{r})(\rho^\downarrow(\mathbf{r}))$ is spin-up(spin-down) charge densities, respectively and $R$ is the radius of the sphere centering the atom. We find that moments are very small (nearly zero) for highly coordinated central atoms, whereas large ($\sim 3.5\ \mu_B$) for those of the surface atoms[3].

*As-doped manganese clusters:* It is interesting to see, what happens if one single arsenic atom is added to the $\mathrm{Mn}_x$ cluster? Binding energy enhances substantially due to the addition of a single arsenic through $s - p$ hybridization.Binding energy of $\mathrm{Mn}_x\mathrm{As}$ clusters are plotted in the Figure 2(a) with a comparison to $\mathrm{Mn}_x$ clusters. Now at this point we can define two kinds of energy gains: The energy gain in adding a single As to $\mathrm{Mn}_x$ cluster $\triangle^1 = E(\mathrm{Mn_xAs}) - E(\mathrm{Mn_x}) - E(\mathrm{As})$ and the energy gain in adding a Mn atom to $\mathrm{Mn}_{1-x}\mathrm{As}$ cluster $\triangle^2 = E(\mathrm{Mn_xAs}) - E(\mathrm{Mn_{1-x}}) - E(\mathrm{Mn})$. The $\triangle^1$ and $\triangle^2$ are plotted in the Figure 2(b). Large positive values of both of these energy gains indicate that the clustering of Mn around is favourable and therefore we argue in favour of possible formation of Mn clusters in the low temperature grown (Ga,Mn)As and (In,Mn)As samples[6]. All the ground states and their respective isomers are magnetically stable i.e. both the spin gaps are positive: the lowest unoccupied molecular orbital of the minority (majority) spin lies above the highest occupied molecular orbital of the majority (minority) spin. These two spin gaps, $\delta_1$ and $\delta_2$ [11], for $\mathrm{Mn_2As}$ (0.83 and 1.34 eV) and $\mathrm{Mn_4As}$ (0.89 and 1.14 eV) are the highest among all clusters. As $x$ increases, $\delta_1$ and $\delta_2$ decrease to a value 0.47 and 0.35 eV, respectively, for $\mathrm{Mn_{10}As}$[6].

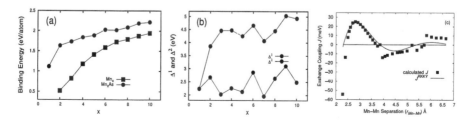

Figure 2: (a) Plots of binding energies for $\mathrm{Mn}_x$ and $\mathrm{Mn}_x\mathrm{As}$ clusters. (b) Two types of energy gains are plotted as a function of cluster size. (c) Plot of exchange coupling as a function of Mn-Mn spatial separation and compared with RKKY form with $k_F$=1.02 Å$^{-1}$.

Now we look into the magnetism in these $\mathrm{Mn}_x\mathrm{As}$ clusters. Generally, inclusion of a single As atom quenches the net moment of the $\mathrm{Mn}_x$ clusters and are given in the table 1. We found that the Mn-Mn coupling in $\mathrm{Mn_2As}$ and $\mathrm{Mn_4As}$ clusters remains ferromagnetic, whereas in all other clusters this coupling is ferrimagnetic. The local magnetic moments at Mn sites $\mathcal{M}_{\mathrm{Mn}}$ is large and basically arise from the localized Mn $3d$ electrons. A negative polarization is induced at As sites due to the strong $p - d$ interaction. For an example, in $Mn_2As$ cluster, $\mathcal{M}_{Mn}$'s are 3.72 $\mu_B$ each with $\mathcal{M}_{As} = -0.26$ $\mu_B$. Polarized neutron diffraction study found a local magnetic moment of $0.23 \pm 0.05$ $\mu_B$ at the As site for NiAs-type MnAs.

To determine the exchange interaction $J_{ij}$'s, the magnetic energy is mapped onto a classical Heisenberg Hamiltonian: $\mathcal{H} = \sum_{ij} J_{ij}(r) \mathbf{S}_i \cdot \mathbf{S}_j$. Variation of exchange coupling with $r_{Mn-Mn}$ for $Mn_2As$ is shown in the Figure 2(c) and compared with the simplest RKKY-type analytic form $J^{RKKY} \propto r^{-3}\cos(2k_F r)$, where $k_F$ is the fermi wave vector. We see that $J$ oscillates and dies down as $1/r^3$ — typical RKKY-type behaviour[6]. However, exchange coupling behaves quite anomalously as cluster size increases: generally, $J$ decreases with cluster size with an exception for $Mn_4As$, where it's value is anomalously high , as well as it has a strong environment dependency. The possible presence of $Mn_2As$ and $Mn_4As$ clusters in the (Ga,Mn)As might be responsible for an increase in the Curie temperature as particularly these two clusters have large Mn-Mn exchange coupling.

## Acknowledgment

This work has been done under the Indian Department of Science and Technology contract SR/S2/CMP-25/2003.

## References

[1] M. B. Knickelbein, Experimental Observation of Superparamagnetism in Manganese Clusters, *Phys. Rev. Lett.* **86**, 5255 (2001).

[2] M. B. Knickelbein, Magnetic ordering in manganese clusters, *Phys. Rev. B* **70**, 14424 (2004).

[3] Mukul Kabir, Abhijit Mookerjee and D. G. Kanhere, Structure, electronic properties and magnetic transition in manganese clusters, *To be published*.

[4] H. Ohno *et al.*, (Ga,Mn)As: A new diluted magnetic semiconductor based on GaAs, *Appl. Phys. Lett.* **69**, 363 (1996).

[5] H. Ohno *et al.*, Making Nonmagnetic Semiconductors Ferromagnetic, *Science* **281**, 951 (1998).

[6] Mukul Kabir, D. G. Kannhere and Abhijit Mookerjee, Large Magnetic Moments of Arsenic-Doped Mn Clusters and their Possible Relevance to Mn-Doped III-V Semiconductor Ferromagnetism, *http://xxx.lanl.gov/abs/physics/0503009*.

[7] P. E. Blöchl, Projector augmented-wave method, *Phys. Rev. B* **50**, 17953 (1994).

[8] G. Kresse and D. Joubert, From ultrasoft pseudopotentials to the projector augmented-wave method, *Phyys. Rev. B* **59**, 1758 (1999).

[9] John P. Perdew, Kieron Burke, and Matthias Ernzerhof, Generalized Gradient Approximation Made Simple, *Phys. Rev. Lett.* **77**, 3865 (1996).

[10] G. Kresse and J. Furthmüller, Efficient iterative schemes for ab initio total-energy calculations using a plane-wave basis set, *Phys. Rev. B* **54** 11169 (1996).

[11] $\delta_1 = - (\epsilon_{HOMO}^{majority} - \epsilon_{LUMO}^{monority})$ and $\delta_2 = - (\epsilon_{HOMO}^{minority} - \epsilon_{LUMO}^{majority})$

Brill Academic Publishers
P.O. Box 9000, 2300 PA Leiden,
The Netherlands

*Lecture Series on Computer*
*and Computational Sciences*
Volume 4, 2005, pp. 1022-1025

# Determination of Atomic Cluster Structure with Cluster Fusion Algorithm

O. I. Obolensky,[1] I. A. Solov'yov, A. V. Solov'yov,[2] W. Greiner

Frankfurt Institute for Advanced Studies,
Johann Wolfgang Goethe-University,
Max-von-Laue str.1, 60438 Frankfurt am Main, Germany

Received 28 June, 2005; accepted in revised form 22 July, 2005

*Abstract:* We present an efficient scheme of global optimization, called cluster fusion algorithm, which has proved its reliability and high efficiency in determination of the structure of various atomic clusters.

*Keywords:* Atomic clusters, global optimization methods, atomic cluster structures, metal clusters, Lennard-Jones clusters

We present an efficient scheme of global optimization, called cluster fusion algorithm. The scheme has been designed within the context of determination of the most stable cluster geometries and applicable for various types of clusters [1].

We have applied the scheme in search for global minima of the multidimensional potential energy surface of metal and noble gas clusters [1, 2, 3, 4, 5]. With this scheme we were able to determine the most stable cluster geometries for up to 150 atoms for noble gas clusters and for up to 22 atoms for metal (sodium, magnesium) clusters. While the global energy optimization for noble gas clusters is a relatively simple problem and optimization could easily be done for larger clusters, the calculations with metal clusters present a serious challenge and require significant computational resources. The principal difference between these two cases consists in the different nature of the atomic interactions in the clusters. For noble gas clusters the interactions have a pair-wise character and can be very well approximated by the Lennard-Jones type of potential, which has a simple analytical form. Metal clusters cannot be accurately described by a pair-wise potential because of delocalization of valence electrons. For these clusters one has to use *ab initio* quantum mechanical methods (Hartree-Fock-based many-body theory or density functional methods) in order to obtain a good agreement with the experimental data. For both types of calculations our algorithm has proven to be reliable and effective tool in multidimensional global optimization.

The proposed algorithm belongs to the class of genetic (also called evolutionary) global optimization methods [6, 7, 8, 9, 10]. In applications to clusters the genetic methods are based on the idea that the larger clusters evolve to low energy states by mutation and/or by mating smaller structures with low potential energy. The success of this procedure reflects the fact that in nature clusters in their ground (i.e., the minimum energy) states often emerge in the cluster fusion process. Numerous versions of the genetic strategies has been adapted to global energy optimization in atomic clusters [11, 12, 13, 14, 15]. One of the strategies within the generic approach assumes that the global energy minimum structure for $N$ atoms can be found on the basis of the global

---

[1]On leave from the A.F. Ioffe Institute, St. Petersburg, Russia. E-mail: o.obolensky@fias.uni-frankfurt.de
[2]On leave from the A.F. Ioffe Institute, St. Petersburg, Russia. E-mail: solovyov@fias.uni-frankfurt.de

energy minimum for $N-1$ atoms. This strategy is implemented, e.g., in the so-called seed growth method [12, 13]. In this method in order to get the most stable cluster of $N$ atoms one atom is added to the most stable cluster of $N-1$ atoms in stochastic manner, near the boundary of the cluster and then optimization of the structure with a given optimization method is performed.

Our method also uses the strategy of adding one atom to a cluster of size $N-1$. There are, however, two important improvement which allow for a much faster convergence. The first one is the fact that the atom is not added at a random place on the surface of the initial cluster. Rather, we use the deterministic approach and the new atom is added to the certain places of the cluster surface, such as the midpoint of a face. The second important feature of our method is that we add the new atom not only to the ground state isomer of size $N-1$, but also to the other, energetically less favorable, isomers. This insures that we do not miss sizes at which smooth evolution within one family of clusters (say, with the same type of lattice) is interrupted and the global energy minimum of the next cluster size lies within another cluster growth branch.

To illustrate that we show in Figure 1 the global energy minimum geometries for Lennard-Jones (LJ) clusters of sizes 28 through 33. It is seen that the clusters $LJ_{30}$ and $LJ_{31}$ belong to different branches of cluster growth (icosahedral and decahedral).

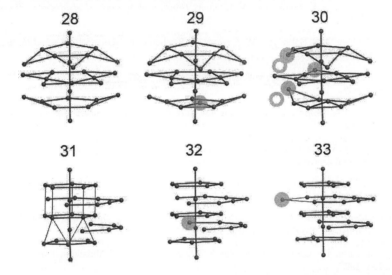

Figure 1: Global energy minimum geometries for Lennard-Jones clusters of sizes 28 through 33. The new atoms added to the cluster are marked by grey circles, while grey rings show the atoms removed.

Another important novel feature of our method is the global optimization technique. Our experience shows that, at least for deterministic addition of one atom to a stable cluster isomer, it is more effective to perform a three-dimensional optimization of coordinates of the newly added cluster rather than applying multidimensional optimization techniques which optimize the coordinates of all atoms in the cluster simultaneously. Therefore, we have employed the following algorithm for optimization of the cluster structure. At each step of the calculation we consider the motion of one atom only, the one, which is the subject to the action of the maximum force. Then we scan the potential energy surface along the direction of the force, keeping the coordinates of all other atoms fixed. At the point in which the kinetic energy of the selected atom is maximum, we set the absolute value of its velocity to zero. This point corresponds to the minimum of the potential valley in which the selected atom moves. When the selected atom is brought to the (local) minimum energy position, the next atom is selected to move and the procedure of the kinetic energy absorption repeats. The calculation stops when all the atoms are at equilibrium.

## Acknowledgment

This work is partially supported by the European Commision within the Network of Excellence project EXCELL, by INTAS under the grant 03-51-6170 and by the Russian Foundation for Basic Research under the grant 03-02-18294-a.

## References

[1] I.A. Solov'yov, A.V. Solov'yov, W. Greiner, *International Journal of Moderm Physics E* **13** 697-736(2004).

[2] I.A. Solov'yov, A.V. Solov'yov, W. Greiner, A. Koshelev, A. Shutovich *Physical Review Letters* **90** 053401-053404(2003).

[3] I.A. Solov'yov, A.V. Solov'yov and W. Greiner 2002 *Physical Review A* **65** 053203-053222(2002).

[4] A. Lyalin, I.A. Solov'yov, A.V. Solov'yov and W. Greiner *Physical Review A* **67** 063203-063216(2003).

[5] A. Lyalin, A.V. Solov'yov, C. Brechignac and W. Greiner, *Journal of Physics B: Atomic Molecular and Optical Physics* **38** L129-L135(2005).

[6] Z. Michalewicz, *Genetic Algorithms + Data Structures = Evolution Programs.* Springer, Berlin, Heidelberg, New York. (3rd Ed.), 1996.

[7] I.H. Osman, and J.P. Kelly (Eds): Meta-Heuristics: Theory and Applications, Kluwer Academic Publishers, Dordrecht, Boston, London, 1996.

[8] F. Glover and M. Laguna, *Tabu Search.* Kluwer Academic Publishers, Dordrecht, Boston, London, 1997.

[9] S. Voss, S. Martello, I.H. Osman and C. Roucairol (Eds): Meta-Heuristics: Advances and Trends in Local Search Paradigms for Optimization, Kluwer Academic Publishers, Dordrecht, Boston, London, 1999.

[10] J.D. Pinter, *Computational Global Optimization in Nonlinear Systems.* Lionheart Publishing, Inc., USA, 2001.

[11] D.E. Goldberg, *Genetic Algorithms in Search, Optimization, and Machine Learning.* Addison-Wesley, Reading, MA, 1989.

[12] J.A Niesse and H.R. Mayne, *Journal of Chemical Physics* **105** 4700(1996).

[13] S.K Gregurick, M.H. Alexander and B. Hartke, *Journal of Chemical Physics* **104** 2684(1996).

[14] D.M Deaven, N. Tit, J.R Morris and K.M. Ho, *Chemical Physics Letters* **256** 195(1996).

[15] D. Romero, C. Barron, S. Gomez, *Computational Physics Communications* **123** 87(1999).

Brill Academic Publishers
P.O. Box 9000, 2300 PA Leiden,
The Netherlands

*Lecture Series on Computer*
*and Computational Sciences*
Volume 4, 2005, pp. 1026-1031

# Structural and Electronic Properties of Metal Clusters

Michael Springborg,[1] Valeri G. Grigoryan[2] Yi Dong,[3] Denitsa Alamanova,[4]
Habib ur Rehman,[5] and Violina Tevekeliyska[6]

Physical and Theoretical Chemistry,
University of Saarland,
66123 Saarbrücken,
Germany

Received 10 July, 2005; accepted in revised form 10 July, 2005

*Abstract:* In order to study the properties of a whole series of metal clusters, for which
the structure has been optimized in an optimized way, we have used various parameterized methods for the calculation of the total energy for a given structure in combination
with different methods for determining the structure of the global total-energy minimum.
Specifically, we have used the embedded-atom method as well as a parameterized density-
functional method. The calculations give first of all the total energy and the structure at
the global total-energy minimum for clusters with up to 150 atoms. In order to extract information from these numbers we have constructed a number of different descriptors from
which stability, overall shape, radial distribution of the atoms, growth patterns, similarity
with finite pieces of the crystal, etc. can be analysed. We present these for clusters of Na,
Cu, Ni, and Ag atoms and in some cases we also compare the results from the different
types of calculations.

*Keywords:* Metal clusters, structure, stability, density-functional calculations, embedded-
atom calculations

*PACS:* 36.40.-c, 36.90.+f, 61.46.+w, 73.22.-f

## 1 Introduction

Clusters are intermediates between small, finite molecules and infinite, periodic crystals. The
finite size or, equivalently, the relative large number of surface atoms compared with the total
number of atoms makes it possible to tune their properties simply through the variation of the size.
Experimental studies on these systems are non-trivial due to difficulties in uniquely identifying the
size of a given clusters, in obtaining sufficiently large amounts of a specific (monodisperse) cluster
size, and in identifying the structure of a given cluster size (compared to a distribution of clusters
with the same size but different structures). On the other hand, theoretical studies are non-trivial
due to the fact that the clusters are large but finite together with the essentially exponential growth
of the number of meta-stable structures as a function of cluster size. Ultimately, this means that
theoretical studies have to consider either single clusters with an essentially known structure when

[1]Corresponding author. e-mail: m.springborg@mx.uni-saarland.de
[2]e-mail: vg.grigoryan@mx.uni-saarland.de
[3]e-mail: y.dong@mx.uni-saarland.de
[4]e-mail: deni@springborg.pc.uni-sb.de
[5]e-mail: haur001@rz.uni-saarland.de
[6]e-mail: vili@springborg.pc.uni-sb.de

using accurate methods or, alternatively, to use less accurate methods on a whole class of clusters and cluster sizes.

The purpose of this contribution is to present some of our results on metal clusters where we have used the second strategy, i.e., we have studied a whole class of clusters for which we have, for each single cluster size, optimized the structure using some unbiased approach. In some cases we have used two different parameterized methods, making it possible to address the accuracy of the two. The calculations yield first of all the total energy as a function of cluster size as well as the coordinates of the nuclei at the structure of the lowest total energy. In order to extract useful information from these data we have constructed various descriptors that we also shall present and discuss.

## 2 Total-Energy Methods

As one method we used the embedded-atom method (EAM) in the parameterization of Voter and Chen [1, 2, 3]. According to this method, the total energy for a system of $N$ atoms is written as a sum of atomic components, each being the sum of two terms. The first term is the energy that it costs to bring the atom of interest into the electron density provided by all other atoms, and the second term is a pair-potential term. Both terms are assumed depending only on the distances between the neighbouring atoms, and do therefore not include any directional dependence. Accordingly, the EAM emphasizes geometrical effects, whereas electronic effects are included only very indirectly.

Furthermore, we used the density-functional tight-binding method (DFTB) as developed by Seifert and coworkers [4, 5]. With this method, the binding energy is written as the difference in the orbital energies of the compound minus those of the isolated atoms augmented with pair potentials. In calculating the orbital energies we need the Hamilton matrix elements $\langle \chi_{m_1 n_1} | \hat{H} | \chi_{m_2 n_2} \rangle$ and the overlap matrix elements $\langle \chi_{m_1 n_1} | \chi_{m_2 n_2} \rangle$. Here, $\chi_{mn}$ is the $n$th atomic orbital of the $m$th atom. The Hamilton operator contains the kinetic-energy operator as well as the potential. The latter is approximated as a superposition of the potentials of the isolated atoms, $V(\vec{r}) = \sum_m V_m(|\vec{r} - \vec{R}_m|)$, and subsequently we assume that the matrix element $\langle \chi_{m_1 n_1} | V_m | \chi_{m_2 n_2} \rangle$ vanishes unless at least one of the atoms $m_1$ and $m_2$ equals $m$. Finally, the pair potentials $U_{m_1, m_2}$ are obtained by requiring that the total-energy curves from parameter-free density-functional calculations on the diatomics are accurately reproduced.

## 3 Structure Optimization

In the EAM calculations we optimized the structure using our own *Aufbau/Abbau* method [6, 7, 8]. The method is based on simulating experimental conditions, where clusters grow by adding atom by atom to a core. By repeating this process **very** many times and in parallel also removing atoms from larger clusters, we can identify the structures of the lowest total energy.

Finally, in the DFTB calculations we used two different approaches. In one approach the structures of the EAM calculations were used as input for a local relaxation, i.e., only the nearest local-total-energy minimum was identified. In another set of calculations, we optimized the structures using the so-called genetic algorithms [9, 10, 11]. Here, from a set of structures we generate new ones through cutting and pasting the original ones. Out of the total set of old and new clusters those with the lowest total energies are kept, and this process is repeated until the lowest total energy is unchanged for a large number of generations.

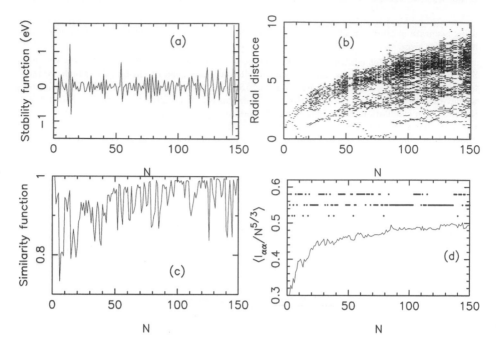

Figure 1: Properties of Au$_N$ clusters from the EAM calculations. The four panels show (a) the stability function, (b) the radial distribution of atoms, (c) the similarity function, and (d) the shape-analysis parameters, respectively. Lengths and energies are given in Å and eV, respectively. In (d) the upper rows show whether the clusters have an overall spherical shape (lowest row), an overall cigar-like shape (middle row), or an overall lens-like shape (upper row).

## 4 An Example: Au Clusters

Gold clusters constitute on of the most intensively studied classes of clusters [12]. Therefore, we shall in this short contribution present results for only this class of clusters.

Figs. 1, 2, and 3 show various properties from the EAM and DFTB calculations on Au$_N$ clusters. With the EAM method in combination with our *Aufbau/Abbau* method in optimizing the structure we studied clusters with $N$ up to 150, giving the results of Fig. 1. Moreover, we used these optimized structures as input for the DFTB calculations for $N$ up to 40, giving the results of Fig. 2. Finally, we also optimized the structures for $N$ up to 40 with the DFTB method in combination with the genetic algorithms, giving the results of Fig. 3.

In order to identify the particularly stable clusters we have introduces the stability function,

$$\Delta_2 E(N) = E_{\text{tot}}(N+1) + E_{\text{tot}}(N-1) - 2E_{\text{tot}}(N) \tag{1}$$

where $E_{\text{tot}}(K)$ is the total energy of the Au$_K$ system. $\Delta_2 E(N)$ has local maxima when Au$_N$ is particularly stable and it is shown in Figs. 1(a), 2(a), and 3(a) and shows clearly different behaviours from the three approaches. In particular, inclusion of electronic effects leads to much larger oscillations in $\Delta_2 E(N)$.

Information on the structure can be obtained from the radial distances of the atoms, defined

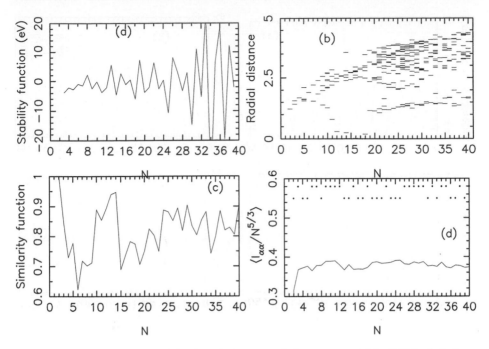

Figure 2: As Fig. 1, but from the DFTB calculations with the structures of the EAM calculations.

as follows. First, we define the center of the $Au_N$ cluster, $\vec{R}_0 = \frac{1}{N}\sum_{i=1}^{N}\vec{R}_i$ and, subsequently, we define for each atom its radial distance $r_i = |\vec{R}_i - \vec{R}_0|$. In Figs. 1(b), 2(b), and 3(b) we show the radial distances for all atoms and all cluster sizes. Each small line shows that at least one atom for the given value of $N$ has exactly that radial distance. The figure shows that somewhere around $N = 10$ a second shell of atoms is being built up, with a central atom for $N = 13$. Around $N = 54$, there are only few values of the radial distance, i.e., the clusters have a high symmetry. Around $N = 75$ we see that a third atomic shell is being formed.

We have earlier found [8] that it was useful to monitor the structural development of the isomer with the lowest total energy through the so-called similarity functions. We shall study how clusters grow and, in particular, if the cluster with $N$ atoms can be derived from the one with $N-1$ atoms simply by adding one extra atom. In order to quantify this relation we consider first the structure with the lowest total energy for the $(N-1)$-atom cluster. For this we calculate and sort all interatomic distances, $d_i$, $i = 1, 2, \cdots, \frac{N(N-1)}{2}$. Subsequently we consider each of the $N$ fragments of the $N$-cluster that can be obtained by removing one of the atoms and keeping the rest at their positions. For each of those we also calculate and sort all interatomic distances $d'_i$, and calculate, subsequently, $q = \left[\frac{2}{N(N-1)}\sum_{i=1}^{N(N-1)/2}(d_i - d'_i)^2\right]^{1/2}$. Among the $N$ different values of $q$ we choose the smallest one, $q_{min}$, and calculate the similarity function $S = \frac{1}{1+q_{min}/u_l}$ ($u_l = 1$ Å) which approaches 1 if the $Au_N$ cluster is very similar to the $Au_{N-1}$ cluster plus an extra atom. This function is shown in Figs. 1(c), 2(c), and 3(c). We see that the structural development is very irregular over the whole range of $N$ that we have considered here, with, however, some smaller

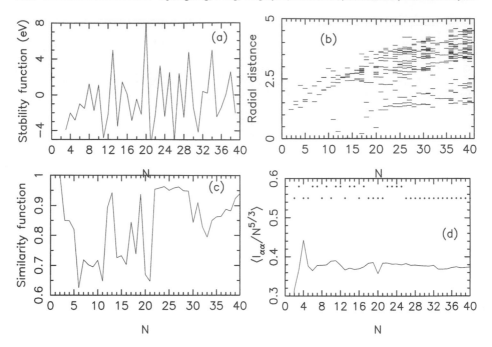

Figure 3: As Fig. 1, but from the DFTB calculations with the genetic-algorithm optimization of the structures.

intervals where $S$ is relatively large, for instance for $N$ slightly above 20.

Finally, we shall consider the overall shape of the clusters. As we showed in our earlier report on Ni clusters [8], it is convenient to study the $3 \times 3$ matrix containing the elements $I_{st} = \frac{1}{u_t^2} \sum_{n=1}^{N} (R_{n,s} - R_{0,s})(R_{n,t} - R_{0,t})$ with $s$ and $t$ being $x$, $y$, and $z$. The three eigenvalues of this matrix, $I_{\alpha\alpha}$, can be used in separating the clusters into being overall spherical (all eigenvalues are identical), more cigar-like shaped (one eigenvalue is large, the other two are small), or more lens-shaped (two large and one small eigenvalue). The average of the three eigenvalues, $\langle I_{\alpha\alpha} \rangle$, is a measure of the overall extension of the cluster. For a homogeneous sphere with $N$ atoms, the eigenvalues scale like $N^{5/3}$. Hence, we show in Figs. 1(d), 2(d), and 3(d) quantities related to $I_{\alpha\alpha}$ but scaled by $N^{-5/3}$. In the figures we also mark the overall shape of the clusters through the upper points with the lowest row meaning spherical, the middle row meaning cigar-shaped, and the upper row meaning lens-shaped clusters. Some clusters with an overall spherical shape can be recognized, that, simultaneously, are clusters of particularly high stability according to Figs. 1(a), 2(a)3(a).

## 5   Conclusions

This short presentation shows clearly that it is possible to address the properties of a whole class of clusters using theoretical methods without making severe approximations on the structure of the clusters but, on the other hand, only when using approximate theoretical methods. Therefore, comparing the results of different theoretical approaches, as done here, is important when estimating the accuracy of the results.

All methods yield similar results with, however, a clear difference depending on whether electronic effects are included directly or indirectly. The former case leads to much more pronounced peaks in the stability function than the latter case.

As we shall in more detail directly at the conference, interesting general trends for different metals but also material-specific differences can be identified from our studies on several different types of metal clusters.

## Acknowledgment

We gratefully acknowledge *Fonds der Chemischen Industrie* for very generous support. This work was supported by the SFB 277 of the University of Saarland and by the German Research Council (DFG) through project Sp439/14-1.

## References

[1] A. F. Voter and S. P. Chen, in *Characterization of Defects in Materials*, edited by R. W. Siegal, J. R. Weertman, and R. Sinclair, MRS Symposia Proceedings No. 82 (Materials Research Society, Pittsburgh, 1987), p. 175.

[2] A. Voter, Los Alamos Unclassified Technical Report No LA-UR 93-3901 (1993).

[3] A. F. Voter, in *Intermetallic Compounds*, edited by J. H. Westbrook and R. L. Fleischer (John Wiley and Sons, Ltd, 1995), Vol. 1, p. 77.

[4] G. Seifert and R. Schmidt, New J. Chem. **16**, 1145 (1992).

[5] G. Seifert, D. Porezag, and Th. Frauenheim, Int. J. Quant. Chem **58**, 185 (1996).

[6] V. G. Grigoryan and M. Springborg, Phys. Chem. Chem. Phys. **3**, 5125 (2001).

[7] V. G. Grigoryan and M. Springborg, Chem. Phys. Lett. **375**, 219 (2003).

[8] V. G. Grigoryan and M. Springborg, Phys. Rev. B **70**, 205415 (2004).

[9] J.-O. Joswig, M. Springborg, and G. Seifert, Phys. Chem. Chem. Phys. **3**, 5130 (2001).

[10] J.-O. Joswig and M. Springborg, Phys. Rev. B **68**, 085408 (2003).

[11] Y. Dong and M. Springborg, in *Proceedings of 3rd International Conference "Computational Modeling and Simulation of Materials"*, Ed. P. Vincenzini *et al.*, Techna Group Publishers, p. 167 (2004).

[12] F. Baletto and R. Ferrando, Rev. Mod. Phys. **77**, 371 (2005).

Brill Academic Publishers

P.O. Box 9000, 2300 PA Leiden

The Netherlands

*Lecture Series on Computer*

*and Computational Sciences*

Volume 4, 2005, pp. 1032-1033

# Symposium
# Computational Economics and Data Analysis

**Chih-Young Hung**

## Preface

Quantitative methodologies are inherently computational, often substantially so. Existing algorithms, however, do not always embody the best of computational techniques, either for efficiency stability, or conditioning. Likewise, environments for doing econometrics and data analysis are inherently computer- based and simulation-based. Computational economics, then, is a natural field that is ever ready to receive new efforts.

Six papers were selected from the seven submitted. These papers represent a diverse area of interests that reflect the wide application of computing on resolving managerial issues. Topics of these papers range from the application of fuzzy MCDM on corporate finance and government's industry policy to the modeling and simulation of various traffic flow problems.

I warmly welcome and invite you to join me in this session and kindly offer your views and comments.

**Profile of Organizer**

Dr Chih-Young Hung serves as the director and associate professor of the Institute of Management of Technology in the National Chiao Tung University, Taiwan. He earned his PhD in finance from the Texas Tech University, U.S.A. He received his bachelor's degree in 1979 from the Department of Electrical and Control Engineering at the National Chiao Tung University, Taiwan. He has published a series of papers in studying industry evolution from finance perspective, and also several professional books on financial related areas.

Current research interests of Dr. Hung include: financial strategies, business valuation, value-based management, technology valuation and venture capital. He is accredited as a valuation analyst by the National Association of Certified Valuation Analyst (NACVA), U.S.A. He also serves as the chairman of the Chinese Society of Business Valuation which promotes academic studies and professional practices of business valuation in the greater China area.

Brill Academic Publishers
P.O. Box 9000, 2300 PA Leiden
The Netherlands

Lecture Series on Computer
and Computational Sciences
Volume 4, 2005, pp. 1034-1038

# Multiple Objective Compromise Optimization Method to Analyze the Development Strategies of Nanotechnology in Taiwan

Wen-Han Tang[1], Jia-Ren Yang[1], Hua-Kai Chiou[2]

[1]Institute of Management of Technology, National Chiao Tung University. Taiwan, R.O.C
[2]Department of Statistics, National Defense Management College, NDU, Taiwan, R.O.C.

Received 3 July, 2005; accepted in revised form 29 July, 2005

*Abstract:* In this research, we proposed a multi-objective compromise optimization method integrating AHP with TOPSIS. This method was then applied to evaluate alternative development strategies of nanotechnology intend for enhancing the global competitive position of Taiwan. Firstly, fuzzy hierarchy analytic process was used to derive the relative weights of the considered criteria from participated evaluators, and then the synthetic value were aggregated by criteria weights with performance score which assigned by participant's subjective judgment. This proposed method assumed that the optimal solution had the shortest distance from the positive ideal solution as well as the farthest distance from the negative ideal solution.

*Keywords:* Multi-objective Compromise Optimization, Nanotechnology, Fuzzy Analytic Hierarchy Process

## 1.   Introduction

Given the bright prospect of applying nanotechnology in various industries, Taiwanese government has undertaken several measures to take on this opportunity. Facing with strong competitions from Mainland China and Korea as well as several other developed countries, Taiwan is eyeing on various strategies of integrating nanotechnologies with her traditional industries in an effort to maintain their current competitive position in the global manufacturing sectors. Given that the stake is high, it is crucial for the government to choose the most proper strategy for developing the nation's nanotechnologies.

## 2.   Fuzzy Analytic Hierarchy Process

AHP is a popular technique used to model subjective decision-making processes based on multiple attributes (Saaty, [9,10]). Here, we employ Buckley's method [2] to derive analytic hierarchy process by allowing fuzzy numbers for the pairwise comparisons, and find the fuzzy weights and fuzzy performance. Hwang and Yoon [3,4] summarized various methods employed to determine weights. Buckley [2] considered a fuzzy positive reciprocal matrix $\tilde{A} = [\tilde{a}_{ij}]$, extending the geometric mean technique to define the fuzzy geometric mean of each row $\tilde{r}_i$ and fuzzy weight $\tilde{w}_i$ corresponding to each criterion as follows:

$$\tilde{r}_i = (\tilde{a}_{i1} \otimes \tilde{a}_{i2} \otimes \cdots \otimes \tilde{a}_{im})^{1/m}; \; \tilde{w}_i = \tilde{r}_i \otimes (\tilde{r}_1 \oplus \tilde{r}_2 \oplus \cdots \oplus \tilde{r}_m)^{-1} \tag{1}$$

where $\oplus$ and $\otimes$ express the addition and multiplication operation of fuzzy numbers, respectively [14].

In addition, the evaluators were required to choose a performance value for each feasible strategies corresponding to considered criteria based on their subjective judgments. In this study, we utilize the fuzzy geometric mean method to determine the criteria weights and integrate the performance value through participated evaluators. In order to provide easy-to-follow process for MCDM problems, here we simplify the process to conduct the synthetic values employing simple additive weighted method to integrate the criteria weights with performance values for each strategy.

The result of the fuzzy synthetic decision reached by each strategy is a fuzzy number. Therefore, it needs to defuzzify the fuzzy numbers for getting the preferred order of the strategies. In previous works the procedure of defuzzification has been to locate the best nonfuzzy performance (BNP) value.

Methods of such defuzzified fuzzy ranking generally include three kinds of method, mean of maximal, center of area (COA), and $\alpha$-cut [12,14]. Utilizing the COA method to determine the BNP is a simple and practical method, the BNP value of the triangular fuzzy number $(LR_i, MR_i, UR_i)$ can be found by as follows:

$$BNP_i = [(UR_i - LR_i) + (MR_i - LR_i)]/3 + LR_i, \forall i \tag{2}$$

[1] Corresponding author, Ph.D. student, E-mail: yjr.mt93g@cc.nctu.edu.tw.

## 3.  Multi-Objective Compromise Optimization Method

With a given reference point, the MCDM problem can then be solved by locating the alternatives or decision points that are the closest to the reference point. Therefore, the problem becomes how to measure the distance to the reference point. Generally, the global criteria method measures the distance by using Minkowski's $L_p$-metric. The $L_p$ metric defines the distance between two points, $f_j$ and $f_j^*$ (the reference point), in $n$-dimensional space as:

$$L_p = \left\{ \sum_{j=1}^{k} (f_j^* - f_j)^p \right\}^{1/p}, \text{ where } p \geq 1 \tag{3}$$

Distance $L_p$ $(p = 1, 2, ..., \infty)$ are especially operationally important. Specifically, for $p = 1$, it implies equal weights for all these deviations, $L_1$ called the Manhattan distance; for $p = 2$, it implies that these deviations are weighted proportionately with the largest deviation having the largest weight, $L_2$ called the Euclidean distance. Ultimately, while $p = \infty$, it implies the largest deviation completely dominates the distance determination, $L_\infty$ usually called the Tchebycheff metric, is the shortest distance in the numerical sense [11]. That is,

$$L_\infty = \max_j \left\{ \left| f_j^* - f_j \right| \middle| j = 1, ..., k \right\} \tag{4}$$

Considering the incommensurability nature among objectives or criteria, Yu and Zeleny [13] normalized the distance family of Eq.(8) to remove the effects of the incommensurability by using the reference point. The distance family then becomes as follows:

$$L_p = \left\{ \sum_{j=1}^{k} \left( \frac{f_j^* - f_j}{f_j^*} \right)^p \right\}^{1/p}, \text{ where } p \geq 1 \tag{5}$$

Hwang and Yoon [3,4] proposed TOPSIS approach to solve multiple attribute decision making problems (MADM) using the concept of optimal compromise solution. Lai et al. [6] further extended the concept of TOPSIS for MADM problems and developed a methodology for solving multiple objective decision making (MODM) problems. In their study using the normalized distance family from Eq.(10) with the ideal solution being the reference point, the problem as Eq.(8) becomes how to solve the following auxiliary problem:

$$\min_{x \in X} L_p = \left\{ \sum_{j=1}^{k} \left( \frac{f_j^* - f_j}{f_j^* - f_j^-} \right)^p \right\}^{1/p} \text{ where } p = 1, 2, ..., \infty \tag{6}$$

where $f_j^*$ is the best value of corresponding $j$-th criterion, $f_j^-$ is the worst value of corresponding $j$-th criterion, and $f^* = \left( f_1^*, ..., f_j^*, ..., f_k^* \right)$ is the vector of positive ideal solution, $f^- = \left( f_1^-, ..., f_j^-, ..., f_k^- \right)$ is the vector of negative ideal solution, respectively. The value chosen for $p$ reflects the way of achieving a compromise by minimizing the weighted sum of the deviations of criteria from their respective reference points.

With the concept of optimal compromise solution, the best alternative or decisions of TOPSIS method are those that have the shortest distance from the positive ideal solution as well as have the farthest distance from the negative ideal solution.

The TOPSIS procedure consists six major steps that rely heavily on computation.

## 4.  Empirical Case

According to the report of technology commercial capability which issued by 3I Group [1] (Table 1), the results indicated that US, Japan, Germany and UK are on the top four in terms of nano-technology industrialization. In Asia, other than Japan, both Taiwan and Korea are among the top five in electronics and manufacturing sectors. Korea is superior to Taiwan in nano-electronics.

Table 1. Nanotechnology Commercial Capability

| Ranking | Pharmaceuticals & Medicines | Materials | Chemical Technologies | Electronics | Manufacturing |
|---|---|---|---|---|---|
| 1 | US (West Coast) | US (West Coast) | Germany | Japan | US (West Coast) |
| 2 | US (East Coast) | US (East Coast) | US (West Coast) | US (West Coast) | US (East Coast) |
| 3 | UK | Japan | US (East Coast) | US (East Coast) | Japan |
| 4 | Germany | Germany | UK | S. Korea | Germany |
| 5 | Switzerland | UK | Japan | Taiwan/ Germany | Taiwan/S. Korea |

In this study, we introduce fuzzy hierarchical analytic process with simple additive weighted method to derive the synthetic values with respect to criteria of development strategies for Taiwan's nanotechnology industry. Furthermore, we employ TOPSIS method to evaluate these proposed strategies.

The development strategies of nanotechnology in Taiwan can therefore be structured into three phases. The first phase is the overall objective. The second phase includes nine aspects of industrialized technology for analysis. The last phase includes eight strategic action plans in order to implement the evaluated aspects The hierarchical frame of the evaluation model is shown in Figure 1.

Figure 1. Hierarchical Frame of Evaluation Model

26 evaluators were involved in this study, nine from industry sector, six from governmental sector, six from academia and five from research institutes. Fuzzy AHP technique utilized to determine the relative weights of considered criteria from evaluators' subjective judgment. The weights from each criterion in the proposed strategy can therefore be determined. We aggregate their own subjective judgments by fuzzy geometric mean method and then conduct the final fuzzy weights, we further utilize fuzzy AHP to derive the aggregated weights of industrialized technology that from subjective judgment of evaluators and then rank the development priority of these technologies, computing the defuzzified BNP values followed that proposed by Opricovic and Tzeng [7] as shown in Table 2.

Table 2. Normalized BNP Values of Criteria

| | Industry Sector | | Government Sector | | Academic Sector | | Research Sector | | Aggregation | |
|---|---|---|---|---|---|---|---|---|---|---|
| | Weigh | Rank | Weigh | Rank | Weigh | Rank | Weigh | Rank | Weigh | Rank |
| T1 | 0.159 | 2 | 0.204 | 1 | 0.204 | 1 | 0.195 | 1 | 0.189 | 1 |
| T2 | 0.113 | 4 | 0.093 | 7 | 0.117 | 4 | 0.060 | 9 | 0.096 | 6 |
| T3 | 0.167 | 1 | 0.106 | 6 | 0.114 | 5 | 0.158 | 2 | 0.138 | 2 |
| T4 | 0.100 | 5 | 0.120 | 2 | 0.067 | 8 | 0.065 | 8 | 0.089 | 8 |
| T5 | 0.119 | 3 | 0.066 | 9 | 0.107 | 6 | 0.073 | 7 | 0.092 | 7 |
| T6 | 0.090 | 6 | 0.116 | 4 | 0.132 | 2 | 0.111 | 5 | 0.111 | 3 |
| T7 | 0.081 | 9 | 0.112 | 5 | 0.052 | 9 | 0.109 | 6 | 0.088 | 9 |
| T8 | 0.084 | 8 | 0.066 | 8 | 0.123 | 3 | 0.116 | 3 | 0.096 | 5 |
| T9 | 0.085 | 7 | 0.118 | 3 | 0.084 | 7 | 0.113 | 4 | 0.099 | 4 |

To determine the performance value of each strategy, evaluators can define their own individual range for the linguistic variables employed in this study according to their subjective judgments within a fuzzy scale. The nonfuzzy performance value of strategies with respect to evaluated criteria are shown in Table 3.

Table 3. BNP Values of Performance Values

|    | S1 | S2 | S3 | S4 | S5 | S6 | S7 | S8 |
|----|--------|--------|--------|--------|--------|--------|--------|--------|
| T1 | 84.256 | 72.551 | 73.333 | 76.435 | 58.710 | 57.306 | 72.180 | 79.512 |
| T2 | 71.856 | 74.384 | 69.519 | 75.791 | 61.588 | 50.872 | 63.031 | 74.872 |
| T3 | 74.713 | 70.235 | 71.706 | 75.460 | 62.195 | 49.996 | 65.544 | 77.543 |
| T4 | 73.312 | 69.786 | 66.578 | 74.904 | 61.093 | 50.850 | 65.271 | 75.273 |
| T5 | 70.035 | 70.588 | 66.734 | 75.129 | 60.388 | 50.009 | 60.736 | 70.155 |
| T6 | 71.910 | 73.533 | 66.482 | 76.486 | 61.617 | 50.832 | 63.129 | 72.533 |
| T7 | 67.319 | 64.571 | 63.821 | 70.361 | 58.176 | 51.748 | 65.189 | 66.413 |
| T8 | 71.484 | 68.737 | 74.092 | 72.768 | 62.253 | 56.919 | 71.113 | 78.195 |
| T9 | 77.707 | 73.914 | 70.887 | 74.619 | 67.289 | 53.867 | 70.133 | 71.813 |

The next step is to integrate the synthetic utility for each strategy. Here we employ simple additive weighted method to integrate the nonfuzzy criteria weights with nonfuzzy performance values for each strategy, the results as shown in Table 4.

Table 4. Nonfuzzy Synthetic Values of Strategies

|    | S1 | S2 | S3 | S4 | S5 | S6 | S7 | S8 |
|----|--------|--------|--------|--------|--------|--------|--------|--------|
| T1 | 15.738 | 13.552 | 13.698 | 14.277 | 10.966 | 10.704 | 13.483 | 14.852 |
| T2 | 7.115  | 7.365  | 6.884  | 7.505  | 6.098  | 5.037  | 6.241  | 7.414  |
| T3 | 10.389 | 9.767  | 9.971  | 10.493 | 8.649  | 6.952  | 9.114  | 10.783 |
| T4 | 6.645  | 6.325  | 6.034  | 6.789  | 5.537  | 4.609  | 5.916  | 6.822  |
| T5 | 6.661  | 6.714  | 6.347  | 7.146  | 5.743  | 4.756  | 5.777  | 6.672  |
| T6 | 7.882  | 8.060  | 7.287  | 8.383  | 6.754  | 5.572  | 6.919  | 7.950  |
| T7 | 5.863  | 5.624  | 5.559  | 6.128  | 5.067  | 4.507  | 5.678  | 5.784  |
| T8 | 6.778  | 6.518  | 7.026  | 6.900  | 5.903  | 5.397  | 6.743  | 7.415  |
| T9 | 7.605  | 7.234  | 6.937  | 7.303  | 6.585  | 5.272  | 6.863  | 7.028  |

Following the procedure as mentioned in Section 3, we derive distance from the positive ideal solution $(S^+)$ and from the negative ideal solution $(S^-)$ of each strategy. Finally, we further compute the relative closeness as Eq.(14) and then assign the preferred order to each strategy according to their closeness index (Table 5).

Table 5. Preferred Order Derived by TOPSIS

|    | $S^+$ | $S^-$ | Closeness | Rank |
|----|---------|---------|-----------|------|
| S1 | 0.00315 | 0.01614 | 0.83664   | 3    |
| S2 | 0.00460 | 0.01485 | 0.76328   | 4    |
| S3 | 0.00595 | 0.01301 | 0.68635   | 5    |
| S4 | 0.00199 | 0.01781 | 0.89940   | 1    |
| S5 | 0.01142 | 0.00773 | 0.40361   | 7    |
| S6 | 0.01878 | 0.00000 | 0.00000   | 8    |
| S7 | 0.00834 | 0.01082 | 0.56467   | 6    |
| S8 | 0.00282 | 0.01674 | 0.85595   | 2    |

From Table 5, we conduct the preferred order of development strategies of industrialized nanotechnology of Taiwan as follows: Emphasize on Intellectual Property Rights (S4) ≻ Information Supply for Opportunity Development (S8) ≻ R&D Supporting (S1) ≻ Professional Incubation (S2) ≻ Tax Incentives (S3) ≻ Finance (S7) ≻ Establish Science Zone (S5) ≻ Government Procurements (S6).

## References

[1]   3I Company *Nanotechnology-size matters: Building a successful nanotechnology company*. UK. 2002.

[2]   J.J. Buckley, "Fuzzy Hierarchical Analysis," *Fuzzy Sets and Systems*, Vol. 17, No. 3, pp. 233-247, 1985.

[3]   S.J. Chen and C.L. Hwang, *Fuzzy Multiple Attribute Decision Making: Methods and Applications*. Springer-Verlag, Berlin, 1992..

[4]   C.L. Hwang, and K. Yoon, *Multiple Attribute Decision Making*, Lecture Notes in Economics and Mathematical Systems, Springer-Verlag, Berlin, 1981.

[5]   R.L. Keeney and H. Raiffa, *Decisions with Multiple Objectives: Preferences and Value Tradeoffs*, New York: John Wiley and Sons, 1976.

[6]   Y.J. Lai, T.Y. Liu, and C.L. Hwang, "TOPSIS for MODM," *European Journal of Operational Research*, 76(3), 486-500, 1994.

[7]   S. Opricovic and G.H. Tzeng, "Defuzzification for a fuzzy multicriteria decision model," *International Journal of Uncertainty, Fuzziness and Knowledge-Based Systems,* Vol.11, No.5, pp.635-652, 2003

[8]   M.C. Roco, *NSTC Subcommittee on Nanoscale Science.* Engineering and Technology Publications, National Science Foundation, USA, 2001.

[9]   T.L. Saaty, "A Scaling Method for Priorities in Hierarchical Structures," *Journal of Mathematical Psychology*, Vol. 15, No. 3, pp. 234-281, 1977.

[10]   T.L. Saaty, *Analytic Hierarchy Process*, McGraw-Hill, New York, 1980.

[11]   R.E. Steuer, *Multiple Criteria Optimization Theory, Computation, and Applications*, Wiley, New York, 1986.

[12]   S.H. Tsaur, G.H. Tzeng and G.C. Wang, "The Application of AHP and Fuzzy MCDM on the Evaluation Study of Tourist Risk," *Annals of Tourism Research*, Vol. 24, No. 4, pp. 796-812, 1997.

[13]   P.L. Yu and M. Zeleny, "The Set of All Nondominated Solutions in Linear Cases and A Multicriteria Simplex Method," *Journal of Mathematical Analysis and Applications*, Vol.49, No.3, pp.430-448, 1975.

[14]   L.A. Zadeh, "Fuzzy Sets," *Information and Control*, Vol. 8, No. 2, pp. 338-353, 1965.

Brill Academic Publishers
P.O. Box 9000, 2300 PA Leiden
The Netherlands

*Lecture Series on Computer
and Computational Sciences*
Volume 4, 2005, pp. 1039-1043

# Using Fuzzy MCDM Approach on A/R Collection Instruments selection in Taiwan's Hsinchu Science Park

Chih-Young Hung[1*], Yi-Hui Chiang [1,2]

[1] Institute of Management of Technology, National Chiao-Tung University, Taiwan
[2] Department of International Trade, Ta-Hwa Institute of Technology, Taiwan

Received 8 July, 2005; accepted in revised form 8 July, 2005

*Abstract:* This paper presents a Fuzzy Multi-Criteria Decision Making methodology (MCDM) model for selecting of A/R collection instruments. The Fuzzy Analytic Hierarchy Process (FAHP) method is used to determine the weightings for evaluation criteria among decision makers, including the managers of the six categories of companies in the Hsinchu Science Park (HSP). The subjectivity and vagueness in the alternatives selection process is dealt with by using fuzzy numbers for linguistic terms. Incorporated the decision makers' attitude towards preference, a crisp overall performance value is obtained for each alternatives based on the concept of FMCDM. This FAHP approach can provide managers with a suitable way to determine the optimum A/R collections instruments.

*Keywords:* Account Receivables Collection Instrument, High-tech Industry, HSP, FAHP, FMCDM

*Mathematics Subject Classification:* AMS-MOS
*AMS-MOS:* 03E72 Fuzzy set theory

## 1. Introduction

Export-oriented industries in Taiwan have contributed greatly to Taiwan's economy and are the main driving force behind Taiwan's "economic miracle". One major contributor to rising exports in Taiwan is the high-tech sector. Established in 1980, the Hsinchu Science Park (HSP) was the first science park of its kind in Taiwan. With its mission being to establish a high quality R&D base for the high-tech industry, the HSP has continuously expanded its infrastructure and facilities with total government investment to date of 1,121 million US dollars Following years of effort, the HSP has become the most successful science park in Taiwan and thus has attracted a considerable attention worldwide. Additionally, the HSP has evolved to become a representative area in Taiwan's high-tech industry.

Generally, four alternative instruments exist for collecting account receivables (A/R). Each instrument has its own merits. For firms doing business internationally, commercial letter of credit (L/C) have been the preferred method of payment. However, according to Central Bank data, the percentage of usage of L/C among exporting firms has declined from 80% in 1980 to11.4% in 2004. At the same time, the percentage of remittance, including Telegraphic Transfer (T/T) and Open Account (O/A), has increased from 10% to 80%. These dramatic changes deserve our concern.

A decline in L/C volume doesn't only means that the collection behavior of exporting firms is changing, but also that the profit potential of the banks involved is declining. Since evaluation of A/R collection instruments is an important task for managers of financial and promotion departments, performance estimate criteria are necessary for an organization's internal audits. Companies in HSP are representative high-tech industries of Taiwan. In recent years, these companies have been facing competitive price-cutting and lower margins, so they have a great need to solve the problem of account receivables (A/R) collection for the purpose of corporate performance and global competitiveness. Therefore, providing a decision-making (DM) model to assist in choosing the ideal collecting instruments should be an important requirement of the industries.

Fuzzy Analytic Hierarchy Process (FAHP) or Fuzzy Multi Criteria Decision Making (FMCDM) analysis has been widely used to deal with DM problems involving multiple criteria evaluation/selection of alternatives. The practical applications reported in the literature [3,4,6-9] have shown advantages in handling unquantifiable/qualitative criteria and obtained quite reliable result. This

---

* Corresponding author, Director, Associate Professor, E-mail: cyhung@cc.nctu.edu.tw

study combines fuzzy sets theory [1,2,10,11] and Analytic Hierarchy Process, in order to help companies in Taiwan HSP select the most appropriate alternative for A/R collecting instruments.

## 2. Alternative A/R collection instruments evaluation model

In general, a seller/exporter and buyer/importer can choose from four methods of payment in international transactions. These options include prepayment, payment by L/C, payment by documentary collections (including D/A and D/P), and O/A. As for collection, the instruments adopted by sellers can be classified into four categories: T/T Advance, L/C, D/A, D/P and O/A. Each type of A/R collection instrument has its own respective advantages and disadvantages, which should be evaluated on an open and fair basis.

In our study, a systematic Fuzzy AHP model for multi-criteria A/R collection instruments strategies is proposed. The steps can be summarized as follows:

Step 1: Form a committee of decision-makers and select the evaluation criteria.

Step 2: Develop a hierarchical structure.

Step 3: Use an AHP method to obtain the subjective criteria weight for each level.

Step 4: Tabulate the linguistic evaluation value for A/R collection instruments, then transform them into triangular fuzzy numbers.

Step 5: Calculate and transform the ratio of each criterion into triangular fuzzy numbers and compute the centroid.

Step 6: Utilize the subjective weight obtained by Steps 3 to determine the integrated weight of the criteria presented above the alternative level.

Step 7: Synthesize the scores and obtain the priorities of A/R collection instruments.

Criteria used to evaluate the A/R collection instruments are from literature reviews and consultation with experts, including one professor in international trade, one professor in financial, two manager of commercial bank and five experienced staff of the HSP Works Bureau. These individuals were asked to rate the accuracy, adequacy and relevance of the criteria and dimensions. Synthesizing the literature review the expert opinions provided the basis for developing the hierarchical structure used in this study. The factors affecting the performance of A/R collection instruments can be classified into three goals and twelve criteria. A multi-criteria decision-making model with three first-tier goals, each with four second-tier evaluation criteria was constructed as illustrated in Figure 1.

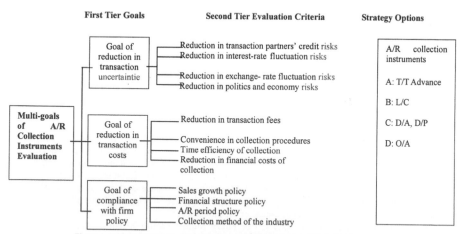

Figure 1. A Multi-Criteria Decision-Making model for Evaluating A/R Collection Instruments

## 3. Empirical study of A/R collection instruments selection

The survey of this study intends to be comprehensive. It covers most perspectives by surveying almost every field of expert in A/R collection. A total of 115 responses were collected of which 12 were discarded because their answers were incomplete. In total, there were 103 complete and acceptable survey responses, which account for a 38.1% return rate. Among the 103 surveyed respondents, 21 were from integrated circuits manufacturers; 25 were from computer & peripherals manufacturers; 12 were from telecommunications manufacturers; 21 were from Opto-electronics manufacturers; 8 were from automation manufacturers, and the remaining 16 surveys were from biotechnology manufacturers and others. In addition to assessing the preferred A/R collection instrument for each one of the six survey groups, the preferences of all of the surveyed respondents as a whole were also studied.

Based on the responses from the survey, a Fuzzy MCDM model was constructed. The model was then used to derive the weighting factors of the three first-tier goals and the twelve second-tier evaluation criteria for four A/R collection instruments. The evaluation results are listed in the Table 1.

The aggregate Consistency Index (C.I.) of AHP is 0.042. The preferential relationship of AHP is found upon the supposed condition of transitivity existence, so it has to satisfy the transitive law of outranking relationship. Consistency test can be used on the judgments of the evaluation decision-makers as well as on the entire hierarchical structure. C.I.= 0.042 means the entire hierarchical structure is acceptable.

The utility scores of the four collection instruments by these groups are summarized in Table 2 and Analysis of variance (ANOVA) is in Table 3. When using the 0.05 level of significance to test for differences among the A/R collection instruments, F =38.36824 > $F_U$=4.458968, or the P-value=7.94E-05 < 0.05. We can conclude that there is evidence of a difference in the average groups among the different A/R collection instruments. F =3.642832 < $F_U$=3.837854, or the P-value=0.056531 > 0.05. We can conclude that there is no evidence of a difference in the average A/R collection instruments among the different groups. It means that four A/R collection instruments have profound impacts on six groups of High-Tech companies, However, there is no significance difference on the selection of instruments.

Table 1. Weighting factors by the Group of All Survey Respondents

| Goals/Evaluation Criteria | Weighting Factors of Goals | Weighting Factors of Evaluation Criteria Within a Goal | Weighting Factors of Evaluation Criteria Across Goals |
|---|---|---|---|
| **Goal of reduction in transaction risks** | **0.344** | | |
| Reduction in transaction partners' credit risks | | 0.404 (1) | 0.139 (2) |
| Reduction in interest-rate fluctuation risks | | 0.201 (3) | 0.069 (8) |
| Reduction in exchange- rate fluctuation risks | | 0.180 (4) | 0.062 (9) |
| Reduction in political and economic risks | | 0.218 (2) | 0.075 (7) |
| **Goal of reduction in transaction costs** | **0.218** | | |
| Reduction of in transaction fees | | 0.211 (3) | 0.046 (11) |
| Convenience of collection procedures | | 0.156 (4) | 0.034 (12) |
| Time efficiency of collection | | 0.349 (1) | 0.076 (6) |
| Reduction in financial costs of collection | | 0.284 (2) | 0.062 (9) |
| **Goal of compliance with firm policy** | **0.438** | | |
| Sales growth policy | | 0.331 (1) | 0.145 (1) |
| Financial structure policy | | 0.240 (2) | 0.105 (3) |
| A/R period policy | | 0.228 (3) | 0.100 (4) |
| Collection method of the industry | | 0.201 (4) | 0.088 (5) |

Table 2. Utility Scores of Four A/R Collection Instruments by Six Survey Groups

| Survey Groups/ A/R Collection Instruments | T/T Advance | L/C | D/A, D/P | O/A |
|---|---|---|---|---|
| Integrated Circuits Manufacturers Group | 62.32 (1) | 43.89 (3) | 24.66 (4) | 45.36 (2) |
| Computer & Peripherals Manufacturers Group | 66.31 (1) | 48.02 (2) | 35.20 (4) | 45.75 (3) |
| Telecommunications Manufacturers Group | 68.81 (1) | 46.77 (2) | 33.39 (4) | 41.88 (3) |
| Electro-Optical manufacturers Group | 65.79 (1) | 44.13 (3) | 35.50 (4) | 49.92 (2) |
| Automation Manufacturers Group | 66.73 (1) | 50.35 (3) | 40.81 (4) | 51.74 (2) |
| Biotechnology and other Manufacturers Groups | 68.55 (1) | 50.35 (3) | 38.20 (4) | 50.41 (2) |
| Group of All Survey Respondents | 66.06 (1) | 46.58 (3) | 33.80 (4) | 47.26 (2) |

Note: The numbers in the parentheses are rankings within the same survey group.

Table 3. ANOVA: Two-Factor Without Replication

| Source of Variation | SS | df | MS | F | P-value | F crit |
|---|---|---|---|---|---|---|
| Rows (Survey Groups) | 77.51524 | 4 | 19.37881 | 3.64283 | 0.056531 | 3.837854 |
| Columns (Collection Instruments) | 408.2159 | 2 | 204.1079 | 38.36824 | 7.94E-05 | 4.458968 |
| Error | 42.55768 | 8 | 5.31971 | | | |
| Total | 528.2888 | 14 | | | | |

## 4. Discussions and conclusion

According to the results, there exists a consensus of the ranking preference among four of the six groups, namely the IC, the telecom, the Opto-electronics and the Biotech, regarding the priorities of the three first-tier goals. The rank preferred by these groups is as follows: (1) the goal of compliance with firm policy; (2) the goal of reduction in transaction risks; and (3) the goal of reduction in transaction costs. In most cases, the importance of compliance with firm policy is much higher than that of the other two goals. "Convenience of collection procedures" under the category of "reduction in transaction costs" is considered the least important by almost all groups. The weighting factor is as low as 0.034 by all surveyed respondents. All survey groups except one that ranked it 11th, ranked it the 12th among twelve evaluation criteria. The results clearly indicate that there is a consensus among firms in the HSIP of Taiwan that the "convenience of collection procedures" is not as relevant an objective for A/R collection instruments.

On the other hand, all groups most favor the T/T Advance and least favor the D/A and D/P to be their choice for A/R collection instruments. Choosing the T/T Advance as their most favored instrument is an indication that all survey groups tried to avoid transaction risks and costs. In other words, the time efficiency of collection is these exporters' main concern. The reasons that D/A and D/P were ranked last could be that the reduction in transaction risks and costs provided by these instruments is very limited, and the sales growth opportunity provided is smaller than that of O/A. Three groups, namely the Computer & Peripherals, the Telecom, and the Biotech, stressed the importance of reduction of trading partners' credit risks in the selection of an A/R collection instrument. They favored the L/C because it is supposed to be able to reduce the credit risks of the trading partners. The IC group and the Opto-electronics group are more concerned about the sales growth policy of their company. Thus, they favored the O/A because it is believed to be helpful to a company's sales growth.

The selection of A/R collection instruments really has profound impact on not only high-tech sellers/exporters and buyers/importers, but also the revenue of the banking industry. With L/C and documentary collection volumes declining, some banks have devoted time to developing new activities—factoring. For example, Chinatrust Commercial Bank, Taishin Commercial Bank, Bank SinoPac and The International Commercial Bank of China (ICBC) etc. They are not only joining the Factors Chain International (FCI), but also adjusting their credit policy and convincing their customs of accepting the new concept. However, the skills and the risks are higher than before. Only few of the banking in Taiwan have the desire to involve at present.

In this paper, we demonstrated the applicability of a multi-criteria decision-making model in the selection of a firm's A/R collection instrument. The paper also revealed the concerns and preferences of those high-tech and export-oriented firms in the HSP of Taiwan. The results of this study might be of interest to authorities in the banking sector or government agencies.

## Acknowledgments

The authors wish to thank the anonymous referees for their careful reading of the manuscript and their fruitful comments and recommendations

## References

[1] Bellman, R. E. and Zadeh, L. A. Decision-making in a fuzzy environment. *Management Science*, 17(4), 141-146. (1970).

[2] Buckley, J. J. Ranking alternatives using fuzzy numbers. *Fuzzy Sets and Systems*, 15(1), 21-31. (1985).

[3] Chou, T.Y., Luang, G. S. Application of a fuzzy multi-criteria decision-making model for shipping company performance evaluation, *Maritime Policy & Management*, 28(4), 375-392. (2001).

[4] Hsieh, T. Y., Lu, S. T., Tzeng, G. H. Fuzzy MCDM approach for planning and design tenders selection in public office buildings, *International Journal of Project Management*, 22, 573-584. (2004).

[5] Saaty, T. L.. A scaling method for priorities in hierarchical structures. *Journal of Mathematical Psychology*, 15(2), 234-281. (1977)

[6] Sheng-Hshiung Tsaur, Te-Yi Chang and Chang-Hua Yen The evaluation of airline service quality by fuzzy MCDM. *Tourism Management, 23*, 107-115. (2002).

[7] Tang, M.T., Tzeng, G.H., Wang SW. A hierarchy fuzzy MCDM method for studying electronic marketing strategies in the information service industry. *Journal of International Information Management*, 8(1), 1-22. (1999).

[8] Tzeng, G. H., Teng, M.H., Chen, J.J., Serafim Opricovic  Multicriteria selection for a restaurant location in Taipei. *Hospitality Management*, 21, 171-187. (2002).

[9] Wu, F. G., Lee, Y. J., Lin, M. C. Using the fuzzy analytic hierarchy process on optimum spatial allocation. International, *Journal of Industrial Ergonomics*, 33, 553-569. (2004).

[10] Zhau, R. and Goving, R. Algebraic characteristics of extended fuzzy numbers. *Information Science*, 54(1), 103-130. (1991).

[11] Zadeh, L. A. Fuzzy set. *Information and Control*, 8(2): 338-353. (1965).

Brill Academic Publishers
P.O. Box 9000, 2300 PA Leiden
The Netherlands

*Lecture Series on Computer*
*and Computational Sciences*
Volume 4, 2005, pp. 1044-1047

# Microscopic Analysis of Desired-Speed Car-Following Phenomena

Hsun-Jung Cho [1], Yuh-Ting Wu

Department of Transportation Technology and Management,
National Chiao Tung University,
Hsinchu, Taiwan, R.O.C.

Received 15 July, 2005; accepted in revised form 12 August, 2005

*Abstract:* A car-following model is proposed in this paper. The proposed model presents a new concept to describe car-following process. Repulsion and thrust act on a following vehicle, and the following vehicle chooses an appropriate speed under the repulsion and the thrust. Illustrative simulations are presented. The simulation result indicates that the proposed model is reasonable, and it can reflect some traffic flow phenomena, such as local stability, asymptotically stability, closing-in, and shying-away. It is worth mentioning that the proposed model employ driver desired speed as model variable so that the proposed model can explain why drivers keep different speeds or different gaps under the same condition. The proposed microscopic model can be the research basis of macroscopic traffic flow model.
*Keywords:* Desired Speed Car-following Model; Traffic Flow; Microscopic Traffic Simulation,

*Mathematics Subject Classification:* 90B06, 90B20, 60K30

## 1. Introduction

Microscopic traffic flow includes car-following and lane-changing. Car-following is the research topic of this study. There are at least four types of car-following models: safe-distance models, stimulus-response model, psycho-physical spacing model, and fuzzy models. The safe-distance models, such as Pipes [1] and PITT models [2], describe that a following car keeps a safe gap between the lead car and the following car. Stimulus-response model [3] expresses the concept that a driver of a vehicle responds to a given stimulus according to the stimulus and sensitivity. Fuzzy models [4] consist of a set of fuzzy inference rules that relate to a particular driving environment. Psycho-physical spacing model [5] divides the car-following process into some behavior zones. Each zone has its behavior rules. It seems that the deficiencies of psycho-physical spacing model and fuzzy models are less. They can describe most traffic phenomena since they have different behavior rules for different traffic conditions. The intelligent driver model [6] has only a few intuitive parameters with realistic values; it reproduces a realistic collective dynamics, and also leads to the plausible microscopic acceleration and deceleration behavior of single drivers. Those simple models (such as: stimulus-response model) can't describe some traffic phenomena (such as: closing-in and shying-away [4]). The models which have different rules for different conditions describe the traffic flow better, but their computing is more complicated.

Microscopic traffic flow models consider the behavior of an individual driver. Under the same condition, some drivers may keep longer distances at a lower speed, and some drivers may keep shorter distances at a higher speed. A microscopic traffic flow model should reflect the traffic phenomenon. Some models mentioned above haven't taken the driver difference into account. Some models employ sensitivity or aggressiveness index to describe the driver difference, but the index can't be measured directly. The intelligent driver model [8] employ driver's desired speed as model variable, but it hasn't discussed the influence of different driver's desired speed.

Microscopic traffic flow model needs a lot of computing time. A simple model needs less computing time and can be extended to a macroscopic traffic flow model easier. The purpose of this research is to develop a simple car-following model that has one or few functions. The proposed model can overcome the above-mentioned shortcomings of those simple models, and it can reflect the difference between different drivers.

---

[1] Corresponding author, Professor, E-mail: hjcho@cc.nctu.edu.tw.

## 2. The Proposed Desired Speed Car-following Model

If there is no lead vehicle, the following vehicle will run at its desired speed. If there is a lead vehicle, the following vehicle may slow down its speed so that it cannot run at its desired speed. The proposed model assumes that repulsion and thrust act on the following vehicle, and the following vehicle will decide its appropriate velocity under the thrust and the repulsion. The concept of the proposed model is shown as Figure 1. The model assumptions are listed below:

1. *Thrust*: Each vehicle has its own desired speed, and the desired speed is the thrust. The thrust makes the following vehicle move forward. If there is no lead vehicle, the following vehicle will run at its desired speed.
2. *Repulsion:* If there is a lead vehicle, the following vehicle should run at a lower speed. Since the lead vehicle makes the following vehicle unable to run at its desired speed, the lead vehicle is considered to be giving repulsion to the following vehicle. The repulsion is related to the speed of lead vehicle, the speed of following vehicle, and the distance between the two vehicles.
3. *Velocity decision:* The following vehicle will decide its appropriate velocity under the thrust and repulsion. The appropriate velocity equals thrust minus repulsion.
4. *Aggressiveness:* The proposed model assumes that the higher the desired speed is, the more aggressive the driver is. The driver with the higher desired speed will keep a higher speed or a shorter gap under the same condition.
5. *Safety:* Since some drivers' behaviors are not safe, the proposed model assumes that drivers do not take the safe-distance into consideration. They only consider the standstill distance headway.

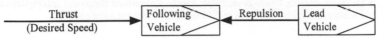

Figure 1: An Illustration of the Car-following Concept.

The proposed model is shown as Equation 1 to Equation 4. If both the lead vehicle and the following vehicle are running, the following vehicle will choose an appropriate speed under the thrust and repulsion, and the appropriate speed equals thrust minus repulsion (shown as Equation 1). If the speed of the lead vehicle is zero (shown as Equation 2), the following vehicle will decelerate its speed so that it can stop before collision. If the lead vehicle is moving and the following vehicle stops, the following vehicle can start to move at next time step and the acceleration is its desired start acceleration (shown as Equation 3). Finally, if both the lead vehicle and the following vehicle stop, the following vehicle will stop at next time step (shown as Equation 4).

$$\tilde{V}_{n,t+1} = v_{n,d}\left(1 - e^{-\lambda\frac{(V_{n-1,t})^{\alpha}}{(V_{n,t})^{\beta}}\left(\frac{x_{n-1,t}-x_{n,t}-S_n}{L}\right)^{\gamma}}\right), \qquad for\ V_{n-1,t} \neq 0\ \&\ V_{n,t} \neq 0 \tag{1}$$

$$\tilde{V}_{n,t+1} = V_{n,t} - \frac{(V_{n,t})^2}{2(x_{n-1,t} - x_{n,t} - S_n)}\Delta t, \qquad for\ V_{n-1,t} = 0\ \&\ V_{n,t} \neq 0 \tag{2}$$

$$\tilde{V}_{n,t+1} = a_{n,d}\Delta t, \qquad for\ V_{n-1,t} \neq 0\ \&\ V_{n,t} = 0 \tag{3}$$

$$\tilde{V}_{n,t+1} = 0, \qquad for\ V_{n-1,t} = 0\ \&\ V_{n,t} = 0 \tag{4}$$

Where

$\tilde{V}_{n,t+1}$ :  the speed of the following vehicle at time step t+1

$v_{n,d}$ :  the desired speed of the following vehicle

$V_{n,t}$ :  the speed of the following vehicle n at time step t

$V_{n-1,t}$ :  the speed of the lead vehicle n-1 at time step t

$x_{n,t}$ :  the position of the following vehicle n at time step t

$x_{n-1,t}$ :  the position of the lead vehicle n-1 at time step t

$S_n$ :  the safe standstill distance headway of the following vehicle n

$\lambda, \alpha, \beta, \gamma, L$ :  Parameters

$\Delta t$ :  the length of a time interval equals to reaction time of drivers

$a_{n,d}$ :  The desired start acceleration of the following vehicle n

## 3. Results and Discussion

In this section, some single lane car-following simulations are presented. First, the traffic stability and the reason of different drivers keep different stable gaps under the same condition are discussed in part A. Then, part B presents the model that can reflect the closing-in and shying-away phenomena.

### 3.1. Traffic Stability

The distance headway between the lead vehicle and the following vehicle reaches a particular value after perturbation caused by the actions of the lead vehicle is referred to as the stability in car-following behavior [1, 4]. Researchers identified two types of traffic stability: local stability and asymptotic stability. Local stability is concerned with the car-following behavior of just two vehicles. Asymptotic stability is concerned with the car-following behavior of a line of vehicles [1, 4].

An example of four vehicles movement process is illustrated below. The example indicates that the proposed model does achieve local and asymptotic stability. The desired speeds of the first vehicle, the second one, the third one, and the last one are 50km/hr, 60km/hr, 70km/hr, and 80km/hr, respectively. Figure 2(a) is the X-T diagram of these four vehicles. Since there is no vehicle in front of the first vehicle, the first vehicle runs at its desired speed (i.e. 50km/hr). The following vehicles run at their desired speed initially. Since their lead vehicle runs at 50km/hr, they decelerate later and keep their speed at 50km/hr finally. This is asymptotic stability.

Figure 2(b) is the distance headways between these vehicles. At first, the distance headways of all vehicles are 50m. Finally, the car-following process is stable, and as the driver's desired speed increases the stable distance headway decrease. Figure 2(b) reflects the model assumptions mentioned above, and it can explain why drivers keep different gaps under the same condition.

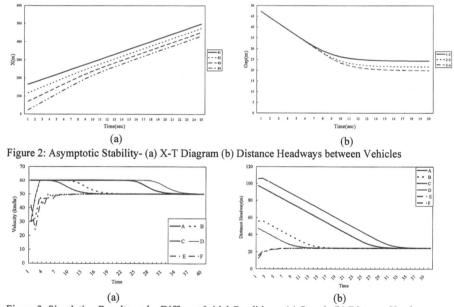

(a)  (b)

Figure 2: Asymptotic Stability- (a) X-T Diagram (b) Distance Headways between Vehicles

(a)  (b)

Figure 3: Simulation Results under Different Initial Conditions- (a) Speeds (b) Distance Headways

The stable distance headway is only dependent on the final speed (or stable speed) and not on anything else [4]. The following example shows that the same lead vehicle and the same following vehicle will result in the same stable distance headway under different initial conditions. The initial conditions include initial gap and initial speed of the following vehicle. The illustrative example presented in Figure 3 includes 6 simulation scenarios with different initial conditions.

Figure 3 is the simulation result. It indicates that the same lead vehicle and the same following vehicle will result in the same stable speed and the same distance headway under different initial conditions. It means that the proposed model can describe the traffic phenomenon: the stable distance headway is only dependent on the final speed and not on initial condition.

### 3.2. Closing-in and shying-away

Sometimes the following vehicle accelerates even though the speed of lead vehicle is slower than its speed (i.e. closing-in) and vice versa (i.e. shying-away) [4]. The following examples indicate that the proposed model can describe the closing-in and shying-away phenomena.

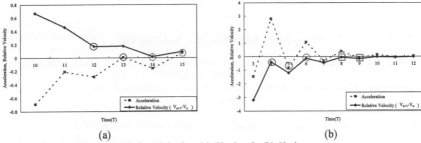

(a)                                                         (b)

Figure 4: Acceleration and Relative Velocity- (a) Closing-in (b) Shying-away

Figure 4 are the simulation results of two simulation examples. Figure 4(a) shows closing-in phenomenon. At T=12, the lead vehicle is slower than the following vehicle, but the following vehicle accelerates its speed at next time step. The same situation occurs at T=14 and T=15, and the phenomenon is closing-in. A shying-away example is illustrated in Figure 4(b). The lead vehicle is faster than the following vehicle at T=4, but the following vehicle still decelerates at next time step. At T=6 and T=8, the same situations occur. Although the lead vehicle is faster than the following vehicle, the following vehicle decelerates its speed at the next time step. This is the so-called shying-away phenomenon.

### 4. Conclusions

In this paper we have computationally investigated the traffic phenomena of car-following. Our model formulation is mainly based on an estimation of speed with considering the repulsion and thrust. Numerical results have confirmed the validity of the proposed model in describing the car-following process. It successfully reflected local stability, asymptotically stability, closing-in, and shying-away. It can explain why drivers keep different speeds or different gaps under the same condition. This approach provides an alternative way to modern traffic flow simulation and is promising in exploring the dynamics of macroscopic traffic flow. We are currently calibrating the simulation results with the field data.

### Acknowledgments

This work is supported in part by Ministry of Education of Taiwan, R.O.C. under Grant EX-91-E-FA06-4-4 and in part by the National Science Council of Taiwan, R.O.C. under Grant NSC 92-2211-E-009-049.

### References

[1]  A.D. May, *Traffic Flow Fundamentals*, Prentice Hall, Englewood Cliffs, 1990.

[2]  M.F. Aycin and R.E. Benekohal, "Comparison of Car-Following Models for Simulation", *Transportation Research Record* **1678** 116-127(1999).

[3]  D.C. Gazis, R. Herman, and R. B. Potts, "Car-Following Theory of Steady-State Traffic Flow", *Operations Research* **7** 499-505(1959).

[4]  P. Chakroborty, and S. Kikuchi, "Evaluation of the General Motors Based Car-Following Models and A Proposed Fuzzy Inference Model", *Transportation Research Part C* **7** 209-235(1999).

[5]  R. Widemann, *Simulation de Stranssenverkehrsflusses*, Schriftenreihe des Instituts fur Verkehrswesen, Heft 8, Universitat Karlsruhe, 1974.

[6]  M. Treiber, A. Hennecke, D. Helbing, "Congested Traffic States in Empirical Observations and Microscopic Simulations", *Physical Review E* **62** 1805-1824(2000).

Brill Academic Publishers
P.O. Box 9000, 2300 PA Leiden
The Netherlands

*Lecture Series on Computer
and Computational Sciences*
Volume 4, 2005, pp. 1048-1052

# Estimating O/D Matrix from Traffic Count with Network Sensitivity Information

[1]Terry Friesz, [2]Hsun-Jung Cho, [3]Pei-Ting Yeh

[1]Harold and Inge Marcus Chaired Professor of Industrial Engineering305 Leonhard Building
University Park, PA 16802
[2,3]Department of Transportation Technology and Management, National Chiao Tung University,
Hsinchu, Taiwan, 30050, ROC

Received 11 July, 2005; accepted in revised form 13 August, 2005

*Abstract:* Under the network user equilibrium principle, the O/D matrix estimation method with network sensitivity information is proposed. The key part of this method is that the rows and the columns of arc/path incidence matrix are partitioned into matrices. In this paper, the sensitivity analysis method is used to find the step direction and the step size of the reassignment. Finally, the example shows the good results.
*Keywords:* Generalized inverse matrix, Network user equilibrium, Wardrop's first behavior principle.

*Mathematics Subject Classification*: 15A09, 34A30, 90B10, 90C31

## 1. Introduction

Under the assumption of the user equilibrium behavior principle [1], the network sensitivity analysis [2, 3, and 4] has been proposed. The O/D matrix estimation methods are proposed by many researches [5, 6, 7, 8], such as bi-level programming, maximum Entropy method. In this research, an O/D matrix estimation method is proposed, which combines the network sensitivity information and the generalized inverse matrix method. The objective is that, under the general condition, we develop a new research method estimating the travel distribution matrix. The assumption of the travel behavior includes user equilibrium principle of the traffic assignment and the travel distribution presented by entropy model [9].

## 2. The Data and Model

The follow notation, which is used throughout the paper, is summarized here for convenience.

$N$ : The set of all nodes in the network.

$I$ : The set of all origin nodes in the network.

$J$ : The set of all destination nodes in the network.

$i, j \in N$ : Node i or node j in the network.

$A$ : The set of all arc in the network.

$a \in A$ : Arc a in the network.

$P_{ij}$ : The set of all paths that the origin node is i, and destination node is j in the network.

$p \in P_{ij}$ : Path p that the origin node is i, and destination node is j in the network.

$\delta_{apij}$ : The arc/path incidence variable, $\delta_{apij}$ =1, where the arc a $\in P_{ij}$ , 0 otherwise.

$\Delta$ : The arc/path incidence matrix, $(..., \delta_{apij} ,...)$ .

$\Lambda$ : The OD/path incidence matrix, $\Lambda$ =1 where the path belongs to some OD pair, 0 otherwise.

$T_{ij}$ : The Demand that the origin node is i, and destination node is j in the network.

---

[1] Corresponding Author: Professor :tlf13@psu.edu

| | |
|---|---|
| $h_{pij}$ | : The flow of the path p that the origin node is i, and destination node is j in the network. |
| $h$ | : The flow vector of all paths in the network. |
| $f_a$ | : The flow of the arc a. |
| $f$ | : The flow vector of all arcs in the network. |
| $c_a(f)$ | : The cost function of arc a. |
| $c(f)$ | : The vector of the arc cost function. |
| $c_{pij}(h)$ | : The cost function of path p that the origin node is i, and the destination node is j. |
| $c(h)$ | : The vector of the path cost function. |
| $u_{ij}$ | : The minimum travel cost function of the origin node i and destination node j in the |

network.

| | |
|---|---|
| $u$ | : The vector of the minimum travel cost function. |
| $R^n$ | : The set of all real number in n space. |

### 2.1. The feasible solution space of equilibrium network O/D matrix

Given the arc flow and arc cost functions, the path will be deleted if the cost is greater than the lowest cost. The relationship is written as:

$$\Delta^0 h' = f \quad \text{where } \Delta^0 \text{ is the } \Delta \text{ deleted the path} \tag{1}$$

$$\Lambda^0 h' = T \quad \text{where } \Lambda^0 \text{ is the } \Lambda \text{ deleted the path} \tag{2}$$

By generalized inverse matrix method, the formulation of the equilibrium network in terms of arc flow. Then the feasible solution space of equilibrium network O/D matrix can be solved when the observed.

### 2.2. Establish the perturbation parameter calibration model

The travel distribution behavior assumption is presented by entropy model. Under this assumption, let estimated O/D matrix to approximate the feasible solution space of the equilibrium network O/D matrix. Then the fittest perturbation parameter can be gained. The method of estimating O/D matrix by the arc flow data can be offered.

Using entropy model presented by Murchland [9], as follow:

$$\min_T \gamma \sum_{ij} c_{ij} T_{ij} + \sum_{ij} T_{ij} (\ln T_{ij} - 1) \tag{3}$$

$$s.t. \quad \sum_j T_{ij} = O_{ij} \quad \forall i \quad (\alpha_i) \tag{4}$$

$$\sum_i T_{ij} = D_{ij} \quad \forall_j \quad (\beta_j) \tag{5}$$

Given $c_{ij}, O_i$ and $D_j$, any of a perturbation parameter can obtain an O/D matrix $T(\gamma)$.

### 2.3. Sensitivity analysis of O/D matrix to perturbation parameter

Let $\varepsilon$ be a parameter, according to Tobin's [10] theorem, the parameter after perturbation is $\gamma(\varepsilon) = \gamma^0 + \varepsilon$. In Murchland's model, in the condition of $\varepsilon = 0$ can be presented as follow:

$$\gamma(\varepsilon)c + \ln T(\varepsilon) - \Lambda_0^T \alpha(\varepsilon) - \Lambda_d^T \beta(\varepsilon) = 0 \tag{6}$$

$$\Lambda_0 T(\varepsilon) - O = 0 \tag{7}$$

$$\Lambda_d T(\varepsilon) - D = 0 \tag{8}$$

Where $\gamma = \begin{bmatrix} \gamma & 0 \\ 0 & \gamma \end{bmatrix}$, $\ln T = \begin{bmatrix} \ln T_{11} \\ \vdots \\ \ln T_{ij} \end{bmatrix}$. According to Tobin's implicit function theorem, the gradient vectors

of the constraint function are required to be linearly independent in sensitivity analysis. Such that the constraint functions have to delete one constraint function. Solve it by Jacobian matrix to get $\varepsilon$.

## 3. Algorithm

The following algorithm is to solve the O/D model.

Step1: $n=1$, choose any $\gamma_n^0$.

Step2: Solve $T_n(\gamma^0)$ from functions (3), (4), and (5) as $T_n(0)$

Step3: Solve $T_n^0$ from following mathematical programming, where $T_n^0$ is the projection of $T_n(0)$ in the feasible solution space of O/D matrix

$$\min_\varepsilon (T - T_n(0))^T (T - T_n(0)) \tag{9}$$

$$s.t. \quad [-M_2 I]T = M_1 \bar{f}^P \tag{10}$$

To get

$$T_n^0 = T_n(0) + [-M_2 \ I]^T ([-M_2 \ I][-M_2 \ I]^T)^{-1} (M_1 f^P - [-M_2 \ I]T_n(0)) \tag{11}$$

Step4: Solve $T_n(0)$ to get the solution $\nabla_\varepsilon T^*$, and let

$$T(\varepsilon) = T_n^0 + \nabla_\varepsilon T^* \cdot \varepsilon \tag{12}$$

Step5: Solve the pace $\varepsilon_n$ from the following method:

Find $\varepsilon_n$ such that $\nabla_\varepsilon T^* \perp (T_n(\varepsilon) - T_n^0)$, and

$$T_n(\varepsilon) - T_n^0 = \nabla_\varepsilon T_n^* \cdot \varepsilon_n - [-M_2 \ I]^T ([-M_2 \ I][-M_2 \ I]^T)^{-1} (M_1 \bar{f}^P - [-M_2 \ I]T_n(0)) \tag{13}$$

The problem solving pace $\varepsilon_n$ is equal to solving the following problem

$$(\nabla_\varepsilon T^*)^T (T_n(\varepsilon) - T_n^0) = 0 \tag{14}$$

The result is

$$\varepsilon_n = (\nabla_\varepsilon T_n^{*T} \cdot \varepsilon_n T_n^*)^{-1} \nabla_\varepsilon T_n^{*T} [-M_2 \ I]^T ([-M_2 \ I][-M_2 \ I]^T)^{-1} (M_1 \bar{f}^P - [-M_2 \ I]T_n(0)) \tag{15}$$

Step6: If $|\varepsilon_n| < \Delta$, $\gamma^* = \gamma_n^0$, and solve $T(\gamma^*)$, stop; otherwise $\gamma_{n+1}^0 = \gamma_n^0 + \varepsilon_n$

Step7: Solve $T_{n+1}(0)$ from functions (3), (4), and (5)

Step8: Find $T_{n+1}^0$ from step3

Step9: If $|T_{n+1}^0 - T_{n+1}(0)| < |T_n^0 - T_n(0)|$, go to step10; otherwise, $\varepsilon = 0.5\varepsilon_n$, go to step6

Step10: $n = n+1$, Take $\gamma_{n+1}^0$, $T_{n+1}(0)$, and $T_{n+1}^0$ go back to step4.

## 4. Example

The saturation flow is 1740, and the cost function of arc is as follow,

$$C_a = a + b(f_a / V_a)^4. \tag{16}$$

Where

$C_a$ : cost function of arc

$f_a$ : arc flow

$V_a$ : arc capacity

$a$ : travel time when arc flow equal to zero

$b$ : calibration value equal to 134.95 (min.)

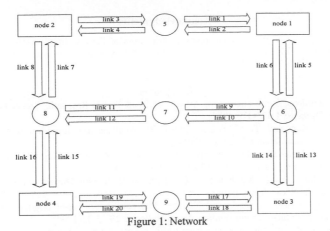

Figure 1: Network

Table 1 : Data for Network

| Travel Type | Traffic flow |
|---|---|
| $O_1,...O_4$ | 242,121,251,81 |
| $D_1,...D_4$ | 184,158,233,120 |
| $f_1,...f_{20}$ | 129, 166, 129, 166, 159, 176, 167, 170, 146, 154, 146, 154, 161, 169, 163, 174, 141, 165, 141, 165 |

The results are presented in Table2. The initial perturbation parameter value is 1, convergent standard value is 0.001, and through six iterations is stable.

Table 2 : Estimating $T_{ij}$

| O \ D | 1 | 2 | 3 | 4 | $O_i$ |
|---|---|---|---|---|---|
| 1 | 0 | 82.165 | 126.944 | 32.891 | 242 |
| 2 | 47.995 | 0 | 36.056 | 36.949 | 121 |
| 3 | 131.005 | 69.835 | 0 | 50.160 | 251 |
| 4 | 5.0 | 6.0 | 70.0 | 0 | 81 |
| $D_j$ | 184 | 158 | 233 | 120 | 695 |

## 5. Conclusion

Under the assumption of user equilibrium, the feasible solution space of O/D matrix is estimated. By compared with the traditional method and assignment method, the network congestion influence is considered. The primary question in this method is that the objective of parameter calibration model is not exactly strictly concave function. So in this method we might find a local solution. Matrix partitioned and generalized inverse matrix approach are employed in this research to transfer feasible solution space of path to feasible solution space of arc, to show the unique solution of arc flow and to get sensitivity information for equilibrium network flow with respect to arc flow.

In addition to arc traffic flow, other suitable data can be used to estimate O/D matrix to make the feasible solution space smaller and to make the estimation more accurate. The result of this research is greatly influenced by the cost function. To develop a cost function for domestic research is the important course.

## Acknowledgments

This work is supported in part by the Ministry of Education of Taiwan, ROC, under grant EX-91- E-FA06-4-4 and in part by the National Science Council of Taiwan under contracts: NSC-92-2622- E-009-011-CC3 and NSC-93-2211-E-009-032.

## References

[1] J. C. Wardrop, Some Theoretical Aspects of Road Traffic Research, *Proceeding Institute of Civil Engineers* **1** 325-378(1952).

[2] R. L. Tobin and T. L. Friesz, Sensitivity Analysis for Equilibrium Network Flow, *Transportation Science* **22** 242-250(1988).

[3] H. J. Cho, Generalized Inverse Approach to Sensitivity Analysis of Equilibrium network flow, *Transportation Planning Journal* **20** 1-14, 1991 (in Chinese).

[4] H. J. Cho, Tony E. Smith and T. L. Friesz, A Ruduction Method for Local Sensitivity Analysis of Network Equilibrium Arc Flows, *Transportation Research Part B* **34** 31-51(2000).

[5] H. Yang, Sasaki. T. Sasaki and Y. Iida., The equilibrium-based origin-destination matrix estimation problem, *Transportation Research Part B* **28** 23-33(1994).

[6] J. Oh, Estimation of trip matrices from traffic counts: An equilibrium approach, Mathematics in Transport Planning and Control 35-44(1992).

[7] M. Florian, and Y. Chen, A coordinate descent method for the bilevel O-D matrix adjustment problem, Preprints of the 7th IFAC/IFORS Symposium on Transportation Systems: Theory and Application of Advanced Technology 1029-1034(1994).

[8] H. Yang, T. Sasaki, Y. Iida and Y. Asakura, Estimation of origin-destination matrices from link traffic counts on congested network, *Transportation Research Part B* **26** 417-434(1992).

[9] J. D. Murchland, *Some Remarks on the Gravity Model of Traffic Distribution and an Equivalent Maximization Formulation*, LSE- TNT-38. Transport Network Theory Unit, London School of Economics, London 1966.

[10] R. L. Tobin, Sensitivity Analysis for Variational Inequalities, *Journal of Optimization Theory and Applications* **48** 191-204(1986).

Brill Academic Publishers
P.O. Box 9000, 2300 PA Leiden
The Netherlands

*Lecture Series on Computer
and Computational Sciences*
Volume 4, 2005, pp. 1053-1056

# User-Oriented Travel Time Prediction
# Using Grey System and Real-time Vehicle Detector Data

Yow-Jen Jou[1], Chien-Lun Lan[2]

[1] Institute of Statistics, National Chiao Tung University, Hsinchu, Taiwan
[2] Department of Transportation, National Chiao Tung University, Hsinchu, Taiwan

Received 5 July, 2005; accepted in revised form 3 August, 2005

*Abstract:* Real-time travel information is increasingly important for ITS. Many high-valued time-sensitive applications can take advantage from this real-time information. This paper describes an approach to estimate travel time with Grey System. Combined with real-time vehicle detector data, wireless communication ability and global positioning system, individual driving preference is then measured and been used to adjust the estimation model. An empirical result is shown at the end of this paper and shows a satisfying result.

*Keywords:* Grey System, Travel Time Prediction,

*Mathematics Subject Classification:* 60J20, 62-07, 90B06

## 1. Introduction

Real-time travel information plays an important role in Intelligent Transportation Systems (ITS). With real-time travel information, travelers can make choice of departure time, route and mode to minimize their travel time. Time-sensitive applications e.g., Just-in-Time delivery and emergency vehicle routing, can take great benefits from the real-time information. There are mainly two approaches used to collect real-time travel information. One approach relies on instrumented vehicles traveling in the traffic, also known as probe vehicles [1, 2]; the other approach relies on collecting information by vehicle detectors (VD) [3]. The focus of this paper is the analysis of real-time VD data and the estimation of freeway travel time. A modified grey system theory is introduced to tackle the problem of having not enough data for statistic analysis. The remainder of this paper is organized as follows, the markov bias corrected grey model is introduced in section 2, and the dynamic freeway travel time estimation model is addressed in section 3. In section 4, an empirical test is shown; section 5, shows the general conclusions of the study.

## 2. Markov Bias Corrected Grey Model

Grey system theory brings up an idea of using small quantity (at least four outputs) of system output to establish a grey model, and approximating the dynamic behavior of the origin system. The grey model, consisting of a series of differential equations, denoted as $GM(n, h)$ where $n$ stands for the order of the differential equations and $h$ stands for the number of variables. The simplest model is the $GM(1, 1)$ model, i.e.,

$$y^{(0)}(k) + az^{(1)}(k) = u$$
$$z^{(1)}(k) = \frac{1}{2}\left[y^{(1)}(k) + y^{(1)}(k-1)\right]$$
$$y^{(1)}(k) = \left(\sum_{m=1}^{k} y^{(0)}(m)\right)$$

(1)

---

[1]Corresponding author, Associate Professor, yjjou@stat.nctu.edu.tw

where $y^{(0)}(k)$ indicates the observation set, $a$ is the development coefficient, $u$ is the grey input. Least square method is been used to calculate $a$ and $u$. The prediction model of *GM(1, 1)* is

$$\hat{y}^{(0)}(n+p) = \left( y^{(0)}(1) - \frac{u}{a} \right) \cdot (1 - e^a) \cdot e^{-a(n+p-1)} \tag{2}$$

where $n$ represent the observation count and $p$ indicates the prediction step.

In order to improve the accuracy of the original grey model, markov process is introduced to deal with the sign change on bias sequence $q^{(0)}(k)$, where

$$q^{(0)}(k) = y^{(0)}(k) - \hat{y}^{(0)}(k) \quad k = 2,3,...,n$$

and also define $\varepsilon^{(0)}(k) = \left| q^{(0)}(k) \right|$.

A *GM(1, 1)* is established for $\varepsilon^{(0)}(k)$, then we get

$$\hat{\varepsilon}^{(0)}(k) = \left( \varepsilon^{(0)}(1) - \frac{u_\varepsilon}{a_\varepsilon} \right) (1 - e^{a_\varepsilon}) e^{-a_\varepsilon(k-1)} \qquad k = 2,3,...n \tag{3}$$

Markov chain is developed to capture the sign change of the bias sequence. Construct the sign change transition matrix $R$ of bias sequence and let the $n^{\text{th}}$ data as initial condition, then the condition probability vector $\pi^{(p)}$ after $p$ steps can be computed by

$$\pi^{(p)} = \pi^{(0)} \hat{R}^p \quad p = 1,2,...$$

Let $\delta(k)$ denotes the sign of $k^{\text{th}}$ bias sequence, then

$$\delta(k) = \begin{cases} +1, & \text{if it is positive} \\ -1, & \text{if it is negative} \end{cases} \quad k = 1,2,...,n,... \tag{4}$$

Combine equation 2, equation 3 and equation 4, we get markov bias corrected grey model, denoted as *MBC-GM(1, 1)*, as follows

$$\hat{y}'^{(0)}(n+p) = \left( y^{(0)}(1) - \frac{u}{a} \right) (1 - e^a) e^{-a(n+p-1)}$$
$$+ \delta(n+p) \left( \varepsilon^{(0)}(1) - \frac{u_\varepsilon}{a_\varepsilon} \right) (1 - e^{a_\varepsilon}) e^{-a_\varepsilon(n+p-1)} \tag{5}$$

## 3. Dynamic Freeway Travel Time Estimation Model

The dynamic travel time estimation provides information about travel time that is going to be experienced by traveler during the entire journey, and that is different from the current travel time. The current travel time of a road section is equal to the road length divided by the average speed on the section, and current travel time of a journey is the summation of current travel time of all section. The estimation of travel time by current travel time ignores the traffic condition changing with time during the total journey, and it might lead to a biased estimation. The dynamic travel time estimation is then proposed to overcome the problem by providing predicted travel time information.

This study mainly focuses on the travel time estimation between interchange pairs on freeway. In reality, it is impossible to obtain actual travel time for individuals, thus the average speed data from vehicle detector (VD) data is been used to estimate the travel time. The average speed data from VD updates every 5 minutes. Being analyzed in this study, the day-of-week average speed is different, thus the estimation time domain had been separated into day-of-week and the 5 minutes VD updating interval had been applied to time-of-day aspect. Hence, there exist 288 time intervals a day. Due to the division of time domain, data sequence will not be able to satisfy the statistics requirement. Grey system theory, which can deal with the issue, is then been introduced to tackle the problem. In this paper, *MBC-GM(1, 1)* is been applied to the speed estimation. The latest 8 speed data is been used to estimate the speed of next time period e.g. , when it come to estimate the speed of 8:00~ 8:05 on Monday morning, the last 8 weeks of speed data on the same time on Monday will be treated as the input data of estimation model. The output of the estimation model will be the speed of 8:00~8:05 this Monday.

Individuals act different on roads. Aggressive drivers tense to keep higher speed than average; while less aggressive drivers tense to keep lower speed. Even the performance of vehicle makes the speed different; when it comes to the road with grades, vehicles with larger weight-to-horsepower ratio may not be able to keep their desire speed. In order to capture the behavior of individuals, a computer with Global Positioning System (GPS) and General Packet Radio Service (GPRS) communication set is installed on the test vehicle. With the GPS, vehicle location can be retrieve; with the GPRS, real-time information can be passed to the vehicle. The individual speed for a single road section is derived from the road section length divided by travel time, and it is then compared with the average speed for the section by real-time communication with real-time VD database. Difference between individual and average will be stored in the individual behavior database. When individuals driving down the same section of the road, the average of difference will be deduct from the *MBC-GM(1, 1)* estimation. The speed estimation after individual behavior adjustment will be used to estimate the travel time. The travel time on each road section of the estimate time period can be computed by $time = distance/speed$.

The traveling *x-t* locus can be represented as figure 1. The steeper the locus curve is, the slower the speed is. As shown in figure X travel time of a journey cannot be simply obtained by adding travel time of each road section in the same time period, the speed of each section will be change as time passed. The speed change with time must be considered.

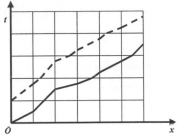

Figure 1: The locus of vehicles,

## 4. Empirical Result

In this empirical test, the target road is a major freeway in Taiwan, the freeway No.1. The total length of the road is 372.7 km, consists of 65 inter-changes and 10 toll gates. The design speed of the freeway is 120 km/h and the speed limitation is 100km/h. Most area of the freeway has 3 lanes each direction. VD is installed over the entire freeway, with intervals of approximately 2 km; there exists at least 1 VD between each interchange pair. The real-time average speed data been used in this study, updates every few minutes, is the average speed between interchange. This data is derived from the average of speed data of all VD on the road section. The toll gates were built across the road and cause a noticeable impact on travel time, because vehicles must slow down to pay. The delay cause by toll gates is estimated by the model developed by Institute of Transportation [7].

The following test is done on the section where the VD data most stable. The length of the test section is 170.4KM; the test vehicle is an ordinary passenger car and the weather is sunny. The individual behavior database contains the data of the same driver driving on the same road section for 10 times. The test results is shown as figure 2, the accumulative estimation bias between actual travel time and predict travel time after 170.4 KM is less then 3%, which shows an impressive result.

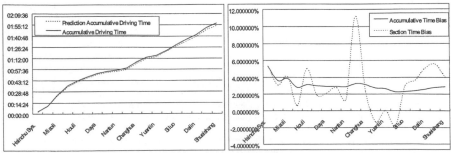

Figure 2: The test results.

## 5. Conclusion

This study implements freeway travel time estimation by combining markov bias corrected grey system theory and individual behavior adjustment. Compare the empirical result with real travel time data obtained from electronic toll system, the result is quite satisfying [8]. Navigation system on vehicles can implement the idea proposed in this paper and provide high accuracy individualized information.

## Acknowledgments

This research was supported in part by the Ministry of Education of Taiwan, ROC under Grants EX-91-E-FA06-4-4 and in part by the National Science Council of Taiwan, ROC under Grants NSC-93-2218-E-009-042 and NSC-93-2218-E-009-043 respectively.

## References

[1] Shawn M. Turner and Douglas J. Holdener, Probe Vehicle Sample Sizes for Real-Time Information: The Houston Experience, *Proceedings of the IEEE 6th International Vehicle Navigation and Information System Conference* 6 3-10(1995).

[2] Yanying Li and Mike McDonald, Link Travel Time Estimation Using Single GPS Equipped Probe Vehicle, *Proceedings of the IEEE 5th International Conference on Intelligent Transportation Systems* 5 932-937(2002).

[3] John Rice and Erik van Zwet, A Simple and Effective Method for Predicting Travel Times on Freeways, *IEEE Transactions on Intelligent Transportation System* 5 200-207(2004).

[4] D. E. Boyce, J. Hicks and A. Sen, In-Vehicle Navigation Requirements for Monitoring Link Travel Times in a Synamic Route Guidance System, *ADVANCE WORKING PAPER SERIES* 2 Urban Transportation Center, University of Illinois, 1991.

[5] John H. Wu, Rolling Error in GM(1, 1) Modeling, *Journal of Grey System* 70-80(2001).

[6] Yow-Jen Jou, Tsu-Tian Lee, Chien-Lun Lan and Chien-Hao Hsu, The Implementatio Markov Bias Corrected Grey System in Freeway Travel Time Prediction, *Proceedings c 6th International Conference on Intelligent Transportation Systems* 6 832-835(2003).

[7] *Highway Travel Time Information and Management System Planning (2/4)*, Institute of Transportation, Ministry of Transportation and Communication, R.O.C, 2004.

[8] Yoshikazu Ohba, Hideki Ueno and Masao Kuwahara, Travel Time Calculation Method for Expressway Using Toll Collection System Data, *Proceedings of the IEEE International Conference on Intelligent Transportation Systems* 471-475(1999).